Chrysomeloidea II (Orsodacnidae, Megalopodidae, Chrysomelidae)

Volume 6/2/1

Catalogue of Palaearctic Coleoptera

VOLUME 6/2/1

The titles published in this series are listed at *brill.com/cpc*

Pachybrachis fimbriolatus Suffrian, 1844

Chrysomeloidea II (Orsodacnidae, Megalopodidae, Chrysomelidae)

Updated and Revised Second Edition

VOLUME 6/2/1

Edited by

J. Bezděk
L. Sekerka

BRILL

LEIDEN | BOSTON

Cover illustration: *Pachybrachis fimbriolatus* Suffrian, 1844 (© Jan Bezděk)

Date of publication: 23 May 2024.

The index of the species-group names is available on site of the Koninklijke Brill BV, Leiden (10.6084/m9.figshare.25540735).

The Library of Congress Cataloging-in-Publication Data is available online at https://catalog.loc.gov
LC record available at https://lccn.loc.gov/2015018977

Typeface for the Latin, Greek, and Cyrillic scripts: "Brill". See and download: brill.com/brill-typeface.

ISBN 978-90-04-70810-5 (hardback, set)
ISBN 978-90-04-44315-0 (hardback, vol. 1)
ISBN 978-90-04-44330-3 (e-book, vol. 1)
DOI 10.1163/9789004443303

Contents

Volume 1

Preface IX
Taxonomic Information XI
Distributional Information XV
Bibliographic Information XXIII
Dating of Publications XXV
Acknowledgements XXXVIII
Editors and Authors XLII

New Nomenclatural and Taxonomic Acts, Comments and Distributional Records 1

Catalogue 169
- superfamily Chrysomeloidea 169
  - family Orsodacnidae 169
    - subfamily Orsodacninae 169
  - family Megalopodidae 170
    - subfamily Zeugophorinae 170
    - subfamily Megalopodinae 173
  - family Chrysomelidae 175
    - subfamily Sagrinae 175
    - subfamily Bruchinae 176
    - subfamily Donaciinae 206
    - subfamily Criocerinae 216
    - subfamily Cassidinae 233
    - subfamily Chrysomelinae 273
    - subfamily Galerucinae 378
    - subfamily Lamprosomatinae 581
    - subfamily Cryptocephalinae 582
    - subfamily Eumolpinae 688
    - subfamily Synetinae 728

Appendix 729
Index to Family-group Names 731
Index to Genus-group Names 733

Volume 2

References 751

Preface

It has been fourteen years since the publication of the sixth volume of the Catalogue of Palaearctic Coleoptera, which contained the entire superfamily Chrysomeloidea. Since then, numerous taxa have been described or changed status, and many genera have undergone modern revisions. There is also a substantial number of new or overlooked data on distribution. Last but not least many mistakes were found that needed to be corrected. Due to the enormous size of the entire superfamily, the sixth volume was split into two parts in the new series. The first part containing Cerambycidae *sensu lato* was published in 2020. In this part, the remaining families of Chrysomeloidea are treated, following in this respect traditional classification, but it must be stated that Orsodacnidae and Megalopodidae have recently been found to belong to the cerambycoid clade, robustly supported by adult and larval morphology as well as molecular data. However, since they were not covered in the first part of Volume Six, they are included here.

This part somewhat deviates from all hitherto published volumes of the series as we made some additions. The general concept of the catalogue is still the same, but we also decided to include all infrasubspecific names, i.e. all names ever published within Palaearctic Orsodacnidae, Megalopodidae, and Chrysomelidae are listed (except for *nomina nuda*). Sometimes it is clear that a name is an infrasubspecific entity by definition (e.g. aberrations). On the other hand, the infrasubspecific status of many names is disputable, which might cause some difficulties in the future, i.e. in homonymies with new taxa. Also, even when the name used is an infrasubspecific entity, there is a distributional record behind it that is still valid. Thus, we decided that all such names should be included to give the reader possibility to find out the present status of each name and to which taxon it refers, as they are not always associated with the original maternal taxon to which they were assigned.

We also put a special emphasis on the distributional data as the previous edition contained many records that were never published, or erroneous or simply repeated from other works despite obvious biogeographic discrepancies. Therefore, the present Edition is mainly based on published records and many suspicious or otherwise problematic distribution records are commented on, to give the reader the most up to date and accurate overview. For this reason, we included new faunistic records in the New Acts chapter.

Finally, the major change is that we decided to include the publication date for each reference cited as accurately as possible and provided the sources of such information. This is an essential basis for the principles of the zoological nomenclature to establish correct synonymies and confirm validity of taxa, and we believe this will be beneficial to all insect taxonomists, as many works contain multiple groups, or at least the cited source provides potentially useful evidence for other volumes/issues within the series. We are aware that many of the dates we give might be specified or changed in the future, because research on this topic is going on and we certainly have overlooked some of the sources. However, we would be satisfied if the data we present were adopted as a basis for further research.

The present part of the Catalogue covers three families, namely Megalopodidae, Orsodacnidae, and Chrysomelidae. The last is the most species rich, and is divided into 11 subfamilies. In total this part contains 596 genera, 165 subgenera, 9398 species and subspecies, and 100 *nomina dubia*. Following new nomenclatural acts are introduced: one new tribe, one new subtribe, one new species, two new subspecies, ten new substitute names, 37 new combinations, 149 new synonymies, ten *nomina protecta*, and nine *nomina oblita*. For the purpose of the stability of nomenclature 15 lectotypes are herein designated. In addition, new distributional data for 351 species are recorded.

Taxonomic Information

The present Catalogue includes all available names, both valid and invalid (synonyms of family group names excepted), of extant beetle taxa described before 1st January 2023 and known to occur in the Palaearctic Region as defined in the Catalogue (see chapter Distribution Information). Taxa described in 2023 are included as far as we were aware of them. Unjustified emendations and names rejected by the Commission are also included, with appropriate indications. Fossil taxa, misspellings, misidentifications and *nomina nuda* are not included in the body of the Catalogue.

The higher classification follows the *Handbook of Zoology* (Leschen & Beutel 2014). The families Orsodacnidae and Megalopodidae, traditionally classified "close" to Chrysomelidae, were recently recovered as belonging to the cerambycoid lineage (Haddad et al. 2018) and resolved as either basal to other cerambycoid lineages or forming a clade with Vesperidae sister to Disteniidae+Cerambycidae. We fully concur with this classification, but both families are included here because they were not covered in the first part of Volume 6 of the Catalogue, and it would be superfluous to create a special volume for them especially because they share most of their references with chrysomelids. The higher classification of Megalopodinae follows Rodríguez-Mirón et al. (2021), though a phylogeny based on molecular data is urgently needed to verify their results. Classification of Galerucinae follows Douglas et al. (2023) with modification that the three tribes are treated here as supertribes, i.e. Alticitae, Galerucitae, and Serraticollitae to prevent diminishing the internal tribal classification of true galerucines. Genera transferred from alticines are listed as unassigned to tribes as their tribal position within Galerucitae was not resolved. The classification of Cryptocephalinae follows Gómez-Zurita & Cardoso (2021) who established the tribes Pachybrachini and Mylassini to solve the unsustainable situation of paraphyletic Cryptocephalini. However, the analysis did not include another crucial taxon – *Acolastus* Gerstaecker, 1855, which share the unique structure of the rectal apparatus with Clytrini. Accordingly, a new subtribe Acolastina of Clytrini is established here.

The present volume covers 761 valid genus-group names and 9498 valid species-group names. The family-group names currently in use are provided with their author and year when published; family synonyms or references to their original descriptions are not included. Type species are given in original combination for all genus-group taxa; in case of synonymy of the type species, the respective valid name is given in parentheses. All genus- and species-group taxa are provided with full reference, i.e. author, year and page where their description appeared. The lowercase prepositions such as French "de" and German "von" are not given with the author's name in the list of taxa. In cases of descriptions in Chinese/Japanese with English translation, both respective pages where the descriptions appeared are given. Species-group taxa subsequently assigned to other genera are provided with their original genus in parentheses, following the page of publication. For species-group taxa proposed as entities lower than species (i.e. ssp., var., ab., forma, morpha, etc.), the original category is given in parentheses following the page and accompanied with the original combination where applicable. Unavailable and infrasubspecific names are listed alphabetically among other available names but not in italics. We find it useful for the reader to have all the names (both available and unavailable) summarized in one place. Taxa considered *incertae sedis* and *nomina dubia* are listed separately at the end of the nearest applicable taxon.

In the past, the same description was occasionally published twice, or even more times, in separate publications. In such cases, the earliest publication is referred to as indicated above, followed by an equals sign (=), and the year and page of the subsequent description(s) given in square brackets. The same style is used in cases where the name was available from the plate and the description appeared later. However, dual descriptions that appeared in the *Bulletins* of the French Entomological Society are ignored and only the oldest one is cited, for more details see the last paragraph in the Bibliographic Information chapter.

The following abbreviations, all given in square brackets following the page of publication, or the original combination when applicable, are used to provide taxonomic information:

DA – doubtful assignment to genus-level taxa
HN – junior homonym
IN – infrasubspecific name
JE – justified emendation
NO – nomen oblitum
NP – nomen protectum
QS – questionable status (denotes names, including infrasubspecific, of uncertain position, e.g. doubtful synonyms or subspecies status)
RN – replacement name
UE – unjustified emendation
UN – unavailable name

All names that are deemed infrasubspecific based on the content of the work i.e., the same publication contained taxa given the status of subspecies as well as varieties, morphs etc., are infrasubspecific entities by definition (ICZN 1999, Art 45.6.4) and each infrasubspecific taxon is commented on in New Acts to avoid any further confusion or misapplication. However, a special case is Julius Weise's monumental book on the Chrysomelidae of Germany, published in six consecutive parts (Lieferungen) within the series *Naturgeschichte der Insecten Deutschlands* (Weise 1881, 1882b, 1884d, 1886a, 1888b, 1893a). There are many varieties established in this book, however, in the third part (Lieferung 3) there is also a subspecies concept applied to the genus *Oreina* (see pp. 444, 447, 451, 462, 466, 476, 481) thus rendering all varieties as unavailable infrasubspecic taxa. For practical reasons and in the interests of stability, we decided to consider each of these parts as an independent publication for nomenclatural purposes and thus only the varieties published in the third part (pp. 369–568, Weise 1884d) are infrasubspecic, while those that appeared in other parts are available names. We believe that this better serves the stability of nomenclature, particularly considering that some of these variety names are currently used as valid species.

The New Acts chapter contains all the novel information with taxonomic and nomenclatural relevance presented in the Catalogue as well as additional comments. Each comment is numbered and cross-referenced with the respective place in the main text of the Catalogue. The New Acts chapter contains: all taxonomic and nomenclatural changes (new taxa, new substitute names, new synonymies, changes in rank, new combinations, *nomina protecta*, *nomina oblita*, type species and lectotype designations, corrections of authorship and spellings), information on infrasubspecific taxa, comments on distribution including new geographic records, and additional general comments when necessary.

The following abbreviations of institutions are used in the section New Nomenclatural and Taxonomic Acts, Comments and Distribution:

IPAE – Institute of Plant and Animal Ecology, Yekaterinburg, Russia
IZAY – Academy of Science of Armenia, Institute of Zoology, Erevan, Armenia
IZCAS – Chinese Academy of Sciences, Institute of Zoology, Beijing, China
JMCB – Jihočeské muzeum v Českých Budějovicích, České Budějovice, Czech Republic
HNHM – Hungarian Natural History Museum, Budapest, Hungary
MHNG – Muséum d'Histoire Naturelle, Geneva, Switzerland
MLUH – Martin-Luther-Universität, Zentralmagazin Naturwissenschaftlicher Sammlungen, Halle (Saale), Germany
MNHN – Muséum National d'Histoire Naturelle, Paris, France
MRSN – Museo Regionale di Scienze Naturali, Torino, Italy

MTD – Museum für Tierkunde, Dresden, Germany
MZH – Finnish Museum of Natural History, Helsinki, Finland
NHMB – Naturhistorisches Museum, Basel, Switzerland
NHMUK – Natural History Museum, London, United Kingdom
NHRS – Naturhistoriska Riksmuseet, Stockholm, Sweden
NMEG – Naturkundesmuseum, Erfurt, Germany
NMPC – National Museum, Praha, Czech Republic
NMSO – Naturmuseum Solothurn, Solothurn, Switzerland
NMW – Naturhistorisches Museum Wien, Wien, Austria
RMCA – Musée Royal de l'Afrique Centrale, Tervuren, Belgium
SDEI – Senckenberg Deutsches Entomologisches Institut, Müncheberg, Germany
SMF – Forschungsinstitut und Naturmuseum Senckenberg, Frankfurt am Main, Germany
SMNS – Staatliches Museum für Naturkunde, Stuttgart, Germany
TARI – Taiwan Agricultural Research Institute, Taichung City, Taiwan
TLMF – Tiroler Landesmuseum Ferdinandeum, Hall in Tirol, Austria
UUZM – Uppsala University, Museum of Evolution, Uppsala, Sweden
ZFMK – Zoologisches Forschungsmuseum "Alexander Koenig", Bonn, Germany
ZIN – Zoological Institute, Russian Academy of Sciences, St. Petersburg, Russia
ZMHUB – Museum für Naturkunde, Leibniz-Institut für Evolutions- und Biodiversitätsforschung, Berlin, Germany
ZMSZ – Zemaljski muzej Bosne i Hercegovine, Sarajevo, Bosnia Hercegovina
ZMUH – Museum der Natur Hamburg, Germany
ZMUK – Universität Kiel, Zoologisches Museum, Kiel, Germany
ZMUM – Moscow State University, Moscow, Russia
ZSM – Zoologische Staatssammlung, München, Germany

Distributional Information

The limits of the Palaearctic Region, like those of other biogeographical regions, are arbitrarily defined (Map 1). For practical reasons, the boundaries of the Palaearctic Region usually follow national boundaries. The region includes Europe, Africa north of the Sahara, and Asia except for the part that is arbitrarily defined as belonging to the Oriental Region.

For each species and subspecies, an outline of its present distribution is given using symbols. The Palaearctic Region is further subdivided into three main parts, Europe (letter **E**), North Africa (letter **N**) and Asia (letter **A**).

Europe includes the Azores, Iceland and Turkey west of the Bosporus. The eastern boundaries are a matter of controversy. In the present Catalogue, Europe includes Russia west of the main range of the Ural Mountains (Permsky Krai, Bashkortostan Republic and Orenburg Oblast entirely, in Sverdlovsk Oblast and Chelyabinsk Oblast only the parts west of 60°E) and a small part of Kazakhstan west of the Ural River. The south-eastern boundaries are the northern political boundaries of Georgia, Azerbaijan and the Caspian and Black seas. The limits of Ukraine reflect the position in 2021; Luhansk Oblast, Donetsk Oblast, Zaporizhzhia Oblast and Kherson Oblast are parts of Ukraine. Crimea is listed separately, mainly due to its unique fauna. Abkhazia is for practical reasons given under Georgia.

In the previous Edition the abbreviation YU designated Serbia and Montenegro (at that time constituting the Federal Republic of Yugoslavia) only, however, many authors adopted it for records from entire former Yugoslavia. During the preparation phase of the Catalogue Serbia and Montenegro were split, what was not reflected in the published version. Subsequently also Kosovo was declared as independent therefore in the present edition each of the countries of former Yugoslavia has its own abbreviation to avoid any further confusions and misapplications.

North Africa includes Morocco (incl. Western Sahara), Algeria, Tunisia, Libya, Egypt west of the Suez Canal, and the Canary, Madeira and Salvage islands.

Asia includes Sinai and the Arabian Peninsula (including Socotra), Turkey east of the Bosporus, the Caucasian republics (Georgia, Armenia and Azerbaijan), the Middle East and Central Asian countries, Russia east of the main range of the Ural Mountains, Korea, Japan (including the Ryukyu Archipelago and the Japanese Pacific Islands), the entire People's Republic of China, Taiwan, Bhutan, Nepal, North India along the base of the Himalayas (Arunachal Pradesh, Uttarakhand (north-western area of former Uttar

Pradesh), Himachal Pradesh, Jammu and Kashmir) and all of Pakistan. Thus, India is the only state for which strict political boundaries were not followed.

Two-letter symbols provided for countries are arranged in alphabetical order within each subdivision. The two-letter symbols for major areas of Russia, North Indian states, and three-letter symbols for provinces of mainland China and for Taiwan (see Map 3) are placed in brackets after the respective country code. Russia is subdivided into seven major sub-regions (Map 2). In previous editions of the Palaearctic Catalogue Siberia was divided into West Siberia and East Siberia only. Now, Siberia is divided into three units: East Siberia, South Siberia and West Siberia, which aim to better describe its zoogeographical diversity. South Siberia includes the entire republics of Altai, Tuva and Khakassia, Kemerovo Region without the part north of Kemerovo, the extreme south of Altai Province, the southern part of Krasnoyarsk Province to the latitude of Krasnoyarsk, the South-West of Irkutsk Region and the western part of the Republic of Buryatia to Selenga River in the east. The symbols RU for Russia and CH for the People's Republic of China, without specification in parentheses, are used only in cases where precise information was not available.

Manchuria (also Mandzhuria or Manchukuo) was a geographical region in Northeast Asia often referred to by various authors, in particular by G. G. Jacobson, until 1949. It is presently divided between China (Heilongjiang, Jilin and Liaoning and eastern part of Inner Mongolia) and Russia (south-eastern part of Russian Far East including Sakhalin); e.g. https://en.wikipedia.org/wiki/Manchuria. In many recent publications, old Manchurian records from the Chinese part were automatically considered to be from Heilongjiang (compare, for example, the record of *Parnops atriceps* Pic, Jacobson (1910b) and its interpretation by Li et al. (2008) through the citation in Tan et al. (2005)). However, at least since 1905, Jacobson also had material from Jilin and Liaoning, and many of his Manchurian records thus belong to these provinces, which can be verified by checking locality labels under respective specimens. The following localities are situated in Jilin (collector Serikov): "Цайцзягоу, ю. Харбина" (Tsaitzyagou, S of Harbin) [= Caijiagou, 45°06′55″N, 126°08′29″E]; and Liaoning (collector Bortkevich, mostly in surroundings of "Мамайкая" (Mamaikaya)) [= Mamajecun, 41°25′36″N 122°51′25″E]): "Чендятунь, з. Мамайкая, Мугден" (Chendyatun', W of Mamaikaya, Mukden) [= Zhengjiatun, 41°28′38″N 122°45′10″E], "Худяпуза, з. Мамайкая, Мугден" (Hudyapuza, W of Mamaikaya, Mukden) [= Houjia Gangzi, 41°24′09″N, 122°47′34″E], "Сышагоу, ю.з. Мамайкая, Мугден" (Syshagou, SW of Mamaikaya, Mukden) [approximately 41°21′N 122°46′E], "Лангаши" (Langashi) [this locality was not found but is certainly situated in the same area]. We thank Alexey Moseyko for providing this essential information about Manchuria.

Some, usually older, distributional records cannot be accommodated within the structure of the Catalogue (e.g., Arabia, Caucasus, North India, Siberia). Such information is given in quotation marks (e.g., "Caucasus") after the last symbol of the subdivision. Distributional records of species-group taxa with isolated occurrence, either endemic to very small geographic area (e.g. islands, mountains etc.) or widely distributed in one area but restricted in another area, may be specified in parentheses.

The extralimital distribution of Palaearctic species is indicated by three letter symbols in boldface located at the end of the respective geographical information. The extralimital regions for the purposes of the Catalogue are defined as follows (see Map 1): Nearctic (**NAR**): north of Mexico; Neotropical (**NTR**): south of the United States; Afrotropical (**AFR**): south of the North African states included in the Palaearctic Region; Oriental (**ORR**): areas south of the People's Republic of China and Taiwan, areas south of Himalayan India, the Philippines, Malaysia and Indonesia west of the Lydekker line; Australian (**AUS**): south and east of the Lydekker line, and Pacific (**PAC**). Introductions are indicated by the letter "i" (e.g., Ei: GB).

In general, the presented distributional information is based on published records scattered in an enormous number of publications. Therefore, the second-level geographic information is not necessarily exhaustive, it should rather be considered as a baseline for future faunal research. The authors of some groups in the current catalogue were unable to find published sources or voucher specimens for some countries or areas listed in the previous Edition. These distribution data are included in the present catalogue but also listed in the Comments. In addition, to present distributional information as accurate as possible, new faunistic records are presented in this Edition, in New Acts chapter separately for each subfamily or tribe.

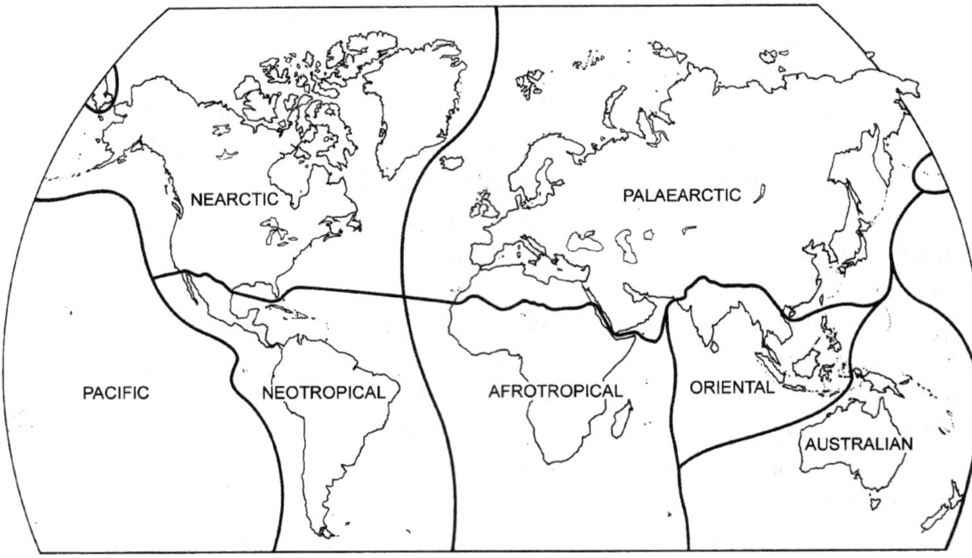

MAP 1 The limits of the geographic Regions as defined for the purpose of this Catalogue

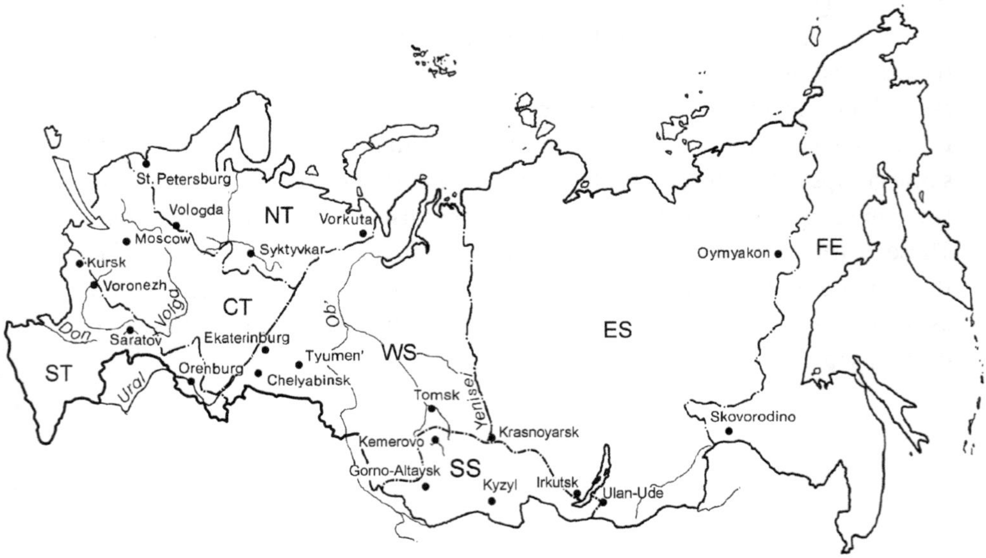

MAP 2 Subdivisions of Russia

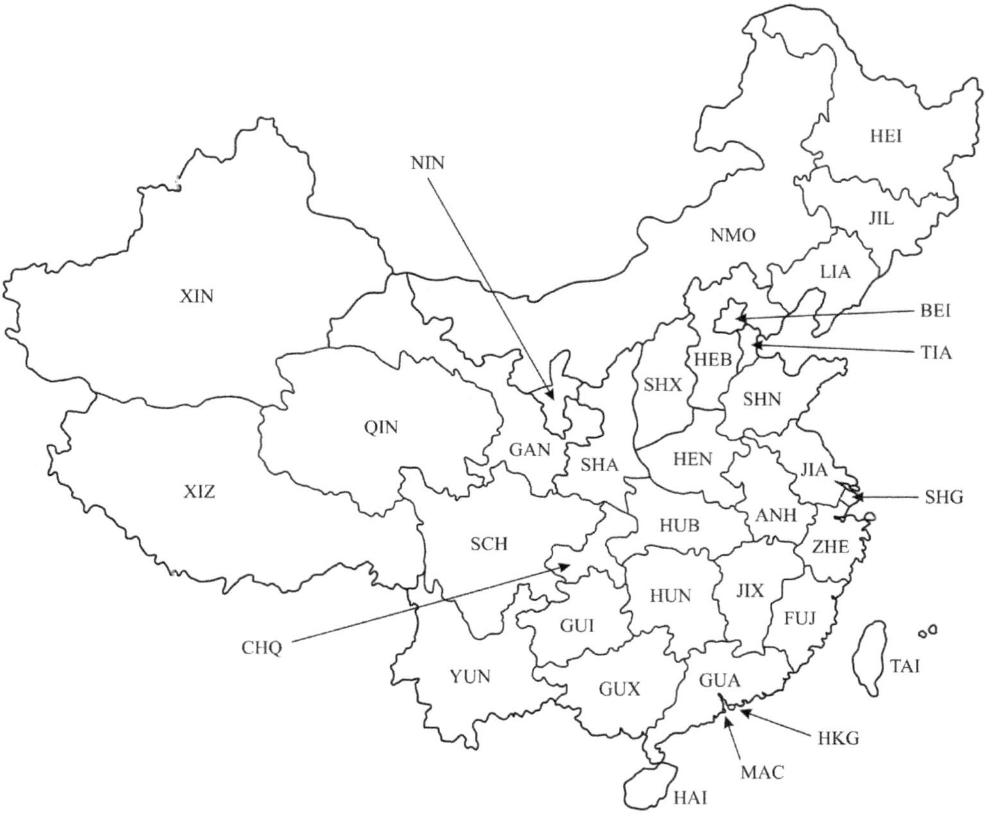

MAP 3 Subdivisions and provinces of the People's Republic of China, and Taiwan

Geographical Symbols

E	**Europe**		
AL	Albania	KR	Crimea
AN	Andorra	KZ	Kazakhstan
AU	Austria	LA	Latvia
AZ	Azores	LS	Liechtenstein
BE	Belgium	LT	Lithuania
BH	Bosnia Herzegovina	LU	Luxembourg
BU	Bulgaria	MA	Malta
BY	Belarus	MD	Moldavia
CR	Croatia	ME	Montenegro
CZ	Czech Republic	NL	The Netherlands
DE	Denmark	NM	North Macedonia
EN	Estonia	NR	Norway
FA	Faeroe Islands	PL	Poland
FI	Finland	PT	Portugal
FR	France (incl. Corsica, Monaco)	RO	Romania
GB	Great Britain (incl. Channel Islands)	RU	Russia
		SB	Serbia
GE	Germany	SK	Slovakia
GR	Greece (incl. Crete)	SL	Slovenia
HU	Hungary	SP	Spain (incl. Gibraltar)
IC	Iceland	SV	Sweden
IR	Ireland	SZ	Switzerland
IT	Italy (incl. Sardinia, Sicily, San Marino)	TR	Turkey
		UK	Ukraine
KO	Kosovo		

N	**North Africa**		
AG	Algeria	MO	Morocco (incl. Western Sahara)
CI	Canary Islands	MR	Madeira Archipelago
EG	Egypt	SG	Salvage Islands
LB	Libya	TU	Tunisia

A	**Asia**		
AB	Azerbaijan	BT	Bhutan
AE	United Arab Emirates	CH	China
AF	Afghanistan	CY	Cyprus
AR	Armenia	GG	Georgia
BA	Bahrain	ID	India

IN	Iran	PA	Pakistan
IQ	Iraq	QA	Qatar
IS	Israel (incl. Palestine)	RU	Russia
JA	Japan	SA	Saudi Arabia
JO	Jordan	SC	South Korea
KI	Kyrgyzstan	SI	Egypt: Sinai
KU	Kuwait	SY	Syria
KZ	Kazakhstan	TD	Tajikistan
LE	Lebanon	TM	Turkmenistan
MG	Mongolia	TR	Turkey
NC	North Korea	UZ	Uzbekistan
NP	Nepal	YE	Yemen (incl. Socotra)
OM	Oman		

Russia: European part

CT	Central European Territory
NT	North European Territory
ST	South European Territory

Russia: Asian part

ES	East Siberia
FE	Far East
SS	South Siberia
WS	West Siberia

China: Provinces, Autonomous Regions or Municipalities, and Taiwan

ANH	Anhui (Anhwei)	JIX	Jiangxi (Kiangsi)
BEI	Beijing (Peking or Peiping)	LIA	Liaoning
CHQ	Chongqing	MAC	Macao
FUJ	Fujian (Fukien)	NIN	Ningxia (Ningsia)
GAN	Gansu (Kansu)	NMO	Nei Mongol (Inner Mongolia)
GUA	Guandong (Kwantung)	QIN	Qinghai (Tsinghai)
GUI	Guizhou (Kweichow)	SCH	Sichuan (Szechwan)
GUX	Guangxi (Kwangsi)	SHA	Shaanxi (Shensi)
HAI	Hainan	SHG	Shanghai
HEB	Hebei (Hopeh)	SHN	Shandong (Shantung)
HEI	Heilongjiang (Heilungkiang)	SHX	Shanxi (Shansi)
HEN	Henan (Honan)	TAI	Taiwan (Formosa)
HKG	Hong Kong	TIA	Tianjin (Tsiensin)
HUB	Hubei (Hupeh)	XIN	Xinjiang (Sinkiang)
HUN	Hunan	XIZ	Xizang (Tibet)
JIA	Jiangsu (Kiangsu)	YUN	Yunnan
JIL	Jilin (Kirin)	ZHE	Zhejiang (Chekiang)

India: Provinces

AP	Arunachal Pradesh
HP	Himachal Pradesh
KA	Jammu and Kashmir, Ladakh
SD	Sikkim, Darjeeling District
UT	Uttarakhand (Uttaranchal, former part of Uttar Pradesh)

World Zoogeographical Regions

AFR	Afrotropical Region
AUS	Australian Region
NAR	Nearctic Region
NTR	Neotropical Region
ORR	Oriental Region
PAC	Pacific Region

Bibliographic Information

The total number of references in the present volume is 5020. They are listed in alphabetic order by the author and then chronologically within each author. The references cover all primary descriptions of the genus-group and species-group names listed in the Catalogue, including unavailable and infra-subspecific names and works supporting the New Acts and other statements published in the present volume. Bibliographical references are not provided for the family-group names, designation of type-species, and any secondary nomenclatural and taxonomic acts.

Names of all authors are given for each reference. Names published in two or more alternative spellings that seem to be equally correct, are spelled according to Schmitt et al. (1998). In such cases the alternative spelling follows the original one in brackets. The original name is used for authors who subsequently adopted additional names, or honorary epithets (e.g., A. Semenov, and not A. Semenov-Tian-Shanskiy). Laporte de Castelnau is referred to as Laporte in recent major works and this is used in the present Catalogue. Jan Bechyně (1920–1973), after his escape from former Czechoslovakia to Germany in 1948, changed his name to Bechyné, which he used consistently for the rest of his life. Because the variant Bechyné significantly predominates in the Chrysomelidae literature, we also use it in the current Catalogue. The name of Georgiy Georgievich Jacobson appears through the literature in several variants. In his publications using Latin alphabet he is consistently spelled as "Jacobson" (except the years 1892–1894 when his name was written as "Jacobsohn"). The Cyrillic alternative is the phonetic transcription always written as "Jakobson", which was used in the previous Edition. In the present Edition we use the spelling Jacobson, which is also the most widely used one. Initials are given for all authors, even if they are missing in the original publication.

Titles of non periodical publications are given in full, i.e. not abbreviated form, with subtitles, as they originally appeared. Titles given in Cyrillic are transliterated and followed by either original translation in parentheses or editorial one in square brackets if no translation was available from the original publication. Titles of periodicals are given in italics in full, only the articles at the beginning of the title were left out. The names or numbers of series are listed as a part of the title. The number of each volume is given in bold and is followed by the respective issue number in parentheses when available. When the year of the volume differs from the actual date of publication, it is given in square brackets after the volume. If periodicals have no volume number, the year of the volume is given in boldface instead. Unnumbered pages are indicated in square brackets. All references are supplemented by the earliest and most accurate date of publication and the source. See the next chapter on Dating of publications for further details.

The *Bulletin de la Société Entomologique de France* was published from 1873 to 1894, in addition to the well-known quarterly version, also in a bi-mensuel version "*Bulletin Bimensuel*". The publication of the bi-mensuel version preceded the quarterly one and the paginations of the two versions are not identical (for details see Evenhuis 2002). All taxa and references in the Catalogue strictly follow the bi-mensuel version, with the exception of Fairmaire (1878), which was published only in the quarterly version. Also the title of *Bulletin de la Société Entomologique de France* changed almost continuously for more than 60 years, appearing usually as *Bulletin des séances*, *Bulletin trimestriel* or *Bulletin entomologique*. In addition, the first numbered volume, 37, was published 100 years after the creation of the *Annales* and the *Bulletin* of the French Entomological Society in 1832. To avoid confusion, this periodical is referred to as *Bulletin de la Société Entomologique de France*, and, from 1832 to 1931, is identified by the year in which it was published. From the year 1932 onwards the volume number is given.

Dating of Publications

One of the fundamental principles of nomenclature, the Principle of priority, is based on the actual date when a publication with a given description appeared, which does not necessarily agree with the date printed in the publication itself. For this reason we decided to verify date of publication for all references and give the earliest date we could find. There have been a few complex treatments on the dating of the books, and some focused on selected journals. Therefore our major objective was to complete the available information and supplement it with new data where it was not summarized. Although we tried to provide information as accurate as possible, this should not be considered as a final version but merely a starting point for others as more research, mainly in libraries, could have been done. However, such research would take much longer than the production of the catalogue itself, so we preferred to close these studies at the present point. Moreover, the global pandemic in 2020–2021 prevented us from visiting large institutional libraries that would have certainly been beneficial for clarifying publishing dates.

Each reference is followed by the date of publication (DP), which is the earliest date we were able to find to prove the existence of the work; but again as mentioned above it should not be considered definitive. Ideally the date is given complete (i.e. day, month and a year), but in many cases only year or year and month were available, so the last day of the respective month or year is placed in square brackets in accordance with the Code (ICZN 1999). Societies and bibliographic journals, journal websites, as well as additional works reporting on dates of publication consulted for dating are listed at the end of this chapter. The source of the date of publication for each reference is given in parentheses following the actual date. They are mostly self-explanatory but we would like to clarify some terms we are using: Cover – stands for all kind of covers and wrappers, permanent or those that tended to be removed when the final volume was bound. Colophon – stands either for bibliographic summary given in the books (typical verso of one of the first pages or at the end) or for that given in journals. Preprint – a work published, with its own specified date of publication (imprint date), in advance of its later reissue as part of a collective or cumulative work. Reprint – a copy of a work contained in a periodical, book or other larger work, intended for distribution detached from the larger work which contains it but without its own specified date of publication (imprint date).

Important proofs of existence provided by the entry stamps or additional important information we were able to trace in the libraries of Natural History Museum, London, UK (NHMUK), Institute of Vertebrate Biology, Czech Academy of Sciences, Brno, Czech Republic (IVB CAS), National Museum, Praha, Czech Republic (NMPC) and Zoological Institute of Russian Academy of Sciences (ZIN). Additional excellent sources of information are freely accessible electronic libraries like *Biodiversity Heritage Library* (BHL) or *Gallica*. In some Russian journals we used the registration date in Vsesoyuznaya Knizhnaya Palata (VKP, All-Union Book House) where all editions in the former USSR were registered. In addition to the date of publication, many Russian publications are accompanied also by the date when they were signed for print (SFP). It is the date when the corrected last version of proofs was ready and approved for printing. The printing started after this date, however, sometimes with a considerable delay, so that the actual date of publication is from about two weeks to several months (exceptionally, a year or more) later than the date when the issue is signed for print. The publications provided with both SFP and DP were issued in the interval between the two dates, but only DP should be understood as a proven date. In several issues of *Entomologicheskoe Obozrenie* the date when the editor-in-chief received the proof-print (now deposited in ZIN) from the printer is accepted as publication date. Moreover, many Russian publications had their date of publication in the Julian calendar, which was for the purpose of this Catalogue converted to Gregorian calendar. For several French publications we use *dépôt legal* (DL) as publication date, which is obligatory to be given for all publicly sold or distributed publications. It typically consists of month and year when a copy of the publication was deposited in the Bibliothèque nationale de France. Some further special cases in dating are discussed below.

For taxonomic papers that were published online first in the years 2000–2011 (Baviera 2007; D.-Y. Ge et al. 2008, 2010; Yus Ramos 2010b; S.-Q. Ge et al. 2011), the dates of publication are given from the printed versions, because the electronic versions available online did not meet the provisions of the Art. 8.6 of the Code (ICZN 1999).

Papers by Bezděk (2017), Biondi & D'Alessandro (2017a), De La Rosa et al. (2012), S.-Q. Ge et al. (2012), Gómez-Zurita & Cardoso (2021), Kolov (2017), Moseyko (2021), Nadein (2013c), Orlova-Bienkowskaya (2015), Özdikmen et al. (2020b), Vela & Bastazo (2012) were published online first, however these papers do not contain a link with registration in ZooBank and thus they do not meet the provisions of the amended Article 8.5 (ICZN 2012). Therefore, the dates from printed versions of respective journals were adopted as the publication dates.

Buysson's (1908a) paper "Une espèce nouvelle du genre Thyamis Steph." was published in *Miscellanea Entomologica* 15 [1907] (5/8): 1–3. On page 3 there is a mention of Martin Jacoby's death on 24 December 1907, so it is probable that the issue was published during 1908. Buysson (1908b) on page 20 explicitly mentioned "Juillet 1908". The copy in ZIN library bears the stamp 16 July 1908 (very probably a stamp from France, as it is not similar to any of the Russian stamps used in ZIN library at that time); we accept this as the date of publication.

Buysson's (1908b) paper "Recherches sur l'epoque d'éclosion des altises ..." (Buysson 1908b) was published as a supplement to *Miscellanea Entomologica* 15 [1907]. According to Le Moult & Bernardi (1943), this supplement was divided into two parts (pp. 1–16 and 17–39), which suggests that it was associated with two different issues of *Miscellanea Entomologica*. We were unable to trace which issue was associated with the first part of the supplement, but the second part was undoubtedly associated with issue 10/11 as stated on page 17. The issue 10/11 deposited in ZIN library bears the stamp 23 September 1908 (very probably a stamp from France, as it is not similar to any of the Russian stamps used in ZIN library at that time); we accept this as the date of publication of Buysson (1908b).

Papers by Daniel J. (1914) and Heikertinger (1915) were intended to appear in the 4th volume of the *Münchener Koleopterologische Zeitschrift*. This was ready for publication in 1914, authors received separata of their papers, but the volume itself was never published due to the outbreak of World War I and also due to the mental illness of the publisher Karl Daniel (see Heikertinger's introduction to Wingelmüller 1937). However, the separata are considered as published works based on Art. 21.8 of the Code (ICZN 1999). The year is not mentioned on the wrappers of the separata, however, the year 1914 is generally accepted, for example in Wingelmüller's revision of the tribe Cionini (Alonso-Zarazaga 2017) or in Bernhauer's works on Staphylinidae (Löbl & Löbl 2015). For J. Daniel's (1914) paper *Der Formenkreis der Phytodecta variabilis Oliv.* we accept 1914 as the year of publication in accordance with other works. The separata of Heikertinger's *Monographie der Koleopterengattung Derocrepis Weise* were printed in 1914, but the author received and distributed them at the beginning of 1915 (Heikertinger 1925b), so we accept January 1915 as the publication date. However, he was probably not sure if the work would be considered as published just by distribution of separata therefore he "reissued" it in an slightly awkward manner. In 1922 he printed a small piece of paper with the following printed text "Wien 1922 | Im Selbstverlage des Verfassers | Franz Heikertinger | Wien XII. Thunhofgasse 8 [vertical bar separates rows]" and glued it on the cover of the remaining stock of the original separata he received in 1915. As a result, two versions of the identical separata exist in libraries. An updated version of the entire revision was published in 1925 (Heikertinger 1925b; Strouhal 1955).

The dating of papers published in *Entomological Review of Japan* within the years 1960–1985 is based on sheets entitled "General index of Entomological review of Japan" irregularly inserted into issues of the journal.

Dating of *Fragments Entomologiques*: Julien Achard published nine issues of his journal *Fragments Entomologiques* during the period 1922–1923, and the last issue (No. 10) was published posthumously in 1926 as a supplement of *Annales et Bulletin de la Société Entomologique de Belgique*. The exact publication dates are written as footers on the first page of issues 2–6. The months of publication of issues 1–3 are also listed on the last page of the wrappers, obviously printed after the printing of issue 3 and subsequently used for each issue. The NMPC library has issues 1–9 as a bound volume. The dates of issue of volumes 7–9 are handwritten as footers on first issue page by Achard himself. Known publication dates for individual issues are summarized as follows: 1 (pp. 1–16, DP: [30] September 1922), 2 (pp. 17–32, DP: 30 October 1922), 3 (pp. 33–48, DP: 30 November 1922), 4 (pp. 49–64, DP: 15 March 1923), 5 (pp. 65–80, DP: 30 April 1923), 6 (pp. 81–96, DP: 31 May 1923), 7 (pp. 97–112, DP: 30 June 1923), 8 (pp. 113–128, DP: 15 August 1923), 9 (pp. 129–144, DP: 30 October 1923), 10 (pp. 145–156, DP: [31 December] 1926).

Gebler (1830) published the paper "Bemerkungen über die Insekten Sibiriens, vorzüglich des Altai" in the second part of Ledebour's *Reise durch das Altai-Gebirge und die soongorische Kirgisen-Steppe*. Schmitt (2017) pointed out that while most copies of the second part have the title page dated 1830, some have been dated as 1829. He went on to add that the most plausible explanation for the two dates is that a first small print run of the Zweiter Theil was produced in 1829, and a second, larger print run was made in 1830 with a change in title page but without stating in the work that it was a subsequent printing. This opinion was opposed by Bousquet (2017), who gave arguments supporting 1830 as the year of publication. In the present Catalogue we follow Bousquet (2017), therefore, Gebler's taxa are assigned to the year 1830.

Graells (1851): Volume 1 of *Memorias de la Real Academia de Ciencias de Madrid* contains three parts and Graells' paper is contained in the second part. There are no original covers available to us, only the title page of the volume with the year 1850. However, Graells' paper was written in 1851 (as stated on page 163). Since the year 1851 is listed consistently in the Chrysomelidae literature for the species described in this work, we accept [31 December] 1851 as the publication date.

Grenier's (1863) *Catalogue des coléoptères de France et matériaux pour servir à la faune des coléoptères français* contains two different release dates: 21 July 1863 on the title page and 6 August 1863 on the cover. A year later Grenier (1864) himself mentioned the publication of the catalogue in August 1863, therefore we accept 6 August 1863 as the publication date.

Hope (1841): Although the title page of the third part of Hope's *Coleopterist's manual* is dated 1840, we accept the publication date April 1841 given by Low (1853). See also the discussion in Bousquet (2016).

Illiger (1794) was published in *Neuestes Magazin für die Liebhaber der Entomologie* 1 (5): 593–620, which was included in the Leipzig fair catalogue *Allgemeines Verzeichniss der Bücher* in 1793 (*Ostermesse* p. 84). However, because the lists of books presented at the fairs were compiled before the event, some of the anticipated books were actually only issued at a later date (Bousquet 2017). Because the wrapper of issue 5 bears the year 1794, we accept [31 December] 1794 as the publication date.

Laboissière's (1932b) paper was published in volume 20 (for the years 1914–1931) of *Annales de l'Association des Naturalistes de Levallois-Perret* but the volume does not contain any specific date or year of publication. However, the volume contains a list of members as of 31 March 1932. The year 1932 is also mentioned in the Zoological Record (for the year 1933) and has been consistently accepted in the Chrysomelidae literature. The publication date [31 December] 1932 is accepted.

Kocher (1963): The publication of the journal *Bulletin de la Société des Sciences Naturelles et Physiques du Maroc* in the 1960's was delayed by one to two years. Kocher's paper was published in the third issue of the volume 42 [1962] but the date of publication was not stated. We can only deduce that the third issue was published between 5 November 1963 (issue 1/2 of volume 42 – see *Comptes Rendus de la Société des Sciences Naturelles du Maroc* 1963: 111) and 30 June 1964 (issue 1/2 of volume 43 – see last page of issue). Because Kocher's paper was listed in the Zoological Record for 1963 (albeit with the year 1962), for the present purposes we accept the date [31 December] 1963, but further research on the date of publication is necessary.

Konstantinov (1992b) and Lopatin (1992e) described several new taxa in the book *Nasekomye Kavkaza* ... edited by G. M. Abdurakhmanov. The year of publication was stated inconsistently in the literature. The book title page states 1991, but in the previous Edition of the catalogue the book was dated 1996 (evidently because it was registered in the ZIN library on 31st January 1996). However, one article that appeared in the book was cited by Belousov & Zamotaylov (1993), thus the book must have existed that year. Finally, Abdurakhmanov (2017) in his biography listed this book with the year 1992, which we accept.

Kraatz (1879a, b) published two papers in the first issue of the volume 23 of *Deutsche Entomologische Zeitschrift*. The title page of the volume and footer on p. 3 has May 1879 on it, while the wrapper of the issue 1 bears January 1879. We assume that January 1879 is correct because the issue was listed in the session from 26 March 1879 of the French Entomological Society (*Bull. Soc. Ent. Fr.* 1879: 77).

Papers published in *Munis Entomology & Zoology* are only provided with month of publication. However, according to the journal website and the general information on the wrappers, the journal is always published on the first day of January and June, which we follow here.

Pérez Arcas (1865a, 1865b) published two papers in *Revista de los Progresos de las Ciencias Exactas, Fisicas y Naturales*. The volume contains nine issues. The heading of the first page of each issue is provided with a month. These months cannot be considered as publication dates, as some works are dated later than the respective month, e.g. Pérez Arcas (1865a) was published in issue 3 announced as "Marzo de 1865" but the paper was finished in May 1865 (p. 186). The exact dates of publication of both of Pérez Arcas' papers remains unclear and we therefore consider them as published on [31 December] 1865, although they appeared in two different volumes.

J. R. Sahlberg (1903): Bouchard et al. (2021) dated this paper as 5th April 1903 on the basis of the minutes of the society meetings (*Öfversigt af Finska Vetenskaps-Societetens Förhandlingar* 1903, 45: 13, *Öfversigt af förhandlingarna*). However, the minutes state that Sahlberg wrote several works and submitted them for publication, thus the date cannot be accepted for the published paper. The exact date of publication is not known and we therefore retain it as [31 December] 1903.

Schneider & Leder's (1878) paper named "Beiträge zur Kenntniss der kaukasischen Käferfauna" was published in two parts in the journal *Verhandlungen des Naturforschenden Vereins in Brünn*: pp. 3–258 and pls 1–4 in volume 16 [1877] (published in 1878) and pp. 3–104 and pls 5–6 in volume 17 [1878] (published in 1879). The separatum includes both parts together with continuous pagination 3–359 and is dated 1878 (for details see Döberl 2009). The exact publication date of the separatum is not known, however, in Gustav Kraatz's review published in *Deutsche Entomologische Zeitschrift* 1879: 424, there is a note that he received a copy after New Year 1879. Therefore, it is highly probable that the separatum was published at the end of 1878, which we accept [31 December] as the date of publication.

Paper by Wiedemann & Germar (1821) was published in the volume 4 of *Magazin der Entomologie*, which was announced in the Leipzig fair catalogue

Allgemeines Verzeichniss der Bücher in spring 1821 (*Ostermesse* p. 155). As explained by Bousquet (2017), some of the anticipated books were actually issued only at a later date. The entry in Leipzig fair catalogue is marked with notice "genannt Sommer" what agrees with *Vorrede* in volume 4 dated 1st June 1821 which clearly proves that the volume was published after the spring trade fair in Leipzig. As we have no further information, the publication date [31 December] 1821 is adopted.

Societies and Recording Journals

Allg. Bibl. Deutsch. – *Allgemeine Bibliographie für Deutschland*

Allg. Verz. Bücher ... Ostermesse/Michaelismesse – *Allgemeines Verzeichniss der Bücher, welche in der Frankfurter und Leipziger Ostermesse/Michaelismesse*

Ann. Soc. Ent. Belg. – *Annales de la Société Entomologique de Belgique*

Ber. Bekanntmach. Verh. Akad. Wiss. Berlin – *Bericht über die zur Bekanntmachung geeigneten Verhandlungen der Königl. Preuss. Akademie der Wissenschaften zu Berlin*

Bibl. France – *Bibliographie de la France*

Bibl. Zool. – *Bibliographia Zoologica*

Bibl. Imp. Contr. Amer. Econ. Ent. – *Bibliography of the more important contributions to American economic entomology*

Bol. Soc. Ent. Arag. – *Boletín de la Sociedad Entomológica Aragonesa*

Bull. Class. Phys.-math. Acad. Imp. Sci. St.-Pétersbourg, Bull. Bibl. – *Bulletin de la Classe Physico-mathématique de l'Académie impériale des Sciences de St.-Pétersbourg, Bulletin Bibliographique*

Bull. Sci. Acad. Imp. Sci. Saint-Pétersbourg – *Bulletin Scientific de l'Académie Impériale des Sciences de Saint-Pétersbourg*

Bull. Soc. Ent. Fr. – *Bulletin de la Société Entomologique de France*

Bull. Soc. Zool. Fr. – *Bulletin de la Société Zoologique de France*

Compt. Rend. Hebdom. Séances Acad. Sci. – *Comptes Rendus Hebdomadaires des Séances de l'Académie des Sciences*

Deutsche Ent. Zeitschr. – *Deutsche Entomologische Zeitschrift*

Ent. Current Lit. – *Entomology Current Literature*

Ent. Litteraturbl. – *Entomologische Litteraturblätter*

Ent. Nachricht. – *Entomologische Nachrichten*

Ent. Zeit. – *Entomologische Zeitung* (Stettin)

Leipzig. Jahrb. Neuest. Lit. – *Leipziger Jahrbuch der Neuesten Literatur*

Lit. Centralbl. Deutschl. – *Literarisches Centralblatt für Deutschland*

Lit. Zeit. – *Literarische Zeitung*

Lotos – *Lotos: Zeitschrift für Naturwissenschaften*
Mitth. Geogr. Ges. – *Mittheilungen der kaiserlich-königlichen geographischen Gesellschaft* (Wien)
Monatsber. Königl. Preuss. Akad. Wiss. Berlin – *Monatsberichte der königlichen Preussische Akademie des Wissenschaften zu Berlin*
Natur. Novit. – *Naturae Novitates*
Öfv. Fin. Vetens.-Soc. Förhand. – *Öfversigt af Finska Vetenskaps-Societetens Förhandlingar*
Öfv. Kongl. Vetens.-Akad. Forhand. – *Öfversigt af Kongliga Vetenskaps-Akademiens Förhandlingar*
Pet. Nouv. Ent. – *Petites Nouvelles Entomologiques*
Proc. Ent. Soc. London – *Proceedings of the Entomological Society of London*
Sitzungsber. Ges. Naturforsch. Freunde Berlin – *Sitzungsberichte der Gesellschaft Naturforschender Freunde zu Berlin*
Trans. Ent. Soc. London (Proc.) – *Transactions of the Entomological Society of London (Proceedings)*
Verh. Zool.-Bot. Ges. Wien – *Verhandlungen der kaiserlich-königlichen zoologisch-botanischen Gesellschaft in Wien*
Zool. Anz. – *Zoologischer Anzeiger*
Ztschr. Ges. Naturwiss. – *Zeitschrift für die gesammten Naturwissenschaften*

Journal Websites used for Dating

Annals of the Entomological Society of America – https://academic.oup.com/aesa/
Atti della Reale Accademia delle Scienze di Torino – http://emeroteca.braidense.it/beic_attacc/scheda_testata.php?IDTestata=911&CodScheda=000R
Biodiversity and Conservation – https://www.springer.com/journal/10531
Caucasian Entomological Bulletin – http://www.ssc-ras.ru/en/journal/caucasian_entomological_bulletin/
Cladistics – https://onlinelibrary.wiley.com/journal/10960031
Elytra, New series – http://kochugakkai.sakura.ne.jp/publication/elytra/elytra-bibliography.html
Entomologia Sinica – https://onlinelibrary.wiley.com/journal/17447917
Entomological Review – https://www.springer.com/journal/11474
Euroasian Entomological Journal – http://kmkjournals.com/journals/EEJ/EEJ_Publ_Dates
Fresenius Environmental Bulletin – https://www.prt-parlar.de/
Insect Molecular Biology – https://onlinelibrary.wiley.com/journal/13652583

Insect Systematics & Evolution – https://brill.com/view/journals/ise/ise-overview.xml
Italian Journal of Zoology – https://www.tandfonline.com/journals/tizo20
Journal of Asia-Pacific Entomology – https://www.sciencedirect.com/journal/journal-of-asia-pacific-entomology/
Journal of Natural History – https://www.tandfonline.com/journals/tnah20
Journal of the New York Entomological Society – https://bioone.org/journals/entomologica-americana
Journal of Zoological Systematics and Evolutionary Research – https://onlinelibrary.wiley.com/journal/14390469
Memoirs of the Entomological Society of Canada – http://www.esc-sec.ca/pdf/ESC-MemoirList.pdf
Microscopy Research & Technique – https://analyticalsciencejournals.onlinelibrary.wiley.com/journal/10970029
Molecular Phylogenetics and Evolution – https://www.journals.elsevier.com/molecular-phylogenetics-and-evolution
Natural History Sciences, Atti della Società Italiana di Scienze Naturali e del Museo Civile di Storia Naturale di Milano – https://sisn.pagepress.org/index.php/nhs/
Pacific Insects – http://hbs.bishopmuseum.org/pi/
Pacific Insects Monograph – http://hbs.bishopmuseum.org/pim/
Polish Journal of Entomology – https://pje-journal.com/resources/html/cms/MAINPAGE
Russian Entomological Journal – https://kmkjournals.com/journals/REJ/REJ_Exact_Publ_Dates
Systematic Entomology – https://resjournals.onlinelibrary.wiley.com/journal/13653113
Terrestrial Arthropod Reviews – https://brill.com/view/journals/tar/tar-overview.xml
Zoological Research – https://www.zoores.ac.cn/
Zoological Science – https://bioone.org/journals/zoological-science
Zoological Systematics – http://www.zootax.com.cn/EN/volumn/home.shtml
Zoology in the Middle East – https://www.tandfonline.com/journals/tzme20
Zoosystematica Rossica – https://www.zin.ru/journals/zsr/
Bieńkowski (2011) – https://www.lap-publishing.com/catalog/
Gómez-Zurita (2008) and Kergoat et al. (2008) – https://brill.com/view/title/13027
Li X.-P. (2013) – https://www.hceis.com/
Moseyko (2013b) – https://brill.com/view/title/24151
Warchałowski (2010b) – https://www.miiz.waw.pl/pl/wydawnictwa/spis-artykulow/37-wydawnictwa/290-the-palaearctic-chrysomelidae-identification-keys
Zampetti & Ricci (2012) – https://pandion-edizioni.blogspot.com/

Works Reporting Dates of Publication

Abdurakhmanov, G. M. 2017: *"Vremya sobirat' kamni ..."* (*Bibliografiya nauchnykh trudov i shtrikhi k portretu*). [*"Time to collect the stones ..."* (*Bibliography of scientific works and strokes to the portrait*)]. Makhachkala: Eko-Press, 304 pp.

Alonso-Zarazaga, M. A. 2001: Antonio Cobos Sánchez (1922–1998). *Graellsia* 57 (2): 191–210.

Alonso-Zarazaga, M. A. & Evenhuis, N. L. 2017: Revised dating of Panzer's Faunae insectorum Germanicae initia. Deutschlands Insecten (1792–1810), based primarily on the Leipzig Book Fair Catalogs. *Sherbornia* 4 (1): 1–8.

Asanovich, T. A., Grozdilova, L. P. & Kerzhner, I. M. 2007: Publications of the Zoological Institute, St.Petersburg. 1. "Ezhegodnik" and "Trudy". *Zoosystematica Rossica* 15 [2006] (2): 195–213.

Barendrecht, G. 1957: A propos du centenaire du Tijdschrift voor Entomologie. *Tijdschrift voor Entomologie* 100 (1): 1–4.

Belousov, I. A. & Zamotaylov, A. S. 1993: On new carabids of the genus Carabus L. (Coleoptera, Carabidae) from the Caucasus. 4th contribution. *Entomologica Basiliensia* **16**: 37–57.

Bezděk, J. 2020: Review of the genus-level names proposed by Johannes Gistel in Chrysomelidae (Coleoptera). *Acta Entomologica Musei Nationalis Pragae* 60 (1): 173–188.

Bouchard, P., Bousquet, Y., Davies, A. E., Alonso-Zarazaga, M. A., Lawrence, J. F., Lyal, C. H. C., Newton, A. F., Reid, C. A. M., Schmitt, M., Ślipiński, S. A. & Smith, A. B. T. 2011: Family-group names in Coleoptera (Insecta). *ZooKeys* **88**: 1–972.

Bousquet, Y. 2012: Catalogue of Geadephaga (Coleoptera, Adephaga) of America, north of Mexico. *ZooKeys* **245**: 1–1722.

Bousquet, Y. 2016: Litteratura Coleopterologica (1758–1900): a guide to selected books related to the taxonomy of Coleoptera with publication dates and notes. *ZooKeys* **583**: 1–776.

Bousquet, Y. 2017: Book catalogues issued for the Leipzig trade fairs and the dating of taxonomic works. *Zoological Bibliography* 4 (5) : 93–126.

Bousquet, Y. 2018: The dating of the fourth volume of Guillaume-Antoine Olivier's "Entomologie, ou histoire naturelle des insectes". *ZooKeys* **734**: 137–148.

Bousquet, Y. & Davies, A. E. 2021: On the dating of the publications (1805–1885) of the Société Impériale des Naturalistes de Moscou. *Sherbornia* 7 (2): 5–20.

Brown, F. M. 1964: The dates of publication of the first ten volumes of the Transactions of the American Entomological Society. *Transactions of the American Entomological Society* 90 (3): 313–321.

Chûjô, M. 1972: A list of entomological publications (1930–1972). *Bulletin of the Biological Society of Kagawa* 5: 1–18.

Dickinson, E. C. 2005: Some bibliographical findings on Notes from the Leyden Museum. *Zoologische Mededeelingen* 79 (2): 1–7.

Dickinson, E. C. & Daneliya, M. E. 2011: A re-examination of the sixth series of the "Mémoires de l'Académie Impériale des Sciences de Saint-Pétersbourg". *Russian Journal of Ornithology* 20 (630): 223–229.

Dickinson, E. C., Overstreet, L. K., Dowsett, R. J. & Bruce, M. D. (eds.) 2011: *Priority! The dating of scientific names in ornithology. A directory to the literature and its reviewers.* Northampton: Aves Press, 319 pp. + CD-ROM.

Döberl, M. 2009: Nomenclatorial notes on Palaearctic Coleoptera (Curculionidae, Cerambycidae, Chrysomelidae). *Entomologische Blätter für Biologie und Systematik der Käfer* 105: 19–23.

Duncan, F. M. (1937): On the dates of publication of the Society's 'Proceedings,' 1859–1926. With an Appendix containing the dates of publication of 'Proceedings,' 1830–1858, compiled by the late F. H. Waterhouse, and of the 'Transactions,' 1833–1869, by the late Henry Peavot, originally published in P. Z. S. 1893, 1913. *Proceedings of the Zoological Society of London* 1937: 71–83.

Evenhuis, N. L. 1994: The publication and dating of P. A. Wytsman's Genera Insectorum. *Archives of Natural History* 21 (1): 49–66.

Evenhuis, N. L. 2002: Publication and dating of the two "Bulletins" of the Société Entomologique de France (1873–1894). *Zootaxa* 70: 1–32.

Evenhuis, N. L. 2003: Publication and dating of the journals forming the Annals and Magazine of Natural History and the Journal of Natural History. *Zootaxa* 385: 1–68.

Evenhuis, N. L. 2011: Publishing and dating of the Bulletin of the Brooklyn Entomological Society (1878–1966). *Zoological Bibliography* 1 (3): 105–115.

Evenhuis, N. L. 2012: Publication and dating of the Exploration Scientifique de l'Algérie: Histoire Naturelle des Animaux Articulés (1846–1849) by Pierre Hippolyte Lucas. *Zootaxa* 3448: 1–61.

Evenhuis, N. L. 2014: Dates of the Leipzig book fairs (1758–1860), with notes on the book catalogs. *Sherbornia* 1 (1) : 1–4 + 1 table.

Evenhuis, N. L. 2015: Dating of the English-language Lingnan University Science Journals (1922–1952). *Sherbornia* 2 (1): 1–7.

Evenhuis, N. L. 2019a: Revised dates of publication for the Zoologie portion of Webb & Berthelot's Histoire Naturelle des Îles Canaries (1836–1844). *Sherbornia* 5: 157–165.

Evenhuis, N. L. 2019b: Charles D. V. d'Orbigny's "Dictionnaire Universel d'Histoire Naturelle": proposed dates for livraisons, with itemized contents of each. *Bishop Museum Occasional Papers* 130: 1–8.

Evenhuis, N. L. 2022: Natural History Journal Dating. Available at: http://hbs.bishop museum.org/dating/sherbornia/journals.html.

Evenhuis, N. L. 2023: Dating of the Verhandlungen der Zoologisch-Botanischen Gesellschaft in Wien (1851–1900). *Sherbornia* 8 (3): 99–107.

Evenhuis, N. L., Pape, T. & Pont, A. C. 2008. The problems of subsequent typification in genus-group names and use of the Zoological Record: a study of selected post-1930 Diptera genus-group names without type species designations. *Zootaxa* **1912**: 1–44.

Gaedike, R. & Behne, L. 2008: Die Zeitschrift "Le Frelon", – Datierung und Inhalt. *Entomologische Blätter* **103/104**: 87–122.

Grenier, A. 1864: [Communications]. *Bulletin de la Société Entomologique de France* **1864** (1): xii–xiii.

Griffin, F. J. 1932: A note on an almost unknown work on Coleoptera. *Entomologist's Monthly Magazine (Third series)* **18**: 65–66.

Griffin, F. J. 1937: The "Archiv der Insectengeschichte" of J. C. Fuessly. Heft 1–8, 1781–1786. *Journal of the Society for the Bibliography of Natural History* **1** (3): 83–85.

Heyden, L. F. J. D. von 1896: *Catalog der Coleopteren von Siberien, mit Einschluss derjenige des östlichen Caspi-Gebietes, von Turcmenien, Turkestan, Nord-Thibet und des Amur-Gebietes. Nachtrag II*. [1896–1898]. Berlin: Deutsche entomologische Gesellschaft, 84 pp.

Kappel, A. W. 1896: *General index to the first twenty volumes of the Journal (Zoology), and the zoological portion of the Proceedings, November 1838 to 1890, of the Linnean Society*. London: Longmans, Green and Co., and Williams and Norgate, viii + 437 pp.

Kerzhner, I. M. 1984: Daty publikacii izdaniya "Trudy Russkogo Entomologicheskogo Obshchestva" i "Horae Societatis Entomologicae Rossicae", 1861–1932. (Dates of publication of "Trudy Russkogo Entomologicheskogo Obshchestva" and "Horae Societatis Entomologicae Rossicae", 1861–1932). *Entomologicheskoe Obozrenie* **63** (4): 849–857.

Kerzhner, I. M. 1994: "Nasekomye Mongolii": dates of publication. *Zoosystematica Rossica* **2** (2) (1993): 246.

Le Moult, E. & Bernardi, G. 1943: Tableau de parution de Miscellanea Entomologica et de ses suppléments et annexes depuis 1892 jusqu'en septembre 1940. *Miscellanea Entomologica* 40 (8): 73–85.

Lefèvre, E. 1885: *Annales de la Société Entomologique de France. Tables générales de 1861 a 1880 inclusivement*. Paris: Lucien Buquet, 286 + [5] pp.

Lefèvre, E. 1895: *Annales de la Société Entomologique de France. Tables générales de 1881 a 1890 inclusivement*. Paris: Société Entomologique de France, 239 pp.

Low, S. 1853: *The British catalogue of books published from October 1837 to December 1852; containing the date of publication, size, price, publisher's name, and edition. Vol. I. – General alphabet*. Sampson Low and Son, London, [2] + 408 + 50 + 58 + 64 pp.

Merkl, O., Grabant, A., Makra, S., Peregovits, L. & Soltész, Z. 2008: Complete list of papers published in the Annales Historico-Naturales Musei Nationalis Hungarici between 1903 and 2007. *Annales Historico-Naturales Musei Nationalis Hungarici* 100: 95–244.

Nagel, P. & Grieder, L. 2019: Herrich-Schäffer's Deutschlands Insecten … (1829–1844), the associate publication of Koch's Deutschlands Crustaceen … (1835–1844), and Heft 110 (1823) of Panzer's Faunae Insectorum Germanicae … : bibliographic notes and dating. *Sherbornia* 5 (1): 1–156.

Nagel, P. & Schmidlin, L. 2014: Silbermann's "Revue entomologique": publication dates for nomenclatural purposes and bibliographic notes (Insecta, mainly Coleoptera). *Zootaxa* **3794** (1): 87–107.

Nolan, E. J. 1913: *An index to the scientific contents of the Journal and Proceedings of the Academy of Natural Sciences of Philadelphia 1812–1912*. Philadelphia: The Academy of Natural Science, xiv + 1419 pp.

Poggi, R. 2008: Le date di stampa dei periodici editi dalla Società Entomologica Italiana dal 1869 al 2007. *Memorie della Società Entomologica Italiana* **87**: 157–188.

Poggi, R. 2010: Gli "Annali" pubblicati dal Museo Civico di Storia Naturale "Giacomo Doria" di Genova: storia del periodico ed indice generali dei primi cento volumi (1870–2009). *Annali del Museo Civico di Storia Naturale "Giacomo Doria"* **101**: 1–529.

Raphael, S. 1970: The publication dates of the Transactions of the Linnean Society of London, Series I, 1791–1875. *Biological Journal of the Linnean Society* **2**: 61–76.

Sherborn, C. D. 1934: Dates of publication of catalogues of natural history (post 1850) issued by the British Museum. *Annals and Magazine of Natural History, Tenth Series* **13** (74): 308–312.

Schmitt, M. 2017: Case 3725 – Lema decempunctata Gebler, 1829 (Insecta, Coleoptera, Chrysomelidae): proposed conservation of the emended specific name. *Bulletin of Zoological Nomenclature* **74**: 28–33.

Standfuss, K. & Kerzhner, I. M. 2004: Publikationsdaten der "Trudy Russkogo Entomologicheskogo Obshchestva" und "Horae Societatis Entomologicae Rossicae", 1861–1932. *Entomofauna, Zeitschrift für Entomologie* **25** (14): 237–248.

Wheeler, G. 1912: On the dates of the publications of the Entomological Society of London. *Transactions of the Entomological Society of London* [**1911**] (3/4): 750–767.

Williams, R. B. 2017: T. V. Wollaston's Coleoptera Hesperidum: correction of its spurious 1867 publication date to 1868. *Zoological Bibliography* **4** (4): 89–92.

Ziani, S., Bezděk, A., Krell, F.-T. & Sommer, D. 2021: Authorship and date of publication of the name Scarabaeus stercorosus (currently Anoplotrupes stercorosus) (Coleoptera: Scarabaeoidea: Geotrupidae). *Zootaxa* **5067** (4): 593–599

Zilli, A. 2021: On the date of publication of Linnaeus' second edition of "Fauna Svecica". *Nota Lepidopterologica* **44**: 69–71.

Acknowledgements

The Editors and Authors of the present volume are indebted to following colleagues and friends who commented on various parts of the Catalogue, pointed to and discussed nomenclatural issues, searched or photographed the type specimens, found or copied the literature, completed references, added distributional information, and helped in various ways: Daniel AGUAYO, Colecciones Científicas de Zoología, Universidad de Granada, Spain; Dirk AHRENS, Zoologisches Forschungsmuseum Alexander Koenig, Bonn, Germany; Miguel Á. ALONSO-ZARAZAGA, Museo Nacional de Ciencias Naturales, Madrid, Spain; Ulf and Kathrin ARNOLD, Berlin, Germany; Vardan ASATRYAN, Institute of Hydroecology and Ichthyology of SCZHE NAS RA, Yerevan, Armenia; Ingolf ASKEVOLD, Tallahassee, Florida, USA; the late Martin BAEHR, Zoologische Staatssammlung, Munich, Germany; Michael and Ditta BALKE, Zoologische Staatssammlung, Munich, Germany; Maxwell V. L. BARCLAY, Natural History Museum, London, UK; Gloria BASTAZO, Málaga, Spain; Johannes BERGSTEN, Naturhistoriska Riksmuseet, Stockholm, Sweden; Aleš BEZDĚK, Institute of Entomology, Biology Centre of the Czech Academy of Sciences, České Budějovice, Czech Republic; Andreas BICK, University of Rostock, Rostock, Germany; Andrzej O. BIEŃKOWSKI, Moscow, Russia; Stephan BLANK, Senckenberg Deutsches Entomologisches Institut, Müncheberg, Germany; Matthias BORER, Naturhistorisches Museum, Basel, Switzerland; Michaela BROJER, Naturhistorisches Museum Wien, Wien, Austria; Harald BRUCKNER, Naturhistorisches Museum Wien, Wien, Austria; Wolfgang BRUNNBAUER, Naturhistorisches Museum Wien, Wien, Austria; Andris BUKEJS, Daugavpils University, Daugavpils, Latvia; Berta CABALLERO, Museu de Ciències Naturals de Barcelona, Barcelona, Spain; Hee-Wook CHO, Nakdonggang National Institute of Biological Resources, Sangju, South Korea; Albert F. DAMAŠKA, Charles University, Praha, Czech Republic; Alex DELOBEL, Fontenay-aux-Roses, France; Ali N. EKIZ, Uşak University, Uşak, Turkey; Mario ELGUETA, Museo Nacional de Historia Natural, Santiago, Chile; Johannes FRISCH, Museum für Naturkunde, Leibniz-Institut für Evolutions- und Biodiversitätsforschung, Berlin, Germany; Michael F. GEISER, Natural History Museum, London, UK; Gloria GEISER, Salzburg, Austria; Irmgard GEISER, Salzburg, Austria; Remigius GEISER sen., Salzburg, Austria; Christoph GERMANN, Naturhistorisches Museum, Basel, Switzerland; Matthew L. GIMMEL, Santa Barbara Museum of Natural History, Santa Barbara, USA; Ali GÖK, Süleyman Demirel University, Isparta, Turkey; Patrick

GROS, Haus der Natur, Salzburg Austria; Alexey A. GUSAKOV, Zoological Museum of Moscow State University, Moscow, Russia; Fritz GUSENLEITNER, Oberösterreichische Landesmuseen, Linz, Austria; Jiří HÁJEK, National Museum, Praha, Czech Republic; Stefan HANOT, Musée Royal de l'Afrique Centrale, Tervuren, Belgium; Michaela HANOUSKOVÁ, Masaryk University, Brno, Czech Republic; Matthias HARTMANN, Naturkundemuseum Erfurt, Erfurt, Germany; Andrea HASTENPFLUG-VESMANIS, Senckenberg Museum Frankfurt, Frankfurt am Main, Germany; Masakazu HAYASHI, Hoshizaki Green Foundation, Izumo, Japan; Markéta HEJČOVÁ, Mendel University in Brno, Brno, Czech Republic; Martin HUSEMANN, Zoologisches Museum Hamburg, Hamburg, Germany; Manfred JÄCH, Naturhistorisches Museum Wien, Wien, Austria; Birgit JAENICKE, Center of Natural History of the University of Hamburg, Hamburg, Germany; Bernd JÄGER, Museum für Naturkunde, Leibniz-Institut für Evolutions- und Biodiversitätsforschung, Berlin, Germany; Vladimír JANSKÝ, Slovak National Museum, Bratislava, Slovakia; Eric JIROUX, Andrésy, France; Mats JONSELL, Swedish University of Agricultural Sciences, Uppsala, Sweden; Ute KACZINSKI, Senckenberg Deutsches Entomologisches Institut, Müncheberg, Germany; Manfred KAHLEN, Tiroler Landesmuseum Ferdinandeum, Hall in Tirol, Austria; Denis G. KASATKIN, Rostov Branch of FSI "VNIIKR", Rostov-on-Don, Russia; Petr KMENT, National Museum, Praha, Czech Republic; Damir KOVAC, Forschungsinstitut Senckenberg, Frankfurt am Main, Germany; Michael KUHLMANN, Zoologisches Museum der Christian-Albrechts-Universität, Kiel, Germany; Leonardo LATELLA, Museo Civico di Storia Naturale, Verona, Italy; Chi-Feng LEE, Taiwan Agricultural Research Institute, Taichung City, Taiwan; Hong-Bin LIANG, Institute of Zoology, Chinese Academy of Sciences, Beijing, China; Andreas LINK, Ansfelden, Austria; Håkan LJUNGBERG, Swedish University of Agricultural Sciences, Uppsala, Sweden; Ivan LÖBL, Muséum d'Histoire Naturelle, Geneva, Switzerland; Michael MALICKY, Oberösterreichische Landesmuseen, Linz, Austria; Antoine MANTILLERI, Museum National d'Histoire Naturelle, Paris, France; Glòria MASÓ, Museu de Ciències Naturals de Barcelona, Barcelona, Spain; Jaakko MATTILA, Finnish Museum of Natural History, Helsinki, Finland; Hans MEJLON, Museum of Evolution, Uppsala University, Uppsala, Sweden; the late Ottó MERKL, Hungarian Natural History Museum, Budapest, Hungary; Peter MICHALIK, University of Greifswald, Greifswald, Germany; Heinz MITTER, Steyr, Austria; Matteo MONTAGNA, University of Naples Federico II, Naples, Italy; Geoffrey E. MORSE, University of San Diego, San Diego, USA; Hans MÜHLE, Zoologische Staatssammlung, München, Germany; Tamás NÉMETH,

Hungarian Natural History Museum, Budapest, Hungary; Rui-E NIE, Institute of Zoology, Chinese Academy of Sciences, Beijing, China; Nikolay B. NIKITSKIY, Zoological Museum of Moscow State University, Moscow, Russia; Esther OCKERMÜLLER, Biologiezentrum Linz, Linz, Austria; Raimundo OUTERELO, Colección de Entomología, Universidad Complutense de Madrid, Spain; Mercedes PARIS, Museo Nacional de Ciencias Naturales, Madrid, Spain; Eduard PETITPIERRE, University of the Balearic Islands, Mallorca, Spain; Isidor PLONSKI, Wien, Austria; Stefan RICHTER, University of Rostock, Rostock, Germany; Alexander RIEDEL, Staatliches Museum für Naturkunde, Karlsruhe, Germany; Christophe RIVIER, Museum National d'Histoire Naturelle, Paris, France; Eduardo RUIZ, Colección de Entomología, Universidad Complutense de Madrid, Spain; Roberta SALMASO, Museo Civico di Storia Naturale, Verona, Italy; Davide SASSI, Castelmarte, Italy; Michail Yu. SAVITSKIY, Moscow State University, Moscow, Russia; Ronald SCHILLER, Naturkundemuseum Leipzig, Leipzig, Germany; Harald SCHILLHAMMER, Naturhistorisches Museum Wien, Wien, Austria; Jürgen SCHMIDL, Friedrich-Alexander-Universität Erlangen, Germany; Thomas SCHMITT, Senckenberg Deutsches Entomologisches Institut, Müncheberg, Germany; Karla SCHNEIDER, Zentralmagazin Naturwissenschaftlicher Sammlungen der Martin-Luther-Universität, Halle-Wittenberg, Germany; the late Wolfgang SCHÖNLEITHNER, Naturhistorisches Museum Wien, Wien, Austria; Mandy SCHRÖTER, Senckenberg Deutsches Entomologisches Institut, Müncheberg, Germany; Martin SCHWARZ, Biologiezentrum Linz, Linz, Austria; Martin SEČANSKÝ, Slovak National Museum, Bratislava, Slovakia; Helena SHAVERDO, Naturhistorisches Museum Wien, Wien, Austria; Shigehiko SHIYAKE, Osaka Museum of Natural History, Osaka, Japan; Dana ŠÍDOVÁ, National Museum, Praha, Czech Republic; Marianna SIMÕES, Senckenberg Deutsches Entomologisches Institut, Müncheberg, Germany; Alexey SOLODOVNIKOV, Danish Natural History Museum, Copenhagen, Denmark; Amoret SPOONER, Oxford University Museum of Natural History, Oxford, UK; Eva SPRECHER, Naturhistorisches Museum, Basel, Switzerland; Takuya TAKEMOTO, Hokkaido University, Sapporo, Japan; Dmitry TELNOV, Natural History Museum, London, UK; Alberto TINAUT, Colecciones Científicas de Zoología, Universidad de Granada, Spain; Luca TOLEDANO, Verona, Italy; Godard TWEEHUYSEN, Dutch Entomological Society, Naturalis, Leiden, The Netherlands; Tomi TRILAR, Slovenian Museum of Natural History, Ljubljana, Slovenia; Karin ULMEN, Zoologisches Forschungsmuseum Alexander Koenig, Bonn, Germany; Peter VĎAČNÝ, Comenius University, Bratislava, Slovakia; Andreas WEIGEL, Wernburg, Germany; Hella WENDT, former Museum für Naturkunde der Humboldt-Universität zu Berlin, Berlin, Germany; Joachim

WILLERS, Museum für Naturkunde, Leibniz-Institut für Evolutions- und Biodiversitätsforschung, Berlin, Germany; Lothar ZERCHE, Senckenberg Deutsches Entomologisches Institut, Müncheberg, Germany; Hong-Zhang ZHOU, Institute of Zoology, Chinese Academy of Sciences, Beijing, China; Stefano ZOIA, Milano, Italy; Isabelle ZUERCHER, Naturhistorisches Museum, Basel, Switzerland.

And above all we would like to thank our families, without whose support, tolerance and understanding this catalogue would not have ever appeared: Monika EILHARDT-ANTON; Petra, Ingeborg and Astrid BEENEN; Eliška, Anna, Antonín and Jaromír BEZDĚK; Olga BEZMAN-MOSEYKO and Ekaterina MOSEYKO; Birgit, Cora, Justus and Klara SCHÖLLER; Katka ŠTAJEROVÁ and Matyáš SEKERKA.

The examination of Chrysomelidae in the Hungarian Natural History Museum, Budapest by Jan BEZDĚK was supported by the Synthesys Project HU-TAF-1388 (http://www.synthesys.info/) financed by the European Community – Research Infrastructure Action under the Seventh Framework Programme.

The research by Jan BEZDĚK in the Finnish Museum of Natural History, Helsinki (FI-TAF-TA4–003), received the support from the SYNTHESYS+ project http://www.synthesys.info/ financed by European Community Research Infrastructure Action under the H2020 Integrating Activities Programme, Project number 823827.

The research by Elisabeth GEISER in the Slovak National Museum, Bratislava, was funded by the Operational Program of Research and Development and co-financed with the European Fund for Regional Development (EFRD) ITMS 26230120004: Building of Research and Development Infrastructure for investigation of genetic biodiversity of organisms and joining the IBOL initiative.

The research of Alexey MOSEYKO was carried out within the framework of ZIN RAS State research assignment No. 122031100272–3.

The examination of type specimens in National Museum, Praha (Czech Republic) carried out by José-Miguel VELA in March 2017 was supported by the Synthesys project CZ-TAF-7039 (http://www.synthesys.info/), funded by the European Community – Research Infrastructure Action under the Seventh Framework Programme.

The examination of Chrysomelidae in the Swedish Natural History Museum, Stockholm by Matthias SCHÖLLER was supported by the Synthesys Project SE-TAF-3924 (http://www.synthesys.info/) financed by the European Community – Research Infrastructure Action.

Lukáš SEKERKA received financial support from the Ministry of Culture of the Czech Republic (DKRVO 2024–2028/5.I.a, National Museum, 00023272).

Editors and Authors

The Editors

Jan BEZDĚK
Mendel University in Brno
Brno, Czech Republic

Lukáš SEKERKA
National Museum
Praha, Czech Republic

Authors

family ORSODACNIDAE C. G. Thomson, 1859	J. BEZDĚK
family MEGALOPODIDAE Latreille, 1802	L. SEKERKA
family CHRYSOMELIDAE Latreille, 1802	
subfamily Sagrinae Leach, 1815	L. SEKERKA
subfamily Bruchinae Latreille, 1802	K.-W. ANTON
subfamily Donaciinae Kirby, 1837	E. GEISER
subfamily Criocerinae Latreille, 1804	J. BEZDĚK, M. SCHMITT
subfamily Cassidinae Gyllenhal, 1813	L. SEKERKA, J. ŚWIĘTOJAŃSKA
subfamily Chrysomelinae Latreille, 1802	
tribe Chrysomelini Latreille, 1802	H. KIPPENBERG, Y. E. MIKHAILOV
tribe Timarchini Motschulsky, 1860	M. DACCORDI, J. M. VELA
subfamily Galerucinae Latreille, 1802	
supertribe Galerucitae Latreille, 1802	R. BEENEN
supertribe Alticitae Newman, 1835	J. BEZDĚK, A. S. KONSTANTINOV
supertribe Serraticollitae B. White, 1942	A. S. KONSTANTINOV
subfamily Lamprosomatinae Lacordaire, 1848	A. S. KONSTANTINOV
subfamily Cryptocephalinae Gyllenhal, 1813	
tribe Clytrini Kirby, 1837	J. BEZDĚK, M. SCHÖLLER
tribe Cryptocephalini Gyllenhal, 1813	M. SCHÖLLER
tribe Pachybrachini Chapuis, 1874	M. SCHÖLLER
tribe Fulcidacini Jacobson, 1924	J. BEZDĚK
subfamily Eumolpinae Hope, 1841	A. G. MOSEYKO
subfamily Synetinae LeConte & Horn, 1883	J. BEZDĚK

Addresses of the Authors

ANTON Klaus-Werner, Grünewaldstraße 13, D-79312 Emmendingen, Germany. E-mail: bruchitax@gmx.de

BEENEN Ron, Martinus Nijhoffhove 51, NL-3437 ZP Nieuwegein, The Netherlands. E-mail: r.beenen@wxs.nl

BEZDĚK Jan, Department of Zoology, Fisheries, Hydrobiology and Apiculture, Mendel University in Brno, Zemědělská 1, CZ-613 00 Brno, Czech Republic. E-mail: bezdek @mendelu.cz

DACCORDI Mauro, c/o Museo di Storia Naturale, Lungadige Porta Vittoria 9, I-37129 Verona, Italy. E-mail: daccordimauro@gmail.com

GEISER Elisabeth, Saint-Julien-Straße 2/314, A-5020 Salzburg, Austria. E-mail: elisa beth.geiser@gmx.at

KIPPENBERG Horst, Langer Platz 21, D-91074 Herzogenaurach, Germany. E-mail: horst@kippenbergs.de

KONSTANTINOV Alexander S., Systematic Entomology Laboratory, USDA, c/o Smithsonian Institution, P. O. Box 37012, National Museum of Natural History, Washington, DC 20013–7012, USA. E-mail: alex.konstantinov@usda.gov

MIKHAILOV Yuri E., Ural State Forest Engineering University, Sibirsky trakt 37, RUS-620100 Yekaterinburg, Russia. E-mail: yuemikhailov@gmail.com

MOSEYKO Alexey G., Zoological Institute, Russian Academy of Sciences, Universitetskaya nab. 1, St. Petersburg, RUS-199034 Russia / All-Russian Institute of Plant Protection, Podbelskogo 3, St. Petersburg – Pushkin, RUS-196608, Russia. E-mail: chrysolesha@mail.ru

SCHMITT Michael, Universität Greifswald, Allgemeine & Systematische Zoologie, Loitzer Straße 26, D-17489 Greifswald, Germany. E-mail: michael.schmitt@uni -greifswald.de

SCHÖLLER Matthias, Humboldt-Universität zu Berlin, Faculty of Life Sciences, Thaer-Institute of Agricultural and Horticultural Sciences, Division Urban Plant Ecophysiology, Lentzeallee 55/57, D-14195 Berlin, Germany. E-mail: Matthias.Schoeller @hu-berlin.de

SEKERKA Lukáš, Department of Entomology, National Museum, Cirkusová 1740, CZ-193 00, Praha – Horní Počernice, Czech Republic. E-mail: sagrinae@gmail.com

ŚWIĘTOJAŃSKA Jolanta, Department of Biodiversity and Evolutionary Taxonomy, University of Wroclaw, Przybyszewskiego 65, PL-51–148 Wrocław, Poland. E-mail: jolanta.swietojanska@uwr.edu.pl

VELA José-Miguel, Instituto Andaluz de Investigación y Formación Agraria y Pesquera (IFAPA), Entomología Agraria, Cortijo de la Cruz, E-29140 Málaga, Spain. E-mails: josem.vela@juntadeandalucia.es, josemiguel.vela@gmail.com

The present work is based upon data provided by the following authors of the first edition of the Catalogue:

family MEGALOPODIDAE Latreille, 1802	
subfamily Zeugophorinae Boving & Craighead, 1931	H. SILFVERBERG
subfamily Megalopodinae Latreille, 1802	L. N. MEDVEDEV
family ORSODACNIDAE C. G. Thomson, 1859	H. SILFVERBERG
family CHRYSOMELIDAE Latreille, 1802	
subfamily Sagrinae Leach, 1815	A. S. KONSTANTINOV, L. SEKERKA
subfamily Bruchinae Latreille, 1802	K.-W. ANTON
subfamily Donaciinae Kirby, 1837	H. SILFVERBERG
subfamily Criocerinae Latreille, 1804	M. SCHMITT
subfamily Cassidinae Gyllenhal, 1813	L. BOROWIEC, L. SEKERKA
subfamily Chrysomelinae Latreille, 1802	H. KIPPENBERG
genus Timarcha Latreille, 1829	J. GÓMEZ-ZURITA, H. KIPPENBERG
subfamily Galerucinae Latreille, 1802	R. BEENEN
subfamily Alticinae Newman, 1835	M. DÖBERL
subfamily Lamprosomatinae Lacordaire, 1848	A. S. KONSTANTINOV
subfamily Cryptocephalinae Gyllenhal, 1813	
tribe Clytrini Kirby, 1837	R. REGALIN, L. N. MEDVEDEV
tribe Cryptocephalini Gyllenhal, 1813	
genus Cryptocephalus Geoffroy, 1762	I. K. LOPATIN, A. SMETANA, M. SCHÖLLER
remaining Cryptocephalini	M. SCHÖLLER, I. LÖBL, I. K. LOPATIN
tribe Fulcidacini Jacobson, 1924	J. BEZDĚK
subfamily Eumolpinae Hope, 1840	A. G. MOSEYKO, E. SPRECHER-UEBERSAX
subfamily Synetinae LeConte & Horn, 1883	H. SILFVERBERG

New Nomenclatural and Taxonomic Acts, Comments and Distributional Records

Orsodacnidae
J. Bezděk

Availability of varieties and forms

001 The following variety and form are deemed infrasubspecific according to Article 45.6.1 (ICZN 1999), since a subspecies concept was used in the same work, and according to available information, these names were not validated prior to 1985 (ICZN 1999, Art 45.6.4.1):
Orsodacne lineola var. *anatolica* Pic, 1916b: 11
Orsodacne lineola forma *marginalis* Roubal, 1941: 6

Excluded taxon

002 El Aalaoui et al. (2021) recorded the Nearctic species *Orsodacne atra* (Ahrens, 1810) from Morocco. However, the photographs in their paper show *Crioceris macilenta* Weise, 1881. Therefore, *Orsodacne atra* is not included as Palaearctic species.

Megalopodidae
L. Sekerka

Comments

003 The present concept of *Zeugophora flavicollis* (Marsham, 1802), which has been used since Suffrian (1840), was questioned and the taxon is probably conspecific with *Z. frontalis* Suffrian, 1840 (Harrison & Harrison 2022). Currently a revision of relevant type material as well as additional material from the whole of Europe is in process to resolve the situation and to clarify the distibution of the two taxa.

Availability of varieties and forms

004 The following variety and forms are deemed infrasubspecific according to Article 45.6.1 (ICZN 1999), since a subspecies concept was used in the same work, and according to available information, these names were not validated prior to 1985 (ICZN 1999, Art 45.6.4.1):
Zeugophora (*Pedrillia*) *annulata* forma *inannulata* Chûjô, 1959: 1
Zeugophora (*Pedrillia*) *annulata* forma *melanaria* Chûjô, 1959: 1
Zeugophora flavicollis var. *notatipes* Pic, 1925b: 9

 | DOI:10.1163/9789004443303_002

Comments on distribution

005 *Zeugophora gracilis unicolor* Chûjô, 1958. The occurence in Taiwan is removed as it was listed erroneously.

006 *Zeugophora indica* Jacoby, 1903. The occurrence in Kashmir (India) is removed because the respective specimens were later described as *Z. wittmeri*.

007 *Zeugophora tricolor* S.-H. Chen & Pu, 1962. It was mistakenly considered as occurring in Yunnan, but the species was described and is so far known only from Gansu.

008 *Zeugophora xanthopoda* Bezděk & Silfverberg, 2010. Its type locality was wrongly interpreted as being situated in Sichuan, while it is situated in Chongqing.

009 *Temnaspis bengalensis* Westwood, 1864. The occurence in Sikkim (India) is removed as it was based on an incorrect interpretation of the type locality "Bengal".

010 Occurrences of *Temnaspis humeralis* Jacoby, 1890 in Hebei and Gansu (China) are removed as the first was confused with Hubei and the other was based on wrong interpretation of data. Occurrences of *T. septemmaculatus* (Hope, 1831) and *T. vitalisi* (Pic, 1922) in Xizang (China) are removed as they were based on wrong interpretation of material from the Tibetan plateau.

011 The following distributional data appeared in the previous version of the Catalogue, however, I was unable to find any published sources. Nevertheless, they are retained here as the occurences are possible.
Zeugophora annulata Baly, 1873: 79 (*Pedrillia*) – **A:** CH (JIX)
Zeugophora dimorpha Gressitt, 1945a: 138 – **A:** CH (JIA)
Zeugophora scutellaris Suffrian, 1840: 99 – **A:** CH (GAN) RU (ES) SY
Temnaspis nigriceps Baly, 1859 – **A:** ID (SD)
Temnaspis septemmaculatus (Hope, 1831) – **A:** ID (SD)

Chrysomelidae: Sagrinae

L. Sekerka

Spelling

012 Pic (1953a) made a lapsus calami when naming a variety "*farrei*" of *Sagra jansoni*, as he misspelled the name of the collector Favre. Accordingly, the name is changed here to *favrei* (ICZN 1999, Art. 32.5.1).

New geographical records

013 *Sagra carbunculus* (Hope, 1842). INDIA: Arunachal Pradesh: Anini vicinity, 28°54'N, 95°56'E, 1700 m, 30.–31.v.2007, 1 spec., L. Dembický leg. (NHMB); Dirang vicinity, 27°21'N, 92°13'E, 1800 m, 8.–22.v.2006, 3 spec., L. Dembický leg. (NHMUK); Mishmi Hills, 10.000 ft., 14.vi.1928, 1 spec., P. Sladen leg. (NHMUK). All L. Sekerka det. Occurence mentioned in the previous Edition was based on the abovementioned specimens but the respective faunistic data were never formally published, thus the abovementioned records are the first for Arunachal Pradesh (India).

014 *Sagra femorata* (Drury, 1773). CHINA: Guizhou: "Kwei Tchéou", 1 spec., L. Sekerka det. (NMPC). New species for the province of Guizhou (China).

015 *Sagra fulgida* Weber, 1801. CHINA: Shanghai, 1 ♀, L. Sekerka det. (NMPC). New species for Shanghai (China).

Comments on distribution

016 *Sagra carbunculus* (Hope, 1842). The occurence in Bhutan, mentioned in the previous Edition, is removed as it was erroneously based on the interpretation of the title of Kimoto's (2005) work while there were no records mentioned; on the other hand, the species quite likely occurs there. Also, the occurence in China is questionable and records of Chen & Pu (1962a) perhaps refer to *S. odontopus* because they interpreted *S. carbunculus* as having a horseshoe-shaped mesosternal process (typical for *S. odontopus*) while the mesosternal process of *S. carbunculus* is normally convex, unmodified. Moreover, it seems that *S. carbunculus* is restricted to the Himalayas and other connected ranges in Myanmar; within China it may reach NE Yunnan and Xizang but I have not seen any material to confirm this hence its occurrence in China is not adopted.

017 Occurrences of *Sagra femorata* (Drury, 1773) in Arunachal Pradesh and Nepal and of *S. jansoni* Baly, 1860 in Arunachal Pradesh mentioned in the previous Edition were based on assumptions only. As there is no published record nor available material from respective areas these records are not accepted in the present Edition.

Chrysomelidae: Bruchinae

K.-W. Anton

Comments

018 The species concept and the synonyms of *Spermophagus abdominalis* (Fabricius, 1781) in the present Edition follow Kingsolver & Borowiec (1988) and Borowiec (1991). The study of photographs of the type specimen (UUZM) of *Bruchus convolvuli* Thunberg, 1816, kindly provided by Jan Bezděk, confirms the synonymy of *S. convovuli* with *S. abdominalis*. All distributional data of *S. rufiventris* Boheman, 1833, currently considered synonym of *S. abdominalis* and listed as valid in the previous Edition (Anton 2010b) are transferred to the latter species.

019 *Spermophagus caricus* Decelle, 1982 was synonymised with *S. turanicus* Lukjanovitch & Ter-Minassian, 1957 by Anton (1999a). Savitskiy (2000) restored the species status of *S. turanicus*, redescribed it and provided differences between the two species. Both, *S. caricus* and *S. turanicus*, are treated as valid species in the present Edition.

020 *Bruchidius fischeri* (Hummel, 1827) was listed as distinct species in the previous Edition (Anton 2010b). In the present Edition the opinion of Schilsky (1905), who treated *B. fischeri* synonym of *B. astragali* (Boheman, 1828), is followed. However,

the type material has not been examined by recent authors, thus the placement of *B. fischeri* is to be considered as questionable [QS].

021 *Bruchidius caeruleus* (Champion, 1919) was transferred to the genus *Conicobruchus* by Singal & Pajni (1986). Nevertheless, it was listed in *Bruchidius* in the previous Edition (Anton 2010b) without explanation. The reasons for this placement are morphological characters of male genitalia indicating a close relationship to the *Bruchidius japonicus* species-group defined by Delobel (2010). Thus the placement in *Bruchidius* is maintained also in the present Edition.

022 *Bruchidius varipictus* (Motschulsky, 1874) was erroneously listed twice (as a valid species and also as synonym of *B. calabrensis*) in the previous Edition (Anton 2010b). In the present Edition, it is correctly listed as synonym of *B. calabrensis* (Blanchard, 1845).

023 Aldridge & Pope (1986) synonymized *Bruchidius canus* (Germar, 1823) with *B. cisti* (Fabricius, 1775), and this was accepted by most of the subsequent authors. The lectotype of *B. canus* (MLUH), designated by Zampetti (1982), and three paralectotypes, all females, were examined and the synonymy is confirmed. Nevertheless, the species name *B. canus* has been used by some authors (e.g. Strejček 1990, Zampetti & Ricci 2012) for different species of the taxonomically complicated *B. lutescens-poupillieri* complex within the *Bruchidius cisti* species-group.

024 *Bruchidius corsicus* (Baudi di Selve, 1890) was usually treated as a synonym of *Bruchidius pauper* (Boheman, 1828). The identity of *B. pauper*, described from "Khersoma Thraciae" [= Kherson, Ukraine], is unclear and the type material is probably lost. No specimens from Ukraine are available to establish a neotype, and no other record from Ukraine was published. Accordingly, *B. pauper* was listed as species incertae sedis and *B. corsicus* was treated as a valid species in the previous Edition (Anton 2010a, b). This arrangement was widely accepted in Bruchinae literature and is also followed in the present Edition.

025 *Bruchidius fulvicornis* (Motschulsky, 1874) was listed as species incertae sedis in the previous Edition (Anton 2010a, b). A syntype (♂, ZMUM) was examined by M. Yu. Savitskiy (Moscow, Russia, pers. comm.) who considers it a junior synonym of *B. imbricornis* (Panzer, 1795). This opinion is maintained in the present Edition.

026 *Bruchidius algiricus* (Allard, 1884) was listed as synonym of *B. incarnatus* (Boheman, 1833) in the previous Edition (Anton 2010b) without explanation. Morphological characters of *B. algiricus* described by Luca (1960) are fully within the variability of *B. incarnatus*, including the structure of the median lobe of aedeagus with armature of sclerites inside the internal sac.

027 *Bruchidius khnzoriani* Karapetjan, 1976 was listed as species incertae sedis in the previous Edition (Anton 2010a, b). Based on the description and photograph of the holotype (♂, IZAY), made by T. Ghrejyan (Yerevan, Armenia) and kindly provided to me by D. Kasatkin (Rostov-on-Don, Russia), it is a member of the *Bruchidius*

astragali species-group and closely related to *B. astragali* (Boheman, 1828) itself. Therefore, it is treated as distinct species in the present Edition.

028 Yus Ramos & François (2015) considered *Bruchidius leprieuri* (E. Jacquet, 1886) as valid species but without examination of type material. The status is confirmed here by the examination of two syntypes (♀♀, MNHN, coll. Pic ex Jacquet) labelled "La Calle". It is a member of the *B. astragali* species-group and closely related to *B. myobromae* (Motschulsky, 1874) and *B. virgatus* (Fåhraeus, 1839).

029 *Bruchidius strictus* Iablokoff-Khnzorian, 1974 was listed as species incertae sedis in the previous Edition (Anton 2010a, b). In the present Edition, it is treated as synonym of *B. longulus* Schilsky, 1905 based on opinion of M. Yu. Savitskiy (Moscow, Russia, pers. comm.).

030 The *Bruchidius lutescens-poupillieri* complex (belonging to the *B. cisti* species-group) includes several undescribed species, partially identified by various authors as *B. canus* or *B. poupillieri*. This complex is in need of a comprehensive revision based on a large series of specimens reared from different host plants, localities and seasons. For *B. lutescens* (Blanchard, 1845) the countries north of the Mediterranean are not listed in the present Edition.

031 *Bruchidius misellus* (Boheman, 1833) was synonymised with *B. debilis* (Gyllenhal, 1933) (currently synonym of *B. cisti* Fabricius, 1775) by Hoffmann (1945) but Zampetti (1982) considered *B. misellus* as valid species. However, Yus Ramos (2010a) treated *B. misellus* again as a synonym of *B. cisti*. The validity of *B. misellus* is confirmed here based on the examination of its syntype (♂, NHRS) labelled "unicolor var., Dejean. Dalm.".

032 Zampetti (1982) considered *Bruchidius olivaceus* (Germar, 1823) as a valid species. Yus Ramos (2010a) erroneously listed *B. olivaceus* as synonym of *B. cisti* (Fabricius, 1775) and provided reference to A. Delobel & B. Delobel (2003), who however explicitly mentioned "*B. olivaceus* auctt.". The validity of *B. olivaceus* is confirmed here based on the description and examination of the lectotype (♂, MLUH) designated by Zampetti (1982).

033 *Bruchidius poupillieri* (Allard, 1868) might be confused with small specimens of *B. lutescens* (Blanchard, 1845). Records from countries outside of the Iberomaghrebian Region are listed but should be verified to make sure to which species they belong.

034 *Bruchus dispar* var. *atricornis* Pic, 1931 was described from "Angleterre: Manchester (British Museum et coll. Pic)" (Pic 1931a) but its status is unknown at present as the type specimens were not traced in NHMUK (M. Geiser, 2023, pers. comm.) nor in MNHN (A. Mantilleri, 2023, pers. comm.). However, it is quite unlikely that it represents *B. dispar* (Gyllenhal, 1833) as the latter species does not occur in Great Britain, and the very similar *B. varius* (Olivier, 1800) was only established in Great Britain as recently as 1994 (Hodge 1997). Moreover, the *B. varius* species-group is taxonomically complicated and includes a few undescribed species, sometimes named by

various authors as "*B. fulvicornis*" or "*Bruchidius* spec. prope *varius*". This entire group needs a comprehensive revision based on large series of specimens reared from various host plants, diverse localities and also across various seasons. For the above stated reasons *Bruchus dispar* var. *atricornis* is here treated as a nomen dubium and its status can only be elucidated if the type material is recovered.

035 *Bruchidius consobrinus* (Allard, 1868) was described from Algeria and listed as synonym of *B. astragali* (Boheman, 1828) by Schilsky (1905). However, *B. astragali* is a species of Asian steppes not occurring in North Africa. Therefore, *Bruchidius consobrinus* is treated as nomen dubium in the present Edition.

036 Motschulsky (1874) noted with doubts that *Bruchidius subalbovittis* (Motschulsky, 1874) is a synonym of *B. fischeri*, however, Schilsky (1905) listed it as synonym of the "var. a" of *B. astragali* (Boheman, 1828). Type material of *B. subalbovittis* has not been examined by recent authors and its identity remains quite unclear. Therefore, *B. subalbovittis* is listed as nomen dubium in the present Edition.

037 Anton (2010a) synonymized *Callosobruchus utidai* Tuda, 2003 with *C. nigritus* Anton, 2000 but this synonymy was omitted from the text part of the Catalogue (Anton 2010b).

038 Yus Ramos (2009) synonymized *Bruchidius endotubercularis* Arora, 1980, with *Megabruchidius dorsalis* (Fåhraeus, 1839). This synonymy was rejected by Fritzsche & A. Delobel (2012), and this view is followed in the present Edition. *Bruchidius endotubercularis* is a south Indian species and thus extralimital to this Catalogue. *Megabruchidius dorsalis* was designated as type species of the genus *Megabruchidius* by Borowiec (1987), however, Zampetti & Ricci (2012) treated it as a species of *Bruchidius* without comment.

039 Although *Megabruchidius sophorae* Tuda & Morimoto, 2004 is within the variability of size and colour of *M. dorsalis* (Fåhraeus, 1839), it is clearly distinguished by the dorsal valve broadly rounded at sides, with apex subtruncate (dorsal valve triangular and arcuate in *M. dorsalis*) and by a different host plant association. Genetic study of *M. sophorae* conducted by Kergoat et al. (2007) also confirmed its species rank.

040 *Mylabris semicarnea* Reitter, 1884 was erroneously listed as synonym of *Acanthoscelides pusillimus* (Sharp, 1885) in the previous Edition (Anton 2010b). However, Baudi (1887) synonymized *M. semicarnea* with *M. plagiata* (currently *Paleobruchidius plagiatus* (Reiche & Saulcy, 1858)). The synonymy is confirmed by examination of holotype (♂, HNHM) of *M. semicarnea*.

041 *Bruchidius saxicola* Iablokoff-Khnzorian, 1965 was listed as a species incertae sedis in the previous Edition (Anton 2010a, b). In the present Edition *B. saxicola* is treated as a synonym of *Paleobruchidius turanus* (Lukjanovitch & Ter-Minassian, 1954) based on the opinion of M. Yu. Savitskiy (Moscow, Russia, pers. comm.).

042 Borowiec (1987) synonymized *Pygobruchidius* Pic, 1951 and *Tuberculobruchus* Decelle, 1951 with *Sulcobruchus* Chûjô, 1937. Later, Anton (1999e) redefined

Sulcobruchus and excluded *Pygobruchidius* and *Tuberculobruchus* without further comments. In the previous Edition species of *Pygobruchidius* were listed within *Bruchidius*. Recent studies based on molecular phylogenetics (Kergoat & Silvain 2004, Kergoat et al. 2008, 2011, Kergoat 2011) show that clades with *Pygobruchidius*/*Tuberculobruchus* also contain morphologically similar species of *Bruchidius*. In the present Edition *Pygobruchidius* is treated as valid genus with *Tuberculobruchus* as synonym (precedence clarified by Borowiec (1987)). *Pygobruchidius* is defined by the following combination of characters: the shape of body elongate to ovoid, usually oblong-ovoid; integument black to yellowish-brown; vestiture recumbent, with dense setae covering the integument completely; head of moderate length, eyes flat or bulging, antennae short, scarcely reaching to elytral callus, with barely visible sexual dimorphism, antennomeres IV(V)–X serrate; pronotum without a lateral carina, pronotal sides convex to conical; scutellum bifid, simple or double-hooked; elytra with protuberance at base of striae III–IV, disc more or less flattened; hind femora incrassate, bicarinate ventrally and with a deep sulcus towards apex between margins, mesoventral margin with more or less developed subapical denticle, hind tibiae with three carinae, mucro varying from shorter to longer than lateral coronal denticle; abdomen massive, abdominal ventrite I simple or with bundle of protruding long setae or with spot of erect setae in the middle of apical half in one or both sexes; last visible tergite enlarged and oblique, aedeagus short and broad, ventral valve square or as long as wide, barely surpassing dorsal valve, lateral lobes apically spatulate, internal sac with less sclerotised needles and microplates, with or without hinge sclerites near orifice. With the exception of the *Bruchidius albosparsus* species-group defined by A. Delobel et al. (2015), *Bruchidius* can be easily distinguished from *Pygobruchidius* by the ventral side of hind femora without sulcus, and by the abdomen not massive. Species of *B. albosparsus* species-group have hind femur with sulcus similar to *Pygobruchidius*, but differ in hind tibiae with four carinae (for detailed description see Anton & A. Delobel 2003) and in aedeagus more or less oblong with ventral valve distinctly surpassing dorsal valve.

043 Anton (2010a), in New acts etc., erroneously synonymized *Bruchidius wittmeri* Decelle, 1977 with *B. urbanus* (Sharp, 1886a), while in the Catalogue part (Anton 2010b) it was correctly listed as synonym of *B. bilineatopygus* (Pic, 1938). The latter synonymy is here confirmed and both taxa are transferred to *Pygobruchidius*.

044 Description and figures of *Bruchus occidentalis* given by Yus Ramos (2021) are confused with *B. sibiricus* Germar, 1823. In comparison with the latter species *B. occidentalis* is on average smaller (length 1.8–2.7 mm), antennomeres VI–XI in males are almost black and wider, elytral pubescence forms small whitish spots similar to *B. luteicornis* Illiger, 1794, and the velum at the apex of the lateral lobes is slimmer, in ventral aspect not exceeding lateral margin of the lobe.

045 Legalov (2011) erroneously listed *Kytorhinus* (*Kytorhinus*) *kergoati* A. Delobel & Legalov, 2009 under the name "*K. rushanensis* Delobel & Legalov, 2009: 72", which is a nomen nudum.

046 *Caryedon kizilkumensis* Ter-Minassian, 1977 was synonymised with *C. prosopidis* Arora, 1977 by Borowiec (1990b). Nevertheless, *C. kizilkumensis* was listed as valid species in the previous Edition (Anton 2010b) without explanation. The reason for resurrection were significant differences in the primary descriptions of the two taxa. This opinion is maintained also in the present Edition.

047 Aldridge & Pope (1986) explicitly removed *Bruchus fasciatus* Olivier, 1800 from the synonymy of *Bruchidius villosus* (Fabricius, 1792), however, Yus Ramos (2019) listed this synonymy again without any comment. The identity of *Bruchus fasciatus* is enigmatic and deposition of its type material is unknown, therefore it is listed as "Bruchinae, nomen dubium".

Availability of varieties and forms

048 The following varieties are deemed infrasubspecific according to Article 45.6.1 (ICZN 1999), since a subspecies concept was used in the same work, and according to available information, these names were not validated prior to 1985 (ICZN 1999, Art 45.6.4.1):

Bruchus dilutus var. *obscuriceps* Pic, 1916b: 10

Bruchus pallidicornis var. *muticus* Fiori, 1906: 206

New synonymy

049 *Bruchidius amurensis* Pic, 1953, **syn. nov.** of *Bruchus loti* Paykull, 1800, based on the study of type material of *B. amurensis*: syntype (♂), "Amur: | Chabarofka || Coll. Koltze || Bruchus | sp || Type || Bruchidius | amurensis | n sp || TYPE || Museum Paris | Coll. M. Pic || SYNTYPE || SYNTYPE | Bruchidius | amurensis Pic, 1953 || MNHN, Paris | EC12511" (MNHN). The photographs of the syntype were kindly provided by C. Rivier (MNHN) and G. Morse (San Diego University, USA). Additional specimens of *B. loti* including those from eastern Siberia were also examined (collection of K.-W. Anton).

050 *Bruchidius gavirai* Yus Ramos, 2013, **syn. nov.** of *Bruchidius lateobscurus* (Pic, 1904), based on descriptions, examination of type material: *Laria lateobscura*, syntype (♀), "Poupillieri var. || v. lateobscurus | Pic || Bou Saada | 18 Mai 95" (MNHN), and study of additional specimens of *B. lateobscurus*, including those from Iberian Peninsula (collection of K.-W. Anton).

051 *Bruchidius horshensis* Zampetti & Toma, 2020, **syn. nov.** of *Bruchidius ochraceus* (Baudi di Selve, 1886), based on descriptions, examination of lectotype (♂, MRSN) of *Mylabris ochracea*, as well as study of additional specimens of *B. ochraceus*, including those from the Levant Region (collection of K.-W. Anton).

052 *Bruchidius migliaccioi* Zampetti, 1998, **syn. nov.** of *Bruchidius lineatus* (Allard, 1868), based on descriptions and redescription (Zampetti & Ricci 2012), as well as study of additional specimens of *B. lineatus* (collection of K.-W. Anton). The curious figure "a" of the male genitalia given by Zampetti & Ricci (2012) does not fit to the description of median lobe in the same publication.

053 *Bruchidius pardoi* Yus Ramos, 2007, **syn. nov.** of *Bruchidius nudus* (Allard, 1868), based on descriptions and study of additional specimens of *B. nudus*, including those from Iberian Peninsula (collection of K.-W. Anton). *Bruchidius pardoi* was listed as synonym of *B. nudus* by Yus Ramos & Labrique (2015) without appropriate comment. In his paper on Iberian bruchids Yus Ramos (2019) noted *B. pardoi* as a synonym of *B. nudus* indicating "(Delobel com. pers.)" without additional details.

054 *Bruchidius salinaensis* Zampetti & Toma, 2020, **syn. nov.** of *Bruchidius nalandus* (Pic, 1927), based on descriptions, examination of type material of *B. nalandus*, paratype (♂), "Ceylon, Nalanda, 2de trim. 89, I. Z. Kannegieter" (MNHN), and study of additional specimens of *B. nalandus* (collection of K.-W. Anton).

055 *Bruchidius samai* Zampetti & Toma, 2020, **syn. nov.** of *Bruchidius astragali* (Boheman, 1828), based on descriptions, examination of lectotype (♂, NHRS) of *Bruchus astragali* designated by Anton (1999a), and study of additional specimens of *B. astragali*, including those from Turkey (collection of K.-W. Anton). The holotype of *B. samai* differs from the typical form of *B. astragali* in darkened reddish parts of antennae as well as pro- and mesotibiae. However, such differences are within the intraspecific variability of *B. astragali*.

056 *Bruchus asiaticus* Iablokoff-Khnzorian, 1964, **syn. nov.** of *Bruchus sibiricus* Germar, 1823, based on descriptions, examination of type material: *B. sibiricus* syntype (♂), "Gb. [triangular yellow label, handwritten with black ink] || coll. Kraatz [printed] || Schilsky det. || sibiricus Gebl. [handwritten with black ink] || Umprepariert: J. Strejček 1959 [handwritten with black ink by J. Strejček] || sibiricus Germ., Sr., 21* Altai [yellow label, handwritten with black ink by F. A. Gebler]" (SDEI); and the photograph of holotype (♂) of *B. asiaticus* (collection of Mark Kalashian, Yerevan, Armenia) made by T. Ghrejyan (Yerevan, Armenia) and kindly provided by D. Kasatkin (Rostov-on-Don, Russia).

057 *Longebruchus* Pic, 1953, **syn. nov.** of *Pygobruchidius* Pic, 1951, based on descriptions and study of additional specimens of the genus *Pygobruchidius* (collection of K.-W. Anton). The combination of characters of *Longebruchus* is the same as in *Pygobruchidius*.

058 *Protobruchus* Zampetti & Toma, 2020, **syn. nov.** of *Callosobruchus* Pic, 1902, based on descriptions and study of additional specimens of the genus *Callosobruchus*, especially those of the *C. maculatus* species-group (collection of K.-W. Anton). The combination of characters defining the genus *Protobruchus* are also present within *Callosobruchus*, especially in *C. subinnotatus* (Pic, 1914) (member of the *C. maculatus* species-group). Absence of carinae on metatibiae mentioned in description of

Protobruchus is incorrect as carinae are hidden under relatively dense setation but barely visible on the photograph provided by Zampetti & Toma (2020).

059 *Protobruchus togoensis* Zampetti & Toma, 2020, **syn. nov.** of *Callosobruchus subinnotatus* (Pic, 1914), based on descriptions and study of additional specimens of *C. subinnotatus* (collection of K.-W. Anton), including examination of male genitalia. The photograph of the holotype of *P. togoensis* shows a relatively short mucro, which usually varyies in *C. subinnotatus* from double to triple the length of the coronar denticle of metatibial apex.

060 *Serratobruchidius* Yus Ramos, 2013, **syn. nov.** of *Bruchidius* Schilsky, 1905, based on the descriptions. The type species of the genus *Bruchidius* is *Bruchus quinqueguttatus* Olivier, 1800, designated by Bridwell (1932), and belonging to the *Bruchidius serraticornis* species-group defined by B. Delobel & A. Delobel (2006) and Yus Ramos (2007a). Later, Yus Ramos (2013a) proposed a new genus *Serratobruchidius* for the *B. serraticornis* species-group and selected *Bruchus rubiginosus* Desbrochers des Loges, 1869 as the type species. Both, *Bruchus quinqueguttatus* and *Bruchus rubiginosus*, belong to the *B. serraticornis* species-group and thus *Serratobruchidius* is the same as *Bruchidius* and is here synonymized with it.

061 *Caryedon cyprus* C. D. Johnson, Southgate & A. Delobel, 2004, **syn. nov.** of *Caryedon germari* (Küster, 1845a), based on descriptions and examination of lectotype (♂, SDEI) of *Caryoborus germari* Küster, 1845, as well as study of additional specimens of *Caryedon germari*, including those from Cyprus (collection of K.-W. Anton).

062 *Caryedon halperini* Anton & A. Delobel, 2004, **syn. nov.** of *Caryedon mesra* C. D. Johnson, Southgate & A. Delobel, 2004, based on descriptions and examination of type series of *C. halperini* (RMCA, collection of K.-W. Anton). Figures of the male genitalia of *C. mesra* provided in the original description are not sufficiently detailed, however, the photograph of the same structures kindly provided by A. Delobel confirms the established synonymy.

Resurrections from synonymy

063 *Bruchidius eupatorius* Arora, 1980 was synonymized with *Bruchidius nalandus* (Pic, 1927) by Anton (2010a). *Bruchidius eupatorius* is restored to species rank based on descriptions, examination of type material of *B. nalandus*: paratype (♂), "Ceylon, Nalanda | 2de trim. 89 I. Z. Kannegieter" (MNHN), as well as study of additional specimens in the collection of K.-W. Anton. *Bruchidius eupatorius* can be easily distinguished from *B. nalandus* by the quadrate scutellum and simply hooked preapical tubercle at elytral stria IV (oblong scutellum and double-hooked tubercle at elytral striae III–IV in *B. nalandus*). In addition, host plants of both species are different, *Eupatorium* spec. in *B. eupatorius* and *Tephrosia purpurea* in *B. nalandus*.

064 *Conicobruchus alpinus* Arora & Singal, 1979 was synonymized with *Conicobruchus kashmiricus* (Pic, 1929b) by Singal & Pajni (1986). *Conicobruchus alpinus* is restored

to species rank based on descriptions and examination of type material of *Bruchus kashmiricus*: syntype (♂), "Islamabad | Kashmir" (MNHN). *Conicobruchus alpinus* can be easily distinguished from *C. kashmiricus* (data in parentheses) by its on average larger size, 2.6–3.5 mm from apex of pronotum to apex of elytra (vs. on average smaller size, 2.3–2.9 mm), and vestiture less dense with short, thin, greyish setae (vs. moderately dense with whitish, long and thick setae), eyes in males strongly bulging (vs. moderately bulging), maximum width of the head including eyes about 1.7 times (vs. 1.2 times) as wide as the width behind eyes, and front tibiae in males slender (vs. broadened), tibial width 0.5–0.6 times (vs. 0.8–0.9 times) the length of first protarsomere. In the previous Edition both discussed taxa were listed in *Bruchidius* (Anton 2010b), while in the present Edition they are placed in *Conicobruchus*.

065 *Kytorhinus* (*Kytorhinus*) *hauseri* Pic, 1906 was synonymized with *Kytorhinus* (*Kytorhinus*) *karasini* Fischer von Waldheim, 1809 by Hoffmann (1965). *Kytorhinus hauseri* is restored to species rank based on descriptions, examination of relevant type material as well as study of additional specimens in the collection of K.-W. Anton. Type material studied: *K. hauseri* syntype (♂), "DSUNGARIA, Borochoro-Gb., coll. Hauser 6.05 || M. Hauseri, Pic. Type. (ZMHUB)"; *Kytorhinus reitteri* Schilsky, 1905 holotype (♂), "Reitteri, *Schils. || Mongolia bor., Reitter" (ZMHUB). *Kytorhinus reitteri* was correctly considered a junior synonym of *K. karasini* by Legalov (1999). *Kytorhinus hauseri* can be easily distinguished from *K. karasini* (data in parentheses) by the following characters: colouration always completely black with large reddish apical patch on elytral apex and very rarely with small indifferent and elongate reddish patch at humeral callus (vs. colouration varying: legs from completely black to predominantly black with tibiae yellowish-reddish, elytra from black to predominantly yellowish-reddish with extreme base, praescutellar region, suture and first interval black), eyes not protuding, distance between eyes about 0.4 times as wide as the width of the head including eyes (vs. less than 0.3), pronotal posterior angles broadly rounded (vs. rectangular), elytra shorter, about 1.2 times as long as their width (vs. about 1.4 times), with greatest width at apical third (vs. at basal third), ventral valve broader, armature of internal sac with slender denticles distinctly longer than their base (vs. slender denticles as long as or shorter than their base).

066 *Megabruchidius tsinensis* (Pic, 1923) was synonymized with *M. dorsalis* (Fåhraeus, 1839) by Yus Ramos (2009) without examination of the type material. *Megabruchidius tsinensis* is restored to species rank based on descriptions, examination of relevant type material and study of additional specimens from the collection of K.-W. Anton. Type material examined: *Bruchus tsinensis* syntype (♂), "Pe Yen Tsin | Yunnan" (MNHN); *Laria tonkinea* Pic, 1904, syntype (♂), "Tonkin" (MNHN). Having the dorsum more or less reddish brown, *M. tsinensis* is more similar to *M. tonkineus* but can be easily distinguished by following characters: hind tibia with

mucro of common size, twice as long as coronar denticles at hind tibial apex (elongated to one third of tibial length in *M. tonkineus*), male genitalia with apex of dorsal valve subtruncate, at sides broadly rounded (apex arcuate in *M. tonkineus*). The latter character separates *M. tsinensis* also from *M. dorsalis*, in addition to the colouration of the elytra.

New combinations

067 *Bruchidius caeruleus* (Champion, 1919), **comb. nov.** from *Conicobruchus*, based on morphological characters, including male genitalia.

068 *Bruchidius postremus* (Gebler, 1839), **comb. nov.** from *Bruchus*, based on morphological characters, including male genitalia, structures of antennae, male mesotarsomere III, shape of body, elytral vestiture, and examination of the syntype (♂) of its synonym, *B. maculatus* Gebler, 1830: "Gebler. || Mont Altai || Bruchus | postremus | Schönh. | maculatus | Gebler" (MZH).

069 *Conicobruchus arcuatipes* (Decelle, 1977), **comb. nov.** from *Bruchidius*, based on morphological characters, including male genitalia.

070 *Conicobruchus letourneuxi* (Caillol, 1894), **comb. nov.** from *Bruchidius*, based on morphological characters, including male genitalia. Based on the structure of antennae, pronotum, shape of body and male genitalia, it is closely related to *C. skaifei* (Pic, 1928).

071 *Conicobruchus steveni* (Gyllenhal, 1839), **comb. nov.** from *Bruchidius*, based on morphological characters, including male genitalia.

072 *Pygobruchidius aureus* (Arora, 1977), **comb. nov.** from *Bruchidius*, based on morphological characters, including male genitalia.

073 *Pygobruchidius baronii* (Decelle, 1977), **comb. nov.** from *Bruchidius*, based on morphological characters, including male genitalia.

074 *Pygobruchidius bilineatopygus* (Pic, 1938), **comb. nov.** from *Bruchidius*, based on morphological characters, including male genitalia.

075 *Pygobruchidius bimaculatopygus* (Anton, 2000), **comb. nov.** from *Bruchidius*, based on morphological characters, including male genitalia.

076 *Pygobruchidius cadenati* (Pic, 1951), **comb. nov.** from *Bruchidius*, based on morphological characters, including male genitalia.

077 *Pygobruchidius densevestitus* (Anton, 2000), **comb. nov.** from *Bruchidius*, based on morphological characters, including male genitalia.

078 *Pygobruchidius devexus* (Fåhraeus, 1839), **comb. nov.** from *Bruchidius*, based on morphological characters, including male genitalia.

079 *Pygobruchidius dimorphus* (Arora, 1977), **comb. nov.** from *Bruchidius*, based on morphological characters, including male genitalia.

080 *Pygobruchidius flavovirens* (Arora, 1977), **comb. nov.** from *Bruchidius*, based on morphological characters, including male genitalia.

081 *Pygobruchidius lineolatus* (Motschulsky, 1874), **comb. nov.** from *Bruchidius*, based on morphological characters, including male genitalia.

082 *Pygobruchidius maculipygus* (Champion, 1919), **comb. nov.** from *Bruchidius*, based on morphological characters, including male genitalia.

083 *Pygobruchidius mimosae* (Arora, 1977), **comb. nov.** from *Bruchidius*, based on morphological characters, including male genitalia.

084 *Pygobruchidius natalensis* (Pic, 1903), **comb. nov.** from *Sulcobruchus*, based on morphological characters, including male genitalia.

085 *Pygobruchidius planophthalmus* (Anton, 1994), **comb. nov.** from *Bruchidius*, based on morphological characters, including male genitalia.

086 *Pygobruchidius ptiniloides* (Fåhraeus, 1839), **comb. nov.** from *Bruchidius*, based on morphological characters, including male genitalia.

087 *Pygobruchidius pygomaculatus* (Arora, 1977), **comb. nov.** from *Bruchidius*, based on morphological characters, including male genitalia.

088 *Pygobruchidius saundersi* (Jekel, 1855), **comb. nov.** from *Bruchidius*, based on morphological characters, including male genitalia.

089 *Pygobruchidius schilskyi* (Arora, 1977), **comb. nov.** from *Bruchidius*, based on morphological characters, including male genitalia.

090 *Pygobruchidius schroderi* (Pic, 1930), **comb. nov.** from *Bruchidius*, based on morphological characters, including male genitalia.

091 *Pygobruchidius senensis* (Pic, 1923), **comb. nov.** from *Bruchidius*, based on morphological characters, including male genitalia.

092 *Pygobruchidius sinaitus* (K. Daniel, 1908), **comb. nov.** from *Bruchidius*, based on morphological characters, including male genitalia.

093 *Pygobruchidius solanensis* (Arora, 1977), **comb. nov.** from *Bruchidius*, based on morphological characters, including male genitalia.

094 *Pygobruchidius sparsemaculatus* (Pic, 1913), **comb. nov.** from *Bruchidius*, based on morphological characters, including male genitalia.

095 *Pygobruchidius terrenus* (Sharp, 1886), **comb. nov.** from *Bruchidius*, based on morphological characters, including male genitalia.

096 *Pygobruchidius urbanus* (Sharp, 1886), **comb. nov.** from *Bruchidius*, based on morphological characters, including male genitalia.

Changes in rank

097 *Bruchidius lateobscurus* (Pic, 1904) **stat. nov.**, formely variety of *B. poupillieri* (Allard, 1868), is raised to species rank based on descriptions, examination of relevant type material as well as study of additional specimens. Type material studied: *Laria poupillieri* var. *lateobscura* (syntype, ♀, MHNH), "Bou Saada | 18 Mai 95"; *Bruchus poupillieri* (syntype, ♂, NHRS): "Algeria | Poupillier", (syntype, ♂, NHRS):

"Oran | [illegible]"; *Bruchidius poupillieri* var. *atramentarius* Schilsky, 1905 (three syntypes, unsexed, ZMHUB): "v. atramentarius Schils.".

098 *Butiobruchus* Prevett, 1966 is downgraded to a subgenus of *Pachymerus* Thunberg, 1805. Nilsson & C. D. Johnson (1993) synonymized *Butiobruchus* with *Pachymerus* Thunberg, 1805, but recently *Butiobruchus* was resurrected as valid genus by Yus Ramos (2018). Based on homogenity of characters of male and female genitalia within *Pachymerus*, *Butiobruchus* is downgraded to a subgenus of *Pachymerus*.

Lectotype designations

099 *Bruchus canescens* Motschulsky, 1874. Lectotype (by present designation): ♂, "Bruchus | canescens | Mus. [illegible] | Mesopotamia" (ZIN). Designated in order to preserve the stability of the nomenclature.

100 *Mylabris paupera* var. *corsica* Baudi di Selve, 1890. Lectotype (by present designation): ♂, "Porto Vecchio | Croissandeau. || Tibiis 4 anteriori | testaceis || M. paupera | v. corsica m. | Corsica [all labels handwritten by Baudi di Selve]" (MRSN), paralectotype (♂), "♂" (MRSN), two paralectotypes (♀♀), "♀" (MRSN). Designated in order to preserve the stability of the nomenclature.

101 *Bruchus suratus* Motschulsky, 1874. Lectotype (by present designation): ♂, "Turcia || Bruchus | suratus | Turcia Motsch." (ZMUM). Designated in order to preserve the stability of the nomenclature.

102 *Bruchidius gombo* Peyerimhoff, 1915. Lectotype (by present designation): ♂, "Bou Saada || Museum Paris | Coll. P. de Peyerimhoff | 1950 || Bruchidius | gombo Peyrh." (MNHN), paralectotype: ♂, same data (MNHN). Designated in order to preserve the stability of the nomenclature.

103 *Bruchus tesserulus* Motschulsky, 1874. Lectotype (by present designation): ♀, "Rus. mer. or. || Bruchus | tesserulus | Motsch. | Rus. Mer.or." (ZIN). Designated in order to preserve the stability of the nomenclature.

104 *Bruchus autumnalis* Motschulsky, 1874. Lectotype (by present designation): ♂, "Bruchus | autumnalis | Motsch. | Transcauc" (ZIN). Designated in order to preserve the stability of the nomenclature.

105 *Bruchus myobromae* Motschulsky, 1874. Lectotype (by present designation): ♂, "Taur [very small, red, handwritten in black ink] || Ex Myobro | ma utri- | gera Taur. || Bruchus | myobromae | Motsch. | Tauria [both labels obscured white, handwritten in black ink]" (ZIN); paralectotype: ♀, "Tauria [very small, red, handwritten in black ink]" (ZIN). Designated in order to preserve the stability of the nomenclature.

106 *Mylabris ochracea* Baudi di Selve, 1886. Lectotype (by present designation): ♂, "Syria" (MRSN). The specimen was provided with a lectotype label by M. F. Zampetti 1988, but to my knowledge the designation was never published. Designated in order to preserve the stability of the nomenclature.

107 *Bruchus incurvatus* Motschulsky, 1874. Lectotype (by present designation): ♀, "Cauc. || Bruchus | incurvatus | Motsch. | Alp. Caucas." [both labels obscured white,

handwritten with black ink]" (ZIN). Designated in order to preserve the stability of the nomenclature.

108 *Caryoborus germari* Küster, 1845. Lectotype (by present designation): ♂, "Acaciae Klug. | Gr. Germari | Küster pallidus | ?Ol Dalmat [pink label with black frame, handwritten by E. F. Germar]" (SDEI). Designated in order to preserve the stability of the nomenclature.

109 *Caryopemon luteonotatus* Pic, 1898. Lectotype (by present designation): ♂, "Inde || Bull. S. Zool. Fr. | 98 p. 173 || Type || luteonotatum | Pic" (MNHN); paralectotype: ♀, "ex Staud" (MNHN). All labels were handwritten by M. Pic. Designated in order to preserve the stability of the nomenclature.

110 *Bruchus schroderi* Pic, 1930. Lectotype (by present designation): ♂, "5/513. | Erfurt | M." (MNHN); paralectotype: ♀, "Erfurt | Allemagne | graines" (MNHN). Designated in order to preserve the stability of the nomenclature.

111 *Bruchus senensis* Pic, 1923. Lectotype (by present designation): ♂, "Yunnan || Bruchus | senensis Pic" (MNHN); paralectotype: ♀, "Yunnan | Fou" (MNHN), two paralectotypes: ♂ and ♀, "Yunnan | Sen" (MNHN). Designated in order to preserve the stability of the nomenclature.

Spellings

112 *Bruchidius dialensis* A. H. Ali & A. Z. M. Ali, 1987. The description of *B. dialensis* contains two different spellings: *dialensis* and *dialeusis*. The species was described from Diala and presumably the name was meant to be derived from the type locality. Hence the spelling *dialensis* is selected here as the correct original spelling in accordance with the principle of the first reviser (ICZN 1999, Art. 24.2.3).

113 *Bruchus brisouti* Kraatz, 1868. The spelling *brisouti* is an incorrect subsequent spelling of the original spelling *brisoutii*. Since the spelling *brisouti* is in prevailing usage it is deemed to be the correct original spelling (ICZN 1999, Art. 33.3.1).

114 *Bruchus perezi* Kraatz, 1868. The spelling *perezi* is an incorrect subsequent spelling of the original spelling *perezii*. Since the spelling *perezi* is in prevailing usage it is deemed to be the correct original spelling (ICZN 1999, Art. 33.3.1).

115 *Pygobruchidius saundersi* (Jekel, 1855). The spelling *saundersi* is an incorrect subsequent spelling of the original spelling *saundersii*. Since the spelling *saundersi* is in prevailing usage it is deemed to be the correct original spelling (ICZN 1999, Art. 33.3.1).

116 *Specularius erythraeus* (Pic, 1908). The spelling *erythraeus* is an incorrect subsequent spelling of the original spelling *erytraeus*. Since the spelling *erythraeus* is in prevailing usage it is deemed to be the correct original spelling (ICZN 1999, Art. 33.3.1).

117 The description of *Spermophagus mesopotameca* A. H. Ali & Kadhim, 1986 contains three different spellings: *mesopotameca*, *mespotameca* and *mesoptameca*. The

spelling *mesopotameca* is selected here as the correct original spelling in accordance with the principle of the first reviser (ICZN 1999, Art. 24.2.3).

118 The description of *Spermophagus swartukensis* A. H. Ali & Kadhim, 1986 contains two different spellings: *swaratukensis* and *swartukensis*. The spelling *swartukensis* is selected here as the correct original spelling in accordance with the principle of the first reviser (ICZN 1999, Art. 24.2.3).

New geographical records

Note. All specimens were identified by K.-W. Anton and deposited in his collection if not stated otherwise.

119 *Bruchidius annulicornis* (Allard, 1868). ALBANIA: Korce, Bamash (S of Erseke), 40°16.766′N 020°38.380′E, 940 m, 9.v.2017, 1 ♀, H. Mühle leg. New species for Albania.

120 *Bruchidius chloroticus* (Dalman, 1833). AUSTRALIA: W. Australia, 80 M beach env., viii.1991, 1 ♂, J. Roppel leg. New adventive species for Australia.

121 *Bruchidius comptus* (Sharp, 1886a). CHINA: Anhui: Dabieshan, 65 km SW of Huoshan, 1400 m, 21.–24.vi.1998, 1 ♂, Bolm leg.; Guizhou: Leigongshan, Xijang, 1200–1900 m, 29.v.–2.vi.1997, 1 ♂, Bolm leg. New species for provinces Anhui and Guizhou (China).

122 *Bruchidius halodendri* (Gebler, 1830). ARMENIA: Yerevan env., Razdan river valley, 1100 m, 27.v.1989, 4 spec., J. Strejček leg. (part of the series NMPC, ex. coll. J. Strejček). New species for Armenia.

123 *Bruchidius lateobscurus* (Pic, 1904). ALGERIA: Kherrata, 8.v.1987, 1 ♂, A. Warchałowski leg. SPAIN: Malaga, Ronda, El Burgo, Sierra de las Nieves, 1000 m, 16.v.1994, 1 ♂, D. Siede leg.; Madrid, Castilla la Nueva, Morata de la Tajuna, 4.v.1985, 1 ♀, D. Siede leg.; Guadaljara, Sacedon, 4.–5.v.1985, 1 ♀, D. Siede leg. First precise data for Algeria and Spain.

124 *Bruchidius lekefensis* Zampetti & Toma, 2020. ALGERIA: Djurdjura, 16.vi.1971, 1 ♂, Hoffer & Horák leg. New species for Algeria.

125 *Bruchidius lespedezae* (Iablokoff-Khnzorian, 1974). CHINA: Anhui: Dabieshan, 65 km SW of Huoshan, 1400 m, 21.–24.vi.1998, 1 ♂, Bolm leg.; Guizhou: Leigongshan, Xijang, 1200–1900 m, 29.v.–2.vi.1997, 8 spec., Bolm leg.; Hunan: Wulungshan, Tienzishan N. R., 800 m, 16.–18.vi.1997, 1 ♂, Bolm leg. New species for provinces of Anhui, Guizhou, and Hunan (China).

126 *Bruchidius lineatus* (Allard, 1868). SWITZERLAND: Ticino: Borello, Contrada Baronessa, 18.v.2008 (photographs by Vincenzo, www.naturamediterraneo.com/forum/topic.asp?TOPIC_ID=308530). New species for Switzerland.

127 *Bruchidius misellus* (Boheman, 1833). SPAIN: Tarragona, Sierra de Prades, 22.v.1989, 1 ♀, A. Elbert leg. First precise data for Spain.

128 *Bruchidius ochraceus* (Baudi di Selve, 1886). AZERBAIJAN: Nakhichevan, Achaibusova env., 20.–25.v.1986, 1 ♂, P. Romantsov leg. (collection of M. Yu. Savitskiy, Moscow, Russia). New species for Azerbaijan.

129 *Bruchidius olivaceus* (Germar, 1823). MONTENEGRO: Sutomoro env., Polenica, 28.v.1974, 2 ♂♂, J. Strejček leg. ARMENIA: Garni prov., Gocht, Azat Riv. valley, 1600 m, 27.v.1989, 1 ♂, Jindra leg. GEORGIA: Tbilisi, vi.1957, 1 ♂ 1 ♀, Dlabola leg. (all specimens NMPC, ex coll. J. Strejček). New species for Armenia, Georgia and Montenegro.

130 *Bruchidius postremus* (Gebler, 1839). KAZAKHSTAN: 60 km E of Dzhambul, Mojunkum Desert, Akirtobe, 1.–9.v.1994, 1 ♂, P. Čechovský leg.; Tekeli prov., Taldi Kurgan, 7.–17.v.1991, 3 ♀♀, N. Tselikov leg. (NHMB; collection of K.-W. Anton); S. Tarbagatay Mt. Range, Blagodamoye, 20.–26.v.1991, 1 ♂ 1 ♀, N. Tselikov leg., (NHMB; collection of K.-W. Anton). KYRGYZSTAN: Kara-Archa River, Kirgizky Mt. Range, 9.v.1994, 1 ♂; Arslanbad [=Arslanbob], 72.9°E 41.3°N, 12. –13.vi.1995, 1 ♀, J. Halada leg.; 130 km N of Namangan, Chatkal Ridge, Jylgyr, 2000 m, 5.vi.1995, 1 ♀, P. Čechovský leg.; N. Inter-Tian-Shan, Kirghizsky Mt., Ala-Artacha, 2900 m, 19.vi.1995, 1 ♀, V. Shchurov leg. (NHMB). UZBEKISTAN: Chimgan, Chatkal Ridge, 26.–28.vi.1989, 2 ♀♀, K. Majer leg. (NHMB; collection of K.-W. Anton); Chimgan, 1750 m, 9.vi.1982, 1 ♀, J. Jelínek leg. New species for Kazakhstan, Kyrgyzstan and Uzbekistan.

131 *Bruchidius pubicornis* Lukjanovitch & Ter-Minassian, 1957. ARMENIA: Sevan, 2500 m, 17.vi.1979, 1 ♂ 3 ♀♀, V. Švihla leg.; Lake Sevan, 16.–21.vi.1981, 1 ♂ 2 ♀♀, J. Visa leg.; Sevan env., 2100–2200 m, 2.vi.1989, 1 ♂ 1 ♀, J. Strejček leg. (NMPC, ex coll. J. Strejček; some specimens in collection of K.-W. Anton). New species for Armenia.

132 *Bruchidius siliquastri* A. Delobel, 2007. GREECE: Ionnina, Konitsa, 39°58.102′N 020°39.752′E, 420 m, 11.v.2011, 1 ♀, H. Mühle leg. New species for Greece.

133 *Bruchidius spathopus* Iablokoff-Khnzorian, 1960. TURKEY: Kars, Sukhoi Fontan, 21.vi.1911, 8 spec., H. Kulzer leg. New species for Turkey.

134 *Bruchidius trifolii* (Motschulsky, 1874). CYPRUS: Paphos district, 2 km SE of Gialia, 9.v.2006, 1 ♂ 1 ♀, H. Mühle leg. New species for Cyprus.

135 *Bruchidius varipes* (Boheman, 1839). SWITZERLAND: Valais: s / Leuk, 30.v.1966, 1 ♂, P. Scherler leg. (collection of C. Germann, Basel, Switzerland). New species for Switzerland.

136 *Bruchus dentipes* (Baudi di Selve, 1886). SPAIN: Huesca, Pyrenées Occidentales, Garganta Mt., 2200 m, 3.vi.1988, 1 ♂, A. Warchałowski leg. First precise data for Spain.

137 *Bruchus ibericus* Anton, 1999. ALGERIA: Blida env., Chrea, 10.v.1987, 1 ♂, V. Kubáň leg., K.-W. Anton det. (NHMB). MOROCCO: Azrou env., 1650 m, 28.iv.1998, 1 ♀, M. Döberl leg. New species for Morocco and first precise data for Algeria.

138 *Bruchus laticollis* Boheman, 1833. CYPRUS: Paphos District, Akamas env., iii. 2013, 1 ♀, J. Roppel leg. New species for Cyprus.

139 *Bruchus lentis* Frölich, 1799. NORTH MACEDONIA: Ohrid, Velestovo, 41°05.508′N 020°49.139′E, 890 m, 5.v.2017, 1 ♂, H. Mühle leg. New species for North Macedonia.

140 *Bruchus libanensis* Zampetti, 1993. SPAIN: Gerona, W of San Hilary Sacalm, 900 m, 7.vi.2000, 1 ♂ 4 ♀♀, K.-W. Anton leg. The species was so far known only from the east

Mediterranean area and the abovementioned record from Spain is thus curious and is considered as introduced. New adventive species for Spain.

141 *Bruchus loti* Paykull, 1800. CHINA: Shaanxi: Qingling Shan Mts., rd. Baoji to Talibai vill., pass 35 km S of Baoji, 21.–23.vi.1998, 1 ♀, O. Šafránek & M. Trýzna leg. New species for Shaanxi (China).

142 *Bruchus luteicornis* Illiger, 1794. GREAT BRITAIN: Nottinghamshire, New Ollerton, x.2012, 1 ♂ (photographs identified as *B. rufipes* and available at http://www.eakringbirds.com/eakringbirds5/insectinfocusbruchusrufipes.htm). New species for Great Britain.

143 *Bruchus occidentalis* Lukjanovitch & Ter-Minassian, 1957. SPAIN: Gerona, Vidreras, Puig Ventos, 19.v.1993, 13 spec., M. Balkenohl leg.; Catalunya, Puig Ventos E, 14.v.1997, 1 ♂, Eifler leg. First precise records for Spain.

144 *Callosobruchus antennifer* Singal & Pajni, 1990. INDIA: Uttarakhand, Gangnani, 1250 m, 13.-20.vi.1981, 1 ♀, M. Brancucci leg. New species for Uttarakhand (India).

145 *Callosobruchus bhutanicus* Decelle, 1977. INDIA: Arunachal Pradesh, betw. Dirang & Bomdila pass, 27°19′N 92°22′E, 2200 m, 15.vi.2004, 1 ♀, L. Dembický leg. (NHMB). New species for Arunachal Pradesh (India).

146 *Callosobruchus chinensis* (Linnaeus, 1758). INDIA: Arunachal Pradesh, Dirang env., 1400–1700 m, 27°21′N 92°13–16′E, 1.–9.vi.2004, 37 spec., L. Dembický leg. (NHMB). New species for Arunachal Pradesh (India).

147 *Caryedon germari* (Küster, 1845). CYPRUS: Paphos district, Prastio, 290 m, 31.vii.2020, 1 ♂, K.-W. Anton leg.; Paphos district, Nikokleia, 5.v.2006, 1 ♀, H. Mühle leg. New species for Cyprus.

148 *Caryedon kizilkumensis* Ter-Minassian, 1978. KYRGYZSTAN: Tash-Kumyr env., 3.vii.2000, 2 ♀♀, V. G. Dolin leg. New species for Kyrgyzstan.

149 *Caryobruchus gleditsiae* (Linnaeus, 1763). SPAIN: Zaragoza, Nuevo Leon, 1800 m, ex sem. Palm., 1 ♀. (NMPC, ex coll. J. Strejček). First precise record for Spanish mainland.

150 *Conicobruchus impubens* (Pic, 1927). NEPAL: Arun valley, Chichila, 2050 m, 7.vi.1992, 1 ♀, J. Probst leg. (NMW). New species for Nepal.

151 *Conicobruchus letourneuxi* (Caillol, 1894). IRAQ: Bagdad, 1 ♀, Kálalová leg., (NMPC, ex coll. J. Strejček). New species for Iraq.

152 *Kytorhinus* (*Kytorhinoides*) *thermopsis* Motschulsky, 1874. RUSSIA: South Siberia, Tuva, Tandinskiy distr., nr. Chadyn Lake, 20.–28.v.1998, 1 ♀, S. N. Vashchenko leg. UZBEKISTAN: 50 km NE of Tashkent, Aktash, 26.–30.vi.1977, 1 ♀, Muche leg. New species for Uzbekistan and first record from the South Siberian Region (Russia).

153 *Kytorhinus* (*Kytorhinus*) *mongolicus* Ter-Minassian, 1973. RUSSIA: South Siberia, Tuva, Khorumnung-Tayga Mts., Buren River valley, 1000 m, 8.–15.vi.1999, 1 ♀, A. Klimenko leg.; Tuva, near Kyzyl, Ka-Chem River, 4.–6.vi.1997, 1 ♀, S. Vashchenko leg. First records from the South Siberian Region (Russia).

154 *Kytorhinus* (*Pygobruchus*) *senilis* (Solsky, 1870). CHINA: Heilongjiang: 160 km N Harbin, Taching City, Red Flag Forest, 9.vi.1994, 1 ♀, V. Kuznetsov leg. New species for Heilongjiang (China).

155 *Pseudopachymerina spinipes* (Erichson, 1834). GREECE: Argolis, Nauplia, 37°34′N 22°48′E, ex seeds of *Acacia* sp., 14.vi.1993, 2 spec., D. Borisch leg., (collection of D. Borisch, Stockholm, Sweden); Rhodes, Archangelos, 28.iv.1993, 1 ♀, A. Elbert leg.; Crete, Chania distr., Georgiopolis, 35°21′37.6″N 24°15′05.9″E, ex seeds of *Acacia farnesiana*, 4.–8.ix.2016, 185 spec., R. Jakob leg. New adventive species for Rhodes and first precise records for Greek mainland and island of Crete.

156 *Pygobruchidius bilineatopygus* (Pic, 1938). FRANCE: Corsica, 1 ♀. (NMPC, ex coll. J. Strejček). QATAR: Doha Zoo, 29.ix.2001, 1 ♂, É. Pierre leg. New adventive species for France and Qatar.

157 *Pygobruchidius sparsemaculatus* (Pic, 1913). GERMANY: "Erfurt | Allemagne | (graines)" [handwritten by M. Pic], 1 spec. (MNHN). New adventive species for Germany.

158 *Salviabruchus retusus* (Baudi di Selve, 1886). IRAN: Bakhtiari prov., Gardameh-ye-Cheri (Samsami), 3000 m, 12.vii.2004, 1 ♀, H. Mühle leg. New species for Iran.

159 *Spermophagus abdominalis* (Fabricius, 1781). CHINA: Chongqing: Jinfo Shan, 7.–13.v.2018, 1 ♀, E. Kučera leg. New species for Chongqing (China).

160 *Spermophagus altaicus* Karapetjan, 1973. RUSSIA: Tuva, nr. Kyzyl, Ka-Chem river, 4.–6.vi.1997, 1 ♂, S. Vashchenko leg. First record from the South Siberian Region (Russia).

161 *Spermophagus calystegiae* (Lukjanovitch & Ter-Minassian, 1957). QATAR: Al Muraydah, Wadi Sadria Farm, 25.iii.2002, 1 ♂, M. Martinez leg.; Rawdat Al Faras University Farm, 24.iii.2002, 1 ♀, M. Martinez leg. RUSSIA: Tuva, Tandinsky distr., nr. Shadyn lake, salines and sands, 20.–28.v.1998, 56 spec., S. Vashchenko leg. New species for Qatar and first record from the South Siberian Region (Russia).

162 *Spermophagus pubiventris* Baudi di Selve, 1886. CYPRUS: Larnaka district, Hala Sultan Tekkesi, 34°53′11.9″N 33°36′28.0″E, 29.iii.2017, 1 ♂ 3 ♀♀, K.-W. Anton leg. New species for Cyprus.

163 *Spermophagus shamszadehi* A. Delobel & Sadeghi, 2013. IRAN: Teheran province, Golhak, 1700 m, 9.–23.vi.1961, 13 spec., J. Klapperich leg. (SMNS; ZFMK; HNHM; collection of K.-W. Anton). Additional records for Iran.

164 *Stator limbatus* (Horn, 1873). OMAN: nr. Barka farm, 50 m, 23°40′50″N 57°50′40″E, 4.–5.viii.1994, 1 ♂, M. D. Gallagher leg. YEMEN: Al Kadan, ex seeds of *Albizzia lebbek*, 3.viii. 2003, 1 ♂ 1 ♀, A. S. Al Absi leg. This species was listed for Oman and Yemen in the first Edition (Anton 2010b) based on abovementioned specimens but without providing any data. Specified records for Oman and Yemen.

Comments on distribution

165 *Bruchidius albopictus* (Allard, 1884). Ricci & Zampetti (2007) noted three specimens from Ifrane, Morocco. This species is a strictly eastern Mediterranean species, therefore the record from Morocco is considered as adventive.

166 *Bruchidius annulicornis* (Allard, 1868) is an east Mediterranean species. The record from Italy by Zampetti & Ricci (2012) is based on misidentification. Table 28 in their publication shows a female of very variable *B. varius* (Olivier, 1800) with completely red antennae, however females of *B. annulicornis* have at least antennomeres V–VIII black. Accordingly, occurrence in Italy is not accepted in the present Edition.

167 *Bruchidius astragali* (Boheman, 1828) is a steppe species distributed from Central Asia westwards to Ukraine and Turkey, southwards to Iran and Iraq. The record of *B. astragali* from Bulgaria (A. Delobel & B. Delobel 2007) was based on misidentification as the figures of male genitalia in their paper refer to *B. varipes* (Boheman, 1839). The records of *B. astragali* from Czech Republic and Slovakia by Benedikt & Mantič (2006) also refer to *B. varipes*. The record from North Macedonia, which appeared in the first edition (Anton 2010b) was based on a single misidentified specimen, therefore is not adopted in the present Edition.

168 *Bruchidius bituberculatus* Schilsky, 1905 is an east Mediterranean species. The record from Italy by Zampetti & Ricci (2012) is surely based on misidentification. Table 44 in their publication shows a female specimen with completely pubescent pygidium, while females of *B. bituberculatus* have pygidium with large shiny and glabrous area without setae. Accordingly, occurrence in Italy is not accepted in the present Edition.

169 *Bruchidius corsicus* (Baudi di Selve, 1890). Kasatkin (2000) listed *B. pauper* (= *B. corsicus*) from Kabardino-Balkar Republic in Northern Caucasus. This record was based on misidentified specimens of a species of the *Bruchidius astragali* species-group (D. Kasatkin, Rostov-on-Don, Russia, pers. comm.) and thus is not accepted here.

170 *Bruchidius halodendri* (Gebler, 1830) is a Central Asian species reaching westwards to Armenia and Iran. The occurrence in Moldavia (Bacal et al. 2013) is quite likely a misidentification of *B. glycyrhizae* (Gyllenhal, 1839). Therefore, the record from Moldavia is not listed accepted in the present Edition.

171 *Bruchidius jocosus* (Gyllenhal, 1833) is a strictly west Mediterranean species. The records from Greece and Rhodes (Ricci & Zampetti 2006) as well as Syria (Ricci & Zampetti 2007) are probably based on Zampetti's misidentification of *B. jocosus* with *B. rubiginosus* (Desbrochers des Loges, 1869) mentioned by Yus Ramos (2010c). The records from France and Italy were considered questionable by Yus Ramos & Labrique (2015) and need confirmation. All the abovementioned records are not accepted in the present Edition.

172 *Bruchidius lekefensis* Zampetti & Toma, 2020. The record of *Bruchidius varipes* (Boheman, 1839) from Magador (= Essaouira), Morocco by Luca (1961) refers very

probably to *B. lekefensis*. However, this record requires verification, thus occurrence in Morocco is not accepted in the present Edition.

173 *Bruchidius martinezi* (Allard, 1868) is a strictly west Mediterranean species. Records from Bulgaria (Stojanova & György 2011), Georgia (Batiashvili & Elerdashvili 1969), Greece (Ricci & Zampetti 2006, Yus Ramos & Angelini 2018) and Turkey (Yus Ramos & Angelini 2017) are surely based on misidentifications, therefore not adopted in the present Edition. To date the author has not seen any specimens of the true *B. martinezi* from the eastern part of the Mediterranean Region but only those of a very similar but yet unknown species, with antennomeres less enlarged and elytra distinctly shorter than *B. martinezi*.

174 *Bruchidius meleagrinus* (Gené, 1839) is a strictly west Mediterranean species. The occurrence in Syria mentioned by Ricci & Zampetti (2006) without any details was undoubtedly based on misidentified specimen(s). Therefore, occurrence in Syria is not accepted in the present Edition.

175 *Bruchidius misellus* (Boheman, 1833) was described from "Dalmatia". Croatia is listed in the present Catalogue, but further confirmation is needed because no additional specimens from this country are known. Yus Ramos (2019) listed *B. misellus* without further data for Iberian Peninsula. Therefore, the first accurate locality data for Spain are provided in the New geographical records chapter.

176 *Bruchidius mulsanti* (Brisout de Barneville, 1863). Record from Austria (Anton 1991) was based on misidentified aberrant specimen of *B. seminarius* (Linnaeus, 1767) and hence is not accepted in the present Edition.

177 *Bruchidius murinus* (Boheman, 1828) is a circummediterranean species reaching northwestwards to central France and Valais Region of Switzerland, and northeastwards to Croatia. Yus Ramos & Hadulla (2019) listed two specimens from Wahner Heide, situated right next to the International Airport Cologne Bonn, Germany. Further confirmation of establishment is needed, and the record is considered adventive in the present Catalogue.

178 *Bruchidius myobromae* (Motschulsky, 1874) is a steppe species distributed in Ukraine and South European Territory of Russia, reaching eastwards to extreme West Kazakhstan. The record from Iran listed in the previous Catalogue (Anton 2010b) was based on misidentification of a single female specimen of *B. virgatus* (Fåhraeus, 1839) and thus is rejected in the present Edition.

179 *Bruchidius niger* Anton, 2004 is an Iberomaghrebian species, not endemic to the Iberian Penisula as noted by Yus Ramos (2019).

180 *Bruchidius nudus* (Allard, 1868). The curious record of a single specimen from Wahner Heide in Central Germany, close to the International Airport Cologne Bonn (Yus Ramos & Hadulla 2019) is considered as adventive.

181 *Bruchidius olivaceus* (Germar, 1823). The type locality of *B. olivaceus* is "Halae Saxonum" (= Halle, Germany) which is highly probably erroneous. This species

is very common in Croatia where Germar travelled and collected during 1811. Yus Ramos (2012a) tentatively listed *B. olivaceus* for Iberian Peninsula but the occurence was never confirmed. *Bruchidius olivaceus* was recorded as adventive species for Great Britain by Aldridge & Pope (1986), however Fig. 8 in their publication clearly shows the aedeagus of closely related *B. unicolor* (Olivier, 1800). Consequently, occurences in Germany, Great Britain and Spain are not included in the present Edition.

182 *Bruchidius prolongatus* Lukjanovitch & Ter-Minassian, 1954. The occurrence in Iran mentioned by Ghahari & Borowiec (2017) was based on a single female specimen identified "*Bruchidius* prope *prolongatus*" by Anton (1999a). This record is not accepted in the present Edition.

183 *Bruchidius pubicornis* Lukjanovitch & Ter-Minassian, 1957 is a Caucasian species. The record from Hungary by György & Merkl (2005) was based on misidentification of *B. varipes* (Boheman, 1839) and is not accepted here.

184 *Bruchidius quinqueguttatus* (Olivier, 1800). A single specimen of this strictly east Mediterranean species was recorded from Tenerife, Canary Islands by Ricci & Zampetti (2007). It is considered as adventive in the present Edition.

185 *Bruchidius reitteri* Schilsky, 1906 is a Caucasian species. A single specimen recorded from Tunisia by Ricci & Zampetti (2007) is undoubtedly a misidentification of *B. lekefensis* Zampetti & Toma, 2020 and is not accepted in the present Edition.

186 *Bruchidius rufisurus* (Allard, 1884) is an east Mediterranean species. Citations from Algeria and Italy (Calabria, Sicily) by Zampetti & Ricci (2012) are misidentifications. Those authors mentioned specimens with partially red elytra, which however are always completely black in *B. rufisurus*, and the female specimen shown on table 45 of their publication is a misidentified male of *B. albolineatus*. Occurrences in Algeria and Italy are therefore not accepted in the present Edition.

187 *Bruchidius seminarius* (Linnaeus, 1767). The records of *Bruchidius seminarius* from Canary Islands by Ricci & Zampetti (2007) were considered to be misidentified specimens of *B. taorminensis* by Yus Ramos (2008), therefore they are not accepted in the present Edition.

188 *Bruchidius sericatus* (Germar, 1823) is a circum-Mediterranean species reaching northwestwards to South France and northeastwards to southern Slovakia. The record of a single specimen from an unknown locality named "Rheinland. Altenehg (Germany)" by Yus Ramos & Hadulla (2019) needs further confirmation of establishment. It is considered as adventive in the present Edition.

189 *Bruchidius villosus* (Fabricius, 1792). A single specimen recorded from Canary Islands by Ricci & Zampetti (2007) is considered as adventive in the present Edition.

190 *Bruchidius virgatoides* Lukjanovitch & Ter-Minassian, 1957. The occurrence in Turkey listed in the first Edition (Anton 2010b) is based on misidentification of a single male specimen, which belongs to an undescribed species close to *B. virgatoides*. Thus, the record is not accepted in the present Edition.

191 *Bruchus brachialis* Fåhraeus, 1839. An old record from Gran Canaria, Canary Islands mentioned by Machado & Oromí (2000) is based on a misidentification of *B. canariensis* Decelle, 1975 or *B. signaticornis* Gyllenhal, 1833 (Yus Ramos 2012a) and is not accepted here.

192 *Bruchus brisouti* Kraatz, 1868 is a rare west Mediterranean species but recently it was recorded from Greece by Yus Ramos & Angelini (2018). The record needs confirmation and is not considered in the present Edition. It was erroneously considered as an endemic species to Sardinia by Zampetti & Ricci (2012).

193 *Bruchus lentis* Frölich, 1799. The records from the Canary Islands by Oromí & García (1995) are based on misidentification of *B. signaticornis* Gyllenhal, 1833 (Yus Ramos 2012b) and hence are not accepted here.

194 *Bruchus perezi* Kraatz, 1868 is a west Mediterranean species. However, Schilsky (1905) mentioned also "Kaukasus", which was repeated by Lukjanovitch & Ter-Minassian (1957). Decelle & Lodos (1989) mentioned Turkey without providing exact locality and with incomplete reference to Lukjanovitch. Finally, "Syrie" was listed by Hoffmann (1945). All those records were accepted by Ricci & Zampetti (2007). Apart from that, Kergoat & Alvarez (2008) noted "Tu." for Turkey as the origin of specimen(s) used for morphometric studies (but "Tu." may even be Tunisia). All abovementioned records were not provided with further details on collecting data or deposition of used specimens, thus they are not accepted in the present Edition. The records from Georgia (Batiashvili & Elerdashvili 1969) and from Greece (Borowiec 1986c) are surely based on misidentifications, thus are not accepted here as well.

195 *Bruchus rufipes* Herbst, 1783 is a circum-Mediterranean species reaching northwards to Central France, Switzerland and the Czech Republic. All records from Great Britain, Germany, including recent record by Yus Ramos & Hadulla (2019), and Alsace Region of France probably pertain to the very similar specimens of *B. occidentalis*. These records are not accepted in the present Edition.

196 *Bruchus sibiricus* Germar, 1823 is a montane species distributed in Central Asia, with western limits of distribution in the Anatolian Highlands in Northeast Turkey. It is not an east Mediterranean species as stated by Yus Ramos & Angelini (2018). Those authors also recorded a single specimen from Skaloti, NE Greece. This locality does not fit the habitat preference of *B. sibiricus* and is considered as misidentification not accepted in the present Edition.

197 *Caryedon kizilkumensis* Ter-Minassian, 1978. The record of *Caryedon prosopidis* Arora, 1977 from Iran (Ghahari & Borowiec 2017) is attributed to *C. kizilkumensis* in the present Edition.

198 *Caryedon serratus* (Olivier, 1790). Most of European records of *C. acaciae* (Gyllenhal, 1833) by various authors refer to *C. serratus*, especially if association with *Arachis hypogea* is mentioned.

199 *Kytorhinus* (*Kytorhinus*) *quadriplagiatus* (Motschulsky, 1839). The occurrence in Russian East Siberia and Turkmenistan listed in the First Edition (Anton 2010b) was based on misidentifications of specimens of *K. karasini* Fischer von Waldheim, 1809 and are not accepted in the present Edition.

200 *Megabruchidius dorsalis* (Fåhraeus, 1839). The occurrence in Mongolia is based on the type locality of *Bruchus rufopygalis* Motschulsky, 1874 (synonym of *M. dorsalis*) only. However, it must be considered as a dubious record because its host plant (*Gleditsia* spp.) does not naturally occur there.

201 *Megacerus eulophus* (Erichson, 1847) was reported from Hamburg (Germany) by Zacher (1949). This record is confirmed here based on examination of two specimens labelled: "Hamburg Stadt | 27.VIII.37 | Lebensmittel Lgr. || leg. Nehls" (ex collection P. Brandl, Kolbermoor, Germany, now in the collection of K.-W. Anton).

202 *Mimosestes mimosae* (Fabricius, 1781) was erroneously listed from La Gomera by Krátký (2021). The figure in his publication refers to *Acanthoscelides macrophthalmus* (Schaeffer, 1907), firstly recorded from the Canary Islands (Gran Canaria and La Palma) by Yus Ramos & García Becerra (2019).

203 *Pseudopachymerina spinipes* (Erichson, 1834). This species was recorded from the Czech Republic based on uncertain data from seed warehouses (Strejček 1990). The record was followed by Strejček (1993), Anton (2010b) and Yus Ramos et al. (2014). On the contrary, this species is not listed as present in the Czech Republic by Strejček (2012). According to Strejček (2017, pers. comm.), there are no verified records from the Czech Republic and thus the occurrence is not listed in the present Edition.

204 *Spermophagus klapperichi* Borowiec, 1985. A single specimen reported from Algeria (Ricci & Zampetti 2007) needs verification as it was quite likely misidentified because it is a rare species so far known only from Levante and Anatolia. Thus, the record is not accepted in the present Edition.

205 *Spermophagus kuesteri* Schilsky, 1905. Three specimens reported from Canary Islands by Decelle (1975c) were collected in the 19th century by G. R. Crotch. Since that time *S. kuesteri* was not found in Canary Islands again, thus the species is considered as imported in the present Edition.

206 *Spermophagus maafensis* Borowiec, 1985 is a strictly Iberomaghrebian species. Two specimens from Rhodes (Greece) recorded by Ricci & Zampetti (2006) are surely misidentified and thus are not accepted in the present Edition.

207 *Spermophagus turanicus* (Lukjanovitch & Ter-Minassian, 1957). The occurrence in Iran (Ghahari & Borowiec 2017) was based on the record by Anton (1999a), which actually refer to *S. caricus* Decelle, 1982.

208 *Spermophagus variolosopunctatus* Gyllenhal, 1833 is an Oriental species. The lectotype is mislabelled as from Ukraine (Borowiec 1991). Previous European records are misidentifications not adopted in the present Edition.

Chrysomelidae: Donaciinae

E. Geiser & J. Bezděk

Comments

209 *Donacia* Fabricius, 1775. For this genus, subgenera are not applied in the Catalogue. Indeed, a genus with about 110 species worldwide is to some extent heterogenous and contains groups of more closely related species. Therefore, several authors tried to assign individual species to subgenera. The most recent study was published by Kippenberg (2015) who split *Donacia* into 15 subgenera based on morphology, host plants and molecular analysis. However, molecular analysis has been performed so far for only one-third of the species, therefore it is certain that with further research a lot of changes will happen in the subgeneric classification of *Donacia*. Furthermore, the fragmentation of *Donacia* into many subgenera collides with the purpose of the catalogue which should be of practical use and should be based on well established and supported facts. In the present Edition all the subgenera are listed as synonyms and all the species names are listed alphabetically.

210 Zschach (1788) described seven nominal taxa of *Donacia* from the collection of Nathanael Gottfried Leske. After Leske's death in 1786 his extensive mineral and natural history collection was sold to the Dublin Society and later on transferred to the National Museum of Ireland. Although much of the Leske's collection has survived in a very good condition to present time, the Coleoptera part was nearly completely destroyed in the 19th century due to the collapse of the museum's roof (Nash 1983). During Jan Bezděk's visit to the National Museum of Ireland in 2018, no types of *Donacia* were found. Out of the seven taxa described by Zschach (1788), four are well established in Donaciinae system either as valid species or synonyms (*D. bicolora* Zschach, 1788, *D. crassipes* Zschach, 1788, *D. fusca* Zschach, 1788, *D. vulgaris* Zschach, 1788). An additional three names (*D. aenea* Zschach, 1788, *D. caerulea* Zschach, 1788, *D. sericea* Zschach, 1788) have disappeared from the literature, their identity is unknown, and they are placed as nomina dubia in the present Catalogue.

211 The description of *Donacia marginata* Hoppe, 1795 was published in April 1795 (Bousquet 2016). As there is no more precise date, the last day of the month must be accepted as the publication date (Art. 21.3.1, ICZN 1999). According to Alonso-Zarazaga & Evenhuis (2017), the descriptions of two of its synonyms, *D. limbata* Panzer, 1795 and *D. vittata* Panzer, 1795, were published on 26 April 1795 (not in 1796 as was previously believed). Therefore, both *D. limbata* and *D. vittata* have priority over the common and widely used *D. marginata*. However, *D. vittata* Panzer, 1795 is also a junior homonym of *D. vittata* Olivier, 1791, now a synonym of *D. aquatica* (Linnaeus, 1758). *Donacia limbata* Panzer, 1795 cannot be regarded as a nomen oblitum based on Art. 23.9.2 (ICZN 1999) because it has been used as a valid name

after 1900 (e.g. W. F. Johnson & Halbert 1902). The case requires an application to the Commission. Meanwhile the usage of *D. marginata* as valid name is to be maintained.

212 *Plateumaris* C. G. Thomson, 1859. The classification of *Plateumaris* with two subgenera has traditionally been applied since Reitter (1920) who introduced the subgenus *Juliusina* Reitter, 1920 to which he assigned *P. bracata* Scopoli, 1772, *P. consimilis* Schrank, 1781, *P. affinis* Kunze, 1818, *P. sulcifrons* Weise, 1900, and *P. rustica* Kunze, 1818. In *Plateumaris* (s. str.) Reitter (1920) placed *P. sericea* (Linnaeus, 1758), *P. discolor* (Panzer 1795), and the newly described *P. annularis* Reitter, 1920. Goecke (1957) synonymized *P. annularis* with *P. roscida* Weise, 1912. All other subsequently described Palaearctic species since 1920 were assigned to *Juliusina* with the exception of *P. shirahatai* Kimoto, 1971, which was placed into *Plateumaris* (s. str.). Askevold (1991) pointed out that the morphological characters do not support the classification into two subgenera. Later on, studies on molecular phylogenetics by Sota et al. (2007) and Hayashi & Sota (2013) revealed that whereas *P. sericea* and *P. shirahatai* are very closely related but clearly distinct species, *P. roscida* is always distant from them in phylograms. Also, numerous Palaearctic species are more closely related to some Nearctic ones, therefore, the establishment of the two subgenera is not justified (Geiser 2023) and thus not accepted in the present Catalogue.

New and reconfirmed synonyms

213 *Donacia bicolora meridionalis* Weise, 1886 was described as a variety. Nethertheless, it was widely used as a subspecies for specimens collected in the southern part of Italy. Studies based on material and description gave no hints that it is a distinct subspecies, and it was synonymized with the nominotypical *D. bicolora* by Geiser & Jäch (2021).

214 *Donacia fukiensis* Goecke, 1944 was restored to species rank from synonymy of *D. clavareaui* Jacobson, 1906 by Cong & Yu (1997) which was confirmed by Geiser (2019) based on a study of the type material.

215 *Donacia mediohirsuta* S.-H. Chen, 1966 is a synonym of *D. clavareaui* Jacobson, 1906 according to Geiser (2019).

216 *Donacia kweilina* S.-H. Chen, 1966 is not a synonym of *D. clavareaui* Jacobson, 1906 as stated in Geiser (2019) but a distinct species (Hayashi & Pütz 2008).

217 *Donacia melanocephala* Marsham, 1802 was treated as a synonym of *D. dentata* Hoppe, 1795 for decades. Study of the type material clearly showed that it is conspecific with *D. semicuprea* Panzer, 1795 (Geiser & Jäch 2021).

218 *Donacia jacobsoni* Semenov & Reichardt, 1927 is a synonym of *D. delagrangei* Pic, 1896 according to Kippenberg (2010c).

219 *Donacia vulgaris issykensis* Jacobson, 1901 is a synonym of *D. vulgaris* Zschach, 1788 according to Goecke (1960) and Warchałowski (1990c) which was also confirmed by E. Geiser based on a study of comparative material and description.

220 *Macroplea skomorokhovi* L. N. Medvedev, 2006 is a synonym of *M. japana* (Jacoby, 1885) according to Lou et al. (2011).

221 *Plateumaris affinis* (Kunze, 1818) is a synonym of *P. rustica* (Kunze, 1818) according to Askevold (1991) which was also confirmed based on a study of comparative material and descriptions (Geiser 2023, Geiser & Geiser 2023).

222 *Plateumaris caucasica* F. A. Zaytsev, 1930 is a synonym of *P. sericea* (Linnaeus, 1758) based on a study of comparative material, description and of biogeographical research (Geiser 2023, Geiser & Geiser 2023).

223 *Plateumaris lacordairii* (Perris, 1864) is a synonym of *P. sericea* (Linnaeus, 1758) according to Askevold (1991) which was also confirmed based on study of comparative material and description (Geiser 2023, Geiser & Geiser 2023).

224 *Plateumaris sibirica* Solsky, 1871 is a synonym of *P. sericea* (Linnaeus, 1758) according to Goecke (1960) and Hayashi (2020) which was also confirmed based on a study of comparative material and description (Geiser 2023, Geiser & Geiser 2023).

225 *Plateumaris sulcifrons* Weise, 1900 is a synonym of *P. rustica* (Kunze, 1818) according to Askevold (1991) which was also confirmed based on a study of description and biogeographical research (Geiser 2023, Geiser & Geiser 2023).

226 *Plateumaris tenuicornis* Balthasar, 1934 is a synonym of *P. consimilis* (Schrank, 1781) according to Bechyné (1942) which was also confirmed based on a study of type material and description (Geiser 2023, Geiser & Geiser 2023).

Taxa of questionable status

For the following taxa, the status reported in the literature is temporarily accepted. However, further study is required to confirm their status.

227 *Donacia fedtschenkoae* Jacobson, 1899 is currently listed as a valid species. Only the holotype is known. It may prove to be a synonym of *D. bactriana* Weise, 1887.

228 *Donacia kirgizkaisaka* Jacobson, 1925 is currently listed as a valid species. Only the type series exists. It may prove to be a synonym of *D. thalassina thalassina* Germar, 1811.

229 *Donacia mannerheimii* Lacordaire, 1845 is currently listed as a valid species. Only the holotype is known. It may prove to be a synonym of *D. bactriana* Weise, 1887, *D. thalassina* Germar, 1811, or *D. marginata* Hoppe, 1795.

230 *Macroplea ranina* Lou & Yu, 2011 is currently listed as a valid species. It may prove to be a synonym of *Neohaemonia voronovae* Medvedev, 1977.

231 *Plateumaris discolor* (Panzer, 1795) is currently listed as a synonym of *P. sericea* (Linnaeus, 1758). It may prove to be a subspecies or even a valid species (Geiser 2023).

232 *Plateumaris obsoleta* Jacobson, 1894 is known from the holotype (female) only and is conspecific either with *P. shirahatai* Kimoto, 1971 or with *P. sericea* (Linnaeus, 1758). However, the only reliable morphological character to discriminate the two taxa is the structure of the endophallus. In the present Catalogue *P. obsoleta* is listed

as a questionable synonym of *P. shirahatai*. For more comprehensive comment see Geiser (2023).

Availability of varieties and forms

233 The following names are deemed infrasubspecific because they were published as a fourth name after a trinomen (ICZN 1999, Art. 45.5):
Donacia thalassina ssp. *rufovariegata* var. *coerulea* Jacobson, 1901b: 105
Plateumaris discolor ssp. *discolor* forma *coelicolor* Bechyné, 1945b: 89
Plateumaris discolor ssp. *discolor* forma *cupraria* Bechyné, 1945b: 89
Plateumaris discolor ssp. *discolor* forma *pseudoviolacea* Bechyné, 1945b: 89
Plateumaris discolor ssp. *discolor* forma *purpuricena* Bechyné, 1945b: 89
Plateumaris discolor ssp. *kratochvili* forma *isocoelicolor* Bechyné, 1945b: 89
Plateumaris discolor ssp. *kratochvili* forma *isocupraria* Bechyné, 1945b: 89
Plateumaris discolor ssp. *kratochvili* forma *isolacordairei* Bechyné, 1945b: 89
Plateumaris discolor ssp. *kratochvili* forma *isopurpuricena* Bechyné, 1945b: 89
Plateumaris discolor ssp. *kratochvili* forma *isoviolacea* Bechyné, 1945b: 89

234 *Donacia sahlbergi* var. *coeruleoviolacea* Jacobson, 1901b: 105 is deemed infrasubspecific according to the Article 45.6.1 (ICZN 1999) as a subspecies concept was used in the same work and to the available knowledge, the name was not validated before 1985 (ICZN 1999, Art 45.6.4).

235 The following forms were proposed for infrasubspecific entities as is unambiguously revealed by the content of the publication. According to available information, these names were not validated prior to 1985 (ICZN 1999, Art 45.6.4.1):
Plateumaris consimilis forma *aerea* Bechyné, 1942: 234
Plateumaris consimilis forma *coerulea* Bechyné, 1942: 234
Plateumaris consimilis forma *flavipes* Bechyné, 1942: 234
Plateumaris consimilis forma *nigripes* Bechyné, 1942: 234
Plateumaris consimilis forma *violacea* Bechyné, 1942: 234
Plateumaris consimilis forma *viridis* Bechyné, 1942: 234

Nomen dubium

236 *Donacia simplex* Thunberg, 1792 (31.xii.1792). We found in Thunberg's collection seven specimens pinned in three different boxes under the name *D. simplex*. Locating and interpreting which specimens are true original types is quite complicated in Thunberg's collection because it is not arranged systematicaly as integrated collection but is partly organized by his publications, and as a result one species can appear more than once in his collection. The study of the seven specimens revealed that it is in fact a mixture of three species: three specimens of *D. vulgaris* Zschach, 1888, one specimen of *D. clavipes* Fabricius, 1792 (30.ix.1792), and three specimens of

D. semicuprea Panzer, 1796. The first two are older names and the last one would not be threatened by precedence because *D. simplex* Thunberg is also a junior primary homonym of *Donacia simplex* Fabricius, 1775. However, none of the seven specimens exactly matches the primary description, thus, *Donacia simplex* Thunberg, 1792 is treated as nomen dubium in the present Catalogue.

Nomina nuda

The following names occur in literature and collections but were never described properly:

237 *Donacia alpina* Apfelbeck in Goecke 1941: 47 is conspecific with *D. freyi* Goecke, 1940.

238 *Donacia trojana* Reitter in Geiser & Jäch 2021: 166 is conspecific with *D. marginata* Hoppe, 1795.

239 *Donacia elegans* Dahl in Geiser & Jäch 2021: 167 is conspecific with *D. polita* Kunze, 1818.

240 *Donacia polita* var. *graecorinsularis* Daniel in Geiser & Jäch 2021: 167 is conspecific with *D. polita* Kunze, 1818.

241 *Donacia imitatrix* Apfelbeck in Goecke 1942: 220 is conspecific with *Plateumaris sericea* (Linnaeus, 1758).

242 *Plateumaris sericea slovacica* Balthasar in Geiser (2023) is conspecific with *P. sericea* (Linnaeus, 1758).

243 *Plateumaris borealis* Mannerheim in Geiser (2023) is conspecific with *P. weisei* (Duvivier, 1885).

Spelling

244 *Donacia bicolora* Zschach, 1788. It was described as "*Donacia bicolora*". Because in classical Latin *bicolor* is used for the male, female and neuter form (single-ended adjective of the third declension) this species is sometimes spelled as "*bicolor*". This would be justified if the author of the original description made an inadvertent (grammatical) error that must be corrected (ICZN 1999, Art. 32.5.1.), but this is not the case. In classical Latin the forms bicolorus, bicolora, bicolorum were also used by Flavius Vopiscus who wrote about 300 AD a biography of emperor Aurelianus (p. 236 in: Lewis C. T. & Short C. 1975: A Latin Dictionary. Oxford University Press, London). Therefore, it is not necessary to correct the spelling of Zschach because it also conforms to classical Latin. So, following the original description, *D. bicolora* is the correct spelling.

245 *Donacia tuberfrons* Goecke, 1934. In the previous Edition of the Catalogue the name was spelled as *D. tuberifrons* and listed as synonym of *D. transversicollis* Fairmaire, 1887. In the original description it was spelled as "*tuberfrons*" without a connecting

vowel. According to the Art. 32.5.1. (ICZN 1999) the "use of an inappropriate connecting vowel", which is the case here, is not to be considered inadvertent error thus the original spelling is to be maintained.

246 *Plateumaris bracata* (Scopoli, 1772). In most publications the species name is spelled "braccata". However, the spelling in the original description of Scopoli (1772) is "Prionus Bracatus" with single "c" (Geiser & Geiser 2023). Scopoli used this epithet presumably because "bracae" means "trousers" or "pants" due to the clubbed shape of the hind femora. There exists also a variant of this word in later medieval Latin spelled "bracca", which Schenkling (1917) used in his explanation of the scientific names of beetles. Because the spelling of the original description is linguistically correct, it has to be preserved unaltered (ICZN 1999, Art. 32.2.1).

Type species

247 *Haemonia* Dejean, 1821. The type species is *Haemonia equiseti* Fabricius, 1798 by monotypy. Dejean (1821) listed under *Haemonia* only one valid species *H. equiseti* and var. *zosterae* as its synonym. See also Bouchard et al. (2024).

New geographical records

248 *Donacia aquatica* (Linnaeus, 1758). KAZAKHSTAN: Temir (Aktjubinsk), 16.v.1999, 1 spec., E. Geiser det. (NMPC). New species for Kazakhstan.

249 *Donacia clavipes clavipes* Fabricius, 1792. CHINA: Xinjiang prov., East-Turkestan, Bagratsch-Kul. [= Bosten-lake], Kurla, v.1902, 9 spec., E. Geiser det. (NMPC). New species for Xinjiang (China).

250 *Donacia dentata* Hoppe, 1795. TURKEY: Ankara prov., Kızılcahamam, 15.vi.1968, 16 spec., G. Wewalka leg., E. Geiser det. (collection of G. Wewalka deposited in NMW); "Turcia", 2 spec., E. Geiser det. (collection of I. Frivaldsky deposited in HNHM). New species for Turkey.

251 *Donacia marginata* Hoppe, 1795. ALBANIA: South-Albania, Liqeni i Butrintit [= Butrintit Lake], 1958, 8 spec., Smetana leg., R. Mlejnek det. (NMPC). CHINA: Xinjiang prov., East-Turkestan, Tian Shan, Yulduz mountains, 1 spec., E. Geiser det. (collection of A. Winkler deposited in NMW). New species for Albania and Xinjiang (China).

252 *Donacia polita* Kunze, 1818. RUSSIA (Southern Territory): Volgograd Oblast, SW of Lake Elton, 11.v.2001, 1 ♂ 1 ♀, A. Nilsson leg., E. Geiser det. (NHRS). New species for Russia.

253 *Donacia semicuprea* Panzer, 1795. GREECE: Corfu, 3 spec., Kyselý leg., R. Mlejnek det. (NMPC). New species for Greece.

254 *Donacia simplex* Fabricius, 1775. ALBANIA: Gjirokastra, Tepelena, Vjosa, 40°17.02′N 20°02.13′E, 132 m, 3.vi.2021, 1 spec., W. Paill, J. Guncy & G. Kunz leg., E. Geiser det. (NMW). GREECE: Peloponnes, Kalavryta, iv.1936, 19 spec., J. Mařan & K. Táborský

leg., R. Mlejnek 2015 det. (NMPC); East-Makedonia, near Drama, 1.–3.v.1989, 1 spec., J. Ježek leg., R. Mlejnek 2015 det. (NMPC). New species for Albania and Greece.

255 *Donacia vulgaris* Zschach, 1788. GREECE: Peloponnes, Argos, 1935, 2 spec., J. Mařan & O. Štěpánek leg., R. Mlejnek det. (NMPC). New species for Greece.

256 *Sominella longicornis* (Jacoby, 1890). CHINA: Guangdong: Danxia Shan NP, Wo Long Gang Forest Walkway 25°01.3′N 113°44.5′E, 100 m, 23.–26.iv.2013, 8 spec., J. Hájek & J. Růžička leg., J. Bezděk det. (NMPC). New species for Guangdong (China).

257 *Macroplea pubipennis* (Reuter, 1875). KAZAKHSTAN: Kazalinsk, 1 spec., L. Sekerka det. (NMPC). New species for Kazakhstan.

Comments on distribution

Concerning the country records numerous additions and other changes have been made in comparison with the previous Edition. For additional details see Geiser & Jäch (2021) and Geiser (2023).

258 *Donacia antiqua* Kunze, 1818 and *D. brevitarsis* C. G. Thomson, 1884 were confused for decades. Therefore, most of the country records in literature are not reliable. Hence only reliable or verified distribution was accepted in the present Edition.

259 *Donacia aquatica* (Linnaeus, 1758) does not occur in Japan (Hayashi 2020).

260 *Donacia clavareaui* Jacobson, 1906 was confused with *D. fukiensis* Goecke, 1944 for decades. The distribution of *D. clavareaui* has been recently verified after the resurrection of *D. fukiensis*. The distribution is adopted from Geiser (2019).

261 *Donacia malinovskyi* Ahrens, 1810 does not occur south of 45° latitude, therefore "Spain" is intentionally omitted in the Catalogue. The records presumably refer to misidentified specimens of *D. dentata* Hoppe, 1795 or *D. semicuprea* Panzer, 1795.

262 *Donacia polita* Kunze, 1818 does not occur in the present-day territory of Hungary.

263 *Donacia reticulata* Gyllenhal, 1817 does not occur in Poland (Bezděk & Mlejnek 2016).

264 *Donacia sparganii sparganii* Ahrens, 1810 does not occur in the present-day territory of Hungary.

265 *Donacia splendens splendens* Jacobson, 1894 does not occur in West Siberia. It is distributed only from Kemerovo Region eastwards to the Pacific coast.

266 *Donacia thalassina* Germar, 1811 does not occur in Japan (Hayashi 2020). In Xinjiang (China), the nominotypical subspecies occurs and not *D. thalassina intermedia* Jacobson, 1899 (see Geiser & Jäch 2021).

267 *Plateumaris consimilis* (Schrank, 1781). Shavrov (1948) described a new subspecies *P. consimilis orientalis* which was latter synonymized with *P. weisei* (Duvivier, 1885). It created confusion as specimens identified as "*P. consimilis*" and published from East Russia and Japan in fact belonged to *P. weisei*. There is no evidence that *P. consimilis* occurs east of Lake Baikal (Geiser 2023).

Chrysomelidae: Criocerinae

J. Bezděk & M. Schmitt

Comments

268 The subgeneric arrangement of *Lema* Fabricius, 1798 is not consistent in the recent Chrysomelidae literature. Moreover, the monophyly of subgenera *Lema*, *Petauristes* Latreille, 1829, *Quasilema* Monrós, 1960, *Microlema* Pic, 1932, and *Bradyceris* Chûjô, 1951 was not confirmed by Matsumura et al. (2014). In the present Catalogue we treat the subgenera *Petauristes*, *Quasilema* and *Microlema* as valid but further comprehensive study on subgeneric arrangement of *Lema* is necessary.

Unavailable name

269 Bogachev & Antsupova (1980) proposed a new replacement name *Asparagocrioceris* to acommodate *Crioceris asparagi*, *C. duodecimpunctata*, *C. quatuordecimpunctata* and *C. bicruciata*. *Asparagocrioceris* is not available name because it is neither a new replacement name for an existing preoccupied name (ICZN 1999, Art. 13.1.3 and 72.7) nor a description or indication of a new genus.

Availability of varieties and forms

270 The following names are deemed infrasubspecific because they were published as a fourth name after a trinomen (ICZN 1999, Art. 45.5):

Crioceris duodecimpunctata ssp. *hypolachna* var. *goetria* Jacobson, 1907c: 26
Crioceris quatuordecimpunctata ssp. *sibirica* forma *confluenta* Heinze, 1943b: 101
Crioceris quatuordecimpunctata ssp. *sibirica* forma *staudingeri* Heinze, 1943b: 101
Lema decempunctata ssp. *japonica* var. *brunneipennis* Kuwayama, 1932: 80
Lema diversa var. *doii* forma *morii* Yuasa, 1939: 163
Lema diversa var. *lewisii* forma *akaheri* Yuasa, 1939: 163

271 The following varieties and forms are deemed infrasubspecific according to Article 45.6.1 (ICZN 1999), since a subspecies concept was used in the same work, and according to available information, these names were not validated prior to 1985 (ICZN 1999, Art 45.6.4.1):

Crioceris duodecimpunctata var. *augustoduni* Pic, 1916a: 6
Crioceris duodecimpunctata var. *semijuncta* Pic, 1916b: 10
Crioceris quatuordecimpunctata var. *donceeli* Pic, 1916a: 6
Crioceris quatuordecimpunctata var. *montandoni* Pic, 1916a: 6
Crioceris quatuordecimpunctata var. *nigripes* J. Müller, 1948: 76
Crioceris quatuordecimpunctata var. *russica* Pic, 1916a: 6
Crioceris quatuordecimpunctata var. *semireducta* Pic, 1916a: 6
Crioceris quatuordecimpunctata var. *turcica* Pic, 1916a: 6

Crioceris quinquepunctata var. *nigerrima* Jacobson, 1901b: 106
Crioceris orientalis var. *choi* Chûjô, 1941c: 452
Lema concinnipennis var. *ventralis* Kuwayama, 1932: 69
Lema decempunctata forma *kiautschauana* Heinze, 1943b: 106
Lema diversa forma *cyaneohumeralis* Heinze, 1943b: 106
Lema diversa var. *doii* Kuwayama, 1932: 78
Lema diversa forma *quadriplagiata* Heinze, 1943b: 106
Lema diversa forma *suturalis* Heinze, 1943b: 107

New geographical records

272 *Crioceris bicruciata* (C. R. Sahlberg, 1823). ARMENIA: Araxesthal, 1 spec., Leder & Reitter leg., J. Bezděk det. (NMPC). IRAQ: Bagdad, 2 spec., Jedlička leg. (NMPC). New species for Armenia and Iraq.

273 *Crioceris macilenta* Weise, 1881. CYPRUS: Salamis, v.1973, 1 spec., Pfeffer leg., J. Bezděk det. (NMPC). New species for Cyprus.

274 *Lema* (*Lema*) *tibialis* Laporte, 1840. SAUDI ARABIA: Asir, Wadi Marabah, WSW of Abha, 18°10′18″N 42°22′12″E, 1197 m, 1.–4.iv.2017, 1 spec., P. Rapuzzi leg., J. Bezděk det. (NMPC). New species for Saudi Arabia.

275 *Lilioceris scapularis* (Baly, 1859). CHINA: Jiangxi: Jinggang Shan Mts., Xiangzhou vill. env., 26°35.5′N 114°16.0′E, 374 m, 26.iv.2011, 2 spec., M. Fikáček & J. Hájek leg., J. Bezděk det. (NMPC). New species for Jiangxi (China).

Comments on distribution

276 The following distributional data appeared in the previous version of the Catalogue, however, we were unable to find any published sources. Nevertheless, they are retained here as the occurences are possible.
Crioceris asparagi (Linnaeus, 1758) – **N:** EG **A:** TD
Crioceris bicruciata (C. R. Sahlberg, 1823) – **E:** NM
Crioceris duodecimpunctata (Linnaeus, 1758) – **E:** AL
Crioceris macilenta Weise, 1881 – **E:** MA
Crioceris nigroornata Clark, 1866 – **A:** ID (HP)
Crioceris paracenthesis (Linnaeus, 1767) – **E:** AN NM **N:** CI
Crioceris quatuordecimpunctata (Scopoli, 1763) – **E:** BH
Crioceris quinquepunctata (Scopoli, 1763) – **E:** BU FR NM
Lema (*Lema*) *bohemani* Clark, 1866 – **A:** CH (GUI)
Lema (*Lema*) *cyanella* (Linnaeus, 1758) – **E:** AN
Lema (*Lema*) *nigrofrontalis* Clark, 1866 – **A:** CH (YUN)
Lema (*Lema*) *obscuritarsis* Jacoby, 1892 – **A:** ID (AP)
Lema (*Lema*) *scutellaris* Kraatz, 1879 – **A:** RU (ES)

Lema (Microlema) decempunctata (Gebler, 1830) – **A:** RU (ES)
Lema (Petauristes) adamsii Baly, 1865 – **A:** CH (SCH YUN)
Lema (Petauristes) jansoni Baly, 1861 – **A:** ID (SD)
Lema (Petauristes) lacordairii Baly, 1865 – **A:** BT
Lema (Petauristes) praeclarior Clark, 1866 – **A:** ID (UT)
Lema (Petauristes) quadripunctata Olivier, 1808 – **A:** ID (SD)
Lema (Petauristes) viridipennis Pic, 1924 – **A:** CH (YUN)
Lilioceris (Lilioceris) adonis (Baly, 1860) – **A:** ID (SD)
Lilioceris (Lilioceris) impressa (Fabricius, 1787) – **A:** ID (AP)
Lilioceris (Lilioceris) lilii lilii (Scopoli, 1763) – **A:** AB AR GG IN
Lilioceris (Lilioceris) lilii stercoraria (Linnaeus, 1767) – **N:** TU
Lilioceris (Lilioceris) merdigera merdigera (Linnaeus, 1758) – **A:** CH (ZHE)
Lilioceris (Lilioceris) semihirsuta (Jacoby, 1908) – **A:** ID (UT)
Lilioceris (Lilioceris) semipunctata (Fabricius, 1801) – **A:** ID (AP UT)
Oulema antonii (Clavareau, 1905) – **A:** BT
Oulema duftschmidi (L. Redtenbacher, 1873) – E: AL AN
Oulema erichsonii (Suffrian, 1841) – E: LS LU
Oulema hoffmannseggii (Lacordaire, 1845) – E: AN
Oulema melanopus (Linnaeus, 1758) – **A:** KI
Oulema obscura (Stephens, 1831) – E: AN
Oulema tristis tristis (Herbst, 1786) – E: RU (CT)

Chrysomelidae: Cassidinae

L. Sekerka

Comments

277 The identity of *Aspidomorpha dorsata* was changed recently due to an incorrect interpretation of the type material (see Sekerka & Barclay 2014). The name *A. dorsata* applies to the taxon that was previously named as *A. fuscopunctata* Boheman, 1854 while *A. dorsata* sensu auctt. (e.g. Świętojańska 2001a) is currently *A. calligera* Boheman, 1854.

278 *Pseudocassis* Steinhausen, 2002, currently considered a synonym of *Cassida* Linnaeus, 1758, is also a junior homonym of *Pseudocassis* Pictet & Campiche, 1863 in Mollusca. A new replacement name is not necessary to be proposed as the taxon is a synonym.

279 *Cassida catenata* Boheman, 1855 was described from Java and is a widespread and common species in Indonesia, reaching Asian mainland in Malaysia and southern Thailand only. While its counterpart in continental Asia is *Cassida circumdata* Herbst, 1799. Nevertheless, there is an isolated population in the Himalaya resembling *C. catenata* and it was assigned to the latter species by Maulik (1919). Such

disjunct occurrence is quite improbable, on the other hand there is no other species to which the population can presently be assigned. The entire group is in a need of a complex revision including genetic data. Therefore, I provisionally, follow Maulik's interpretation.

280 *Cassida fausti* Spaeth, 1926 was proposed as "lokalform" of *C. vibex* but in the same work, a concept of subspecies was also used, and therefore the name should become infrasubspecific. However, it has been accepted as a valid species since Zaytsev (1938) thus it is valid from its original description (ICZN 1999, Article 45.6.4.1).

281 The genus *Laccoptera* was until now considered to have been described in 1855. The name was first defined in key to genera by Boheman (1854) but without any species included. Boheman (1855) provided a detailed description and included numerous species to the genus. The name is valid from its definition in 1854 (ICZN 1999, Articles 12.1 and 67.2.2) and the year is changed accordingly.

282 The identity of *Cassida chinensis* Fabricius, 1798 (previously classified under *Basiprionota* Chevrolat, 1836) was changed recently due to an incorrect interpretation of the type material (see Sekerka in prep.). The name applies to the taxon that was previously named as *Laccoptera nepalensis* Boheman, 1855 (e.g. Świętojańska 2001a). *Basiprionota satrapa* (Boheman, 1862) is a valid name for taxon known until recently as *B. chinensis* sensu auctt.

283 *Sinispa* Uhmann, 1938 is valid from its description although no type species was designated in the original publication because it is a combined description of a new genus and a single new species (ICZN 1999, Article 13.4).

Type species

284 *Agonia* was until now considered as proposed by Weise (1905c) and accordingly Maulik (1919) designated *Gonophora wallacei* Baly, 1859 as the type species. However, the name is valid from Weise (1904e), who described two species under the name *Agonia*, namely *A. femoralis* Weise, 1904 and *A. nigripennis* Weise, 1904. This makes Maulik's (1919) designation invalid as *G. wallacei* was not originally included. *Agonita nigripennis* is nowadays placed in the genus *Klitispa* Uhmann, 1940 while *A. femoralis* is a representative of *Agonita* (= *Agonia*) as presently used. Therefore, I hereby designate *A. femoralis* as the type species of the genus. The name *Agonia* is a junior homonym and was replaced by *Agonita* Strand, 1942.

285 *Hispellinus* Weise, 1897. In the previous Edition (Borowiec & Sekerka 2010b) *Hispa multispinosa* Germar, 1848 was erroneously listed as the type species. Weise (1897d) created the genus and included three species: *Hispellinus callicanthus* (Bates, 1866), *H. moestus* (Baly, 1888), and *H. multispinosus* but did not designate any of them as the type species. Weise (1905f) synonymized it with *Monochirus* and mentioned that *H. multispinosus* is the type of the latter, however this does not apply to the name *Hispellinus*. The genus was again revalidated due to homonymy and Gressitt

(1950) designated *H. callicanthus* as the type species, which is to my knowledge the first correct designation.

286 *Klitispa* Uhmann, 1940 was first described by Uhmann (1939b). He included four species, however, did not designate any of them as the type species, thus did not meet provisions of the Article 13.3 (ICZN 1999) and the name was nomen nudum. *Klitispa* was validated by Uhmann (1940a) as he designated *Agonia nigripennis* Weise, 1904 as the type species and provided reference to the original description.

287 *Triplispa* Weise, 1897 was introduced in a catalogue of species of *Dactylispa* Weise, 1897 and comprised five species but was not provided with description or type species designation (Weise 1897d). Chen et al. (1986: 285) stated in the Chinese text that the type species is *Hispa confluens* Baly, 1890. Because this species was originally included by Weise (1897d) it constitutes a valid type species designation.

New synonymy

288 The subgenera of *Cassida* Linnaeus, 1758 are not considered valid in the present Catalogue in accordance with Borowiec (2007). In general, true cassidines are strongly adapted to their environment and it can be said that they are characterised by reductions with low number of apomorphies. As a result, the generic system, including nowadays hundreds of taxa, is based on a combination of a few dozen characters (mainly homoplasies). This is particularly true in the case of the tribe Cassidini and namely the genus *Cassida*. The subgenera were proposed for the Palaearctic and partly Oriental taxa only while the rest of the diversity of this vast genus (currently 470 species) remained unclassified. The subgenera occuring in the western part of the Palaearctic Region are mainly based on subtle characteristics, usually results of adaptive convergence, and thus do not bear phylogenetic signal. This is multiplied by the simple fact that most workers using and proposing these subgenera were only familiar with the European fauna and thus created and promoted a system taken out of context. Moreover, all these authors were always looking only for supportive characters to separate the subgenera without defining any broader phylogenetic concept. Consequently, the subgenera are either artificial or monotypic, based on a few characters, frequently just a single one, and of disproportional significance. If such approach was applied to species inhabiting tropical regions, for instance Madagascar, then almost every species would deserve to have its own subgenus. Such atomization would inevitably create more disorder. Instead, the use of species-groups is in place, as they do not create ballast in nomenclature and despite often being intuitive, are monophyletic or associate phylogenetically related species. For the reasons stated above also the recently proposed subgenera by Özdikmen & Bal (2021), which were purely based on the structure of spermathecae, are formally synonymized with the nominotypical genus: *Cassida* Linnaeus, 1758 = *Diversivascula* Özdikmen & Bal, 2021 **syn. nov.** = *Longiampulla* Özdikmen &

Bal, 2021 **syn. nov.** = *Reliquacassida* Özdikmen & Bal, 2021 **syn. nov.** Özdikmen et al. (2023) suggested that *Pilemostoma* Desbrochers des Loges, 1891 should be classified as subgenus of *Cassida* based on the structure of genitalia. This view is not accepted here in accordance with reasons stated above and *Pilemostoma* is treated as genus.

289 *Cassida circumdata* Herbst, 1799 = *C. nigroramosa* S.-H. Chen & Zia, 1964 **syn. nov.** based on study of relevant type material in IZCAS. *Cassida nigroramosa* was proposed for extreme colour morph of *C. circumdata* with black disc of pronotum and elytra and basal and posterolateral spots on the explanate margin of elytra. This form was described also as *C. nilgiriensis* Borowiec & Takizawa, 1991.

290 *Cassida fausti* Spaeth, 1926 = *C. circassica* L. N. Medvedev, 1962 **syn. nov.** The holotype of *C. circassica* represents only a pale-coloured form of *C. fausti* agreeing with it in all principal external morphological characters (i.e. the shape of body, punctation and pubescence of elytra, the structure of clypeus, and the colouration of ventral parts) including the structure of spermatheca.

291 *Cassida liquefacta* Spaeth, 1912. Borowiec (1999) synonymized *C. praetimida* Spaeth, 1912, *C. sincera* Spaeth, 1912, and *C. semipallens* Spaeth, 1912 with *C. liquefacta* Spaeth, 1912. All four taxa were described in the same publication (Spaeth 1912b). Borowiec & Sekerka (2010a) restored the name *C. praetimida* as valid because it appeared first in the respective publication and the other three names were synonymized with it. This was followed in the recent revision of the African species of *Cassida* (Borowiec & Świętojańska 2022). However, there is no page priority in the present edition of the Code (ICZN 1999) and the principle of the First Revising Author is to be maintained (ICZN 1999, Article 24.2.4). Consequently, *C. liquefacta* is here restored as valid and the following synonymy is proposed: *C. liquefacta* Spaeth, 1912 **stat. res.** = *C. chapuisi* Spaeth, 1912 **syn. nov.** = *C. praetimida* Spaeth, 1912 **syn. nov.** = *C. semipallens* Spaeth, 1912 **syn. nov.** = *C. sincera* Spaeth, 1912 **syn. nov.** = *C. concolor* Spaeth, 1926 **syn. nov.**

292 *Cassida murraea caucasica* Roubal, 1931 was omitted from all chrysomelid literature known to me until now. Roubal (1931c) described it without giving it any explicit status within the name itself but wrote in its description "Diese neue Rasse ...". There are several other taxa proposed in the same paper as forma and variety and the appropriate shortcut (f. or var.) was placed in the name what suggests that *C. murraea caucasica* was of higher rank for him and therefore is considered as an available subspecific name. *Cassida murraea* is a widespread species with many local forms, some of which are presently considered as subspecies. On the other hand, *C. murraea* is in need of a complex study of diverse material and based also on molecular data to analyze affinities among individual populations. The population described as *C. murraea caucasica* fits within the present-day concept of the nominotypical species and is most similar to the form described as *C. pseudomurraea* Gruev, 1978 from the Balkans. Therefore, the following formal

synonymy is established: *C. murraea murraea* Linnaeus, 1758 = *C. murraea caucasica* Roubal, 1931 **syn. nov.**

293 *Cassida physodes* (Boheman, 1855) = *C. viridiguttata* (Chen & Zia, 1964) **syn. nov.** based on study of relevant type material in IZCAS. The holotype of *C. viridiguttata* nicely matches the lectotype of *C. physodes* differing only in having all spots on the apical half of elytral disc isolated while partly coalescent in *C. physodes*. However, this character is of no significance as the pattern and size of the spots is quite variable even within a single population.

294 *Dactylispa parbatya* Maulik, 1919 = *D. fabricii* L. N. Medvedev, 2009 **syn. nov.** The holotype of the latter species is morphologically identical with that of *D. parbatya*, just the frontal spines of prothorax are bifurcate while more or less trifurcate in *D. parbatya*. However, this character is variable within the type series of *D. parbatya* and even some specimens have a bifurcate spine on one side and a trifurcate one on the other.

295 In the previous Edition (Borowiec & Sekerka 2010b) the concept of subspecies of *Dicladispa armigera* was adopted from Chen et al. (1986). Subsequently, I had opportunity to study extensive material from the entire Oriental Region and it is very common that within a single population several of these nominal taxa occur together. Therefore, the following synonymy is proposed: *Dicladispa armigera* (Olivier, 1808) = *D. boutani* (Weise, 1904) **syn. nov.** = *D. similis* (Uhmann, 1927) **syn. nov.** = *D. armigera yunnanica* Chen & Sun, 1962 **syn. nov.**

296 *Platypria hystrix* (Fabricius, 1798) = *P. alces* Gressitt, 1938 **syn. nov.** based on study of respective type material and additional specimens from China. The latter taxon was separated based on longer and slenderer spines, however, these characters are very variable among populations of *P. hystrix* and the holotype of *P. alces* is fully within its intraspecific variability.

New names

297 *Callispa yuae* **nom. nov.** is proposed for *C. quadricollis* Yu & Li, 2002, primary junior homonym of *C. quadricollis* Weise, 1897 (currently in synonymy of *C. dimidiatipennis* Baly, 1859), and named for Yu Peiyu, one of the authors of the original name.

298 *Dactylispa cheni* **nom. nov.** is proposed for *D. similis* Chen & Tan, 1985, primary junior homonym of *D. similis* Achard, 1917 (currently in synonymy of *D. spinulosa* (Gyllenhal, 1817)), and named for Chen Sicien, one of the authors of the original name.

New combinations

299 Basu (1999) described two new species of *Gonophora* Chevrolat, 1836 having narrow body, truncate apex of elytra and not serrate lateral margin of pronotum, which

are typical characters for the genus *Sinagonia* S.-H. Chen & Tan, 1962. Despite, I did not have opportunity to examine type material of the two taxa I propose the following new combinations: *Sinagonia pitambara* (Basu, 1999) **comb. nov.** and *Sinagonia raktava* (Basu, 1999) **comb. nov.** Original descriptions are sufficient for the transfer, however, the identity of the species remains enigmatic.

Spellings

300 *Aspidomorpha* Hope, 1841. Most authors used the spelling *Aspidomorpha* until Seeno & Wilcox (1982) who considered *Aspidimorpha* as the correct spelling and attributed (erroneously) the name *Aspidomorpha* to Agassiz (1846). Since then, *Aspidimorpha* is used, however, old authors were correct as Hope (1841) used both spellings and explicitly wrote in the Errata, which appeared in the same book: "Aspidimorpha read Aspidomorpha, and elsewhere". Therefore, Hope is the first revising author (ICZN 1999, Articles 24.2.4 and 32.5.1.1) and *Aspidomorpha* is the original correct spelling.

301 Aspidomorphina. Borowiec (1994a) emended the name Aspidomorphini to Aspidimorphini based on Seeno & Wilcox's (1982) erroneous interpretation of the name *Aspidomorpha* (see the previous note). Consequently, the name is here reversed to the original spelling.

302 *Basiprionota westermanni* (Mannerheim, 1845). Original spelling "*westermanei*" was changed to "*westermanni*" by Boheman (1850) and is in use since then. Although Boheman did not clarify this change the name was correctly emended (ICZN 1999, Article 32.5.1).

303 Bothryonopini. The genus *Bothryonopa* was proposed by Guérin-Méneville (1840c) but subsequent authors cited it with different spellings and authorship, i.e. *Bothrionopa* (Chevrolat 1842) or *Botryonopa* (Baly 1859a). The latter name was used in the last published catalogue of the subfamily (Uhmann 1958) as well as revision of the entire tribe (Würmli 1976a), however with authorship erroneously attributed to Blanchard (1845b). Staines (2010) clarified the situation and confirmed *Bothryonopa* as the original spelling, which is in use since that time (e.g. Sekerka 2015) and accordingly the spelling of the tribe was adjusted to Bothryonopini (ICZN 1999, Article 32.5.3.2).

304 *Cassida justa* Spaeth, 1914. Original spelling of the name was "*iusta*" (Spaeth 1914d), however, Spaeth (1914e) changed it to "*justa*" (incorrect subsequent spelling). Nevertheless, the latter spelling is consistently used since then and thus is deemed as valid (ICZN 1999, Article 33.3.1).

305 *Dactylispa fabriziae* L. N. Medvedev, 2009. Medvedev, (2009b) wrote in the etymology: "The species is dedicated to one of its collectors, Mrs. Silvia Fabrici" and used the spelling "*fabricii*". However, the name was incorrectly spelled as the collector was Silvia Fabrizi, and the surname is spelled the same on the locality label

attached to the holotype; moreover, a wrong masculine ending was used. The spelling is emended here to "*fabriziae*" in accordance with the article 32.5.1 (ICZN 1999).

306 *Oncocephala* Agassiz, 1846. The name first appeared as "*Onchocephala*" in the second edition of Dejean's catalogue, but no valid species was included hence the name is unavailable. Subsequently it was validated by Guérin-Méneville (1844), who described two new species of *Onchocephala* and the name retains his authorship. Agassiz (1846b: 259) unjustifiably emended the name as he wrote: "Onchocephala Chevr[olat]. Col[eoptera]. 1834 (Oncocephala)". Bouchard & Bousquet (2020) stated that *Onchocephala* Guérin-Méneville, 1844 is a junior homonym of Onchocephala de Blainville, 1828, which however refers to an order level name thus the principle of priority cannot be applied. Even though Agassiz's spelling is used with a few exceptions as valid since it was introduced it cannot be deemed as justified emendation (ICZN 1999, Art. 33.2.3.1) because it was not attributed to the author of the original name. Moreover, the name cannot be conserved using *nomen protectum* as *Onchocephala* as it was used as valid by Maulik (1916). Therefore, *Onchocephala* Guérin-Méneville, 1844 has a priority and an application to the Commission is necessary to conserve the present usage. Meanwhile the usage of *Oncocephala* Agassiz, 1846 as valid name is to be maintained.

307 *Rhadinosa laghua* Maulik, 1915. Original spelling of the name was "*laghu*", however, Maulik (1919a) changed it to "*laghua*" (incorrect subsequent spelling). Nevertheless, the latter spelling has been in prevailing usage since then and thus is deemed as valid (ICZN 1999, Article 33.3.1).

308 Two different spellings appeared in the original descriptions of the four following taxa: *Aspidomorpha yunnana* and "*yunnanica*" (Chen & Zia 1964), *Taiwania simanica* and "*simauica*" (Chen & Zia 1961), *Dactylispa carinata* and "*carinate*" (Chen & Tan 1961), and *D. omeia* and "*omei*" (Chen & Tan 1961). In the subsequent publication (Chen et al. 1986) the authors used only one spelling, which is hence deemed as the correct one (*A. yunnana, T. simanica, D. carinata*, and *D. omeia*) (ICZN 1999, Article 24.2.4).

Protected names

309 *Cassida sanguinolenta* Schrank, 1776 (published on 21st April 1776) is a senior primary homonym of *C. sanguinolenta* Müller, 1776 (published on 28th April 1776) and senior synonym of *C. canaliculata* Laicharting, 1781. The latter two are well-known Palaearctic species that have been consistently treated as valid for more than 130 years, while *C. sanguinolenta* Schrank was synonymized with *C. austriaca* Fabricius, 1792 (= *C. canaliculata*) already by Herbst (1799). Fortunately, the present usage can be maintained according to the Art. 23.9.1. (ICZN 1999). To my knowledge the name *C. sanguinolenta* Schrank was never used as valid after 1899 (provision of the Art. 23.9.1.1). Below are given 25 references for both species to be maintained, published by more than 10 authors during the past 50 years and during a span greater

than 10 years (provision of the Art. 23.9.1.2). As both required provisions are met *Cassida sanguinolenta* Schrank is nomen oblitum, and *C. sanguinolenta* Müller and *C. canaliculata* are nomina protecta. References for *Cassida sanguinolenta* Müller (nomen protectum): Aslan et al. (2018), Bordy (1997, 2000), Bordy & Doguet (1987), Borowiec (1999), Brovdij (1983), Gruev (1992), Gruev & Tomov (1987, 1998, 2007), Kismali & Sassi (1994), Lopatin (1977b), Lopatin & Kulenova (1986), Maican & Serafim (2017), Recalde & Vicente (2005), Sekerka (2004, 2006b, 2010), Steinhausen (1994, 2002), Świętojańska (2005), Warchałowski (1978, 2000a, 2003, 2010b). References for *Cassida canaliculata* (nomen protectum): Aslan et al. (2018), Bordy (1997, 2000), Bordy & Doguet (1987), Borowiec (1999, 2012), Brovdij (1983), Gruev (1992), Gruev & Tomov (1987, 1998, 2007), Heisswolf et al. (2007), Kismali & Sassi (1994), Lopatin (1977b), Lopatin & Kulenova (1986), Maican & Serafim (2017), Sekerka (2004, 2006b, 2010), Steinhausen (1994, 2002), Warchałowski (1978, 2000a, 2003, 2010b).

Nomina dubia

310 *Cassida alpina* Comolli, 1837. This taxon was completely overlooked and to my knowledge it was never cited in the subsequent literature, certainly not after 1900. The name is a senior primary homonym of *C. alpina* Bremi-Wolf, 1855, which applies to a well-known species distributed in the Alps. The description of Comolli is not informative enough to allow me to decide about its identity and I was not able to trace the type material yet therefore the name is considered a nomen dubium for now. In case it would prove to be a different taxon then *C. alpina* in its current use the latter should be fixed by nomen protectum.

311 *Cassida angusticollis* Hagenbach, 1822 was synonymized with *C. vibex* by Spaeth (1914e), nevertheless, the description and figure suggest that the two taxa are not conspecific. However, the original figure is quite peculiar, and it is not possible to determine which taxon it represents. The type material was not traced so far and thus the name is considered nomen dubium.

312 *Dactylispa maculosa* (Fairmaire, 1889). Chen et al. (1986) considered this taxon a nomen nudum, however, Fairmaire (1889b) provided several morphological characters for it thus met the criteria of the Code (ICZN 1999) and the taxon is valid. However, its identity is unknown and it is here treated as nomen dubium.

Change in rank

313 Aspidomorphina. The tribe Aspidomorphini was recently recognized as paraphyletic within Cassidini based on cladistic analysis of morphological characters and synonymized with the latter tribe (López-Pérez et al. 2017). Nevertheless, *Aspidomorpha* and related genera forms quite likely monophyletic coherent group supported by larval and adult morphology as well as host plant association. Therefore, they are considered here a subtribe of Cassidini.

Autorship

314 *Cassida fausti* Spaeth, 1926. According to the preface of the Cassidinae volume in the series *Bestimmungs-Tabellen* (Spaeth & Reitter 1926), Reitter wrote a letter to Spaeth in 1919 and asked him to finish the manuscript and proposed a joint authorship as he did not have enough material and was feeling unwell. Several months after that Reitter died and his son Emmerich sent the manuscript to Spaeth. Spaeth realized that it was much less complete than he thought, and he wrote most of the text. All taxa proposed in this work are credited specifically to Reitter or Spaeth with the exception of *C. fausti*, which was until now credited to both authors. However, 1) the description is written in the "ich" form, 2) all species level taxa proposed in this work were proposed by Spaeth; 3) and the holotype of *C. fausti* is provided with Spaeth's own identification label on which is stated "Spaeth det." only. Therefore, it is obvious that only Spaeth was responsible for description and the authorship is changed accordingly.

315 *Cassida palaestina* Reiche & Saulcy, 1858. Was until now considered as described only by Reiche. However, in the primary description (Reiche & Saulcy 1858b) "R. et S." is mentioned after the taxon name, clearly indicating that both authors are responsible for it.

Generic and subgeneric assignment

316 The following species of *Dactylispa* Weise, 1897 are assigned to the subgenera based on the primary descriptions only and these assignments must be understood as preliminary until type material is available for examination: *D. chanchala* Basu & Saha, 1977, *D. malabikae* Basu & Saha, 1977, *D. manii* Vazirani, 1972, and *D. molina* Basu & Saha, 1977 to the nominotypical subgenus; *D. amala* Basu & Saha, 1977, *D. bhaumiki* Basu & Saha, 1977, *D. jonathani* Basu & Saha, 1977, and *D. shira* Basu & Saha, 1977 to *Triplispa* Weise, 1897.

317 *Dactylispa bulbifera* Medvedev, 1993 is assigned to the nominotypical subgenus based on examination of the holotype.

318 *Hispa filicornis* Motschulsky, 1860 is tentatively assigned to *Monohispa* Weise, 1897 based on the primary description and a note published by Medvedev (2007a) but the type material is at present not available for study.

Availability of varieties and forms

319 The following varieties and form are deemed infrasubspecific according to Article 45.6.1 (ICZN 1999), since a subspecies concept was used in the same work, and according to available information, these names were not validated prior to 1985 (ICZN 1999, Art 45.6.4.1):

Callispa bowringii var. *atripes* Pic, 1924g: 27

Cassida pannonica forma *fraudulenta* Spaeth, 1914c: 141

Cassida pupillata var. *olcesei* Pic, 1908b: 46
Estigmena chinensis var. *atricollis* Pic, 1924g: 29
Estigmena chinensis var. *nigripennis* Pic, 1924g: 29
Estigmena chinensis var. *ruficollis* Pic, 1924g: 30
Lasiochila gestroi var. *ruficolor* Pic, 1924g: 31

Excluded species

320 *Agonita pitava* Basu, 1999 was included in the previous Edition (Borowiec & Sekerka 2010b) because of erroneous interpretation of the type locality – Calcutta in West Bengal.

321 *Chaeridiona metallica* Baly, 1869. Würmli (1976b) incorrectly synonymized *C. cupreoviridis* Gressitt, 1950 with this species and both species erroneously appeared in the previous Edition (Borowiec & Sekerka 2010b) with identical distribution. *Chaeridiona metallica* is restricted to West and South India and does not occur in the Palaearctic Region as defined for the purpose of the present Catalogue.

322 *Downesia auberti* (Fairmaire, 1889). Uhmann (1955a) recorded *Downesia latenigra basipennis* Pic, 1924 from Fujian and Zhejiang, which was latter synonymized with *D. auberti*. However, Uhmann overlooked that the population from the same locality (Tianmushan in Zhejiang) was described as *D. vandykei* Gressitt, 1939 and latter also recorded from the same localities in Fujian (Wuyishan Mts). Chen et al. (1986) attributed Uhmann's records to *D. vandykei* thus excluding *D. latenigra* Pic, 1924 from fauna of China. However, this was overlooked, and the species was erroneously introduced to the previous Edition of the Catalogue in Errata published in the volume 7 (Löbl & Smetana 2011). *Downesia auberti* is so far known only from Vietnam.

323 *Neodownesia sita* Maulik, 1919 was erroneously mentioned in the previous Edition (Borowiec & Sekerka 2010b) based on Uhmann's (1940b) record; the population was latter on described as a distinct species *N. rubra* Gressitt, 1953 (see Sekerka 2015).

324 *Oocassida schultzei* Spaeth, 1917. This species was included in the previous Edition (Borowiec & Sekerka 2010b) and occurrence in Iran was mentioned. Borowiec & Ghahari (2021) catalogued Cassidinae fauna of Iran and stated that the occurrence of this species is based on the Palaearctic Catalogue only. However, I was not able to trace any published record or material in collections and the record must have been included in error.

New geographical records

Note. In the previous edition of the Catalogue many new distributions were given, however, without being mentioned as such and without being provided with a faunistic record. This created quite a lot of confusion, hence all such cases are reported here as new species to the respective countries and provided with full faunistic records along with additional new records. All specimens were identified by L. Sekerka.

325 *Anisodera fraterna* Baly, 1888. CHINA: Hainan: Wuchihshan, 26.iv.2011, 1 spec., Y.-T. Chung leg. (TARI). New species for Hainan (China).

326 *Aspidomorpha* (s. str.) *calligera* Boheman, 1854. INDIA: Uttakharand: Dehra Dun, viii.1942, 5 spec., K. Benner leg. (ZMUH). New species for Uttakharand (India).

327 *Aspidomorpha* (s. str.) *miliaris* (Fabricius, 1775). CHINA: Guizhou: "Kouy-Tchéou" [= Guizhou], 1903, 7 spec., J. R. Chaffanjon leg. (MNHN, 2 spec. in collection of L. Sekerka, Praha, Czech Republic). INDIA: Arunachal Pradesh: Roing vicinity, 28°08′N 95°50′E, 500 m, 23.–28.v.2007, 1 spec., P. Pacholátko leg. (NHMB); Himachal Pradesh: Renuka Lake, 30.60997°N 077.44978°E, 690 m, 8.v.2016, 6 spec., R. Ambrus leg. (collection of R. Ambrus, Praha, Czech Republic, 2 spec. in collection of L. Sekerka, Praha, Czech Republic); Uttarakhand: Bhimtal, 2.x.1972, 2 spec., 1976, 2 spec., F. Smetacek leg. (ZMUH); Haridwar-Chilla, 5.–14.viii.1994, 18 spec., M. Valenta leg. (NMPC). New species for Guizhou (China) and Arunachal Pradesh, Himachal Pradesh and Uttarakhand (India).

328 *Basiprionota bisignata* (Boheman, 1862). CHINA: Shanghai: "Ho-Chan", 1 ♀, Mouton leg. (MNHN). According to Théry (1895: cxi) the locality Ho-Chan is situated in the environs of Shanghai, however, I was not able to find its precise location. New species for Shanghai (China).

329 *Basiprionota gressitti* L. N. Medvedev, 1957. CHINA: Shaanxi: Liziping env., nr. Shimian, 200 km SW of Yaan, 1.vii.1991, 1 ♂, 1 ♀, Z. Kejval leg. (NHMUK); Qinling Mts., Xunyangba env., 1200 m, 20.v.–10.vi.2000, 2 ♂♂ (NHMUK, collection of L. Sekerka, Praha, Czech Republic); Tabaishan Mts., 1800 m, v.–vi.2000, 6 ♂♂, 4 ♀♀, A. Plutenko leg. (collection of L. Sekerka, Praha, Czech Republic). Specimens from Xunyangba were previously erroneously identified and published as *B. bisignata* (Boheman, 1862) by Borowiec (2001b). New species for Shaanxi (China).

330 *Basiprionota satrapa* (Boheman, 1862). CHINA: Heilongjiang: Suiling, 1 ♂, W. A. Maw leg. (NHMUK). New species for Heilongjiang (China).

331 *Basiprionota sexmaculata* (Boheman, 1850). INDIA: West Bengal: Darjeeling without further data, 1 ♂ (NHMUK); Darjeeling Distr., Mahanuddi Valley-Nurbong, 1914, 1 ♂, H. Stevens leg. (NHMUK). New species for Darjeeling District of West Bengal (India).

332 *Basiprionota whitei* (Boheman, 1857). CHINA: Shanghai: "Ho-Chan", 5 ♂♂, 5 ♀♀, Mouton leg. (MNHN); Sichuan: "Tá-tsien-lou" [= Kangding], 1894, 7 ♂♂, 14 ♀♀ (MNHN, 4 spec. in collection of L. Sekerka, Praha, Czech Republic). According to Théry (1895: cxi) the locality Ho-Chan is situated in the environs of Shanghai, however, I was not able to find its precise location. New species for Shanghai and Sichuan (China).

333 *Cassida aurora* Weise, 1907. CROATIA: Mrkviste, 1276 m, 26.vi.1910, 1 spec., Meusel leg. (collection of L. Sekerka, Praha, Czech Republic). New species for Croatia.

334 *Cassida azurea* Fabricius, 1801. BOSNIA AND HERZEGOVINA: Mostar, 1 spec., J. Daniel leg. (HNHM). LUXEMBOURG: Zolver, Galgenberg [= Soleuvre, Gaalgebierg], 29.vii.1917, 1 spec., Heymes leg. (collection of F. Fritzlar, Jena, Germany). MONTENEGRO: Prokletje Mts., Vusenje, 8.vi.2012, 12 spec., J. Stanovský leg. (collection of J. Stanovský, Ostrava, Czech Republic, 4 spec. in collection of L. Sekerka, Praha, Czech Republic); Herceg Novi env., 10.vii.1986, 1 spec., V. Beneš leg. (collection of L. Sekerka, Praha, Czech Republic). New species for Bosnia and Herzegovina, Luxembourg, and Montenegro.

335 *Cassida bergeali* Bordy, 1996. BOSNIA AND HERZEGOVINA: Pazarič, 2.vi.1935, J. Fodor leg. (HNHM). BULGARIA: Michajlovgrad-Vraca env., 20.v.1994, J. Strejček leg. (NMPC). SERBIA: Braničevo: Mustapič, 44°33′41″N 21°31′57″E, 7.v.2015, 5 spec., R. Udržal leg., 3 spec., J. Jelínek leg. (collection of L. Sekerka, Praha, Czech Republic). New species for Bosnia and Herzegovina, Bulgaria and Serbia.

336 *Cassida berolinensis* Suffrian, 1844. NORTH KOREA: Myohyang [= Mjohjang (Myohyangsan)], 8.viii.1959, 1 spec., Pisarski & Prószuński leg. (collection of J. Bezděk, Brno, Czech Republic). New species for North Korea.

337 *Cassida canaliculata* Laicharting, 1781. TURKEY: Afyon prov., 3 km NW of Basören, 26.v.1998, 1 spec., J. Voříšek leg. (collection of L. Sekerka, Praha, Czech Republic). The species was until now reported from Turkey based on vaguely labelled material without precise locality data besides "Turkey". Occurrence in Turkey confirmed.

338 *Cassida circumdata* Herbst, 1799. CHINA: Macao: Macao, without further data, 1 spec. (MNHN). New species for Macao (China).

339 *Cassida deflorata* Suffrian, 1844. LIBYA: N of Bengazi, Merg, 12.iv.1926, 8 spec., T. Chaworth Musters leg. (NHMUK, 2 spec. in collection of L. Sekerka, Praha, Czech Republic). New species for Libya.

340 *Cassida exilis* Boheman, 1854. INDIA: Uttarakhand: Narendra Nagar, 21.iii.1993, S. B. Singh leg. (NHMUK). New species for Uttarakhand (India).

341 *Cassida fausti* Spaeth, 1926. BULGARIA: Harmanli, Marica River, 10.v.1985, 1 spec., Kadlec & Voříšek leg. (collection of L. Sekerka, Praha, Czech Republic); Jasna Poljana, 23.–24.vi.2008, 2 spec., J. Plecháč leg. (collection of L. Sekerka, Praha, Czech Republic); Nesebar, viii.1957, 1 spec. (NMPC); N of Nesebar, 2 km E of Banja, 15.v.2016, 1 spec., Z. Švec leg. (collection of L. Sekerka, Praha, Czech Republic); Rožen env., pr. Melnik, 41°31′44″N 23°25′53″E, 500–550 m, 5.v.2009, 1 spec., 17.vi.2011, 4 spec., J. Pelikán leg. (collection of J. Pelikán, Hradec Králové, Czech Republic, 1 spec. in collection of L. Sekerka, Praha, Czech Republic); Rožeň, 5.vi.2013, 1 spec., E. Ezer leg. (collection of E. Ezer, Zlín, Czech Republic); Primorsko env., 42°16.24′N 27°44.30′E, 21–42 m, 21.–27.vi.2017, 1 spec., P. Pacholátko leg. (NHMUK). CRIMEA: Massandra Park,

7.vi.2007, 1 spec., J. Mertlik leg. (collection of L. Sekerka, Praha, Czech Republic). NORTH MACEDONIA: Rakle, Babuna, 5.–7.vi.1978, 2 spec., I. Rozner leg. (collection of L. Sekerka, Praha, Czech Republic). ROMANIA: Iaşi County, S of Suhuleţ, 46°53′19.0″N 27°16′12.7″E, 4.vi.2018, 1 spec., J. Pelikán leg. (collection of J. Pelikán, Hradec Králové, Czech Republic). SERBIA: Mustapič env., 3.vi.2007, 1 spec., R. Udržal leg. (collection of L. Sekerka, Praha, Czech Republic). New species for Bulgaria, Crimea, North Macedonia Romania and Serbia.

342 *Cassida ferruginea* Goeze, 1777. RUSSIA: Stavropol Reg.: ca. 10 km W of Kislovodsk, 950–1200 m, 29.v.1969, Muche leg. (collection of L. Sekerka, Praha, Czech Republic). To my knowledge, first record for South European Territory of Russia.

343 *Cassida flaveola* Thunberg, 1794. ARMENIA: Cachkadzor env., 1600 m, 11.vi.1988, J. Strejček leg. (collection of L. Sekerka, Praha, Czech Republic). MONGOLIA: Chentei Aimak, Mörön gol, vii.1984, 1 spec., Malec leg. (collection of L. Sekerka, Praha, Czech Republic). MONTENEGRO: "Jezero Susice" [= Sušičko Jezero], 12.–24.VIII.1933, 1 spec., J. Fodor leg. (HNHM). New species for Armenia, Mongolia, and Montenegro.

344 *Cassida hemisphaerica* Herbst, 1799. TURKEY: prov. Balikesir, Kapidagi Yarimadasi, Cakilkoy, 1.vi.2001, 2 spec., G. & V. Rozner leg. (collection of I. Rozner, Budapest, Hungary, collection of L. Sekerka, Praha, Czech Republic). The occurrence in Turkey, mentioned in the previous Edition of the Catalogue, was based on the abovementioned specimen but without providing data or source. Therefore, the species is here formally recorded as new for Turkey.

345 *Cassida humeralis* Kraatz, 1874. PORTUGAL: Faro: Serro da Cima, 37°14′26.6″N 8°04′45.3″E, 261 m, 2.v.2013, L. Sekerka leg. (collection of L. Sekerka, Praha, Czech Republic); Santarém: 10 km W of Coruche (Rd. 114–3), 38°57′52.6″N 8°37′51.6″E, 58 m, 8.v.2013, 18 spec., L. Sekerka leg. (collection of L. Sekerka, Praha, Czech Republic); 9 km N of Monchique (Rd. N 266), 37°21′57.0″N 8°31′17.8″E, 426 m, 29.iv.2013, 10 spec., L. Sekerka leg. (collection of L. Sekerka, Praha, Czech Republic); Setúbal: 2 km W of Muda, 38°16′52.2″N 8°42′15.2″E, 53 m, 27.iv.2013, 12 spec., L. Sekerka leg. (collection of L. Sekerka, Praha, Czech Republic). New species for Portugal.

346 *Cassida icterica* Boheman, 1854. INDIA: Himachal Pradesh: Shimla, vii.1909, C. M. C. leg. (NHMUK). New species for Himachal Pradesh (India).

347 *Cassida inopinata* Sassi & Borowiec, 2006. BOSNIA AND HERZEGOVINA: Pazarič, 2.vi.1935, 1 spec., J. Fodor leg. (collection of L. Sekerka, Praha, Czech Republic). MONTENEGRO: Herceg Novi env., 10.vii.1986, 9 spec., V. Beneš leg. (collection of L. Sekerka, Praha, Czech Republic); Mojkovac, 17.vii.2006, 1 spec., J. Řehounek leg. (collection of L. Sekerka, Praha, Czech Republic); Sutorina, 16.vi.1911, 1 spec., Spaney-Schumacher leg. (collection of L. Sekerka, Praha, Czech Republic). New species for Bosnia and Herzegovina, and Montenegro.

348 *Cassida inquinata* Brullé, 1833. BOSNIA AND HERZEGOVINA: Žepče-Žepačko-polje, 200 m, 26.vii.1911, 1 spec., Spaney-Schumacher leg. (collection of L. Sekerka,

Praha, Czech Republic). SERBIA: Kopaonik, 1 spec., Rambousek leg. (NMPC). New species for Bosnia and Herzegovina, and Serbia.

349 *Cassida jacobsoni* Spaeth, 1914. CHINA: Jiangxi: Tianlangshan Mts., Xin-Huan, 1600 m, vii.2004, 5 spec. (collection of L. Sekerka, Praha, Czech Republic). New species for Jiangxi (China).

350 *Cassida japana* Baly, 1874. CHINA: Hong Kong: "Hong Kong", without further data, 2 spec. (NHMUK); Yunnan: "Yunnan", without further data, 1902–1903, 3 spec., Courtellemont leg. (MNHN, 1 spec. in collection of L. Sekerka, Praha, Czech Republic). New species for Hong Kong and Yunnan (China).

351 *Cassida justa* Spaeth, 1914. INDIA: Uttarakhand: Dehra Dun, Mussoorie, Kemptee Falls, 800 m, viii.1961, 9 spec., G. Scherer leg. (NHMB); Dehra Dun, Mussoorie, 2006 m, 7 spec., G. Scherer leg. (NHMB, 3 spec. in collection of L. Sekerka, Praha, Czech Republic); Naini Tal, Bhim Tal, 609 m, 3 spec., G. Scherer leg. (NHMB, 1 spec. in collection of L. Sekerka, Praha, Czech Republic). New species to Uttarakhand (India).

352 *Cassida leucanthemi* Bordy, 1995. MONTENEGRO: Belasica Mts., Lazovi, above Biogradsko jezero, 42°53′48″N 19°36′51″E, 1390 m, 15.vi.2019, 1 spec., P. Jansa leg. (collection of L. Sekerka, Praha, Czech Republic). New species for Montenegro.

353 *Cassida lineola* Creutzer, 1799. CHINA: Tianjin: "Tientsin", 25.vi.1906, 1 spec., F. M. Thomson leg. (NHMUK). CROATIA: Sisak, 1 spec. (NMPC). GREECE: Centr. Macedonia, near Karydochori, 41°18′43.6″N 23°34′20.28″E, 9.vi.2011, 1 spec., J. Krátký leg. (collection of L. Sekerka, Praha, Czech Republic). New species for Croatia, Greece, and Tianjin (China).

354 *Cassida margaritacea* Schaller, 1783. TURKEY: Tekirdag vil., Hayrabolu, 5.vi.2001, Gy. Rozner leg. (collection of I. Rozner, Budapest, Hungary). New species for European part of Turkey.

355 *Cassida mongolica* Boheman, 1854. CHINA: Heilongjiang: "Mandschourie, Ourga á Tsitsikhar", 1896, 2 spec., J. Chaffanjon leg. (MNHN). New species for Heilongjiang (China).

356 *Cassida murraea halophila* Spaeth, 1926. KAZAKHSTAN: Djambul, vi.1964, 11 spec., J. Gottwald leg. (NMPC, 4 spec. in collection of L. Sekerka, Praha, Czech Republic). UKRAINE: "Falzfeinowo a. Dniepr" [= Falz-Fein Chateau, ca. 4 km SE of Novooleksandrivka, 47°14′N, 33°54′E], 12.v.–7.vi.1914, 2 spec., W. Ramme leg. (ZMHUB, collection of L. Sekerka, Praha, Czech Republic). New species for Kazakhstan and Ukraine.

357 *Cassida murraea murraea* Linnaeus, 1767. SERBIA: Belgrad, 1919, 1 spec., C. Koch leg. (NMPC); Resavica, 30.v.1988, 2 spec., F. Kantner leg. (NMPC); Sarajevo, 2 spec. (HNHM). New species for Serbia.

358 *Cassida nebulosa* Linnaeus, 1758. ARMENIA: Kuybishev, Riv. Agatev, 1500 m, 19.vii.1977, 12 spec., Zombori leg. (HNHM). CHINA: Jiangxi: "Kiang-Si", 1 spec. (MNHN). KYRGYZSTAN: Issykkul, Annaevo, 4.vi.1989, 1 spec., Z. Kletečka leg. (JMCB). SLOVENIA: Bikič, 4 spec., V. Zoufal leg. (NMPC); Borkasovo, 1.vi.1915,

1 spec., J. Fodor leg. (HNHM); Ljublana, 1 spec., Grabowski leg. (HNHM); Morovič, 7 spec. (HNHM, 4 spec. in NMPC), 3 spec., Štípek leg. (NMPC), 5 spec., V. Zoufal leg. (2 spec. in HNHM, 3 spec. in NMPC). New species for Armenia, Jiangxi (China), Kyrgyzstan, and Slovenia.

359 *Cassida nobilis* Linnaeus, 1758. TURKEY: Edirne, 3.vii.1993, 1 spec., F. Kantner leg. (NMPC). New species for European part of Turkey.

360 *Cassida palaestina* Reiche & Saulcy, 1858. LEBANON: Amiom, 2.vi.1944, 1 spec., (NHMUK); Beyrouth, 1 spec. (MNHN); Djezin, 5 spec. (MTD, collection of L. Sekerka, Praha, Czech Republic). SAUDI ARABIA: Hedjaz, 5 spec. (NHMUK). New species for Lebanon and Saudi Arabia.

361 *Cassida pallidicollis* Boheman, 1857. CHINA: Tianjin: "Tientsin", v.1905, 2 spec., vi.1905, 1 spec., F. M. Thomson leg. (NHMUK). New species for Tianjin (China).

362 *Cassida pannonica* Weise, 1892. ARMENIA: Gocht (pr. Garni), Azat River valley, 1600 m, 12.vi.1988, 2 spec., J. Strejček leg. (NMPC); ca 10 km NO Erivan Dzhrwezh-Schlucht, 4.vii.1970, 1 spec., H. Muche leg. (MTD); Ararat Reg., Khosrov res., Vedi area, 1.vi.2002, 2 spec., M. Kalashian leg. (collection of L. Sekerka, Praha, Czech Republic); Bjurakan, 16.-26.vi.1987, 1 spec., Kadlec & Voříšek leg. (NMPC). RUSSIA: Tunkinskie Goltsy, r. Shumak, Perevalnyj, 21.viii.2009, 1 spec., K. Orszulik leg. (collection of K. Orszulik, Frýdek-Místek, Czech Republic). New species for Armenia and first record from Southern Siberia (Russia) as defined in the present Catalogue.

363 *Cassida panzeri* Weise, 1907. SERBIA: Sveta Petka, ix.1902, 1 spec., Horváth leg. (HNHM). New species for Serbia.

364 *Cassida parvula* Boheman, 1854. TURKMENISTAN: Kopet-Dagh Mts., 6km W of Germob, Kurkulab, 850 m, 19.iv.1993, 2 spec., M. Hreblay, Gy. László & A. Podlussány leg. (HNHM, collection of L. Sekerka, Praha, Czech Republic). New species for Turkmenistan.

365 *Cassida prasina* Illiger, 1798. ARMENIA: Little Caucasus Mts., Cachkadzor env. 2300 m, 8.–11.vi.1998, 1 spec., Z. Jindra leg. (NMPC); Dilidjan env., Kujbušev, 2 spec., Z. Jindra leg. (NMPC); Kotayk Region, 6–8 km N of Arzakan, 18.vii.2002, 1 spec., M. Kalashian leg. (collection of J. Bezděk, Brno, Czech Republic); Erivan Umgebung, 60 km NO, Zachkadsor, 1800–2050m, 12.viii.1970, 2 spec., H. Muche leg. (MTD); Cakhkadzor, 13.–17.viii.1979, 2 spec., T. Vásárhely (HNHM). KOSOVO: Peč, Radavac, 17.v.1971, 3 spec., Papp & Horvatovich leg. (HNHM); Žleb Mts., Peč, Rugovo, 16.v.1971, 1 spec., Papp & Horvatovich leg. (HNHM); Šar Mts., Brezovica, 20.–23.v.1971, 1 spec., Papp & Horvatovich leg. (HNHM). SYRIA: Krak des Chevaliers, 15.vii.1997, 1 spec., R. Sehnal leg. (collection of R. Sehnal, Velenice, Czech Republic). New species for Armenia, Kosovo, and Syria.

366 *Cassida pusilla* Waltl, 1835. MONTENEGRO: Bečiči pr. Budva, without date, 1 spec. (NMPC), 4.v.1975, 2 spec., J. Strejček leg. (NMPC), 23.–29.vi.2012, 7 spec., T. Staněk

leg. (collection of T. Staněk, Hradec Králové, Czech Republic); Bar, 9.vi.1978, 1 spec., Necker leg. (ZSM). New species for Montenegro.

367 *Cassida rubiginosa rubiginosa* O. F. Müller, 1776. ANDORRA: Pal, Xixerella, Pyrreneus, 26.vi.1987, 1 spec., A. Podlussány leg. (collection of L. Sekerka, Praha, Czech Republic). SYRIA: Akbés, 1900, 1 spec., Ch. Delagrange leg. (MNHN). New species for Andora and Syria.

368 *Cassida sanguinolenta* O. F. Müller, 1776. GREECE: Thessalia: Spilia env., Ossa Mt., 1600 m, 23.–25.v.2006, J. Plecháč leg. (collection of J. Pelikán, Hradec Králové, Czech Republic), 820 m, 1.vi.2016, 1 spec., B. Zbuzek leg. (NMPC). MONTENEGRO: Belasica Mts., Lazovi, Dolovi Lalaviča-Bendovač Mt., 42°53′42″N 19°37′38″E, 763 m, 15.vi.2019, 2 spec., P. Jansa leg. (collection of L. Sekerka, Praha, Czech Republic). SYRIA: Krak des Chevaliers, 15.vii.1997, 1 spec., R. Sehnal leg. (collection of R. Sehnal, Velenice, Czech Republic). TURKEY: Çanakkale: "Port Baklar" [= Bolayir], 1 spec. (NHMUK). New species for Greece, Montenegro, Syria, and European part of Turkey.

369 *Cassida sareptana* Kraatz, 1874. KAZAKHSTAN: "Uralsk" [= Oral], 2 spec., Reitter leg. (HNHM). First record for European part of Kazakhstan.

370 *Cassida seladonia* Gyllenhal, 1827. ALBANIA: Kula Lums, 2.vii.1918, 1 spec., Csiki leg. (HNHM). LIBYA: Cyrenaica, N of Bengazi, Merg, iv.1926, 1 spec., T. C. Musters leg. (NHMUK). NORTH MACEDONIA: Prilep prov., Mazuciste, 31.v.1998, 1 spec., A. Orosz leg. (HNHM). TURKEY: Besika Bay, 1 spec. (NHMUK). New species for Albania, Libya, North Macedonia, and European part of Turkey.

371 *Cassida seraphina* Ménétriés, 1836. TURKEY: Istanbul vil., Silivri, 23.v.1984, 1 spec., D. Szalóki leg. (HNHM). New species for European part of Turkey.

372 *Cassida stigmatica* Suffrian, 1844. TURKEY: Edirne, 3.vii.1993, 1 spec., F. Kantner leg. (NMPC). TURKMENISTAN: Badchyz Schutz-Geb., Akar-Tseschme, 21.iv.1993, 2 spec., W. Dolin leg. (NHMB). New species for European part of Turkey and Turkmenistan.

373 *Cassida subreticulata* Suffrian, 1844. SYRIA: Akbés, 1890, 1 spec., Ch. Delagrange leg. (MNHN). TURKEY: Çanakkale: "Port Baklar" [= Bolayir], 1 spec. (NHMUK). New species for Syria and European part of Turkey.

374 *Cassida transcaucasica* Borowiec & Świętojańska, 2001. TURKEY: Bitlis: 10 km W of Tatvan, El Aman Hanı, 1850 m, 22.–24.vi.2002, 1 spec., T. Rybárik leg. (collection of L. Sekerka, Praha, Czech Republic). New species for Turkey. From Turkey was described *C. strejceki* Sekerka, 2006, which differs in larger and stouter body and particularly in the explanate margin of elytra being broad and forming an obvious gutter. However, the abovementioned specimen has the shape of body of the typical specimens of *C. transcaucasica*. Nevertheless, more specimens are required to confirm that it truly belongs to *C. transcaucasica* and not being an extreme specimen of *C. strejceki*. In general, the entire *C. undecimnotata* species-group is rather difficult for identification/interpretation and also studies based on molecular data should be involved.

375 *Cassida varians* Herbst, 1799. CHINA: Yunnan: Xiaguan env., 25.36°N 100.12°E, 2000 m, 18.vi.1994, 1 spec., J. Farkač & D. Král leg. (NHMB). INDIA: Arunachal Pradesh: Dirang vicinity, 27°21'–23'N 92°13'–16'E, 1550 m, 1.–9.vi.2004, 2 spec., L. Dembický leg. (NHMB); Etalin vicinity, 28°36'56"N 95°53'21"E, 700 m, 12.–25.v.2012, 27 spec., O. Šauša leg. (NHMUK, 5 spec. in collection of L. Sekerka, Praha, Czech Republic). New species for Yunnan (China) and Arunachal Pradesh (India).

376 *Cassida vespertina* Boheman, 1862. CHINA: Yunnan: "Recu de Lou-Nan" [= Lounan County, currently Shilin County], 1931, 2 spec. (MNHN). New species for Yunnan (China).

377 *Cassida virguncula* Weise, 1889. CHINA: Henan: Denfeng, Umgebung Henan, 16.iv.1983, 12 spec., W. Wittmer leg. (HNHM, collection of L. Sekerka, Praha, Czech Republic). New species for Henan (China).

378 *Cassida vittata* Villers, 1789. MALTA: "Malta", 1901, 3 spec., M. Cameron leg. (NHMUK); Luga, 27.vi.1956, 1 spec., G. V. P. Sewell leg. (NHMUK); Saptan, 28.ii.1956, 2 spec., G. V. P. Sewell leg. (collection of L. Sekerka, Praha, Czech Republic); Wied Babu, 28.ii.1956, 1 spec., G. V. P. Sewell leg. (NHMUK). SERBIA: Sar planina, 1 spec. (collection of L. Sekerka, Praha, Czech Republic); Fruška Góra, 1 spec. (collection of L. Sekerka, Praha, Czech Republic). New species for Malta and Serbia.

379 *Chiridopsis bowringii* (Boheman, 1855). CHINA: Guizhou: "Kouy-Tchéou", 1903, 1 spec., J. R. Chaffanjon leg. (MNHN). New species for Guizhou (China).

380 *Chiridopsis mimica* (Weise, 1905). INDIA: Arunachal Pradesh: 8 km S of Jamiri-Sessa vicinity, 27°07'-09'N 92°34'E, 350 m, 26.v.–4.vi.2005, 2 spec., L. Dembický leg. (NHMUK); West Bengal: Nurbong, Mahanuddi Vy., Darjiling, 1914, 4 spec., H. Stevens leg. (NHMUK); Sikkim, Gopaldhara, Rungbong Valley, 4 spec., H. Stevens leg. (NHMUK). The material from Gopaldhara was published by Maulik (1919) as *C. scalaris* (Weber, 1801). New species for Arunachal Pradesh, Sikkim and Darjeeling District of West Bengal (India).

381 *Craspedonta levis* Borowiec, 1993. CHINA: Guangxi: Rongshui Miao Autonomous Co., Yuan Bao Mts., 25°23'14.3"N 109°09'17.8"E, 1510 m, v.2007, 14 spec., M. Hackel & R. Sehnal leg. (NHMUK, 6 spec. in collection of L. Sekerka, Praha, Czech Republic). New species for Guangxi (China).

382 *Dicladispa testacea* (Linnaeus, 1767). JORDAN: Jerash: Burma env., Al Huna, 15.v.2010, 7 spec., M. Snížek leg. (collection of L. Sekerka, Praha, Czech Republic). MONTENEGRO: Herceg Novi, vii.1986, 16 spec., V. Beneš leg. (collection of L. Sekerka, Praha, Czech Republic); Virpazar, 4.vii.2010, 22 spec., D. Čagánek leg. (collection of D. Čagánek, Otrokovice, Czech Republic, 2 spec. in collection of E. Ezer, Zlín, Czech Republic). New species for Jordan and Montenegro.

383 *Epistictina viridimaculata* (Boheman, 1850). INDIA: Uttarakhand: Kumaon Division, C. Almora, vii.1920, 2 spec., H. G. Champion leg. (NHMUK). New species for Uttarakhand (India).

384 *Estigmena chinensis* Hope, 1841. CHINA: Sichuan: Daxue Shan Mts., N of San Ya, 28°45.3'N 101°57.9'E, 3040 m, 10.vi.2005, 1 spec., Trýzna leg. (collection of L. Sekerka, Praha, Czech Republic). New species for Sichuan (China).

385 *Glyphocassis* (s. str.) *trilineata szechuana* Gressitt, 1952. CHINA: Jiangxi: Wumeishan, Quing-An Reg., v.–vi.2004, 2 spec., Wang leg. (collection of L. Sekerka, Praha, Czech Republic). New species for Jiangxi (China).

386 *Glyphocassis* (s. str.) *trilineata trilineata* (Hope, 1831). INDIA: Arunachal Pradesh: Dirang vicinity, 27°21'–23'N 92°13'–16'E, 1500–1800 m, 1.–10.vi.2004, 1 spec., R. Businský leg. (NHMB). New species for Arunachal Pradesh (India).

387 *Gonophora pulchella* Gestro, 1888. CHINA: Guangxi: Yulin Distr., Liuwan Mts., SW Yulin, 350–400 m, 16.xi.1993, 1 spec., H. Schillhammer leg. (NMW). Additional record for Guangxi (China). The species was recorded as new to Guangxi by Yang et al. (2023).

388 *Hypocassida cornea* (Marseul, 1868). ALBANIA: Lezhë distr., 2.5 km NE of Ndreaj, 10.vi.2018, 1 spec., T. Németh, Cs. & V. Szénasi leg. (HNHM). BULGARIA: Arkutino, 15.v.1985, 1 spec., Kadlec & Voříšek leg. (collection of L. Sekerka, Praha, Czech Republic). CROATIA: Baško Pole env., pr. Baška Voda, 1.–6.vi.2008, 12 spec., T. Kopecký leg. (collections of T. Kopecký, Ústí nad Orlicí, Czech Republic, and L. Sekerka, Praha, Czech Republic). CZECH REPUBLIC: Moravia: NP Podyjí, Havraníky, 17.vi.2002, 1 spec., 19.vi.2002, 2 spec., 22.iv.2002, 2 spec., Z. Kraus leg. (collections of Z. Kraus, Znojmo, Czech Republic, and L. Sekerka, Praha, Czech Republic). MONTENEGRO: Rumija Mts., NE of Stegvaši, 42°03'51.5"N, 19°22'22.0"E, 482 m, 17.vi.2019, 7 spec., P. Jansa leg. (collection of L. Sekerka, Praha, Czech Republic). NORTHERN MACEDONIA: Skopje, Sv. Nikita, 560 m, 20.vi.1968, Necker leg. (collection of L. Sekerka, Praha, Czech Republic). POLAND: Bieszczadzki PN, Gernik env., 29.vi.2013, 1 spec., R. Škoda leg. (collection of L. Sekerka, Praha, Czech Republic). ROMANIA: Cheile Turzii NR, 2 km SW of Săndulești, 15.vi.2005, 3 spec., L. Sekerka & K. Štajerová leg. (collection of L. Sekerka, Praha, Czech Republic). SERBIA: Deliblatska peščara, Šušara env., 8.–9.vi.2010, 7 spec., V. Zieris leg. (collections of J. Pelikán, Hradec Králové, Czech Republic, and L. Sekerka, Praha, Czech Republic). UKRAINE: Lugansk distr., Severodonetsk env., Provalskaya step NR, 5.–10.iv.2002, 2 spec., V. Rybak leg. (collection of L. Sekerka, Praha, Czech Republic). New species for Albania, Bulgaria, Croatia, Czech Republic, Montenegro, Northern Macedonia, Poland, Romania, Serbia and Ukraine. Bordy (2008) recorded it from Bulgaria based on a single specimen, however the locality is actually situated in Romania.

389 *Ischyronota desertorum* (Gebler, 1833). ROMANIA: Histria, 15.vii.1995, 1 spec., M. Mikát leg. (collection of J. Bezděk, Brno, Czech Republic); Istria-NR Grindul Saile, 3.–4.vi.2008, 7 spec., R. Škoda leg. (collection of R. Škoda, Liberec, Czech Republic), 26.–28.vi.2010, 45 spec., L. Sekerka leg. (collection of L. Sekerka, Praha, Czech Republic), 19.–20.ix.2010, 2 spec., P. Kresl leg. (collection of P. Kresl, Spůle, Czech Republic), 22.ix.2010, 20 spec., J. Krátký leg. (collection of J. Krátký, Hradec

Králové, Czech Republic, 8 spec. in collection of L. Sekerka, Praha, Czech Republic); Lacu-Sarat, 3 spec., A. L. Montadon leg. (HNHM). TURKEY: Çanakkale: "Port Baklar" [= Bolayir], 6 spec. (NHMUK). New species for Romania and European part of Turkey.

390 *Laccoptera* (*Sindiola*) *vigintisexnotata* Boheman, 1855. CHINA: Guizhou: "Kouy-Tchéou" [= Guizhou], 1903, 3 spec., J. R. Chaffanjon leg. (MNHN, 1 spec. in collection of L. Sekerka, Praha, Czech Republic). New species for Guizhou (China).

391 *Lasiochila cylindrica* (Hope, 1831). INDIA: Arunachal Pradesh: Etalin vicinity, 28°36′56″N 95°53′21″E, 700 m, 12.–25.v.2012, 4 spec., O. Šauša leg. (NHMUK). New species for Arunachal Pradesh (India).

392 *Macromonycha apicalis* (Gebler, 1844). KYRGYZSTAN: Issik-Kul: Terskej Alatau Mts., 2665 m, 3.vii.1998, 1 spec., Gurko leg. (collection of L. Sekerka, Praha, Czech Republic). New species for Kyrgyzstan.

393 *Nabathaea pygmaea* Spaeth, 1911. EGYPT: SE Sinai, ca 8 km SW of Dahab, 28°27′00″N 34°27′30″E, 17.iv.1997, 5 spec. (NHMB, 2 spec. in collection of L. Sekerka, Praha, Czech Republic). New species for Sinai (Egypt).

394 *Notosacantha maculipennis* (Boheman, 1857). BHUTAN: Domphu-Gopani, 1400 m, 1983, 1 spec., B. Bhakta leg. (NHMB). New species for Bhutan.

395 *Oxylepus deflexicollis* (Boheman, 1862). CROATIA: Ugljan Is., 19.vii.1966, 4 spec., A. Hoffer leg. (NMPC, 1 spec. in collection of L. Sekerka, Praha, Czech Republic). New species for Croatia.

396 *Pilemostoma fastuosum* (Schaller, 1783). ARMENIA: Cachcadzor, 8.vi.1988, 1 spec., Netušil leg. (collection of V. Šilha, Nové Strašecí, Czech Republic). New species for Armenia.

Comments on distribution

Distributional data presented in this Edition are based on published records only. All occurrences mentioned in the previous Edition (Borowiec & Sekerka 2010b) were thoughtfully verified whether they are based on reliable information and all questionable cases are commented in this and following chapters dealing with the distribution. I have found a large number of records that have been overlooked in the previous Edition and are incorporated here; these are not discussed further. However, all occurrences that appeared in the previous Edition and are not accepted here are commented on.

397 *Agonita parvula* (Gestro, 1890). Würmli (1975) recorded this species from Bhutan and in distribution also mentioned Fujian and Taiwan without any reference or explanation. This was because *A. tricolor* (Chûjô, 1933) was considered a subspecies of *A. parvula* by Gressitt (1950) and Gressitt & Kimoto (1963a), while it is a valid species (Chen et al. 1986, Lee 2015b). Thus, the records from Fujian and Taiwan are not accepted for *A. parvula*.

398 *Cassida flaveola* Thunberg, 1794. Occurrence in Croatia given in the previous Edition (Borowiec & Sekerka 2010b) was based on specimens generally labelled as "Dalmacia". No other published record or more precisely labelled material is known to me.

399 *Cassida nobilis* Linnaeus, 1758. Occurrence in Israel is doubtful as *C. pfefferi* Sekerka, 2006 was described from this region, and so far all specimens I have seen from Israel belonged to that species.

400 *Cassida piperata* Hope, 1843. Shi (2008) published a record of *C. versicolor* (Boheman, 1855) from Anhui but based on the photo presented in the publication it pertains to *C. piperata*. Accordingly, the record is transferred to the latter species.

401 *Cassida sanguinosa* Suffrian, 1844. Occurrence in Austria is doubtful (see Sekerka 2010).

402 *Cassida undecimnotata* Gebler, 1841 is a typically central Asian species. In the past the species was broadly defined with records from E Turkey to W China, however, it was subsequently divided into several species based on external morphology. Although the newly defined species are partly difficult to be recognized morphologically, particularly in the eastern part of their range, they are supported also geographically. Populations inhabiting western part were classified as *C. transcaucasica* Borowiec & Świętojańska, 2001 (distributed in the Caucasus and eastern Iran) and *C. strejceki* Sekerka 2006 found in eastern Turkey. Records of *C. undecimnotata* from Armenia and Georgia certainly belong to *C. transcaucasica*. There are also two records from Turkey, known to me: Malatya (Heyden 1888) and Buglan Gecidi, 7 km E of Solhan (Medvedev 2015b) but I have had no opportunity to examine the respective material. Medvedev (2015b) did not provide any details, however, Heyden (1888) mentioned a black spot on the pronotum, which is present in both recently described taxa as well as in typical specimen(s) of *C. undecimnotata*. However, in the latter species it is in general a very rare feature present usually in specimens from eastern part of its range and towards mountains (e.g. Tian-Shan, Pamir etc.) while most of the specimens have red pronotum. On the contrary, *C. transcaucasica* and *C. strejceki* have always a black spot on the pronotum. Moreover, Buglan Gecidi is the type locality of *C. strejceki* and I have seen some recent material from Malayta belonging also to the latter species. It is very unlikely that two species of *C. undecimnotata* species-group would occur together, therefore I assume that the specimens published by Heyden (1888) and Medvedev (2015b) belong to *C. strejceki* and consequently distribution of *C. undecimnotata* in Turkey is rejected.

403 *Chiridula semenowi* Weise, 1889. Distribution in China and North Korea is adopted after summary distribution for this species in the book *Fauna of the beetles from Ningxia* (Ren 2019), however, there are no precise published records known to me besides those from Xinjiang.

404 *Estigmena chinensis* Hope, 1841. Pic (1924g) mentioned in the description of the var. *atricollis* Pic, 1924 distribution "Chine, Indes". Nevertheless, the original specimen from China is labelled as "Kouy-Tchéou, Koúy Yàng Foú" which refers to Guiyang in the province of Guizhou which is added to the distribution of *E. chinensis*.

405 *Hypocassida cornea* (Marseul, 1868). The species was reported from Cyprus by Bordy (2008) but the record should be verified as all specimens that I have seen from Cyprus always undoubtedly belonged to *H. subferruginea* (Schrank, 1776).

406 *Ischyronota desertorum* (Gebler, 1833). The record from Egypt (Alfieri 1976) should be verified whether it belongs to this or another species. It is the unique record of this species from North Africa, where typically *I. brisouti* (Reitter, 1889) occurs.

407 *Laccoptera tredecimpunctata* (Fabricius, 1801). Occurrence in China is doubtful as records from all mentioned provinces are based on old and vaguely labelled specimens with province name only. The species is distributed on islands in Indonesia, Malaysia, and the Philippines.

408 *Platypria echidna* (Guérin-Méneville, 1840). Occurrence in Japan (Gressitt 1939c) is doubtful as no more records exist (Kimoto 1966b).

409 Borowiec (2009) recorded the following species from Xizang based on material collected by Li Jingke: *Aspidomorpha dorsata* (Fabricius, 1787) (as *A. fuscopunctata* Boheman, 1854), *A. miliaris* (Fabricius, 1775), *A. sanctaecrucis* (Fabricius, 1794), *Basiprionota pudica* (Spaeth, 1925), *B. sexmaculata* (Boheman, 1850), *B. westermanni* (Mannerheim, 1845), *Cassida inciens* Spaeth, 1926, and *C. truncatipennis* (Spaeth, 1914). It is well known that the material collected by Li Jingke is not trustworthy as it was often mislabelled. In this case most of the specimens are morphologically identical to populations known from northern Laos (Mt. Phu Phan) and not to those from adjacent tropical forests in Arunachal Pradesh. Although at least part of the species might occur in Xizang all records are considered erroneous and thus not adopted in the Catalogue. Świętojańska & Borowiec (2006) described *Aspidomorpha tibetana* based on the same material, which is consequently excluded from the Palaearctic Region. Reliably labelled material, identical to the holotype is known from northern Thailand.

Distributional data that appeared in the previous edition of the Catalogue but were erroneous and are rejected

410 *Acmenychus planus* Maulik, 1919. The occurrence in Saudi Arabia was based on interpretation of the type locality "Arabia".

411 *Agonita laeta* L. N. Medvedev & Sprecher-Uebersax, 1999. The occurrence in the Oriental Region was based on erroneous interpretation of locality data from Darjeeling.

412 *Agonita picea* Gressitt, 1953 and *A. pilipes* (S.-H. Chen & Sun, 1962). Guandong appeared erroneously in the English key in Chen et al. (1986) as in the Chinese text is mentioned material from Guangxi instead.

413 *Basiprionota decemmaculata* (Boheman, 1850). The occurrence in Hainan refers to *B. multipunctata* (Gressitt, 1938); the occurrence in Pakistan refers to *B. decemstillata* (Boheman, 1857).

414 *Basiprionota sexmaculata* (Boheman, 1850). The occurrence in Hainan was based on *B. reducta* (Gressitt, 1938), which is synonym of *B. opima* (Spaeth, 1925) and not *B. sexmaculata*.

415 *Callispa dimidiatipennis* Baly, 1859, *C. karena* Maulik, 1919, and *C. sundara* Maulik, 1919. Their occurrence in Guandong was based on incorrect interpretation of Hainan.

416 *Callispa ruficollis* Fairmaire, 1889. Guangxi was erroneously mentioned in the English key in Chen et al. (1986) as the Chinese text does not contain any information on Guangxi. It was reported from Guangxi by Yang et al. (2023) but according to the photograph of the voucher specimen it does not conform the species presently recognized as *C. ruficollis* and the record is not accepted here.

417 *Cassida algirica* Boheman, 1847. The occurrence in Albania was erroneously given as there is no record published. Occurrences in Armenia and Turkey are doubtful and quite likely based on misidentifications.

418 *Cassida atrata* Fabricius, 1787. The occurrence in Italy was erroneously given as there is no record published; the occurrence in Latvia was rejected by Bukejs et al. (2009).

419 *Cassida berolinensis* Suffrian, 1844. The occurrence in Hebei was based on incorrect interpretation of Beijing.

420 *Cassida brevis* Weise, 1884. European part of Turkey was mentioned by Gruev (2005) in a summary table without comment or published record; subsequently it was not accepted in the catalogue of Turkish leaf beetles (Ekiz et al. 2013). The occurrence in Lebanon was mentioned by mistake.

421 *Cassida deltoides* Weise, 1889. The occurrence in Mongolia was based on erroneous interpretation of the type locality Ordos, which is located nowadays in Chinese province of Inner Mongolia.

422 *Cassida ginpinica* (S.-H. Chen & Zia, 1961). The occurrence in Nepal was based on misidentification of *C. varians* Herbst, 1799 (see Kimoto & Takizawa 1981).

423 *Cassida inquinata* Brullé, 1833. The occurrence in Slovakia was based on misidentification of *C. rufovirens* Suffrian, 1844 (see Sekerka 2010).

424 *Cassida lineola* Creutzer, 1799. The occurrence in Bosnia Herzegovina was based on erroneous interpretation of locality data and is replaced with Serbia.

425 *Cassida liquefacta* Spaeth, 1912. Lopatin (2008) recorded this species as new to the United Arab Emirates. However, the species was misidentified based on the photograph given in the paper as it represents *Seminabathaea arabica* (Spaeth, 1911). Therefore the record is here transferred to the latter species.

426 *Cassida margaritacea* Schaller, 1783. The occurrence in Great Britain was based on erroneous interpretation of *C. spergulae* Marsham, 1802, which is conspecific with

C. hemisphaerica Herbst, 1799. The occurrence in Azerbaijan was erroneously given as no published record exists.

427 *Cassida nebulosa* Linnaeus, 1758. The occurrence in European Turkey was erroneously given as no record exists (e.g. Ekiz et al. 2013).

428 *Cassida nigriventris* Boheman, 1854 and *C. occursans* Spaeth, 1914. Occurrence of these species in Bhutan was adopted in the previous Edition (Borowiec & Sekerka 2010b) after Borowiec (1990a), who mentioned it among their general distribution but did not provide any formal record. I assume this is an error as I was unable to find any published record of these species from Bhutan, although their occurrence there is quite probable. Nevertheless, both species were mentioned in the paper entitled "The Cassidinae of Nepal, Bhutan and Norther Territories of India ..." (Kimoto 1981c) and the title is most likely the reason, why Bhutan was mentioned in their distribution, although they were not recorded from there.

429 *Cassida pannonica* Weise, 1892. The occurrence in Latvia was rejected by Bukejs et al. (2009). In the previous Edition the species was listed for West Siberia, however, the locality of the respective specimen is situated in South Siberia as defined in the present Edition.

430 *Cassida panzeri* Weise, 1907. The occurrence in Japan was based on misidentification of *C. ferruginea* Goeze, 1777 (see Suenaga 2013).

431 *Cassida parvula* Boheman, 1854. The occurrence in Uzbekistan was mentioned in error. Although the species possibly occurs there, there is no reliable record to my knowledge.

432 *Cassida rothschildi* Spaeth, 1922. The occurrence in United Arab Emirates was given erroneously and is here corrected to Saudi Arabia.

433 *Cassida rubiginosa rubiginosa* O. F. Müller, 1776. All records from China, Japan, Russian Far East, North Korea and South Korea belong to the subspecies *rugosopunctata* Motschulsky, 1866.

434 *Cassida sareptana* Kraatz, 1874. The record from Mongolia was rejected and belongs to *C. spaethi* Weise, 1901 (see Medvedev 1979c).

435 *Cassida seraphina* Ménétriés, 1836. The occurence in Armenia, Kazakhstan and Russian South European territory was erroneously mentioned and pertains to *C. hablitziae* Motschulsky, 1838, which was in the past considered a subspecies of *C. seraphina*.

436 *Cassida subreticulata* Suffrian, 1844. The occurrence in Latvia was rejected by Bukejs et al. (2009).

437 *Cassida sussamyrica* Spaeth, 1926. The occurrence in Turkmenistan was based on erroneous interpretation of one of the type localities, which are situated in the present-day Kyrgyzstan.

438 *Cassida vespertina* Boheman, 1862. The species was in the past recorded as from "Korea" (Spaeth & Reitter 1926) and this record was followed by subsequent authors.

However, there are no exact data for present-day North Korea, although the species quite likely occurs there.

439 *Cassida vittata* Villers, 1789. Occurrence in Japan was mentioned by Spaeth (1914c), besides that there is no other known record, and the specimen(s) was possibly mislabeled.

440 *Cassidispa bipuncticollis* S.-H. Chen, 1941. The occurrence in Shaanxi was based on wrong interpretation of the type locality, replaced by Shanxi.

441 *Chaeridiona cupreoviridis* Gressitt, 1950. The occurrence in the Oriental Region was erroneously given as there is no existing record.

442 *Chiridopsis scalaris* (Weber, 1801). The occurrence in Sikkim/Darjeeling District was based on misidentification and belongs to *C. mimica* (Weise, 1905).

443 *Chiridula semenowi* Weise, 1889. The occurrence in Armenia was based on unpublished and misidentified specimen similar to *C. reitteri* Weise, 1892; no published record exists from Pakistan, nor I was able to trace any faunistic material and thus the record is not accepted.

444 *Dactylispa albopilosa* (Gestro, 1888). The occurrence in Sikkim/Darjeeling District and Pakistan was interpreted after records published for *D. soror* Weise, 1897, which is currently a synonym of *D. nigripennis* (Motschulsky, 1860).

445 *Dactylispa corpulenta* Weise, 1897. The occurrence in Sikkim/Darjeeling District and Nepal was based on interpretation of "Himalaya" published by Uhmann (1958) and no further precise records exist.

446 *Dactylispa paucispina* Gressitt, 1939. The occurrence in Jiangsu was an error, corrected to Jiangxi.

447 *Dactylispa subquadrata subquadrata* (Baly, 1874). Yunnan was erroneously given and refers to the subspecies *australis* S.-H. Chen & Tan, 1964.

448 *Dactylispa tewfiki* Pic, 1939. The occurrence in the Afrotropical Region was erroneously mentioned as there is no existing record.

449 *Downesia javana ginpinica* S.-H. Chen & C.-C. Tan, 1962 was erroneously mentioned from the Oriental Region, which refers to the nominotypical subspecies from Java.

450 *Downesia sasthi* Maulik, 1923. The occurrence in Sikkim/Darjeeling District was based on erroneous interpretation of the type locality, replaced by Uttarakhand; also, occurrence in the Oriental Region was erroneously given as there is no existing record.

451 *Downesia tarsata* Baly, 1869. The occurrence in Guandong was based on misidentification and refers to *D. balyi* Gressitt, 1950.

452 *Hispa ramosa* Gyllenhal, 1817. The Chinese provinces Anhui, Guandong, Guangxi, Hainan, Sichuan and Yunnan were erroneously mentioned, all records belong to *H. andrewesi* (Weise, 1897); the occurrence in Sikkim/Darjeeling District was based on erroneous interpretation of West Bengal.

453 *Hypocassida cornea* (Marseul, 1868). The occurrence in Libya was erroneously given as there is no existing record.

454 *Hypocassida subferruginea* (Schrank, 1776). The occurrence in Sweden was erroneously given as the species was not recorded from there (e.g. Silfverberg 2010).

455 *Lasiochila bicolor* Pic, 1924. Occurrence in Hunan (Tan & Yu 1992) is doubtful as the species is restricted to Taiwan. In continental China another similar species, *L. fukiena* Gressitt, 1950 is known (previously considered a subspecies of *L. insulana* Uhmann, 1929 = *L. bicolor*), which probably has wider distribution, but no reliable material was published so far. Despite I did not examine the material published by Tan & Yu (1992) as *L. bicolor* I do not accept this record due to its unlikely geographic distribution.

456 *Leptispa abdominalis abdominalis* Baly, 1859. The occurrence in Hebei was based on erroneous interpretation of locality data and is replaced by Beijing.

457 *Leptispa godwini* Baly, 1869. The occurrence in Jiangsu was based on erroneous interpretation of locality data and is replaced by Shandong.

458 *Leptispa pici* Uhmann, 1958. The occurrence in Yunnan was erroneously mentioned in the English key as it did not appear in the Chinese text of the book (Chen et al. 1986).

459 *Macrispa saundersi* Baly, 1859. The occurrence in Bhutan was based on incorrect interpretation locality data "British Bootang: Maria Basti" by Würmli (1975); the locality it is situated in the Darjeeling District of West Bengal.

460 *Macromonycha apicalis* (Gebler, 1844). The occurrence in Armenia and Azerbaijan is based on interpretation of Reitter's old locality "Araxesthal", however this seems to refer more specifically to Ordubad as Reitter himself in publications often wrote "Araxesthal bei Ordubad" but only "Araxesthal" was printed on the labels (e.g. Zerche 2003).

461 *Notosacantha darjeelingensis* Borowiec & Takizawa, 1991. The occurrence in the Oriental Region was erroneously mentioned as there is no existing record.

462 *Notosacantha fumida* (Spaeth, 1913). The occurrence in Guandong was mentioned erroneously as the respective records belong to *N. moderata* Chen & Zia, 1964 (see Chen & Zia 1964).

463 *Oncocephala atratangula* Gressitt, 1938. The occurrence in Guandong was based on erroneous interpretation of Hainan.

464 *Oxylepus deflexicollis* (Boheman, 1862). The occurrence in Turkey was erroneously mentioned as there is no existing record (e.g. Ekiz et al. 2013).

465 *Pilemostoma fastuosum* (Schaller, 1783). The occurrence in Italy was based on erroneous interpretation of locality data and refers to Montenegro; also the occurrence in the Netherlands was erroneously given as there is no published record.

466 *Platypria hystrix* (Fabricius, 1798). The occurrence in Guandong was based on erroneous interpretation of Hainan.

467 *Rhoptrispa dilaticornis* (Duvivier, 1891). The occurrence in Hebei was based on wrong interpretation of locality data and is replaced by Beijing.
468 *Seminabathea arabica* (Spaeth, 1911). The occurrence in Saudi Arabia was based on erroneous interpretation of the type locality "Arabien".
469 *Sinispa tayana* (Gressitt, 1939). The occurrence in Guandong was based on erroneous interpretation of Hainan.
470 *Wallacea dactyliferae* Maulik, 1919. The population from Taiwan proved to represent a distinct species *W. yuae* Lee & Sekerka, 2018.

Unverified distributional data

471 The following distributional data appeared in the previous Edition (Borowiec & Sekerka 2010b), however, I was unable to find any published sources. Nevertheless, they could have been based on unpublished specimens identified by Lech Borowiec and are retained here as the occurrences are quite possible.
Cassida aurora Weise, 1907 – E: GR
Cassida azurea Fabricius, 1801 – **A**: RU (ES)
Cassida berolinensis Suffrian, 1844 – **A**: CH (SHX)
Cassida denticollis Suffrian, 1844 – E: AL PT
Cassida flaveola Thunberg, 1794 – E: GR
Cassida inquinata Brullé, 1833 – E: SL **A**: AR IS LE
Cassida linnavuorii Borowiec, 1986 – **A**: LE
Cassida nebulosa Linnaeus, 1758 – E: AL
Cassida parvula Boheman, 1854 – E: RU (CT)
Cassida subreticulata Suffrian, 1844 – **A**: JA
Hypocassida subferruginea (Schrank, 1776) – **A**: RU (FE)
Leptispa filiformis Germar, 1842 – **N**: TU
Pilemostoma fastuosum (Schaller, 1783) – E: SL SP

Chrysomelidae: Chrysomelinae: Chrysomelini

H. Kippenberg & Yu. E. Mikhailov

Comments

472 The position of the genus *Timarcha* Samouelle, 1819 in Chrysomelidae system is not satisfactorily resolved. Although preferred by some authors (e.g. Gómez-Zurita et al. 2008, Nie et al. 2020, Zhang et al. 2022), an independent subfamily Timarchinae is not yet generally accepted by the scientific community. In the current Catalogue we have decided to divide Chrysomelinae into two tribes Chrysomelini and Timarchini, which reflects the relative isolation of the genus *Timarcha* and such arrangement is widely accepted in the current Chrysomelidae literature (e.g. Daccordi 1994, Reid 2014). However, this arrangement forced some changes in Chrysomelini compared to the previous Edition. The former subtribes Gastrophysina, Phratorina and

Prasocurina are incorporated into the subtribe Chrysomelina, tribe Doryphorini is downgraded to subtribe Doryphorina, the former subtribe Chrysolinina is incorporated into subtribe Doryphorina, and tribes Entomoscelini and Gonioctenini are downgraded as subtribes Entomoscelina and Gonioctenina.

473 Subtribe Sclerophaedonina Steinhausen, 1996. Species belonging to the subtribe Entomoscelina have adults with closed anterior coxal cavities and pupae with two urogomphi. The adults of species from subtribe Chrysomelina have anterior coxal cavities open and pupae without urogomphi. The subtribe Sclerophaedonina was re-established for the genera *Apterocuris* Jacobson, 1901, *Colaphellus*, Weise, 1916, *Colaspidema*, Laporte, 1833, and *Sclerophaedon* Weise, 1882 whose adults have open anterior coxal cavities and pupae with two urogomphi (Mikhailov 2023). Steinhausen (1996) also introduced the subtribe Colaphina Steinhausen, 1996, explicitly based on "*Colaphus* Dahl". The catalogue by Dahl (1823) was suppressed for the purposes of nomenclature (ICZN 1964, Opinion 710), hence the name *Colaphus* is available from L. Redtenbacher (1845). Since the name *Colaphus* is a junior homonym, the valid genus name is *Colaphellus*. Although Steinhausen (1996) reported the incorrect authorship, he considered the name *Colaphus* to be valid and therefore the derived family-level name Colaphina is available. As both names, Sclerophaedonina and Colaphina, were published in the same work, we fix the precedence of Sclerophaedonina in accordance with the principle of the First Reviser (ICZN 1999, Art. 24.2.2), and Colaphina is treated its synonym. Finally, Petitpierre (2019) proposed the subtribe Colaspidemina for *Colaspidema*, *Colaphellus* and *Sclerophaedon*. Because it contains the same genera as Sclerophaedonina we treat Colaspidemina its synonym.

474 *Ambrostoma superbum* (Thunberg, 1787). Thunberg's collection in UUZM contains one syntype of this species. Bieńkowski (2019) disputed its authenticity and commented on the discrepancies of this specimen with the original description. We had a possibility to examine this specimen and in our opinion it is a true syntype pertaining to the genus *Ambrostoma* Motschulsky, 1845. Bieńkowski (2019) designated the neotype of *Chrysomela superba* Thunberg, 1787 from Ethiopia. We treat this neotype designation as invalid and in direct conflict with Art. 75.1 (ICZN 1999) because a syntype of *Chrysomela superba* exists.

475 *Chrysolina* Motschulsky, 1860. In the subfamily Chrysomelinae, the genus *Chrysolina* is the most controversial construct. At the moment, *Chrysolina* combines nearly 500 species worldwide divided into about 70 subgenera and a number of species groups, as their members often differ markedly in morphology. A comprehensive revision is lacking and will be possible based on complex analysis using also molecular data. Bieńkowski (2019) published a meritorious and extensive study on *Chrysolina* based on external morphology. It is used here as a basis,

however, there are many details that are critically addressed and discussed further. The second important source containing nomenclature acts and comments is Fauna Iberica (Petitpierre 2019). A special problem is the large number of species described from the eastern Palaearctic Region, which are currently not assigned to any subgenus.

476 *Chrysolina* (*Allohypericia*) *aeruginosa* (Faldermann, 1835) has currently six recognized subspecies, but a proper revision has never been published and their distribution is not well documented. Five subspecies (except *Ch. a alpina* L. N. Medvedev, 1980 from Gobi Altai Mts. in Mongolia) have been described or recorded from China (Yang et al. 2015, Bieńkowski 2019). Nominotypical subspecies with dorsum shining or dull (surface finely but clearly shagreened) is common in South Siberia and Mongolia but it also has been recorded from many provinces of China (Yang et al. 2015). However, we excluded provinces from which other subspecies were reliably recorded. The remaining provinces are inhabited by *Ch. aeruginosa*, but the exact subspecies is unclear, so they are listed under nominotypical subspecies only provisionally.

477 *Chrysolina* (*Anopachys*) *sundukovi* Mikhailov, 2006. Although Bieńkowski (2007, 2008, 2019) persistently considered *Ch. sundukovi* a synonym of *Ch. watanabei* Takizawa, 1970, differences in the shape of aedeagus, punctation of the dorsal side of body and allopatry proved that these taxa are two distinct species. *Chrysolina sundukovi* is hence considered a valid species in the present Catalogue.

478 *Chrysolina* (*Arctolina*) *poretzkyi* (Jacobson, 1897) and *Ch.* (*Arctolina*) *tundralis* (Jacobson, 1910) are not synonyms. Mikhailov (2018) demonstrated clear differences between both taxa after the examination of the holotype of *Ch. poretzkyi* which was considered lost for the long time. Moreover, examination of the type of *Ch.* (*Arctolina*) *kuznetzowi* (Jacobson, 1897) showed that it is a senior synonym of *Ch. tundralis*. Despite well-documented differences, Bieńkowski (2019) synonymized all these taxa, but his point of view is not followed in the present Catalogue.

479 *Chrysolina* (*Bechynia*) *montana* (Gebler, 1848). Bieńkowski (2005) placed *Ch. montana* into subgenus *Sibiriella* L. N. Medvedev, 1999 based on external characters. However, the representatives of the subgenus *Bechynia* are characterised by the shape of aedeagus and flagellum very different from *Ch. montana*. Provisionally, we leave *Ch. montana* in *Bechynia* as a doubtful assignment [DA].

480 *Chrysolina* (*Centoptera*) *bicolor* (Fabricius, 1775). Petitpierre & Alonso-Zarazaga (2019b) pointed out that *Chrysomela bicolor* Fabricius, 1775 is a junior homonym of *Chrysomela bicolor* Linnaeus, 1767 (now *Alagoasa bicolor*, Galerucinae) and the Fabrician name cannot be conserved based on Art. 23.9.1 (ICZN 1999). The case should be presented to the Commission for a ruling under the plenary power and meanwhile the use of *Chrysomela bicolor* Fabricius, 1775 is to be maintained.

481 *Chrysolina* (*Chalcoidea*) *interstincta* (Suffrian, 1851). Fleury & Vitali (2019) treated *Chrysolina interstincta depressa* (Fairmaire, 1854) as a valid subspecies distributed in France and distinct from *Ch. i. subseriata* (Suffrian, 1851) from Central and North Italy. The only reason, they provided was the vicariant geographical distribution due to the Alps, where *Ch. interstincta* does not occur. But it was not supported by any additional morphological or molecular characters. Furthermore, Fleury & Vitali (2019) were probably unsure of the status of *Ch. i. depressa*, as they mentioned it as "sous-espèces", "(sous-)espèces" and "espèces". We follow Bourdonné (2012) and Bieńkowski (2019) who considered *Ch. i. depressa* a synonym of *Ch. i. subseriata* based on the study of respective type material.

482 *Chrysolina* (*Chalcoidea*) *marginata* (Linnaeus, 1758) is a variable species widespread through the western Palaearctic Region and North America. A reliable revision is lacking, therefore some forms are used here provisionally as subspecies. *Chrysolina m. glacialis* (Weise, 1884) without or nearly without red elytral margin from the high elevation of the Alps and *Ch. m. dierythra* (Rottenberg, 1872) with brownish red elytra from Sicily are treated here as subspecies due to their more or less unique and isolated populations, despite similar colour forms occur also in other subspecies. Two described Pyrenean subspecies, *Ch. m. sculpticollis* Bechyné, 1948 and *Ch. m. iniussa* Bechyné, 1950, are provisionally synonymized here (see comment 544). *Chrysolina m. bodemeyeri* (Weise, 1910) is treated as subspecies based on the study of lectotype (present designation, see comment 562). Bieńkowski (2019) regarded *Ch. sanguineocincta* (Crotch, 1872) and *Ch. waldheimi* Kippenberg, 2010 as subspecies of *Ch. marginata*. Due to their distinct morphology of aedeagus (and due to the different external characters of *Ch. waldheimi*) both taxa are treated here as valid species.

483 *Chrysolina* (*Chalcoidea*) *marginata sculpticollis* Bechyné, 1948, described from the Central Pyrenees, is a valid subspecies. Bieńkowski (2019: 754) examined three female syntypes (collected by Coiffat in 1946 in the subalpine zone) and synonymized it with *Ch.* (*Chalc.*) *interstincta subseriata* (Suffrian, 1851), a taxon inhabiting Italy and France outside the Pyrenees. A verification from photographs showed that all three syntypes (deposited in NMPC) pertain to *Ch. marginata* as well as a male from the same series but not labelled as syntype. Based on the description characteristic features of the ssp. *sculpticollis* are: dark bluish surface, primary rows of elytral punctures rather coarse and irregular, and partly undulating intervals (especially in females). The three syntypes have rather smooth and regular elytral punctation and do not entirely agree with Bechyné's description. However, they were labelled as "cotypes" by Bechyné himself, therefore we accept them as syntypes. The mentioned characters overlap with *Ch. m. iniussa* Bechyné, 1950 which is provisionally synonymized with ssp. *sculpticollis* (see comment 544). In the High Pyrenees, there is a form of *Ch. marginata* corresponding exactly the Bechyné's description

of *Ch. sculpticollis* but additionally characterised by the lateral red margin limited to epipleura or at most extending also to the ultimate interval (Petitpierre 2019), which may represent a separate high-alpine subspecies. The occurrence of distinct subspecies in the same area but at different altitudes is well known, e.g. in species of *Oreina* and even in *Ch. marginata* (see *Ch. marginata glacialis* (Weise, 1884) from the High Alps). However, further studies are necessary.

484 *Chrysolina* (*Chrysolina*). Current assessment of the subspecies of *Ch. staphylaea* (Linnaeus, 1758) has to be revised. In the present Catalogue we keep only *Ch. staphylaea staphylaea* and *Ch. staphylaea daurica* (Gebler, 1832). Bieńkowski (2019) transferred Ch. costalis (Olivier, 1807) and *Ch. wollastoni* (Bechyné, 1957) from *Chrysolina* (s. str.) to subgenus *Rhyssoloma* Wollaston, 1854. In the present Catalogue both species are again listed in *Chrysolina* (s. str.) based on Petitpierre (2023).

485 *Chrysolina* (*Chrysomorpha*) *cerealis* (Linnaeus, 1767) is a polymorphic species forming several subspecies. However, neither a comprehensive revision, nor a satisfactory delimitation of its subspecies was published. The key given by Bieńkowski (2019) does not work reliably and the classification of subspecies traditionally used in European faunistic papers is chaotic. The arrangement given in the present Catalogue has to be understood as provisional. The nominotypical form (which meets Linnaeus' description) occurs in Spain and France and, contrary to the opinion of Bieńkowski (2019), also in Central Europe (e.g. in Germany, Austria, Switzerland, Czech Republic), in parts of the Balkan Peninsula and possibly in North Greece. *Chrysolina c. mixta* (Küster, 1844), differs in its mostly bluish-violet colouration, lives probably allopatrically in France, Switzerland, Italy, the Balkan Peninsula and North Greece. A smaller form, mainly without bright colours, striped or non-striped, is known as *Ch. c. megerlei* (Fabricius, 1801). Its exact distribution is uncertain, the typical specimens are known from E Austria, Croatia, Czech Republic, Hungary, Romania (Transsylvania), Bulgaria (Stara planina), Poland, Slovakia and West Ukraine. Bieńkowski (2019) added most part of eastern Europe, Asia Minor, Caucasus, Kazakhstan and West Siberia as a part of distribution area *Ch. c. megerlei*. However, we prefer to regard these populations as *Ch. c. rufolineata* (Motschulsky, 1860), which is defined mainly by less bright and peculiar greenish testaceous colouration but poorly characterised by structural morphology. *Chrysolina c. livonica* (Motschulsky, 1860) described from the Baltic area is a somewhat doubtful form which can be very similar to small specimens of *Ch. c. mixta* and *Ch. c. megerlei*. The distribution of this form is uncertain, especially its dispersion to Scandinavia and NW European Russia. *Chrysolina c. cyaneoaurata* (Motschulsky, 1860) is known from South Siberia, probably it occurs also in Mongolia and even in the Far East, on average it is more slender than *Ch. cerealis* s. str. and with a somewhat different shape of aedeagus. Larger females can be very similar to *Ch. cerealis* s. str. In addition, Jurado-Rivera & Petitpierre (2015) found the genetic distances between

Ch. cerealis s. str. and *Ch. c. cyaneoaurata* (from Altai) unusually high for an intraspecific taxa, suggesting that they do not belong to the same species. Therefore, the status of Siberian populations remains unclear and requires further study.

486 *Chrysolina* (*Colaphodes*) *haemoptera* (Linnaeus, 1758). Based on Bieńkowski (2019) we consider all described subspecies of *Ch. haemoptera* as synonyms.

487 *Chrysolina* (*Colaphoptera*) *globosa bigorrensis* (Fairmaire, 1865). Bieńkowski (2019) listed *Ch. bigorrensis* as valid species. In the present Catalogue we follow the revision of European *Colaphoptera* by Kippenberg (2017) and *Ch. bigorrensis* is treated as subspecies of *Ch. globosa* (Panzer, 1802).

488 *Chrysolina* (*Colaphoptera*) *purpurascens crassimargo* (Germar, 1823). Bieńkowski (2019) listed *Ch. purpurascens allobrogorum* Bechyné, 1950, *Ch. crassimargo ominosa* Bechyné, 1950 and *Ch. purpurascens stoeckleini* Bechyné, 1950 as synonyms of *Ch. purpurascens purpurascens* (Germar, 1822). In the present Catalogue we follow the revision of European *Colaphoptera* by Kippenberg (2017) and all mentioned taxa are treated as synonyms of *Ch. purpurascens crassimargo*.

489 *Chrysolina* (*Colaphosoma*) *sturmi* (Westhoff, 1882). The greenish form of *Ch. sturmi*, described as var. *polonica* (Weise, 1883), distributed predominantly in the Ukraine, is regarded here as colour aberration only and not as subspecies because it lives sympatrically with the specimens of typical colour.

490 *Chrysolina* (*Erythrochrysa*) *kirigaminensis* Suzuki & Saito, 2011 was downgraded to a subspecies of *Ch. polita* (Linnaeus, 1758) by Bienkowski (2019). However, the drawing of aedeagus in the original description of *Ch. kirigaminensis* indicates a remarkable difference from *Ch. polita*. Therefore, *Ch. kirigaminensis* is regarded as a valid species in the present Catalogue.

491 *Chrysolina* (*Erythrochrysa*) *ogloblini* (Ter-Minassian, 1950) is a valid species based on its peculiar flagellum (see Bieńkowski 2019) but no reliable difference in the external morphology of *Ch. polita* (Linnaeus, 1758) and *Ch. ogloblini* is known. Also it is unknown whether both species live sympatrically or not, therefore the distribution of *Ch. polita* in the Caucasus area and in Near East needs to be verified.

492 *Chrysolina* (*Heliostola*) *gibbipennis* (Faldermann, 1835) was placed in the subgenus *Sibiriella* by Bieńkowski (2019) with some doubts. Due to external characters as well as the typical form of aedeagus we assigned *Ch. gibbipennis* to the subgenus *Heliostola* in the present Catalogue.

493 *Chrysolina* (*Heliostola*) *katonica* Lopatin, 1988 was placed in the subgenus *Sibiriella* by Bieńkowski (2005, 2019). Due to external characters as well as the typical form of aedeagus we assigned *Ch. katonica* to the subgenus *Heliostola* in the present Catalogue.

494 *Chrysolina* (*Hypericia*) *ohoi* Chûjô, 1958, described from Taiwan, was synonymised by Bieńkowski (2019) with *Ch. nikkoensis* (Jacoby, 1885) described from Japan. This conclusion was based on comparison of the original description and photo of the

type of *Ch. ohoi* with the specimens considered by Bieńkowski (2019) as *Ch. nikkoensis*. The aedeagus of *Ch. ohoi* necessary for the precise identification was not examined. Therefore, we treat *Ch. ohoi* as a distinct species not conspecific with *Ch. nikkoensis*.

495 *Chrysolina (Jeanclaudia) tibialis* (Jacobson, 1896), described as *Chrysomela tibialis*, is a junior primary homonym of *Chrysomela tibialis* Suffrian, 1851 (now classified in *Phratora* Chevrolat, 1836). Kippenberg (2010a) replaced *tibialis* Jacobson with a new name *Ch. neotibialis* Kippenberg, 2010 which is in disagreement with Art. 23.9.5 (ICZN 1999). We follow Bieńkowski (2019) who suggested maintaining the usage of *Ch. tibialis* (Jacobson, 1896) until a case could be presented to the Commission for a ruling under the plenary power.

496 *Chrysolina (Jeanclaudia) undulata* (Gebler, 1833), described as *Chrysomela undulata*, is a junior primary homonym of *Chrysomela undulata* Linnaeus, 1763 (now classified in *Phyllocharis* Dalman, 1824). The two taxa were not considered congeneric after 1899. We concur with Bieńkowski (2019) that the prevailing usage of both names is to be maintained and the case should be referred to the Commission for a ruling under the plenary power (ICZN 1999, Art. 23.9.5).

497 *Chrysolina (Lopatinica) curvata* Kippenberg, 2012. The description of *Ch. kataevi* Lopatin, 2000 has an inaccurate drawing of the aedeagus which led to a misidentification of this species in Kippenberg (2012b). As shown by Bienkowski (2019), the description of *Ch. daccordiana* Kippenberg, 2012 refer to true *Ch. kataevi* while *Ch. kataevi* sensu Kippenberg (2012b: 324–325, Figs. 6a, b, 13a, b) represents another taxon. Examination of type material and additional material showed the existence of transitions in the aedeagus shape between *Ch. kataevi* sensu Kippenberg (2012b) and *Ch. curvata*. Therefore, *Ch. kataevi* sensu Kippenberg (2012b) is considered identical to *Ch. curvata*.

498 *Chrysolina (Melasomoptera) grossa grossa* (Fabricius, 1792). In agreement with Bieńkowski (2019) *Ch. grossa* var. *tingitana* (Escalera, 1914) is regarded as synonym of *Ch. grossa grossa*.

499 *Chrysolina (Melasomoptera) lucida* (Olivier, 1807). Following Petitpierre (2019) *Ch. lucida torresi* Bechyné, 1950 is regarded here as synonym of *Ch. lucida*.

500 *Chrysolina (Ovosoma)*. A comprehensive revision of this subgenus does not exist. The species of *Ovosoma* are characterised by a remarkably complicated structure of the aedeagus. Most species have individualy developed apical part of the aedeagus. The described subspecies are based on external characters (surface structure, punctation, colour) only.

501 *Chrysolina (Ovosoma) cretica* (Olivier, 1807) has surprisingly the same apical part of the aedeagus as *Ch. vernalis* (Brullé, 1833), despite both taxa are quite different in their external features. They are allopatric: *Ch. cretica* is restricted to the island of

Crete (Kreta), whereas *Ch. vernalis* is widespread from Near East to Spain. It is possible that both are conspecific (with *Ch. cretica* as a local island population).

502 *Chrysolina* (*Ovosoma*) *orientalis* (Olivier, 1807) has almost the same apical part of the aedeagus as *Ch. sahlbergii* (Ménétriés, 1832). Bieńkowski (2019) characterised *Ch. orientalis* as being strongly brachypterous (alae length ½ of elytra) and separated it from *Ch. sahlbergii* which is indeed brachypterous too but has an alae length of about elytral length (or sometimes more). However, besides this artificial character there is no other morphogical support to discriminate the two taxa. Studies proved that there is an immense variability in the size of alae in different populations of *Ch. sahlbergii* and in other features as well. As long as a study using other scientific methods is lacking *Ch. orientalis* and *Ch. sahlbergii* ought to be treated correctly as a complex. Therefore, *Ch. sahlbergii* is marked here as a questionable species [QS].

503 *Chrysolina* (*Ovosoma*) *vernalis* (Brullé, 1833) has many different forms. Two western subspecies, ssp. *cantabrica* (Heyden, 1870) and ssp. *pyrenaica* (Dufour, 1843), were well characterised by Petitpierre (2019). Dahlgren (1984) studied the other forms of *Ch. vernalis* and found populations of different colouration, with fine and sparse or strong and deep punctation, and intermediate forms. Moreover, such similar populations have mosaic distribution. Our own studies confirm these results. A well-founded classification of subspecies of *Ch. vernalis* is not yet available, therefore all these forms ar regarded here provisionally as synonyms of the nominotypical form, with exception of ssp. *herii* (Herrich-Schäffer, 1838) (sensu Bieńkowski 2019), an island population from Corfu having distinct colouration.

504 *Chrysolina* (*Ovostoma*) *globipennis* (Suffrian, 1851). This taxon is inconsistently treated in Chrysomelidae literature, either as species or as subspecies of *Ch. olivieri* Bedel, 1892. There are arguments for both options. In the present Catalogue we follow Bieńkowski (2019) who considered it a subspecies. However, it was neglected that *Ch. globipennis* is older and thus has priority over *Ch. olivieri*. Therefore, the nominal taxon is changed here to *Ch. globipennis* and further, *Ch. olivieri* including its former subspecies are listed as subspecies *Ch. globipennis*.

505 *Chrysolina* (*Pleurosticha*) *novozhenovi* Mikhailov, 2007 (**stat. res.**) is a sibling species of *Ch. sylvatica* (Gebler, 1823) and they are sympatric in West and East Sayans Mts. The main distinguishing feature of *Ch. novozhenovi* is the shape of its entirely inflated endophallus clearly different from that of *Ch. sylvatica*. Bienkowski (2019) synonymised *Ch. novozhenovi* with *Ch. sylvatica* claiming that this character is not sufficiently studied in *Chrysolina*. However, *Pleurosticha* is the only subgenus where the shape of endophallus was studied in almost all known species (Mikhailov 2006b, 2007c; Mikhailov & Guskova 2013a). Therefore, the decision to synonymise *Ch. novozhenovi* and *Ch. sylvatica* is rejected and both species are treated as valid in the present Catalogue.

506 *Chrysolina* (*Pleurosticha*) *romani* Mikhailov, 2007 was described with two subspecies: nominotypical one from SE Tuva and ssp. *burjatica* Mikhailov, 2007 from the Sayans Mts. in Buryatia. The latter was described based on two specimens (male and female) only and Bienkowski (2019) synonymised it with the nominotypical form. However, clear differences in the shapes of aedeagus and fore tarsi in males, indicated in the original description (Mikhailov 2007b), brings us to the opposite conclusion and we restore *Ch. burjatica* to the species level.

507 *Chrysolina* (*Pleurosticha*) *pirka* Takizawa, 1970 was described from Hokkaido (Japan). Bienkowski (2012, 2019) treated it as subspecies *Ch. cavigera pirka* and indicated its distribution also in SE Siberia, Khabarovsk Krai and Sakhalin. Our studies of a series of topotypes (7 males, 4 females) from Hokkaido and the specimens from eastern Siberia, Russian Far East and Sakhalin demonstrated that the true *Ch. pirka* and the specimens from Russia considered as *Ch. cavigera pirka* by Bienkowski (2019) are not conspecific. *Chrysolina pirka* from Hokkaido is relatively large, flattened and having flat elytral intervals while the specimens from the mountains of East Siberia, Sikhote-Alin and Sakhalin are smaller, convex and have ridge-shaped elytral intervals. Therefore, we treat only Japanese specimens as *Ch. pirka* and the populations from the mountains of Siberia eastwards from Baikal, Russian Far East and Sakhalin are treated here as *Ch. cavigera kodarensis* Mikhailov, 2006 (resurrected from synonymy with *Ch. pirka*).

508 *Chrysolina* (*Stichoptera*). External characters traditionally used to delimit species of *Stichoptera* are rather variable (geographically or even individually) and therefore not always reliable. The aedeagal structures (especially the tip of the flagellum) allow more precise identification (Kippenberg 2020b). *Chrysolina gypsophilae* (Küster, 1845), *Ch. limitata* (Küster, 1845), *Ch. lucidicollis* (Küster, 1845), *Ch. rossia* (Illiger, 1802), *Ch. grossepunctata* (Har. Lindberg, 1950), and *Ch. variolosa* (Petagna, 1819) share the same type of the tip of flagellum. Three of them (*Ch. gypsophilae*, *Ch. limitata* and *Ch. lucidicollis*) show nearly uniform habitus. The study of their type material showed that the morhological differences are not sufficient to consider them as distinct species and a new classification is proposed. Three subspecies of *Ch. gypsophilae* are recognized here: widespreaded *Ch. gypsophilae* s. str. (incl. *Ch. limitata* **syn. conf.**), *Ch. gypsophilae lucidicollis* (**stat. nov.**) from Sardinia, Spain, Maghreb, and *Ch. gypsophilae grossepunctata* (**stat. nov.**) from the Canary Islands. The latter shows certain differences in external morphology and somewhat deviating chromosome numbers (Bourdonné et al. 2013). *Chrysolina limitata* is a synonym of *Ch. gypsophilae* s. str. because of corresponding morphology and sympatric distribution. *Chrysolina rossia* is regarded here as valid species which can be distinguished from *Ch. gypsophilae* by more or less conical shape of the pronotum and smaller elytral punctation. Both species are distributed sympatrically in some

countries but yet they are distinct in our opinion. *Chrysolina lucidicollis sierrana* Franz, 1958 was synonymized with nominotypical taxon by Bieńkowski (2019) and the synonymy is accepted here. *Chrysolina variolosa* is clearly different from *Ch. gypsophilae* by external characters.

509 *Chrysolina* (*Stichoptera*) *gypsophilae* (Küster, 1845). Méquignon (1945) described *Chrysomela gypsoides* var. *morvennensis* Méquignon, 1945. Bieńkowski (2019) regarded it as synonym of *Ch. sanguinolenta* (Linnaeus, 1758). In the present Catalogue we follow Petitpierre (2019) who regarded it conspecific with *Ch. gypsophilae gypsophilae*.

510 *Chrysolina* (*Stichoptera*) *latecincta* (Demaison, 1896) has been often confused with *Ch. kuesteri* (Helliesen, 1912) and *Ch. lucidicollis* (Küster, 1845), especially in Spain, however, *Ch. latecincta* has a different shape of the flagellum. Consequently, ssp. *epipleurica* (Reitter, 1913) from Spain is classified as a subspecies of *Ch. latecincta* and not of *Ch. kuesteri*. *Chrysolina* (*Stichoptera*) *kuesteri friderici* (H. Wagner, 1927) is not a synonym of *Ch. latecincta epipleurica* (Reitter, 1913) and is restored here as a valid subspecies. *Chrysolina* (*Stichoptera*) *oceanoripensis* Bourdonné, Doguet & Petitpierre, 2013 has the same type of the tip of flagellum as *Ch. latecincta*, but is not brachypterous. Provisionally we maintain its species status.

511 *Chrysolina* (*Stichoptera*) *variolosa* (Petagna, 1819). The name *Chrysomela melanostigma* Kollar was mentioned in old catalogues and subsequently attributed to Herrich-Schäffer (e.g. Weise 1916b). Bieńkowski (2019) doubted the existence of the name as he did not find it in works of Herrich-Schäffer. Nevertheless, Herrich-Schäffer (1838b) trully mentioned this name in synonymy in the key in as "*melanostigma* Koll. *sicula* Dej." and consequently neither of the two names is available with his authorship. The name *melanostigma* was made available by Sturm (1843) who used it as an unnecessary replacement name for *Ch. sicula* (Lefebvre, 1827).

512 *Chrysolina* (*Synerga*). Numerous different forms (aberrations, varieties, subspecies) have been described due to the extensive colour variability in some species of *Synerga*. We follow Bieńkowski (2019), who united all forms of *Ch. herbacea* (Duftschmid, 1825), and classified *Ch. viridana* (Küster, 1844) in the two traditionally recognized geographical subspecies *Ch. viridana* s. str. and *Ch. viridana chloris* (P. H. Lucas, 1847). Forms of *Chrysolina coerulans* (Scriba, 1791) were reduced by Bieńkowski (2019) to three geographical subspecies. Whereas *Ch. coerulans bella* (Jacoby, 1890) is defined by its distinctive aedeagus, *Ch. coerulans* s. str. and *Ch. coerulans angelica* (Reiche & Saulcy, 1858) differ only in the colouration. Within the same publication *Ch. coerulans angelica* was characterised as a form with bright variable colours and a conspicuous lateral elytral vitta, and *Ch. coerulans* s. str. as a blue coloured form. This is, however, not quite correct. There are populations of *Ch. coerulans* s. str. with bright green-golden surface distributed in its entire range, but also populations with dark olive-green (Czech Republic) or pure red (South

Anatolia) surface, and other populations with a clear lateral elytral vitta. Aedeagal differences between the two subspecies have not been determined yet. In the present Catalogue we use *Ch. coerulans angelica* as a valid subspecies, however, its relationship to nominal subspecies requires further investigation. Suffrian (1851) described *Chrysomela palustris* Suffrian, 1851 from Sardinia, Sicily and Maghreb which suggests that he had a mixture of *Ch. viridana* s. str. and *Ch. viridana chloris* at his disposal. In the present Catalogue, *Ch. palustris* is preliminary regarded as synonym of *Ch. viridana* s. str.

513 *Chrysolina* (*Taeniosticha*) *pseudolurida samarensis* Bechyné, 1950. Based on the study of one syntype (female) and additional specimens, Bieńkowski (2019) synonymized *Ch. samarensis* with *Ch. pseudolurida* (Roubal, 1917). We examined the syntype (male) of *Ch. samarensis* deposited in NMPC. Its aedeagus is distinctly narrower and smaller than in typical *Ch. pseudolurida*. Because of still unclear subspecific classification of *Ch. pseudolurida*, especially in the eastern part of European Russia and West Siberia, and lack of material from these regions we provisionally treat *Ch. samarensis* as a valid subspecies of *Ch. pseudolurida*.

514 *Chrysolina* (*Taeniosticha*) *tuvensis* L. N. Medvedev, 1976 was originally placed near *Ch.* (*Pezocrosita*) *sahlbergiana* (Jacobson, 1901) based on similarity of colour pattern. Subsequently, Bourdonné (2005) attributed apterous *Ch. tuvensis* to *Taeniosticha* as the only representative in South Siberia close to *Ch. reitteri* (Weise, 1884) from the Caucasus. *Chrysolina tuvensis* was described from the Tuvinian Hollow but later it was found also in West Sayans (Mikhailov 2009a). The indication of *Ch. pseudolurida* (= *Ch. lurida* (Linnaeus, 1767)) for the Sayans by Medvedev & Dubeshko (1992) has not been confirmed and can be attributed also to the alpine form of *Ch. tuvensis*. The recent transfer of *Ch. tuvensis* to the subgenus *Jeanclaudia* Mikhailov, 2009 by Bieńkowski (2019) is completely unreasonable and is rejected here.

515 *Chrysolina* (*Threnosoma*) *fimbrialis hungarica* (Fuss, 1861). Following Daccordi & Ruffo (1979) and our own studies *Ch. fimbrialis hungarica* is to be regarded as a subspecies and not a synonym of *Ch. fimbrialis* (Küster, 1845).

516 *Chrysolina* (*Timarchoptera*). Mikhailov (2003) synonymised the subgenera *Paraheliostola* Medvedev, 1992 and *Timarchoptera* Motschulsky, 1860 based on intermediate morphological characters of *Ch. lomakini* Mikhailov, 2003 and *Ch. soiota khakassa* Mikhailov, 2003. The recent finding of one more species, *Ch. olegkabakovi* Bieńkowski, 2023, supported this point of view (Bieńkowski 2023a). However, the results of molecular phylogeny by Jurado-Rivera & Petitpierre (2015) showed *Ch. haemochlora* (Gebler, 1823) in extraordinary position within a clade together with two species of the genus *Oreina* Motschulsky, 1860 and distant from all other studied species of *Chrysolina* including *Ch. soiota* (Jacobson, 1924). Jurado-Rivera & Petitpierre (2015) suggested to resurrect the subgenus *Paraheliostola* from synonymy and later on Petitpierre (2021) clearly indicated its resurrection. A recent phylogenetic analysis (Gauthier et al. 2023) did not confirm the alternate position

of *Ch. haemochlora* and placed it among other *Chrysolina* clades far from *Oreina*. Therefore, we follow here the morphological evidence and regard the subgenera *Paraheliostola* and *Timarchoptera* as synonyms.

517 Jakob (1952b, 1953a, 1955, 1958b) described several Asian forms of *Chrysomela populi* Linnaeus, 1758, *Chr. saliceti* (Weise, 1884) and *Chr. tremulae* Fabricius, 1787. Some of them were already synonymized. However, the status *Chr. saliceti quadricollis* Jakob, 1958 and *Chr. tremula selengensis* Jakob, 1953 need verification, both taxa are provisionally listed as subspecies marked with [QS]. Similarly, *Chr. saliceti turcestanica* (Reineck, 1937) is probably a synonym of *Chr. saliceti afghanica* (Reineck, 1937) but a definitive decision requires a more detailed study.

518 *Chrysomela collaris* Linnaeus, 1758 is widely distributed in the boreal zone of Eurasia. Medvedev & Khruleva (2011) treated the arctic forms *alpina* Zetterstedt, 1828, *blaisdelli* Van Dyke, 1938, and *hyperborea* L. N. Medvedev, 2011 as subspecies of *Ch. collaris*, and two very close taxa, *Ch. taimyrensis* L. N. Medvedev, 1969 and *Ch. wrangeliana* L. N. Medvedev, 1973, as separate species. To support this arrangement and due to extensive variability of elytral sculpture, they used colour pattern of inner surface of elytra, but its taxonomic value is questionable. An alternative treatment of *Ch. collaris* as a separate boreal-montane species and the Arctic forms (except *alpina* Zetterstedt, 1828) as subspecies of *Ch. blaisdelli* Van Dyke was presented in previous Edition (Kippenberg 2010b) and also supported by O. A. Khruleva, who coauthored the ecological part of the paper with L. N. Medvedev (Medvedev & Khruleva 2011). In the present Catalogue ssp. *blaisdelli* and ssp. *alpina* are treated as valid and ssp. *hyperborea* is synonymized with ssp. *blaisdelli*, however, further studies are necessary.

519 *Colaphellus hoeftii* (Ménétriés, 1832) is treated here as a valid species. The flagellum of the aedeagus in *C. hoeftii* is distinctly different from that of *C. sophiae* (Schaller, 1783) (Machatschke 1954). The distribution areas of *C. hoeftii* and *C. sophiae* are overlapping at least in Crimea and South Urals but transitional forms were not found (Bieńkowski 2011).

520 *Crosita* Motschulsky, 1860. A phylogenetic analysis by Gauthier et al. (2023) placed *Crosita altaica* (Gebler, 1823) among the species of the genus *Chrysolina*, but the authors did not propose to downgrade *Crosita* as a subgenus of *Chrysolina*. We agree with this decision and keep treating *Crosita* as a separate genus because the result based on one species only is not enough to indicate the position of *Crosita* within *Chrysolina*.

521 *Crosita altaica* (Gebler, 1823). The study of well-defined subspecies of *Crosita altaica* (Gebler, 1823) in Kazakhstan demonstrated the importance of the structure of aedeagus in addition to the dorsal sculpture. Traditionally, *Cr. clementzae* Jacobson, 1899 has been readily distinguished from *Cr. altaica* by the elytral sculpture tending

to form rows of punctures. However, the elytral sculpture of *Cr. altaica* is variable and the rows or ridges may appear among completely irregular punctation. The shapes of aedeagus of both, *Cr. altaica* and *Cr. clementzae*, are similar but with clearly visible slight specific differences. Consequently, the subspecies *Cr. clementzae atasica* L. N. Medvedev, 1976 and *Cr. clementzae barkulica* L. N. Medvedev, 1976 are transferred as subspecies of *Cr. altaica* based on the shape of aedeagus. *Crosita altaica barkulica* was described from Xinjiang but occurs also in Mongolia and the records of subspecies other than *C. altaica altaica* from Mongolia (e.g. Medvedev 1982) likely belong to this subspecies. In general, the subspecific classification of *C. altaica* in China and Mongolia is not clear yet. *Crosita bogutensis* Lopatin, 1996 has unusual almost black colour but in other characters is very close to *C. altaica* and therefore treated here as the subspecies *Cr. altaica bogutensis*. On the contrary, *Cr. urumchiana* S.-H. Chen, 1934 was treated as a subspecies of *Cr. altaica* by Chinese authors (Chen & Wang 1962, Yang et al. 2014) but it has a peculiar aedeagus shape and a different host plant and we confirm it as a valid species.

522 *Entomoscelis adonidis* (Pallas, 1771). Recently, Bieńkowski & Orlova-Bienkowskaya (2023) made special research to show that at least in SW Russia and the Caucasus only *E. adonidis* s. str. exists without any subspecies, and *E. dorsalis bashkiriae* Kippenberg, 2020 is also its synonym. Their conclusion is not accepted for several reasons. Neither aedeagus shape nor endophallus were carefully examined. What they illustrate as the "flagellum" is in fact an uninflated endophallus. The everted endophalli of *Entomoscelis* spp. at our disposal have a very small spiral flagellum similar to that of *Oreomela* Jacobson, 1895 and *Cystocnemis* Motschulsky, 1860. Statistical analysis is absent in this paper but significant methodological problems present that caused the confirmation bias. First of all, as it was checked from the data presented by Bieńkowski & Orlova-Bienkowskaya (2023), almost all the characters have strong correlations with one another, therefore the analysis of separate single features is useless. The calculations were made for normal (Gaussian) distribution of the characters by default although almost all characters significantly deviate from normal distribution. The so-called "75% rule" taken from Mayr & Ashlock (1991) is not a rule at all and not even commonly used to separate subspecies in insects. Further, this rule was not intended and therefore not valid for different, although presumably closely related, species. To resolve questions of this kind, also DNA analysis is essential. The host plant situation is still questionable. The neotype was described from a series collected at the type locality, but the authors did not state if the neotype was collected from Brassicaceae or from *Adonis* as indicated by Pallas. It cannot be excluded that both *Entomoscelis* species in question occur in this locality. To designate a neotype of *E. adonidis* in this situation was a premature decision.

523 *Fasta* Petitpierre & Alonso Zarazaga, 2019 was elevated to genus level by Gauthier et al. (2023) and it is followed in the present Catalogue. *Fasta fastuosa ventricosa* (Suffrian, 1858) from Spain and Portugal is distinguished from the nominal form only by a more rounded habitus. Similar specimens are known also from the Apennines, therefore the subspecies rank of *ventricosa* is doubtful and marked here as a questionable status [QS]. *Fastuolina* Warchałowski, 1991 is unavailable name according to Petitpierre & Alonso Zarazaga (2019a).

524 Cho (2016) treated *Cercyonops* Jacobson, 1901 and *Machomena* Dubois, 1887 as valid genera. In the present Catalogue we prefer to regard them as subgenera of *Gonioctena* as in the previous Edition (Kippenberg 2010b).

525 Cho (2016), in his Ph.D. Thesis, synonymized subspecies of *G. aegrota* (Fabricius, 1798). However, we recognize them as defined by Kippenberg (2001) and accepted by Petitpierre (2019).

526 *Gonioctena linnaeana* (Schrank, 1781) is widespread in the Palaearctic Region and has clear intraspecific differences in aedeagal shape. The nominal subspecies described by Schrank (1781) from Austria and distributed in the Alps, the Pyrenees and Spain is characterised by a long apical process of the aedeagus in dorsal view (Cho 2016: figs. 238–239, Petitpierre 2019: fig. 33c). The main part of Europe and Turkey is inhabited by *G. linnaeana calcarata* Fabricius, 1792 with a somewhat shorter aedeagal process (Cho 2016: figs. 240–241, 243–245). The populations from southern and eastern parts of European Russia (Orenburg, Samara, Chelyabinsk regions, Republics of Bashkortostan and Komi), Siberia, Mongolia but also from Norway have a short process of aedeagus (Cho 2016: 242, 246–247) and belong to *G. linnaeana orientalis* (Weise, 1884) or alternatively to undescribed forms. Unfortunately, the distribution of these subspecies cannot be presented here as detailed as desirable because reliable information about aedeagus proportions of *G. linnaeana* is available for a limited number of countries only.

527 *Gonioctena nivosa* (Suffrian, 1851). Cho et al. (2016) treated *G. arctica* Mannerheim, 1853 conspecific with *G. nivosa*. In the present Catalogue we prefer the opinion of Wilcox (1972) and E. G. Riley et al. (2003), who treated *G. nivosa* and *G. arctica* as subspecies.

528 *Gonioctena charitonowi* (Palmén, 1946) has been listed several times as a synonym of *G. norvegica* (A. Strand, 1936) (e.g. Cho 2016, Cho et al. 2016, Kippenberg 2010b). However, Palmén (1946) discussed the existing specimens with Ogloblin's label "charitonowi" but he was unsure of the validity of this taxon. Because he was not working with *G. charitonowi* as with a valid one, it cannot be considered available name (ICZN 1999, Art. 11.5).

529 *Gonioctena viminalis* (Linnaeus, 1758). The eastern form of *G. viminalis* is characterised by red legs, pronotum widest before base, and the tip of aedeagus slightly wider than in the nominate form. It is regarded here as a subspecies *G. viminalis pallidipes* (Weise, 1893).

530 *Gonioctena* (*Goniomena*) *pallida* (Linnaeus, 1758). Bechyně (1948a) introduced six subspecies and 21 aberrations of *G. pallida*. In his PhD Thesis, Cho (2016) examined types of almost all subspecies and stated that they are conspecific with *G. pallida*. All aberrations are listed under *G. pallida* according to Bechyně (1948a) as original specimens were not examined.

531 *Gonioctena* (*Goniomena*) *quinquepunctata* (Fabricius, 1787). As pointed out by Petitpierre & Alonso-Zarazaga (2019b), the well-known *G. quinquepunctata* (Fabricius, 1787), described in *Chrysomela*, is a junior primary homonym of *Chr. quinquepunctata* Linnaeus, 1767 (now *Leucocera quinquepunctata*). Its senior synonym *Ch. undata* Thunberg, 1784 is a homonym of *Ch. undata* DeGeer, 1775 (now *Platyphora undata*), but *Ch. quinquepunctata* Fabricius, 1787 was conserved as nomen protectum by Kippenberg (2010a). Petitpierre & Alonso-Zarazaga (2019b) suggested *Ch. dispar* Paykull, 1799 as the oldest available name for this taxon. However, the conservation of *Ch. quinquepunctata* Fabricius, 1787 under the Article 23.9.5 of the Code was proposed to the Commission. Until a decision is made under plenary powers, the usage of *Ch. quinquepunctata* Fabricius, 1787 is to be maintained.

532 As a result of their phylogenomic analysis of the genus *Oreina* Chevrolat, 1836 Gauthier et al. (2023) proposed to downgrade the monotypic subgenera *Frigidorina* Kühnelt, 1984 and *Virgulatorina* Kühnelt, 1984 as synonyms of the subgenus *Chrysochloa* Hope, 1841. However, the phylogenetic results do not agree with the morphological characters and therefore H. Kippenberg, the co-author of Gauthier et al. (2023), does not support these new synonymies. Both species, *O. frigida* (Weise, 1883) and *O. virgulata* (Germar, 1823), differ from all members of subgenus *Chrysochloa* essentially in the shape of aedeagus, moreover, *O. virgulata* differs also in the last maxillary palpomere not distinctly narrower than the penultimate one. We propose to keep the subgenera *Frigidorina* and *Virgulatorina* valid in accordance with the previous Edition (Kippenberg 2010b).

533 *Oreina* (*Allorina*) *caerulea* (Olivier, 1791). Bontems (1984, 2010) used the name *O. luctuosa* (Olivier, 1807) instead of *O. caerulea* due to missing type material of *O. caerulea*. With respect to the stability of nomenclature we prefer the usage of *O. caerulea*.

534 *Oreina* (*Chrysochloa*) *redikortzevi* (Jacobson, 1924). No specific differences between *O. redikortzevi* from South Siberia and *O. speciosissima* (Scopoli, 1763) from European mountains are known. Some subspecies of *O. speciosissima*, especially ssp. *drenskii* (Gruev, 1974) or ssp. *troglodytes* (Kiesenwetter, 1861) cannot be reliably distinguished from *O. redikortzevi* either by external characters or by the aedeagus shape (Bontems 2001). Although similar disjunct distribution in the mountains of Europe and Siberia is described in *Chrysolina relucens* (Rosenhauer, 1847) (see Bieńkowski 1997), we avoid to downgrade *O. redikortzevi* as a subspecies of *O. speciosissima* or even its synonymy until a more comprehensive study is available.

535 *Oreina (Chrysochloa) speciosissima carpathica* Kraatz, 1859 was synonymized with *O. speciosissima fuscoaenea* (Schummel, 1844) by Bontems (2009). This synonymy was not supported by zoogeography as the specimens of *carpathica* from an unknown Carpathian locality were compared and equalized to specimens from a very distant Sudetes Mountains. The subspecific classification of *O. speciosissima* in the Carpathians has not been satisfactorily studied until now. From the Southern Carpathians *O. speciosissima juncorum* Suffrian, 1851 was described. This subspecies proved to be a form of the subalpine zone, while in the alpine zone another (unnamed) form exists. The situation is similar to the distribution of nominotypical subspecies (subalpine) and ssp. *troglodytes* (alpine) in the Swiss Alps. Until the status of the Carpathian subspecies of *O. speciosissima* is not evaluated we regard *carpathica* as a valid subspecies.

536 *Oreina (Oreina) alpestris bannatica* Weise, 1884. This taxon was usually attributed to Suffrian (1851a) who mentioned it under var. β of *Ch. speciosa* Linnaeus, 1767, and thus the name is unavailable (see comment 1115). Weise (1884d) treated it as variety but the book contained also subspecies and thus varieties are deemed infrasubspecific. However, Bechyně (1958b) adopted the variety *bannatica* Weise, 1884 as subspecies and thus it is deemed available from Weise (1884) based on the Art. 45.6.4.1 (ICZN 1999).

537 *Oreina (Oreina) viridis* (Duftschmid, 1825). Bontems (2010) proposed the usage of *O. nivalis* (Suffrian, 1851) instead of *O. viridis* due to Duftschmid's not completely convincing description. With respect to the stability of nomenclature we prefer the usage of *O. viridis*.

538 *Oreomela yunnana* Lopatin, 2004, the type species of the subgenus *Oreomelina* Lopatin, 2004, proved to be *Suinzona* (see Ge et al. 2011). Consequently, *Oreomelina* is a synonym of *Suinzona*. Other species classified until now in *Oreomelina* are listed as *Oreomela* species unassigned to subgenera in the present Catalogue.

539 *Phaedon (Phaedon) cochleariae graecus* Apfelbeck, 1912. Two specimens of this variety from the type locality Myli in Apfelbeck's collection (ZMSZ) not labelled as types but probably belonging to the type series were examined. The variety is characterised by a fine and dispersed punctation of pronotum. All examined specimens of *P. cochleariae* from its entire range have stronger and denser punctation on pronotum (especially near base). Provisionally, we accept the var. *graecus* as subspecies of *P. cochleariae* but a detailed study is necessary to clarify its status.

540 The status of *Plagiosterna placida* (S.-H. Chen, 1934) is unclear. Gressit & Kimoto (1963a) treated *P. placida* as a possible subspecies of *P. adamsi*. In the present Catalogue we follow Yang et al. (2014, 2015) who considered *P. placida* and *P. adamsi* as distinct species.

541 *Prasocuris suturella* (Reiche & Saulcy, 1858) has been regarded until now as synonym of *P. distincta* (P. H. Lucas, 1846). It is raised here to subspecies of *P. distincta* based on the distinctive colouration.

542 *Chrysomela betulae* Linnaeus, 1758, *Galleruca morbillosa* Fabricius, 1792, and *Galleruca punctulata* Fabricius, 1792 were proposed to be nomina oblita by Wilcox (1975), apparently based on Art. 23b of the Code. However, Art. 23b was in use in the preriod 1961–1973 (ICZN 1985a, 1999) and therefore the Wilcox's statement, published in 1975, has to be treated as invalid act.

New and confirmed synonyms

543 *Chrysolina* (*Arctolina*) *tundralis* (Jacobson, 1910), **syn. nov.** of *Ch.* (*A.*) *kuznetzowi* (Jacobson, 1897), based on examination of the type material.

544 *Chrysolina* (*Chalcoidea*) *marginata iniussa* Bechyné, 1950, **syn. nov.** of *Ch.* (*Chalc.*) *m. sculpticollis* Bechyné, 1948. *Chrysolina marginata iniussa* was described from the Eastern Pyrenees and was distinguished from Central Pyrenean *Ch. m. sculpticollis* by its different colouration and sculpture. The examination of a series of *Ch. m. iniussa* from Mt. Canigou (Eastern Pyrenees) showed a large scale of variability including characteristic dark-bronzy coloured *iniussa*-form as well as blackish *sculpticollis*-forms. Examined type material of *Ch. m. sculpticollis* proved a surface variability overlapping with *Ch. m. iniussa*. Mixed populations have also been observed by Eduard Petitpierre (2023, pers. comm.) in other Pyrenean localities.

545 *Chrysolina* (*Chalcoidea*) *sanguineocincta pinarbasiense* Özdikmen, Şahin & Bal, 2020, **syn nov.**, and *Ch. marginata unificans* Bechyné, 1950, **syn nov.** of *Ch.* (*Chalc.*) *marginata bodemeyeri* (Weise, 1910), based on photographs of the aedeagus in the original description of *Ch. sanguineocincta pinarbasiense* and based on the drawing of the aedeagus of a syntype of *Ch. marginata unificans* in Bieńkowski (2019, Fig. 2: 1).

546 *Chrysolina* (*Cyrtochrysolina*) *dissipabilis* Bechyné, 1950 is a synonym of *Ch. marcasitica marcasitica* (Germar, 1823) based on Kippenberg (2017).

547 *Chrysomela collaris* ssp. *hyperborea* L. N. Medvedev, 2011, **syn. nov.** of *Chr. collaris* ssp. *blaisdelli* Van Dyke, 1938, based on examination of a series of specimens from Chukotka. The difference from the ssp. *blaisdelli* given by L. N. Medvedev (2011b) in the diagnosis of ssp. *hyperborea* and in the key based on elytral sculpture and colour pattern of inner surface of elytra proved insufficient even within a sample from one locality (Aiopechan Island in Chaunskaya Guba Bay), where specimens shared the characters of both taxa.

548 *Chrysomela cyaneoviridis* Gruev, 1994, **syn. nov.** of *Chr. lapponica* Linnaeus, 1758, based on the examination of holotype (male) of *Ch. cyaneoviridis* deposited in HNHM. The holotype of *Chr. cyaneoviridis* represents an unicolorous specimen of *Ch. lapponica*.

549 *Colaphellus grouvellei* Achard, 1923 is confirmed here as a synonym of *Colaphellus bowringii* (Baly, 1865) based on the study of the type series of *C. grouvellei* deposited in NMPC. Bieńkowski (2019) confused *Colaphellus bowringii* (Baly, 1865) and

Chrysolina (*Pierryvettia*) *bowringii* (Baly, 1860) and erroneously treated *Colaphellus grouvellei* as a synonym of *Chrysolina bowringii*.

550 *Colaphellus sophiae* ssp. *transsylvanicus* Machatschke, 1954, **syn. nov.**, and *C. sophiae* ssp. *amasiae* Machatschke, 1954, **syn. nov.** of *C. sophiae* (Schaller, 1783). Both subspecies show no difference in the shape of aedeagus and flagellum in comparison to the nominal subspecies, therefore they are synonymised with *C. sophiae*.

551 *Gastrophysa viridula* ssp. *lenta* (Weise, 1887), **syn. nov.**, and *G. viridula* ssp. *pennina* (Weise, 1882), **syn. nov.** of *G. viridula* (DeGeer, 1775). Both subspecies are synonymized with *G. viridula* based on topotypic material from Switzerland and Siberia which do not show any characters supporting the separation of subspecies.

552 *Oreina* (*Allorina*) *smaragdina* Weise, 1883, **stat. nov.** = *O. luctuosa* var. *auricollis* Stierlin, 1887, **syn. nov.** = *O. bidentata* Bontems, 1981, **syn. nov.** = *O. luctuosa* var. *cuprina* Weise, 1883, **syn. nov.** Orlova-Bienkowskaya (2007) raised var. *auricollis* to species level and used it as valid name instead of *O. bidentata*. However, there are two older available names of varieties proposed by Weise (1883c). Following principle of the First Reviser (ICZN 1999, Art. 24.2.3) we select *O. luctuosa* var. *smaragdina* as valid name and the taxa *O. luctuosa* var. *auricollis*, *O. bidentata* Bontems, 1981, and *O. luctuosa* var. *cuprina* are proposed as its new synonyms.

553 *Oreina* (*Chrysochloa*) *tristis* (Fabricius, 1792) is treated as synonym of *O.* (*Chrysochloa*) *cacaliae cacaliae* (Schrank, 1785) in the present Catalogue. Bontems (2021) resurrected *O. tristis* as subspecies of *O. cacaliae* and characterised it by dark bluish colour and by the punctuation of elytra being somewhat less wrinkled than in the generally greenish nominal form from the North-Eastern Alps. However, the specimens of *O. cacaliae* from all over the Alps show a variable elytral punctation which does not allow a clear separation. Although the populations of *O. cacaliae* distributed in great parts of E- and W-Alps are always bluish coloured, the colouration itself is not sufficient to consider them as distinct geographic subspecies. Moreover, at least one dark blue population lives in the Eastern Bavarian Alps (surrounded by greenish populations), which cannot be distinguished from the dark populations from Southern Alps by external characters. Therefore, *O. tristis* is considered here as a colour aberration only and listed in synonymy with *O. cacaliae cacaliae*. Generally, it cannot be excluded that *O. cacaliae* from the Alps and the northern Balkan Peninsula is a conglomerate of different subspecies, which however at the moment cannot be reliably recognized as the characters are not defined.

554 *Oreina* (*Intricatorina*) *intricata* (Germar, 1823). The study of *Oreina intricata* specimens from its entire range proved that recognition of the two geographic subspecies (a more eastern and a more western one) should not be maintained: neither the external characters nor the shape of aedeagus support these two geographic forms reliably. The colouration is not a criterion as the dark bluish populations occur in the Alps as well as in the Carpathians. Several populations in the Alps have the

same structure of surface as populations from the Carpathians so they cannot be discriminated. Therefore, ssp. *anderschii* (Duftschmid, 1825) is synonymized with *O. intricata*.

New taxa

555 *Chrysolina* (*Colaphoptera*) *purpurascens scarisoarae* Kippenberg, **ssp. nov.** Reitter (1913c) proposed a name for an aberration *Ch. crassimargo* ab. *plumbeonigra*, which is an unavailable infrasubspecific name (Art. 45.6.2, ICZN 1999). Nevertheless, Bechyné (1950b) attributed the name to Reitter, treated it as a subspecies of *Ch. purpurascens* (Germar, 1822) based on two females from Romania (Banat and Retezat) and accompanied it with a short description, hence he validated the name but the authorship must be attributed to him (Art. 45.5.1, ICZN 1999). At the same he used two spellings and Bieńkowski (2019) fixed *plumbeonigra* as the correct one. Two years later, Bechyné (1952b) reported this taxon also from Bihor Mts. (as *plumbeoingra* sic!) without more precise record and compared it with *Ch. hemisphaerica banatica* Bechyné, 1950. Kippenberg (2017) considered the population from Bihor Mts. a distinct subspecies, however, he mistakenly evaluated the note by Bechyné (1952b) as an independent taxonomic act and considered it a junior homonym of *Ch. purpurascens plumbeonigra* Bechyné 1950. Due to the pressumed homonymy he proposed a new name *Ch. purpurascens scarisoarae* for the Bihor population and considered the original Bechyné's specimen its holotype. However, nothing indicates that Bechyné (1952b) changed his opinion about identity of *Ch. purpurascens plumbeonigra*, which can be supported by the fact that he attributed the taxon again to Reitter, thus it is just a subsequent mention of the name, which cannot be considered as the second validation of it (e.g. see Example under the Article 72.7, ICZN 1999). Therefore, *Ch. purpurascens scarisoarae* is here formally validated. The description was given by Kippenberg (2017: 119). Type material: Holotype ♂ (NHMB): Bihar-Gebirge / Apusenen-Gebirge: Scarisoara, Casa de Piatra, Juni 1922, leg. Jeannel & Winkler. Paratypes: 3 ♂♂ (collection of F. Fritzlar, Jena, Germany, collection of H. Kippenberg, Herzogenaurach, Germany), Munti Bihor, Garda de Sus, 25.V.2013, leg. H. Müller. Distribution: Romania (Bihor Mountains). Differential diagnosis: *Chrysolina purpurascens scarisoarae* is similar to the nominal subspecies but its body is black, elytra strongly punctate and the apex of aedeagus more elongate (Fig. 63 in Kippenberg 2017). Nominal subspecies has the body dark coppery, elytra less strongly punctate and the apex of aedeagus less elongate with rounded sides (Fig. 46 in Kippenberg 2017).

556 *Oreina* (*Oreina*) *speciosa jurassica* Kippenberg, **ssp. nov.** The population of *O. speciosa* from the French/Swiss Jura mountains was traditionally, but erroneously, referred to *O. pretiosa* (Suffrian, 1851) (e.g. Weise 1884d, Bechyné 1958b). However, *O. pretiosa* was explicitly proposed for "flat" form and based on the examination

of the type material it is a synonym of *O. speciosa speciosa* (Linnaeus, 1767) (see Bontems 2010). In the previous Edition, Kippenberg (2010b) listed the population from Jura mountains as *O. speciosa discolor* (Weise, 1884). The variety *discolor* was described in the work containing also taxa in status of subspecies and thus is deemed infrasubspecific. It was neither adopted as the valid name of a species or subspecies before 1985 and it cannot be used for the name of subspecies (ICZN 1999, Art. 45.6.4). Due to the lack of available names, the population from Jura mountains is described here as *Oreina* (*Oreina*) *speciosa jurassica* Kippenberg, **ssp. nov.** for *Orina gloriosa* var. *pretiosa* sensu Weise (1884d: 450). Type material: Holotype, ♂, Schweizer Jura, Delémont, Birseschlucht, 400 m, 47°23'N 7°21'E, 21.6.1984, leg. Kippenberg. Paratypes: 53 spec., data as in holotype; 6 spec., Schweizer Jura, 5 km NW Delevier, 500 m, 20.6.1984, leg. Kippenberg; 4 spec., CH (Jura), St. Brais, 25 km SW Delémont, 17.vi.1982, leg. Simon; 1 ♀, CH (Jura), Lützeltal, 10 km N Delémont, 20.v.1984, leg. Simon; 2 spec., F (Jura), Dept. Doubs, St. Hippolyte, 9.vi.1957, leg. Daickler (all specimens deposited in TLMF). The subspecies is characterised by short and stout body, elytra in lateral view distinctly rounded, especially in females (length/width ratio of elytra in females of *O. speciosa jurassica* **ssp. nov.** is about 1.30; in females of *O. speciosa speciosa* about 1.45). The subspecies is restricted to the Jura Mountains in eastern France and western Switzerland where it inhabits lower localities up to 500 m a.s.l. (*O. speciosa speciosa* prefers subalpine to alpine zones).

New name

557 *Chrysolina* (*Paracrosita*) *persica* (Jakob, 1960), a junior homonym of *Ch. haemoptera persica* Jolivet, 1951, was regarded for many years as a synonym of *Ch. armeniaca* (Faldermann, 1837). Bieńkowski (2019) provisionally divided *Ch. armeniaca* in two subspecies, *Ch. armeniaca* s. str. and *Ch. armeniaca iranica* L. N. Medvedev, 1975 (= *persica* Jakob, 1960). However, also *Ch. armeniaca iranica* L. N. Medvedev, 1975 is a junior homonym of *Ch. coerulans iranica* Jakob, 1954 and thus cannot be used as valid name. Therefore, a substitute name for *Ch. persica* (Jakob, 1960) is proposed here: *Chrysolina armeniaca neopersica* Kippenberg & Mikhailov, **nom. nov.**

Type species

558 *Allorina* Weise, 1902 (subgenus of *Oreina*) was introduced without designation of a type species. *Chrysomela luctuosa* Olivier, 1807 treated as the type species by Bontems (1984) and Kippenberg (2010b) can not be used as it was not an originally included species. The correct type species is *Oreina tristis* sensu Weise (1902f) designated by Kühnelt (1984: 173), which refers to *Oreina smaragdina* (Weise, 1893) (see Bontems 1981).

559 *Plagiodera* Chevrolat, 1836. As pointed out by Bousquet & Bouchard (2013) the type species of *Plagiodera* is *Chrysomela armoraciae* Linnaeus, 1758 designated

by Thomson (1859) which is also the type species of *Phaedon* Latreille, 1829 and is currently classified in this genus. To promote the stability, an application to the Commission to designate *Chrysomela versicolora* Laicharting, 1781 as the type species of *Plagiodera* as suggested by Bousquet & Bouchard (2013) should be proposed. For the purpose of the Catalogue the current usage is to be maintained.

560 *Pseudotaeniochrysea* Daccordi, 1980 is a synonym of *Chrysolina* (*Taeniochrysea*). Daccordi (1980) designated *Chrysomela superba* Thunberg, 1787 as the type species and applied it to the African *Chrysolina* species. However, the syntype deposited in the Thunberg's collection (UUZM) is not *Chrysolina* but East Asian *Ambrostoma*. The oldest available name for the African *Chrysolina* is *Ch. limbolata* (Reiche, 1850). Consequently, the type species of *Pseudotaeniochrysea* is *Chrysomela superba* Thunberg, 1787 sensu Daccordi (1980) (= *Chrysomela americana* var. *limbolata* Reiche, 1850).

561 *Romalorina* Weise, 1906 (syn. of *Oreina* s. str.) was introduced without designation of type species. Bontems (1978: 78) designated "*Chrysomela speciosa* Fabricius, 1775" as type species, however, this taxon was not originally included. Kühnelt (1984) published "*speciosa* Linné, 1767 (= *gloriosa* var. *excellens* Weise)" as type species, which is also an invalid designation because Weise (1906e) did not mention *speciosa* Linné, and *excellens* was listed as aberration. Because the previous designations are invalid *O. alpestris* Schummel, 1844 is designated here as a type species of *Romalorina*.

Lectotype designation

562 *Chrysomela bodemeyeri* Weise, 1910 was described from "Persien: Luristan, Sultanabad". In the Weise's collection in ZMHUB there is a series of ten specimens under the name *Chrysomela bodemeyeri*, but only two of them represent syntypes with certainty (the other specimens were obviously added to the series later). One syntype is a male, labelled "Sultanabad", its aedeagus was extracted and described by Weise (1910c) and belongs to the *Chrysolina marginata* complex and is here designed as lectotype. The second syntype (female), designed here as paralectotype and provided with handwritten label by J. Weise "*Ch. bodemeyeri* m.", belongs to *Ch. sanguineocincta* (Crotch, 1872). The other eight specimens in the series belong to *Ch. sanguineocincta*, as well, and are labelled as lectotype and paralectotypes respectively, by Michel Rapilly. The designation was never published and even if so it would still be invalid because these specimens do not belong to the type series.

563 *Galleruca calcarata* Fabricius, 1792. The Fabricius collection in ZMUK houses two male syntypes. The first one, labelled "G: calcarata | D: de Lövenskiold || TYPE || Gonioctena | linneana | Sch. | rev. Daccordi 80 || zmuc | 00026761" is designated here as lectotype and the second, labelled "TYPE || Gonioctena | linneana | Sch. | rev. Daccordi 80 || zmuc | 00026762" becomes a paralectotype. The lectotype is designated in order to conserve status of thus taxon and avoid its further misinterpretation.

Species unassigned to subgenera

564 *Chrysolina caspica* (Weise, 1892), formerly classified in the subgenus *Colaphoptera*, and *Chrysolina lehri* (Lopatin, 1970), formerly classified in the subgenus *Chalcoidea*, are unassigned to subgenus as morphology of their aedeagi do not conform that present within the subgenera *Colaphoptera* and *Chalcoidea*, respectively.

Nomina dubia

565 *Chrysomela betulae* Linnaeus, 1758: 369. The identity of this taxon is unknown and thus it is placed as nomen dubium under *Chrysomela*.

566 *Chrysolina dahlii* (Herrich-Schäffer, 1838) is sometimes supposed to be conspecific with *Ch.* (*Sphaerochrysolina*) *umbratilis* (Weise, 1887) (e.g. Bieńkowski 2019). However, its identity remains unknown and therefore it is listed as nomen dubium under *Chrysolina*.

567 *Chrysolina distincta* (Küster, 1844). Bieńkowski (2019) listed this taxon as synonym of *Ch. marcasitica marcasitica* (Germar, 1823). However, its identity is unclear, therefore we placed it as as nomen dubium under *Chrysolina*.

568 *Chrysolina subincrassata* (Duftschmid, 1825). Bieńkowski (2019) placed *Ch. subincrassata* as unclear taxon within subgenus *Cyrtochrysolina*. However, the identity of *Ch. subincrassata* is enigmatic and it is not possible to reasonably determine to which subgenus it belongs, therefore we placed it as as nomen dubium under *Chrysolina*.

569 *Chrysolina tagenii* (Herrich-Schäffer, 1838) was regarded as a doubtful taxon by Petitpierre (2019).

Spelling

570 *Gonioctena* (*Sinomela*) *nigroplagiata* Baly, 1862. The original spelling *nigroplagata* is obviously an incorrect original spelling (printer's error) which must be corrected to *nigroplagiata* based of the Art. 32.5.1 (ICZN 1999). Apparently the intention was "plagiata" (= spotted) and not "plagata" (= with wounds).

571 *Gonioctena linnaeana calcarata* (Fabricius, 1792). The original spelling *Galleruca calearata* is obviously an incorrect original spelling. The lectotype has the handwritten label "*G. calcarata*". The printing error was already corrected to *calcarata* by Fabricius (1801a: 431).

572 *Plagiosterna adamsi* (Baly, 1864) was described as *Lina adamsii* in honour of A. Adams. The prevailing usage of the name is *adamsi* that in fact is an incorrect subsequent spelling (ICZN 1999, Art. 33.4). However, according to Art. 33.3.1 of the Code the spelling *adamsi* is in prevailing usage, is attributed to the publication of the original spelling, and thus is deemed to be a correct original spelling.

Protected name

573 As shown by Petitpierre & Alonso-Zarazaga (2019b: 548) *Helodes* Paykull, 1799 is an available name which threatens the well-established name *Prasocuris* Latreille, 1802. In order to preserve stability, we propose to activate the Art. 23.9 of ICZN (1999), considering *Helodes* nomen oblitum and *Prasocuris* nomen protectum. To our knowledge, *Helodes* was not treated as a valid name after 1900 and, thus, the provisions of the Article 23.9.1.1 are met. To meet the conditions of Art. 23.9.1.2, we hereby provide supporting references of 25 papers, published by at least 10 authors in the immediately preceeding 50 years, and encompassing a span of not less than 10 years, where *Prasocuris* is used as valid genus name: Bezděk (2021a), Bieńkowski (2004), Biondi et al. (1995), Borowiec et al. (2011), Daccordi & Gruev (1991), Debreuil (2014), Geiser (2001), Gruev (1992), Gruev & Tomov (2007), Jolivet & Hawkeswood (1995), Kimoto & Takizawa (1994), Kippenberg (1994, 2010b), Lopatin (2010), Lopatin & Kulenova (1986), Lopatin et al. (2004), Medvedev (1982), Petitpierre (2019), Rheinheimer & Hassler (2018), Vig (2002), Warchałowski (2003, 2010b), Winkelman & Debreuil (2008), Yang et al. (2014, 2015).

Availability of varieties and forms

574 Gradl (1881) presented many variations of *Chrysomela lapponica* from Eger (Cheb, Czech Republic) explicitly characterised by their colour pattern only ("Farbverschiedenheiten und Färbungsentwicklungen"). Therefore, in our opinion, the content of the work unambiguously reveals that these names were proposed for an infrasubspecific entities (ICZN 1999, Art. 45.6.4):

Melasoma lapponica var. *bimaculata* Gradl, 1881: 305
Melasoma lapponica var. *bipustulata* Gradl, 1881: 305
Melasoma lapponica var. *coerulea* Gradl, 1881: 306
Melasoma lapponica var. *marginata* Gradl, 1881: 306
Melasoma lapponica var. *quadrimaculata* Gradl, 1881: 305
Melasoma lapponica var. *quadrinotata* Gradl, 1881: 305
Melasoma lapponica var. *rubrocoerulea* Gradl, 1881: 305
Melasoma lapponica var. *sexmaculata* Gradl, 1881: 305

575 Letzner (1853) proposed 13 new varieties explicitly for only colour forms of *Oreina speciosissima* (Scopoli, 1763). All are conspecific with the taxon presently recognized as *O. speciosissima fuscoaenea* (Schummel, 1844). Letzner (1853) was inconsistent through his paper in application of the latter name as he mentioned it as a distinct form or variety and also as conspecific with *O. speciosissima*. In the present Catalogue we consider these varieties unavailable because, in our opinion, the content of the work unambiguously reveals that these names were proposed for infrasubspecific entities (ICZN 1999, Art. 45.6.1 and 45.6.4):

Chrysomela speciosissima var. *coerulescens* Letzner, 1853: 89
Chrysomela speciosissima var. *cupreoviridis* Letzner, 1853: 89
Chrysomela speciosissima var. *fuscocuprea* Letzner, 1853: 89
Chrysomela speciosissima var. *ignita* Letzner, 1853: 89
Chrysomela speciosissima var. *nigrescens* Letzner, 1853: 89
Chrysomela speciosissima var. *nigroaenea* Letzner, 1853: 89
Chrysomela speciosissima var. *nigrocoerulea* Letzner, 1853: 89
Chrysomela speciosissima var. *pretiosa* Letzner, 1853: 89
Chrysomela speciosissima var. *tristis* Letzner, 1853: 89
Chrysomela speciosissima var. *violacea* Letzner, 1853: 89
Chrysomela speciosissima var. *viridescens* Letzner, 1853: 89
Chrysomela speciosissima var. *viridiaenea* Letzner, 1853: 89
Chrysomela speciosissima var. *viridis* Letzner, 1853: 89

576 The following names are deemed infrasubspecific because they were published as a fourth name after a trinomen (ICZN 1999, Art. 45.5):
Chrysolina cerealis ssp. *mixta* natio *fratercula* Bechyné, 1952b: 364
Chrysolina cerealis ssp. *plorans* natio *opulentiformis* Bechyné, 1948b: 13
Chrysolina cerealis ssp. *plorans* natio *podolica* Bechyné, 1948b: 13
Chrysolina cerealis ssp. *plorans* natio *relucentiformis* Bechyné, 1948b: 13
Orina cacaliae var. *lissahorensis* var. *janaceki* Weise, 1896c: 110
Orina speciosissima ssp. *convergens* var. *pyrenaica* Weise, 1884d: 481
Orina viridis ssp. *merkli* var. *transsylvanica* Weise, 1884d: 466
Phytodecta linneana ssp. *bergrothi* var. *correspondens* Jacobson, 1901b: 128
Phytodecta linneana ssp. *bergrothi* var. *simplex* Jacobson, 1901b: 128
Phytodecta variabilis ssp. *atlasica* var. *secsaouia* Kocher, 1953c: 100

577 The following varieties and forms are deemed infrasubspecific according to Article 45.6.1 (ICZN 1999), since a subspecies concept was used in the same works, and according to available information, these names were not validated prior to 1985 (ICZN 1999, Art 45.6.4.1):
Ambrostoma quadriimpressum var. *purpureocupreum* Jacobson, 1901b: 127
Ambrostoma quadriimpressum var. *viridicyaneum* Jacobson, 1901b: 127
Apterocuris sibiricus var. *aurichalcea* Jacobson, 1901a: 86
Apterocuris sibiricus var. *cyanea* Jacobson, 1901a: 86
Apterocuris sibiricus var. *purpurea* Jacobson, 1901a: 86
Apterocuris sibiricus var. *violacea* Jacobson, 1901a: 86
Cercyonops caraganae var. *hieroglyphica* Jacobson, 1901a: 84 [= 1901b: 129]
Cercyonops caraganae var. *limbata* Jacobson, 1901a: 84 [= 1901b: 129]
Cercyonops caraganae var. *sesquialtera* Jacobson, 1901a: 84 [= 1901b: 129]
Cercyonops caraganae var. *simplex* Jacobson, 1901a: 84 [= 1901b: 129]
Chrysochloa basilea var. *aeneoviridis* Jacobson 1901a: 100

Chrysochloa basilea var. *cuprescens* Jacobson 1901a: 100
Chrysochloa basilea var. *nigrocoerulea* Jacobson 1901a: 100
Chrysochloa basilea var. *virens* Jacobson 1901a: 100
Chrysochloa cacaliae var. *resplendens* J. Daniel, 1903a: 183
Chrysochloa ganglbaueri forma *nigra* Jakob, 1953b: 157
Chrysochloa ganglbaueri forma *viridis* Jakob, 1953b: 157
Chrysochloa intricata var. *carnica* J. Müller, 1948: 96
Chrysochloa speciosissima forma *gabrielis* Depoli, 1926b: 103
Chrysochloa vittigera var. *maritima* J. Daniel, 1903a: 184
Chrysomela aeneipennis var. *subcoerulea* J. R. Sahlberg, 1913c: 248
Chrysomela americana var. *lesinae* Weise, 1884d: 411
Chrysomela aurichalcea var. *nigricans* Jacobson, 1901a: 100
Chrysomela bankii var. *chlorizans* Weise, 1884d: 379
Chrysomela bicolor var. *marseuli* Kocher, 1959: 58
Chrysomela confossa var. *dorsalis* Weise, 1884d: 395
Chrysomela crassipes var. *porphyropus* Peyerimhoff, 1915a: 32. Bieńkowski (2001, 2019) regarded it as a subspecies of *Chrysolina crassipes* (P. H. Lucas, 1847), which however does not validate the name.
Chrysomela femoralis var. *laeta* Weise, 1884d: 372
Chrysomela grossa var. *illita* Weise, 1884d: 432
Chrysomela guttifera var. *sericata* Jacobson, 1901b: 125
Chrysomela gypsoides var. *morvennensis* Méquignon, 1945: 29
Chrysomela hyperici var. *ambigua* Weise, 1884d: 407
Chrysomela hyperici var. *privigna* Weise, 1884d: 407
Chrysomela marginalis forma *richteri* Roubal, 1934: 38
Chrysomela marginata var. *solitaria* Weise, 1884d: 391
Chrysomela menthastri var. *croatica* Weise, 1884d: 426
Chrysomela quadrigemina var. *indigena* Weise, 1884d: 405
Chrysomela rufilabris var. *perfecta* Jacobson, 1901b: 124
Chrysomela rufoaenea var. *fallaciosa* Weise, 1884d: 401
Chrysomela sahlbergi var. *venefica* Weise, 1884d: 369
Chrysomela staphylaea var. *palliata* Jacobson, 1901b: 125
Chrysomela superstes var. *altiatlantica* Kocher, 1959: 53
Chrysomela superstes var. *antoinei* Kocher, 1959: 53
Chrysomela superstes var. *crebieri* Kocher, 1959: 53
Chrysomela superstes var. *ifranensis* Kocher, 1959: 52
Chrysomela varians var. *pratensis* Weise, 1884d: 429
Chrysomela viridana var. *sardea* Weise, 1884d: 426
Entomoscelis cornea var. *viberti* Bedel, 1915: 207
Entomoscelis rumicis var. *mediojuncta* Pic, 1939d: 284

Gastrolina thoracica var. *cupricollis* Chûjô, 1941c: 459
Gastrolina peltoidea var. *cimex* Jacobson, 1901b: 134
Melasoma collaris var. *thoracica* Weise, 1884d: 560
Melasoma lapponica var. *altaica* Weise, 1884d: 557
Melasoma lapponica var. *cruciata* Jacobson, 1901b: 133
Melasoma lapponica var. *multipunctata* Jacobson, 1901b: 133
Melasoma lapponica var. *quadripunctata* Jacobson, 1901b: 133
Melasoma lapponica var. *quadripustulata* Jacobson, 1901a: 90
Melasoma lapponica var. *violaceipennis* Jacobson, 1901b: 133
Melasoma vigintipunctata var. *pustulata* Weise, 1884d: 559
Orina bifrons var. *aerosa* Weise, 1884d: 461
Orina cacaliae var. *nubigena* Weise, 1884d: 476
Orina frigida var. *kiesenwetteri* Weise, 1884d: 487
Orina frigida var. *rhaetica* Weise, 1884d: 487
Orina gloriosa var. *discolor* Weise, 1884d: 451
Orina gloriosa var. *nubila* Weise, 1884d: 450
Orina gloriosa var. *virgo* Weise, 1884d: 450
Orina splendidula var. *fairmairei* Weise, 1884d: 475
Orina virgulata var. *serena* Weise, 1884d: 468
Orina vittigera var. *glacialis* Weise, 1884d: 459
Paropsides nigrofasciata forma *bivittata* Reineck, 1922b: 371
Paropsides duodecimpustulata forma *sexmaculata* Reineck, 1922b: 367
Phaedon cochleariae var. *obesus* Weise, 1884d: 546
Phaedon laevigatus var. *caucasicus* Weise, 1884d: 543
Phaedon laevigatus var. *violaceus* Weise, 1884d: 543
Phaedon segnis var. *obscurus* Weise, 1884d: 540
Phyllodecta laticollis var. *purpurascens* Pic, 1925b: 9
Phyllodecta viennensis var. *coerulea* Weise, 1884d: 517
Phyllodecta vulgatissima var. *aestiva* Weise, 1884d: 515
Phyllodecta vulgatissima var. *obscura* Weise, 1884d: 515
Phytodecta fornicata var. *picea* Weise, 1884d: 505
Phytodecta fornicata var. *innotata* Weise, 1884d: 505
Phytodecta nivosa var. *eppelsheimi* Weise, 1884d: 501
Phytodecta nivosa var. *funesta* Weise, 1884d: 501
Phytodecta nivosa var. *personata* Weise, 1884d: 501
Phytodecta olivacea var. *nigricans* Weise, 1884d: 507
Phytodecta olivacea forma *umbriceps* Roubal, 1941: 51
Phytodecta pallida var. *decipiens* Weise, 1884d: 510
Phytodecta pallida var. *borealis* Weise, 1884d: 510
Phytodecta quinquepuncata var. *aucupariae* Jacobson, 1901b: 128

Phytodecta quinquepunctata var. *sorbi* Weise, 1884d: 508
Phytodecta quinquepunctata var. *unicolor* Weise, 1884d: 508
Phytodecta sexpunctata var. *berytensis* Pic, 1916b: 10
Phytodecta sexpunctata var. *binotaticollis* Pic, 1916b: 10
Phytodecta sexpunctata var. *obliquemaculata* Pic, 1916b: 10
Phytodecta sibirica var. *fraterna* Jacobson, 1901a: 82
Phytodecta viminalis var. *cincta* Weise, 1884d: 492
Phytodecta viminalis var. *munda* Weise, 1884d: 491
Phytodecta viminalis var. *transitoria* Jacobson, 1901b: 129
Prasocuris vicina var. *hispanica* Weise, 1884d: 534
Sclerophaedon carniolicus var. *obscurus* Weise, 1884d: 536
Sternoplatys fausti var. *completus* Jacobson, 1901b: 131
Sternoplatys fulvipes var. *piceipes* Jacobson, 1901b: 133

578 The following varieties were proposed in works also using a concept of subspecies, however, they were validated before 1985 and thus are deemed to be subspecific from their original publication (ICZN 1999, Art. 45.6.4.1):

Chrysochloa cacaliae var. *senilis* J. Daniel, 1903a: 183 was adopted by Bechyné (1950b) as subspecies.

Chrysomela fuliginosa var. *galii* Weise, 1884d: 394 was adopted by Bechyné (1950b) as subspecies.

Chrysomela marginata var. *glacialis* Weise, 1884d: 391 was adopted by Bechyné (1950b) as subspecies.

Chrysomela rufilabris var. *discriminata* Jacobson, 1901b: 124 was adopted by Medvedev (1982) as species.

Orina alpestris var. *bannatica* Weise, 1884d: 447 was adopted by Bechyné (1958b) as subspecies.

Orina alpestris var. *umbrosa* Weise, 1884d: 447 was adopted by Bechyné (1958b) as subspecies.

Orina cacaliae var. *macera* Weise, 1884d: 476 was adopted by Bechyné (1958b) as subspecies.

Orina bifrons var. *stussineri* Weise, 1884d: 461 was adopted by Bechyné (1958b) as subspecies.

Orina virgulata var. *candens* Weise, 1884d: 468 was adopted by Bechyné (1958b) as subspecies.

Orina virgulata var. *praefica* Weise, 1884d: 468 was adopted by Bechyné (1958b) as subspecies.

Phytodecta linnaeana var. *orientalis* Weise, 1884d: 498 was adopted by Bechyné (1948a) as species.

Phytodecta sibirica var. *sundmani* Jacobson, 1901b: 128 was adopted by Medvedev (1982) as species.

Unavailable names

579 *Chrysomela* (*Microlina*) was introduced by L. N. Medvedev in 1973 in his Doctoral Thesis (Riley et al. 2003), which was only submitted to a committee and was not part of any edition. Such thesis is an unpublished work as it does not meet provisions of the Article 8 of the Code (ICZN 1999, 2012b); see also example under the Article 9 in the Amendment. Consequently, the taxa proposed in the thesis are unavailable. The name *Microlina* was mentioned by Lopatin (1977b), Medvedev (1982), Lopatin & Kulenova (1986) and Medvedev & Dubeshko (1992) but neither of the authors made it available because the type species was not designated (ICZN 1999, Art. 13.3).

580 The following names are unavailable as the types were not explicitly fixed in the original description (ICZN 1999, Art. 16.4):

Oreomela crassicornis Lopatin, 2010: 267
Oreomela mushketovi Lopatin, 2010: 255
Zygogramma suturalis ssp. *volata* Kovalev, 2002: 76

581 The following names, which appeared in the Chrysomelinae volume of the Fauna Sinica series (Yang et al. 2014), are unavailable because they were not explicitly indicated as new (ICZN 1999, Art. 16.1), only provided with references to non-existent descriptions:

Chrysolina (*Lithopteroides*) *pseudolaeviguttata* Daccordi & S.-Q. Ge, 2014: 480
Gonioctena (*Asiphytodecta*) *brunnea* Daccordi & S.-Q. Ge, 2014: 382
Gonioctena (*Brachyphytodecta*) *bedzeki* Yang, S.-Q. Ge, Wang, Li & Cui, 2014: 373
Gonioctena (*Gonioctena*) *kalganensis* S.-Q. Ge & Daccordi, 2014: 363
Phaedon germinates Daccordi & S.-Q. Ge, 2014: 237
Phaedon huizuensis Daccordi & S.-Q. Ge, 2014: 241
Phaedon igori Daccordi & S.-Q. Ge, 2014: 241
Phaedon prosternalis Daccordi & S.-Q. Ge, 2014: 247
Phratora cuiae Daccordi & S.-Q. Ge, 2014: 337
Phratora deqinensis S.-Q. Ge & Wang, 2014: 340
Phratora wangi Daccordi & S.-Q. Ge, 2014: 356

582 Descriptions of following taxa first appeared in a Ph.D. Thesis (Cho 2016), which was only submitted to a committee and was not part of any edition. Such thesis is an unpublished work as it does not meet provisions of the Article 8 of the Code (ICZN 1999, 2012b); see also example under the Article 9 in the Amendment. Consequently, the taxa proposed in the thesis are unavailable, however, most of them were subsequently validated by a properly published descriptions in scientific journals; remaining ones (denoted by an asterisk) are still undescribed. Although some of the taxa are extralimital to the present Catalogue, we list all of them here.

**Gonioctena* subgenus *Tumidocnema* Cho, 2016: 331

Gonioctena (*Asiphytodecta*) *allardi* Cho, 2016: 26 [validated by Cho (2021: 82), extralimital: Vietnam]

Gonioctena (*Asiphytodecta*) *jindrai* Cho, 2016: 46 [validated by Cho & Borowiec (2016c: 762)]

Gonioctena (*Asiphytodecta*) *mantillerii* Cho, 2016: 49 [validated by Cho (2021: 84)]

**Gonioctena* (*Asiphytodecta*) *napolovi* Cho, 2016: 54 [extralimital: Laos, Vietnam]

Gonioctena (*Asiphytodecta*) *truncaticornis* Cho, 2016: 74 [validated by Cho & Borowiec (2016c: 766), extralimital: Laos]

Gonioctena (*Brachyphytodecta*) *brancuccii* Cho, 2016: 84 [extralimital: Laos; described by Sprecher-Uebersax & Daccordi (2016) under the same name and from the same type locality but based on different type material]

Gonioctena (*Brachyphytodecta*) *davidi* Cho, 2016: 85 [validated by Cho (2017a: 446)]

Gonioctena (*Brachyphytodecta*) *kangdingensis* Cho, 2016: 101 [validated by Cho (2017a: 449)]

**Gonioctena* (*Brachyphytodecta*) *klapperichi* Cho, 2016: 104

Gonioctena (*Brachyphytodecta*) *kuatunensis* Cho, 2016: 106 [validated by Cho (2017b: 175)]

Gonioctena (*Brachyphytodecta*) *longshengensis* Cho, 2016: 110 [validated by Cho (2017b: 176)]

Gonioctena (*Gonioctena*) *amurensis* Cho, 2016: 132 [validated by Cho & Borowiec (2016b: 90)]

Gonioctena (*Gonioctena*) *jani* Cho, 2016: 175 [validated by Cho & Borowiec (2016b: 97)]

**Gonioctena* (*Gonioctena*) *taebaekensis* Cho, 2016: 250

**Gonioctena* (*Gonioctena*) *thibetana* Cho, 2016: 257

**Gonioctena* (*Tumidocnema*) *xinjiangana* Cho, 2016: 332

New geographical records

583 *Chrysomela cuprea* Fabricius, 1775. GREECE: Xanthi, Stavroupoli, 5.–9.vi.2018, 1 ♀, M. Egger leg., H. Kippenberg det. (collection of H. Kippenberg, Herzogenaurach, Germany). New species for Greece.

584 *Chrysomela saliceti turcestanica* (Reineck, 1937). UZBEKISTAN: Fergana, Chamsaabad, 2000 m, 15.vii.1984, 1 ♀, L. Behne leg., H. Kippenberg det. (collection of H. Kippenberg, Herzogenaurach, Germany). New species for Uzbekistan.

585 *Chrysolina* (*Allochrysolina*) *lepida gastonis* (Fairmaire, 1875). TUNISIA: Zaghouan, 1.vi.2009, 1 ♂, W. Ziegler leg., H. Kippenberg det. (collection of H. Kippenberg, Herzogenaurach, Germany). New species for Tunisia.

586 *Chrysolina* (*Allohypericia*) *aeruginosa poricollis* (Motschulsky, 1860). CHINA: Nei-Mongol, Jining, 1.viii.1997, 2 ♂♂ 2 ♀♀, E. Kučera leg., Yu. Mikhailov det. (collection

of Yu. Mikhailov, Yekaterinburg, Russia); Ningxia Autonomous Region, Helan Shan, rd. to Suyukou National Forest Park, 28 km WNW Yinchuan, 1404 m, 38°37′46,6″N 105°57′05,7″E, semi-steppe, under stones, 19.vi.2011, 1 ♂ 3 ♀♀, D. W. Wrase leg., Yu. Mikhailov det. (collection of Yu. Mikhailov, Yekaterinburg, Russia). New species for provinces Inner Mongolia and Ningxia (China).

587 *Chrysolina* (*Bittotaenia*) *aeneipennis* (Reiche & Saulcy, 1858). SAUDI ARABIA: El Riyadh, 5.ix.–9.xii.1962, 1 ♀, E. Diehl leg., M. Daccordi det. (collection of H. Kippenberg, Herzogenaurach, Germany). New species for Saudi Arabia.

588 *Chrysolina* (*Bittotaenia*) *mellyi* (Stål, 1857). INDIA: Uttarakhand, Rishikesh, Uttar Kashi env., 1400 m, 6.vii.1989, 1 ♀, A. Riedel leg., H. Kippenberg det. (collection of H. Kippenberg, Herzogenaurach, Germany). New species for Uttarakhand (India).

589 *Chrysolina* (*Chalcoidea*) *marginata marginata* (Linnaeus, 1758). TURKEY: Kırklareli prov., Istrançadağ near Demirköy, 700 m, 15.iv.1979, 1 ♀, W. Heinz leg., H. Kippenberg det. (collection of H. Kippenberg, Herzogenaurach, Germany). New species for European part of Turkey.

590 *Chrysolina* (*Chalcoidea*) *waldheimi* Kippenberg, 2010. CHINA: Xinjiang prov., Urumqui, Tianchi Bogda Shan, 23.vii.1990, 1♂ 1♀, J. Kolibáč leg., H. Kippenberg det. (collection of H. Kippenberg, Herzogenaurach, Germany); Xinjiang, Urumqui See, 43°54,369′N 88°07,104′E, 1800 m, 16.vi.1997, 1 ♂, H. Burtscher leg., H. Kippenberg det. (collection of Andreas Kapp, Götzis, Austria); Xinjiang prov., Karlit Mt., 80 km NW of Hami city, 3300 m, 18.–19.vii.1998, 1 ♀, H. Kippenberg det. (collection of H. Kippenberg, Herzogenaurach, Germany). New species for Xinjiang (China).

591 *Chrysolina* (*Heliostola*) *lichenis lichenis* (Richter, 1820). GERMANY: Sachsen, Glauchau, 6.viii.1978, 1 ♀, Jorrath leg., H. Kippenberg det. (collection of H. Kippenberg, Herzogenaurach, Germany). New species for Germany.

592 *Chrysolina* (*Hypericia*) *cuprina* (Duftschmid, 1825). ALBANIA: Vlore, Drimadhe env., 3.–4.v.2015, 1 ♂, D. & L. Kopr leg., J. Bezděk det. (collection of D. Kopr, Brno, Czech Republic). GREECE: Evros, 6 of km N Dadia, 8.vi.2014, 1 ♂, H. Mühle leg., H. Kippenberg det. (collection of H. Kippenberg, Herzogenaurach, Germany); NNE Seres Orini E, 1000 m, 30.vi.1997, 1 ♂, J. Messutat leg., H. Kippenberg det. (collection of H. Kippenberg, Herzogenaurach, Germany); Voria Pindos W. Pirsogiani, 1200 m, 28.vi.1997, 1 ♂, J. Messutat leg., H. Kippenberg det. (collection of H. Kippenberg, Herzogenaurach, Germany); Thessalien, Stomion, Mt. Ossia, 300 m, 17.v.1997, 1 ♂, I. Wolf leg., H. Kippenberg det. (collection of H. Kippenberg, Herzogenaurach, Germany); Florina, 24.v.2008, 1 ♀, M. Egger leg., H. Kippenberg det. (collection of H. Kippenberg, Herzogenaurach, Germany); Thessalien, Larissa, Ossa/Oros, 25.v.1988, 1 ♀, Hiermeier leg., H. Kippenberg det. (collection of H. Kippenberg, Herzogenaurach, Germany); Phthiotis, Almyrôs, 20.v.1968, 1 ♀, A. Senglet leg., H. Kippenberg det. (collection of H. Kippenberg, Herzogenaurach, Germany). New species for Albania and Greece.

593 *Chrysolina* (*Hypericia*) *difficilis* (Motschulsky, 1860). CHINA: Jilin: 25 km SSE of Erdaobaihe Town, Naitoushan Mts., upper of Naidaohe River, 1100 m, 42.22°N 128.27°E, 12.vii.2012, 1 ♂, R. Dudko leg., Yu. Mikhailov det. (collection of Yu. Mikhailov, Yekaterinburg, Russia); 33 km SSE Erdaobaihe Town, upstream of Erdaojian River, 1100 m, 42.19°N 128.35°E, 13.vii.2012, 1 ♂, R. Dudko leg., Yu. Mikhailov det. (collection of Yu. Mikhailov, Yekaterinburg, Russia). New species for Jilin (China).

594 *Chrysolina* (*Hypericia*) *geminata* (Paykull, 1799). RUSSIA: Sverdlovsk Reg., Achit Distr., env. of Sazhina village, 11.–29.viii.1998, 2 ♂♂ 1 ♀, P. Rudoiskatel' leg., Yu. Mikhailov det. (collection of Yu. Mikhailov, Yekaterinburg, Russia); Sverdlovsk Reg., Achit Distr., env. of Sarga village, 29.ix.1984, 1 ♀, E. Zinoviev leg., Yu. Mikhailov det. (collection of Yu. Mikhailov, Yekaterinburg, Russia). New species for Central European Territory of Russia.

595 *Chrysolina* (*Minckia*) *impavida* Bechyné, 1949. LEBANON: Bqaatouta, NW of Beskinta, 250 m, 1.iv.1997, 3 spec., W. Heinz leg., H. Kippenberg det. (collection of H. Kippenberg, Herzogenaurach, Germany). New species for Lebanon.

596 *Chrysolina* (*Ovosoma*) *turca* (Fairmaire, 1865). GREECE: Thraki Prov., Evros env., Avdela, 41°25,840'N 26°13,21'E, 100 m, 1.iv.2015, 1 ♂, L. Nádai, D. Szaloki & A. Podlussány leg., J. Bezděk det. (HNHM). New species for Greece.

597 *Chrysolina* (*Sibiriella*) *kholsunica* Mikhailov, 2001. RUSSIA (South Siberia): Altai, NE spurs of Kholzun Mt. Range, upper right tributary of Poperechnaya Krasnoyarka River, 1750–1800 m, 50.176°N 84.921°E, 1.vii.2021, 1♂ 1♀, R. & E. Dudko leg., Yu. Mikhailov det. (collection of Yu. Mikhailov, Yekaterinburg, Russia); Altai, NE spurs of Kholzun Mt. Range, upper Poperechnaya Krasnoyarka River, 1950 m, 50.176°N 84.928°E, 2.vii.2021, 1 ♂, R. & E. Dudko leg., Yu. Mikhailov det. (collection of Yu. Mikhailov, Yekaterinburg, Russia). New species for Russia.

598 *Chrysolina* (*Synerga*) *coerulans angelica* (Reiche & Saulcy, 1858). JORDAN: 20 km SW of Madaba, 31°38.063'N 35°41.883'E, 400 m, 26.v.2007, 49 spec., J. Bezděk leg. and det. (collection of J. Bezděk, Brno, Czech Republic); Madaba Prov., Wadi Hisban, 31°49.718'N 35°47.673'E, 681 m, 1.iv.2016, 1 spec., W. Ziegler leg., H. Kippenberg det. (collection of H. Kippenberg, Herzogenaurach, Germany). New species for Jordan.

599 *Chrysolina* (*Synerga*) *coerulans coerulans* (Scriba, 1791). GREECE: Korfu, Agios Georgios, 18.ix.2001, 1 ♀, O. Blochwitz leg., H. Kippenberg det. (collection of H. Kippenberg, Herzogenaurach, Germany). New species for Greece.

600 *Chrysolina* (*Volosatik*) *liqingzhaoae* Daccordi & S.-Q. Ge, 2011. CHINA: Gansu, Mts. 38 km SE of Longnan, 3130 m, 33°11'16"N 105°14'24"E, 7.viii.2012, 1 ♀, D. W. Wrase leg., M. Daccordi det., (collection of H. Kippenberg, Herzogenaurach, Germany). New species for Gansu (China).

601 *Colaphellus hoeftii* (Ménétriés, 1832). UZBEKISTAN: Amirsoy, 2200 m, 3.vi.2022, 2 spec., J. Stanovský leg., J. Bezděk det. (collection of J. Stanovský, Ostrava, Czech Republic). New species for Uzbekistan.

602 *Entomoscelis dorsalis bashkiriae* Kippenberg, 2020. ARMENIA: Dzhervedz, 18.vi.1982, 1 ♂, M. Kalashian leg., Yu. Mikhailov det. (collection of P. Romantsov, St.-Petersburg, Russia); Vedi, 1.vi.1926, 1 ♂, A. Schelkovnikov leg., H. Kippenberg & Yu. Mikhailov det. (collection of A. Bieńkowski, Moscow, Russia); Erivan, 1 ♀, Maljushenko leg., H. Kippenberg & Yu. Mikhailov det. (coll. A. Bieńkowski, Moscow, Russia). CRIMEA: South Bank Territory [Greater Yalta], v.2015, 1 ♂, P. S. Sitnikov leg., Yu. Mikhailov det. (collection of V. Stolbov, Tyumen', Russia). RUSSIA (Central Territory): Tatarstan, Chistopol Distr., Utjakovo vill., steppe, 24.vi.2014, 1 ♂, S. V. Dedyukhin leg., H. Kippenberg & Yu. Mikhailov det. (collection of A. Bieńkowski, Moscow, Russia). RUSSIA (West Siberia): Chelyabinsk Reg., Ilmensky Nature Reserve, 30.vi.1986, 2 ♂♂ 1 ♀, P. A. Kosintsev leg., same locality, 4.vii.1990, 1 ♂, B. V. Krasutsky leg., both Yu. Mikhailov det. (IPAE Yekaterinburg, Russia). Additional records from Russia (Central Territory) and new species for Armenia, Crimea and Russia (West Siberia).

603 *Fasta fastuosa fastuosa* (Scopoli, 1763). MONTENEGRO: Durmitor, Dolina Susice, 7.–27.vii.1933, 1 spec., J. Fodor leg., J. Bezděk det. (HNHM); Durmitor, Susica jezero, 12.–24.vii.1933, 1 spec., J. Fodor leg., J. Bezděk det. (HNHM); Durmitor, Škrcko jezero, 13.vii.1933, 7 spec., J. Fodor leg., J. Bezděk det. (HNHM); Durmitor, Dobri Do, 1900 m, vii.1980, Krätschmer leg., H. Kippenberg det. (collection of H. Kippenberg, Herzogenaurach, Germany). New species for Montenegro.

604 *Gastrophysa* (*Exiguipenna*) *analis* Reitter, 1890. MONTENEGRO: Bjelasica Mts., 6.vi.2012, 4 spec., J. Stanovský leg., J. Bezděk det. (collection of J. Bezděk, Brno, Czech Republic). New species for Montenegro.

605 *Gastrophysa* (*Gastrophysa*) *polygoni* (Linnaeus, 1758). UZBEKISTAN: 7 km S of Yiangol, 22.v.2022, 1 spec., J. Stanovský leg., J. Bezděk det. (collection of J. Stanovský, Ostrava, Czech Republic); Syr Darya, Jomasho´y env., 5.vi.2022, 1 spec., J. Stanovský leg., J. Bezděk det. (collection of J. Stanovský, Ostrava, Czech Republic). New species for Uzbekistan.

606 *Gastrophysa* (*Gastrophysa*) *viridula* (DeGeer, 1775). ARMENIA: Sevan Lake W of Tsovagyugh, 15.vi.2018, 1 ♂, W. Ziegler leg., H. Kippenberg det. (collection of W. Ziegler, Rondeshagen, Germany). New species for Armenia.

607 *Gonioctena* (*Gonioctena*) *flexuosa* (Baly, 1860). CHINA: Hunan: Wugonshan Mts., An-Fu, 1650 m, viii.2004, 1 ♂ 1♀, local collectors leg., H. Kippenberg det. (collection of H. Kippenberg, Herzogenaurach, Germany). New species for Hunan (China).

608 *Gonioctena* (*Gonioctena*) *linnaeana calcarata* (Fabricius, 1792). ARMENIA: Kotayk Reg., 6–8 km N of Arzakan, 18.vii.2002, 1 ♂, M. Kalashian leg., J. Bezděk det. (collection of J. Bezděk, Brno, Czech Republic). New species for Armenia.

609 *Gonioctena* (*Goniomena*) *intermedia* (Helliesen, 1913). MONTENEGRO: Durmitor, Dolina Susice, 7.–27.vii.1933, 1 ♂ 3 ♀♀, J. Fodor leg., J. Bezděk det. (HNHM). New species for Montenegro.

610 *Gonioctena* (*Goniomena*) *pallida* (Linnaeus, 1758). GREECE: Makedonia, Drama Reg., Western Rodopi, E of Kanandere, Forest of Elatia, 1350 m, 13.–14.vi.2002, 2 ♀♀, I. Wolf leg., H. Kippenberg det. (collection of H. Kippenberg, Herzogenaurach, Germany). New species for Greece.

611 *Oreina* (*Allorina*) *caerulea* (Olivier, 1791). MONTENEGRO: Durmitor, Sedlo, 1800 m, viii.1976, 3 spec., B. Drovenik leg., H. Kippenberg det. (collection of H. Kippenberg, Herzogenaurach, Germany). New species for Montenegro.

612 *Oreina* (*Chrysochloa*) *speciosissima binaghii* Daccordi & Ruffo, 1976. FRANCE: Alpes-Maritimes, St. Martin-Vesubie, Lac de Madonne de la Fenestre, 2200–2270 m, 12.vii.1977, 5 spec., H. Kippenberg leg. and det. (collection H. Kippenberg, Herzogenaurach, Germany); Col de Vars, 2300–2500m, 24.vii.1977, 12 spec., H. Kippenberg leg. and det. (collection H. Kippenberg, Herzogenaurach, Germany). New subspecies for France.

613 *Oreina* (*Virgulatorina*) *virgulata ljubetensis* (Apfelbeck, 1912). GREECE: Drama Prov., Elatia forest, 26.vii.2009, 2 ♂♂, Hengmith & Ziegler leg., H. Kippenberg det. (collection of H. Kippenberg, Herzogenaurach, Germany). New species for Greece.

614 *Phaedon* (*Hemiphaedon*) *subtilis subtilis* Weise, 1900. TURKMENIA: Repetek Nat. Res., 23.–29.v.1992, 1 ♀, A. Napolov leg., H. Kippenberg det. (collection of H. Kippenberg, Herzogenaurach, Germany). New species for Turkmenia.

615 *Phaedon* (*Phaedon*) *armoraciae* (Linnaeus, 1758). GEORGIA: Trialetskiy Chrebet, Bakuriani, 1800–2000 m, 4.–7.vii.1980, 1 ♀, Wrase & Schülke leg., Kippenberg det. (collection of H. Kippenberg, Herzogenaurach, Germany). New species for Georgia.

616 *Phaedon* (*Phaedon*) *cochleariae cochleariae* (Fabricius, 1792). ALBANIA: Bistrice, NE of Sarande, 18.iv.2016, 2 spec., K. Orszulik leg., J. Bezděk det. (collection of K. Orszulik, Frýdek-Místek, Czech Republic). New species for Albania.

617 *Phaedon* (*Phaedon*) *laevigatus* (Duftschmid, 1825). ALBANIA: Peshtan env., Vjoses riv., 7 km E of Tepelene, 20.iv.2016, 14 spec., K. Orszulik leg., J. Bezděk det. (collection of K. Orszulik, Frýdek-Místek, Czech Republic, 2 spec. in collection of J. Bezděk, Brno, Czech Republic). New species for Albania.

618 *Phratora* (*Phratora*) *alternata* Lopatin, 2007. CHINA: Sichuan: Min Shan, pass 17 km NE of Songpan, 3920 m, 32°44′23″N 103°44′31″E, 10.viii.2012, 3 spec., D. W. Wrase leg., M. Daccordi det. (collection of H. Kippenberg, Herzogenaurach, Germany). New species for Sichuan (China).

619 *Phratora* (*Phratora*) *atrovirens* (Cornelius, 1857) SLOVENIA: Sanntaler Alpen, Logarska dolina, 1100 m, 27.vi.1988, 1 ♂, H. Kippenberg leg. and det. (collection of H. Kippenberg, Herzogenaurach, Germany). New species for Slovenia.

620 *Phratora* (*Phratora*) *coerulescens* Küster, 1848. ARMENIA: Gegharkunik prov., Dilijan W of Semyonovka, 15.vi.2018, 4 spec., W. Ziegler leg., H. Kippenberg

det. (collections of W. Ziegler, Rondeshagen, Germany and H. Kippenberg, Herzogenaurach, Germany). GEORGIA: Zekari Pass, Sairme env., 2200 m, 1.vii.2017, 1 ♀, K. Orszulik leg., J. Bezděk det. (collection of K. Orszulik, Frýdek-Místek, Czech Republic). New species for Armenia and Georgia.

621 *Phratora* (*Phratora*) *polaris polaris* (J. S. Schneider, 1886). ICELAND: Landmannalaugar, 1.viii.2021, 14 spec., G. Bastazo & J. M. Vela leg., J. M. Vela & M. Daccordi det. (collections of J.-M. Vela, Málaga, Spain and M. Daccordi, Verona, Italy). New species for Iceland.

622 *Phratora* (*Phratora*) *vitellinae* (Linnaeus, 1758). GREECE: Thessalia, Monodendri, 4.v.1999, 1 ♂, M. Szatyor leg., J. Bezděk det. (HNHM). NORTH MACEDONIA: Mavrovo, 16.vii.1937, 1 ♂, J. Fodor leg., J. Bezděk det. (HNHM); Galičnik, Bistra Planina, 7.–14.vii.1937, 3 ♀♀, J. Fodor leg., J. Bezděk det. (HNHM). CHINA: Sichuan: 20 km N of Shuangjingsi, 3400 m, 22.vi.2007, 32°06′38.3″N 102°23′02.3″E, 1 ♂, Croy leg., M. Daccordi det. (collection of H. Kippenberg, Herzogenaurach, Germany). New species for Greece, North Macedonia and Sichuan (China).

623 *Plagiodera maculicollis* S.-H. Chen, 1934. CHINA: Sichuan prov., ca. 220 km SW of Chengdu, Emei Shan, 600 m, 16.–18.vi.1990, 4 spec., Kleinfeld leg., H. Kippenberg det. (collection of H. Kippenberg, Herzogenaurach, Germany). New species for Sichuan (China).

624 *Plagiodera rufescens* (Gyllenhal, 1808). INDIA: Uttarakhand: Rishikesh, 2.–4.vii.1989, 1 spec., A. Riedel leg., M. Daccordi det. (collection of H. Kippenberg, Herzogenaurach, Germany). New species for Uttarakhand (India).

625 *Plagiosterna* (*Linaeidea*) *aenea aenea* (Linnaeus, 1758). GREECE: Macedonia Prov., Kerkini Mts., Ano Poroia, 520–660 m, 13.vi.2013, 3 spec., Z. Malinka leg., J. Bezděk det. (collection of Z. Malinka, Opava, Czech Republic). New species for Greece.

Comments on distribution

626 *Chrysolina* (*Chalcoidea*) *marginata bodemeyeri* (Weise, 1910). The records of *Ch. marginata* (Linnaeus, 1758) from the Caucasus known to us pertain to *Ch. marginata bodemeyeri*, therefore, we refer all records of *Ch. marginata* from the Caucasian countries to *Ch. marginata bodemeyeri*.

627 *Chrysolina* (*Chrysomorpha*) *cerealis* (Linnaeus, 1767). The classification of subspecies within this taxon is unsatisfactory (see comment 485). The species was registered for some countries (Belarus, Finland, Great Britain, Moldavia, Norway, Slovenia, Sweden) but subspecies were ignored. In addition, specimens from some other countries (e.g. Greece, North Macedonia, Montenegro, Bosnia Hercegovina) cannot be reliably associated with any of the currently recognized subspecies. The situation is further complicated by the presence of more than one form in some countries of Central and South Europe. For these reasons following distributional data are adopted with some doubts and require further confirmation: *Ch. cerealis cerealis* (Bulgaria, Belarus, Croatia, Great Britain, Greece, Hungary, North

Macedonia, Montenegro, Norway, Poland, Slovenia and Sweden); *Ch. cerealis cyaneoaurata* (Far East and West Siberia); *Ch. cerealis livonica* (Finland, Norway, Poland and Sweden); *Ch. cerealis megerlei* (Moldavia and Poland); and *Ch. cerealis mixta* (Austria).

628 Al-Nadawi (2019) published *Chrysolina herbacea* (Duftschmid, 1825) as new to Iraq, however the taxon was mistakenly named as "*Chrysolina herbacea alnadawi* (Duftschmidt, 1825)". The name is nomen nudum as it was done unintentionally and not explicitly indicated as new (ICZN 1999, Art. 16.1). Based on the photograph presented in Al-Nadawi (2019) the Iraqi specimens were misidentified and belong to *Ch. coerulans angelica* (Reiche & Saulcy, 1858) and not *Ch. herbacea*, which hence does not occur in Iraq.

629 *Crosita altaica faldermanni* Krynicki, 1832 is distributed in Central Kazakhstan between Aral Sea and Balkhash Lake and is well characterised morphologically. The occurrence in southern territory of European Russia is based on a single female with the label "Astrakhan, V. Yakovlev" deposited in ZIN. This record is questionable and was never confirmed later. Also, the occurrence in Mongolia (e.g. Medvedev 1982) is cannot be confirmed by available material. Thus, these records are not accepted here.

630 *Crosita kowalewskii kowalewskii* Gebler, 1836 was described from SE Russian Altai (type locality: "Gebirgsbach Topolewka des Altai einige Werste von seinem Ausflusse in den Argut"). Since the description it has never been found again in Russian Altai but only in Mongolia and China (mainly Mongolian Altai). However, the locality at the Argut River is hardly accessible, which could explain lack of contemporary material.

631 *Cystocnemis* (*Cystocnemis*) *concolor* (Kraatz, 1879). *Oreomela mohri* Daccordi, 1975, synonym of the latter taxon, was described from Iran evidently based on mislabelled specimens. *Cystocnemis concolor* is distributed in eastern Kazakhstan only and Iran is thus not listed in the present Catalogue.

632 *Entomoscelis rumex* Kippenberg, 2020. The type locality "Hispaniae" stated in the original description of *Chrysomela rumicis* Fabricius, 1787 is an evident error. The syntypes deposited in Fabrician collection in Kiel (Germany) are provided with the label "Rumicis | e Barb: Vahl" referring to North African "Barbaria" which corresponds to the known distribution of this species. Therefore, Spain is not listed in the present Catalogue.

633 *Gonioctena linnaeana calcarata* (Fabricius, 1792). Records of *G. linnaeana* from Croatia are tentatively assigned to this subspecies, but it requires further verification of respective voucher specimens, as we have not seen any material.

634 *Plagiosterna* (*Linaeidea*) *adamsi* Baly, 1864 and *P.* (*Linaeidea*) *placida* (S.-H. Chen, 1934). In the present Edition we follow the distribution presented by Yang et al. (2014, 2015), however, most records need verification because the published data may refer to both taxa.

635 *Sternoplatys clementzi* Jacobson, 1901. The record from Kazakhstan by Gabdullina (2016) is based on misidentification and refers to *Oreothassa weisei* (Reitter, 1901).

Chrysomelidae: Chrysomelinae: Timarchini

M. Daccordi & J. M. Vela

Comments

636 *Timarcha goettingensis* (Linnaeus, 1758) is distinct from other species in the unique morphology of aedeagus, aedeagal sclerite and endophallus. However, it contains numerous forms with different degree of differentiation of both, external and endophallus characters. Consequently, it is difficult to be sure which taxa deserve to be retained at subspecies rank. The subspecific classification of *T. goettingensis* used for the purpose of this Catalogue cannot be considered definitive until further study of the species and its intra- and inter-population variability is made.

New synonyms, confirmed synonyms and changes in rank

637 *Timarcha corinthia* Fairmaire & Allard, 1873 = *T. corinthia* var. *albanica* Apfelbeck, 1912, **syn. conf.**, based on a study of topotypic and additional comparative specimens.

638 *Timarcha metallica* (Laicharting, 1781) = *T. gibba* Hagenbach, 1825, **syn. nov.**, based on a study of topotypic specimens of both taxa.

639 *Timarcha brachydera* Fairmaire & Allard, 1873 = *T. brachydera* ssp. *distans* Bechyné, 1944, **syn. conf.** = *T. brachydera* ssp. *figurata* Bechyné, 1944, **syn. conf.** = *T. brachydera* ssp. *limitata* Bechyné, 1944, **syn. conf.** = *T. brachydera* ssp. *pumilio* Bechyné, 1944, **syn. conf.**, based on examination of the type material of all taxa.

640 *Timarcha calceata* Pérez Arcas, 1865 = *T. geniculata* (Germar, 1823), **syn. nov.**, based on the examination of the type material of both taxa. *Timarcha geniculata* (Germar, 1823) is a junior homonym of *T. geniculata* (C. R. Sahlberg, 1823). See comment 681 for priority of names.

641 *Timarcha catalaunensis* Fairmaire, 1884, **stat. res.**, based on the examination of the type material and additional comparative specimens.

642 *Timarcha corallipes* Fairmaire & Allard, 1873 = *T. corpulenta* Bechyné, 1944, **syn. nov.** = *T. insignis* ssp. *inornata* Bechyné, 1944, **syn. nov.** (topotypic specimens examined) = *T. melas* Bechyné, 1944, **syn. nov.** = *T. nitidipes* Bechyné, 1944, **syn. nov.** = *T. opacicollis* Bechyné, 1944, **syn. nov.**, based on the examination of the type material of all taxa if not stated otherwise.

643 *Timarcha crassaticollis* Fairmaire & Allard, 1873 = *T. henonii* Fairmaire & Allard, 1873, **syn. nov.**, based on the examination of the type material of both taxa.

644 *Timarcha dislocata* Bechyné, 1948, **stat. res.**, based on the examination of the type material and additional comparative specimens.

645 *Timarcha elliptica* Fairmaire & Allard, 1873, **stat. res.**, based on the examination of the type material and additional comparative specimens.

646 *Timarcha endora* Buquet, 1840 (type material examined) = *T. brevipennis* Pic, 1919, **syn. nov.** = *T. endora* var. *demoflysi* Normand, 1950, **syn. nov.**, based on examination of comparative specimens.

647 *Timarcha fallax* Pérez Arcas, 1865 = *T. aitanae* Petitpierre & Anichtchenko, 2018, **syn. nov.** = *T. tenuicornis* Fairmaire, 1880, **syn. conf.**, based on the examination of the type material of all taxa and additional comparative specimens.

648 *Timarcha goettingensis* ssp. *chloropus* (Germar, 1823) (type material examined) = *T. asturiensis* Kraatz, 1879, **syn. nov.** (type material examined) = *T. convexifrons* Fairmaire, 1880, **syn. nov.** (based on the description) = *T. gallaeciana* Chevrolat, 1840, **syn. conf.** (based on the description) = *T. gougeletii* Fairmaire, 1860, **syn. nov.** (based on the description and additional comparative specimens) = *T. janthinipes* Fairmaire, 1880, **syn. nov.** (type material examined) = *T. sphaeroptera* Fairmaire & Allard, 1873, **syn. conf.** (type material examined) = *T. trapezicollis* Fairmaire & Allard, 1873, **syn. nov.** (type material examined).

649 *Timarcha goettingensis* ssp. *globipennis* Fairmaire & Allard, 1873 = *T. cyanescens* ssp. *aequata* Bechyné, 1953, **syn. nov.** = *T. cyanescens* ssp. *guipuzcoana* Bechyné, 1948, **syn. nov.** (based on description and additional comparative specimens) = *T. cyanescens* ssp. *vizcayensis* Bechyné, 1953, **syn. nov.** = *T. leseleuci* Marseul, 1883, **syn. nov.**, based on the examination of the type material of all taxa if not otherwise stated.

650 *Timarcha humberti* Peyerimhoff, 1925, **stat. res.** = *T. mairei* ssp. *tichuktensis* Kocher, 1957, **syn. nov.**, based on the examination of the type material of both taxa.

651 *Timarcha iblanensis* Antoine, 1939 = *T. mairei* ssp. *lecerfi* Kocher, 1953, **syn. nov.** = *T. mairei* ssp. *peyerimhoffi* Kocher, 1953, **syn. nov.** = *T. theryi* Kocher, 1959, **syn. nov.**, based on the examination of the type material of all taxa.

652 *Timarcha lusitanica* (Fabricius, 1781) (photographs of type material examined) = *T. occidentalis* Herrich-Schäffer, 1838, **syn. conf.** (based on original description) = *T. scortea* (Germar, 1823), **syn conf.** (type material examined).

653 *Timarcha maroccana* Weise, 1882 = *T. freyi* Bechyné, 1953, **syn. nov.** = *T. maroccana* ssp. *liebmanni* Bechyné, 1953, **syn. conf.**, based on the examination of the type material of all taxa.

654 *Timarcha monserratensis* Bechyné, 1962, **stat. res.**, based on the examination of the type material and additional comparative specimens.

655 *Timarcha nicaeensis* A. Villa & G. B. Villa, 1835 [NP] (topotypic specimens examined) = *T. affinis* Leach & Risso, 1827 [NO], **syn. nov.** (based on original description) = *T. apuana* Daccordi & Ruffo, 1990, **syn. nov.** (type material examined) = *T. fremyi* Jolivet, 1944, **syn. conf.** (type material examined) = *T. fremyi* ssp. *ovata*

Jolivet, 1944, **syn. conf.** (based on original description) = *T. validicornis* Fairmaire & Allard, 1873, **syn. nov.** (type material examined).

656 *Timarcha pimelioides* ssp. *pimelioides* Herrich-Schäffer, 1838 (topotypic specimens examined) = *T. aguilari* Bechyné, 1946b, **syn. conf.** = *T. calabrica* Bechyné, 1944, **syn. conf.** = *T. chalcosoma* Fairmaire, 1868, **syn. nov.** = *T. gravis* Rosenhauer, 1856, **syn. nov.** = *T. insularis* Bechyné, 1944, **syn. conf.** = *T. insularis* ssp. *martialis* Bechyné, 1953, **syn. conf.** = *T. insularis* ssp. *ruffoi* Bechyné, 1953, **syn. conf.**, based on the examination of the type material of all taxa if not otherwise stated.

657 *Timarcha pimelioides* ssp. *turbida* Erichson, 1841, **stat. nov.** = *T. amethystipes* Chevrolat, 1873, **syn. nov.** = *T. camoensii* Fairmaire & Allard, 1873, **syn. nov.** = *T. marani* Bechyné, 1944, **syn. nov.** = *T. pfefferi* Bechyné, 1945, **syn. nov.** = *T. laevigata* ssp. *robustior* Bechyné, 1944, **syn. nov.** = *T. rugosa* var. *rugosoidea* Jolivet, 1944, **syn. nov.** = *T. staneki* Bechyné, 1944d: 366, **syn. nov.** = *T. taborskyi* Bechyné, 1944, **syn. nov.** = *T. tarsata* Bechyné, 1944, **syn. nov.** = *T. gravis* ssp. *tunisea* Pic, 1919, **syn. nov.** (based on original description and approximate topotypic specimens) = *T. variolosa* Bechyné, 1944, **syn. nov.**, based on the examination of the type material of all taxa if not stated otherwise.

658 *Timarcha piochardi* Fairmaire, 1874, **stat. res.** = *T. aurichalcea* Bechyné, 1953, **syn. nov.**, based on the examination of the type material of both taxa.

659 *Timarcha punctella* Marseul, 1871 = *T. djerbensis* Pic, 1919, **syn. nov.** = *T. inaequalis* Pic, 1924, **syn. nov.**, based on the examination of the type material of all taxa.

660 *Timarcha rugosa* ssp. *rugosa* (Linnaeus, 1767) = *T. angusta* ssp. *algirica* Bechyné, 1947, **syn. nov.** = *T. angusta* Bechyné, 1944, **syn. nov.** = *T. araneiformis* Bechyné, 1947, **syn. nov.** = *T. rugosa* ssp. *aterrima* Bechyné, 1944, **syn. nov.** = *T. djelfaica* Bechyné, 1947, **syn. nov.** = *T. espanoli* Bechyné, 1948, **syn. nov.** = *T. rugosa* ssp. *kourili* Bechyné, 1944, **syn. nov.** = *T. gnaptoroides* Bechyné, 1944, **syn. nov.** = *T. generosa* ssp. *khalouensis* Jolivet, 1966, **syn. nov.** = *T. rugosa* var. *kobelti* Heyden, 1883, **syn. nov.** = *T. mesatlantica* Kocher, 1953, **syn. nov.** (based on original description and topotypic specimens) = *T. rugosa* ssp. *minutissima* Bechyné, 1944, **syn. nov.** = *T. rugosa* var. *rufipes* Weise, 1882, **syn. nov.** (based on original description and approximate topotypic specimens) = *T. generosa* var. *rugata* Kocher, 1953, **syn. nov.** = *T. generosa* ssp. *rungsi* Jolivet, 1966, **syn. nov.** = *T. spuria* Bechyné, 1947, **syn. nov.** = *T. generosa* ssp. *subscabra* Jolivet, 1966, **syn. nov.** (based on the original description). The type material examined if not stated otherwise.

661 *Timarcha rugosa* ssp. *ventricosa* Weise, 1882, **stat. nov.** = *T. atlantica* Bechyné, 1944, **syn. nov.** = *T. rugosa* var. *aurulenta* Kocher, 1957, **syn. nov.** (based on original description and comparative specimens) = *T. atlantica* ssp. *mima* Bechyné, 1947, **syn. nov.** = *T. tangeriana* Bechyné, 1944, **syn. nov.**, based on the examination of the type material if not stated otherwise.

662 *Timarcha rugulosa* Herrich-Schäffer, 1838 = *T. globata* Fairmaire & Allard, 1873, **syn. nov.** = *T. lomnickii* Miller, 1867, **syn. nov.** = *T. rugulosa* ssp. *sinaiana* Bechyné, 1947, **syn. nov.** = *T. rugulosa* ssp. *stichai* Bechyné, 1947, **syn. nov.**, based on original descriptions and additional comparative specimens.

663 *Timarcha scabripennis* Fairmaire, 1868 = *T. scabripennis* var. *bechynei* Kocher, 1959, **syn. nov.** = *T. scabripennis* ssp. *ifranensis* Jolivet, 1968, **syn. nov.** = *T. scabripennis* var. *infoveolata* Kocher, 1953, **syn. nov.** = *T. scabripennis* ssp. *vazquezi* Jolivet, 1968, **syn. nov.**, based on the examination of the type material of all taxa and additional comparative specimens.

664 *Timarcha sobrina* Fairmaire, 1884 = *T. arragonica* Balbi, 1892, **syn. nov.**, based on the examination of the type material of both taxa.

665 *Timarcha spectacula* Bechyné, 1962, **stat. nov.**, based on the examination of the type material.

666 *Timarcha strangulata* Fairmaire, 1862 (based on the description and comparative specimens) = *T. strangulata* ssp. *aubryi* Jeanne, 1967, **syn. nov.** (based on the description and comparative specimens) = *T. strangulata* ssp. *breiti* Bechyné, 1953, **syn. nov.** (type material examined) = *T. strangulata* ssp. *decens* Bechyné, 1953, **syn. conf.** (type material examined) = *T. strangulata* ssp. *epipleuralis* Bechyné, 1953, **syn. conf.** (type material examined) = *T. strangulata* ssp. *sculptipennis* Bechyné, 1948, **syn. nov.** (based on the description and comparative specimens).

667 *Timarcha teluetica* Escalera, 1914, **stat. res.** (type material examined) = *T. achardi* Bechyné, 1945, **syn. nov.** (type material examined) = *T. punctella* ssp. *antiatlantica* Jolivet, 1968, **syn. nov.** (type material examined) = *T. apicicornis* Bechyné, 1948, **syn. nov.** (type material examined) = *T. punctella* ssp. *sarhroensis* Jolivet, 1968, **syn. nov.** (type material examined) = *T. punctella* var. *schrammi* Kocher, 1957, **syn. nov.**, based on the description and additional comparative specimens.

Protected name

668 *Timarcha affinis* Leach & Risso, 1827 (senior synonym of *Timarcha nicaeensis* A. Villa & G. B. Villa, 1835) was not used as valid species name after 1900. The following 25 papers, published by more than ten authors in the immediately preceeding 50 years, and encompassing a span of not less than ten years allow to propose *T. nicaeensis* as nomen protectum and *T. affinis* as nomen oblitum based on Art. 23.9.1. and 23.9.2. (ICZN 1999): Biondi et al. (1995), Chevin (1985, 1986, 1994a, 1994b), Cox (1994), Daccordi & Ruffo (1988, 1990), Daccordi et al. (2020), Gómez-Zurita (2004, 2008), Gómez-Zurita & Kippenberg (2010), Gómez-Zurita et al. (2000a, 2000b, 2004), Jolivet & Hawkeswood (1995), Jolivet & Petitpierre (1973), Jolivet et al. (2014), Leonardi & Sassi (1997), Minelli & Vittorelli (1976), Petitpierre & Cuadras (1977), Tiberghien (2014), Warchałowski (2003, 2010b), Winkelman & Debreuil (2008).

Availability of varieties and forms

669 The following names are deemed infrasubspecific because they were published as a fourth name after a trinomen (ICZN 1999, Art. 45.5):

Timarcha generosa ssp. *mesatlantica* var. *rungsi* Kocher, 1959: 77

Timarcha generosa ssp. *mesatlantica* var. *subscabra* Kocher, 1959: 77

Timarcha tenebricosa ssp. *submontana* natio *helvetica* Bechyné, 1945a: 9

670 The following varieties and forms are deemed infrasubspecific according to Article 45.6.1 (ICZN 1999), since a subspecies concept was used in the same works, and according to available information, these names were not validated prior to 1985 (ICZN 1999, Art 45.6.4.1):

Timarcha balearica var. *martini* Pic, 1919c: 20

Timarcha balearica var. *violacea* Pic, 1919c: 20

Timarcha normanna var. *bivillensis* Jolivet, 1943a: 69

Timarcha pimelioides var. *aenescens* Pic, 1919c: 20

Timarcha pratensis forma *goidanichi* Depoli, 1926b: 99

Timarcha pratensis forma *subrugosa* Depoli, 1926b: 99

671 The following varieties were proposed in works using also concept of subspecies, however, they were validated before 1985 and thus are deemed to be subspecific from their original publication (ICZN 1999, Art. 45.6.4.1):

Timarcha laevigata var. *rugosoidea* Jolivet, 1944: 100 was adopted by Bechyné (1948d) as subspecies.

Timarcha punctella var. *schrammi* Kocher, 1957: 135 was adopted by Jolivet (1968a) as subspecies.

Timarcha generosa var. *rugata* Kocher, 1953b: 319 was adopted by Bechyné (1953) as subspecies.

Timarcha scabripennis var. *bechynei* Kocher, 1959: 78 was adopted by Jolivet (1968b) as subspecies.

Timarcha scabripennis var. *infoveolata* Kocher, 1953b: 328 was adopted by Jolivet (1968b) as subspecies.

Timarcha ventricosa var. *aurulenta* Kocher, 1957: 131 was adopted by Jolivet (1966b) as subspecies.

Spelling

672 Kocher (1953b: 331) described *Timarcha generosa* ssp. *mediatlantis*. Later, Kocher (1959) incorrectly spelled this name as *mesatlantica* and this name was used also by Jolivet (1966a), Warchałowski (2003, 2010b) and Gómez-Zurita & Kippenberg (2010). As the incorrect subsequent spelling *mesatlantica* is in prevailing usage and is attributed to the publication of the original spelling, it is to be preserved and deemed to be a correct original spelling (ICZN 1999, Art. 33.3.1).

Applications to the Commission

673 *Tenebrio caeruleus* Linnaeus, 1758 is a senior synonym of *Timarcha balearica* Gory, 1833 as stated by Vela et al. (2020). An application to the Commission has been submitted to conserve the name *Timarcha balearica* Gory, 1833, by giving it precedence over the misinterpreted senior synonym *Tenebrio caeruleus* Linnaeus, 1758 (presently the type species of genus *Helops* Fabricius, 1775, in Tenebrionidae). In the meanwhile, prevailing usage is to be maintained.

674 *Chrysomela lusitanica* Fabricius, 1781: 116 (now in *Timarcha*) is a primary homonym of *C. lusitanica* Linnaeus 1767: 1066 (now *Exosoma lusitanicum*, Galerucinae). The application to conserve the name *Timarcha lusitanica* should be referred to the Commission. Meanwhile prevailing usage of both names is to be maintained (ICZN 1999, Art. 23.9.5.).

Nomina dubia

675 The type material of *Timarcha angulicollis* Motschulsky, 1849: 151 (surroundings of Cartagena, Southeastern Spain) and *T. immarginata* Herrich-Schäffer, 1838a: 24c (type locality not given) have not been found to date. Because identity of these two taxa cannot be assessed based on the original description they are treated as nomina dubia.

676 The extremely short description of *Timarcha apricaria* Gistel, 1832 was discovered at a final stage of preparing of the Catalogue. As the travelling to museums was almost impossible during pandemy we were unable to check Gistel's collection in Zoologische Staatssammlung Munich (Germany). *Timarcha apricaria* Gistel, 1832 is a senior homonym of *T. apricaria* Waltl, 1835 (type material examined). At the moment we treat *Timarcha apricaria* Gistel, 1832 as nomen dubium and *T. apricaria* Waltl, 1835 as valid name, until we have the opportunity to study the respective type material.

Nomina nuda

The following names are nomina nuda, however, their original material was examined and their identities are as follows:

677 *Timarcha hesperica* Rambur in Dejean, 1836: 399 is conspecific with *T. apricaria apricaria* Waltl, 1835

678 *Timarcha rugosula* Rambur in Dejean 1836: 399 is conspecific with *T. insparsa* Rosenhauer, 1856

679 *Timarcha parnassia* Rambur in Dejean, 1836: 399 is conspecific with *T. intermedia lugens* Rosenhauer, 1856

Homonymy

680 *Chrysomela scabra* Olivier, 1807: 507 (a junior synonym of *T. rugosa rugosa* (Linnaeus, 1767)) is a junior homonym of *Chrysomela scabra* Geoffroy 1785: 111.

681 *Chrysomela geniculata* Germar, 1823: 582, published on 5 October 1823, is a junior homonym of *Chrysomela geniculata* C. R. Sahlberg, 1823b: 73 (now a junior synonym of *Timarcha goettingensis* ssp. *goettingensis* (Linnaeus, 1758)), published on 16 June 1823. The oldest available name for *T. geniculata* (Germar, 1823) is *T. calceata* Pérez Arcas, 1865. The type specimens of *Chrysomela geniculata* Germar, 1823 and *C. geniculata* C. R. Sahlberg, 1823 were examined (the latter from photographs).

Patria falsa

The following type locaties are considered to be patria falsa:

682 "Algeria" for *T. fossulata* Bechyné, 1944, a synonym of *T. strophium strophium* Weise, 1888, distributed in Spain.

683 "Syria, Tripoli" for *T. amethystipes* Chevrolat, 1873, a synonym of *T. pimelioides turbida* Erichson, 1841, distributed in Tunisia and Algeria.

684 "Iles Baléares" for *T. elliptica* Fairmaire & Allard, 1873: 180 distributed in northeastern Spain.

Chrysomelidae: Galerucinae: Galerucitae

R. Beenen

New synonymy

685 *Trichelytron* Apfelbeck, 1912 was established as a subgenus of *Luperus* and included *L. apollinis* Apfelbeck, 1912 and *L. cyanipennis* Küster, 1848. Beenen (2010a) designated *L. apollinis* as the type species of *Trichelytron*. At that time *L. apollinis* was classified in *Calomicrus* and also in the previous Edition *Trichelytron* was treated as synonym of *Calomicrus*. However, the examination of photographs of a single male syntype in the collection of the late Andrzej Warchałowski (now NHMUK) proved this species to be a true *Luperus*; the third antennal segment is about four times as long as the second. This syntype is labelled: Apfelbeck | Veluchi || subalpin || specimen | de Dom. Karnožicki | receptum || Cotypus || Luperus apollinus APFELB. | det. A. Warchałowski. Therefore, *Calomicrus apollinis* is transferred back to the genus *Luperus* and *Trichelytron* is synonymized here with *Luperus*.

686 *Sphenoraia nigripennis* Clark, 1865 **stat. res.** = *Galleruca bicolor* Hope, 1831 **syn. nov.** = *S. hopei* Beenen, 2005 **syn. nov.** Beenen (2005a) proposed *S. hopei* as substitute for the homonym *Galleruca bicolor* Hope, however, overlooked the available synonyms, of which *S. nigripennis* Clark, 1865 is the oldest.

Restored synonymy

687 *Galerucella medvedevi* Beenen, 2008 = *G. rishwani* Makhan, 2012 (**syn. reconf.**). Ghahari & Beenen (2020) synonymized *G. rishwani* Makhan, 2012 with

G. medvedevi Beenen, 2008 and presented arguments for their proposal. Makhan et al. (2020) refuted these arguments and depicted again the aedeagi of both taxa showing diminutive differences. Extensive studies of aedeagi of *Galerucella* species from all parts of the Palaearctic Region revealed that the assumed differences by Makhan et al. (2020) are subject to intraspecific variation and do not justify existence of *G. rishwani*. Accordingly, the synonymy is reestablished. A single sentence in the paper by Makhan et al. (2020) "*Galerucella pusilla* has to be regarded as a junior synonym of *G. medvedevi*" cannot be accepted because *G. medvedevi* was described in 2008 while *G. pusilla* (Duftschmid, 1825) is much older and thus has priotity.

Type species designation

688 *Sphenoraia* Clark, 1865. Clark (1865b) originally included two species, *S. flavicollis* Clark, 1865 and *S. nigripennis* Clark, 1865, but the type species was not designated. The type species designation of *Galleruca bicolor* Hope, 1831 by Gressitt & Kimoto (1963a) is invalid as it was not originally included nominal taxon. *Sphenoraia nigripennis* Clark, 1865 is here designated as type species of *Sphenoraia*.

New replacement name

689 *Ochralea* Clark, 1865 with type species *Ochralea nigricornis* Clark, 1865 is a junior homonym of *Ochralea* Chevrolat, 1836 (synonym of *Oides* Weber, 1801) with type species *Adorium flavum* Olivier, 1807. Although *Ochralea* Chevrolat has not been used as a valid name after 1899, the conditions of Art. 23.9.1.2 (ICZN 1999) for reversal of precendence are not met and usage of *Ochralea* Clark cannot be maintained. *Ochralepta* Beenen, **nom. nov.** is proposed for *Ochralea* Clark, 1865, type species *Ochralea nigricornis* Clark, 1865, gender feminine (nec *Ochralea* Chevrolat, 1836).

New combinations

690 *Charaea tomokunii* (Takizawa, 1990), **comb. nov.** from *Exosoma*, based on the original description and study of additional specimens.

691 *Erganoides ictericus* (Weise, 1889), **comb. nov.** from *Charaea*, based on the photograph of one of the three syntypes deposited in ZIN.

Primary homonym

692 *Aplosonyx ancora* Laboissière, 1935 is a junior primary homonym that may be preserved because its senior homonym *Aplosonyx ancora* L. Redtenbacher, 1868 (now in *Aulacophora*) was placed in another genus before 1900 (ICZN 1999, Art. 23.9.5). The case requires an application to the Commission.

Doubtful assignment

693 *Mimastra brancuccii* L. N. Medvedev, 2011. The female holotype, deposited in NHMB and examined by J. Bezděk, has the elytra densely covered with short pubescence and almost certainly cannot be classified in the genus *Mimastra* Baly, 1865. However, the male remains unknown, which significantly complicates its proper placement. Therefore, *M. brancuccii* is temporarily left in *Mimastra* and marked as doubtful assignment [DA].

Spellings

694 *Agetocera* Hope, 1841. When Hope (1841) introduced the name *Agetocera* he stated that the name *Aegelocerus* Hope, 1831 was a print error. However, since the words differ completely, the provisions of article 32.5 (ICZN 1999) are not met and the name *Aegelocerus* is listed as a synonym, although unavailable because it is to be seen as an inadvertent error.

695 *Decarthrocera* Laboissière, 1937. Laboissière (1937a) used two different, though allied, names for a new genus he proposed: *Decarthrocera* and *Decarthrocerina*. The last name he also used for a new subtribe. In his description of the genus, he used the name *Decarthrocera* and only in his description of the species *D. viberti* he used *Decarthrocerina*. Because it is just a few lines below the introduction of the subtribe, it likely is an error (lapsus calami). Since Laboissière's publication, the name "Decarthrocerina" has never been used at the genus level. It has been used as name at the tribal level only (Wilcox 1975, Beenen, 2010b). The spelling *Decarthrocera* was fixed by Bouchard et al. (2024) as the correct original spelling in accordance with the principle of the First Reviser (ICZN 1999, Art. 24.2.3).

696 *Pyrrhalta qianana* Nie & Yang, 2013. Nie & Yang (2013) used two different spellings of this species: *qianana* in the key on page 43 and in the formal description on page 52, and *quinana* in the legend to figures on page 54. The last spelling is likely to be an error (lapsus calami). The spelling *qianana* is fixed here as the correct original spelling in accordance with the principle of the First Reviser (ICZN 1999, Art. 24.2.3).

697 *Yunaspes medogensis* Jiang, 1988. Jiang (1988) used two different spellings of this species: *modogensis* in the formal description in Chinese on page 392, and *medogensis* in the English description on page 395. The species name is derived from the type locality Mêdog (spelling according to *The Times Atlas of the World*, comprehensive edition) in Xizang, China, so it is likely that the spelling *medogensis* was intended. Although the name spelled as *modogensis* is used by Beenen (2010b) and Yang et al. (2015), the spelling *medogensis* is fixed here as the correct original spelling in accordance with the principle of the First Reviser (ICZN 1999, Art. 24.2.3).

Availability of varieties and forms

698 The following varieties are deemed infrasubspecific according to Article 45.6.1 (ICZN 1999), since a subspecies concept was used in the same work, and according to available information, these names were not validated prior to 1985 (ICZN 1999, Art 45.6.4.1):

Aplosonyx chalybeus var. *jeanvoinei* Laboissière, 1935b: 109
Diorhabda elongata var. *bipustulata* Normand, 1937: 126
Euliroetis lameyi var. *obscuripes* Ogloblin, 1936: 200, 405
Galeruca barbara var. *atlasica* Kocher, 1959: 62
Galerucella distincta var. *jureceki* Pic, 1921a: 2
Galerucella luteola var. *bicarinata* Ogloblin, 1936: 108, 391
Galerucida ornatipennis var. *decolora* Laboissière, 1935b: 120
Galerucida ornatipennis var. *violacea* Laboissière, 1935b: 120
Galerucida pectoralis var. *humeralis* Laboissière, 1935b: 124
Luperus circumfusus var. *luteovittatus* Pic, 1934h: 27
Luperus longicornis var. *tenebricosus* Ogloblin, 1936: 291, 426
Phyllotreta funesta var. *collaris* Chûjô, 1940b: 117
Sphenoraia micans var. *cyanella* Laboissière, 1935b: 132

New geographical records

699 *Agelasa nigriceps* Motschulsky, 1861. NORTH KOREA: South Hamgyong Prov., Mukbang Mt., 1.–12.vii.2010, 1 spec., Jingke Li leg., R. Beenen det. (collection of H. Hebauer, Rain, Gemany); Chonmassan Mt., Pyonganbuk-Do, 20.–27.vii.2007, 2 spec., Jingke Li leg., R. Beenen det. (collection of H. Hebauer, Rain, Germany). New species for North Korea.

700 *Anadimonia latifascia* (Gressitt & Kimoto, 1963). CHINA: Fujian: Ziyungdongshan, NW slopes, 25°46′N, 117°20′E, 700–1100 m, 29.iv.2008, 11 spec., J. Turna leg., R. Beenen det. (9 spec. in collection of H. Hebauer, Rain, Germany; 2 spec. in collection of R. Beenen, Nieuwegein, Netherlands). New species for Fujian (China).

701 *Aplosonyx flavipennis* S.-H. Chen, 1964. CHINA: Guangxi: Tangwanshan Mt., Baise city, 20.–28.vi.2011, 4 spec., Jingke Li leg., R. Beenen det. (collection of H. Hebauer, Rain, Germany); Xizang: Ma-ma yuo, Zayu-county, 15.–26.vi.2010, 3 spec., Jingke Li leg., R. Beenen det. (collection of H. Hebauer, Rain, Germany). New species for Guangxi and Xizang (China).

702 *Aplosonyx orientalis* Jacoby, 1892. CHINA: Fujian: Wuyishan Mts. NNR, Dazhulan vill., 27°41.8–42.1′N 117°38.6–9′E, 720 m, 27.v. + 1.vi.2018, 8 spec., J. Hájek, D. Král, J. Růžička & L. Sekerka leg., J. Bezděk det. (NMPC). New species for Fujian (China).

703 *Apophylia dellacasai* Bezděk, 2006. SAUDI ARABIA: Asir, 6 km N of Ainah, 200 m, 3.iv.2017, 2 ♂♂ 1 ♀, P. Rapuzzi leg., J. Bezděk det. (NMPC); Bahah, 5 km N of Al Makhwah, 464 m, 30.iii.2017, 1 ♂ 1 ♀, P. Rapuzzi leg., J. Bezděk det. (NMPC). New species for Saudi Arabia.

704 *Apophylia nigriceps* Laboissière, 1927. CHINA: Sichuan: way between Xaijin and Danba, Taiping Qiao Xiang env., 2160 m, 26.vii.–6.viii.2006, 1 spec., Puchner leg., R. Beenen det. (collection of R. Schuh, Wiener Neustadt, Austria). New species for Sichuan (China).

705 *Arthrotus chinensis* (Baly, 1879). CHINA: Henan: Funiu Shan, 33°31′N 111°56′E, Baotianman, 1500–1750 m, 16.–17.v.2009, 2 spec., J. Turna leg., R. Beenen det. (collection of H. Hebauer, Rain, Germany). New species for Henan (China).

706 *Calomicrus fallax* (Joannis, 1865). TUNISIA: 15 km E of Ain Draham, 12.vi.2011, 1 spec., M. Egger leg., R. Beenen det. (collection of M. Egger, Wattens, Austria). New species for Tunisia.

707 *Cassena bicolor* (Gressitt & Kimoto, 1963). CHINA: Hubei: Guanmenshan, 1350–1600 m, 31°265′N 110°22′E, 24.v.2008, 5 spec., J. Turna leg., R. Beenen det. (3 spec. in collection of H. Hebauer, Rain, Germany; 2 spec. in collection of R. Beenen, Nieuwegein, Netherlands). New species for Hubei (China).

708 *Charaea iniquum* (Weise, 1889). INDIA: Jammu and Kashmir: Gulmarg, 2700 m, 27.vii.1977, 2 ♂♂ 2 ♀♀, Richter leg., J. Bezděk det. (collection of J. Bezděk, Brno, Czech Republic); Sikkim: Teesta Vy., Tungu, 1. –15.vii.1903, 1 ♂ 1 ♀, Tibet Expedition leg., J. Bezděk det. (NHMUK); Uttarakhand: Chamoli District, Gghangaria, 3000–3300 m, 19.–20.vii.1989, 5 ♂♂ 4 ♀♀, Heinz leg., J. Bezděk det. (collection of J. Bezděk, Brno, Czech Republic). New species for Jammu and Kaschmir, Sikkim, and Uttarakhand (India).

709 *Charaea kelloggi* (Gressitt & Kimoto, 1963). CHINA: Hainan: Changjiang County, Heiling Mt., 1560 m, 7.–16.vi.2008, 1 ♂ 1 ♀, Li Jingke leg., J. Bezděk det. (♂ in collection of H. Hebauer, Rain, Germany; ♀ in collection of R. Beenen, Nieuwegein, Netherlands). New species for Hainan (China).

710 *Charaea tomokunii* (Takizawa, 1990). INDIA: Uttarakhand: 15 S of Badrinath, 30 km N of Joshimat, 2800 m, 10.vii.1994, 4 ♂♂ 10 ♀♀, M. Snížek leg., J. Bezděk det. (collection of J. Bezděk, Brno, Czech Republic). New species for Uttarakhand (India).

711 *Cneorane braeti* Duvivier, 1892. NEPAL: Kakara env., 25.–26.vii.2012, 1 spec., E. Kučera leg., R. Beenen det. (collection of U. Heinig, Berlin, Germany). New species for Nepal.

712 *Cneorane orientalis* Jacoby, 1892. CHINA: Hainan: Jianfengling Mt, Sanya City, 10.–27.vi.2011, 1 spec., Jingke Li leg., R. Beenen det. (collection of H. Hebauer, Rain, Germany). New species for Hainan (China).

713 *Cneoranidea parasinica* L.-J. Zhang & Yang, 2005. CHINA: Fujian: Wuyishan Mts. NNR, Taoyuanyu valley, 27°43.7′N 117°42.8′E, 575 m, 1 ♂ 1 ♀, J. Hájek, D. Král, J. Růžička & L. Sekerka leg., J. Bezděk det. (NMPC). New species for Fujian (China).

714 *Erganoides similis* S.-H. Chen, 1942. CHINA: Gansu: Wenxian, 7.–8.vi.2014, 1 ♂, E. Kučera leg., R. Beenen det. (collection of U. Heinig, Berlin, Germany). New species for Gansu (China).

715 *Euluperus pinicola* (Duftschmid, 1825). GREECE: Epirus Prov., Pindos Mts., Milia, 1360 m, 11.vi.2013, 3 spec., Z. Malinka leg., J. Bezděk det. (collection of Z. Malinka, Opava, Czech Republic). ROMANIA: Bizac, Potoci env., 500–600 m, 16.–23.vi.1996, 4 spec., P. Průdek leg., J. Bezděk det. (collection of J. Bezděk, Brno, Czech Republic). New species for Greece and Romania.

716 *Euluperus pseudoaltaicus* (Medevedv, 1975). GEORGIA: Svanetia, Lalkhori, W of Ushguli, 1800 m, 6.vii.2015, 1 spec., W. Ziegler leg., R. Beenen det. (collection of W. Ziegler, Rondeshagen, Germany). New species for Georgia.

717 *Furusawaia konstantinovi* (Lopatin, 2009). CHINA: Sichuan: S of Yanyuan, Chuandongzi massiv, 3200 m, 8.–19.vi.2017, 1 ♀, C. Reuter leg., J. Bezděk det. (NMEG). New species for Sichuan (China).

718 *Galeruca angusta* Küster, 1849. ALGERIA: Ain el Beida (Timgad ruins in Aurès Mountains), 1700 m, 5.iii.1991, 5 spec., W. Heinz leg., R. Beenen det. (4 spec. in collection of D. Erber, in ZMHUB, 1 spec. in collection of R. Beenen, Nieuwegein, The Netherlands). MOROCCO: Mutawassit Ksabi, NNE of Midelt, 27.iii.1990, 1 spec., H. Teunissen leg., R. Beenen det. (collection of A. P. J. A. Teunissen, Eindhoven, The Netherlands); Moyen Atlas, Mischliffen, 1800–2000 m, 2.iv.1991, 1 spec. W. Heinz leg., R. Beenen det. (collection of D. Erber, in ZMHUB); Santarém, Fatima, 30.iv.1992, 3 spec., J. Teunissen leg., R. Beenen det. (collection of A. P. J. A. Teunissen, Eindhoven, The Netherlands); SE Marrakesh, Pass Tizi-n-Tichka, Hoher Atlas, 2200 m, 17.ii.2004, 3 spec., Assmann leg., R. Beenen det. (collection of W. A. Starke, Warendorf, Germany); Marrakesh, Haut Atlas, 2200 m, Tizi-n-Tichka pass, 17.ii.2004, 36 spec., D. W. Wrase leg., R. Beenen det. (collection of U. Heinig, Berlin, Germany); Azrou, Ifrane, 22.v.2007, swept from low vegetation, M. A. Hielkema leg., R. Beenen det. (collection of R. Beenen, Nieuwegein, The Netherlands). TUNISIA: Ferme Shitta Djebel Eddyr, 7 km NE El Kef, 29.iii.1977, 1 spec., R. Beenen det. (HNHM); El Kesra, vi.1982, 1 spec. J. Picka leg., R. Beenen det. (collection of J. Pelikán, Hradec Králové, Czech Republic); Haffiz, 5.vi.1982, 1 spec., J. Picka leg., R. Beenen det. (collection of J. Pelikán, Hradec Králové, Czech Republic); El Kef, 31.v.–2.vi.1999, 8 spec., P. A. A. Teunissen leg., R. Beenen det. (4 spec. in collection of A. P. J. A. Teunissen, Eindhoven, The Netherlands, 4 spec. in collection of R. Beenen, Nieuwegein, The Netherlands). New species for Algeria, Morocco and Tunisia.

719 *Galeruca daurica* (Joannis, 1865). CHINA: Heilongjiang: Harbin env. [45°45′N, 126°38′E], 5 spec., V. Ya. Tolmacheva leg., D. A. Ogloblin det. (ZIN); Liaoning: Zhengjiatun, 8.v.–16.vi.1905, 25 spec., Bortkevich leg., D. A. Ogloblin det. (ZIN). New species for Heilongjiang and Liaoning (China).

720 *Galeruca melanocephala* (Ponza, 1805). GEORGIA: Kakheti Prov., Koreti, w Telavi, Flusstal Hochwasser, 42°06′01.65″N 45°14′38.25″E, 561 m, 12.vi.2016, 2 spec.,

W. Ziegler leg., R. Beenen det. (collections of W. Ziegler, Rondeshagen, Germany and R. Beenen, Nieuwegein, Netherlands). New species for Georgia.

721 *Galeruca reichardti* Jacobson, 1925. CHINA: Heilongjiang: Imyanpo Railway Station [= Yimianpozhen, 45°03′N 128°04′E], 9.vi.1911, 1 spec., A. Emelyanov leg., D. A. Ogloblin det. (ZIN). New species for Heilongjiang (China).

722 *Galeruca rufa* Germar, 1823. UZBEKISTAN: Tashkent, Habib Abdullaev street, 450 m, 1.–18.iv.2009, 9 spec. L. & E. Hendrich leg., R. Beenen det. (6 spec. in collection of U. Heinig, Berlin, Germany, 3 spec. in collection of R. Beenen, Nieuwegein, Netherlands). New species for Uzbekistan.

723 *Galerucella calmariensis* (Linnaeus, 1767). CHINA: Liaoning: Langashi, 20.vi.–11. vii.1905, 1 spec., D. A. Ogloblin det. (ZIN). New species for Liaoning (China).

724 *Galerucella placida* Baly, 1878. CHINA: Gansu: W Qinling Shan, 43 km N Chengdan, 1750 m, 34°06′24″N 105°46′43″E, moist valley with creek and ponds, moist meadow with tall herbaceous vegetation, 28.vii.2012, 2 spec., D. W. Wrase leg., R. Beenen det. (collection of U. Heinig, Berlin, Germany); Jiangsu: Yangzhou, Slender West lake, 21.viii.2005, 2 spec., on *Lagerstroemia indica*, L. Zhang leg., R. Beenen det. (collection of L. Zhang, Yangling, China); Sichuan: Guanxian, 31.00N 103.36E, 700 m, 20.v.1989, 1 spec., J. Kolibáč leg., R. Beenen det. (NHMB); Guanxian, 31.00N 103.36E, 700 m, 8.–20.vii.1990, 2 spec., J. Kolibáč leg., R. Beenen det. (NHMB); Emei Mt., 29.30N 103.2036E, 500–1200 m, 4.–18.v.1989, 1 spec., J. Kolibáč leg., R. Beenen det. (NHMB); Yunnan: Dali Bai Nat. Aut. Pref., 4 km E Dali old town, shore terrain of Er Hai lake, 25.42N 100.01.5E, 2020 m, young willows with knot grass under vegetation in roots, 27.viii.–2.ix.2003, 23 spec., D. W. Wrase leg., R. Beenen det. (collection of U. Heinig, Berlin, Germany); Lincang Pref., Mekong Valley, small creek cleft 38 km SSE Lincang, 23°33′13,3″N 100°09′56,8″E, 854 m, field edge in small cleft, on vegetation, 11.ix.2009, 12 spec., D. W. Wrase leg., R. Beenen det. (collection of U. Heinig, Berlin, Germany); Baoshan pref., Gaoligung, 24°51′22″N 98°45′36″E, 2100–2200 m, pasture, under stones, plants, moss, 31.v.2007, 1 spec., D. W. Wrase leg., R. Beenen det. (collection of U. Heinig, Berlin, Germany). New species for China (Gansu, Jiangsu, Sichuan, and Yunnan).

725 *Gallerucida gloriosa* (Baly, 1861). CHINA: Henan: Wanxianshan, 35°42′N 113°36′E, 840 m, 15.v.2006, 1 spec., J. Turna leg., R. Beenen det. (collection of U. Heinig, Berlin, Germany). New species for Henan (China).

726 *Gallerucida rufometallica* Gressitt & Kimoto, 1963. CHINA: Yunnan: Baoshan, 8.vi.1993, 1 spec., O. Šauša, R. Beenen det. (collection of U. Arnold, Berlin, Germany). New species for Yunnan (China).

727 *Luperomorpha arabica* Doguet, 1979. OMAN: Dhofar Prov. Rakhyut env., 25–50 m, 1.–8.viii.1999, 1 ♂, R. Červenka leg., J. Bezděk det. (collection of J. Bezděk, Brno, Czech Republic); Dhofar Prov., Jabal al Qamar, 3 km W of Rakhyut, 16°45′19″N

53°23′48″E, 65 m, 11.iv.2013, 20 spec., R. Fouqué leg., J. Bezděk det. (collection of J. Bezděk, Brno, Czech Republic); Dhofar Prov., Wadi Nashib, 20 km E of Salalah, 0–50m, 16.–22.ix.2006, 3 spec., S. Jákl leg., R. Beenen det. (collections of M. Schöller, Berlin, Germany, and R. Beenen, Nieuwegein, Netherlands). New species for Oman.

728 *Luperus flaviceps* Apfelbeck, 1912. ALBANIA: Korce Prov., Radanj env., Lengatices valley, 2.vi.2013, 1 ♂ 1 ♀, P. Vonička leg., J. Bezděk det. (collection of J. Bezděk, Brno, Czech Republic). New species for Albania.

729 *Luperus flavus* Rosenhauer, 1856. PORTUGAL: Lusitania (B. J.), Alvito, iv.2000, 1 spec., P. Foot leg., R. Beenen det. (collection of R. Beenen, Nieuwegein, Netherlands); Guarda, 10 km W of Vilar Formoso, 21.v.1988, 2 spec., M. A. Hielkema leg., R. Beenen det. (collection of R. Beenen, Nieuwegein, Netherlands). New species for Portugal.

730 *Luperus semiflavus* Ogloblin, 1936. NORTH KOREA: South Hamgyong Prov., Mukbang Mt., 1.–12.vii.2010, 1 spec., Jingke Li leg., R. Beenen det. (collection of H. Hebauer, Rain, Germany). New species for North Korea.

731 *Menippus beeneni* C.-F. Lee, Bezděk & Suenaga, 2012. CHINA: Henan: Funiu Shan, Baotianman, 33°31′N 111°56′E, 1500–1750 m, 5.vi.2009, 2 spec., J. Turna leg., R. Beenen det. (collections of H. Hebauer, Rain, Germany and R. Beenen, Nieuwegein, Netherlands). New species for Henan (China).

732 *Morphosphaera albipennis* Allard, 1889. CHINA: Guangxi: Shangsi County, Shiwandashan Nat. Forest Park, 265 m, 29.iii.2015, 2 spec., J. Kontschán, J.-N. Li, S. Li, W.-H. Li, D. Murányi & G.-Q. Wang leg., J. Bezděk det. (HNHM). New species for Guangxi (China).

733 *Paraplotes clavicornis* Gressitt & Kimoto, 1963. CHINA: Jiangxi: Jinggang Shan Mts., Xiangzhou vill. env., 26°35.5′N 114°16.0′E, 374 m, 26.iv.2011, 1 ♂, M. Fikáček & J. Hájek leg., J. Bezděk det. (collection of J. Bezděk, Brno, Czech Republic). New species for Jiangxi (China).

734 *Paridea costata* (Chûjô, 1935). CHINA: Hunan: ca. 100 km S of Huaihua, Huitong, 7.–12.vii.1992, 1 ♀, C. Holzschuh leg., J. Bezděk det. (NMEG). New species for Hunan (China).

735 *Pyrrhalta huangshana* S.-H. Chen, 1964. CHINA: Guangxi: 4.5 km NE of Jinxiu, 24°09′36″N 110°13′28″E, 1210 m, 1.vii.2018, 1 ♀, D. Rédei & J.-F. Tsai leg., J. Bezděk det. (HNHM); Fujian: Wuyishan Mts., Sangang vill., 27°45.0′N 117°40.7′E, 720 m, 23.v. + 3.vi.2018, 8 spec., J. Hájek, D. Král, J. Růžička & L. Sekerka leg., J. Bezděk det. (NMPC). New species for Fujian and Guangxi (China).

736 *Radymna persica* (Fadermann, 1837). UZBEKISTAN: Zaamin N. P., 39°43.072′N 68°26.657′E, 1450 m, 16.v.2019, 1 ♂, Z. Malinka leg., J. Bezděk det. (collection of Z. Malinka, Opava, Czech Republic); 11 km NE of Boysun, 1811 m, 7.v.2019, 1 ♂ 1 ♀, R. Ambrus leg., J. Bezděk det. (collection of R. Ambrus, Praha, Czech Republic). New species for Uzbekistan.

737 *Sermylassa halensis* (Linnaeus, 1767): ARMENIA: Syunik Reg., 2 km E of Gorhayk, 15.viii.2002, 3 spec., M. Kalashian leg., J. Bezděk det. (collection of J. Bezděk, Brno, Czech Republic). New species for Armenia.

738 *Strobiderus nigriceps* Laboissière, 1936. CHINA: Yunnan: Mengla, Wangtianshu Nature Reserve, 21°37′09.6″N 101°35′17.2″E, 740 m, 28.vii.2017, 1 spec., D. Rédei & J.-F. Tsai leg., J. Bezděk det. (HNHM); 15 km SE of Mohan, Miaozhai env., 21°08′57.2″N 101°45′30.7″E, 1236 m, 30.–31.vii.2017, 1 spec., D. Rédei & J.-F. Tsai leg., J. Bezděk det. (HNHM). New species for Yunnan (China).

739 *Xanthogaleruca aenescens* Fairmaire, 1878. CHINA: Heilongjiang: Mudanjiang [44°33′N 129°37′E], 15.viii.1900, 1 spec., Korolkov leg., D. A. Ogloblin det. (ZIN); Liaoning: Zhengjiatun, 8.–15.vi.1905, 14 spec., Bortkevich leg., D. A. Ogloblin det. (ZIN). NORTH KOREA: South Hamgyong Prov., Mukbang Mt., 1.–12.vii.2010, 1 spec., Jingke Li leg., R. Beenen det. (collection of H. Hebauer, Rain, Germany). New species for Heilongjiang and Liaoning (China), and North Korea.

740 *Xanthogaleruca luteola* (O. F. Müller, 1766). ARMENIA: Kchozrou, 23.–24.vi.2006, 1 spec., Dr. Louda leg., R. Beenen det. (collection of H. Hebauer, Rain, Gemany). CYPRUS: Paphos Prov., Kidasi, 25.iv.2016, W. Ziegler leg., R. Beenen det. (collection of W. Ziegler, Rondeshagen, Germany). PAKISTAN: Azad Jammu and Kashmir, Neelum valley, Sharda env., 1900–2100 m, 24.–29.viii.1988, 1 spec., Heinz leg., J. Bezděk det. (collection of J. Bezděk, Brno, Czech Republic). SYRIA: 10 km N of Jablah, 10.ix.1998, 1 spec., A. Bezděk leg., J. Bezděk det. (collection of J. Bezděk, Brno, Czech Republic). TUNISIA: Ain Draham, 10.vii.2008, 1 spec., M. Mantič leg., J. Bezděk det. (collection of M. Mantič, Hlučín, Czech Republic). New species for Cyprus, Pakistan, Syria, and Tunisia, confirmed occurrence in Armenia.

Comments on distribution

741 *Arthrotus flavocinctus* (Hope, 1831). Lee & Bezděk (2013) listed only Nepal and some provincies of India as Palaearctic countries for *Arthrotus flavocinctus*. Yang et al. (2015) were aware of this work but also listed many Chinese provinces. Although it is uncertain if they are correct, all these provinces are listed in the present Catalogue.

742 *Atysa diversicornis* (Pic, 1931). Distribution of this species is uncertain. In the original description, Pic (1931b) stated that although it was labelled as from Turkey it most probably originally came from India ("Acquis comme provenant d'Asie Mineure, mais originaire vraisemblablement des Indes"). The type specimen in Pic's collection (MNHN) bears the labels "Asia minor" and "? Indes". The species certainly does not come from Turkey and resembles members of the Oriental fauna, but it is absolutely impossible to determine whether it came from regions considered in the present Catalogue or is extralimital. Therefore, it is listed only with generalised distribution information.

743 *Galeruca angusta* Küster, 1849 was recorded from northern Africa ("Tout le nord africain jusque dans l'Atlas marocain") by Laboissière (1934b), but no African countries were listed for this species in the first Edition (Beenen 2010b).

744 *Hoplasoma philippinense* Jacoby, 1894 is not included as, according to Bezděk (2012a), the single Palaearctic record from Korea is either mislabelled or accidentally introduced.

745 *Strobiderus nigripennis* (Jacoby, 1900) is not included in the present Catalogue although Yang et al. (2015) listed it for China without detailed information. It is unlikely that this south Indian species occurs also in China.

Chrysomelidae: Galerucinae: Galerucitae

J. Bezděk & A. S. Konstantinov

New names

746 *Luperomorpha bryanti* Bezděk & Konstantinov, **nom. nov.** for *Luperomorpha caerulea* Bryant 1939, a junior homonym of *Luperomorpha caerulea* (Jacoby 1900).

747 *Luperomorpha deorali* Bezděk & Konstantinov, **nom. nov.** for *Luperomorpha nepalensis* L. N. Medvedev, 1990, a junior homonym of *Luperomorpha nepalensis* L. N. Medvedev, 1984.

748 *Luperomorpha medvedevi* Bezděk & Konstantinov, **nom. nov.** for *Luperomorpha schmidti* L. N. Medvedev, 2009, a junior homonym of *Luperomorpha schmidti* L. N. Medvedev, 2004.

Chrysomelidae: Galerucinae: Alticitae

J. Bezděk & A. S. Konstantinov

Comments

749 Approximately 40 available family-group names exist within Alticitae (Bouchard et al. 2011). However, most of them have never been properly diagnosed. Published phylogenetic studies of various scopes that include Alticitae have been plagued by small and geographically limited taxon sampling and poor choice of morphological characters, which resulted in poor resolution of deeper nodes of the flea beetle phylogeny (e.g. Nie et al. 2018, Nadein 2013a, 2013c, 2015). Because of these uncertainties, recent authors (e.g. Riley et al. 2003, Biondi and D'Alessandro 2010, Döberl 2010b) have not used available family-group names. We here follow this approach.

750 *Aphthona agrostidis* Gistel, 1857 is an available name treated conspecific with *Neocrepidodera cyanescens concolor* (K. Daniel, 1900) (see Monrós & Bechyné 1956). In the errata to the volume 6 of Catalogue of Palaearctic Coleoptera (Löbl & Smetana 2011) *A. agrostidis* was added as a nomen oblitum by Manfred Döberl (2012, pers. comm.). However, the nomen oblitum was not established in accordance with the Code (ICZN 1999) and must therefore be considered an invalid act and thus

A. agrostidis still threatens the name *N. cyanescens concolor*. Unfortunately, since the conditions of Art. 23.9.1.2 are not met, nomen oblitum and nomen protectum cannot be applied. The case requires an application to the Commission. Meanwhile the usage of *N. cyanescens concolor* as valid name is to be maintained.

751 A cosmopolitan genus *Chaetocnema* Stephens, 1831 with its 496 valid species has one of the most complicated nomenclatural history among flea beetles. *Chaetocnema* species are distributed through biogeographic regions as follows: Afrotropical – 147 species, Australian – 27, Nearctic – 61, Neotropical – 102, Oriental – 92, Palearctic – 90. Nearctic, Oriental, and Palearctic faunas were recently revised (White 1996, Konstantinov et al. 2011, Ruan et al. 2019). Several genus-level names have been proposed for *Chaetocnema* species, some of which have been treated as synonyms and some as subgenera. Of the latter, two subgenera are most widely used with some consensus on the ways to separate them: *Chaetocnema* Stephens, 1831 and *Udorpes* Motschulsky, 1845 (Konstantinov et al. 2011, Ruan et al. 2019). Zhang et al. (2023) made significant progress in circumscribing *Chaetocnema* and *Udorpes*. However, as has been mentioned repeatedly (e.g. Konstantinov et al. 2011), it is impossible to adequately classify *Chaetocnema* species worldwide without comprehensive analyses of a variety of data, morphological and especially molecular. Unfortunately, a recently published attempt (Özdikmen 2021) to completely reclassify *Chaetocnema* species wordwide only complicates the matter. One of the most dramatic outcomes of the new classification was the erection of ten more subgenera. Proposed changes are made entirely based on characters of the spermatheca. As the result, characters traditionally used to separate two well established subgenera are in contradiction with characters of the spermatheca. Spermathecal characters are inadequeately divided into states (which might be altogether impossible considering the relative simplicity of spermatheca and the number of taxa it expected to reliably diagnose) and various subgenera share many of them. In addition many *Chaetocnema* species are not placed in any given subgenus. Taken altogether, these considerations make using the proposed classification highly problematic, therefore, it is not implemented here and the names of all generic level taxa are listed under *Chaetocnema* as synonyms in the present Catalogue without comments on their status, as is done for other genera treated here.

752 *Longitarsus pallens* (Stephens, 1831). Heikertinger & Csiki (1939) recorded the name *L. pallens* two times in their world catalogue of flea beetles: *L. pallens* Foudras 1860 and *L. pallens* Kutschera, 1863, but, surprisingly, *L. pallens* (Stephens, 1831: 313) is missing there. However, neither Foudras nor Kutschera described such taxon. According to Weise (1893), *L. pallens* sensu Foudras is a misidentification of *L. pallens* (Stephens, 1831) and Weise named those specimens as *L. foudrasi* Weise, 1893. *Longitarsus pallens* Kutschera, 1863 is a non-existent name, because Kutschera

(1863b) only placed *L. pallens* sensu Foudras in synonymy with *L. verbasci* (Panzer, 1794) (now syn. of *L. tabidus* (Fabricius, 1775)). In the previous Edition (Döberl 2010b) *L. pallens* (Stephens, 1831) was mistakenly placed in synonymy with *L. foudrasi* Weise, 1893. Because the identity of *L. pallens* is unknown we placed it as a nomen dubium in the present Catalogue.

753 *Oedionychis* Latreille, 1829. Two spellings and attributions of this generic name were interchangeably used in leaf beetle literature: *Oedionychus* Berthold, 1827 and *Oedionychis* Latreille, 1829. A version of the name was first used by Latreille (1825: 405) in French transcription "Oedionyque" and therefore is not available. Berthold (1827: 401) published German translation of Latreille (1825) and latinized all names, including *Oedionychus*. However, the name was not accompanied with description or valid species and, therefore is unavailable. Latreille (1829: 154) used *Oedionychis* and included *Altica marginella* Olivier, 1808 and thus made the name available. Therefore, the correct name is *Oedionychis* Latreille, 1829.

754 *Psylliodes* Latreille, 1829. During its long nomenclatural history *Psylliodes* has been usually treated as feminine (see for example Heikertinger 1916). Leonardi & Gruev (1993) and Doguet (1994) changed the gender to masculine because the genus names ending with *-odes* were strictly stated as masculine without any exception in the Code that was valid at that time (ICZN 1985a, Art. 30). However, the current version of the Code (ICZN 1999) introduced the possibility of exception if the author combined the generic names with a female adjective (Art. 30.1.4.2.). Based on this article Döberl (2009a) suggested to treat the gender of *Psylliodes* as feminine. Latreille (1829) included the following species names when introduced *Psylliodes*: *chrysocephala, napi, hyosciami* [sic!], *dulcamarae* and *affinis*. The names *napi, hyosciami* [sic!] and *dulcamarae* cannot determine the gender as they are genitives. The name *affinis* is a Latin compositum with masculine and feminine termination *-is*, and neuter *-e*, so this also does not indicate whether Latreille intended *Psylliodes* as masculinum or feminine (only eliminates neuter). Last, and crucial, is *chrysocephala* which is a latinised feminine adjective. Therefore, in the present Catalogue we follow Döberl (2009a) and treat *Psylliodes* as feminine.

New synonyms

755 *Hartmannia* L. N. Medvedev, 2015, **syn. nov.** of *Liprus* Motschulsky, 1861. Upon examination of the type specimens, it became obvious that *Hartmannia matthiasi* Medvedev, 2015 is congeneric with *Liprus*. It shares all the characters of the head, legs, elytra, and other body parts. The only difference is shape of the pronotal base where antebasal transverse impression (also present in *Liprus*) is interrupted on both sides of the middle. Moreover, *Hartmannia* Medvedev, 2015 is a junior homonym of *Hartmannia* Newton, 1891 (Gastropoda, Architaenioglossa, Megalomastomatidae).

756 *Sphaerochabria* L. N. Medvedev, 1999, **syn. nov.** of *Sphaeroderma* Stephens, 1831. Upon examination of the holotype, it became obvious that *Sphaerochabria nepalica* Medvedev is undoubtedly congeneric with *Sphaeroderma. Sphaerochabria nepalica* shares with *Sphaeroderma testaceum* (Fabricius, 1775) all details of the head (with all the grooves, furrows, and setae), pronotum, including the shape and structure of the anterolateral callosity, elytra, and legs (including claws and spurs). There is not a single character that may be used to separate *Sphaerochabria* from *Sphaeroderma*, so the synonymy is straight forward.

Primary homonym

757 *Chrysomela dorsalis* Fabricius, 1781: 135 (now in *Longitarsus*) is a primary junior homonym of *Chrysomela dorsalis* Fabricius, 1777: 220 (now *Entomoscelis dorsalis*, Chrysomelinae) but the names were not considered congeneric after 1899. An application to conserve the name *Longitarsus dorsalis* (Fabricius, 1781) should be referred to the Commission. Meanwhile prevailing usage of both names is to be maintained (ICZN 1999, Art. 23.9.5.).

New combinations

758 *Liprus matthiasi* (L. N. Medvedev, 2015), **comb. nov.** from *Hartmannia.*

759 *Sphaeroderma nepalicum* (L. N. Medvedev, 1999), **comb. nov.** from *Sphaerochabria.*

Type species

760 *Plectroscelis* Chevrolat, 1836. Konstantinov et al. (2011) designated *Altica dentipes* sensu Oliver (1808) (= *Altica chlorophana* Duftschmid 1825), fixed under Article 70.3 (ICZN 1999), because *Altica dentipes* was misidentified in the first subsequent designation by Chevrolat (1845: 6). Bousquet & Bouchard (2013) questioned this designation with the explanation that Chevrolat (1847: 267) treated "*dentipes* Ol." and "*dentipes* Koch" as two different species and therefore his type species designation of *Altica dentipes* sensu Olivier is invalid since Olivier's taxon is not originally included in Dejean's catalogue (Dejean 1836). In the present Catalogue we treat the designation made by Konstantinov et al. (2011) as correct. The opinion of Chevrolat (1847) that "*dentipes* Ol." and "*dentipes* Koch" are two different species is irrelevant as it was published two years after his first type species designation and in our opinion it cannot be taken into consideration.

Change of authorship

761 *Crepidodera obscuripes* Bechyné, 1956 was previously attributed to Heikertinger (1912g). However, Heikertinger described it as an aberration and thus the name is unavailable. Bechyně (1956b) treated it as a subspecies, without a description, but

he gave a full bibliographic reference. Therefore, based on Arts. 13.1.2. and 45.6.2 (ICZN 1999) the author of *obscuripes* is Bechyné (1956b).

762 *Neocrepidodera spectabilis* Döberl, 1995 was previously attributed to J. Daniel (1904b), who introduced it as *Crepidodera rhaetica* var. *spectabilis*. In the same paper, J. Daniel used also subspecies category, which makes varieties infrasubspecific. From the same reason, var. *spectabilis* is unavailable also in Porta (1934) and Heikertinger (1948). Döberl (1995b) was the first who explicitly elevated *Crepidodera spectabilis* to species level, gave the description and mentioned the type material, therefore, the authorship must be credited to him.

Protected names

763 *Crepidodera arbustorum* Gistel, 1857 is an available name treated conspecific with *Longitarsus curtus* (Allard, 1861) (see Monrós & Bechyné 1956). In the errata to volume 6 of Catalogue of Palaearctic Coleoptera (Löbl & Smetana 2011) *C. arbustorum* was added as a nomen oblitum by Manfred Döberl (2012, pers. comm.). However, the nomen oblitum was not established in accordance with the Code and must therefore be considered an invalid act and thus the name still threatens the well-established name *L. curtus*. In order to preserve the stability, we proposed to use Art. 23.9 of ICZN (1999), considering *C. arbustorum* as nomen oblitum and *L. curtus* as nomen protectum. To our knowledge, *C. arbustorum* was not treated as a valid name after 1900 and, thus, the conditions in Art. 23.9.1.1 are met. The mention of a name in the list of species described by Gistel (Strand 1917) cannot, in our opinion, be taken as use of a valid name in the sense of the Code. To meet the conditions of Art. 23.9.1.2, we hereby provide supporting references of 25 papers, published by at least 10 authors in the immediately preceding 50 years, and encompassing a span of not less than 10 years, where *Longitarsus curtus* is used as valid species name: Bezděk (2021a), Bieńkowski (2004), Biondi (1989), Biondi et al. (1995), Borowiec et al. (2011), Čížek (2006), Čížek & Doguet (2008), Döberl (1995b, 2010b), Doguet (1994, 2014), Farina (2021), Geiser (2001), Gruev (1979, 1992), Gruev & Döberl (1997, 2005), Gruev & Tomov (2007), Gruev et al. (1993), Rheinheimer & Hassler (2018), Silfverberg (2004), Schmitt & Rönn (2011), Warchałowski (1996, 2003, 2010b).

764 *Haltica aerata* Dalman, 1823 is an available name which threatens the well-established name *Dibolia femoralis* L. Redtenbacher, 1848. In order to preserve the stability, we propose to use Art. 23.9 of ICZN (1999), considering *Haltica aerata* as nomen oblitum and *Dibolia femoralis* as nomen protectum. To our knowledge, *Haltica aerata* was not treated as a valid name after 1900 and, thus, the conditions in Article 23.9.1.1 are met. To meet the conditions of Art. 23.9.1.2, we hereby provide supporting references of 25 papers, published by at least 10 authors in the immediately preceding 50 years, and encompassing a span of not less than 10 years, where

Dibolia femoralis is used as valid species name: Aslan et al. (1999), Bezděk (2021a), Bieńkowski (2004), Biondi et al. (1995), Borowiec et al. (2011), Čížek (2006), Čížek & Doguet (2008), Döberl (1995b, 2010b), Doguet (1994, 2014), Farina (2015), Geiser (2001), Gruev (1979, 1992), Gruev & Döberl (1997, 2005), Gruev & Tomov (2007), Gruev et al. (1993), Mohr (1981), Nadein (2015), Rheinheimer & Hassler (2018), Schmitt & Rönn (2011), Warchałowski (2003, 2010b).

765 *Thyamis foudrasi* Crotch, 1871 (usually placed in synonymy of *Longitarsus ganglbaueri ganglbaueri* Heikertinger, 1912) is an available name which threatens the well-established name *L. ganglbaueri ganglbaueri* Heikertinger, 1912 and, due to homonymy, also *L. foudrasi* Weise, 1893. In order to preserve the stability, we proposed to activate the Art. 23.9 of ICZN (1999), considering *Thyamis foudrasi* Crotch, 1871 as nomen oblitum and both *L. ganglbaueri* Heikertinger, 1912 and *L. foudrasi* Weise, 1893 as nomina protecta. To our knowledge, *Thyamis foudrasi* Crotch, 1871 was not treated as a valid species name after 1900 and, thus, the conditions in Article 23.9.1.1 are met. To meet the conditions of Art. 23.9.1.2, we hereby provide supporting references of 25 papers, published by at least 10 authors in the immediately preceding 50 years, and encompassing a span of not less than 10 years, where *L. ganglbaueri* Heikertinger, 1912 and *L. foudrasi* Weise, 1893 are used as valid species names. The references for *L. ganglbaueri*: Aslan et al. (1999), Bezděk (2021a), Bieńkowski (2004), Biondi et al. (1995), Borowiec et al. (2011), Brelih et al. (2003), Čížek (2006), Čížek & Doguet (2008), Cox (2007), Döberl (1995b, 2010b), Doguet (1994, 2014), Gruev (1979, 1992), Gruev & Döberl (1997, 2005), Gruev & Tomov (2007), Gruev et al. (1993), LeSage (1988), Lopatin et al. (2004), Schmitt & Rönn (2011), Warchałowski (1996, 2003, 2010b). The references for *L. foudrasi*: Aslan et al. (1999), Baselga & Novoa (2006), Bezděk (2021a), Bieńkowski (2004), Biondi et al. (1995), Borowiec et al. (2011), Čížek (2006), Čížek & Doguet (2008), Döberl (1995b, 2010b), Doguet (1994, 2014), Gruev (1979, 1992), Gruev & Döberl (1997, 2005), Gruev & Tomov (2007), Gruev et al. (1993), Lopatin et al. (2004), Salvi et al. (2019), Schmitt & Rönn (2011), Vig (2002), Warchałowski (1996, 2003, 2010b).

Taxa removed from the catalogue

766 *Crepidodera teotiai* Kacker, 1969, listed in the previous Edition (Döberl 2010b) as Alticinae incertae sedis, was described from Kanpur in Uttar Pradesh state and thus it is extralimital.

767 *Mniophila muscorum* var. *seriatopunctata* Roubal, 1932: 130 was listed in the previous Edition (Döberl 2010b) as a synonym of *M. muscorum* (Koch, 1803). However, Roubal (1932) listed only the name without any description. We believe the description of this taxon does not exist, therefore we remove var. *seriatopunctata* from the Catalogue.

768 *Monomacra* Chevrolat, 1836. The only *Monomacra* species mentioned in the previous Edition (Döberl 2010b) is *M. nisotroides* (S.-H. Chen, 1939). However it was synonymized with *Hermaeophaga perraudieri* (Allard, 1889) by Medvedev (2009e). Therefore, the genus is removed from the current Catalogue.

Resurrection

769 *Longitarsus scrophulariae* Ohno, 1968 was treated as a homonym of *L. suturatus* var. *scrophulariae* Normand, 1937 and replaced by *L. okushirensis* Gruev, 1996. In the present Catalogue, *L. scrophulariae* Ohno, 1968 **stat. res.** is considered a valid name as *L. suturatus* var. *scrophulariae* Normand, 1937 was described in a work also containing taxa with the status of subspecies, and therefore is infrasubspecific.

Availability of varieties and forms

770 The following names are deemed infrasubspecific because they were published as a fourth name after a trinomen (ICZN 1999, Art. 45.5):
Batophila acutangula kamikochiana forma *hira* Nakane, 1958: 313
Mniophila muscorum var. *wroblewskii* forma *fortepunctata* Horion, 1939: 142
Oedionychus limbata lateralis var. *cruciata* Kocher, 1959: 67
Podagrica menetriesii discedens forma *minor* J. R. Sahlberg, 1913b: 100

771 The following varieties and form are deemed infrasubspecific according to Article 45.6.1 (ICZN 1999), since a subspecies concept was used in the same work, and according to available information, these names were not validated prior to 1985 (ICZN 1999, Art 45.6.4.1):
Aphthona crassipes var. *alata* Jacobson, 1901a: 93
Aphthona crassipes var. *heringi* Heikertinger, 1944: 48
Aphthona depressa var. *subparallela* Normand, 1937: 132
Aphthona vaulogeri var. *cyrenaica* Heikertinger, 1944: 80
Chaetocnema concinna var. *nitidicollis* Jacobson, 1901a: 91
Chalcoides aurata var. *subunicolor* Pic, 1918d: 18
Chalcoides aurea var. *purpurascens* Pic, 1918d: 18
Chalcoides fulvicornis var. *libanicola* Pic, 1918d: 18
Chalcoides gemmata var. *semicyanea* Pic, 1918d: 19
Chalcoides gemmata var. *semiviridis* Pic, 1918d: 19
Chalcoides plutus var. *punctatithorax* Pic, 1918d: 18
Crepidodera nobilis var. *interstitialis* J. Daniel, 1904b: 281
Crepidodera rhaetica var. *spectabilis* J. Daniel, 1904b: 273
Longitarsus lateralis var. *unionis* J. R. Sahlberg, 1913c: 257
Longitarsus suturatus var. *scrophulariae* Normand, 1937: 132
Oedionychis limbata var. *schrammi* Kocher, 1959: 79

Podagrica malvae var. *cedricola* J. R. Sahlberg, 1913c: 251
Podagrica malvae var. *convexicollis* J. R. Sahlberg, 1913c: 251
Podagrica malvae forma *subaptera* J. R. Sahlberg, 1913c: 251
Psylliodes inflata var. *completa* Normand, 1937: 137
Sangariola fortunei var. *unicoloripennis* Chûjô, 1941b: 173
Sphaeroderma apicale var. *immaculithorax* Nakane, 1958: 311

772 The following varieties were proposed in works using also concept of subspecies, however, they were validated before 1985 and thus are deemed to be subspecific from their original publication (ICZN 1999, Art. 45.6.4.1):

Crepidodera impressa var. *obtusangula* J. Daniel, 1904b: 249 was adopted by Biondi (1982b) as subspecies.

Longitarsus nigrocillus forma *atriplaga* Heikertinger, 1913a: 29 was adopted by Kocher (1958) as subspecies.

Minota obesa var. *minima* Heikertinger, 1912e: 60 was adopted by Bechyné (1956b) as subspecies.

Orestia alpina var. *hungarica* Heikertinger, 1910c: 43 was adopted by Bechyné (1956b) as subspecies.

Phyllotreta chinensis forma *shirahatai* Jind. Madar, 1959: 5 was adopted by Chûjô & Kimoto (1961) as subspecies.

Psylliodes chrysocephala var. *angulicollis* Heikertinger, 1926: 119 was adopted by Bechyné (1956b) as subspecies.

Psylliodes cuprea forma *testaceoconcolor* Heikertinger, 1926: 122 was adopted as subspecies by Bechyné (1956b) and Leonardi (1971), and as species by Furth (1984).

New geographical records

773 *Aphthona lutescens* (Gyllenhal, 1813). UNITED ARAB EMIRATES: Wadi Shawkah, 9.iv.–24.vi.2007, water trap, 1 ♀, A. van Harten leg., M. Döberl det. (collection of J. Bezděk, Brno, Czech Republic). New species for United Arab Emirates.

774 *Chaetocnema afghana* Gruev, 1988. UNITED ARAB EMIRATES: Wadi Bih dam, 19.ii.–30.iv.2008, light trap, 1 spec., A. van Harten leg., M. Biondi det. (collection of J. Bezděk, Brno, Czech Republic); Wadi Bih dam, 16.xii.2009–8.iii.2010, water trap, 1 ♂ 4 ♀♀, A. van Harten leg., M. Biondi det. (collection of J. Bezděk, Brno, Czech Republic). New species for United Arab Emirates.

775 *Chaetocnema ljuba* Bechyné, 1955. OMAN: Darbaat, 18.–21. vi. 2003, 1 ♂ 1 ♀, leg. R. Červenka, M. Biondi det. (collection of J. Bezděk, Brno, Czech Republic). New species for Oman.

776 *Chaetocnema tarsalis* Wollaston, 1860. OMAN: 10 km W of Al Mughsayl, 2.ix.2018, 1 ♀, K. Orszulik leg., J. Bezděk det. (collection of J. Stanovský, Ostrava, Czech Republic). New species for Oman.

777 *Epitrix abeillei* (Bauduer, 1874). YEMEN: Haraz Mts., S slope of Al-lan Mt., 2600 m, 23.vi.2010, 1 ♀, V. Hula & J. Niedobová leg., M. Döberl det. (collection of J. Bezděk, Brno, Czech Republic). New species for Yemen.

778 *Longitarsus* (*Longitarsus*) *buettikeri* Doguet, 1985. UNITED ARAB EMIRATES: Hatta, 30.i.–26.ii.2006, light trap, 1 ♂, A. van Harten leg., M. Döberl det. (collection of J. Bezděk, Brno, Czech Republic); Hatta, 24°49′N 56°07′E, 17.–24.v.2006, light trap, 1 ♂, A. van Harten leg., M. Döberl det. (collection of J. Bezděk, Brno, Czech Republic). YEMEN: Sana'a env. Bait Bows dam, 15°16,398′N 44°11,634′E, 2300 m, 21. vi. 2010, 1 spec., V. Hula & J. Niedobová leg., M. Döberl det. (collection of J. Bezděk, Brno, Czech Republic). New species for United Arab Emirates and continental Yemen.

779 *Longitarsus* (*Testergus*) *hittita* Biondi, 1996. ARMENIA: Yerevan env., Nubarashen, 13.v.1988, 3 spec., A. S. Konstantinov leg. and det. (collection of A. S. Konstantinov, Washington, USA). New species for Armenia.

780 *Phyllotreta fallaciosa* Heikertinger, 1941. YEMEN: Kawr Sayban Mt., NW Al Mukalla, 575 m, 14°37′N 49°03′E, at light, 29.iii.2007, 1 ♂, M. Rejzek leg., M. Döberl det. (collection of J. Bezděk, Brno, Czech Republic). New species for Yemen.

781 *Phyllotreta procera* (L. Redtenbacher, 1848). YEMEN: Sana'a Gov., Wadi Dhahr (wasted gardens), 15°26.4′N 44°07.5′E, 2255 m, 2 spec., 2.xi.2010, J. Bezděk leg. M. Döberl det. (collection of J. Bezděk, Brno, Czech Republic); Wadi as Sudd, 10 km W of Ma'rib, 15°24′N 45°16′E, 1117 m, 8.x.2005, 1 ♂, P. Kabátek leg., M. Döberl det. (collection of J. Bezděk, Brno, Czech Republic). New species for continental Yemen.

782 *Psylliodes hospes* Wollaston, 1854. OMAN: Al Batinah S. Gov., Al Jabal al Akhdar, Wadi Bani Habib, 23°04,9′N 57°36,5′E, 1900 m, 1.xi.2019, 1 spec., J. Hájek & A. Reiter leg., J. Bezděk det. (NMPC). New species for Oman.

Comments on distribution

783 *Aphthona potanini* Konstantinov, 1998 was described from two specimens collected in Fu-byan-ho valley but the Chinese province was not specified. The locality refers to Fubian River valley placed between the villages Kuangjiawan and Fubianxiang in Sichuan province.

784 *Chaetocnema constricta* Ruan, Konstantinov & Yang, 2014 is widely distributed in China. Takizawa (2021) published many specimens from Japan but pointed out the difference in aedeagus of Japanese specimens. The identity of Japanese populations requires more comprehensive study, and Japan is not listed in the present Catalogue.

Chrysomelidae: Cryptocephalinae: Clytrini

M. Schöller

Newly established taxon

785 Acolastina Schöller, **subtrib. nov.**

Type genus: *Acolastus* Gerstäcker, 1855.

Description. Body cylindrical, glabrous or covered with short hairs. Head hypognathous, eyes large, strongly convex and protruding, canthus shallow but distinct. Frontoclypeus broad and short, more or less separated from frons, labrum transverse, apically concave. Mandibles in male simple or strongly enlarged. Antennae filiform. Prothorax transverse, pronotum with rounded angles, posterior border of pronotum with short median truncate lobe, hind margin of pronotum bordered, untoothed. Pronotum in most species setose, with or without lateral margin. Mesoscutellum large, not elevated, triangular, with rounded apex. Prosternal process narrow and elevated between coxae, basally pointed. Procoxal cavities closed behind. Elytron with base turned upward forming ridge, in most species with distinct denticle. Elytral punctation confused, in most species elytra setose. Epipleurae indistinct. Mesothoracal intercoxal projection narrow. Hind wings developed, transparent, with costales, distal and median veins, and radiale weakly chitinised, radial triangle subtriangular. Abdomen with simple pygidium and ventrites. Rectal apparatus composed of five sclerites, ventrally two apodemes, dorsally three sclerites, two dorsal apodemes narrow and oriented dorso-ventrally, without broadened terminations, dorsal central plate transverse to square. Male femora simple or strongly enlarged, tibiae with apical spur. Tarsi with claws simple. Male genitalia with Y-shaped tegmen, tegminal manubrium present, tergal apodeme Y-shaped and simple, median lobe ventrally prolonged in most species. Female genitalia with hook-shaped spermatheca.

Diagnosis. Head with large protruding eyes. Prothorax transverse, hind margin of pronotum bordered, untoothed, with short median truncate lobe. Mesoscutellum large, not elevated, triangular. Prosternal process narrow and elevated between coxae, basally pointed. Rectal apparatus with three dorsal and two ventral sclerites, dorsal apodemes narrow, not wider than rectum, dorsal and ventral sclerotisations of the lateral fold present.

Comparison. This subtribe takes a basal position within Clytrini and can be easily distinguished from all other Clytrini tribes by the filiform antennae and the narrow dorsal dorso-ventrally oriented apodemes of the rectal apparatus. In the related tribes Mylassini Gómez-Zurita & Cardoso, 2021, and Pachybrachini Chapuis, 1874 the antennae are also filiform, these tribes differ from Acolastina by the broad prosternal process and a single or two dorsal sclerites of the rectal apparatus, respectively (see Monrós 1949, Schöller 2008). In Pachybrachini the mesoscutellum is elevated and truncate, more rectangular. In Mylassini the eyes are weakly protruding only, the posterior border of pronotum has a long median truncate lobe and the elytra are with regular rows of punctures.

Discussion. The systematics of the Cryptocephalinae were recently studied by Gómez-Zurita & Cardoso (2021), who concluded the Cryptocephalini are paraphyletic due to the inclusion of the genera *Mylassa* and *Acolastus*. Gómez-Zurita & Cardoso (2021) proposed the tribe Mylassini but did not study the genus *Acolastus*. The unique structure of the rectal apparatus is uniform in Clytrini and is also present in *Acolastus* (Schöller 2008). The characters given under the description paragraph and the results of a previous, unpublished cladistic analysis by the author, suggest a position within the Clytrini sensu the classification of Cryptocephalinae by Gómez-Zurita & Cardoso (2021).

Chrysomelidae: Cryptocephalinae: Clytrini
J. Bezděk

New taxon

786 *Labidostomis* (*Labidostomis*) *luristanica* Bezděk, **sp. nov.** This name was first published as *L. elegans* var. *luristanica* Pic, 1920 but is deemed infrasubspecific according to Article 45.6.1 (ICZN 1999), since a subspecies concept was used by Pic (1920a), and the name was not validated prior to 1985 (ICZN 1999, Art 45.6.4.1). Rapilly (1984) treated var. *luristanica* as synonym of *L. shirazica* Lopatin, 1979. Warchałowski (2004b) used *L. luristanica* as valid species in the key to species of the *Labidostomis elegans* species-group. Bezděk & Regalin (2015) and Bezděk (2018a) mistakenly interpreted the Article 45.5.1 (ICZN 1999) and assigned the authorship to Warchałowski (2004b). However, Warchałowski (2004b) cannot be considered the author of the species because in his work *L. luristanica* is not explicitly indicated as intentionally new (Article 16.1, ICZN 1999) and the name-bearing types are not explicitely fixed (Article 16.4, ICZN 1999). So this species is formally described only now. The detailed description was given by Bezděk (2018a). Type material: Holotype ♂ (NMPC): "IRAN, Kordestan prov. | Sanandaj env. | 29.iv.2006 | D. Farbiak leg. || ex coll. F. Kantner | National Museum | Prague, Czech Republic || Labidostomis (s. str.) | shirazica Lopatin | det. F. Kantner 2007". Paratypes: 2 ♂♂ 2 ♀♀ (NMPC), same data as in holotype; 1 ♂ (collection of J. Bezděk, Brno, Czech Republic): "IRAN, Kordestan prov. | Sanandaj env. | 29.iv.2006 | D. Farbiak leg."; 1 ♂ (NMPC): "IRAN, Zanjan prov., 11.-12.V.2006 | 7 km E Sa'id Abad | (steppe, stream; at light) | 36°23,3'N 48°11,7'E; 1635 m | Jiří Hájek & Pavel Chvojka leg."; 2 ♂♂ 1 ♀ (collection of D. Frank, Praha, Czech Republic): "IR, Kordestan prov., 30km | S Sanandaj, Danan env. | 1350 mnm | 35°01'18"N, 46°59'06"E, | 14 V 2016 leg. David Frank"; 1 ♀ (MNHN): "v.

Bodemeyer | Persien | Luristan || elegans || v. luristanica | Pic || v. luristanica Pic || type || TYPE || Museum Paris | Pic || Labidostomis | shirazicus Lopatin | M Rapilly dét. 1981"; 1 ♀ (NMEG): "IRAN, Prov. Chahar | Mahali-o-Bakhtiyari | 32°23′57″N, 50°34′17″E | Haruni, Bachtal | 10.V.2012, 2300 m | leg.: J. Weipert"; 1 ♀ (collection of F. Fritzlar, Jena, Germany): "IRAN, Pr. Chahar Mahali- | o-Bakhtiyari, Harunie, | stream, dry slopes, 2300m | 32°23′57″N, 50°34′17″E | 10.V.2012, leg. D. Frenzel". Distribution: Iran (Lorestan, Chaharmahal and Bakhtiari, Kurdistan, Zanjan provinces). Differential diagnosis: *Labidostomis luristanica* has brownish labrum medially darkened and large elytral spot always reaching the scutellum. Similar species *L. shirazica* Lopatin, 1979 and *L. regalini* Bezděk, 2018 differ in the structure of internal sclerites of aedeagus (Figs 22–25 in Bezděk 2018a). The females of *L. luristanica* have very long ductus spermathecae, ca six times as long as vasculum (Fig. 10 in Bezděk 2018a), while ductus spermathecae in females of *L. shirazica* and *L. regalini* is much shorter, ca 2–3 times as long as vasculum (Figs 11–12 in Bezděk 2018a).

Type species

787 *Melitonoma* Chevrolat, 1836. The type species is *Cryptocephalus* pallens Fabricius, 1787 (now classified in *Diapromorpha* Lacordaire, 1848) by monotypy as it is the only available name listed in Dejean's (1836) catalogue. Monrós (1953) designated *Clytra decempunctata* Olivier, 1808 as type species of *Melitonoma*, but this taxon is listed as a species inquirendum in Dejean's (1836) catalogue and is deemed not to be originally included (ICZN 1999, Art. 67.2.5). A request to the Commission is necessary to change the type species of *Melitonoma*. Meanwhile the current usage is to be maintained.

New synonymy

788 *Clytra atraphaxidis atraphaxidis* (Pallas, 1773) = *C. atraphaxidis maculifrons* Zubkov, 1833 **syn. nov.** Both taxa can be distinguished only by the colouration of the head, which, in view of colour variability throughout the entire range, cannot be used as a character to delimit subspecies. Moreover, the distributional areas of both taxa are largely overlapping. The syntype of *C. maculifrons* Zubkov, 1833 deposited in MZH was examined: "♀ || Karelin. || Astrabad. || Coll. Mannerh. || [small red blank label]".

789 *Clytra laeviuscula* ab. *biinterrupta* Reitter, 1913 is listed under *Clytra valeriana valeriana* Ménétriés, 1832 based on the examination of two relevant Reitter's specimens (females) deposited in HNHM.

790 *Macrolenes dentipes* (Olivier, 1808) = *Cyaniris picturata* Normand, 1950 **syn. nov.** = *Smaragdina tunisea* Warchałowski, 2000 **syn. nov.** *Cyaniris picturata* was described based on one female from Tunisia (Normand 1950a). The deposition of the holotype is unknown, however, the description undoubtedly refers to *Macrolenes dentipes*.

Due to homonymy with *Cyaniris picturata* Jacoby, 1901, Warchałowski (2000c) replaced Normand's name with *Smaragdina tunisea*.

791 *Clythra sennariensis* Lacordaire 1848: 224 (now in *Paraclytra*), used as valid name in recent publications (e.g. Bezděk & Kantner 2010), was proposed under a variety denoted by a letter and is thus unavailable (see comment 1126). The name *P. sennariensis* was validated by Medvedev (1971b) who treated it as a subspecies. The oldest available name for the taxon is *P. rufitarsis* (Lacordaire, 1848). Consequently, the following synonymy is proposed: *Paraclytra rufitarsis* (Lacordaire, 1848) = *P. aethiopica* (L. N. Medvedev, 1971) **syn. nov.** = *P. plicatipennis* (Pic, 1939) **syn. nov.** = *P. sennariensis* (L. N. Medvedev, 1971) **syn. nov.**

Primary homonyms

The following names are junior homonyms that may be preserved because their senior homonyms were placed in other genera before 1900 (ICZN 1999, Art. 23.9.5). These cases require an application to the Commission. Meanwhile the usage of those names is to be maintained.

792 *Clytra bodemeyeri arabica* L. N. Medvedev, 1961, junior primary homonym of *Clythra arabica* Lacordaire, 1848 (now in *Tituboea* Lacordaire, 1848).

793 *Clythra* (*Labidostomis*) *limbata* Lacordaire, 1848 (now in *Labidostomis* Chevrolat, 1836), junior primary homonym of *Clytra limbata* Steven, 1806 (now in *Smaragdina* Chevrolat, 1836).

Spelling

794 Roubal (1948a) described new form of *Coptocephala unifasciata* using two different spellings: *triornata* on p. 33 and *trinotata* on p. 34. I hereby fix *triornata* as the correct original spelling in accordance with the principle of the First Reviser (ICZN 1999, Art. 24.2.3).

Availability of varieties and forms

795 The following varieties and forms are deemed infrasubspecific according to Article 45.6.1 (ICZN 1999), since a subspecies concept was used in the same works, and according to available information, these names were not validated prior to 1985 (ICZN 1999, Art 45.6.4.1):

Antipa biguttata var. *legionis* Kocher, 1959: 12
Chilotoma musciformis var. *orientalis* Jacobson, 1901b: 107
Clytra arida var. *ehnbergi* Jacobson, 1901a: 94
Clytra arida var. *ehnbergi* Jacobson, 1901b: 110
Clytra atraphaxidis var. *defecta* Jacobson, 1901b: 110
Clytra atraphaxidis var. *quinquemaculata* Pic, 1920a: 7
Clytra nigrocincta var. *cyprica* Pic, 1918b: 11

Clytra nigrocincta var. *multipunctata* Pic, 1920a: 7
Clytra nigrocincta var. *subinterrupta* Pic, 1920a: 7
Clytra novempunctata var. *juncta* Pic, 1920a: 7
Clytra valerianae var. *drurei* Pic, 1920a: 7
Clytra valerianae var. *subjuncta* Pic, 1920a: 7
Coptocephala kerimi var. *rubriceps* Pic, 1916b: 10
Coptocephala melanocephala var. *andalusiaca* Pic, 1918d: 17
Coptocephala melanocephala var. *espanoli* Pic, 1933c: 15
Coptocephala melanocephala var. *theryi* Pic, 1918d: 17
Coptocephala pallens var. *octonotata* Chûjô, 1952b: 89
Coptocephala rubicunda forma *bimaculata* Polentz, 1939: 10
Coptocephala rubicunda var. *dalmatina* Pic, 1918d: 17
Coptocephala scopolina forma *bipunctata* Depoli, 1926b: 92
Coptocephala scopolina forma *destituta* Depoli, 1926b: 92
Coptocephala scopolina forma *lateremaculata* Depoli, 1926b: 92
Coptocephala scopolina forma *scutellaris* Depoli, 1926b: 92
Coptocephala scopolina forma *separata* Depoli, 1926b: 92
Coptocephala sexstigma var. *impressiceps* Pic, 1918d: 18
Coptocephala unifasciata forma *reducta* J. Müller, 1948: 80
Cyaniris laboissierei var. *binhana* Pic, 1928c: 34
Gynandrophthalma octomaculata var. *melania* Chûjô, 1952b: 82
Labidostomis arcuata var. *arisi* Pic, 1920a: 7
Labidostomis centromaculata var. *lineata* Pic, 1920a: 6
Labidostomis centromaculata var. *obliterata* Pic, 1920a: 6
Labidostomis centromaculata var. *suturella* Pic, 1920a: 6
Labidostomis elegans var. *inhumeralis* Pic, 1920a: 6
Labidostomis elegans var. *luristanica* Pic, 1920a: 6
Labidostomis hybrida var. *atlasica* Kocher, 1959: 7
Labidostomis rufomarginata var. *reymondi* Kocher, 1959: 7
Labidostomis sibirica var. *fallacissima* Jacobson, 1901b: 106
Labidostomis sibirica var. *transitoria* Jacobson, 1901b: 106
Labidostomis taxicornis forma *latina* Ulrich, 1924: 154
Macrolenes dentipes var. *binotaticollis* Pic, 1916a: 6
Titubaea chobauti var. *semijuncta* Pic, 1918d: 18
Titubaea macropus var. *armeniaca* Pic, 1918d: 18

796 The following varieties were proposed in works using also concept of subspecies, however, they were validated before 1985 and thus are deemed to be subspecific from their original publication (ICZN 1999, Art. 45.6.4.1):

Clytra atraphaxidis var. *asiatica* Chûjô, 1941c: 455 was adopted as species by Gressitt & Kimoto (1961), and as subspecies by Medvedev (1961) and Tiberghien (1969).
Gynandrophthalma aurita var. *hammarstroemi* Jacobson, 1901b: 108 was adopted as subspecies by Medvedev (1982).
Lachnaea lucidicollis var. *peyerimhoffi* Kocher, 1953c: 86 was adopted as species by Codina Padilla (1958).
Titubaea octopunctata var. *hypomelaena* Bedel, 1921: 61 was adopted as species by Kocher (1958).
Labidostomis sibirica var. *transitoria* Jacobson, 1901a: 94 was adopted as subspecies by Medvedev (1982).

Unavailable names

797 X.-P. Li (2013) described *Labidostomis rubimacula* in his MSc. Thesis, which was only submitted to a committee and was not part of any edition. Such thesis is an unpublished work as it does not meet provisions of the Article 8 of the Code (ICZN 1999, 2012b); see also example under the Article 9 in the Amendment. Consequently, *Labidostomis rubimacula* must be considered as unpublished name.

798 The electronic file of an unpublished book by I. K. Lopatin 2005: Zhuki-listoedy podsemeystva Cryptocephalinae (Coleoptera, Chrysomelida) Rosii i sopredelnykh territoriy. [Leaf beetles of the subfamily Cryptocephalinae (Coleoptera, Chrysomelidae of Russia and adjacent territories], written in Russian, exists only in electronic version. It was neither printed nor an ebook-number is available and therefore this publication does not meet the criteria of amended Art. 8.5 (ICZN 2012b). Consequently, *Acolastus gurjevae unicolor* Lopatin, 2005: 118 and *Acolastus pallidus tianshanicus* Lopatin, 2005: 127 described in this book are unavailable. Both taxa were correctly described by Lopatin (2009c).

New geographical records

799 *Acolastus* (*Acolastus*) *x-signatus* (Lopatin, 1985). TURKEY: Kars, Çıldır Gölü, 26.v.1967, 1 ♂ 2 ♀♀, M. Schöller det. (NHMUK). New species for Turkey.

800 *Acolastus* (*Lapsionus*) *praevius* (Suffrian, 1857). MOROCCO: S. Zagora, Qued Draa, 30.312N 5.841W, 25.iv.2018, 1 ♀, Schmid-Egger leg., M. Schöller det. (M. Schöller collection, Berlin, Germany). New species for Morocco.

801 *Aetheomorpha seminigra pumilio* (Lacordaire, 1848). OMAN: Taqah – Nashib rd., 150–500 m, 26.–28.ix.2003, 3 spec., R. Červenka leg., J. Bezděk det. (collection of J. Bezděk, Brno, Czech Republic). Confirmed occurrence in Oman.

802 *Cheilotoma erythrostoma erythrostoma* Faldermann, 1837. ARMENIA: Armenia, Ararat prov., Khosrov res., 2.–3.vi.2004, 1 ♀, J. Bezděk det. (collection of J. Bezděk,

Brno, Czech Republic). HUNGARY: Budafok, 30.v.1922, 2 ♂♂; Budafok, 1 ♂ 1 ♀; Érdliget, 25.v.1936, 1 ♀, E. Csiki leg.; Budapest, Rószadomb, 6.vi.1918, 1 ♀, E. Csiki leg. All specimens from Hungary were identified by J. Bezděk and are deposited in HNHM. Confirmed occurrence in Armenia, new species for Hungary.

803 *Clytra appendicina* (Lacordaire, 1848). ARMENIA: Ararat prov., Khosrov forest reserve, 18.vi.2004, 1 ♀, M. Kalashian leg., J. Bezděk det. (NMPC); Vayots Dzor prov., 4–5 km S of Agarakadzor, 12.viii.2004, 1 ♀, M. Kalashian leg., J. Bezděk det. (NMPC). BOSNIA HERZEGOVINA: Pazarič, 2.vi.1933, 1 ♀, J. Fodor leg., J. Bezděk det. (HNHM). MONTENEGRO: Plevija mun., 5.5 km SE Glibaci, 13.vi.2018, 1 ♀, T. Németh, Cs. & V. Szénási leg., J. Bezděk det. (HNHM). PORTUGAL: Serra de Estrela, NW of Torre, 1600 m, 29.v.2019, 1 ♀, K. Orszulik leg., J. Bezděk det. (collection of K. Orszulik, Frýdek-Místek, Czech Republic). SWITZERLAND: Grisons: Bregaglia, Val Bondasca, 19.vii.1984, 1 ♂, C. Besuchet leg., R. Regalin det. (MHNG); Ticino: Rovio, 1100 m, 17.v.1982, 1 ♀, C. Besuchet leg., R. Regalin det. (MHNG); Valais: Marécottes, 1100 m, 1 ♀, vii.1980, G. Toumayeff leg., R. Regalin det. (MHNG). New species for Armenia, Bosnia Herzegovina, Montenegro and Portugal, confirmed occurrence in Switzerland.

804 *Clytra atraphaxidis atraphaxidis* (Pallas, 1773). KYRGYZSTAN: Issyk-kul, vi.1968, 1 ♂ 1 ♀, Pokorný leg., J. Bezděk det. (NMPC); Aflatun, 31.v.1995, 1 ♂, M. Múčka leg., J. Bezděk det. (NMPC). Confirmed occurrence for Kyrgyzstan.

805 *Clytra medvedevi* Lopatin, 1966. PAKISTAN: Baluchistan, Gawal forest, 16.iii.1930, 2 ♂♂, W. N. Evans leg., J. Bezděk det. (NHMUK); Gawal, x.1931, 1 ♂ 1 ♀, D. Harrison leg., J. Bezděk det. (NHMUK); Waziristan, iii.–iv.1930, 3 ♂♂ 2 ♀♀, rev. G. Palacios leg., J. Bezděk det. (NHMUK). New species for Pakistan.

806 *Clytra novempunctata* Olivier, 1808. CYPRUS: Panagia vill., 50 km N of Pafos, 1100 m, 1.vi.1999, 2 ♂♂ 2 ♀♀, F. & L. Kantner leg., J. Bezděk det. (NMPC). Confirmed occurrence in Cyprus.

807 *Clytra quadripunctata quadripunctata* (Linnaeus, 1758). IRAN: Ostan-e Khorasan prov., Elburs Mts., 60 km E of Minudasht, 800–1200 m, 22.–23.v.2007, 1 ♂, Z. Košťál leg., J. Bezděk det. (NMPC). Confirmed occurrence in Iran.

808 *Clytra weisei* Monrós, 1953. SYRIA: Halabiyah, Eufrat River, 50 km N of Davr az Zawr, 24.iv.2000, 4 ♂♂, F. & L. Kantner leg., J. Bezděk det. (NMPC). Confirmed occurrence in Syria.

809 *Coptocephala destinoi* Fairmaire, 1884. CYPRUS: "Cypern", 1 ♂ (MZLU). The occurence of *C. destinoi* in Cyprus published in the previous Edition was based on the abovementioned specimen (R. Regalin, 2016, pers. comm.).

810 *Coptocephala scopolina kuesteri* Kraatz, 1872. SWITZERLAND: Ticino: Biasca, 15.vii.1951, 1 ♂, J. P. Wolf leg., R. Regalin det. (MHNG). Confirmed occurrence in Switzerland.

811 *Coptocephala scopolina scopolina* (Linnaeus, 1767). SWITZERLAND: Ticino: Salorino, viii.1956, 1 ♂, Lautner lgt. et det. (NHMB); Arzo, 30.vi.1988, 1 ♂ 1 ♀, P. Scherler leg., J. Bezděk det. (NMSO); Rovio, 3.viii.1971, 1 ♂, P. Scherler leg., J. Bezděk det. (NMSO); Castel S. Pietro, 21.vii.1977, 1 ♀, P. Scherler leg., J. Bezděk det. (NMSO). Confirmed occurrence in Switzerland.

812 *Coptocephala simillima* Lodewyckx, 1995. NORTH MACEDONIA: Shar Planina, Popova Shapka, 25.vii.1937, 1 ♂ 2 ♀♀, J. Fodor leg., J. Bezděk det. (HNHM); Skoplje, vii.1937, 1 ♂ 2 ♀♀, J. Fodor leg., J. Bezděk det. (HNHM). New species for North Macedonia.

813 *Coptocephala unifasciata unifasciata* (Scopoli, 1763). SWITZERLAND: Ticino: Salorino, viii.1956, 1 ♂ 1 ♀, Lautner lgt. et det., J. Bezděk rev. (NHMB) Confirmed occurrence in Switzerland.

814 *Diapromorpha dejeanii dejeanii* (Lacordaire, 1848). INDIA: Arunachal Pradesh: Dirang vicinity, 1.–10.vi.2004, 1 ♂, R. Businský leg., J. Bezděk det. (collection of J. Bezděk, Brno, Czech Republic). New species for Arunachal Pradesh (India).

815 *Diapromorpha pallens* (Fabricius, 1787). CHINA: Shaanxi: Qin Ling Shan Mts., road Baoji – Taibai vill., pass 40 km S of Baoji, 21.–23.vi.1998, 1 ♀, Z. Jindra leg., J. Bezděk det. (NMPC). INDIA: Uttaranchal: Rishikesh env., 7.viii.1989, 2 ♂♂, Hiermeier leg., J. Bezděk det. (NMPC). New species for Shaanxi (China), confirmed occurrence for Uttaranchal (India).

816 *Epimela viridicollis* (Jacoby, 1899). INDIA: Uttaranchal: Kumaon, Deoprayak, 11.vii.1991, 1 ♂, Richter leg., J. Bezděk det. (NMPC). New species for Uttaranchal (India).

817 *Labidostomis axillaris* (Lacordaire, 1848). KOSOVO: Mitrovica, Rudnik, 4.vi.1917, 3 ♂♂, E. Csiki leg., J. Bezděk det. (HNHM); "Albania, Ipek" [= Kosovo, Peć], 27.vi.1917, 5 ♂♂, E. Csiki leg., J. Bezděk det. (HNHM). SWITZERLAND: Ticino: „Tessin", 1 ♂ 2 ♀♀, R. Regalin det. (MHNG); Mendrisio, 1 ♂ 2 ♀♀, R. Regalin det. (MHNG). New species for Kosovo, confirmed occurrence in Switzerland.

818 *Labidostomis brevipennis* Faldermann, 1837. JORDAN: S of Tafila, 27.–30.iii.2013, 1 ♀, M. Snížek leg., J. Bezděk det. (collection of F. Fritzlar, Jena, Germany). New species for Jordan.

819 *Labidostomis lucida* (Germar, 1823). SWITZERLAND: Ticino: Alpe Brogoldone, s/Claro, 1900 m, 17.vii.1990, 1 ♂, C. Besuchet leg., R. Regalin det. (MHNG); P. S. Giacomo, 2300 m, 21.vii.1994, 1 ♂, C. Besuchet leg., R. Regalin det. (MHNG); Valais: Foret d'Aletsch, 1900–2000 m, 10.vii.1991, 1 ♂ 2 ♀♀, C. Besuchet leg., R. Regalin det. (MHNG). Confirmed occurrence in Switzerland.

820 *Labidostomis tridentata* (Linnaeus, 1758). SWITZERLAND: Genéve: L'Allondon, 5.v.1961, 2 ♀♀, J. Rappo leg., R. Regalin det. (MHNG). Confirmed occurrence in Switzerland.

821 *Macrolenes dentipes* (Olivier, 1808). EUROPEAN TURKEY: "Turquie d'Europe, Constantinople, Frère Charles Collection & Cordier", 2 ♂♂ 2 ♀♀, J. Bezděk det. (NMPC). New species for European Turkey.

822 *Miochira gracilis* (Lacordaire, 1848). INDIA: Uttaranchal: Haldwani-Kathgodam, 21.–22.vii.2003, 2 ♂♂ 1 ♀, Z. Kejval & M. Trýzna leg., J. Bezděk det. (collection of J. Bezděk, Brno, Czech Republic). New species for Uttaranchal (India).

823 *Physosmaragdina nigrifrons* (Hope, 1843). CHINA: Hongkong, 1 ♀, J. Bezděk det. (NMPC). New species for Hongkong (China).

824 *Smaragdina concolor concolor* (Fabricius, 1792). ANDORRA: Andorra, 5.vi.1961, 4 ♀♀, Hellén leg., J. Bezděk det. (MZH). SWITZERLAND: Ticino: Comano, v.1968, 2 ♀♀, R. Köstlin det. (MHNG); Comano, 15.v.1968, 6 ♂♂ 2 ♀♀, R. Regalin det. (MHNG). New species for Andorra, confirmed occurrence in Switzerland.

825 *Smaragdina discolor discolor* (Solsky, 1881). KYRGYZSTAN: Shamaldy-Sai, 30.v.1995, 1 ♂, M. Múčka leg., J. Bezděk det. (NMPC). Confirmed occurrence in Kyrgyzstan.

826 *Smaragdina diversipes* (Letzner, 1840). SWITZERLAND: Bern: Justistal, 1300–1500m, 22.vii.1979, 1 ♂, Puthz leg., D. Erber det. (MHNG); Solothurn: Weissenstein, 16.viii.1970, 1 ♀, Pesarini leg., J. Bezděk det. (MHNG). Confirmed occurrence in Switzerland.

827 *Smaragdina flavicollis* (Charpentier, 1825). GREECE: Taygetos, Kefisia, 25.iv.1935, 1 ♀, J. Fodor leg., J. Bezděk det. (HNHM); Kephallenia, 1 ♀, Paganetti leg., J. Bezděk det. (HNHM). Confirmed occurrence for Greece.

828 *Smaragdina graeca* (Kraatz, 1872). KOSOVO: Peč, Pečka Banja, 600 m, 18.v.1971, 1 ♀, Papp & Horvatovitch leg., J. Bezděk det. (HNHM). New species for Kosovo.

829 *Smaragdina salicina* Scopoli, 1763. ARMENIA: Bjurakan, v.–vi.1973, 1 ♂, J. Šmelhaus leg., J. Bezděk det. (NMPC). Confirmed occurrence in Armenia.

830 *Smaragdina unipunctata* (Olivier, 1808). ARMENIA: Meghri, 17.iv.2001, 2 ♂♂ 2 ♀♀, Aghababian leg., J. Bezděk det. (collection of J. Bezděk, Brno, Czech Republic). Confirmed occurrence in Armenia.

831 *Tituboea macropus* (Illiger, 1800). MONTENEGRO: Petrovac, 14.vii.1975, 1 ♂, J. Bezděk det. (collection of J. Bezděk, Brno, Czech Republic). Confirmed occurrence in Montenegro.

832 *Tituboea nigriventris* Lefèvre, 1872. KYRGYZSTAN: Uch-Kurgan, Alai Mt., 21.vii.1993, 1 ♂, L. Nádai leg., J. Bezděk det. (HNHM). New species for Kyrgyzstan.

Comments on distribution

833 *Aetheomorpha seminigra pumilio* (Lacordaire, 1848). Specimens from Afghanistan and Jordan published by Lopatin (1966c) and deposited in HNHM pertain to

Melixanthus melanocephalus (Suffrian, 1857). The occurrence of *A. seminigra pumilio* in Afghanistan and Jordan is removed.

834 *Smaragdina concolor concolor* (Fabricius, 1792) was newly recorded from Turkey by Özdikmen et al. (2020) based on one specimen from Osmaniye Prov. I had a possibility to examine the photographs of this specimen and it is a female of some another *Smaragdina* species. Therefore, *Smaragdina concolor concolor* is not listed for Turkey.

835 The following distributional data appeared in the previous version of the Catalogue, however, I was unable to find any published sources. Nevertheless they are retained here as the occurences are possible.

Aetheomorpha hakka Gressitt & Kimoto, 1961 – **A:** CH (SCH)
Aetheomorpha virgula Jacoby, 1908 – **A:** ID (SD)
Aspidolopha thoracica Jacoby, 1892 – **A:** CH (GUA GUI)
Clytra (*Clytra*) *espanoli* Daccordi & Petitpierre, 1977 – **E:** IT (Sardegna)
Clytra (*Clytra*) *quadripunctata puberula* Weise, 1898 – **E:** AN
Clytra (*Clytra*) *quadripunctata quadripunctata* (Linnaeus, 1758) – **A:** TM
Coptocephala bistrinotata (Fabricius, 1803) – **N:** AG
Coptocephala scopolina scopolina (Linnaeus, 1767) – **E:** AN
Coptocephala unifasciata australis L. N. Medvedev, 1965 – **A:** TM
Coptocephala unifasciata unifasciata (Scopoli, 1763) – **E:** MD ME **A:** AR
Diapromorpha pallens (Fabricius, 1787) – **A:** CH (GUA) ID (AP)
Diapromorpha pinguis (Lacordaire, 1848) – **A:** CH (YUN)
Exomis degenevei (Pic, 1932) – **A:** CH (GUX)
Exomis peplopteroides Weise, 1889 – **A:** CH (SHA)
Labidostomis (*Labidostomis*) *centrisculpta centrisculpta* Pic, 1920 – **A:** CH (XIN)
Labidostomis (*Labidostomis*) *curta* Lopatin, 1961 – **A:** TM UZ
Labidostomis (*Labidostomis*) *cyanicornis* (Germar, 1822) – **E:** SL
Labidostomis (*Labidostomis*) *deserta* L. N. Medvedev, 1971 – **A:** TD
Labidostomis (*Labidostomis*) *lucida* (Germar, 1823) – **E:** BH
Labidostomis (*Labidostomis*) *lusitanica* (Germar, 1823) – **E:** IT (Sardegna)
Labidostomis (*Labidostomis*) *metallica metallica* Lefèvre, 1872 – **E:** UZ
Labidostomis (*Labidostomis*) *propinqua* Faldermann, 1837 – **A:** SY
Labidostomis (*Labidostomis*) *rugicollis* Lefèvre, 1872 – **A:** KI UZ
Labidostomis (*Labidostomis*) *schneideri* Weise, 1900 – **A:** AF
Labidostomis (*Labidostomis*) *tridentata* (Linnaeus, 1758) – **E:** MD
Lachnaia (*Lachnaia*) *paradoxa* (Olivier, 1808) – **N:** EG LB
Lachnaia (*Lachnaia*) *sexpunctata* (Scopoli, 1763) – **E:** RU (ST)
Macrolenes dentipes (Olivier, 1808) – **E:** SL

Miochira signaticollis signaticollis (Kollar & L. Redtenbacher, 1844) – **A:** ID (SD)
Physosmaragdina crassipes (Duvivier, 1891) – **A:** ID (UT)
Smaragdina aurita auritoides (Achard, 1923) – **A:** AR
Smaragdina aurita hammarstroemi (Jacobson, 1901) – **A:** CH (LIA NMO)
Smaragdina flavicollis (Charpentier, 1825) – E: LS RU (ST)
Smaragdina flavifrons Gressitt & Kimoto, 1961 – **A:** CH (XIZ)
Smaragdina graeca (Kraatz, 1872) – E: TR
Smaragdina stenroosi (Jacobson, 1901) – **A:** UZ
Smaragdina thoracica thoracica (Fischer von Waldheim, 1842) – **A:** TD
Smaragdina unipunctata (Olivier, 1808) – **A:** JO
Smaragdina xanthaspis (Germar, 1823) – E: RU (CT)
Tituboea macropus (Illiger, 1800) – E: RU (CT)
Tituboea sexmaculata (Fabricius, 1781) – **A:** SY

Chrysomelidae: Cryptocephalinae: Cryptocephalini, Pachybrachini

M. Schöller

New synonymy

836 *Cryptocephalus* subgenus *Sopidus* Jacobson, 1901 = *Aryana* Berti & Rapilly, 1973 **syn. nov.** = *Asionus* Lopatin, 1988 **syn. nov.** = *Asiopus* Lopatin, 1965 **syn. nov.** The subgenus *Sopidus* Jacobson, 1901 was proposed for *Cryptocephalus gamma* Herrich-Schäffer, 1835 (type species by monotypy). Six years later, Jacobson (1907a) added *C. apicalis* Gebler, 1830 to *Sopidus*. Lopatin (1963c) described *C. arnoldii sogdianus* in *Sopidus* and included a total of five species in *Sopidus*. For unknown reasons, in the *Cryptocephalus* subgeneric revision (Lopatin 1965a) the subgenus *Sopidus* was not mentioned and almost all *Sopidus* species were included into the newly established subgenus *Asiopus* Lopatin, 1965.

New replacement names

837 *Cryptocephalus furcatus* Schöller, **nom. nov.** for *Cryptocephalus bifurcatus* L. Medvedev, 1978 (nec *Cryptocephalus triangularis* var. *bifurcatus* Pic, 1917, now in synonymy with *Cryptocephalus triangularis* Hope, 1831).

838 *Cryptocephalus macrodactylus macrocyaneus* Schöller, **nom. nov.** for *Cryptocephalus cyaneus* Ballion, 1878 (nec *Cryptocephalus cyaneus* Olivier, 1791, now in synonymy with *Cryptocephalus violaceus violaceus* Laicharting, 1781).

Spelling

839 Burlini (1970) used two spellings in the original description of a subspecies of *Cryptocephalus* (*Burlinius*) *pygmaeus*: *C. pygmaeus atlanticus* (p. 38) and *C.*

pygmaeus atlasicus (p. 39). The spelling *atlanticus* is fixed here as the original spelling in accordance with the principle of the First Reviser (ICZN 1999, Art. 24.2.3).

840 Tan (1992b) used two spellings *Cryptocephalus lactinus* and *C. lactineus* in the original description. The spelling *lactineus* is fixed here as the original spelling in accordance with the principle of the First Reviser (ICZN 1999, Art. 24.2.3). Lactineus is the correct term for milky-white.

841 Tan (1992b) used two spellings *Cryptocephalus lanpingensis* and *C. langpingensis* in the original description. The spelling *lanpingensis* is fixed here as the original spelling in accordance with the principle of the First Reviser (ICZN 1999, Art. 24.2.3). Lanping is the site in Yunnan to which the specific epithet refers.

842 Lopatin (1991a) used the two spellings *Pachybrachis abditus* and *P. additus* in the original description. Lopatin (2010) used the spelling *abditus* and therefore he is deemed to be the First Reviser according to Art. 24.2.4 (ICZN 1999).

843 Romantsov (2012) used two spellings *Pachybrachis parthicus* and *P. parthiacus* in the original description. The spelling *parthicus* is fixed here as the original spelling in accordance with the principle of the First Reviser (ICZN 1999, Art. 24.2.3).

Protected names

844 For many years *Cryptocephalus gamma* Herrich-Schäffer, 1835 was assumed to have been described in 1829. The year of publication was corrected to 1835 (see Nagel & Grieder 2019). Two previous synonyms *C. sesquistriatus* Krynicki, 1832 and *C. maculipes* Zubkov, 1833 were not used as valid species names after 1900 and are threatening now the well-established name *C. gamma*. The following 25 publications, published by more than 10 authors in the immediately preceeding 50 years, and encompassing a time span of not less than 10 years allow to state *C. gamma* as nomen protectum and *C. sesquistriatus* and *C. maculipes* as nomina oblita based on Art. 23.9.1.2 (ICZN 1999): Bacal et al. (2013), Bardin & Timraleev (2007), Bezděk (2021a), Bieńkowski (2004), Crişan (1995), Dedyukhin (2016a, 2016b, 2016c), Gruev (2005), Gruev & Tomov (1998, 2007), Guskova (2010), Japoshvili & Aslan (2020), Lopatin (1977b), Lopatin et al. (2004, 2010), Maican & Serafim (2012), Moseyko et al. (2019), Schmitt & Rönn (2011), Schöller (2020), Serafim & Maican (2011), Sergeeva & Dedyukhin (2021), Warchałowski (2003, 2010b), Zaytsev et al. (2013).

845 *Cryptocephalus quadriguttatus* O. F. Müller, 1776, a synonym of *Cryptocephalus moraei* (Linnaeus, 1758) is a senior homonym of the well-established name *C. quadriguttatus* C. F. W. Richter, 1820, and was not used as valid species name after 1900. The following 25 publications, published by more than 10 authors in the immediately preceeding 50 years, and encompassing a time span of not less than 10 years allow to state *C. quadriguttatus* C. F. W. Richter, 1820 as nomen protectum and *C. quadriguttatus* O. F. Müller, 1776 as nomen oblitum based on Art. 23.9.1.2 (ICZN

1999): Bezděk (2021a), Bieńkowski (2004), Fritzlar et al. (2022), Gavrilović & Ćurčić (2011), Gruev (2005), Gruev & Tomov (1998, 2007), Guskova et al. (2018), Jakob (1979), Japoshvili & Aslan (2020), Kippenberg (2003), Lopatin (1977b), Lopatin et al. (2004, 2010), Maican & Serafim (2012), Medvedev & Roginskaya (1988), Mohr (1977), Moseyko et al. (2019), Özdikmen (2011), Özdikmen et al. (2014), Samin et al. (2019), Sassi & Kismali (2000), Serafim & Maican (2011), Warchałowski (1991e, 2003, 2010b).

Comments

846 Gmelin (1790: 1714) listed *Cryptocephalus caerulescens* citing Fabricius (1787: 85, n. 10) and copying some details like "habitat in Barbariae plantis, *rubricollis* statura et magnitudine". However, Fabricius (1787: 85) described *Cistela coeruleus* with almost the same details ("habitat in Barbariae plantis Dom Vahl, statura et magnitudo praecedentis [= *ruficollis*]"). The genus *Cistela* is directly following *Cryptocephalus* in Fabricius' book. Therefore, *Cryptocephalus caerulescens* was not described by Gmelin but the diagnosis is a reference to Fabricius (1787), and the name was incorrectly spelled. Consequently, *C. caerulescens* C. R. Sahlberg, 1839 is not a homonym of *C. caerulescens* Gmelin, 1790 but a valid name. Recently, *C. androgyne* Marseul, 1874 was considered the valid name because of a suspected homonymy of *C. caerulescens* C. R. Sahlberg, 1839 with Gmelin's name (e.g. Warchałowski 2010c). In the present Catalogue *C. caerulescens* C. R. Sahlberg, 1839 is treated as valid species and *C. androgyne* as its synonym.

847 *Cryptocephalus etruscus* Heyden, 1892. Sassi (1995) attributed this name to Suffrian (1847b). However, the name was not available because Suffrian (1847b) introduced it in synonymy with *C. ilicis* Olivier, 1808. The name *etruscus* was made available by Heyden (1892a) and consequently the name has his authorship.

848 *Cryptocephalus nigrolimbatus* Jacoby, 1890 was transferred from the subgenus *Burlinius* to *Cryptocephalus* (s. str.) by Duan et al. (2021b). Because this transfer was not based on the study of relevant types, *C. nigrolimbatus* is still classified in the subgenus *Burlinius* in the present Catalogue.

849 *Cryptocephalus turcicus* Suffrian, 1847 is an unavailable name introduced as aberration and Suffrian (1847b) stated explicitly that the taxon is not a proper species. The name *Cryptocephalus turcicus* is available from Weise (1882b) who used it as species and provided its description. However, the oldest available name for this taxon is *C. pistaciae* Suffrian, 1853 and *C. turcicus* Weise, 1882 is proposed as its new synonym. The reversal of precedence based on Art. 23.9.1 (ICZN 1999) is not possible because *C. pistaciae* was used as an available name and was used after 1900 by Gozis (1907: 68, 97, 98).

850 *Buprestis lunatus* Scopoli, 1763 was used by Gozis (1908: 40) as a variety of *Pachybrachis hieroglyphicus* (Laicharting, 1781) and then it was listed with a

question mark by Clavareau (1913) among synonyms of *P. hieroglyphicus*. Scopoli's collection is considered lost, however, the combination of characters given in the original description, i.e. pronotum and elytra completely black and red legs, as well as the information given on the biology (in flowers of *Scabiosa*) is not in accordance with *P. hieroglyphicus* and with the genus *Pachybrachis* at all. For these reasons, *B. lunatus* Scopoli, 1763 is included in the present Catalogue as nomen dubium in Cryptocephalini.

851 The subgenus *Microstilus* was mentioned by Pic (1950) in the description of *Stylosomus marocanus* Pic, 1950. As *S. marocanus* was compared with *S. minutissimus* Germar, 1823 in the differential diagnosis, it is assumed as incorrect subsequent spelling of *Microstylus* Rey, 1883 (which includes *S. minutissimus*) and not a proposal of a new subgenus.

Primary homonym

852 *Cryptocephalus sexmaculatus* Olivier, 1792, junior primary homonym of *Cryptocephalus sexmaculatus* Fabricius, 1781 (now in *Tituboea*), is preserved because its senior homonym was placed in another genus before 1900 (ICZN 1999, Art. 23.9.5). For the purpose of stability of nomenclature, the junior primary homonym is not to be automatically replaced in this case, and the case would require an application to the Commission. Consequently, in the present Catalogue, the name *Cryptocephalus sexmaculatus* Olivier, 1792 is used as valid with *Cryptocephalus obliteratifer* Pic, 1900 as its synonym.

Unavailable names

853 The following taxa were proposed in a MSc. Thesis (X.-P. Li 2013), which was only submitted to a committee and was not part of any edition. Such thesis is an unpublished work as it does not meet provisions of the Article 8 of the Code (ICZN 1999, 2012b); see also example under the Article 9 in the Amendment. Consequently, all taxa proposed in the thesis are unavailable:

Cryptocephalus apressus X.-P. Li, 2013: 38

Cryptocephalus maculatus X.-P. Li, 2013: 44

Cryptocephalus heilongjiangensis X.-P. Li, 2013: 47 (unnecessary replacement name for of *C. manchuricus* Gressit & Kimoto, 1961)

854 The electronic file of an unpublished book by I. K. Lopatin 2005: Zhuki-listoedy podsemeystva Cryptocephalinae (Coleoptera, Chrysomelida) Rosii i sopredelnykh territoriy. [Leaf beetles of the subfamily Cryptocephalinae (Coleoptera, Chrysomelidae of Russia and adjacent territories], written in Russian, exists only in electronic version. It was neither printed nor an ebook-number is available and therefore this publication does not meet the criteria of amended Art. 8.5 (ICZN

2012b), consequently *Cryptocephalus* (*Burlinius*) *dilutellus badakshanicus* Lopatin, 2005: 326 is an unavailable name.

Availability of varieties and forms

855 The following names are deemed infrasubspecific because they were published as a fourth name after a trinomen (ICZN 1999, Art. 45.5):

Cryptocephalus octopunctatus ssp. *apotmetus* var. *bicrux* Jacobson, 1901a: 78
Cryptocephalus raddei ssp. *rosinae* var. *obliteratithorax* Pic, 1904a: 27
Cryptocephalus raddei ssp. *rosinae* var. *raddensis* Pic, 1904a: 27
Cryptocephalus sericeus ssp. *monticola* forma *lydiae* Depoli, 1926a: 14

856 The following varieties and forms are deemed infrasubspecific according to Article 45.6.1 (ICZN 1999), since a subspecies concept was used in the same work, and according to available information, these names were not validated prior to 1985 (ICZN 1999, Art 45.6.4.1):

Cryptocephalus abdominalis var. *limbaticeps* Pic, 1920a: 6
Cryptocephalus alboscutellatus var. *escalerai* Pic, 1908j: 93
Cryptocephalus aureolus forma *coeruleotinctus* Depoli, 1926b: 93
Cryptocephalus coerulescens forma *quadripunctata* Polentz, 1938: 59
Cryptocephalus cordiger var. *completus* Jacobson, 1901b: 111
Cryptocephalus crassus var. *susicus* Kocher, 1959: 25
Cryptocephalus cribratus var. *syriacus* Pic, 1908j: 92
Cryptocephalus cristula forma *suturalis* Depoli, 1926a: 15
Cryptocephalus cristula forma *viridicollis* Depoli, 1926a: 15
Cryptocephalus crux var. *incompletus* Jacobson, 1901b: 112
Cryptocephalus decemmaculatus var. *champenoisi* Pic, 1925b: 9
Cryptocephalus decemmaculatus var. *externesignatus* Pic, 1925b: 9
Cryptocephalus decemmaculatus var. *gardanus* Pic, 1927l: 11
Cryptocephalus decemmaculatus var. *germanicus* Pic, 1927l: 11
Cryptocephalus decemmaculatus var. *goulettei* Pic, 1927l: 11
Cryptocephalus decemmaculatus var. *marionensis* Pic, 1927l: 11
Cryptocephalus decemmaculatus var. *miscerus* Pic, 1927l: 11
Cryptocephalus decemmaculatus var. *morvandicus* Pic, 1925b: 9
Cryptocephalus decemmaculatus var. *prixi* Pic, 1927l: 11
Cryptocephalus decemmaculatus var. *teutonicus* Pic, 1927l: 11
Cryptocephalus dilutellus var. *fedtschenkoi* Jacobson, 1901b: 114
Cryptocephalus frenatus var. *lapponicus* Pic, 1939f: 30
Cryptocephalus fulvus var. *desgozisi* Pic, 1908j: 94
Cryptocephalus hypochoeridis var. *frigidus* Jacobson, 1901b: 113
Cryptocephalus instabilis var. *kiotosinus* Pic, 1908j: 92
Cryptocephalus instabilis var. *multiconnexus* Pic, 1908j: 92
Cryptocephalus japanus var. *posticedepictus* Pic, 1908j: 92

Cryptocephalus limbellus var. *obscurior* Jacobson, 1901b: 112
Cryptocephalus luteosuturalis var. *biquinquemaculatus* Pic, 1928c: 34
Cryptocephalus luteosuturalis var. *fouanus* Pic, 1928c: 34
Cryptocephalus luteosuturalis var. *humeromaculatus* Pic, 1928c: 34
Cryptocephalus mannerheimi var. *medioflavus* Pic, 1904a: 27
Cryptocephalus mannerheimi var. *medioniger* Pic, 1904a: 27
Cryptocephalus marginatus var. *albopectoralis* Pic, 1908j: 93
Cryptocephalus marginatus var. *devaulxi* Méquignon, 1948: 79
Cryptocephalus marginatus var. *sabaudus* Méquignon, 1948: 79
Cryptocephalus multiplex var. *savioi* Pic, 1928c: 33
Cryptocephalus ocellatus var. *montanellus* J. Müller, 1948: 88
Cryptocephalus pallidocinctus var. *latecinctus* Pic, 1908j: 93
Cryptocephalus parvulus var. *devillei* J. Müller, 1948: 85
Cryptocephalus pusillus var. *arcuatetestaceus* Pic, 1926e: 10
Cryptocephalus pusillus var. *ernesti* Pic, 1926e: 10
Cryptocephalus pusillus var. *multinigronotatus* Pic, 1926e: 10
Cryptocephalus pusillus var. *parisiensis* Pic, 1926e: 10
Cryptocephalus pusillus var. *planchezi* Pic, 1926e: 10
Cryptocephalus quadripunctatus var. *martini* Pic, 1908j: 92
Cryptocephalus regalis var. *cuprescens* Jacobson, 1901b: 113
Cryptocephalus reitteri var. *fiumensis* Pic, 1908j: 93
Cryptocephalus rufipes var. *subflavoapicalis* Pic, 1955d: 6
Cryptocephalus sarafschanensis var. *solskyanus* Jacobson, 1901b: 112
Cryptocephalus sarafschanensis var. *stenroosi* Jacobson, 1901b: 112
Cryptocephalus scapulitarsis var. *apicimaculatus* Jacobson, 1901b: 111
Cryptocephalus scapulitarsis var. *correspondens* Jacobson, 1901b: 111
Cryptocephalus sericeus forma *bicolor* Depoli, 1926b: 94
Cryptocephalus sibiricus var. *flavomixtus* Jacobson, 1901b: 113
Cryptocephalus sinuatus var. *anticeincisus* Pic, 1908j: 92
Cryptocephalus stschukini var. *externopunctatus* Jacobson, 1901b: 112
Cryptocephalus taiwanus var. *innotaticollis* Chûjô, 1954d: 207
Cryptocephalus tappesi var. *missisiensis* Pic, 1908j: 92
Cryptocephalus tarsalis var. *banghaasi* Pic, 1908j: 93
Cryptocephalus tarsalis var. *kaschgarensis* Pic, 1908j: 93
Cryptocephalus tarsalis var. *subexquisitus* Pic, 1908j: 94
Cryptocephalus tataricus var. *ephippiatus* Jacobson, 1901b: 111
Cryptocephalus tschimganensis var. *ahngeri* Jacobson, 1901b: 114
Cryptocephalus zambanellus forma *viridiaureus* Depoli, 1926b: 94
Dioryctus longicornis var. *nigrobasalis* Chûjô, 1954d: 170
Dioryctus nigrofasciatus var. *melanopus* Chûjô, 1954d: 158
Dioryctus rufonigrus var. *nigrescens* Chûjô, 1954d: 153

Pachybrachus (sic!) *exclusus* var. *guilhelmi* Pic, 1955d: 6
Pachybrachus (sic!) *hyppophaes* var. *pratsensis* Pic, 1955d: 6
Pachybrachus (sic!) *pallidulus* var. *molloensis* Pic, 1955d: 6
Pachybrachys scripticollis var. *arisi* Pic, 1908b: 45
Pachybrachys simius var. *favarcqi* Pic, 1907k: 186
Pachybrachys tesselatus var. *melanopygus* J. Müller, 1948: 84

857 *Cryptocephalus zubovskyi* var. *flavoirroratus* Jacobson, 1901a: 97 was described in a work also using a concept of subspecies and thus would be infrasubspecific. However, it was adopted as valid name of a subspecies by Chen (1940c) and thus is deemed to be subspecific from its original publication (ICZN 1999, Art. 45.6.4.1).

New geographical records

858 *Adiscus hauseri* (Weise, 1895). BHUTAN: Phuntsholing, 87 km of Thimphu, 1680 m, 22.v.1972, 1 ♂, Nat. Hist. Museum Basel expedition leg., M. Schöller det. (ZMHUB). New species for Bhutan.

859 *Adiscus niger* (S.-H. Chen, 1941). CHINA: Shaanxi: Daba Shan, Quanyuan, 32°2′N 106°7′E, 28.vi.2001, 1 ♀, O. Šafránek leg., M. Schöller det. (collection of J. Hebauer, Rain, Germany). New species for Shaanxi (China).

860 *Adiscus variabilis* (Jacoby, 1890). CHINA: Fujian: Emei Feng, 27°01′N 117°04′E, 8.–9.v.2008, 1 ♂ 2 ♀♀, J. Turna leg., M. Schöller det. (collection of J. Hebauer, Rain, Germany); Guizhou: Leigongshan, Xijang, 1200–1900 m, 29.v.–2.vi. 1997, 1 ♂ 1 ♀, Bolm leg., M. Schöller det. (ZMHUB; collection of Uwe Heinig, Berlin, Germany); Henan: Dalaoshan forest park, 31°05′N 110°95′E, 9.–10.vi.2004, 2 ♂♂ 2 ♀♀, J. Turna leg., M. Schöller det. (collection of J. Hebauer, Rain, Germany). New species for Fujian, Guizhou and Henan (China).

861 *Coenobius piceipes* Gressitt, 1942. CHINA: Yunnan: Xishan Mts., 24.57N 102.38E, 19.v.1993, 2300 m, 1 ♂, Bolm leg. (ZMHUB). New species for Yunnan (China).

862 *Cryptocephalus* (*Burlinius*) *aphthonoides* S.-H. Chen, 1942. CHINA: Yunnan: Weishan Mt., 25.10N 100.21E, 1800–2500 m, 22.–25.vi.1992, 1 ♂, V. Kubáň leg., M. Schöller det. (collection of M. Schöller, Berlin, Germany). New species for Yunnan (China).

863 *Cryptocephalus* (*Burlinius*) *chrysopus* Gmelin, 1790. NORTH MACEDONIA: 30 km N of Veles, v.1969, 1 ♂ 2 ♀♀, G. Wewalka leg., M. Schöller det. (NMW). New species for North Macedonia.

864 *Cryptocephalus* (*Burlinius*) *exiguus amiculus* Baly, 1873. CHINA: Sichuan: Jiushaigou env., 2400 m, 29.vi.1996, 2 ♂♂ 1 ♀, D. Erber leg., M. Schöller det. (ZMHUB). New species for Sichuan (China).

865 *Cryptocephalus* (*Burlinius*) *exiguus variceps* Weise, 1884. GEORGIA: Abchasia, Mt. Gagrensis, vi.1968, 1 ♀, Novotný leg., M. Schöller det. (NMPC). New species for Georgia.

866 *Cryptocephalus* (*Burlinius*) *gangaridus* Lopatin, 1979. NEPAL: Modi Khola, Pothana, 1900 m, 8.vi.1984, 1 ♂, C. Holzschuh leg., M. Schöller det. (ZMHUB). New species for Nepal.

867 *Cryptocephalus* (*Burlinius*) *labiatus* Linnaeus, 1761. GREECE: Corfu, Gouvia, 3.vi.1981, 1 ♂, J. Olsson leg., M. Schöller det. (NHRS); Thrakien, Stravoupoli Komnina, Hotel Nemesis, 41°10′26″N 24°43′27″E, 126 m, 23.–28.v.2016, 1 ♀, A. Link leg., M. Schöller det. (collection of A. Link, Haid, Austria). New species for Greece.

868 *Cryptocephalus* (*Burlinius*) *macellus* Suffrian, 1860. SWITZERLAND: Valais: „Wallis", 1 ♂ 2 ♀♀, Stierlin leg., J. Bezděk det. (SDEI); s/Hohtenn, 1000 m, 21.vi.1991, 1 ♂, C. Besuchet leg., D. Sassi det. (MHNG); Ausserberg, 1000 m, 26.vi.1991, 1 ♂, C. Besuchet leg., D. Sassi det. (MHNG); Branson, 7.viii.1966, 1 ♂, D. Sassi det. (MHNG); Visperterminen, 1310 m, 21.vii.1992, 1 ♂, B. Merz leg., D. Sassi det. (MHNG). New species for Switzerland.

869 *Cryptocephalus* (*Burlinius*) *politus sindonicus* Marseul, 1874. TUNISIA: Serashan, Aman-Kutan, 2000 m, 1.vi.1996, 1 ♂, S. Bajdak leg., M. Schöller det. (collection of M. Schöller, Berlin, Germany). New species for Tunisia.

870 *Cryptocephalus* (*Burlinius*) *pusus* Schöller, 2009. CHINA: Sichuan: Liziping env., 28.vi.–3.vii.1991, 1 ♂, R. Dunda leg., M. Schöller det. (ZMHUB). New species for Sichuan (China).

871 *Cryptocephalus* (*Burlinius*) *pygmaeus vittula* Suffrian, 1848. CRIMEA: Litt. Merid, Massandra, vii.–ix.1951, 2 ♂♂, W. Sachin leg., M. Schöller det. (collection of M. Schöller, Berlin, Germany). Confirmed occurrence in Crimea.

872 *Cryptocephalus* (*Burlinius*) *querceti* Suffrian, 1848. SWITZERLAND: Genève: Corsier, vi.–vii.1990, 1 ♀, C. Besuchet leg., D. Sassi det. (MHNG). Confirmed occurrence in Switzerland.

873 *Cryptocephalus* (*Burlinius*) *scapularis* Suffrian, 1848. SWITZERLAND: Vaud: Grangettes, 25.v.1901, 1 ♂, A. Gaud leg., D. Sassi det. (MHNG); Ticino: Gudo, vi.1993, 1 ♀, C. Besuchet leg., D. Sassi det. (MHNG). New species for Switzerland.

874 *Cryptocephalus* (*Cryptocephalus*) *ajeschae* Schöller, 2006. SAUDI ARABIA: Asir, Wadi Marabah, WSW of Abha, 18°10′18″N, 42°22′12″E, 1197 m, 1 ♀, P. Rapuzzi leg., M. Schöller det. (NMPC). New species for Saudi Arabia.

875 *Cryptocephalus* (*Cryptocephalus*) *albolineatus* Suffrian, 1847. SWITZERLAND: Graubünden: Zermaz, Bergwiese, 2400 m, 3.vii.2014, 1 ♂, V. Gollkowski leg., M. Schöller det. (collection of M. Langer, Niederwiesa, Germany). Confirmation of occurrence in Switzerland.

876 *Cryptocephalus* (*Cryptocephalus*) *anticus* Suffrian, 1848. GREECE: Corfu, North, Roda env., 39°47′46″N 19°46′18″E, 19.v.–2.vi.2012, 1 ♂, Bahr & Winkelmann leg., M. Schöller det. (collection of M. Schöller, Berlin, Germany). SWITZERLAND: Genève: Mategnin, 29.vii.1954, 1 ♂, J. Simonet leg., J. Bezděk det. (MHNG). New species for Greece, confirmation of occurrence in Switzerland.

877 *Cryptocephalus* (*Cryptocephalus*) *bameuli* Duhaldeborde, 1999. RUSSIA: South Tuva, E. Tanu-Ola Mts., Tcha-Ova-Art pass, 1500–1700 m, 15.–24.vi.2004, 4 ♂♂ 4 ♀♀, S. N.Vashchenko leg., M. Schöller det. (collection of J. Hebauer, Rain, Germany); Khadzhiko, Ashe, vi. 1973, 1 ♂, Novotný leg., M. Schöller det. (NMPC); Ossetia, near Mrtuyen, 20.vi.1978, 1 ♂, J. Walda leg., M. Schöller det. (NMPC). CRIMEA: Aluschta, 25.vii.1969, 1 ♂, M. Schöller det. (ZSM). New species for Crimea, confirmed occurrence in Russian Southern European Territory and Southern Siberia.

878 *Cryptocephalus* (*Cryptocephalus*) *biledjekensis* Pic, 1909. ARMENIA: Tavush Province, M4 ca. 7 km S Dilijan, ca. 1700 m, 15.vi.2018, 1 ♀, R. Lukat leg., M. Schöller det. (NMEG). GEORGIA: Telavi Distr., 5 km N Kobadze village, Turlo River valley, 1100 m, 5.–8.vi.2008, 1 ♀, Putchkov leg., M. Schöller det. (collection of J. Hebauer, Rain, Germany). New species for Armenia and Georgia.

879 *Cryptocephalus* (*Cryptocephalus*) *bipunctatus cautus* Weise, 1893. CHINA: Shandong: Tsingtau, 1 ♀, M. Schöller det. (SMF). New species for Shandong (China).

880 *Cryptocephalus* (*Cryptocephalus*) *brancuccii* Lopatin, 1984. INDIA: West Bengal: Darjeeling district, Kalimpong, 1.viii.1984, 1 ♀, C. Holzschuh leg., M. Schöller det. (ZMHUB). New species for West Bengal: Darjeeling district (India).

881 *Cryptocephalus* (*Cryptocephalus*) *caerulescens caerulescens* R. C. Sahlberg, 1839. SYRIA: Slinfeh, Jabal al-Asariyah Mts., 1500 m, 11.–12.vi.1997, 1 ♀, P. Kabátek leg., M. Schöller det. (NMPC). New species for Syria.

882 *Cryptocephalus* (*Cryptocephalus*) *caerulescens pelleti* Marseul, 1874. PORTUGAL: Silves N. E., Barragem do Arade, 13.v.1994, 1 ♂ 1 ♀, C. Bayer & H. Winkelmann leg., M. Schöller det. (collection of M. Schöller, Berlin, Germany). New species for Portugal.

883 *Cryptocephalus* (*Cryptocephalus*) *dimidiatipennis* Jacoby, 1895. INDIA: Uttarakhand: Rishikesh, 23 km from Uttar Kashi, 1400 m, 6.vii.1989, 1 ♂ 1 ♀, A. Riedel leg., M. Schöller det. (collection of H. Kippenberg, Herzogenaurach, Germany). New species for Uttarakhand (India).

884 *Cryptocephalus* (*Cryptocephalus*) *duplicatus* Suffrian, 1847. ARMENIA: Aragats Mt., N. Biurakan, 2050 m, 24.vi.2019, 1 ♂, Skoupý leg., M. Schöller det. (collection of M. Schöller, Berlin, Germany). GREECE: North-East Greece, Didimotichon, 3.vi.1980, 1 ♂ 1 ♀, H. Roehr leg., M. Schöller det. (ZFMK); Evros Province, Koutóvouno, 12.v.2001, 1 ♂, J. Schneider leg., M. Schöller det. (NMPC). Confirmation of occurrence in Armenia and Greece.

885 *Cryptocephalus* (*Cryptocephalus*) *ensifer* Hope, 1831. INDIA: Uttarakhand: Kumaon hills, Nainital, 1900–2200 m, 25.–30.vi.1989, 1 ♂ 1 ♀, R. Schuh leg., M. Schöller det. (collection of M. Schöller, Berlin, Germany). New species for Uttarakhand (India).

886 *Cryptocephalus* (*Cryptocephalus*) *eous* Lopatin, 1952. CHINA: Henan: Funiu Shan, Shirenshan, 15.vi.2007, 1 ♂, L. Mikulenka leg., M. Schöller det. (collection of J. Hebauer, Rain, Germany). New species for Henan (China).

887 *Cryptocephalus* (*Cryptocephalus*) *fatuus* Suffrian, 1860. INDIA: Jammu and Kashmir: Ladakh, viii.1986, 1 ♀, Werner leg., L. N. Medvedev det. (ZMHUB). New species for Jammu and Kashmir (India).

888 *Cryptocephalus* (*Cryptocephalus*) *globicollis* Suffrian, 1847. SWITZERLAND: Freiburg: Lossy, 25.v.1957, 1 ♂, D. Sassi det. (MHNG); Ticino: Morcote, vi.1960, 1 ♀, D. Sassi det. (MHNG). New species for Switzerland.

889 *Cryptocephalus* (*Cryptocephalus*) *hyacinthinus* Suffrian, 1860. CHINA: Shaanxi: Micang Shan, Liping forest park, 32°43′N 106°34′E, 1700–1850 m, 9.–10.vi.2009, 2 ♂♂ 1 ♀, J. Turna leg., M. Schöller det. (collection of J. Hebauer, Rain, Germany); Sichuan: East of Zhangla, 3500 m, 27.vi.1996, 1 ♂, D. Erber leg., M. Schöller det. (ZMHUB). New species for Shaanxi and Sichuan (China).

890 *Cryptocephalus* (*Cryptocephalus*) *imperialis* Laicharting, 1781. CRIMEA: Suma, 29.vi.1956, 1 ♀, M. Schöller det. (collection of M. Schöller, Berlin, Germany). New species for Crimea.

891 *Cryptocephalus* (*Cryptocephalus*) *jansoni* Baly, 1877. CHINA: Beijing: Temple of Heaven, 1 ♀, Tizhurist leg., M. Schöller det. (ZMHUB). New species for Beijing (China).

892 *Cryptocephalus* (*Cryptocephalus*) *luteosignatus* Pic, 1922. CHINA: Beijing: "Peking", 1934, 1 ♂ 1 ♀, N. Filippov leg., M. Schöller det. (collection of M. Schöller, Berlin, Germany). New species for Beijing (China).

893 *Cryptocephalus* (*Cryptocephalus*) *messutati* Kippenberg, 2011. GREECE: Chalkidiki, NE Megáli, Panayia, 150 m, 40°25′39″N 23°42′45″E, 17.vii.2003, 1 ♀, Bahr, Bayer & Winkelmann leg. (collection of M. Schöller, Berlin, Germany). New species for Greece.

894 *Cryptocephalus* (*Cryptocephalus*) *multiplex multiplex* Suffrian, 1860. CHINA: Gansu: Xiahe env., 3000–3200 m, 24.vi.–26.vi.1993, 1 ♀, Heinz leg. (ZMHUB). New species for Gansu (China).

895 *Cryptocephalus* (*Cryptocephalus*) *nitidus* (Linnaeus, 1758). ANDORRA: La Vella, 30.vi.1975, 1 ♀, M. Schöller det. (collection of M. Schöller, Berlin, Germany). GREECE: Thrakien, Katafigio Livaditi, 41°18′46″N 24°40′15″E, 1230 m, 25.v.2016, 1 ♂, A. Link leg., M. Schöller det. (collection of A. Link, Haid, Austria); Kiknos, 41°15′48″N 24°51′57″E, 538 m, 27.v.2016, 1 ♀, A. Link leg., M. Schöller det. (collection of A. Link, Haid, Austria); Kavalla, Monte Lekani, 4.vii.2015, 1 ♀, M. Egger leg. (collection of M. Egger, Bad Enns, Austria). New species for Andorra and Greece.

896 *Cryptocephalus* (*Cryptocephalus*) *nobilis* Kraatz, 1879. CHINA: Henan: Funiu Shan, Baotianman, 33°31′N 111°56′E, 1500–1750 m, 16.–17.v.2009, 1 ♂, J. Turna leg., M. Schöller det. (collection of J. Hebauer, Rain, Germany). New species for Henan (China).

897 *Cryptocephalus* (*Cryptocephalus*) *parvulus* O. F. Müller, 1776. GREECE: Larissa env., 2 km SW of Stomio, Ossa Mt., 300 m, 24.v.2011, 2 ♂♂, A. Kotán, A. Márkus & T. Németh leg., J. Bezděk det. (HNHM). NORTH MACEDONIA: Prespasee, 800 m, 28.vi.1968, 1 ♀, Necker leg., M. Schöller det. (ZSM). New species for Greece and North Macedonia.

898 *Cryptocephalus* (*Cryptocephalus*) *peyroni* Marseul, 1874. EGYPT: Cairo, 1 ♂, M. Schöller det. (SMF). New species for Egypt.

899 *Cryptocephalus* (*Cryptocephalus*) *pistaciae* Suffrian, 1853. SWITZERLAND: Ticino: Stabio, 18.vi.1971, 1 ♂, A. Spälti leg., D. Sassi det. (MHNG). GREECE: Evros Province, Koutóvouno, 12.v.2001, 1 ♀, J. Schneider leg., M. Schöller det. (NMPC). New species for Switzerland, confirmed occurrence in Greece.

900 *Cryptocephalus* (*Cryptocephalus*) *pustulipes* Ménétriés, 1837. CHINA: Henan: Funiu Shan, Baotianman, 33°31′N 111°56′E, 1500–1750 m, 9.–10.vi.2008, 1 ♂, M. Schöller det. (collection of J. Hebauer, Rain, Germany). Sichuan: Jiushaigou env., 2400 m, 29.vi.1996, 1 ♀, D. Erber leg., M. Schöller det. (ZMHUB). New species for Henan and Sichuan (China).

901 *Cryptocephalus* (*Cryptocephalus*) *quadripustulatus* Gyllenhal, 1813. LIECHTENSTEIN: Schaan, 1500 m, 26.vii.2021, 1 ♀, Bureš & Weyrová leg., J. Bezděk det. (collection of L. Bureš, Brno, Czech Republic). New species for Liechtenstein.

902 *Cryptocephalus* (*Cryptocephalus*) *regalis regalis* Gebler, 1830. CHINA: Shandong: Tsingtau, 1 ♀, S. Glaue leg., M. Schöller det. (ZMHUB). New species for Shandong (China).

903 *Cryptocephalus* (*Cryptocephalus*) *sericeus* Linnaeus, 1758. ARMENIA: Aygut env., 1420 m, 40.6752N 45.2042E, 1.–4.vi.2016, 1 ♂, S. Murzin leg., M. Schöller det. (collection of M. Schöller, Berlin, Germany). RUSSIA: Rostov-on-Don reg., Pukhhijakovsky env., 100 m, 47.53N 40.61E, 1.–2.vii.2016, 1 ♂, E. Khachikov leg., M. Schöller det. (collection of M. Schöller, Berlin, Germany). SWITZERLAND: Bern: Kiental, 22.vii.1953, 1 ♀, O. Lundblad leg., M. Schöller det. (NHRS). Confirmed occurrence in Armenia, Russia (European Southern Territory) and Switzerland.

904 *Cryptocephalus* (*Cryptocephalus*) *sexsignatus* Fabricius, 1801. PAKISTAN: Islamabad, Lake Rawal, on *Cotoneaster*, 2.viii.1986, 1 ♂, M. Schöller det. (collection of R. Beenen, Nieuwegein, Netherands). New species for Pakistan.

905 *Cryptocephalus* (*Cryptocephalus*) *signatifrons* Stierlin, 1896. GREECE: Ossa-mountain, Karitsa, 39°51′04″N 22°46′06″E, 50 m, 11.v.2008, 1 ♀, A. Kopetz leg., M. Schöller det. (collection of M. Schöller, Berlin, Germany). New species for Greece.

906 *Cryptocephalus* (*Cryptocephalus*) *solivagus* Leonardi & Sassi, 2001. RUSSIA: Tuva, 50 km north from Kyzyl Ujukski Range, 1650 m, 15.–16.vii.2000, 1 ♂ 1 ♀, R. A. Saldaitis & J. Krueger leg., M. Schöller det. (collection of J. Hebauer, Rain, Germany); Tuva, Akademica Obrucheva Mt. Range, Ulug-Oo River, 52′04″N 95′46″E, 1680 m,

14.vii.1998, 1 ♀, A. Saldaitis leg., M. Schöller det. (collection of J. Hebauer, Rain, Germany). Confirmed occurrence in Southern Siberia (Russia).

907 *Cryptocephalus* (*Cryptocephalus*) *sonani* Chûjô, 1934. CHINA: Sichuan: Gongga Shan, 2600–2900 m, 23.vii.1994, 1 ♂, K.-W. Anton leg., M. Schöller det. (ZMHUB); Moxi-Hailougou, 29.37N 102.07E, 1200–1900 m, 1.–7.vii.1994, 1 ♀, D. Král & J. Farkač leg., M. Schöller det. (collection of M. Schöller, Berlin, Germany). New species for Sichuan (China).

908 *Cryptocephalus* (*Cryptocephalus*) *tricinctus* Kollar & L. Redtenbacher, 1844. CHINA: Yunnan: Xiaguan env., 2100 m, 28.vii.2002, 1 ♂ 1 ♀, S. Murzin & I. Shokhin leg., M. Schöller det. (collection of M. Schöller, Berlin, Germany); Yunnan, 1 ♀, M. Schöller det. (NMPC). New species for Yunnan (China).

909 *Cryptocephalus* (*Cryptocephalus*) *trimaculatus* P. Rossi, 1790. CRIMEA: Litt. Merid, Masandra, vii.–ix. 1951, 1 ♂ 1 ♀, M. Sashin leg., M. Schöller det. (collection of M. Schöller, Berlin, Germany). New species for Crimea.

910 *Cryptocephalus* (*Cryptocephalus*) *violaceus violaceus* Laicharting, 1781. TURKEY: Kaçkar-Dağı, Kavron-valley, ca. 2200 m, 21.viii.1996, 1 ♂, C. Reuter leg., M. Schöller det. (collection of M. Schöller, Berlin, Germany). Confirmation of occurrence in Asian Turkey by male specimen.

911 *Cryptocephalus* (*Homalopus*) *coryli* (Linnaeus, 1758). GEORGIA: Tbilisi area, Telovani env., 41°48.633′N 44°42.755′E, 1.–2.vi.2019, 1 ♀, Z. Malinka leg., J. Bezděk det. (collection of Z. Malinka, Opava, Czech Republic). New species for Georgia.

912 *Cryptocephalus* (*Homalopus*) *lividimanus* Suffrian, 1852. ANDORRA: 7 km from Canillo, along mountainside, 27.vi.1976, 1 ♂, M. Schöller det. (collection of M. Schöller, Berlin, Germany). New species for Andorra.

913 *Cryptocephalus* (*Homalopus*) *quadripunctatus* Olivier, 1808. ANDORRA: Andorra, 5.vi.1961, 2 ♀♀, Hellén leg., D. Sassi det. (MZH). New species for Andorra.

914 *Cryptocephalus* (*Sopidus*) *bidentulus bidentulus* Suffrian, 1854. RUSSIA: Tuva: near Kyzyl, right bank of Biy-Khem River, slopes of the hills, 10.–19.v.2002, 1 ♂ 1 ♀, S. N. Vashchenko leg., M. Schöller det. (collection of J. Hebauer, Rain, Germany). Confirmed occurrence in Russia: Southern Siberia.

915 *Cryptocephalus* (*Sopidus*) *bohemius* Drapiez, 1819. RUSSIA: Tuva: right bank of the Biy-Khem River, Ujukskyi mountains, dry slopes of the hills, 800 m, 22.v.–06.vi.1999, 1 ♂ 2 ♀♀, S. N. Vashchenko leg., M. Schöller det. (collection of J. Hebauer, Rain, Germany). Confirmed occurrence in Russia: Southern Siberia.

916 *Cryptocephalus* (*Sopidus*) *diadochus* Lopatin, 1963. KYRGYZTAN: Chreb.-Magbal, 7.v.1981, 1 ♂ 1 ♀, J. Jelínek leg., M. Schöller det. (NMPC). New species for Kyrgyztan.

917 *Cryptocephalus* (*Sopidus*) *koltzei* Weise, 1887. CHINA: Beijing: Beijing, 4.vi.1989, 1 ♀, 15.iv.1989, 1 ♀, 6.v.1989, 1 ♂ 1 ♀, O. Majzlan leg., M. Schöller det. (NMW). New species for Beijing (China).

918 *Cryptocephalus* (*Sopidus*) *sarafschanensis iliensis* Weise, 1900. CHINA: Xinjiang: bei Tihra, Urumtsi, 2.ii.–19.v.1926, 1 ♂, S. G. Beick leg., M. Schöller det. (ZSM). New species for Xinjiang (China).

919 *Cryptocephalus* (*Sopidus*) *spurius* Lopatin, 1956. AZERBAIJAN: 60 km S of Baku, Aljat, 13.–16.vi.1983, 1 ♀, V. Kubáň leg., M. Schöller det. (NMPC). New species for Azerbaijan.

920 *Cryptocephalus* (*Sopidus*) *stchukini* Faldermann, 1835. RUSSIA: Eastern Tuva, Naryn river, 50°11′N 95°39′E, 1540 m, 22.vi.1996, 1 ♀ R. A. Saldaitis & J. Krueger leg., M. Schöller det. (collection of J. Hebauer, Rain, Germany). Confirmed occurrence in Russia: Southern Siberia.

921 *Melixanthus melanocephalus* (Suffrian, 1857). AFGHANISTAN: Jalalabad, 500 m, 30.iii.1953, 1 spec., J. Klapperich leg., J. Bezděk det. (HNHM). This specimen was published by Lopatin (1966c) as *Aetheomorpha seminigra pumilio* (Lacordaire, 1848). New species for Afghanistan.

922 *Pachybrachis albicans* Weise, 1882. ARMENIA: Vedi-Uchcaozor, 30.v.1989, 1 ♂, V. Šilha leg., M. Schöller det. (NMPC). New species for Armenia.

923 *Pachybrachis distictopygus* Jacobson, 1901. CHINA: Heilongjiang: Harbin, 1.viii.1994, 1 ♀, V. Kuznetsov leg., M. Schöller det. (collection of M. Schöller, Berlin, Germany). New species for Heilongjiang (China).

924 *Pachybrachis excisus* Weise, 1897. GREECE: Thessalia, Kalomira, 39°45′N 21°22′E, 800 m, 19.vii.2014, 1 ♂, Z. Kraus leg., J. Bezděk det. (collection of J. Bezděk, Brno, Czech Republic); Thessalia, Matoneri, 39°45′N 21°22′E, 820 m, 6.-10.vii.2016, 1 ♂ 2 ♀♀, Z. Kraus leg., J. Bezděk det. (collection of Z. Kraus, Mikulovice, Czech Republic). New species for Greece.

925 *Pachybrachis hieroglyphicus* (Laicharting, 1781). SWITZERLAND: Ticino: Bellinzona, 12.vii.1953, 3 ♂♂ 3 ♀♀, O. Lundblad leg., M. Schöller det. (NHRS). Confirmed occurrence in Switzerland.

926 *Pachybrachis jordanicus* Lopatin, 1984. SYRIA: Burqush env., 25 km W of Damascus, 5.v.2000, 1 ♂ 1 ♀, F. & L. Kantner leg., M. Schöller det. (NMPC); 3 km NNW of Rabba, 31°18,045′N 35°43,534′E, 900 m, 27.v.2007, 1 ♀, F. & L. Kantner leg., M. Schöller det. (NMPC); Slenfeh, 1300 m, 2.vi.2000, 1 ♂ 1 ♀, Preiss, Sama, Magnani & Baiocchi leg., M. Schöller det. (collection of M. Langer, Niederwiesa, Germany). New species for Syria.

927 *Pachybrachis mendax mendax* Suffrian, 1860. ARMENIA: Vayotsdzor reg., 7.vi.2000, 1 ♂, Kalashian leg., M. Schöller det. (collection of M. Schöller, Berlin, Germany); Dzhrwezh-Schlucht, ca. 10 km nordöstlich Jeriwan, 28.v.1971, 1 ♂, W. H. Muche leg., M. Schöller det. (MTD); Eriwan, 23.vi.1898, 3 ♂, M. Schöller det. (ZSM). New species for Armenia.

928 *Pachybrachis merkensis* Lopatin, 1968. UZBEKISTAN: Ak-Taš, 4.vii.1976, 1 ♂ 1 ♀, J. Jelínek leg., M. Schöller det. (NMPC). New species for Uzbekistan.

929 *Pachybrachis scripticollis* Faldermann, 1837. AZERBAJIAN: Baku, 20.vii.1979, 1 ♀, L. Medvedev leg., M. Schöller det. (collection of M. Schöller, Berlin, Germany). New species for Azerbajian.

930 *Pachybrachis scriptidorsum* Marseul, 1874. CHINA: Xinjiang: Tian Shan, Urumqui 70 km Houxia, 2000–3000 m, 16.–21.VII.1991, 1 ♂, M. Snížek leg., M. Schöller det. (collection of M. Schöller, Berlin, Germany). New species for Xinjiang (China).

931 *Stylosomus* (*Stylomicrus*) *minutissimus* Germar, 1823. TUNISIA: Ain el Hamaraya, 30 km SO of Ain Draham, 7.vi.1994. 2 ♂♂ 2 ♀♀, F. Kantner leg., M. Schöller det. (NMPC). New species for Tunisia.

932 *Stylosomus* (*Stylosomus*) *bipartitus* Fairmaire, 1867. TUNISIA: Monastir env., 18.vi.1994, 2 ♂♂ 2 ♀♀, F. Kantner leg., M. Schöller det. (NMPC). New species for Tunisia.

933 *Stylosomus* (*Stylosomus*) *flavus flavus* Marseul, 1874. CRIMEA: Crimea, 1 ♂, M. Schöller det. (collection of M. Schöller, Berlin, Germany). New species for Crimea.

934 *Stylosomus* (*Stylosomus*) *x-signum* Pic, 1899b. SPAIN: Mallorca, Alcudia, 1.–14.iv.2012, 3 ♂♂ 2 ♀♀, H. Winkelmann leg., M. Schöller det. (collection of M. Schöller, Berlin, Germany). New species for Balearic Islands.

Chrysomelidae: Cryptocephalinae: Fulcidacini

J. Bezděk

Type species

935 The first valid typification of the genus *Chlamys* Knoch, 1801 was published by Chen (1940a), who listed *Ch. gibbosus* as type species and placed *Ch. tuberosa* as its only synonym and, thus, constituted fixation of *Ch. tuberosa* as type species of *Chlamys* based on Art. 69.2.2 (ICZN 1999). Therefore, the subsequent designation of *Ch. foveolata* Knoch, 1801 by Navajas (1946) is invalid. However, the identity of *Ch. tuberosa* is uncertain being treated as nomen dubium in Chlamisini (Riley et al. 2003), species with uncertain identity possibly belonging to *Neochlamisus* Karren, 1972 (Clark et al. 2004), or a synonym of *Neochlamisus gibbosus* (Fabricius, 1777) (Shin et al. 2012). An application to the Commission to designate *Chlamys foveolata* Knoch, 1801 as the type species should be proposed to conserve the concept of *Chlamys* and its replacement name. See also Bouchard et al. (2024).

Chrysomelidae: Eumolpinae

J. Bezděk, A. G. Moseyko & M. Schöller

Identity of *Cryptocephalus vitis* Fabricius, 1775

936 In the 20th century, *Cryptocephalus vitis* Fabricius, 1775 was usually treated as synonym, aberration, or even subspecies of *Bromius obscurus* (Linnaeus, 1758). However, the placement of *C. vitis* is not consistent in Chrysomelidae literature. For example, Panzer (1799c, 68: 5) depicted the male of *C. coryli* under the name *C. vitis*. Clavareau (1913) listed *C. vitis* as synonym of *C. coryli* (Linnaeus, 1758) in the catalogue of Cryptocephalinae, and Clavareau (1914) in the catalogue of Eumolpinae wrote "*vitis* auctor (nec F.)" under *Bromius obscurus*. Fabricius (1775) referred *C. vitis* to Geoffroy (1762). Geoffroy's book was rejected for nomenclatorial purposes (ICZN 1954, Opinion 228), but subsequently some names of genera were made available (ICZN 1994, Opinion 1754). Geoffroy's description is too short and confusing, and it is impossible to state the correct identity of his species. Fabricius (1775) was the first who made *Cryptocephalus vitis* available. The crucial key to solve the identity of *C. vitis* is to find the original type specimens. The Fabricius' collection in Kiel (Germany) contains three specimens of *C. vitis* which are conspecific with *B. obscurus* (aberration with brown elytra). However, another potential syntype of *C. vitis*, deposited in the Hunterian collection, University of Glasgow, is a male conspecific with *Cryptocephalus coryli* (Linnaeus, 1758). In all five most significant books of Fabricius (1775, 1781, 1787, 1792a, 1801a), he consistently repeated the description of *C. vitis* as "niger, glaber, elytris rufis" which is in agreement with a male of *C. coryli* but contradictory to *B. obscurus* whose dorsum is densely covered by setae. Based on the compilation of various available information, it seems to be clear that the specimens deposited in Kiel are not original type specimens while the specimen in Hunterian collection is with a high degree of probability a syntype. It is well known that Fabricius travelled Great Britain, personally met Banks, Drury and Hunter, used their collections, and many Fabrician types are deposited now in the Natural History Museum in London (Banks collection) and in University of Glasgow (Hunter collection). Zimsen (1964) wrote: "Then, in 1775, his Systema Entomologicae was published, the main part of which was based on his own collectings in England and on English collections in general". Fabricius often mentioned the collections from which the specimens he described originated. Unfortunately, this information is missing in the description of *C. vitis*, which can be explained by the fact that Fabricius attributed this species to Geoffroy and apparently did not intend to describe this species as new to science. Until 1787, *C. vitis* is always mentioned in his books in close proximity to *C. coryli* (Fabricius 1775, 1781, 1787). In addition, the original description is accompanied by a note "An satis distinctus a *C. coryli*?" [= is it quite distinct from *C. coryli*?]. A similar sentence "An a sequenti satis distinctus?" [= is it quite distinct from the following?] can be found in the second book (Fabricius 1781). In his last two books, Fabricius (1792a, 1801a) changed the placement and *C. vitis* is placed next to *B. obscurus*, although the description still includes the word "glaber". Considering all available information summarized above, we are able to reconstruct the genesis as follows. For the description of *C. vitis*, Fabricius most likely used a specimen from Hunter's collection (apparently only one male specimen) and this specimen subsequently remained in Hunter's collection. After that, Fabricius came back to continental Europe and started to build his own collection. Over the years, he mistakenly placed specimens of *B. obscurus* under the name *C. vitis* in his collection. Probably based on that, he placed *C. vitis* very close to *B. obscurus* in his books from 1792 and 1801, respectively. The Hunter collection was not available for study for decades, and the potential syntype of *C. vitis* was studied as late as 2008 by M. Schöller. Subsequent authors who examined the more accessible Kiel collection found under *C. vitis* specimens of *B. obscurus* and thus considered it synonym of the latter. However, in the present Catalogue, *C. vitis* is treated as synonym of *C. coryli*. *Cryptocephalus vitis* is the type species of both *Eumolpus* Illiger, 1798 and *Adoxus* Kirby, 1837, now synonyms of *Bromius* Chevrolat, 1836. The name *Eumolpus* Illiger, 1798 was placed in the Official Index of Rejected and Invalid Generic Names in Zoology (ICZN 2012a, Opinion 2298). *Adoxus* Kirby, 1837 is problematic as it is the type genus of Adoxini Baly, 1863 and *Cryptocephalus vitis* could threaten this usage. If authors discover that a type species was misidentified, they may select, and thereby fix as type species, the species that will best serve stability and universality (ICZN 1999, Art. 70.3). In accordance with Art. 70.3.2 we thereby select *Bromius obscurus* (Linnaeus, 1758) (= *Cryptocephalus vitis* sensu Kirby (1837)) as the type species of *Adoxus* Kirby, 1837.

Chrysomelidae: Eumolpinae

A. G. Moseyko

Newly established taxon

937 Floricolini Moseyko, **trib. nov.**

= Eupalini Verma, Gomez-Zurita, Jolivet & Vig, 2005: 164, nomen nudum

Type genus: *Floricola* Gistel, 1848

Description. Body elongate (in type genus) or elongate-oval, slightly flattened, covered with short hairs. Head without constriction behind eyes. Frontoclypeus not separated from frons. Prothorax transverse, with developed lateral edging, angulated (in type genus) or slightly sinuate laterally. Fore edge of prothorax ("propleurae") ventro-laterally concave. Procoxal cavities transverse, teardrop-shaped, closed behind, with trochantine well-visible in any position of coxa. Prothoracal intercoxal projection narrow, absence of lateral convexities, procoxae without emarginations near it. Mesothoracal intercoxal projection narrow, without lateral convexities, mesocoxae globular, without emarginations. Metacoxae transverse. Abdomen with simple pygidium and ventrites. Wings developed (type genus) or wingless. Femora simple, without tooths, tibiae with two small apical spurs. Tarsi with tarsomere III divided to lobes almost at base. Claws simple. Male genitalia with well developed but short tegmen and well-developed Y-shaped ventral sclerite, reaching (in type genus) or connected with abdominal tergite VIII. Female genitalia with moderately long ovipositor, styli absent (in type genus) or present. Spermatheca J-shaped, with twisted duct and gland, placed on separate projection.

Diagnosis. Head without constriction behind eyes. Prothorax transverse, edged. Procoxal cavities transverse, teardrop-shaped, closed behind, with trochantine well-visible in any position of coxa. Prothoracal intercoxal projection narrow, absence of lateral convexities, procoxae without emarginations near it. Hind femora not increased. Tarsomere III deeply bilobed.

Comparison. This tribe seems to be the most primitive within Eumolpinae and can be easily distinguished from all other eumolpinae tribes by transverse procoxal cavities with well visible trochantine. This character is similar to those in related subfamilies (with controversial position), Synetinae, Megascelinae (Megascelidinae) and Spilopyrinae. Synetinae have the head constricted behind eyes, conic procoxae and open procoxal cavities, Megascelinae has narrow and not edged pronotum, very narrow intercoxal prosternal process, subconic procoxae and increased hind legs, Spilopyrinae has partly connated abdominal ventrites, in many cases with one apical spur on protibia, spermatheca with duct and gland duct placed close to each other and other larval morphology (see Reid 2000).

Discussion. Taxonomic position and characters of the genus *Eupales* Lefèvre, 1885 were discussed by Reid (2000), Verma et al. (2005) and also some characters were shortly discussed later (Moseyko 2017). Verma et al. (2005) proposed the tribe name Eupalini for the genus *Eupales* (now a junior synonym of *Floricola*), but the tribe remained undescribed (Löbl 2010) and is validated here as Floricolini **trib. nov.** The genus *Rhodopaea* Gruev & Tomov, 1968 is also placed to Floricolini sharing most of important characters with *Floricola*.

Comments

938 Names of tribes and their authorships are used in accordance with Moseyko & Kirejtshuk (2013). The tribe Adoxini Baly, 1863 is considered valid with Bromiini Chapuis, 1874 as its junior synonym. The tribe Typophorini Baly, 1865 is considered valid with Nodinini Chen, 1940 as its junior synonym. The tribe Euryopini Lefèvre, 1885 is considered valid with Colasposomini Springlova, 1960 as its junior synonym.

Change of authorship

939 *Aloria* Bryant, 1939 was proposed for two included species without designation of type species and, thus, it does not meet the criteria of availability for genus-group names proposed after 1930 (ICZN 1999, Art. 13.3). Later, *Aloria* was cited in Zoological

Record (Imperial Institute of Entomology, 1940) as monotypic genus, containing only one species, *Aloria purpurea* Bryant, 1939. According to opinion of Evenhuis et al. (2008), such citation is a validation of this taxon with anonymous authorship. Thus, *Aloria* Anonymous, 1940 with type species *Aloria purpurea* Bryant, 1939, by monotypy, is a valid genus.

Spelling

940 *Nodina obliquocostata* L. N. Medvedev, 2011. The original description (Medvedev 2011c) includes two different spellings, *obliquocostata* in the description and *obliquostriata* in the abstract. The spelling *obliquocostata* is selected here as the correct original spelling in accordance with the principle of the First Reviser (ICZN 1999, Art. 24.2.3).

Availability of varieties and forms

941 The following names are deemed infrasubspecific because they were published as a fourth name after a trinomen (ICZN 1999, Art. 45.5):
Lypesthes ater ater forma *tibialis* Ohno, 1958b: 174
Macrocoma cylindrica mesatlantica var. *fauconnieri* Kocher, 1959: 40
Xanthonia placida hachijoensis forma *obscura* Ohno, 1960b: 66
Xanthonia placida hachijoensis forma *scutellata* Ohno, 1960b: 66

942 The following varieties are deemed infrasubspecific according to Article 45.6.1 (ICZN 1999), since a subspecies concept was used in the same work, and according to available information, these names were not validated prior to 1985 (ICZN 1999, Art 45.6.4.1):
Aphilenia parvula var. *nigrescens* Jacobson, 1901b: 117
Basilepta elegans var. *melanura* Chûjô, 1956b: 40
Chloropterus versicolor var. *infuscatus* J. R. Sahlberg, 1913c: 246
Demotina elegans var. *futamon* Nakane, 1958: 306
Eupales ulema var. *viridis* Pic, 1916a: 10
Macrocoma henoni var. *nigrita* Kocher, 1959: 30
Macrocoma henoni var. *reymondi* Kocher, 1959: 30
Macrocoma henoni var. *sarroensis*, Kocher, 1959: 30

943 *Macrocoma henoni* var. *atlasica* Kocher, 1959: 30 was described in a work using also concept of subspecies and thus would be infrasubspecific. However, it was adopted as valid name of a subspecies by Kocher (1969b) and is deemed to be subspecific from its original publication (ICZN 1999, Art. 45.6.4.1).

944 The following varieties and forms are deemed to be infrasubspecific as were first published after 1960 (ICZN 1999, Art. 45.6.3):
Acrothinium gaschkevitschii [sic!] forma *miyakensis* Ohno, 1968b: 35
Acrothinium gaschkevitschii [sic!] forma *obscurum* Ohno, 1968b: 35
Macrocoma bolivari var. *siruensis* Kocher, 1969a: 113

Macrocoma henoni occidentalis var. *lindbergi* Kocher, 1967: 292
Macrocoma henoni occidentalis var. *pardoi* Kocher, 1967: 291
Macrocoma henoni occidentalis var. *susica* Kocher, 1967: 290

New name

945 *Aulexis gracilicornis* Gressitt & Kimoto, 1961: 264 from Hainan is a primary homonym of *Aulexis gracilicornis* Weise, 1922: 455 from the Philippines. A new replacement name *Aulexis gressitti* Moseyko, **nom. nov.** is proposed for the younger name.

New synonymy and new combinations

946 *Abirus fortunei* (Baly, 1861) = *Abirus sinensis* var. *guerryi* Pic, 1927 **syn. nov.** = *Abirus latus* Pic, 1944 **syn. nov.** = *Abirus latus* var. *semicyaneus* Pic, 1944 **syn. nov.** based on study of extensive material and opinion of Eroshkina (1992).

947 *Chrysochus goniostoma* Weise, 1889 = *Chrysochus singularis* var. *caerulescens* Pic, 1927 **syn. nov.** transferred from the synonymy of *Ch. chinensis* Baly, 1859 based on the study of type material.

948 *Colaspinella* Weise, 1893. Classification used in the present Catalogue follows revision based on the study of all relevant type specimens, which will be published later this year (Moseyko & Bezděk, in prep.).

949 *Cleorina* Lefèvre, 1885 = *Colposcelis* Dejean, 1836 **syn. nov.** *Colposcelis* is transferred from the synonymy with *Pagria* Lefèvre, 1884 and synonymized with *Cleorina*, because its type species, *Colaspis viridiaenea* Gyllenhal, 1808 (extralimital to the present Catalogue), is here transferred to the genus *Cleorina* (**comb. nov.**) based on study of the type specimen. *Colposcelis* Dejean, 1836 is junior homonym of *Colposcelis* Dejean, 1834 in Tenebrionidae.

950 *Demotina fasciata* Baly, 1874 = *Demotina pseudoimasakai* J. Park, 2013 **syn. nov.** based on the study of type material of *D. fasciata* and original description and illustrations of *D. pseudoimasakai*.

951 *Fidia phoebicola* (Tan, 1983) = *Lypesthes vittatus* Zhou & Tan, 1997 **syn. nov.** based on the study of type material of both taxa.

952 *Basilepta cribricollis* (Motschulsky, 1860) = *Nodostoma bicoloripes* Pic, 1930 **syn. nov.** = *Nodostoma bicoloripes* var. *guerryi* Pic, 1930 **syn. nov.** = *Nodostoma chinensis* Lefèvre, 1877 **syn. nov.** = *Nodostoma oberthuri* Lefèvre, 1877 **syn. nov.** transferred from the synonymy with *B. fulvipes* (Motschulsky, 1860) based on the study of type material.

Taxa excluded from Palaearctic fauna

Two monotypic genera were erroneously described from the Palaearctic Region based on mislabelled specimens.

953 *Phortus* Weise, 1899 was established for *P. creticus* Weise, 1899, described from Crete. Daccordi (1977c) considered Crete as an incorrect locality and synonymized

P. creticus with *Graphops curtipennis* (Melsheimer, 1847), a Nearctic taxon, and the genus *Phortus*, accordingly, with *Graphops* Leconte, 1884.

954 *Weiselina* Reitter, 1913 was established for *Nodostoma lenkorana* Reitter, 1890, described from one specimen provided with the label "Lenkoran, Leder" (nowadays in Azerbaijan). The label was considered wrong and taxon was excluded from Palaearctic fauna by Moseyko & Sprecher-Uebersax (2010). The holotype fully corresponds to Afrotropical genus *Paraivongius* Pic, 1936.

New geographical records

955 *Bromius obscurus* (Linnaeus, 1758). GREECE: Morea [= Peloponnisos], Cumani, 1 spec., Brenske leg., A. G. Moseyko det. (ZIN). TADJIKISTAN: "Karategen" [= Rasht Valley], 25.iii.1899, 1 spec., A. G. Moseyko det. (ZIN, ex coll. Semenov-Tian-Shansky). NORTH KOREA: Ryanggang Prov., Muduvan Mt., near Baekdusan (Cheonjji) Lake, 41°57′N 128°11′E, 20.viii.1950, 1 spec., N. Borchsenius leg., A. G. Moseyko det. (ZIN). New species for Greece and Tadjikistan, confirmed for North Korea.

956 *Basilepta cribricollis* (Motschulsky, 1860). CHINA: Jilin: Caijiagou, 9.v.1905, 1 spec., Serikov leg., A. G. Moseyko det. (ZIN); Liaoning: Houjia Gangzi, 3.vii.1905, 1 spec., Bortkevich leg., A. G. Moseyko det. (ZIN); Sichuan: Ho-Dzi-Gou Canyon near Lun Ngan Fou [= Duobu River Valley N of Long'an Zhen], 1893, 1 spec., Berezovsky leg., A. G. Moseyko det. (ZIN). New species for Jilin, Liaoning and Sichuan (China).

957 *Chloropterus lefevrei arabicus* Lopatin, 2008. OMAN: Ash Sharqiyah N. Gov., Masirah Island, Hiql vill. env., 20°21,9′N 58°47,6′E, 26.x.2019, 1 spec., J. Hájek & A. Reiter leg., J. Bezděk det. (NMPC). New species for Oman.

958 *Chloropterus ornatus* Lopatin, 1984. SYRIA: Dayr as Zawr, SE of Euphrates River bank, 26.v.2009, 1 ♂ 1 ♀, F. Houška leg., A. G. Moseyko det. (collection of J. Bezděk, Brno, Czech Republic). New species for Syria.

959 *Colasposoma* (*Falsonerissus*) *coracinum* (Lopatin, 1996). OMAN: Al Batinah North Gov., near Ar Rajmi, wadi, 23.–24.iii.2019, 11 spec., D. Frank leg., J. Bezděk det. (collections of D. Frank, Praha, and J. Bezděk, Brno, Czech Republic). New species for Oman.

960 *Macrocoma brunnipes brunnipes* (Olivier, 1808). JORDAN: 10 km N of Ajlun, 22.v.2007, 1 ♂, J. Bezděk leg. and det. (collection of J. Bezděk, Brno, Czech Republic); Jarash, Burma env., Al Huna, 15.v.2010, 3 ♂♂, M. Snížek leg., J. Bezděk det. (collection of J. Bezděk, Brno, Czech Republic). New species for Jordan.

961 *Macrocoma zarudnii* Lopatin, 1985. OMAN: Batinah Prov., Rahbah, oasis in wadi ca NW, 4.iv.2012, 2 spec., A. Reiter leg. (collection of J. Bezděk, Brno, Czech Republic); Fanjah, 14.iii.2019, 1 spec., K. Orszulik leg. (collection of K. Orszulik, Frýdek-Místek, Czech Republic); Ad Dakhiliyah Gov., N of Subayb, wadi, 28.–29.iii.2019, 19 spec., D. Frank leg., J. Bezděk det. (collection of D. Frank, Praha, Czech Republic). SAUDI

ARABIA: Riyadh, Ibex Res. Prot. Area (W of Hotat Bani Tamim), 11.iv.2017, 11 spec., P. Rapuzzi leg., J. Bezděk det. (NMPC). New species for Oman and Saudi Arabia.

962 *Pagria ussuriensis* Moseyko & L. N. Medvedev, 2005. CHINA: Jilin: Caijiagou, 9.v.1905, 1 spec., Serikov leg., A. G. Moseyko det. (ZIN); Liaoning: Zhengjiatun, 29.v.1905, 1 spec., Bortkevich leg., A. G. Moseyko det. (ZIN). New species for Jilin and Liaoning (China).

963 *Parnops atriceps* Pic, 1903. CHINA: Liaoning: Zhengjiatun, 3.vi.1905, 1 spec., Bortkevich leg., A. G. Moseyko det. (ZIN). New species for Liaoning (China).

964 *Rhyparida bimaculata* Jacoby, 1889. CHINA: Yunnan: Jinping env., ., 25.V.1956, 1 ♂, Huang Ke Ren leg., A. G. Moseyko det. (IZCAS). New species for Yunnan (China).

Comments on distribution

965 *Basilepta fulvipes* (Motschulsky, 1860). The record from Mongolia (Lopatin 1975b) is confirmed. *Basilepta cribricollis* (Motschulsky, 1860) was stil not found in Mongolia.

966 *Basilepta kaszabi* Lopatin, 1962 is distributed in China (Shanghai), according to the label of one specimen labelled as a syntype but not included in the description.

967 *Basilepta pectoralis* Pic, 1931 is distributed in China (Yunnan), according to the original locality label of the type.

968 *Cleoporus cuprescens* (Baly, 1878). This species was interpreted and assigned to the genus *Cleoporus* in the previous Edition (Moseyko & Sprecher-Uebersax 2010) based on study of a specimen from Khasi Hills (Meghalaya, India) considered as the type. However, it proved not to be a part of the type series as the true type material is from Jhellam Valley (Azad Kashmir, Pakistan). Since the true type material was not studied and is probably lost, the placement in *Cleoporus* must be considered as doubtful assignment.

969 *Cleorina instriata* Pic, 1937 is distributed in China (Yunnan), according to the original locality label of the type.

970 *Colaspoides tenenbaumi* Pic, 1942 is distributed in China (Yunnan), according to the original locality label of the type.

971 *Colasposoma dauricum* (Mannerheim, 1849). The records from Jilin, based on Zhengjiatun and Langashi localities (Moseyko et al. 2018), refer to Liaoning.

972 *Pachnephorus cylindricus* P. H. Lucas, 1846. All records from Eastern Europe and Turkey are preliminarily referred to *P. lateralis* Reitter, 1901. Also, some records of *P. cylindricus* may belong to *P. corinthius* Fairmaire, 1862, especially from Albania and Greece.

973 *Pachnephorus lateralis* Reitter, 1901. The label data of syntype deposited in HNHM are: "Caucasus Mlok. 97. [= Caucasus, L. Mlokosiewicz leg., 1897]". According to material from ZIN, in the second half of the 1890s, L. Mlokosiewicz collected insects only in Lagodekhi valley in Georgia. So, the type locality of *P. lateralis* is specified here to Georgia.

974 *Pagria signata* (Motschulsky, 1860). Records of this species from China, except Yunnan, belong to other species. In case of North-East China mostly to *P. ussuriensis* Moseyko & L. Medvedev, 2005, in other cases, including Xizang and Chongqing, mostly to *P. grata* (Baly, 1867).

975 *Parheminodes mouhoti* (Baly, 1864). Records from China (as *Chysochus pulcher* Baly, 1864), given by Gressitt & Kimoto (1961) are doubtful and can be a confusion with *P. collaris* Chen, 1940, described from the same region and not studied by these authors.

976 *Parnops atriceps* Pic, 1903. Jacobson (1910b) published this species from Mandzhuria, which was misinterpreted as Heilongjiang (Li et al. 2008, Tan et al. 2005). However, the voucher specimen was collected in Liaoning. Therefore, Heilongjiang was removed from the present Catalogue.

Editorial comments

J. Bezděk & L. Sekerka

General comment

In the final stage of preparation of the Catalogue we noticed that some of the names that are currently considered available with their original authorship, and were never previously questioned, were actually published in synonymy in the original publication, or mentioned under varieties denoted by Greek or Latin letters and thus in most of the cases are not available, although they have been listed in the previous Edition and used by many authors. This mainly concerns works by Suffrian (and several of his contemporaries), who used either Greek or Latin letters for their varieties and sometimes also mentioned one or more manuscript names under which they received material from various colleagues. However, these names are to be considered as mentioned for documentary purposes only, as in all cases the Greek or Latin letter is mentioned in discussions instead of the other names, making it obvious that the authors did not intend to use these names, and sometimes also proposed an entirely new name for these varieties in subsequent publications, e.g. *Cryptocephalus pistaciae* Suffrian, 1853 for the var. γ of *C. flavipes* Fabricius, 1781 of Suffrian (1847b). This issue was discussed with the members of the International Commission on Zoological Nomenclature, who confirmed that in the case of varieties, these names do not exist within the framework of the Code. Nevertheless, some of the names used for these varieties were made available by subsequent authors with their authorship. In most cases the first source was J. Weise's monograph on Chrysomelidae in the series *Naturgeschichte der Insekten Deutschlands*, which however also listed varieties denoted by Latin letters. Nevertheless, Weise had different styles of how these varieties were presented, as the name itself either 1) immediately followed the denoting letter (this is the case for new varieties he proposed and also for some of the old ones) and then followed by a diagnosis; or 2) the denoting letter is followed by the diagnosis and the name is listed afterwards, often several of them, for instance as they appeared in Suffrian's works. The first case is to be considered as an intended usage of a name, as in this way are mentioned all new names proposed by Weise or unavailable names of earlier authors that he adopted and intentionally used for the respective taxon, and consequently all such mentioned names are to be treated as available with Weise's authorship. The second style is again mentioning name(s) for documentary purposes and the name(s) remain unavailable until such time as they are mentioned in a way to be made available by a subsequent author, at which point they would take the authorship and date of the author and publication that did so.

Names mentioned in synonymy could be made available if published in accordance with Article 11.6.1. (ICZN 1999) before 1961. Suffrian most likely again mentioned these names for documentary purposes and did not intend to use them, as in some cases in the same publication a taxon with the same name is properly described, e.g. Suffrian (1851a: 7) mentioned "*Chrysomela carbonaria* Dahl" in synonymy with *Ch. atra* Herrich-Schäffer, 1838, while on p. 49 of the same work, *Ch. carbonaria* Suffrian, 1851 is properly described. There are other similar cases. Although Suffrian had no intention of using these names that he mentioned in synonymy, according to the Code, they must still be considered as potentially available. The earliest authors to use the names listed in synonymy as valid were Gistel (1856), Calwer (1857) and Compayo (1862), however, they provided only lists of names and did not provide a reference to Suffrian's original work, and hence the names cannot be linked with certainty and remain unavailable. On the contrary, J. Weise provided extensive taxonomic treatment on Chrysomelidae, where these names in question were listed under his varieties denoted with Latin letters; the names are to be considered unavailable, mentioned only for documentary purposes unless immediately following the denoting letter of the respective variety and referring to Suffrian's work. Validations by other authors are clear and do not require further comments.

To avoid further confusion or misapplication we summarized all names in question, mentioned by Suffrian and his contemporaries, and provide their subsequent citations if ever mentioned. This list is restricted to Palaearctic taxa. The names are divided into three categories: 1) unavailable names proposed in synonymy (we provide mentions of these names in subsequent publication alhough these do not represent validations for documentary purposes); 2) invalid names mentioned under varieties (provided with additional comments where necessary); and 3) names that were made available by actions of subsequent authors, which are commented on in detail and clarified. All unavailable names are excluded from the Catalogue.

Unavailable names proposed in synonymy, not validated before 1961 to our knowledge

977 *Sagra festiva* (Lacordaire 1845: 35, in synonymy with *S. chrysochlora* Lacordaire, 1845).

978 *Cassida rosea* Ziegler (Suffrian 1844d: 214, in synonymy with *C. azurea* Fabricius, 1801).

979 *Cassida graminis* Dahl and *C. immutabilis* Ziegler (Suffrian 1844a: 97–98, in synonymy with *C. austriaca* Fabricius, 1792). Weise (1893a: 1085) listed *C. graminis* under var. a of *C. canaliculata* in a way that cannot be considered as a validation. *Cassida immutabilis* was listed as a species by Calwer (1857: 675) in a way that cannot be considered as a validation.

980 *Cassida sardoa* Heyden and *C. scolymi* Küster (Suffrian 1844b: 139–140, in synonymy with *C. deflorata* Suffrian, 1844).

981 *Cassida silesiaca* Megerle (Suffrian 1844f: 276, in synonymy with *C. ferruginea* Goeze, 1777).

982 *Cassida pectoralis* Parreyss, *C. punctata* Andersch, *C. retifera* Ziegler and *C. sulfurea* Megerle (Suffrian 1844a: 94, in synonymy with *C. hemisphaerica* Herbst, 1799).

983 *Cassida herbacea* Dejean (Suffrian 1844b: 141, in synonymy with *C. hexastigma* Suffrian, 1844). Listed as a species by Calwer (1857: 638) and Companyo (1862: 806) in a way that cannot be considered as a validation.

984 *Cassida millefolii* Dahl (Suffrian 1844e: 248, in synonymy with *C. nobilis* Linnaeus, 1758).

985 *Cassida fenestrata* Megerle and *C. signaticollis* Megerle (Suffrian 1844c: 191, in synonymy with *C. chloris* Suffrian, 1844).

986 *Cassida fallax* Ziegler (Suffrian 1844b: 137, in synonymy with *C. rubiginosa* O. F. Müller, 1776).

987 *Cassida fenestrata* Megerle (Suffrian 1844d: 207, in synonymy with *C. sanguinolenta* O. F. Müller, 1776).

988 *Cassida anceps* Ziegler, *C. cruentata* Megerle and *C. pallipes* Ziegler (Suffrian 1844a: 107, in synonymy with *C. sanguinosa* Suffrian, 1844).

989 *Cassida podolica* Megerle (Suffrian 1844b: 148, in synonymy with *C. vibex* Linnaeus, 1767).

990 *Cassida variabilis* Dahl (Suffrian 1844a: 100), in synonymy with *C. vittata* Villers, 1789). Listed as a species by Gistel (1856: 156) in a way that cannot be considered as a validation.

991 *Cassida nigriventris* Waltl and *C. testacea* Dahl (Suffrian 1844e: 256, in synonymy with *C. oblonga* Illiger, 1798).

992 *Chrysomela zetterstedtii* Dejean (Suffrian 1851a: 194, in synonymy with *Chrysomela alpina* Zetterstedt, 1828). Listed as *Lina zetterstedtii* by Gistel (1856: 155) in a way that cannot be considered as a validation.

993 *Gastrophysa chalybea* Dejean (Suffrian 1851a: 241, in synonymy with *Chrysomela janthina* Suffrian, 1851). Listed as a species by Calwer (1857: 662) in a way that cannot be considered as a validation.

994 *Phaedon chalybeum* Sturm (Suffrian 1851a: 253, in synonymy with *Chrysomela cochleariae* Fabricius, 1792). Listed as a species by Calwer (1857: 663) in a way that cannot be considered as a validation.

995 *Chrysomela salicus* Kiesenwetter and *Ch. triglochinis* Schaum (Suffrian 1851a: 257, in synonymy with *Ch. concinna* Stephens, 1831).

996 *Phaedon remotissimus* Imhoff (Suffrian 1851a: 256, in synonymy with *Chrysomela salicina* Heer, 1845). Listed as a species by Calwer (1857: 663) in a way that cannot be considered as a validation.

997 *Chrysomela peregrina* Parreyss (Suffrian 1851a: 266, in synonymy with *Ch. flavocincta* Brullé, 1833). Listed as a species by Calwer (1857: 664) in a way that cannot be considered as a validation.

998 *Chrysomela cyanipes* Ahrens, *Ch. genuensis* Dejean and *Ch. nigroaenea* Sturm (Suffrian 1851a: 59, in synonymy with *Ch. molluginis* Suffrian, 1851). *Chrysomela genuensis* and *Ch. nigroaenea* were listed as a species by Calwer (1857: 657–658) in a way that cannot be considered as a validation.

999 *Chrysomela betulae* Latreille (Suffrian 1851a: 133, in synonymy with *Ch. lepida* Olivier, 1807).

1000 *Chrysomela inflata* Kunze, *Ch. lata* Märkel and *Ch. villae* De Christophori (Suffrian 1851a: 188, in synonymy with *Ch. asclepiadis* A. Villa & G. B. Villa, 1833).

1001 *Chrysomela irregularis* Eversmann (Suffrian 1851a: 74, in synonymy with *Ch. besseri* Krynicki, 1832).

1002 *Chrysomela circumcincta* Dejean (Suffrian 1851a: 79, in synonymy with *Ch. circumducta* Suffrian, 1851). Listed as a species by Calwer (1857: 658) in a way that cannot be considered as a validation.

1003 *Chrysomela fimbriata* Hoffmannsegg (Suffrian 1851a: 77, in synonymy with *Ch. coerulescens* Suffrian, 1851).

1004 *Chrysomela modesta* Sturm (Suffrian 1851a: 19, in synonymy with *Ch. aethiops* Olivier, 1807). Listed as a species by Calwer (1857: 658) in a way that cannot be considered as a validation.

1005 *Chrysomela alpina* Dahl (Suffrian 1851a: 52, in synonymy with *Ch. crassicollis* Suffrian, 1851). Listed as a species by Calwer (1857: 657) in a way that cannot be considered as a validation.

1006 *Chrysomela rutilans* (Suffrian 1851a: 31, in synonymy with *Ch. opulenta* Suffrian, 1851).

1007 *Chrysomela bithynica* Rosenhauer (Suffrian 1851a: 55, in synonymy with *Ch. haemoptera* Linnaeus, 1758).

1008 *Chrysomela globata* Dahl and *Ch. rufocuprea* Dejean (Suffrian 1851a: 41, in synonymy with *Ch. hemisphaerica* Germar, 1817). *Chrysomela rufocuprea* was listed by Calwer (1857: 657) in a way that cannot be considered as a validation. Kippenberg (1998) considered the name valid and wrote: "für die Rasse *hemisphaerica* (Germ.) s.str. [junior primary homonym] tritt der Name *rufocuprea* (Suffrian) ein.". This cannot be considered as a validation of the name nor as a new act, as new replacement names already had to be explicitly proposed as new at that time (ICZN 1985a, Art. 13(a)(iii)). Kippenberg (2010b) also used the name as a valid subspecies of *Ch. purpurascens*. This was followed in a revision of European species of *Colaphoptera* Motschulsky, 1860 (Kippenberg 2017) but although *Ch. purpurascens rufocuprea* was provided with a description, this cannot be considered as a validation of the name as it was not explicitly indicated as intentionally new (ICZN 1999, Art. 16.1). The oldest available name for the taxon is *Chrysomela hemisphaerica* var. *laeta* Weise, 1882.

1009 *Chrysomela dahlii*, *Ch. germari* Tischer and *Ch. guentheri* Sturm (Suffrian 1851a: 24–25, in synonymy with *Ch. purpurascens* Germar, 1822). In the same publication was also properly described *Ch. dahlii* Suffrian, 1851 on p. 28. Calwer (1857: 658) listed the name "dahlii Knoch", and thus it is not clear to which of the two names it could apply. However, *Ch. dahlii* (p. 24) remains unavailable. *Chrysomela guentheri* was listed as a species by Calwer (1857: 657) in a way that cannot be considered as a validation.

1010 *Chrysomela alpina* and *Ch. crassicollis* Dahl (Suffrian 1851a: 23, in synonymy with *Ch. marcasitica* Germar, 1823).

1011 *Chrysomela tanaceti* Klingelhöfer (Suffrian 1851a: 99, in synonymy with *Ch. graminis* Linnaeus, 1758).

1012 *Chrysomela femorata* Dejean (Suffrian 1851a: 17, in synonymy with *Ch. baetica* Suffrian, 1851). Listed as a species by Calwer (1857: 658) in a way that cannot be considered as a validation.

1013 *Chrysomela heteropunctata* Megerle (Suffrian 1851a: 120, in synonymy with *Ch. lamina* Fabricius, 1792).

1014 *Chrysomela patruelis* Dejean (Suffrian 1851a: 15, in synonymy with *Ch. cretica* Olivier, 1807).

1015 *Chrysomela bicolor* var. *turcica* (Suffrian 1851a: 45, in synonymy with *Ch. orientalis* Olivier, 1807).

1016 *Chrysomela atroaenea* Steven, *Ch. impressa* Eversmann, *Ch. maura* Dejean and *Ch. sulcicollis* Sturm (Suffrian 1851a: 119–120, in synonymy with *Ch. morio* Krynicki, 1832). *Chrysomela sulcicollis* and *Ch. maura* were listed as a species by Calwer (1857: 657–658) in a way that cannot be considered as a validation.

1017 *Chrysomela barbara* Jan and *Ch. bithynica* Küster (Suffrian 1851a: 44, in synonymy with *Ch. herii* Herrich-Schäffer, 1838). *Chrysomela barbara* was listed as a species by Gistel (1856: 167) and Calwer (1857: 657) in a way that cannot be considered as a validation.

1018 *Chrysomela chalcipennis* Sturm (Suffrian 1851a: 42, in synonymy with *Ch. vernalis* Brullé, 1833). Listed as a species by Gistel (1856: 167) and Calwer (1857: 657) in a way that cannot be considered as a validation.

1019 *Chrysomela marginepunctata* Géné and *Ch. rufomarginata* (Suffrian 1851a: 65, in synonymy with *Ch. gypsophilae* Küster, 1845).

1020 *Chrysomela zenkeri* Dahl (Suffrian 1851a: 39, in synonymy with *Ch. rossia* Illiger, 1802). Listed as a species by Calwer (1857: 657) in a way that cannot be considered as a validation.

1021 *Chrysomela escheri* Lasserre and *Ch. melanosticta* Kollar (Suffrian 1851a: 64, in synonymy with *Ch. sparshalli* Stephens, 1835).

1022 *Chrysomela auronitens* Escher-Zoll (Suffrian 1851a: 94, in synonymy with *Ch. ignita* Olivier, 1807).

1023 *Chrysomela reticulata* Dahl (Suffrian 1851a: 11, in synonymy with *Ch. cribrosa* Ahrens, 1812).

1024 *Chrysomela carbonaria* Dahl and *Ch. coeruleipes* Rollet (Suffrian 1851a: 7, in synonymy with *Ch. atra* Herrich-Schäffer, 1838). *Chrysomela carbonaria* was listed as a species by Gistel (1856: 247) and Calwer (1857: 657) in a way that cannot be considered as a validation.

1025 *Chrysomela rufitarsis* Sturm (Suffrian 1851a: 101, in synonymy with *Ch. fastuosa* (Scopoli, 1763)).

1026 *Chrysomela alpicola* Hoppe (Suffrian 1851a: 183, in synonymy with *Ch. luctuosa* Olivier, 1807). Listed as a species by Gistel (1856: 96) in a way that cannot be considered as a validation.

1027 *Chrysomela alpicola* Heer (Suffrian 1851a: 155, in synonymy with *Ch. tristis* Fabricius, 1792).

1028 *Chrysomela lapidicola* Märkel (Suffrian 1851a: 142, in synonymy with *Ch. fuscoaenea* Schummel, 1844). Listed as a species by Calwer (1857: 658) in a way that cannot be considered as a validation.

1029 *Chrysomela laeta* Schmidt and *Chr. punctata* Dahl (Suffrian 1851a: 142, in synonymy with *Ch. speciosissima* (Scopoli, 1763)).

1030 *Chrysomela cyanipes* Märkel and *Chr. subrugosa* Sturm (Suffrian 1851a: 181, in synonymy with *Ch. intricata* Germar, 1823). *Chrysomela subrugosa* was listed as a species by Calwer (1857: 659) in a way that cannot be considered as a validation.

1031 *Chrysomela vittata* Dahl (Suffrian 1851a: 161, in synonymy with *Ch. aenescens* Suffrian, 1851).

1032 *Chrysomela diversa* Sturm (Suffrian 1851a: 166, in synonymy with *Ch. gloriosa* Fabricius, 1782). Listed as a species by Calwer (1857: 659) in a way that cannot be considered as a validation.

1033 *Chrysomela vittata* Dahl (Suffrian 1851a: 167, in synonymy with *Ch. vittigera* Suffrian, 1851).

1034 *Chrysomela alpina* Lasserre and *Chr. nivalis* Kunze (Suffrian 1851a: 223, in synonymy with *Ch. nivosa* Suffrian, 1851).

1035 *Clythra quinquemaculata* Dejean (Lacordaire 1848: 235, in synonymy with *Clythra* (*Diapromorpha*) *dejeanii* Lacordaire, 1848).

1036 *Cryptocephalus spitzyi* Friwaldszki (Suffrian 1848: 59, in synonymy with *C. bilineatus* (Linnaeus, 1767)).

1037 *Cryptocephalus dimidiatus* and *C. dimidiaticornis* Ziegler (Suffrian 1848: 58, in synonymy with *C. tesselatus* Germar, 1813).

1038 *Cryptocephalus wawarius* Koy (Suffrian 1848: 65, in synonymy with *C. vittula* Suffrian, 1848). Listed as a species by Gistel (1856: 177) in a way that cannot be considered as a validation.

1039 *Cryptocephalus coecus* Kunze (Suffrian 1848: 97, in synonymy with *C. querceti* Suffrian, 1848).

1040 *Cryptocephalus strigatus* Creutzer (Suffrian 1848: 105, in synonymy with *C. strigosus* Germar, 1823).

1041 *Cryptocephalus octonotatus* Dahl and *C. octopustulatus* Megerle (Suffrian 1848: 42, in synonymy with *C. anticus* Suffrian, 1848).

1042 *Cryptocephalus auratus* Megerle (Suffrian 1847b: 137, in synonymy with *C. aureolus* Suffrian, 1847). Listed as a species by Gistel (1856: 380) in a way that cannot be considered as a validation.

1043 *Cryptocephalus bisbipustulatus* Sturm (Suffrian 1853b: 93, in synonymy with *C. bimaculatus* Fabricius, 1781). Listed as a species by Calwer (1857: 676) in a way that cannot be considered as a validation.

1044 *Cryptocephalus octopustulatus* Sturm (Suffrian 1848: 46, in synonymy with *C. gravidus* Herrich-Schäffer, 1836).

1045 *Cryptocephalus uralensis* Mannerheim (Suffrian 1853b: 110, in synonymy with *C. elongatus* Germar, 1823).

1046 *Cryptocephalus chloromanus* Megerle, *C. chlorophanus* Ullrich, *C. frontalis*, *C. furcatus* Megerle and *C. giloipes* Ziegler (Suffrian 1847b: 174, in synonymy with *C. flavipes* Fabricius, 1781).

1047 *Cryptocephalus callifer* Germar and *C. nigriventris* Parreyss (Suffrian 1848: 9, in synonymy with *C. flavescens* D. H. Schneider, 1792). Weise (1882b: 209) listed *C. callifer* under var. a of *C. frenatus* Laicharting, 1781 in a way that cannot be considered as a validation.

1048 *Cryptocephalus blockii* Dahl and *C. quadrinotatus* Sturm (Suffrian 1847b: 38, in synonymy with *C. ilicis* Olivier, 1808 sensu Suffrian (1847b)). *Cryptocephalus blockii* was subsequently listed by Marseul (1874: 39) as a variety but together with

C. binotatus Dahl, and thus the name remains in synonymy. Finally, it was listed by Heyden (1891: 368) in a nomenclatural note clarifying that the name *C. blockii* Dahl is also a junior homonym of *C. blockii* Rossi and that the name *C. etruscus* should apply to this taxon, thus again it was listed as unavailable name. *Cryptocephalus quadrinotatus* was listed as a species by Calwer (1857: 675) in a way that cannot be considered as a validation. Marseul (1874) mentioned it in synonymy with *C. ilicis*. Finally, Weise (1881: 162) mentioned it under *C. ilicis* together with *C. etruscus* not giving priority to either of the names, and moreover they were mentioned as variability but without being given an explicit category (i.e. var., ab., forma etc.); therefore, both names remain unavailable.

1049 *Cryptocephalus cyaneus* Dejean, *C. picticeps* Ziegler and *C. punctiger* Gyllenhal (Suffrian 1848: 15, in synonymy with *C. janthinus* Germar, 1823). *Cryptocephalus cyaneus* was listed as a species by Calwer (1857: 675) in a way that cannot be considered as a validation.

1050 *Cryptocephalus punctifrons* Sturm (Suffrian 1847b: 165, in synonymy with *C. nitens* (Linnaeus, 1761)).

1051 *Cryptocephalus signaticollis* Sturm (Suffrian 1854b: 78, in synonymy with *C. petraeus* Suffrian, 1854).

1052 *Cryptocephalus confluens* Mannerheim and *C. transversomaculatus* Motschulsky (Suffrian 1853b: 97–98, in synonymy with *C. transversalis* Suffrian, 1853).

1053 *Cryptocephalus fallax* Eversmann (Suffrian 1848: 109, in synonymy with *C. quadriguttatus* C. F. W. Richter, 1820).

1054 *Cryptocephalus bigeminus* Illiger and *C. similis* Sturm (Suffrian 1847b: 177, in synonymy with *C. quadripustulatus* Gyllenhal, 1813). Weise (1882b: 212) listed *C. similis* together with another name under var. a of *C.* in a way that cannot be considered as a validation.

1055 *Cryptocephalus comptus* (Suffrian 1853b: 89, in synonymy with *C. rugicollis* Olivier, 1792).

1056 *Cryptocephalus pratorum* Megerle (Suffrian 1847b: 131, in synonymy with *C. sericeus* (Linnaeus, 1758)). Listed as a species by Gistel (1856: 380) in a way that cannot be considered as a validation.

1057 *Cryptocephalus grandis* Dejean (Suffrian 1847b: 33, in synonymy with *C. sexmaculatus* Olivier, 1792). Listed as a species by Gistel (1856: 225), Calwer (1857: 675) and Compayo (1862: 819) in a way that cannot be considered as a validation.

1058 *Cryptocephalus pictus* Ziegler and *C. separandus* Ziegler (Suffrian 1847b: 87, in synonymy with *C. sexpunctatus* (Linnaeus, 1758)). Weise (1881: 152) listed *C. pictus* and *C. separandus* under var. f and var. h of *C. sexpunctatus* in a way that cannot be considered as a validation.

1059 *Cryptocephalus punctulatus* Dejean (Suffrian 1848: 50, in synonymy with *C. rossii* Suffrian, 1848). Listed as a species by Calwer (1857: 676) and Compayo (1862: 820) in a way that cannot be considered as a validation.

1060 *Cryptocephalus humeralis* Sturm (Suffrian 1847b: 80, in synonymy with *C. variegatus* Fabricius, 1782). Weise (1881: 156) listed *C. humeralis* under var. c of *C. distinguendus* D. H. Schneider, 1792 in a way that cannot be considered as a validation.

1061 *Cryptocephalus cerabrae* Motschulsky (Suffrian 1853b: 117, in synonymy with *C. pini* (Linnaeus, 1758) and *C. cembrae* Motschulsky (Suffrian 1854b: 122, in synonymy with *C. pini* (Linnaeus, 1758)).

1062 *Cryptocephalus porrii* Villa (Suffrian 1847b: 161, in synonymy with *C. abietis* Suffrian, 1847).

1063 *Cryptocephalus gigas* Heyden (Suffrian 1847b: 65, in synonymy with *C. loreyi* Solier, 1837).

1064 *Cryptocephalus aemulus* Friwaldszki (Suffrian 1847b: 93, in synonymy with *C. laevicollis* Gebler, 1830), *C. octosignatus* Eversmann (Suffrian 1848: 105, in synonymy with *C. laevicollis* Gebler, 1830) and *C. eversmanni* Mannerheim (Suffrian 1853b: 99, in synonymy with *C. laevicollis* Gebler, 1830).

1065 *Cryptocephalus pilosus* Faldermann (Faldermann 1837: 384 and Suffrian 1853b: 114, in synonymy with *C. lobatus* Fabricius, 1792).

1066 *Cryptocephalus clythroides* Motschulsky (Suffrian 1854b: 131, in synonymy with *C. bivulneratus* Faldermann, 1835).

1067 *Cryptocephalus pertinax* Eversmann (Suffrian 1848: 107, in synonymy with *C. ypsilon* Suffrian, 1847).

1068 *Cryptocephalus brachialis* Mannerheim (Suffrian 1854b: 136, in synonymy with *C. hirtipennis* Faldermann, 1835).

1069 *Cryptocephalus sellatus* Mannerheim (Suffrian 1854b: 121, in synonymy with *C. semiviridis* Suffrian, 1854).

1070 *Pachybrachis lignosus* Dahl (Suffrian 1848: 128, in synonymy with *P. hippophaes* Suffrian, 1848). Weise (1882b: 260) listed *P. lignosus* under var. a of *P. hippophaes* in a way that cannot be considered as a validation.

1071 *Cryptocephalus galegae* (Suffrian 1854b: 154, in synonymy with *Pachybrachis nigropunctatus* Suffrian, 1854).

Unavailable names mentioned under varieties denoted with a letter

1072 *Bruchus ineptus* Schönherr (Fåhraeus 1839: 90, under var. β of *B. varicolor* Boheman, 1833).

1073 *Cassida aurea* Boeber (Suffrian 1844d: 213, under var. β of *C. lineola* Creutzer, 1799).

1074 *Cassida albiventris* Germar (Suffrian 1844e: 243, under var. γ of *C. margaritacea* Schaller, 1783).

1075 *Cassida marmorata* Kollar (Suffrian 1844a: 103, under var. α of *C. murraea* Linnaeus, 1767).

1076 *Cassida suturalis* Ziegler and *C. distinguenda* Ziegler (Suffrian 1844b: 147, under var. β of *C. vibex* Linnaeus, 1767).

1077 *Chrysomela subaenea* Megerle was first mentioned by Duftschmid (1825: 194) in a text with observations on *Chrysomela hottentotta* Fabricius, 1792 (nowadays synonym of *Ch. haemoptera* (Linnaeus, 1758)). Duftschmid mentioned variety β as "nigro-aenea" not associating it with any name. In the text is a description and is also mentioned "schwarzlich erzfarbig" specimen(s) that were under the name *Ch. subaenea* but it is not directly linked to the variety β. Therefore, there are two interpretations possible, the name was either published in synonymy or based on assumption under variety β but in the latter case only for documentary purposes. According to our knowledge the name was not used again with reference to Duftschmidt and therefore is unavailable, and accordingly it does not threaten the name *Ch. subaenea* Suffrian, 1851, which was considered its junior homonym in the previous Edition of the Catalogue.

1078 *Chrysomela gibba* Dahl and *Chr. globata* Dahl (Suffrian 1851a: 26, under var. β of *Ch. purpurascens* Germar, 1822). *Chrysomela globata* was mentioned several times in species lists, e.g. Calwer (1857: 657) and Compayo (1862: 812), but neither of these mentions represent a validation of this name. Kippenberg (2010b) considered the name a valid subspecies of *Ch. purpurascens*, overlooking that the name is actually unavailable. It was followed in a revision of European species of *Colaphoptera* Motschulsky, 1860 (Kippenberg 2017) and although *Ch. globata* was provided with description it cannot be considered as a validation of the name as it was not explicitly indicated as intentionally new (ICZN 1999, Art. 16.1). The taxon has three other available names that however have been described in the same publication: *Ch. banatica* Bechyné, 1950 (junior homonym), *Ch. plumbeonigra* Bechyné, 1950, and *Ch. sutilis* Bechyné, 1950, and priority among them was not established before. Following the Principle of the first revising author (Art. 24.2, ICZN 1999) the name *Ch. sutilis* Bechyné, 1950 is selected as the valid name for this taxon.

1079 *Chrysomela aurolimbata* Besser and *Chr. rutilans* Dahl (Suffrian 1851a: 98, under var. δ of *Ch. graminis* Linnaeus, 1758).

1080 *Chrysomela gleissneri* Zawadzky (Suffrian 1851a: 151, under var. α of *Ch. cacaliae* Schrank, 1785).

1081 *Chrysomela venusta* Villa, *Chr. aenea* Sturm and *Oreina cyanea* Sturm (Suffrian 1851a: 147, under var. α and var. β of *Ch. elongata* Suffrian, 1851). *Oreina cyanea* was mentioned as a species by Calwer (1857: 657) in a way that cannot be considered as a validation. According to our knowledge the name was never made available although it was listed among the synonyms of *Ch. elongata* in the previous version of the Catalogue. This was probably due to Bechyné (1958b), who listed the name with Suffrian's authorship as a valid subspecies of *Ch. elongata*, but failed to provide a description or bibliographic reference to it and therefore the name remains unavailable.

1082 *Chrysomela pretiosa* Sturm (Suffrian 1851a: 143, under var. ε of *Ch. speciosissima* (Scopoli, 1763)).

1083 *Chrysomela alpina* Heer (Suffrian 1851a: 164, under var. β of *Ch. nivalis* Suffrian, 1851).

1084 *Chrysomela avellanae* Germar and *Chr. ruficornis* Sturm (Suffrian 1851a: 235, under var. α and var. γ of *Ch. pallida* Linnaeus, 1758).

1085 *Plectroscelis insolita* Dejean (Foudras, 1860a: 230, under var. b of *Chaetocnema sahlbergii* (Gyllenhal, 1827)). Subsequent authors (Allard, 1860e: 570, Kutschera 1864e: 349) only repeated the original text. As far as we know, the name was never made available.

1086 *Cyaniris fuscitarsis* Dejean (Lacordaire 1848: 297, under var. A of *Clythra* (*Gynandrophthalma*) *nigritarsis* Lacordaire, 1848). Mentioned as a species by Calwer (1857: 672) and Compayo (1862: 819) in a way that cannot be considered as a validation.

1087 *Clythra subfasciata* Dejean (Lacordaire 1848: 201, under var. B of *Clythra* (*Clythra*) *succincta* Lacordaire, 1848). The name was erroneously considered as available by Regalin & Medvedev (2010) and Bezděk (2012b, 2016), but actually was never used in a way that would represent validation and hence remains unavailable. The oldest available name for this taxon is *C. subopacipennis* Pic, 1943.

1088 *Labidostomis laticollis* Dahl (Lacordaire 1848: 69, under var. B of *Clythra* (*Labidostomis*) *axillaris* Lacordaire, 1848). Weise (1881: 94) mentioned *L. laticollis* together with *L. axillaris* under var. a of *L. lucida* (Germar, 1823) for documentary purposes only and thus the name remains unavailable.

1089 *Labidostomis taurica* Dejean (Lacordaire 1848: 62, under var. A of *Clythra* (*Labidostomis*) *brevipennis* (Faldermann, 1837)).

1090 *Clythra nigronotata* Chevrolat (Lacordaire 1848: 167, under var. A of *Clythra* (*Barathroea*) *cerealis* Olivier, 1808).

1091 *Lachnaea triguttata* Dejean (Lacordaire 1848: 184, under var. A of *Clythra* (*Lachnaea*) *tristigma* Lacordaire, 1848).

1092 *Clythra ilicis* Germar (Lacordaire 1848: 288, under var. A of *Clythra* (*Gynandrophthalma*) *hypocrita* Lacordaire, 1848).

1093 *Clythra octopunctata* Klug (Lacordaire 1848: 162, under var. A of *Clythra* (*Tituboea*) *paykullii* Lacordaire, 1848).

1094 *Cryptocephalus fulvicollis* Dahl (Suffrian 1848: 74, under var. β of *C. minutus* Fabricius, 1792). Weise (1882b: 237) mentioned *C. fulvicollis* under var. a of *C. fulvus* (Goeze, 1777) for documentary purposes only and thus the name remains unavailable.

1095 *Cryptocephalus paradoxus* Friwaldszki (Suffrian 1848: 35, under var. ε of *C. bipunctatus* (Linnaeus, 1758)). Subsequently mentioned by Weise (1893a: 1119) as a variety but together with *C. quadrinotatus* Herrich-Schäffer, 1838 and thus not made available.

1096 *Cryptocephalus alpinus* Dahl and *C. furcatus* Dahl (Suffrian 1847b: 172, under var. β of *C. flavipes* Fabricius, 1781). *Cryptocephalus furcatus* was mentioned as a species by Gistel (1856: 380) and Calwer (1857: 676) in a way that cannot be considered as a validation. Weise (1881: 214) mentioned *C. alpinus* and *C. furcatus* under var. h of *C. flavipes* Fabricius, 1781 for documentary purposes only and thus the name remains unavailable.

1097 *Cryptocephalus fasciatus* Sturm, *C. interruptofasciatus* Ziegler oder Megerle and *C. interruptus* Ziegler oder Megerle (Suffrian 1847b: 188, under var. β of *C. moraei* (Linnaeus, 1758)). *Cryptocephalus interruptofasciatus* was subsequently mentioned together with *C. cruciatus* Dahl under a variety by Marseul (1874: 165) and Weise (1888: 225), however, for documentary purposes only and thus remains unavailable.

1098 *Cryptocephalus rufolimbatus* Reiche (Suffrian 1847b: 51, under var. γ of *C. imperialis* Fabricius, 1798). Weise (1881: 163) mentioned *C. rufolimbatus* under var. c of *C. primarius* Harold, 1872 for documentary purposes only and thus the name remains unavailable.

1099 *Cryptocephalus caelatus* Reiche and *C. verrucosus* Kunze (Suffrian 1847b: 43, under var. η of *C. rugicollis* Olivier, 1792). *Cryptocephalus verrucosus* with Suffrian's authorship was mentioned by Weise (1881: 178) among the colour variability of *C. rugicollis* but not as a variety or form and thus the name is infrasubspecific.

1100 *Cryptocephalus alpinus* Stenz, *C. bicolor* Megerle, *C. centaurii* Megerle, *C. chlorodius* Megerle, *C. coeruleus* Periol., *C. melanarius* Stenz, *C. smaragdinus* Ziegler, *C. subtilis* Dahl, *C. unicolor* Ziegler and *C. viridis* Dahl (Suffrian 1847b: 124, under var. β of *C. violaceus* Laicharting, 1781). Weise (1881: 190) mentioned *C. smaragdinus* under var. c of *C. violaceus* for documentary purposes and thus the name remains unavailable.

1101 *Cryptocephalus dispar* Dahl (Suffrian 1847b: 101, under var. γ of *C. coryli* (Linnaeus, 1758)).

1102 *Cryptocephalus cyanipennis* Dejean (Suffrian 1852b: 72, under var. β of *Diachus auratus* (Fabricius, 1801)).

Names in synonymy or as a variety denoted by a letter, that were made available by actions of subsequent authors

1103 *Spermophagus glabratus* Schönherr (Gyllenhal 1839: 141, under var. β of *S. convolvuli* (Thunberg, 1816)). The name is available from Lukjanovitch & Ter-Minassian (1957: 195) who used it for a species (in genus *Euspermophagus*) and provided description.

1104 *Lema collaris* Dahl (Lacordaire 1845: 577, under var. B of *Crioceris brunnea* Fabricius, 1792). The name was made available by Weise (1881: 71) who adopted it as a variety and provided description.

1105 *Cassida melanocephala* Ziegler (Suffrian 1844e: 243, under var. γ of *C. margaritacea* Schaller, 1783). The name *C. melanocephala* was made available by Weise (1893a: 1091) who used it as a name of a variety and gave its description.

1106 *Cassida pannonica* Dahl (Suffrian 1844b: 147, under var. β of *C. vibex* Linnaeus, 1767). The name *C. pannonica* was made available by Weise (1892f) who used it as a name of a variety and gave its description.

1107 *Chrysomela geniculata* Ziegler (Duftschmid, 1825: 185, under var. γ of *Ch. collaris* Linnaeus, 1758). Weise (1884d: 560) adopted it as a name of variety but the respective part of the book contains also taxa classified as subspecies and thus the varieties are deemed infrasubspecific. The name is available from Stierlin (1897) who used it for a variety provided with a short description.

1108 *Chrysomela cupraria* Eversmann (Suffrian 1851a: 189, in synonymy with *Chrysomela aurichalcea* Mannerheim, 1825). Kolbe (1886) explicitly described *Ch. cupraria* Eversmann as a new species without reference to Suffrian, thus it has to be understood as an independent nomenclatural act and not validation of name published in synonymy. Consequently, the name *Ch. cupraria* in Suffrian (1851a) remains unavailable but is available from Kolbe (1886).

1109 *Chrysomela ericae* Dahl (Suffrian 1851a: 110, under var. η of *Ch. cerealis* Linnaeus, 1767). The name *C. ericae* was made available by Motschulsky (1860b: 227) who provided its description.

1110 *Chrysomela rufolineata* Motschulsky (Suffrian 1851a: 111, under var. λ of *Ch. cerealis* Linnaeus, 1767). The name *Ch. rufolineata* was made available by Motschulsky (1860b: 227) who provided its description.

1111 *Chrysomela sturmii* Ullrich (Suffrian 1851a: 36, under var. β of *Ch. goettingensis* Linnaeus, 1761). The name, with the spelling *sturmi*, was made available by Westhoff (1882) who provided its description. *Chrysomela goettingensis* Linnaeus, 1761 is a primary junior homonym of *Ch. goettingensis* Linnaeus, 1758 (now in *Timarcha*), consequently *Ch. sturmi* is in usage being the oldest available name for this taxon.

1112 *Chrysomela hungarica* Friwaldszky (Suffrian 1851a: 8, under var. γ of *Ch. fimbrialis* Küster, 1845). The name *Ch. hungarica* was made available by Fuss (1861a).

1113 *Chrysomela inflata* Herbst (Suffrian 1851a: 7, in synonymy with *Ch. atra* Herrich-Schäffer, 1838). Because *Chrysomela atra* Herrich-Schäffer, 1838 is a junior primary homonym of *Chrysomela atra* Goeze, 1777 (now synonym of *Chrysolina haemoptera* (Linnaeus, 1758)), Weise (1916b) explicitly proposed *Ch. inflata* as a new replacement name for *Ch. atra* Herrich-Schäffer, 1838. The latter has to be treated as independent nomenclatural act and not as a validation of a name published in synonymy. Consequently, the name *Ch. inflata* in Suffrian (1851a) remains unavailable, but is available from Weise (1916b).

1114 *Chrysomela viridescens* Kollar (Suffrian 1851a: 143, under var. γ of *Ch. speciosissima* (Scopoli, 1763)). The name *Ch. viridescens* Kollar was used by Letzner (1853) as a variety provided with a description, but varieties in Letzner (1853) are deemed infrasubspecific in the present catalogue (see comment 575). The name *Ch. viridescens* is available from Weise (1883a).

1115 *Chrysomela bannatica* Sturm (Suffrian 1851a: 173, under var. β of *Ch. speciosa* Linnaeus, 1767). Weise (1884d) treated it as a variety but the respective part of the book contains also taxa classified as subspecies and thus the varieties are deemed infrasubspecific. However, Bechyné (1958b) adopted the variety *bannatica* Weise, 1884 as a subspecies and thus it is deemed available with Weise's authorship (Art. 45.6.4.1, ICZN 1999). In the present Catalogue it is listed as *Oreina alpestris bannatica* (Weise, 1884).

1116 *Chrysomela heerii* Bremi (Suffrian 1851a: 164, under var. α of *Ch. nivalis* Suffrian, 1851). The name *Ch. heerii* is available from Weise (1883c) who used it for a variety provided with a description.

1117 *Chrysomela splendens* Jurine (Suffrian 1851a: 164, under var. δ of *Ch. nivalis* Suffrian, 1851). The name *Ch. splendens* is available from Weise (1883c) who used it for a variety provided with a description.

1118 *Chrysomela decostigma* Andersch (Duftschmid, 1825: 203, under var. ζ of *Ch. tibialis* Duftschmid, 1825). Weise (1884d: 498) adopted it as a name of variety but the respective part of the book contains also taxa classified as subspecies and thus the varieties are deemed infrasubspecific. The name, with the spelling *decastigma*, is available from Stierlin (1896b) who used it for a variety provided with a short description.

1119 *Chrysomela flavicollis* Dahl (Duftschmid, 1825: 206, under var. κ of *Ch. dispar* Paykull, 1798). Weise (1884d: 508) adopted it as a name of variety but the respective part of the book contains also taxa classified as subspecies and thus the varieties are deemed infrasubspecific. The name is available from Stierlin (1896b) who used it for a variety provided with a short description.

1120 *Timarcha angusticollis* Géné (Fairmaire & Allard 1873: 170 and Fairmaire 1884b: 93, under var. C of *T. tenebricosa* (Fabricius, 1775)). Weise (1882b: 322) listed *T. angusticollis* within the synonyms of *T. tenebricosa*. The name *angusticollis* was subsequently adopted as a valid name of subspecies by Bechyné (1945a: 8) and thus validated with Weise's authorship (ICZN 1999, Art. 11.6). It is also secondary homonym of *Chrysomela angusticollis* Duftschmid, 1825, now in synonymy with *T. goettingensis goettingensis* (Linnaeus, 1758).

1121 *Podagrica tristicula* Chevrolat (Allard 1860e: 543, under var. b of *P. saracena* (Reiche & Saulcy, 1858)). The name was made available by Weise (1886a: 684) who adopted it for a species and provided description.

1122 *Clythra ovata* Klug (Lacordaire 1848: 200, under var. A of *Clythra* (*Clythra*) *nigrocincta* Lacordaire, 1848). The name is available from Weise (1881: 117) who used it for a variety provided with a description.

1123 *Labidostomis syriaca* Dejean (Lacordaire 1845: 75, under var. B of *Clythra* (*Labidostomis*) *centromaculata* Gené, 1839). The name was made available by Weise (1881: 95) who adopted it as a variety and provided description.

1124 *Labidostomis dalmatina* Dejean (Lacordaire 1848: 71, under var. C of *Clythra (Labidostomis) longimana* (Linnaeus, 1761)). Weise (1881: 96) mentioned *L. dalmatina* under var. a of *L. longimana*, for documentary purposes only, that cannot be considered as a validation. Depoli & Goidanich (1926) treated it as subspecies of *L. longimana* but without any description which also does not make the name available. Therefore, it is available from Medvedev (1990a) with his authorship as he treated it a subspecies of *L. longimana* and provided a description.

1125 *Melitonoma scita* Dejean (Lacordaire 1848: 377, under var. E of *Clythra (Melitonoma) sobrina* Lacordaire, 1848). In our opinion, the authorship can be credited to Weise (1909: 164), who mentioned the name under a variety, where it immediately followed the denoting letter.

1126 *Clythra sennariensis* Kollar (Lacordaire 1848: 224, under var. D of *Clythra (Clythra) notata* Klug, 1835). The name *Ch. sennariensis* was validated by Medvedev (1971b) who treated it as a subspecies. The oldest available name for the taxon is *C. rufitarsis* Lacordaire, 1848.

1127 *Cheilotoma manicata* Reiche (Lacordaire 1848: 304, under var. A of *Clythra (Gynandrophthalma) affinis* Illiger, 1794). The name *manicata* was made available by Lefèvre (1872b) who treated it as a species and provided it with a description.

1128 *Clytra aegyptiaca* Motschulsky (Lefèvre, 1872b: 337, in synonymy with *Gynandrophthalma menetriesii* (Ménétriés, 1832)). It was subsequently adopted as a valid name of a variety of *G. menetriesii* by Weise (1881: 125) and thus validated with Lefèvre's authorship (ICZN 1999, Art. 11.6).

1129 *Clytra venusta* Lacordaire (Lefèvre, 1872b: 337, in synonymy with *Gynandrophthalma menetriesii* (Ménétriés, 1832)). It was subsequently adopted as a valid name of a variety of *G. menetriesii* by Weise (1881: 125) and thus validated with Lefèvre's authorship (ICZN 1999, Art. 11.6).

1130 *Cryptocephalus rufilabris* Sturm (Suffrian 1848: 19, in synonymy with *C. janthinus* Germar, 1823). Later on, Suffrian (1853b) explicitly described *C. rufilabris* as a new species. The description by Suffrian (1853b) has to be treated as independent nomenclatural act and not as adoption of a name published in synonymy. Consequently, the name *C. rufilabris* remains unavailable from Suffrian (1848) but is available from Suffrian (1853b).

1131 *Cryptocephalus etruscus* Dej. (Suffrian 1847b: 38, in synonymy with *C. ilicis* Olivier, 1808 sensu Suffrian (1847b)). It was listed as a species by Calwer (1857: 675) and Compayo (1862: 820) in a way that cannot be considered as a validation. Weise (1881: 162) mentioned it under *C. illicis* among variability together with *C. quadrinotatus* Sturm without giving a priority to one of them, thus they remained unavailable. The name *etruscus* was made available by Heyden (1892a) and consequently the name has his authorship.

1132 *Cryptocephalus binotatus* Dahl (Suffrian 1847b: 38, in synonymy with *C. ilicis* Olivier, 1808 sensu Suffrian (1847b)). The name was subsequently listed by Marseul (1874: 39) as a variety, but together with *C. blockii* Dahl, which cannot be considered as a validation. Gozis (1907: 72) used it as a valid name for a variety and provided it with a description, thus making it available with his authorship.

1133 *Cryptocephalus cruciatus* Dahl (Suffrian 1847b: 188, under var. β of *C. moraei* (Linnaeus, 1758)). The name was subsequently mentioned together with *C. interruptofasciatus* Ziegler under a variety by Marseul (1874: 165) and Weise (1888: 225), but for documentary purposes only, and thus remains unavailable. The name *C. cruciatus* is available from Gozis (1907: 151) who used it for a variety and provided its description.

1134 *Cryptocephalus turcicus* Heyden (Suffrian 1847b: 173, under var. γ of *C. flavipes* Fabricius, 1781). The name *C. turcicus* is available from Weise (1882b: 215) who used it as a species and provided a description. However, the oldest available name for this taxon is *C. pistaciae* Suffrian, 1853, which was explicitly proposed for var. γ of *C. flavipes*. Accordingly, *C. turcicus* Weise, 1882 is synonymized with *C. pistaciae*. A reversal of precedence based on Art. 23.9.1 (ICZN 1999) is not possible because *C. pistaciae* was used as a valid name after 1900 by Gozis (1907: 68, 97, 98).

1135 *Cryptocephalus maurus* Friwaldszky (Suffrian 1848: 108, under var. β of *C. quadriguttatus* C. F. W. Richter, 1820). Weise (1882b: 211) mentioned *C. maurus* together with two other names under var. c of *C. quadriguttatus* for documentary purposes only, what cannot be considered as a validation. The name is available from Stierlin (1896a) who used it for a variety provided with a description.

1136 *Cryptocephalus apicalis* Ziegler (Suffrian 1847b: 177, in synonymy with *C. quadripustulatus* Gyllenhal, 1813). Marseul (1874: 161) used the name for a variety without reference to Suffrian (1847b), however, in the index it is listed as "*apicalis* Mars.", which hence makes the name available with his authorship. Consequently, *C. apicalis* in Suffrian (1847b) remains unavailable.

1137 *Cryptocephalus bisignatus* Megerle (Suffrian 1847b: 177, under var. β of *C. quadripustulatus* Gyllenhal, 1813). The name was made available by Marseul (1874: 161) who used it for a variety and provided its description.

1138 *Cryptocephalus signatifrons* Sturm (Suffrian 1847b: 172, under var. β of *C. flavipes* Fabricius, 1781). Mentioned as a species by Calwer (1857: 676) in a way that cannot be considered as a validation. Weise (1881: 214) mentioned *C. signatifrons* together with two other names under var. h of *C. flavipes* Fabricius, 1781 for documentary purposes only, that cannot be considered as a validation. *Cryptocephalus signatifrons* is available from Stierlin (1896a) who used it as a variety provided with a description, and it takes his authorship.

1139 *Cryptocephalus intrusus* Megerle (Suffrian 1847b: 137, in synonymy with *C. aureolus* Suffrian, 1847). Weise (1881: 183) adopted *C. intrusus* for a variety, provided it with a description and thus made it available with his authorship as there is no reference to Suffrian. Consequently, the name *C. intrusus* in Suffrian (1847b) remains unavailable. In the present Catalogue *C. intrusus* is listed as synonym of *C. zambanellus* Marseul, 1874 in agreement with Weise (1881, 1916b).

1140 *Cryptocephalus temesiensis* Megerle (Suffrian 1847b: 101, under var. γ of *C. coryli* (Linnaeus, 1758)). The name *C. temesiensis* was made available by Marseul (1874: 55) who used it as a name of a variety and provided its description.

1141 *Cryptocephalus sinuatus* Sturm (Suffrian 1847b: 97, in synonymy with *C. fasciatus* Herrich-Schäffer, 1835). Mentioned as a species by Calwer (1857: 676) in a way that cannot be considered as a validation. The name *C. sinuatus* is available from Harold (1872) who proposed it as a new replacement name for *C. fasciatus*, which should be considered as an independent nomenclatural act and not as a validation of the name published in synonymy. Consequently, the name *C. sinuatus* remains unavailable from Suffrian (1847b) but is available from Harold (1872).

1142 *Pachybrachis ater* Dahl (Suffrian 1848: 119, in synonymy with *P. piceus* Suffrian, 1848) but with comment that Dahl had under this name also unicoloured specimens of *P. maculatus* Parr. (described by Suffrian as a distinct species), thus to avoid confusion he established a new taxon *P. piceus*. However, Weise (1882b: 264) used the name *P. ater* Dahl for a variety of *P. limbatus* (including *P. maculatus* Suffrian used for another variety of it) but did not referred to Suffrian hence the name cannot be associated with certainty and is unavailable from Suffrian (1848) but available with Weise's (1882b) authorship.

1143 *Pachybrachis erycinus* Géné (Suffrian 1848: 122, in synonymy with *P. scriptus* (Herrich-Schäffer, 1838)). The name was made available by Weise (1882b: 254) who used it for a variety and provided description. As the reference to Suffrian (1848) was not mentioned, the name retains Weise's authorship.

1144 *Pachybrachis hellwigii* Dahl (Suffrian 1848: 122, in synonymy with *P. scriptus* (Herrich-Schäffer, 1838)). The name was made available by Weise (1882b: 254) who used it for avariety and provided description. As the reference to Suffrian (1848) was not mentioned, the name retains Weise's authorship.

New distribution records based on photographs presented on iNaturalist

1145 We noticed that some of the photographs presented on iNaturalist website (www.inaturalist.org) actualy represent new faunistic records for respective countries but at the same time they remain formally unpublished and largely not reflected in the literature. We decided to use these records and present them here in order to present as precise distribution of palaearctic Chrysomelidae as possible. On the other hand, we have picked up only photographic records clearly showing all necessary characters and thus allowing unambiguous identification of the respective species.

Abiromorphus anceyi Pic, 1924 – **A:** CH (SHG) – https://www.inaturalist.org/observations/166821725

Abirus fortunei (Baly, 1861) – **A:** CH (HKG) – https://www.inaturalist.org/observations/25051747

Aoria bowringii (Baly, 1860) – **A:** CH (SHG) – https://www.inaturalist.org/observations/166542107

Aoria nigripennis Gressitt & Kimoto, 1961 – **A:** CH (ZHE) – https://www.inaturalist.org/observations/163286335

Aspidomorpha orientalis Boheman, 1857 – **A:** ID (HP) – https://www.inaturalist.org/observations/125586897

Basilepta cribricollis (Motschulsky, 1860) – **A:** CH (HEN) – https://www.inaturalist.org/observations/169603342

Basilepta cribricollis (Motschulsky, 1860) – **A:** CH (SHA) – https://www.inaturalist.org/observations/172395437

Basiprionota bisignata (Boheman, 1862) – **A:** CH (LIA) – https://www.inaturalist.org/observations/180695533

Callisina rufa Tan, 1992 – **A:** CH (HKG) – https://www.inaturalist.org/observations/173930141

Callisina rufa Tan, 1992 – **A:** CH (MAC) – https://www.inaturalist.org/observations/57338934

Cassida circumdata Herbst, 1799 – **A:** CH (SHG) – https://www.inaturalist.org/observations/134659750

Cassida inflata Gressitt, 1952 – **A:** CH (JIA) – https://www.inaturalist.org/observations/142420301

Cassida lineola Creutzer, 1799 – **A:** CH (SHN) – https://www.inaturalist.org/observations/166363661

Cassida murraea murraea Linnaeus, 1767 – **E:** LU – https://www.inaturalist.org/observations/48153102

Cassida nobilis Linnaeus, 1758 – **E:** KO – https://www.inaturalist.org/observations/49941093

Cassida nobilis Linnaeus, 1758 – **E:** LU – https://www.inaturalist.org/observations/111298194

Cassida pallidicollis Boheman, 1857 – **A:** KZ – https://www.inaturalist.org/observations/63955368

Cassida piperata Hope, 1843 – **A:** CH (CHQ) – https://www.inaturalist.org/observations/171709518

Cassida rati Maulik, 1923 – **A:** CH (HAI) – https://www.inaturalist.org/observations/148819986

Cassida rati Maulik, 1923 – **A:** CH (XIZ) – https://www.inaturalist.org/observations/97732017

Cassida sanguinolenta O. F. Müller, 1776 – E: LU – https://www.inaturalist.org/observations/21924226

Cassida sauteri (Spaeth, 1913) – **A**: CH (ANH) – https://www.inaturalist.org/observations/161791602

Chrysochus asclepiadeus (Pallas, 1773) – E: AL – https://www.inaturalist.org/observations/144241888

Chrysolina americana (Linnaeus, 1758) – E: LU – https://www.inaturalist.org/observations/168472200

Chrysolina americana (Linnaeus, 1758) – **A**: AB – https://www.inaturalist.org/observations/158493464

Chrysolina cerealis cerealis (Linnaeus, 1767) – E: LU – https://www.inaturalist.org/observations/162109524

Chrysolina coerulans coerulans (Scriba, 1791) – E: LU – https://www.inaturalist.org/observations/165495549

Chrysolina herbacea (Duftschmid, 1825) – E: LU – https://www.inaturalist.org/observations/128667286

Chrysolina hyperici (Forster, 1771) – E: LU – https://www.inaturalist.org/observations/26074490

Chrysolina oricalcia (O. F. Müller, 1776) – E: LU – https://www.inaturalist.org/observations/167718073

Chrysolina polita (Linnaeus, 1758) – E: LU – https://www.inaturalist.org/observations/123267565

Chrysolina staphylaea staphylaea (Linnaeus, 1758) – E: LU – https://www.inaturalist.org/observations/99082162

Chrysomela cuprea Fabricius, 1775 – E: LU – https://www.inaturalist.org/observations/13328353

Chrysomela populi Linnaeus, 1758 – E: LU – https://www.inaturalist.org/observations/168251789

Chrysomela vigintipunctata (Scopoli, 1763) – E: LU – https://www.inaturalist.org/observations/168352262

Cleoporus badius Lefèvre, 1889 – **A**: CH (HKG) – https://www.inaturalist.org/observations/167854934

Cleorina janthina Lefèvre, 1885 – **A**: CH (MAC) – https://www.inaturalist.org/observations/108696241

Clytra novempunctata Olivier, 1808 – E: KO – https://www.inaturalist.org/observations/48118367

Colasposoma viridicoeruleum Motschulsky, 1860 – **A**: CH (HKG) – https://www.inaturalist.org/observations/6710940

Colasposoma viridicoeruleum Motschulsky, 1860 – **A**: CH (MAC) – https://www.inaturalist.org/observations/78523360

Cryptocephalus octopunctatus (Scopoli, 1763) – E: AN – https://www.inaturalist.org/observations/78977311

Fidia phoebicola (Tan, 1983) – **A:** CH (HKG) – https://www.inaturalist.org/observations/87064680

Gastrophysa viridula (DeGeer, 1775) – E: AN – https://www.inaturalist.org/observations/15226121

Glyphocassis lepida Spaeth, 1914 – **A:** CH (CHQ) – https://www.inaturalist.org/observations/47959820

Gonioctena decemnotata (Marsham, 1802) – E: LU – https://www.inaturalist.org/observations/161273177

Gonioctena viminalis viminalis (Linnaeus, 1758) – E: LU – https://www.inaturalist.org/observations/114389535

Gonophora pulchella Gestro, 1888 – **A:** ID (UT) – https://www.inaturalist.org/observations/92871046, https://www.inaturalist.org/observations/91928778

Laccoptera chinensis (Fabricius, 1798) – **A:** CH (SHA) – https://www.inaturalist.org/observations/127615640

Parascela cribrata (Schaufuss, 1871) – **A:** CH (JIA) – https://www.inaturalist.org/observations/115077313

Plagiosterna aenea aenea (Linnaeus, 1758) – E: LU – https://www.inaturalist.org/observations/127473118

Platycorynus undatus (Olivier, 1792) – **A:** CH (HKG) – https://www.inaturalist.org/observations/126303455

Platycorynus undatus (Olivier, 1792) – **A:** CH (MAC) – https://www.inaturalist.org/observations/24924445

Prasocuris glabra (Herbst, 1783) – E: LU – https://www.inaturalist.org/observations/121395279

Prasocuris hannoveriana (Fabricius, 1775) – E: LU – https://www.inaturalist.org/observations/157584531

Prasocuris junci (Brahm, 1790) – E: LU – https://www.inaturalist.org/observations/80722456

Sagra femorata (Drury, 1773) – **A:** BT – https://www.inaturalist.org/observations/58937924

Spitiella collaris (Baly, 1878) – E: ID (AP) – https://www.inaturalist.org/observations/154706134

Timarcha metallica (Laicharting, 1781) – E: LU – https://www.inaturalist.org/observations/127894535

Timarcha tenebricosa (Fabricius, 1775) – E: LU – https://www.inaturalist.org/observations/161756457

Xanthogaleruca luteola (O. F. Müller, 1766) – **A:** LE – https://www.inaturalist.org/observations/29056257

Catalogue

superfamily CHRYSOMELOIDEA
family ORSODACNIDAE C. G. Thomson, 1859
subfamily **Orsodacninae** C. G. Thomson, 1859

genus *Orsodacne* Latreille, 1802: 223 type species: *Chrysomela cerasi* Linnaeus, 1758 [ICZN (2002), Opinion 1989]

Orsodacna Latreille, 1804a: 350 [incorrect subsequent spelling]

arakii Chûjô, 1942d: 33 **A:** JA

kurosawai Chûjô, 1949b: 24

nakanei Chûjô, 1942d: 34

cerasi Linnaeus, 1758: 376 (*Chrysomela*) **E:** AN AU BE BH BU BY CR CZ DE EN FI FR GB GE HU IT LA LS LT LU MD NL NR PL RO RU (CT NT ST) SB SL SP SV SK SZ UK **A:** AB IN KZ RU (ES FE SS WS) TR

apicalis Csiki, 1953: 121 (ab.) [IN]

atra Gistel 1857b: 527, 545 (*Lema*)

baudii Pic, 1913i: 154 (var.)

bohemica Roubal, 1914: 92 (ab.) [IN]

budensis Csiki, 1953: 120 (ab.) [IN]

cantharoides Fabricius, 1775: 120 (*Crioceris*)

chlorotica Olivier, 1791b: 203 (*Crioceris*)

duftschmidi Weise, 1891a: 355 (var.) [RN]

fulvicollis Fabricius, 1792b: 5 (*Crioceris*)

glabrata Panzer, 1795a: 170 (*Crioceris*)

horvathi Laczó, 1912: 5 (ab.) [IN]

humerata Csiki, 1953: 121 (ab.) [IN]

johannis Csiki, 1941: 55 (ab.) [IN]

lacordairei Pic, 1913i: 154 (var.) [RN]

limbata Olivier, 1808: 754 (*Orsodacna*)

limbatipennis Pic, 1913i: 154 (var.)

lineola Lacordaire, 1845: 76 (*Orsodacna*) [HN]

melanura Fabricius, 1792b: 22 (*Galleruca*)

nigroculata Moll, 1784b: 197 (*Chrysomela*)

palaseki Zoufal, 1926: 49 (ab.) [IN]

pallida Geoffroy, 1785: 96 (*Crioceris*)

pulchra Csiki, 1953: 121 (ab.) [IN]

simplex Csiki, 1953: 120 (ab.) [IN]

suturalis Jacobson, 1922b: 54 (*Orsodacna*, ab.) [IN]

tenuecincta Csiki, 1953: 121 (ab.) [IN]

 | DOI:10.1163/9789004443303_003

theresae Pic, 1913i: 154 (var.)

humeralis Latreille, 1804a: 350 (*Orsodacna*) E: AL AU BH BU CR CZ FR GB GE GR HU IT MD ME NM PL PT RO RU (ST) SB SK SL SP SZ TR UK **N:** MO **A:** AB AF CY GG IN IQ IS LE SY TR

anatolica Pic, 1916b: 11 (var.) [IN] [# 001]

brancsiki Laczó, 1909: 57 (ab.) [IN]

coerulescens Duftschmid, 1825: 248

croatica Weise, 1883d: 251 (*Orsodacna*, var.)

delagrangei Pic, 1896b: 30 (var.)

flava Csiki, 1899: 93 (var.)

graeca Pic, 1941e: 13 (*Orsodacna*, var.)

hispanica Pic, 1913i: 154 (var.)

innotatipennis Pic, 1941d: 15 (*Orsodacna*, var.)

kraatzi Pic, 1913i: 154 (var.)

lineola Panzer, 1795a: 170 (*Crioceris*) [HN]

luteoapicalis Pic, 1941d: 15 (*Orsodacna*, var.)

luteonotata Pic, 1941e: 13 (*Orsodacna*, var.)

marginalis Roubal, 1941: 6 (forma) [IN] [# 001]

marginata Csiki, 1899: 94 (var.)

marginella Duftschmid, 1825: 248

mespili Lacordaire, 1845: 73 (*Orsodacna*)

nematoides Lacordaire, 1845: 84 (*Orsodacna*)

nigriceps Latreille, 1807: 44 (*Orsodacna*)

nigricollis Olivier, 1808: 755 (*Orsodacna*)

picipennis Duftschmid, 1825: 250

ruficollis Pic, 1894g: cclxxxiv

suturata Csiki, 1953: 121 (ab.) [IN]

syriaca Pic, 1896f: 202 (var.)

variabilis Baly, 1877f: 377 (*Orsodacna*)

violacea Chevrolat, 1833: pl. 47 [= 1844: 257] (*Orsodacna*)

mesopotamica Legalov, 2022: 82 **A:** IQ

yunnanica Legalov, 2021: 75 **A:** CH (YUN)

genus *Orsodacne*, excluded taxon

atra Ahrens, 1810: 46 (*Donacia*) **NAR** [# 002]

family MEGALOPODIDAE Latreille, 1802

subfamily **Zeugophorinae** Böving & Craighead, 1931

genus *Zeugophora* Kunze, 1818b: 71 [conserved name, ICZN (1986), Opinion 1382]

type species: *Crioceris subspinosa* Fabricius, 1781

Auchenia Thunberg, 1792: 95, 116 [suppressed name, ICZN (1986), Opinion 1382] type species: *Crioceris subspinosa* Fabricius, 1781
Austrolema Oke, 1932: 164 type species: *Austrolema vitinea* Oke, 1932
Bruchomima Achard, 1916a: 47 type species: *Bruchomima chloropelta* Achard, 1916
Macrozeugophora Achard, 1914b: 288 type species: *Macrozeugophora ornata* Achard, 1914
Papuleptura Gressitt, 1959: 80 type species: *Papuleptura alticola* Gressitt, 1959
Pedrillia Westwood, 1864: 280 type species: *Pedrillia longicornis* Westwood, 1864
Pedrillimorpha Papp, 1946d: 25 [UN]
Pedrilliomorpha Pic, 1917a: 9 type species: *Pedrilliomorpha atrosuturalis* Pic, 1917
Pedrilonga Papp, 1946d: 26 [UN]
Pedrinella Papp, 1946d: 26 [UN]
Taraxis LeConte, 1850: 237 type species: *Taraxis abnormis* LeConte, 1850

ancora Reitter, 1900a: 164 **A:** CH (GAN NIN NMO QIN SHA)
pseudancora Reitter, 1900a: 165 (var.)

annulata Baly, 1873: 79 (*Pedrillia*) **A:** CH (HEI JIL JIX LIA NIN NMO) JA NC RU (FE) SC [# 011]
biguttata Kraatz, 1879a: 119 (*Pedrillia*)
disconotata Pic, 1906b: 27 (*Pedrillia*, var.)
inannulata Chûjô, 1959: 1 (forma) [IN] [# 004]
melanaria Chûjô, 1959: 1 (forma) [IN] [# 004]
theresae Pic, 1945c: 13 (*Pedrillia*, var.)

atrosuturalis Pic, 1917a: 9 (*Pedrilliomorpha*) **A:** ID (SD)
bicolor Kraatz, 1879a: 120 (*Pedrillia*) **A:** CH (FUJ GAN LIA) NC RU (FE) SC
bifasciata Gressitt & Kimoto, 1961: 25 **A:** CH (ANH FUJ HEN ZHE)
bimaculata Kraatz, 1879b: 129 **A:** CH (HEI) RU (FE)
brancuccii L. N. Medvedev, 1993c: 360 **A:** NP
chinensis L. N. Medvedev, 1998e: 65 **A:** CH (SCH)
chujoi Ohno, 1961b: 36 **A:** JA (Honshu)
cribrata S.-H. Chen, 1974a: 43, 47 **A:** CH (QIN)
cupka Takemoto, 2019: 55 **A:** JA (Hokkaido) RU (FE: Sakhalin)
cyanea S.-H. Chen, 1974a: 44, 47 **A:** CH (QIN SHA SCH YUN)
decorata Chûjô, 1937d: 146 (*Pedrillia*) **A:** CH (TAI)
dimorpha Gressitt, 1945a: 138 **A:** CH (GUA HUN JIA JIX TAI) [# 011]
emeica K.-Q. Li & H.-B. Liang, 2018: 135 **A:** CH (SCH)
euonymorum K.-Q. Li & H.-B. Liang, 2020: 57 **A:** CH (GUI NIN)

flavicollis Marsham, 1802: 217 (*Auchenia*) **E:** AU BE BU BY CR CZ DE FR GB GE HU IT LA LU NL PL RO RU (CT NT) SK SL SP SZ UK **A:** RU (ES FE SS WS) [# 003]
australis Weise, 1881: 58 (var.)
notatipes Pic, 1925b: 9 (var.) [IN] [# 004]

flavitarsis L. N. Medvedev, 1998e: 68 **A:** ID (SD)

flavithorax K.-Q. Li & H.-B. Liang, 2020: 61 **A:** CH (GUI)
flavonotata Chûjô, 1935a: 69 (*Pedrillia*) **A:** CH (TAI) JA (Iriomote-jima, Ishigaki-jima)
formosana Gressitt, 1945a: 140 [RN] **A:** CH (TAI)
  ornata Chûjô, 1937d: 148 (*Pedrillia*) [HN]
frontalis Suffrian, 1840: 100 E: AU BE BY CZ DE EN GE IT LA LT PL RU (CT) SV **A:** RU (ES FE) [# 003]
gracilis gracilis Chûjô, 1937d: 149 (*Pedrillia*) **A:** CH (TAI)
gracilis unicolor Chûjô, 1958c: 1 (ssp.) **A:** JA (Okinawa) [# 005]
himalayana L. N. Medvedev, 1998e: 67 **A:** ID (UT)
hozumii Chûjô, 1953a: 1 **A:** JA RU (FE)
impressa S.-H. Chen & Pu, 1962b: 116, 119 **A:** CH (YUN)
indica Jacoby, 1903a: 81 **A:** CH (YUN) **ORR** [# 006]
  flavipes Jacoby, 1908: 14 (*Pedrillia*)
japonica Chûjô, 1951c: 119 **A:** JA (Hokkaido)
longicornis Westwood, 1864: 280 (*Pedrillia*) **A:** CH (YUN) ID (SD UT) NP **ORR**
  andrewesi Jacoby, 1903a: 82
  bicoloriventris Pic, 1911h: 144 (*Pedrillia*)
maai Kimoto & Gressitt, 1979: 206 **A:** NP **ORR**
maculata Chûjô, 1941d: 463 **A:** CH (GUI TAI)
multisignata Pic, 1944a: 7 (*Pedrillia*) **A:** CH (SCH)
  interrupta Pic, 1944a: 8 (*Pedrillia*, var.)
nepalica L. N. Medvedev, 1998e: 67 **A:** NP
nigricollis Jacoby, 1885b: 195 (*Pedrillia*) **A:** CH (NIN) JA RU (FE) SC
nigroaerea Lopatin, 2008d: 831 **A:** CH (GAN)
nigroapica K.-Q. Li & H.-B. Liang, 2018: 143 **A:** CH (YUN)
nitida Chûjô, 1932b: 340 (*Pedrillia*) **A:** CH (TAI)
ornata Achard, 1914b: 289 (*Macrozeugophora*) **A:** CH (JIA JIX YUN)
pallidicincta Gressitt, 1945a: 139 [RN] **A:** ID (SD) NP
  unifasciata Pic, 1917a: 9 [HN]
parva Crowson, 1946: 96 **A:** ID (UT) NP
ruficollis Chûjô, 1932b: 342 (*Pedrillia*) **A:** CH (TAI)
  grandis Chûjô, 1937d: 147 (*Pedrillia*)
scutellaris Suffrian, 1840: 99 E: AU BE BH BY CZ DE FI FR GE HU IT LA LT NL NR PL RO RU (NT ST) SB SK SV SZ UK **A:** CH (GAN HEB HEI JIL LIA NMO QIN SHA XIN) GG IN IS KI KZ RU (ES FE SS WS) SY TD TR **NARi** [# 011]
subspinosa Fabricius, 1781: 155 (*Crioceris*) E: AU BE BH BU BY CR CZ DE EN FI FR GB GE HU IT KR LA LT LU MD NL NR PL RO RU (CT NT ST) SK SL SV SZ TR UK **A:** KZ MG RU (ES FE SS WS)
tetraspilota L. N. Medvedev, 1998e: 66 **A:** CH (YUN)
tricolor S.-H. Chen & Pu, 1962b: 116, 118 **A:** CH (GAN) [# 007]
trifasciata K.-Q. Li & H.-B. Liang, 2020: 69 **A:** CH (YUN)

trisignata An & Kwon, 2002: 273 **A**: SC
turneri Power, 1863: 8735 **E**: BY CZ DE EN FI GB GE LA LT NR PL RO RU (CT NT) SV **A**: CH (BEI) MG RU (ES SS WS)
rufotestacea Kraatz, 1872a: 162
unifasciata Jacoby, 1885b: 197 (*Pedrillia*) **A**: JA
varipes Jacoby, 1885b: 196 (*Pedrillia*) **A**: JA
weisei Reitter, 1889b: 43 **A**: AR IN
wittmeri L. N. Medvedev, 1993c: 359 **A**: ID (KA)
xanthopoda Bezděk & Silfverberg, 2010: 61 [RN] **A**: CH (CHQ) [# 008]
flavipes P.-Y. Yu, 1997a: 838, 845 (*Pedrillia*) [HN]
yuae K.-Q. Li & H.-B. Liang, 2020: 72 **A**: CH (YUN)
yunnanica S.-H. Chen & Pu, 1962b: 116, 119 **A**: CH (XIZ YUN) NP **ORR**

subfamily **Megalopodinae** Latreille, 1802
tribe Megalopodini Latreille, 1802
subtribe Temnaspidina Rodríguez-Mirón, 2021

genus *Clythraxeloma* Kraatz, 1879b: 143 type species: *Clythraxeloma cyanipennis* Kraatz, 1879
Clytraxeloma Clavareau, 1913: 18 [UE] type species: *Clythraxeloma cyanipennis* Kraatz, 1879
assamensis assamensis Jacoby, 1908: 91 (*Temnaspis*) **A**: CH (HAI) **ORR**
assamensis yunnana S.-H. Chen & Pu, 1962b: 115, 118 (*Poecilomorpha*, ssp.) **A**: CH (YUN)
cyanipennis Kraatz, 1879b: 143 **A**: CH (BEI FUJ GAN HEB HEI JIA JIX LIA MNO SHA ZHE) NC RU (FE) SC
discolineata Pic, 1938g: 356 (*Temnaspis*) **A**: CH (YUN) **ORR**
downesii Baly, 1859b: 205 (*Temnaspis*) **A**: CH (GUI HUB SCH YUN) ID ("North India") NP **ORR**
inlineata Pic, 1922f: 7 (*Temnaspis*, var.)
penae Gressitt, 1942a: 285 (*Poecilomorpha*)
laosensis Pic, 1922f: 7 (*Temnaspis*) **A**: CH (YUN) **ORR**
maculata Pic, 1926b: 9 (*Temnaspis*) **A**: CH (GUX SCH YUN) **ORR**
mediofasciata Pic, 1955e: 9 (*Temnaspis*, var.)
mouhoti Baly, 1864e: 435 (*Temnaspis*) **A**: CH (YUN) **ORR**
oberthuri Jacoby, 1895b: 253 (*Temnaspis*)
pretiosa Reineck, 1923: 609 (*Poecilomorpha*) **A**: CH (FUJ GUA GUX HAI HUB HUN JIX TAI YUN ZHE)
elegantula Gressitt, 1942a: 284 (*Poecilomorpha*)

genus *Temnaspis* Lacordaire, 1845: 716 type species: *Megalopus javanus* Guérin-Méneville, 1844
Colobaspis Fairmaire, 1894a: 225 type species: *Colobaspis flavonigra* Fairmaire, 1894

ashlocki Kimoto & Gressitt, 1979: 208 **A:** NP **ORR**
atrithorax Pic, 1934g: 87 (*Colobapsis* [sic!]) **A:** CH (HUN SCH)
bengalensis Westwood, 1864: 276 **A:** NP **ORR** [# 009]
centromaculatus L. N. Medvedev & Sprecher-Uebersax, 1997: 206 **A:** NP
elegans Chûjô, 1951d: 64 **A:** CH (TAI)
femoratus Gressitt, 1942a: 287 (*Colobaspis*) **A:** CH (FUJ GUX SCH)
flavicornis Jacoby, 1892b: 875 **A:** CH (GUI GUX SCH YUN) **ORR**
bidentatus Pic, 1922e: 27
flavoniger Fairmaire, 1894a: 225 (*Colobaspis*) **A:** CH (SCH XIZ)
formosanus Reineck, 1923: 606 (*Colobaspis*) **A:** CH (TAI)
fraxini Komiya, 1986: 6 (*Colobaspis*) **A:** CH (TAI)
humeralis Jacoby, 1890a: 86 **A:** CH (CHQ HUB SHA) [# 010, # 011]
humerosus Lopatin, 2013: 765 **A:** CH (SCH)
insignis Baly, 1859b: 208 **A:** CH (HAI YUN) ID ("North India") **ORR**
brunneipennis Pic, 1926b: 9
japonicus Baly, 1873: 78 **A:** JA RU (FE)
kwangtungensis Gressitt, 1942a: 289 (*Colobaspis*) **A:** CH (GUA GUI)
nankineus Pic, 1914f: 20 (*Colobapsis* [sic!]) **A:** CH (HEB HEN HUB JIA JIX SHN SHX TAI ZHE) RU (FE) SC
coreanus Chûjô, 1934d: 34
nigriceps Baly, 1859b: 207 **A:** CH (YUN) ID (SD) NP **ORR** [# 011]
bicoloriceps Pic, 1922f: 8 (*Colobaspis*)
nigropunctatus Pic, 1896a: 36 (*Clytraxeloma*) **A:** SY TR
quadrimaculatus Reitter, 1908: 135 (*Poecilomorpha*)
omeiensis Gressitt, 1942a: 290 (*Colobaspis*) **A:** CH (SCH)
pallidus Gressitt, 1942a: 290 (*Colobaspis*) **A:** CH (FUJ HUN JIX SHA ZHE)
puae K.-Q. Li & H.-B. Liang, 2013: 380 **A:** CH (YUN) **ORR**
pulcher Baly, 1859b: 206 **A:** CH (FUJ HUN JIX ZHE) **ORR**
diversesignatus Pic, 1955c: 22
quadriplagiatus Bryant, 1934: 225 **A:** ID (UT) NP
regalis Achard, 1920: 48 (*Colobaspis*) **A:** CH (GUI SCH XIZ YUN)
rubripennis Kraatz, 1879b: 143 **A:** CH ("North-east") RU (FE) SC
bonneuili Pic, 1947c: 15
sanguinicollis S.-H. Chen & Pu, 1962b: 115, 118 **A:** CH (YUN)
sauteri Reineck, 1923: 607 (*Colobaspis*) **A:** CH (TAI)
rubus Chûjô, 1932a: 313 (*Colobaspis*)
septemmaculatus Hope, 1831: 28 (*Megalopus*) **A:** CH (SCH TAI YUN ZHE) ID (HP SD) NP **ORR** [# 011]
nigroplagiatus Jacoby, 1892b: 876
quinquemaculatus Baly, 1859b: 206
shirakii Chûjô, 1932a: 315 **A:** CH (TAI)

speciosus Baly, 1859b: 204 **A:** ID (SD) NP
syringa K.-Q. Li & H.-B. Liang, 2013: 389 **A:** CH (BEI NIN SHA)
testaceus Gressitt & Kimoto, 1961: 30 (*Poecilomorpha*) **A:** CH (YUN)
vitalisi Pic, 1922f: 8 (*Colobaspis*) **A:** CH (YUN) **ORR** [# 011]
pulcherrimus Reineck, 1923: 605 (*Colobaspis*)

family CHRYSOMELIDAE Latreille, 1802
subfamily **Sagrinae** Leach, 1815

genus *Sagra* Fabricius, 1792b: 51 type species: *Tenebrio femoratus* Drury, 1773
Orthosagra Crowson, 1946: 106 type species: *Sagra bicolor* Lacordaire, 1845
Prosagra Crowson, 1946: 106 type species: *Sagra jansoni* Baly, 1860
Sagrania Rafinesque, 1815: 117 [UE] type species: *Tenebrio femoratus* Drury, 1773
Sagrina Crowson, 1946: 106 [HN] type species: *Sagra carbunculus* Hope, 1842
Sagrinola Monrós & Bechyné, 1956: 1120 [RN] type species: *Sagra carbunculus* Hope, 1842
Tinosagra Weise, 1905a: 33 type species: *Sagra tristis* Fabricius, 1798
carbunculus Hope, 1842b: 128 [= 1842c: 248] **A:** ID (AP SD UT) NP **ORR** [# 013, # 016]
femorata Drury, 1773: 64 (*Tenebrio*) **A:** BT CH (FUJ GUA GUI GUX HAI HKG HUB HUN JIX SCH TAIi YUN ZHE) ID (SD UT) JAi PA **AUS ORR** [# 014, # 017, # 1145]
abdominalis Jacoby, 1895a: 52
aenea Olivier, 1807: 500
andamanensis Jolivet, 1951b: 2 (ssp.)
atricolor Pic, 1953a: 8 (var.)
borneoensis Jacoby, 1898b: 352
brevipes Jacoby, 1889a: 148
chrysochlora Lacordaire, 1845: 35 [# 977]
dentipes Fabricius, 1787: 66 (*Alurnus*)
druryi Lacordaire, 1845: 32
empyrea Lacordaire, 1845: 31
fabricii Lacordaire, 1845: 52
formosa Lacordaire, 1845: 49
heterodera Lacordaire, 1845: 50
ignita Lacordaire, 1845: 47
jeanvoinei Pic, 1953a: 8 (var.)
longicollis Lacordaire, 1845: 35
longipes Baly, 1878g: 337
mutabilis Baly, 1864d: 433
nigrita Olivier, 1807: 500
olivieri Weise, 1913a: 18 (var.)
papuana Jacoby, 1889b: 203
pfeifferi Baly, 1860c: 239
puncticollis Jacoby, 1884b: 188

purpurea Lichtenstein, 1796: 60 [conserved name, ICZN (1995b), Opinion 1820]
quadraticollis Lacordaire, 1845: 40
queenslandica Mjöberg, 1917: 14
speciosa Lacordaire, 1845: 38
splendida Weber, 1801: 61
superba Lacordaire, 1845: 39
tonkinensis Kuntzen, 1914: 125 (ssp.)
tridentata Weber, 1801: 62
viridis Sulzer, 1776: 64 (*Tenebrio*)
weberi Lacordaire, 1845: 50

fulgida Weber, 1801: 62 **A:** CH (FUJ GUA GUI GUX HKG HUB JIX SCH SHA XIZ YUN) **ORR** [# 015]
insuturalis Pic, 1953a: 8 (var.)
janthina S.-H. Chen, 1942b: 107 (ssp.)
leechi Jacoby, 1888: 339
minuta Pic, 1930d: 37
subalutacea Pic, 1953a: 8

humeralis Jacoby, 1904a: 294 **A:** CH (GUI YUN ZHE)

jansoni Baly, 1860c: 243 **A:** CH (FUJ YUN) **ORR** [# 017]
favrei Pic, 1953a: 8 (var.) [# 012]
multipunctata Jacoby, 1889a: 149

moghanii S.-H. Chen & Pu, 1962a: 107, 112 **A:** CH (HEN HUB YUN)

mouhoti Baly, 1861: 193 **A:** CH (YUN) **ORR**
suturalis Pic, 1953a: 7

odontopus Gistel, 1831: 309 **A:** CH (GUA YUN) NP **ORR** [# 016]
cyanescens Pic, 1953a: 8 (var.)
impressipennis Pic, 1953a: 7
petelii Lacordaire, 1845: 44
pygmea Lacordaire, 1845: 53

subfamily **Bruchinae** Latreille, 1802
tribe Amblycerini Bridwell, 1932
subtribe Amblycerina Bridwell, 1932

genus *Amblycerus* Thunberg, 1815: 121 type species: *Bruchus robiniae* Fabricius, 1781 **[Appendix]**

robiniae Fabricius, 1781: 75 (*Bruchus*) Ei: HU RO **NAR**

subtribe Spermophagina Borowiec, 1987

genus *Spermophagus* Schoenherr, 1833: 102 type species: *Spermophagus titivilitius* Boheman, 1833
Euspermophagus Zacher, 1930: 237 type species: *Mylabris sericea* Geoffroy, 1785

Spermatophagus Gistel, 1856: 375 [UE] type species: *Spermophagus titivilitius* Boheman, 1833

abdominalis Fabricius, 1781: 76 (*Bruchus*) Ei: GB UK A: CH (CHQ FUJ HEI HKG HUN SCH YUN) ID (SD) JA NC NP RU (FE) SC YE **ORR AFR** [# 018, # 159]

albonotatus Chûjô, 1937e: 198

atriceps Pic, 1924f: 456

atromaculatus Pic, 1917d: 11

burgeoni Pic, 1924f: 457 (var.)

complectus Sharp, 1886a: 37

convolvuli Thunberg, 1816: 46 (*Bruchus*) [# 018]

diversipes Pic, 1922c: 16

diversipes Pic, 1941c: 2 [HN]

elisabethae Pic, 1924f: 457

glabratus Lukjanovitch & Ter-Minassian, 1957: 195 (*Euspermophagus*) [# 1103]

japonicus Schilsky, 1906: 94

kiotensis Pic, 1918a: 8

notatipennis Pic, 1932a: 332

notatithorax Pic, 1917d: 9 (var.)

rufipennis Pic, 1917d: 9

rufiventris Boheman, 1833: 107 [# 018]

rufopygidialis Pic, 1929a: 316

sparsemaculatus Pic, 1917d: 10

sublineatus Boheman, 1839: 140

testaceiventris Pic, 1917d: 11

undulatus Chûjô, 1937e: 199

vadoni Pic, 1942d: 10

aeneipennis Pic, 1917d: 9 **A:** ID (SD) NP **ORR**

albosparsus Gyllenhal, 1833: 110 **A:** CH (YUN) ID (SD UT) NP **ORR**

andamensis Pic, 1917d: 10 (var.)

subsignatus Gyllenhal, 1839: 139

tessellatus Motschulsky, 1858: 97

altaicus Karapetjan, 1973: 44 **A:** MG RU (SS) [# 160]

atrispinus Borowiec, 1995: 99 **A:** CH (YUN) ID (SD) NP **ORR**

borowieci Anton, 1996b: 481 **A:** IN PA

brevipes Pic, 1927c: 13 **A:** AE IN OM PA YE **AFR**

braunsi Pic, 1935a: 66

humilis Decelle, 1970: 265

calystegiae Lukjanovitch & Ter-Minassian, 1957: 193 (*Euspermophagus*) **E:** AL AU BE BH BU BY CR CZ FR GE GR HU IT LA LT LU MA NL NM PL PT RO RU (CT ST) SK SL SP SZ TR **N:** AG EG LB MO TU **A:** AB AF AR CY GG IN IQ IS JO KZ LE QA RU (SS: Tuva) SA SY TD TM TR UZ [# 161]

canus Baudi di Selve, 1886: 110 **A:** CH ("W. China") IN KZ PA TD TM

caricus Decelle, 1982a: 31 **E:** GR (Rhodes) **A:** IN JO LE SY TR UZ [# 019]
caucasicus Baudi di Selve, 1886: 111 **E:** RU (ST) **A:** AB AF AR CH (SCH SHA YUN) IN IS KI KZ NC PA RU (FE) TD TM TR UZ **ORR**
eous Lukjanovitch & Ter-Minassian, 1957: 199 (*Euspermophagus*)
confusus Borowiec, 1986a: 163 **E:** BH BU GR HU KR RO RU (ST: Caucasus) TR **A:** AB AR GG SY TR
decellei Borowiec, 1985d: 465 **A:** IN
drak Borowiec, 1991: 53 **A:** CH (YUN) **ORR**
klapperichi Borowiec, 1985a: 19 **A:** IS JO TR [# 204]
kuesteri Schilsky, 1905: 4 **E:** AL BH BU CR FR GEi GR IT KR NM PT RO RU (ST) SP TR **N:** AG CIi LB MO TU **A:** AB AF AR CY GG IN IQ IS JO KI KZ LE SY TM TR UZ [# 205]
albipilis Chobaut, 1898a: 87 (var.)
swartukensis A. H. Ali & Kadhim, 1986: 45 [# 118]
kuskai Borowiec, 1985f: 782 **A:** ID (UT) **ORR**
latithorax Boheman, 1828: 116 (*Bruchus*) **Ei:** GB **AFR**
gossypii Chevrolat, 1872a: 8
latipennis Pic, 1918a: 9
tomentosus Klug, 1835: 42
trogodermoides Fairmaire, 1902: 247 (*Bruchus*)
lindbergorum Decelle, 1975c: 140 **N:** CI (Gran Canaria, La Palma, Tenerife)
longepygus Anton, 1993: 380 **A:** CH (FUJ YUN) ID (SD) NP TD **ORR**
lukjanovitschi Savitskiy, 2000: 556 **A:** IS SY TR
arcis A. Delobel, 2011: 108
maafensis Borowiec, 1985a: 17 **E:** SP (Cadiz, Malaga) **N:** AG MO [# 206]
monardi Decelle, 1975a: 26 **A:** YE (Socotra) **AFR**
multilineolatus Pic, 1918a: 7 **A:** CH (HUN YUN) RU (FE)
negligens Pic, 1917d: 10 **A:** CH (HUN YUN) **ORR**
javanus Pic, 1918a: 8
niger Motschulsky, 1866b: 405 **A:** BT CH (GUA HAI TAI YUN) ID (SD) NP **AFRi ORR**
bifasciolatus Motschulsky, 1874: 250
formosanus Pic, 1917d: 9
immaculatus Pic, 1922c: 16 (var.)
tonkineus Pic, 1917d: 8
pubiventris Baudi di Selve, 1886: 110 (var.) **E:** GR (Crete, Cyclades, Rhodes) **A:** AE CY IS JO LE SY TR [# 162]
punjabensis Borowiec, 1991: 103 **A:** CH (YUN) ID (SD) **ORR**
rufipes Ter-Minassian, 1975: 247 (*Euspermophagus*) **A:** KI MG

sericeus Geoffroy, 1785: 112 (*Mylabris*) **E:** AL AU BE BH BU BY CR CZ DE FI FR GB GE GR HU IT KR LA LT LU MA MD NL NM PL PT RO RU (CT NT ST) SK SL SP SV SZ TR UK **N:** AG EG MO TU **A:** AB AF AR CH (NMO TIA XIN) CY GG IN IQ IS JO KI KZ LE MG RU (ES FE SS WS) SA SY TD TM TR UZ

afghanicus Borowiec, 1985a: 7 (ssp.)

cardui Boheman, 1828: 117 (*Bruchus*)

euphorbiae Küster, 1848c: 54

mesopotamecus A. H. Ali & Kadhim, 1986: 45 [# 117]

subdenudatus Motschulsky, 1874: 251

sulcifrons Küster, 1848c: 52

shamszadehi A. Delobel & Sadeghi, 2013: 310 **A:** IN (Teheran Yazd) [# 163]

sinensis Pic, 1918a: 9 **A:** BT CH (YUN) ID (SD) NP **ORR**

kurseongensis Decelle, 1977: 405

sophorae Fåhraeus, 1839: 136 **A:** YE **AFR ORR**

fulvopubens Pic, 1918a: 9

uniformis Pic, 1932a: 332

stemmleri Decelle, 1977: 407 **A:** AF BT CH (YUN) ID (HP KA UT) NP PA **ORR**

turanicus Lukjanovitch & Ter-Minassian, 1957: 195 (*Euspermophagus*) **A:** KI UZ [# 019, # 207]

variolosopunctatus Gyllenhal, 1833: 110 **Ei:** UK **A:** CH (HAI TAI YUN) ID (SD) NP **ORR** [# 208]

maculatus Schilsky, 1905: 5 (var.)

subundulatus Pic, 1917d: 10 (var.)

wittmeri Borowiec, 1985d: 463 **A:** IS

genus *Spermophagus*, nomina dubia

testaceicolor Pic, 1917d: 11 **A:** JA

tigridis A. H. Ali & Kadhim, 1986: 50 **A:** IQ

genus *Zabrotes* Horn, 1885: 156 type species: *Zabrotes cruciger* Horn, 1885

subfasciatus Boheman, 1833: 111 (*Spermophagus*) **Ei:** AL AU AZ BE BU CZ FR GB GE GR HU IT NL PL PT RO SK SP SZ UK **Ni:** AG CI EG MO **Ai:** CH ("China") ID (UT) IN IS JA RU (FE) SY TR YE **AFRi NARi NTR ORRi**

cingulatus Suffrian, 1870a: 169 (*Bruchus*)

dorsopictus Lepesme, 1941: 201 (*Spermophagus*)

musculus Boheman, 1833: 112 (*Spermophagus*)

nesapius Fåhraeus, 1839: 108 (*Bruchus*)

pectoralis Sharp, 1885: 492 (*Spermophagus*)

semicinctus Horn, 1894: 411 (*Spermophagus*)

semifasciatus Boheman, 1839: 137 (*Spermophagus*)

tribe Bruchini Latreille, 1802

subtribe Acanthoscelidina Bridwell, 1946

genus *Acanthobruchidius* Borowiec, 1980a: 127 type species: *Mylabris spinigera* Baudi di Selve, 1886

spiniger Baudi di Selve, 1886: 47 (*Mylabris*) E: GR (Lesbos, Rhodes) **A:** IQ IN IS JO LE SY TR

discipennis Allard, 1868: 94 (*Bruchus*)

rubrithorax Pic, 1900b: 31 (*Bruchus*, var.)

genus *Acanthoscelides* Schilsky, 1905: C type species: *Bruchus irresectus* Fåhraeus, 1839 (= *Bruchus obtectus* Say, 1831) **[Appendix]**

griseolus Fall, 1910: 175 (*Bruchus*) **Ai:** JA (Honshu) **NAR NTR**

kiotensis Pic, 1913h: 14 (*Bruchus*)

macrophthalmus Schaeffer, 1907: 300 (*Bruchus*) **Ei:** SP **Ni:** CI EG MR **Ai:** CH (HAI TAI YUN) CY JA SI **AFRi AUSi NAR NTR ORRi PACi**

obtectus Say, 1831: 1 (*Bruchus*) **Ei:** AL AN AU AZ BE BH BU BY CR CZ DE EN FI FR GB GE GR HU IC IR IT KR LA LS LT LU MA MD NL NM NR PL PT RO RU (CT NT ST) SB SK SL SP SV SZ TR UK **Ni:** AG CI EG MO MR SG TU **Ai:** AB AE AR CH (TAI) CY GG ID (HP KA UT) IN IS JA LE RU (ES FE WS) SA SY TR YE **AFRi AUSi NARi NTR ORRi**

fabae Fitch, 1861: 63 (*Bruchus*)

fabae C. V. Riley, 1871: 52 (*Bruchus*) [HN]

gilvipes Motschulsky, 1874: 238 (*Bruchus*)

incretus Walker, 1859b: 261 (*Bruchus*)

irresectus Fåhraeus, 1839: 18 (*Bruchus*)

leguminarius Gyllenhal, 1833: 69 (*Bruchus*)

melanocephalus Fåhraeus, 1839: 87 (*Bruchus*)

pallidipes Fåhraeus, 1839: 91 (*Bruchus*)

subellipticus Wollaston, 1854: 420 (*Bruchus*)

tetricus Gyllenhal, 1839: 22 (*Bruchus*)

varicornis Motschulsky, 1874: 239 (*Bruchus*)

pallidipennis Motschulsky, 1874: 210 (*Bruchus*) **Ei:** AN AU BH BU CR CZ FR GE GR HU IT LU MD ME NM PL RO RU (ST) SB SK SL SZ UK **Ai:** AB AR CH (JIX NIN SHA XIN) JA KZ NC RU (FE) TD **NAR NTR**

collusus Fall, 1910: 176 (*Bruchus*)

perplexus Fall, 1910: 177 (*Bruchus*)

tarnawskii Borowiec, 1980b: 167

pusillimus Sharp, 1885: 479 (*Bruchus*) **Ai:** JA (Honshu) **NAR NTR**

genus *Algarobius* Bridwell, 1946: 54 type species: *Bruchus prosopis* LeConte, 1858

prosopis LeConte, 1858: 77 (*Bruchus*) **Ni:** EG **Ai:** AE IN KU OM QA SA SI YE **AFRi AUSi NAR NTR ORRi PACi**

desertorum LeConte, 1858: 78 (*Bruchus*)
uniformis LeConte, 1858: 77 (*Bruchus*)

genus *Borowiecius* Anton, 1994b: 108 type species: *Bruchus ademptus* Sharp, 1886
ademptus Sharp, 1886a: 36 (*Bruchus*) **A:** CH (FUJ SHA TAI YUN) ID (SD) JA NC SC **NARi ORR**
fusculus Anton, 1994b: 110 **A:** OM YE **AFR**
siamensis Anton, 1999b: 67 **A:** ID (UT) **ORR**
varicolor Boheman, 1833: 69 (*Bruchus*) Ei: IT **AFR** [# 1072]
sparsepubens Pic, 1929g: 25 (*Bruchus*, var.)

genus *Bruchidius* Schilsky, 1905: B type species: *Bruchus quinqueguttatus* Olivier, 1800
Serratobruchidius Yus Ramos, 2013a: 18 type species: *Bruchus rubiginosus* Desbrochers des Loges, 1869 [# 060]
Sparteus Bridwell, 1946: 55 type species: *Bruchus villosus* Fabricius, 1792
albolineatus Blanchard, 1845a: lxxxii (*Bruchus*) **E:** FR (Corsica) GR IT MA PT SP **N:** AG EG MO TU **A:** CY IS JO TR
croccipennis Motschulsky, 1874: 209 (*Bruchus*)
flavescens P. H. Lucas, 1846b: pl. 35 [= 1847c: 403] (*Bruchus*)
gracilis Blanchard, 1845a: lxxxii (*Bruchus*)
hoyeri Tenenbaum, 1915: 842 (*Kytorrhinus*)
albopictus Allard, 1884a: 11 (*Bruchus*) **E:** GR **N:** MOi **A:** CY IS JO LE SY TR [# 165]
albosparsus Fåhraeus, 1839: 52 (*Bruchus*) **A:** IS **N:** EG **AFR**
advena Wollaston, 1870: 26 (*Bruchus*)
spadiceus Fåhraeus, 1839: 54 (*Bruchus*)
anderssoni Decelle, 1975b: 188 **A:** ID (UT) NP **ORR**
compositus Arora, 1977: 43
annulicornis Allard, 1868: 107 (*Bruchus*) **E:** AL BH BU GR NM TR **A:** AB AR CY GG IN IS JO LE SY TR [# 119, # 166]
anobioides Baudi di Selve, 1886: 72 (*Mylabris*) **E:** GR TR **A:** CY IN IQ IS JO LE TR
zampettii Borowiec, 1985e: 774
antennatus Wollaston, 1864: 383 (*Bruchus*) **N:** CI (Gran Canaria, La Palma, Tenerife)
apicipennis Heyden, 1892b: 101 (*Mylabris*) **A:** CH (NIN XIN) IN KI KZ MG RU (WS) TM
arabicus Decelle, 1979a: 319 **A:** IS SA SI YE
armeniacus Ter-Minassian, 1970: 273 **A:** AR TR
asiricus Decelle, 1979a: 323 **A:** OM SA YE
astragali Boheman, 1828: 107 (*Bruchus*) **E:** KR RU (ST) UK **A:** IN IQ KZ TR [# 167]
atbasaricus Lukjanovitch & Ter-Minassian, 1954: 67
fischeri Hummel, 1827: 9 (*Bruchus*) [QS] [# 020]
samai Zampetti & Toma, 2020: 560 [# 055]

aurivillii Blanc, 1889: xliii (*Mylabris*) **N:** AG MO TU **A:** SA YE **AFR**
mimus Gyllenhal, 1833: 43 (*Bruchus*) [HN]
senegalensis Pic, 1912l: 92 (*Bruchus*) [RN]
baharicus Decelle, 1979a: 324 **A:** SA
bcharreensis Zampetti & Toma, 2020: 558 **A:** LE
beauprei Pic, 1905c: 122 (*Laria*) **N:** EG
bernardi A. Delobel & Anton, 2004: 178 E: IT
biguttatus Olivier, 1800: [79]: 20 (*Bruchus*) E: AL BH BU CR FR GR IT KR MA MD NM PT RU (ST) SP TR **N:** AG MO TU **A:** CY GG IN IS JO LE SY TR
denticornis Allard, 1868: 113 (*Bruchus*)
fulvipennis Germar, 1823: 180 (*Bruchus*)
juvenilis Schilsky, 1905: 43 (var.)
mendicans Weise, 1891a: 332 (*Mylabris*, var.) [RN]
palaestinus Pic, 1904c: 40 (*Laria*)
tuberculifer Caillol, 1894: cclxix (*Mylabris*)
bimaculatus Olivier, 1800: [79]: 18 (*Bruchus*) E: AL BH BU CR FR GR HU IT KO KR MA ME NM PL PT RU (ST) SB SL SP TR **N:** AG CI EG MO TU **A:** AB AR CY IN IQ IS JO LE RU SY TM (Kopet Dag) TR **ORRi**
dispergatus Gyllenhal, 1833: 46 (*Bruchus*)
megriensis Ter-Minassian, 1976: 301
variegatus Germar, 1823: 181 (*Bruchus*)
bituberculatus Schilsky, 1905: 90 (var.) E: AL BH CR GR TR **N:** EG **A:** CY IN IQ IS JO LE SY TR **[# 168]**
borowieci Anton, 1998: 3 E: BU FR GR IT (Sicily, Sardinia) PT SP **N:** AG MO **A:** IS JO TR
buettikeri Decelle, 1979a: 321 **A:** AE IS JO OM SA SI YE **AFR**
bythinocerus Reitter, 1890c: 197 (*Mylabris*) **A:** AB AR IN JO TR
caeruleus Champion, 1919: 244 (*Bruchus*) **A:** BT CH (XIZ) ID (HP KA SD UT) NP **[# 021, # 067]**
calabrensis Blanchard, 1845a: lxxxii (*Bruchus*) E: AL BH BU CR FR GR IT KR MA NM PT RU (ST) SB SP TR **N:** AG MO TU **A:** CY GG IN IS LE SY TR
selvei Pic, 1913b: 110 (*Bruchus*) [RN]
sordidus Baudi di Selve, 1886: 84 (*Mylabris*, var.)
stylophorus K. Daniel, 1904: 87 (*Laria*)
varipictus Motschulsky, 1874: 240 (*Bruchus*) **[# 022]**
canescens Motschulsky, 1874: 231 (*Bruchus*) E: BU GR ME NM SB TR **A:** CY IN IQ IS JO LE SY TR **[# 099]**
parumpunctatus Baudi di Selve, 1886: 82 (*Mylabris*)
caninus Kraatz, 1870b: 334 (*Bruchus*) E: BU CR FR GR IT NM SP TR **N:** AG MO TU **A:** CY IS JO LE SY TR
angustulus Motschulsky, 1874: 218 (*Bruchus*)

sordidatus Allard, 1884a: 6 (*Bruchus*)
uniformis Brisout de Barneville, 1866: 385 (*Bruchus*) [HN]
centromaculatus Allard, 1868: 107 (*Bruchus*) **N:** EG **A:** IN IS PA SA YE **AFR AUSi**
albonotatus Pic, 1930d: 13 (*Bruchus*)
lanceolatus Motschulsky, 1874: 211 (*Bruchus*)
sahlbergi Schilsky, 1905: 94
chloroticus Dalman, 1833: 41 (*Bruchus*) **N:** EG **A:** CH (YUN) ID (HP) IN IQ JO PA YE **AFR ORR AUSi** **[# 120]**
amplicollis Wollaston, 1868: 141 (*Bruchus*)
angustifrons Schilsky, 1905: 52
celebensis Pic, 1922c: 15 (*Bruchus*)
diversiventris Pic, 1928a: 299 (*Bruchus*, ssp.)
vectabilis Gyllenhal, 1839: 100 (*Bruchus*)
vinsoni Pic, 1936b: 299 (*Bruchus*, var.)
cinerascens Gyllenhal, 1833: 55 (*Bruchus*) **E:** AU BU CR CZ FR GE GR HU IT KR MA RU (CT ST) SK SP TR UK **N:** AG MO TU **A:** AB AR CY GG IN IS JO LE SY TR
eryngii Brisout de Barneville, 1863: 93 (*Bruchus*)
musculus Solsky, 1866: 95 (*Bruchus*)
cisti Fabricius, 1775: 65 (*Bruchus*) **E:** AL AU BE BU CR CZ FR GB GE GR HU IT KR LT NL NM PL PT SK SL SZ SP SV SZ TR **N:** AG MO TU **A:** CY IN IQ LE TR **NAR**i
canus Germar, 1823: 183 (*Bruchus*) **[# 023]**
debilis Gyllenhal, 1833: 53 (*Bruchus*)
nugax Motschulsky, 1874: 226 (*Bruchus*)
pulverulentus Motschulsky, 1874: 218 (*Bruchus*)
rostratus Motschulsky, 1874: 232 (*Bruchus*)
comptus Sharp, 1886a: 36 (*Bruchus*) **A:** CH (ANH BEI GUI HUB) JA NC RU (FE) SC **[# 121]**
mandschuricus Pic, 1913b: 110 (*Bruchus*) [RN]
modestus Schilsky, 1906: 97
convexicollis Lukjanovitch & Ter-Minassian, 1957: 137 **E:** KR UK **A:** KI KZ TD UZ
onobrychidis Lukjanovitch & Ter-Minassian, 1957: 138
coreanus Chûjô, 1937e: 194 **A:** NC SC
corsicus Baudi di Selve, 1890: 344 (*Mylabris*, var.) **E:** BU CR CZ FR GR HU IT PT SK SP **N:** AG MO TU **[# 024, # 100, # 169]**
crassicornis Lukjanovitch & Ter-Minassian, 1957: 141 **A:** KZ
danilevskyi Anton, 1999c: 443 **A:** KZ
desmodei Arora, 1980: 42 **A:** CH (HKG) ID (HP UT) **ORR**
dilutus Motschulsky, 1874: 209 (*Bruchus*) **E:** GRi (Rhodes) RU (ST) **N:** EG LB **A:** AB AE AR CH (XIN) IN IQ IS JO KZ PA SI TD TM TR UZ **ORR**
fulvus Allard, 1884a: 5 (*Bruchus*)

obscuriceps Pic, 1916b: 10 (*Bruchus*, var.) [IN] [# 048]
pallidulus Reitter, 1895b: 159 (*Mylabris*)
dispar Gyllenhal, 1833: 45 (*Bruchus*) **E:** AL AU BH BU CR CZ FR GR HU IT KO KR NM PL PT RU (ST) SK SL SP SZ TR **N:** AG EG MO TU **A:** AB AR CY GG IN IQ IS JO LE RU SY TR
albescens Allard, 1868: 112 (*Bruchus*)
braccatus Gyllenhal, 1833: 50 (*Bruchus*)
delicatulus Motschulsky, 1874: 211 (*Bruchus*)
palliatus Baudi di Selve, 1886: 86 (*Mylabris*, var.)
elnairensis Pic, 1931d: 35 (*Bruchus*) **N:** EG **A:** SA YE **AFR**
acaciaephilus Anton, 2000b: 246
fallaciosus Iablokoff-Khnzorian, 1960: 73 **A:** AR IN
formosanus Pic, 1913g: 144 (*Bruchus*) **A:** CH (TAI) **ORR**
foveolatus Gyllenhal, 1833: 81 (*Bruchus*) **E:** AL AU AZ BH BU CR CZ FR GR IT ME NM PT SK SP TR **N:** AG CI LB MO MR TU **A:** AR CY IN IS JO LE SY TM (Nebit Dag) TR
concolor Blanchard, 1845a: lxxxiv (*Bruchus*)
costatus Blanchard, 1845a: lxxxiv (*Bruchus*)
grandicornis Blanchard, 1845a: lxxxiii (*Bruchus*)
ignarium Allard, 1868: 106 (*Bruchus*)
maurus Motschulsky, 1874: 233 (*Bruchus*)
miser Fåhraeus, 1839: 70 (*Bruchus*)
oblongus Blanchard, 1845a: lxxxiv (*Bruchus*)
ovalis Blanchard, 1845a: lxxxiv (*Bruchus*)
republicanus Jekel, 1855: 10 (*Bruchus*)
sardous Motschulsky, 1874: 218 (*Bruchus*)
suratus Motschulsky, 1874: 233 (*Bruchus*) [# 101]
fulvescens Baudi di Selve, 1886: 68 (*Mylabris*, var.) **E:** GR **N:** EG **A:** CY IN IQ IS JO LE SY TR
riedeli Borowiec, 1985e: 776
gardneri Pic, 1938e: 400 (*Bruchus*) **A:** ID (UT) NP **ORR**
glycyrhizae Gyllenhal, 1839: 55 (*Bruchus*) **E:** GEi HU KR MD GRi RO RU (ST) UK **A:** AB AR GG IN IQ KI KZ MG SY TM TR UZ **NARi** **ORRi**
halimodendri Motschulsky, 1874: 209 (*Bruchus*)
obscuripennis Lukjanovitch & Ter-Minassian, 1957: 140 (ssp.)
peregii Hajóss, 1938: 128
silacialis Motschulsky, 1874: 226 (*Bruchus*)
gombo Peyerimhoff, 1915a: 50 **N:** AG MO [# 102]
gracilicollis Chûjô, 1937e: 195 **A:** CH (TAI)
guanchorum Decelle, 1975c: 132 **N:** CI (Gran Canaria, La Palma, Tenerife)

halodendri Gebler, 1830: 143 (*Bruchus*) **A:** AR IN KI KZ RU (FE WS) UZ **[# 122, # 170]**

reichardti Lukjanovitch & Ter-Minassian, 1954: 69

holosericeus Schoenherr, 1832: 213 (*Bruchus*) E: AL BH BU CR FR GR IT NM RU (ST) TR **A:** AB AR IN IQ IS LE SY TM TR

canaliculatus Mulsant & Rey, 1858: 10 (*Bruchus*)

imbricornis Panzer, 1795a: 292 (*Bruchus*) E: AL AU BE BU CZ FR GBi GE GR HU IT KR MA NM PT RO RU (ST) SK SL SP SZ TR **N:** MO **A:** AB AR CY GG LE TR

fulvicornis Motschulsky, 1874: 215 (*Bruchus*) [# 025]

fulviventris Blanchard, 1845a: lxxxii (*Bruchus*)

galegae P. Rossi, 1794: 82 (*Bruchus*)

nugarius Motschulsky, 1874: 215 (*Bruchus*)

incarnatus Boheman, 1833: 72 (*Bruchus*) E: FRi GBi IT (Sicily) PLi PT SP **N:** AG CIi EG LB MO TU **AFR**

algiricus Allard, 1884a: 9 (*Bruchus*) [# **026**]

mocquerysii Kraatz, 1870b: 333 (*Bruchus*)

segmentatus Caillol, 1894: cclxx (*Mylabris*)

jakuticus Ter-Minassian & Egorov, 1981: 23 **A:** RU (ES FE)

japonicus Harold, 1878: 87 (*Mylabris*) **A:** CH (CHQ FUJ HUN JIX NE) JA NC RU (FE) SC

fulvipes Roelofs, 1879: lv (*Bruchus*)

jocosus Gyllenhal, 1833: 73 (*Bruchus*) E: PT SP **N:** AG MO TU **[# 171]**

discipennis Fåhraeus, 1839: 95 (*Bruchus*)

germari Schilsky, 1905: 61 (*Bruchidius*)

histrio Boheman, 1828: 112 (*Bruchus*)

longicornis Germar, 1823: 179 (*Bruchus*)

kamtschaticus Motschulsky, 1874: 224 (*Bruchus*) **A:** RU (FE: Kamchatka)

kashmirensis Arora, 1980: 48 **A:** AF ID (KA) PA

kaszabi Ter-Minassian, 1973: 79 **A:** AF CH (NIN NMO) IN KZ MG RU (ES) UZ

khnzoriani Karapetjan, 1976: 189 **A:** AR [# 027]

kieneri Zampetti, 1992: 249 **A:** AB AR CY IQ IN IS JO TR

kiritschenkoi Ter-Minassian & Egorov, 1981: 21 **A:** NC RU (FE)

koenigi Schilsky, 1906: 96 E: KR UK **A:** AR GG IN TR

lateobscurus Pic, 1904c: 40 (*Laria*, var.) E: SP **N:** AG **A:** TR [# 097, **# 123**]

gavirai Yus Ramos, 2013b: 58 [# 050]

latior Pic, 1932a: 330 (*Bruchus*) **A:** NP

lautus Sharp, 1886a: 37 (*Bruchus*) **A:** CH (FUJ) JA NC RU (FE) SC

chujoi Pic, 1953g: 16

diversimembris Pic, 1953g: 16 (var.) [HN]

rubellus Ter-Minassian & Egorov, 1981: 20

lekefensis Zampetti & Toma, 2020: 555 **N:** AG TU **[# 124, # 172]** *leprieuri* E. Jacquet, 1886: clxxi (*Bruchus*) **N:** AG MO TU [# 028]

lespedezae Iablokoff-Khnzorian, 1974: 126 (*Acanthoscelides*) **A:** CH (ANH FUJ GUI HUN) JA NC RU (FE) SC [# 125]

lichenicola Wollaston, 1854: 421 (*Bruchus*) **N:** CI MR

floricola Wollaston, 1864: 383 (*Bruchus*)

lindbergi Hoffmann, 1961: 38 **A:** AF IN KZ TD TM UZ

lineatus Allard, 1868: 117 (*Bruchus*) **E:** AL BH BU CR FR (Corsica) GR IT NM SZ TR **N:** AG **A:** IQ LE SY TR [# 126]

migliaccioi Zampetti, 1998: 55 [# 052]

lividimanus Gyllenhal, 1833: 68 (*Bruchus*) **E:** AL AU AZ BH BU BY CR CZ FI FR GBi GE GR HU IT MD NM PL PT RU (CT ST) SK SL SP SZ TR **N:** AG CIi EG MO MR TU **A:** CY IS JO LE SY TR

clavatus Motschulsky, 1874: 239 (*Bruchus*)

commodus Motschulsky, 1874: 236 (*Bruchus*)

divisus Motschulsky, 1874: 237 (*Bruchus*)

fulvicornis Rey, 1893c: 37 (*Bruchus*, var.) [HN]

implicitus Motschulsky, 1874: 216 (*Bruchus*)

inaequalis Rey, 1893c: 37 (*Bruchus*, var.)

laticornis Blanchard, 1845a: lxxxiii (*Bruchus*)

latus Blanchard, 1845a: lxxxiii (*Bruchus*)

obsoletus Blanchard, 1845a: lxxxiii (*Bruchus*)

planeticus Motschulsky, 1874: 216 (*Bruchus*)

plumbeus P. H. Lucas, 1846b: pl. 35 [= 1847c: 402] (*Bruchus*)

reichei Allard, 1868: 111 (*Bruchus*)

retamae Vogel, 1861: 49 (*Bruchus*)

subconicus Rey, 1893c: 37 (*Bruchus*, var.)

teneriffae Gyllenhal, 1839: 105 (*Bruchus*)

velaris Fåhraeus, 1839: 90 (*Bruchus*)

loebli Borowiec, 1985e: 777 **A:** IN IS TR

longulus Schilsky, 1905: 79 **E:** FR GR IT NM PT SP TR UK **N:** MO **A:** AB AF AR CY IN IQ IS JO KI KZ TD TM TR UZ

longus Pic, 1913b: 110 (*Bruchus*) [RN]

strictus Iablokoff-Khnzorian, 1974: 126 [# 029]

tibiellus Baudi di Selve, 1886: 93 (*Mylabris*)

lucifugus Boheman, 1833: 47 (*Bruchus*) **E:** FRi RU (ST) **A:** AB AR TR

tesserulus Motschulsky, 1874: 214 (*Bruchus*) [# 103]

lutescens Blanchard, 1845a: lxxxiv (*Bruchus*) **E:** BU GR IT NM PT SP TR **N:** MO **A:** AB AR CY GG IN IQ IS JO KI KZ LE SY TD TM TR UZ [# 030]

anatolicus Pic, 1904c: 40 (*Laria*, var.)

autumnalis Motschulsky, 1874: 226 (*Bruchus*) [# 104]

marginalis Fabricius, 1777: 212 (*Bruchus*) **E:** AL AU BE BH BU BY CR CZ FI FR GE HU IT LA LT NL NM PL RU (ST) SB SK SL SP SZ UK **A:** AB AR GG RU (SS) TR
marginellus Fabricius, 1792b: 371 (*Bruchus*)
persicae Walckenaer, 1802: 230 (*Bruchus*)
martinezi Allard, 1868: 112 (*Bruchus*) **E:** FR IT PT SP **N:** AG MO TU [# 173]
melanocerus Ter-Minassian, 1973: 79 **A:** KI KZ MG RU (ES)
meleagrinus Gené, 1839: 75 (*Bruchus*) **E:** FR IT MA PT SP **N:** AG MO TU [# 174]
albomaculatus Graells, 1858a: 78 [= 1858b: 116] (*Bruchus*)
eximius Chevrolat, 1882: 212 (*Spermophagus*)
paracentesis Motschulsky, 1874: 229 (*Bruchus*)
mendosus Gyllenhal, 1839: 72 (*Bruchus*) **A:** BT ID (HP KA SD UT) IN NP YE **ORR**
blemeris Arora, 1980: 22
decretus Walker, 1859b: 261 (*Bruchus*)
minimus Motschulsky, 1858: 97 (*Bruchus*)
vulgaris Arora, 1977: 45
microminutus Arora, 1980: 59 **A:** ID (UT) **ORR**
misellus Boheman, 1833: 56 (*Bruchus*) **E:** CR FR IT PT SP **N:** MO [# 031, # 127, # 175]
monstrosicornis Pic, 1904c: 40 (*Laria*) **E:** GR **A:** IN TR
carthami Hoffmann, 1956: 249 (ssp.)
mordelloides Baudi di Selve, 1886: 94 (*Mylabris*) **E:** BU GR KR NM RU (CT ST) TR UK **A:** CY IS JO LE SY TR
muehlei Anton, 2000b: 251 **A:** SA YE **AFR**
mulsanti Brisout de Barneville, 1863: 50 (*Bruchus*) **E:** AL BH BU CR FR GR HU IT KR MD NM PT RU (ST) SK SP TR **A:** AB AR CY GG IS JO LE TD (Gissarskiy Khrebet) TR [# 176]
murinus Boheman, 1828: 113 (*Bruchus*) **E:** AL BH BU CR FR GEi GR IT KR NM PT RO RU (ST) SP SZ (Valais) TR **N:** AG CIi LB MO TU **A:** CY IS SY TR [# 177]
alni Fåhraeus, 1839: 113 (*Bruchus*)
grisescens Rey, 1893b: 26 (*Bruchus*, var.)
macrocerus Rey, 1893b: 26 (*Bruchus*, var.)
nigritarsis Fåhraeus, 1839: 113 (*Bruchus*)
mussooriensis Arora, 1980: 54 **A:** ID (HP SD UT) NP **ORR**
myobromae Motschulsky, 1874: 212 (*Bruchus*) **E:** IT (South Tirol) KR RU (CT ST) UK **A:** KZ [# 105, # 178]
scutulatus Baudi di Selve, 1890: 342 (*Mylabris*, var.)
nalandus Pic, 1927c: 12 (*Bruchus*) **Ei:** IT **A:** AE ID (HP SD UT) IN YE (incl. Socotra) **AFR ORR** [# 063]
indigoferae Singh & Saini, 1978: 129
salinaensis Zampetti & Toma, 2020: 557 [# 054]
tephrosiae Arora, 1977: 47

nangalensis Arora, 1980: 57 **A:** ID (HP) **ORR**

nanus Germar, 1823: 182 (*Bruchus*) **E:** AL BH BU CR FR GR IT KR MD NM PT SL RU (ST) SP TR UK **N:** AG MO **A:** AB AF AR CY GG IN JO TR

griseus Motschulsky, 1874: 219 (*Bruchus*)

nepalensis Anton & A. Delobel, 2017: 163 **A:** NP

niger Anton, 2004: 381 **E:** PT SP **N:** AG MO **[# 179]**

nudus Allard, 1868: 114 (*Bruchus*) **E:** BU CR FR GEi GR IT ME NM PT SL SP **N:** AG MO TU **A:** IN IS LE TM TR **[# 180]**

pardoi Yus Ramos, 2007b: 39 [# 053]

obscuripes Gyllenhal, 1839: 21 (*Bruchus*) **E:** AL BU CR FR GR HU IT MD NM PT RU (ST) SP TR **N:** AG MO TU **A:** AF AR CY GG IN IQ IS JO LE SY TR

obscurus Arora, 1977: 48 **A:** ID (UT)

ochraceus Baudi di Selve, 1886: 61 (*Mylabris*) **A:** AB AR JO LE SY TR **[# 106, # 128]**

elegans Iablokoff-Khnzorian, 1959: 73

horshensis Zampetti & Toma, 2020: 554 [# 051]

olivaceus Germar, 1823: 183 (*Bruchus*) **E:** AL BU CR CZ GR HU IT ME RU (ST) SK TR **N:** AG MO **A:** AB AR GG IS JO TR **[# 032, # 129, # 181]**

calabrus Baudi di Selve, 1886: 75 (*Mylabris*, var.)

orientalis Arora, 1980: 20 **A:** ID (HP) **ORR**

osellai Zampetti, 1979a: 391 **A:** ID (KA) PA

pardellus A. Delobel, Anton, Le Ru & Kergoat, 2013: 53 **A:** YE **AFR**

picipes Germar, 1823: 182 (*Bruchus*) **E:** AU BH BU CR CZ FR GR HU IT KR NM PT RO SK SP SZ TR **N:** AG MO TU **A:** AF IN TR "Caucasus"

albopunctatus Blanchard, 1845a: lxxxiii (*Bruchus*)

ganglbaueri Schilsky, 1905: 78

hoffmanni Tempère, 1954: 50

tarsalis Gyllenhal, 1833: 49 (*Bruchus*)

planicornis Anton & A. Delobel, 2017: 168 **A:** IN TR

poecilus Germar, 1823: 180 (*Bruchus*) **E:** BH BU CR GR IT KR ME RO SP TR UK **N:** AG LB MO TU **A:** AB AR GG IN IS JO LE SY TR

decorus Boheman, 1828: 106 (*Bruchus*)

postremus Gebler, 1839: 46 (*Bruchus*) **A:** KI KZ RU (SS) UZ **[# 068, # 130]**

maculatus Gebler, 1833: 291 (*Bruchus*) [HN]

poupillieri Allard, 1868: 116 (*Bruchus*) **E:** GR IT MA PT SP **N:** AG EG MO TU **A:** AB AR CY GG IN IQ IS JO TR [# 033]

atramentarius Schilsky, 1905: 48 (var.)

prolongatus Lukjanovitch & Ter-Minassian, 1954: 68 **A:** KI KZ UZ **[# 182]**

pubicornis Lukjanovitch & Ter-Minassian, 1957: 159 **A:** AB AR TR **[# 131, # 183]**

punctoterminalis Arora, 1980: 46 **A:** ID (HP KA SD UT) NP PA **ORR**

pusillus Germar, 1823: 181 (*Bruchus*) **E:** AL BH BU CR CZ FR GE GR HU IT KR NM PL RU (ST) SK SL SP SZ TR **N:** AG MO TU **A:** AR CY IN IQ IS JO LE TR

discrepans Iablokoff-Khnzorian, 1964: 170
stierlini Allard, 1868: 117 (*Bruchus*)
pygmaeus Boheman, 1833: 80 (*Bruchus*) **E:** AL BH BU CR FR GR IT KO KR NM PT RU (ST) SP SZ TR **N:** AG EG MO TU **A:** CY IS JO LE TR
anxius Fåhraeus, 1839: 72 (*Bruchus*)
conicus Motschulsky, 1874: 219 (*Bruchus*)
perparvulus Boheman, 1839: 68 (*Bruchus*)
quinqueguttatus Olivier, 1800: [79]: 15 (*Bruchus*) **E:** BEi BU CR GBi GEi GR IT (Sicily) NM RU (ST) TR **N:** CIi EG **A:** AB AR CY IN IQ IS JO LE SY TR **AUSi** [# 184]
laterufus Pic, 1904c: 40 (*Laria*, var.)
leucophaeus Allard, 1884a: 12 (*Bruchus*)
novemguttatus Motschulsky, 1874: 231 (*Bruchus*)
quadriguttatus Boheman, 1828: 111 (*Bruchus*)
quinquepunctatus Motschulsky, 1874: 230 (*Bruchus*)
septemguttatus Motschulsky, 1874: 230 (*Bruchus*)
signatus L. Redtenbacher, 1843: 987 (*Bruchus*)
suratus Schilsky, 1905: 63 (*Bruchidius*, var.)
raddianae Anton & A. Delobel, 2003: 178 **E:** IT (Lampedusa) PT SP **N:** AG EG LB MO TU **A:** AE IN IS JO OM SA YE **AFR ORR**
reitteri Schilsky, 1906: 95 **A:** AB AR IN TR [# 185]
richteri Lukjanovitch & Ter-Minassian, 1954: 70 **A:** AB AR IN TR
robustus Lukjanovitch & Ter-Minassian, 1957: 134 **A:** AB AR IN IQ JO LE SY TM (Kopet Dag) TR
popovi Ter-Minassian, 1976: 304
rubiginosus Desbrochers des Loges, 1869: 399 (*Bruchus*) **E:** FR IT MA PT SP **N:** AG MO TU
lamellicornis Baudi di Selve, 1886: 49 (*Mylabris*)
ruficolor Pic, 1942d: 11 **A:** YE **AFR**
rufisurus Allard, 1884a: 6 (*Bruchus*) **E:** GR (Crete, Rhodes) NM TR **N:** EG **A:** CY IN IQ IS JO LE SY TR [# 186]
saudicus Decelle, 1979a: 326 **A:** SA YE **AFR**
savitskyi Anton, 1999c: 445 **A:** TD UZ
seminarius Linnaeus, 1767: 605 (*Bruchus*) **E:** AL AU BH BU CR CZ FR GE GR HU IT KR MA MD ME NM PL PT RO RU (CT ST) SB SK SL SP SV SZ TR UK **N:** AG LB MO TU **A:** AB AR CY GG IN IQ IS JO LE SY TR [# 187]
affinis Stephens, 1831: 214 (*Bruchus*) [HN]
basalis Gyllenhal, 1833: 52 (*Bruchus*)
cribicollis Motschulsky, 1874: 236 (*Bruchus*)
diversipes Rey, 1893c: 37 (*Bruchus*, var.)
femoralis Gyllenhal, 1833: 51 (*Bruchus*)
fuscipes Rey, 1893c: 37 (*Bruchus*, var.)

sanremi O. Schneider, 1893: 36 (*Bruchus*, var.)

siculus Fåhraeus, 1839: 58 (*Bruchus*)

varicornis Rey, 1893c: 37 (*Bruchus*, var.)

xanthicornis Rey, 1893c: 37 (*Bruchus*, var.)

sericatus Germar, 1823: 184 (*Bruchus*) E: AL BH BU CR FR GEi GR HU IT KR NM SB SK SL SP TR **N:** AG **A:** AB AR GG CY IN IS JO LE SY TR **[# 188]**

horvathi Hoffmann, 1964: 419

minimus Blanchard, 1845a: lxxxiv (*Bruchus*)

serraticornis Fabricius, 1775: 65 (*Bruchus*) **A:** AR IS TR

serricollis Morimoto, 1990: 93 **A:** JA

siliquastri A. Delobel, 2007: 266 Ei: BE BU CZ FR GB GE GR HU IT KR NL RO RU (ST) SB SK SL SP TR **A:** CH (SHA FUJ) GGi SC **[# 132]**

sivasensis Zampetti, 1984: 399 **A:** AB AR IN TR

siwalikus Arora, 1980: 38 **A:** ID (HP UT) **ORR**

snizeki A. Delobel, Anton, Le Ru & Kergoat, 2013: 56 **A:** YE **AFR**

spathopus Iablokoff-Khnzorian, 1960: 74 **A:** AR IN TR **[# 133]**

talyshensis Ter-Minassian, 1970: 275 **A:** AB AR IN SY TR

taorminensis Blanchard, 1845a: lxxxiii (*Bruchus*) E: CR FR GR IT PT SP **N:** AG CI MO MR TU **A:** TR

tenerifensis Anton, 1999d: 399 **N:** CI (Gran Canaria, Tenerife)

thunbergi Southgate, Howe & Brett, 1957: 81 [RN] **A:** CH ("China")

chinensis Thunberg, 1816: 46 (*Bruchus*) [HN]

tibialis Boheman, 1828: 114 (*Bruchus*) E: AL BH CR FR GEi GR IT KR PT RU (ST) SP TR **N:** AG EG MO TU **A:** AB AR CY GG IN IS JO LE SY TM TR

basifasciatus Motschulsky, 1874: 214 (*Bruchus*)

incanescens Motschulsky, 1874: 216 (*Bruchus*)

lubricus Gyllenhal, 1839: 46 (*Bruchus*)

nigrolineatus Motschulsky, 1874: 213 (*Bruchus*)

tragacanthae Olivier, 1800: [79]: 15 (*Bruchus*) **A:** IN

tricolor Anton & A. Delobel, 2017: 165 **A:** ID (UT) **ORR**

trifolii Motschulsky, 1874: 235 (*Bruchus*) E: AL BH BU CR FR GR IT NM PT SL SP TR UK **N:** AG EG LB MO TU **A:** AB CY IN IS JO LE SY TR **ORR**i **[# 134]**

alfierii Pic, 1923f: 104 (*Bruchus*)

tuberculatus Hochhut, 1847: 453 (*Bruchus*) E: AL BH BU CR FR GBi GR HU IT KR NM RU (ST) SP TR **N:** AG MO TU **A:** AB AF AR CY GG ID (KA) IN IQ IS JO KI KZ LE SY TD TM TR UZ

annulipes Allard, 1868: 111 (*Bruchus*)

radulus Desbrochers des Loges, 1869: 400 (*Bruchus*)

tuberculicauda Lukjanovitch & Ter-Minassian, 1954: 70 **A:** CH (XIN) KI KZ MG RU (ES)

turkmenicus Anton, 1996a: 475 **A:** IN TD TM

uberatus Fåhraeus, 1839: 40 (*Bruchus*) **N:** EG TU **A:** AE IN QA YE **AFR ORRi**
baudoni Caillol, 1908: viii (*Laria*)
unicolor Olivier, 1800: [79]: 17 (*Bruchus*) **E:** AL AU BEi BU BY CZ FR GBi HU IT KR LA LT MD PL PT RO SK RU (ST) SP SZ UK **N:** MO **A:** AB AR GG IN KZ TR RU (WS) **NARi**
decellei Zampetti, 1982: 392
varipes Boheman, 1839: 63 (*Bruchus*) **E:** BU CR CZ HU RU (ST "Ural") SK SZ (Valais) **A:** AB AR IN TR [# 135]
aestuosus Motschulsky, 1874: 214 (*Bruchus*)
bagdasarjani Lukjanovitch & Ter-Minassian, 1954: 68
forticornis Motschulsky, 1874: 217 (*Bruchus*)
lukjanovitschi Ter-Minassian, 1970: 274
varius Olivier, 1800: [79]: 18 (*Bruchus*) **E:** AL AN BE BH BU CR CZ FR GB GE GR HU IT KR LU NL NM PL PT RU (ST) SK SL SP SZ TR UK **N:** AG LB MO MR TU **A:** AB AF AR CY GG IN IS LE SY TM TR
assimilis Rey, 1893b: 26 (*Bruchus*, var.)
inspergatus Gyllenhal, 1833: 48 (*Bruchus*)
magnicornis Küster, 1845a: 36 (*Bruchus*)
nebulosus Olivier, 1800: [79]: 20 (*Bruchus*)
ruficornis Motschulsky, 1874: 213 (*Bruchus*)
submaculatus Rey, 1893b: 26 (*Bruchus*, var.)
vestitialis Anton, 1996a: 478 **A:** TM
villosus Fabricius, 1792b: 373 (*Bruchus*) **E:** AL AU BE BH BU CR CZ DE FI FR GB GE GR HU IR IT KO KR LA LT LU ME NL NM PL PT RO RU (CT ST) SB SK SL SP SV SZ TR UK **N:** AG CIi EG MO **A:** GG TR **NARi AUSi** [# 189]
antennarius Motschulsky, 1874: 232 (*Bruchus*)
ater Marsham, 1802: 236 (*Bruchus*)
obscuricornis Blanchard, 1845a: lxxxii (*Bruchus*)
pubescens Germar, 1823: 184 (*Bruchus*)
virgatoides Lukjanovitch & Ter-Minassian, 1957: 168 **A:** TD UZ [# 190]
virgatus Fåhraeus, 1839: 20 (*Bruchus*) **A:** AB AR IN LE SY TR
brignolii Zampetti, 1979b: 204
varipubens Pic, 1953g: 16
wollastoni Decelle, 1975c: 127 **N:** CI MR
wuermlii Decelle, 1977: 402 **A:** BT ID (HP KA SD UT) NP PA **ORR**
cassiae Arora, 1977: 46

genus *Bruchidius*, nomina dubia
atricornis Pic, 1931a: 1 (*Bruchus*, var.) **E:** GB [# 034]
cavicollis Motschulsky, 1874: 239 (*Bruchus*) **N:** EG
consobrinus Allard, 1868: 110 (*Bruchus*) **N:** AG [# 035]

dialensis A. H. Ali & A. Z. M. Ali, 1987: 2 **A:** IQ **[# 112]**
gagliardii Pic, 1933c: 14 **E:** IT
holoniger A. H. Ali & A. Z. M. Ali, 1987: 4 **A:** IQ
incipiens Kolenati, 1858: 116 (*Bruchus*) **A:** "Caucasus et Transcaucasia"
infectus Motschulsky, 1874: 234 (*Bruchus*) **E:** GR (Crete) **N:** EG
inops Motschulsky, 1874: 221 (*Bruchus*) **N:** EG
pauper Boheman, 1828: 115 (*Bruchus*) **E:** "Khersoma Thraciae" [# 024]
subalbovittis Motschulsky, 1874: 221 (*Bruchus*) **E:** RU (ST: Daghestan) [# 036]
zangezuricus Karapetjan, 1989: 178 **A:** AR

genus *Callosobruchus* Pic, 1902e: 6 type species: *Curculio chinensis* Linnaeus, 1758
Protobruchus Zampetti & Toma, 2020: 562 type species: *Protobruchus togoensis* Zampetti & Toma, 2020 (= *Callosobruchus subinnotatus* Pic, 1914) [# 058, # 059]
aethiopicus Decelle, 1971b: 243 **A:** YE **AFR**
analis Fabricius, 1781: 75 (*Bruchus*) **Ei:** BE EN FI FR GB RU (ST) **Ni:** AG EG **A:** AE CH ("Chine") ID (HP KA UT) IN JA RU (FE) SA SY TR **ORR AFRi**
ciceri Rondani, 1877: 174 (*Bruchus*)
glaber Allibert, 1847: 16 (*Bruchus*)
jekelii Allibert, 1847: 15 (*Bruchus*)
mauritiensis Pic, 1932b: 53 (*Bruchus*)
obliquus Allibert, 1847: 14 (*Bruchus*) [HN]
antennifer Singal & Pajni, 1990: 769 **A:** CH (HAI TAI YUN) ID (UT) NP **ORR** **[# 144]**
bhutanicus Decelle, 1977: 393 **A:** BT ID (AP SD UT) NP **[# 145]**
cherenensis Pic, 1939d: 2 (*Bruchus*) **A:** AE OM YE **AFR**
omanicus Anton, 1994b: 107
chinensis Linnaeus, 1758: 386 (*Curculio*) **Ei:** AL AN AU AZ BE BH BU BY CR CZ DE EN FI FR GB GE GR HU IC IR IT LA LS LT LU MA MD NL NM NR PL PT RO RU (ST) SB SK SL SP SV SZ UK **Ni:** AG CI EG MO MR TU **A:** BT CH (BEI CHQ FUJ HUN LIA NIN SCH TAI YUN XIN) CYi ID (AP HP KA SD UT) IN IQ IS JA JO KU LE NC NP RU (FE) SC SY TR YE **AFRi AUSi NARi NTRi ORR PACi** **[# 146]**
barbicornis Fabricius, 1801b: 403 (*Bruchus*)
biguttatus Fabricius, 1801b: 402 (*Bruchus*)
biguttellus Schoenherr, 1833: 42 (*Bruchus*) [HN]
bistriatus Fabricius, 1801b: 402 (*Bruchus*)
pecticornis Linnaeus, 1767: 605 (*Bruchus*)
rubens Boheman, 1833: 34 (*Bruchus*)
rufobrunneus Wollaston, 1870: 25 (*Bruchus*)
rufus DeGeer, 1775: 281 (*Bruchus*)
scutellaris Fabricius, 1792b: 372 (*Bruchus*)
dolichosi Gyllenhal, 1839: 5 (*Bruchus*) **A:** ID (UT) NP **ORR**
albocallosus Pic, 1932a: 329 (*Bruchus*)

imitator Kingsolver, 1999: 209 **A:** CH ("China") NP **ORR**
latealbus Pic, 1926f: 2 (*Bruchus*) **A:** CH (ANH FUJ TAI) **ORR**
albobasalis Chûjô, 1937e: 191
latitarsis Decelle, 1969: 288 **A:** YE **AFR**
maculatus Fabricius, 1775: 65 (*Bruchus*) **Ei:** AL AZ BE BU BY CZ DE FI FR GB GE GR HU IT NR PL PT RU (ST) SK SP SV SZ **Ni:** AG CI EG MO **A:** AE AF BT CH (XIN) ID (HP KA SD UT) IN IQ IS JA NP PA RU (FE) SA TD TR YE **AFR AUSi NARi NTRi ORRi PACi**
ambiguus Gyllenhal, 1839: 11 (*Bruchus*)
calcaratus Wollaston, 1868: 140 (*Bruchus*)
millingeni Pic, 1900b: 31 (*Bruchus*)
ornatus Boheman, 1828: 103 (*Bruchus*)
quadrimaculatus Fabricius, 1792b: 371 (*Bruchus*)
sinuatus Fåhraeus, 1839: 8 (*Bruchus*)
trabuti Caillol, 1919: 54 (*Acanthoscelides*)
vicinus Gyllenhal, 1833: 36 (*Bruchus*)
montanus Anton, 2000a: 25 **A:** ID (UT) NP **ORR**
multinigrus Singal & Pajni, 1990: 767 **A:** ID (HP)
nigripennis Allard, 1895: 226 (*Bruchus*) **A:** ID (HP SD UT) NP **ORR**
nigritus Anton, 2000a: 20 **A:** NP **ORR**
utidai Tuda, 2003: 197 [# 037]
phaseoli Gyllenhal, 1833: 37 (*Bruchus*) **Ei:** AL CZ FR GB GR IT PL SP **Ni:** AG EG **A:** CH ("China") ID (UT) IS JAi KU KZi RUi (FE) SA YE **AFR AUSi NARi NTRi ORRi PACi**
conicicollis Fairmaire, 1898b: 247 (*Bruchus*)
figuratus Gyllenhal, 1839: 12 (*Bruchus*)
rhodesianus Pic, 1902e: 6 (*Bruchus*) **Ei:** GB **A:** YE (Socotra) **AFR ORRi**
marshalli Pic, 1902e: 6 (*Bruchus*)
theobromae Linnaeus, 1767: 605 (*Bruchus*) **A:** CH (TAI) ID (UT) **ORR**
cajanis Arora, 1977: 24
madurensis Pic, 1927c: 12 (*Bruchus*)
maindroni Pic, 1914h: 6 (*Bruchus*)
taiwanensis Chûjô, 1937e: 191
theobromatis Fabricius, 1801b: 397 (*Bruchus*)

genus *Caryedes* Hummel, 1827: 11 type species: *Bruchus faldermanni* Mannerheim, 1827
Falsobruchus Pic, 1913b: 110 type species: *Bruchus cristatus* Fåhraeus, 1839
brasiliensis Thunberg, 1816: 45 (*Bruchus*) **Ei:** FR **NTR**
faldermanni Mannerheim, 1827: 10 (*Bruchus*)

genus *Conicobruchus* Decelle, 1951: 181 type species: *Bruchus strangulatus* Fåhraeus, 1839

Cornutobruchus Decelle, 1975b: 181 type species: *Cornutobruchus veddarus* Decelle, 1975

albopubens Pic, 1931b: 26 (*Bruchus*) Ei: GB **A:** ID (UT) PA YE **AFR NARi ORR**

alpinus Arora & Singal, 1979: 55 **A:** ID (HP KA UT) PA [# 064]

arcuatipes Decelle, 1977: 395 (*Bruchidius*) **A:** BT CH (YUN) NP [# 069]

impubens Pic, 1927c: 10 (*Bruchus*) **A:** CH (FUJ YUN) ID (SD UT) NP **ORR** [# 150]

kashmiricus Pic, 1929b: 4 (*Bruchus*) **A:** ID (HP KA UT) NP PA [# 064]

letourneuxi Caillol, 1894: cclxx (*Mylabris*) **N:** EG **A:** IQ YE [# 070, # 151]

mellyi Pic, 1904c: 40 (*Laria*)

medaniensis Decelle, 1982b: 282 (*Bruchidius*) **A:** AE SA YE (incl. Socotra) **AFR**

pilosus Boheman, 1828: 108 (*Bruchus*) **A:** YE **AFR**

skaifei Pic, 1928d: 20 (*Bruchus*, var.) **A:** AE **AFR**

steveni Gyllenhal, 1839: 99 (*Bruchus*) **E:** RU (ST) UK **A:** AB AF AR IN IS KI KZ TD TM TR UZ [# 071]

cachrydis Lukjanovitch & Ter-Minassian, 1957: 149 (ssp.)

orchesioides Heyden, 1892c: 109 (*Mylabris*)

genus *Decellebruchus* Borowiec, 1987: 149 type species: *Bruchus walkeri* Pic, 1912

atrolineatus Pic, 1921e: 15 (*Bruchus*) Ei: BE GB GE FR IT **N:** AG CIi EG **A:** AE IQi ISi LEi SA SYi TRi YE **AFR NTRi ORRi**

semiflabellatus Pic, 1931d: 34 (*Bruchus*) [HN]

walkeri Pic, 1912l: 92 (*Bruchus*) [RN] **A:** ID (UT) **AFRi ORR**

figuratus Walker, 1859b: 261 (*Bruchus*) [HN]

manillensis Pic, 1927c: 12 (*Callosobruchus*, var.)

genus *Horridobruchus* Borowiec, 1984b: 116 type species: *Horridobruchus corniger* Borowiec, 1984 (= *Pseudopachymerus quadridentatus* Pic, 1923)

quadridentatus Pic, 1923g: 15 (*Pseudopachymerus*) **A:** CH (GUA HKG SHG ZHE)

corniger Borowiec, 1984b: 116

genus *Kingsolverius* Borowiec, 1987: 147 type species: *Kingsolverius gibbicollis* Borowiec, 1987 (= *Bruchus malaccanus* Pic, 1913)

malaccanus Pic, 1913c: 8 (*Bruchus*) **A:** BT CH (HAI YUN) ID (SD) **ORR**

gibbicollis Borowiec, 1987: 147

genus *Megabruchidius* Borowiec, 1984b: 121 type species: *Megabruchidius bifoveolatus* Borowiec, 1984 (= *Bruchus dorsalis* Fåhraeus, 1839)

dorsalis Fåhraeus, 1839: 98 (*Bruchus*) Ei: AL AU BE BH BU CR CZ FR GE HU IT KR MD ME NL PL RO RU (ST) SK SL SP SZ TR UK **A:** CH (CHQ FUJ GUI

HKG SCH TAI XIN YUN) GGi JA KIi KZi MG SC TMi TRi UZi **NARi NTRi ORR** [# 038, # 200]

bifoveolatus Borowiec, 1984b: 121

rufopygalis Motschulsky, 1874: 243 (*Bruchus*)

sophorae Tuda & Morimoto, 2004: 105 **A**: JA (Honshu) [# 039]

tonkineus Pic, 1904d: 42 (*Laria*) Ei: BE BU CR FR GE GR HU ME RO RU (ST) SB SK SL SP SZ TR **A**: CH (YUN) GGi TRi **AFRi AUSi NARi NTRi ORR**

tsinensis Pic, 1923g: 14 (*Bruchus*) **A**: CH (GUI SCH YUN) **ORR** [# 066]

genus *Merobruchus* Bridwell, 1946: 54 type species: *Bruchus julianus* Horn, 1894

julianus Horn, 1894: 410 (*Bruchus*) Ei: IT (Sicily) **NAR**

ochreolineatus Fall, 1910: 186 (*Bruchus*)

genus *Mimosestes* Bridwell, 1946: 54 type species: *Bruchus sallaei* Sharp, 1885

Cercidiestes Bridwell, 1946: 55 type species: *Bruchus ulkei* Horn, 1873

amicus Horn, 1873: 337 (*Bruchus*) **Ni**: EG **Ai**: IS SI **NAR NTR ORRi**

mimosae Fabricius, 1781: 76 (*Bruchus*) Ei: AZ DE FR GE HU IT NL PL SP TR **Ni**: AG CI EG MO **Ai**: IN TR **NAR NTR** [# 202]

breweri Crotch, 1867: 389 (*Bruchus*)

caliginosus Baudi di Selve, 1890: 341 (*Mylabris*)

dominicanus Jekel, 1855: 12 (*Bruchus*)

immunis Sharp, 1885: 474 (*Bruchus*)

innotatus Pic, 1912l: 92 (*Bruchus*) [RN]

inornatus Horn, 1873: 333 (*Bruchus*) [HN]

strigatus Motschulsky, 1874: 237 (*Bruchus*)

subrufus Motschulsky, 1874: 225 (*Bruchus*)

genus *Paleoacanthoscelides* Borowiec, 1985c: 457 type species: *Bruchus gilvus* Gyllenhal, 1839

gilvoides Lukjanovitch & Ter-Minassian, 1957: 177 (*Acanthoscelides*) **A**: AB AR TR

gilvus Gyllenhal, 1839: 30 (*Bruchus*) **E**: AL BU CR FR IT GR MA PL RU (ST) SP TR **N**: AG TU **A**: AB AF AR CY GG IN IQ KZ TD TM TR UZ **ORRi**

genus *Paleobruchidius* Egorov, 1989: 750 type species: *Acanthoscelides turanus* Lukjanovitch & Ter-Minassian, 1954

chesneyae Lukjanovitch & Ter-Minassian, 1957: 175 (*Acanthoscelides*) **A**: TD

colutae Lukjanovitch & Ter-Minassian, 1957: 180 (*Acanthoscelides*) **A**: TD

gussakovskii Lukjanovitch & Ter-Minassian, 1954: 72 (*Acanthoscelides*) **A**: IN TD UZ

plagiatus Reiche & Saulcy, 1858a: 649 (*Bruchus*) **E**: GR IT KR RU (ST) **A**: AR CY IN IS JO LE SY TR

semicarnea Reitter, 1884a: 99 (*Mylabris*) [# 040]

semenovi Lukjanovitch & Ter-Minassian, 1954: 73 (*Acanthoscelides*) **A:** AF KZ TD TM UZ
sugonjaevi Ter-Minassian, 1976: 302 (*Bruchidius*) **A:** AF
turanus Lukjanovitch & Ter-Minassian, 1954: 72 (*Acanthoscelides*) **A:** KI KZ TD UZ
saxicola Iablokoff-Khnzorian, 1965: 23 (*Bruchidius*) [# 041]

genus *Penthobruchus* Kingsolver, 1973: 142 type species: *Pachymerus germaini* Pic, 1894
germaini Pic, 1894b: 65 (*Pachimerus*) Ei: FR IT (Sicily) **A**i: IN **AFRi NTR AUSi**

genus *Pseudopachymerina* Zacher, 1952: 467 type species: *Bruchus lallemanti* Marseul, 1875 (= *Bruchus spinipes* Erichson, 1833)
spinipes Erichson, 1834: 252 (*Bruchus*) Ei: BE DE FR GE GR IT KR PL SP TR **N**i: AG CI EG MO TU **A**i: IS JO LE SY TR **AFRi NTR** [# 155, # 203]
aculeatus Motschulsky, 1874: 236 (*Bruchus*)
iaspideus Erichson, 1847: 125 (*Bruchus*)
lallemanti Marseul, 1875: xxxix (*Bruchus*)

genus *Pygobruchidius* Pic, 1951b: 427 type species: *Pygobruchidius latiorithorax* Pic, 1951 [# 042]
Longebruchus Pic, 1953b: 4 type species: *Bruchus clermonti* Pic, 1953 [# 057]
Tuberculobruchus Decelle, 1951: 179 type species: *Bruchus natalensis* Pic, 1903
aureus Arora, 1977: 60 (*Bruchidius*) **A:** ID (HP UT) **ORR** [# 072]
baronii Decelle, 1977: 400 (*Bruchidius*) **A:** BT ID (HP SD) NP **ORR** [# 073]
lineolatus Arora, 1977: 57 (*Bruchidius*) [HN]
bilineatopygus Pic, 1938e: 399 (*Bruchus*) Ei: FR (Corsica) GE **A:** AE AF BT ID (HP KA SD UT) NP PA QA YE **ORR** [# 043, # 074, # 156]
albizziae Arora, 1977: 55 (*Bruchidius*)
cingalicus Decelle, 1975b: 183 (*Bruchidius*)
variegatus Arora, 1980: 26 (*Bruchidius*)
wittmeri Decelle, 1977: 398 (*Bruchidius*) [# 043]
bimaculatopygus Anton, 2000b: 248 (*Bruchidius*) **A:** SA [# 075]
cadenati Pic, 1951d: 1100 (*Bruchidius*, var.) **A:** SA YE **AFR** [# 076]
densevestitus Anton, 2000b: 249 (*Bruchidius*) **A:** SA **AFR** [# 077]
devexus Fåhraeus, 1839: 112 (*Bruchus*) **A:** CH ("China boreali") [# 078]
dimorphus Arora, 1977: 53 (*Bruchidius*) **A:** ID (HP UT) [# 079]
flavovirens Arora, 1977: 56 (*Bruchidius*) **A:** ID (HP SD UT) NP **ORR** [# 080]
lineolatus Motschulsky, 1874: 225 (*Bruchus*) **A:** BT ID (HP SD) NP **ORR** [# 081]
maculipygus Champion, 1919: 245 (*Bruchus*) **A:** ID (HP UT) [# 082]
mimosae Arora, 1977: 54 (*Bruchidius*) **A:** BT ID (HP SD UT) NP **ORR** [# 083]

natalensis Pic, 1903f: 170 (*Bruchus*) Ei: GB **AFR** [# 084]
planophthalmus Anton, 1994a: 101 (*Bruchidius*) **A:** SA YE [# 085]
ptilinoides Fåhraeus, 1839: 103 (*Bruchus*) **A:** CH (GUA HKG NIN TIA) MG NC RU (FE) SC [# 086]
pygomaculatus Arora, 1977: 62 (*Bruchidius*) **A:** ID (HP KA UT) **ORR** [# 087]
saundersi Jekel, 1855: 6 (*Bruchus*) **Ni:** EG **A:** ID (HP UT) **ORR** [# 088, # 115]
chevrolati Allard, 1868: 108 (*Bruchus*)
schilskyi Arora, 1977: 61 (*Bruchidius*) **A:** ID (UT) **ORR** [# 089]
schroderi Pic, 1930c: 6 (*Bruchus*) Ei: GE **A:** ID (UT) **ORR** [# 090, # 110]
importatus Pic, 1930c: 6 (*Bruchus*, var.)
senensis Pic, 1923g: 14 (*Bruchus*) **A:** CH (YUN) [# 091, # 111]
sinaitus K. Daniel, 1908: 40 (*Laria*) **N:** AG EG MO TU **A:** IN IS JO SA SI OM YE **AFR** [# 092]
rabinovitchi Pic, 1939h: 148 (*Bruchus*)
solanensis Arora, 1977: 52 (*Bruchidius*) **A:** ID (HP) NP [# 093]
sparsemaculatus Pic, 1913g: 144 (*Bruchus*) Ei: GE **A:** ID (HP UT) **AFRi ORR** [# 094, # 157]
multilineolatus Arora, 1977: 58 (*Bruchidius*)
terrenus Sharp, 1886a: 35 (*Bruchus*) Ei: BU CR GR FR HU IT KR PT RO RU (ST) SB SK SP TR **A:** CH ("China" TAI) GGi JA UZi **NARi** [# 095]
notatus Chûjô, 1937e: 196 (*Bruchidius*)
urbanus Sharp, 1886a: 35 (*Bruchus*) **A:** CH (FUJ HP HUN YUN) ID (SD UT) JA NC NP SC **ORR** [# 096]
stevensi Pic, 1929a: 315 (*Bruchus*, var.)

genus *Salviabruchus* Decelle, 1983: 244 type species: *Mylabris retusa* Baudi di Selve, 1886
retusus Baudi di Selve, 1886: 102 (*Mylabris*) **A:** IN IQ IS LE SY TR [# 158]

genus *Sennius* Bridwell, 1946: 55 type species: *Bruchus cruentatus* Horn, 1873
lebasi Fåhraeus, 1839: 25 (*Bruchus*) **Ai:** JA **NAR NTR**
atriceps Pic, 1927c: 12 (*Bruchus*)
celatus Sharp, 1885: 449 (*Bruchus*)
rufescens Motschulsky, 1874: 222 (*Bruchus*)

genus *Specularius* Bridwell, 1938: 71 type species: *Specularius erythrinae* Bridwell, 1938
albus Pic, 1928a: 298 (*Bruchus*) Ei: IT **A:** IN SA YE **AFR ORRi**
bridwelli Arora, 1977: 37
erythraeus Pic, 1908h: 86 (*Laria*) Ei: GB **AFR** [# 116]
impressithorax Pic, 1932c: 36 (*Bruchus*) Ei: GB NL **Ni:** MR **AFR NTRi ORRi**

erythrinae Bridwell, 1938: 71

minor Zacher, 1952: 467 (var.)

genus *Stator* Bridwell, 1946: 55 type species: *Bruchus pruininus* Horn, 1873

limbatus Horn, 1873: 326 (*Bruchus*) Ei: FR (Corsica) IT (Sardinia) **Ai**: AE CY IN OM SY YE **AFRi NAR NTR** [# 164]

interruptus Sharp, 1885: 470 (*Bruchus*)

genus *Sulcobruchus* Chûjô, 1937e: 189 type species: *Bruchus sauteri* Pic, 1927

griseosuturalis Pic, 1932a: 331 (*Bruchus*) **A**: ID (HP SD UT) CH (YUN) **ORR**

kingsolveri Arora, 1977: 86

maculatithorax Pic, 1928a: 298 (*Bruchus*) **A**: ID (UT) **ORR**

sauteri Pic, 1927i: 248 (*Bruchus*) **A**: CH (TAI ZHE) JA **ORR**

discus S.-F. Zhang & Liu, 1987: 191, 192

subtribe Bruchina Latreille, 1802

genus *Bruchus* Linnaeus, 1767: 604 [conserved name, ICZN (1995a), Opinion 1809] type species: *Dermestes pisorum* Linnaeus, 1758

Acanthobruchus Yus Ramos, 1978: 317 type species: *Bruchus venustus* Fåhraeus, 1839

Bruchoides Yus Ramos, 1978: 317 type species: *Bruchus tessellatus* Mulsant & Rey, 1858

affinis Frölich, 1799: 55 **E**: AU BE BH BU BY CR CZ DE EN FI FR GB GE GR HU IT KR LA LT LU MD NL NR PL RO RU (ST) SK SL SP SV SZ TR UK **N**: AG EG **A**: AB AF AR CH (XIN) GG IN KI KZ NC RU (WS) SY TD TR **NARi ORRi**

bedeli Hoffmann, 1945: 44 (var.)

flavimanus Boheman, 1833: 59

monticola Bedel, 1901: 348 (*Laria*, var.)

mulkaki Lukjanovitch & Ter-Minassian, 1957: 98

obscuritarsis Motschulsky, 1874: 244

ruthenicus Becker, 1892: 69

altaicus Fåhraeus, 1839: 80 **E**: RU (ST) UK **A**: AB AR GG IN TM TR UK

incurvatus Motschulsky, 1874: 243 [# 107]

anatolicus Anton, 1999f: 655 **A**: TR

atomarius Linnaeus, 1761: 183 (*Curculio*) **E**: AN AU BE BH BU BY CR CZ DE EN FI FR GB GE GR HU IR IT KR LA LT LU MD NL NM NR PL RO RU (CT NT ST) SK SL SP SV SZ TR UK **N**: AG MO **A**: AR CH (XIN) GG IN KI KZ LE MG NC RU (ES FE WS SS) SY TR **NARi**

fahraei Gyllenhal, 1839: 85

granarius Linnaeus, 1767: 605

pilifer Olivier, 1791a: 562 (*Curculio*)

sarothamni Hoffmann, 1938: 13 (var.)

troglodytes Fåhraeus, 1839: 78

barimanii A. Delobel & Sadeghi, 2014: 437 **A:** IN

brachialis Fåhraeus, 1839: 79 **E:** AL AU AZ BE BH BU CR CZ FR GBi GE GR HU IT KR LU MD ME NM PL PT RO RU (ST) SK SL SP SZ TR UK **N:** AG MO **A:** AB AR CH ("China") GG IN LE SCi TR **NARi ORRi** [# 191]

rugicollis Motschulsky, 1874: 238

brisouti Kraatz, 1868: 322 **E:** FR IT (Sardinia) SP **N:** MO [# 113, # 192]

canariensis Decelle, 1975c: 118 **N:** CI (Gran Canaria, La Gomera, Tenerife)

dentipes Baudi di Selve, 1886: 33 (*Mylabris*) **E:** AL BEi CR FR GBi GR IT KR RU (ST) SP SZ (Valais) TR UK **N:** AG EG TU **A:** AB AF AR CH (XIN) CY IN IQ IS JO KZ LE SY TD TM TR UZ **ORRi AUSi** [# 136]

diversepubens Pic, 1930e: 14 (var.)

ochraceosignatus Heyden, 1894: 65 (*Mylabris*)

emarginatus Allard, 1868: 104 **E:** BU CR FR GBi GR IT KR MD NM RO RU (ST) SP TR **N:** AG MO TU **A:** AB AR GG IN IS JO LE SY TR **ORRi**

ervi Frölich, 1799: 56 **E:** AUi BEi BU CZi FR GBi GEi GR HUi IT MA PLi SKi SPi SVi SZi TR "Yugoslavia" **N:** AG CI EG MO TU **A:** AF CY IN IQ IS JO LE SY TR **AFRi AUSi NTRi ORRi**

bipunctatus Sulzer, 1776: 36

sertatus Illiger, 1805: 131

griseomaculatus Gyllenhal, 1833: 66 **E:** BE BU CR FR GE GR IT LU PT SP **N:** AG MO TU

adeps Vogel, 1875: 339

exiguus Rosenhauer, 1856: 236

hamatus Miller, 1881: 228 **E:** BH BU CR FRi GR KR RO RU (ST) TR UK **A:** AB AR IN LE SY TM (Kopet Dag) TR

hierroensis Decelle, 1979b: 143 **N:** CI (Hierro)

ibericus Anton, 1999f: 656 **E:** PT SP **N:** AG MO [# 137]

laticollis Boheman, 1833: 71 **E:** AL BH BU CR FR GR IT KR LAi MA NM RU (ST) SK SP TR UK **N:** AG MO **A:** AB AF AR CY GG IN IS KZ SY TD TM TR UZ [# 138]

lentis Frölich, 1799: 57 **E:** AL AUi BEi BU CR CZi FR FIi GBi GEi GR HUi IT KR MD NM PLi SBi SKi RUi (ST) SP SVi TR UK SZi **N:** AG EG MO TU **A:** AB AF AR CY GG IDi (HP KA UT) IN IQ IS JO KU LE SY TD TR UZ **NARi** [# 139, # 193]

libanensis Zampetti, 1993: 215 **E:** BU GR HU KR SPi **A:** AB AR GG IN LE TR [# 140]

lorestanus Kasatkin, 2018: 41 **A:** IN

loti Paykull, 1800: 158 **E:** AU BE BU BY CR CZ DE EN FI FR GE GB GR HU IT KR LA LT LU MD NL NR PL RO RU (CT NT ST) SK SL SP SV SZ UK **A:** CH (SHA XIN) GG IN JA KZ RU (ES FE WS SS) TR [# 141]

amurensis Pic, 1953g: 16 (*Bruchidius*) [# 049]

lathyri Stephens, 1831: 214

maculatipes Pic, 1927g: 153

muelleri Schilsky, 1905: 16

oxytropidis Gebler, 1830: 143

wasastjernii Fåhraeus, 1839: 77

lugubris Fåhraeus, 1839: 84 **A**: AR AF GG IN RU (ST: Dagestan) TR

luteicornis Illiger, 1794: 619 **E**: AU BE BU BY CR CZ DE FR GB GE GR HU IT KO KR LU MD ME NL NM PL RO RU (CT ST) SK SL SP SV SZ TR UK **N**: EG MO **A**: AB AR GG IN TR **ORR**i **[# 142]**

mirabilicollis Ter-Minassian, 1968: 181 **A**: TM (Kopet Dag)

nikdeli A. Delobel & Sadeghi, 2014: 434 **A**: IN (Azarbayjan) TR (Erzurum, Kars)

occidentalis Lukjanovitch & Ter-Minassian, 1957: 109 (ssp.) **E**: AU BE BU CZ FR GB GE HU IT PL RU (CT ST) SK SP SV SZ UK **A**: KZ **[# 044, # 143]**

pavlovskii Lukjanovitch & Ter-Minassian, 1954: 71 **A**: KZ TD

perezi Kraatz, 1868: 321 E: FR IT (Sicily) SP **N**: AG MO TU **[# 114, # 194]**

meridionalis Allard, 1868: 103

pisorum Linnaeus, 1758: 356 (*Dermestes*) E: AL AN AU AZ BE BH BU BY CR CZ DE EN FI FR GB GE GR HU IC IR IT KO KR LA LS LT LU MA MD NL NM NR NT PL PT RO RU (CT NT ST) SB SK SL SP SV SZ TR UK **N**: AG CI EG MO MR TU **A**: AB AF AR CY GG IN IS JO KZ LE SY TD TR UZ **Ai**: CH (CHQ NIN SCH XIN) ID (HP KA SD UT) JA NC RU (FE WS) SC **AFR**i **AUS**i **NAR**i **NTR**i **ORR**i

cruciger Geoffroy, 1785: 112 (*Mylabris*)

intermedius Motschulsky, 1854b: 10

lunaris Rey, 1893a: 3 (var.)

obscurus Philippi & Philippi, 1864: 360

pisi Linnaeus, 1767: 604

salicis Scopoli, 1763: 22 (*Laria*)

sparsus Fabricius, 1801b: 398

rufimanus Boheman, 1833: 58 **E**: AL AN AU BE BH BU BY CR CZ DE EN FI FR GB GE GR HU IC IR IT KO KR LA LS LT LU MA MD NL NM NR PL PT RO RU (CT NT ST) SB SK SL SP SV SZ TR UK **N**: AG AZ CI EG MO MR TU **A**: AB AF AR CY GG IN IS JO LE SY TR **Ai**: CH (CHQ NIN SCH SHG XIN) JA NC RU (FE WS) SC **AFR**i **NAR**i **ORR**i

ecalcaratus K. Daniel, 1906: 180 (*Laria*)

fabae Motschulsky, 1854b: 6

velutinus Mulsant & Rey, 1858: 27 (var.)

rufipes Herbst, 1783: 29 **E**: AL AU BH BU CR CZ FR GR HU IT KR MA MD NM PL PT RO RU (ST) SB SK SL SP SZ TR UK **N**: AG AZ CI MO MR TU **A**: AB AR GG IN IS LE SY TR **ORR**i **[# 195]**

apicatus Rey, 1893a: 3 (var.)

ervi Motschulsky, 1854b: 14 [HN]

flavicornis Rey, 1893a: 3 (var.)

guttalis Rey, 1893a: 3 (var.)

minor Rey, 1893a: 3 (var.)
nubilus Boheman, 1833: 60
sibiricus Germar, 1823: 179 E: RU (ST) **A:** AB AR CH (XIN) GG IN KI KZ MG RU (ES FE SS WS) TD TR UZ [# 044, # 196]
asiaticus Iablokoff-Khnzorian, 1964: 169 [# 056]
transcaucasicus Lukjanovitch & Ter-Minassian, 1957: 109 (ssp.)
signaticornis Gyllenhal, 1833: 64 E: AUi AZ BH BU CZi FR GBi GEi GR HUi IT MA MDi PLi PT SKi SP SZi TR "Yugoslavia" **N:** AG CI EG MO TU **A:** CY IN IS LE SY TM TR **NARi ORRi**
inornatus Küster, 1849c: 72
lugens E. Jacquet, 1888: 8 (var.)
muticus Fiori, 1906: 206 (var.) [IN] [# 048]
pallidicornis Boheman, 1833: 65
ruficornis Allard, 1868: 103
tetragonus Baudi di Selve, 1886: 28 (*Mylabris*) **A:** SY TR
tristiculus Fåhraeus, 1839: 81 E: AL AZ BH BU CR FR GR HU IT MA NM PT RO RU (ST) SB SK SP TR **N:** AG CI EG MO MR TU **A:** AB CY GG IN IS SY TM (Kopet Dag) TR **ORRi**
azoricus Crotch, 1867: 390
gyllenhalii Hochhut, 1847: 457
oblongus Rosenhauer, 1856: 235
tristis Boheman, 1833: 63 E: BEi BU CR FR GBi GR HU IT KR MA SP TR **N:** AG CIi MO EG TU **A:** CY IN IS JO LE SY TR **NARi ORRi**
ulicis Mulsant & Rey, 1858: 33 E: BEi BU FR GR HU IT MA NM PT RO SB SP TR **N:** AG CIi EG MO **A:** AB AF CY IN IS JO SY TM (Kopet Dag) TR **NARi**
terminatus Wollaston, 1864: 381
vavilovi Bekman, 1929: 160 (ssp.)
venustus Fåhraeus, 1839: 75 E: AU BU CR CZ FR GE GR HU IT PL RU (ST) SK SP UK **A:** AB AR TR
elegans Motschulsky, 1874: 243
viciae Olivier, 1800: [79]: 12 E: AL AU BH BU CR CZ DE FR GE GR HU IT KO KR NM PL RO RU (ST) SB SK SL SP SV SZ TR UK **N:** MO **A:** AB AF AR GG IN IS LE PA SY TR
nigripes Gyllenhal, 1833: 66
punctellus Boheman, 1828: 109

genus *Bruchus*, nomina dubia
adustus Motschulsky, 1874: 228 **A:** JA
atratus Motschulsky, 1874: 218 **A:** RU ("Sibir. or.")
biovalis Marseul, 1888b: 458 [RN] **A:** TR
ovalis Motschulsky, 1874: 217

brunnipes Allard, 1868: 116 (*Bruchus*) **A:** SY
calamitosus Motschulsky, 1874: 242 **A:** RU ("Sib. or.")
campanulatus Motschulsky, 1874: 242 **A:** "Caucas."
diversicornis Motschulsky, 1874: 213 **E:** RU (ST)
glaucus Motschulsky, 1874: 220 **A:** RU ("Sibir. or.")
insitivus Motschulsky, 1874: 241 **E:** RU (ST)
latiusculus Motschulsky, 1874: 242 **A:** "Des. Kirgh. occ."
longicornis Thunberg, 1816: 47 **A:** CH ("Sinensia")
  litteratus Schoenherr, 1833: 102 [RN]
maculosus Motschulsky, 1874: 212 **A:** GG
obliquus Grimmer, 1841: 40 **A:** AU
pachycerus Motschulsky, 1874: 242 **A:** "Transcauc."
pinetorum Motschulsky, 1874: 240 **E:** "Sib. Eur. bor."
rubrimanus Motschulsky, 1874: 217 **A:** GG
septentrionalis Motschulsky, 1874: 240 **E:** RU (NT)
tardus Motschulsky, 1874: 219 **A:** GG
tessellatus Mulsant & Rey, 1858: 38 **E:** FR
transversoguttatus Motschulsky, 1874: 223 [distribution not stated]
umbellatarum Fabricius, 1787: 41 **N:** AG
umbellatarum Gistel, 1857b: 518 [HN] **E:** GE
venustulus Motschulsky, 1874: 243 **A:** GG
vitticollis Motschulsky, 1874: 214 **A:** GG

subtribe Megacerina Bridwell, 1946

genus *Megacerus* Fåhraeus, 1839: 34 type species: *Bruchus pescaprae* Fåhraeus, 1839
subgenus *Megacerus* Fåhraeus, 1839: 34 type species: *Bruchus pescaprae* Fåhraeus, 1839
eulophus Erichson, 1847: 124 (*Bruchus*) Ei: GE **NTR** [# 201]
  laticornis Blanchard, 1851: 288 (*Bruchus*)
  leucaster Blanchard, 1851: 293 (*Bruchus*)

tribe Kytorhinini Bridwell, 1932

genus *Kytorhinus* Fischer von Waldheim, 1809: 298 type species: *Kytorhinus karasini* Fischer von Waldheim, 1809
subgenus *Kytorhinoides* Legalov, 2011: 262 type species: *Kytorhinus thermopsis* Motschulsky, 1874
thermopsis Motschulsky, 1874: 208 **A:** CH (SCH XIN) KZ MG RU (ES FE SS) UZ **[# 152]**
  aridus Motschulsky, 1874: 208
subgenus *Kytorhinus* Fischer von Waldheim, 1809: 298 type species: *Kytorhinus karasini* Fischer von Waldheim, 1809
hauseri Pic, 1906f: 50 (*Kytorrhinus*) **A:** CH (GAN QIN SCH XIN) **[# 065]**

immixtus Motschulsky, 1874: 208 **A:** CH (NIN XIN) MG RU (ES FE)
karasini Fischer von Waldheim, 1809: 302 **A:** KI KZ MG RU (ES SS WS) TD [# 065]
posticus Hoffmann, 1970: 2 (*Kytorrhinus*, var.) [IN]
reitteri Schilsky, 1905: 11
uniformis Hoffmann, 1965: 70 (*Kytorrhinus*, var.) [IN]
kaszabi Hoffmann, 1965: 64 (*Kytorrhinus*) **A:** MG RU (ES)
nigricornis Hoffmann, 1965: 65 (*Kytorrhinus*, var.) [IN]
kergoati A. Delobel & Legalov, 2009: 72 **A:** CH (XIN XIZ) TD (Pamir) [# 045]
mongolicus Ter-Minassian, 1973: 82 (*Kytorrhinus*) **A:** MG RU (ES SS) [# 153]
pectinicornis Melichar, 1912b: 170 (*Kytorrhinus*) **E:** AU GE (Allgäu) RU (CT NT ST: Caucasus, Ural) **A:** GG KI KZ RU (ES FE SS WS: Ural) **NAR**
kerzhneri Egorov, 1996: 156 (*Kytorrhinus*)
lygaeus Iablokoff-Khnzorian, 1974: 123
obscurus Lukjanovitch & Ter-Minassian, 1957: 185
prolixus Fall, 1926: 204 (*Mylabris*)
zherichini Egorov, 1996: 156 (*Kytorrhinus*)
piptanthi Decelle, 1971a: 110 **A:** BT NP
quadriplagiatus Motschulsky, 1839: 57 (*Bruchus*) **E:** RU (CT ST) UK **A:** KZ RU (WS: South Ural) [# 199]
luteipennis Baudi di Selve, 1890: 344 (*Mylabris*, var.)
tibetanus Decelle, 1971a: 112 **A:** CH (XIZ) ID (UT) NP
subgenus *Pygobruchus* Sharp, 1886a: 38 type species: *Pygobruchus scutellaris* Sharp, 1886
caraganae Ter-Minassian, 1960: 417 (*Kytorrhinus*) **A:** RU (FE: South Primorie)
lefevrei Pic, 1924i: 31 **A:** CH (SCH XIZ YUN)
sericeus Decelle, 1971a: 107
senilis Solsky, 1870: 310 (*Bruchus*) **A:** CH (BEI HEI) JA NC RU (FE) [# 154]
scutellaris Sharp, 1886a: 38 (*Pygobruchus*)
sharpianus Bridwell, 1932: 106
tenebrosus Baudi di Selve, 1886: 105 (*Mylabris*)

tribe Pachymerini Bridwell, 1929

subtribe Caryedontina Bridwell, 1929

genus ***Caryedon*** **Schoenherr, 1823: 1134** type species: *Bruchus serratus* Olivier, 1790
acaciae Gyllenhal, 1833: 97 (*Bruchus*) **Ei:** HU SP **N:** EG MO **A:** IN JO OM SA YE **AFR ORRi**
capicola Motschulsky, 1874: 248 (*Caryaborus*)
angeri Semenov, 1896b: 383 (*Caryoborus*) **Ei:** CR **N:** AG EG **A:** AE AF IN IQ IS JO LE OM PA SA SY TM TR **AFR**
palaestinicus Southgate, 1977: 195 (ssp.)

cassiae Gyllenhal, 1833: 95 (*Bruchus*) **A:** YE **AFR**
minutus Pic, 1902d: 146 (*Caryoborus*)
ceylonicus Pic, 1924g: 25 (*Pachymerus*) **A:** ID (SD UT) NP **ORR**
crineus Arora, 1977: 102 **A:** ID (HP UT)
femoralis Anton & A. Delobel, 2004: 72 **N:** AG MO **AFR**
furcatus Anton & A. Delobel, 2004: 76 **A:** OM SA YE **AFR**
germari Küster, 1845a: 37 (*Caryoborus*) **E:** AL BH BU CR GR KR NM RU (ST) TR **A:** AB AR CY GG IN IQ IS LE SY TR [# 108, # 147]
cyprus C. D. Johnson, Southgate & A. Delobel, 2004: 35 [# 061]
lisaeae Southgate, 1971: 413
gonagra Fabricius, 1798: 159 (*Bruchus*) **Ei:** CZ FR RU (ST) **N:** CIi EG **A:** AE ID (HP KA SD) IN IQ IS JO KU OM PA RU SA YE (incl. Socotra) **ORR AUSi**
irakensis A. H. Ali & A. Z. M. Ali, 1988: 3
notativentris Pic, 1924g: 24 (*Pachymerus*)
tamarindi Decaux, 1894: 129 (*Caryoborus*)
indus Motschulsky, 1858: 98 (*Caryoborus*) **A:** BT ID (SD) NP **ORR**
punjabensis Pajni & Singh, 1977: 225
arori Singh & Saini 1978: 131
kizilkumensis Ter-Minassian, 1978: 55 **A:** AB AF IN KI KZ TD TM UZ [# 046, # 148, # 197]
lagonychii Motschulsky, 1874: 247 (*Caryaborus*) **A:** IN TM (Caspian coast)
languidus Gyllenhal, 1839: 129 (*Bruchus*) **A:** BT ID (HP SD UT) NP **ORR**
longipennis Pic, 1898d: 174 (*Caryoborus*) **Ei:** GB **AFR**
combreti Prevett, 1965: 533
longivertex Borowiec, 1984a: 519 **A:** CH (HKG)
mesra C. D. Johnson, Southgate & A. Delobel, 2004: 57 **A:** IN IS TR
halperini Anton & A. Delobel, 2004: 81 [# 062]
montanus Arora, 1977: 99 **A:** ID (HP UT) NP **ORR**
nongoniermai Anton & A. Delobel, 2004: 84 **N:** EG LB MO **A:** AE IS JO OM SA YE **AFR**
omanensis Yus Ramos, 2010b: 259 **A:** OM (Dhofar)
opacus Arora, 1977: 103 **A:** ID (HP UT) **ORR**
pallidus Olivier, 1790: 199 (*Bruchus*) **Ei:** CZ FR IT **AFR**
annulicornis Pic, 1950f: 211 (*Pachymerus*, var.)
prosopidis Arora, 1977: 105 **A:** ID (HP) **ORR**
saudicus Anton, 1994a: 99 **N:** EG **A:** IS JO SA
serratus Olivier, 1790: 199 (*Bruchus*) **Ei:** BE CZ FR GE GR IT PT SP SV **Ai:** CY IN KU OM SA **AFR NARi** [# 198]
sibutensis Pic, 1924g: 25 (*Pachymerus*)
sudanensis Southgate, 1971: 411 **Ei:** GB **N:** AG EG **A:** YE (Socotra) **AFR**

yemenensis Decelle, 1979a: 328 **N:** EG **A:** AE IN IQ IS JO OM PA SA SI YE

subtribe Caryopemina Bridwell, 1929

genus *Caryopemon* Jekel, 1855: 25 type species: *Caryopemon hieroglyphicus* Jekel, 1855

Protocaryopemon Borowiec, 1987: 53 type species: *Protocaryopemon archetypus* Borowiec, 1987 (= *Caryopemon giganteus* Pic, 1909)

cruciger Stephens, 1839: 265 (*Caryoborus*) Ei: CZ FR GB **AFR ORRi**

giganteus Pic, 1909b: 34 **A:** CH (YUN) ID (SD) NP **ORR**

archetypus Borowiec, 1987: 53 (*Protocaryopemon*)

hieroglyphicus Jekel, 1855: 27 **A:** CH (YUN) ID (UT) **ORR**

centronotatus Pic, 1924i: 31

lajoyei Pic, 1924i: 31 (var.)

quadriguttatus Chevrolat, 1877: 116

lhostei Pic, 1924i: 30 Ei: FR **ORR**

luteonotatus Pic, 1898d: 173 **A:** CH (YUN) ID ("Inde") NP **ORR** [# 109]

subtribe Pachymerina Bridwell, 1929

genus *Caryobruchus* Bridwell, 1929: 148 type species: *Dermestes gleditsiae* Linnaeus, 1763

gleditsiae Linnaeus, 1763: 9 (*Dermestes*) Ei: SP **Ni:** CI **NTR** [# 149]

arthriticus Fabricius, 1801b: 398 (*Bruchus*)

fuscus Goeze, 1777: 332 (*Bruchus*)

genus *Pachymerus* Thunberg, 1805: 282 type species: *Dermestes bactris* Linnaeus, 1763

subgenus *Butiobruchus* Prevett, 1966b: 183 [RN] type species: *Pachymeroides bridwelli* Prevett, 1966 [# 098]

Pachymeroides Prevett, 1966a: 81 [HN] type species: *Pachymeroides bridwelli* Prevett, 1966

bridwelli Prevett, 1966a: 81 (*Pachymeroides*) **Ai:** IS **NTR**

subgenus *Pachymerus* Thunberg, 1805: 282 type species: *Dermestes bactris* Linnaeus, 1763

cardo Fåhraeus, 1839: 127 (*Bruchus*) Ei: GB **NTR**

tribe Rhaebini Blanchard, 1845

genus *Rhaebus* Fischer von Waldheim, 1823: pl. XLVII [= 1824: 178] type species: *Rhaebus gebleri* Fischer von Waldheim, 1823

gebleri Fischer von Waldheim, 1823: pl. XLVII [= 1824: 180] **E:** RU (CT ST) **A:** CH (NIN NMO) IN IS KI KZ MG RU (ES SS) TR

amnoni Lopatin & Chikatunov, 2000b: 32

beckeri Suffrian, 1867a: 143

komarovi Lukjanovitch, 1939: 551

mannerheimii Motschulsky, 1845a: 108
sagroides Solsky, 1867: 181
lukjanovitschi Ter-Minassian, 1973: 76 **A**: CH (NMO) MG
solskyi Kraatz, 1879g: 277 E: RU (CT ST) **A**: CH (NIN NMO QIN XIN) KI KZ MG RU (ES SS) TM (Kopet Dag)

subfamily Bruchinae, nomina dubia
Bruchus bipunctatus Roemer, 1789: 6 **E**: "Rhaetia"
Bruchus capsularius Schrank, 1789: 70 **E**: AU
Bruchus fasciatus Olivier, 1800: [79]: 20 **E**: FR [# 047]

subfamily **Donaciinae** Kirby, 1837
tribe Donaciini Kirby, 1837

genus *Donacia* Fabricius, 1775: 195 type species: *Donacia crassipes* Fabricius, 1775 [# 209]
Arundonacia Kippenberg, 2015: 257 type species: *Donacia clavipes* Fabricius, 1792
Askevoldia Kippenberg, 1994: 20 type species: *Donacia reticulata* Gyllenhal, 1817
Brevidonacia Kippenberg, 2015: 258 type species: *Donacia semicuprea* Panzer, 1795
Crassodonacia Kippenberg, 2015: 258 type species: *Donacia polita* Kunze, 1818
Cyphogaster Goecke, 1934: 215 type species: *Donacia provostii* Fairmaire, 1885
Donaciella Reitter, 1920: 38 type species: *Donacia tomentosa* Ahrens, 1810
Donaciomima L. N. Medvedev, 1973b: 876 type species: *Donacia clavareaui* Jacobson, 1906
Donacocia Gistel, 1856: 377 type species: *Donacia brevicornis* Ahrens, 1810
Flavodonacia Kippenberg, 2015: 262 type species: *Rhagium fennica* Paykull, 1800
Mergodonacia Kippenberg, 2015: 263 type species: *Leptura versicolorea* Brahm, 1790
Plateumaroides Iablokoff-Khnzorian, 1962: 116 type species: *Donacia fastuosa* Iablokoff-Khnzorian, 1962 (= *Donacia kraatzi* Weise, 1881)
Pseudodonacia Reitter, 1920: 27 type species: *Donacia kraatzi* Weise, 1881
Smaragdonia Kippenberg, 2015: 264 type species: *Donacia dentata* Hoppe, 1795
Sotaiana Kippenberg, 2015: 265 type species: *Donacia sparganii* Ahrens, 1810
aequidorsis Jacobson, 1894b: 152 **E**: KZ RU (ST) **A**: RU (WS) TR
akiyamai Komiya, 2001: 41 **A**: JA
antiqua Kunze, 1818a: 21 **E**: FI RU (CT NT) SV [# 258]
gracilis Suffrian, 1845: 366
simplicifrons Lacordaire, 1845: 135
apricans Lacordaire, 1845: 158 **E**: IT (Sicily) SP **N**: AG MO
andalusiaca Kraatz, 1870a: 271 (var.)
inermis Fuente, 1904: 389 (var.)

aquatica Linnaeus, 1758: 397 (*Leptura*) E: AL AU BE BH BY CR CZ DE EN FI FR GB GE HU IR IT KR LA LS LT LU NL NR PL RO RU (CT NT ST) SB SK SL SP SV SZ UK **A:** CH (HEI) GG IN KZ RU (ES FE SS WS) TR **[# 248, # 259]**

coccineofasciata Harrer, 1784: 226 (*Leptura*)

concinna Weise, 1898: 179 (var.)

cyaneicollis Olsoufieff, 1903: 303 (var.)

dentipes Fabricius, 1792b: 116

fasciata Herbst, 1784: 100

latonae Gistel, 1857b: 586

vittata Olivier, 1791b: 292

waldaica Olsoufieff, 1903: 303 (var.)

aureocincta J. R. Sahlberg, 1921: 36 E: FI RU (NT) SV

bactriana Weise, 1887d: 325 **A:** CH (XIN) KI KZ RU (ES SS) TD UZ

biformis Reitter, 1920: 36 (var.)

coeruleoviolacea Jacobson, 1901b: 105 (var.) [IN] [# 234]

cyanea Shavrov, 1948: 50 (ab.) [IN]

rubra Shavrov, 1948: 51 (ab.) [IN]

sahlbergi Jacobson, 1901b: 103

bicolora Zschach, 1788: 27 E: AL AU BE BH BU BY CR CZ DE EN FI FR GB GE GR HU IR IT KZ LA LT LU MD ME NL NR PL PT RO RU (CT NT ST) SB SK SL SP SV SZ UK **A:** AB AR CH (XIN) GG IN IS KI KZ RU (ES SS WS) TR UZ **[# 210, # 244]**

aurea Hoppe, 1795: 43

collaris Panzer, 1795a: 216

meridionalis Weise, 1886d: 250 (var.) [# 213]

sagittariae Fabricius, 1792b: 117

bicoloricornis S.-H. Chen, 1941a: 10 **A:** CH (BEI HEB HEI HUB JIA JIX TAI ZHE) JA RU (FE) SC

subcylindrica S.-H. Chen, 1941a: 11

brevicornis Ahrens, 1810: 26 E: AU BE BH BY CR CZ DE EN FI FR GE HU IT LA LT MD NL NR PL RO RU (CT NT ST) SB SK SL SP SV SZ UK **N:** AG **A:** KZ RU (WS)

platysterna C. G. Thomson, 1866: 118

noesskei Hänel, 1912: 91 (ab.) [IN]

brevitarsis C. G. Thomson, 1884: 202 E: AU BE BH BY CR CZ FI FR GE HU IT LA LT NR PL RU (CT NT) SK SL SV SZ UK **A:** RU (WS) [# 258]

breviuscula Jacobson, 1899b: 3 **A:** RU (FE)

cinerea Herbst, 1784: 100 E: AL AU BE BH BU BY CR CZ DE EN FI FR GB GE GR HU IR IT KR LA LT LU MD NL NR PL PT RO RU (CT NT ST) SB SK SL SP SV SZ UK **A:** AB CH (XIN) GG IN KZ RU (SS WS) TR UZ

hydrochaeridis Fabricius, 1801b: 129

hydrocharis Fabricius, 1792b: 118

tersata Panzer, 1795b: 16

clavareaui Jacobson, 1906b: 311 **A:** CH (HEI NMO SCH YUN) JA MG RU (ES FE) SC [**# 260**]

mediohirsuta S.-H. Chen, 1966: 145, 147 [**# 215**]

clavipes clavipes Fabricius, 1792b: 117 **E:** AU BE BH BU BY CR CZ DE EN FI FR GB GE HU IR IT KZ LA LS LT LU MD ME NL NM NR PL PT RO RU (CT NT ST) SK SL SP SV SZ TR UK **A:** CH (XIN) IN KI KZ RU (SS WS) TR [**# 249**]

aerea Schrank, 1798: 684

crassipes Zschach, 1788: 27 [HN] [**# 210**]

menyanthidis Gyllenhal, 1813: 662

menyanthis Fabricius, 1801b: 129

mutica C. G. Thomson, 1866: 109

clavipes glabrata Solsky, 1871: 245 **A:** CH (BEI HEB HEI JIL LIA) MG RU (ES FE)

crassipes Fabricius, 1775: 195 **E:** AU BE BH BU BY CR CZ DE EN FI FR GB GE HU IR IT KR KZ LA LT LU ME NL NR PL RO RU (CT NT ST) SB SK SL SP SV SZ UK **A:** CH (GAN HEB SHA SHX) KZ RU (ES SS WS) TR

anthracina Everts, 1901: 399 (ab.) [IN]

micans Hoppe, 1795: 39

spinosa C. G. Thomson, 1866: 110

striata Panzer, 1795b: 1

delagrangei Pic, 1896a: 35 **A:** AR GG TR

jacobsoni Semenov & Reichardt, 1927: 218 [**# 218**]

dentata Hoppe, 1795: 40 **E:** AU BE BH BU BY CR CZ DE EN FI FR GB GE HU IR IT KR KZ LA LT MD NL NR PL RO RU (CT NT ST) SB SK SL SP SV SZ UK **A:** GG KZ MG RU (ES SS WS) TR [**# 250**]

angustata Kunze, 1818a: 8

autumnalis Westhoff, 1882: 253 (var.)

chrysochlora Westhoff, 1882: 253 (var.)

diabolus Westhoff, 1882: 253 (var.)

indentata Kolosov, 1925: 3 (ab.) [IN]

phellandrii C. R. Sahlberg, 1839: 271

fedtschenkoae Jacobson, 1899b: 1 **A:** UZ [**# 227**]

fennica Paykull, 1800: 70 (*Rhagium*) **E:** BY EN FI LA LT RU (CT NT) **A:** RU (ES FE SS WS)

flemola Goecke, 1944: 7 **A:** CH (JIL) NC JA RU (FE) SC

jacobsoniana Shavrov, 1946: 36

freyi Goecke, 1940: 852 **E:** AL [**# 237**]

frontalis Jacoby, 1893b: 261 **A:** CH (BEI CHQ FUJ GUI GUX HAI HEB HGK HUB JIA JIX SCH SHA SHG SHX TAI YUN) JA

fukiensis Goecke, 1944: 9 **A:** CH (FUJ) [**# 214**]

galaica Báguena, 1959: 209 **E:** SP PT

gracilicornis Jacobson, 1899b: 6 **E:** RU (ST) **A:** AB AR GG IN SY TM
  ornata Shavrov, 1948: 50 (ab.) [IN]
  transcaucasica Sumakov, 1901b: 454
hirtihumeralis Komiya & Kubota, 1987: 41 **A:** JA
humilis Weise, 1912: 76 **A:** CH (JIX)
hungi Hayashi & Pütz, 2008: 243 **A:** CH (SCH)
impressa Paykull, 1799a: 193 **E:** AU BE BH BU BY CR CZ DE EN FI FR GB GE GR HU IR IT KR LA LT LU NL NR PL PT RO RU (CT NT ST) SB SK SL SP SV SZ UK **N:** AG MO **A:** AB GG IN KZ RU (ES WS) TR
  armillata Shavrov, 1948: 50 (ab.) [IN]
japana Chûjô & Goecke, 1956: 60 **A:** JA
katsurai Kimoto, 1981a: 24 **A:** JA
kirgizkaisaka Jacobson, 1925c: 7 **A:** KZ **[# 228]**
knipowitschi Jacobson, 1927: 45 **A:** CH (HEI) RU (FE)
koenigi Jacobson, 1899b: 4 **A:** GG
kraatzi Weise, 1881: 38 **A:** AR TR
  fastuosa Iablokoff-Khnzorian, 1962: 116
kweilina S.-H. Chen, 1966: 145, 147 **A:** CH (GUX) **[# 216]**
lenzi Schönfeldt, 1888: 33 **A:** CH (ANH GUX HEN HUB HUN JIA JIX SCH TAI) JA RU (FE) SC **ORR**
  formosana Chûjô, 1934g: 526
  wiepkeni Weise, 1898: 178
lungtanensis Hayashi & C.-F. Lee, 2009: 153 **A:** CH (TAI)
lusow Hayashi & C.-F. Lee, 2007: 551 **A:** CH (TAI)
malinovskyi Ahrens, 1810: 39 **E:** AU BH BY CZ FI FR GE HU LT LU NL PL RO RU (CT) SK SZ UK **A:** RU (WS) **[# 261]**
  ararensis Thierriat, 1943: 22 (var.)
  arundinis Ahrens, 1810: 41
  caroli Thierriat, 1943: 22 (var.)
  ladonensis Thierriat, 1943: 22 (var.)
  mequignoni Thierriat, 1943: 22 (var.)
mannerheimii Lacordaire, 1845: 159 **A:** IN **[# 229]**
marginata Hoppe, 1795: 42 **E:** AL AU BE BH BU BY CR CZ DE EN FI FR GB GE GR HU IR IT KR LA LT LU NL NM PL PT RO RU (CT NT ST) SB SK SL SP SV SZ TR UK **N:** MO **A:** AB CH (XIN) GG IN IS KI KZ RU (WS) TD TR UZ **[# 211, # 238, # 251]**
  australis Jacobson, 1892a: 428 (var.)
  chlamydata Balthasar, 1934: 129 (ab.) [IN]
  lateralis Bonelli, 1812: 171
  lemnae Fabricius, 1801b: 128
  limbata Panzer, 1795b: 12 **[# 211]**

unicolor Westhoff, 1882: 255 (var.)

vittata Panzer, 1795b: 11 [HN] [# 211]

microcephala J. Daniel, 1904a: 89 **A:** IN IQ TR

testaceipes Pic, 1908d: 52

mistshenkoi Jacobson, 1910b: 53 **A:** GG TR

nitidior Nakane, 1963a: 18 (*Plateumaris*) **A:** JA

obscura Gyllenhal, 1813: 654 **E:** AU BE BU BY CZ DE EN FI FR GB GE HU IR IT LA LT LU NL NR PL RO RU (CT NT) SK SL SP SV SZ UK "Yugoslavia" **A:** KZ RU (WS)

barovskyi Jacobson, 1922b: 54 (var.)

impressa Ahrens, 1810: 23 [HN]

ochroleuca Weise, 1912: 76 **A:** CH (HEI) RU (ES FE SS)

flavidula Reitter, 1913a: 124

ozensis Nakane, 1954: 739 **A:** JA

polita Kunze, 1818a: 29 **E:** AL BH BU CR FR GR IT ME PT RU (ST) SB SL SP TR UK **N:** AG MO TU **A:** TR [# 239, # 240, # 252, # 262]

provostii Fairmaire, 1885a: lxiv **A:** CH (ANH BEI FUJ GAN GUA GUI HAI HEB HEI HEN HUB HUN JIA JIX LIA SCH SHA SHN SHX TAI TIA ZHE) JA NC RU (FE) SC **ORR**

brevicollis Weise, 1898: 177

nitidicollis Weise, 1898: 177

sikanga Gressitt, 1942a: 277

yuasai Nakane, 1963a: 18

recticollis Jacoby, 1893b: 261 **A:** NP **ORR**

flava Goecke, 1936: 228 (ab.) [IN]

reticulata Gyllenhal, 1817: 37 **E:** AU BE CR CZ FR GE HU IT LU ME NL PT SL SP SZ **N:** AG MO [# 263]

appendiculata Ahrens, 1810: 34 [HN]

coeruleans Shavrov, 1948: 50 (ab.) [IN]

semicuprea Panzer, 1795b: 14 **E:** AU BE BH BU BY CZ DE EN FI FR GB GE GR HU IR IT LA LT LU NL NR PL RO RU (CT NT ST) SB SK SL SP SV SZ UK **A:** GG RU (ES WS) [# 253]

iris Westhoff, 1882: 259 (var.)

melanocephala Marsham, 1802: 348 (*Leptura*) [# 217]

moerens Kunze, 1818a: 48

tenebrans Westhoff, 1882: 259 (var.)

simplex Fabricius, 1775: 195 **E:** AL AU BE BH BU BY CR CZ DE EN FI FR GB GE GR HU IR IT KR LA LS LT LU MD NL NR PL PT RO RU (CT NT ST) SB SK SL SP SV SZ TR UK **N:** AG MO **A:** CH (NMO SHX) GG IN MG RU (ES WS) SC SY TR [# 254]

atrocoerulea Westhoff, 1882: 259 (var.)

aeruginosa Westhoff, 1882: 259 (var.)

aurichalcea Westhoff, 1882: 259 (var.)
linearis Hoppe, 1795: 46
pulcherrima Weise, 1893a: 1118 (var.)
sanguinea Westhoff, 1882: 259 (var.)
sparganii gracilipes Jacoby, 1885b: 191 **A:** CH (NMO) JA MG RU (ES FE)
tschitscherini Semenov, 1895a: 141
sparganii sparganii Ahrens, 1810: 20 **E:** AU BE BY CZ DE EN FI FR GB GE IT LA LT NL NR PL RU (CT NT) SK SL SV SZ UK **A:** RU (WS) **[# 264]**
coelestis Weise, 1893a: 1117 (var.)
tridens Jacobson, 1892a: 425 (var.)
splendens hiurai Kimoto, 1983a: 11 **A:** JA (Honshu)
splendens splendens Jacobson, 1894f: 22 **A:** JA (Hokkaido) MG RU (ES FE SS) **[# 265]**
springeri J. Müller, 1916: 95 **E:** AU CZ FR GE IT LA SV SZ
thalassina intermedia Jacobson, 1899b: 7 **A:** CH (HEB HEI NMO QIN XIZ) MG RU (ES FE SS)
coerulea Jacobson, 1901b: 105 (var.) [IN] [# 233]
rufovariegata Jacobson, 1901b: 105 (ssp.)
semenovi Jacobson, 1907b: 5
thalassina thalassina Germar, 1811: 29 **E:** AU BE BH BU BY CR CZ DE EN FI FR GB GE GR HU IR IT LA LT LU ME NL NR PL RO RU (CT NT ST) SB SK SL SP SV SZ TR UK **A:** AB AR CH (XIN) GG IN KZ MG RU (WS) TR **[# 266]**
extensa Shavrov, 1948: 51 (ab.) [IN]
inermipes Shavrov, 1948: 51 (ab.) [IN]
porphyrogenita Westhoff, 1882: 255 (var.)
viridis Shavrov, 1948: 51 (ab.) [IN]
tomentosa Ahrens, 1810: 42 **E:** AU BE BH BU BY CR CZ EN FI FR GE GR HU IT KR KZ LA LT LU NL PL RO RU (CT NT ST) SK SL SV SZ UK **A:** AB IN IS KZ RU (SS WS) TM TR UZ
boelensis Wiel, 1956: 18 (ab.) [IN]
tominagai Hayashi, 2000: 6 **A:** JA
transversicollis Fairmaire, 1887b: 135 **A:** CH (SCH YUN)
tuberfrons Goecke, 1934: 221 [# 245]
ussuriensis L. N. Medvedev, 1973b: 876 **A:** NC RU (FE)
versicolorea Brahm, 1790: 135 (*Leptura*) **E:** AU BE BU BY CR CZ DE EN FI FR GB GE HU IR IT KR LA LT LU MD NL NR PL PT RO RU (CT NT ST) SB SK SL SP SV SZ UK **A:** GG KZ RU (ES FE SS WS)
bidens Olivier, 1791b: 291
cincta Germar, 1810: 15
lusatica Hänel, 1911: 125 (ab.) [IN]
suffriani Westhoff, 1882: 253 (var.)

vulgaris Zschach, 1788: 27 **E:** AU BE BH BU BY CZ DE EN FI FR GB GE GR HU IR IT LA LS LT LU NL NR PL PT RO RU (CT NT ST) SB SK SL SP SV SZ UK **A:** AB AF AR CH (BEI HEB HEI JIL XIN) GG IN JA KI KZ MG RU (ES FE SS WS) TD TR UZ **[# 210, # 255]**

concolor Westhoff, 1882: 260 (var.)

issykensis Jacobson, 1901b: 105 (ssp.) **[# 219]**

typhae Ahrens, 1810: 37

viridula J. R. Sahlberg, 1870b: 438

genus *Donaciasta* Fairmaire, 1901b: 233 [RN] type species: *Donacilla perrieri* Fairmaire, 1901

Donacilla Fairmaire, 1901a: 127 [HN] type species: *Donacilla perrieri* Fairmaire, 1901

Donaciocrioceris Pic, 1936a: 10 type species: *Donaciocrioceris dentata* Pic, 1936

Prodonacia S.-H. Chen, 1966: 144, 146 type species: *Prodonacia shishona* S.-H. Chen, 1966

assama Goecke, 1936: 224 (*Donacia*) **A:** CH (YUN) **ORR**

shishona S.-H. Chen, 1966: 144, 146 (*Prodonacia*)

vietnamensis Kimoto & Gressitt, 1979: 202 (*Donacia*)

genus *Sominella* Jacobson, 1908: 622 type species: *Donacia macrocnemia* Fischer von Waldheim, 1823

longicornis Jacoby, 1890a: 84 (*Donacia*) **A:** CH (CHQ FUJ GUA HUB HUN SCH ZHE) **ORR** **[# 256]**

macrocnemia Fischer von Waldheim, 1823: pl. XLVII [= 1824: 235] (*Donacia*) **A:** CH (HEI HUB JIL LIA SHN) MG NC RU (ES FE) SC

excisipennis Jacobson, 1894d: 243 (*Plateumaris*)

tribe Haemoniini S.-H. Chen, 1941

genus *Macroplea* Samouelle, 1819: 211 type species: *Donacia zosterae* Fabricius, 1801 (= *Rhagium muticum* Fabricius, 1792)

Apelma Billberg, 1820: 53 type species: *Donacia zosterae* Fabricius, 1801 (= *Rhagium muticum* Fabricius, 1792)

Calamobia Gistel, 1856: 378 type species: *Donacia equiseti* Fabricius, 1798 (= *Donacia appendiculata* Panzer, 1794)

Haemonia Dejean, 1821: 114 type species: *Donacia equiseti* Fabricius, 1798 (= *Donacia appendiculata* Panzer, 1794) **[# 247]**

appendiculata Panzer, 1794b: 17 (*Donacia*) **E:** AU BE BY CZ DE EN FI FR GB GE HU IR KR LA LT NL NR PL RO RU (CT NT ST) SK SP SV SZ UK **A:** CH (XIN) KZ RU (SS WS)

chevrolatii Lacordaire, 1845: 215 (*Haemonia*)

equiseti Fabricius, 1798: 128 (*Donacia*)

flavicollis Bellevoye, 1870: 93 (*Haemonia*, var.)
lineata Chevrolat, 1844: 258 (*Crioceris*)
mosellae Bellevoye, 1870: 92 (*Haemonia*)
mucronata Hoppe, 1795: 47 (*Donacia*)
rugipennis J. R. Sahlberg, 1870a: 65 (*Haemonia*)

huaxiensis Lou & H.-B. Liang, 2011: 3 **A:** CH (GUI)

japana Jacoby, 1885b: 190 (*Haemonia*) **A:** CH (GUI HEB HEI HUB HUN JIA LIA) JA RU (FE)

skomorokhovi L. N. Medvedev, 2006: 385 [# 220]

mutica Fabricius, 1792b: 306 (*Rhagium*) **E:** BE DE EN FI FR GB GE HU IT KR LA NL NR PL RO RU (NT ST) SV **N:** AG **A:** AB CH (HEI TIA) JA KI KZ MG RU (ES FE) TM TR UZ

balatonica Székessy, 1941b: 21 (*Haemonia*, ssp.)
baltica Seidlitz, 1875: 504 (*Haemonia*, var.)
curtisii Lacordaire, 1845: 214 (*Haemonia*)
gyllenhalii Lacordaire, 1845: 218 (*Haemonia*)
incisa J. R. Sahlberg, 1870a: 67 (*Haemonia*)
kraussei Reitter, 1910: 116
lapponica Hellén, 1937: 4 (*Haemonia*, var.)
ruppiae Germar, 1831: 9 (*Donacia*)
sahlbergii Lacordaire, 1845: 220 (*Haemonia*)
schiodtei Guérin-Méneville, 1844: 259 (*Crioceris*)
zosterae Fabricius, 1801b: 127 (*Donacia*)

pubipennis Reuter, 1875: 326 (*Haemonia*) **E:** FI SV **A:** CH (GAN HEI NIN XIN) KI KZ RU (ES) [# 257]

incostata Pic, 1907a: 100 (*Haemonia*)
piligera Weise, 1889g: 576 (*Haemonia*)

ranina Lou & Yu, 2011: 7 **A:** CH (SCH) [# 230]

genus *Neohaemonia* Székessy, 1941a: 148 type species: *Haemonia nigricornis* Kirby, 1837

voronovae L. N. Medvedev, 1977b: 489 **A:** MG RU (ES)

tribe Plateumarini Böving, 1922

genus *Plateumaris* C. G. Thomson, 1859: 154 type species: *Donacia nigra* Fabricius, 1792 (= *Prionus bracatus* Scopoli, 1772) [# 212]

Euplateumaris Iablokoff-Khnzorian, 1966: 121 type species: *Leptura sericea* Linnaeus, 1758
Juliusina Reitter, 1920: 41 type species: *Prionus bracatus* Scopoli, 1772

akiensis Tominaga & Katsura, 1984: 25 **A:** JA

amurensis Weise, 1898: 179 **A:** RU (ES FE)

bracata Scopoli, 1772: 100 (*Prionus*) E: AU BE BH BU BY CR CZ DE EN FI FR GB GE HU IR IT LA LT LU MD ME NL NR PL RO RU (CT NT ST) SB SK SL SV SZ UK A: AB CH (XIN) GG IN KI KZ RU (SS WS) **[# 246]**

abdominalis Olivier, 1800: [75]: 9 (*Donacia*)

fairmairi LeGrand, 1861: 265 (var.)

nigra Fabricius, 1792b: 117 (*Donacia*)

palustris Herbst, 1784: 100 (*Donacia*)

violacea Pallas, 1773: 724 (*Leptura*)

consimilis Schrank, 1781: 155 (*Leptura*) E: AL AU BE BH BU BY CR CZ DE FR GE HU IT LS LT LU NL NM PL RO RU (CT) SB SK SL SP SV SZ UK **A:** GG RU (WS) TR **[# 267]**

aerea Bechyné, 1942: 234, 236 (forma) [IN] **[# 235]**

assimilis Schrank, 1781: 156 (*Leptura*)

coerulea Bechyné, 1942: 234, 236 (forma) [IN] **[# 235]**

discolor Hoppe, 1795: 45 (*Donacia*) [HN]

flavipes Bechyné, 1942: 234, 236 (forma) [IN] **[# 235]**

nigripes Bechyné, 1942: 234, 236 (forma) [IN] **[# 235]**

rufipes Olivier, 1791b: 292 (*Donacia*)

tenuicornis Balthasar, 1934: 128 **[# 226]**

variabilis Kunze, 1818a: 39 (*Donacia*)

violacea Bechyné, 1942: 234, 236 (forma) [IN] **[# 235]**

viridis Bechyné, 1942: 234, 236 (forma) [IN] **[# 235]**

constricticollis babai Chûjô, 1959: 2 (ssp.) **A:** JA

constricticollis constricticollis Jacoby, 1885b: 192 (*Donacia*) **A:** JA RU (FE: Kuriles)

kurilensis L. N. Medvedev, 1978d: 83 (ssp.)

constricticollis toyamensis Tominaga & Katsura, 1984: 27 (ssp.) **A:** JA

chugokuensis Tominaga & Katsura, 1984: 28 (ssp.)

roscida Weise, 1912: 77 **A:** CH (HEI NMO) RU (ES FE)

annularis Reitter, 1920: 41

rustica Kunze, 1818a: 31 (*Donacia*) E: AU BE BH BU BY CR CZ DE EN FI FR GB GE HU IR IT KR LA LS LT LU ME NL NR PL RO RU (CT NT ST) SB SK SL SP SV SZ UK **N:** AG **A:** IN RU (WS) TR

affinis Kunze, 1818a: 37 (*Donacia*) **[# 221]**

forojulensis Gortani, 1906: 20 (ab.) [IN]

fusca Zschach, 1788: 27 (*Donacia*) [NO] **[# 210]**

pallipes Kunze, 1818a: 35 (*Donacia*)

picipes Weise, 1898: 180 (var.)

planicollis Kunze, 1818a: 34 (*Donacia*)

sulcifrons Weise, 1900b: 267 **[# 225]**

sericea Linnaeus, 1758: 397 (*Leptura*) E: AU BE BH BU BY CR CZ DE EN FI FR GB GE GR HU IR IT LA LS LT LU KR MD ME NL NM NR PL RO RU (CT NT ST) SB SK SL SP SV SZ TR UK N: AG A: AB AR CH (BEI HEB ZHE) GG IN JA KZ MG NC RU (ES FE SS WS) SC TR **[# 241, # 242]**

aenea Olivier, 1791b: 292 (*Donacia*)

armata Paykull, 1799a: 194 (*Donacia*)

asiatica Faldermann, 1837: 322 (*Donacia*)

atropurpurea Westhoff, 1882: 256 (*Donacia*, var.)

caucasica F. A. Zaytsev, 1930: 111 **[# 222]**

coelicolor Bechyné, 1945b: 89 (forma) [IN] **[# 233]**

comari Suffrian, 1846: 84 (*Donacia*)

cupraria Bechyné, 1945b: 89 (forma) [IN] **[# 233]**

discolor Panzer, 1795a: 216 (*Donacia*) **[# 231]**

festucae Fabricius, 1792b: 116 (*Donacia*)

geniculata C. G. Thomson, 1866: 123 (*Donacia*)

intermedia Apfelbeck, 1912: 239

isocoelicolor Bechyné, 1945b: 89 (forma) [IN] **[# 233]**

isocupraria Bechyné, 1945b: 89 (forma) [IN] **[# 233]**

isolacordairei Bechyné, 1945b: 89 (forma) [IN] **[# 233]**

isopurpuricena Bechyné, 1945b: 89 (forma) [IN] **[# 233]**

isoviolacea Bechyné, 1945b: 89 (forma) [IN] **[# 233]**

kratochvili Bechyné, 1945b: 89 (ssp.)

lacordairii Perris, 1864: 300 (*Donacia*) **[# 223]**

laevicollis C. G. Thomson, 1866: 125 (*Donacia*)

levigata Csiki, 1953: 120 (ab.) [IN]

luctuosa Westhoff, 1882: 256 (*Donacia*, var.)

micans Panzer, 1795b: 9 (*Donacia*)

nigrita Schilsky, 1908: 603 (var.)

nipponensis Nakane, 1963a: 18

nymphaeae Fabricius, 1792b: 116 (*Donacia*)

palustris Schilling, 1838: 99 (*Donacia*) [HN]

proteus Kunze, 1818a: 23 (*Donacia*)

pseudoviolacea Bechyné, 1945b: 89 (forma) [IN] **[# 233]**

purpuricena Bechyné, 1945b: 89 (forma) [IN] **[# 233]**

sibirica Solsky, 1871: 245 (*Donacia*) **[# 224]**

socia S.-H. Chen, 1941a: 9

tatrica Balthasar, 1934: 130 (ab.) [IN]

tenebricosa Westhoff, 1882: 256 (*Donacia*, var.)

violacea Hoppe, 1795: 44 (*Donacia*) [HN]

viridis Csiki, 1953: 120 (ab.) [IN]

shirahatai Kimoto, 1971b: 1 **A**: CH (JIL) JA MG RU (FE) SC
macropenis Nakane, 1999: 45
obsoleta Jacobson, 1894d: 243 [QS] [# 232]
weisei Duvivier, 1885c: cxvi (*Donacia*) **E**: FI RU (CT NT) SV **A**: CH (HEI NMO) JA MG RU (ES FE SS WS) SC [# 243]
hirashimai Kimoto, 1963: 13
mongolica Semenov, 1895b: 267 (*Donacia*)
morimotoi Kimoto, 1963: 13
orientalis Shavrov, 1948: 49 (ssp.)
sachalinensis L. N. Medvedev, 1973b: 876

Donaciinae, nomina dubia
aenea Gistel, 1857b: 524, 563 (*Donacocia*) E: GE
aenea Zschach, 1788: 27 (*Donacia*) E: "Europa" [# 210]
caerulea Zschach, 1788: 27 (*Donacia*) E: "Europa" [# 210]
sericea Zschach, 1788: 27 (*Donacia*) E: "Europa" [# 210]
simplex Thunberg, 1792: 118 (*Donacia*) [HN] E: SV [# 236]

subfamily **Criocerinae** Latreille, 1804
tribe Criocerini Latreille, 1804

genus *Crioceris* Geoffroy, 1762: 237 [conserved name, ICZN (1994), Opinion 1754] type species: *Chrysomela asparagi* Linnaeus, 1758 [ICZN (1970), Opinion 908]
Anthurga Gistel, 1848c: [9] type species: *Chrysomela asparagi* Linnaeus, 1758
Asparagocrioceris Bogachev & Antsupova, 1980: 57 [UN] [# 269]
afghana L. N. Medvedev, 1978c: 859 **A**: AF IN
asparagi Linnaeus, 1758: 376 (*Chrysomela*) **E**: AL AU BE BH BU BY CR CZ DE FI FR GB GE GR HU IT KR LA LS LT LU MD ME NL NM PL PT RO RU (CT ST) SK SL SP SV SZ UK **N**: AG EG **A**: AB AR CY GG IN IS LE RU (WS) SI SY TD TM TR **AFRi NARi NTRi** [# 276]
anticeconjuncta Pic, 1900d: 65 (var.)
apiceconjuncta Pic, 1900d: 65 (var.)
brancsiki Laczó, 1912: 5 (ab.) [IN]
campestris Linnaeus, 1767: 602 (*Chrysomela*)
cruciata Schuster, 1905: 213 (forma)
crucifera Tunkl, 1929: 164 (*Lema*, var.)
dingleri Báguena, 1958: 22 (ab.) [IN]
fauconneti Pic, 1912f: 50 (var.)
heydeni Pic, 1906d: 123 (var.)
impupillata Pic, 1900d: 65 (var.)
incrucifera Pic, 1900d: 65 (var.)
integreconjuncta Bodor, 1965: 301 (ab.) [IN]

linnei Pic, 1895d: 88 (var.)
maculipes Krynicki, 1834: 171 (*Lema*)
malleti Pic, 1934j: 31 (ab.) [IN]
mediobipunctata Bodor, 1965: 301 (ab.) [IN]
moguntiaca Schuster, 1906: 152 (forma)
multiplicata Pic, 1912f: 50 (var.)
normalis Schuster, 1905: 212 (forma)
obliterata Pic, 1906h: 96 (var.) [HN]
octomaculata Tunkl, 1929: 164 (var.) [HN]
paganettii Pic, 1934j: 30 (ab.) [IN]
pici Heyden, 1906: 125 (var.)
pupillata Ahrens, 1812b: 30 (*Lema*)
quadripunctata Schuster, 1905: 213 (forma)
quadristigma Wiel, 1956: 19 (ab.) [IN]
schusteri Heyden, 1906: 124 (var.)
sexmaculata Tunkl, 1929: 164 (var.) [HN]
sexsignata Heyden, 1906: 126 (var.)
tau Tunkl, 1929: 164 (*Lema*, var.)
trifasciata Schuster, 1905: 213 (forma)
turcestanica L. N. Medvedev, 1955: 1177 (ssp.)
verdcourti Báguena, 1958: 22 (ab.) [IN]

atrolateralis Pic, 1932g: 32 **A**: CH (YUN)

bicruciata C. R. Sahlberg, 1823a: 54 [= 1823c: 54, = 1829: 25] (*Lema*) E: BU GR NM RU (ST) TR **A**: AB AR CY GG IN IQ IS JO KZ TD TR UZ **[# 272, # 276]**
decorata Morawitz, 1860: 299
distincta Lacordaire, 1845: 587
merkli Pic, 1909e: 138 (var.)

duodecimpunctata Linnaeus, 1758: 376 (*Chrysomela*) E: AL AU BE BH BU BY CR CZ DE FR GE GR HU IT KR LA LT LU MD NL NM PL RO RU (CT NT ST) SB SK SL SP SV SZ TR UK **A**: AB AR GG IN KI KZ MG RU (ES SS WS) SY TD TM TR **NARi** [# 276]
abeillei Pic, 1894g: cclxxxiv
augustoduni Pic, 1916a: 6 (var.) [IN] [# 271]
baguenai Fuente, 1927: 33 (ab.) [IN]
bisbiconneza Pic, 1909a: 99 (var.)
bisquadripunctata Kaszab, 1962: 28 (ab.) [IN]
delagrangei Pic, 1912f: 51 (var.)
dodecastigma Suffrian, 1841a: 40 (*Lema*)
fauconneti Pic, 1919e: 19 (var.) [HN]
gallica Pic, 1919e: 19 (var.)
gisellae Kaszab, 1962: 28 (ab.) [IN]

guranyii Kaszab, 1962: 28 (ab.) [IN]
heydeni Wiel, 1956: 19 (ab.) [IN]
macgillavryi Wiel, 1956: 19 (ab.) [IN]
mediofaciata Laczó, 1928a: 8 (ab.) [IN]
kaszabi Bodor, 1965: 301 (ab.) [IN]
laboissierei Vitale, 1930: 137 (ab.) [IN]
peyroni Pic, 1919e: 19 (var.)
revyi Kaszab, 1962: 27 (ab.) [IN]
ruffi Kaszab, 1962: 28 (ab.) [IN]
semijuncta Pic, 1916b: 10 (var.) [IN] [# 271]
steinmanni Bodor, 1965: 302 (ab.) [IN]
subannulipes Pic, 1919e: 19 (var.)
hypopsila demokidovi Semenov, 1909: 298 **A:** TD TM UZ
submaculata Semenov, 1909: 299 (ab.) [IN]
hypopsila hypopsila Jacobson, 1907c: 25 **A:** TM
iliensis Weise, 1900b: 268 **A:** CH (XIN) KZ MG RU (ST) TM
macilenta Weise, 1881: 79 **E:** CR FR GBi GR IT MA PT SP **A:** CY **N:** AG LB MO TU [# 273, # 276]
algirica Chobaut, 1907: 179 (ab.) [IN]
bimaculata Fuente, 1908b: 389 (var.)
bisbipunctata Pic, 1934j: 31 (ab.) [IN]
carthagonovae Báguena, 1958: 21 (ab.) [IN]
corsica Pic, 1897d: 3 (var.)
henoni Pic, 1912f: 51 (var.)
hipponensis Pic, 1897d: 3 (var.)
hispanica Weise, 1892a: 373 (var.)
iberica Heyden, 1906: 126 (var.)
jacqueti Pic, 1897d: 3 (var.)
jordai Fuente, 1908a: 121 (var.)
kabyliana Chobaut, 1907: 179 (ab.) [IN]
lineata Pic, 1897d: 3 (var.)
mediodisjuncta Pic, 1906c: 34 (var.)
quadrimaculata Chobaut, 1907: 179 (ab.) [IN]
quadrimaculata Fuente, 1908b: 389 (var.)
sefrensis Pic, 1912f: 50 (var.)
simoni Weise, 1883d: 251 (var.)
subreducta Pic, 1908i: 91 (var.)
tournieri Pic, 1897d: 3 (var.)
vittigera Costa, 1842: 125 (*Lema*)
weisei Heyden, 1906: 126 (var.) [HN]
nigroornata Clark, 1866a: 40 [RN] **A:** ID (HP UT) NP **ORR** [# 276]

multimaculata Jacoby, 1898a: 185
nigropicta Clark, 1866b: 70 [HN]
nigropicta Wollaston, 1864: 394 **N:** CI
orientalis orientalis Jacoby, 1885b: 195 **A:** CH (BEI HEI JIL SHA SHN TAI) JA MG NC RU (FE) SC
biparticeps Pic, 1932g: 32 (var.)
choi Chûjô, 1941c: 452 (var.) [IN] [# 271]
goetria Jacobson, 1907c: 26 (var.) [IN] [# 270]
hypolachna Jacobson, 1907c: 26 (ssp.)
hutahosi Yuasa, 1938: 42 (ab.) [IN]
paracruusa Jacobson, 1907c: 26 (ab.) [IN]
sakahati Yuasa, 1938: 41 (ab.) [IN]
ussurica Pic, 1929c: 120 (var.)
yotubosi Yuasa, 1938: 41 (ab.) [IN]
orientalis tshingisana Lopatin, 1966a: 237 **A:** MG
oschanini Dohrn, 1883: 40 **A:** CH (XIN) KI KZ MG TD TM UZ
egregia Weise, 1900b: 269
koltzei Weise, 1892d: 131
paracenthesis Linnaeus, 1767: 1066 (*Chrysomela*) **E:** AL AN BH BU CR FR GR IT MA ME NM PT SL SP TR **A:** SI TR **N:** AG CI MO TU [# 276]
cruciata Bonelli, 1812: 170
dahlii Lacordaire, 1845: 589
interrupta Pic, 1907b: 112 (var.)
pallidicolor Pic, 1907b: 112 (var.)
suturalis Gravenhorst, 1807: 139 (*Lema*)
tetrastigma Báguena, 1958: 22 (ab.) [IN]
quatuordecimpunctata Scopoli, 1763: 36 (*Attelabus*) **E:** AL AU BH BU BY CR CZ FR GE HU KR MD PL RO RU (CT ST) SB SK SL UK **A:** AB CH (BEI FUJ GUI GUX HEB HEI HEN JIA JIL NIN NMO SHN TAI YUN ZHE) GG IN JA KZ MG NC RU (ES FE SS WS) SC TR [# 276]
blatnyi Roubal, 1949: 46 (ab.) [IN]
confluenta Heinze, 1943b: 101 (forma) [IN] [# 270]
csikii Kaszab, 1962: 29 (ab.) [IN]
deficiens Roubal, 1949: 46 (ab.) [IN]
donceeli Pic, 1916a: 6 (var.) [IN] [# 271]
externepunctata Csiki, 1953: 120 (ab.) [IN]
imitatrix Lopatin, 1967d: 169 (ab.) [IN]
innotaticeps Pic, 1916c: 17
internepunctata Csiki, 1953: 120 (ab.) [IN]
issikii Chûjô, 1951a: 79

lichtneckerti Kaszab, 1962: 29 (ab.) [IN]
mandschurica L. N. Medvedev, 1958: 107 (ssp.)
montandoni Pic, 1916a: 6 (var.) [IN] [# 271]
moravica Roubal, 1946: 44 (ab.) [IN]
nigripes J. Müller, 1948: 76 (var.) [IN] [# 271]
parumpunctata Roubal, 1949: 46 (ab.) [IN]
russica Pic, 1916a: 6 (var.) [IN] [# 271]
semireducta Pic, 1916a: 6 (var.) [IN] [# 271]
sibirica Weise, 1887h: 165 (var.)
staudingeri Heinze, 1943b: 101 (forma) [IN] [# 270]
tschangana Pic, 1932g: 32 (var.)
turcica Pic, 1916a: 6 (var.) [IN] [# 271]
quinquepunctata Scopoli, 1763: 36 (*Attelabus*) **E:** AU BH BU BY CR CZ FR GE GR HU IT MD NM PL RO RU (CT ST) SB SK SL UK **A:** AB RU (WS) [# 276]
kaufmanni Kaszab, 1962: 31 (ab.) [IN]
nigerrima Jacobson, 1901b: 106 (var.) [IN] [# 271]
quinquepunctata Schrank, 1781: 97 (*Chrysomela*) [HN]
thoracica Weise, 1893a: 1118 (var.) [HN]
signatifrons Pic, 1920d: 20 **A:** CH (QIN YUN) **ORR**
obliteraticollis Pic, 1920d: 20 (var.)
sokolowi Jacobson, 1894e: 270 **A:** TR "Caucasus"

genus *Lilioceris* Reitter, 1913c: 79 type species: *Chrysomela merdigera* Linnaeus, 1758 [ICZN (1979), Opinion 1130]
subgenus *Bradyceris* Chûjô, 1951a: 82 type species: *Crioceris lewisi* Jacoby, 1885
lewisi Jacoby, 1885b: 194 (*Crioceris*) **A:** JA
subgenus *Chujoita* Monrós, 1960: 148 type species: *Crioceris camelus* Duvivier, 1884
dromedarius Baly, 1861b: 279 (*Crioceris*) **A:** CH (HAI) **ORR**
foveolata Pic, 1921b: 3 (*Crioceris*)
rouyeri Pic, 1916d: 18 (*Crioceris*)
gibba Baly, 1861b: 280 (*Crioceris*) **A:** CH (FUJ GUA GUI GUX HUB JIA JIX SCH TAI YUN ZHE) NP "Korea" **ORR**
formosensis Chûjô, 1933a: 48 (*Crioceris*)
lemoulti Pic, 1916c: 17 (*Crioceris*, var.)
semifoveolata Jacoby, 1908: 76 (*Crioceris*)
subgenus *Lilioceris* Reitter, 1913c: 79 type species: *Chrysomela merdigera* Linnaeus, 1758 [ICZN (1979), Opinion 1130]
adonis Baly, 1860a: 150 (*Crioceris*) **A:** CH (GUX HAI SCH YUN) ID (SD "Northern India") **ORR** [# 276]
superba Pic, 1916c: 16 (*Crioceris*)
vitalisi Pic, 1930d: 2 (*Crioceris*)

antevallata Jakob, 1954c: 45 **A:** IN
apicalis P.-Y. Yu, 1992a: 527, 531 **A:** CH (HUN)
atrilateralis Kimoto & Takizawa, 1973: 171 **A:** NP
balyi Chûjô, 1962c: 1 **A:** JA
bipartiticollis Pic, 1934e: 22 **A:** CH (SCH)
brancuccii L. N. Medvedev, 1992f: 379 **A:** ID (SD) NP
cantonensis Heinze, 1943b: 104 **A:** CH (ANH FUJ GUA GUI GUX HEN JIX SCH SHA SHG ZHE)
cheni Gressitt & Kimoto, 1961: 46 **A:** CH (FUJ GUA GUX HAI HUN SCH TAI XIZ YUN) NP **ORR NARi**
chodjaii Berti & Rapilly, 1976: 56 **A:** IN
consentanea Lacordaire, 1845: 561 (*Crioceris*) **A:** CH (FUJ) **ORR**
cupreosuturalis Gressitt, 1942b: 298 (*Crioceris*) **A:** CH (GAN SCH)
cyanicollis Pic, 1916c: 16 (*Crioceris*) **A:** CH (CHQ FUJ GUA GUI GUX HUB HUN SCH TAI YUN ZHE) NP **ORR**
cyaneitarsis Pic, 1921c: 137 (*Crioceris*)
guerryi Pic, 1921b: 2 (*Crioceris*)
shirakii Chûjô, 1944: 570
subviridicollis Pic, 1932c: 10 (*Crioceris*, var.)
tonkinea Pic, 1916c: 16 (*Crioceris*)
dentifemoralis Long, 1988: 231, 233 **A:** CH (HAI)
egena Weise, 1922b: 41 (*Crioceris*) **A:** CH (ANH CHQ FUJ GUA GUI GUX HAI HEB HKG HUN SCH TAI YUN ZHE) ID (SD UT) NP **ORR**
coomani Pic, 1928e: 88 (*Crioceris*)
faldermanni Guérin-Méneville, 1844: 264 (*Crioceris*) [RN] **E:** BU GR RO RU (ST) TR **A:** AB AR CY GG IQ IN IS JO LE SY TR
cornuta Faldermann, 1837: 323 (*Lema*) [HN]
notaticollis Pic, 1934d: 21 (var.)
semirufa Marseul, 1868: 204 (*Crioceris*)
flavipennis Baly, 1860a: 151 (*Crioceris*) **A:** CH (XIZ) ID ("Northern India") NP
formosana Heinze, 1943b: 104 (ssp.) **A:** CH (TAI) JA (Ryukyus)
loochooana Nakane, 1956: 170 (ssp.)
fouana Pic, 1932c: 11 (*Crioceris*) **A:** CH (GAN HUB SCH SHA YUN)
glabra Jakob, 1961: 61 **A:** CH (YUN)
grahami Gressitt & Kimoto, 1961: 49 **A:** CH (SCH)
gressitti L. N. Medvedev, 1958: 111 **A:** CH (GUI SCH YUN)
hainanensis Gressitt, 1942b: 299 (*Crioceris*) **A:** CH (HAI)
impressa Fabricius, 1787: 88 (*Crioceris*) **A:** CH (CHQ FUJ GUA GUI GUX HAI HKG HUB HUN SCH TAI YUN ZHE) ID (AP SD UT) NP **ORR** **[# 276]**
castanea Lacordaire, 1845: 564 (*Crioceris*)
crassicornis Olivier, 1808: 731 (*Crioceris*)

inflaticornis Gressitt & Kimoto, 1961: 50
laticornis Gressitt, 1942b: 300 (*Crioceris*)
maai Gressitt & Kimoto, 1961: 53
omophloides Lacordaire, 1845: 564 (*Crioceris*)
ruficornis Pic, 1921c: 136 (*Crioceris*)
subcostata Pic, 1921b: 2 (*Crioceris*)
iridescens Pic, 1916d: 17 (*Crioceris*) **A:** CH (GUI YUN) **ORR**
jakobi R. E. White, 1981: 31 [RN] **A:** CH (FUJ JIA)
minima Jakob, 1961: 61 [HN]
jianfenglingensis Long, 1988: 231, 233 **A:** CH (HAI)
klapperichi Pic, 1955c: 23 (*Crioceris*) **A:** CH (FUJ)
laosensis Pic, 1916c: 16 (*Crioceris*) **A:** CH (FUJ GUX HAI HUB XIZ YUN ZHE) ID (SD) NP **ORR**
bechynei L. N. Medvedev, 1958: 109
lateritia Baly, 1864a: 613 (*Crioceris*) **A:** CH (ANH CHQ FUJ GUA GUI GUX HKG HUB HUN JIA JIX SCH SHG ZHE)
potens Weise, 1922b: 39 (*Crioceris*)
smilacis Gressitt, 1942b: 304 (*Crioceris*)
lianzhouensis Long, 2000b: 262, 264 **A:** CH (GUA)
lilii laeviuscula Weise, 1881: 69 (*Crioceris*) (var.) E: PT SP
lilii lilii Scopoli, 1763: 36 (*Attelabus*) E: AL AN AU BE BH BU BY CR CZ DE EN FI FR GB GE GR HU IR IT KO KZ LA LS LT LU MA MD ME NL NM NR PL PT RO RU (CT NT ST) SB SK SL SP SV SZ UK **N:** CI **A:** AB AR CH (HEI JIL NMO XIN) GG IN IS KI KZ MG RU (ES FE SS WS) SY TR **NARi** [# 276]
amurensis Pic, 1916e: 14 (*Crioceris*)
biguttata Münzner, 1934: 31 (ab.) [IN]
liliorum C. G. Thomson, 1866: 138 (*Crioceris*)
melanocephala Say, 1826: 294 (*Lema*)
regeli Ballion, 1878: 371 (*Crioceris*)
rubra liliorum DeGeer, 1775: 339 (*Chrysomela*) [UN]
schepmanni Everts, 1920a: 231 (ab.) [IN]
lilii stercoraria Linnaeus, 1767: 600 (*Chrysomela*) **N:** AG MO TU [# 276]
locuples Clark, 1866b: 64 (*Crioceris*) **A:** NP **ORR**
luteohumeralis Pic, 1923b: 10 (*Crioceris*) **A:** CH (GUX HAI) **ORR**
major Pic, 1916c: 15 (*Crioceris*) **A:** CH (GUA GUI GUX HAI HKG YUN) **ORR**
jeanvoinei Pic, 1927k: 21 (*Crioceris*, var.)
medogensis Xu & Liang, 2024: 59 **A:** CH (XIZ)
merdigera latior Pic, 1916f: 18 (*Crioceris*) **A:** CH ("Tchao Pin Lo") NP
latiorithorax Pic, 1954b: 12 [= 1954c: 7] (*Crioceris*, var.)
merdigera merdigera Linnaeus, 1758: 375 (*Chrysomela*) E: AL AU BE BH BU BY CR CZ DE EN FI FR GE GR HU IT KO KR LA LS LT LU MD ME NL NR PL RO RU (CT

NT ST) SB SK SL SP SV SZ UK **A:** CH (BEI CHQ FUJ GUX HEB HEI HUB JIL LIA NMO SCH SHN SHX TAI ZHE) GG JA KZ MG NC NP RU (ES FE SS WS) SC TR [# 276]

abdominalis Comolli, 1837: 45 (*Lema*)

brunnea Fabricius, 1792b: 6 (*Crioceris*)

collaris Weise, 1881: 71 (*Crioceris*, var.) [# 1104]

convallaria Harrer, 1791: 142 (*Crioceris*)

forojuliensis J. Müller, 1948: 77 (*Crioceris*) (ssp.)

mediciana Lacordaire, 1845: 579 (*Crioceris*) [RN]

nigritarsis Doi, 1928: 372 (*Crioceris*)

rufipes Herbst, 1783: 67 (*Crioceris*)

similis Brahm, 1790: 79 (*Chrysomela*) [RN]

suffrianii F. J. Schmidt, 1842: 27 (*Lema*)

minima Pic, 1933d: 12 (*Crioceris*) **A:** CH (FUJ GAN SCH ZHE)

miwai Chûjô, 1951a: 83 **A:** CH (TAI)

multimaculata Jacoby, 1893b: 266 **A:** CH (HKG)

nepalensis Takizawa, 1989a: 327 **A:** CH (XIZ) ID (SD) NP

neptis Weise, 1922b: 40 (*Crioceris*) **A:** CH (FUJ GUA GUX HAI HKG HUN JIX TAI ZHE) NP

melli Heinze, 1943b: 104

nigropectoralis Pic, 1928e: 88 (*Crioceris*) **A:** CH (FUJ GUI HUB HUN JIX SCH TAI YUN ZHE) NP **ORR**

ochracea Gressitt, 1942b: 302 (*Crioceris*)

yunnanica L. N. Medvedev, 1958: 110 (ssp.)

nobilis L. N. Medvedev, 1958: 112 **A:** CH (GAN GUI HUB SCH)

parvicollis Baly, 1873: 76 (*Crioceris*) **A:** JA

pulchella Baly, 1860a: 152 (*Crioceris*) **A:** CH (XIZ) ID ("Northern India")

quadripustulata Fabricius, 1787: 88 (*Crioceris*) **A:** CH (HAI YUN) ID (SD) **ORR**

speciosa Fabricius, 1792b: 384 (*Attelabus*)

rufimembris Pic, 1921b: 2 (*Crioceris*) **A:** CH (GUX YUN)

rufometallica Pic, 1923b: 10 (*Crioceris*) **A:** CH (HAI YUN) **ORR**

rugata Baly, 1865d: 154 (*Crioceris*) **A:** JA

scapularis Baly, 1859b: 195 (*Crioceris*) **A:** CH (FUJ GUA GUI GUX HAI HUB HUN JIA JIX SHA SHN ZHE) JA (Tsushima) NC RU (FE) SC **ORR** [# 275]

bilineata Pic, 1916f: 18 (*Crioceris*, var.)

coreana Pic, 1916f: 18 (*Crioceris*, var.)

quelpaerti Pic, 1954a: 3 (*Crioceris*, var.)

reducta Pic, 1916f: 18 (*Crioceris*, var.)

schneideri abnobae Pankow & Schmitt, 2017: 194 **E:** AU GE FR IT SL

schneideri schneideri Weise, 1900b: 268 (*Crioceris*) **E:** IT

semicostata Jacoby, 1908: 77 (*Crioceris*) **A:** CH (XIZ) ID (SD) **ORR**

semihirsuta Jacoby, 1908: 75 (*Crioceris*) **A**: ID (UT) **ORR** **[# 276]**
semimetallica Gressitt & Kimoto, 1961: 56 **A**: CH (SCH)
semipunctata Fabricius, 1801a: 472 (*Lema*) **A**: CH (FUJ GUA GUX HAI SCH YUN) ID (AP SD UT) NP **ORR** **[# 276]**
dehaanii Guérin-Méneville, 1844: 261 (*Crioceris*)
sieversi Heyden, 1887d: 271 (*Crioceris*) **A**: CH (BEJ FUJ GUI HEB HEI HEN HUB HUN JIL JIX SHA SHN ZHE) JA (Tsushima) RU (FE) SC
rubricollis R. E. White, 1981: 41 [RN]
ruficollis Baly, 1865d: 155 (*Crioceris*) [HN]
sinica Heyden, 1887d: 270 (*Crioceris*) **A**: CH (BEI CHQ FUJ GAN GUI GUX HAI HEB HEI HEN HUB HUN JIA JIL JIX SCH SHA SHN YUN ZHE) NC SC
chinensis Jacoby, 1888: 340 (*Crioceris*)
sparsipunctata L. N. Medvedev, 1958: 111 (ssp.)
subpolita Motschulsky, 1861a: 22 (*Crioceris*) **A**: JA
theana Reitter, 1898a: 22 (*Crioceris*) **A**: CH (BEI FUJ HEB HEI LIA ZHE) RU (FE) SC
thibetana Pic, 1916f: 18 (*Crioceris*) **A**: CH (YUN XIZ)
tibialis A. Villa & G. B. Villa, 1839: 63 (*Lema*) **E**: AU FR GE IT SZ
alpina L. Redtenbacher, 1848: 517 (*Lema*)
nigripes Pic, 1891: 51 (*Crioceris*, var.)
triplagiata Jacoby, 1888: 340 (*Crioceris*) **A**: CH (FUJ JIX ZHE)
trinotata Pic, 1955c: 22 (*Crioceris*)
unicolor Hope, 1831: 28 (*Crioceris*) **A**: CH (YUN) NP **ORR**
badia Lacordaire, 1845: 560 (*Crioceris*)
seminigra Jacoby, 1889b: 153 (*Crioceris*)
vietnamica L. N. Medvedev, 1985b: 66 **A**: CH (YUN XIZ) **ORR**
wagneri Jacobson, 1893a: 122 (*Crioceris*) **A**: MG NC RU (ES FE)
laticollis Reitter, 1893: 302 (*Crioceris*)
xinglongensis Long, 1988: 232, 233 **A**: CH (HAI)
yuae Long, 2000a: 416, 418 **A**: CH (HAI)
yunnana Weise, 1913d: 220 (*Crioceris*) [RN] **A**: CH (HUB SCH YUN) ID (SD) **ORR**
crassicornis Fairmaire, 1887b: 136 (*Crioceris*) [HN]
zayuensis Xu & Liang, 2024: 61 **A**: CH (XIZ)
zhentangensis Xu & Liang, 2024: 56 **A**: CH (XIZ)

genus *Manipuria* Jacoby, 1908: 84 type species: *Manipuria dohertyi* Jacoby, 1908
dohertyi Jacoby, 1908: 84 **A**: CH (YUN) **ORR**
yuae Y. Xu, Bi & H.-B. Liang, 2021: 31 **A**: CH (XIZ YUN)

genus *Mecoprosopus* Chûjô, 1951a: 75 type species: *Mecoprosopus fulvus* Chûjô, 1951
minor Pic, 1916d: 18 (*Crioceris*) **A**: CH (FUJ GUA GUI GUX HUB TAI YUN ZHE)
fulvus Chûjô, 1951a: 76
bicoloricollis L. N. Medvedev, 1958: 108 (ssp.)

tribe Lemini Gyllenhal, 1813

genus *Lema* Fabricius, 1798: 90 type species: *Lema cyanea* Fabricius, 1798 [ICZN (1970), Opinion 908]

subgenus *Lema* Fabricius, 1798: 90 type species: *Lema cyanea* Fabricius, 1798 [ICZN (1970), Opinion 908] **[# 268]**

Atactolema Heinze, 1927b: 163 type species: *Lema australis* Lacordaire, 1845

Sulcatolema Pic, 1928b: 96 type species: *Leptura coromandeliana* Fabricius, 1798

Trichonotolema Heinze, 1927b: 165 type species: *Lema coelestina* Klug, 1835

assamensis Jacoby, 1891a: 31 **A:** ID (SD) **ORR**

atkinsoni Jacoby, 1908: 28 **A:** ID (SD)

balyana Jacoby, 1908: 20 **A:** NP **ORR**

becquarti Gressitt, 1942b: 308 **A:** CH (ZHE)

bifoveipennis Pic, 1934i: 1 **A:** CH (SCH YUN)

bimaculaticeps Pic, 1924b: 12 **A:** CH (XIZ)

birmanica Jacoby, 1892b: 872 **A:** NP **ORR**

bohemani Clark, 1866a: 24 [RN] **A:** CH (FUJ GUA GUI HAI HKG JIX SCH ZHE) **[# 276]**

forticornis Pic, 1924b: 14

russula Boheman, 1859: 153 (*Crioceris*) [HN]

castanea Jacoby, 1908: 37 **A:** CH (GUA HAI) ID (AP) NP **ORR**

cephalotes Lacordaire, 1845: 328 **A:** YE **AFR**

cirsicola Chûjô, 1959: 2 **A:** JA

concinnipennis Baly, 1865d: 157 **A:** CH (BEI CHQ FUJ GAN GUI GUX HAI HEN HEB HUB HUN JIA JIX SHA SCH TAI YUN ZHE) JA NC RU (FE) SC

atriventris Pic, 1924b: 13

haemorrhoidalis Weise, 1889g: 576

kiotoensis Pic, 1924b: 12

constrictofasciata Jacoby, 1908: 46 **A:** ID (SD UT) NP **ORR**

epipleuralis Jacoby, 1908: 21

nepalensis Bryant, 1952: 602

coomani Pic, 1924b: 13 **A:** CH (GUX TAI YUN) JA (Ryukyus) **ORR**

paagai Chûjô, 1933a: 25

rangoonensis R. E. White, 1981: 40 [RN]

rugifrons Jacoby, 1889b: 151 [HN]

coreensis Monrós, 1960: 182 [RN] **A:** SC

coreana Chûjô, 1933a: 30 [HN]

peninsularis Gressitt & Kimoto, 1961: 70 [RN]

coronata Baly, 1873: 72 **A:** CH (FUJ GUA GUX HUB HUN JIA SCH TAI ZHE) JA SC

formosana Kuwayama, 1932: 75

sagensis Heinze, 1943b: 107

spinipes Pic, 1938b: 16

takanabensis Nakane, 1950: 53 (ab.) [IN]

cyanea Fabricius, 1798: 92 **A:** CH (FUJ GUA HKG HUN TAI XIZ YUN) ID (HP SD) NP **ORR**

angustepunctata Gressitt, 1942b: 307

bicolorithorax Pic, 1924b: 13

lacosa Pic, 1924b: 12

persicariae Chûjô, 1933a: 38

wallacei Jacoby, 1884b: 10

cyanella Linnaeus, 1758: 376 (*Chrysomela*) **E:** AL AN AU BE BH BU BY CR CZ DE EN FI FR HU GB GE GR HU IR IT LA LT LU MD NL NM NR PL RO RU (CT NT ST) SB SK SL SP SV SZ UK **A:** AB CH (GAN HEB HEI JIL LIA NMO SCH TAI XIN) GG IN JA KI KZ MG NC RU (ES FE SS WS) SC TR **AUSi NARi** **[# 276]**

inaequalicollis Pic, 1924b: 13

kuwayamae Monrós, 1960: 182 (ssp.) [UN] **[Appendix]**

nigricans Jacobs, 1926: 166 (ab.) [IN]

obscurior Pic, 1897d: 3 (var.)

puncticollis Curtis, 1830: 323 (*Crioceris*)

rugicollis Suffrian, 1841b: 97

ventralis Kuwayama, 1932: 69 (var.) [IN] **[# 271]**

delicatula Baly, 1873: 75 **A:** CH (FUJ GUA HUB HUN JIA ZHE) JA SC

demangei Pic, 1924b: 14 **A:** CH (HAI XIZ) **ORR**

corpulentina Gressitt, 1942b: 311

dilecta Baly, 1873: 74 **A:** JA SC

hiranoi Suzuki, 2005: 93 (forma) [IN]

michioi Suzuki, 2005: 88

diversa Baly, 1873: 71 **A:** CH (ANH BEI CHQ FUJ GUA GUI GUX HEB HEI HEN HUB HUN JIA JIL JIX LIA SCH SHA SHN ZHE) JA NC RU (FE) SC

akaheri Yuasa, 1939: 163 (forma) [IN] **[# 270]**

chinensis Jacoby, 1890a: 85

coreana Pic, 1924b: 13

cyaneohumeralis Heinze, 1943b: 106 (forma) [IN] **[# 271]**

doii Kuwayama, 1932: 78 (var.) [IN] **[# 271]**

lewisii Baly, 1873: 72

morii Yuasa, 1939: 163 (forma) [IN] **[# 270]**

quadriplagiata Heinze, 1943b: 106 (forma) [IN] **[# 271]**

suturalis Heinze, 1943b: 107 (forma) [IN] **[# 271]**

diversipes Pic, 1921b: 3 **A:** CH (TAI YUN)

diversitarsis Pic, 1927m: 2 **A:** CH (FUJ) **ORR**

djoui Gressitt, 1942b: 312 **A:** CH (GUA)

esakii Chûjô, 1951a: 110 **A:** CH (TAI)

feae feae Jacoby, 1892b: 873 **A:** CH (SCH YUN) **ORR**
fulvofrontalis Jacoby, 1900b: 97 **A:** ID (SD) **ORR**
fuscitarsis Jacoby, 1897a: 240 **A:** "Arabia" **AFR**
impotens Lacordaire, 1845: 335 **A:** ID (SD) **ORR**
inconspicua Gressitt, 1942b: 315 **A:** CH (HAI)
infranigra Pic, 1924b: 11 **A:** CH (CHQ FUJ GUA GUI GUX HAI HUB HUN JIX SCH YUN ZHE)
pulchrula Gressitt, 1942b: 317
kimotoi L. N. Medvedev & Sprecher-Uebersax, 1999: 358 **A:** NP **ORR**
lacertosa Lacordaire, 1845: 339 **A:** CH (FUJ HAI SCH TAI) ID (SD) JA (Ryukyus) NP **ORR**
jeanvoinei Pic, 1932c: 11
phungi Pic, 1924b: 13
lauta Gressitt & Kimoto, 1961: 69 **A:** CH (FUJ)
maheensis Jacoby, 1908: 39 **A:** ID (HP) **ORR**
malabarica Jacoby, 1908: 18 **A:** NP **ORR**
marginalis Gressitt, 1942b: 316 **A:** CH (HAI)
minutissima Pic, 1923g: 19 **A:** JA
nigricollis Jacoby, 1891a: 31 **A:** CH (XIZ) ID (AP SD) **ORR**
nigrofrontalis Clark, 1866b: 40 **A:** CH (HAI YUN) ID (SD "North India") NP [# 276]
nigrosignata Pic, 1927b: 393 **A:** CH (HAI) **ORR**
fanheanga Gressitt, 1942b: 314
obscuritarsis Jacoby, 1892b: 872 **A:** ID (AP) **ORR** [# 276]
omanica L. N. Medvedev, 2012e: 415 **A:** OM YE (Socotra)
palpalis Lacordaire, 1845: 315 **A:** CH (XIZ) ID (SD) NP **ORR**
pallidipes Pic, 1916f: 18 (*Crioceris*)
paradoxa Jacoby, 1904a: 380 **A:** NP **ORR**
pauperata Lacordaire, 1845: 336 **A:** "Arabia" **AFR**
perplexa Baly, 1890b: 487 **A:** CH (GUA GUX HAI HKG YUN) **ORR**
chujoi Gressitt & Kimoto, 1961: 64
piceocastanea Gressitt & Kimoto, 1961: 71 **A:** CH (SCH)
praeusta Fabricius, 1792b: 8 (*Crioceris*) **A:** CH (FUJ GUA GUI GUX HAI HKG SCH TAI XIZ YUN) ID (HP SD UT) NP PA **ORR**
allardi Baly, 1890b: 486 (var.)
atropectoralis Pic, 1943a: 3
binghami Jacoby, 1908: 61
binhana Pic, 1924b: 11
bisulcata Baly, 1890b: 486
bretinghami Baly, 1861b: 278
brevesuturalis Pic, 1943b: 13

coimbatorensis Pic, 1943a: 3 (var.)
commelinae Gressitt, 1942b: 310
coromandeliana Fabricius, 1798: 154 (*Leptura*)
cyaneosuturalis Pic, 1923b: 10
cyanipennis Fabricius, 1801a: 472
dichroa Blanchard, 1846: pl. 18 [= 1853: 310] (*Crioceris*)
fleutiauxi Baly, 1890b: 487
fulvula Lacordaire, 1845: 338
gangetica Weise, 1903a: 20 (var.)
lateralis Jacoby, 1893b: 265
malayana Lacordaire, 1845: 378
mediofasciata Heinze, 1942: 56
melanocera Lacordaire, 1845: 374
melanura Fabricius, 1801a: 474
nodieri Pic, 1943a: 3
obscuriventris Pic, 1935f: 16 (var.)
philippina Weise, 1922c: 424 (var.)
rufipes Weise, 1922c: 424 (ab.) [IN]
semifulva Jacoby, 1889b: 152
semiviridiceps Pic, 1943a: 3
spinotibialis Pic, 1949b: 12
stevensi Baly, 1878e: 309
takara Chûjô, 1933a: 23
tibiella Weise, 1903a: 20
trifasciata Jacoby, 1908: 44
trisulcata Gressitt, 1942b: 321
viridenotata Pic, 1943a: 3

psyche Baly, 1865d: 157 **A:** ID (SD UT "Northern India") NP **ORR**
pygmaea Kraatz, 1879b: 129 **A:** CH (HEI) RU (FE) SC
rufofrontalis Jacoby, 1908: 62 **A:** ID (SD) NP **ORR**
rufolineata Pic, 1924b: 12 **A:** CH (YUN)
rufotestaea Clark, 1866b: 29 **A:** CH (CHQ FUJ GUA GUI GUX HAI HKG HUB HUN JIX SCH TAI YUN ZHE) ID (SD) NP **ORR**

annulitarsis Pic, 1924b: 13
nitobei Chûjô, 1933a: 28
palonensis Jacoby, 1892b: 873
saigonensis Pic, 1923b: 10

scutellaris Kraatz, 1879b: 130 (*Crioceris*) **A:** CH (HEI) JA NC RU (ES FE) SC [# 276]
sikanga Gressitt, 1942b: 319 **A:** CH (SCH)
spoliata Jacoby, 1908: 50 **A:** CH (XIZ) **ORR**

szechuana Gressitt & Kimoto, 1961: 72 **A:** CH (SCH)
terminata Lacordaire, 1845: 341 **A:** NP **ORR**
tibialis Laporte, 1840: 509 (*Crioceris*) **A:** SA **AFR** [# 274]
vanharteni L. N. Medvedev, 2012e: 415 **A:** OM YE
subgenus *Microlema* Pic, 1932g: 33 [RN] type species: *Lema akinini* Heyden, 1887 [# 268]
Pseudolema Pic, 1928b: 96 type species: *Lema akinini* Heyden, 1887
decempunctata Gebler, 1830: 196 (*Lema*) [conserved spelling, ICZN (2019), Opinion 2439] **A:** CH (BEI FUJ GAN GUA HEB HEN HUB HUN JIA JIX NIN NMO QIN SCH SHA SHN SHX XIN XIZ ZHE) JA KZ MG NC RU (ES FE SS) SC [# 276]
brunneipennis Kuwayama, 1932: 80 (var.) [IN] [# 270]
japonica Weise, 1889g: 562 (var.)
kiautschauana Heinze, 1943b: 106 (forma) [IN] [# 27-]
nadari Pic, 1945c: 13 (var.)
nankinea Pic, 1945c: 13 (var.)
ropunctata Gebler, 1830: 196 (*Lema*) [incorrect original spelling, ICZN (2019), Opinion 2439]
semiobliterata Pic, 1907b: 112 (var.)
quadrimaculata Gebler, 1844: 105 **A:** CH (XIN) KI KZ TD UZ
akinini Heyden, 1887c: 323
bimaculata Weise, 1900b: 268 (var.)
distigma Weise, 1913d: 220 (ab.) [IN]
margelanica Weise, 1894a: 67
subgenus *Petauristes* Latreille, 1829: 136 type species: *Lema crassipes* Olivier, 1808 [# 268]
Bradylema Weise, 1901c: 146 type species: *Bradylema rusticella* Weise, 1901
Bradylemoides Heinze, 1930: 28 type species: *Crioceris grossa* J. Thomson, 1858
Enoplolema Heinze, 1943a: 23 type species: *Lema adhaerens* Weise, 1913
adamsii Baly, 1865d: 156 **A:** CH (FUJ GUA HUN JIX SCH YUN ZHE) JA SC [# 276]
quadripunctata Swartz, 1808: 281 [HN]
swartzii Clark, 1866a: 25 [RN]
trapezinotata Gressitt, 1942b: 320
crioceroides Jacoby, 1893b: 271 [RN] **A:** CH (GUA GUI YUN) **ORR**
robusta Jacoby, 1892b: 869 [HN]
semiopaca Pic, 1931e: 139
duplicata Gressitt, 1942b: 314 **A:** CH (HAI)
femorata Guérin-Méneville, 1844: 259 (*Crioceris*) **A:** CH (XIZ YUN) ID (AP SD) **ORR**
anticejuncta Pic, 1929d: 15 (var.)
assamica Heinze, 1941: 208 (*Bradylema*, ssp.)
darjeelingensis Heinze, 1941: 208 (*Bradylema*, ssp.)

fortunei Baly, 1860a: 148 **A:** CH (ANH BEI CHQ FUJ GUA GUI GUX HEB HUB HUN JIA SCH SHA SHN TAI XIN YUN ZHE) JA SC

postrema Bates, 1866: 353

gahani Jacoby, 1899b: 67 **A:** ID (SD) **ORR**

honorata Baly, 1873: 73 **A:** CH (BEI CHQ FUJ GUI GUX HEB HUB HUN JIX LIA SCH SHN TAI YUN ZHE) JA NP SC **ORR**

infracyanea Pic, 1924b: 11

indicola Takizawa & Basu, 1987: 273 **A:** ID (SD)

jansoni Baly, 1861b: 277 **A:** CH (YUN ZHE) ID (HP SD) NP **ORR** [# 276]

histrio Clark, 1866b: 29

koshunensis Chûjô, 1934f: 282 **A:** CH (TAI)

kurosuzi Chûjô, 1951a: 99 (var.)

lacordairii Baly, 1865a: 23 **A:** BT CH (GUA) NP **ORR** [# 276]

mouhoti Baly, 1878e: 311 **A:** CH (FUJ HAI) **ORR**

rufonubes Gressitt, 1942b: 318

nepalica L. N. Medvedev, 2003a: 317 **A:** NP

pectoralis unicolor Clark, 1866b: 23 **A:** CH (GUA GUI GUX HAI HKG HUB TAI YUN) JA (Ryukyus) NP **ORR**

major Pic, 1924b: 14

nigrotibialis L. N. Medvedev, 1958: 110 (*Lilioceris*)

vandae Gressitt, 1942b: 321

praeclarior Clark, 1866a: 27 [RN] **A:** CH (YUN) ID (SD UT) **ORR** [# 276]

praeclara Clark, 1866b: 38 [HN]

quadripunctata Olivier, 1808: 731 (*Crioceris*) [conserved name, ICZN (2011a), Opinion 2271] **A:** CH (YUN) ID (SD) NP **ORR** [# 276]

ceylonica Heinze, 1941: 208 (*Bradylema*, ssp.)

maindroni Pic, 1929d: 15 (var.)

malaccana Pic, 1929d: 15 (var.)

philippinica Heinze, 1941: 208 (*Bradylema*, ssp.)

quadrisignata Boisduval, 1835: 534

subinnotata Pic, 1929d: 15

rufobrunnea Jacoby, 1908: 28 **A:** NP **ORR**

singularis Jacoby, 1908: 32 **A:** CH (HAI) ID (SD) **ORR**

tenebrosipennis Heinze, 1943b: 105 (*Bradylema*) **A:** KI

viridipennis Pic, 1924b: 11 **A:** CH (YUN) [# 276]

subgenus *Quasilema* Monrós, 1951: 468 type species: *Lema apicalis* Lacordaire, 1845 [# 268]

bilineata bilineata Germar, 1823: 527 (*Crioceris*) Ei: IT **Ni:** CI **AFRi AUSi NTR**

atrithorax Pic, 1947b: 4 (var.)

flavipennis Heinze, 1928: 61 (var.)

nigricollis Heinze, 1928: 61 (var.)
suturalis Pic, 1941c: 8 (var.)
solani Fabricius, 1798: 93 **Ai**: CH (TAI) **NAR**
trivittata trivittata Say, 1824a: 429 **Ai**: CH (TAI) JA (Ryukyu Islands) **NAR NTR**
immaculicollis Chevrolat, 1835b: 112
trivirgata LeConte, 1859b: 22 [HN]
lecontei Clark, 1866a: 31 [RN]

genus *Ortholema* Heinze, 1943b: 108 type species: *Ortholema abnormis* Heinze, 1943 (= *Lema mandarensis* Jacoby, 1897)
gracilenta Chûjô, 1951a: 113 (*Lema*) **A**: CH (TAI)
mandarensis Jacoby, 1897d: 421 (*Lema*) **A**: CH (FUJ GUA GUI GUX HAI HUB) **ORR**
abnormis Heinze, 1943b: 108
elongatior Pic, 1929h: 138 (*Lema*)
funerea Jacoby, 1908: 69 (*Lema*)
punctaticeps Pic, 1929h: 138 (*Lema*, var.)

genus *Oulema* Gozis, 1886: 33 type species: *Chrysomela melanopus* Linnaeus, 1758
subgenus *Oulema* Gozis, 1886: 33 type species: *Chrysomela melanopus* Linnaeus, 1758
Conradsia Pic, 1936d: 10 type species: *Conradsia suturalis* Pic, 1936
Hapsidolema Heinze, 1927b: 162 type species: "*Lema lichenis* Voet, 1806" (= *Crioceris obscura* Stephens, 1831)
Incisophthalma Heinze, 1929: 289 type species: *Lema infima* Lacordaire, 1845
Ulema Bedel, 1891: 116 [UE] type species: *Chrysomela melanopus* Linnaeus, 1758
Xoidolema Heinze, 1931: 206 type species: *Xoidolema rhodesiana* Heinze, 1931
algerica Pic, 1897d: 3 (*Lema*) **N**: AG TU
lacordairei Desbrochers des Loges, 1875: 149 (*Lema*) [HN]
antonii Clavareau, 1905: 6 (*Lema*) [RN] **A**: BT **ORR** [# 276]
duvivieri Jacoby, 1900b: 96 (*Lema*) [HN]
atrosuturalis Pic, 1923g: 18 (*Lema*) **A**: CH (FUJ GUA GUI GUX HAI HUB JIA JIX SCH SHN TAI YUN) JA SC **ORR**
melania Chûjô, 1951a: 119 (*Lema*, var.)
dilutipes Fairmaire, 1888b: 149 (*Lema*) **A**: CH (BEI HUB LIA ZHE) JA RU (FE) SC
kajimurai Ohno, 1962a: 99
downesii Baly, 1865d: 156 (*Lema*) **A**: CH (SCH TAI) NP **ORR**
nigrosuturalis Clark, 1866b: 37 (*Lema*)
duftschmidi L. Redtenbacher, 1873: 446 (*Lema*) [RN] **E**: AL AN AU BE BH BU BY CZ DE FI FR GB GE GR HU IT KO KR LA LS LT LU MA ME NL NM NR PL PT RO RU (CT ST) SB SK SL SP SV SZ TR UK **A**: AB AF AR CH (XIN) CY IN IS JO KI KZ LE SY TD TM TR UZ **N**: AG CI MO MR TU [# 276]

cyanipennis Duftschmid, 1825: 243 [HN] (*Lema*)

erichsonii Suffrian, 1841b: 104 (*Lema*) **E:** AU BE BH BU BY CZ EN FI FR GB GE HU IR IT LA LS LT LU MD NM PL RO RU (CT NT ST) SB SK SP SZ UK **A:** AB CH (HEB HEI HUB NMO XIN) GG JA KZ MG RU (ES FE SS WS) SC **[# 276]**

lipperti Gredler, 1866: 405 (*Lema*, var.)

sapporensis S. Matsumura, 1911b: 140 (*Lema*)

weisei Seidlitz, 1891a: 726 [= 1891b: 818] (*Lema*)

hayashii Ohno, 1962e: 45 **A:** JA

hoffmannseggii Lacordaire, 1845: 396 (*Lema*) **E:** AN FR IT PT SP **N:** AG MO TU **[# 276]**

obscura Normand, 1937: 120 (*Lema*, ab.) [IN]

purpuricollis Reiche, 1861b: 92 (*Lema*)

infima Lacordaire, 1845: 337 (*Lema*) **A:** SA **AFR**

magistrettiorum Ruffo, 1964: 74 (*Lema*) **E:** IT

mauroi Bezděk & Baselga, 2015: 291 **E:** IT

melanopus Linnaeus, 1758: 376 (*Chrysomela*) **E:** AL AN AU BE BH BU BY CR CZ DE EN FI FR GB GE GR HU IR IT KR LA LS LT LU MA MD NL NM NR PL PT RO RU (CT NT ST) SB SK SL SP SV SZ TR UK **N:** MO SG **A:** AB AR GG IN IQ IS JO KI KZ LE MG RU (ES FE SS WS) SY TR **NAR**i **[# 276]**

atrata Waltl, 1835: 81 (*Lema*, var.) [HN]

azurea Voet, 1806: 37 (*Crioceris*) [UN]

hordei Geoffroy, 1785: 95 (*Crioceris*)

melanopoda O. F. Müller, 1776: 85 (*Crioceris*)

nigricans Westhoff, 1882: 261 (*Lema*, var.)

waltli Heinze, 1927a: 141 (*Lema*, var.) [RN]

nigrofrontalis Takizawa & Basu, 1987: 274 **A:** ID (HP)

obscura Stephens, 1831: 281 (*Crioceris*) **E:** AL AN AU BE BH BU BY CR CZ DE EN FI FR GB GE GR HU IR IT KO LA LS LT LU MD NL NM NR PL RO RU (CT NT ST) SB SK SL SP SV SZ UK **A:** AB AR GG KZ MG RU (ES FE SS WS) TR **[# 276]**

gallaeciana Heyden, 1870: 164 (*Lema*)

laevicollis Baudi di Selve, 1889: 195 (*Lema*, var.)

lichenis Voet, 1806: 42 (*Chrysomela*) [UN]

nigra Delherm de Larcenne, 1885: 120 (*Lema*)

oryzae Kuwayama, 1931: 155 (*Lema*) **A:** CH (FUJ GUA GUI GUX HEI HEN HUB HUN JIA JIX JIL LIA SCH SHA TAI YUN ZHE) JA NC RU (FE) SC

jakobi Monrós, 1960: 183 (*Lema*) [RN]

rugifrons Jakob, 1954a: 102 (*Lema*) [HN]

suvorovi Jacobson, 1931: 255 (*Lema*)

rufocyanea Suffrian, 1847a: 100 (*Lema*) **E:** AU BE BH BU CR CZ FR GE GR HU IT KO ME NL NM RO SB SK SL SZ UK **A:** TR

septentrionis Weise, 1880d: 158 (*Lema*) **E:** AU CZ DE EN FR GB GE HU IR LA LT MD NL NM NR PL RO RU (CT NT ST) SB SK SV SZ UK **A:** RU (FE SS WS)
tristis kuwayamai Ohno, 1962e: 43 (ssp.) **A:** JA RU (FE)
tristis tristis Herbst, 1786: 165 (*Crioceris*) **E:** AU BH BU BY CR CZ FR GE HU IT LT MD PL PT RO RU (CT ST) SB SK SL SZ UK **A:** AB CH (BEI GAN HEB HEI HEN HUB JIL LIA NIN NMO SHA SHN) GG IN JA KZ MG NC RU (ES FE SS WS) SC UZ [# 276]
flavipes Suffrian, 1841b: 100 (*Lema*)
verae Bezděk & Baselga, 2015: 295 **E:** PT SP
viridula Gressitt, 1942b: 322 **A:** CH (HEB HUB JIL JIX SHN) NC SC
subgenus *Gracilema* Chûjô, 1964: 257 type species: *Oulema iwatai* Chûjô, 1964
globicollis Baly, 1865d: 158 (*Lema*) **A:** CH (GUA) **ORR**
bengalensis Duvivier, 1891a: xxvi (*Lema*)
langana Pic, 1923b: 11 (*Lema*)
subelongata Pic, 1924b: 10 (*Lema*) **A:** CH (SCH YUN)
yunnana Pic, 1923b: 10 (*Lema*) **A:** CH (YUN)

subfamily **Cassidinae** Gyllenhal, 1813
tribe Anisoderini Chapuis, 1875

genus *Anisodera* **Chevrolat, 1836: 363** type species: *Alurnus ferrugineus* Fabricius, 1801
Lissochila Weise, 1911b: 40 type species: *Alurnus ferrugineus* Fabricius, 1801
fraterna Baly, 1888c: 655 **A:** CH (GUX HAI YUN) ID (AP) NP **ORR** [# 325]
brevipennis Pic, 1924g: 29
sinuaticollis Pic, 1926b: 10
guerinii Baly, 1859a: 168 **A:** CH (GUX YUN) ID (AP SD) **ORR**
foveolata Pic, 1924g: 29
propinqua Baly, 1888c: 657 **A:** CH (YUN) **ORR**
barbicornis Weise, 1897d: 119
rugulosa S.-H. Chen & P.-Y. Yu, 1964: 106, 116 **A:** CH (YUN)

genus *Estigmena* **Hope, 1841: 174** type species: *Estigmena chinensis* Hope, 1841
chinensis Hope, 1841: 175 **A:** CH (GUA GUI GUX HAI SCH YUN) ID (AP SD UT) NP PA **ORR** [# 384, # 404]
atricollis Pic, 1924g: 29 (var.) [IN] [# 319]
latior Pic, 1924g: 30 (ssp.)
nigripennis Pic, 1924g: 29 (var.) [IN] [# 319]
ruficollis Pic, 1924g: 30 (var.) [IN] [# 319]

genus *Lasiochila* **Weise, 1916a: 38** [RN] type species: *Anisodera excavata* Baly, 1859
Anisodera Baly, 1859a: 101 [HN] type species: *Anisodera excavata* Baly, 1859

Anisoderopsis Maulik, 1916: 570 type species: *Anisodera excavata* Baly, 1859

anthracina P.-Y. Yu, 1985: 1, 5 **A:** CH (GUX)

bicolor Pic, 1924g: 30 **A:** CH (TAI) [# 455]

balli Uhmann, 1930a: 87

formosana Pic, 1924g: 30

insulana Uhmann, 1929: 149 [RN]

insularis Uhmann, 1927b: 110 [HN]

cylindrica Hope, 1831: 27 (*Trogosita*) **A:** CH (GUX XIZ YUN) ID (AP SD) NP PA **ORR** [# 391]

convexicollis Gestro, 1920: 390 (*Anisoderopsis*)

mausonica Uhmann, 1930a: 91

subparallela Pic, 1924g: 30

dimidiatipennis S.-H. Chen & P.-Y. Yu, 1962: 122, 132 **A:** CH (YUN)

estigmenoides S.-H. Chen & P.-Y. Yu, 1962: 122, 132 **A:** CH (YUN)

excavata Baly, 1859a: 105 (*Anisodera*) **A:** CH (GUX XIZ YUN) ID (SD) NP PA **ORR**

alternata Maulik, 1919c: 383 (*Anisoderopsis*)

fukiena Gressitt, 1950: 65 (ssp.) **A:** CH (FUJ)

gestroi Baly, 1888c: 656 (*Anisodera*) **A:** CH (YUN) **ORR**

ruficolor Pic, 1924g: 31 (var.) [IN] [# 319]

rufipes Pic, 1924g: 31

imitans Uhmann, 1951a: 74 **A:** BT NP **ORR**

latior P.-Y. Yu, 1985: 2, 5 **A:** CH (GUX)

longipennis Gestro, 1906: 474 (*Anisodera*) **A:** CH (YUN) **ORR**

apicalis Pic, 1927n: 132

krishna Maulik, 1919c: 382 (*Anisoderopsis*)

rufipennis Pic, 1924g: 30

monticola S.-H. Chen & P.-Y. Yu, 1964: 106, 116 **A:** CH (GUX)

tribe Basiprionotini Gressitt, 1952 (1914)

genus *Basiprionota* **Chevrolat, 1836: 367** type species: *Cassida octopunctata* Fabricius, 1787

Monax Gistel, 1848b: xi [RN] type species: *Cassida octopunctata Fabricius*, 1787

Prioptera Hope, 1841: 152 type species: *Cassida octopunctata* Fabricius, 1787

Stenoprioptera Spaeth, 1914c: 132 type species: *Stenoprioptera tibetana* Spaeth, 1914

andrewesi Weise, 1897d: 99 (*Prioptera*) **A:** NP **ORR**

angusta Spaeth, 1914a: 17 (*Prioptera*) **A:** CH (TAI)

fulva Chûjô, 1934c: 147 (*Prioptera*, ab.) [IN]

bimaculata Thunberg, 1789: 86 (*Cassida*) **A:** CH (SCH YUN) ID (SD) **ORR**

impustulata Boheman, 1850: 46 (*Prioptera*)

bisignata Boheman, 1862: 22 (*Prioptera*) **A**: CH (ANH BEI FUJ GAN GUA GUI GUX HAI HEB HEN HKG HUB HUN JIA JIX LIA SCH SHA SHG SHN SHX YUN ZHE) **ORR** **[# 328, # 1145]**

pallida Wagener, 1881: 25 (*Prioptera*)

decemmaculata Boheman, 1850: 60 (*Prioptera*) **A**: BT CH (YUN) ID (AP SD UT) NP **ORR** **[# 413]**

cardoni Spaeth, 1925: 430 (*Prioptera*, ab.) [IN]

decemsignata Boheman, 1850: 62 (*Prioptera*)

fuscicornis Weise, 1897d: 101 (*Prioptera*, var.)

macularia Spaeth, 1925: 430 (*Prioptera*, ssp.)

nigricornis Baly, 1863a: 9 (*Prioptera*)

pallidicornis Boheman, 1850: 61 (*Prioptera*)

decemstillata Boheman, 1857: 9 (*Prioptera*) **A**: ID (SD) PA

gressitti L. N. Medvedev, 1957d: 554 **A**: CH (HUB SCH SHA) **[# 329]**

laotica Spaeth, 1933a: 496 (*Prioptera*) **A**: CH (YUN) **ORR**

lata S.-H. Chen & Zia, 1964: 123, 133 **A**: CH (YUN)

multipunctata Gressitt, 1938a: 186 (*Prioptera*) **A**: CH (HAI)

omeia S.-H. Chen & Zia, 1964: 123, 133 **A**: CH (GUI SCH ZHE)

opima Spaeth, 1925: 406 (*Prioptera*) **A**: CH (HAI)

reducta Gressitt, 1938a: 185 (*Prioptera*, ssp.)

prognata Spaeth, 1925: 433 (*Prioptera*) **A**: CH (YUN) ID (SD) **ORR**

pudica Spaeth, 1925: 403 (*Prioptera*) **A**: CH (GAN GUI GUX HUB HUN SCH YUN) **ORR** **[# 409]**

satrapa Boheman, 1862: 17 (*Prioptera*) **A**: CH (FUJ GUA GUI GUX HEI HEN HKG HUB HUN JIA JIX SCH SHA SHG YUN ZHE) **ORR** **[# 282, # 330]**

sexmaculata Boheman, 1850: 49 (*Prioptera*) **A**: BT CH (YUN) ID (AP SD) NP PA **ORR** **[# 331, # 409, # 414]**

maculipennis Boheman, 1850: 50 (*Prioptera*)

punctipennis Wagener, 1877: 59 (*Prioptera*)

rugosa Baly, 1863a: 8 (*Prioptera*)

tibetana Spaeth, 1914c: 132 (*Stenoprioptera*) **A**: CH (GAN HUB SCH XIZ YUN)

westermanni Mannerheim, 1845: 864 (*Prioptera*) **A**: CH (GUI HUN YUN) **ORR** **[# 302, # 409]**

whitei Boheman, 1857: 11 (*Prioptera*) **A**: CH (ANH FUJ GUA GUI GUX HUB HUN JIA JIX SCH SHG ZHE) **[# 332]**

trabeata Fairmaire, 1888c: 46 (*Prioptera*)

genus *Craspedonta* Chevrolat, 1836: 367 type species: *Imatidium leayanum* Latreille, 1807

Calopepla Hope, 1841: 152 type species: *Imatidium leayanum* Latreille, 1807

leayana insulana Gressitt, 1938a: 187 (*Calopepla*, ssp.) **A:** CH (HAI) **ORR**
leayana leayana Latreille, 1807: 50 (*Imatidium*) **A:** CH (YUN) ID (AP SD) NP PA **ORR**
atritarsis Pic, 1927k: 21 (*Calopepla*)
dentata Pic, 1927k: 21 (*Calopepla*, var.)
nigriventris Weise, 1897d: 98 (*Calopepla*)
notaticeps Pic, 1927k: 22 (*Calopepla*, var.)
levis Borowiec, 1993: 79 **A:** CH (FUJ GUX) [# 381]

genus *Epistictina* **Hincks, 1950: 509** [RN] type species: *Epistictia viridimaculata* Boheman, 1850
Epistictia Boheman, 1850: 12 [HN] type species: *Epistictia viridimaculata* Boheman, 1850
fulvonigra Maulik, 1913: 107 (*Epistictia*) **A:** CH (YUN) **ORR**
perplexa Baly, 1863a: 7 (*Epistictia*) **A:** CH (YUN) **ORR**
reicheana Guérin-Méneville, 1844: 286 (*Calopepla*) **A:** NP PA **ORR**
matronula Boheman, 1850: 14 (*Epistictia*)
selecta Boheman, 1850: 13 (*Epistictia*)
trivandrumensis Maulik, 1913: 108 (*Epistictia*, var.)
viridimaculata Boheman, 1850: 15 (*Epistictia*) **A:** CH (GUI GUX HAI YUN) ID (AP SD UT) NP PA **ORR** [# 383]
collaris Weise, 1897d: 99 (*Epistictia*, var.)
marginata Kirsch, 1875: 56 (*Epistictia*)
parryi Baly, 1863a: 8 (*Epistictia*)

tribe Bothryonopini Chapuis, 1875 [# 303]

genus *Bothryonopa* **Guérin-Méneville, 1840c: 332** type species: *Bothryonopa sanguinea* Guérin-Méneville, 1840
Hispopria Baly, 1859a: 94 type species: *Hispopria foveicollis* Baly, 1859
bicolor Uhmann, 1927b: 109 (*Botryonopa*) **A:** CH (TAI)
sheppardi Baly, 1859a: 92 (*Botryonopa*) **A:** BT ID (AP SD) **ORR**

genus *Macrispa* **Baly, 1859a: 90** type species: *Macrispa saundersi* Baly, 1859
saundersi Baly, 1859a: 91 **A:** ID (SD) **ORR** [# 459]
krishnalohita Maulik, 1915: 369

genus *Neodownesia* **Gressitt, 1953: 121** type species: *Neodownesia rubra* Gressitt, 1953
rubra Gressitt, 1953: 122 **A:** CH (FUJ GUX) [# 323]

genus *Neodownesia*, taxon excluded from Palaearctic fauna
sita Maulik, 1919a: 109 (*Wallacea*) **ORR** [# 323]

genus *Wallacea* **Baly, 1859a:** 97 type species: *Wallacea bowringii* Baly, 1859

Wallaceana Maulik, 1928: 159 [RN] type species: *Wallacea bowringii* Baly, 1859

abscisa Uhmann, 1939a: 4 (*Wallaceana*) **A:** CH (YUN) **ORR**

dactyliferae Maulik, 1919a: 107 Ei: FR **A:** CH (YUN) PA **ORR** [# 470]

gorbunovi L. N. Medvedev, 1997e: 264 (*Pistocia*) **A:** NP

yuae C.-F. Lee & Sekerka, 2018: 299 **A:** CH (TAI)

tribe Callispini Chapuis, 1875

genus *Amblispa* **Baly, 1859a:** 10 type species: *Amblispa dohrnii* Baly, 1859

laevigata Guérin-Méneville, 1844: 278 (*Microrhopala*) **A:** ID (HP UT) NP PA **ORR**

purpurascens Maulik, 1919a: 73 (var.)

viridis Maulik, 1919a: 73 (var.)

genus *Callispa* **Baly, 1859a:** 4 type species: *Callispa fortunii* Baly, 1859

subgenus *Callispa* Baly, 1859a: 4 type species: *Callispa fortunii* Baly, 1859

ajaya Basu, 1999: 156 **A:** ID (SD)

almora Maulik, 1923: 599 **A:** ID (UT)

amabilis Gestro, 1911b: 123 **A:** CH (YUN) **ORR**

angusta Gressitt, 1950: 67 **A:** CH (FUJ)

apicalis Pic, 1924d: 99 **A:** CH (FUJ GUX HUN JIA JIX ZHE)

biarcuata S.-H. Chen & P.-Y. Yu, 1961: 460, 477 **A:** CH (GUA)

bijaya Basu, 1999: 154 **A:** ID (SD)

bipartita Kung & P.-Y. Yu, 1961: 461, 477 **A:** CH (YUN)

bowringii Baly, 1859a: 5 **A:** CH (FUJ GUA GUI GUX HAI HKG HUB JIA JIX SCH SHG YUN) **ORR**

atripes Pic, 1924g: 27 (var.) [IN] [# 319]

proxima Baly, 1869: 364

brettinghami Baly, 1869: 365 **A:** CH (GUX YUN) ID (AP SD) **ORR**

brevicornis Baly, 1869: 365 **A:** CH (YUN) **ORR**

brihata Basu, 1999: 155 **A:** ID (SD)

cyanea cyanea S.-H. Chen & P.-Y. Yu, 1961: 461, 476 **A:** CH (GUX YUN) **ORR**

cyanea fukienica S.-H. Chen & P.-Y. Yu, 1964: 108, 117 (ssp.) **A:** CH (FUJ)

cyanipennis Pic, 1924g: 27 **A:** CH (FUJ)

debilis Gressitt & Kimoto, 1963b: 929 **A:** CH (HUB)

dimidiatipennis Baly, 1859a: 7 **A:** CH (GUX HAI YUN) ID ("Northern India") **ORR** [# 415]

hypoenops Maulik, 1919a: 60

jeanvoinei Pic, 1927k: 25

latithorax Pic, 1927k: 25 (var.)

quadricollis Weise, 1897d: 113
recticollis Gressitt, 1938b: 321 (spp.)
vicina Pic, 1927k: 25
donckieri Pic, 1924d: 99 **A:** CH (FUJ GUX SCH)
duodecimmaculata Chapuis, 1876: xvii **A:** PA **ORR**
elliptica Gressitt, 1939b: 198 **A:** CH (JIX)
feae Baly, 1888c: 654 **A:** CH (YUN) ID (SD) NP **ORR**
flaveola Uhmann, 1931: 193 **A:** CH (YUN) **ORR**
fortunii Baly, 1859a: 6 **A:** CH (ANH FUJ GUA HAI JIX SHN YUN ZHE)
emarginata Gressitt, 1938b: 322 (spp.)
ruficeps Pic, 1929h: 139
frontalis L. N. Medvedev, 1992c: 147 **A:** CH (GUX) **ORR**
fulvescens S.-H. Chen & P.-Y. Yu, 1961: 461, 477 **A:** CH (YUN)
himalayana L. N. Medvedev, 1993c: 375 **A:** ID (SD) NP **ORR**
horni Uhmann, 1927a: 58 **A:** ID (SD)
houjayi C.-F. Lee, Świętojańska & Staines, 2012: 833 **A:** CH (TAI)
jaya Basu, 1999: 152 **A:** ID (SD)
kabakovi L. N. Medvedev, 1992c: 145 **A:** CH (YUN) **ORR**
karena Maulik, 1919a: 62 **A:** CH (HAI YUN) **ORR** [# 415]
limbifera P.-Y. Yu & Kung, 1961: 462, 477 **A:** CH (SCH)
submarginata L. N. Medvedev, 1993c: 374
marginicollis L. N. Medvedev, 2012d: 508 **A:** CH (YUN)
montivaga Maulik, 1919a: 63 **A:** NP **ORR**
mungphua Maulik, 1919a: 59 **A:** ID (SD) **ORR**
latipennis Pic, 1924d: 99
nigricollis S.-H. Chen & P.-Y. Yu, 1961: 462, 476 **A:** CH (YUN)
nigrimembris S.-H. Chen & P.-Y. Yu, 1964: 107, 117 (ssp.) **A:** CH (FUJ SCH YUN)
nigripennis S.-H. Chen & P.-Y. Yu, 1964: 109, 117 **A:** CH (GUX)
obliqua S.-H. Chen & P.-Y. Yu, 1964: 108, 117 **A:** CH (FUJ ZHE)
paharia Basu, 1999: 158 **A:** ID (SD)
pallida Gestro, 1888a: 130 **A:** CH (YUN) **ORR**
curtipennis Pic, 1926a: 324
fleutiauxi Baly, 1890b: 489
popovi S.-H. Chen & P.-Y. Yu, 1961: 463, 476 **A:** CH (YUN)
procedens Uhmann, 1939a: 1 **A:** CH (YUN) **ORR**
pseudapicalis P.-Y. Yu, 1985: 4, 5 **A:** CH (JIA YUN)
ruficollis Fairmaire, 1889b: 84 **A:** CH (FUJ GUI HUB HUN SCH ZHE) [# 416]
specialis P.-Y. Yu, 1985: 2, 5 **A:** CH (YUN)
sundara Maulik, 1919a: 61 **A:** CH (GUX HAI YUN) **ORR** [# 415]

tsoui C.-F. Lee, Świętojańska & Staines, 2012: 846 **A:** CH (TAI)

uhmanni S.-H. Chen & P.-Y. Yu, 1961: 463, 476 **A:** CH (YUN)

weigeli L. N. Medvedev, 2009d: 412 **A:** NP

yuae Sekerka, **nom. nov.** **A:** CH (HAI) [# 297]

quadricollis P.-Y. Yu & Li, 2002: 461, 462 [HN]

tribe Cassidini Gyllenhal, 1813

subtribe Aspidomorphina Chapuis, 1875 [# 301, # 313]

genus *Aspidomorpha* **Hope, 1841: 158** type species: *Cassida miliaris* Fabricius, 1775 [# 300]

subgenus *Aspidomorpha* Hope, 1841: 158 type species: *Cassida miliaris* Fabricius, 1775

Aspidimorpha Hope, 1841: 158 [incorrect original spelling]

birmanica Spaeth, 1914d: 544 **A:** NP **ORR**

calligera Boheman, 1854: 297 **A:** BT CH (GUX HAI YUN) ID (AP SD UT) NP PA **ORR** [# 277, # 326]

chandrika Maulik, 1918: 322 **A:** CH (HAI SHA YUN) ID (AP SD) **ORR**

renidens Spaeth, 1926d: 118

difformis Motschulsky, 1861a: 27 (*Deloyala*) **A:** CH (BEI FUJ GAN GUI HEB HEI HUB HUN JIL LIA SCH SHA TAI ZHE) JA NC RU (FE) SC

japonica Spaeth, 1926b: 10 (ab.) [IN]

dorsata Fabricius, 1787: 64 (*Cassida*) **A:** CH (GUA GUX HAI YUN) ID (AP SD) NP **ORR** [# 277, # 409]

fuscopunctata Boheman, 1854: 298

rubrodorsata Boheman, 1854: 310

fraterna Baly, 1863a: 11 **A:** CH (YUN) **ORR**

furcata Thunberg, 1789: 87 (*Cassida*) **A:** BT CH (FUJ GUA GUI GUX HAI HKG JIA JIX MAC SCH SHG TAI YUN ZHE) ID (AP SD UT) NP PA **ORR**

micans Fabricius, 1801a: 398 (*Cassida*)

gruevi Borowiec, 1985b: 451 **A:** SA YE **AFR**

indica Boheman, 1854: 318 **A:** CH (FUJ GUI HAI HUB HUN SCH TAI XIZ YUN) ID (AP SD UT) NP PA **ORR**

egena Boheman, 1854: 317

miliaris Fabricius, 1775: 91 (*Cassida*) **A:** CH (CHQ GUA GUI GUX HAI HKG TAI YUN) ID (AP HP SD UT) NP PA **AUS ORR** [# 327, # 409]

amplissima Boheman, 1854: 260

celebensis Blanchard, 1853: 316

flaveola Weise, 1910b: 143 (ab.) [IN]

inundata Weise, 1910b: 143 (ab.) [IN]

quatuordecimpunctata Olivier, 1808: 943 (*Cassida*)

orientalis Boheman, 1857: 107 **A:** ID (AP HP SD UT) **ORR** [# 1145]

olivacea Wagener, 1881: 49

sanctaecrucis Fabricius, 1794: 446 (*Cassida*) **A:** BT CH (FUJ GUA GUI GUX HAI HUN SCH YUN) ID (AP SD UT) NP PA **ORR** [# 409]

bajula Boheman, 1854: 288

insularis Spaeth, 1912a: 118 (var.)

snizeki Świętojańska, 2001a: 173 (*Aspidimorpha*) **A:** CH (GUI)

stevensi Baly, 1863a: 11 **A:** CH (YUN) **ORR**

vietnamica L. N. Medvedev & Eroshkina, 1982: 113

yunnana S.-H. Chen & Zia, 1964: 132, 138 [# 308]

transparipennis Motschulsky, 1861b: 41 (*Coptocycla*) **A:** CH (BEI GUI HEB HEI HUB JIL LIA SHN) JA NC RU (FE) SC

elliptica Gorham, 1885: 280

vetula Weise, 1900b: 295 (var.)

subgenus *Dianaspis* S.-H. Chen & Zia, 1984a: 80, 82 type species: *Dianaspis bifoveolata* S.-H. Chen & Zia, 1984 (= *Aspidomorpha denticollis* Spaeth, 1932)

denticollis Spaeth, 1932b: 200 **A:** CH (YUN) **ORR**

bifoveolata S.-H. Chen & Zia, 1984a: 80, 82 (*Dianaspis*)

genus *Aspidomorpha*, taxon excluded from Palaearctic fauna *tibetana* Świętojańska & Borowiec, 2006: 542 (*Aspidimorpha*) ORR [# 409]

genus *Conchyloctenia* Spaeth, 1902: 449 type species: *Cassida hybrida* Boheman, 1854

nigrovittata Boheman, 1854: 341 (*Cassida*) **A:** PA **ORR**

genus *Laccoptera* **Boheman, 1854: 152** type species: *Laccoptera excavata* Boheman, 1855 [# 281]

subgenus *Laccopteroidea* Spaeth, 1952: 345 type species: *Cassida tredecimpunctata* Fabricius, 1801

cheni Świętojańska, 2001a: 209 **A:** CH (GUI)

chinensis Fabricius, 1798: 84 (*Cassida*) **A:** BT CH (ANH BEI FUJ GAN GUA GUI GUX HAI HEB HEN HKG HUB HUN JIA JIX SCH SHA SHG TAI XIZ YUN ZHE) ID (AP HP SD UT) JA NP PA **ORR** [# 282, # 1145]

bohemani Weise, 1910d: 42

nepalensis Boheman, 1855: 76

quadrimaculata Thunberg, 1789: 86 (*Cassida*) [HN]

thunbergi Spaeth, 1914a: 15 [RN]

fruhstorferi Spaeth, 1905: 117 **A:** CH (YUN) **ORR**

plagiograpta Maulik, 1919a: 349 (var.)

prominens S.-H. Chen & Zia, 1964: 131, 138 **A:** CH (GUI YUN)

tredecimpunctata Fabricius, 1801a: 398 (*Cassida*) **A:** CH (GUA HAI HKG) **ORR** [# 407]

philippinensis Blanchard, 1846: pl. 18 [= 1853: 321] (*Aspidomorpha*)
yunnanica Spaeth, 1914b: 226 **A:** CH (SCH YUN)
subgenus *Sindiola* Spaeth, 1903a: 111 type species: *Sindiola parallelipennis* Spaeth, 1903
burmensis Spaeth, 1938: 231 (*Sindiola*, ssp.) **A:** CH (YUN) **ORR**
hospita Boheman, 1855: 68 **A:** CH (GUA GUI GUX HAI SCH YUN) **ORR**
puncticollis Gressitt, 1938a: 189 (ssp.)
vigintisexnotata Boheman, 1855: 66 **A:** CH (GUI GUX YUN) **ORR** [# 390]
subgenus *Sindiolina* Świętojańska, 2001a: 294 type species: *Cassida sedecimmaculata* Boheman, 1857
sedecimmaculata Boheman, 1857: 119 (*Cassida*) **A:** CH (GUI YUN) ID (AP SD) NP **ORR**

genus *Nilgiraspis* **Spaeth, 1932a: 2** type species: *Aspidomorpha andrewesi* Spaeth, 1914
andrewesi Spaeth, 1914d: 546 (*Aspidomorpha*) **A:** CH (YUN) **ORR**

subtribe Cassidina Gyllenhal, 1813

genus *Cassida* **Linnaeus, 1758: 362** type species: *Cassida nebulosa* Linnaeus, 1758 [# 288]
Alledoya Hincks, 1950: 508 [RN] type species: *Cassida seraphina* Ménétriés, 1836
Betacassida Steinhausen, 2002: 26 type species: *Cassida nebulosa* Linnaeus, 1758
Cassidula Weise, 1889h: 260 [HN] type species: *Cassida nobilis* Linnaeus, 1758
Cassidulella E. Strand, 1928: 2 [RN] type species: *Cassida nobilis* Linnaeus, 1758
Crepidaspis Spaeth, 1912a: 119 type species: *Crepidaspis varicornis* Spaeth, 1912
Cyclocassida S.-H. Chen & Zia, 1961: 442, 448 type species: *Taiwania variabilis* S.-H. Chen & Zia, 1961
Cyrtonocassis S.-H. Chen & Zia, 1961: 446, 451 type species: *Cyrtonocassis tumidicollis* S.-H. Chen & Zia, 1961
Deloyala L. Redtenbacher, 1858: 952 [HN] type species: *Cassida seraphina* Ménétriés, 1836
Diversivascula Özdikmen & Bal, 2021: 5285 type species: *Cassida sanguinolenta* Müller, 1776 [# 288]
Dolichocassida Günther, 1958: 568 type species: *Cassida veselyi* Günther, 1958
Eremocassis Spaeth, 1926b: 15 type species: *Eremocassis transcaspica* Spaeth, 1926
Evaspistes *Gistel*, 1856: 381 type species: *Cassida viridis* Linnaeus, 1758
Lasiocassis Gressitt, 1952: 485 type species: *Cassida vespertina* Boheman, 1862
Longiampulla Özdikmen & Bal, 2021: 5285 type species: *Cassida vibex* Linnaeus, 1767 [# 288]
Lordicassis Reitter, 1926c: 23 type species: *Cassida undecimnotata* Gebler, 1841
Lordiconia Reitter, 1926c: 23 type species: *Cassida canaliculata* Laicharting, 1781
Mionycha Weise, 1891e: 204 type species: *Cassida azurea* Fabricius, 1801
Mionychella Spaeth, 1952: 346 type species: *Cassida hemisphaerica* Herbst, 1799
Odontionycha Weise, 1891e: 204 type species: *Cassida viridis* Linnaeus, 1758

Onychocassis Spaeth, 1926b: 23 type species: *Cassida brevis* Weise, 1884

Pseudocassida Desbrochers des Loges, 1891: 15 type species: *Cassida murraea* Linnaeus, 1768

Pseudocassis Steinhausen, 2002: 28 [HN] type species: *Cassida flaveola* Thunberg, 1794 [# 278]

Reliquacassida Özdikmen & Bal, 2021: 5285 type species: *Cassida rubiginosa* Müller, 1776 [# 288]

Taiwania Spaeth, 1913a: 47 type species: *Taiwania sauteri* Spaeth, 1913

Tylocentra Reitter, 1926: 24 type species: *Cassida turcmenica* Weise, 1892

Yunocassis S.-H. Chen & Zia, 1961: 442, 448 type species: *Cassida appluda* Spaeth, 1926

achardi Spaeth, 1926e: 93 (*Taiwania*) **A:** CH (YUN) **ORR**

algirica Boheman, 1847a: pl. 44 [= 1847b: 513] **E:** FR GR IT SP **N:** AG TU **A:** AR TR [# 417]

alidagiensis Özdikmen, Coral Şahin & Bal, 2020b: 1172 **A:** TR

alpina Bremi-Wolf, 1855: 197 **E:** AU FR IT SL SP SZ

alticola S.-H. Chen & Zia, 1984b: 383, 384 **A:** CH (SCH)

amurensis Kraatz, 1879b: 141 (*Coptocycla*) **A:** CH (GAN HEI JIL SHA) NC RU (FE) SC

andrewesi Weise, 1897d: 111 **A:** ID (UT) NP PA **ORR**

angusta Marseul, 1876: 35 **N:** AG LB MO TU

olcesei Pic, 1908b: 46 (var.) [IN] [# 319]

angustifrons Weise, 1891e: 205 **E:** PT SP **N:** AG

oranensis Desbrochers des Loges, 1891: 39

appluda Spaeth, 1926e: 86 **A:** CH (YUN) **ORR**

atrata Fabricius, 1787: 62 **E:** AL AU BH BU BY CR CZ FR GE GR HU KO KR MD ME NM PL RO SB SK SL TR UK **A:** AB IN TR [# 418]

atrofemorata Borowiec & Sassi, 2002: 145 **A:** CH (QIN SCH)

atrosignata Sekerka & Borowiec, 2008: 615 **A:** ID (AP)

aurora Weise, 1907: 14 [RN] **E:** AU BH CR CZ GR HU IT PL RO RU (CT ST) SK SL UK [# 333, # 471]

rosea Boheman, 1854: 387 [HN]

australica Boheman, 1855: 257 (*Coptocycla*) **A:** CH (SCH SHA YUN XIZ) ID (KA SD UT) NP PA **ORR**

nigridorsis Spaeth, 1926c: 56 (*Metriona*, ab.) [IN]

azurea Fabricius, 1801a: 389 **E:** AU BE BH BU BY CR CZ FR GE HU IT LU MD ME PL RO SK SP SZ UK **N:** AG **A:** IN KZ RU (ES SS WS) TR [# 334, # 471, # 978]

lucida Suffrian, 1844d: 215

ornata Creutzer, 1799: 118 [HN]

femorata Csiki, 1953: 134 (ab.) [RN]

basicollis S.-H. Chen & Zia, 1964: 125, 134 (*Taiwania*) **A:** CH (YUN)

bella Faldermann, 1837: 325 **E:** RU (ST) **A:** AR GG TR

belliformis Maulik, 1919a: 396 **A:** ID (SD)

bergeali Bordy, 1996: 244 **E:** AU BH BU CZ FR GE HU PL RO SB SK SZ [# 335] *berolinensis* Suffrian, 1844f: 270 E: AU BU CZ GE HU KR PL RO RU (CT ST) SK SV UK **A:** AB AR CH (BEI HEI NMO SHA SHX XIN) GG KI KZ MG NC RU (ES FE SS WS) TR UZ **[# 336, # 419, # 471]**

daurica Boheman, 1854: 457

guentheri Bechyné, 1944a: 77 (ab.) [IN]

pallidiventris Reitter, 1913c: 215 (ab.) [IN]

pectoralis Weise, 1896a: 81 (var.)

brevis Weise, 1884a: 159 E: GR **A:** AF AR IN IQ IS SY TM TR [# 420]

cerata Desbrochers des Loges, 1891: 40

gibbosula Kraatz, 1888: 77

canaliculata Laicharting, 1781: 109 [NP] E: AU BH BU CR CZ FR GE HU IT KR MD PL RO RU (CT ST) SB SK SL SZ UK **A:** IN KZ TR [# 309, # 337, # 979]

austriaca Fabricius, 1792a: 293

illyrica Boheman, 1854: 357 (var.)

sanguinolenta Schrank, 1776: 61 [NO] [# 309]

speciosa Brahm, 1790: 74

splendida Böber, 1795: 136

catenata Boheman, 1855: 262 (*Coptocycla*) **A:** BT ID (AP SD UT) **ORR** [# 279]

callosa Kirsch, 1875: 56 (*Coptocycla*)

ferrarii Spaeth, 1904: 77 (*Metriona*)

fulgida Boheman, 1855: 347 (*Coptocycla*)

cherrapunjiensis Maulik, 1919a: 397 **A:** CH (GUI JIX XIZ YUN) **ORR**

circumdata Herbst, 1799: 268 **A:** CH (FUJ GUA GUI GUX HAI HKG HUB HUN JIA JIX MAC SCH SHG TAI XIZ YUN ZHE) ID (AP SD UT) JA NP PA **ORR** [# 279, # 338, # 1145]

baeri Spaeth, 1903b: 130 (ab.) [IN]

cuticula Gressitt, 1938a: 191

dentata Maulik, 1913: 114 (*Metriona*, var.)

effusa Boheman, 1854: 320 (*Aspidomorpha*)

nilgiriensis Borowiec & Takizawa, 1991: 644

nigroramosa S.-H. Chen & Zia, 1964: 127, 135 (*Taiwania*) [# 289]

pescadorensis Chûjô, 1934c: 162 (*Metriona*, ab.) [IN]

u-fuscum Wiedemann, 1823: 74

trivittata Fabricius, 1801a: 397

concha Solsky, 1871: 264 **A:** CH (GUA JIA LIA TAI ZHE) JA NC RU (FE) SC

juglans Gressitt, 1942d: 4

conchyliata Spaeth, 1914d: 564 (*Metriona*) **A:** CH (GUX YUN) ID (SD) **ORR**

corallina Boheman, 1862: 306 E: FR SP **N:** AG

artemisiae Brisout de Barneville, 1880b: 234

corbetti Weise, 1897d: 108 (*Metriona*) **A:** CH (YUN) **ORR**

crucifera Kraatz, 1879f: 271 (*Coptocycla*) **A:** JA CH (HAI TAI) **ORR** *culminis* S.-H. Chen & Zia, 1964: 129, 136 (*Taiwania*) **A:** CH (YUN) **ORR**

deflorata Suffrian, 1844b: 139 **E:** FR IT PT SP **N:** AG LB MO TU **[# 339, # 980]**

herbea Boheman, 1847a: pl. 44 [= 1847b: 513]

limbaticollis Fairmaire, 1856: 531

dehradunensis Borowiec & Takizawa, 1991: 641 **A:** ID (UT)

deltoides Weise, 1889g: 644 **A:** CH (BEI GAN HUN JIA JIX NIN NMO SHA SHG XIN ZHE) **[# 421]**

klapperichi Spaeth, 1940: 37

denticollis Suffrian, 1844c: 186 **E:** AL AU BE BH BU BY CR CZ DE EN FI FR GB GE GR HU IT KI LA LS LT LU MD NL NM NR PL PT RO RU (CT NT ST) SB SK SL SP SV SZ UK **A:** AB AR CH (XIN) GG KZ MG RU (ES FE SS WS) TR **[# 471]**

fuscicollis Weise, 1893a: 1096 (var.)

mongolensis L. N. Medvedev, 1957d: 556 (ssp.)

desultrix Spaeth, 1914d: 566 (*Metriona*) **A:** CH (YUN) ID (SD) NP **ORR**

florea Spaeth, 1914d: 567 (*Metriona*, var.)

diops S.-H. Chen & Zia, 1964: 125, 134 (*Taiwania*) **A:** CH (YUN)

discalis Gressitt, 1938c: 386 **A:** CH (GUA JIX)

dorsonotata Boheman, 1854: 409 **A:** ID (SD) NP **ORR**

aspectabilis Spaeth, 1914d: 557

elongata Weise, 1893a: 1108 (var.) **E:** BU RU (CT ST) UK **A:** KZ TR

eoa Spaeth, 1928: 46 (*Metriona*) **A:** CH (YUN) **ORR**

exilis Boheman, 1854: 407 **A:** ID (UT) PA **ORR** **[# 340]**

expansa Gressitt, 1952: 492 **A:** CH (HAI) **ORR**

smaragdocruciata L. N. Medvedev & Eroshkina, 1982: 114

expressa Spaeth, 1914b: 230 (*Metriona*) **A:** CH (HUB SCH YUN)

fausti Spaeth, 1926b: 39 (forma) **E:** BU KR MD NM RO RU (ST) SB TR UK **A:** AB TR **[# 280, # 314, # 341]**

circassica L. N. Medvedev, 1962a: 389 [# 290]

feae Spaeth, 1904: 71 **A:** CH (YUN) **ORR**

ferruginea Goeze, 1777: 213 **E:** AU BH BU BY CR CZ EN FR GE GR HU IT LA LT PL RO RU (NT ST) SB SK SL SP SV SZ UK **N:** AG **A:** JA RU (ES FE SS WS) SC TR **[# 342, # 981]**

thoracica Geoffroy, 1785: 141

tincta Weise, 1891e: 205

flaveola Thunberg, 1794: 103 **E:** AL AU BE BH BU BY CR CZ DE EN FI FR GB GE GR HU IR IT KO KR LA LT LU MD ME NL NM NR PL RO RU (CT NT ST) SB SK SL SP SV SZ UK **N:** AG **A:** AR CH (HEI) IN KZ MG RU (ES FE SS WS) TR **NAR** **[# 343, # 398, # 471]**

atrata Gerhardt, 1891: 28 (var.) [HN]

atrodorsalis Spaeth, 1914c: 142 [RN]

dorsalis Desbrochers des Loges, 1891: 27 (var.) [HN]

exsculpta Charpentier, 1825: 231

kusnetzovi Matis, 1974: 1101

littoralis Weise, 1893a: 1111 (var.)

obsoleta Illiger, 1798: 484

pallida Thunberg, 1794: 103

pallida Paykull, 1799a: 50 [HN]

flavoscutata Spaeth, 1914d: 556 **A:** CH (YUN) ID (SD UT) NP PA **ORR**

fumida Spaeth, 1914d: 562 **A:** CH (YUN) **ORR**

fuscorufa Motschulsky, 1866a: 178 **A:** CH (BEI FUJ GAN GUA GUI GUX HAI HEB HEI HEN HUB HUN JIA JIL JIX LIA QIN SCH SHA SHN SHX TAI ZHE) JA MG NC RU (FE) SC

consociata Baly, 1874: 213

laticollis Gressitt, 1952: 510

russata Fairmaire, 1887d: 335

sikanga Gressitt, 1952: 518

gansuica S.-H. Chen & Zia, 1964: 124, 133 **A:** CH (GAN GUI)

ginpinica S.-H. Chen & Zia, 1961: 442, 449 (*Taiwania*) **A:** CH (YUN) **[# 422]**

hablitziae Motschulsky, 1838: 182 **E:** RU (ST) **A:** AR GG IN KZ TR

hainanensis P.-Y. Yu, 2002: 465, 466 (*Taiwania*) **A:** CH (HAI)

hemisphaerica Herbst, 1799: 226 **E:** AU BE BH BU BY CR CZ DE EN FI FR GB GE GR HU IR IT KR LA LT MA NL NR PL PT RO RU (NT) SK SL SP SV SZ UK **N:** AG CI MO MR TU **A:** GG RU (WS) TR **[# 344, # 982]**

anglica Curtis, 1826: 127

bicolor Bechyné, 1944a: 77 (ab.) [IN]

concinna Stephens, 1832: 371

corii Fairmaire, 1876a: 25

heydeni Spaeth, 1914c: 136 (ab.) [IN]

nigriventris Heyden, 1877: 364 (var.) [HN]

peculiosa Weise, 1895d: 224 (var.)

rossii Wollaston, 1857: 130

spergulae Marsham, 1802: 144

hexastigma Suffrian, 1844b: 140 **E:** FR IT PT SP SZ **N:** AG MO TU **[# 983]**

humeralis Kraatz, 1874a: 103 **E:** FR PT SP **N:** AG **[# 345]**

hyalina Weise, 1891e: 205 **E:** SP

icterica Boheman, 1854: 400 **A:** CH (GUX XIZ YUN) ID (HP SD UT) NP PA **ORR** **[# 346]**

petulans Spaeth, 1914d: 550

ictericiformis Borowiec, 2009: 476 **A:** ID (AP) **ORR**

immaculicollis S.-H. Chen & Zia, 1961: 445, 450 (*Taiwania*) **A:** CH (YUN)

imparata Gressitt & Kimoto, 1963a: 971 [RN] **A:** CH (GUX YUN)
imitatrix Gressitt, 1952: 493 [HN]
inciens Spaeth, 1926e: 85 **A:** CH (GUX YUN) **ORR** **[# 409]**
indochinensis Spaeth, 1919: 204 (*Metriona*, ssp.) **A:** CH (GUX YUN) **ORR**
inflata Gressitt, 1952: 521 **A:** CH (HUB JIA) **[# 1145]**
inopinata Sassi & Borowiec, 2006: 547 **E:** AU BH CR GR IT ME SL SZ **[# 347]**
inquinata Brullé, 1833: 266 **E:** AL AU BH BU CR CZ FR GR HU IT KR NM PT RO RU (ST) SB SL SP SZ TR **N:** AG MO TU **A:** AB AR GG IS LE SY TM TR **[# 348, # 423, # 471]**
bohemani Brisout de Barneville, 1860: 346
depressa Suffrian, 1844b: 142
ovoidea Fairmaire, 1904: 271
plana Charpentier, 1825: 232
sordida Brisout de Barneville, 1860: 348
tangeriana Pic, 1908e: 58
insulana Gressitt, 1952: 495 **A:** CH (TAI)
jacobsoni Spaeth, 1914c: 138 **A:** CH (FUJ HUB HUN JIX SCH SHA ZHE) **[# 349]**
japana Baly, 1874: 212 **A:** CH (ANH FUJ GAN GUA GUI GUX HKG HUB HUN JIA JIX SCH SHA SHG TAI YUN ZHE) JA SC **ORR** **[# 350]**
annamita Spaeth, 1919: 197
rugifera Kraatz, 1879f: 274
justa Spaeth, 1914d: 554 **A:** ID (SD UT) **ORR** **[# 304, # 351]**
koreana Borowiec & Cho, 2011: 447 **A:** RU (FE) SC
kunminica S.-H. Chen & Zia, 1964: 126, 134 (*Taiwania*) **A:** CH (YUN) **ORR**
leucanthemi Bordy, 1995: 377 **E:** AU BY CR CZ FR GE HU IT ME PL RO SK SL SP SZ UK **[# 352]**
lineola Creutzer, 1799: 119 **E:** AU BU BY CR CZ GR HU KR LA PL RO RU (CT ST) SB SK UK **A:** AB CH (BEI FUJ GAN GUA GUX HEB HEI HUB JIA JIX NMO SCH SHA SHN SHX TAI TIA YUN ZHE) JA KI KZ MG NC RU (ES FE SS WS) SC TR UZ **[# 353, # 424, # 1073, # 1145]**
bicostata Fischer von Waldheim, 1842: 24
formosana Chûjô, 1934c: 176 (ab.) [IN]
japonica Chûjô, 1951b: 46 (var.)
nigroguttata Gorham, 1885: 281
nigrostrigata Fairmaire, 1888b: 157
russica Herbst, 1799: 232
sibirica Gebler, 1833: 306
signata Herbst, 1799: 234
suturalis Fischer von Waldheim, 1842: 24
linnavuorii Borowiec, 1986b: 573 **A:** IQ LE TR **[# 471]**

liquefacta Spaeth, 1912b: 499 **A:** YE **AFR** **[# 291, # 425]**
chapuisi Spaeth, 1912b: 502 **[# 291]**
concolor Spaeth, 1926a: 11 **[# 291]**
praetimida Spaeth, 1912b: 497 **[# 291]**
semipallens Spaeth, 1912b: 500 **[# 291]**
sincera Spaeth, 1912b: 498 **[# 291]**
major Kraatz, 1874a: 103 (var.) **E:** IT (Sicily) PT SP **N:** AG MO TU
corrosa Desbrochers des Loges, 1891: 39
ellipsodes Marseul, 1876: 35
tingitana Escalera, 1914: 535
mandli Spaeth, 1921: 85 **A:** CH (BEI HEB HEI HUB JIA JIL LIA NIN NMO SHN SHX TIA) MG NC RU (FE) SC
manipuria Maulik, 1923: 606 **A:** CH (YUN) **ORR**
margaritacea Schaller, 1783: 259 **E:** AL AU BE BH BU BY CR CZ DE EN FR GE GR HU IT KR LA LT LU MD ME NL NM PL RO RU (CT NT ST) SB SK SL SP SZ TR UK **N:** AG **A:** AR GG IN RU (WS) SY TR **[# 354, # 426, # 1074]**
albiventris Heyden, 1877: 365 (var.)
melanocephala Weise, 1893a: 1091 (var.) **[# 1105]**
mutabilis Villers, 1789: 93
nigrescens Weise, 1893a: 1091 (var.)
superba Gmelin, 1790: 1644
medvedevi Lopatin, 1965c: 40 **A:** AF KI KZ TD UZ
mishmiensis Sekerka & Borowiec, 2008: 612 **A:** ID (AP)
mongolica Boheman, 1854: 449 **A:** CH (BEI GUA HEB HEI HUB JIA NMO SHA SHN TIA) JA MG RU (FE) SC **[# 355]**
moori Boheman, 1857: 124 **A:** CH (XIN) KI KZ
murraea halophila Spaeth, 1926b: 29 (ssp.) **E:** RU (ST) UK **A:** IN KZ **[# 356]**
murraea murraea Linnaeus, 1767: 575 **E:** AL AU BE BH BU BY CR CZ DE EN FI FR GB GE GR HU IT KR KZ LA LT LU MD ME NL NM PL RO RU (CT NT ST) SB SK SL SP SV SZ TR UK **A:** AB CH (BEI XIN) GG JA KI KZ MG RU (ES FE SS WS) TR **[# 357, # 1075, # 1145]**
caucasica Roubal, 1931c: 184 (ssp.) **[# 292]**
dorsalis Weise, 1893a: 1092 (var.) [HN]
immaculata Desbrochers des Loges, 1891: 16 (var.)
inundata Weise, 1907: 14 (var.) [RN]
maculata Linnaeus, 1767: 575
pseudomurraea Gruev, 1978: 171
subobliterata Pic, 1902h: 80 (var.)
varia Latreille, 1804b: 28 [RN]
variegata Geoffroy, 1785: 141 [HN]

murraea ussuriensis Spaeth, 1921: 84 (ssp.) **A:** RU (FE)

nebulosa Linnaeus, 1758: 363 **E:** AL AU BE BH BU BY CR CZ DE EN FI FR GB GE GR HU IT KO KR LA LT LU MD NL NM NR PL RO RU (CT NT ST) SB SK SL SP SV SZ UK **A:** AB AR CH (BEI GAN GUA GUI HAI HEB HEI HUB JIA JIL JIX LIA NIN NMO QIN SCH SHA SHG SHN SHX TIA XIN YUN) CY GG IN JA KI KZ MG NC RU (ES FE SS WS) SC TD TR UZ **[# 358, # 427, # 471]**

affinis Fabricius, 1775: 88

nigra Herbst, 1799: 258

tigrina DeGeer, 1775: 168

nepalica L. N. Medvedev, 1997e: 265 **A:** ID (UT) NP

nigriventris Boheman, 1854: 410 **A:** CH (GUX XIZ YUN) ID (AP HP SD UT) NP PA **ORR** **[# 428]**

hopei Spaeth, 1914e: 113 [RN]

pallida Hope, 1831: 30 [HN]

nigrocastanea S.-H. Chen & Zia, 1964: 129, 136 (*Taiwania*) **A:** CH (YUN)

nobilis Linnaeus, 1758: 363 **E:** AL AU BE BH BU BY CR CZ DE EN FI FR GB GE GR HU IR IT KO KR LA LT LU MD ME NL NR PL RO RU (CT NT ST) SB SK SL SP SV SZ TR UK **N:** AG **A:** AB AR CH (HEI XIN) GG IN IS JA KI KZ MG NC RU (ES FE SS WS) SC TR UZ **[# 359, # 399, # 984, # 1145]**

dollmanni Donisthorpe, 1913: 294 (var.)

laevis Herbst, 1799: 250

obscura Weise, 1893a: 1113 (var.)

obscurella Weise, 1907: 14 (var.) [RN]

pulchella Panzer, 1796b: 15

rosea Illiger, 1798: 486

splendidula Marsham, 1802: 147

urticae Brahm, 1790: 109

viridula Paykull, 1799a: 49

nucula Spaeth, 1914b: 228 **A:** CH (GUX YUN) **ORR**

obtusata Boheman, 1854: 405 **A:** CH (FUJ GUA GUX HAI HKG MAC TAI YUN ZHE) ID (SD UT) NP PA **ORR**

interstitialis Spaeth, 1901a: 12 (*Cassidula*)

juno Boheman, 1862: 324

picifrons Weise, 1908c: 259

pusillula Boheman, 1862: 327

residua Weise, 1901b: 53

occursans Spaeth, 1914d: 560 **A:** CH (YUN) ID (AP SD) NP PA **ORR** **[# 428]**

olympica Sekerka, 2005: 285 **E:** GR

ovalis Spaeth, 1914c: 143 **E:** FR **N:** AG TU

palaestina Reiche & Saulcy, 1858b: 55 **E:** RU (ST) **A:** AB AF AR CY GG IN IQ IS JO KI KZ LE SA SY TD TM TR UZ **[# 315, # 360]**

nigrofemorata Spaeth, 1926b: 42 (ab.) [IN]

incompta Weise, 1887d: 334

paliji Matis, 1970: 31 **A**: KI

pallidicollis Boheman, 1857: 138 **A**: CH (ANH BEI GUX HEB HEI JIA JIL LIA NMO SCH SHN SHX TIA ZHE) KZ MG NC RU (ES FE SS) SC [# 361, # 1145]

diabolica Kraatz, 1879b: 142

morawitzi Jacobson, 1894d: 245

pannonica Weise, 1892f: 208 (var.) **E**: AU BH BU BY CZ FR GE GR HU KR LT MD NM PL RO RU (CT ST) SB SK SZ UK **A**: AB AR CH (XIN) GG IN IS KI KZ LE RU (SS) SY TM TR UZ [# 362, # 429, # 1106]

fraudulenta Spaeth, 1914c: 141 (forma) [IN] [# 319]

panzeri Weise, 1907: 14 [RN] **E**: AU BE BU BY CR CZ EN FI FR GE HU IT KR LA LT MD NR PL RO RU (CT NT ST) SB SK SL SP SV SZ UK **A**: KZ RU (ES FE SS WS) [# 363, # 430]

thoracica Panzer, 1796a: 24 [HN]

parvula Boheman, 1854: 428 **E**: BU KR RO RU (CT ST) UK **A**: AB CH (GAN HEB JIA JIL NMO QIN SHA SHN TIA XIN) GG KI KZ MG NC RU (ES FE SS WS) TM TR [# 364, # 431, # 471]

comparata Rybakow, 1889: 289

navicula Boheman, 1854: 429

pellegrini Marseul, 1868: 213 **A**: CY IS LE SA **N**: TU

tectiformis Normand, 1950a: 50

perplexa S.-H. Chen & Zia, 1961: 445, 450 (*Taiwania*) **A**: CH (GUX YUN) **ORR**

persica Spaeth, 1926b: 59 **A**: IN IQ

persicana Borowiec, 1999: 273 [RN] **A**: IN

persica L. N. Medvedev, 1956d: 898 [HN]

pfefferi Sekerka, 2006a: 253 **A**: CY IS

physodes Boheman, 1855: 266 (*Coptocycla*) **A**: CH (YUN) **ORR**

viridiguttata S.-H. Chen & Zia, 1964: 129, 137 (*Taiwania*) [# 293]

piperata Hope, 1843a: 64 [= 1843b: 62] **A**: CH (ANH BEI CHQ FUJ GUA GUI GUX HEB HEI HEN HUB JIA JIL JIX LIA QIN SCH SHA SHG SHN TAI TIA YUN ZHE) JA NC RU (FE) SC **ORR** [# 400, # 1145]

biguttulata Kraatz, 1879f: 275

labilis Boheman, 1854: 402

sparsa Gorham, 1885: 284 (*Coptocycla*)

plausibilis Boheman, 1862: 395 (*Coptocycla*) **A**: CH (HAI YUN) **ORR**

objecta Spaeth, 1919: 203 (*Metriona*)

postarcuata S.-H. Chen & Zia, 1964: 128, 136 (*Taiwania*) **A**: (FUJ GUI HUB SCH ZHE)

prasina Illiger, 1798: 481 **E**: AL AU BE BH BU BY CR CZ DE FI FR GE GR HU IR IT KO KR LA LT MD ME NL NM PL RO RU (CT NT ST) SB SK SL SP SV SZ TR **A**: AB AR CH (JIL XIN) GG IN KZ RU (ES FE SS WS) SY TR [# 365]

chloris Suffrian, 1844c: 188 [# 985]

languida Cornelius, 1851: 88

sanguinolenta O. F. Müller, 1776: 65 [NP] **E**: AL AU BE BH BU BY CR CZ DE EN FI FR GB GE GR HU IT KO KR KZ LA LT LU MD ME NL NM NR PL RO RU (CT NT ST) SB SK SL SP SV SZ TR UK **N**: AG **A**: AB AR GG IN KZ RU (FE SS WS) SY TR [# 309, # 368, # 987, # 1145]

cruentata Donovan, 1793: 70

flaviventris Kraatz, 1874b: 127

sanguinosa Suffrian, 1844a: 105 **E**: AU BE BH BY CR CZ DE EN FI FR GB GE HU IR IT LA LT MD NL NR PL RO RU (NT CT ST) SK SL SP SV UK **N**: AG MO **A**: KZ RU (ES FE SS WS) TR [# 401, # 988]

viridissima Reitter, 1913c: 217 (ab.) [IN]

sareptana Kraatz, 1874b: 126 **E**: KR KZ RU (CT ST) UK **A**: KZ [# 369, # 434]

saginata Spaeth, 1914d: 553 **A**: PA **ORR**

saucia Weise, 1889h: 260 **A**: AB AR GG IN TR

sauteri Spaeth, 1913a: 48 (*Taiwania*) **A**: CH (ANH FUJ GUI GUX HUB JIX SCH TAI XIZ YUN ZHE) JA (Okinawa) **ORR** [# 1145]

schawalleri L. N. Medvedev, 1990b: 46 **A**: NP

seladonia Gyllenhal, 1827: 644 **E**: AL AU BE BH BU BY CZ DE FR GE GR HU IT KR LA LT NL NM PL PT RO RU (CT NT) SK SP SV SZ TR UK **N**: AG LB MO TU **A**: TR [# 370]

filaginis Perris, 1856: lxxix

rotundicollis Brisout de Barneville, 1860: 348

semipunctata S.-H. Chen & Zia, 1964: 124, 134 **A**: CH (SCH)

seraphina Ménétriés, 1836: 151 **E**: BU GR TR **A**: IN TR [# 371, # 435, # 471]

testudo Suffrian, 1844f: 284

sigillata Gorham, 1885: 284 (*Coptocycla*) **A**: CH (FUJ GUA GUI JIA JIX TAI ZHE) JA NC SC

simanica S.-H. Chen & Zia, 1961: 446, 450 (*Taiwania*) **A**: CH (XIZ YUN) **ORR** [# 308]

sodalis S.-H. Chen & Zia, 1964: 130, 137 (*Taiwania*) **A**: CH (YUN)

spaethi Weise, 1901a: 458 [RN] **A**: CH (BEI HEB HEI NMO SHA) MG NC RU (FE)

kraatzi Weise, 1900b: 295 [HN]

mandschukuoensis Spaeth, 1942: 13 (ssp.)

spaethiana Gressitt, 1945b: 147 [RN] **A**: CH (FUJ GUA HUB HUN ZHE)

spaethi Gressitt, 1942d: 2 (*Taiwania*) [HN]

stevensi Sekerka, 2011: 499 **A**: ID (SD) **ORR**

stigmatica Suffrian, 1844d: 206 **E**: AU BE BH BU BY CR CZ DE EN FI FR GE GR HU IT KR LA LT MD NL NM PL RO RU (CT NT ST) SB SK SP SZ TR UK **N**: TU **A**: AF CH (XIN) GG IN KI KZ RU (ES SS WS) TM TR [# 372]

strejceki Sekerka, 2006c: 562 **A**: TR [# 402]

subprobata S.-H. Chen & Zia, 1964: 126, 135 (*Taiwania*) **A**: CH (YUN) **ORR** *subreticulata* Suffrian, 1844e: 244 **E:** AL AU BE BH BU CR CZ FR GE GR HU IT KR MD ME NM PL RO RU (CT ST) SB SK SL SP SZ TR UK **A:** AB AR CH (XIN) GG IN JA KI KZ MG RU (ES FE SS WS) SY TM TR [# 373, # 436, # 471]

helva Spaeth, 1914c: 137

hincksi L. N. Medvedev, 1957d: 555

splendidula Suffrian, 1844d: 223 [HN]

ventralis Weise, 1893a: 1090 (var.)

subtilis Weise, 1897d: 111 **A:** PA **ORR**

sussamyrica Spaeth, 1926b: 60 **A:** CH (XIN) KI KZ [# 437]

tenasserimensis Spaeth, 1926c: 56 (*Metriona*) **A:** CH (GUX YUN) **ORR**

binorbis S.-H. Chen & Zia, 1961: 444, 449 (*Taiwania*)

tianshanica Borowiec & Świętojańska, 2001a: 70 **A:** KI KZ

transcaucasica Borowiec & Świętojańska, 2001a: 72 **A:** AB AR GG IN TR [# 374, # 402]

truncatipennis Spaeth, 1914d: 565 (*Metriona*) **A:** CH (YUN) **ORR** [# 409]

tsinlinica S.-H. Chen & Zia, 1964: 124, 133 **A:** CH (SHA)

tumidicollis S.-H. Chen & Zia, 1961: 446, 451 (*Cyrtonocassis*) **A:** CH (YUN)

turcmenica Weise, 1892g: 238 **A:** AF CH (XIN) IN KI KZ MG TD TM UZ

arata Lopatin, 1970a: 258

hauseri Weise, 1894a: 72

undecimnotata Gebler, 1841b: 615 **A:** AF CH (XIN) IN KI KZ MG RU (SS) TD TM UZ [# 402]

impicticollis Weise, 1894f: 96 (*Chiridula*, var.)

undecimguttata Reitter, 1890e: 265 (var.)

uniorbis S.-H. Chen & Zia, 1961: 443, 449 (*Taiwania*) **A:** CH (JIX YUN) **ORR**

variabilis S.-H. Chen & Zia, 1961: 443, 449 (*Taiwania*) **A:** CH (JIX YUN) **ORR**

varians Herbst, 1799: 269 **A:** BT CH (YUN) ID (AP HP SD UT) NP PA **ORR** [# 375]

consors Spaeth, 1938: 234

purulenta Boheman, 1862: 436 (*Coptocycla*)

rugulosa Boheman, 1857: 129

syrtica Boheman, 1857: 129

velaris Weise, 1896b: 28 **A:** CH (HEI JIL QIN SHA XIZ) JA NC RU (FE) SC

versicolor Boheman, 1855: 414 (*Coptocycla*) **A:** CH (FUJ GUA GUI GUX HAI HEI HKG HUB HUN JIX SCH TAI YUN ZHE) JA NC RU (FE) SC [# 400]

thais Boheman, 1862: 463 (*Coptocycla*)

vespertina Boheman, 1862: 357 **A:** CH (ANH BEI FUJ GAN GUA GUI GUX HEB HEI HUB HUN JIA JIX NIN NMO SCH SHA SHX TAI XIZ YUN ZHE) JA MG RU (FE) SC [# 376, # 438]

vibex Linnaeus, 1767: 575 **E:** AL AU BE BH BU BY CR CZ DE EN FI FR GB GE GR HU IR IT KO KR LA LS LT LU MD ME NL NM NR PL PT RO RU (CT NT ST) SB SK SL SP SV SZ TR UK **A:** CH (BEI HEB JIL XIN) GG JA KZ MG RU (ES FE SS WS) TR [# 989, # 1076]

discoidea Weise, 1893a: 1108 (var.)
dorsalis Herbst, 1799: 227
liriophora Kirby, 1797: 8
significata Csiki, 1953: 134 (ab.) [IN]

virguncula Weise, 1889g: 645 **A:** CH (BEI GAN HEB HEN JIA JIX LIA NIN QIN SHA SHX XIN) [# 377]
lenis Spaeth, 1926b: 59

viridis Linnaeus, 1758: 362 **E:** AL AU BE BH BU BY CR CZ DE EN FI FR GB GE GR HU IR IT KO KR KZ LA LS LT LU MD ME NL NM NR PL PT RO RU (CT NT ST) SB SK SL SP SV SZ TR UK **N:** MO **A:** AB AR CH (HEI JIL XIN) GG IN JA KZ LE NC RU (ES FE SS WS) SC SY TR UZ **NARi**
cardui DeGeer, 1775: 174
equestris Fabricius, 1787: 62
flaviceps Marseul, 1876: 35
japonica Yasutomi, 1952: 93, 98 (ssp.) [HN]
nigriceps Fairmaire, 1851b: 528
nigroconcolor Heikertinger, 1918: 5 (forma)

vitalisi Spaeth, 1928: 46 (*Metriona*) **A:** CH (GUX YUN) **ORR**

vittata Villers, 1789: 93 **E:** AL AU BE BH BY CR CZ DE EN FR GB GE GR HU IT KR LA LT LU MA NL PL PT RO RU (CT NT ST) SB SK SP SV SZ UK **N:** AG EG MO TU **A:** CH (SCH) IN IS KZ RU (ES FE SS WS) SC TM TR UZ [# 378, # 439, # 990]
atlantica Escalera, 1914: 535 (*Cassidula*)
dimbakar Maulik, 1923: 607
oblonga Illiger, 1798: 485 [# 991]
salicorniae Curtis, 1826: 127

weisei Jacobson, 1894d: 246 **A:** KZ TD TM UZ
transcaspica Spaeth, 1926b: 16 (*Eremocassis*)

genus *Cassida*, nomina dubia

alpina Comolli, 1837: 45 E: IT [# 310]

angusticollis Hagenbach, 1822: 8 E: SZ [# 311]

genus *Chiridopsis* **Spaeth, 1922: 1003** type species: *Coptocycla nigrosepta* Fairmaire, 1891

bipunctata Linnaeus, 1767: 578 (*Cassida*) **A:** ID (SD) PA **ORR**
bipunctipennis Boheman, 1855: 118 (*Coptocycla*)
promiscua Boheman, 1855: 130 (*Coptocycla*)
sexmaculata Boheman, 1855: 114 (*Coptocycla*)
sexnotata Fabricius, 1798: 82 (*Cassida*)
singularis Weise, 1901b: 53 (*Chirida*, var.)

bistrimaculata Boheman, 1855: 112 (*Coptocycla*) **A:** CH (YUN) PA **ORR**

bistrinotata Boheman, 1855: 113 (*Coptocycla*)
bowringii Boheman, 1855: 123 (*Coptocycla*) **A:** CH (GUA GUI GUX HAI HKG YUN) NP **ORR** [# 379]
mimica Weise, 1905d: 210 (*Chirida*) **A:** ID (AP SD) **ORR** [# 380]
punctata indochinensis Spaeth, 1919: 199 (*Chirida*, ssp.) **A:** CH (GUA GUX HAI HUB YUN) **ORR**
scalaris Weber, 1801: 51 (*Cassida*) **A:** CH (YUN) **ORR** [# 442]
selecta Weise, 1905c: 128 (*Chirida*) **A:** PA **ORR**
binduta Maulik, 1919a: 423 (*Chirida*)
septemnotata Boheman, 1855: 133 (*Coptocycla*) **A:** BT ID (SD) **ORR**

genus *Chiridula* **Weise, 1889g: 647** type species: *Chiridula semenowi* Weise, 1889
semenowi Weise, 1889g: 647 **A:** AB AF CH (GAN HEB HEI JIL NIN NMO QIN SHX XIN) IN KI KZ NC TD TM [# 403, # 443]
fovangula Reitter, 1901d: 164 (*Cassida*)

genus *Glyphocassis* **Spaeth, 1914d: 547** type species: *Cassida trilineata* Hope, 1831
Hebdomecosta Spaeth, 1915: 361 type species: *Hebdomecosta reitteri* Spaeth, 1915 (= *Coptocycla spilota* Gorham, 1885)
lepida Spaeth, 1914c: 145 (*Coptocycla*) **A:** CH (CHQ HUB JIX SCH XIZ ZHE) [# 1145]
tetrasticta Gressitt, 1942d: 3
spilota gansuica L. N. Medvedev, 1957d: 555 (ssp.) **A:** CH (GAN)
spilota spilota Gorham, 1885: 285 (*Coptocycla*) **A:** CH (BEI FUJ GUI HEB HUN JIA JIX SCH SHA SHG SHN ZHE) JA NC RU (FE) SC **ORR**
reitteri Spaeth, 1915: 362 (*Hebdomecosta*)
shirahatai Chûjô, 1949a: 9 (*Hebdomecosta*)
trilineata szechuana Gressitt, 1952: 480 (ssp.) **A:** CH (HUB JIX SCH) [# 385]
trilineata trilineata Hope, 1831: 30 (*Cassida*) **A:** CH (GAN GUI GUX YUN) ID (AP HP SD UT) NP PA **ORR** [# 386]
melanosticta Spaeth, 1914d: 548 (ab.) [IN]

genus *Hypocassida* **Weise, 1893a: 1074** type species: *Cassida subferruginea* Schrank, 1776
convexipennis Borowiec, 2000: 601 **A:** IN
cornea Marseul, 1868: 214 (*Cassida*) **E:** AL AU BU CR CZ FR (Corsica) GR HU IT ME NM PL PT RO RU (ST) SB SK UK **N:** AG EG MO TU **A:** AE CY LE SY TR **AFR** [# 388, # 405, # 453]
grossepunctata Bordy, 2008: 52 **E:** FR IT SP
meridionalis Suffrian, 1844f: 276 (*Cassida*) **E:** FR IT PT SP **N:** AG MO **A:** IS TR
viturati Pic, 1905h: 179 (*Cassida*, var.)

subferruginea Schrank, 1776: 62 (*Cassida*) **E:** AL AU BE BH BU BY CR CZ DE EN FI FR GE GR HU IT KO KR LA LT LU MD ME NL NM PL PT RO RU (CT NT ST) SB SK SL SP SZ TR UK **N:** AG EG MO TU **A:** AF AR CH (BEI HEB HEI NIN SHA XIN) CY GG IN IQ IS KI KZ MG RU (ES FE SS WS) SI TD TM TR UZ [# 454, # 471]

costata Boheman, 1857: 140 (*Cassida*)

ferruginea Fabricius, 1781: 108 (*Cassida*)

fusca Laicharting, 1781: 112 (*Cassida*)

marginata Bechyné, 1944a: 76 (ab.) [IN]

sobrina Weise, 1893a: 1083 (*Cassida*, var.)

genus ***Ischyronota*** **Weise, 1891e: 204** type species: *Cassida desertorum* Gebler, 1833

brisouti Reitter, 1889c: 288 (*Cassida*) **N:** AG MO

nitidula Desbrochers des Loges, 1891: 44 (*Cassida*) [HN]

conicicollis Weise, 1890a: 487 (*Cassida*) **A:** CH (NIN NMO XIN) IN KZ MG TD TM UZ

desertorum Gebler, 1833: 305 (*Cassida*) **E:** KR RO RU (CT ST) TR UK **N:** EG **A:** AB AF AR CH (XIN) GG IN IS KI KZ MG PA RU (SS) SY TD TM TR UZ [# 389, # 406]

araxicola Reitter, 1889c: 288 (*Cassida*)

basimargo Reitter, 1901b: 108 (var.)

gibbula Boheman, 1854: 430 (*Cassida*)

jakowleffi Reitter, 1889c: 288 (*Cassida*)

salsolae Desbrochers des Loges, 1891: 43 (*Cassida*)

elevata Reitter, 1890b: 175 (*Cassida*) **E:** RU (ST) **A:** AB AR IN IS SI KZ SI TM UZ

nitidula Weise, 1890g: 367 (*Cassida*)

jordanensis Borowiec, 1986b: 575 (ssp.) **A:** JO IS TR

schusteri Spaeth, 1914c: 144 **A:** CH (XIN) IN KZ MG TM

spaethi Reitter, 1901b: 108 **E:** RU (ST) **A:** IN KZ TM

deserticola Reitter, 1901b: 108 (var.)

genus ***Macromonycha*** **Spaeth, 1911: 271** type species: *Cassida apicalis* Gebler, 1845

Kari Maulik, 1923: 602 type species: *Kari brunnea* Maulik, 1923 (= *Cassida apicalis* Gebler, 1845)

anatolica Weise, 1900c: 139 [= 1900d: 166] (*Chiridula*) **A:** TR

apicalis Gebler, 1844: 105 (*Cassida*) **E:** KZ **A:** AB AF AR IN IQ IS KI KZ SY TD TM TR UZ [# 392, # 460]

brunnea Maulik, 1923: 602 (*Kari*)

discoidalis Reitter, 1891: 35 (*Cassida*, var.)

olivieri Boheman, 1854: 332 (*Cassida*)

suberosa Weise, 1889m: 132 (*Cassida*)

kantnerorum Sekerka, 2008: 99 **A:** JO TR ("Anatolia")

genus *Nabathaea* Spaeth, 1911: 272 type species: *Nabathaea pygmaea* Spaeth, 1911

Falsocassidula Pic, 1939h: 149 type species: *Cassida rabinovitchi* Pic, 1939

pygmaea Spaeth, 1911: 272 **N:** EG **A:** SI YE **[# 393]**

rabinovitchi Pic, 1939h: 149 (*Cassida*)

genus *Oocassida* Weise, 1897d: 110 type species: *Cassida pudibunda* Boheman, 1857

cruenta Fabricius, 1792a: 293 (*Cassida*) **A:** ID (UT) NP PA **ORR**

obscura Fabricius, 1792a: 295 (*Cassida*)

unicolor Fabricius, 1801a: 391 (*Cassida*)

pudibunda Boheman, 1857: 133 (*Cassida*) **A:** ID (UT) NP PA **ORR**

tunisiensis Boheman, 1854: 419 (*Cassida*) **N:** AG MO TU

biskrensis Desbrochers des Loges, 1884: 169 (*Cassida*)

koechlini Marseul, 1866: lxviii (*Cassida*)

genus *Oocassida*, excluded taxon

schultzei Spaeth, 1917: 437 **AFR** **[# 324]**

genus *Oxylepus* Desbrochers des Loges, 1884: 170 type species: *Cassida deflexicollis* Boheman, 1862

Oxylepis Desbrochers des Loges, 1891: 6 [UE] type species: *Cassida deflexicollis* Boheman, 1862

boroveci Borowiec, 2001a: 350 **N:** TU

deflexicollis Boheman, 1862: 333 (*Cassida*) **E:** CR FR GR IT SP **N:** AG CI EG MO TU **A:** IN IS **[# 395, # 464]**

capucinus Desbrochers des Loges, 1884: 170

excelsus Desbrochers des Loges, 1891: 38 (*Cassida*)

involutus Fairmaire, 1880b: 419 (*Cassida*)

palaestinus Desbrochers des Loges, 1884: 171

sahlbergi Spaeth, 1901b: 755 (var.)

vagepunctatus Weise, 1891e: 203 (*Chelysida*)

kossmati Spaeth, 1901b: 752 **A:** OM YE **AFR**

genus *Pilemostoma* Desbrochers des Loges, 1891: 14 type species: *Cassida vittata* Fabricius, 1798 (= *Cassida fastuosa* Schaller, 1783) **[# 288]**

fastuosum Schaller, 1783: 259 (*Cassida*) **E:** AL AU BE BH BU BY CR CZ DE EN FR GB GE GR HU KR LA LU MD ME NM PL RO RU (CT NT ST) SB SK SL SP SV SZ TR UK **A:** AR GG IN KI KZ MG RU (ES SS WS) TR UZ **[# 396, # 465, # 471]**

bucharicum Spaeth, 1914c: 136

delicatulum Weise, 1893a: 1084 (*Cassida*, var.)

sanguineum Böber, 1795: 135 (*Cassida*)

nigrinum Weise, 1893a: 1084 (*Cassida*, var.)
ocellatum Herbst, 1799: 244 (*Cassida*)
vittatum Fabricius, 1798: 81 (*Cassida*)

genus *Rhytidocassis* **Spaeth, 1941: 316** type species: *Cassida limbiventris* Boheman, 1854
Chloocassis Spaeth, 1952: 347 type species: *Odontionycha indicola* Duvivier, 1892
indicola Duvivier, 1892b: 448 (*Cassida*) **A:** PA **ORR**
hina Maulik, 1919a: 415 (*Chirida*)
iranella Lopatin, 1984a: 92 (*Cassida*) **A:** IN
lopatini Borowiec & Świętojańska, 2001b: 325 **A:** IN

genus *Seminabathea* **Borowiec, 1994a: 14** type species: *Nabathaea pygmaea* sensu Borowiec (1994a) (= *Nabathaea arabica* Spaeth, 1911)
Neonabathaea Borowiec, 1994b: 158 type species: *Nabathaea arabica* Spaeth, 1911
arabica Spaeth, 1911: 273 (*Nabathaea*) **A:** "Arabia" AE OM **[# 425, # 468]**

genus *Thlaspida* **Weise, 1899b: 272** type species: *Coptocycla cribrosa* Boheman, 1855
biramosa biramosa Boheman, 1855: 408 (*Coptocycla*) **A:** CH (ANH FUJ GAN GUA GUI GUX HAI HEN HUB HUN JIA JIX SCH SHA TAI YUN ZHE) JA NC SC **ORR**
biramosa Weise, 1899b: 273 [HN]
chinensis Spaeth, 1926b: 64
formosae Spaeth, 1913a: 46
immaculatipennis Chûjô, 1934c: 157 (ab.) [IN]
japonica Spaeth, 1914a: 17 [RN]
tristis Weise, 1899b: 273
biramosa omeia S.-H. Chen & Zia, 1964: 131, 137 (ssp.) **A:** CH (SCH)
cribrosa Boheman, 1855: 404 (*Coptocycla*) **A:** CH (SCH TAI XIZ YUN) **ORR**
lewisii Baly, 1874: 214 (*Coptocycla*) **A:** CH (FUJ HEI LIA) JA NC RU (FE) SC
testacea Rybakow, 1884b: 136 (*Coptocycla*)
pygmaea L. N. Medvedev, 1957d: 555 **A:** CH (SCH)
tsoui Borowiec & C.-F. Lee, 2009: 350 **A:** CH (TAI)

genus *Thlaspidosoma* **Spaeth, 1901a: 4** type species: *Thlaspidosoma dohrni* Spaeth, 1901
breve S.-H. Chen & Zia, 1964: 131, 138 **A:** CH (YUN)

tribe Coelaenomenoderini Weise, 1911

genus *Javeta* **Baly, 1859a: 108** type species: *Javeta pallida* Baly, 1859
foveicollis Gressitt, 1939a: 168 (*Wallaceana*) **A:** CH (HAI)
maculata Sun, 1985: 7, 8 **A:** CH (HAI)

tribe Cryptonychini Chapuis, 1875

genus *Brontispa* Sharp, 1904: 924 type species: *Brontispa froggatti* (= *Oxycephala longissima* Gestro, 1885)

Bronthispa Weise, 1911b: 46 [UE] type species: *Brontispa froggatti* (= *Oxycephala longissima* Gestro, 1885)

Planispa Chûjô, 1937f: 223 type species: *Planispa castaneipennis* Chûjô, 1937 (= *Bronthispa mariana* Spaeth, 1937)

longissima Gestro, 1885: 162 (*Oxycephala*) **Ai**: CH (FUJ HAI TAI YUN) JA **AUS ORR**i

froggatti Sharp, 1904: 924

javana Weise, 1922a: 62 (*Bronthispa*)

longipennis Gestro, 1892: 1017 (*Oxycephala*)

reicherti Uhmann, 1929: 145 (*Bronthispa*)

selebensis Gestro, 1923: 16 (*Bronthispa*)

simmondsi Maulik, 1927: 105

genus *Octodonta* Chapuis, 1875: 289 type species: *Octodonta depressa* Chapuis, 1875

subgenus *Octodonta* Chapuis, 1875: 289 type species: *Octodonta depressa* Chapuis, 1875

depressa Chapuis, 1875: 290 **Ai**: CH (TAI) **ORR**

anceyi Pic, 1924b: 5 (*Downesia*)

nipae Maulik, 1921: 451 (*Plesispa*) **Ai**: CH (HAI FUJ) CY **ORR**

tribe Gonophorini Chapuis, 1875

genus *Agoniella* **Weise, 1911b: 57** type species: *Agonia manilensis* Weise, 1910

sonani Chûjô, 1933b: 317 (*Agonia*) **A**: CH (TAI)

genus *Agonita* **E. Strand, 1942: 391** [RN] type species: *Agonia femoralis* Weise, 1904 [# 284]

Agonia Weise, 1904e: 99 [HN] type species: *Agonia femoralis* Weise, 1904 [# 284]

apicata S.-H. Chen & Sun, 1964: 111, 119 **A**: CH (YUN)

carbunculus Maulik, 1919a: 139 (*Agonia*) **A**: CH (GUA XIZ) **ORR**

castanea Tan & Sun, 1962: 125, 134 (*Gonophora*) **A**: CH (YUN)

chinensis Weise, 1922a: 75 (*Agonia*) **A**: CH (FUJ GUA GUI GUX HAI HKG HUB HUN JIA JIX NIN SCH SHA SHN YUN ZHE) **ORR**

atripennis Pic, 1926b: 10 (*Agonia*)

dangali L. N. Medvedev, 2009b: 408 **A**: ID (SD)

darjeelingensis Basu, 1999: 167 **A**: ID (SD)

discrepans Uhmann, 1954: 509 [= 1955a: 188] **A**: CH (FUJ HAI)

himalayensis Maulik, 1919a: 127 (*Agonia*) **A**: ID (SD) "Himalayas" **ORR**

immaculata Gestro, 1888b: 175 (*Gonophora*) **A**: CH (GUX YUN) NP **ORR**

nigrimembris Pic, 1926b: 11 (*Agonia*)

indenticulata Pic, 1924b: 6 (*Agonia*) **A:** CH (GUX JIX YUN) **ORR**
klapperichi Uhmann, 1955b: 172 **A:** AF ID (UT) NP
kunminensis Tan & Sun, 1962: 125, 134 (*Gonophora*) **A:** CH (YUN)
laeta L. N. Medvedev & Sprecher-Uebersax, 1999: 369 **A:** ID (SD) NP [# 411]
laticeps Gressitt, 1939b: 199 (*Agonia*) **A:** CH (JIX)
lohita Basu, 1999: 169 **A:** ID (SD)
metasternalis Tan & Sun, 1962: 125, 135 (*Gonophora*) **A:** CH (YUN)
nanpinensis S.-H. Chen & Sun, 1964: 111, 119 **A:** CH (FUJ)
nepalica L. N. Medvedev, 2003a: 322 **A:** NP
nigra Tan & Sun, 1962: 126, 135 (*Gonophora*) **A:** CH (YUN)
omeia S.-H. Chen & Sun, 1964: 112, 119 **A:** CH (SCH)
omoro Takizawa, 1975a: 57 **A:** CH (TAI) JA (Isigaki, Okinawa Is.)
pallidipennis Maulik, 1919a: 135 (*Agonia*) **A:** ID (SD) NP **ORR**
parvula Gestro, 1890: 237 (*Gonophora*) **A:** BT **ORR** [# 397]
picea Gressitt, 1953: 124 **A:** CH (FUJ GUX HUN ZHE) [# 412]
pictipes S.-H. Chen & Sun, 1964: 112, 119 **A:** CH (FUJ)
pilipes S.-H. Chen & Sun, 1962: 126, 135 (*Gonophora*) **A:** CH (FUJ GUX SCH) [# 412]
ruficollis S.-H. Chen & Sun, 1964: 113, 120 **A:** CH (SCH)
saundersii Baly, 1859a: 110 (*Gonophora*) **A:** ID (SD) NP **ORR**
sculpturata Gressitt, 1953: 124 **A:** CH (FUJ HAI JIX SCH YUN) **ORR**
seminigra Tan & Sun, 1962: 126, 135 (*Gonophora*) **A:** CH (YUN)
tricolor Chûjô, 1933b: 314 (*Agonia*) **A:** CH (FUJ GUX JIX TAI)
tricostata P.-Y. Yu & H. Huang, 2002: 456, 457 **A:** CH (HAI)
tristis S.-H. Chen & Sun, 1962: 127, 136 (*Gonophora*) **A:** CH (HUB SCH)
unicolor Chûjô, 1933b: 316 (*Agonia*) **A:** CH (TAI)
yuae C.-F. Lee, 2015b: 265 **A:** CH (TAI)

genus *Agonita*, excluded taxon
pitava Basu, 1999: 169 **ORR** [# 320]

genus *Downesia* **Baly, 1859a:** 107 type species: *Downesia insignis* Baly, 1859
Hanoia Fairmaire, 1889a: 375 type species: *Hanoia auberti* Fairmaire, 1889
abdominalis Weise, 1922a: 72 **A:** ID (SD)
atrata Baly, 1869: 377 **A:** CH (GUX YUN) ID (SD) NP **ORR**
balyi Gressitt, 1950: 83 **A:** CH (GUA)
fulvipennis Baly, 1888c: 660 **A:** CH (GUX YUN) **ORR**
gestroi Baly, 1888c: 660 **A:** BT CH (FUJ GUX YUN) ID (SD) NP **ORR**
atripes Pic, 1928c: 35
donckieri Pic, 1924d: 99
ruficolor Pic, 1924b: 4

viridana Herbst, 1799: 224

probata Spaeth, 1914b: 227 **A**: CH (GUA HUB HUN JIX YUN) NP **ORR**

purpuricollis Spaeth, 1914b: 229 (*Metriona*) **A**: CH (GUI HUB SCH YUN) **ORR**

pusilla Waltl, 1835: 82 **E**: AL CR FR GR IT ME PT SP SZ **N**: AG MO TU **A**: TR **[# 366]**

disticta Boheman, 1862: 336

mollis Boheman, 1854: 437

puncticollis Suffrian, 1844e: 257

pupillata Boheman, 1854: 439

veselyi Günther, 1958: 568

pyrenaea Weise, 1893a: 1082 **E**: FR PT SP

lusitanica Sassi, 1993: 331

quadriramosa Gressitt, 1952: 499 **A**: CH (FUJ GUA HUB HUN JIX SCH TAI ZHE)

quinaria S.-H. Chen & Zia, 1964: 128, 135 (*Taiwania*) **A**: CH (GAN GUX SCH YUN) **ORR**

quinqueasteriza L. N. Medvedev & Eroshkina, 1988: 133

rati Maulik, 1923: 605 **A**: CH (FUJ GAN GUA GUI GUX HAI HUB HUN JIX SCH TAI XIZ YUN ZHE) **ORR** **[# 1145]**

ratina S.-H. Chen & Zia, 1964: 127, 135 (*Taiwania*) **A**: CH (GUX)

reitteri Weise, 1892g: 238 **A**: AR GG TR

reticulicosta S.-H. Chen & Zia, 1964: 130, 137 (*Taiwania*) **A**: CH (YUN) **ORR**

rothschildi Spaeth, 1922: 1002 **A**: SA YE **AFR** **[# 432]**

rubiginosa rubiginosa O. F. Müller, 1776: 65 **E**: AL AN AU BE BH BU BY CR CZ DE EN FI FR (incl. Corsica) GB GE GR HU IR IT KO KR LA LS LT LU MD ME NL NM NR PL PT RO RU (CT NT ST) SB SK SL SP SV SZ TR UK **N**: AG MO **A**: AB AR GG IS KZ MG RU (ES SS WS) SY TR **AUSi NARi** **[# 367, # 433, # 986]**

fuliginosa Weise, 1893a: 1104 (var.)

graeca Kraatz, 1874a: 104

lata Suffrian, 1844b: 139

melanosceles Schrank, 1798: 520

similis Marsham, 1802: 144

singularis Stephens, 1832: 369

rubiginosa rugosopunctata Motschulsky, 1866a: 177 **A**: CH (FUJ HUB JIA JIL JIX NIN QIN SHA SHX TAI XIN XIZ ZHE) JA NC RU (FE) SC

babai Kimoto, 1986: 128 (ssp.)

erudita Baly, 1874: 212

taiwana Gressitt, 1952: 517 (ssp.)

rufovirens Suffrian, 1844b: 144 **E**: AU BE BH BU CR CZ FR GE HU KR PL RO RU (CT ST) SK SP SZ UK **A**: AR GG TR

rhilensis Weise, 1891e: 205

ruralis Boheman, 1862: 461 (*Coptocycla*) **A**: CH (TAI YUN) NP **ORR**

gracilis Uhmann, 1955a: 188 **A:** CH (FUJ)
javana ginpinica S.-H. Chen & Tan, 1962: 123, 133 (ssp.) **A:** CH (YUN) [# 449]
kwangtunga Gressitt, 1950: 84 **A:** CH (GUA)
marginicollis Weise, 1922a: 71 **A:** CH (FUJ)
nigripennis S.-H. Chen & Tan, 1962: 123, 133 **A:** CH (YUN)
puncticollis S.-H. Chen & Tan, 1962: 123, 133 **A:** CH (YUN)
sasthi Maulik, 1923: 600 **A:** CH (YUN) ID (UT) NP [# 450]
simulans S.-H. Chen & Sun, 1964: 110, 118 **A:** CH (YUN)
strandi Uhmann, 1943: 121 **A:** CH (ZHE)
strigicollis Baly, 1876b: 128 **A:** CH (YUN) **ORR**
tarsata Baly, 1869: 377 **A:** CH (GUX HKG JIX MAC) [# 451]
thoracica S.-H. Chen & Sun, 1964: 110, 118 **A:** CH (GUA GUX YUN)
vandykei Gressitt, 1939c: 133 **A:** CH (FUJ GUA JIX ZHE) **ORR** [# 322]

genus *Downesia*, excluded taxon
auberti Fairmaire, 1889a: 375 (*Hanoia*) **ORR** [# 322]

genus *Gonophora* **Chevrolat, 1836: 366** type species: *Hispa haemorrhoidalis* Weber, 1801
Distolaca Baly, 1859a: 116 type species: *Distolaca whitei* Baly, 1859
Gonophora Baly, 1859a: 108 [HN] type species: *Hispa haemorrhoidalis* Weber, 1801
pulchella Gestro, 1888b: 176 **A:** CH (GUA GUX HAI YUN) ID (SD UT) NP PA **ORR** [# 387, # 1145]
annamita Pic, 1930d: 4
bengalensis Weise, 1908b: 214
clermonti Pic, 1927k: 25 (var.)
vitalisi Pic, 1930d: 4 (var.)

genus *Klitispa* **Uhmann, 1940a: 143** type species: *Agonia nigripennis* Weise, 1905 [# 286]
mutilata S.-H. Chen & Sun, 1964: 113, 120 **A:** CH (GUX JIX)
rugicollis Gestro, 1890: 238 (*Gonophora*) **A:** CH (GUX YUN) **ORR**

genus *Micrispa* **Gestro, 1897a: 65** type species: *Gonophora sinuata* Gestro, 1885
dentatithorax Pic, 1924b: 4 (*Gonophora*) **A:** CH (GUX YUN) **ORR**
yunnanica S.-H. Chen & Sun, 1962: 127, 136 (*Gonophora*)

genus *Sinagonia* **S.-H. Chen & Tan, 1962: 124, 133** type species: *Gonophora maculigera* Gestro, 1888
angulata S.-H. Chen & Tan, 1962: 124, 133 (*Gonophora*) **A:** CH (YUN)

foveicollis S.-H. Chen & Tan, 1962: 124, 134 (*Gonophora*) **A:** CH (FUJ GUA GUX JIX YUN)
maculigera Gestro, 1888a: 131 (*Gonophora*) **A:** CH (FUJ GUA HAI YUN) NP **ORR**
juncta Pic, 1924b: 5 (*Agonia*, var.)
multimaculata Pic, 1924b: 5 (*Agonia*)
pitambara Basu, 1999: 172 (*Gonophora*) **A:** ID (SD) [# 299]
raktava Basu, 1999: 171 (*Gonophora*) **A:** ID (SD) [# 299]

tribe Hispini Gyllenhal, 1813

genus *Acmenychus* **Weise, 1905f: 318** type species: *Hispa inermis* Zubkov, 1833
caucasicus Heyden, 1878: 343 (*Monochirus*) **A:** AB AR IN
inermis Zubkov, 1833: 337 (*Hispa*) E: RU (ST) **A:** CH (NMO XIN) IN KI KZ MG TD TM UZ
potanini Weise, 1890a: 485 (*Monochirus*)
planus Maulik, 1919b: 409 **A:** IS JO "Arabia" "Kurdestan" "Turcomania" [# 410]

genus *Asamangulia* **Maulik, 1915: 378** type species: *Asamangulia cuspidata* Maulik, 1915 (= *Hispa tuberculosa* Motschulsky, 1860)
horni Uhmann, 1927b: 112 **A:** CH (TAI)
longispina Gressitt, 1950: 104 **A:** CH (FUJ GUA GUX HAI JIX YUN ZHE)
tuberculosa Motschulsky, 1860b: 239 (*Hispa*) **A:** AF NP PA **ORR**
cuspidata Maulik, 1915: 378
dreesi Uhmann, 1955b: 174
yonakuni Kimoto & Gressitt, 1966: 570 **A:** JA (Yonaguni-jima)

genus *Cassidispa* **Gestro, 1899a: 174** type species: *Cassidispa mirabilis* Gestro, 1899
bipuncticollis S.-H. Chen, 1941c: 196 **A:** CH (NIN SHX) [# 440]
femoralis S.-H. Chen & P.-Y. Yu, 1976: 218, 224 **A:** CH (NIN XIZ)
maderi Uhmann, 1938: 225 **A:** CH (YUN)
mirabilis Gestro, 1899a: 175 **A:** CH (FUJ HEB NIN SCH SHX ZHE)
relicta L. N. Medvedev, 1957a: 295, 296 **A:** CH (NMO) RU (FE)

genus *Chrysispa* **Weise, 1897d: 139** type species: *Hispa viridicyanea* Kraatz, 1895
acanthina Reiche, 1850: 398 (*Hispa*) **A:** SA **AFR**

genus *Dactylispa* **Weise, 1897d: 137** type species: *Dactylispa andrewesi* Weise, 1897 (= *Hispa longicornis* Motschulsky, 1860)
subgenus *Dactylispa* Weise, 1897d: 137 type species: *Dactylispa andrewesi* Weise, 1897 (= *Hispa longicornis* Motschulsky, 1860)
albopilosa Gestro, 1888b: 181 (*Hispa*) **A:** CH (JIX) NP **ORR** [# 444]

discicollis Gestro, 1890: 254 (*Hispa*)
kamarupa Maulik, 1919a: 223
maculata Gestro, 1890: 252 (*Hispa*)
anula Maulik, 1919a: 243 **A:** NP **ORR**
approximata Gressitt, 1939b: 201 **A:** CH (FUJ GUA GUX JIX SCH XIZ ZHE)
atkinsonii Gestro, 1897c: 132 (*Hispa*) **A:** BT CH (XIZ) ID (SD UT) NP **ORR**
balyi Gestro, 1890: 250 (*Hispa*) [RN] **A:** CH (GUA GUX HAI JIX YUN) **ORR**
discoidalis Baly, 1888c: 664 (*Hispa*) [HN]
basalis Gestro, 1897a: 86 (*Hispa*) **A:** CH (YUN) **ORR**
binotaticollis S.-H. Chen & Tan, 1964: 415, 422 **A:** CH (GUX YUN)
bulbifera Medvedev, 1993c: 375 **A:** ID (UT) **[# 317]**
burmana Uhmann, 1939a: 7 **A:** CH (YUN) **ORR**
chanchala Basu & Saha, 1977: 383 [DA] **A:** ID (SD) **ORR** **[# 316]**
chiayiana Kimoto, 1978b: 71 **A:** CH (TAI)
corpulenta Weise, 1897d: 132 **A:** CH (GUX HAI YUN) "Himalaya" **ORR** **[# 445]**
delicatula Gestro, 1888b: 182 (*Hispa*) **A:** CH (GUA GUX HAI JIX YUN) NP **ORR**
longula Maulik, 1919a: 204
dohertyi Gestro, 1897c: 133 (*Hispa*) **A:** CH (YUN) ID (KA) **ORR**
doriae Gestro, 1890: 256 (*Hispa*) **A:** CH (YUN) ID (AP SD) NP PA **ORR**
elegantula Duvivier, 1892b: 447 (*Hispa*) **A:** ID (SD UT) NP PA **ORR**
bilasa Maulik, 1919a: 216
feae Gestro, 1888b: 183 (*Hispa*) **A:** CH (FUJ GUA GUX HAI JIX) PA **ORR**
filiola Weise, 1897d: 135
ferrugineonigra Maulik, 1919a: 208 **A:** CH (YUN) **ORR**
flavoapicalis Tan, 1993b: 352, 355 **A:** CH (HUN)
flavomaculata Uhmann, 1930b: 133 **A:** CH (GUX HAI YUN) **ORR**
foveiscutis S.-H. Chen & Tan, 1961: 466, 479 **A:** CH (YUN)
fukienica S.-H. Chen & Tan, 1964: 416, 423 **A:** CH (FUJ GUX)
humeralis Weise, 1905c: 120 **A:** NP **ORR**
inaequalis S.-H. Chen & Tan, 1964: 416, 423 **A:** CH (YUN)
issiki Chûjô, 1938f: 12 **A:** CH (FUJ GUA GUX JIX ZHE) JA
kaulina Gressitt, 1950: 118
kantakita Maulik, 1919a: 239 **A:** ID (UT) **ORR**
lameyi Uhmann, 1930b: 134 **A:** CH (GUX YUN) **ORR**
latifrons S.-H. Chen & Tan, 1961: 467, 479 **A:** CH (YUN)
lohita Maulik, 1919a: 197 **A:** BT CH (XIZ) ID (HP SD UT) NP **ORR**
longicornis Motschulsky, 1860b: 240 (*Hispa*) **A:** ID (UT) **ORR**
andrewesi Weise, 1897d: 129
severinii Gestro, 1897c: 129 (*Hispa*)
longispina Gressitt, 1938b: 328 **A:** CH (FUJ GUA GUX HAI HUB JIX YUN) **ORR**

malabikae Basu & Saha, 1977: 384 [DA] **A:** ID (SD) [# 316]
malaisei Uhmann, 1939a: 5 **A:** CH (YUN) **ORR**
manii Vazirani, 1972: 145 [DA] **A:** ID (AP SD) [# 316]
mauliki Gressitt, 1950: 123 **A:** CH (GUA)
melanocera S.-H. Chen & Tan, 1961: 467, 479 **A:** CH (YUN)
mendica Weise, 1897d: 131 **A:** CH (YUN) **ORR**
mixta Kung & Tan, 1961: 468, 479 **A:** CH (YUN)
miyamotoi Kimoto, 1970a: 178 **A:** CH (TAI)
molina Basu & Saha, 1977: 385 [DA] **A:** ID (SD) [# 316]
montivaga Gestro, 1898: 261 (*Hispa*) **A:** ID (SD) **ORR**
nigripennis Motschulsky, 1860b: 239 (*Hispa*) **A:** ID (SD) NP **ORR**
soror Weise, 1897d: 134
nigromaculata Motschulsky, 1860b: 239 (*Hispa*) **A:** CH (FUJ GUA GUX HAI HUN JIX YUN) **ORR**
xanthospila Gestro, 1890: 261 (*Hispa*)
parva S.-H. Chen & Tan, 1961: 468, 479 **A:** CH (YUN)
pectinata L. N. Medvedev, 1997e: 264 **A:** NP
pilosa Tan & Kung, 1961: 469, 478 **A:** CH (GUA GUX YUN) **ORR**
polita S.-H. Chen & Tan, 1961: 469, 478 **A:** CH (YUN)
pradhana Maulik, 1919a: 244 **A:** ID (SD) NP
praefica Weise, 1897d: 135 **A:** ID (SD) NP PA **ORR**
pubescens S.-H. Chen & Tan, 1962: 120 [RN] **A:** CH (YUN)
aureopilosa S.-H. Chen & Tan, 1961: 464, 480 [HN]
pugnax Gestro, 1897c: 134 (*Hispa*) **A:** ID (SD) **ORR**
puwena S.-H. Chen & Tan, 1961: 470, 479 **A:** CH (YUN)
sauteri Uhmann, 1927b: 115 **A:** CH (FUJ GUA GUI GUX HUB HUN JIX SCH TAI YUN ZHE) **ORR**
piceomaculata Gressitt, 1939a: 179
setifera atra S.-H. Chen & Tan, 1961: 470, 479 (ssp.) **A:** CH (YUN)
setifera setifera Chapuis, 1877c: 54 (*Hispa*) **A:** CH (GUI GUX HAI JIX)
sjostedti Uhmann, 1928a: 102 **A:** CH (GUA GUI HUB JIX YUN) **ORR**
spinosa Weber, 1801: 65 (*Hispa*) **A:** CH (YUN) **ORR**
cariana Gestro, 1890: 251 (*Hispa*)
sumatrana Weise, 1911a: 173
tonkinea Uhmann, 1930b: 135
superspinosa Gressitt & Kimoto, 1963a: 933 **A:** CH (HUB SCH)
tewfiki Pic, 1939h: 148 **N:** EG [# 448]
variabilis Maulik, 1919a: 211 **A:** ID (SD) NP **ORR**
vulnifica Gestro, 1908: 186 **A:** CH (YUN)
wittmeri Würmli, 1975: 374 **A:** BT ID (SD UT)

xanthopus Gestro, 1898: 262 (*Hispa*) **A:** CH (JIX TAI) ID (SD) NP **ORR**
xisana S.-H. Chen & Tan, 1961: 471, 479 **A:** CH (YUN)
subgenus *Platypriella* S.-H. Chen & Tan, 1961: 459, 478 type species: *Hispa excisa* Kraatz, 1879
badia S.-H. Chen & Tan, 1961: 474, 481 **A:** CH (YUN)
cheni Sekerka, **nom. nov.** **A:** CH (YUN) **[# 298]**
similis S.-H. Chen & Tan, 1985: 269, 271 [HN]
digitata Uhmann, 1955a: 202 **A:** CH (FUJ)
excisa excisa Kraatz, 1879b: 140 (*Hispa*) **A:** CH (ANH FUJ GUA GUI GUX HEI HUB HUN JIA JIL JIX SCH SHA SHN TAI YUN ZHE) NC RU (FE) SC
repanda Weise, 1922a: 81 (var.)
excisa meridionalis S.-H. Chen & Tan, 1961: 475, 481 (ssp.) **A:** CH (YUN)
latipennis Chûjô, 1933b: 329 **A:** CH (TAI)
latispina Gestro, 1899d: 228 (*Hispa*) **A:** CH (FUJ GUA GUX HAI JIX YUN) **ORR**
saurus Uhmann, 1955a: 194
planispina Gressitt, 1950: 125 **A:** CH (FUJ YUN)
platyacantha Gestro, 1897c: 126 (*Hispa*) **A:** CH (FUJ GUA HUN JIX YUN) NP **ORR**
platycanthoides Kimoto, 2001: 69 **A:** NP
spiniloba S.-H. Chen & Tan, 1964: 420, 426 **A:** CH (YUN) **ORR**
subquadrata australis S.-H. Chen & Tan, 1964: 421, 426 (ssp.) **A:** CH (FUJ HUN YUN)
subquadrata subquadrata Baly, 1874: 216 (*Hispa*) **A:** CH (HUB JIA JIX SHN) JA SC **[# 447]**
adstricta Weise, 1922a: 81
tienmuensis S.-H. Chen & Tan, 1964: 421, 427 **A:** CH (FUJ ZHE)
subgenus *Triplispa* Weise, 1897d: 150 type species: *Hispa confluens* Baly, 1890 [# 287]
amala Basu & Saha, 1977: 381 [DA] **A:** ID (SD) **[# 316]**
angulosa Solsky, 1871: 262 (*Hispa*) **A:** CH (ANH BEI FUJ GAN GUA GUI GUX HEB HEI HEN HUB HUN JIA JIL JIX LIA SCH SHA SHG SHN SHX TAI TIA YUN ZHE) JA NC RU (FE) SC
chujoi Shirôzu, 1957: 55
flavomarginata Shirôzu, 1957: 53
japonica Baly, 1874: 215 (*Hispa*)
masonii Gestro, 1923: 9
rufescens Shirôzu, 1957: 55
ussurina Uhmann, 1928b: 35
asoka Maulik, 1919a: 218 **A:** BT CH (YUN) ID (SD) NP **ORR**
bhaumiki Basu & Saha, 1977: 380 [DA] **A:** ID (SD) **[# 316]**
brevispina S.-H. Chen & Tan, 1964: 417 , 424 **A:** CH (ZHE)
brevispinosa brevispinosa Chapuis, 1877c: 56 (*Hispa*) **A:** BT CH (XIZ) ID (AP HP SD UT) NP PA **ORR**

srnkae Weise, 1897d: 133
peregrina Maulik, 1919a: 184
brevispinosa yunnana S.-H. Chen & Tan, 1961: 465, 479 (ssp.) **A:** CH (YUN)
carinata S.-H. Chen & Tan, 1961: 472, 480 **A:** CH (YUN) [# 308]
cervicornis Gressitt, 1950: 113 **A:** CH (FUJ GUA GUI ZHE)
chaturanga Maulik, 1919a: 237 **A:** CH (GUA GUX HKG HUN) **ORR**
chinensis Weise, 1904e: 102 **A:** CH (FUJ GUA GUI GUX HAI HUB HUN JIX SCH TAI YUN ZHE) ID (AP SD UT) NP PA **ORR**
bindusara Maulik, 1919a: 193
gestroi Gressitt, 1938b: 327
insulicola Chûjô, 1933b: 326
confluens Baly, 1890b: 490 (*Hispa*) **A:** CH (YUN) ID (UT) NP **ORR**
vestita Maulik, 1919a: 214
crassicuspis Gestro, 1908: 184 **A:** CH (FUJ GUA GUI HKG HUB HUN JIX SCH SHA SHX YUN)
fleutiauxii Gestro, 1923: 12 **A:** CH (GUX YUN) **ORR**
fumida S.-H. Chen & Tan, 1964: 417, 424 **A:** CH (YUN)
gonospila Gestro, 1897c: 130 (*Hispa*) **A:** CH (YUN) **ORR**
gressitti Uhmann, 1955a: 205 **A:** CH (FUJ)
harsha Maulik, 1919a: 188 **A:** ID (UT) NP **ORR**
higoniae higoniae Lewis, 1896: 342 (*Hispa*) **A:** CH (FUJ GUA GUI GUX HAI HUN JIX SCH TAI XIZ YUN ZHE) JA NP **ORR**
corpulentina Uhmann, 1927b: 113
higoniae szechuanensis S.-H. Chen & Tan, 1961: 472, 480 (ssp.) **A:** CH (SCH)
intermedia S.-H. Chen & Tan, 1961: 473, 480 **A:** CH (FUJ GUX YUN)
jonathani Basu & Saha, 1977: 379 [DA] **A:** ID (SD UT) **ORR** **[# 316]**
kambaitica Uhmann, 1939a: 6 **A:** CH (YUN) **ORR**
klapperichi Uhmann, 1955a: 192 **A:** CH (FUJ GUX HUB JIX TAI YUN)
koreana An, Kwon & S. M. Lee, 1985: 5 **A:** SC
kumatai Kimoto, 1996: 50 **A:** CH (TAI)
leonardi Ritsema, 1876: 48 (*Hispa*) **A:** NP **ORR**
maculithorax Gestro, 1908: 189 **A:** CH (FUJ GUA GUI GUX HAI HUN JIX SCH YUN)
sinuispina Gressitt, 1938b: 331
multifida Gestro, 1890: 263 (*Hispa*) **A:** CH (YUN) **ORR**
nigrodiscalis Gressitt, 1938b: 329 **A:** CH (FUJ GUA GUX HAI JIX YUN)
omeia S.-H. Chen & Tan, 1961: 473, 481 **A:** CH (SCH) [# 308]
pallidicollis Gressitt, 1938b: 330 **A:** CH (HAI)
papilla Tan, 1993b: 353, 355 **A:** CH (GUI)
parbatya Maulik, 1919a: 234 **A:** CH (FUJ YUN) ID (SD UT) NP **ORR**
fabricii L. N. Medvedev, 2009b: 410 [incorrect original spelling] **[# 294, # 305]**

fabriziae L. N. Medvedev, 2009b: 410 [JE] [# 305]
paucispina Gressitt, 1939a: 178 **A:** CH (FUJ GUA GUX JIX SCH) **[# 446]**
pici borealis L. N. Medvedev, 1973b: 884 (ssp.) **A:** RU (FE)
pici pici Uhmann, 1934: 28 **A:** CH (FUJ GUA GUX JIX YUN) **ORR**
marginicollis Gressitt, 1939b: 203
protuberance Tan, 1982b: 39, 41 **A:** CH (GUX)
pungens Boheman, 1859: 202 (*Hispa*) **A:** CH (GUA HAI HKG JIX) **ORR**
atracumina Gressitt, 1938b: 326
quinquespina Tan, 1982b: 40, 41 **A:** CH (YUN)
ramuligera Chapuis, 1877c: 53 (*Hispa*) **A:** CH (JIX YUN) **ORR**
reitteri Spaeth, 1933b: 136 **A:** CH (JIX SCH)
scutellaris S.-H. Chen & Tan, 1961: 474, 481 **A:** CH (YUN)
serrulata S.-H. Chen & Tan, 1964: 419, 425 **A:** CH (HEI JIA TAI)
shira Basu & Saha, 1977: 382 [DA] **A:** ID (SD) **[# 316]**
spectabilis Gestro, 1914: 75 **A:** CH (GUX HUB YUN)
sternalis S.-H. Chen & Tan, 1964: 418, 424 **A:** CH (GUX)
stoetzneri diannana S.-H. Chen & Tan, 1986: 328, 612 (ssp.) [RN] **A:** CH (YUN)
yunnana S.-H. Chen & Tan, 1964: 419, 425 (ssp.) [HN]
stoetzneri stoetzneri Uhmann, 1955a: 204 **A:** CH (FUJ JIX)
taiwana Takizawa, 1978b: 598 **A:** CH (TAI)
tientaina S.-H. Chen & Tan, 1964: 419, 426 **A:** CH (ZHE)
uhmanni Gressitt, 1950: 132 **A:** CH (FUJ GUI GUX HUB SCH YUN)

genus *Dactylispa*, nomen dubium
maculosa Fairmaire, 1889b: 84 (*Hispa*) **A:** CH (JIX) RU (FE) **[# 312]**

genus *Dicladispa* Gestro, 1897a: 81 type species: *Hispa testacea* Linnaeus, 1767
Cirrispa Uhmann, 1936: 123 type species: *Cirrispa conradsi* Uhmann, 1936 (= *Hispa comata* Weise, 1922)
armigera Olivier, 1808: 763 (*Hispa*) **A:** CH (FUJ GUA GUI GUX HAI HEN HUB HUN JIA JIX LIA SCH SHA SHG TAI YUN ZHE) ID (SD) JA NP PA **ORR**
aenescens Baly, 1887: 412 (*Hispa*)
boutani Weise, 1904e: 101 (*Hispa*) **[# 295]**
cyanipennis Motschulsky, 1860b: 238 (*Hispa*)
semicyanea Pic, 1932g: 26 (*Hispa*)
similis Uhmann, 1927b: 116 (*Hispa*) **[# 295]**
yunnanica S.-H. Chen & Sun, 1962: 130, 137 (ssp.) **[# 295]**
yunusi Abdullah & Qureshi, 1969a: 103 (*Hispa*)
comata Weise, 1922a: 84 (*Hispa*) **N:** EG **A:** SY **AFR**
conradsi Uhmann, 1936: 123 (*Cirrispa*)

dama Chapuis, 1877c: 52 (*Hispa*) **A:** ID (UT) **ORR**

abdominalis Baly, 1888c: 664 (*Hispa*)

iranica Lopatin, 1984a: 91 **A:** IN

megacantha Gestro, 1890: 249 (*Hispa*) **A:** BT CH (YUN) ID (SD) **ORR**

birendra Maulik, 1919a: 250 (*Hispa*)

occator Brullé, 1839: 73 (*Hispa*) **N:** CI

pallescens Guérin-Méneville, 1841: 13 (*Hispa*) **N:** EG **A:** PA SA YE **AFR ORR**

holtzi Uhmann, 1928c: 88 (*Hispa*)

pallida Guérin-Méneville, 1841: 13 (*Hispa*)

pavida Weise, 1902b: 234 (*Hispa*)

testacea Linnaeus, 1767: 603 (*Hispa*) **E:** AL BU CR FR GBi GR IT ME PT SP SZ TR **N:** AG CI EG LB MO TU **A:** CY IS JO LE SY TR [# 382]

algeriana Guérin-Méneville, 1841: 12 (*Hispa*)

numida Guérin-Méneville, 1841: 14 (*Hispa*)

genus *Hispa* **Linnaeus, 1767: 603** type species: *Hispa atra* Linnaeus, 1767

Hispella Chapuis, 1875: 334 type species: *Hispa atra* Linnaeus, 1767

andrewesi Weise, 1897d: 126 (*Hispella*) **A:** CH (ANH GUA GUX HAI SCH XIZ YUN) ID (UT) NP **ORR**

donckieri Pic, 1924g: 28

singhalensis Maulik, 1919a: 161 (*Hispella*, var.)

atra Linnaeus, 1767: 603 **E:** AL AN AU BE BH BU BY CR CZ DE FR GE GR HU IT KO KR LS LT LU MD ME NL NM PL PT RO RU (CT ST) SB SK SL SP SV SZ TR UK **N:** AG MO TU **A:** AB AF AR CH (HEB HEI SHX XIN) GG IN IS KI KZ MG RU (ES SS WS) SY TD TR UZ

algira Motschulsky, 1860b: 238

aptera Motschulsky, 1860b: 238

balbii Donckier de Donceel, 1900: 602 [RN]

cariosa Reiche & Saulcy, 1858b: 54

minuta Balbi, 1892b: 195 [HN]

brachycera Gestro, 1897c: 123 (*Hispella*) **A:** ID (HP) PA **ORR**

echinata S.-H. Chen & Sun, 1964: 115, 121 **A:** CH (YUN)

ramosa Gyllenhal, 1817: 6 **A:** AF ID (HP UT) NP **ORR** [# 452]

ceylonica Motschulsky, 1860b: 238

tarsata Świętojańska, 2001b: 479 **A:** IN

genus *Hispellinus* **Weise, 1897d: 144** type species: *Hispa callicantha* Bates, 1866 [# 285]

Monochirus Chapuis, 1875: 330 [HN] type species: *Hispa multispinosa* Germar, 1848

callicanthus Bates, 1866: 354 (*Hispa*) **A:** CH (ANH FUJ GUA GUI GUX HAI HUB HUN JIA JIX TAI YUN) PA **ORR**

moestus Baly, 1888c: 662 (*Monochirus*)

chinensis Gressitt, 1950: 96 **A:** CH (GUA HUN JIX SCH SHN) NC SC
formosanus Uhmann, 1927b: 111 (*Monochirus*) **A:** CH (FUJ TAI) **ORR**
minor Maulik, 1919a: 155 (*Monochirus*) **A:** ID (AP) **ORR**
moerens Baly, 1874: 215 (*Hispa*) **A:** CH (HEB HEI JIA JIX LIA SHG SHN TAI YUN) JA RU (FE) SC

genus ***Jambhala*** **Würmli, 1975: 377** type species: *Jambhala nekula* Würmli, 1975
nekula Würmli, 1975: 377 **A:** BT NP

genus ***Monohispa*** **Weise, 1897d: 147** type species: *Hispa singularis* Gestro, 1888
filicornis Motschulsky, 1860b: 239 (*Hispa*) [DA] **A:** NP **ORR** **[# 318]**
singularis Gestro, 1888b: 179 (*Hispa*) **A:** CH (GUA YUN) **ORR**
tuberculata Gressitt, 1950: 131 (*Dactylispa*)

genus ***Platypria*** **Guérin-Méneville, 1840b: 139** type species: *Hispa echidna* Guérin-Méneville, 1840
Lobacantha Kirby, 1837: 227 [suppressed name, ICZN (1955), Opinion 376] type species: *Hispa erinacea* Fabricius, 1801
subgenus *Dichirispa* Gestro, 1890: 229 type species: *Platypria abdominalis* Chapuis, 1877
arabica L. N. Medvedev, 2012e: 418 **A:** YE
subgenus *Platypria* Guérin-Méneville, 1840b: 139 type species: *Hispa echidna* Guérin-Méneville, 1840
acanthion Gestro, 1890: 245 **A:** CH (YUN) **ORR**
erethizon Gestro, 1897b: 112
aliena S.-H. Chen & Sun, 1962: 130, 138 **A:** CH (YUN)
chiroptera Gestro, 1899a: 172 **A:** BT CH (YUN) ID (SD) **ORR**
garthwaitei Bhasin, 1943: 9
echidna Guérin-Méneville, 1840b: 139 (*Hispa*) **A:** CH (YUN) ID (SD) JA NP PA **ORR** **[# 408]**
erinacea Fabricius, 1801b: 59 (*Hispa*) **A:** ID (UT) NP PA **ORR**
andrewesi Weise, 1906b: 404
bengalensis Gestro, 1897b: 112 (var.)
hystrix Fabricius, 1798: 116 (*Hispa*) **A:** BT CH (GUX HAI JIX SCH YUN) ID (HP SD UT) NP PA **ORR** **[# 466]**
alces Gressitt, 1938b: 334 **[# 296]**
digitata Gestro, 1888b: 178
echinogale Gestro, 1897b: 115 [= 1897d: 71]
ericulus Gestro, 1890: 247
latenigra Pic, 1927k: 26 (var.)

sumatrensis Csiki, 1900: 196

testaceicollis Heller, 1916: 116 (ssp.)

melli Uhmann, 1955a: 211 **A:** CH (ANH FUJ GUA GUX HUB HUN JIX ZHE) JA

paracanthion S.-H. Chen & Sun, 1962: 131, 138 **A:** CH (YUN)

parva S.-H. Chen & Sun, 1964: 115, 121 **A:** CH (YUN)

yunnana Gressitt, 1939a: 183 **A:** CH (YUN)

genus ***Polyconia*** **Weise, 1905f: 318** type species: *Hispella spinicornis* Kraatz, 1895

caroli Leprieur, 1883: 115 (*Hispa*) E: IT (Sicily) SP **N:** AG

genus ***Rhadinosa*** **Weise, 1905f: 318** type species: *Hispa nigrocyanea* Motschulsky, 1860

fleutiauxi Baly, 1890b: 491 (*Hispa*) **A:** CH (FUJ GUA GUX HEN HUB HUN JIX YUN) **ORR**

girija Maulik, 1915: 377 **A:** NP

laghua Maulik, 1915: 376 **A:** ID (SD UT) NP PA **ORR** [# 307]

lebongensis Maulik, 1919a: 168 **A:** CH (FUJ JIX YUN) ID (HP SD UT) NP **ORR**

nigrocyanea Motschulsky, 1860b: 237 (*Hispa*) **A:** CH (ANH FUJ GUA HEB HEI HUB JIA JIX NMO SCH SHX XIN YUN ZHE) JA NC RU (FE) SC

parvula Motschulsky, 1860b: 238 (*Hispa*) **A:** BT **ORR**

nigritula Boheman, 1859: 203 (*Hispa*)

reticulata Baly, 1888c: 665 (*Hispa*) **A:** NP **ORR**

yunnanica S.-H. Chen & Sun, 1962: 129, 137 **A:** CH (YUN)

genus ***Rhoptrispa*** **S.-H. Chen & Tan, 1964: 414, 422** type species: *Dactylispa luhi* Uhmann, 1951 (= *Hispa dilaticornis* Duvivier, 1891)

clavata Weise, 1902b: 235 **A:** SA YE **AFR**

clavicornis S.-H. Chen & Tan, 1964: 415, 422 (*Dactylispa*) **A:** CH (YUN)

dilaticornis Duvivier, 1891a: xlviii (*Hispa*) **A:** AE AF BT CH (BEI GUA SHN TAI YUN) ID (UT) IN NP PA **AFR ORR** [# 467]

abnormis Gressitt & Kimoto, 1963a: 917 (*Rhadinosa*)

arisana Chûjô, 1933b: 324 (*Dactylispa*)

assamensis Weise, 1904d: 451 (*Dactylispa*)

luhi Uhmann, 1951b: 19 (*Dactylispa*)

perpusilla Gestro, 1911a: 28 (*Dactylispa*)

pusilla Weise, 1904e: 103 (*Dactylispa*)

genus ***Sinispa*** **Uhmann, 1938: 224** type species: *Sinispa yunnana* Uhmann, 1938 [# 283]

tayana Gressitt, 1939a: 172 (*Rhadinosa*) **A:** CH (HAI: Taya Is.) [# 469]

yunnana Uhmann, 1938: 224 **A:** CH (YUN)

tribe Leptispini Fairmaire, 1868

genus *Leptispa* **Baly, 1859a: 1** type species: *Leptomorpha filiformis* Germar, 1842

Lepthispa Weise, 1911b: 44 [UE] type species: *Leptomorpha filiformis* Germar, 1842

Leptomorpha Germar, 1842: 10 [HN] type species: *Leptomorpha filiformis* Germar, 1842

Paradownesia Gestro, 1899c: 220 type species: *Downesia longipennis* Gestro, 1890

Parallelispa Fairmaire, 1885b: 238 type species: *Parallelispa quadratricollis* Fairmaire, 1885

abdominalis abdominalis Baly, 1859a: 3 **A:** CH (BEI FUJ GUA HKG JIX MAC TAI: Kinmen Is. ZHE) [# 456]

atripes Pic, 1925a: 16

conicicollis Voronova & Zaytsev, 1982: 121

abdominalis meridiana S.-H. Chen & P.-Y. Yu, 1964: 109, 118 (ssp.) **A:** CH (GUA GUX HAI)

allardi Baly, 1890b: 489 **A:** CH (YUN) **ORR**

oberthurii Gestro, 1899c: 221 (*Downesia*)

bengalensis Takizawa, 1989a: 330 **A:** ID (SD)

collaris S.-H. Chen & P.-Y. Yu, 1962: 464, 477 **A:** CH (GUX YUN)

filiformis Germar, 1842: 10 (*Leptomorpha*) E: GR IT PT SP **N:** AG MO TU [# 471]

filiformis Guérin-Méneville, 1844: 277 (*Leptomorpha*) [HN]

godwini Baly, 1869: 364 **A:** CH (FUJ HUB HUN JIX SHG YUN ZHE) **ORR** [# 457]

fruhstorferi Gestro, 1906: 483 (*Paradownesia*)

impressa Uhmann, 1939a: 3 **A:** CH (YUN) **ORR**

kanistha Basu, 1999: 159 **A:** ID (SD)

krishna Basu, 1999: 161 **A:** ID (SD)

longipennis Gestro, 1890: 243 (*Downesia*) **A:** BT CH (FUJ GUA GUX) **ORR**

magna S.-H. Chen & P.-Y. Yu, 1962: 464, 477 **A:** CH (YUN)

miwai Chûjô, 1933b: 310 **A:** CH (TAI)

bicolor Chûjô, 1933b: 311 [HN]

formosana Chûjô, 1934b: 249 [RN]

miyamotoi Kimoto, 1957: 77 **A:** JA (Amami-Ôshima)

parallela parallela Gestro, 1899c: 220 (*Downesia*) **A:** ID (SD)

parallela yunnana S.-H. Chen & P.-Y. Yu, 1964: 109, 118 (ssp.) **A:** CH (YUN)

pici Uhmann, 1958: 188 [RN] **A:** CH (GUX) **ORR** [# 458]

bicoloripennis Voronova & Zaytsev, 1982: 123

perroti Pic, 1939b: 17 [HN] (*Lepthispa*)

pygmaea Baly, 1859a: 2 **A:** CH (FUJ GUX JIX YUN ZHE) NP PA **ORR**

viridis Gressitt, 1950: 76

rufithorax Maulik, 1919a: 78 **A:** ID (UT) PA **ORR**

taguchii Chûjô, 1956c: 18 **A:** JA

genus *Ovotispa* **L. N. Medvedev, 1992c: 149** type species: *Leptispa atricolor* Pic, 1928
atricolor Pic, 1928g: 23 **A:** CH (YUN) **ORR**
angustior Pic, 1951b: 11 (*Lepthispa*)

tribe Notosacanthini Gressitt, 1952 (1929)

genus *Notosacantha* **Chevrolat, 1836: 367** type species: *Cassida echinata* Fabricius, 1801
Festa Gistel, 1848b: xi [RN] type species: *Cassida echinata Fabricius*, 1801
Hoplionota Hope, 1841: 152 type species: *Cassida echinata* Fabricius, 1801
Notocantha Agassiz, 1846b: 252 [UE] type species: *Cassida echinata* Fabricius, 1801
arisana Chûjô, 1934c: 149 (*Hoplionota*) **A:** CH (TAI)
castanea Spaeth, 1913b: 507 (*Hoplionota*) **A:** CH (GUX HAI TAI YUN) **ORR**
centinodia Spaeth, 1913b: 514 (*Hoplionota*) **A:** CH (YUN XIZ) **ORR**
circumdata Wagener, 1881: 17 (*Hoplionota*) **A:** CH (YUN) **ORR**
darjeelingensis Borowiec & Takizawa, 1991: 646 **A:** ID (SD) **[# 461]**
fumida Spaeth, 1913b: 513 (*Hoplionota*) **A:** CH ("South China") **[# 462]**
ginpinensis S.-H. Chen & Zia, 1961: 441, 447 **A:** CH (YUN)
ihai Chûjô, 1958c: 19 (ssp.) **A:** JA (Akusekijima, Amami-Ôshima, Ishigakijima, Okinawa, Tokara Isls)
jammuensis Borowiec & Takizawa, 1991: 647 **A:** ID (KA)
lenta Spaeth, 1913b: 519 (*Hoplionota*) **A:** ID (AP) **ORR**
loochooana Chûjô, 1961: 91 (ssp.) **A:** JA (Amami-Ôshima Is.)
maculipennis Boheman, 1857: 5 (*Hoplionota*) **A:** BT ID (AP SD UT) NP PA **ORR** **[# 394]**
moderata S.-H. Chen & Zia, 1964: 122, 132 **A:** CH (GUA) **ORR**
nigrodorsata S.-H. Chen & Zia, 1961: 442, 448 **A:** CH (YUN) **ORR**
nishiyamai Komiya, 2002: 215 **A:** JA (Okinawa-Hontô Is.)
oblongopunctata Gressitt, 1938d: 573 (*Hoplionota*) **A:** CH (HAI YUN) **ORR**
pakistanica Borowiec, Sultan & Rafi, 2008: 288 **A:** PA
sauteri Spaeth, 1914f: 295 (*Hoplionota*) **A:** CH (FUJ GUA GUX JIX TAI YUN ZHE) **ORR**
marginalis Gressitt, 1942d: 1 (*Hoplionota*)
vietnamica L. N. Medvedev, 1992c: 154
shibatai Kimoto, 1981b: 3 **A:** CH (TAI)
shishona S.-H. Chen & Zia, 1964: 122, 132 **A:** CH (YUN) **ORR**
sinica Gressitt, 1952: 449 **A:** CH (FUJ HKG ZHE)
tenuicula Spaeth, 1913b: 505 (*Hoplionota*) **A:** CH (YUN) ID (SD) NP **ORR**
nepalensis Borowiec & Takizawa, 1991: 650
trituberculata Gressitt, 1952: 450 **A:** CH (GUX TAI)
warchalowskii Świętojańska & Sekerka, 2007: 681 **A:** ID (AP)

tribe Oncocephalini Chapuis, 1875

genus *Chaeridiona* **Baly, 1869: 380** type species: *Chaeridiona metallica* Baly, 1869

Choeridiona Gemminger & Harold, 1876: 3609 [UE] type species: *Chaeridiona metallica* Baly, 1869

cupreoviridis Gressitt, 1950: 77 (*Choeridiona*) **A:** CH (FUJ) **[# 441]**

picea Baly, 1869: 382 **A:** PA **ORR**

pseudometallica Basu, 1999: 163 **A:** ID (SD)

genus *Chaeridiona*, taxon excluded from Palaearctic fauna

metallica Baly, 1869: 381 **ORR** **[# 321]**

genus *Oncocephala* **Agassiz, 1846b: 259** type species: *Onchocephala quadrilobata* Guérin-Méneville, 1844 **[# 306]**

Nepius J. Thomson, 1858: 225 type species: *Nepius corrosus* J. Thomson, 1858 (= *Onchocephala senegalensis* Guérin-Méneville, 1844)

Onchocephala Guérin-Méneville, 1844: 280 type species: *Onchocephala quadrilobata* Guérin-Méneville, 1844 [# 306]

atratangula Gressitt, 1938b: 324 (*Oncocephalus*) **A:** CH (GUX HAI YUN) **[# 463]**

formosana Chûjô, 1933b: 312 **A:** CH (TAI)

grandis S.-H. Chen & P.-Y. Yu, 1962: 128, 136 **A:** CH (YUN)

hemicyclica S.-H. Chen & P.-Y. Yu, 1962: 128, 137 **A:** CH (GUX YUN)

quadrilobata Guérin-Méneville, 1844: 281 (*Onchocephala*) **A:** CH (GUX YUN) **ORR**

tuberculata Olivier, 1793: 99 (*Hispa*) **A:** NP PA **ORR**

weisei Gestro, 1899b: 324 **A:** CH (YUN) **ORR**

yunnanica S.-H. Chen & P.-Y. Yu, 1962: 129, 137 (ssp.)

genus *Prionispa* **Chapuis, 1875: 337** type species: *Prionispa subopaca* Chapuis, 1875

champaka Maulik, 1919a: 94 **A:** CH (JIX YUN) **ORR**

cheni Staines, 2007: 20 [RN] **A:** CH (YUN)

tuberculata S.-H. Chen & P.-Y. Yu, 1964: 114, 121 (*Chaeridiona*) [HN]

clavata P.-Y. Yu, 1992b: 609, 610 (*Chaeridiona*) **A:** CH (ḤUN)

dentata Pic, 1938g: 354 **A:** CH (YUN) **ORR**

himalayensis Maulik, 1915: 371 **A:** ID (SD) **ORR**

cuneata Uhmann, 1954: 510

houjayi C.-F. Lee, Świętojańska & Staines, 2009: 559 **A:** CH (TAI)

laeta L. N. Medvedev, 1990b: 43 **A:** ID (SD) NP

longicornis Gestro, 1906: 485 **A:** ID (SD) **ORR**

opacipennis S.-H. Chen & P.-Y. Yu, 1962: 127, 136 **A:** CH (YUN)

sinica Gressitt, 1950: 78 **A:** CH (FUJ GUX)

subfamily **Chrysomelinae** Latreille, 1802
tribe Chrysomelini Latreille, 1802 [# 472]
subtribe Chrysomelina Latreille, 1802

genus *Agasta* Hope, 1841: 177 type species: *Agasta formosa* Hope, 1841

formosa Hope, 1841: 177 **A**: CH (GUA GUX YUN) ID (SD UT) NP **ORR**

fleutiauxi Laboissière, 1932b: 130

genus *Chrysomela* Linnaeus, 1758: 368 type species: *Chrysomela populi* Linnaeus, 1758

Eleia Gistel, 1848a: viii, xiv [RN] type species: *Chrysomela populi* Linnaeus, 1758

Ernobia Gistel, 1856: 379 type species: *Chrysomela populi* Linnaeus, 1758

Gymnota Gistel, 1848a: 403 [= 1848b: 123] [HN] type species: *Chrysomela populi* Linnaeus, 1758

Lina Latreille, 1829: 151 type species: *Chrysomela populi* Linnaeus, 1758

Macrolina Motschulsky, 1860b: 198 type species: *Coccinella vigintipunctata* Scopoli, 1763

Melasoma Dillwyn, 1829: 56 type species: *Chrysomela populi* Linnaeus, 1758

Microdera Stephens, 1839: 307 [HN] type species: *Coccinella vigintipunctata* Scopoli, 1763

Microlina Lopatin, 1977b: 166 [UN] [# 579]

Pachylina L. N. Medvedev & Chernov, 1969: 537 type species: *Chrysomela collaris* Linnaeus, 1758

Strickerus R. Lucas, 1920: 413 [RN] type species: *Coccinella vigintipunctata* Scopoli, 1763

collaris alpina Zetterstedt, 1828: 414 **E**: FI NR SV [# 518, # 992]

connexa Weise, 1921: 4 (*Melasoma*, ab.) [IN]

collaris blaisdelli Van Dyke, 1938: 48 (*Chrysolina*) **A**: RU (ES: northeastern Yakutia FE: Chukotka) **NAR** [# 518]

engelhardti Hatch, 1939: 49 (*Chrysolina*)

hyperborea L. N. Medvedev, 2011b: 579 (ssp.) [# 518, # 547]

collaris collaris Linnaeus, 1758: 371 **E**: AL AU BH BU BY CR CZ DE EN FI FR GE GR HU IT KO LA LT NL NM NR PL RO RU (CT NT) SB SK SV SZ UK **A**: CH (HEI) KZ MG RU (ES FE SS WS) TR [# 518]

alnetorum Gistel, 1857b: 520 (*Lina*)

custodians Gistel, 1857b: 542 (*Lina*)

cyanipennis Csiki, 1953: 130 (*Melasoma*, ab.) [IN]

daurica Motschulsky, 1860b: 224 (*Lina*)

escheri Heer, 1836: 31

geniculata Stierlin, 1897: 578 (*Lina*, var.) [# 1107]

salicis Fabricius, 1792a: 319

thoracica Weise, 1884d: 560 (*Melasoma*, var.) [IN] [# 577]

cuprea Fabricius, 1775: 96 **E**: AU BH BU BY CR CZ EN FI FR GE GR HU IT LA LT LU ME NL NR PL RO RU (CT NT) SB SK SP SV SZ UK **A**: IN NC PA RU (ES FE SS WS) SC [# 583, # 1145]

coeruleovirens Gabriel, 1928: 8 (*Melasoma*, ab.) [IN]
cupreobrunnea Csiki, 1953: 130 (*Melasoma*, ab.) [IN]
humeralis Eschscholtz, 1818: 465
ruficaudis DeGeer, 1775: 305
sarmatica Weise, 1887h: 187 (*Melasoma*, var.)

lapponica Linnaeus, 1758: 371 **E:** AU BH BU BY CR CZ EN FI FR GE HU IT LA LT NL NR PL RO RU (CT NT ST) SB SK SL SP SV SZ UK **A:** CH (JIL HEI NMO XIN) JA KI KZ MG NC RU (ES FE SS WS)

altaica Weise, 1884d: 557 (*Melasoma*, var.) [IN] [# 577]
ammosovi L. N. Medvedev, 1978e: 121 (ab.) [IN]
bimaculata Gradl, 1881: 305 (*Melasoma*, var.) [IN] [# 574]
bipustulata Gradl, 1881: 305 (*Melasoma*, var.) [IN] [# 574]
bulgharensis Fabricius, 1798: 88 (*Chrysomela*)
coerulea Gradl, 1881: 306 (*Melasoma*, var.) [IN] [# 574]
cruciata Jacobson, 1901b: 133 (*Melasoma*, var.) [IN] [# 577]
curvilinea DeGeer, 1775: 302
cyaneoviridis Gruev, 1994: 77 [# 548]
litua Marseul, 1888a: 13 [= 311] (*Lina*, var.)
marginata Gradl, 1881: 306 (*Melasoma*, var.) [IN] [# 574]
multipunctata Jacobson, 1901b: 133 (*Melasoma*, var.) [IN] [# 577]
quadrimaculata Gradl, 1881: 305 (*Melasoma*, var.) [IN] [# 574]
quadrinotata Gradl, 1881: 305 (*Melasoma*, var.) [IN] [# 574]
quadripunctata Suffrian, 1858b: 389 (*Chrysomela*)
quadripunctata Jacobson, 1901b: 133 (*Melasoma*, var.) [IN] [# 577]
quadripunctata Lengerken, 1913: 124 (*Melasoma*, forma) [HN]
quadripustulata Jacobson, 1901a: 90 (*Melasoma*, var.) [IN] [# 577]
rubrocoerulea Gradl, 1881: 305 (*Melasoma*, var.) [IN] [# 574]
sexmaculata Gradl, 1881: 305 (*Melasoma*, var.) [IN] [# 574]
unicolor Marseul, 1888a: 13 (*Lina*, var.)
violaceipennis Jacobson, 1901b: 133 (*Melasoma*, var.) [IN] [# 577]

populi Linnaeus, 1758: 370 **E:** AL AU BE BH BU BY CR CZ DE EN FI FR GB GE GR HU IT KR LS LA LT LU MD ME NL NM NR PL PT RO RU (CT NT ST) SB SL SK SP SZ TR UK **A:** AB AF AR CH (AHN BEI CHQ FUJ GAN GUA GUI GUX HEB HEI HUB HUN JIA JIL JIX QIN LIA NIN NMO SCH SHA SHN SHX XIN XIZ YUN ZHE) GG ID (AP SD UT) IN JA MG NC NP PA RU (ES FE SS WS) SC TR [# 1145]

asiatica Jakob, 1952b: 105 (*Melasoma*, ssp.)
janaceki Reitter, 1892b: 240 (*Melasoma*, var.)
kitaica Jakob, 1952b: 105 (*Melasoma*, ssp.)
nigricollis Jakob, 1952b: 105 (*Melasoma*, ssp.)
parvicollis Jakob, 1955: 231 (*Melasoma*)

rufina Brancsik, 1914: 61 (*Melasoma*, ab.) [IN]

violaceicollis Bechyné, 1954b: 671 (ssp.)

saliceti afghanica Reineck, 1937: 186 (*Melasoma*, ssp.) **A:** AF

saliceti quadricollis Jakob, 1958b: 38 (*Melasoma*) [QS] **A:** "Central Asia" [# 517]

saliceti saliceti Weise, 1884d: 565 (*Melasoma*) **E:** AL AU BE BH BU BY CR CZ DE EN FI FR GE GR HU IT KO KR LA LS LT MD NL PL RU (CT NT ST) SB SK SL SP SV SZ UK **A:** CH (NMO XIN) IN KZ MG RU (ES FE SS WS) TR

bouvieri Achard, 1916b: 5 (*Melasoma*)

nigripennis Reitter, 1894b: 192 (*Melasoma*, var.)

saliceti turcestanica Reineck, 1937: 187 (*Melasoma*, ssp.) [QS] **A:** AF CH (XIN) KI KZ (South) TD UZ [# 517, # 584]

salicivorax Fairmaire, 1888c: 40 (*Lina*) **A:** CH (ANH BEI CHQ GAN GUI HEB HEN JIX LIA NIN SHA SCH SHN ZHE) SC

completa Jacobson, 1903: 15 (*Melasoma*, var.)

deflorata Jacobson, 1903: 15 (*Melasoma*, var.)

hajeki Breit, 1913: 297 (*Melasoma*)

imperfecta Breit, 1913: 298 (*Melasoma*, ab.) [IN]

octodecimpunctata Jacoby, 1888: 346 (*Melasoma*) [HN]

succinea Breit, 1913: 298 (*Melasoma*, ab.) [IN]

taimyrensis L. N. Medvedev, 1969: 534 **A:** RU (ES: Taymyr peninsula) [# 518]

tremula selengensis Jakob, 1953a: 35 (*Melasoma*) [QS] **A:** RU (ES) [# 517]

tremula tremula Fabricius, 1787: 69 **E:** AL AU BE BU BY CR CZ DE EN FI FR GB GE GR HU IT KR LA LT MD NL NR PL RO RU (CT NT ST) SB SP SK SL SV SZ TR UK **A:** AF CH (ANH BEI GUI HEB HEI JIL LIA NMO QIN SCH YUN XIZ ZHE) IN JA KZ MG NC PA RU (ES FE SS WS) SC TR

longicollis Suffrian, 1851a: 199 [QS]

vigintipunctata alticola S.-Y. Wang, 1992b: 635 (ssp.) **A:** CH (QIN SCH YUN)

vigintipunctata vigintipunctata Scopoli, 1763: 78 (*Coccinella*) **E:** AL AU BE BH BU BY CR CZ EN FR GE GR HU IT KZ LA LS LT LU MD ME NL NM PL RO RU (CT NT ST) SB SK SL SV SZ TR UK **A:** AB CH (ANH CHQ FUJ GAN GUI HEB HUB HUN JIL LIA NIN SCH SHA TAI YUN ZHE) GG IN JA NC RU (ES FE SS WS) SC TR [# 1145]

costella Marseul, 1888a: 10 [= 308] (*Lina*, var.)

diluta Marseul, 1888a: 10 [= 308] (*Lina*, var.)

incontaminata Heyden, 1887d: 263 (*Melasoma*, var.)

miniata Auel, 1909: 318 (*Melasoma*, var.)

octodecimpunctata Linnaeus, 1767: 582 (*Coccinella*)

pustulata Weise, 1884d: 559 (*Melasoma*, var.) [IN] [# 577]

viciesmaculata Bergsträsser, 1778: 87

wrangeliana L. N. Medvedev, 1973b: 881 **A:** RU (FE: Wrangel Island) [# 518]

genus *Chrysomela*, nomen dubium
Chrysomela betulae Linnaeus, 1758: 369 **A**: SV [# 542, # 565]

genus *Gastrolina* Baly, 1859c: 61 type species: *Gastrolina depressa* Baly, 1859
Linastica Motschulsky, 1860b: 200 type species: *Chrysomela peltoidea* Gebler, 1832
depressa Baly, 1859c: 61 **A**: CH (AHN FUJ GAN GUA GUI GUX HEB HEN HUB HUN JIA SCH SHA ZHE) JA
cupricollis Chûjô, 1941c: 459 (var.) [IN] [# 577]
immaculicollis S.-H. Chen, 1936c: 83 (var.)
pallipes S.-H. Chen, 1974b: 196, 197 (ssp.) **A**: CH (YUN)
peltoidea Gebler, 1832: 74 (*Chrysomela*) **A**: CH (HEI HUB SHX) JA RU (ES FE)
cimex Jacobson, 1901b: 134 (var.) [IN] [# 577]
japana Jacoby, 1885b: 210
thoracica Baly, 1864d: 228 **A**: CH (GAN HEB HEI JIL LIA SCH SHA) JA NC RU (FE) SC
tonkinea S.-H. Chen, 1931b: 128 **A**: CH (YUN) **ORR**

genus *Gastrolinoides* Chûjô & Kimoto, 1960: 7 type species: *Melasoma japonica* Harold, 1877
japonica Harold, 1877a: 362 (*Melasoma*) **A**: CH (HEI HEB HUB GAN SCH SHA) JA NC RU (FE) SC
nigripes Kimoto, 1969: 21 **A**: CH (TAI) JA

genus *Gastrophysa* Chevrolat, 1836: 405 type species: *Chrysomela polygoni* Linnaeus, 1758
subgenus *Exiguipenna* Jolivet, 1951a: 3 type species: *Chrysomela unicolor* sensu Jolivet (1951) (= *Gastrophysa janthina* Suffrian, 1851)
analis Reitter, 1890d: 396 (*Gastroidea*) **E**: BH ME **[# 604]**
janthina Suffrian, 1851a: 240 (*Chrysomela*) **E**: PT SP **[# 993]**
naceirii Graells, 1858a: 100 [1858b: 138] (*Colaphus*)
subgenus *Gastrophysa* Chevrolat, 1836: 405 type species: *Chrysomela polygoni* Linnaeus, 1758
Gasterophysia Gistel, 1856: 380 [UE] type species: *Chrysomela polygoni* Linnaeus, 1758
Gastroeidea Hope, 1841: 164 type species: *Chrysomela polygoni* Linnaeus, 1758
Gastroidea Agassiz, 1846b: 160 [UE] type species: *Chrysomela polygoni* Linnaeus, 1758
atrocyanea Motschulsky, 1860b: 222 **A**: CH (ANH BEI CHQ FUJ GAN HEB HEI HEN HUB HUN JIA JIX LIA NIN NMO QIN SCH SHA SHG TAI YUN ZHE) JA NC RU (ES FE) SC **ORR**
tonkinea Achard, 1923d: 134 (*Gastroidea*)

mannerheimi Stål, 1858: 252 (*Chrysomela*) **A:** CH (BEI HEI HEB JIL LIA NIN NMO XIN) JA MG RU (ES FE) **ORR**

amoena Weise, 1889g: 602 (*Gastroidea*)

popovi Marseul, 1888a: 78 [= 376]

suturalis Motschulsky, 1860b: 222

suturella Fairmaire, 1887d: 330 (*Colaspidema*)

polygoni Linnaeus, 1758: 370 (*Chrysomela*) **E:** AU BE BH BU BY CR CZ DE EN FI FR GB GE GR HU IT KO KR LA LS LT LU MD ME NL NM NR PL PT RO RU (CT NT ST) SB SK SL SP SV SZ TR UK **A:** AB AF AR CH (GAN GUA HEB HEI LIA NIN NMO XIN) GG IN KZ LE MG NC RU (ES FE SS WS) SC SY TR UZ **NAR** **[# 605]**

caeruleipennis Say, 1826: 296 (*Chrysomela*)

elongata Jolivet, 1951a: 13 (ssp.)

nigrocornis Jacobs, 1926: 166 (*Gastroidea*, ab.) [IN]

obtusa O. F. Müller, 1776: 82 (*Chrysomela*)

rubripes Philippi & Philippi, 1864: 390 (*Phaedon*)

ruficollis Fabricius, 1775: 820 (*Chrysomela*)

ruficeps Gebler, 1844: 106 **A:** CH (XIN)

viridula DeGeer, 1775: 311 (*Chrysomela*) **E:** AL AN AU BE BH BU BY CZ DE EN FI FR GB GE HU IT LA LS LT MD ME NL NR PL RO RU (CT NT ST) SB SK SL SP SV SZ UK **A:** AB AR CH (XIN) GG KZ MG RU (ES FE SS WS) TR **NAR** **[# 606, # 1145]**

caucasica Jolivet, 1951a: 22 (ssp.)

cyanescens Weise, 1898: 189 (*Gastroidea*, var.)

cyanicollis Mader, 1931: 20 (*Gastroidea*, ab.) [IN]

formosa Say, 1824a: 451 (*Chrysomela*)

fulgurans Achard, 1923d: 133 (*Gastroidea*, var.)

lenta Weise, 1887h: 174 (*Gastroidea*) [# 551]

pennina Weise, 1882b: 307 (*Gastroidea*, var.) [# 551]

raphani Herbst, 1783: 59 (*Chrysomela*)

rumicis Schrank, 1785: 318 (*Chrysomela*)

genus *Gastrophysa*, nomina dubia

amphibia Motschulsky, 1860b: 222 **A:** KI

duodecimpunctata Motschulsky, 1861b: 41 (*Colaphus*) **A:** RU (FE)

malvae Geoffroy, 1785: 109 (*Chrysomela*) **E:** FR

russica Fabricius, 1792a: 323 (*Chrysomela*) **E:** "Russia merid."

ruficollis Fabricius, 1787: 71 (*Chrysomela*) [HN]

virescens Gebler, 1842: 39 **A:** KZ

genus *Neophaedon* Jacobson, 1901a: 90 [NP] type species: *Chrysomela pyritosa* P. Rossi, 1792

Alitene Gistel, 1856: 380 [NO] type species: *Chrysomela pyritosa* P. Rossi, 1792

balangshanensis S.-Q. Ge, S.-Y. Wang & Yang, 2002a: 321, 324 (*Phaedon*) **A:** CH (SCH)

poneli Bergeal, 2001: 153 **E:** GR (Crete)

pseudopyritosus Codina Padilla, 1963b: 79 (*Phaedon*) **E:** SP **N:** MO

pusillus S.-Q. Ge & Daccordi, 2013a: 137 **A:** CH (GAN SCH)

pyritosus P. Rossi, 1792: 31 (*Chrysomela*) **E:** AU BE BH BU BY CR CZ FR GE GR HU IT KR ME NL NM PL RO RU (ST) SB SK SL SZ TR UK **A:** AB AF GG IN KZ TM TR UZ

alutaceus Fleischer, 1909c: 271 (*Phaedon*, var.)

coerulescens Weise, 1900b: 286 (*Phaedon*, var.)

graminicola Drapiez, 1819a: 53 (*Chrysomela*)

graminicola Duftschmid, 1825: 214 (*Chrysomela*) [HN]

graminis Panzer, 1799a: 4 (*Altica*)

segnis Weise, 1884d: 540 (*Phaedon*) **E:** AU BH CZ PL RO SK SL UK

deubeli Weise, 1900b: 285 (*Phaedon*) [QS]

obscurus Weise, 1884d: 540 (*Phaedon*, var.) [IN] [# 577]

pyritosus Duftschmid, 1825: 211 (*Chrysomela*) [HN]

szmolayi Marseul, 1889: 103 (*Phaedon*)

tatrensis Marseul, 1889: 102 (*Phaedon*)

tatricus Weise, 1916b: 124 [UE]

sichuanicus Lopatin, 2006c: 597 **A:** CH (SCH)

genus *Odontoedon* S.-Q. Ge & Daccordi, 2013b: 201 type species: *Phaedon fulvescens* Weise, 1922

chinensis Gressitt & Kimoto, 1963a: 335 (*Phaedon*) **A:** CH (CHQ HUB)

cupreus S.-Y. Wang, 1992a: 176, 177 (*Phaedon*)

fulgidus Ge & S.-Y. Wang, 2004b: 77 (*Phaedon*)

fulvescens Weise, 1922b: 55 (*Phaedon*) **A:** CH (GUA GUI CUX HUN JIX TAI YUN ZHE) **ORR**

globulus Daccordi & S.-Q. Ge, 2013a: 207 **A:** CH (GAN)

impressus Daccordi & Yang, 2013: 209 **A:** CH (SCH)

kippenbergi Daccordi & S.-Q. Ge, 2013a: 210 **A:** CH (SCH)

limbatus Lopatin, 2002a: 115 (*Phaedon*) **A:** CH (SCH)

lopatini S.-Q. Ge & Daccordi, 2013b: 212 **A:** CH (YUN)

maculicollis S.-H. Chen, 1974a: 45, 48 (*Phaedon*) **A:** CH (SCH)

potentillae S.-Y. Wang, 1992b: 633, 644 (*Phaedon*) **A:** CH (SCH)

rufulus S.-Q. Ge & Daccordi, 2013b: 216 **A:** CH (YUN)

sericeus Daccordi & S.-Q. Ge, 2013a: 217 **A:** CH (SCH)
sichuanus Daccordi & S.-Q. Ge, 2013a: 218 **A:** CH (SCH)
taiwanus S.-Q. Ge & Daccordi, 2013b: 220 **A:** CH (TAI)
thibetanus S.-Q. Ge & Yang, 2013: 220 **A:** CH (XIZ)

genus *Oreothassa* Jacobson, 1901a: 86 type species: *Oreothassa martjanowi* Jacobson, 1901
martjanowi Jacobson, 1901a: 88 **A:** KZ (Altai) RU (SS)
weisei Reitter, 1901d: 163 (*Prasocuris*) **A:** KZ (Altai) RU (SS)

genus *Phaedon* Latreille, 1829: 151 type species: *Chrysomela armoraciae* Linnaeus, 1758
subgenus *Hemiphaedon* Jacobson, 1901b: 130 type species: *Phaedon subtilis* Weise, 1900
magnificus Lopatin, 1985a: 47 **A:** KZ (Aral Sea region)
subtilis opacus Lopatin, 1976a: 114 (ssp.) **A:** KZ (south-east)
subtilis subtilis Weise, 1900b: 285 **A:** KI TD TM UZ **[# 614]**
subgenus *Paraphaedon* Sharp, 1910: 4 type species: *Chrysomela tumidula* Germar, 1823
tumidulus Germar, 1823: 588 (*Chrysomela*) **E:** FR GB SP
ignitus Reiche, 1864: 246 (*Phaedon*)
subgenus *Phaedon* Latreille, 1829: 151 type species: *Chrysomela armoraciae* Linnaeus, 1758
Emmetrus Motschulsky, 1860b: 221 type species: *Chrysomela betulae* sensu Fabricius (1775) (= *Chrysomela armoraciae* Linnaeus, 1758)
Poebia Gistel, 1856: 380 type species: *Chrysomela armoraciae* Linnaeus, 1758
armoraciae Linnaeus, 1758: 369 (*Chrysomela*) **E:** AU BE BH BU BY CR CZ DE EN FI FR GB GE GR HU IT LA LT ME NL NR PL PT RO RU (CT NT ST) SB SK SL SP SV SZ UK **A:** AB CH (BEI HEB NIN XIN) GG IN KZ MG RU (ES FE SS WS) TR **NARi** **[# 615]**
gomphoceros Stephens, 1831: 333 [QS]
nasturtii Gistel, 1857b: 531, 547 (*Alitene*)
parvulus Duftschmid, 1825: 213 (*Chrysomela*)
plantaginis DeGeer, 1775: 322 (*Chrysomela*)
veronicae Bedel, 1892b: 153
chaukiensis Sprecher-Uebersax & Daccordi, 2012: 416 **A:** NP
cochleariae cochleariae Fabricius, 1792a: 330 (*Chrysomela*) **E:** AL AU BE BH BU BY CR CZ DE EN FI FR GB GE HU IT KO KR LA LS LT MD NL NM NR PL RO RU (CT NT ST) SB SK SL SP SV SZ TR UK **A:** AB CH (FUJ GUX HEI HEN JIA SHA XIN YUN ZHE) GG IN JA KZ MG NC RU (ES FE SS WS) SC TR **N:** MO **ORR NARi** **[# 616, # 994]**
auratus Motschulsky, 1860b: 231 (*Emmetrus*)

baolacensis Achard, 1923e: 138
brassicae Baly, 1874: 174
egenus Gyllenhal, 1827: 649 (*Chrysomela*)
galeopsis Letzner, 1849: 97
hederae Suffrian, 1851a: 253 (*Chrysomela*)
incertus Baly, 1874: 175
neglectus R. F. Sahlberg, 1838: 204 (*Chrysomela*)
nigritus Chûjô, 1951b: 38
obesus Weise, 1884d: 546 (var.) [IN] [# 577]
omissus R. F. Sahlberg, 1838: 203 (*Chrysomela*)
regnianus Tottenham, 1941: 14 [QS]
cochleariae graecus Apfelbeck, 1912: 257 (var.) E: GR **[# 539]**
concinnus Stephens, 1831: 334 **E:** DE EN FI FR GB GE IC IR NL NR PL RU (NT) SV **A:** RU (ES FE WS) MG **[# 995]**
commutatus Kontkanen, 1933: 73 (ssp.)
contemptus Kontkanen, 1933: 73 (ssp.)
inauratus Kontkanen, 1933: 73 (ssp.)
ivalianus Sprecher-Uebersax & Daccordi, 2012: 415 **A:** NP
laevigatus Duftschmid, 1825: 214 (*Chrysomela*) E: AL AU BH BU BY CR CZ FR GE GR HU IT LA LS LT MD ME NM PL RO RU (CT) SB SK SL SP SZ UK **A:** AB GG **NARi** **[# 617]**
aeruginosus Suffrian, 1858b: 395 (*Chrysomela*) [QS]
caucasicus Weise, 1884d: 543 (var.) [IN] [# 577]
gramicus Duftschmid, 1825: 215 (*Chrysomela*)
sabulicola Suffrian, 1851a: 248 (*Chrysomela*)
violaceus Weise, 1884d: 543 (var.) [IN] [# 577]
mellyi Achard, 1923e: 137 **A:** CH (HKG)
menthae Wollaston, 1864: 404 **N:** CI
rufipes Brullé, 1839: 74 (*Chrysomela*) [QS]
salicinus Heer, 1845: 17 (*Chrysomela*) E: AU FR IT SP SZ **[# 996]**
tristis Achard, 1923e: 137 (var.)
tashigaonensis Sprecher-Uebersax & Daccordi, 2012: 416 **A:** NP

genus *Phaedon*, species unassigned to subgenera
alpinus S.-Q. Ge, S.-Y. Wang & Yang, 2002a: 321, 323 **A:** CH (QIN SCH YUN)
alticola S.-H. Chen, 1974a: 45, 48 **A:** CH (GAN QIN SCH YUN)
apterus S.-H. Chen & S.-Y. Wang, 1984a: 172, 175 **A:** CH (SCH YUN)
cheni Daccordi, 1979b: 451 **A:** ID (SD)
chujoi Daccordi, 1979b: 446 **A:** ID (SD) **ORR**
flavotibialis Lopatin, 2005d: 876 **A:** CH (SCH)

geminatus Daccordi & S.-Q. Ge, 2015: 17 **A**: CH (SCH)
germinates Daccordi & S.-Q. Ge, 2014: 237 [UN] [# 581]
gressitti Daccordi, 1979b: 448 **A**: CH (XIZ) NP
huizuensis Daccordi & S.-Q. Ge, 2015: 20 **A**: CH (QIN)
huizuensis Daccordi & S.-Q. Ge, 2014: 241 [UN] [# 581]
igori Daccordi & S.-Q. Ge, 2015: 20 **A**: CH (SCH)
igori Daccordi & S.-Q. Ge, 2014: 241 [UN] [# 581]
insolitus S.-Q. Ge & Daccordi, 2015: 20 **A**: CH (SCH)
kabakovi Lopatin, 1998b: 833 **A**: RU (FE: Amur Region)
kimotoi Daccordi, 1979b: 449 **A**: NP
nepala Gruev, 1989: 52 (*Gonioctena*)
lesagei Daccordi, 1984: 326 **A**: NP
prosternalis Daccordi & S.-Q. Ge, 2015: 25 **A**: CH (SCH)
prosternalis Daccordi & S.-Q. Ge, 2014: 247 [UN] [# 581]
thomsoni Daccordi, 1978b: 214 **A**: NP ID (SD)
wittmeri Daccordi, 1979b: 451 **A**: BT
wumingshanensis S.-Q. Ge, S.-Y. Wang & Yang, 2002a: 322, 325 **A**: CH (SCH)
yodai Chûjô, 1966a: 553 **A**: NP

genus *Phaedon*, nomen dubium
bogdanovikatjkovi Jacobson, 1921: 20 **A**: KZ (Aktobe Region)

genus *Phaedonia* Weise, 1898: 210 type species: *Chrysomela circumcincta* C. R. Sahlberg, 1823
indica S.-H. Chen, 1933e: 381 (*Phaedon*) **A**: BT CH (XIZ) NP **ORR**

genus *Phratora* Chevrolat, 1836: 405 type species: *Chrysomela vitellinae* Linnaeus, 1758 **[Appendix]**
subgenus *Chaetoceroides* E. Strand, 1935: 285 [RN] type species: *Chrysomela vulgatissima* Linnaeus, 1758
Chaeroceta Mroczkowski, 1990: 214 [RN] type species: *Chrysomela vulgatissima* Linnaeus, 1758
Chaetocera Weise, 1884d: 514 [HN] type species: *Chrysomela vulgatissima* Linnaeus, 1758
vulgatissima Linnaeus, 1758: 370 (*Chrysomela*) **E**: AU BE BH BU BY CR CZ DE FI FR GB GE GR HU IT LA LS LT LU NL NR PL PT RO RU (CT NT) SB SK SL SP SV SZ TR UK **A**: AB CH (HEI NMO XIN) GG KZ MG RU (ES FE SS WS) SC TR **NAR**
aestiva Weise, 1884d: 515 (*Phyllodecta*, var.) [IN] [# 577]
inhonesta Weise, 1884d: 514 (*Phyllodecta*)
interstitialis Mannerheim, 1853: 259 [QS]
longula Motschulsky, 1860b: 229
obscura Weise, 1884d: 515 (*Phyllodecta*, var.) [IN] [# 577]

obtusicollis Motschulsky, 1860b: 230
striata Motschulsky, 1860b: 231 [QS]
unicolor Marsham, 1802: 185 (*Chrysomela*)

subgenus *Phratora* Chevrolat, 1836: 405 type species: *Chrysomela vitellinae* Linnaeus, 1758 **[Appendix]**
Phyllodecta Kirby, 1837: 216 type species: *Chrysomela vitellinae* Linnaeus, 1758

abdominalis abdominalis Baly, 1878h: 375 **A:** CH (XIZ) NP **ORR**
abdominalis pakistana Daccordi, 1982a: 398 (ssp.) **A:** PA
aenea S.-Y. Wang, 1992b: 636, 644 **A:** CH (SCH)
alternata Lopatin, 2007: 180 **A:** CH (GAN SCH) **[# 618]**
atrovirens Cornelius, 1857: 396 (*Chrysomela*) E: AL AU BE BH BY CR CZ DE EN FI FR GE HU LA LT NL NR PL RU (CT NT) SK SL SV SZ UK **A:** AB CH (HEI) MG RU (ES FE WS) "Caucasus" **[# 619]**
belousovi Lopatin, 2009b: 436 **A:** CH (YUN)
bicolor Gressitt & Kimoto, 1963a: 367 **A:** CH (QIN SCH XIZ YUN)
purpurea S.-Q. Ge, S.-Y. Wang & Yang, 2002b: 327 [HN]
bispinula S.-Y. Wang, 1992b: 637, 645 **A:** CH (SCH YUN)
biuncinata S.-Q. Ge, S.-Y. Wang & Yang, 2002b: 327 **A:** CH (YUN)
caperata S.-Q. Ge, S.-Y. Wang & Yang, 2002b: 328 **A:** CH (YUN)
cheni S.-Q. Ge, S.-Y. Wang & Yang, 2002b: 326 **A:** CH (YUN)
coerulescens Küster, 1848b: 100 **A:** AR GG TR **[# 620]**
horioni Mohr, 1968a: 48 (*Phyllodecta*)
costipennis S.-H. Chen, 1965: 221, 223 **A:** CH (SHA)
cuprea S.-Y. Wang, 1992b: 637, 645 **A:** CH (YUN)
daccordii S.-Q. Ge & S.-Y. Wang, 2004a: 134 **A:** CH (SCH YUN)
flavipes S.-H. Chen, 1963: 452, 456 **A:** BT CH (XIZ) NP
gracilis S.-H. Chen, 1965: 222, 223 **A:** CH (SCH YUN XIZ)
grandis Chûjô, 1956c: 6 (*Phyllodecta*) **A:** JA
hartmanni Medvedev, 2009b: 407 **A:** ID (SD)
jinchuanensis S.-Q. Ge & S.-Y. Wang, 2004a: 135 **A:** CH (SCH)
zhouzhiensis S.-Q. Ge & Yang, 2005: 1
koreana Takizawa, 1985a: 4 **A:** JA (Honshu) SC
laticollis Suffrian, 1851a: 262 (*Chrysomela*) E: AL AU BE BH BU BY CR CZ DE EN FR GB GE GR HU IT KO KR LA LT LU MD ME NL NR PL RO RU (CT NT) SB SK SP SZ UK **A:** AB CH (CHQ GAN GUI HUB NIN SCH YUN) JA KZ LE MG NC RU (ES FE SS WS) SC TR
cavifrons C. G. Thomson, 1866: 278 (*Phyllodecta*)
diversipes Pic, 1924e: 26 (*Phyllodecta*, var.)
multipunctata Jacoby, 1890b: 117 (*Phyllodecta*)
purpurascens Pic, 1925b: 9 (*Phyllodecta*, var.) [IN] **[# 577]**
viridiaurata Pic, 1924e: 26 (*Phyllodecta*, var.)

maya Daccordi, 1979b: 453 **A:** BT
mirabilis Lopatin, 2002a: 117 **A:** CH (SCH)
moha Daccordi, 1977a: 348 **A:** BT CH (SCH YUN)
parva Gressitt & Kimoto, 1963a: 370 **A:** CH (HEI)
phaedonoides occidentalis S.-H. Chen, 1965: 222, 224 (ssp.) **A:** CH (SCH YUN)
kabaki Lopatin, 2002f: 875
phaedonoides phaedonoides S.-H. Chen, 1934d: 78 (*Phyllodecta*) **A:** CH (BEI)
polaris leederi Steinhausen, 1993: 27 (ssp.) E: AU GE
polaris polaris J. S. Schneider, 1886: 66 (*Phyllodecta*) **E:** FI GB IC NR RU (NT) SV [# **621**]
polaris reitteri Lopatin, 1962c: 325 (*Phyllodecta*) **A:** MG NC RU (ES FE)
quadrithoracilis S.-Q. Ge & Yang, 2005: 4 **A:** CH (SHA)
ryanggangensis Gruev, 1994: 79 **A:** CH (HEB) NC RU (FE)
similis Chûjô, 1958b: 76 (*Phyllodecta*) **A:** CH (TAI)
tamangi L. N. Medvedev, 2009d: 409 **A:** NP
tibialis antennaria Apfelbeck, 1912: 256 (*Phyllodecta*) [QS] **E:** GR NM
tibialis tibialis Suffrian, 1851a: 259 (*Chrysomela*) [conserved name, ICZN (1984a), Opinion 1270] **E:** AL AU BE BH BU BY CR CZ DE EN FR GE HU IT LA LS LT ME NL NM NR PL RO RU (CT) SB SK SL SP SZ UK **A:** IN KZ TR
altaica Motschulsky, 1860b: 230
coerulea Weise, 1884d: 517 (*Phyllodecta*, var.) [IN] [# 577]
cornelii Weise, 1882a: 158 (*Phyllodecta*, var.)
vitellinae Linnaeus, 1758: 370 (*Chrysomela*) **E:** AL AU BE BH BU BY CR CZ DE EN FI FR GB GE GR HU IC IT KR LA LS LT ME NL NM NR PL PT RO RU (CT NT ST) SB SK SL SP SV SZ TR UK **A:** AB CH (GAN HEI JIL LIA NMO QIN SCH XIN) GG IN MG NC NP RU (ES FE SS WS) SC TR **NAR** [# **622**]
angusticollis Motschulsky, 1860b: 230
brevicollis Motschulsky, 1860b: 231
funesta Faldermann, 1837: 363 [QS]
latipennis Motschulsky, 1860b: 230
major Stierlin, 1863b: 65
nigrica Motschulsky, 1860b: 230
riparia Scopoli, 1763: 72 (*Coccinella*)
sinensis S.-H. Chen, 1934a: 4 (*Phyllodecta*)

genus *Phratora*, nomen dubium
laticollis Motschulsky, 1860b: 231 [HN]

genus *Phratora*, unavailable names
cuiae Daccordi & S.-Q. Ge, 2014: 337 [UN] [# 581]
deqinensis S.-Q. Ge & S.-Y. Wang, 2014: 340 [UN] [# 581]
wangi Daccordi & S.-Q. Ge, 2014: 356 [UN] [# 581]

genus *Plagiodera* Chevrolat, 1836: 404 type species: *Chrysomela armoraciae* Linnaeus, 1758 [# 559]

Pseudoparopsis Blackburn, 1899: 99 type species: *Paropsis nitidipennis* Boisduval, 1835

atmanama Daccordi, 1982a: 403 **A:** ID (UT) **ORR**

bicolor bicolor Weise, 1889g: 606 **A:** CH (GAN SCH YUN)

bicolor hengduanica S.-H. Chen & S.-Y. Wang, 1984a: 172, 174 (ssp.) **A:** CH (YUN)

borealis Gressit & Kimoto, 1963: 344 **A:** CH (LIA)

cupreata S.-H. Chen, 1934d: 59 **A:** CH (GUI SCH YUN)

maculicollis S.-H. Chen, 1934d: 59 **A:** CH (SCH) [# 623]

rufescens Gyllenhal, 1808: 267 (*Chrysomela*) **A:** ID (SD UT) NP **ORR** [# 624]

cinctipennis Baly, 1862a: 26 [QS]

septemvittata Stål, 1858: 252 **A:** CH (HAI HKG TAI) **ORR**

subglobosa Hope, 1831: 31 (*Coccinella*) **A:** NP

versicolora Laicharting, 1781: 148 (*Chrysomela*) **E:** AL AU BE BH BU BY CR CZ DE EN FI FR GB GE GR HU IT KO KR LA LT LU MD ME NL NM NR PL PT RO RU (CT NT ST) SB SK SL SP SV SZ TR UK **N:** MO **A:** AB AF AR CH (ANH BEI FUJ GAN GUI HEB HEI HEN HKG HUB HUN JIA JIL JIX LIA NIN NMO SCH SHA SHN TAI TIA XIN YUN ZHE) CY GG ID (KA UT) IN JA LE MG NC NP PA RU (ES FE SS WS) SC SI TD TM TR UZ **NARi ORR**

chinensis Weise, 1898: 212

clavicornis Stephens, 1831: 332 (*Phaedon*)

celestina Baly, 1864d: 229

cognata Gmelin, 1790: 1690 (*Chrysomela*)

distincta Baly, 1874: 174

hanoiensis S.-H. Chen, 1934d: 58

laevicollis Lopatin, 1962c: 325 (ssp.)

orientalis S.-H. Chen, 1934d: 56 (var.)

rufithorax S.-H. Chen, 1934d: 57 [= 1936b: 169] (var.)

rufivestis Bechyné, 1954b: 668 (ab.) [IN]

salicis Panzer, 1795a: 172 (*Galleruca*)

salicis C. G. Thomson, 1866: 271 [HN]

similis Herbst, 1783: 58 (*Chrysomela*)

virescens Hope, 1831: 31 (*Coccinella*) **A:** NP

yunnanica S.-H. Chen, 1934d: 60 (var.) **A:** CH (SCH YUN) **ORR**

genus *Plagiosterna* Motschulsky, 1860b: 196 type species: *Plagiosterna rufolimbata* Motschulsky, 1860 (= *Plagiosterna egregia* Gerstaecker, 1855)

Linamorpha Motschulsky, 1860b: 197 type species: *Lina erythroptera* Blanchard, 1851

Melasomida Schaeffer, 1920b: 117 [RN] type species: *Chrysomela californica* Rogers, 1856

Paraplagiomorpha Daccordi, 1987: 445 type species: *Plagiodera prasinipennis* Erichson, 1847

Plagiomorpha Motschulsky, 1860b: 200 type species: *Chrysomela californica* Rogers, 1856
Plagioschema Daccordi, 1987: 446 type species: *Chrysomela encausta* Klug, 1829
Pseudolina Schaeffer, 1920a: 330 [HN] type species: *Chrysomela californica* Rogers, 1856
subgenus *Linaeidea* Motschulsky, 1860b: 199 type species: *Chrysomela aenea* Linnaeus, 1758
adamsi Baly, 1864d: 229 (*Lina*) **A:** CH (GUA GUI LIA SCH YUN ZHE) NC SC **ORR** [# 540, # 572, # 634]
aenea aenea Linnaeus, 1758: 369 (*Chrysomela*) **E:** AL AU BE BH BU BY CR CZ DE EN FI FR GE GB GR HU IT LA LS LT LU ME NL NM NR PL PT RO RU (CT NT) SB SK SL SP SV TR UK **A:** CH (HEI JIL LIA TAI) JA NC RU (ES FE SS WS) SC TR [# 625, # 1145]
amethystina Gmelin, 1790: 1688 (*Chrysomela*)
bicolor Schilsky, 1908: 603 (*Melasoma*, var.)
caeruleoviolacea DeGeer, 1775: 316 (*Chrysomela*)
fuscicornis Fabricius, 1781: 126 (*Chrysomela*) [HN]
haemorrhoidalis O. F. Müller, 1776: 81 (*Chrysomela*)
purpureotincta Pic, 1924e: 26 (*Melasoma*, var.)
violacea Olivier, 1791a: 718 (*Chrysomela*) [HN]
aenea insularis Chûjô, 1940a: 78 (*Chrysomela*, ssp.) [QS] **A:** JA
aenea tsutsuii Nakane, 1955: 38 (*Linaeidea*, ssp.) [QS] **A:** JA
chlorina Maulik, 1926: 69 (*Chrysomela*) **A:** ID (SD UT) NP
formosana Bates, 1866: 354 (*Lina*) **A:** CH (TAI) JAi (Okinawa) **ORR**
placida S.-H. Chen, 1934d: 63 [= 1936b: 176] (*Chrysomela*, var.) **A:** CH (CHQ SCH XIN XIZ YUN) NP [# 540, # 634]
emmerichi Pic, 1934g: 86 (*Melasoma*, var.)
metalla Bechyné, 1954b: 670 (*Chrysomela*, ab.) [IN]
minutior Bechyné, 1954b: 669 (*Chrysomela*, ssp.)
ornaticollis S.-H. Chen, 1936c: 80 (*Chrysomela*, var.) [RN]
praecox Bechyné, 1954b: 669 (*Chrysomela*, ab.) [IN]
viridana Bechyné, 1954b: 670 (*Chrysomela*, ab.) [IN]
subgenus *Macromela* Chûjô, 1958b: 31 type species: *Melasoma maculicollis* Jacoby, 1890
maculicollis Jacoby, 1890b: 117 (*Melasoma*) **A:** CH (FUJ GUI GUX HUB HUN SCH YUN ZHE) JA
insularis Chûjô, 1958b: 32 (*Chrysomela*, ssp.) [HN]
subgenus *Xenolina* Jacoby, 1905a: 481 type species: *Xenolina marginata* Jacoby, 1905
aeneipennis Baly, 1859c: 61 (*Lina*) **A:** CH (AHN CHQ FUJ GUA GUI HUB HUN JIN SCH YUN ZHE)
siemsseni Weise, 1922b: 56 (*Plagiodera*)
miniaticollis Hope, 1831: 30 (*Chrysomela*) **A:** NP **ORR**

genus ***Prasocuris*** **Latreille, 1802: 224** [NP] type species: *Chrysomela phellandrii* Linnaeus, 1758 **[# 573]**

subgenus *Hydrothassa* C. G. Thomson, 1859: 158 type species: *Chrysomela marginella* Linnaeus, 1758

Agrostithassa Jacobson, 1921: 24 type species: *Chrysomela aucta* Fabricius, 1787 (= *Chrysomela glabra* Herbst, 1783)

Alonsina Warchałowski, 1995a: 459 type species: *Phratora burakowskii* Warchałowski, 1995

Eremosis Gozis, 1882: 207 type species: *Chrysomela glabra* Herbst, 1783

anatolica Coral Şahin & Özdikmen, 2019: 327 **A:** TR

bicolora Rapilly, 1981: 143 **A:** IN

fairmairei Brisout de Barneville, 1866: 423 (*Phratora*) E: PT SP

burakowskii Warchałowski, 1995a: 460 (*Phratora*)

planiuscula Weise, 1887b: 188 (*Hydrothassa*)

flavocincta Brullé, 1833: 271 (*Chrysomela*) E: AL BH BU CR GR HU KO ME SB TR **A:** TR **[# 997]**

glabra Herbst, 1783: 60 (*Chrysomela*) E: AU BE BH BU BY CR CZ DE EN FI FR GB GE GR HU IT LA LS LT LU MD NL NR PL PT RO RU (CT NT ST) SK SL SP SV SZ UK **N:** MO **A:** GG TR **[# 1145]**

aucta Fabricius, 1787: 72 (*Chrysomela*)

egena Gyllenhal, 1827: 649 (*Chrysomela*)

hispanica Graells, 1851: 152 (*Phaedon*)

umhangii Wencker, 1866: 133 (var.)

hannoveriana Fabricius, 1775: 102 (*Chrysomela*) E: BE BY CZ DE EN FI FR GB GE LA LT LU NL PL RU (NT ST) SK SV UK **A:** RU (ES WS) **[# 1145]**

calthae Weise, 1883d: 252 (*Hydrothassa*, var.)

degenerata Mäklin, 1881: 16 (var.)

germanica Weise, 1883d: 252 (*Hydrothassa*, var.)

potentillae Herbst, 1783: 55 (*Chrysomela*)

ranunculi Herbst, 1783: 55 (*Chrysomela*)

marginella marginella Linnaeus, 1758: 371 (*Chrysomela*) E: AU BE BY CR CZ DE EN FI FR GB GE HU IT LA LT LU MD NL NR PL RO RU (CT: Kaliningrad region NT) SB SK SL SV SZ UK **A:** RU (ES SS WS)

devillei Bullock, 1928: 104 (*Hydrothassa*, var.)

marginella eoa Lopatin, 1962c: 326 (ssp.) **A:** RU (FE)

oblongiuscula Fairmaire, 1884a: lxix **N:** AG TU

suffriani Küster, 1852: 100 (*Helodes*) E: FR (Corsica) IT (Sardinia)

pumila Reiche, 1862a: 298 (*Phratora*)

subgenus *Prasocuris* Latreille, 1802: 224 [NP] type species: *Chrysomela phellandrii* Linnaeus, 1758 **[# 573]**

Helodes Paykull, 1799a: 84 [NO] type species: *Chrysomela phellandrii* Linnaeus, 1758 **[# 573]**

distincta distincta P. H. Lucas, 1846b: pl. 46 [= 1849: 539] (*Helodes*) E: FR IT **N:** AG

marginicollis Suffrian, 1851a: 270 (*Chrysomela*)
notaticollis Pic, 1913k: 179 (var.)
pallidithorax Pic, 1901e: 59 (var.)
subapicalis Pic, 1913k: 178 (var.)

distincta suturella Reiche & Saulcy, 1858b: 38 (*Chrysomela*) **A:** IS LE SY [# 541]
insuturella Pic, 1913e: 129 (var.)

gressitti Daccordi & Gruev, 1991: 379 **A:** CH (HEI)

junci Brahm, 1790: 62 (*Chrysomela*) **E:** AU BE BH BU BY CR CZ DE EN FI FR GB GE GR HU IT KR LA LT LU NL NM NR PL PT RO RU (CT NT ST) SB SK SL SP SV SZ UK **A:** AB GG IN TR **N:** AG [# 1145]
atra Duftschmid, 1825: 237 (*Helodes*)
beccabungae Illiger, 1794: 600 (*Chrysomela*)
olivaceus Pic, 1934d: 22 (var.)
parallelicollis Jakob, 1954c: 49
violacea Fabricius, 1792b: 11 (*Crioceris*)

phellandrii Linnaeus, 1758: 376 (*Chrysomela*) **E:** AU BE BH BU BY CR CZ DE EN FI FR GB GE GR HU IT KR LA LT MD NL NR PL PT RO RU (CT incl. Kaliningrad Region NT ST) SB SK SL SP SV SZ TR UK **N:** MO **A:** JA RU (ES WS) TR **NAR**
cicutae Weise, 1883d: 252 (var.)
sii Weise, 1883d: 252 (var.)

vicina P. H. Lucas, 1846b: pl. 46 [= 1849: 539] (*Helodes*) **E:** FR (Corsica) IT PT SP **N:** AG MO TU
chalybea Suffrian, 1851a: 271 (*Chrysomela*)
hispanica Weise, 1884d: 534 (var.) [IN] [# 577]
litigiosa Rosenhauer, 1856: 323 (*Chrysomela*)

genus *Sternoplatys* Motschulsky, 1860b: 220 type species: *Sternoplatys fulvipes* Motschulsky, 1860

subgenus *Austrosternoplatys* Jacobson, 1921: 24 type species: *Sternoplatys longulus* Weise, 1890

longulus Weise, 1890a: 480 (*Phaedon*) **A:** CH (XIN) KI TD

subgenus *Sternoplatys* Motschulsky, 1860b: 220 type species: *Sternoplatys fulvipes* Motschulsky, 1860

clementzi Jacobson, 1901b: 131 **A:** CH (NMO) MG RU (ES FE: upper Kolyma River SS) [# 635]
tolli Jacobson, 1901b: 132
weisei Csiki, 1901a: 118

fausti Weise, 1884d: 540 (*Phaedon*) **A:** RU (FE)
completus Jacobson, 1901b: 131 (var.) [IN] [# 577]

fulvipes baicalicus Weise, 1900b: 285 (*Phaedon*) **A:** RU (ES FE)
motschulskyi Jacobson, 1901b: 133

fulvipes fulvipes Motschulsky, 1860b: 220 **A:** CH (HEI JIL LIA) RU (FE)
hemisphaericus Weise, 1887h: 187 (*Phaedon*)
piceipes Jacobson, 1901b: 133 (var.) [IN] [# 577]

genus *Yulongedon* Daccordi & S.-Q. Ge, 2012b: 634 type species: *Yulongedon jambhalai* Daccordi & S.-Q. Ge, 2012
formosus Lopatin, 2012: 635 **A:** CH (GAN SCH)
jambhalai Daccordi & S.-Q. Ge, 2012b: 637 **A:** CH (XIZ)

subtribe Doryphorina Motschulsky, 1860

genus *Agrosteella* L. N. Medvedev, 1987: 72 type species: *Agrosteomela fallaciosa* Stål, 1862
biconvexa S.-Q. Ge, S.-Y. Wang, Yang & W.-Z. Li, 2002: 84 **A:** CH (XIZ)
cheni S.-Q. Ge, S.-Y. Wang, Yang & W.-Z. Li, 2002: 82 **A:** CH (XIZ)
fallaciosa Stål, 1862: 5 (*Paralina*) **A:** CH (XIZ) ID (AP SD) NP **ORR**
jini S.-Q. Ge, S.-Y. Wang, Yang & W.-Z. Li, 2002: 86 **A:** CH (XIZ)
medvedevi Daccordi, 2001: 89 (*Agrosteomela*) **A:** NP
oligotricha S.-Q. Ge, S.-Y. Wang, Yang & W.-Z. Li, 2002: 85 **A:** CH (XIZ)
punctata S.-Q. Ge, S.-Y. Wang, Yang & W.-Z. Li, 2002: 85 **A:** CH (XIZ)
violaceicollis S.-Q. Ge, S.-Y. Wang, Yang & W.-Z. Li, 2002: 86 **A:** CH (YUN)

genus *Agrosteomela* Gistel, 1857b: 605 type species: *Chrysomela caschmirensis* Kollar & L. Redtenbacher, 1844 (= *Chrysomela indica* Hope, 1831)
Paralina Baly, 1860a: 155 type species: *Chrysomela indica* Hope, 1831
chinensis Weise, 1922b: 54 (*Paralina*) **A:** CH (FUJ GUI JIA HUB HUN SCH TAI XIZ)
flavipennis S.-Q. Ge, Yang, S.-Y. Wang & W.-Z. Li, 2004: 281 **A:** CH (XIZ)
impressiuscula Fairmaire, 1878b: 135 (*Paralina*) **A:** CH (GUI SCH YUN XIZ) **ORR**
indica Hope, 1831: 29 (*Chrysomela*) **A:** CH (SCH TAI XIZ YUN) ID (AP SD) NP **ORR**
caschmirensis Kollar & L. Redtenbacher, 1844: 558 (*Chrysomela*)
elata Stål, 1857: 60 (*Lina*)
nigrita S.-Q. Ge, Yang, S.-Y. Wang & W.-Z. Li, 2004: 282 **A:** CH (SCH)

genus *Ambrostoma* Motschulsky, 1845a: 109 type species: *Chrysomela quadriimpressa* Motschulsky, 1845 (= *Chrysomela superba* Thunberg, 1787)
subgenus *Ambrostoma* Motschulsky, 1845a: 109 type species: *Chrysomela quadriimpressa* Motschulsky, 1845 (= *Chrysomela superba* Thunberg, 1787)
chinkinyui Kimoto & Osawa, 1995: 15 **A:** CH (TAI)
fasciatum S.-H. Chen, 1936e: 721 **A:** CH (ZHE)
fortunei Baly, 1860c: 94 (*Chrysomela*) **A:** CH (ANH FUJ GUA GUI HEI HEN HUN JIA JIX SHA ZHE)

chinense Motschulsky, 1860b: 228
chusanicum Gruev, 1980b: 83 (ssp.)
fulgurans Achard, 1922b: 15 **A:** CH (HUN YUN) **ORR**
koreanum Cho & Borowiec, 2013: 3 **A:** SC
leigongshanum S.-Y. Wang, 1992a: 175, 177 **A:** CH (GUI)
omeishanum Gressitt & Kimoto, 1963a: 328 **A:** CH (SCH)
rugosopunctatum S.-H. Chen, 1936e: 718 **A:** CH (HUB YUN) **ORR**
daccordii L. N. Medvedev, 2007d: 434
laosense Kimoto & Gressitt, 1981: 372
superbum Thunberg, 1787b: 44 (*Chrysomela*) **A:** CH (BEI HEB HEI JIL LIA SHN) MG NC RU (ES FE SS WS) [# 474, # 560]
quadriimpressum Motschulsky, 1845a: 109
purpureocupreum Jacobson, 1901b: 127 (var.) [IN] [# 577]
viridicyaneum Jacobson, 1901b: 127 (var.) [IN] [# 577]
subgenus *Parambrostoma* S.-H. Chen, 1936e: 718 type species: *Ambrostoma sublaevis* S.-H. Chen, 1934
ambiguum S.-H. Chen, 1936e: 728 **A:** NP
kippenbergi Daccordi & S.-Q. Ge, 2012a: 341 (*Parambrostoma*) **A:** NP
mahesa Hope, 1831: 30 (*Chrysomela*) **A:** CH (XIZ) NP
nepalense Motschulsky, 1860b: 228
medvedevi Daccordi & S.-Q. Ge, 2012a: 342 (*Parambrostoma*) **A:** NP
montanum L. N. Medvedev, 1990b: 11 **A:** NP
shuteae Daccordi, 1977b: 133 **A:** NP
sublaeve S.-H. Chen, 1934b: 66 **A:** NP

genus *Calligrapha* Chevrolat, 1836: 398 type species: *Chrysomela polyspila* Germar, 1821: 176
subgenus *Calligrapha* Chevrolat, 1836: 398 type species: *Chrysomela polyspila* Germar, 1821: 176
Boliographa Motschulsky, 1860b: 198 type species: *Chrysomela philadelphica* Linnaeus, 1758
Metallographa Motschulsky, 1860b: 198 type species: *Calligrapha marmorata* Fabricius, 1798
Phyllis Gistel, 1848a: 403 [= 1848b: 123] [HN] type species: *Phyllis fulvipes* Gistel, 1848
Polyspila Hope, 1841: 165 type species: *Chrysomela philadelphica* Linnaeus, 1758
polyspila Germar, 1821: 176 (*Chrysomela*) Ei: AZ **AUS**i **NTR PAC**i
plagata Achard, 1923a: 70 (*Polyspila*, var.)
bilineolata Achard, 1923c: 120 (*Polyspila*, var.)

genus *Chrysolina* Motschulsky, 1860b: 210 [NP] [conserved name, ICZN (1984c), Opinion 1279] type species: *Chrysomela staphylaea* Linnaeus, 1758 [# 475]

subgenus *Allochrysolina* Bechyné, 1950b: 133 type species: *Chrysomela fuliginosa* Olivier, 1807

confossa Fairmaire, 1865: 76 (*Chrysomela*) **N:** AG

dorsalis Weise, 1884d: 395 (*Chrysomela*, var.) [IN] [# 577]

semiopaca Fairmaire, 1873c: 357 (*Chrysomela*)

fuliginosa coriacea Suffrian, 1851a: 59 (*Chrysomela*) E: PT SP

rugipennis Harold, 1875c: 140 (*Chrysomela*)

fuliginosa fuliginosa Olivier, 1807: 514 (*Chrysomela*) [NP] E: FR IT SP

espanoli Bechyné, 1950b: 134 (ssp.)

microsticha Bechyné, 1949a: 54 (ssp.)

opaca Suffrian, 1853a: 124 (*Chrysomela*)

fuliginosa galii Weise, 1884d: 394 (*Chrysomela*, var.) [NP] E: BE CZ FR GE NL SZ [# 578]

gendreaui Bechyné, 1949a: 55 (ssp.)

molluginis Suffrian, 1851a: 57 (*Chrysomela*) [HN] [# 998]

violaceocoerulea Geoffroy, 1785: 106 (*Chrysomela*) [NO]

fuliginosa subtilipunctata Petitpierre, 2019: 428 (ssp.) E: SP

lepida gastonis Fairmaire, 1875b: 538 (*Chrysomela*) **N:** AG MO TU [# 585]

gentianea Bechyné, 1952b: 379 (ab.) [IN]

lepida lepida Olivier, 1807: 522 (*Chrysomela*) **E:** FR PT SP **A:** CY [# 999]

dotalis Bechyné, 1950b: 155 (ab.) [IN]

semigrans Bechyné, 1950b: 155 (ab.) [IN]

opacicollis Fairmaire, 1865: 77 (*Chrysomela*) **N:** MO

ruginosa Fairmaire, 1873c: 356 (*Chrysomela*) **N:** AG

subgenus *Allohypericia* Bechyné, 1950b: 159 type species: *Chrysomela lobicollis* Fairmaire, 1887 (= *Taeniosticha poricollis* Motschulsky, 1860)

Bourdonneana Kippenberg, 2010a: 68 type species: *Chrysomela aeruginosa* Faldermann, 1835

aeruginosa aeruginosa Faldermann, 1835: 440 (*Chrysomela*) **A:** CH (ANH FUJ GAN HEI JIL LIA NIN NMO QIN XIZ) KZ MG RU (ES SS FE) [# 476]

dimidiata Ménétriés, 1837: 181 (*Chrysomela*)

distans Csiki, 1901a: 116 (*Chrysomela*)

instructa Motschulsky, 1860b: 228 (*Taeniosticha*)

regularis Motschulsky, 1860b: 228 (*Taeniosticha*)

tarda Motschulsky, 1860b: 228 (*Taeniosticha*)

urbana Csiki, 1901a: 117 (*Chrysomela*)

aeruginosa alpina L. N. Medvedev, 1980a: 317 (ssp.) **A:** MG (Mongolian Altai) [# 476]

aeruginosa chingana Bechyné, 1952b: 382 (ssp.) **A:** CH (NMO) [# 476]

aeruginosa kabana Lopatin, 2009b: 435 (ssp.) **A:** CH (GAN) [# 476]

aeruginosa muralis Csiki, 1901a: 116 (*Chrysomela*) **A:** CH (HEB) [# 476]

aeruginosa poricollis Motschulsky, 1860b: 228 (*Taeniosticha*) **A:** CH (BEI NMO NIN SHX) [# 476, # 586]

lobicollis Fairmaire, 1887d: 331 (*Chrysomela*)
mandarina Achard, 1922b: 16 (*Chrysomela*)
arctica L. N. Medvedev, 1980b: 93 **A:** RU (FE: Wrangel Island)
centralasiae Lopatin, 1970a: 257 (*Chrysomela*, ssp.) **A:** CH (NMO QIN) MG RU (ES: Irkutsk Region)
koltzei brunneipennis S. Matsumura, 1911b: 141 (*Chrysomela*) **A:** RU (FE: Sakhalin)
koltzei koltzei Weise, 1887h: 179 (*Chrysomela*) **A:** RU (FE ES WS)
koltzei lamii Takizawa, 1970: 118 **A:** JA RU (FE)
mauroi Lopatin, 2005b: 571 **A:** CH (SCH)
orochi Takemoto, 2022: 459 **A:** JA (Honshu)
peninsularis Bechyné, 1952b: 382 (ssp.) **A:** CH (HEN SHA ZHE) RU (FE) SC
perforata perforata Gebler, 1830: 216 (*Chrysomela*) **A:** MG RU (SS)
perforata simillima Mohr, 1966: 96 (*Chrysomela*, ssp.) **A:** MG RU (SS)
changaiensis L. N. Medvedev, 1980a: 318 (ssp.)
pallidipes L. N. Medvedev, 1980a: 318 (ssp.)
perforata turczaninoffi Harold, 1875a: 176 (*Chrysomela*) [RN] **A:** MG RU (ES SS)
variolosa Ménétriés, 1854: 40 (*Chrysomela*) [HN]
pubitarsis Bechyné, 1950b: 160 **A:** CH (BEI)
purpurata Faldermann, 1833: 70 (*Chrysomela*) **A:** KZ MG RU (ES WS)
campestris Weise, 1912: 84 (*Chrysomela*) [HN]
subgenus *Anopachys* Motschulsky, 1860b: 202 type species: *Chrysomela asclepiadis* A. Villa & G. B. Villa, 1833
asclepiadis asclepiadis A. Villa & G. B. Villa, 1833: 36 (*Chrysomela*) **E:** AU FR IT SZ [# 1000]
thurntaxisi Schatzmayr, 1927: 151 (*Chrysomela*, ssp.)
mesauca Allenspach, 1945: 341 (*Chrysomela*, forma)
asclepiadis bohemica J. Müller, 1948: 95 (*Chrysomela*, ssp.) **E:** AU CR CZ GE IT KZ PL RO RU (CT ST) SK SL UK **A:** AB KI KZ
problematica Kaszab, 1962: 54 (*Chrysomela*, ssp.)
viridisplendens Bechyné, 1950b: 148 (ab.) [IN]
aurichalcea Mannerheim, 1825: 39 (*Chrysomela*) **E:** BY LA LT NT RU (CT ST) **A:** CH (CHQ FUJ GAN GUI GUX HEB HEI HEN NIN SCH SHA SHN SHX TAI YUN ZHE) JA KI KZ MG NC RU (ES FE SS WS) SC **ORR**
amethystina Kolbe, 1886: 228 (*Chrysomela*) [HN]
collaris Weise, 1916b: 59 (*Chrysomela*, var.) [RN]
cupraria Kolbe, 1886: 229 (*Chrysomela*) [# 1108]
elevata Suffrian, 1851a: 189 (*Chrysomela*)
fokiensis Bechyné, 1950b: 147 (ssp.)
japana Marseul, 1886: 41 (*Chrysomela*, var.) [HN]
kiotensis Bechyné, 1950b: 147 (ab.) [IN]
kwanghsiensis Bechyné, 1950b: 147 (ssp.)

nigricans Jacobson, 1901a: 100 (*Chrysomela*, var.) [IN] [# 577]
omisiensis Bechyné, 1950b: 147 (ssp.)
pekinensis Fairmaire, 1887d: 331 (*Chrysomela*)
recticollis Weise, 1887h: 182 (*Chrysomela*, var.) [HN]
relator Bechyné, 1950b: 147 (ab.) [IN]
uvida Bechyné, 1950b: 147 (ab.) [IN]
vagesplendens Bechyné, 1950b: 148 (ssp.)
violaceicollis Motschulsky, 1862: 21 (*Anopachys*)
wallacei Baly, 1862a: 21 (*Chrysomela*)
yunnanica Bechyné, 1950b: 148 (ssp.)
eurina Frivaldszky, 1883: 17 (*Chrysomela*) **E:** AU CZ HU PL RO RU (CT) SK **A:** RU (SS WS)
perplexa Breit, 1920: 86 (*Chrysomela*)
gensanensis Weise, 1900b: 282 (*Chrysomela*) **A:** CH (GUA GUI HUB LIA) NC SC
koreana Chûjô, 1941a: 68 **A:** NC SC
lineella Weise, 1887h: 182 (*Chrysomela*, var.) **A:** RU (FE)
lineigera Jacobson, 1901b: 125 (*Chrysomela*) **A:** CH (HEI NMO) RU (ES FE)
neglecta Bieńkowski, 1998: 133 **A:** RU (FE)
pala Bieńkowski, 1998: 135 **A:** RU (FE)
quadrangulata Motschulsky, 1860b: 226 (*Chrysomorpha*) **A:** MG RU (ES FE)
linaeides Weise, 1896a: 80 (*Chrysomela*)
omoka Jacobson, 1924b: 83 (*Chrysomela*)
relucens Rosenhauer, 1847: 62 (*Chrysomela*) **E:** AU IT RU (CT: South and North Ural NT: Murmansk Reg.) **A:** KZ (Altai) RU (ES FE SS WS: Berezovo)
aquilina Bechyné, 1948b: 11 (ab.) [IN]
schatzmayri J. Müller, 1916: 96 (*Chrysomela*) **E:** IT
stalii Baly, 1860c: 95 (*Chrysomela*) **A:** CH (JIX)
micans Jacoby, 1893c: 105 (*Chrysomela*)
sundukovi Mikhailov, 2006a: 135 **A:** RU (FE) [# 477]
watanabei Takizawa, 1970: 121 **A:** JA RU (FE: South Sakhalin)
subgenus *Apterosoma* Motschulsky, 1861a: 23 type species: *Apterosoma angusticollis* Motschulsky, 1861
Caudatochrysa Bechyné, 1950b: 149 type species: *Apterosoma angusticollis* Motschulsky, 1861
aino Takizawa, 1970: 117 **A:** JA
angusticollis Motschulsky, 1861a: 23 (*Apterosoma*) **A:** CH (HEI ZHE) JA RU (FE)
japana Baly, 1874: 171 (*Chrysomela*)
porosirensis Takizawa, 1970: 120 **A:** JA
subgenus *Arctolina* Kontkanen, 1959: 31 type species: *Chrysomela birulai* Jacobson, 1910 (= *Chrysomela subsulcata* Mannerheim, 1853)
ballioni Lopatin, 1968c: 549 (*Chrysomela*) **A:** KZ

boeberi Harold, 1874c: 3415 (*Chrysomela*) [RN] **A**: RU (ES FE)

sulcata Germar, 1823: 589 (*Chrysomela*) [HN]

boreosinica Lopatin, 2004a: 616 **A**: CH (XIN: Tien-Shan)

borochorensis Lopatin, 2000c: 133 **A**: CH (XIN: Tien-Shan)

naratica Lopatin, 2000c: 134

bungei Jacobson, 1910a: 62 (*Chrysomela*) **A**: RU (ES FE: Arctic territories)

cyanella Gebler, 1830: 213 (*Chrysomela*) **A**: KZ (Altai) RU (SS: Altai)

teleuta Jacobson, 1922a: 521 (*Chrysomela*)

dolini Lopatin, 1999b: 891 **A**: KZ (Dzhungar Alatau)

kaikana Lopatin, 1993b: 5 **A**: KZ (Dzhungar Alatau)

kryzhanovskii Lopatin, 1968c: 550 (*Chrysomela*) **A**: KZ (Dzhungar Alatau)

kuznetzowi Jacobson, 1897a: 434 (*Chrysomela*) **E**: RU (CT NT) **A**: RU (ES WS) [# 478]

tundralis Jacobson, 1910a: 65 (*Chrysomela*, var.) [# 543]

magniceps J. R. Sahlberg, 1885a: 38 (*Chrysomela*) **A**: RU (ES FE: Arctic territories) **NAR**

novosibirica Jacobson, 1910a: 60 (*Chrysomela*, var.)

wollosowiczi Jacobson, 1910a: 59 (*Chrysomela*, var.)

octocosta Jacobson, 1924b: 78 (*Chrysomela*) **A**: KZ (Dzhungar Alatau)

oirota Lopatin, 1990c: 50 **A**: KZ (West Altai) RU (SS: Altai)

poretzkyi olschwangi Mikhailov, 2018: 294 (ssp.) **E**: RU (CT: South Ural)

poretzkyi poretzkyi Jacobson, 1897a: 434 (*Chrysomela*) **E**: RU (CT: Bashkortostan, Samara Reg., Lipetsk Reg.) **A**: RU (WS: Chelyabinsk Reg.) [# 478]

saurica Jacobson, 1924b: 78 (*Chrysomela*) **A**: KZ (Saur Mts.)

septentrionalis Ménétriés, 1851: 73 (*Chrysomela*) **E**: RU (NT) **A**: RU (ES FE WS: Arctic territories) **NAR**

caurina Brown, 1962: 64

sculpturata Jacobson, 1896a: 548 (*Chrysomela*)

subsulcata Mannerheim, 1853: 254 (*Chrysomela*) **A**: RU (ES FE: Arctic territories) **NAR**

birulai Jacobson, 1910a: 56 (*Chrysomela*)

glacialis Jacobson, 1910a: 59 (*Chrysomela*, var.) [HN]

tastavica Lopatin, 1993b: 6 **A**: CH (XIN) KZ (Dzhungar Alatau)

tshingilica Mikhailov, 2021b: 12 **A**: CH (XIN: Altai)

valichanovi Lopatin, 1990c: 53 **A**: KZ (Dzhungar Alatau)

subgenus *Atlasiana* Bourdonné & Doguet, 1991: 57 type species: *Chrysomela seriatopora* Fairmaire, 1867

Mimophaedon Bourdonné, 1997: 349 type species: *Chrysolina pourtoyi* Bourdonné, 1997

abchasica Weise, 1892j: 405 (*Chrysomela*) **A**: TR

circassicola Reitter, 1913c: 112 (*Chrysomela*)

pourtoyi Bourdonné, 1997: 343 **E**: FR

seriatopora Fairmaire, 1867: 415 (*Chrysomela*) **N**: AG LB TU

edoughensis Fairmaire, 1873c: 357 (*Chrysomela*)

subgenus *Bechynia* Bourdonné, 1977: 330 type species: *Chrysolina platypoda* Bechyné, 1950

milleri Weise, 1884b: 162 (*Chrysomela*) **E**: BH CR IT SL

montana Gebler, 1848: 23 (*Chrysomela*) [DA] **A**: RU (SS) [# 479]

substrangulata Bourdonné, 1986: 237

philotesia Daccordi & Ruffo, 1980: 357 **E**: GR

platypoda Bechyné, 1950b: 62 **E**: FR

subgenus *Bittotaenia* Motschulsky, 1860b: 206 type species: *Chrysomela salviae* Germar, 1823

aeneipennis Reiche & Saulcy, 1858b: 34 (*Chrysomela*) **N**: AG EG LB **A**: IS JO SA SI SY TR [# 587]

tripolitanica Bechyné, 1950b: 178 (*Crosita*)

compuncta Weise, 1898: 191 (*Chrysomela*) **A**: CY IS LE SY TR

grata brancuccii Daccordi, 1982b: 406 (*Crosita*) **A**: AE OM

grata derelicta Marseul, 1887: 84 (*Chrysomela*) **E**: GR (Rhodes) **A**: SY

subcoerulea J. R. Sahlberg, 1913c: 248 (*Chrysomela*, var.) [IN] [# 577]

grata grata Faldermann, 1837: 361 (*Chrysomela*) **A**: AB AF AR GG IN IQ TM TR

megriensis Ter-Minassian, 1950: 133 (*Chrysomela*)

pseudoturanica Bechyné, 1950b: 177 (*Crosita*, ab.) [IN]

grata leonardii Daccordi, 1976b: 79 (*Crosita*) **A**: SI

grata nigrolucens Lopatin, 1985b: 770 (ssp.) **A**: IN

mellyi Stål, 1857: 60 (*Chrysomela*) **A**: AF CH (XIZ) ID (KA SD UT) IN PA **ORR** [# 588]

caelestina Baly, 1879c: 193 (*Crosita*)

salviae catalonica Bechyné, 1950b: 178 (*Crosita*, ssp.) **E**: SP

salviae salviae C. R. Sahlberg, 1823b: 76 [= 1823c: 76, = 1829: 30] (*Chrysomela*) **E**: AL BH BU CR GR IT ME NM RO SL RU (ST) TR **A**: AB AR CY GG IN IS LE SY TR

connectens Bechyné, 1950b: 178 (*Crosita*, ab.) [IN]

cyanea Papp, 1946c: 23 (*Chrysomela*, ab.) [IN]

salviae Germar, 1823: 586 (*Chrysomela*) [HN]

sculptipennis Faldermann, 1837: 359 (*Chrysomela*)

solida Papp, 1946c: 23 (*Chrysomela*, ab.) [IN]

turanica Reitter, 1889e: 131 (*Chrysomela*) **A**: AB AR IN TD TM

subgenus *Cecchiniola* Jacobson, 1908: 624 type species: *Chrysomela platyscelidina* Jacobson, 1898

platyscelidina Jacobson, 1898: 201 (*Chrysomela*) **E**: KR

subgenus *Centoptera* Motschulsky, 1860b: 207 type species: *Chrysomela regalis* Olivier, 1807 (= *Chrysomela bicolor* Fabricius, 1775)

bicolor Fabricius, 1775: 95 (*Chrysomela*) [HN] **E**: IT SP **N**: AG CI EG LB MO SY TU **A**: IS LE SI [# 480]

canariensis Brullé, 1839: 73 (*Chrysomela*)

consularis Erichson, 1841: 190 (*Chrysomela*) [HN]

dolorosa Fairmaire, 1873c: 359 (*Chrysomela*, var.)
lusitanica Gmelin, 1790: 1668 (*Chrysomela*) [HN]
marseuli Kocher, 1959: 58 (*Chrysomela*, var.) [IN] [# 577]
mima Marseul, 1887: 99 (*Chrysomela*, var.)
nigropunctata Reitter, 1872: 175 (*Chrysomela*)
regalis Olivier, 1807: 538 (*Chrysomela*)
viridicaerulea Forskål, 1775: 77 (*Chrysomela*)
subgenus *Chalcoidea* Motschulsky, 1860b: 209 type species: *Chrysomela marginata* Linnaeus, 1758
Craspeda Motschulsky, 1860b: 191 type species: *Chrysomela besseri* Krynicki, 1832
alaiensis Lopatin, 1998b: 833 (ssp.) **A:** KI
analis Linnaeus, 1767: 592 (*Chrysomela*) E: AU BY CZ DE EN FI FR GE HU KR LA LT NL NR PL RO RU (CT NT ST) SK SP SZ UK **A:** AB GG TR **N:** AG
brunneomicans Roubal, 1942b: 63 (*Chrysomela*, ab.) [IN]
lomata Herbst, 1783: 54 (*Chrysomela*)
prasina Suffrian, 1851a: 85 (*Chrysomela*)
schach Fabricius, 1792a: 326 (*Chrysomela*)
tristis Tenenbaum, 1927: 151 (*Chrysomela*, ab.) [IN]
besseri Krynicki, 1832: 171 (*Chrysomela*) E: RU (ST) UK **A:** AB GG KI KZ TD UZ [# 1001]
cinctipennis Harold, 1874c: 3416 (*Chrysomela*) [RN]
circumducta Suffrian, 1851a: 77 (*Chrysomela*) [HN] [# 1002]
brahma Takizawa, 1980b: 51 **A:** ID (HP)
brunnicornis bermani L. N. Medvedev, 1978e: 119 **A:** RU (ES)
brunnicornis brunnicornis Weise, 1887h: 175 (*Chrysomela*) **A:** MG RU (SS)
koshantschikovi Jacobson, 1925a: 51 (*Chrysomela*)
brunnicornis wrangeliani Voronova, 1982: 125 (ssp.) **A:** RU (FE: Wrangel Is.)
carnifex burdigalensis Bechyné, 1949b: 86 (ssp.) **E:** FR
carnifex carnifex Fabricius, 1792a: 325 (*Chrysomela*) E: AU BY CR CZ DE EN FR GE HU LA LT PL RO RU (CT ST) SK SZ UK **A:** KZ RU (WS)
konowi Weise, 1889c: 127 (*Chrysomela*, var.)
planitiae Kaszab, 1962: 53 (*Chrysomela*, ab.) [IN]
carnifex coerulescens Suffrian, 1851a: 76 (*Chrysomela*) E: FR GE SZ [# 1003]
carnifex cruentata Suffrian, 1851a: 74 (*Chrysomela*) [QS] E: PT SP
allofossulata Bechyné, 1950b: 165 (ab.) [IN]
fossulata Suffrian, 1853a: 101 (*Chrysomela*)
isofossulata Bechyné, 1950b: 165 (ab.) [IN]
pseudocruentata Bechyné, 1950b: 164 (ab.) [IN]
pseudofossulata Bechyné, 1950b: 165 (ab.) [IN]
toletensis Bechyné, 1949b: 86 (ab.) [IN]
carnifex hollandei Bourdonné, 2012: 117 (*Chalcoidea*) [QS] **N:** AG

carnifex melanaria Suffrian, 1851a: 115 (*Chrysomela*) **E:** BH CR FR GE IT
aethiops Olivier, 1807: 519 (*Chrysomela*) [# 1004]
metallica Sirgue, 1929: 35 (*Chrysomela*, var.) [HN]
provincialis Harold, 1874b: 152 (*Chrysomela*) [RN]
dieckmanni Mohr, 1966: 94 (*Chrysomela*) **A:** CH (XIN)
immarginata Rybakow, 1884a: 135 (*Chrysomela*) **A:** CH (XIN) KI KZ
unicolor Gebler, 1844: 105 (*Chrysomela*) [HN]
instabilis Mäklin, 1878: 30 (*Chrysomela*) **A:** RU (ES FE WS)
interstincta graellsii Pérez Arcas, 1872: 124 (*Chrysomela*) **E:** PT SP
escorialensis Bechyné, 1949b: 85 (ab.) [IN]
interstincta interstincta Suffrian, 1851a: 79 (*Chrysomela*) **E:** IT
interstincta subseriata Suffrian, 1851a: 80 (*Chrysomela*) **E:** FR IT **[# 481, # 483]**
coiffaiti Bechyné, 1949b: 85 (ssp.)
depressa Suffrian, 1854a: 317 (*Chrysomela*) [# 481]
ivanovi Mikhailov, 2022: 280 **A:** KZ
janbechynei Cobos, 1953: 129 **E:** SP
curvilinea Weise, 1884b: 164 (*Chrysomela*) [HN]
levi Okhrimenko, 1990: 64 **E:** RU (ST: Taman' Peninsula) **A:** RU (WS: Omsk Region)
lia Jacobson, 1896b: 551 (*Chrysomela*) **A:** AF IN KI TD TM UZ
freyi Bechyné, 1950b: 167
haarlovi Jakob, 1962: 194 (*Chrysomela*)
lopatini Mohr, 1966: 103 (*Chrysomela*) **A:** MG RU (ES)
luchti Lopatin, 2000c: 130 **A:** CH (XIN)
luteocincta Fairmaire, 1864: 647 (*Chrysomela*) **N:** AG
marginata bodemeyeri Weise, 1910c: 36 (*Chrysomela*) **A:** AB AR GG IN TR **[# 482, # 562, # 626]**
pinarbasiensis Özdikmen, Coral Şahin & Bal, 2020a: 86 (ssp.) [# 545]
unificans Bechyné, 1950b: 166 (ssp.) [# 545]
marginata dierythra Rottenberg, 1872a: 243 (*Chrysomela*) **E:** IT (Sicily) **[# 482]**
convexior Bechyné, 1948b: 15 (ssp.)
marginata finitima Brown, 1962: 67 **A:** RU (ES FE WS: Arctic territories) **NAR**
borealis L. N. Medvedev, 1980b: 94 (ssp.)
marginata glacialis Weise, 1884d: 391 (*Chrysomela*, var.) **E:** AU IT SZ **[# 482, # 578]**
marginata marginata Linnaeus, 1758: 371 (*Chrysomela*) **E:** AL AU BE BH BU BY CR CZ DE EN FI FR GE GR HU IT KO KR LA LS LT MD ME NL NM NR PL RO RU (NT ST) SB SK SL SV SZ TR UK **A:** TR **[# 482, # 589]**
amata Bechyné, 1949b: 86 (ab.) [IN]
caroniensis Bechyné, 1956a: 86 (ab.) [IN]
cinctella Gyllenhal, 1827: 650 (*Chrysomela*, var.)
diachalca Bechyné, 1946a: 107, 111 (ab.) [IN]

lugdunensis Bechyné, 1949b: 86 (ab.) [IN]
marginicollis Derenne, 1949: 171 (*Chrysomela*, var.)
parisiensis Bechyné, 1949b: 86 (ab.) [IN]
portai Bechyné, 1948b: 15 (ssp.)
purini Jacobson, 1896a: 554 (*Chrysomela*, var.)
roubali Bechyné, 1946a: 110 (ssp.) [RN]
rugosopunctata Roubal, 1917: 3 (*Chrysomela*, var.)
solitaria Weise, 1884d: 391 (*Chrysomela*, var.) [IN] [# 577]
trebinjensis Roubal, 1917: 3 (*Chrysomela*, var.)
marginata sculpticollis Bechyné, 1948b: 15 (ssp.) **E:** FR SP **[# 482, # 483]**
iniussa Bechyné, 1950b: 165 (ssp.) [# 544]
pusa Lopatin, 1962c: 323 (*Chrysomela*) **A:** KZ (Altai) MG RU (ES SS)
purkynei Mohr, 1966: 103 (*Chrysomela*, ssp.)
rufilabris Faldermann, 1835: 443 (*Chrysomela*) **A:** MG RU (ES)
perfecta Jacobson, 1901b: 124 (*Chrysomela*, var.) [IN] [# 577]
sajanica Jacobson, 1925a: 52 (*Chrysomela*) **A:** RU (SS)
sanguineocincta Crotch, 1872: 268 (*Chrysomela*) **A:** IS JO LE SI SY **[# 482, Appendix]**
sellata Weise, 1894d: 92 (*Chrysomela*) **A:** TR
bruneli Demaison, 1896: 12 (*Chrysomela*)
cilissa Jacobson, 1924b: 81 (*Chrysomela*)
concolor Demaison, 1896: 12 (*Chrysomela*, var.)
nebulosa Demaison, 1896: 12 (*Chrysomela*, var.)
songarica Gebler, 1842: 39 (*Chrysomela*) **E:** RU (ST) TR **A:** AB AR GG IN IS KI KZ TM TR UZ
amasiensis Weise, 1894d: 92 (*Chrysomela*)
brullei Portevin, 1934: 232 (*Chrysomela*) [RN]
circumducta Ménétriés, 1848: 52 (*Chrysomela*)
chalybea Brancsik, 1899: 103 (*Chrysomela*, var.)
cyanescens Jacobson, 1894b: 159 (*Chrysomela*, var.)
hyrcana Weise, 1884d: 389 (*Chrysomela*)
lepida Brullé, 1833: 270 (*Chrysomela*) [HN]
superstes roulleaui Kocher, 1959: 55 (*Chrysomela*, ssp.) **N:** MO
superstes siruense Kocher, 1953c: 95 (*Chrysomela*, ssp.) **N:** MO
superstes superstes Bedel, 1921: 61 (*Chrysomela*, ssp.) **N:** MO
altiatlantica Kocher, 1959: 53 (*Chrysomela*, var.) [IN] [# 577]
antoinei Kocher, 1959: 53 (*Chrysomela*, var.) [IN] [# 577]
crebieri Kocher, 1959: 53 (*Chrysomela*, var.) [IN] [# 577]
ifranensis Kocher, 1959: 52 (*Chrysomela*, var.) [IN] [# 577]
itoensis Kocher, 1959: 74 (*Chrysomela*, ab.) [IN]
taygetana Bechyné, 1952b: 374 **E:** GR

tesari elbursica Lopatin, 1981a: 373 **A**: IN TM
tesari tesari Roubal, 1936a: 68 (*Chrysomela*) **A**: AB AR GG TR
kulzeri Bechyné, 1950b: 166
turgaica Jacobson, 1910b: 59 (*Chrysomela*) **A**: KZ
vagecincta sarroensis Kocher, 1959: 56 (*Chrysomela*) **N**: MO
obscuriventris Codina Padilla, 1960: 66 (*Chrysomela*, ssp.)
vagecincta vagecincta Fairmaire, 1875b: 538 (*Chrysomela*) **N**: AG MO TU
vishnu Hope, 1831: 30 (*Chrysomela*) **A**: BT ID (HP SD) NP **ORR**
cingulata Baly, 1860c: 97 (*Chrysomela*)
parvati Daccordi, 1982a: 396
waldheimi Kippenberg, 2010a: 69 [RN] **A**: CH (XIN) KI KZ MG RU (SS) TD UZ [# 482, # 590]
sulcata Fischer von Waldheim, 1842: 25 (*Chrysomela*) [HN]
zamotajlovi L. N. Medvedev & Okhrimenko, 1992: 866 **E**: RU (ST: Lagonaki Plateau)
zangana S.-H. Chen & S.-Y. Wang, 1981b: 512, 516 **A**: CH (XIZ)
nyalamana S.-H. Chen & S.-Y. Wang, 1981b: 512, 516
subgenus *Chrysocrosita* Bechyné, 1950b: 90 type species: *Heliostola spectabilis* Motschulsky, 1860
Bechynea L. N. Medvedev, 1966: 40 type species: *Chrysolina kabakovi* L. N. Medvedev, 1966
Lithocrosita L. N. Medvedev, 1982: 79 type species: *Chrysolina rugulosa* sensu Medvedev (1982) (= *Crosita concinna* Weise, 1894)
alaschanica Jacobson, 1898: 196 (*Crosita*) **A**: CH (GAN NMO)
filchnerae Weise, 1908a: 92 (*Crosita*)
concinna Weise, 1894i: 154 (*Crosita*) **A**: MG RU (SS)
fuyunica alta S.-H. Chen, 1961: 431, 434 (ssp.) **A**: CH (XIN)
fuyunica fuyunica S.-H. Chen, 1961: 430, 434 **A**: CH (XIN)
jakowlewi Weise, 1894i: 155 (*Crosita*) **A**: RU (SS)
josefi Daccordi & S.-Q. Ge, 2009b: 806 **A**: CH (HEN SHA)
nikolskyi adzhalamica L. N. Medvedev, 1970c: 162 (ssp.) **A**: RU (FE)
nikolskyi nikolskyi Jacobson, 1898: 200 (*Chrysomela*) **A**: JA (Hokkaido) RU (FE: Sakhalin Isl., South Kuriles)
kabakovi L. N. Medvedev, 1966: 41
kawakamii S. Matsumura, 1911b: 141 (*Timarcha*)
nikolskyi sutschanica L. N. Medvedev, 1970c: 161 (ssp.) **A**: RU (FE) SC
sogdiorum Weise, 1892d: 138 (*Chrysomela*) **A**: KI
spectabilis Motschulsky, 1860b: 223 (*Heliostola*) **A**: MG RU (FE SS)
polychroma L. N. Medvedev, 1975b: 182 (ssp.)
viridipurpurea L. N. Medvedev, 1975b: 182 (ssp.)
sulcicollis przewalskyana Kippenberg, 2010a: 70 (ssp.) [RN] **A**: CH (NMO)
przewalskyi Jacobson, 1898: 197 (*Crosita*) [HN]

sulcicollis solida Weise, 1898: 207 (*Chrysomela*) **A:** NC SC
sulcicollis sulcicollis Fairmaire, 1887d: 330 (*Chrysomela*) **A:** CH (BEI HEB LIA NIN NMO SHX)
teichophila Csiki, 1901a: 117 (*Chrysomela*)
subgenus *Chrysolina* Motschulsky, 1860b: 210 [NP] [conserved name, ICZN (1984c), Opinion 1279] type species: *Chrysomela staphylaea* Linnaeus, 1758
Chloemela Gistel, 1856: 379 [NO] type species: *Chrysomela staphylaea* Linnaeus, 1758
Stichosoma Motschulsky, 1860b: 208 type species: *Chrysomela banksii* [sic!] Fabricius, 1775
bankii Fabricius, 1775: 95 (*Chrysomela*) **E:** AZ CR FR GB IT MA PT SP SZ **N:** AG CI MO MR TU **NAR**i
ausonia Schatzmayr, 1942a: 162 (*Chrysomela*, ssp.)
caesariensis Jolivet, 1951b: 4 (ssp.)
chlorizans Weise, 1884d: 379 (*Chrysomela*, var.) [IN] [# 577]
corcyraea Jolivet, 1951b: 4 (ssp.)
ibizensis Bechyné, 1950b: 93 (ssp.)
interior Jolivet, 1951d: 194 (ssp.) [RN]
maderensis Jolivet, 1951b: 4 (ssp.)
mediterranea Jolivet, 1951b: 4 (ssp.) [HN]
peloritana Schatzmayr, 1942a: 162 (*Chrysomela*, ssp.)
phaeaca Jolivet, 1951d: 194 (ssp.) [RN]
vitalei Schatzmayr, 1942a: 162 (*Chrysomela*, ssp.)
blanchei Fairmaire, 1865: 75 (*Chrysomela*) **A:** CY IS LE SY TR **N:** EG
nilotica Bechyné, 1950b: 97 (ssp.)
rufa biharica Breit, 1919: 15 (*Chrysomela*) **E:** RO
rufa crassicollis Suffrian, 1851a: 50 (*Chrysomela*) **E:** BU RO [# 1005]
pachysomoides Bechyné, 1950b: 100 (ssp.)
temeraria Bechyné, 1950b: 100 (ab.) [IN]
rufa frieseri Bechyné, 1950b: 99 (ssp.) **E:** AU CZ GE SZ
bohumilae Bechyné, 1950b: 99 (ssp.)
rufa rementina Bechyné, 1950b: 101 (ssp.) **E:** RO
rufa robusta Breit, 1919: 15 (*Chrysomela*, ssp.) **E:** CR SL
rufa rufa Duftschmid, 1825: 186 (*Chrysomela*) **E:** AU BH CR IT SL
macromela Bechyné, 1952b: 366 (ssp.)
rufa squalida Suffrian, 1851a: 31 (*Chrysomela*) **E:** CZ GE HU PL RO SK UK
dahlii Suffrian, 1851a: 28 (*Chrysomela*) [HN]
dichotoma Bechyné 1950: 99 (ab.) [IN]
diminuta Bechyné, 1950b: 99 (ssp.)
lapidaria Bechyné, 1950b: 100 (ssp.) [RN]
opulenta Suffrian, 1851a: 30 (*Chrysomela*) [HN] [# 1006]
staphylaeoides Bechyné, 1950b: 99 (ssp.)

sulcicollis Gerhardt, 1911: 337 (*Chrysomela*, var.) [HN]

staphylaea daurica Gebler, 1832: 73 (*Chrysomela*) **A**: CH (NMO XIN) KI KZ MG RU (ES FE) SC [# 484]

palliata Jacobson, 1901b: 125 (*Chrysomela*, var.) [HN] [IN] [# 577]

staphylaea staphylaea Linnaeus, 1758: 370 (*Chrysomela*) **E**: AU BE BH BU BY CR CZ DE EN FA FI FR GB GE GG HU IC IT KR LA LS LT LU MD NL NM NR PL PT RO RU (CT NT ST) SB SK SL SP SZ UK **N**: LB **A**: AB GG TR RU (SS WS) **NARi** **[# 484, # 1145]**

arthritica Bechyné, 1950b: 93 (ssp.)

cuprea DeGeer, 1775: 294 (*Chrysomela*) [HN]

lederi Weise, 1878: 332 (*Chrysomela*)

lepida Stephens, 1831: 341 (*Chrysomela*) [HN]

rubrocuprea Geoffroy, 1785: 108 (*Chrysomela*)

sharpi Fowler, 1890: 304 (*Chrysomela*, var.)

subferruginea Suffrian, 1851a: 21 (*Chrysomela*)

subgenus *Chrysolinopsis* Bechyné, 1950b: 82 type species: *Chrysomela gemina* Brullé, 1839

gemina Brullé, 1839: 73 (*Chrysomela*) **N**: CI

nitens Brullé, 1839: 74 (*Chrysomela*)

subgenus *Chrysomorpha* Motschulsky, 1860b: 204 type species: *Chrysomela cerealis* Linnaeus, 1767

cerealis cerealis Linnaeus, 1767: 588 (*Chrysomela*) **E**: AU BE BH BU BY CR CZ DE FR GB GE GR HU LU ME NL NM NR PL SB SL SP SV SZ **[# 485, # 627, # 1145]**

bivittata Schrank, 1789: 66 (*Chrysomela*)

costata G. Dufour, 1902: 84 (*Chrysomela*, var.)

ericae Motschulsky, 1860b: 227 (*Chrysomorpha*) [# 1109]

fasciata Geoffroy, 1785: 108 (*Chrysomela*) [HN]

fastidita Bechyné, 1948b: 12 (ab.) [IN]

fulgens L. J. M. Dufour, 1851: 353 (*Chrysomela*)

kiesenwetteri Motschulsky, 1860b: 227 (*Chrysomorpha*)

laminula Herrich-Schäffer, 1838b: [12] (*Chrysomela*)

luxurians Olivier, 1807: 546 (*Chrysomela*)

nominata Bechyné, 1948b: 12 (forma) [IN]

ornata Ahrens, 1812b: 13 (*Chrysomela*)

octovittata Schrank, 1789: 66 (*Chrysomela*)

opulentiformis Bechyné, 1948b: 13 (natio) [IN]

pseudocerealis Bechyné, 1948b: 13 (ab.) [IN]

purkynei Bechyné, 1948b: 12 (ab.) [IN]

uzeli Bechyné, 1948b: 13 (ab.) [IN]

cerealis cyaneoaurata Motschulsky, 1860b: 227 (*Chrysomorpha*) [QS] **A:** KI MG RU (ES FE SS WS) **[# 485, # 627]**

cerealis livonica Motschulsky, 1860b: 227 (*Chrysomorpha*) [QS] **E:** EN FI LA LT NR PL RU (NT) SV **[# 485, # 627]**

cerealis megerlei Fabricius, 1801a: 439 (*Chrysomela*) **E:** AU (East) BU (Stara planina) CR CZ HU MD PL RO SK UK **[# 485, # 627]**

alternans Panzer, 1799b: 16 (*Chrysomela*) [HN]

hibernacea Bechyné, 1948b: 13 (ab.) [IN]

plorans Bechyné, 1948b: 13 (ssp.)

podolica Bechyné, 1948b: 13 (natio) [IN] [# 576]

relucentiformis Bechyné, 1948b: 13 (natio) [IN] [# 576]

tincticollis Bechyné, 1948b: 13 (ab.) [IN]

cerealis mixta Küster, 1844: 87 (*Chrysomela*) **E:** AL AU BH CR FR GR IT KO ME NM SB SZ (Ticino) **[# 485, # 627]**

fratercula Bechyné, 1952b: 364 (natio) [IN] [# 576]

mixtoidea Bechyné, 1952b: 364 (ab.) [IN]

opulentiformis Bechyné, 1948b: 13 (natio) [IN] [# 576]

violacea Schaller, 1783: 270 (*Chrysomela*) [HN, QS]

cerealis rufolineata Motschulsky, 1860b: 227 (*Chrysomorpha*) [QS] **E:** KR RU (CT ST) UK **A:** AB AR GG KZ RU (WS) TR (Black Sea area) **[# 485, # 1110]**

parumsplendens Bechyné, 1952b: 365 (ab.) [IN]

pretiosissima Bechyné, 1948b: 13 (ab.) [IN]

subgenus *Colaphodes* Motschulsky, 1860b: 212 type species: *Chrysomela hottentotta* Fabricius, 1792 (= *Chrysomela haemoptera* Linnaeus, 1758)

haemoptera Linnaeus, 1758: 369 (*Chrysomela*) **E:** AL AU BE BH BU BY CR CZ DE EN FI FR GB GE GR FR HU IT KO KR LA LS LT LU ME MD NL NM PL PT RO RU (CT NT ST) SB SK SL SP SV SZ TR UK **A:** AB GG IN TR **[# 486, # 1007]**

acuticollis Fairmaire, 1878a: clxxix (*Chrysomela*)

atra Goeze, 1777: 301 (*Chrysomela*)

atrata Gmelin, 1790: 1688 (*Chrysomela*)

byzantia Jolivet, 1951d: 194 (ssp.) [RN]

camerani Piolti, 1880: 378 (*Chrysomela*)

corvina Weise, 1916b: 75 (*Chrysomela*, var.) [RN]

erythroptera Schrank, 1781: 70 (*Chrysomela*)

hottentotta Fabricius, 1792a: 309 (*Chrysomela*) [# 1077]

molluginis Brahm, 1790: 226 (*Chrysomela*)

nigra Geoffroy, 1785: 106 (*Chrysomela*)

ottomana Jolivet, 1951b: 6 (ssp.) [HN]

persica Jolivet, 1951b: 6 (ssp.)

unicolor Suffrian, 1851a: 55 (*Chrysomela*) [HN]

subgenus *Colaphoptera* Motschulsky, 1860b: 215 type species: *Chrysomela hemisphaerica* Germar, 1817 (= *Chrysomela hemisphaerica* var. *laeta* Weise, 1882)

apsilaena Silfverberg, 1977: 93 [RN] E: RU (ST) A: GG

fuscicornis Weise, 1892j: 406 (*Chrysomela*) [HN]

kubanensis L. N. Medvedev & Okhrimenko, 1992: 873 (ssp.)

globosa banatica Csiki, 1940c: 920 (*Chrysomela*, var.) E: RO

peripherica Bechyné, 1952b: 367 (ssp.)

globosa bigorrensis Fairmaire, 1865: 77 (*Chrysomela*) E: FR SP [# 487]

globosa friulica Kippenberg, 2017: 125 (ssp.) E: AU IT

globosa globosa Panzer, 1802: 6 (*Chrysomela*) E: AU BH CR CZ GE RO SL

aerea Duftschmid, 1825: 187 (*Chrysomela*) [QS]

globosa kahleni Kippenberg, 2017: 126 (ssp.) E: BH CR IT (Friulia) SL

globosa reprehensa Bechyné, 1950b: 103 (ssp.) E: BH BU SB

planicollis Breit, 1919: 14 (*Chrysomela*) A: AR SY TR

pliginskii Reitter, 1913b: 291 (*Chrysomela*) E: KR

taurica Breit, 1919: 19 (*Chrysomela*)

purpurascens cartolarii Kippenberg, 2017: 103 (ssp.) E: IT

purpurascens crassimargo Germar, 1823: 584 (*Chrysomela*) E: AU CZ GE HU IT SL SZ [# 488]

allobrogorum Bechyné, 1950b: 109 (ssp.)

bavarica Bechyné, 1950b: 108 (ssp.)

ominosa Bechyné, 1950b: 108 (ssp.)

stoeckleini Bechyné, 1950b: 109 (ssp.)

purpurascens franzi Bechyné, 1952b: 368 (ssp.) [RN] E: CR SL

croatica Franz, 1952: 5 (*Chrysomela*, ssp.) [HN]

viridis Weise, 1882b: 359 (*Chrysomela*, var.) [HN]

purpurascens goricae Kippenberg, 2017: 112 (ssp.) E: IT SL

purpurascens laeta Weise, 1882b: 360 (*Chrysomela*, var.) E: CR SL [# 1008]

fallaciosa J. Müller, 1948: 92 (*Chrysomela*) [HN]

hemisphaerica Germar, 1817: 204 (*Chrysomela*) [HN] [# 1008]

muelleri Kippenberg, 2010a: 70 (ssp.) [RN]

rufocuprea Suffrian, 1851a: 41 (*Chrysomela*) [UN] [# 1008]

purpurascens ligurica Kippenberg, 2017: 113 (ssp.) E: IT

purpurascens notiophila Bechyné, 1952b: 368 E: BU BH KO ME

purpurascens praealpina Kippenberg, 2017: 114 (ssp.) E: AU IT SL

purpurascens purpurascens Germar, 1822a: 6 (*Chrysomela*) E: BE CZ FR GE PL SK [# 1009, # 1078]

avia Weise, 1887a: 185 (*Chrysomela*, var.)

carpathica Papp, 1946c: 23 (*Chrysomela*, ab.) [IN]

dahlii Matzek, 1843: 155 (*Chrysomela*)

purpurascens scarisoarae Kippenberg, **ssp. nov.** E: RO [# 555]

scarisoarae Kippenberg, 2017: 119 (ssp.) [RN, UN]

purpurascens sutilis Bechyné, 1950b: 109 (ssp.) E: BH RO [# 1078]

banatica Bechyné, 1950b: 110 (ssp.) [HN]

bechyneana Kaszab, 1962: 78 (*Chrysomela*) [RN]

globata Suffrian, 1851a: 26 (*Chrysomela*) [UN] [# 1078]

plumbeonigra Bechyné, 1950b: 115 (ssp.)

plumbeonigra Reitter, 1913c: 114 (*Chrysomela*, ab.) [IN]

rosti Weise, 1892j: 406 (*Chrysomela*) E: RU (ST) A: GG

subgenus *Colaphosoma* Motschulsky, 1860b: 216 type species: *Chrysomela goettingensis* Linnaeus, 1761 (= *Chrysomela sturmi* Westhoff, 1882)

sturmi Westhoff, 1882: 268 (*Chrysomela*, var.) E: AL AU BE BH BU BY CR CZ DE EN FI FR GB GE FR HU IT KR KZ LA LS LT LU MD NL NM NR PL RO RU (CT NT ST) SB SK SL SV SZ SP TR UK A: AB GG KZ RU (ES SS WS) [# 489, # 1111]

diversipes Bedel, 1892b: 147 (*Chrysomela*) [RN]

fuscipes Gmelin, 1790: 1672 (*Chrysomela*) [HN]

goettingensis Linnaeus, 1761: 160 (*Chrysomela*) [HN]

nigrita Townson, 1797: 459 (*Chrysomela*) [HN]

polonica Weise, 1883d: 253 (*Chrysomela*, var.)

subgenus *Crositops* Marseul, 1883: 105 type species: *Chrysomela pedestris* Gebler, 1823

kabaki Lopatin, 1988b: 587 A: KZ (South Altai)

pedestris Gebler, 1823: 118 (*Chrysomela*) A: KZ RU (SS WS)

pterosticha Fischer von Waldheim, 1842: 25 (*Chrysomela*)

roddi misha Bieńkowski, 2019: 367 (ssp.) E: RU (CT: Lipetsk Reg.)

roddi roddi Jacobson, 1897a: 432 (*Chrysomela*) E: RU (CT: Samara Reg., Bashkortostan) A: RU (WS: Chelyabinsk Reg.)

subgenus *Cyrtochrysolina* Kippenberg, 2012c: 146 type species: *Chrysomela marcasitica* Germar, 1823

Cyrtochrysolina Kippenberg, 1994: 59 [UN]

marcasitica marcasitica Germar, 1823: 585 (*Chrysomela*) E: AU BH CR HU ME SL [# 1010]

dissipabilis Bechyné, 1950b: 106 (ssp.) [# 546]

marcasitica turgida Weise, 1882b: 355 (*Chrysomela*, var.) E: CZ PL RO SK UK

bokori Csiki, 1953: 129 (*Chrysomela*, ab.) [IN]

cupreopurpurea Gerhard, 1909: 420 (*Chrysomela*, ab.) [IN]

gerhardi Weise, 1916b: 80 (*Chrysomela*, ab.) [IN]

pannonica Weise, 1882b: 355 (*Chrysomela*, var.)

subgenus *Diachalcoidea* Bechyné, 1955a: 349 type species: *Chrysomela sacarum* Weise, 1890

Paradiachalcoidea Daccordi, 1978a: 745 type species: *Chrysolina vignai* Daccordi, 1978 *aegyptiaca* Olivier, 1807: 528 (*Chrysomela*) **A:** IN IQ IS JO LE SI SY TR **N:** EG LB

aleppensis Bechyné, 1955a: 350 (ssp.)

assurensis Bechyné, 1955a: 350 (ssp.)

palmyrensis Bechyné, 1955a: 350

rufomarginata Baly, 1879c: 191 (*Chrysomela*) [HN]

rufomarginata Marseul, 1887: 127 (*Chrysomela*) [HN]

dohrnii Fairmaire, 1865: 81 (*Chrysomela*) **A:** LE SY

peyroni Ancey, 1876: 94 (ssp.)

sacarum embiensis Lopatin, 1996: 641 (ssp.) **A:** KZ (West)

sacarum sacarum Weise, 1890a: 479 (*Chrysomela*) **A:** AF IN KI KZ (South and South-East) TD TM TR UZ

badakhshanica Iablokoff-Khnzorian, 1978: 119 (*Chrysomela*)

immatura Jacobson, 1893b: 244 (*Chrysomela*, var.)

subgenus *Erythrochrysa* Bechyné, 1950b: 91 type species: *Chrysomela polita* Linnaeus, 1758

kirigaminensis Suzuki & Saito, 2011: 132 **A:** JA (Honshu) **[# 490]**

ogloblini Ter-Minassian, 1950: 134 (*Chrysomela*) **A:** AB AR GG LE SY TR **[# 491]**

polita Linnaeus, 1758: 370 (*Chrysomela*) **E:** AL AU BE BH BU BY CR CZ DE EN FI GB GE GR HU IT FR KO KZ LA LT LU MD NL NM NR PL RO RU (CT NT ST) SB SK SL SP SV SZ TR UK **A:** CH (NMO SHN XIN) CY IN IS KI KZ LE MG RU (ES FE WS) SY TR UZ **[# 1145]**

adamsi Baly, 1879c: 190 (*Chrysomela*)

epipleuralis Jacobson, 1896a: 554 (*Chrysomela*, var.)

kafkana Reitter, 1898c: 359 (*Chrysomela*, var.)

menthae Schrank, 1776: 63 (*Chrysomela*)

subgenus *Euchrysolina* Bechyné, 1950b: 83 type species: *Chrysomela graminis* Linnaeus, 1758

graminis auraria Motschulsky, 1860b: 225 (*Dlochrysa*) **A:** CH (HEI JIL NMO TAI) MG RU (FE) SC

falsocerealis Bechyné, 1950b: 85 (ab.) [IN]

graminis graminis Linnaeus, 1758: 369 (*Chrysomela*) **E:** AU BE BH BU BY CR CZ DE EN FI FR GB GE GR HU IT KR KZ LA LT NL NM NR PL RO RU (CT NT ST) SB SK SL SV SZ UK **A:** AB GG KZ MG RU (ES SS WS) TR **[# 1011, # 1079]**

artemisiae Motschulsky, 1860b: 225 (*Chrysomela*)

christianae Mallet, 1933: 75 (*Chrysomela*, ab.) [IN]

fulgida Fabricius, 1801a: 432 (*Chrysomela*)

isofulgida Bechyné, 1950b: 85 (ab.) [IN]

nigrocuprea Mallet, 1924: 77 (*Chrysomela*, var.)

ponoriana Marcu, 1957: 544 (*Chrysomela*, ab.) [IN]

rugulosa Mallet, 1933: 75 (*Chrysomela*, var.)

rutilans Gravenhorst, 1807: 134 (*Chrysomela*)

semiviolacea Mallet, 1924: 77 (*Chrysomela*, ab.) [IN]
taupini Mallet, 1924: 77 (*Chrysomela*, var.)
graminis santonici Contarini, 1847: 189 (*Chrysomela*) E: IT FR (Corsica)
edomita Bechyné, 1950b: 85 (ab.) [IN]
mediterranea Bechyné, 1950b: 85 (ssp.)
schallehni Reineck, 1915a: 240 (*Chrysomela*, var.)
virgata Motschulsky, 1860b: 224 (*Dlochrysa*) **A**: CH (HEI JIL NMO) JA NC RU (FE) SC
eximia Baly, 1862a: 20 (*Chrysomela*)
obscurofasciata Jacoby, 1885b: 208 (*Chrysomela*)
subgenus *Gnathomela* Jacobson, 1895b: 297 [RN] type species: *Orina dolens* Kraatz, 1885
Pachygnatha Weise, 1892h: 352 [HN] type species: *Chrysomela dolens* Kraatz, 1885
dolens Kraatz, 1885: 298 (*Orina*) **A**: TD UZ
chrysomeliformis Bogachev, 1961: 112 (*Lobodera*)
glasunovi Bechyné, 1962: 182 (*Gnathomela*) **A**: UZ
nuratavica Lopatin, 1972: 349 (*Gnathomela*, ssp.)
igorlopatini Kippenberg, 2010a: 70 [RN] **A**: UZ
meridionalis Lopatin, 1972: 350 (*Gnathomela*) [HN]
laevigata kirghizica Lopatin, 1990d: 51 (*Gnathomela*, ssp.) **A**: KI
laevigata laevigata Lopatin, 1972: 349 (*Gnathomela*) **A**: KI UZ
mutica Lopatin, 1972: 351 (*Gnathomela*) **A**: UZ
ovczinnikovi Lopatin, 1990d: 51 (*Gnathomela*) **A**: KI
praestans Lopatin, 1972: 350 (*Gnathomela*) **A**: TD UZ
valida Lopatin, 1972: 349 (*Gnathomela*) **A**: KZ UZ
subgenus *Heliostola* Motschulsky, 1860b: 190 type species: *Chrysomela islandica* Germar, 1823 (= *Chrysomela lichenis* Richter, 1820)
carpathica carpathica Fuss, 1856: 31 (*Chrysomela*) E: RO UK
hopffgarteni Weise, 1890c: 30 (*Chrysomela*, var.)
internitens Bechyné, 1947a: 61 (ab.) [IN]
jaspidea Bechyné, 1947a: 61 (ab.) [IN]
lucifuga Bechyné, 1947a: 61 (ab.) [IN]
pseudotypica Kaszab, 1962: 77 (*Chrysomela*, ab.) [IN]
carpathica gabrieli Weise, 1903c: 163 (*Chrysomela*, var.) E: CZ PL SK (Tatra)
adulterata Bechyné, 1947a: 62 (ab.) [IN]
gibbipennis Faldermann, 1835: 441 (*Chrysomela*) **A**: KZ RU (ES SS WS) [# 492]
despecta Kontkanen, 1958: 208
schewyrewi Jacobson, 1896a: 547 (*Chrysomela*)
katonica Lopatin, 1988b: 590 **A**: KZ (Altai) [# 493]
lichenis ahena Germar, 1823: 586 (*Chrysomela*) E: AU SZ
cunctabunda Bechyné, 1947a: 63 (ab.) [IN]
neta Bechyné, 1947a: 63 (ab.) [IN]

lichenis lichenis Richter, 1820: 6 (*Chrysomela*) E: CZ (Krkonoše) GE (Sachsen) PL [# 591]
februa Bechyné, 1947a: 62 (ab.) [IN]
hiemata Bechyné, 1947a: 62 (ab.) [IN]
islandica Germar, 1823: 585 (*Chrysomela*)
mendosa Bechyné, 1947a: 62 (ab.) [IN]
nigerrima Weise, 1882b: 350 (*Chrysomela*, var.)
prunicea Bechyné, 1947a: 62 (ab.) [IN]
quadricollis Weise, 1882b: 350 (*Chrysomela*, var.)
lichenis moraviaca Weise, 1882b: 350 (*Chrysomela*, var.) E: CZ PL SK
atypica Kaszab, 1962: 77 (*Chrysomela*, ab.) [IN]
conducta Bechyné, 1947a: 63 (ab.) [IN]
pervilis Bechyné, 1947a: 63 (ab.) [IN]
plesienis Bechyné, 1947a: 63 (ab.) [IN]
tatrica Bechyné, 1947a: 63 (ab.) [IN]
lichenis rhipaea Weise, 1898: 204 (*Chrysomela*, var.) E: RO UK
jugosa Bechyné, 1947a: 64 (ab.) [IN]
urgens Bechyné, 1947a: 64 (ab.) [IN]
suturalis Bechyné, 1947a: 63 (ssp.) [HN]
schneideri Weise, 1882b: 349 (*Chrysomela*) E: PL SK (Tatra)
subgenus *Hypericia* Bedel, 1899: 258 type species: *Chrysomela hyperici* Forster, 1771
anatolica Dahlgren, 1984: 42 (*Chrysomela*) A: BU TR
brunsvicensis Gravenhorst, 1807: 135 (*Chrysomela*) E: AU BE CZ DE FR GB GE LU NL PL PT SP SZ
duplicata Megerle von Mühlfeld, 1815: 148 (*Chrysomela*)
subseriepunctata Dietrich, 1857: 135 (*Chrysomela*)
corcyria Suffrian, 1851a: 133 (*Chrysomela*) E: GR IT
brenskei Bechyné, 1952b: 379 (ab.) [IN]
corsica Bechyné, 1949a: 52 (ab.) [IN]
extempulans Bechyné, 1949a: 52 (ab.) [IN]
falsocuprina Bechyné, 1952b: 379 (ab.) [IN]
falsogeminata Bechyné, 1952b: 379 (ab.) [IN]
patibilis Bechyné, 1949a: 52 (ab.) [IN]
plausibilis Bechyné, 1949a: 52 (ab.) [IN]
reuleauxi Brenske, 1890: 14 (*Chrysomela*)
cuprina Duftschmid, 1825: 177 (*Chrysomela*) E: AL AU BU BY CR CZ FR GR HU IT PL RO RU (ST) SB SK SL UK A: AB GG IN TR [# 592]
aeneola Šprinqlová, 1945: 143 (*Chrysomela*, ab.) [IN]
bechynei Šprinqlová, 1945: 142 (*Chrysomela*, ab.) [IN]
daghestanica Reitter, 1913c: 111 (*Chrysomela*)

hypericicola Šprinlová, 1945: 143 (*Chrysomela*, ab.) [IN]
nigra Reitter, 1913: 112 (*Chrysomela*, ab.) [IN]
nigritula Bechyné, 1949a: 52 (ab.) [IN]
nigritula Bechyné, 1954a: 90 (ssp.)
perviridis Reitter, 1913c: 112 (*Chrysomela*, ab.) [IN]
semicoerulea Šprinlová, 1945: 143 (*Chrysomela*, ab.) [IN]
semicuprea Šprinlová, 1945: 143 (*Chrysomela*, ab.) [IN]
semiviridis Šprinlová, 1945: 143 (*Chrysomela*, ab.) [IN]
staneki Bechyné, 1949a: 52 (ssp.)
vulgaris Šprinlová, 1945: 142 (*Chrysomela*, ab.) [IN]

didymata Scriba, 1791: 294 (*Chrysomela*) **E:** AL BH BU CR FR GR HU IT ME NM UK **A:** AR CY IN KZ LE SY TM TR
elongatior Bechyné, 1952b: 379 (ssp.)
euviridis Bechyné, 1952b: 379 (ab.) [IN]
quinquejugis Marsham, 1802: 173 (*Chrysomela*)

difficilis Motschulsky, 1860b: 228 (*Taeniosticha*) **A:** CH (JIL NMO) JA KZ MG RU (ES FE SS) SC **[# 593]**
dilecta Bechyné, 1952b: 380 (ssp.)
exgeminata Bechyné, 1952b: 380 (ssp.)
nikinoja Bechyné, 1950b: 155
pseudogeminata Bechyné, 1950b: 156
shikokensis Nakane, 1963: 19
sibirica Weise, 1887h: 177 (*Chrysomela*) [HN]
ussuriensis Jacobson, 1901b: 126 (*Chrysomela*)
yezoensis S. Matsumura, 1911b: 142 (*Chrysomela*)

fricata Bechyné, 1950b: 158 **A:** CH (FUJ GUI)

geminata Paykull, 1799a: 65 (*Chrysomela*) **E:** AL AU BE BH BU BY CR CZ DE EN FR GE GR IT NL NM NR LA LT LU PL PT RO RU (CT NT) SB SK SL SP SV SZ UK **A:** AB GG TR **[# 594]**
approximata Megerle von Mühlfeld, 1815: 148 (*Chrysomela*)
lugubrina Csiki, 1953: 129 (*Chrysomela*, ab.) [IN]
werneri Šprinlová, 1945: 142 (*Chrysomela*, ab.) [IN]
utens Bechyné, 1949a: 53 (ab.) [IN]

gracilis Bechyné, 1950b: 157 **A:** CH (GUA GUI GUX HUB JIX SCH YUN) **ORR**

hyperici Forster, 1771: 20 (*Chrysomela*) **E:** AL AU BE BH BU BY CR CZ DE EN FI FR GB GE GR HU IT KR LA LT LU ME NL NM NR PL PT RO RU (CT NT ST) SB SK SP SV SZ TR UK **A:** AB AR CY GG IN KZ RU (WS) TR UZ **N:** AG TU **AUSi NARi** **[# 1145]**
ambigua Weise, 1884d: 407 (*Chrysomela*, var.) [IN] [# 577]
anglica Gmelin, 1790: 1689 (*Chrysomela*)

cillerana Moll, 1784b: 191 (*Chrysomela*)
fucata Fabricius, 1781: 126 (*Chrysomela*)
gemellata Geoffroy, 1785: 110 (*Chrysomela*)
privigna Weise, 1884d: 407 (*Chrysomela*, var.) [IN] [# 577]
viridula Laicharting, 1781: 147 (*Chrysomela*)

medogana S.-H. Chen & S.-Y. Wang, 1981b: 511, 516 **A:** CH (XIZ)

nikkoensis Jacoby, 1885b: 207 (*Chrysomela*) **A:** JA

ohoi Chûjô, 1958b: 50 **A:** CH (TAI) **[# 494]**

quadrigemina Suffrian, 1851a: 125 (*Chrysomela*) **E:** AL AU AZ BE CR DE FR GE GR IT MA NL NM PT SP SZ **N:** AG MO MR TU **AFRi AUSi NARi NTRi**
alternata Suffrian, 1851a: 127 (*Chrysomela*)
conversaria Bechyné, 1952b: 380 (ssp.)
indigena Weise, 1884d: 405 (*Chrysomela*, var.) [IN] [# 577]
isidis Bechyné, 1952b: 380 (ssp.)
normandi Bechyné, 1949a: 51 (ssp.)

syriaca Weise, 1884d: 408 (*Chrysomela*) **A:** IS SY

subgenus *Hypochalcoidea* Bourdonné, 2012: 123 type species: *Chrysomela rufilabris* var. *discriminata* Jacobson, 1901

discriminata Jacobson, 1901b: 124 (*Chrysomela*, var.) **A:** CH (NMO) MG RU (ES) **[# 578]**
fuscipes Weise, 1890a: 479 (*Chrysomela*) [HN]
infuscipes Weise, 1916b: 76 (*Chrysomela*) [RN]

mongolensis Lopatin, 1966a: 238 (*Chrysomela*) **A:** MG

subgenus *Jeanclaudia* Mikhailov, 2009b: 183 type species: *Chrysomela ordinata* Gebler, 1823

asperata Lopatin, 1990c: 54 (ssp.) **A:** KZ (South Altai) RU (SS: Altai)

baicalica Mohr, 1966a: 96 (*Chrysomela*) **A:** RU (ES: Yakutia)
cyaneovinosa L. N. Medvedev, 1978e: 120

ordinata Gebler, 1823: 119 (*Chrysomela*) **E:** RU (CT: Ulyanovsk Reg.) **A:** KZ RU (SS WS)

tatianae Mikhailov, 2010: 606 **A:** KZ (Altai)

tibialis Jacobson, 1896a: 548 (*Chrysomela*) [HN] **A:** RU (SS) **[# 495]**
neotibialis Kippenberg, 2010a: 70 [RN]

undulata Gebler, 1833: 308 (*Chrysomela*) [HN] **A:** RU (SS) **[# 496]**

subgenus *Lithopteroides* E. Strand, 1935: 295 [RN] type species: *Chrysomela musiva* Gebler, 1830 (= *Chrysomela exanthematica* Wiedemann, 1821)
Lithoptera Motschulsky, 1860b: 210 [HN] type species: *Chrysomela musiva* Gebler, 1830 (= *Chrysomela exanthematica* Wiedemann, 1821)
Neopotanina Abdullah & Qureshi, 1969b: 119 type species: *Neopotanina hamidi* Abdullah & Qureshi, 1969 (= *Chrysomela exanthematica* Wiedemann, 1821)

exanthematica exanthematica Wiedemann, 1821: 178 (*Chrysomela*) **A:** AF CH (ANH CHQ FUJ GAN GUA GUI JIA JIL HEB HEI HEN HUB HUN LIA NIN QIN SCH SHA SHX TAI XIN XIZ YUN ZHE) ID (KA UT) JA NC NP PA RU (ES FE) SC **ORR**
consimilis Baly, 1874: 172 (*Chrysomela*) [RN]
hamidi Abdullah & Qureshi, 1969b: 119 (*Neopotanina*)
laeviguttata Chûjô, 1958b: 52
laevipunctata Lewis, 1879: 28 (*Chrysomela*) [RN]
nepalensis Hope, 1831: 30 (*Chrysomela*)
speculifera Kollar & L. Redtenbacher, 1844: 558 (*Chrysomela*)
subaenea Motschulsky, 1860b: 229 (*Lithoptera*) [HN]
exanthematica gemmifera Motschulsky, 1860b: 229 (*Lithoptera*) **A:** MG RU (ES SS WS: Polar Ural)
guttifera Motschulsky, 1860b: 229 (*Lithoptera*)
nigrogemmata Motschulsky, 1860b: 229 (*Lithoptera*)
marseuli Weise, 1898: 200 (*Chrysomela*, var.) [RN]
sericata Jacobson, 1901b: 125 (*Chrysomela*, var.) [IN] [# 577]
exanthematica musiva Gebler, 1830: 215 (*Chrysomela*) [RN] **A:** KI KZ MG RU (SS)
guttata Gebler, 1817: 316 (*Chrysomela*) [HN]
subgenus *Lopatinica* Kippenberg, 2012b: 218 type species: *Chrysomela porphyrea* Faldermann, 1837
adzharica adzharica Lopatin, 1988b: 589 **A:** GG TR
adzharica excavata Kippenberg, 2012b: 323 (ssp.) **A:** TR
adzharica heinzi Kippenberg, 2012b: 322 (ssp.) **A:** TR
boluensis Kippenberg, 2012b: 326 **A:** TR
curvata Kippenberg, 2012b: 320 **A:** GG TR [# 497]
kabalaki Şen, 2015: 252 **A:** TR
kataevi Lopatin, 2000b: 858 **A:** TR [# 497]
daccordiana Kippenberg, 2012b: 325
porphyrea kubanica Bechyné, 1952b: 370 (ssp.) **E:** RU (ST) **A:** GG
exsul Bechyné, 1952b: 371 (ssp.)
minutior Bechyné, 1952b: 370 (ssp.)
porphyrea porphyrea Faldermann, 1837: 354 (*Chrysomela*) **E:** RU (ST) **A:** GG TR
differens Franz, 1952: 6 (*Chrysomela*)
diga Bechyné, 1952b: 370 (ssp.)
erivanicola Bechyné, 1952b: 370 (ssp.)
kutaisa Bechyné, 1952b: 371 (ssp.)
trapezicollis Bechyné, 1952b: 371
subgenus *Maenadochrysa* Bechyné, 1950b: 116 type species: *Chrysomela femoralis* Olivier, 1791
affinis affinis Fabricius, 1787: 67 (*Chrysomela*) **N:** AG TU
xanthophryna Bechyné, 1950b: 118 (ssp.)

affinis atlantica Escalera, 1914: 524 (*Chrysomela*) N: MO
affinis baetica Suffrian, 1851a: 16 (*Chrysomela*) E: SP [# 1012]
chalcomelaena Bechyné, 1950b: 119 (ab.) [IN]
vectigalis Bechyné, 1950b: 119 (ab.) [IN]
affinis hecateia Bechyné, 1950b: 119 (ssp.) E: SP
indomita Bechyné, 1950b: 119 (ssp.)
affinis hyacinthina Suffrian, 1851a: 45 (*Chrysomela*) E: IT (Calabria, Sicily)
bruttiana Bourdonné, 1999: 45 (ssp.)
cribellata Suffrian, 1851a: 46 (*Chrysomela*)
thalassina Reiche & Saulcy, 1858b: 29 (*Chrysomela*)
affinis ilerdensis Petitpierre, 2019: 450 (ssp.) E: SP
affinis lepineyi Kocher, 1959: 48 (*Chrysomela*) N: MO
affinis mesatlantica Kocher, 1959: 47 (*Chrysomela*) N: MO
affinis nevadensis Cobos, 1952: 4 (*Chrysomela*, ssp.) E: SP
crassipes P. H. Lucas, 1847a: pl. 45 [= 1847c: 533] (*Chrysomela*) N: AG
porphyropus Peyerimhoff, 1915a: 32 (*Chrysomela*, var.) [IN] [# 577]
femoralis aveyronensis Bechyné, 1950b: 117 (ssp.) E: FR
sortilega Bechyné, 1950b: 117 (ab.) [IN]
femoralis caliginosa Olivier, 1807: 521 (*Chrysomela*) E: FR IT
femoralis femoralis Olivier, 1791a: 690 (*Chrysomela*) E: FR SP
anquisita Bechyné, 1950b: 120 (ab.) [IN]
balanyensis Bechyné, 1950b: 121 (ssp.)
bargusiana Bechyné, 1950b: 121 (ssp.)
camena Bechyné, 1950b: 120 (ssp.)
confusa Suffrian, 1851a: 47 (*Chrysomela*)
laeta Weise, 1884d: 372 (*Chrysomela*, var.) [IN] [# 577]
ootypa Bechyné, 1952b: 372 (ssp.)
parumnitens Bechyné, 1952b: 372 (ssp.)
pernitescens Bechyné, 1952b: 372 (ssp.)
putealis Bechyné, 1952b: 372 (ssp.)
varipes Suffrian, 1851a: 17 (*Chrysomela*)
femoralis rufofemorata Heyden, 1870: 171 (*Chrysomela*) E: SP
vicinitatis Bechyné, 1950b: 120 (ssp.)
pseudoaenea Fairmaire, 1865: 74 (*Chrysomela*) N: MO
subgenus *Medvedevlevna* Özdikmen, 2008: 644 [RN] type species: *Chrysolina pudica* L. N. Medvedev, 1970
Jacobsonia L. N. Medvedev, 1970c: 162 [HN] type species: *Chrysolina pudica* L. N. Medvedev, 1970
Bienkowskia L. N. Medvedev, 2009a: 58 [RN] type species: *Chrysolina pudica* L. N. Medvedev, 1970

pudica L. N. Medvedev, 1970c: 162 **A:** CH (SCH)

mikhailovi Lopatin, 2011b: 379 [HN]

yurimikhailovi X.-L. Wang & S.-Q. Ge, 2020: 69 [RN]

subgenus *Melasomoptera* Bechyné, 1950b: 141 type species: *Chrysomela grossa* Fabricius, 1792

grossa chloromaura Olivier, 1807: 553 (*Chrysomela*) E: PT SP

gallega Fairmaire, 1862a: 595 (*Chrysomela*)

grossa grossa Fabricius, 1792a: 317 (*Chrysomela*) E: BH BU CR FR GR IT MA ME SL SP SZ **N:** AG TU MO **[# 498]**

dulcinula Bechyné, 1950b: 143 (ab.) [IN]

illita Weise, 1884d: 432 (*Chrysomela*, var.) [IN] [# 577]

tingitana Escalera, 1914: 525 (*Chrysomela*, var.)

lucida Olivier, 1807: 553 (*Chrysomela*) E: FR IT PT SP **[# 499]**

cheni Bechyné, 1950b: 73 [RN]

chloromaura Charpentier, 1825: 233 (*Chrysomela*) [HN]

dichroa L. J. M. Dufour, 1843: 107 (*Chrysomela*)

lucidula S.-H. Chen, 1934d: 36 [HN]

suarezi Cobos, 1952: 3 (ssp.)

torresi Bechyné, 1950b: 144 (ssp.)

lutea Petagna, 1819: 32 (*Chrysomela*) E: AL BH GR IT

laevipennis Suffrian, 1851a: 202 (*Chrysomela*)

subgenus *Minckia* E. Strand, 1935: 292 [RN] type species: *Chrysomela lamina* Fabricius, 1792 (= *Chrysomela oricalcia* O. F. Müller, 1776)

Hoplosoma Motschulsky, 1860b: 211 [HN] type species: *Chrysomela lamina* Fabricius, 1792 (= *Chrysomela oricalcia* O. F. Müller, 1776)

Sulcicollis J. R. Sahlberg, 1913c: 247 [HN] type species: *Chrysomela chalcites* Germar, 1823

chalcites Germar, 1823: 587 (*Chrysomela*) **E:** BU CR GR HU IT KR MD RO RU (CT ST) SK TR UK **A:** AB AF AR CY GG IN IQ IS SY TD TR RU (WS)

festiva Ménétriés, 1832: 235 (*Chrysomela*)

scovitzii Ménétriés, 1832: 235 (*Chrysomela*)

impavida Bechyné, 1949a: 54 (ssp.) E: GR **A:** CY IS LE SY TR **[# 595]**

oricalcia O. F. Müller, 1776: 82 (*Chrysomela*) **E:** AL AU BE BH BU CR CZ DE FR GB GE GR IT HU KR LA LS LT LU MD NL NM NR PL RO RU (CT: Kaliningrad Region) SB SL SK SV SZ TR UK **A:** TR **[# 1145]**

atrocoerulea Depoli, 1940: 308 (*Chrysomela*, ab.) [IN]

austriaca Olivier, 1791a: 717 (*Chrysomela*)

bicolor Gabriel, 1902: 60 (*Chrysomela*, var.) [HN]

bulgarensis Schrank, 1781: 70 (*Chrysomela*)

dieneri Merkl, 1897: 209 (*Chrysomela*, var.)

hobsoni Stephens, 1831: 343 (*Chrysomela*)

incrassata Marsham, 1802: 186 (*Chrysomela*)
laevicollis Olivier, 1807: 519 (*Chrysomela*)
lamina Fabricius, 1792a: 311 (*Chrysomela*) [# 1013]
nigricollis Weise, 1916b: 84 (*Chrysomela*, ab.) [IN]
olivacea Schaller, 1783: 272 (*Chrysomela*)
pagana Bechyné, 1946a: 107, 111 (*Chrysomela*, ab.) [IN]
protrusa Bechyné, 1946a: 107, 111 (*Chrysomela*, ab.) [IN]
roscida Bechyné, 1946a: 107, 111 (*Chrysomela*, ab.) [IN]
vieta Bechyné, 1946a: 107, 111 (*Chrysomela*, ab.) [IN]
peregrina Herrich-Schäffer, 1838b: [8] (*Chrysomela*) E: CR FR IT MA PT SP **N**: AG MO TU
erythromera P. H. Lucas, 1847a: pl. 45 [= 1847c: 534] (*Chrysomela*)
meridionalis Herrich-Schäffer, 1838b: [14] (*Chrysomela*)
oberndorferi Brenske, 1890: 14 (*Chrysomela*, var.)
schottii Suffrian, 1851a: 39 (*Chrysomela*)
rufoaenea Suffrian, 1851a: 122 (*Chrysomela*) E: AU CZ FR GE SK SP SZ
fallaciosa Weise, 1884d: 401 (*Chrysomela*, var.) [IN] [# 577]
subgenus *Ovosoma* Motschulsky, 1860b: 214 type species: *Chrysomela vernalis* Brullé, 1833 [# 500]
Byrrhiformis J. R. Sahlberg, 1913c: 247 type species: *Chrysomela vernalis* Brullé, 1833
Gemellata J. R. Sahlberg, 1913c: 247 type species: *Chrysomela sahlbergii* Ménétriés, 1832
Parkaniola Bechyné, 1950b: 130 type species: *Chrysolina susterai* Bechyné, 1950
atrovirens atrovirens Frivaldszky, 1876: 338 (*Chrysomela*) E: RO
atrovirens minckwitzi Apfelbeck, 1912: 249 (*Chrysomela*) E: BH
atrovirens winneguthi J. Müller, 1948: 94 (*Chrysomela*, ssp.) E: AL ME
cretica Olivier, 1807: 518 (*Chrysomela*) E: GR (Crete) [# 501, # 1014]
comisabunda Bechyné, 1950b: 126 (ab.) [IN]
extrita Bechyné, 1950b: 126 (ab.) [IN]
insulicola Bechyné, 1950b: 126 (ab.) [IN]
praevia Bechyné, 1950b: 126 (ab.) [IN]
orientalis Olivier, 1807: 512 (*Chrysomela*) E: GR TR **A**: TR [# 502, # 1015]
bythiniensis Bechyné, 1950b: 126 (ab.) [IN]
campicola Gistel, 1857b: 576 (*Chrysomela*)
thraeissa Bechyné, 1950b: 126 (ssp.)
sahlbergii Ménétriés, 1832: 235 (*Chrysomela*) [QS] E: GR **A**: AB CY GG IN IQ IS LE IQ SY TR **N**: EG [# 502]
assyrica Bechyné, 1950b: 127 (ssp.) [QS]
bleusei Bechyné, 1950b: 127 (ab.) [IN]
halysa Bechyné, 1950b: 127 [QS]
cupreopunctata Reiche & Saulcy, 1858b: 30 (*Chrysomela*)

intercalaria Bechyné, 1950b: 127 (ssp.)
palaestina Bechyné, 1950b: 127 (ssp.)
rhodia Bechyné, 1950b: 127
sahlbergii Faldermann, 1837: 358 (*Chrysomela*) [HN]
venefica Weise, 1884d: 369 (*Chrysomela*, var.) [IN] [# 577]
susterai Bechyné, 1950b: 130 E: HU KR RU (CT ST) SK UK A: AB KZ RU (SS WS)
morio Krynicki, 1832: 171 (*Chrysomela*) [HN] [# 1016]
turca Fairmaire, 1865: 74 (*Chrysomela*) E: BU GR TR A: TR IN [# 596]
vernalis cantabrica Heyden, 1870: 170 (*Chrysomela*) E: SP [# 503]
vernalis herii Herrich-Schäffer, 1838c: pl. 2b (*Chrysomela*) E: GR (Corfu) [# 503, # 1017]
vernalis pyrenaica L. J. M. Dufour, 1843: 106 (*Timarcha*) E: FR SP [# 503]
canfrancensis Bechyné, 1952a: 172 (ssp.)
carbonaria Suffrian, 1851a: 49 (*Chrysomela*)
egelida Bechyné, 1952a: 172 (ssp.)
gallica Weise, 1882b: 365 (*Chrysomela*, var.)
vernalis vernalis Brullé, 1833: 269 (*Chrysomela*) [NP] E: AL BH BU CR GR ME NM TR A: AB IN LE TR [# 503, # 1018]
incerta Herrich-Schäffer, 1838b: [10] (*Chrysomela*)
italica Weise, 1882b: 365 (*Chrysomela*, var.) [QS]
moreana Bechyné, 1950b: 125 (ab.) [IN]
muchei Mohr, 1969: 390 (*Chrysomela*, ssp.)
ottomana Bechyné, 1950b: 125 (ssp.) [QS]
ottomana Weise, 1906e: 554 (*Chrysomela*, ab.) [IN]
wittmeri L. N. Medvedev, 1975a: 13 A: TR
guichardi Gruev, 1976: 77 (*Chrysomela*)
subgenus *Ovostoma* Motschulsky, 1860b: 215 type species: *Chrysomela coerulea* Olivier, 1807 (= *Chrysomela olivieri* Bedel, 1892)
globipennis collina Csiki, 1953: 128 (*Chrysomela*, var.) E: RO [# 504]
deubeli Bechyné, 1948b: 14 (ssp.) [HN]
subalpina Csiki, 1953: 129 (*Chrysomela*, var.)
globipennis euminuta Bechyné, 1950b: 64 (ssp.) E: SK (Tatra) [# 504]
globipennis globipennis Suffrian, 1851a: 36 (*Chrysomela*) E: MD PL RO SK UK [# 504]
marmarosensis Reitter, 1913c: 116 (*Chrysomela*, ab.) [IN]
styriaca Bechyné, 1946a: 108, 112 (ab.) [IN]
globipennis olivieri Bedel, 1892b: 148 (*Chrysomela*) [RN] E: AL AU BH BU CR CZ GE HU KO ME MD RO SB SL SK TR UK A: AB GG TR [# 504]
azurea Bechyné, 1946a: 108, 112 (ssp.)
caerula Csiki, 1953: 128 (*Chrysomela*) [RN]
coerulea Olivier, 1807: 515 (*Chrysomela*) [HN]
ehumerosa Bechyné, 1950b: 63 (ssp.)

montanella Bechyné, 1950b: 63 (ssp.)

slovaca Bechyné, 1946a: 108, 112 (ssp.)

globipennis veneta J. Müller, 1948: 91 (*Chrysomela*, ssp.) E: IT SL [# 504]

subgenus *Palaeosticta* Bechyné, 1952b: 383 type species: *Chrysomela diluta* Germar, 1823

diluta Germar, 1823: 591 (*Chrysomela*) E: FR SP PT

nubila Breit, 1920: 83 (*Chrysomela*, ab.) [IN]

kocheri Codina Padilla, 1962: 74 (*Chrysomela*) **N**: MO

numida Reiche, 1864: 245 (*Chrysomela*) **N**: AG LB MO TU

pardoi Codina Padilla, 1962: 78 (*Chrysomela*) **N**: MO TU

ruffoi benjaminica Daccordi, 1972: 515 (*Chrysomela*, ssp.) **A**: IS

ruffoi ruffoi Daccordi, 1972: 512 (*Chrysomela*) **N**: LB

subgenus *Paracrosita* Daccordi, 1982b: 411 type species: *Chrysomela armeniaca* Faldermann, 1837

armeniaca armeniaca Faldermann, 1837: 355 (*Chrysomela*) **A**: AB AR GG

armeniaca neopersica Kippenberg & Mikhailov, **nom. nov.** [RN] **A**: AB AF IN KI TM [# 557]

iranica L. N. Medvedev, 1975a: 17 (ssp.) [HN]

persica Jakob, 1960: 26 (*Chrysomela*) [HN]

subgenus *Pezocrosita* Jacobson, 1901a: 78 [= 1901b: 120] type species: *Crosita kuznetzowi* Jacobson, 1901 (= *Chrysolina sahlbergiana jacobsoni* Lopatin, 1970)

altimontana Rybakow, 1889: 286 (*Chrysomela*) **A**: CH (QIN SCH)

amplicollis Jacobson, 1896a: 550 (*Chrysomela*) **A**: CH (QIN)

brevilata Heyden, 1886c: 277 (*Chrysomela*) **A**: KI KZ

cyrtonastes Weise, 1892k: 413 (*Chrysomela*) [RN]

cyrtonoides Weise, 1892d: 139 (*Chrysomela*) [HN]

convexicollis Jacobson, 1901b: 122 (*Chrysomela*) **A**: MG RU (SS)

cyanopurpurea Ballion, 1878: 379 (*Chrysomela*) **A**: CH (XIN)

belousovi Lopatin, 2000c: 132

dalailamai Lopatin, 1998b: 827 **A**: CH (QIN)

dubeshkoae L. N. Medvedev, 1974: 181 **A**: MG

bannikovae L. N. Medvedev, 1978a: 193

erzinica Mikhailov, 2002: 61 **A**: RU (SS: Tuva)

freyensis Kippenberg, 2010a: 70 [RN] **A**: CH (SCH)

freyi Daccordi, 1976a: 35 (*Semenowia*) [HN]

gruevi Lopatin, 2007: 178 **A**: CH (GAN)

helenae Lopatin, 1968c: 548 (*Chrysomela*) **A**: KZ

glebi Lopatin, 1988b: 585

hohxilensis Daccordi & S.-Q. Ge, 2011a: 34 **A**: CH (QIN)

hyperboreica Mikhailov, 2002: 64 **A**: RU (CT: North Ural: Kytlym Massif)

irinae Lopatin, 2008d: 833 **A:** CH (SCH)
juldusana Lopatin, 1962c: 321 (*Chrysomela*) **A:** CH (XIN) KI KZ
daccordii Lopatin, 2000c: 135 [HN]
oschanini Lopatin, 1965b: 9 (*Chrysomela*)
ketmenica Lopatin, 1970b: 192 (*Chrysomela*) **A:** KZ
koenigi Jacobson, 1896a: 552 (*Chrysomela*) **A:** KI TD
katarinae Lopatin, 1965b: 10 (*Chrysomela*)
kiritshenkoi Lopatin, 1970b: 191 (*Chrysomela*) **A:** KI
korgonica Mikhailov & Reshetnikov, 2021: 1013 **A:** RU (SS: Altai: Korgonsky Range)
kozlovi Lopatin, 1988b: 586 **A:** CH (QIN XIZ)
medvedevi Lopatin, 1970b: 186 (*Chrysomela*) **A:** KZ (Kent Mts.)
mohri Lopatin, 1970b: 191 (*Chrysomela*) **A:** KI
obovata almaatica Lopatin, 1962c: 322 (*Chrysomela*) **A:** KZ
obovata obovata Jacobson, 1896a: 553 (*Chrysomela*) **A:** KI
petrenkoi Lopatin, 1992a: 69 **A:** KZ
przewalskii Jacobson, 1896a: 551 (*Chrysomela*) **A:** CH (GAN XIZ)
burchana Lopatin, 1998b: 829
regeli Lopatin, 2002a: 113 **A:** CH (XIN)
roborowskii Jacobson, 1896a: 550 (*Chrysomela*) **A:** CH (QIN)
fallax Jacobson, 1896a: 551 (*Chrysomela*)
sahlbergiana jacobsoni Lopatin, 1970b: 183 (*Chrysomela*) [RN] **A:** RU (SS)
kuznetzowi Jacobson, 1901a: 78 (*Crosita*) [HN]
sahlbergiana sahlbergiana Jacobson, 1901b: 120 (*Crosita*) **A:** MG RU (SS)
sarcandica Lopatin, 1990d: 49 **A:** KZ
tekessica Lopatin, 2000c: 136 **A:** CH (XIN)
terskeica Romantsov, 2005: 881 **A:** KI (Terskey Ala-Too Mts.)
tshatkalica Lopatin, 1970b: 193 (*Chrysomela*) **A:** KI
ulugkhemica Mikhailov, 2002: 59 **A:** RU (SS: Tuva)
urjanchaica Jacobson, 1925a: 50 (*Chrysomela*) **A:** RU (SS: Tuva)
verestschaginae Lopatin, 1992a: 68 **A:** KI KZ
yupeiyuae Lopatin, 1998b: 829 **A:** CH (QIN)
subgenus *Pierryvettia* Bechyné, 1950b: 68 type species: *Chrysomela stictica* Stål, 1857
aeneomicans S.-H. Chen, 1934d: 35 **A:** CH (YUN)
baronii Daccordi, 1979b: 443 **A:** ID (SD) **ORR**
laotiana Sprecher-Uebersax & Daccordi, 2016: 466
bowringii Baly, 1860c: 96 (*Chrysomela*) **A:** CH (FUJ GUA GUI HKG TAI YUN) **ORR**
niobe Stål, 1861: 463 (*Chrysomela*)
stevensii Baly, 1862a: 23 (*Chrysomela*)
conglomerata Maulik, 1926: 37 **A:** ID (UT) NP **ORR**
helferi Bechyné, 1950b: 72

inconstans Wiedemann, 1823: 74 (*Chrysomela*) **A:** NP PA **ORR**
bonvouloirii Baly, 1862a: 23 (*Chrysomela*)
democratica Duvivier, 1891a: xliii (*Chrysomela*)
jeanneli S.-H. Chen, 1934d: 37 **A:** CH (YUN) **ORR**
karachia Maulik, 1926: 38 **A:** PA
separata Baly, 1860c: 96 (*Chrysomela*) **A:** CH (GUA GUI XIZ YUN) ID (KA UT) BT NP **ORR**
aurata Suffrian, 1851a: 102 (*Chrysomela*) [HN]
grutii Baly, 1862a: 22 (*Chrysomela*)
indosinensis Bechyné, 1950b: 70 (*Chrysolina*, ssp.)
perforata Kollar & L. Redtenbacher, 1844: 557 (*Chrysomela*) [HN]
redtenbacheri Kippenberg, 2010a: 70 [RN]
siamensis Jacoby, 1900b: 119 (*Chrysomela*)
tonkinea Fairmaire, 1889a: 372 (*Chrysomela*)
subgenus *Pleurosticha* Motschulsky, 1860b: 191 type species: *Chrysomela sylvatica* Gebler, 1823
arctoalpina Mikhailov, 2006b: 299 **A:** RU (SS: Kuznetsky Alatau Mts.)
borisi Bieńkowski, 2023b: 76 **A:** RU (SS)
burjatica Mikhailov, 2007b: 684 (ssp.) **A:** RU (SS: East Sayan) **[# 506]**
cavigera cavigera J. R. Sahlberg, 1885a: 35 (*Chrysomela*) **A:** RU (FE) **NAR**
cavigera kodarensis Mikhailov, 2006b: 301 (ssp.) **A:** RU (ES FE) **[# 507]**
cavigera tolli Jacobson, 1910a: 54 (*Chrysomela*) **A:** RU (ES FE WS: Arctic territories)
rufipes Ménétriés, 1851: pl. III (*Chrysomela*) [HN]
gebleri L. N. Medvedev, 1979b: 83 **A:** KZ (Altai) RU (SS)
baicalica L. N. Medvedev, 1992g: 107 (ssp.) [HN]
sajanensis L. N. Medvedev, 1979b: 84 (ssp.)
lagunovi Mikhailov, 2007d: 113 **E:** RU (CT: South and North Urals)
latimargo Weise, 1896a: 80 (*Chrysomela*) **A:** MG RU (SS: East Sayan)
ambulans Faldermann, 1835: 442 (*Chrysomela*) [NO]
changaica Lopatin, 1968b: 219 (*Chrysomela*)
mordkovitshi Mikhailov, 2007c: 258 **A:** RU (SS: West Altai)
novozhenovi Mikhailov, 2007c: 261 **A:** RU (SS) **[# 505]**
pirka Takizawa, 1970: 118 **A:** JA **[# 507]**
reshetnikovi Bieńkowski, 2023b: 76 **A:** RU (SS)
romani Mikhailov, 2007b: 683 **A:** RU (SS: Tuva) **[# 506]**
shapkini Mikhailov & Guskova, 2013a: 126 **A:** MG (Mongolian Altai)
sylvatica Gebler, 1823: 118 (*Chrysomela*) **A:** RU (SS) **[# 505]**
subcostata Gebler, 1848: 27 (*Chrysomela*)
uraltuvensis Mikhailov, 2000b: 130 **A:** RU (SS: West Sayan)

subgenus *Pseudocrosita* Lopatin, 1999b: 891 type species: *Crosita bactriana* Lopatin, 1961

bactriana Lopatin, 1961c: 102 (*Crosita*) **A:** KI TD

globicollis Lopatin, 1970b: 187 (*Chrysomela*)

subgenus *Pseudolithoptera* L. N. Medvedev, 1970c: 159 type species: *Chrysolina interlucea* L. N. Medvedev, 1970

interlucea L. N. Medvedev, 1970c: 159 **A:** NC

subgenus *Rhyssoloma* Wollaston, 1854: 458 type species: *Chrysomela fragariae* Wollaston, 1854

costalis Olivier, 1807: 513 (*Chrysomela*) **N:** CI

obsoleta Brullé, 1839: 73 (*Chrysomela*)

fragariae Wollaston, 1854: 458 (*Chrysomela*) **N:** MR

onychina Wollaston, 1860b: 459 (*Chrysomela*) [RN]

wollastoni Bechyné, 1957: 4 (*Oreina*) [RN] **N:** CI

rutilans Wollaston, 1864: 402 (*Chrysomela*) [HN]

subgenus *Semenowia* Weise, 1889g: 605 type species: *Semenowia chalcea* Weise, 1889

chalcea Weise, 1889g: 605 (*Semenowia*) **A:** CH (GAN QIN)

bienkowskii Lopatin, 2000c: 133

subgenus *Sibiriella* L. N. Medvedev, 1999b: 1014 type species: *Chrysolina paradoxa* L. N. Medvedev, 1999

Altailina Mikhailov, 2000b: 133 type species: *Chrysolina dudkoi* Mikhailov, 2000

capricornus Mikhailov, 2000b: 139 **A:** KZ (Altai) RU (SS: Altai)

dudkoi Mikhailov, 2000b: 134 **A:** KZ (Altai) RU (SS: Altai)

ivanovskiana Mikhailov, 2000b: 136 (ssp.)

kholsunica Mikhailov, 2001: 334 [RN] **A:** KZ (Altai) RU (SS: Altai) [# 597]

ogloblini Mikhailov, 2000b: 137 [HN]

paradoxa L. N. Medvedev, 1999b: 1014 **A:** RU (SS)

subgenus *Sphaerochrysolina* Kippenberg, 2010a: 69 type species: *Chrysomela olivacea* var. *umbratilis* Weise, 1887

Sphaerochrysolina Kippenberg, 1994: 64 [UN]

umbratilis Weise, 1887a: 185 (*Chrysomela*, var.) **E:** CZ GE PL RO SK UK

atra Matzek 1843: 154 (*Chrysomela*) [HN]

erudita Bechyné, 1952b: 367 (ssp.)

olivacea Suffrian, 1851a: 53 (*Chrysomela*) [HN]

olivaceocuprea Bechyné, 1950b: 105 (ab.) [IN]

olivicolor Weise, 1916b: 94 (*Chrysomela*, ab.) [IN]

subgenus *Sphaeromela* Bedel, 1899: 260 type species: *Chrysomela varians* Schaller, 1783

varians Schaller, 1783: 271 (*Chrysomela*) **E:** AU BE BH BU BY CR CZ DE EN FI FR GB GE HU IT LA LS LT LU NL NM NR PL RO RU (CT NT ST) SB SK SL SP SV SZ UK **N:** AG **A:** AB RU (SS WS) TR **NAR**i

aerifera Bechyné, 1950b: 152 (ab.) [IN]
aethiops Fabricius, 1792a: 309 (*Chrysomela*)
centaura Herbst, 1783: 56 (*Chrysomela*)
centaurii Fabricius, 1787: 68 (*Chrysomela*)
havelkai Bechyné, 1949a: 55 (ssp.)
hyperici DeGeer, 1775: 312 (*Chrysomela*) [HN]
malleti J. Jacquet, 1935: 95 (*Chrysomela*, var.)
malletiana Bechyné, 1950b: 152 (ab.) [IN]
margarita Olivier, 1791a: 709 (*Chrysomela*)
marshami Donovan, 1799: 83 (*Chrysomela*)
nigricollis Mallet, 1936: 141 (*Chrysomela*, ab.) [IN]
nigroolivacea Bechyné, 1950b: 152 (ab.) [IN]
ooensis Weise, 1916b: 83 (*Chrysomela*) [RN]
pratensis Weise, 1884d: 429 (*Chrysomela*, var.) [IN] [# 577]
pulchella Fleischer, 1918a: 76 (*Chrysomela*, ab.) [IN]
subaenea Suffrian, 1851a: 60 (*Chrysomela*)
violetta Bechyné, 1950b: 152 (ab.) [IN]
viridiaenea Marsham, 1802: 184 (*Chrysomela*)

subgenus *Stichoptera* Motschulsky, 1860b: 209 type species: *Chrysomela sanguinolenta* Linnaeus, 1758 [# 508]
Cobosorina David, 1953b: 186 type species: *Chrysochloa colasi* Cobos, 1952
Ovomorpha Motschulsky, 1860b: 213 type species: *Chrysomela rossia* Illiger, 1802

bourdonnei Daccordi & Ruffo, 2005: 115 **E**: IT
colasi Cobos, 1952: 5 (*Chrysochloa*) **E**: SP
gypsophilae grossepunctata Har. Lindberg, 1950: 13 (*Chrysomela*) **N**: CI [# 508]
grancanariensis Har. Lindberg, 1953: 11 (*Chrysomela*)
gypsophilae gypsophilae Küster, 1845a: 71 (*Chrysomela*) **E**: AL AU BH BU BY CR CZ DE EN FI FR GE GR HU IT KR KO LA LT ME NL NM NR PL RO RU (NT ST) SB SK SV TR UK **A**: AB AR CY GG IN IS KI KZ SI SY TR [# 508, # 1019]
morvennensis Méquignon, 1945: 29 (*Chrysomela*, var.) [IN] [# 509, # 577]
limitata Küster, 1845a: 72 (*Chrysomela*)
gypsophilae lucidicollis Küster, 1845a: 73 (*Chrysomela*) **E**: FR IT PT SP **N**: MO TU [# 508]
gaubilii P. H. Lucas, 1847a: pl. 45 [= 1847c: 535] (*Chrysomela*)
sierrana Franz, 1958: 127 (*Chrysomela*, ssp.)
kuesteri friderici H. Wagner, 1927: 111 (*Chrysomela*, ssp.) **E**: PT SP [# 510]
kuesteri kuesteri Helliesen, 1912: 12 (*Chrysomela*) **E**: AU BE BH BU BY CZ FR GE HU LA ME NL PL RO SB SK SL SZ UK
latecincta decipiens Franz, 1938a: 207 (*Chrysomela*, ssp.) **E**: FR (Pyrénées)
laboissierei Bechyné, 1952b: 377 (ssp.)

rufohumeralis Pic, 1939f: 26 (*Chrysomela*, var.)
latecincta epipleurica Reitter, 1913c: 110 (*Chrysomela*, var.) E: SP (Asturiás) [# 510]
latecincta hellieseni Silfverberg, 1977: 93 (ssp.) [RN] E: NR
crassicornis Helliesen, 1912: 13 (*Chrysomela*) [HN]
latecincta hustachei Laboissière, 1939: 128 (*Chrysomela*) E: FR IT
latecincta intermedia Franz, 1938b: 250 (*Chrysomela*, ssp.) E: GB IR
latecincta latecincta Demaison, 1896: 12 (*Chrysomela*) E: FR IT SZ [# 510]
graja Franz, 1938a: 208 (*Chrysomela*, ssp.)
vallesiaca Franz, 1949a: 12 (*Chrysomela*, ssp.)
latecincta norica Holdhaus, 1914: 126 (*Chrysomela*) E: AU SZ
holdhausi Franz, 1949a: 11 (*Chrysomela*, ssp.)
raetica Franz, 1938b: 252 (*Chrysomela*, ssp.)
latecincta tarragonensis Bechyné, 1950b: 140 (ssp.) E: PT SP
mactata Fairmaire, 1859c: clii (*Chrysomela*) E: PT SP
insignis Breit, 1920: 81 (*Chrysomela*, ssp.)
oceanoripensis Bourdonné, Doguet & Petitpierre, 2013: 34 E: FR [# 510]
rossia Illiger, 1802: 415 (*Chrysomela*) E: AU BH BU CR CZ FR GR HU IT RO SB SK SL SZ [# 508, # 1020]
mancinii J. Müller, 1924b: 118 (*Chrysomela*)
rossii Suffrian, 1851a: 38 (*Chrysomela*) [UE]
sanguinolenta Linnaeus, 1758: 371 (*Chrysomela*) E: AU BE BH BY BU CZ DE EN FI FR GB GE GR HU IT KR LA LT MD ME NL NM NR PL RO RU (CT NT ST) SB SK SL SV SZ TR UK A: AB AR IN KI KZ MG RU (ES FE SS WS) TR
breiti Franz, 1938b: 254 (*Chrysomela*, ssp.)
distinguenda Stephens, 1839: 309 (*Chrysomela*)
marginalis Duftschmid, 1825: 182 (*Chrysomela*)
porosa Gebler, 1830: 217 (*Chrysomela*)
richteri Roubal, 1934: 38 (*Chrysomela*, forma) [IN] [# 577]
rubromarginata DeGeer, 1775: 298 (*Chrysomela*)
stachydis Gené, 1839: 79 (*Chrysomela*) E: FR (Corsica) IT (Sardinia)
variolosa Petagna, 1819: 32 (*Chrysomela*) E: IT (Sicily) MA [# 508]
melanostigma Sturm, 1843: 290 (*Chrysomela*) [RN] [# 511]
sicula Lefebvre, 1827: 104 (*Chrysomela*)
sparshalli Stephens, 1835: 425 (*Chrysomela*) [# 1021]
subgenus *Synerga* Weise, 1900b: 283 type species: *Chrysomela bella* Jacoby, 1890 [# 512]
Chrysonotum J. R. Sahlberg, 1913c: 247 type species: *Chrysomela viridana* Küster, 1844
Menthastriella Bechyné, 1950b: 74 type species: *Chrysomela herbacea* Duftschmid, 1825
coerulans angelica Reiche & Saulcy, 1858b: 33 (*Chrysomela*) A: AB AR CY GG IN IQ IS JO RU (ST) SY TM N: EG [# 512, # 598]
alnadawi Al-Nadawi, 2019: 839 [UN] [# 628]

islandica Faldermann, 1837: 362 (*Chrysomela*) [HN]
iranica Jakob, 1954c: 47 (*Chrysomela*, ssp.)
mimeomicans Bechyné, 1950b: 79 (ab.) [IN]
pseudostarhoni Bechyné, 1950b: 79 (ab.) [IN]
splendorifera Motschulsky, 1860b: 226

coerulans bella Jacoby, 1890f: 253 (*Chrysomela*) **A**: AF CH (HUB) ID (KA) IN KI PA TD UZ **ORR** **[# 512]**
piffli Lopatin, 1967b: 325 (*Chrysomela*, ssp.)
punjabiensis Abdullah & Qureshi, 1969b: 117
uzbekorum Bechyné, 1950b: 78

coerulans coerulans Scriba, 1791: 286 (*Chrysomela*) **E**: AU BE BH BU BY CR CZ FR GB GE GR (Corfu) HU IT LA LS LT LU NL NM PL RO RU (CT NT ST) SB SL SK SZ TR **A**: GG RU (WS: Chelyabinsk Reg.) TR **[# 512, # 599, # 1145]**
adameki Bechyné, 1950b: 78 (ab.) [IN]
evittata Bechyné, 1950b: 78 (ab.) [IN]
oblonga Duftschmid, 1825: 188 (*Chrysomela*)
obscura Laczó, 1928b: 4 (*Chrysomela*, ab.) [IN]
olivaceonigra Fleischer, 1892: 141 (*Chrysomela*, var.)
relicta L. N. Medvedev, 1977c: 35 (ssp.)
starhoni Reitter, 1913c: 118 (*Chrysomela*, ab.) [IN]

herbacea Duftschmid, 1825: 192 (*Chrysomela*) **E**: AL AU BE BH BU BY CR CZ DE FR GB GE GR HU IT KO KR LS LT LU MD ME NL NM PL PT RO RU (CT NT ST) SB SK SL SP SZ TR UK **A**: AB AR GG IN KZ RU (WS) TM TR **[# 512, # 1145]**
alacris Bechyné, 1950b: 77 (ssp.)
blanda Motschulsky, 1860b: 226 (*Chrysomela*)
caucasica Motschulsky, 1860b: 225 (*Chrysomela*)
cribellata Motschulsky, 1860b: 226 (*Chrysomela*) [HN]
croatica Weise, 1884d: 426 (*Chrysomela*, var.) [IN] [# 577]
cuprea Jolivet, 1943b: 79 (*Chrysomela*, ab.) [IN]
cyaneonigra Jolivet, 1943b: 79 (*Chrysomela*, ab.) [IN]
durabilis Bechyné, 1950b: 76 (ab.) [IN]
elaverensis Buysson, 1924: 89 (*Chrysomela*, var.)
expectata Bechyné, 1950b: 76 (ab.) [IN]
fulgida Motschulsky, 1860b: 226 (*Chrysomela*)
fulminans Suffrian, 1851a: 92 (*Chrysomela*)
mariannae Gistel, 1857b: 595 (*Chrysomela*)
menthastri Suffrian, 1851a: 90 (*Chrysomela*)
meridionalis Jolivet, 1951e: 472 (race) [HN]
paveli Kaszab, 1962: 55 (*Chrysomela*, ab.)
recticollis Motschulsky, 1860b: 225 (*Chrysomela*)
resplendens Suffrian, 1855: 144 (*Chrysomela*)

rugicollis Weidenbach, 1859: 84 (*Chrysomela*)
semiglobosa Reineck, 1922a: 80 (*Chrysomela*, forma)
solistima Bechyné, 1950b: 76 (ab.) [IN]
talyshana Bechyné, 1950b: 77 (ssp.)
thoracica Bechyné, 1950b: 76 (ab.) [IN]
tolosana Buysson, 1924: 89 (*Chrysomela*, var.)

suffriani Fairmaire, 1859b: 282 (*Chrysomela*) **E**: FR (Corsica) IT (Sardinia)

viridana chloris P. H. Lucas, 1847a: pl. 45 [= 1847c: 536] (*Chrysomela*) **E**: IT (Sicily) PT SP **N**: AG MO TU [# 512]
algecirensis Bechyné, 1950b: 75 (ab.) [IN]
ignita Olivier, 1807: 524 (*Chrysomela*) [HN] [# 1022]

viridana viridana Küster, 1844: 85 (*Chrysomela*) **E**: GBi (Isle of Wight) FR (incl. Corsica) IT (Sardinia) [# 512]
aurocuprea Fairmaire, 1859b: 282 (*Chrysomela*)
cupreopurpurea A. Costa, 1884: 49 (*Chrysomela*, var.)
palustris Suffrian, 1851a: 96 (*Chrysomela*) [# 512]
pseudochloris Bechyné, 1950b: 75 (ab.) [IN]
sardea Weise, 1884d: 426 (*Chrysomela*, var.) [IN] [# 577]

subgenus *Taeniochrysea* Bechyné, 1950a: 121 [= 1950b: 87] type species: *Chrysomela americana* Linnaeus, 1758
Pseudotaeniochrysea Daccordi, 1980: 302 type species: *Chrysomela superba* Thunberg, 1787 sensu Daccordi (1980) (= *Chrysomela americana* var. *limbolata* Reiche, 1850) [# 560]

americana Linnaeus, 1758: 372 (*Chrysomela*) **E**: AL AUi BEi CR FR GEi GBi GR IRi IT KRi LAi LUi MA ME NLi NM PLi PT ROi SB SL SP SZi TRi **N**: AG MO MR TU **A**: ABi CYi INi ISi TR [# 1145]
barbarica Gmelin, 1790: 1683 (*Chrysomela*)
decemstriata Goeze, 1777: 301 (*Chrysomela*)
desdoueti Marseul, 1887: 103 (*Chrysomela*, var.)
insulivagans Bechyné, 1950a: 122 (ssp.)
lesinae Weise, 1884d: 411 (*Chrysomela*, var.) [IN] [# 577]
nitidula Geoffroy, 1785: 108 (*Chrysomela*)
striata Voet, 1806: 44 (*Chrysomela*) [UN]
ubertini Marseul, 1887: 104 (*Chrysomela*, var.)

subgenus *Taeniosticha* Motschulsky, 1860b: 207 type species: *Chrysomela lurida* Linnaeus, 1767 (= *Chrysolina pseudolurida* Roubal, 1917)

alatavica Jacobson, 1910b: 58 (*Chrysomela*) **A**: CH (XIN) KZ

bergeali Bourdonné, 2005: 339 (*Craspeda*) **A**: CH (XIN) KZ

dzhungarica Jacobson, 1910b: 58 (*Chrysomela*) **A**: CH (XIN) KI KZ

imperfecta Breit, 1920: 82 (*Chrysomela*) **E**: RU (ST) **A**: AB AF AR GG IN TM TR
bakuensis Bechyné, 1952b: 377 (ssp.)
kaltenbachi Jakob, 1954c: 46 (*Chrysomela*)
plusquamperfecta Bechyné, 1952b: 377 (ssp.)

khalyktavica Lopatin, 2005b: 572 **A:** CH (XIN)
klimenkoi Romantsov, 2008: 842 **A:** KZ (Dzhungar Alatau)
koktumensis Lopatin & Kulenova, 1987: 39 **A:** KZ (Dzhungar Alatau)
kuldzhensis Lopatin, 1976a: 113 **A:** CH (XIN)
petitpierrei Kippenberg, 2004: 19 **E:** SP
pseudolurida lineata Bechyné, 1952b: 376 (ssp.) **E:** CZ HU PL SK
lineata Papp, 1946c: 23 (*Chrysomela*, ab.) [IN]
pappi Bourdonné, 2008: 18 (*Craspeda*, ssp.) [UN] **[Appendix]**
pseudocarnifex Kaszab, 1962: 54 (*Chrysomela*, ab.) [IN]
pseudolurida obscurefacta Bechyné, 1952b: 376 (ssp.) **E:** IT
fageli Bechyné, 1957: 2 (*Oreina*, ssp.)
pseudolurida pseudolurida Roubal, 1917: 3 (*Chrysomela*) **E:** AL BH BU CR GE FR KR KZ NM RO RU (CT ST) SB **A:** GG RU (WS) TR
diluta Krynicki, 1832: 171 (*Chrysomela*) [HN]
jailensis Bechyné, 1952b: 376 (ssp.)
lurida Linnaeus, 1767: 590 (*Chrysomela*) [HN]
mangaliana Bechyné, 1952b: 376 (ssp.)
nevesinjensis Bechyné, 1952b: 376 (ssp.)
saxonica Silfverberg, 1977: 93 (ssp.) [RN]
sequana Silfverberg, 1977: 93 (ssp.) [RN]
striata Geoffroy, 1785: 105 (*Chrysomela*) [HN]
pseudolurida samarensis Bechyné, 1950b: 137 **E:** RU (CT ST) **A:** RU (WS) **[# 513]**
reitteri Weise, 1884b: 163 (*Chrysomela*) **E:** RU (ST: North Ossetia) **A:** GG
sairannurica Romantsov, 2008: 844 **A:** CH (XIN)
tianshanica Jacobson, 1910b: 59 (*Chrysomela*) **A:** CH (XIN) KI KZ
kungeyana Bourdonné, 2005: 322 (*Craspeda*)
tuvensis L. N. Medvedev, 1976b: 243 **A:** RU (SS: Tuva) **[# 514]**
subgenus *Threnosoma* Motschulsky, 1860b: 213 type species: *Chrysomela helopioides* Suffrian, 1851
afra afra Erichson, 1841: 190 (*Chrysomela*) **N:** AG MO
tangeriana Kocher, 1959: 43 (*Chrysomela*)
afra maritima Peyerimhoff, 1938: 53 (*Chrysomela*, ssp.) **N:** AG
anceyi anceyi Marseul, 1868: 211 (*Chrysomela*) **A:** IS LE SY TR
anceyi libanicola Marseul, 1868: 212 (*Chrysomela*) **A:** LE SY
praestantialis Bechyné, 1950b: 52 (ab.) [IN]
winkleri Breit, 1919: 18 (*Chrysomela*, ssp.)
anceyi pluscula Bechyné, 1950b: 52 (ssp.) **A:** IQ
arambourgi Peyerimhoff, 1938: 52 (*Chrysomela*) **N:** AG
avulsa Bechyné, 1946a: 106, 109 (ssp.) **E:** BH CR ME
rufocingulata Csiki, 1953: 129 (*Chrysomela*, var.)

cribrosa cribrosa Ahrens, 1812a: 10 (*Chrysomela*) E: BH CR SL [# 1023]
lucidula Apfelbeck, 1912: 248 (*Chrysomela*, var.)
cribrosa sirentensis Meier, 1900a: 78 (*Chrysomela*) E: IT
maestitialis Bechyné, 1950b: 55
fimbrialis fimbrialis Küster, 1845a: 74 (*Chrysomela*) E: AU CZ IT HU RO SK SL
molluginis L. Redtenbacher, 1848: 544 (*Chrysomela*) [HN]
redtenbacheri Bechyné, 1946a: 106, 109 (ab.) [IN]
fimbrialis hungarica Fuss, 1861a: 151 (*Chrysomela*) E: RO [# 515, # 1112]
manicata Bechyné, 1950b: 53 (ab.) [IN]
fimbrialis langobarda Daccordi & Ruffo, 1979: 313 (ssp.) E: IT
helopioides Suffrian, 1851a: 12 (*Chrysomela*) E: SP
korbi Weise, 1891b: 149 (*Chrysomela*)
inflata inflata Weise, 1916b: 76 (*Chrysomela*) [RN] E: IT (Sicily) [# 1113]
atra Herrich-Schäffer, 1838b: [10] (*Chrysomela*) [HN] [# 1024]
inflata vitiosa Bechyné, 1950b: 57 (ssp.) N: TU
joliveti Bechyné, 1950b: 56 E: FR
mairei mairei Peyerimhoff, 1928: 41 (*Chrysomela*) N: MO
mairei pauliani Peyerimhoff, 1939: 34 (*Chrysomela*, ssp.) N: MO
obenbergeri Bechyné, 1950b: 54 E: RO
obenbergeriana Bechyné, 1958b: 88 (*Oreina*) [RN]
obscurella Suffrian, 1851a: 11 (*Chrysomela*) E: FR IT
pelagica Chevrolat, 1863: 120 (*Chrysomela*)
remigrata Bechyné, 1950b: 57 (ssp.)
osellai Daccordi & Ruffo, 1979: 316 E: IT
serdanensis Jolivet, 1966b: 128 N: MO
solata Fairmaire, 1881: 218 (*Chrysomela*) N: AG
extricata Bechyné, 1950b: 57 (ssp.)
scorodon Marseul, 1886: 12 (*Chrysomela*)
tagana Suffrian, 1851a: 9 (*Chrysomela*) E: PT
atra Gravenhorst, 1807: 133 (*Chrysomela*) [HN]
timarchoides Brisout de Barneville, 1882: 245 (*Chrysomela*) E: FR SP
tortipennis Fairmaire, 1865: 72 (*Chrysomela*) N: AG
pertusa Fairmaire, 1865: 73 (*Chrysomela*)
rubricrus Desbrochers des Loges, 1898: 44 (*Chrysomela*)
vermiculosa Marseul, 1886: 19 (*Chrysomela*) N: AG
weisei Frivaldszky, 1883: 16 (*Chrysomela*) E: RO
kercesorae Kaszab, 1962: 53 (*Chrysomela*, ab.) [IN]
pachysoma Hubenthal, 1911: 192 (*Chrysomela*, var.)

subgenus *Timarchida* Ganglbauer, 1897: 571 type species: *Timarchida deubeli* Ganglbauer, 1897 *deubeli* Ganglbauer, 1897: 572 (*Timarchida*) E: RO

flecki Reitter, 1913c: 124 (*Timarchida*)

subgenus *Timarchomela* Achard, 1922b: 17 type species: *Timarchomela aeneolucens* Achard, 1922

Doeberlia Warchałowski, 2007: 482 type species: *Doeberlia subopaca* Warchałowski, 2007 (= *Chrysolina doeberli* Daccordi & Yang, 2009)

aeneolucens Achard, 1922b: 17 (*Timarchomela*) **A:** CH (YUN)

melanaspis Achard, 1922b: 18 (*Timarchomela*, var.)

baoshanica Lopatin, 2009b: 430 **A:** CH (YUN)

confucii Lopatin, 2007: 176 **A:** CH (SCH YUN)

costulata Achard, 1922b: 18 (*Timarchomela*) **A:** CH (YUN)

dalia S.-H. Chen & S.-Y. Wang, 1984a: 170, 173 **A:** CH (YUN)

doeberli Daccordi & Yang, 2009b: 817 [RN] **A:** CH (HUN)

subopaca Warchałowski, 2007: 482 (*Doeberlia*) [HN]

gansuica Lopatin, 2006c: 595 **A:** CH (GAN)

lii Daccordi & S.-Q. Ge, 2011b: 297 **A:** CH (YUN)

maximi Lopatin, 2011b: 375 **A:** CH (YUN)

nigrorugosa Lopatin, 2005d: 874 **A:** CH (GAN)

nushana S.-H. Chen & S.-Y. Wang, 1984a: 170, 174 **A:** CH (YUN)

oxanae Lopatin, 2013: 767 **A:** CH (SCH)

subgenus *Timarchoptera* Motschulsky, 1860b: 188 type species: *Chrysomela haemochlora* Gebler, 1823 **[# 516]**

Cyrtocnemis Motschulsky, 1868: 82 type species: *Timarcha rubra* Motschulsky, 1845

Paraheliostola L. N. Medvedev, 1992g: 105 type species: *Chrysomela soiota* Jacobson, 1924

haemochlora Gebler, 1823: 120 (*Chrysomela*) **A:** KZ (Altai) MG RU (ES SS WS)

rubra Motschulsky, 1845a: 109 (*Timarcha*)

lomakini Mikhailov, 2003: 220 **A:** RU (SS: Khansyn Mt. Range)

olegkabakovi Bieńkowski, 2023a: 28 **A:** RU (FE: Khabarovsk Krai)

soiota khakassa Mikhailov, 2003: 224 (ssp.) **A:** RU (SS: Khakassia)

soiota soiota Jacobson, 1924b: 80 (*Chrysomela*) **A:** RU (SS: West Sayan)

subgenus *Upseleatlasia* Bourdonné, 2012: 121 type species: *Chrysomela furva* Peyerimhoff, 1926

furva Peyerimhoff, 1926: 96 (*Chrysomela*) **N:** MO

villiersi ruficornis Kocher, 1953c: 95 (*Chrysomela*, ssp.) **N:** MO

villiersi villiersi Peyerimhoff, 1939: 34 (*Chrysomela*) **N:** MO

subgenus *Vittatochrysa* Lopatin, 1977b: 146 type species: *Chrysomela nigrovittata* Ballion, 1878

nigrovittata Ballion, 1878: 380 (*Chrysomela*) **A:** CH (XIN) KI KZ

subgenus *Volosatik* Bieńkowski, 2022: 6 type species: *Chrysolina fascinatrix* Lopatin, 1998

fascinatrix Lopatin, 1998b: 831 **A:** CH (YUN)
genriki Bieńkowski, 2022: 17 **A:** CH (YUN)
igori Bieńkowski, 2022: 14 **A:** CH (YUN)
ilyakabaki Bieńkowski, 2022: 18 **A:** CH (YUN)
liqingzhaoae Daccordi & S.-Q. Ge, 2011b: 300 **A:** CH (GAN YUN) [# 600]
marinae Bieńkowski, 2022: 16 **A:** CH (YUN)
wangi Lopatin, 2005b: 570 **A:** CH (YUN)
subgenus *Zeugotaenia* Motschulsky, 1860b: 206 type species: *Chrysomela limbata* Fabricius, 1775
jacobyi Baly, 1878a: 38 (*Chrysomela*) **A:** CH (SHN) RU (FE)
pavlenkoi Jacobson, 1924b: 82 (*Chrysomela*)
jenisseiensis Breit, 1920: 81 (*Chrysomela*) **E:** RU (ST) **A:** MG RU (ES SS)
limbata discipennis Ménétriés, 1848: 52 (*Chrysomela*) **E:** KZ RU (ST)
limbata hochhuthii Suffrian, 1851a: 72 (*Chrysomela*) **A:** CH (HEI) KZ MG RU (ES SS)
limbata limbata Fabricius, 1775: 101 (*Chrysomela*) **E:** AL AN AU BE BH BU BY CR CZ DE FR GE GR HU KO KR LA MA MD ME NL NM PL RO RU (CT ST) SB SK SL SP SV SZ UK **A:** AB GG KZ RU (WS)
findelii Suffrian, 1851a: 70 (*Chrysomela*)
kavani Bechyné, 1950b: 170 (ssp.)
limbifera Küster, 1846a: 91 (*Chrysomela*)
limbata luigionii Depoli, 1936: 139 (*Chrysomela*, ssp.) **E:** BH IT FR
limbata russiella Bieńkowski & Orlova-Bienkowskaya, 2011: 954 (ssp.) **E:** RU (ST) UK
limbata volodi Bieńkowski & Orlova-Bienkowskaya, 2011: 955 (ssp.) **A:** AR GG TR

genus *Chrysolina*, species unassigned to subgenera
amica Lopatin, 2009b: 431 **A:** CH (SCH)
baimana Lopatin, 2009b: 432 **A:** CH (SCH)
balthazari Bieńkowski, 2023c: 207 **A:** CH (YUN)
boccaccioi Daccordi & Yang, 2011b: 314 **A:** CH (SCH)
caspari Bieńkowski, 2023c: 200 **A:** CH (SCH)
caspica Weise, 1892j: 406 (*Chrysomela*) [RN] **E:** RU (ST) **A:** GG [# 564]
caucasica Weise, 1882b: 352 (*Chrysomela*) [HN]
jdanieli Roubal, 1912b: 27 (*Chrysomela*)
claripes Lopatin, 2002a: 114 **A:** CH (SCH)
cuiae S.-Q. Ge & Daccordi, 2011a: 38 **A:** CH (XIZ)
daccordii L. N. Medvedev & Sprecher-Uebersax, 1999: 366 (*Semenovia*) **A:** NP
davidiani Lopatin, 2002a: 114 **A:** CH (SCH)
dhaulagirica arunensis L. N. Medvedev, 1992f: 386 (ssp.) **A:** NP
dhaulagirica dhaulagirica L. N. Medvedev, 1990b: 12 **A:** NP
dohertyi Maulik, 1926: 27 **A:** CH (YUN) **ORR**
jelineki Daccordi & Yang, 2009b: 809

fortunata Wollaston, 1864: 402 (*Chrysomela*) **N:** CI
foveopunctata Fairmaire, 1888c: 39 (*Chrysomela*) **A:** CH (YUN)
geae Lopatin, 2006c: 595 **A:** CH (SCH)
gyacaensis Daccordi & Yang, 2011a: 35 **A:** CH (XIZ)
hartmanni L. N. Medvedev, 1999c: 182 **A:** NP
hongyuanensis Daccordi & S.-Q. Ge, 2011a: 37 **A:** CH (SCH)
infernalis Lopatin, 2007: 177 **A:** CH (SCH)
jiangi Lopatin, 2006c: 594 **A:** CH (SCH)
jinxiaoae S.-Q. Ge, 2011: 292 **A:** CH (SCH)
kamali Abdullah & Qureshi, 1969b: 117 **A:** PA
kippenbergi Lopatin, 2008d: 833 **A:** CH (SCH)
korotjaevi Lopatin, 2009b: 434 **A:** CH (SCH)
lehri Lopatin, 1970b: 189 (*Chrysomela*) **A:** KZ **[# 564]**
libaii Daccordi & S.-Q. Ge, 2011b: 295 **A:** CH (SCH)
lijieae Daccordi & Yang, 2011b: 298 **A:** CH (SCH)
luobinwangi S.-Q. Ge & Daccordi, 2011b: 304 **A:** CH (SCH)
luyoui Daccordi & S.-Q. Ge, 2011b: 305 **A:** CH (SCH)
matruelis Lopatin, 2011b: 378 **A:** CH (SCH)
melchiori Bieńkowski, 2023c: 205 **A:** CH (YUN)
mikhailovi S.-Q. Ge & Daccordi, 2011b: 308 **A:** CH (XIZ)
  lishangyini S.-Q. Ge & Daccordi, 2011b: 303
  markamensis Daccordi & S.-Q. Ge, 2011a: 32
mirabilis Daccordi, 1976a: 37 (*Semenowia*) **A:** CH (SCH)
murciana Kippenberg, 2012a: 59 (ssp.) **E:** SP
nagaja Daccordi, 1982a: 397 (*Semenowia*) **A:** NP PA
nesterovae Lopatin, 2011b: 376 **A:** CH (SCH)
nigrita Fabricius, 1792a: 309 (*Chrysomela*) **E:** FR
nixiana Lopatin, 2008d: 837 **A:** CH (YUN)
pieli S.-H. Chen, 1936c: 96 **A:** CH (HEN JIX ZHE)
pingchuana Lopatin, 2013: 768 **A:** CH (SCH)
postviolacea Marseul, 1887: 100 (*Chrysomela*) **N:** AG
purpureoviridis Lopatin, 2005d: 875 **A:** CH (SCH)
romandudkoi Mikhailov, 2019: 60 **A:** NP
rotundata Lopatin, 2002a: 113 **A:** CH (SCH)
sergeyi Lopatin, 2009b: 433 **A:** CH (GAN SCH)
seriepunctata Weise, 1887h: 176 (*Chrysomela*) **A:** CH (HUB) JA RU (FE)
  bechynei Gressitt & Kimoto, 1963a: 315 (*Oreina*)
shuyongi S.-Q. Ge & Daccordi, 2011a: 40 **A:** CH (YUN)
sichuanica Lopatin, 2002a: 111 **A:** CH (SCH)

sicieni S.-Q. Ge & Yang, 2011: 310 **A:** CH (SCH)
sinica Lopatin, 2008d: 835 **A:** CH (SCH)
songpana Lopatin, 2007: 179 **A:** CH (SCH)
taibaica S.-H. Chen, 1961: 431, 434 **A:** CH (SHA)
tani Lopatin, 1998b: 830 **A:** CH (GAN HEI SCH)
daosana Lopatin, 2007: 177
tangalaensis Kimoto, 2001: 32 **A:** NP
viridiopaca Lopatin, 2004a: 616 **A:** CH (XIN)
volkovitshi Lopatin, 2009b: 434 **A:** CH (GAN)
wangboi Daccordi & Yang, 2011b: 311 **A:** CH (SCH YUN)
warchalowskii Lopatin, 2005b: 569 **A:** CH (GAN)
yunnana Lopatin, 2008d: 836 **A:** CH (YUN)
zhangi S.-Q. Ge & Daccordi, 2011a: 41 **A:** CH (YUN)
zhongdiana S.-H. Chen & S.-Y. Wang, 1984a: 171, 174 **A:** CH (YUN)

genus *Chrysolina*, nomina dubia
caerulea Gmelin, 1790: 1687 (*Chrysomela*) **E:** FR
dahlii Herrich-Schäffer, 1838b: [8] (*Chrysomela*) **[# 566]**
distincta Küster, 1844: 89 (*Chrysomela*) **E:** FR **[# 567]**
florea Herrich-Schäffer, 1838b: [11] (*Chrysomela*) [NO] [distribution not stated]
subincrassata Duftschmid, 1825: 179 (*Chrysomela*) **E:** AU **[# 568]**
tagenii Herrich-Schäffer, 1838b: [9] (*Chrysomela*) **E:** PT **[# 569]**
collita Bechyné, 1950b: 121 (ab.) [IN]
violacea Goeze, 1777: 301 (*Chrysomela*) **E:** FR

genus *Chrysolina*, unavailable name
pseudolaeviguttata Daccordi & S.-Q. Ge, 2014: 480 [UN] **[# 581]**

genus *Crosita* Motschulsky, 1860b: 190 type species: *Chrysomela altaica* Gebler, 1823 **[# 520]**
afghanica Lopatin, 1962d: 1812 **A:** AF (Badakhshan) TD (Badakhshan)
altaica altaica Gebler, 1823: 117 (*Chrysomela*) **A:** CH (XIN) KI KZ MG **[# 521]**
insignis Fischer von Waldheim, 1842: 20 (*Chrysomela*)
petraea Marseul, 1883: 103 (var.)
altaica atasica L. N. Medvedev, 1976c: 908 (ssp.) **A:** MG (Atas-Ula Mts.) **[# 521]**
altaica barkulica L. N. Medvedev, 1976c: 909 (ssp.) **A:** CH (XIN) MG **[# 521]**
altaica bogutensis Lopatin, 1996: 641 **A:** KZ (Boguty Mts.) **[# 521]**
altaica faldermanni Krynicki, 1832: 170 (*Chrysomela*) **A:** KZ **[# 629]**
altaica heptapotamica Jacobson, 1896a: 547 **A:** KI KZ

altaica kalbensis Mikhailov, 2009b: 187 (ssp.) **A:** KZ (Kalbinsky Range)
altaica maximovitschii Zubkov, 1833: 337 (*Chrysomela*) **E:** RU (ST: Orenburg Region) **A:** KZ RU (SS)
borochorensis L. N. Medvedev, 1976c: 909 **A:** CH (XIN)
clementzae Jacobson, 1899b: 9 **A:** CH (XIN) MG [# 521]
aenea Jacobson, 1899b: 10 (var.)
cuprescens Jacobson, 1899b: 10 (var.)
cyanea Jacobson, 1899b: 11 (var.)
cyanescens Jacobson, 1899b: 11 (var.)
elegans Lopatin, 1968b: 216 **A:** MG
kaszabi Lopatin, 1970a: 256 **A:** MG
kowalewskii kowalewskii Gebler, 1836: 344 (*Chrysomela*) **A:** CH (XIN) MG RU (SS) [# 630]
cupreoviridula Jacobson, 1898: 198 (var.)
gobica Lopatin, 1968b: 217 (ssp.)
kowalewskii kozlovi L. N. Medvedev, 1976c: 910 (ssp.) **A:** CH (GAN)
kowalewskii matronula Weise, 1894i: 155 **A:** MG
cyaneoviridula Lopatin, 1968b: 218 (ab.) [IN]
potanini Jacobson, 1899b: 8
transitoria Lopatin, 1968b: 218 (ab.) [IN]
unicolor Lopatin, 1968b: 218 (ab.) [IN]
viridula Jacobson, 1898: 198 (var.)
pigra Weise, 1894i: 155 **A:** MG RU (SS)
gemellata Lopatin, 1971a: 232
rugulosa Gebler, 1841b: 620 (*Chrysomela*) **A:** MG RU (SS)
longipes Jacobson, 1898: 197
urumchiana S.-H. Chen, 1934a: 2 **A:** CH (XIN) [# 521]

genus *Cyrtonus* Latreille, 1829: 149 [NP] type species: *Chrysomela rotundata* Herrich-Schäffer, 1838
Cyrtonussa E. Strand, 1935: 287 [RN] type species: *Chrysomela rotundata* Herrich-Schäffer, 1838
almeriensis Cobos, 1953: 133 E: SP (Andalucía: Almería)
angusticollis Fairmaire, 1851a: 543 E: SP (Andalucía: Cádiz, Málaga)
arcasi Fairmaire, 1884c: 361 E: SP (Andalucía, Castilla La Mancha: Albacete)
brevis Fairmaire, 1851a: 547 E: SP (Catalonia: Barcelona)
canalisternus Marseul, 1883: 20 E: PT (North-East: Vila Real)
charidemi Vela & Bastazo, 2009: 80 E: SP (Andalucía: Almería)
cobosi Vela & Bastazo, 2009: 78 E: SP (Andalucía: Granada, Málaga)

contractus Fairmaire, 1882: 19 E: SP (Andalucía: Granada, Almería)
coruscans Vuillefroy, 1868: 295 E: PT (North, Central) SP (Asturias, Galicia, Castilla-León: Ávila)
cupreovirens Pérez Arcas, 1872: 134 E: SP
curtulus Fairmaire, 1883: 262 E: PT (North, Central)
curtus Fairmaire, 1851a: 547 E: PT (North) SP (Asturias, Cantabria)
cylindricus Marseul, 1883: 25 E: SP (central and eastern Andalucía)
mateui Cobos, 1954a: 151
denticulatus Chevrolat, 1872b: cxxxii E: SP (Castilla-León: Soria)
dorsolineatus Fairmaire, 1883: 256 E: SP (Andalucía)
dufourii L. J. M. Dufour, 1848: 412 E: FR (Pyrenees) SP (Pyrenees)
coarctatus Mulsant & Wachanru, 1850: 413
ehlersi Fairmaire, 1884c: 361 E: SP (Murcia, Valencia: Alicante)
elegans Germar, 1813: 127 (*Eumolpus*) E: PT (South) SP (Andalucía: Huelva)
eumolpus Fairmaire, 1851a: 545 E: SP (Andalucía, Aragón: Teruel)
fairmairei Rosenhauer, 1856: 315 E: SP (Andalucía: Málaga)
conformis Fairmaire, 1883: 254
gratiosus Rosenhauer, 1856: 314
franzi Cobos, 1954c: 323 E: PT (Vila Real) SP (North-West Castilla-León, Galicia: Pontevedra)
gadorensis Cobos, 1954c: 322 (ssp.) E: SP (Andalucía: Almería)
gibbicollis antoinei Kocher, 1959: 60 (ssp.) N: MO (Sebdou Basin)
gibbicollis gibbicollis Fairmaire, 1866a: 73 N: MO (Tanger)
gibbicollis ouezzani Kocher, 1959: 59 (ssp.) N: MO (Rif Occidental)
heydeni Fairmaire, 1884c: 361 E: SP (Asturias, Cantabria)
majoricensis Breit, 1908: 64 E: SP (Mallorca)
martorellii Fairmaire, 1880a: 38 E: SP (Extremadura: Cáceres, Andalucía: Córdoba)
minor Fairmaire, 1883: 267 E: SP (Andalucía: Málaga)
montanus Fairmaire, 1851a: 542 E: SP (Madrid, Castilla-León: Ávila, Segovia)
oomorphus Fairmaire, 1882: 19 E: SP (Asturias, Castilla-León)
pardoi Cobos, 1953: 131 E: SP (Murcia, Castilla La Mancha: Albacete, Andalucía: Almería, Jaén, Granada)
plumbeus Fairmaire, 1851a: 540 E: SP (Murcia, Valencia: Alicante, Andalucía: Almería, Granada)
major Fairmaire, 1883: 252
pazii Fairmaire, 1883: 254
puncticeps Fairmaire, 1882: 19 E: SP (Aragón: Teruel, Castilla La Mancha: Cuenca, Guadalajara)
punctipennis Fairmaire, 1858: 744 E: FR (Pyrénées Orientales) SP (Catalonia: Barcelona)

punctulatus Fairmaire, 1883: 268 E: PT (Guarda)
rotundatus Herrich-Schäffer, 1838b: [7] (*Chrysomela*) E: FR (Provence) SP (Valencia)
sycophanta Fairmaire, 1883: 263
riffensis Cobos, 1954c: 321 (ssp.) N: MO (North)
ruficornis Fairmaire, 1851a: 544 E: SP (Aragón, Castilla-León, Castilla La Mancha, Madrid, Navarra)
scutellatus Fairmaire, 1883: 264 E: PT (Bragança)
strictus Fairmaire, 1883: 262 E: SP (Castilla La Mancha: Ciudad Real, Toledo)
thoracicus Fairmaire, 1851a: 544 E: SP (Andalucía: Cádiz)
versicolor Marseul, 1883: 22 E: PT (Leiria, Lisboa) SP (Castilla-León: Ávila)

genus *Fasta* Petitpierre & Alonso Zarazaga, 2019a: 406 type species: *Coccinella fastuosa* Scopoli, 1763 [# 523]
Fastuolina Warchałowski, 1991d: 281 [UN]
fastuosa fastuosa Scopoli, 1763: 74 (*Coccinella*) E: AL AN AU BE BH BU BY CR CZ DE EN FI FR GB GE GR HU IT KO KR LA LS LT LU MD ME NL NM NR PL RO RU (CT NT ST) SB SK SL SP SV SZ TR UK **A:** AB AF GG IN KZ RU (ES WS) TR **NARi** [# 603, # 1025]
aenea Geoffroy, 1785: 108 (*Chrysomela*)
andorrensis Bechyné, 1950b: 174 (*Dlochrysa*, ssp.)
biroi Csiki, 1953: 129 (*Chrysomela*, var.)
callichloris Bechyné, 1950b: 174 (*Dlochrysa*, ab.) [IN]
coromandeliana Maulik, 1926: 40 (*Chrysolina*)
cupreonitens Marseul, 1886: 46 (*Chrysomela*, var.)
galeopsidis Schrank, 1798: 532 (*Chrysomela*)
inexplicabilis Brancsik, 1910: 189 (*Chrysomela*)
jodasi Bechyné, 1950b: 174 (*Dlochrysa*, ab.) [IN]
obscura Fleischer, 1917b: 121 (*Chrysomela*, ab.) [IN]
revyi Kaszab, 1962: 55 (*Chrysomela*, ab.) [IN]
rugosicollis Fleischer, 1917b: 121 (*Chrysomela*, ab.) [IN]
subfastuosa Motschulsky, 1860b: 226 (*Chrysomela*)
fastuosa ventricosa Suffrian, 1858a: 539 (*Chrysomela*) [QS] E: PT SP [# 523]

genus *Humba* S.-H. Chen, 1934d: 47 [RN] type species: *Chrysomela cyanicollis* Hope, 1831
Eumela Baly, 1875a: 23 [HN] type species: *Chrysomela cyanicollis* Hope, 1831
Eumelaella E. Strand, 1935: 289 [RN] type species: *Chrysomela cyanicollis* Hope, 1831
cyanicollis Hope, 1831: 29 (*Chrysomela*) **A:** CH (GAN GUA GUI GUX HUB HUN SCH XIZ YUN) ID (AP SD) NP **ORR**

davidis Fairmaire, 1878b: 134 (*Sphaerolina*)
pyrrhopyga Stål, 1857: 60 (*Chrysomela*)
tonkinensis Weise, 1902g: 110 (*Eumela*)

genus *Leptinotarsa* Chevrolat, 1836: 397 [conserved name, ICZN (1985b), Opinion 1290] type species: *Leptinotarsa heydenii* Stål, 1858
Myocoryna Stål, 1859: 316 type species: *Chrysomela juncta* Germar, 1823
Polygramma Chevrolat, 1836: 397 [suppressed name, ICZN (1985b), Opinion 1290] type species: *Chrysomela juncta* Germar, 1823
Thlibocoryna C. V. Riley, 1875: 17 [RN] type species: *Chrysomela juncta* Germar, 1823
decemlineata Say, 1824a: 453 (*Doryphora*) **Ei:** AL AN AU BE BH BU BY CR CZ DE EN FI FR GB GE GR HU IT KO KR KZ LA LS LT LU MD ME NL NM NR PL PT RO RU (CT NT ST) SB SK SL SP SV SZ TR UK **Ni:** LB **Ai:** AB AR CH (HEI JIL XIN) GG RU (ES FE SS WS) IN IQ IS KI KZ SY TD TM TR UZ **NAR NTR**
intermedia Tower, 1906: 7
multilineata Stål, 1859: 316 (*Myocoryna*)
multitaeniata Stål, 1859: 317 (*Myocoryna*)
oblongata Tower, 1906: 6
rubicunda Tower, 1906: 7

genus *Oreina* Chevrolat, 1836: 402 type species: *Chrysomela speciosa* Linnaeus, 1758
subgenus *Allorina* Weise, 1902f: 105 type species: *Orina tristis* Fabricius, 1792 sensu Weise (1902) (= *Oreina smaragdina* Weise, 1883) **[# 558]**
Allorinula Weise, 1916b: 98 [RN] type species: *Orina tristis* Fabricius, 1792 sensu Weise (1902) (= *Oreina smaragdina* Weise, 1883)
caerulea Olivier, 1791a: 718 (*Chrysomela*) **E:** AL AU BE BH BU BY CR DE FR GE GR HU ME NM PL RO RU (CT NT ST) SB SK SL SP UK **A:** RU (WS: Sverdlovsk and Chelyabinsk regions) **[# 533, # 611, Appendix]**
cyanea Hänel, 1937: 26 (*Chrysochloa*, ab.) [IN]
delaunayi Mallet, 1925: 28 (*Chrysochloa*, ab.) [IN]
delaunayi Bechyné, 1958b: 83 (ssp.)
gaertneri Weise, 1894k: 250 (*Orina*, var.)
luctuosa Olivier, 1807: 549 (*Chrysomela*) [# 1026]
marxzellana Bechyné, 1958b: 83 (ssp.)
nigripennis Hänel, 1937: 26 (*Chrysochloa*, ab.) [IN]
nigritula Weise, 1894k: 250 (*Orina*, var.)
olivacea Hänel, 1937: 26 (*Chrysochloa*, ab.) [IN]
onega Bechyné, 1958b: 83 (ssp.)
rareua Bechyné, 1958b: 83 (ssp.)

rugulosa Suffrian, 1851a: 181 (*Chrysomela*)
serbica Apfelbeck, 1912: 253 (*Chrysochloa*, var.)
suffriani Bontems, 1984: 195 (ssp.) [RN]
tenebrosa Weise, 1883c: 243 (*Orina*, var.)
violacea Hänel, 1937: 26 (*Chrysochloa*, ab.) [IN]
canavesei Bontems, 1984: 191 E: IT
collucens J. Daniel, 1903a: 180 (*Chrysochloa*, ssp.) E: FR IT
fenestrellana J. Daniel, 1903a: 180 (*Chrysochloa*, ssp.)
vesulina J. Daniel, 1903a: 180 (*Chrysochloa*, ab.) [IN]
viridifulgens J. Daniel, 1903a: 181 (*Chrysochloa*, ab.) [IN]
smaragdina Weise, 1883c: 243 (var.) E: AU FR IT SL SZ [# 552]
auricollis Stierlin, 1887: 137 (var.) [# 552]
bidentata Bontems, 1981: 108 [# 552]
coeruleoviridis Hänel, 1937: 26 (*Chrysochloa*, ab.) [IN]
cuprina Weise, 1883c: 243 (*Orina*, var.) [# 552]
cyanea Hänel, 1937: 26 (*Chrysochloa*, ab.) [IN]
nigripennis Hänel, 1937: 26 (*Chrysochloa*, ab.) [IN]
nigripennis Burlini, 1942b: 128 (ab.) [IN]
olivacea Hänel, 1937: 26 (*Chrysochloa*, ab.) [IN]
violacea Hänel, 1937: 26 (*Chrysochloa*, ab.) [IN]
subgenus *Chrysochloa* Hope, 1841: 165 type species: *Chrysomela tristis* Fabricius, 1792 (= *Chrysomela cacaliae* Schrank, 1785)
Alpaeixena Motschulsky, 1860b: 203 type species: *Chrysomela senecionis* Schummel, 1844
cacaliae albanica J. Müller, 1948: 96 (*Chrysochloa*, ssp.) [QS] E: AL
cacaliae arvernica Bontems, 2005: 146 (ssp.) E: FR
cacaliae barii Schatzmayr, 1943: 29 (*Chrysochloa*, ssp.) E: IT
cacaliae bearnica Bontems, 2005: 148 (ssp.) E: FR SP
cacaliae bohemica Weise, 1889c: 127 (*Orina*, var.) E: CZ GE
aureopennis Fassati, 1961: 420 (*Chrysochloa*, ab.) [IN]
ignita Hänel, 1937: 31 (*Chrysochloa*, ab.) [IN]
marani Fassati, 1961: 417 (*Chrysochloa*, ssp.)
pseudocacaliae Fassati, 1961: 420 (*Chrysochloa*, ab.) [IN]
subolivacea Fassati, 1961: 420 (*Chrysochloa*, ab.) [IN]
cacaliae cacaliae Schrank, 1785: 317 (*Chrysomela*) E: AU BE FR GE IT LS SL SZ
alpina Weber, 1801: 57 (*Galleruca*)
coeruleolineata Duftschmid, 1825: 169 (*Chrysomela*)
ignita Hänel, 1937: 31 (*Chrysochloa*, ab.) [IN]
macera Weise, 1884d: 476 (*Orina*, var.) [# 578]
nubigena Weise, 1884d: 476 (*Orina*, var.) [IN] [# 577]
nubigena Stierlin, 1896b: 565 (*Orina*, var.)

olivacea Hänel, 1937: 31 (*Chrysochloa*, ab.) [IN]
pseudobohemica Hänel, 1937: 31 (*Chrysochloa*, ab.) [IN]
resplendens J. Daniel, 1903a: 183 (*Chrysochloa*, var.) [IN] [# 577]
sumptuosa L. Redtenbacher, 1848: 550 (*Chrysomela*)
tristis Fabricius, 1792a: 315 (*Chrysomela*) [# 553, # 1027]
violacea Hänel, 1937: 31 (*Chrysochloa*, ab.) [IN]
violaceicollis Hänel, 1937: 31 (*Chrysochloa*, ab.) [IN]

cacaliae dinarica Apfelbeck, 1912: 256 (*Chrysochloa*, var.) [QS] E: BH BU CR ME NM SB SL
gabrielis Depoli, 1926b: 103 (*Chrysochloa*, forma) [IN] [# 577]
violacea Depoli, 1940: 336 (*Chrysochloa*, ab.) [IN]
violaceicollis Hänel, 1937: 31 (*Chrysochloa*, ab.) [IN]

cacaliae feldbergensis Bechyné, 1958b: 91 (ssp.) [QS] E: FR GE

cacaliae magistrettii Schatzmayr, 1942a: 164 (*Chrysochloa*) E: IT

cacaliae senecionis Schummel, 1844: 198 (*Chrysomela*) E: CZ GE PL RO SK UK [# 1080]
coeruleoviridis Hänel, 1937: 32 (*Chrysochloa*, ab.) [IN]
cyanipennis Weise, 1883a: 215 (*Orina*, var.)
fraudulenta Weise, 1883a: 215 (*Orina*, var.)
janaceki Weise, 1896c: 110 (*Orina*, var.) [IN] [# 576]
lissahorensis Weise, 1896b: 110 (*Orina cacaliae*, var.)
nigra Hänel, 1937: 32 (*Chrysochloa*, ab.) [IN]
nigripennis Hänel, 1937: 32 (*Chrysochloa*, ab.) [IN]
olivacea Hänel, 1937: 32 (*Chrysochloa*, ab.) [IN]
pseudoletzneri Hänel, 1937: 31 (*Chrysochloa*, ab.) [IN]
tristicula Weise, 1883a: 215 (*Orina*, var.)
viridicollis Hänel, 1937: 32 (*Chrysochloa*, ab.) [IN]

cacaliae senilis J. Daniel, 1903a: 183 (*Chrysochloa*, var.) E: FR IT [# 578]

cacaliae tussilaginis Suffrian, 1851a: 155 (*Chrysomela*) E: FR SP
violacea Hänel, 1937: 31 (*Chrysochloa*, ab.) [IN]

elongata elongata Suffrian, 1851a: 146 (*Chrysomela*) [conserved name, ICZN (2011b), Opinion 2278] E: IT [# 1081]
tridentina Franz, 1949a: 16 (*Chrysochloa*, ssp.)

elongata occidentalis Ruffo, 1946: 180 (*Chrysochloa*, ssp.) E: BE FR IT SZ
gallica Franz, 1949a: 16 (*Chrysochloa*, ssp.)
violacea Hänel, 1937: 33 (*Chrysochloa*, ab.) [IN]

elongata ruffoi Franz, 1949a: 23 (*Chrysochloa*, ssp.) E: IT SZ
aurocuprea Hänel, 1937: 33 (*Chrysochloa*, ab.) [IN]

elongata siparii Luigioni, 1930: 206 (*Chrysochloa*) E: IT

elongata styriaca Franz, 1949a: 13 (*Chrysochloa*, ssp.) E: AU

elongata zangherii Daccordi & Ruffo, 1987: 16 (ssp.) E: IT

elongata zoiai Daccordi & Ruffo, 1987: 14 (ssp.) E: IT
fairmairiana Gozis, 1882: 206 (*Chrysomela*) [RN] E: FR SP
fairmairei Weise, 1884d: 475 (*Orina*, var.) [IN] [# 577]
splendidula Fairmaire, 1865: 81 (*Chrysomela*) [HN]
genei Suffrian, 1851a: 147 (*Chrysomela*) E: FR IT
carbonaria Binaghi, 1973: 5 (*Chrysochloa*, forma) [IN]
elegans Aragona, 1830: 27 (*Chrysomela*) [HN]
redikortzevi Jacobson, 1924a: 20 [= 1925c: 14] (*Chrysochloa*) A: RU (SS) [# 534]
speciosissima binaghii Daccordi & Ruffo, 1976: 397 (ssp.) E: FR IT [# 612]
speciosissima bisculpta Bechyné, 1958b: 92 (ssp.) [QS] E: GE
speciosissima carpathica Kraatz, 1859: 288 (var.) E: "Carpathes" [# 535]
speciosissima convergens Suffrian, 1851a: 145 (*Chrysomela*) E: FR SP
cyanea Hänel, 1937: 32 (*Chrysochloa*, ab.) [IN]
pyrenaica Weise, 1884d: 481 (*Orina*, var.) [IN] [# 576]
violacea Hänel, 1937: 32 (*Chrysochloa*, ab.) [IN]
speciosissima drenskii Gruev, 1974b: 969 (*Chrysochloa*, ssp.) E: BU GR
speciosissima fuscoaenea Schummel, 1844: 196 (*Chrysomela*) E: CZ GE PL SK [# 1028, # 1082]
aurea Hänel, 1937: 32 (*Chrysochloa*, ab.) [IN]
coeruleoviridis Hänel, 1937: 32 (*Chrysochloa*, ab.) [IN]
coerulescens Letzner, 1853: 89 (*Chrysomela*, var.) [IN] [# 575]
cupreoviridis Letzner, 1853: 89 (*Chrysomela*, var.) [IN] [# 575]
cyanea Hänel, 1937: 32 (*Chrysochloa*, ab.) [IN]
flammea Jakob, 1965: 12 (*Chrysochloa*, ssp.)
fuscocuprea Letzner, 1853: 89 (*Chrysomela*, var.) [IN] [# 575]
guidonis Csiki, 1953: 130 (*Chrysochloa*, ab.) [IN]
ignita Letzner, 1853: 89 (*Chrysomela*, var.) [IN] [# 575]
letzneri Weise, 1883a: 214 (*Orina*, var.) [RN]
nigra Hänel, 1937: 32 (*Chrysochloa*, ab.) [IN]
nigrescens Letzner, 1853: 89 (*Chrysomela*, var.) [IN] [# 575]
nigroaenea Letzner, 1853: 89 (*Chrysomela*, var.) [IN] [# 575]
nigrocoerulea Letzner, 1853: 89 (*Chrysomela*, var.) [IN] [# 575]
nigripennis Hänel, 1937: 32 (*Chrysochloa*, ab.) [IN]
olivacea Hänel, 1937: 32 (*Chrysochloa*, ab.) [IN]
opaca Hänel, 1937: 32 (*Chrysochloa*, ab.) [IN]
pretiosa Letzner, 1853: 89 (*Chrysomela*, var.) [IN] [# 575]
rufipes Hänel, 1937: 32 (*Chrysochloa*, ab.) [IN]
schummeli Weise, 1883a: 214 (*Orina*, var.)
silesiaca Weise, 1884d: 481 (*Orina*, ssp.)
tristis Letzner, 1853: 89 (*Chrysomela*, var.) [IN] [# 575]

violacea Letzner, 1853: 89 (*Chrysomela*, var.) [IN] [# 575]
violacea Stierlin, 1896b: 566 (*Orina*, var.)
violaceicollis Hänel, 1937: 32 (*Chrysochloa*, ab.) [IN]
viridescens Letzner, 1853: 89 (*Chrysomela*, var.) [IN] [# 575]
viridescens Weise, 1883a: 214 (*Orina*, var.) [# 1114]
viridiaenea Letzner, 1853: 89 (*Chrysomela*, var.) [IN] [# 575]
viridicollis Hänel, 1937: 32 (*Chrysochloa*, ab.) [IN]
viridis Letzner, 1853: 89 (*Chrysomela*, var.) [IN] [# 575]
speciosissima juncorum Suffrian, 1851a: 144 (*Chrysomela*) E: RO
coeruleoviridis Hänel, 1937: 32 (*Chrysochloa*, ab.) [IN]
herculis Kaszab, 1962: 56 (*Chrysochloa*, ab.) [IN]
speciosissima natarsia Bechyné, 1958b: 92 (ssp.) [QS] E: AU
speciosissima priela Bechyné, 1958b: 92 (ssp.) [QS] E: SP
speciosissima pyhrgassia Bechyné, 1958b: 92 (ssp.) [QS] E: AU
speciosissima solarii Daccordi & Ruffo, 1976: 398 (ssp.) E: IT
speciosissima speciosissima Scopoli, 1763: 74 (*Coccinella*) E: AU BE BH CR FR GE IT KO LS ME NM SL SZ [# 1029]
alpina Hänel, 1937: 32 (*Chrysochloa*, ab.) [IN]
aurea Hänel, 1937: 32 (*Chrysochloa*, ab.) [IN]
coeruleoviridis Hänel, 1937: 32 (*Chrysochloa*, ab.) [IN]
cupricollis Hänel, 1937: 32 (*Chrysochloa*, ab.) [IN]
fothi Hänel, 1934: 90 (*Chrysochloa*, ab.) [IN]
guidonis Csiki, 1953: 130 (*Chrysochloa*, ab.) [IN]
nigripennis Hänel, 1937: 32 (*Chrysochloa*, ab.) [IN]
opaca Hänel, 1937: 32 (*Chrysochloa*, ab.) [IN]
perinii Jakob, 1953c: 132 (*Chrysochloa*)
stilleri Csiki, 1953: 130 (*Chrysochloa*, ab.) [IN]
violaceicollis Depoli, 1940: 336 (*Chrysochloa*, ab.) [IN]
viridicollis Hänel, 1937: 32 (*Chrysochloa*, ab.) [IN]
speciosissima troglodytes Kiesenwetter, 1861: 391 (*Chrysomela*, var.) E: FR IT SZ
coerulea Hänel, 1937: 33 (*Chrysochloa*, ab.) [IN]
coeruleoviridis Hänel, 1937: 33 (*Chrysochloa*, ab.) [IN]
crocetta Bechyné, 1958b: 92 (ssp.)
cupricollis Hänel, 1937: 32 (*Chrysochloa*, ab.) [IN]
fuscoaenea Hänel, 1937: 32 (*Chrysochloa*, ab.) [IN]
rugosa Hänel, 1937: 33 (*Chrysochloa*, ab.) [IN]
violacea Hänel, 1937: 33 (*Chrysochloa*, ab.) [IN]
viridiaurea Hänel, 1937: 32 (*Chrysochloa*, ab.) [IN]
viridicollis Hänel, 1937: 33 (*Chrysochloa*, ab.) [IN]
viridis Hänel, 1937: 33 (*Chrysochloa*, ab.) [IN]

subgenus *Frigidorina* Kühnelt, 1984: 174 type species: *Orina frigida* Weise, 1883 [# 532]
frigida Weise, 1883c: 247 (*Orina*) E: AU FR GE IT LS SZ
cupreata Weise, 1887e: 364 (*Orina*, var.)
kiesenwetteri Weise, 1884d: 487 (*Orina*, var.) [IN] [# 577]
rhaetica Weise, 1884d: 487 (*Orina*, var.) [IN] [# 577]
subgenus *Intricatorina* Kühnelt, 1984: 174 type species: *Chrysomela intricata* Germar, 1823
intricata Germar, 1823: 596 (*Chrysomela*) E: AL AU BE BH BU CR CZ FR GE IT LS ME NM PL RO SB SK SL SZ UK [# 554, # 1030]
amethystina Weise, 1883a: 211 (*Orina*, var.)
anderschii Duftschmid, 1825: 166 (*Chrysomela*) [# 554]
aurea Hänel, 1937: 27 (*Chrysochloa*, ab.) [IN]
aurulenta Suffrian, 1851a: 179 (*Chrysomela*)
carnica J. Müller, 1948: 96 (*Chrysochloa*, var.) [IN] [# 577]
coerulea Hänel, 1937: 27 (*Chrysochloa*, ab.) [IN]
coeruleoviridis Hänel, 1937: 26 (*Chrysochloa*, ab.) [IN]
collaris Hänel, 1937: 26 (*Chrysochloa*, ab.) [IN]
cyanea Hänel, 1937: 27 (*Chrysochloa*, ab.) [IN]
cyaneicollis Hänel, 1937: 27 (*Chrysochloa*, ab.) [IN]
nigripennis Hänel, 1937: 26 (*Chrysochloa*, ab.) [IN]
olivacea Hänel, 1937: 27 (*Chrysochloa*, ab.) [IN]
seminigra Weise, 1883a: 211 (*Orina*, var.)
violacea Hänel, 1937: 27 (*Chrysochloa*, ab.) [IN]
subgenus *Oreina* Chevrolat, 1836: 402 type species: *Chrysomela speciosa* Linnaeus, 1767
Dlochrysa Motschulsky, 1860b: 203 type species: *Chrysomela speciosa* Linnaeus, 1767
Fastuolina Warchałowski, 1993c: 102 type species: *Chrysomela speciosa* Linnaeus, 1767
Orina Agassiz, 1846b: 263 [UE] type species: *Chrysomela speciosa* Linnaeus, 1767
Romalorina Weise, 1906e: 558 type species: *Chrysomela alpestris* Schummel, 1844 [# 561]
alpestris alpestris Schummel, 1844: 197 (*Chrysomela*) E: CZ HU PL SK UK
alutacea Hänel, 1937: 27 (*Chrysochloa*, ab.) [IN]
bicolora Weise, 1883a: 212 (*Orina*, var.)
bohuslavi Fleischer, 1916a: 100 (*Chrysochloa*, ab.) [IN]
coeruleoviridis Hänel, 1937: 27 (*Chrysochloa*, ab.) [IN]
fontinalis Weise, 1883a: 212 (*Orina*, var.)
liturata Hänel, 1937: 27 (*Chrysochloa*, ab.) [IN]
moesta Weise, 1883a: 212 (*Orina*, var.)
nigripennis Hänel, 1937: 27 (*Chrysochloa*, ab.) [IN]
olivacea Weise, 1883a: 212 (*Orina*, var.)
punctatissima Suffrian, 1851a: 174 (*Chrysomela*)

rivularis Weise, 1883a: 212 (*Orina*, var.)
stejskali Bechyné, 1958b: 84 (ssp.)
stejskali Fleischer, 1914: 65 (*Chrysochloa*, ab.) [IN]
umbrosa Weise, 1884d: 447 (*Orina*, var.) [# 578]
viridicollis Hänel, 1937: 27 (*Chrysochloa*, ab.) [IN]
alpestris balcanica Weise, 1883c: 245 (*Orina*, var.) E: BU GR ME SB
alpestris bannatica Weise, 1884d: 447 (*Orina*, var.) E: RO SK UK [# 536, # 578, # 1115]
aurea Hänel, 1937: 29 (*Chrysochloa*, ab.) [IN]
anthrisci Weise, 1900b: 283 (*Chrysochloa*, var.)
bodemeyeri Weise, 1900b: 283 (*Chrysochloa*, var.)
coeruleoviridis Hänel, 1937: 29 (*Chrysochloa*, ab.) [IN]
cyanicollis Hänel, 1937: 27 (*Chrysochloa*, ab.) [IN]
nigra Hänel, 1937: 27 (*Chrysochloa*, ab.) [IN]
nigripennis Hänel, 1937: 27 (*Chrysochloa*, ab.) [IN]
nigrocoerulea Hänel, 1937: 29 (*Chrysochloa*, ab.) [IN]
nigrovittata Hänel, 1937: 29 (*Chrysochloa*, ab.) [IN]
violacea Hänel, 1937: 27 (*Chrysochloa*, ab.) [IN]
polonina Bechyné, 1958b: 85 (ssp.)
punctatissima Weise, 1900b: 283 (*Chrysochloa*, var.)
alpestris marsicana Luigioni, 1933: 121 (*Chrysochloa*) E: IT
coelestina Luigioni, 1933: 126 (*Chrysochloa*, ab.) [IN]
coeruleovittata Luigioni, 1933: 126 (*Chrysochloa*, ab.) [IN]
pseudoexcellens Luigioni, 1933: 126 (*Chrysochloa*, ab.) [IN]
viridula Binaghi, 1938: 41 (*Chrysochloa*, ab.) [IN]
alpestris meridiana Marchand, 1939: 157 (*Chrysochloa*, var.) E: NM
alpestris nigrina Suffrian, 1851a: 176 (*Chrysomela*) E: AN FR SP
cyanicollis Hänel, 1937: 27 (*Chrysochloa*, ab.) [IN]
navarica David, 1952: 109 (*Chrysochloa*, ssp.)
nigrina Hänel, 1937: 27 (*Chrysochloa*, ab.) [IN]
putoni Weise, 1892b: 374 (*Orina*, var.)
violaceicollis Hänel, 1937: 27 (*Chrysochloa*, ab.) [IN]
viridicollis Hänel, 1937: 27 (*Chrysochloa*, ab.) [IN]
alpestris polymorpha Kraatz, 1880: 35 (var.) E: GE
aurea Hänel 1937: 27 (*Chrysochloa*, ab.) [IN]
bivittata Hänel, 1937: 27 (*Chrysochloa*, ab.) [IN]
cyanicollis Hänel, 1937: 27 (*Chrysochloa*, ab.) [IN]
gaukleri Bechyné, 1958b: 84 (ssp.)
gersfeldiana Bechyné, 1958b: 84 (ssp.)
ignita Hänel, 1937: 27 (*Chrysochloa*, ab.) [IN]

pseudoillustris Hänel, 1937: 27 (*Chrysochloa*, ab.) [IN]
violaceicollis Hänel, 1937: 27 (*Chrysochloa*, ab.) [IN]
viridis Hänel, 1937: 27 (*Chrysochloa*, ab.) [IN]
vinariensis Weise, 1902c: 240 (*Chrysochloa*, var.)
alpestris variabilis Weise, 1883c: 245 (*Orina*, var.) E: AU BH CR FR GE IT LS NM SL SZ
apfelbecki Winkler, 1930: 1286 (*Chrysochloa*, var.) [RN]
croatica Apfelbeck, 1912: 254 (*Chrysochloa*, var.) [HN]
coeruleoviridis Hänel, 1937: 28 (*Chrysochloa*, ab.) [IN]
curtula Apfelbeck, 1912: 254 (*Chrysochloa*, ab.) [IN]
curtula Bechyné, 1958: 85 (ssp.)
cyanea Hänel, 1937: 28 (*Chrysochloa*, ab.) [IN]
cyanicollis Hänel, 1937: 28 (*Chrysochloa*, ab.) [IN]
imitatrix Apfelbeck, 1912: 254 (*Chrysochloa*, var.)
kraussi Bechyné, 1958b: 85 (ssp.) [RN]
smaragdella Marchand, 1938: 208 *Chrysochloa*, ab.) [IN]
straubiana Marchand, 1938: 207 (*Chrysochloa*, ssp.)
violacea Hänel, 1937: 28 (*Chrysochloa*, ab.) [IN]
viridis Hänel, 1937: 28 (*Chrysochloa*, ab.) [IN]
weisei Krauss, 1899: 205 (*Orina*, var.) [HN]
bifrons bifrons Fabricius, 1792a: 314 (*Chrysomela*) E: AU FR GE IT LS SZ
aenescens Suffrian, 1851a: 161 (*Chrysomela*) [# 1031]
aerosa Weise, 1884d: 461 (*Orina*, var.) [IN] [# 577]
bifrons cadorensis Bechyné, 1958b: 87 (ssp.) E: AU IT
bifrons decora Richter, 1820: 7 (*Chrysomela*) E: CZ PL
aurata Weise, 1883a: 213 (*Orina*, var.)
cyanescens Weise, 1883a: 213 (*Orina*, var.)
virescens Weise, 1883a: 213 (*Orina*, var.)
bifrons heterocera Reitter, 1917: 7 (*Chrysochloa*) E: PL RO SK UK
addita Roubal, 1940c: 92 (*Chrysochloa*, ab.) [IN]
diabolica Marchand, 1939: 157 (*Chrysochloa*, ab.) [IN]
obenbergeri Marchand, 1939: 156 (*Chrysochloa*, var.)
pseudoviridis Marchand, 1939: 157 (*Chrysochloa*, ab.) [IN]
bifrons recuperata Bechyné, 1958b: 87 (ssp.) E: AU GE IT LS SZ
gadmena Bechyné, 1958b: 87 (ssp.)
bifrons stussineri Weise, 1884d: 461 (*Orina*, var.) E: AU BH BU CR GE SB SL [# 578]
ganglbaueri Jakob, 1953b: 157 (*Chrysochloa*) E: AN FR SP
auberi David, 1954: 102 (*Chrysochloa*)
casasecai Franz, 1958: 129 (*Chrysochloa*, ab.) [IN]
iberica Franz, 1958: 129 (*Chrysochloa*, ssp.)

nigra Jakob, 1953b: 157 (*Chrysochloa*, forma) [IN] [# 577]
viridis Jakob, 1953b: 157 (*Chrysochloa*, forma) [IN] [# 577]
gloriosa Fabricius, 1782: 497 (*Chrysomela*) E: AU BE FR GE IT LS SZ [# 1032]
aurichalcea Hänel, 1937: 29 (*Chrysochloa*, ab.) [IN]
coerulea Hänel, 1937: 29 (*Chrysochloa*, ab.) [IN]
coeruleoviridis Hänel, 1937: 29 (*Chrysochloa*, ab.) [IN]
cyanea Hänel, 1937: 29 (*Chrysochloa*, ab.) [IN]
glacialis Weise, 1884d: 459 (*Orina*, var.) [IN] [# 577]
glacialis Stierlin, 1896b: 562 (*Orina*, var.)
maritima J. Daniel, 1903a: 184 (*Chrysochloa*, var.) [IN] [# 577]
matheyi Matcha, 1917: 54 (*Chrysochloa*, ab.) [IN]
simplex Burlini, 1942b: 128 (ab.) [IN]
violacea Hänel, 1937: 29 (*Chrysochloa*, ab.) [IN]
vittigera Suffrian, 1851a: 166 (*Chrysomela*) [# 1033]
liturata liturata Scopoli, 1763: 73 (*Coccinella*) E: AU IT SL
aurea Hänel, 1937: 29 (*Chrysochloa*, ab.) [IN]
aurichalcea Hänel, 1937: 29 (*Chrysochloa*, ab.) [IN]
cyanicollis Hänel, 1937: 29 (*Chrysochloa*, ab.) [IN]
phalerata Germar, 1834: 13 (*Chrysomela*)
violaceicollis Hänel, 1937: 29 (*Chrysochloa*, ab.) [IN]
liturata tomasii Daccordi & Ruffo, 1977: 99 (ssp.) E: IT
speciosa bosnica Apfelbeck, 1912: 253 (*Chrysochloa*, var.) [QS] E: BH BU CR ME SB
arrogans Apfelbeck, 1912: 253 (*Chrysochloa*, var.)
coeruleoviridis Depoli, 1940: 335 (*Chrysochloa*, ab.) [IN]
cyanicollis Depoli, 1940: 335 (*Chrysochloa*, ab.) [IN]
depolii Csiki, 1953: 129 (*Chrysomela*, ab.) [IN]
divisivittis Depoli, 1940: 335 (*Chrysochloa*, ab.) [IN]
humeralis Depoli, 1940: 335 (*Chrysochloa*, ab.) [IN]
padewiethi Kaszab, 1962: 55 (*Chrysomela*, ab.) [IN]
viridis Depoli, 1940: 335 (*Chrysochloa*, ab.) [IN]
speciosa huberi Bechyné, 1958b: 86 (ssp.) [QS] E: FR (Vosges)
speciosa jurassica Kippenberg, **ssp. nov.** E: FR SZ (Jura) [# 556]
discolor Weise, 1884d: 451 (*Orina*, var.) [IN] [# 577]
speciosa lugdunensis Weise, 1909: 191 (*Chrysomela*) [RN] E: FR (Massif Central)
superba Olivier, 1791a: 705 (*Chrysomela*) [HN]
speciosa speciosa Linnaeus, 1767: 588 (*Chrysomela*) E: AU BE FR GE IT LS SL SZ
atramentaria Weise, 1892b: 374 (*Orina*, var.)
aurea Hänel, 1937: 28 (*Chrysochloa*, ab.) [IN]
bluehweissi Marchand, 1933: 455 (*Chrysochloa*, ab.) [IN]

coerulea Hänel, 1937: 28 (*Chrysochloa*, ab.) [IN]
coeruleoviridis Hänel, 1937: 28 (*Chrysochloa*, ab.) [IN]
cupricollis Hänel, 1937: 28 (*Chrysochloa*, ab.) [IN]
cyanea Hänel, 1937: 28 (*Chrysochloa*, ab.) [IN]
cyanicollis Hänel, 1937: 28 (*Chrysochloa*, ab.) [IN]
divisa Hänel, 1937: 28 (*Chrysochloa*, ab.) [IN]
excellens Weise, 1884d: 451 (*Orina*, ssp.)
illustris J. Daniel, 1903a: 182 (*Chrysochloa*, ab.) [IN]
nigra Hänel, 1937: 28 (*Chrysochloa*, ab.) [IN]
nigripennis Hänel, 1937: 28 (*Chrysochloa*, ab.) [IN]
nigrovittata Hänel, 1937: 28 (*Chrysochloa*, ab.) [IN]
nubila Weise, 1884d: 450 (*Orina*, var.) [IN] [# 577]
pavonina Marchand, 1933: 455 (*Chrysochloa*, ab.) [IN]
polita Hänel, 1937: 28 (*Chrysochloa*, ab.) [IN]
phalerata L. Redtenbacher, 1848: 549 (*Chrysomela*) [HN]
pretiosa Suffrian, 1851a: 167 (*Chrysomela*)
pseudolitturata Bechyné, 1958b: 86 (ssp.)
pseudoliturata J. Müller, 1916: 98 (*Chrysochloa*, ab.) [IN]
sapphirus Fabricius, 1801a: 432 (*Chrysomela*)
venusta Suffrian, 1851a: 175 (*Chrysomela*) [QS]
violacea Hänel, 1937: 28 (*Chrysochloa*, ab.) [IN]
virgo Weise, 1884d: 450 (*Orina*, var.) [IN] [# 577]

sulcata basilea Gebler, 1823: 122 (*Chrysomela*) A: KZ RU (ES: Yenisei River valley SS WS)
aeneoviridis Jacobson 1901a: 100 (*Chrysochloa*, var.) [IN] [# 577]
coerulea Hänel, 1937: 30 (*Chrysochloa*, ab.) [IN]
cuprescens Jacobson 1901a: 100 (*Chrysochloa*, var.) [IN] [# 577]
nigrocoerulea Jacobson 1901a: 100 (*Chrysochloa*, var.) [IN] [# 577]
violacea Hänel, 1937: 30 (*Chrysochloa*, ab.) [IN]
virens Jacobson 1901a: 100 (*Chrysochloa*, var.) [IN] [# 577]

sulcata sulcata Gebler, 1823: 123 (*Chrysomela*) A: MG RU (SS)
foveolata Faldermann, 1833: 69 (*Chrysomela*)

viridis merkli Weise, 1884d: 466 (*Orina*, ssp.) E: PL RO SK UK
aurichalcea Hänel, 1937: 30 (*Chrysochloa*, ab.) [IN]
coerulea Hänel, 1937: 30 (*Chrysochloa*, ab.) [IN]
cuprea Hänel, 1937: 30 (*Chrysochloa*, ab.) [IN]
cyanipennis Hänel, 1937: 30 (*Chrysochloa*, ab.) [IN]
nigra Hänel, 1937: 30 (*Chrysochloa*, ab.) [IN]
nigripennis Hänel, 1937: 30 (*Chrysochloa*, ab.) [IN]
nigroviolacea Hänel, 1937: 30 (*Chrysochloa*, ab.) [IN]

transsylvanica Weise, 1884d: 466 (*Orina*, var.) [IN] [# 576]
violacea Hänel, 1937: 30 (*Chrysochloa*, ab.) [IN]
viridis Hänel, 1937: 30 (*Chrysochloa*, ab.) [IN]
viridis viridis Duftschmid, 1825: 167 (*Chrysomela*) E: AU BH FR GE IT LS NM SL SP SZ [# 537]
auricollis Hänel, 1937: 30 (*Chrysochloa*, ab.) [IN]
bicolora Papp, 1946c: 24 (*Chrysochloa*, ab.) [IN]
coeruleoviridis Hänel, 1937: 30 (*Chrysochloa*, ab.) [IN]
cupreonigra Hänel, 1937: 30 (*Chrysochloa*, ab.) [IN]
heerii Weise, 1883c: 246 (*Orina*, var.) [# 1116]
ignita Comolli, 1837: 49 (*Lina*) [HN]
lugubris Weise, 1883c: 246 (*Orina*, var.)
mirifica Weise, 1883c: 244 (*Orina*, var.)
nivalis Suffrian, 1851a: 163 (*Chrysomela*) [# 1083]
splendens Weise, 1883c: 246 (*Orina*, var.) [# 1117]
violaceipennis Hänel, 1937: 30 (*Chrysochloa*, ab.) [IN]
viridis Hänel, 1937: 30 (*Chrysochloa*, ab.) [IN]
viscoi Schatzmayr, 1942a: 163 (*Chrysochloa*)
viridis waldenburgi J. Müller, 1925: 27 (*Chrysochloa*, ssp.) E: SL
subgenus *Protorina* Weise, 1894k: 252 type species: *Chrysomela melanocephala* Duftschmid, 1825 (= *Chrysomela melancholica* Heer, 1845)
ludovicae cantabricola Kippenberg, 2008: 370 (ssp.) E: SP
ludovicae ludovicae Mulsant, 1854: 18 [= 1856: 12] (*Chrysomela*) E: AN FR SP
nigriceps Fairmaire, 1857a: 545 (*Chrysomela*)
melancholica melancholica Heer, 1845: 17 (*Chrysomela*) E: AU FR GE IT LS SL SZ
melanocephala Duftschmid, 1825: 170 (*Chrysomela*) [HN]
muriauxi Weill, 1951: 154 (*Chrysochloa*)
neomelanocephala Kippenberg, 1998: 315 (forma) [IN]
solitaria Kippenberg, 1994: 80 (forma) [IN]
melancholica visoi Kippenberg, 2008: 375 (ssp.) E: FR IT
peirolerii Bassi, 1834: 465 (*Chrysomela*) E: IT SZ
picea Hänel, 1937: 30 (*Chrysochloa*, ab.) [IN]
plagiata commutata Suffrian, 1861b: 435 (*Chrysomela*) E: AU GE SL
divisoria Bechyné, 1958b: 89 (ssp.)
kaprunensis Bechyné, 1958b: 89 (ssp.)
plagiata pennina Binaghi, 1938: 39 (*Chrysochloa*, ssp.) E: IT
plagiata plagiata Suffrian, 1861b: 436 (*Chrysomela*) E: PL RO SK UK
distinctissima Ulanowski, 1886: 22 (var.)
dorsalis Weise, 1887a: 186 (*Orina*, var.)
rufipes Weise, 1894k: 256 (*Orina*, var.)

retenta retenta Weise, 1894k: 256 (*Orina*) E: AU
retenta tatrica Kippenberg, 2008: 384 (ssp.) E: RO SK UK
schipkana hercegovinensis Bechyné, 1958b: 89 (ssp.) [RN] E: AL BH CR ME
croatica Weise, 1905b: 80 (*Protorina*, var.) [HN]
schipkana schipkana Jakob, 1953b: 158 (*Chrysochloa*, ssp.) [RN] E: BU SB
balcanica Jakob, 1952a: 99 (*Chrysochloa*, ssp.) [HN]
sibylla Binaghi, 1938: 36 (*Chrysochloa*) E: IT
subgenus *Virgulatorina* Kühnelt, 1984: 174 type species: *Chrysomela virgulata* Germar, 1823 [# 532]
Hyporeina Bourdonné & Doguet, 1986: 19 type species: *Chrysomela virgulata* Germar, 1823
virgulata ljubetensis Apfelbeck, 1912: 255 (*Chrysochloa*, var.) E: AL BH BU GR KO NM SB [# 613]
tricolor Apfelbeck, 1912: 255 (*Chrysochloa*, var.)
virgulata praefica Weise, 1884d: 468 (*Orina*, var.) E: BU CZ PL RO SK UK [# 578]
aurea Hänel, 1937: 31 (*Chrysochloa*, ab.) [IN]
bulgarica Obenberger, 1917: 44 (*Chrysochloa*, ssp.)
cepelarensis Obenberger, 1917: 44 (*Chrysochloa*, ssp.)
coelicolor Obenberger, 1917: 45 (*Chrysochloa*, ssp.)
cuprea Csiki, 1953: 129 (*Chrysomela*, ab.) [IN]
nigra Hänel, 1937: 31 (*Chrysochloa*, ab.) [IN]
orientalis Weise, 1900b: 283 (*Chrysochloa*, var.)
purkynei Obenberger, 1917: 43 (*Chrysochloa*)
violacea Hänel, 1937: 31 (*Chrysochloa*, ab.) [IN]
virgulata virgulata Germar, 1823: 596 (*Chrysomela*) E: AU GE FR IT LS SL SZ
alcyonea Suffrian, 1851a: 156 (*Chrysomela*)
candens Weise, 1884d: 468 (*Orina*, var.) [# 578]
cyanea Hänel, 1937: 31 (*Chrysochloa*, ab.) [IN]
isofastuosa Bechyné, 1958b: 90 (ssp.)
mona Bechyné, 1958b: 90 (ssp.)
papei Weise, 1887e: 364 (*Orina*, var.)
peneaui David, 1953a: 13 (*Chrysochloa*)
serena Weise, 1884d: 468 (*Orina*, var.) [IN] [# 577]
serena Stierlin, 1896b: 564 (*Orina*, var.)

genus *Oreina*, nomina dubia
monticola Duftschmid, 1825: 167 (*Chrysomela*) E: AU
nobilis Waltl, 1839b: 226 (*Chrysomela*) E: GE

genus *Zygogramma* Chevrolat, 1836: 398 type species: *Chrysomela pulcra* Fabricius, 1792 (= *Chrysomela suturalis* Fabricius, 1775)

subgenus *Zygospila* Achard, 1923a: 65 type species: *Calligrapha arcuigera* Stål, 1859
bicolorata Pallister, 1953: 45 **Ai**: BT ID (HP KA UT) NP PA **AUSi ORRi NTR**
suturalis Fabricius, 1775: 95 (*Chrysomela*) **Ei**: LA RU (ST) UK **Ai**: GG RU (ES FE) TR **NAR NTR**
casta Rogers, 1856: 33 (*Chrysomela*)
pulcra Fabricius, 1792a: 313 (*Chrysomela*)
volata Kovalev, 2002: 76 (ssp.) [UN] [# 580]

subtribe Entomoscelina Reitter, 1913

genus ***Cyrtonastes*** **Fairmaire, 1874c: 42** type species: *Cyrtonastes aeneomicans* Fairmaire, 1874
subgenus *Cyrtonastes* Fairmaire, 1874c: 42 type species: *Cyrtonastes aeneomicans* Fairmaire, 1874
aeneomicans Fairmaire, 1874c: 43 **A**: SY
balachowskyi Berti & Daccordi, 1974a: 608 **A**: SY "Mesopotamia"
confusus Berti & Daccordi, 1974a: 607 **A**: SY TR
grandis Lopatin, 1994: 527 **A**: TR
henoni Berti & Daccordi, 1974a: 606 **A**: SY
libanensis Berti & Daccordi, 1974a: 605 **A**: LE
ovatus Berti & Daccordi, 1974a: 603 **A**: SY
phaedonoides Fairmaire, 1874c: 44 **A**: LE
schereri Berti & Daccordi, 1974a: 609 **A**: IS
subgenus *Natocyrstes* Kippenberg, 2010a: 69 type species: *Cyrtonastes seriatoporus* Fairmaire, 1880
Natocyrstes Berti & Daccordi, 1974b: 853 [UN]
lacedaemonis Berti & Daccordi, 1974b: 856 **E**: GR (Peloponnisos)
peloponnesiacus Berti & Daccordi, 1974b: 858 **E**: GR (Peloponnisos)
ruffoi Berti & Daccordi, 1974b: 861 E: AL
seriatoporus Fairmaire, 1880d: 243 **E**: GR (Kythira) **A**: TR
cythereias Pic, 1903h: 147
weisei Reitter, 1884b: 121 E: GR (Korfu)
corcyreus Marseul, 1889: 107
zacynthi Berti & Daccordi, 1974b: 859 **E**: GR (Zakynthos)

genus ***Cystocnemis*** **Motschulsky, 1860b: 179** type species: *Chrysomela discoidea* Gebler, 1830
subgenus *Cystocnemis* Motschulsky, 1860b: 179 type species: *Chrysomela discoidea* Gebler, 1830
concolor Kraatz, 1879e: 266 (*Entomoscelis*) **A**: KZ (Saur-Tarbagatai, Dzhungar Alatau) [# 631]

mohri Daccordi, 1975: 147 (*Oreomela*)
karavaevi Jacobson, 1926a: 234 (ab.) [IN]
discoidea discoidea Gebler, 1830: 213 (*Chrysomela*) **A:** KZ RU (SS)
discoidea gebleri Jacobson, 1897b: 75 **A:** KZ CH (XIN)
discoidea jacobsoniana Mikhailov, 2013: 306 (ssp.) **A:** KZ
discoidea oreas Mikhailov, 2013: 307 (ssp.) **A:** KZ (Altai) RU (SS)
levmedvedevi Mikhailov & Guskova, 2013b: 540 **A:** MG (Mongolian Altai)
zintshenkoi Mikhailov & Guskova, 2013b: 543 **A:** KZ (South Altai)
subgenus *Entomomela* Jacobson, 1926a: 242, 267 type species: *Oreomela oirata* Jacobson, 1926
arnoldii Lopatin, 1974b: 176 (*Oreomela*) **A:** MG (Mongolian Altai)
oirata Jacobson, 1926a: 267 (*Oreomela*) **A:** KZ (Altai) MG (Mongolian Altai) RU (SS: Altai)

genus *Cystocnemis*, species unassigned to subgenus
koslovi Jacobson, 1926a: 233 **A:** CH (NMO)

genus *Entomoscelis* Chevrolat, 1836: 402 type species: *Chrysomela adonidis* Pallas, 1771
adonidis adonidis Pallas, 1771: 463 (*Chrysomela*) E: AU BH BU CZ FR GE GR HU IT KR LT MD NM PL PT RO RU (CT NT: Vologda Region ST) SB SK SP TR UK **A:** AF CH (XIN) IN KI KZ MG PA RU (ES FE SS WS) SY TD TM TR UZ **NAR** [# 522]
americana Brown, 1942: 172
spuria Jacobson, 1896b: 108 (var.)
trilineata Fabricius, 1777: 219 (*Chrysomela*)
varentzowi Jacobson, 1896a: 546 (var.)
adonidis caucasica Kippenberg, 2020a: 262 (ssp.) E: RU (ST) **A:** AB AR GG IN TR [# 522]
adonidis goliath Abeille de Perrin, 1897: 42 **A:** TR
berytensis Reiche & Saulcy, 1858b: 36 (*Chrysomela*) **A:** AR IN IQ IS JO SY TR
cornea Abeille de Perrin, 1897: 43 **N:** AG LB MO TU
mediojuncta Pic, 1939d: 284 (var.) [IN] [# 577]
multipunctata Kocher, 1969a: 113 (ab.) [IN]
occidentalis Escalera, 1914: 526
thalmanni Kocher, 1969a: 113 (ssp.)
viberti Bedel, 1915: 207 (var.) [IN] [# 577]
deserticola Lopatin, 1967c: 953 **A:** CH (NMO)
dorsalis bashkiriae Kippenberg, 2020a: 268 (ssp.) E: KR RU (CT: Bashkortostan, Tatarstan) **A:** AR RU (WS: Chelyabinsk Reg.) [# 522, # 602]
dorsalis dorsalis Fabricius, 1777: 220 (*Chrysomela*) E: AU HU SK UK

dorsalis hammarstroemi Jacobson, 1901b: 118 (ssp.) **A:** RU (ES SS)
dorsalis iranica Kippenberg, 2020a: 270 (ssp.) **A:** IN
dorsalis turkestana Mikhailov, 2020: 270 (ssp.) **A:** KZ
erythrocnema Jacobson, 1893b: 241 **A:** AF TD UZ
nigriventris Daccordi & Yang, 2009a: 422 **A:** CH (BEI GAN SHA SHX)
orientalis Motschulsky, 1860b: 222 **A:** CH (BEI GAN HEB HEI HUB JIA LIA NIN NMO SHA SHN TIA ZHE) MG NC RU (ES SS FE) SC
rufipennis Kraatz, 1879e: 266 (var.)
pilula Lopatin, 1967c: 952 **E:** RU (ST: Dagestan) **A:** AB GG KZ
pulla Daccordi & S.-Q. Ge, 2009a: 421 **A:** CH (BEI)
rumex Kippenberg, 2020a: 272 [RN] **A:** AG LB MO TU **[# 632]**
rumicis Fabricius, 1787: 70 (*Chrysomela*) [HN]
detersa Bedel, 1915: 207 (ab.) [IN]
sacra Linnaeus, 1758: 372 (*Chrysomela*) **E:** BU CR GR KR MD NM RO RU (CT ST) UK **A:** AB AR GG IN IS IQ JO SY TR
sancta Mader, 1943: 44 (ab.) [IN]
suturalis Weise, 1882b: 312

genus *Entomoscelis*, nomen dubium
melanostoma Gmelin, 1790: 1673 (*Chrysomela*)

genus *Oreomela* Jacobson, 1895a: 274 type species: *Xenomela heydeni* Weise, 1885
subgenus *Craspedomela* Jacobson, 1926a: 242 type species: *Oreomela avinovi* Jacobson, 1926
avinovi Jacobson, 1926a: 268 **A:** ID (KA)
subgenus *Leptomela* Jacobson, 1926a: 242 type species: *Oreomela kaznakovi* Jacobson, 1926
grumi Jacobson, 1926a: 264 **A:** CH (GAN)
kaznakovi Jacobson, 1926a: 265 **A:** CH (QIN)
njamtzoana Jacobson, 1926a: 267 (ssp.)
koltzei Weise, 1896a: 81 **A:** CH (QIN)
subgenus *Olcomela* Jacobson, 1926a: 236 type species: *Oreomela suvorovi* Jacobson, 1910
dolini Lopatin, 2005a: 63 **A:** KZ (Saur Mts.)
korolkovi Jacobson, 1926a: 243 **A:** KZ (Dzhungar Alatau)
milae Lopatin, 2007: 182 **A:** CH (XIN)
suvorovi Jacobson, 1910b: 56 **A:** CH (XIN) KZ (Dzhungar Alatau)
subgenus *Oreomela* Jacobson, 1895a: 274 type species: *Xenomela heydeni* Weise, 1885
abramovi Jacobson, 1926a: 253 **A:** KI (Talass Alatau)
andreevi Jacobson, 1926a: 250 **A:** KI (Terskei Alatau)

annapurnae L. N. Medvedev & Sprecher-Uebersax, 1998: 26 **A:** NP
bergi Jacobson, 1910b: 55 **A:** KI TD
borealis Lopatin, 1999b: 892 **A:** KZ
borochorensis Lopatin, 2002f: 878 **A:** CH (XIN)
celyphoides Jacobson, 1897b: 76 **A:** CH (XIN) KZ
clypealis Jacobson, 1901b: 119 **A:** KI
dudkorum Mikhailov, 2007a: 434 **A:** RU (SS)
dungana Jacobson, 1926a: 245 **A:** KI (Sarydzhas Mts.)
dzhungara Jacobson, 1926a: 262 **A:** KZ (Dzhungar Alatau)
fuscipes Weise, 1885c: 407 (*Xenomela*) **A:** KI
fausti Weise, 1890e: 23 (*Xenomela*)
gansuica Lopatin, 2002f: 877 **A:** CH (GAN)
gracilis Lopatin, 1999b: 894 **A:** KI
heydeni Weise, 1885c: 406 (*Xenomela*) **A:** KI
hohlbecki Jacobson, 1926a: 251 **A:** KI
joliveti Lopatin, 1976a: 114 **A:** KZ (Ketmen' Mts.)
juldusana Lopatin, 1962c: 324 **A:** CH (XIN)
kabaki Lopatin, 1995e: 177 **A:** KZ (Dzhungar Alatau)
kaszabi Lopatin, 1962c: 323 **A:** CH (XIZ)
korotjaevi Lopatin, 1983a: 93 **A:** KZ (Trans-Ili Alatau)
kungeica Lopatin, 1976a: 114 **A:** KI (Kungei Alatau)
kutzenkoi Jacobson, 1926a: 259 **A:** KI (Atbashi Mts.)
lopatini Gruev, 1991: 67 **A:** KI (Susamyr Mts.)
medvedevi Lopatin, 1968c: 552 **A:** TD (Badakhshan)
mitjaevi Lopatin, 1996b: 642 [RN] **A:** KZ (Dzhungar Alatau)
dzhungara Lopatin, 1990c: 55 [HN]
montivaga Lopatin, 1987a: 84 **A:** KI (Susamyr Mts.)
multipunctata Lopatin, 2008d: 838 **A:** KZ
muzartea Jacobson, 1926a: 254 **A:** CH (XIN)
pedashenkoi Jacobson, 1926a: 257 **A:** KI (Terskei Alatau)
przewalskii Jacobson, 1897b: 76 **A:** CH (XIN) KZ (Dzhungar Alatau)
puncticollis Lopatin, 1999b: 892 **A:** KZ (Dzhungar Alatau)
radkewiczi Jacobson, 1926a: 260 **A:** KZ
recticollis Lopatin, 1995e: 176 **A:** KI
romantsovi Mikhailov, 2007a: 438 **A:** RU (SS)
rueckbeili Jacobson, 1926a: 248 **A:** KI (Terskei Alatau)
rufipes Lopatin, 2002f: 876 **A:** CH (XIN)
sapozhnikovi Jacobson, 1926a: 255 **A:** KI
sarydzhasea Jacobson, 1926a: 246 **A:** KI (Sarydzhas Mts.)
scutellaris Jacobson, 1926a: 247 **A:** KI

semenovi Jacobson, 1895a: 278 **A:** KZ (Trans-Ili Alatau)
shnitnikovi Jacobson, 1926a: 258 **A:** KI
sussamyrica Lopatin, 1987a: 84 **A:** KI (Susamyr Mts.)
tarantsha Jacobson, 1926a: 249 **A:** KZ (Dzhungar Alatau, Ketmen' Mts.)
tarbagataica Lopatin, 1968c: 552 **A:** KZ (Tarbagatai Mts.)
transalaica Lopatin, 1965b: 11 **A:** KI TD
transiliensis Lopatin, 1976a: 114 **A:** KZ (Trans-Ili Alatau)
trotzinae Jacobson, 1894c: 239 (*Xenomela*) **A:** KI
jacobsoni Semenov, 1894: 546 (*Xenomela*)
tshernavini Jacobson, 1926a: 261 **A:** KI
tuvensis Mikhailov, 2007a: 438 **A:** RU (SS)
weisei Jacobson, 1895a: 280 **A:** KI (Terskei Alatau)
invisa Lopatin, 1962c: 324
zaslavskii Lopatin, 1976a: 114 **A:** KI (Atbashi Mts.)
subgenus *Pseudolina* Jacoby, 1896c: 252 type species: type species: *Pseudolina indica* Jacoby, 1896
Apaksha Maulik, 1926: 95 type species: *Apaksha himalayensis* Maulik, 1926
himalayensis himalayensis Maulik, 1926: 96 (*Apaksha*) **A:** ID (UT)
himalayensis nepalica L. N. Medvedev & Sprecher-Uebersax, 1998: 26 (ssp.) **A:** NP
indica Jacoby, 1896c: 253 (*Pseudolina*) **A:** ID ("North Western Prov.") **ORR**
meridionalis hindukushi L. N. Medvedev, 1985a: 373 (ssp.) **A:** AF
meridionalis ladakhia Daccordi, 1979b: 455 (ssp.) **A:** ID (KA)
meridionalis meridionalis Lopatin, 1967b: 324 **A:** AF PA
rama Maulik, 1926: 92 (*Pseudolina*) **A:** ID (UT)
tianshanica S.-H. Chen, 1961: 429, 433 **A:** CH (XIN)
subgenus *Rhaphomela* Jacobson, 1926a: 242 type species: *Oreomela timarchisca* Jacobson, 1926
timarchisca Jacobson, 1926a: 270 **A:** ID (KA)

genus *Oreomela*, species unassigned to subgenera
caudata S.-H. Chen, 1976: 212, 222 **A:** CH (QIN)
cheni Lopatin 2002f: 879 **A:** CH (XIN)
coerulea S.-H. Chen, 1976: 213, 222 **A:** CH (QIN)
cupreata S.-H. Chen, 1976: 213, 222 **A:** CH (QIN)
dubeshkoae L. N. Medvedev, 1977a: 349 **A:** MG
foveipennis S.-H. Chen & S.-Y. Wang, 1985: 104, 106 [= 1986a: 55, 57] **A:** CH (XIN)
fulvicornis Lopatin, 2008d: 837 **A:** CH (YUN) [# 538]
geae Lopatin, 2011b: 384 **A:** CH (SCH) [# 538]
inflata Lopatin, 2006c: 598 **A:** CH (YUN) [# 538]
laticollis S.-Y. Wang, 1992b: 628, 643 **A:** CH (QIN)

magnifica Lopatin, 2011b: 383 **A:** CH (SCH) [# 538]
melanosoma S.-Y. Wang, 1992b: 628, 643 **A:** CH (XIZ)
mirabilis Lopatin, 1998b: 834 **A:** CH (SCH) [# 538]
nigerrima S.-H. Chen, 1976: 212, 222 **A:** CH (XIZ)
nigroviolacea S.-H. Chen, 1976: 213, 222 **A:** CH (QIN)
nitidicollis S.-Y. Wang, 1992b: 629, 643 **A:** CH (SCH)
obtusa Lopatin, 2013: 769 **A:** CH (SCH) [# 538]
rugipennis S.-H. Chen & S.-Y. Wang, 1981b: 511, 515 **A:** CH (XIZ)
setigera S.-Y. Wang, 1992b: 630, 644 **A:** CH (SCH)
sichuanica Lopatin, 2005b: 573 **A:** CH (SCH) [# 538]
splendens Lopatin, 2013: 771 **A:** SCH [# 538]
striata Lopatin, 2004a: 618 **A:** CH (HUN) [# 538]
violacea S.-Y. Wang, 1992b: 630, 644 **A:** CH (SCH)

genus *Oreomela*, unavailable names
crassicornis Lopatin, 2010: 267 [UN] [# 580]
mushketovi Lopatin, 2010: 255 [UN] [# 580]

genus *Phola* Weise, 1890a: 482 type species: *Phola keyserlingi* Weise, 1890 (= *Chrysomela octodecimguttata* Fabricius, 1775)
Phyllophila Stål, 1857: 59 [HN] type species: *Phyllophila sedecimpustulata* Stål, 1857 (= *Chrysomela octodecimguttata* Fabricius, 1775)
octodecimguttata Fabricius, 1775: 100 (*Chrysomela*) **A:** CH (ANH CHQ FUJ GAN GUA GUX HAI JIX SCH SHA TAI YUN ZHE) JA **AUS ORR**
cybele Stål, 1861: 464 (*Chalcolampra*)
dipa Maulik, 1926: 89 (*Chalcolampra*)
keyserlingi Weise, 1890a: 482
sedecimpustulata Stål, 1857: 59 (*Phyllophila*)
vicitis Fairmaire, 1888a: 39 (*Chalcolampra*)

genus *Potaninia* Weise, 1889g: 603 type species: *Potaninia polita* Weise, 1889
assamensis Baly, 1879a: 437 (*Entomoscelis*) **A:** ID (SD) CH (CHQ GUI HUB HUN SCH XIZ YUN) NP **ORR**
collaris Weise, 1905e: 216
metallica Baly, 1888a: 85 (*Entomoscelis*)
polita Weise, 1889g: 604

genus *Suinzona* S.-H. Chen, 1931b: 130 type species: *Suinzona laboissierei* S.-H. Chen, 1931
Oreomelina Lopatin, 2004a: 617 type species: *Oreomela yunnana* Lopatin, 2004 [# 538]

belousovi Lopatin, 2002a: 117 (*Oreomela*) **A:** CH (SCH)
bergeali Daccordi & S.-Q. Ge, 2011c: 651 **A:** CH (SCH)
bienkowskii Daccordi & S.-Q. Ge, 2011c: 652 **A:** CH (SCH)
borowieci Cho & Kim, 2021: 2 **A:** SC
cheni Daccordi & Yang, 2011c: 652 **A:** CH (SCH)
cuiae S.-Q. Ge & Daccordi, 2011c: 654 **A:** CH (SCH)
cyrtonoides Jacoby, 1885b: 206 (*Chrysomela*) **A:** JA
faldermanni Daccordi & Yang, 2011c: 655 **A:** CH (SCH)
furialis Lopatin, 2013: 766 (*Chrysolina*) **A:** CH (SCH)
gebleri S.-Q. Ge & Daccordi, 2011c: 656 **A:** CH (SCH)
jacobsoni Daccordi & S.-Q. Ge, 2011c: 657 **A:** CH (SCH)
konstantinovi Daccordi & S.-Q. Ge, 2011c: 657 **A:** CH (SCH)
laboissierei S.-H. Chen, 1931b: 130 **A:** CH (SCH) **ORR**
lopatini Daccordi & S.-Q. Ge, 2011c: 658 **A:** CH (SCH)
medvedevi Daccordi & S.-Q. Ge, 2011c: 660 **A:** CH (SCH)
menetriesi Daccordi & S.-Q. Ge, 2011c: 661 **A:** CH (SCH)
mikhailovi Daccordi & S.-Q. Ge, 2011c: 662 **A:** CH (SCH)
monticola S.-H. Chen, 1961: 429, 433 **A:** CH (SCH)
motschulskyi Daccordi & S.-Q. Ge, 2011c: 662 **A:** CH (SCH)
ogloblini Daccordi & S.-Q. Ge, 2011c: 663 **A:** CH (YUN)
pallasi S.-Q. Ge & Daccordi, 2011c: 664 **A:** CH (SCH)
parva Lopatin, 2005d: 876 (*Oreomela*) **A:** CH (SCH)
potanini Lopatin, 2002a: 118 (*Oreomela*) **A:** CH (SCH)
wangi Lopatin, 2002a: 119 (*Oreomela*) **A:** CH (SCH)
yangi Lopatin, 2005d: 877 (*Oreomela*) **A:** CH (SCH)
yunnana Lopatin, 2004a: 617 (*Oreomela*) **A:** CH (YUN) [# 538]

genus *Taipinus* Lopatin, 2007: 181 type species: *Taipinus rotundatus* Lopatin, 2007
convexus Daccordi & S.-Q. Ge, 2011c: 665 **A:** CH (HUB)
elatus Daccordi & S.-Q. Ge, 2011c: 666 **A:** CH (GUI GUX YUN)
globosus Daccordi & S.-Q. Ge, 2011c: 667 **A:** CH (HEN)
kabaki Lopatin, 2011b: 382 **A:** CH (SHA)
magnus Lopatin, 2011b: 382 **A:** CH (SHA)
rotundatus Lopatin, 2007: 182 **A:** CH (GAN)

genus *Xenomela* Weise, 1884c: 226 type species: *Xenomela kraatzi* Weise, 1884
ballioni Lopatin, 1990a: 118 **A:** KI (Kirgiz Alatau)
belousovi Lopatin, 1990a: 118 **A:** KZ UZ
dohrni Solsky, 1881: 81 (*Entomoscelis*) **A:** TD UZ
karatavica Lopatin, 1990a: 117 **A:** KZ (Karatau Mts.)

konstantinovi Lopatin, 1995e: 175 **A:** UZ
kraatzi Weise, 1884c: 227 **A:** KI
laevigata Lopatin, 1990a: 115 **A:** KI
minckwitzae kreutzbergi Lopatin, 1990a: 121 **A:** KZ (Karzhantau Mts.) UZ
minckwitzae minckwitzae Lopatin, 1990a: 121 **A:** KZ (Talas and Ugam Mts.)
minckwitzae Jacobson, 1926a: 232 (morpha) [IN]
minckwitzae pskemica Lopatin, 1990a: 121 **A:** UZ (Pskem Mts.)
ovczinnikovi Lopatin, 1995e: 174 **A:** UZ
ovipennis Ballion, 1878: 384 (*Chrysomela*) **A:** CH (XIN) KI KZ
fulvipes Ballion, 1878: 383 (*Chrysomela*)
marginicollis Ballion, 1878: 382 (*Chrysomela*)
seriata Jacobson, 1926a: 232 (ab.) [IN]
regeli Jacobson, 1897b: 74 **A:** UZ

genus *Yunnaedon* Daccordi & L. N. Medvedev, 2000: 297 type species: *Yunnaedon pankui* Daccordi & L. N. Medvedev, 2000
belousovi Lopatin, 2011b: 381 **A:** CH (SCH)
foveatus Lopatin, 2011b: 380 **A:** CH (SCH)
pankui Daccordi & L. N. Medvedev, 2000: 298 **A:** CH (YUN)

subtribe Gonioctenina Motschulsky, 1860

genus *Centroscelis* Chevrolat, 1836: 403 type species: *Chrysomela notata* Fabricius, 1781
subgenus *Centroscelis* Chevrolat, 1836: 403 type species: *Chrysomela notata* Fabricius, 1781
kadleci Bezděk, Daccordi & Kantner, 2012: 238 **A:** YE

genus *Gonioctena* Chevrolat, 1836: 403 type species: *Chrysomela viminalis* Linnaeus, 1758
subgenus *Asiphytodecta* S.-H. Chen, 1935b: 126 type species: *Phytodecta tredecimmaculata* Jacoby, 1888
cinctipennis Achard, 1924: 33 (*Phytodecta*, var.) **A:** CH (GUI YUN) **ORR**
eburoides Achard, 1924: 34 (*Phytodecta*) **A:** CH (GUI) **ORR**
emeishana Bezděk, 2002: 10 **A:** CH (GUI GUX SCH SHN)
flavoplagiata Jacoby, 1890b: 117 (*Phytodecta*) **A:** CH (GAN GUI HUB SCH) **ORR**
guangxiae Kippenberg, 2010a: 70 [RN] **A:** CH (GUX) **ORR**
oculata S.-Y. Wang & S.-Q. Ge, 2002: 122, 124 [HN]
jindrai Cho & Borowiec, 2016c: 762 **A:** CH (SCH)
jindrai Cho, 2016: 46 [UN] [# 582]
mantillerii Cho, 2021: 84 **A:** CH (SCH)
mantillerii Cho, 2016: 49 [UN] [# 582]

medogana S.-Y. Wang, 2004a: 72, 73 **A:** CH (XIZ)
nigrosparsa Fairmaire, 1889a: 373 (*Paropsides*) **A:** CH (YUN) **ORR**
ornata L. N. Medvedev, 2009d: 411 **A:** NP
oudai Cho & Borowiec, 2016a: 170 **A:** CH (SCH)
bedzeki Yang, S.-Q. Ge, S.-Y. Wang, W.-Z. Li & Cui, 2014: 373 [UN] [# 581]
subgeminata S.-H. Chen, 1934d: 75 (*Phytodecta*) **A:** CH (ANH FUJ GUA HUN JIX SCH TAI ZHE)
tonkinensis S.-H. Chen, 1934d: 76 (*Phytodecta*, var.) **A:** CH (GUX) **ORR**
tredecimmaculata Jacoby, 1888: 347 (*Phytodecta*) **A:** CH (CHQ FUJ GUA GUI GUX HUB HUN JIA JIX SCH SHA TAI YUN ZHE) **ORR**
ohmomoi Cho, Takizawa & Borowiec, 2016: 365
riyuetanensis Cho, Takizawa & Borowiec, 2016: 366
taiwanensis Achard, 1924: 34 (*Phytodecta*, var.)
trilochana Maulik, 1926: 81 (*Phytodecta*) **A:** CH (ANH GUX YUN) **ORR**
subgenus *Brachyphytodecta* Bechyné, 1948a: 101 type species: *Spartophila fulva* Motschulsky, 1861
altimontana S.-H. Chen & S.-Y. Wang, 1984: 173, 175 **A:** CH (YUN)
andrzeji Daccordi & S.-Q. Ge, 2007: 580 **A:** CH (HUB SCH)
arunensis Cho & Borowiec, 2016a: 173 **A:** NP
davidi Cho, 2017a: 446 **A:** CH (SCH)
davidi Cho, 2016: 85 [UN] [# 582]
flavipennis Jacoby, 1888: 347 (*Phytodecta*) **A:** CH (CHQ FUJ GUA HUN JIX ZHE) **ORR**
fulva Motschulsky, 1861b: 41 (*Spartophila*) **A:** CH (CHQ FUJ GAN GUA HEI HUB JIA JIL JIX NIN SCH ZHE) NC RU (ES FE) SC **ORR**
incondita Weise, 1898: 213 (*Phytodecta*) **A:** ID (SD) NP
jamadai Takizawa, 2015: 236 **A:** JA (Honshu)
kangdingensis Cho, 2017a: 449 **A:** CH (SCH)
kangdingensis Cho, 2016: 101 [UN] [# 582]
kidoi Takizawa & Daccordi, 1998: 105 **A:** JA
kuatunensis Cho, 2017b: 175 **A:** CH (FUJ)
kuatunensis Cho, 2016: 106 [UN] [# 582]
lesnei S.-H. Chen, 1931a: 111 (*Phytodecta*) **A:** CH (HUN YUN ZHE) **ORR**
liui C.-F. Lee & Hsieh, 2022: 22 **A:** CH (TAI)
longshengensis Cho, 2017b: 176 **A:** CH (GUX)
longshengensis Cho, 2016: 110 [UN] [# 582]
melanota Kippenberg, 2010a: 70 [RN] **A:** CH (GUX ZHE)
melanoptera S.-H. Chen & Young, 1941: 209 (*Asiphytodecta*) [HN]
rubripennis Baly, 1862a: 28 **A:** JA
plagipennis Achard, 1924: 33 (*Phytodecta*, var.)
tenebrosa Weise, 1910c: 37 (*Phytodecta*, var.)

scutellaris Baly, 1862a: 27 **A:** CH (FUJ JIX SCH ZHE)

dichroa Fairmaire, 1888b: 153

foochowensis Gruev, 1989: 53

issikii Chûjô, 1958b: 69 (*Phytodecta*)

thoracica Baly, 1862a: 27 [HN]

semiglobosa Achard, 1924: 33 (*Phytodecta*) **A:** CH (GUI)

subgenus *Cercyonops* Jacobson, 1901a: 83 type species: *Chrysomela caraganae* Gebler, 1823 **[# 524]**

caraganae Gebler, 1823: 121 (*Chrysomela*) **E:** RU (NT: Komi Republic) **A:** RU (ES SS WS)

hieroglyphica Jacobson, 1901a: 84 [= 1901b: 129] (var.) [IN] [# 577]

limbata Jacobson, 1901a: 84 [= 1901b: 129] (var.) [IN] [# 577]

menetriesii Gebler, 1841b: 621 (*Spartophila*)

sesquialtera Jacobson, 1901a: 84 [= 1901b: 129] (var.) [IN] [# 577]

simplex Jacobson, 1901a: 84 [= 1901b: 129] (var.) [IN] [# 577]

subgenus *Gonioctena* Chevrolat, 1836: 403 type species: *Chrysomela viminalis* Linnaeus, 1758

Goniactis Gistel, 1856: 380 [UE] type species: *Chrysomela viminalis* Linnaeus, 1758

Phytodecta Kirby, 1837: 213 type species: *Chrysomela rufipes* DeGeer, 1775 (= *Chrysomela decemnotata* Marsham, 1802)

Platyphytodecta Bechyné, 1948a: 100 type species: *Phytodecta flexuosa* Baly, 1860

amurensis Cho & Borowiec, 2016b: 90 **A:** MG RU (FE)

amurensis Cho, 2016: 132 [UN] [# 582]

coreana Bechyné, 1948a: 114 (*Phytodecta*) **A:** CH (JIL) MG NC RU (FE) SC

decemnotata Marsham, 1802: 175 (*Chrysomela*) **E:** AU BE BH BU BY CR CZ DE EN FI FR GB GE HU IT LA LT LU NL NR PL RO RU (CT NT ST) SB SK SL SP SV SZ TR UK **A:** CH (HEI JIL) GG KZ MG RU (ES FE WS) TR **AFRi** **[# 1145]**

conjuncta Reineck, 1910: 66 (*Phytodecta*, ab.) [IN]

fulvipes Duftschmid, 1825: 199 (*Chrysomela*)

guranyii Kaszab, 1962: 57 (*Phytodecta*, ab.) [IN]

populi Bechyné, 1948a: 113 (*Phytodecta*, ab.) [IN]

praecox Bechyné, 1948a: 113 (*Phytodecta*, ab.) [IN]

rufipes DeGeer, 1775: 295 (*Chrysomela*) [HN]

sexpunctata Fabricius, 1787: 70 (*Chrysomela*) [HN]

taeniolata Bechyné, 1948a: 113 (*Phytodecta*, ab.) [IN]

tremulae Bechyné, 1948a: 113 (*Phytodecta*, ab.) [IN]

tropica Weise, 1919: 196 (*Phytodecta*, ab.) [IN]

wachsmanni Kaszab, 1962: 57 (*Phytodecta*, ab.) [IN]

filippovi V. L. Medvedev, 2004: 41 **A:** CH (GAN SCH)

flavicornis Suffrian, 1851a: 215 (*Chrysomela*) [conserved name, ICZN (1984a), Opinion 1270] E: AU BE CZ FI FR GE HU LA NR PL RO RU (CT: Ural NT) SK SL SP SV SZ UK **A:** JA MG RU (ES FE SS WS)

bicolor Heyden, 1883a: 53 (*Phytodecta*, var.) [HN]

borealis L. N. Medvedev, 1963b: 114 (*Phytodecta*, ssp.) [HN]

borealicola Kippenberg, 2010a: 70 (ssp.) [RN]

infernalis Penecke, 1901: 20 (*Phytodecta*, var.)

limbatipennis Achard, 1924: 32 (*Phytodecta*, var.)

nigrivestris Reitter, 1913c: 129 (*Phytodecta*, ab.)

flexuosa Baly, 1860a: 156 (*Phytodecta*) **A:** CH (ANH FUJ GAN GUA HUN JIA JIX SCH SHA ZHE) [# 607]

inornata S.-H. Chen, 1934d: 74 [= 1938b: 293] (*Phytodecta*, var.) [HN]

quadriplagiata Fairmaire, 1889b: 75 (*Phytodecta*)

fujiana Cho & Borowiec, 2016a: 178 **A:** CH (FUJ)

gracilicornis Kraatz, 1879b: 135 (*Phytodecta*) **A:** CH (HEI) NC MG RU (ES FE SS) SC

innocens Mader, 1937: 285 (*Phytodecta*, ab.) [IN]

kiberi Chûjô, 1941a: 74 (*Phytodecta*, var.)

manca Mader, 1937: 285 (*Phytodecta*, ab.) [IN]

munaguro Chûjô, 1941a: 75 (*Phytodecta*, var.)

signaticollis Chûjô, 1941a: 75 (*Phytodecta*, var.)

signaticollis Mader, 1937: 285 (*Phytodecta*, ab.) [IN]

sunkangensis Kimoto & Kawase, 1966: 44

hiranoi Takizawa, 1989b: 1 **A:** JA (Honshu)

hoki Takizawa, 2007: 48 **A:** JA (Honshu)

holdhausi Leeder, 1950: 21 (*Phytodecta*) **E:** AU GE IT SL SZ

nigrobrunnea Leeder, 1950: 22 (*Phytodecta*, ab.) [IN]

obscuricollis Leeder, 1950: 22 (*Phytodecta*, ab.) [IN]

honshuensis chujoi L. N. Medvedev, 1966: 41 **A:** JA (Hokkaido) RU (FE: Kurile Islands)

honshuensis honshuensis Nakane, 1963a: 19 **A:** JA (Honshu)

honshuensis ochotensis L. N. Medvedev, 1968: 85 (ssp.) **A:** RU (FE)

honshuensis sachalinensis L. N. Medvedev, 1968: 84 (ssp.) **A:** RU (FE: Sakhalin)

honshuensis yeeni J.-K. Li, 1992: 189 (ssp.) [QS] **A:** CH (JIL)

iyonis Suenaga, 2012: 30 **A:** JA (Shikoku)

jacobsoni Ogloblin & L. N. Medvedev, 1956: 897 (*Phytodecta*) **A:** CH (HEI) MG RU (ES FE) SC

jani Cho & Borowiec, 2016b: 97 **A:** RU (ES FE)

jani Cho, 2016: 175 [UN] [# 582]

japonica japonica Chûjô & Kimoto, 1960: 5 **A:** JA RU (FE: Kurile Islands)

japonica jilinensis J.-K. Li, 1992: 188 (ssp.) [QS] **A:** CH (JIL)

japonica zhangi J.-K. Li, 1992: 189 (ssp.) [QS] **A:** CH (JIL)

kamikawai Chûjô, 1958b: 72 (*Phytodecta*) **A:** CH (TAI)
kanfani S.-H. Chen, 1941c: 191 (*Phytodecta*) **A:** CH (ANH ZHE)
katsuyai Takizawa, 2007: 48 **A:** JA (Kyushu)
kaufmanni Miller, 1880: 7 (*Phytodecta*) **E:** AU IT SL
koryeoensis Cho & C.-F. Lee, 2010: 53 **A:** SC
laeta L. N. Medvedev, 1973b: 882 **A:** NC RU (FE) SC
linnaeana calcarata Fabricius, 1792b: 19 (*Galleruca*) **E:** BE BH BU BY CR CZ EN FR GE GR HU IT LA LT ME NL NM PL RO RU (CT) SB SK SV TR UK **A:** AR GG TR [# 526, # 563, # 571, # 608, # 633]
apfelbecki Kaszab, 1962: 65 (*Phytodecta*, ab.) [IN]
basalis Bechyné, 1948a: 120 (*Phytodecta*, ab.) [IN]
biguttata Bechyné, 1948a: 121 (*Phytodecta*, ab.) [IN]
biroi Kaszab, 1962: 65 (*Phytodecta*, ab.) [IN]
biscutellata Kaszab, 1962: 64 (*Phytodecta*, ab.) [IN]
bohemica Achard, 1924: 31 (*Phytodecta*, var.)
congener Bechyné, 1948a: 120 (*Phytodecta*, ab.) [IN]
continua Bechyné, 1948a: 121 (*Phytodecta*, ab.) [IN]
csikii Kaszab, 1962: 63 (*Phytodecta*, ab.) [IN]
decastigma Stierlin, 1896b: 569 (*Phytodecta*, var.) [# 1118]
deficiens Bechyné, 1948a: 120 (*Phytodecta*, ab.) [IN]
dispersa Bechyné, 1948a: 120 (*Phytodecta*, ab.) [IN]
distans Bechyné, 1948a: 120 (*Phytodecta*, ab.) [IN]
fasciolata Bechyné, 1948a: 121 (*Phytodecta*, ab.) [IN]
flavotibialis Bechyné, 1948a: 121 (*Phytodecta*, ab.) [IN]
gebhardti Kaszab, 1962: 61 (*Phytodecta*, ab.) [IN]
hexagonalis Kaszab, 1962: 63 (*Phytodecta*, ab.) [IN]
hofferi Bechyné, 1948a: 121 (*Phytodecta*, ab.) [IN]
humeralis Bechyné, 1948a: 120 (*Phytodecta*, ab.) [IN]
imperfecta Bechyné, 1948a: 121 (*Phytodecta*, ab.) [IN]
inclinans Bechyné, 1948a: 121 (*Phytodecta*, ab.) [IN]
kanabei Kaszab, 1962: 64 (*Phytodecta*, ab.) [IN]
kraatzi Westhoff, 1882: 273 (*Phytodecta*, var.)
langhofferi Kaszab, 1962: 61 (*Phytodecta*, ab.) [IN]
lunulata Bechyné, 1948a: 121 (*Phytodecta*, ab.) [IN]
media Bechyné, 1948a: 120 (*Phytodecta*, ab.) [IN]
mihoki Kaszab, 1962: 64 (*Phytodecta*, ab.) [IN]
mocsaryi Kaszab, 1962: 64 (*Phytodecta*, ab.) [IN]
mutenicensis Bechyné, 1948a: 121 (*Phytodecta*, ab.) [IN]
nigricollis Westhoff, 1882: 273 (*Phytodecta*, var.)
nigrolocana Czwalina, 1885: 256 (*Phytodecta*, var.)

octonotata Bechyné, 1948a: 120 (*Phytodecta*, ab.) [IN]
parietis Bechyné, 1948a: 121 (*Phytodecta*, ab.) [IN]
picturata Bechyné, 1948a: 121 (*Phytodecta*, ab.) [IN]
postica Bechyné, 1948a: 120 (*Phytodecta*, ab.) [IN]
pragensis Bechyné, 1948a: 120 (*Phytodecta*, ab.) [IN]
privorai Bechyné, 1948a: 121 (*Phytodecta*, ab.) [IN]
proxima Bechyné, 1948a: 120 (*Phytodecta*, ab.) [IN]
punctula Bechyné, 1948a: 121 (*Phytodecta*, ab.) [IN]
quadrangularis Bechyné, 1948a: 120 (*Phytodecta*, ab.) [IN]
reducta Kaszab, 1962: 61 (*Phytodecta*, ab.) [IN]
reversa Bechyné, 1948a: 120 (*Phytodecta*, ab.) [IN]
satanas Westhoff, 1882: 273 (*Phytodecta*, var.)
stilleri Kaszab, 1962: 63 (*Phytodecta*, ab.) [IN]
stredai Kaszab, 1962: 63 (*Phytodecta*, ab.) [IN]
tetragramma Bechyné, 1948a: 121 (*Phytodecta*, ab.) [IN]
tibialis Duftschmid, 1825: 202 (*Chrysomela*) [HN] [rejected name, ICZN (1984a), Opinion 1270]
trapezoidalis Kaszab, 1962: 61 (*Phytodecta*, ab.) [IN]
triandrae Suffrian, 1851a: 216 (*Chrysomela*)
triguttata Bechyné, 1948a: 120 (*Phytodecta*, ab.) [IN]
unireducta Kaszab, 1962: 61 (*Phytodecta*, ab.) [IN]
vadia Bechyné, 1948a: 121 (*Phytodecta*, ab.) [IN]
vigilans Bechyné, 1948a: 121 (*Phytodecta*, ab.) [IN]

linnaeana linnaeana Schrank, 1781: 69 (*Chrysomela*) **E**: AU FR (Pyrenees) GE (Alps) IT (Alps) LS SL SP SZ **[# 526]**

linnaeana orientalis Weise, 1884d: 498 (*Phytodecta*, var.) **E**: FI NR RU (CT: Urals NT ST) **A**: KZ MG RU (ES FE SS WS) **[# 526, # 578]**

alternata Bechyné, 1948a: 122 (*Phytodecta*, ab.) [IN]
ascendens Bechyné, 1948a: 119 (*Phytodecta*, ab.) [IN]
aterrima Bechyné, 1948a: 122 (*Phytodecta*, ab.) [IN]
carbonaria Bechyné, 1948a: 122 (*Phytodecta*, ab.) [IN]
gradualis Bechyné, 1948a: 122 (*Phytodecta*, ab.) [IN]
jureceki Bechyné, 1948a: 122 (*Phytodecta*, ab.) [IN]
melanocnema Jacobson, 1897a: 436 (*Phytodecta*, var.)
minuta Bechyné, 1948a: 122 (*Phytodecta*, ab.) [IN]
repetens Bechyné, 1948a: 122 (*Phytodecta*, ab.) [IN]
samarica Bechyné, 1948a: 122 (*Phytodecta*, ab.) [IN]
tockoensis Bechyné, 1948a: 122 (*Phytodecta*, ab.) [IN]

lukasi Cho, 2022: 114 **A**: CH (ZHE)

melli Gressitt & Kimoto, 1963a: 363 (ssp.) **A**: CH (GUA JIX) **ORR**

metallica Cho & Borowiec, 2014: 298 **A**: CH (SCH)

mongolica Cho & Borowiec, 2016a: 180 A: CH (NMO)
kalganensis S.-Q. Ge & Daccordi, 2014: 363 [UN] [# 581]
morimotoi Kimoto, 1963: 15 A: JA (Honshu)
nivosa arctica Mannerheim, 1853: 257 E: FI NR RU (CT: North Ural NT) SV A: KZ (Altai) RU (ES FE SS WS) NAR [# 527]
affinis Gyllenhal, 1808: 257 (*Chrysomela*) [HN]
bergrothi Jacobson, 1901b: 128 (*Phytodecta*, ssp.)
correspondens Jacobson, 1901b: 128 (*Phytodecta*, var.) [IN] [# 576]
decaspilota Achard, 1924: 32 (*Phytodecta*, var.)
dinah Bechyné, 1948a: 123 (*Phytodecta*)
fennica Bechyné, 1948a: 124 (*Phytodecta*, ab.) [IN]
guttifer Bechyné, 1948a: 123 (*Phytodecta*, ab.) [IN]
hamata Achard, 1924: 32 (*Phytodecta*, var.)
marae Bechyné, 1948a: 119 (*Phytodecta*, ab.) [IN]
mutata Achard, 1924: 31 (*Phytodecta*, var.) [RN]
salicis Motschulsky, 1860b: 223
simplex Jacobson, 1901b: 128 (*Phytodecta*, var.) [IN] [# 576]
sinuatocollis Bechyné, 1948a: 114 (*Phytodecta*, ab.) [IN]
zanaida Bechyné, 1948a: 125 (*Phytodecta*, ab.) [IN]
nivosa nivosa Suffrian, 1851a: 222 (*Chrysomela*) E: AU FR GE IT LS SP SZ [# 527, # 1034]
aethiops Heyden, 1883a: 53 (*Phytodecta*, var.)
apicalis Heyden, 1883a: 53 (*Phytodecta*, var.)
cedehensis Ronchetti, 1922: 89 (*Phytodecta*, var.)
clythroides Gradl, 1882: 331 (*Phytodecta*, var.)
eppelsheimi Weise, 1884d: 501 (*Phytodecta*, var.) [IN] [# 577]
eppelsheimi Stierlin, 1896b: 570 (*Phytodecta*, var.)
excisa Bechyné, 1948a: 125 (*Phytodecta*, ab.) [IN]
funesta Weise, 1884d: 501 (*Phytodecta*, var.) [IN] [# 577]
funesta Stierlin, 1896b: 570 (*Phytodecta*, var.)
hexangularis Bechyné, 1948a: 125 (*Phytodecta*, ab.) [IN]
iablokoffi Cantonnet, 1968: 43 (*Phytodecta*, ab.) [IN]
immarginata Achard, 1924: 33 (*Phytodecta*, var.)
latefasciata Bechyné, 1948a: 126 (*Phytodecta*, ab.) [IN]
limitata Bechyné, 1948a: 126 (*Phytodecta*, ab.) [IN]
marginata Gradl, 1882: 331 (*Phytodecta*, var.)
marginipennis Bechyné, 1948a: 126 (*Phytodecta*, ab.) [IN]
nana Gradl, 1882: 330 (*Phytodecta*, var.)
nigricollis Gradl, 1882: 331 (*Phytodecta*, var.)

octopunctata Gradl, 1882: 330 (*Phytodecta*, var.)
personata Weise, 1884d: 501 (*Phytodecta*, var.) [IN] [# 577]
personata Stierlin, 1896b: 570 (*Phytodecta*, var.)
ruficollis Weise, 1891d: 160 (*Phytodecta*, var.)
rufula Kraatz, 1879c: 55 (*Phytodecta*, var.)
scutellaris J. R. Sahlberg, 1885b: 55 (*Phytodecta*) [HN]
stenomera Dufour, 1851: 353 (*Chrysomela*)
trinotata Bechyné, 1948a: 125 (*Phytodecta*, ab.) [IN]
tyrolensis Gradl, 1882: 331 (*Phytodecta*, var.)
undulata Pic, 1924e: 27 (*Phytodecta*, var.)
vicina Bechyné, 1948a: 126 (*Phytodecta*, ab.) [IN]

norvegica A. Strand, 1936: 104 (*Phytodecta*) **E:** FI NR RU (NT) SV **A:** MG RU (ES SS WS)
charitonowi Palmén, 1946: 231 (*Phytodecta*) [UN] [# 528]
janovskii L. N. Medvedev, 1976a: 234

ogloblini L. N. Medvedev & Dubeshko, 1972: 80 **A:** CH (HEI) MG NC RU (ES FE) SC
suwai Takizawa, 1985a: 7

rufa Kraatz, 1879b: 139 (*Phytodecta*) **A:** JA (Honshu) RU (ES FE) SC
kamiyai Kimoto, 1963: 14
ussuriensis L. N. Medvedev, 1964: 180 (*Phytodecta*)

shibatai Takizawa, 1982: 57 **A:** JA

sibirica Weise, 1893a: 1128 (*Phytodecta*) **E:** RU (CT) **A:** CH (JIL) JA MG NC RU (ES FE SS WS) SC
cheni Bechyné, 1948a: 102 (*Phytodecta*)
fraterna Jacobson, 1901a: 82 (*Phytodecta*, var.) [IN] [# 577]
sorbi Motschulsky, 1860b: 223 [HN]
sobrina Weise, 1916b: 179 (*Phytodecta*) [RN]

sichuana Cho & Borowiec, 2014: 295 **A:** CH (SCH)

simotuke Takizawa, 2007: 45 **A:** JA (Honshu)

springlovae Bechyné, 1948a: 116 (*Phytodecta*) **A:** JA (Hokkaido) RU (FE: Sakhalin)
gradualis Bechyné, 1948a: 116 (*Phytodecta*, ab.) [IN]

sundmani Jacobson, 1901b: 128 (*Phytodecta*, var.) **A:** CH (HEI) MG RU (ES FE SS WS) [# 578]
sundmani Palmén, 1946: 227 (*Phytodecta*) [HN]

takahashii Kimoto, 1963: 15 **A:** JA

tatesinensis Takizawa, 2007: 49 **A:** JA (Honshu)

viminalis pallidipes Weise, 1893a: 1129 (*Phytodecta*, var.) **A:** CH (HEI JIL LIA) MG NC RU (ES FE SS) SC [# 529]
collaris Weise, 1887h: 186 (*Phytodecta*, var.) [HN]
curvungula Bechyné, 1948a: 141 (*Phytodecta*, ab.) [IN]

denominanda Achard, 1924: 32 (*Phytodecta*, var.) [RN]
karafutensis Kôno & Tamanuki, 1926: 294 (*Phytodecta*, ab.) [IN]
transitoria Jacobson, 1901b: 129 (*Phytodecta*, var.) [IN] [# 577]

viminalis viminalis Linnaeus, 1758: 371 (*Chrysomela*) E: AN AU BE BH BU BY CR CZ DE EN FI FR GB GE HU IR IT LA LS LT LU NL NR PL RU (CT NT ST) RO SB SK SL SV SZ TR UK **A:** GG KZ MG RU (ES SS WS) **[# 1145]**

adumbrata Bechyné, 1948a: 110 (*Phytodecta*, ab.) [IN]
approximata Bechyné, 1948a: 110 (*Phytodecta*, ab.) [IN]
baaderi Panzer, 1794b: 14 (*Chrysomela*)
basipunctata Gradl 1882: 329 (*Phytodecta*, var.)
bicolor Kraatz, 1879c: 49 (*Phytodecta*)
bifasciata Bechyné, 1948a: 110 (*Phytodecta*, ab.) [IN]
bisquadripunctata Csiki, 1953: 130 (*Phytodecta*, ab.) [IN]
carneolica Voet, 1806: 41 (*Chrysomela*) [UN]
cincta Weise, 1884d: 492 (*Phytodecta*, var.) [IN] [# 577]
cincta Stierlin, 1896b: 568 (*Phytodecta*, var.)
circumducta Bechyné, 1948a: 110 (*Phytodecta*, ab.) [IN]
c-notata Pic, 1924e: 27 (*Phytodecta*, var.)
collaris Gradl, 1882: 329 (*Phytodecta*, var.)
complicata Bechyné, 1948a: 109 (*Phytodecta*, ab.) [IN]
conjugata Bechyné, 1948a: 109 (*Phytodecta*, ab.) [IN]
cyathigera Bechyné, 1948a: 110 (*Phytodecta*, ab.) [IN]
decempunctata Linnaeus, 1758: 371 (*Chrysomela*)
decemstilata Bechyné, 1948a: 111 (*Phytodecta*, ab.) [IN]
declivis Bechyné, 1948a: 108 (*Phytodecta*, ab.) [IN]
defectiva Bechyné, 1948a: 109 (*Phytodecta*, ab.) [IN]
diffusa Bechyné, 1948a: 107 (*Phytodecta*, ab.) [IN]
distigma Achard, 1924: 31 (*Phytodecta*, var.)
elegantula Bechyné, 1948a: 109 (*Phytodecta*, ab.) [IN]
eximia Bechyné, 1948a: 110 (*Phytodecta*, ab.) [IN]
fantisi Bechyné, 1948a: 109 (*Phytodecta*, ab.) [IN]
flamulata Bechyné, 1948a: 108 (*Phytodecta*, ab.) [IN]
fleischeri Bechyné, 1948a: 110 (*Phytodecta*, ab.) [IN]
florea Bechyné, 1948a: 110 (*Phytodecta*, ab.) [IN]
fridrichsdalensis O. F. Müller, 1764: 8 (*Chrysomela*)
fulva Gradl, 1882: 329 (*Phytodecta*, var.) [HN]
gradli Heyden, 1883b: 53 (*Phytodecta*, var.) [RN]
grata Bechyné, 1948a: 109 (*Phytodecta*, ab.) [IN]
havelkai Bechyné, 1948a: 108 (*Phytodecta*, ab.) [IN]

havliki Bechyné, 1948a: 110 (*Phytodecta*, ab.) [IN]
interversa Achard, 1924: 31 (*Phytodecta*, var.)
kocoureki Bechyné, 1948a: 109 (*Phytodecta*, ab.) [IN]
kodymi Bechyné, 1948a: 110 (*Phytodecta*, ab.) [IN]
labialis Bechyné, 1948a: 104, 106, 111 (*Phytodecta*, ssp.)
lepidoptera Bechyné, 1948a: 110 (*Phytodecta*, ab.) [IN]
marekorum Bechyné, 1948a: 108 (*Phytodecta*, ab.) [IN]
marginalis Gradl, 1882: 329 (*Phytodecta*, var.)
mediolineata Pic, 1924e: 27 (*Phytodecta*, var.)
monostigma Achard, 1924: 31 (*Phytodecta*, var.)
munda Weise, 1884d: 491 (*Phytodecta*, var.) [IN] [# 577]
munda Stierlin, 1896b: 567 (*Phytodecta*, var.)
musili Bechyné, 1948a: 110 (*Phytodecta*, ab.) [IN]
nigerrima Gradl, 1882: 330 (*Phytodecta*, var.)
nigrescens Gradl, 1882: 330 (*Phytodecta*, var.)
nigripes DeGeer, 1775: 296 (*Chrysomela*)
notatithorax Pic, 1924e: 26 (*Phytodecta*, var.)
obenbergeri Bechyné, 1948a: 111 (*Phytodecta*, ab.) [IN]
opposita Bechyné, 1948a: 109 (*Phytodecta*, ab.) [IN]
oculata Gradl, 1882: 329 (*Phytodecta*, var.)
ornata Csiki, 1953: 130 (*Phytodecta*, ab.) [IN]
panzeri Bechyné, 1948a: 108 (*Phytodecta*, ab.) [IN]
paracenthesis Bechyné, 1948a: 108 (*Phytodecta*, ab.) [IN]
pterostigma Bechyné, 1948a: 111 (*Phytodecta*, ab.) [IN]
pudica Bechyné, 1948a: 108 (*Phytodecta*, ab.) [IN]
quadrimaculata Heyden, 1883b: 53 (*Phytodecta*, var.)
quadripunctata Heyden, 1883b: 53 (*Phytodecta*, var.)
reniplagiata Penecke, 1901: 20 (*Phytodecta*, var.)
roubali Bechyné, 1948a: 108 (*Phytodecta*, ab.) [IN]
samsinaki Bechyné, 1948a: 109 (*Phytodecta*, ab.) [IN]
seminigra Weise, 1893a: 1126 (*Phytodecta*, var.)
separata Bechyné, 1948a: 111 (*Phytodecta*, ab.) [IN]
sigmoidea Bechyné, 1948a: 109 (*Phytodecta*, ab.) [IN]
signata Scopoli, 1763: 74 (*Coccinella*)
stictica Bechyné, 1948a: 108 (*Phytodecta*, ab.) [IN]
tacta Bechyné, 1948a: 110 (*Phytodecta*, ab.) [IN]
trijuncta Bechyné, 1948a: 109 (*Phytodecta*, ab.) [IN]
trimaculata Pic, 1924e: 26 (*Phytodecta*, var.)
tripuncticollis Kaszab, 1962: 57 (Phytodecta, ab.) [IN]

uniguttata Bechyné, 1948a: 108 (*Phytodecta*, ab.) [IN]
urbani Bechyné, 1948a: 110 (*Phytodecta*, ab.) [IN]
vittula Bechyné, 1948a: 109 (*Phytodecta*, ab.) [IN]
w-nigrum Bechyné, 1948a: 108 (*Phytodecta*, ab.) [IN]
x-notatus Bechyné, 1948a: 111 (*Phytodecta*, ab.) [IN]
zajiceki Bechyné, 1948a: 109 (*Phytodecta*, ab.) [IN]
zigzag Bechyné, 1948a: 110 (*Phytodecta*, ab.) [IN]
warchalowskii Daccordi & S.-Q. Ge, 2007: 584 **A**: CH (HUB)
yunnana V. L. Medvedev, 1999: 12 **A**: CH (YUN)
subgenus *Goniomena* Motschulsky, 1860b: 180 type species: *Chrysomela pallida* Linnaeus, 1758
intermedia Helliesen, 1913: 5 (*Phytodecta*) **E**: AU BH CR CZ EN FI FR GE HU IT LA LT ME NR PL RO RU (CT: Kaliningrad Region NT) SK SL SZ [# 609]
barsica Kaszab, 1962: 72 (*Phytodecta*, ab.) [IN]
brunnicollis Kaszab, 1962: 73 (*Phytodecta*, ab.) [IN]
deletepunctata Kaszab, 1962: 73 (*Phytodecta*, ab.) [IN]
falsopallida Kaszab, 1962: 72 (*Phytodecta*, ab.) [IN]
herculeana Kaszab, 1962: 73 (*Phytodecta*, ab.) [IN]
interposita Franz & Palmén, 1950: 15 (*Phytodecta*) **E**: AU CR CZ GE HU IT PL RO SK SZ UK **A**: IS
brunnea Kaszab, 1962: 72 (*Phytodecta*, ab.) [IN]
hades Kaszab, 1962: 72 (*Phytodecta*, ab.) [IN]
israelita Lopatin & Friedman, 2003: 343
pectoralis Kaszab, 1962: 71 (*Phytodecta*, ab.) [IN]
plagicollis Kaszab, 1962: 71 (*Phytodecta*, ab.) [IN]
rubriventris Kaszab, 1962: 70 (*Phytodecta*, ab.) [IN]
rufithorax Kaszab, 1962: 71 (*Phytodecta*, ab.) [IN]
pallida Linnaeus, 1758: 370 (*Chrysomela*) **E**: AU BE BH BU BY CZ DE EN FI FR GB GE GR HU IT LA LS LT ME NL NM NR PL RU (CT NT ST) RO SK SL SV SZ UK **A**: KZ MG RU (ES SS WS) **[# 530, # 610, # 1084]**
alpestris Bechyné, 1948a: 96, 130 (*Phytodecta*, ab.) [IN]
apicalis Milani & Ruffo, 1963: 115 (*Phytodecta*, forma) [IN]
atra Bechyné, 1948a: 98, 132 (*Phytodecta*, ab.) [IN]
atrobrunnea Bechyné, 1948a: 97, 131 (*Phytodecta*, ab.) [IN]
balcanica Bechyné, 1948a: 99, 132 (*Phytodecta*, ab.) [IN]
borealis Weise, 1884d: 510 (*Phytodecta*, var.) [IN] [# 577]
borealis Stierlin, 1896b: 571 (*Phytodecta*, var.)
brunneiventris Kaszab, 1962: 70 (*Phytodecta*, ab.) [IN]
carpathica Bechyné, 1948a: 98, 156 (*Phytodecta*, ssp.)

decipiens Stierlin, 1896b: 571 (*Phytodecta*, var.)
decipiens Weise, 1884d: 510 (*Phytodecta*, var.) [IN] [# 577]
distincta Bechyné, 1948a: 97, 131 (*Phytodecta*, ab.) [IN]
erythrura Reitter, 1914a: 46 (*Phytodecta*, ab.) [IN]
frontalis Olivier, 1807: 565 (*Chrysomela*)
grossa Bechyné, 1948a: 98, 132 (*Phytodecta*, ab.) [IN]
guttulosa Bechyné, 1948a: 97, 131 (*Phytodecta*, ab.) [IN]
fuscidorsis Reitter, 1914a: 46 (*Phytodecta*, ab.) [IN]
haemorrhoidalis Linnaeus, 1758: 369 (*Chrysomela*)
melaniventris Milani & Ruffo, 1963: 115 (*Phytodecta*, forma) [IN]
melas Bechyné, 1948a: 96, 130 (*Phytodecta*, ab.) [IN]
nigricolor Reitter, 1914a: 46 (*Phytodecta*, ab.) [IN]
nigricolor Bechyné, 1948a: 97, 155 (*Phytodecta*, ssp.)
nigripennis Weise, 1887e: 364 (*Phytodecta*, var.)
nigriscutis Bechyné, 1948a: 99, 132 (*Phytodecta*, ab.) [IN]
nigrithorax Reitter, 1914a: 46 (*Phytodecta*, ab.) [IN]
nigritincta Achard, 1924: 35 (*Phytodecta*, var.)
nigrobasalis Reitter, 1914a: 46 (*Phytodecta*, ab.) [IN]
occidentalis Bechyné, 1948a: 98 (*Phytodecta*, ab.) [IN]
occidentis Bechyné, 1948a: 131 (*Phytodecta*, ab.) [IN]
oralis Bechyné, 1948a: 97, 131 (*Phytodecta*, ab.) [IN]
padi DeGeer, 1775: 301 (*Chrysomela*)
pectoralis Bechyné, 1948a: 98, 132 (*Phytodecta*, ab.) [IN]
plagiata Bechyné, 1948a: 96, 129 (*Phytodecta*, ab.) [IN]
reticulata Bechyné, 1948a: 99, 156 (*Phytodecta*, ssp.)
sanguinicollis Bechyné, 1948a: 96, 130 (*Phytodecta*, ab.) [IN]
silesiaca Bechyné, 1948a: 96, 129 (*Phytodecta*, ab.) [IN]
straminea Bechyné, 1948a: 97, 131, 155 (*Phytodecta*, ab.) [IN]
sudetica Bechyné, 1948a: 96, 129 (*Phytodecta*, ssp.)
tatrica Bechyné, 1948a: 97, 131, 155 (*Phytodecta*, ab.) [IN]
testacea Bechyné, 1948a: 96, 130 (*Phytodecta*, ab.) [IN]
variolosa Bechyné, 1948a: 98, 131 (*Phytodecta*, ssp.)
vulgaris Bechyné, 1948a: 98, 132 (*Phytodecta*, ssp.)
wadasi Bechyné, 1948a: 98, 131 (*Phytodecta*, ab.) [IN]
xanthomelaena Bechyné, 1948a: 96, 129 (*Phytodecta*, ab.) [IN]

quinquepunctata Fabricius, 1787: 72 (*Chrysomela*) [HN, NP] E: AL AU BE BH BU BY CR CZ DE EN FI FR GE HU IT LA LS LT LU ME NL NM NR PL RO RU (CT NT) SB SK SL SP SV SZ UK A: NP RU (ES FE WS) [# 531]
aucupariae Gistel, 1857b: 525 (*Genioctena*)

aucupariae Jacobson, 1901b: 128 (*Phytodecta*, var.) [IN] [# 577]
bechynei Roubal, 1948b: 125 (*Phytodecta*, ab.) [IN]
dispar Paykull, 1799a: 66 (*Chrysomela*)
flavicollis Stierlin, 1896b: 571 (*Phytodecta*, var.) [# 1119]
flavipennis Krauss, 1900: 241 (*Phytodecta*, var.) [HN]
jenisonii Gistel, 1857b: 525 (*Chrysomela*)
kendii Csiki, 1953: 131 (*Phytodecta*, ab.) [IN]
limbata Csiki, 1953: 131 (*Phytodecta*, ab.) [IN]
melanopa Achard, 1924: 35 (*Phytodecta*, var.) [RN]
melanoptera Penecke, 1898: 254 (*Phytodecta*, var.)
morbillosa Fabricius, 1792b: 19 (*Galleruca*) [# 542]
nepalica V. L. Medvedev, 1999: 14 (ssp.)
nigriventris Penecke, 1898: 254 (*Phytodecta*, var.)
obscura Grimmer, 1841: 48 (*Paropsis*)
ochroptera Achard, 1924: 35 (*Phytodecta*, var.) [RN]
padi Penecke, 1898: 254 (*Phytodecta*, var.) [HN]
peneckei Csiki, 1953: 131 (*Phytodecta*, ab.) [IN]
punctulata Fabricius, 1792b: 19 (*Galleruca*) [# 542]
scutellata Trella, 1930: 194 (*Phytodecta*, ab.) [IN]
sorbi Grimmer, 1841: 47 (*Paropis*)
sorbi Weise, 1884d: 508 (*Phytodecta*, var.) [IN] [# 577]
sorbi Stierlin, 1896b: 571 (*Phytodecta*, var.) [HN]
undata Thunberg, 1784: 12 (*Chrysomela*) [HN, NO]
unicolor Weise, 1884d: 508 (*Phytodecta*, var.) [IN] [# 577]
unicolor Stierlin, 1896b: 570 (*Phytodecta*, var.)

subgenus *Machomena* Dubois, 1887: 3 type species: *Spartophila lineata* Gené, 1839 [# 524]
Acanthodon Weise, 1880b: 76 [HN] type species: *Spartophila lineata* Gené, 1839
lineata Gené, 1839: 80 (*Spartophila*) **E**: FR IT

subgenus *Sinomela* S.-H. Chen, 1935b: 126 type species: *Gonioctena aeneipennis* Baly, 1862
aeneipennis Baly, 1862a: 27 **A**: CH (JIA ZHE) SC
fortunei Baly, 1864d: 228 **A**: CH (CHQ GUI HUN JIA JIX SCH SHG ZHE)
extensa Achard, 1924: 33 (*Phytodecta*, var.)
hikosana Kimoto, 1974a: 144 **A**: JA (Kyushu)
nagaii Nakane, 1963b: 226 **A**: JA (Okinoerabu Island)
nigroplagiata Baly, 1862a: 28 **A**: CH (FUJ TAI: Matsu Islands) JA (Honshu) [# 570]
robusta Jacoby, 1885b: 209 (*Phytodecta*)
osawai Kimoto, 1996: 29 **A**: CH (TAI)

subgenus *Spartomena* Reitter, 1913c: 128 type species: *Gonioctena fornicata* Brüggemann, 1873

Spartiophila Küster, 1848a: 91 [HN] type species: *Chrysomela sexpunctata* Fabricius, 1787 sensu Küster (1848) (= *Gonioctena fornicata* Brüggemann, 1873)

akbesiana Fairmaire, 1884d: 175 **A:** IS SY TR

picea Weise, 1884d: 505 (*Phytodecta*, var.) [IN] [# 577]

fornicata Brüggemann, 1873: 518 **E:** AL AU BH BU CR CZ FR GB GE GR HU IT KO MD NM PL RO RU (CT ST) SB SK SZ TR UK **A:** IN IS LE SY TR

berytensis Pic, 1916b: 10 (*Phytodecta*, var.) [IN] [# 577]

binotaticollis Pic, 1916b: 10 (*Phytodecta*, var.) [IN] [# 577]

bistrimaculata Kaszab, 1962: 66 (*Phytodecta*, ab.) [IN]

bohumilae Bechyné, 1945e: 112 (*Phytodecta*, ab.) [IN]

conjuncta Endrődi, 1957: 347 (*Phytodecta*, ab.) [IN]

diabolica Endrődi, 1957: 348 (*Phytodecta*, ab.) [IN]

dieneri Kaszab, 1962: 66 (*Phytodecta*, ab.) [IN]

extracojuncta Kaszab, 1962: 68 (*Phytodecta*, ab.) [IN]

hexaspilota Bechyné, 1945e: 112 (*Phytodecta*, ab.) [IN]

innotata Weise, 1884d: 505 (*Phytodecta*, var.) [IN] [# 577]

innotatithorax Laczó, 1928a: 8 (*Phytodecta*, ab.) [IN]

inornata Heyden, 1889a: 203 (*Phytodecta*, var.)

kuesteri Csiki, 1953: 131 (*Phytodecta*, ab.) [IN]

nigriceps Csiki, 1953: 130 (*Phytodecta*, ab.) [IN]

obliquemaculata Pic, 1916b: 10 (*Phytodecta*, var.) [IN] [# 577]

rossola Marseul, 1888a: 57 [= 355]

septempunctata Bechyné, 1945e: 111 (*Phytodecta*, ab.) [IN]

seminigra Mikšić, 1954: 81 (*Phytodecta*, ab.) [IN]

sequensi Reitter, 1916: 234 (*Spartophila*, ab.) [IN]

speiseri Kaszab, 1962: 69 (*Phytodecta*, ab.) [IN]

transversa Csiki, 1953: 131 (*Phytodecta*, ab.) [IN]

tetraspilota Bechyné, 1945e: 111 (*Phytodecta*, ab.) [IN]

triangularis Bechyné, 1945e: 111 (*Phytodecta*, ab.) [IN]

tripallida Kaszab, 1962: 69 (*Phytodecta*, ab.) [IN]

vittipennis Csiki, 1953: 131 (*Phytodecta*, ab.) [IN]

subgenus *Spartophila* Chevrolat, 1836: 403 type species: *Chrysomela litura* Fabricius, 1775 (= *Chrysomela olivacea* Forster, 1771)

Amblis Gistel, 1848a: 403 [= 1848b: 123] [HN] type species: *Chrysomela litura* Fabricius, 1775 (= *Chrysomela olivacea* Forster, 1771)

olivacea Forster, 1771: 23 (*Chrysomela*) **E:** AU BE BU BY CR CZ DE FR GB GE HU IT LU NL NM PL PT RU (ST) RO SK SP SZ UK **N:** MO

agnani Pic, 1924e: 27 (*Phytodecta*, var.)

areata Bedel, 1899: 265 (*Phytodecta*, var.) [HN]

bipartita Achard, 1924: 34 (*Phytodecta*, var.)
escorialensis Achard, 1924: 35 (*Phytodecta*, var.)
fauconetti Pic, 1924e: 27 (*Phytodecta*, var.)
flavicans Fabricius, 1787: 72 (*Chrysomela*)
litura Fabricius, 1775: 102 (*Chrysomela*)
nigricans Weise, 1884d: 507 (*Phytodecta*, var.) [IN] [# 577]
nigricans Stierlin, 1896b: 570 (*Phytodecta*, var.)
obscurissima Pic, 1907g: 146 (*Phytodecta*, var.)
quittardi Pic, 1924e: 27 (*Phytodecta*, var.)
rufitarsis Marsham, 1802: 182 (*Chrysomela*)
suturata Geoffroy, 1785: 111 (*Chrysomela*)
thoracica Geoffroy, 1785: 109 (*Chrysomela*)
tulipa Geoffroy, 1785: 111 (*Chrysomela*)
umbriceps Roubal, 1941: 51 (*Phytodecta*, forma) [IN] [# 577]
vernalis Gmelin, 1790: 1689 (*Chrysomela*)
weisei Reitter, 1886: 71 (*Phytodecta*)

subgenus *Spartoxena* Motschulsky, 1860b: 180 type species: *Chrysomela aegrota* Fabricius, 1798

Phytodectella Cantonnet, 1968: 39 type species: *Phytodecta levasseuri* Cantonnet, 1968 (= *Phytodecta irrorata* Weise, 1891)

aegrota aegrota Fabricius, 1798: 87 (*Chrysomela*) E: PT SP **A**: MO [# 525]
capreae Illiger, 1802: 413 (*Chrysomela*)
iberica J. Daniel, 1914: 252 (*Phytodecta*, ab.) [IN]
icterica Weise, 1891d: 160 (*Phytodecta*, var.)
koltzei Weise, 1891d: 160 (*Phytodecta*, var.)
nubila Fuente, 1910b: 446 (*Phytodecta*, var.)
pseudospartii J. Daniel, 1914: 251 (*Phytodecta*, ab.) [IN]
rubra Zulueta, 1925: 206 (*Phytodecta*, var.)
schrammi Pic, 1924e: 27 (*Phytodecta*, var.)
sexnotata Fabricius, 1798: 86 (*Chrysomela*)

aegrota litoralis Kippenberg, 2001: 19 (ssp.) E: PT SP [# 525]

aegrota nanula Kippenberg, 2010a: 70 (ssp.) [RN] E: SP [# 525]
nana Kippenberg, 2001: 18 (ssp.) [HN]

gobanzi Reitter, 1902b: 141 (*Phytodecta*) E: IT SL
lagarinensis Halbherr, 1908: 33 (*Phytodecta*, var.)
montivaga J. Daniel, 1914: 258 (*Phytodecta*, ab.) [IN]
thurnheri Roubal, 1927b: 135 (*Phytodecta*, ab.) [IN]

irrorata irrorata Weise, 1891d: 160 (*Phytodecta*, var.) **N**: AG LB MO TU
atlasica Kocher, 1953c: 98 (*Phytodecta*, ssp.)
levasseuri Cantonnet, 1968: 46 (*Phytodecta*)
maroccana J. Daniel, 1914: 260 (*Phytodecta*, ab.) [IN]

pellax Normand, 1947b: 84 (*Phytodecta*)
quatuordecimlineata J. Daniel, 1914: 260 (*Phytodecta*, ab.) [IN]
quadrinotata Kocher, 1953c: 99 (*Phytodecta*, ab.) [IN]
irrorata lateralis Kocher, 1955: 286 (*Phytodecta*, ssp.) N: MO
leprieuri Pic, 1911b: 98 (*Phytodecta*, var.) E: PT SP
acanthodon J. Daniel, 1914: 244 (*Phytodecta*, ab.) [IN]
bilunata J. Daniel, 1914: 241 (*Phytodecta*)
bilunata Reineck, 1911b: 479 (*Phytodecta*, ab.) [IN]
cantabrica J. Daniel, 1914: 244 (*Phytodecta*, ab.) [IN]
marginata Reineck, 1911b: 480 (*Phytodecta*, ab.) [IN]
marginata Fuente, 1920: 151 (*Phytodecta*) [HN]
paganettii J. Daniel, 1914: 246 (*Phytodecta*, ab.) [IN]
rubromarginata Achard, 1924: 36 (*Phytodecta*, forma) [RN]
tigrina J. Daniel, 1914: 244 (*Phytodecta*, ab.) [IN]
procax Normand, 1947b: 85 (*Phytodecta*) N: AG
universa Normand, 1947b: 85 (*Phytodecta*, ab.) [IN]
pseudogobanzi Kippenberg, 2001: 20 E: SP
secsaouia Bourdonné & Doguet, 1979: 50 N: AG MO
secsaouia Kocher, 1953c: 100 (*Phytodecta*, var.) [IN] [# 576]
theae Baviera, 2007: 390 E: IT (Sicily)
variabilis Olivier, 1791a: 708 (*Chrysomela*) E: FR SP
binotaticeps Pic, 1911b: 98 (*Phytodecta*, var.)
gallica J. Daniel, 1914: 255 (*Phytodecta*, ab.) [IN]
grandini Desbrochers des Loges, 1870c: 132
heterogyna J. Daniel, 1914: 255 (*Phytodecta*, ab.) [IN]
luteonotata Pic, 1912a: 3 (*Phytodecta*, var.)
marginata Brancsik, 1914: 61 (*Phytodecta*, var.) [HN]
navasi Fuente, 1901: 44 (*Phytodecta*, var.)
spartii Olivier, 1807: 569 (*Chrysomela*)
unipunctata Olivier, 1791a: 708 (*Chrysomela*)

genus *Gonioctena*, nomen dubium
leseleuci Marseul, 1888a: 55 [= 353] N: MO

genus *Gonioctena*, unavailable names
subgenus *Tumidocnema* Cho, 2016: 331 [UN] [# 582]
brunnea Daccordi & S.-Q. Ge, 2014: 382 [UN] [# 581]
klapperichi Cho, 2016: 104 [UN] [# 582]
taebaekensis Cho, 2016: 250 [UN] [# 582]
thibetana Cho, 2016: 257 [UN] [# 582]
xinjiangana Cho, 2016: 332 [UN] [# 582]

genus *Paropsides* Motschulsky, 1860b: 192 type species: *Paropsis duodecimpustulatus* Gebler, 1825

bouvieri S.-H. Chen, 1931a: 110 **A:** CH (GUX) **ORR**

chennelli Baly, 1879a: 438 **A:** CH (XIZ) ID (SD)

nigrofasciata Jacoby, 1888: 348 **A:** CH (ANH CHQ GUI GUX HUB HUN JIX SCH YUN ZHE)

bisquadrimaculata Mader, 1943: 44 (ab.) [IN]

bivittata Reineck, 1922b: 371 (forma) [IN] [# 577]

humeralis Mader, 1943: 44 (ab.) [IN]

nigropunctata Jacoby, 1892b: 918 **A:** CH (YUN) ID (SD) NP **ORR**

soriculata Swartz, 1808: 246 (*Chrysomela*) **A:** CH (ANH CHQ FUJ GUA GUI GUX HEB HEN HUB HUN JIA JIL JIX LIA NIN NMO SHC SHX YUN ZHE) JA NC RU (ES FE) SC **ORR**

decempustulata S.-H. Chen, 1934d: 66 (var.)

duodecimpustulata Gebler, 1825: 54 (*Paropsis*)

hieroglyphica Gebler, 1825: 55 (*Paropsis*)

immaculata Mader, 1943: 44 (ab.) [IN]

maculicollis Jacoby, 1890b: 118

melli Reineck, 1922b: 370 (ssp.)

sexmaculata Reineck, 1922b: 367 (forma) [IN] [# 577]

suturalis S.-H. Chen, 1934d: 66 (var.)

yuasai Ohno, 1958a: 34, 40 (ssp.)

genus *Paropsisterna* Motschulsky, 1860b: 193 type species: *Notoclea sexpustulata* Marsham, 1808

Chrysophtharta Weise, 1901d: 166 type species: *Paropsis nobilitata* Erichson, 1842

Niliosoma Motschulsky, 1860b: 194 type species: *Paropsis testacea* Olivier, 1807

Sterromela Weise, 1915b: 436 type species: *Paropsis subcostata* Chapuis, 1877

Xanthogramma Weise, 1923: 63 type species: *Xanthogramma pellucida* Weise, 1923

selmani Reid & De Little, 2013: 396 Ei: GB IR **AUS**

genus *Peltoschema* Reitter, 1880b: 4 type species: *Peltoschema filicornis* Reitter, 1880

Acacicola Lea, 1903: 392 type species: *Acacicola tristis* Lea, 1903

Pyrgo Weise, 1901d: 171 [HN] type species: *Paropsis hamadryas* Stål, 1861

Pyrgoides Aslam, 1967: 129 [UN]

Pyrgoides Kelly & Reid 1999: 264 [RN] type species: *Paropsis hamadryas* Stål, 1861

subgenus *Asiparopsis* S.-H. Chen, 1934d: 69 [= 1938b: 288] type species: *Paropsides pardalis* Jacoby, 1892

convexa Weise, 1902h: 368 (*Pyrgo*) **A:** CH (GUX YUN) **ORR**

mausonensis Achard, 1914a: 72 (*Pyrgo*) [RN]

unicolor S.-H. Chen, 1934d: 70 [= 1938b: 289] (*Asiparopsis*, var.)

pardalis Jacoby, 1892b: 918 (*Paropsides*) **A:** CH (GUX YUN) **ORR**

genus *Trachymela* Weise, 1908d: 8 type species: *Paropsis sloanei* Blackburn, 1897
Chondromela Weise, 1915b: 435 type species: *Chondromela mjoebergi* Weise, 1915
sloanei Blackburn, 1897: 681 (*Paropsis*) **Ei:** GR PT SP **Ai:** CH (GUA TAI) **Ni:** CI **AUS NARi NTRi**

subtribe Sclerophaedonina Steinhausen, 1996 **[# 473]**

genus *Apterocuris* Jacobson, 1901a: 85 type species: *Chrysomela sibirica* Gebler, 1830
brinevi Mikhailov, 2023: 8 **A:** RU (FE)
sibirica Gebler, 1830: 218 (*Chrysomela*) **A:** KZ RU (SS)
aurichalcea Jacobson, 1901a: 86 (var.) [IN] [# 577]
cyanea Jacobson, 1901a: 86 (var.) [IN] [# 577]
puncticollis Gebler, 1833: 308 (*Chrysomela*)
purpurea Jacobson, 1901a: 86 (var.) [IN] [# 577]
violacea Jacobson, 1901a: 86 (var.) [IN] [# 577]

genus *Colaphellus* Weise, 1916b: 113 [RN] type species: *Chrysomela sophiae* Schaller, 1783
Colaphus L. Redtenbacher, 1845: 116 [HN] type species: *Chrysomela sophiae* Schaller, 1783
apicalis Ménétriés, 1848: 54 (*Colaspidema*) **A:** AB AF IN IS KI KU KZ SI SY TM TR UZ
tibialis L. Redtenbacher, 1849: 9 (*Colaphus*)
bowringii Baly, 1865a: 35 (*Colaphus*) **A:** CH (ANH BEI CHQ FUJ GAN HEB HEI HEN HUB HUN FUJ GAN GUA GUI GUX HUB HUN JIA JIL JIX LIA NIN NMO QIN SCH SHA SHN YUN ZHE) **ORR**
grouvellei Achard, 1923d: 130 [# 549]
foveolatus Gebler, 1848: 36 (*Phaedon*) **A:** RU (SS)
hoeftii Ménétriés, 1832: 235 (*Chrysomela*) **E:** KR MD RU (CT ST) UK **A:** AF CH (XIN) GG IN KI KZ MG RU (WS) TM UZ **[# 519, # 601]**
jakutus Machatschke, 1954: 208 **A:** MG RU (ES SS)
alpicola Warchałowski, 2004a: 113 (*Colaphus*) [RN]
alpinus Gebler, 1833: 307 (*Chrysomela*) [HN]
joliveti Bechyné, 1954b: 665 **N:** AG
nitidicollis Weise, 1889g: 601 (*Colaphus*) **A:** CH (QIN)
palaestinus Achard, 1923d: 129 **E:** MA **A:** IN IQ IS JO LE SY
pulchellus arabicus L. N. Medvedev, 1996: 249 (*Colaphus*, ssp.) **A:** SA
pulchellus pulchellus P. H. Lucas, 1846b: 46 [= 1847c: 517] (*Colaspidema*) **N:** AG MO TU **A:** SI
obscuricolor Pic, 1942e: 11 (var.)
sophiae Schaller, 1783: 272 (*Chrysomela*) **E:** AU BE BU BY CZ GE GR HU KR LA LT MD NL NM PL RO RU (CT) SK SP (incl. Mallorca) TR UK **A:** IN TR **[# 519]**
amasiae Machatschke, 1954: 212 (ssp.) [# 550]
erythropus Gmelin, 1790: 1690 (*Chrysomela*)
iconiensis Pic, 1901c: 35 (*Colaphus*)

rufipes Herbst, 1783: 59 (*Chrysomela*)
viennensis Schrank, 1781: 74 (*Chrysomela*) [QS]
transsylvanicus Machatschke, 1954: 210 (ssp.) [# 550]
tenuipes Weise, 1883d: 254 (*Colaphus*) **N:** MO
zarudnyi L. N. Medvedev, 1973b: 882 **A:** IN

genus *Colaspidema* Laporte, 1833: 19, 21 type species: *Chrysomela barbara* Fabricius, 1792
subgenus *Colaphomega* Reitter, 1913c: 124 type species: *Chrysomela rufifrons* Olivier, 1807
dufourii Pérez Arcas, 1865a: 185 (*Colaphus*) E: PT SP
rufifrons Olivier, 1807: 532 (*Chrysomela*) **N:** AG MO
latenigrum Pic, 1928f: 9 (var.)
signatipenne Guérin-Méneville, 1844: 296 **N:** AG MO TU
implagatum Achard, 1923d: 128 (*Colaphomega*, var.)
plagiatum Weise, 1882b: 304 (var.)
reductum Pic, 1928f: 9 (var.)
subgenus *Colaspidema* Laporte, 1833: 19, 21 type species: *Chrysomela barbara* Fabricius, 1792
Colaphus Dejean, 1836: 411 [RN] type species: *Chrysomela barbara* Fabricius, 1792
barbarum Fabricius, 1792a: 326 (*Chrysomela*) E: FR PT SP **N:** AG MO
atrum Olivier, 1791a: 719 (*Chrysomela*) [HN]
maculicolle Lefèvre, 1873b: 8

genus *Sclerophaedon* Weise, 1882b: 303 type species: *Chrysomela carniolica* Germar, 1823
subgenus *Sclerophaedon* Weise, 1882b: 303 type species: *Chrysomela carniolica* Germar, 1823
carniolicus Germar, 1823: 588 (*Chrysomela*) E: AU BH CR CZ GE ME PL RO SK SL SZ UK
obscurus Weise, 1884d: 536 (var.) [IN] [# 577]
transsilvanicus Fuss, 1861b: 169 (*Phaedon*)
carpathicus Weise, 1875: 366 (*Phaedon*, var.) E: PL RO SK UK
kenderesyi Kiesenwetter, 1879: 256 (*Chrysomela*)
daccordii Lopatin, 2006c: 597 **A:** CH (GAN)
fulvicornis S.-H. Chen, 1974a: 45, 48 (*Phaedon*) **A:** CH (SCH)
kabaki Daccordi & S.-Q. Ge, 2013b: 88 **A:** CH (SCH)
murzini Daccordi & S.-Q. Ge, 2013b: 89 **A:** CH (SCH)
orbicularis Suffrian, 1851a: 243 (*Chrysomela*) E: AU BE CR FR GE IT SL SZ
modesta Gravenhorst, 1807: 136 (*Chrysomela*) [HN]
orientalis Daccordi & S.-Q. Ge, 2013b: 91 **A:** CH (GAN)

punctatus Daccordi & S.-Q. Ge, 2013b: 93 **A:** CH (SHA)
rufipes Daccordi & S.-Q. Ge, 2013b: 95 **A:** CH (SCH)
shennongi Daccordi & S.-Q. Ge, 2013b: 96 **A:** CH (HUB)
subgenus *Tantraedon* Daccordi & L. N. Medvedev, 2000b: 220 type species: *Phaedon besucheti* Daccordi, 1984
besucheti Daccordi, 1984: 323 (*Phaedon*) **A:** ID (SD) **ORR**
brendelli Daccordi & L. N. Medvedev, 2000b: 228 **A:** NP
nepalicus Daccordi & L. N. Medvedev, 2000b: 224 **A:** NP
prajapati Daccordi & L. N. Medvedev, 2000b: 226 **A:** NP
takizawai Daccordi & L. N. Medvedev, 2000b: 222 **A:** NP

tribe Timarchini Motschulsky, 1860 [# 472]

genus *Timarcha* Samouelle, 1819: 213 type species: *Chrysomela tenebricosa* Fabricius, 1775
subgenus *Metallotimarcha* Motschulsky, 1860b: 187 type species: *Chrysomela metallica* Laicharting, 1781
corinthia Fairmaire & Allard, 1873: 197 E: AL BH CR GR ME NM SB
albanica Apfelbeck, 1912: 248 (var.) [# 637]
hummelii Faldermann, 1837: 351 E: RU (ST) **A:** AB AR GG IN TR
armeniaca Faldermann, 1837: 352
starcki Bechyné, 1953: 85 (ssp.)
metallica Laicharting, 1781: 144 (*Chrysomela*) E: AU BE BH BU CR CZ FR GE HU IT LU ME NL PL RO SB SK SL SZ UK **[# 1145]**
gibba Hagenbach, 1825: 487 [# 638]
globosa Herrich-Schäffer, 1838a: 23c
portai Bechyné, 1954a: 90 (ssp.)
splendens Matzek, 1843: 154
thiemei Kuntzen, 1919: 231 (ssp.)
wendleri Fleischer, 1909a: 32 (ab.) [IN]
subgenus *Timarcha* Samouelle, 1819: 213 type species: *Chrysomela tenebricosa* Fabricius, 1775
Timarchostoma Motschulsky, 1860b: 187 type species: *Chrysomela coriaria* Laicharting, 1781
apricaria apricaria Waltl, 1835: 83 E: SP PT [# 677]
coarcticollis Fairmaire & Allard, 1873: 147
paulinoi Kraatz, 1879i: 383
nitentula Bechyné, 1953: 92 (ssp.)
apricaria heydeni Weise, 1882b: 321 E: SP
apricaria melitensis Weise, 1882b: 318 **N:** MO
riffensis Fairmaire, 1887c: cxciii
apricaria parvicollis Rosenhauer, 1856: 320 E: SP
elocata Bechyné, 1953: 92 (ssp.)

seidlitzi Kraatz, 1879i: 383
balearica Gory, 1833: pl. 49 [= 1844: 300] E: SP (Balearic Islands) [# 673]
balearica Pérez Arcas, 1865a: 180 [HN]
caerulea Linnaeus, 1758: 418 (*Tenebrio*) [# 673]
coerulescens Bechyné, 1946b: 30 (ab.) [IN]
discolor Bechyné, 1946b: 30 (ab.) [IN]
longicornis Bechyné, 1946b: 30 (ab.) [IN]
martini Pic, 1919c: 20 (var.) [IN] [# 670]
nigriventris Bechyné, 1946b: 30 (ab.) [IN]
olivacea Bechyné, 1946b: 30 (ab.) [IN]
semicoerulea Bechyné, 1946b: 30 (ab.) [IN]
tricolor Bechyné, 1946b: 30 (ab.) [IN]
violacea Pic, 1919c: 20 (var.) [IN] [# 670]
viridipennis Bechyně 1946b: 30 (ab.) [IN]
brachydera Fairmaire & Allard, 1873: 165 N: AG TU [# 639]
distans Bechyné, 1944e: 372 (ssp.) [# 639]
figurata Bechyné, 1944e: 372 (ssp.) [# 639]
limitata Bechyné, 1944e: 372 (ssp.) [# 639]
pumilio Bechyné, 1944e: 371 (ssp.) [# 639]
calceata Pérez Arcas, 1865a: 181 E: PT SP [# 640]
geniculata Germar, 1823: 582 (*Chrysomela*) [HN] [# 640, # 681]
catalaunensis Fairmaire, 1884b: 100 E: SP [# 641]
corallipes Fairmaire & Allard, 1873: 191 N: AG [# 642]
corpulenta Bechyné, 1944d: 289 [# 642]
inornata Bechyné, 1944d: 288 (ssp.) [# 642]
melas Bechyné, 1944d: 288 [# 642]
nitidipes Bechyné, 1944d: 290 [# 642]
opacicollis Bechyné, 1944d: 289 [# 642]
crassaticollis Fairmaire & Allard, 1873: 166 N: AG [# 643]
henonii Fairmaire & Allard, 1873: 167 [# 643]
cyrenaica Peyerimhoff, 1925b: 27 N: LB
dislocata Bechyné, 1948d: 61 (ssp.) E: SP [# 644]
elliptica Fairmaire & Allard, 1873: 180 E: SP [# 645, # 684]
endora Buquet, 1840: 243 N: AG TU [# 646]
brevipennis Pic, 1919c: 19 [# 646]
demoflysi Normand, 1950b: 108 (var.) [# 646]
fallax Pérez Arcas, 1865a: 183 E: SP [# 647]
aitanae Petitpierre & Anichtchenko, 2018: 376 [# 647]
ametella Bechyné, 1962: 185 (ab.) [IN]
tenuicornis Fairmaire, 1880e: 332 [# 647]

goettingensis bruleriei Bellier, 1870: xxvii E: FR [# 636]
dubitabilis Fairmaire & Allard, 1873: 183
obenbergeri Bechyné, 1947b: 8 (ssp.)
goettingensis chloropus Germar, 1823: 583 (*Chrysomela*) E: PT SP [# 636, # 648]
asturiensis Kraatz, 1879i: 384 [# 648]
convexifrons Fairmaire, 1880e: 332 [# 648]
gallaeciana Chevrolat, 1840: 17 [# 648]
gougeletii Fairmaire, 1860: ccxvi [# 648]
janthinipes Fairmaire, 1880e: 333 [# 648]
sphaeroptera Fairmaire & Allard, 1873: 192 [# 648]
trapezicollis Fairmaire & Allard, 1873: 196 [# 648]
goettingensis fracassii Meier, 1900a: 78 (var.) E: IT [# 636]
schenklingi Meier, 1900b: 161
goettingensis globipennis Fairmaire & Allard, 1873: 186 E: SP [# 636, # 649]
aequata Bechyné, 1953: 87 (ssp.) [# 649]
guipuzcoana Bechyné, 1948e: 203 (ssp.) [# 649]
leseleuci Marseul, 1883: 62 [# 649]
vizcayensis Bechyné, 1953: 87 (ssp.) [# 649]
goettingensis goettingensis Linnaeus, 1758: 368 (*Chrysomela*) E: AL AU BE BH BU BY CR CZ DE FR GB GE HU IT KR LT LU ME NL NM PL RO RU (ST) SB SK SP SV UK [# 636]
aerea Herrich-Schäffer, 1838a: 24a
altimontana Bechyné, 1948e: 204
andrzejowskii Krynicki, 1832: 169 (var.)
angusticollis Duftschmid, 1825: 163 (*Chrysomela*)
arvernensis Laboissière, 1939: 128 (race)
belchitensis Bechyné, 1962: 184 (ssp.)
bivillensis Jolivet, 1943a: 69 (var.) [IN] [# 670]
carpathica Bechyné, 1946c: 137 (ssp.)
coiffaiti Bechyné, 1948e: 201
conveniens Bechyné, 1948e: 201 (ab.) [IN]
coriaria Laicharting, 1781: 143 (*Chrysomela*)
cornuta Bechyné, 1944b: 143
corsica Bechyné, 1944b: 145 (ssp.)
cyanescens Fairmaire, 1862a: 593
cylindricollis Bechyné, 1948e: 203 (ssp.)
daillei Laboissière, 1939: 127
duftschmidi Bechyné, 1946c: 136 (morpha) [IN]
franzi Bechyné, 1962: 184 (ssp.)
gallica Fairmaire & Allard, 1873: 182

geniculata C. R. Sahlberg, 1823b: 73 [= 1823c: 73, = 1829: 29] (*Chrysomela*) [# 681]
iopa Marsham, 1802: 170 (*Chrysomela*)
irridescens Bechyné, 1948e: 202 (ab.) [IN]
latipes Linnaeus, 1767: 678 (*Tenebrio*)
macropa Gistel, 1857b: 537
mangaliana Bechyné, 1946c: 138 (ssp.)
maritima Perris, 1856: lxxix
melica Bechyné, 1947b: 6
migratoria Bechyné, 1948e: 203 (ssp.)
moderata Bechyné, 1948e: 202 (ssp.)
monticola Dufour, 1851: 353
normanna Pasquet, 1923: 235
obsoleta Laboissière, 1937b: 164 (ssp.)
orophila Bechyné, 1948e: 204
parisiensis Bechyné, 1946c: 137 (ab.) [IN]
perezii Fairmaire, 1884a: 103
pontavicei Marseul, 1883: 67
praecox Bechyné, 1953: 89 (ssp.)
prunneri Herrich-Schäffer, 1838a: 22e
pseudoperezi Bechyné, 1953: 89 (ssp.)
pygmaeola Bechyné, 1953: 89 (ssp.)
purpurascens Bechyné, 1946c: 136 (ab.) [IN]
recticollis Fairmaire, 1862a: 592
reticulata Laboissière, 1937b: 165
rugosula Rosenhauer, 1856: 321
sardea A. Villa & G. B. Villa, 1835: 50
sicelidis Reiche, 1861a: 736
semilaevis Fairmaire, 1884b: 105
semirufa Pic, 1931a: 2 (var.)
sphaeromorpha Bechyné, 1948e: 203 (ssp.)
splendorifera Bechyné, 1946c: 137 (ssp.)
sublaevis Fairmaire & Allard, 1873: 174
susterai Bechyné, 1944b: 146
temperei Jeanne, 1967: 22
thoracica Bechyné, 1948e: 202 (ssp.)
venosula Weise, 1882b: 325
viator Bechyné, 1948e: 204
violaceonigra DeGeer, 1775: 298 (*Chrysomela*)
vivesi Bechyné, 1962: 184 (ssp.)

goettingensis interstitialis Fairmaire, 1862a: 594 E: AN FR SP [# 636]

affinis Laboissière, 1937b: 166 [HN]
atramentaria Bechyné, 1948e: 200 (ssp.)
cebennica Jeanne, 1967: 13 (ssp.)
colasi Laboissière, 1937b: 164
confinis Bechyné, 1948e: 200 (ab.) [IN]
pseudostrangulata Bechyné, 1953: 85 (ssp.)
regularis Bechyné, 1948e: 204 (ssp.)
sinuatocollis Fairmaire, 1862a: 595
theodoridesi Jolivet, 1950a: 235
goettingensis parnassia Fairmaire, 1868a: 261 E: GR IT TR A: AR TR [# 636]
olivieri Fairmaire, 1868a: 261
goettingensis pratensis Duftschmid, 1825: 162 (*Chrysomela*) E: AL AU BH BU CR GR HU IT ME NM RO SL SB [# 636]
durmitorensis Apfelbeck, 1912: 248 (var.)
globulata Fairmaire, 1874b: 526
goidanichi Depoli, 1926b: 99 (forma) [IN] [# 670]
montana Fairmaire & Allard, 1873: 174
oblongula Fairmaire, 1880e: 332
paulmeyeri Stiller, 1923: 139 (var.)
stepaneki Bechyné, 1946c: 134
subrugosa Depoli, 1926b: 99 (forma) [IN] [# 670]
granadensis Bechyné, 1948d: 60 E: SP
humberti Peyerimhoff, 1925a: 15 (ssp.) N: MO [# 650]
tichuktensis Kocher, 1957: 128 (ssp.) [# 650]
iblanensis Antoine, 1939: 150 N: MO [# 651]
lecerfi Kocher, 1953b: 316 (ssp.) [# 651]
peyerimhoffi Kocher, 1953b: 316 (ssp.) [# 651]
theryi Kocher, 1959: 75 [# 651]
insignis Guérin-Méneville, 1860: cxc N: AG
insparsa Rosenhauer, 1856: 318 E: SP [# 678]
intermedia intermedia Herrich-Schäffer, 1838a: 21c E: SP
ehlersi Weise, 1882b: 319 (var.)
sericea Fairmaire & Allard, 1873: 153
intermedia lugens Rosenhauer, 1856: 319 E: SP [# 679]
nevadensis Fairmaire, 1884b: 81
parnassia Weise, 1882b: 332 [HN]
kiesenwetteri kiesenwetteri Kraatz, 1879i: 382 E: SP
kiesenwetteri sagrensis Kuntzen, 1911: 215 E: SP
lusitanica Fabricius, 1781: 116 (*Chrysomela*) E: PT [# 652, # 674]
occidentalis Herrich-Schäffer, 1838a: 23a [# 652]

scortea Germar, 1823: 583 (*Chrysomela*) [# 652]
mairei Peyerimhoff, 1923: 78 **N:** MO
marginicollis Rosenhauer, 1856: 317 E: SP
escalerae Kuntzen, 1911: 216 (var.)
splendida Pérez Arcas, 1872: 121
violetta Bechyné, 1948: 61 (ab.) [IN]
maroccana Weise, 1882b: 325 **N:** MO [# 653]
freyi Bechyné, 1953: 96 [# 653]
liebmanni Bechyné, 1953: 96 (ssp.) [# 653]
monserratensis Bechyné, 1962: 184 (ssp.) E: SP [# 654]
montsiensis Vela & Daccordi, 2021: 294 E: SP
nicaeensis A. Villa & G. B. Villa, 1835: 49 [NP] E: FR IT SZ [# 655, # 668]
affinis Leach & Risso, 1827: 206 [NO] [# 655, # 668]
apuana Daccordi & Ruffo, 1990: 103 [# 655]
designata Bechyné, 1945a: 10 (ab.) [IN]
fraterna Bechyné, 1945a: 10 (ab.) [IN]
fremyi Jolivet, 1944: 96 [# 655]
italica Herrich-Schäffer, 1838a: 22f
ovata Jolivet, 1944: 97 (ssp.) [# 655]
validicornis Fairmaire & Allard, 1873: 172 [# 655]
peresa Peyerimhoff, 1908: 124 **N:** AG
pimelioides pimelioides Herrich-Schäffer, 1838a: 21a E: IT [# 656]
aenescens Pic, 1919c: 20 (var.) [IN] [# 670]
aguilari Bechyné, 1946b: 27 [# 656]
calabrica Bechyné, 1944e: 369 [# 656]
chalcosoma Fairmaire, 1868a: 261 [# 656]
gravis Rosenhauer, 1856: 316 [# 656]
insularis Bechyné, 1944e: 369 [# 656]
martialis Bechyné, 1953: 93 (ssp.) [# 656]
ruffoi Bechyné, 1953: 93 (ssp.) [# 656]
pimelioides turbida Erichson, 1841: 189 **N:** AG TU [# 657]
aeruginosa Bechyné, 1944e: 366 (ab.) [IN]
amethystipes Chevrolat, 1873: 206 [# 657, # 683]
camoensii Fairmaire & Allard, 1873: 171 [# 657]
marani Bechyné, 1944e: 368 [# 657]
pfefferi Bechyné, 1945c: 64 [# 657]
robustior Bechyné, 1944e: 366 (ssp.) [# 657]
rugosoidea Jolivet, 1944: 100 (var.) [# 657, # 671]
staneki Bechyné, 1944e: 366 [# 657]

taborskyi Bechyné, 1944e: 368 [# 657]
tarsata Bechyné, 1944e: 365 [# 657]
tunisea Pic, 1919c: 19 (ssp.) [# 657]
variolosa Bechyné, 1944e: 363 [# 657]
piochardi Fairmaire, 1874b: 525 E: SP [# 658]
aurichalcea Bechyné, 1948e: 205 (ab.) [IN]
aurichalcea Bechyné, 1953: 90 [# 658]
bruleriei Weise, 1916b: 204 [HN, UE]
brulerii Fairmaire & Allard, 1873: 154 [HN]
nigrocuprea Bechyné, 1948e: 205 (ab.) [IN]
prujai Kocher, 1965: 198 N: MO
punctella Marseul, 1871: 387 N: AG LB MO TU [# 659]
djerbensis Pic, 1919c: 20 [# 659]
inaequalis Pic, 1924c: 22 [# 659]
punica P. H. Lucas, 1847a: pl. 45 [= 1847c: 532] N: AG
rugosa rugosa Linnaeus, 1767: 678 (*Tenebrio*) E: SP N: AG MO TU [# 660]
algirica Bechyné, 1947c: 58 (ssp.) [# 660]
angusta Bechyné, 1944c: 220 [# 660]
araneiformis Bechyné, 1947c: 57 [# 660]
aterrima Bechyné, 1944c: 219 (ssp.) [# 660]
bicolor Bechyné, 1944c: 216 (ab.) [IN]
djelfaica Bechyné, 1947c: 58 [# 660]
espanoli Bechyné, 1948d: 62 [# 660]
generosa Erichson, 1841: 189
gnaptoroides Bechyné, 1944c: 220 [# 660]
idiochroma Bechyné, 1944c: 219 (ab.) [IN]
incolorata Bechyné, 1944c: 216 (ab.) [IN]
khalouensis Jolivet, 1966a: 121 (ssp.) [# 660]
kobelti Heyden, 1883c: 221 (var.) [# 660]
kourili Bechyné, 1944c: 219 (ssp.) [# 660]
laeviventris Bechyné, 1944c: 218 (morpha) [IN]
mediatlantis Kocher, 1953b: 331 [unused original spelling] [# 672]
mesatlantica Kocher, 1953b: 331 [used incorrect subsequent spelling] [# 660, # 672]
minutissima Bechyné, 1944c: 219 (ssp.) [# 660]
nigra Bechyné, 1944c: 216 (ab.) [IN]
obscura Bechyné, 1944c: 215 (ab.) [IN]
pedestris Bechyné, 1944c: 219 (ab.) [IN]
rufipes Weise, 1882b: 326 (var.) [# 660]
rugata Kocher, 1953b: 319 (var.) [# 660, # 671]

rungsi Kocher, 1959: 77 (var.) [IN] [# 669]
rungsi Jolivet, 1966a: 122 (ssp.) [# 660]
scabra Olivier, 1807: 507 (*Chrysomela*) [HN] [# 680]
similis Bechyné, 1944c: 215 (ab.) [IN]
spuria Bechyné, 1947c: 58 [# 660]
subscabra Kocher, 1959: 77 (var.) [IN] [# 669]
subscabra Jolivet, 1966a: 122 (ssp.) [# 660]
violacea Jolivet, 1944: 99 (ab.) [IN]
viridis Jolivet, 1944: 99 (ab.) [IN]

rugosa ventricosa Weise, 1882b: 327 **N**: MO [# 661]
atlantica Bechyné, 1944c: 216 [# 661]
aurulenta Kocher, 1957: 131 (var.) [# 661, # 671]
cuprina Bechyné, 1947c: 57 (ab.) [IN]
mima Bechyné, 1947c: 57 (ssp.) [# 661]
tangeriana Bechyné, 1944c: 217 [# 661]

rugulosa Herrich-Schäffer, 1838a: 22c E: CR CZ HU KR MD PL RO SB SK TR UK [# 662]
globata Fairmaire & Allard, 1873: 177 [# 662]
lomnickii Miller, 1867: 503 [# 662]
sinaiana Bechyné, 1947b: 5 (ssp.) [# 662]
stichai Bechyné, 1947b: 5 (ssp.) [# 662]

scabripennis Fairmaire, 1868a: 261 **N**: MO [# 663]
bechynei Kocher, 1959: 78 (var.) [# 663, # 671]
ifranensis Jolivet, 1968c: 234 (ssp.) [# 663]
infoveolata Kocher, 1953b: 328 (var.) [# 663, # 671]
scabripennis Marseul, 1871: 388 [HN]
vazquezi Jolivet, 1968c: 235 (ssp.) [# 663]

scutellaris Waltl, 1835: 84 E: PT SP
erosa Fairmaire & Allard, 1873: 157
hispanica Herrich-Schäffer, 1838a: 22b
laevisterna Fairmaire, 1880e: 332
rugipennis Pérez Arcas, 1865b: 442
transversicollis Fairmaire, 1884b: 87
vermiculata Fairmaire, 1880e: 332

sobrina Fairmaire, 1884b: 85 E: SP [# 664]
arragonica Balbi, 1892a: 50 [# 664]

spectacula Bechyné, 1962: 185 (ssp.) E: SP [# 665]

strangulata Fairmaire, 1862a: 594 E: AN FR SP [# 666]
aubryi Jeanne, 1967: 23 (ssp.) [# 666]
breiti Bechyné, 1953: 90 (ssp.) [# 666]

decens Bechyné, 1953: 90 (ssp.) [# 666]

epipleuralis Bechyné, 1953: 90 (ssp.) [# 666]

sculptipennis Bechyné, 1948e: 199 (ssp.) [# 666]

strophium carmelenae Petitpierre, 2013: 65 E: SP

strophium strophium Weise, 1888a: 124 E: SP

fossulata Bechyné, 1944c: 221 [# 682]

teluetica Escalera, 1914: 521 **N:** AG MO [# 667]

achardi Bechyné, 1945d: 102 [# 667]

antiatlantica Jolivet, 1968a: 95 [= 1968b: 117] (ssp.) [# 667]

apicicornis Bechyné, 1948d: 62 [# 667]

sarhroensis Jolivet, 1968a: 95 [= 1968b: 117] (ssp.) [# 667]

schrammi Kocher, 1957: 135 (var.) [# 667, # 671]

tenebricosa Fabricius, 1775: 94 (*Chrysomela*) [conserved name, ICZN (2021), Opinion 2474] E: AL AU BE BH BU CR CZ DE FR GB GE GR HU IR IT KR LU ME NL NM PL PT RO RU (ST) SB SK SL SP SZ TR UK **A:** AB GG IN TR [# 1145]

anglica Bechyné, 1945a: 9 (ssp.)

angusticollis Weise, 1882b: 322 [HN] [# 1120]

bulgarica Bechyné, 1945a: 9 (ab.) [IN]

caraboides Geoffroy, 1785: 110 (*Chrysomela*)

chionophila Apfelbeck, 1912: 247 (var.)

circassica Reitter, 1913a: 108 (var.)

coerulea Berkenhout, 1769: 111 (*Tenebrio*) [HN]

crassipes Grimmer, 1841: 47

grossa O. F. Müller, 1776: 81 (*Chrysomela*)

helvetica Bechyné, 1945a: 9 (natio) [IN] [# 669]

heyrovskyi Bechyné, 1945a: 8 (ssp.)

iberica Motschulsky, 1849: 152

laevigata Linnaeus, 1767: 678 (*Tenebrio*) [suppressed name, ICZN (2021), Opinion 2474]

moravica Bechyné, 1945a: 8 (ssp.)

motschulskyi Bechyné, 1945a: 10 (ssp.)

normandiana Bechyné, 1945a: 9 (ssp.)

pauperata Bechyné, 1945a: 8 (ssp.)

semipolita Chevrolat, 1863a: 120

sontiaca J. Müller, 1948: 91

subcyanea Motschulsky, 1849: 152

submontana Bechyné, 1945a: 9 (ssp.)

transitoria Bechyné, 1948e: 199 (ssp.)

unicolor Herbst, 1786: 161 (*Chrysomela*)

tingitana Fairmaire & Allard, 1873: 159 **N:** MO

tortosensis Bechyné, 1948d: 60 (ssp.) E: SP

genus *Timarcha*, nomina dubia
angulicollis Motschulsky, 1849: 151 [# 675]
apricaria Gistel, 1832: 130 [# 676]
immarginata Herrich-Schäffer, 1838a: 24c [# 675]

subfamily **Galerucinae** Latreille, 1802
supertribe Galerucitae Latreille, 1802
tribe Decarthrocerini Laboisièrre, 1937

genus *Decarthrocera* Laboissière, 1937a: 27 type species: *Decarthrocera viberti* Laboissière, 1937 [# 695]
viberti Laboissière, 1937a: 29 **N:** AG

tribe Galerucini Latreille, 1802

genus *Anadimonia* Ogloblin, 1936: 127, 391 type species: *Anadimonia potanini* Ogloblin, 1936
Trichocerophysa Gressitt & Kimoto, 1963a: 471 type species: *Trichocerophysa latifascia* Gressitt & Kimoto, 1963
latifascia Gressitt & Kimoto, 1963a: 472 (*Trichocerophysa*) **A:** CH (FUJ GUI HAI TAI) NP **ORR** [# 700]
hainana Gressitt & Kimoto, 1963a: 473 (*Trichocerophysa*)
potanini Ogloblin, 1936: 128, 392 **A:** CH (SCH SHA)

genus *Apophylia* J. Thomson, 1858: 221 type species: *Apophylia chloroptera* J. Thomson, 1858
Apophylana L. N. Medvedev, 2019a: 167 type species: *Apophylana thoracica* Medvedev, 2019
Bequaertinia Laboissière, 1922c: 263 type species: *Bequaertinia nodicornis* Laboissière, 1922
Galerucesthis Weise, 1897a: 296 type species: *Auchenia thalassina* Faldermann, 1835
Glyptolus Jacoby, 1884b: 62 type species: *Glyptolus viridis* Jacoby, 1884
Malaxia Fairmaire, 1878b: 139 type species: *Malaxia flavovirens* Fairmaire, 1878
Malaxioides Fairmaire, 1888b: 155 type species: *Malaxioides grandicornis* Fairmaire, 1888
aeruginosa Hope, 1831: 29 (*Auchenia*) **A:** ID (SD "North-east India") NP **ORR**
asahinai Chûjô, 1962b: 23 **A:** CH (TAI)
taiwanica Bezděk, 2003d: 505
assamensis Jacoby, 1891a: 34 (*Malaxia*) **A:** BT CH (XIZ) ID (AP SD UT) NP **ORR**
ciliaticornis Pic, 1931b: 21
himalayana L. N. Medvedev, 1993c: 364
beeneni Bezděk, 2003c: 207 **A:** CH (ANH BEI FUJ GUA GUI GUX HAI HEB HEI HKG HUB HUN JIA JIL JIX LIA NMO SCH SHA SHN SHX TAI XIZ ZHE) NC SC **ORR**
bertiae Bezděk, 2003d: 495 **A:** ID (HP UT) PA **ORR**
brancuccii L. N. Medvedev, 1998c: 28 **A:** CH (YUN) ID (UT) NP **ORR**

cervenkai Bezděk, 2005b: 171 **A:** OM YE **AFR**
cheni Bezděk & L.-J. Zhang, 2006: 148 **A:** CH (GUA)
cyanipennis Laboissière, 1927: 59 **A:** CH (XIZ) **ORR**
dellacasai Bezděk, 2006a: 240 **N:** EG **A:** OM SA [# 703]
elongata Jacoby, 1896a: 8 (*Malaxia*) **A:** JA (Ryukyus)
elsholtziae alticola S.-H. Chen, 1976: 208, 220 (ssp.) **A:** CH (XIZ)
elsholtziae elsholtziae S.-H. Chen, 1976: 207, 219 **A:** CH (XIZ)
eoa Ogloblin, 1936: 140, 393 **A:** CH (LIA) NC RU (FE) SC
loukashkini Gressitt & Kimoto, 1963a: 429
epipleuralis Laboissière, 1927: 60 **A:** CH (GUA GUI HAI HUN SCH XIZ YUN) "India" **ORR**
geniculata Pic, 1931b: 21
flavovirens Fairmaire, 1878b: 139 (*Malaxia*) **A:** CH (CHQ FUJ SHA XIZ ZHE)
thoracica Gressitt & Kimoto, 1963a: 433
grandicornis Fairmaire, 1888b: 155 (*Malaxioides*) **A:** CH (BEI GAN GUI HEB HEI SCH SHA SHN SHX) SC **ORR**
shirozui Takizawa, 1985a: 15
hajeki Bezděk, 2003b: 4 **A:** BT ID (AP) NP
halberstadti Bezděk, 2006b: 260 **A:** ID (AP) **ORR**
kimotoi Bezděk, 2003d: 499 **A:** JA
laotica Bezděk, 2005a: 38 **A:** CH (YUN) **ORR**
lebongana Maulik, 1936: 82 **A:** BT CH (SCH YUN) ID (AP SD) NP
maculata Kimoto, 1977: 352
martensi Bezděk, 2003c: 213 **A:** BT ID (UT) NP
melli Gressitt & Kimoto, 1963a: 431 **A:** CH (FUJ HUN YUN)
pavlae Bezděk, 2003a: 92
micheli Bezděk, 2003b: 7 **A:** ID (SD) NP
miyamotoi Kimoto, 1969: 23 **A:** CH (TAI)
nepalica Bezděk, 2003c: 218 **A:** NP
nigriceps Laboissière, 1927: 62 **A:** CH (FUJ GUI HUN SCH TAI YUN) JA **ORR** [# 704]
nilakrishna Maulik, 1936: 83 **A:** ID (HP UT) NP
pectoralis Pic, 1927h: 189 **A:** CH (HKG) **ORR**
poggii Bezděk, 2003c: 221 **A:** PA **ORR**
purpurea Allard, 1889a: 332 (*Malaxia*) **A:** CH (FUJ SCH YUN) **ORR**
saigonensis Pic, 1927m: 2
rugiceps Gressitt & Kimoto, 1963a: 432 **A:** CH (GAN HUN SCH XIZ YUN)
samoderzhenkovi L. N. Medvedev, 1993c: 363 **A:** ID (AP) NP
savioi Pic, 1931b: 22 **A:** CH (FUJ GUX JIA ZHE)
schawalleri L. N. Medvedev, 1992d: 12 **A:** CH (XIZ) NP
securigera Chûjô, 1962a: 105 **A:** CH (YUN) **ORR**

sericea Fabricius, 1798: 69 (*Cantharis*) **A:** BT ID (AP SD UT) **ORR**
crotchi Jacoby, 1887a: 107 (*Galerucella*)
orientalis Jacoby, 1900b: 129 (*Malaxia*)
shuteae Bezděk, 2003d: 500 **A:** ID (HP KA UT) **ORR**
sikkimensis Bezděk, 2003d: 504 **A:** ID (SD)
sprecherae Bezděk, 2003a: 95 **A:** CH (YUN)
thalassina Faldermann, 1835: 437 (*Auchenia*) **A:** CH (GAN GUI HEB HEN JIA JIL LIA NIN NMO SHA SHX) MG RU (ES FE) SC
trinotata Gressitt & Kimoto, 1963a: 435 **A:** CH (SCH)
trochanterina Gressitt & Kimoto, 1963a: 436 **A:** CH (SCH YUN)
variicollis Laboissière, 1927: 61 **A:** CH (SHA SCH YUN)
robustior Pic, 1946b: 14
velai Bezděk, 2003c: 222 **A:** CH (TAI)
kaoi Bezděk & C.-F. Lee, 2009: 430
vietnamica Samoderzhenkov, 1988: 80 **A:** CH (XIZ) ID ("India") **ORR**
viridipennis Jacoby, 1886a: 744 (*Galerucella*) **A:** JA
yangi Bezděk & L.-J. Zhang, 2006: 146 **A:** CH (XIZ)
yunnanica Bezděk, 2003a: 97 **A:** CH (YUN)

genus *Apterogaleruca* Chûjô, 1962b: 26 type species: *Apterogaleruca hirtihumeralis* Chûjô, 1962
hirtihumeralis Chûjô, 1962b: 29 **A:** CH (TAI)
uenoi Kimoto, 1969: 29 **A:** CH (TAI)

genus *Arima* Chapuis, 1875: 217 type species: *Adimonia brevipennis* Illiger, 1805 (= *Meloe marginata* Fabricius, 1781)
brachyptera Küster, 1844: 61 (*Adimonia*) E: IT
buai Havelka, 1960: 130 E: IT
marginata caprai Havelka, 1960: 112 (ssp.) E: IT
marginata freyi Havelka, 1960: 121 (ssp.) E: FR
marginata heyrovskyi Havelka, 1960: 115 (ssp.) E: FR
marginata marginata Fabricius, 1781: 328 (*Meloe*) E: IT
brevipennis Illiger, 1805: 168 (*Galleruca*)
cincta Weise, 1886a: 574 (var.)
savonaensis Havelka, 1960: 111 (natio) [IN]
marginata obenbergeri Havelka, 1960: 113 (ssp.) E: FR
vesubiensis Havelka, 1960: 115 (natio) [IN]
marginata pseudobrachyptera Havelka, 1960: 120 (ssp.) E: FR
maritima brigaensis Havelka, 1960: 127 (ssp.) E: FR
maritima fascensis Havelka, 1960: 126 (ssp.) E: IT
maritima maritima Bua, 1953: 173 E: FR IT

genus *Atysa* Baly, 1864d: 238 type species: *Atysa terminalis* Baly, 1864

Falsoplatyxantha Pic, 1927k: 23 type species: *Falsoplatyxantha aurantiaca* Pic, 1927

Formosogalerucella Pic, 1928c: 32 type specis *Formosogalerucella brevithorax* Pic, 1928

Triaplatarthris Fairmaire, 1878b: 138 type species: *Triaplatarthris pyrochroides* Fairmaire, 1878

brevithorax Pic, 1928c: 32 (*Formosogalerucella*) **A:** CH (TAI) JA

japonica Pic, 1929e: 45 (var.)

collaris Gressitt & Kimoto, 1963a: 408 (*Triaplatarthris*) **A:** CH (FUJ SCH TAI ZHE)

diversicornis Pic, 1931b: 22 (*Falsoplatyxantha*) **A/ORR:** [unknown distribution] [# 742]

gressitti L. N. Medvedev, 2005b: 233 [RN] **A:** CH (FUJ ZHE)

marginata Gressitt & Kimoto, 1963a: 409 (*Triaplatarthris*) [HN]

himalayana L. N. Medvedev, 2005b: 233 **A:** ID (UT)

marginata cinnamomi S.-H. Chen, 1978: 55, 56 **A:** CH (FUJ)

marginata marginata Hope, 1831: 29 (*Auchenia*) **A:** CH (FUJ GAN GUI HUB SCH SHA TAI ZHE) ID (AP KA UT) NP PA **ORR**

sudiyana Maulik, 1936: 248

montivaga Maulik, 1936: 250 **A:** CH (TAI YUN) ID (AP UT) NP **ORR**

mureana Maulik, 1936: 247 **A:** CH (FUJ SCH ZHE) NP PA **ORR**

nepalica L. N. Medvedev, 2005b: 231 **A:** NP

porphyrea Fairmaire, 1889b: 76 (*Galerucella*) **A:** CH (FUJ JIX SCH XIN YUN)

pyrochroides Fairmaire, 1878b: 138 (*Triaplatarthris*) **A:** CH (HUB)

thoracica L. N. Medvedev, 2005c: 286 **A:** CH (YUN)

genus *Beenenia* Bezděk, 2012d: 404 type species: *Beenenia scanticola* Bezděk, 2012

kabateki Bezděk, 2012d: 409 **A:** YE (Socotra)

scanticola Bezděk, 2012d: 406 **A:** YE (Socotra)

genus *Belarima* Reitter, 1913c: 134 type species: *Adimonia violacea* P. H. Lucas, 1847

obliqua Beenen, 2019: 2 **N:** AG

violacea P. H. Lucas, 1847a: pl. 44 [= 1849: 540] (*Adimonia*) **E:** IT **N:** AG MO TU

genus *Chujoa* Gressitt & Kimoto, 1963a: 437 type species: *Atysa uetsukii* Chûjô, 1954

uetsukii Chûjô, 1954a: 2 (*Atysa*) **A:** JA

genus *Clitena* Baly, 1864d: 229 type species: *Clitena limbata* Baly, 1864

Mesodonta Baly, 1865g: 99 type species: *Clitena limbata* Baly, 1864

maculipennis S.-H. Chen, 1942a: 12 **A:** CH (GAN)

genus *Clitenella* Laboissière, 1927: 53 type species: *Galleruca fulminans* Faldermann, 1835

Callopistria Chevrolat, 1836: 378 [HN] type species: *Galleruca fulminans* Faldermann, 1835

fulminans Faldermann, 1835: 438 (*Galleruca*) **A:** CH (CHQ FUJ GUI GUX HEB HUB HUN JIX NMO SCH SHA SHN TAI YUN ZHE) MG **ORR**

coerulea Chûjô, 1938c: 152 (var.)

ignitincta Fairmaire, 1878b: 135 (*Lina*) **A:** CH (FUJ GUA JIA JIX SCH XIZ ZHE)

punctata Laboissière, 1927: 55 **A:** CH (YUN) **ORR**

yunnana Yang & W.-Z. Li, 1997a: 386, 390 (*Galeruca*)

purpureovittata S.-H. Chen, 1942a: 20 **A:** CH (GUI HUN SCH)

genus *Diorhabda* Weise, 1883e: 316 type species: *Galeruca elongata* Brullé, 1833

carinata Faldermann, 1837: 329 (*Galleruca*) **E:** UK **A:** AB AF AR CH (XIN) GG IN IQ KI KZ PA SY TD TR TM UZ **NAR**i

carinulata Desbrochers des Loges, 1870a: 134 (*Galeruca*) **E:** RU (ST) **A:** AB CH (FUJ GAN NIN NMO XIN) IN KI KZ MG TD TM UZ **NAR**i

basicornis Laboissière, 1935a: 324 (ab.) [IN]

deserticola S.-H. Chen, 1961: 431, 435 (ssp.)

elongata Brullé, 1833: 271 (*Galeruca*) **E:** AL BH BU CR GR NM PT RU (ST) SP TR **N:** AG EG **A:** CY LE SY TR **NAR**i

costalis Mulsant & Wachanru, 1852: 176 [= 1853a: 16] (*Galleruca*)

meridionalis Berti & Rapilly, 1973: 883 (ssp.) **A:** IN PA SY

octocostata Gahan, 1896: 460 **A:** SA SY YE **AFR**

rybakowi Weise, 1890a: 484 **A:** CH (GAN NIN NMO SCH SHA XIN) KZ MG

koltzei Weise, 1900b: 288

sublineata P. H. Lucas, 1847a: pl. 44 [= 1849: 542] (*Galleruca*) **E:** FR PT SP **N:** AG EG MO TU **A:** IQ YE **AFR**

bipustulata Normand, 1937: 126 (var.) [IN] [# 698]

tarsalis Weise, 1889g: 623 **A:** CH (GAN HEB LIA NIN NMO QIN SHX YUN XIN) MG RU (ES SS)

trirakha Maulik, 1936: 236 **A:** ID (HP)

genus *Doryxena* Baly, 1861a: 202 type species: *Galleruca grossa* Hope, 1831

geniculata Baly, 1879a: 451 **A:** ID (SD) NP **ORR**

grossa Hope, 1831: 28 (*Galleruca*) **A:** ID ("India") NP **ORR**

minor Kimoto, 2004: 53 **A:** NP

genus *Doryxenoides* Laboissière, 1927: 57 type species: *Doryxenoides tibialis* Laboissière, 1927

hainanus Gressitt & Kimoto, 1963a: 415 **A:** CH (HAI)

siva Maulik, 1936: 124 (*Doryxena*) **A:** ID (SD) **ORR**

tibialis Laboissière, 1927: 58 **A:** CH (HUB SHA YUN) NC NP SC

genus ***Galenaria*** **L. N. Medvedev, 2007d: 438** type species: *Galenaria flava* Medvedev, 2007

flava L. N. Medvedev, 2007d: 438 **A:** CH (YUN) **ORR**

genus ***Galeruca*** **Geoffroy, 1762: 251** [conserved name, ICZN (1994), Opinion 1754] type species: *Chrysomela tanaceti* Linnaeus, 1758

subgenus *Emarhopa* Weise, 1886a: 657 type species: *Galeruca rufa* Germar, 1823

maculaticeps Pic, 1920b: 10 **A:** SY

rufa Germar, 1823: 603 E: AL AU BH BU CR CZ FR GR HU IT KR MD ME NM RO RU (ST) SB SK SL SP (Balearic Islands) UK **A:** AB IN RU (WS) TR UZ **NARi** [# 722]

reticulata Küster, 1844: 65 (*Adimonia*)

subgenus *Fassatia* Havelka, 1955: 115 type species: *Galeruca microptera* Havelka, 1955

microptera Havelka, 1955: 116 **N:** MO

subgenus *Galerima* Reitter, 1903: 133 type species: *Adimonia monticola* Kiesenwetter, 1850

canigoensis Fauvel, 1892: 315 E: FR (Pyrenees) SP (Pyrenees)

miegii Pérez Arcas, 1874: 151 (*Adimonia*) E: FR (Pyrenees) SP (Pyrenees)

monticola Kiesenwetter, 1850: 224 (*Adimonia*) E: FR (Pyrenees) SP (Pyrenees)

villiersi Berti & Rapilly, 1983: 277 E: SP (Pyrenees)

subgenus *Galerotoma* Reitter, 1903: 139 type species: *Adimonia haagi* Joannis, 1865

haagi Joannis, 1865: 63 (*Adimonia*) E: PT SP **N:** MO

hamaticollis Fairmaire, 1868b: 500 (*Adimonia*)

tenebricosa Weise, 1886a: 641 (var.)

subgenus *Galeruca* Geoffroy, 1762: 251 [conserved name, ICZN (1994), Opinion 1754] type species: *Chrysomela tanaceti* Linnaeus, 1758

Adimonia Laicharting, 1781: 190 type species: *Chrysomela tanaceti* Linnaeus, 1758

altissima S.-H. Chen & Jiang, 1981: 459, 482 **A:** CH (XIN XIZ)

angelae Havelka, 1958: 199 E: SP

angulicollis Kocher, 1959: 64 **N:** MO

angusta Küster, 1849a: 91 (*Adimonia*) E: IT (Sicily) FR PT SP **N:** AG MO TU [# 718, # 743]

armeniaca Weise, 1886a: 643 E: BU GR KR RU (ST) UK **A:** AB AR CY GG IN IS JO KI KZ SY TD TM TR UZ

fulvimargo Reitter, 1902a: 186

usitata Jacobson, 1926b: 169 (ab.) [IN]

artemisiae Rosenhauer, 1856: 324 (*Adimonia*) E: SP

baetica Weise, 1891b: 150 E: SP

barbara Erichson, 1841: 191 (*Adimonia*) **N:** AG MO TU

atlasica Kocher, 1959: 62 (var.) [IN] [# 698]

goudoti Joannis, 1865: 42 (*Adimonia*)
montigena Peyerimhoff, 1915a: 34 (ssp.)
circassica Reitter, 1889a: 104 **E:** RU (ST) **A:** AR GG TR
corsica Joannis, 1865: 43 (*Adimonia*) **E:** FR (Corsica)
dahlii dahlii Joannis, 1865: 36 (*Adimonia*) **E:** AU CZ FR HU IT KR MD PL RU (CT ST) SK UK **A:** TR
dahlii vicina Solsky, 1871: 255 (*Adimonia*) **A:** CH (GAN GUI HEB HEI HUN JIL NIN NMO SHA SHX XIN) JA MG NC RU (ES FE SS) SC
japonica Weise, 1894j: 168 (var.)
daurica Joannis, 1865: 52 (*Adimonia*) **A:** CH (GAN HEI LIA NMO) MG RU (ES FE) SC **[# 719]**
mongolica Csiki, 1901a: 119 (*Adimonia*)
sedakovi Joannis, 1865: 54 (*Adimonia*)
extensa Motschulsky, 1862: 22 (*Adimonia*) **A:** CH ("North-east China") JA RU (ES FE) SC
banghaasi Weise, 1894j: 168
heydeni Weise, 1887g: 272 **A:** CH (JIL) JA NC RU (ES FE) SC
hunyadensis Csiki, 1953: 133 **E:** RO
hyrcana L. N. Medvedev & Mirzoeva, 1969: 56 **A:** AB
ida Havelka, 1956: 51 **E:** GR (Crete)
impressicollis Pic, 1934e: 22 **A:** TR
improvisa Havelka, 1956: 50 **N:** AG
indica Baly, 1878h: 381 (*Galleruca*) **A:** ID (HP UT) NP PA **ORR**
holzschuhi Mandl, 1986: 72 (*Pseudadimonia*) [HN]
interrupta Illiger, 1802: 423 (*Galleruca*) **E:** BE BY CZ DE FR GE IT LA LT PL PT SB SK SP SV UK **N:** AG LB MO TU **A:** CY IS LE SY TR
jucunda Faldermann, 1837: 328 (*Adimonia*) **E:** AU BH BU CR CZ FR GE GR HU IT MD NM PL RO RU (CT NT ST) SB SK SV SZ UK **A:** AB AF CH (NMO XIN XIZ) IN JO MG RU (ES SS WS) SY TM TR "Caucasus"
circumdata Duftschmid, 1825: 219 (*Galleruca*) [HN]
circumcincta Mannerheim, 1844: 199 (*Adimonia*)
florentina L. Redtenbacher, 1848: 522 (*Adimonia*)
hungarica Frivaldszky, 1876: 340 (*Adimonia*)
lobata Joannis, 1865: 46 (*Adimonia*)
oelandica Boheman, 1851: 232 (*Adimonia*)
sequensi Reitter, 1903: 138
sicelidis Weise, 1886a: 650 (var.)
laticollis C. R. Sahlberg, 1838: 218 (*Galleruca*) **E:** AU BE BY CZ DE EN FI FR GB GE HU IT LA LT PL RU (CT NT) SK SP SV SZ UK **A:** CH KZ MG RU (SS WS)
confinis Gebler, 1848: 12 (*Adimonia*)
flava Küster, 1854: 86 (*Adimonia*)

fontinalis Boheman, 1853: 93 (*Adimonia*)
villae Comolli, 1837: 46 (*Adimonia*)
littoralis Fabricius, 1787: 74 (*Chrysomela*) **E**: AL BH CR FR (Corsica) GR IT ME NM SB SL **A**: CY TR
atrofusca Geoffroy, 1762: 252 [UN]
brevis Joannis, 1865: 68 (*Adimonia*)
fuliginosa Joannis, 1865: 48 (*Adimonia*)
luctuosa Joannis, 1865: 30 (*Adimonia*) **E**: FR SP
macchoi Joannis, 1865: 28 (*Adimonia*) **E**: PT SP
nebrodensis Ragusa, 1887: 216 **E**: IT (Sicily)
nigrolineata nigrolineata Mannerheim, 1825: 37 (*Galleruca*) **A**: KI KZ MG RU (WS)
nigrolineata rutoga S.-H. Chen & Jiang, 1981: 458, 482 (ssp.) **A**: CH (XIN XIZ)
nigrolineata trubetzkoi Jacobson, 1925b: 51 [= 1926b: 163] **A**: ID (KA)
nigrolineata vittatipennis Baly, 1878h: 380 (*Galleruca*) **A**: AF CH (HEB NMO SHN XIN) KI KZ TD **ORR**
major Jacobson, 1901b: 138 (ssp.)
obscura Joannis, 1865: 31 (*Adimonia*) **E**: BH CR GR ME
pallasia Jacobson, 1925b: 51 [= 1926b: 165] **A**: CH (GAN NIN NMO QIN XIZ)
parallelipennis Beenen, 2002: 23 **A**: CH (QIN)
planiuscula Laboissière, 1937a: 29 **A**: IN TR
pomonae cretica Weise, 1889b: 65 (var.) **E**: GR (Crete)
pomonae pannonica Franz, 1974: 438 (ssp.) **E**: AU (Neusiedl) HU
pomonae petshenega Jacobson, 1925b: 51 [= 1926b: 167] (ssp.) **E**: RU (CT ST)
pomonae pomonae Scopoli, 1763: 83 (*Tenebrio*) **E**: AL AU BE BH BU BY CR CZ DE EN FI FR GE GR HU IT KR LA LS LT LU MD ME NL NM NR PL RO RU (CT NT ST) SB SK SL SV SZ UK **N**: MO **A**: AB CH (QIN SHX XIN) GG IN KI KZ MG RU (FE SS WS) TD TR
abbreviata Joannis, 1865: 66 (*Adimonia*)
alpestris Weise, 1886a: 652 (var.)
anthracina Weise, 1886a: 653 (var.)
aubei Joannis, 1865: 70 (*Adimonia*)
dispar Joannis, 1865: 34 (*Adimonia*)
erratica Joannis, 1865: 69 (*Adimonia*)
gredleri Joannis, 1865: 41 (*Adimonia*)
javeti Joannis, 1865: 64 (*Adimonia*)
rufescens Joannis, 1865: 34 (*Adimonia*)
rustica Schaller, 1783: 274 (*Chrysomela*)
subrubra Reitter, 1903: 137
regularis Beenen & Yang, 2007: 67 **A**: CH (SCH)
reichardti Jacobson, 1925b: 51 [= 1926b: 167] **A**: CH (GAN HEB HEI LIA NMO SCH SHN SHX XIN) NC RU (ES FE) SC **[# 721]**

reichei Joannis, 1865: 40 (*Adimonia*) E: IT (Sicily)
rugosa Joannis, 1865: 37 (*Adimonia*) E: CR
sardoa Gené, 1839: 79 (*Adimonia*) E: FR (Corsica) IT (Sardinia) **N**: AG TU
sicana Reiche, 1861a: 737 (*Adimonia*) E: IT (Sicily)
sinensis Laboissière, 1937a: 30 **A**: CH (SCH XIZ)
spectabilis lacericollis Semenov, 1909: 299 E: RU (ST) **A**: AB IN TM
spectabilis orientalis Osculati, 1844: 72 (*Adimonia*) **A**: IN IS JO LE SY TR
spectabilis spectabilis Faldermann, 1837: 326 (*Adimonia*) E: GR RU (ST) TR **A**: AB AR CH (FUJ GUI HUB HUN LIA SCH SHX XIN XIZ) CY GG IN IS JA LE RU (FE) SY TR "Central Asia" "South Russia"
pelleti Joannis, 1865: 38 (*Adimonia*)
spectabilis ussuriensis L. N. Medvedev, 1973b: 882 (ssp.) **A**: RU (FE)
tanaceti convexa Jacobson, 1925b: 51 [= 1926b: 167] (ssp.) E: RU (ST) **A**: AR
tanaceti incisicollis Motschulsky, 1860b: 232 (*Adimonia*) **A**: CH (HEI NMO XIN) MG NC RU (ES FE) SC
tanaceti tanaceti Linnaeus, 1758: 369 (*Chrysomela*) E: AL AN AU BE BH BU BY CR CZ DE EN FI FR GB GE GR HU IT KR KZ LA LS LT LU MD ME NL NM NR PL PT RO RU (CT NT ST) SB SK SL SP SV SZ UK **N**: AG MO TU **A**: AB AR GG IN JA KI KZ SY RU (SS WS) TR
banghaasi Laboissière, 1912a: 329 (var.) [HN]
bonvouloiri Joannis, 1865: 65 (*Adimonia*)
cicatricosa Chevrolat, 1872b: cxxxi (*Adimonia*)
declivis Joannis, 1865: 49 (*Adimonia*)
gibbosa Reiche & Saulcy, 1858b: 40 (*Adimonia*)
rufifrons Laboissière, 1912b: 375 [RN]
tristis Scopoli, 1763: 83 (*Tenebrio*)
tripoliana Chevrolat, 1873: 206 (*Adimonia*): **A**: LE
weisei Reitter, 1903: 137 **A**: CH (NMO JIL) KZ MG NC RU (ES FE WS)
aetha Jacobson, 1925b: 52 [= 1926b: 168] (ab.) [IN]
subgenus *Haptoscelis* Weise, 1886a: 658 type species: *Galeruca melanocephala* Ponza, 1805
melanocephala Ponza, 1805: 57 (*Chrysomela*) E: AU BU BY CR CZ DE FI FR GE HU IT KR PL RO RU (NT ST) SK SV TR UK **A**: GG RU (WS) TR [# 720]
aptera Bonelli, 1812: 168 (*Galleruca*)
baltica Weise, 1904c: 368 (var.)
haematidia Germar, 1823: 603
kubanensis Formánek, 1900: 127 (*Lochmaea*)
subgenus *Rhabdotilla* Jacobson, 1911: pl. 59 type species: *Rhabdotilla rosti* Jacobson, 1911 (= *Galeruca sexcostata* Jacoby, 1904)
Galemira Beenen, 2003: 2 type species: *Galeruca sexcostata* Jacoby, 1904

gyangzea S.-H. Chen & Jiang, 1987: 53, 57 **A:** CH (XIZ)
himalayensis Jacoby, 1896c: 283 **A:** ID (HP)
holzschuhi Mandl, 1981: 179 (*Pallasiola*, ssp.) **A:** ID (KA)
sexcostata Jacoby, 1904a: 405 **A:** ID (KA) PA **ORR**
rosti Jacobson, 1911: pl. 59 (*Rhabdotilla*)
theoneiformis Mandl, 1970: 167 (*Pallasiola*)
subcostata Beenen, 2003: 3 **A:** PA
zangana S.-H. Chen & Jiang, 1987: 53, 57 **A:** CH (XIZ)

genus *Galeruca*, species unassigned to subgenera
barovskyi Jacobson, 1925b: 51 [= 1926b: 164] **A:** CH (QIN SCH XIZ)
comaica S.-H. Chen & Jiang, 1981: 457, 482 **A:** CH (XIZ)

genus *Galerucella* Crotch, 1873: 55 type species: *Chrysomela nymphaeae* Linnaeus, 1758
subgenus *Galerucella* Crotch, 1873: 55 type species: *Chrysomela nymphaeae* Linnaeus, 1758
Hydrogaleruca Laboissière, 1922b: 33 type species: *Chrysomela nymphaeae* Linnaeus, 1758
birmanica Jacoby, 1889b: 215 (*Lochmaea*) **A:** BT CH (GUA HUB JIA LIA SHN SHG YUN) ID ("India") NP **ORR**
chujoi Komiya, 2005: 95 **A:** JA (Hokkaido)
consentanea Hope, 1831: 29 (*Galleruca*) **A:** NP
grisescens Joannis, 1865: 98 (*Galleruca*) **E:** AU BE BY EN FI FR GB GE HU IT LA LT NL PL RU (CT NT ST) SZ UK **A:** AF CH (ANH CHQ FUJ GAN GUA GUI GUX HAI HEB HEI HEN HKG HUB HUN JIA JIL JIX LIA NMO SCH SHA SHN SHX TAI XIN XIZ YUN ZHE) JA MG NP NC RU (ES FE SS WS) SC TR **ORR**
distincta Baly, 1874: 178 (*Galleruca*)
jureceki Pic, 1921a: 2 (var.) [IN] [# 698]
reducta S.-H. Chen, 1942a: 19
rubi Tamanuki, 1928: 2
vittaticollis Baly, 1874: 178 (*Galleruca*)
yakushimana Nakane, 1958: 308 (*Hydrogaleruca*, ssp.)
nipponensis Laboissière, 1922b: 34 (*Hydrogaleruca*) **A:** CH (FUJ GUA HUB JIX SHA SHN TAI ZHE) FUJ JA RU (ES FE) SC
paludosa Weise, 1922b: 68
nymphaeae Linnaeus, 1758: 376 (*Chrysomela*) **E:** AL AU BE BH BU BY CR CZ DE EN FI FR GB GE GR HU IT KR LA LT MD ME NL NM NR PL RO RU (CT NT ST) SB SK SL SV SZ UK TR **A:** AB CH (HEI HAI) IN IS KZ MG RU (ES FE SS WS) TR **NAR**
aquatica Geoffroy, 1785: 103 (*Galeruca*) [HN]
femoralis Melsheimer, 1847: 161 (*Galeruca*)
fergussoni Fowler, 1910: 228

hydrolapathi Lohse, 1989: 66 (ssp.)

kerstensi Lohse, 1989: 67

luctuosa Mannerheim, 1852: 368 (*Galleruca*)

marginella Kirby, 1837: 220 (*Galleruca*)

sagittariae Gyllenhal, 1813: 511 (*Galleruca*)

stilleri Papp, 1945: 345 (var.)

ohkurai Kimoto & Takahashi, 1992: 99 **A**: JA

ozeana Nakane, 1963a: 20 **A**: JA

placida Baly, 1878h: 381 **A**: AF BT CH (GAN JIA SCH YUN) ID (SD KA UT) NP PA **ORR** [# 724]

angulosa Pic, 1928g: 30

mastersi Blackburn, 1896: 87 (*Adimonia*)

rugosa Jacoby, 1884b: 65

vartiani Lopatin, 1967e: 439

subgenus *Neogalerucella* Chûjô, 1962b: 38 type species: *Chrysomela tenella* Linnaeus, 1761

anserina Ødegaard & Hanssen, 2020: 342 E: NR

calmariensis Linnaeus, 1767: 600 (*Chrysomela*) E: AL AU BE BH BU BY CR CZ DE EN FI FR GB GE GR HU IT KR LA LS LT LU MD ME NL NM NR PL RO RU (CT NT ST) SB SK SL SP SV SZ TR UK **N**: AG MO **A**: AB CH (JIL LIA XIN) GG IN JA KI KZ MG RU (FE SS WS) TM TR **NAR**i [# 723]

aquatica O. F. Müller, 1776: 83 (*Galeruca*)

griseaalni DeGeer, 1775: 325 (*Chrysomela*) [UN]

lythri Gyllenhal, 1813: 513 (*Galleruca*)

nigrina Delahon, 1913: 532 (ab.) [IN]

pallida Geoffroy, 1762: 253 (*Galeruca*) [UN]

lineola lineola Fabricius, 1781: 149 (*Crioceris*) E: AL AU BE BH BU BY CR CZ DE EN FI FR GB GE GR HU IT KR LA LS LT LU MD ME NL NM NR PL PT RO RU (CT NT ST) SB SK SL SP SV SZ TR UK **N**: AG MO TU **A**: AB AR CH (JIL) GG IN JA KZ MG NC RU (ES FE SS WS) SC TD TR

maculicornis Faldermann, 1837: 332 (*Galleruca*)

pici Laboissière, 1913: 69

verna Laicharting, 1781: 194 (*Adimonia*)

lineola solarii Burlini, 1942b: 127 E: IT

medvedevi Beenen, 2008: 74 **A**: IN RU (FE) TR

rishwani Makhan, 2012: 1 [# 687]

pusilla Duftschmid, 1825: 230 (*Galleruca*) E: AL AU BE BH BU BY CR CZ DE EN FI FR GB GE GR HU IT KR LA LS LT MD ME NL NM NR PL RO RU (CT NT ST) SB SK SL SP SV SZ UK **A**: AB CH (GAN GUA HKG JIL LIA SHN SHX XIZ) GG IN KI KZ MG RU (WS) TR **NAR**i

flavidula Reitter, 1913c: 140

semenowi Rybakow, 1889: 288 (*Lochmaea*) **A:** CH (QIN) MG
tenella Linnaeus, 1761: 171 (*Chrysomela*) **E:** AU BE BH BU BY CZ DE EN FI FR GB GE GR HU IT KR LA LS LT LU NL NR PL RU (CT NT ST) SB SK SL SP SV SZ UK **A:** AB CH (GAN HEB NMO QIN) GG IN KI KZ MG RU (FE SS WS) TR
jakowleffi Csiki, 1901a: 118
melanaria Everts, 1920b: 290 (ab.) [IN]
minima Weidenbach, 1859: 84 (*Galeruca*)
parva Herbst, 1783: 66 (*Crioceris*)
sibirica Csiki, 1901a: 118

genus *Geinella* E. Strand, 1935: 289 [RN] type species: *Geina invenusta* Jacobson, 1925
Geina Jacobson, 1925b: 52 [= 1925d: 143] [HN] type species: *Geina invenusta* Jacobson, 1925
crassicornis S.-H. Chen & Jiang, 1987: 55, 58 **A:** CH (XIZ)
cuprea S.-H. Chen & Jiang, 1981: 467, 485 **A:** CH (XIZ)
intermedia S.-H. Chen & Jiang, 1987: 56, 58 **A:** CH (XIZ)
invenusta Jacobson, 1925b: 52 [= 1925d: 145] (*Geina*) **A:** CH (QIN XIZ)
krishna Maulik, 1936: 278 (*Shaira*) **A:** CH (XIZ)
punctipennis S.-H. Chen & Jiang, 1987: 56, 59 **A:** CH (XIZ)
rugosa S.-H. Chen & Jiang, 1987: 56, 59 **A:** CH (XIZ)
splendida S.-H. Chen, Jiang & S.-Y. Wang, 1987: 64, 69 **A:** CH (SCH)
trapezicollis Warchałowski, 2001b: 185 **A:** CH (XIZ)

genus *Geinula* Ogloblin, 1936: 142, 394 type species: *Geinula jacobsoni* Ogloblin, 1936
antennata S.-H. Chen, 1961: 432, 435 **A:** CH (SCH)
cheni Lopatin, 2008d: 839 **A:** CH (SCH)
coeruleipennis S.-H. Chen, Jiang & S.-Y. Wang, 1987: 66, 71 **A:** CH (SCH)
jacobsoni Ogloblin, 1936: 143, 395 **A:** CH (QIN SCH TAI XIZ)
longipilosa S.-H. Chen, Jiang & S.-Y. Wang, 1987: 65, 70 **A:** CH (SCH)
nigra Gressitt & Kimoto, 1963a: 422 **A:** CH (SCH)
rugipennis S.-H. Chen, Jiang & S.-Y. Wang, 1987: 65, 70 **A:** CH (SCH)
similis S.-H. Chen, Jiang & S.-Y. Wang, 1987: 66, 70 **A:** CH (SCH)
trifoveolata S.-H. Chen, Jiang & S.-Y. Wang, 1987: 67, 71 **A:** CH (SCH)

genus *Hirtigaleruca* Chûjô, 1962b: 43 type species: *Hirtigaleruca aptera* Chûjô, 1962
aptera Chûjô, 1962b: 46 **A:** CH (TAI)

genus *Lochmaea* Weise, 1883e: 316 type species: *Chrysomela caprea* Linnaeus, 1758
Lochmaeata E. Strand, 1935: 295 [RN] type species: *Chrysomela caprea* Linnaeus, 1758
caprea Linnaeus, 1758: 376 (*Chrysomela*) **E:** AL AU BE BH BU BY CR CZ DE EN FI FR GB GE GR HU IT KR LA LS LT LU ME NL NM NR PL RO RU (CT NT ST) SB SC

SK SL SP SV SZ TR UK **A:** AB CH (GAN HEB HEI JIL LIA NIN NMO SHX) GG IN JA KZ MG NC RU (ES FE SS WS) SC TR

circumflexa Burlini, 1946: 78 (ab.)

cribrata Solsky, 1871: 257 (*Adimonia*, var.)

griseonitida DeGeer, 1775: 325 (*Chrysomela*)

livida Geoffroy, 1785: 103 (*Galeruca*)

longicornis Gmelin, 1790: 1688 (*Chrysomela*)

luctuosa Weise, 1886a: 611 (var.)

nigra Geoffroy, 1762: 254 (*Galeruca*) [UN]

pallescens Gmelin, 1790: 1724 (*Cryptocephalus*)

pallida Herbst, 1783: 66 (*Crioceris*)

pallidipennis Küster, 1847: 81 (*Adimonia*)

polygonata Laicharting, 1781: 193 (*Adimonia*)

reitteri Laboissière, 1914: 82 (var.)

saturata Stephens, 1831: 288 (*Galeruca*)

cheni C.-F. Lee, 2019: 91 **A:** CH (TAI)

crataegi Forster, 1771: 28 (*Chrysomela*) **E:** AL AU BE BH BU BY CR CZ DE FR GB GE GR HU IT KR LA LT LU MD ME NL NM PL RO RU (CT NT ST) SB SC SK SL SP SV SZ UK **N:** AG TU **A:** CH (GAN HEN SHX) GG MG RU (FE WS) TR

binotata Duftschmid, 1825: 222 (*Galleruca*)

femoralis Duftschmid, 1825: 225 (*Galleruca*)

flava Donisthorpe, 1934: 231 (ab.) [IN]

lineata Pic, 1927e: 7 (var.)

nigronotata Pic, 1912k: 90 (var.)

obscura Laczó, 1928a: 8 (ab.) [IN]

orientalis Weise, 1924a: 80 (var.) [RN]

pallida Joannis, 1865: 77 (*Adimonia*)

sanguinea Fabricius, 1775: 119 (*Crioceris*)

sanguineorubra Goeze, 1777: 326 (*Chrysomela*)

tibialis Duftschmid, 1825: 224 (*Galleruca*)

huanggangana Yang & J.-S. Wang, 1998: 261, 263 (*Lochmaeata*) **A:** CH (FUJ)

jungchani C.-F. Lee, 2019: 95 **A:** CH (TAI)

lesagei Kimoto, 1996: 32 **A:** CH (TAI)

limbata Pic, 1898b: 93 **A:** IN IS JO SY TR

setulosa J. R. Sahlberg, 1913a: 76 (*Galerucella*)

machulkai Roubal, 1926: 246 **E:** RU (ST) **A:** AB AR GG IN TR

maculata Kimoto, 1979b: 466 **A:** ID (SD) NP

nepalica L. N. Medvedev, 2005c: 287 **A:** NP

scutellata Chevrolat, 1840: 17 (*Adimonia*) **E:** PT SP

joliveti Cobos, 1955: 61

singalilaensis Takizawa, 1990a: 284 **A:** ID (SD)
smetanai Kimoto, 1996: 30 **A:** CH (TAI)
suturalis C. G. Thomson, 1866: 151 (*Galeruca*) **E:** AU BE BY CZ DE EN FI FR GB GE HU IT LA LT LU MD NL NR PL PT RU (CT NT) SK SP SV SZ UK **A:** AB RU (WS)
nigrita Weise, 1886a: 614 (var.)
tsoui C.-F. Lee, 2019: 83 **A:** CH (TAI)

genus *Menippus* Clark, 1864: 257 type species: *Menippus cynicus* Clark, 1864
Issikia Chûjô, 1961: 87 type species: *Galeruca issikii* Chûjô, 1961
arnoldi Beenen, 2019: 4 **A:** CH (SCH SHA)
asahinai Chûjô, 1962b: 33 (*Galerotella*) **A:** CH (TAI)
beeneni C.-F. Lee, Bezděk & Suenaga, 2012: 5 **A:** CH (HEN SCH SHA SHX) [# 731]
shaanxianus L. N. Medvedev, 2019a: 166 (*Pyrrhalta*)
brevicornis Jacoby, 1889b: 214 (*Diorhabda*) **A:** NP **ORR**
canellinus Fairmaire, 1889a: 374 **A:** CH (GUX HAI) **ORR**
cervinus Hope, 1831: 29 (*Galleruca*) **A:** CH (HAI GUX) ID (UT) NP **ORR**
dimidiaticornis Jacoby, 1889b: 218 **A:** ID ("India") **ORR**
garoanus Maulik, 1936: 271 (*Galerotella*)
doeberli Suenaga, C.-F. Lee & Bezděk, 2017: 234 **A:** JA (Ryukyus)
gressitti C.-F. Lee, Bezděk & Suenaga, 2012: 7 **A:** CH (HAI) **ORR**
hsuehleeae C.-F. Lee, Bezděk & Suenaga, 2012: 9 **A:** CH (TAI)
issikii Chûjô, 1961: 86 (*Galeruca*) **A:** JA (Ryukyus)
nigricornis Ohno, 1962c: 27 (*Pyrrhalta*)
nepalensis C.-F. Lee, Bezděk & Suenaga, 2012: 11 **A:** NP
sakishimanus Suenaga, C.-F. Lee & Bezděk, 2017: 237 **A:** JA (Ryukyus)
sericeus Weise, 1889g: 622 (*Galerucella*) **A:** CH (FUJ HUN SCH SHA TAI)

genus *Mimastracella* Jacoby, 1903a: 120 type species: *Mimastracella hirsuta* Jacoby, 1903
Eriosarda Jacoby, 1903a: 121 type species: *Eriosarda metallica* Jacoby, 1903
bicolor Kimoto, 1984: 47 **A:** CH (TAI)
flavomarginata Takizawa, 1978a: 123 **A:** CH (TAI)
lateralis S.-H. Chen, 1942a: 21 **A:** CH (SCH)
ochracea S.-H. Chen, 1942a: 22 **A:** CH (SCH)

genus *Monoxia* LeConte, 1865: 221 type species: *Galleruca angularis* LeConte, 1859
obesula Blake, 1939: 167 Ei: FR GR (Santorini) IT MA SP **NAR**

genus *Nyctidromas* Semenov, 1896a: 556 type species: *Leptosonyx hirtus* Weise, 1885
Nyctiphantus Semenov, 1902: 353 [RN] type species: *Leptosonyx hirtus* Weise, 1885

bergi Semenov, 1904: 195 (*Nyctiphantus*) **A:** KZ
bicoloripennis L. N. Medvedev, 2005c: 286 (*Nyctiphantus*) **A:** CH (XIZ)
custos Semenov, 1903a: 181 (*Nyctiphantus*) **A:** KZ
hirtus hirtus Weise, 1885b: 315 (*Leptosonyx*) **A:** KZ MG CH (NMO XIN)
hirtus nocturnus Semenov, 1891: 381 (*Leptosonyx*) **A:** TM UZ

genus *Ophraella* Wilcox, 1965: 43 type species: *Galleruca notata* Fabricius, 1801
communa LeSage, 1986: 43 **Ei:** BH CR HU IT RO SB SL SZ **Ai:** CH (ANH GUA FUJ HEB HUB HUN JIA JIX TAI ZHE) JA SC **NAR NTR**

genus *Pallasiola* Jacobson, 1925b: 51 [RN] type species: *Chrysomela absinthii* Pallas, 1773
Pallasia Weise, 1886a: 577 [HN] type species: *Chrysomela absinthii* Pallas, 1773
absinthii Pallas, 1773: 725 (*Chrysomela*) **E:** RU (CT) **A:** CH (GAN HEB HEI HUB JIL LIA NIN NMO QIN SCH SHA SHX YUN XIN XIZ) ID (KA) KI KZ MG RU (ES FE SS WS) TD

genus *Periclitena* Weise, 1902e: 157 type species: *Galleruca vigorsii* Hope, 1831
cyanea Clark, 1865b: 259 (*Clitena*) **A:** CH (HAI) **ORR**
sinensis Fairmaire, 1888c: 43 (*Agetocera*) **A:** CH (GAN GUA GUI GUX JIA JIX SCH YUN ZHE) **ORR**
tonkinensis Laboissière, 1929a: 266 **A:** CH (GUA GUI GUX JIA JIX SCH YUN ZHE) **ORR**
vigorsii Hope, 1831: 29 (*Galleruca*) **A:** CH (GUA HAI XIZ YUN) ID (AP SD) NP PA **ORR**
coerulans Hope, 1831: 29 (*Galleruca*)
fulgida Laboissière, 1927: 53 (var.)
igneipennis Baly, 1865g: 99 (*Clitena*)
melancholica Baly, 1864d: 231 (*Clitena*)
viridissima Weise, 1922b: 66 (var.)

genus *Pseudadimonia* Duvivier, 1891a: xlvi type species: *Colaspis variolosa* Hope, 1831
debria Maulik, 1936: 95 **A:** ID (SD)
dilatata Jiang, 1991: 85, 87 **A:** CH (XIZ)
femoralis Jiang, 1991: 84, 87 **A:** CH (XIZ)
hirtipes Jiang, 1991: 83, 86 **A:** CH (XIZ)
microphtalma Achard, 1922a: 37 **A:** CH (SCH)
parafemoralis Jiang, 1991: 85, 87 **A:** CH (YUN)
pararugosa Jiang, 1991: 86, 88 **A:** CH (XIZ)
punctipennis Jiang, 1991: 84, 87 **A:** CH (XIZ)
rugosa Laboissière, 1927: 57 **A:** CH (YUN)
variolosa Hope, 1831: 30 (*Colaspis*) **A:** BT CH (XIZ YUN) ID (AP SD) NP **ORR**

genus *Pterophthinus* Gressitt & Kimoto, 1963a: 424 type species: *Pterophthinus viridipennis* Gressitt & Kimoto, 1963
viridipennis Gressitt & Kimoto, 1963a: 425 **A:** CH (SCH XIZ)

genus *Pyrrhalta* Joannis, 1865: 82 type species: *Galeruca viburni* Paykull, 1799
Chapalia Laboissière, 1929a: 269 type species: *Chapalia jeanvoinei* Laboissière, 1929
Decoomanius Laboissière, 1927: 55 type species: *Decoomanius limbatus* Laboissière, 1927
Hoplostines Blackburn, 1890a: 361 type species: *Hoplostines viridipennis* Blackburn, 1890
Tricholochmaea Laboissière, 1933a: 963 type species: *Galerucella semifulva* Jacoby, 1886
alishanensis C.-F. Lee & Bezděk, 2021a: 32 **A:** CH (TAI)
angulaticollis Gressitt & Kimoto, 1963a: 444 **A:** CH (HUB)
annulicornis Baly, 1874: 177 (*Galleruca*) **A:** CH ("North-east China") JA NC RU (ES FE) SC
aurata Maulik, 1936: 219 (*Galerucella*) **A:** BT **ORR**
bhutanensis Kimoto, 1977: 351 **A:** BT
brunneipes Gressitt & Kimoto, 1963a: 446 **A:** CH (CHQ FUJ GAN GUA GUI SCH SHX YUN ZHE)
chinensis Jacoby, 1890c: 166 (*Galeruca*) **A:** CH (GUA HUB)
corpulenta Gressitt & Kimoto, 1963a: 447 **A:** CH (FUJ GAN GUX HUB SHA)
crassipunctata Yang, 1992b: 555, 587 **A:** CH (GUI HUN)
darjeelingensis Kimoto, 1979b: 463 **A:** ID (SD) NP
digambara Maulik, 1936: 221 (*Galerucella*) **A:** BT ID (HP SD UT) NP PA **ORR**
discalis Gressitt & Kimoto, 1963a: 448 **A:** CH (GUI HUB HUN SCH TAI)
dorsalis S.-H. Chen, 1942a: 16 (*Galerucella*, ssp.) **A:** CH (CHQ GAN GUI HUN SCH SHA SHX)
erosa Hope, 1843a: 66 [= 1843b: 64] (*Galleruca*) **A:** CH (GUA)
esakii Kimoto, 1963: 17 **A:** JA
flavescens Weise, 1887h: 190 (*Galerucella*) **A:** JA RU (FE)
konishii Kimoto, 1963: 17
formosanensis C.-F. Lee & Bezděk, 2021a: 52 **A:** CH (TAI)
fossata S.-H. Chen, 1942a: 19 (*Galerucella*) **A:** CH (SCH)
fuscipennis Jacoby, 1886a: 746 (*Galerucella*) **A:** JA NC RU (ES FE) SC
rosinae Pic, 1905a: 99 (*Galerucella*)
gracilicornis S.-H. Chen, 1942a: 16 (*Lochmaea*) **A:** CH (FUJ SCH)
gressitti Kimoto, 1969: 25 **A:** CH (TAI)
griseovillosa Jacoby, 1890c: 165 (*Galeruca*) **A:** CH (FUJ GUX HUB JIA SCH SHN ZHE)
houjayi C.-F. Lee & Bezděk, 2021a: 18 **A:** CH (TAI)
huangshana S.-H. Chen, 1964: 207, 211 **A:** CH (ANH FUJ GUX ZHE) [# 735]
hupehensis Gressitt & Kimoto, 1963a: 451 **A:** CH (HUB)
igai Kimoto, 1981b: 1 **A:** CH (TAI)
indica Laboissière, 1933a: 964 (*Lochmaea*) **A:** ID (SD) NP

ishiharai Kimoto, 1994: 191 **A:** CH (TAI)
jungchani C.-F. Lee & Bezděk, 2021a: 73 **A:** CH (TAI)
kawashimai Kimoto, 1964b: 297 **A:** JA
kobayashii Kimoto, 1974b: 25 **A:** CH (TAI)
kwangtungensis Gressitt & Kimoto, 1963a: 453 **A:** CH (GUA HKG YUN)
laosica L. N. Medvedev, 2013d: 270 **A:** CH (YUN) **ORR**
lineatipes Takei, 1916: 254 **A:** CH (ANH CHQ FUJ GAN GUA GUI GUX HEI HUB HUN JIL JIX LIA NIN SCH SHA SHX TAI YUN ZHE) JA SC
 humeralis S.-H. Chen, 1942a: 17 (*Galerucella*)
longipilosa S.-H. Chen, 1942a: 15 (*Lochmaea*) **A:** CH (SCH SHA)
lopatini Beenen, 2008: 65 [= 2010a: 74] [RN] **A:** CH (SCH)
 ornatipennis Lopatin, 2002f: 879 (*Galerucella*) [HN]
lucka Bezděk & C.-F. Lee, 2019: 528 **A:** CH (YUN) **ORR**
lui C.-F. Lee & Bezděk, 2021a: 77 **A:** CH (TAI)
maculata Gressitt & Kimoto, 1963a: 456 **A:** CH (FUJ TAI)
martensi L. N. Medvedev & Sprecher-Uebersax, 1999: 366 **A:** CH (XIZ) ID (SD) NP
medvedevi Sprecher-Uebersax & Zoia, 2002: 162 **A:** NP
meifena Kimoto, 1976: 4 **A:** CH (TAI)
meihuai C.-F. Lee & Bezděk, 2021a: 42 **A:** CH (TAI)
metallica Gressitt & Kimoto, 1963a: 457 **A:** CH (GUI SCH YUN)
multicostata Pic, 1928c: 32 (*Galerucella*) **A:** CH (YUN) **ORR**
 octocostata L. N. Medvedev, 2000d: 261 (*Atysa*)
nigromaculata Yang, 1992b: 556, 587 **A:** CH (GUI HUN)
ningpoensis Gressitt & Kimoto, 1963a: 459 **A:** CH (CHQ SHA SHX ZHE) JA
ochracea Gressitt & Kimoto, 1963a: 460 **A:** CH (FUJ JIX ZHE)
ohbayashii Kimoto, 1984: 46 **A:** CH (TAI)
pseudolimbata Yang, 2010: 128 [RN] **A:** CH (HUB SHX)
 limbata S.-H. Chen, 1942a: 13 (*Lochmaea*, ssp.)
qianana Nie & Yang, 2013: 52 **A:** CH (GUI SCH) **[# 696]**
ruficollis Gressitt & Kimoto, 1963a: 462 **A:** CH (GUI HUB HUN)
salicicola Wilcox, 1971: 82 [RN] **A:** CH (FUJ HUB)
 salicis Gressitt & Kimoto, 1963a: 463 (*Pyrrhalta*) [HN]
scutellata Hope, 1831: 29 (*Galleruca*) **A:** BT CH (FUJ GUI HUB HUN JIX SHA) ID ("India") NP
 tumida Gressitt & Kimoto, 1963a: 467
semifulva Jacoby, 1886a: 745 (*Galerucella*) **A:** CH (FUJ HEI JIL LIA SHA TAI) JA RU (ES FE) SC
 japonica Weise, 1922b: 67 (*Lochmaea*)
 modesta Jacoby, 1886a: 745 (*Galerucella*)
 signaticeps Weise, 1887h: 191 (*Galerucella*)

shirozui Kimoto, 1969: 26 **A:** CH (TAI)
sikanga Gressitt & Kimoto, 1963a: 465 **A:** CH (SCH)
silfverbergi Lopatin, 2005b: 574 **A:** CH (SCH)
submetallica S.-H. Chen, 1942a: 14 (*Lochmaea*) **A:** CH (FUJ GUA HUB JIA SCH)
sulcatipennis S.-H. Chen, 1942a: 18 (*Galerucella*) **A:** CH (GUI HUB HUN SCH YUN)
tahsiangi C.-F. Lee & Bezděk, 2021a: 22 **A:** CH (TAI)
taiwana Kimoto, 1969: 27 **A:** CH (TAI)
takeii Chûjô, 1950: 114 (*Lochmaea*) **A:** JA
takizawai Kimoto, 1996: 32 **A:** CH (TAI)
tatesuji Kimoto, 2001: 45 **A:** NP
tianmuensis S.-H. Chen, 1964: 207, 211 **A:** CH (ZHE)
tibialis Baly, 1874: 176 (*Galleruca*) **A:** CH (GAN GUI HUB JIX) JA SC
    nigrimembris Fairmaire, 1888c: 41 (*Galerucida*)
tsoui Bezděk & C.-F. Lee, 2019: 531 **A:** CH (TAI) **ORR**
unicostata Pic, 1937a: 20 (*Galerucella*) **A:** CH (HAI SCH YUN) **ORR**
    hainanensis Gressitt & Kimoto, 1963a: 450
ussuriensis Romantsov, 2021: 159 (*Tricholochmaea*) **A:** RU (FE)
viburni Paykull, 1799a: 89 (*Galeruca*) E: AU BE BH BU BY CR CZ DE EN FI FR GB GE HU IT KR LA LT LU MD NL NR PL RO RU (CT NT ST) SK SL SP (Navarra) SV SZ UK **A:** AB AR GG KZ RU (SS WS) **NARi**
    nigrescens Ochs, 1925: 80 (var.)
viridipennis Kimoto, 1981b: 2 **A:** CH (TAI)
wilcoxi Gressitt & Kimoto, 1965: 800 [RN] **A:** CH (CHQ GUI SHA YUN)
    fulva Laboissière, 1929a: 265 (*Clitena*) [HN]
    kimotoi Aslam, 1967: 128 [RN]
wulaiensis C.-F. Lee & Bezděk, 2021a: 69 **A:** CH (TAI)
xizangana S.-H. Chen & Jiang, 1981: 459, 483 **A:** CH (XIZ)
yasumatsui Kimoto, 1964b: 301 **A:** JA (Ryukyus)
yoshimotoi Kimoto & Gressitt, 1966: 521 **A:** JA (Ryukyus)

genus *Radymna* Reitter, 1913c: 135 type species: *Diorhabda rickmersi* Weise, 1900
    Farsogaleruca Lopatin, 1981b: 625 type species: *Farsogaleruca rufina* Lopatin, 1981
    Galerupipla Maulik, 1936: 222 type species: *Galerupipla brunnea* Maulik, 1936
    Prophyllis Reitter, 1913c: 135 type species: *Galleruca persica* Faldermann, 1837
damascena Joannis, 1865: 96 (*Galleruca*) **A:** SY
fischeri Faldermann, 1837: 334 (*Galleruca*) E: KR RU (ST) **A:** AB AR GG IN TM TR
    subnigra Weise, 1878: 335 (*Galerucella*, var.)
insperabilis Lopatin, 1981b: 627 (*Farsogaleruca*) **A:** IN
    latifrons Beenen, 2014: 90
maculicollis Beenen, 2014: 93 **A:** IN IS TR

nigrifrons Laboissière, 1914: 83 (*Diorhabda*) **A:** AR IN TR
persica Faldermann, 1837: 331 (*Galleruca*) E: GR (Samos, Rhodes) KR RU (ST) **A:** AB AF AR CH (HEI XIN) CY GG IN IQ IS KZ PA SY TD TM TR UZ **ORR** [# 736]
brunnea Maulik, 1936: 224 (*Galerupipla*)
disconigra Pic, 1909d: 131 (*Diorhabda*, var.)
flaveola Brancsik, 1899: 104 (*Diorhabda*, var.)
sareptana Stierlin, 1863a: 501 (*Diorhabda*)
turcica Stierlin, 1867: 226 (*Galeruca*)
quadrimaculata L. Redtenbacher, 1849: 9 (*Galleruca*) **A:** IN TR
rickmersi Weise, 1900b: 289 (*Diorhabda*) **A:** KZ TD TM UZ
ornaticollis Reitter, 1900b: 231 (*Lochmaea*)
pamirica Mandl, 1968: 29 (*Pallasiola*)
reitteri Havelka, 1958: 202 (*Galeruca*)
rufina Lopatin, 1981b: 626 (*Farsogaleruca*) **A:** IN

genus *Sastra* Baly, 1865f: 253 type species: *Sastra placida* Baly, 1865
Eriosardella Chûjô, 1935e: 219 type species: *Eriosardella costata* Chûjô, 1935
fulvomarginata Takizawa, 1988b: 540 **A:** ID (SD) NP
medvedevi Bezděk & Beenen, 2010: 74 [RN] **A:** NP
suturalis L. N. Medvedev, 2003a: 320 [HN]

genus *Sastracella* Jacoby, 1900a: 294 type species: *Sastra fulvipennis* Jacoby, 1884
cinnamomea Yang, 1995b: 211, 213 **A:** CH (JIX ZHE)
collaris Kimoto, 2001: 46 **A:** NP
harmandi Laboissière, 1933a: 961 (*Sastra*) **A:** ID (SD)
himalayana L. N. Medvedev, 2007c: 293 **A:** ID (SD) NP
laosensis Kimoto, 1989a: 23 **A:** CH (YUN) **ORR**

genus *Sastroides* Jacoby, 1884c: 218 type species: *Sastroides bimaculata* Jacoby, 1884
birmanica Jacoby, 1889b: 217 **A:** ID (AP) **ORR**
gigantica Maulik, 1936: 248 (*Atysa*)
dohertyi Maulik, 1936: 265 (*Sastra*) **A:** NP **ORR**
hirtipennis Jacoby, 1891a: 33 (*Sastra*) **A:** ID (AP) **ORR**
indica Jacoby, 1894: 315 **A:** ID (SD) NP **ORR**
livida Laboissière, 1935c: 143 (*Sastra*) **A:** CH (HAI YUN) **ORR**
mamaya Maulik, 1936: 257 (*Sastra*)
brunnea Gressitt & Kimoto, 1963a: 412 (*Mimastracella*)
nigriceps Kimoto, 2004: 59 **A:** ID (SD)
purpurascens Hope, 1831: 29 (*Galleruca*) **A:** CH (GUI HUN) ID (SD) NP **ORR**
acutipennis Laboissière, 1933a: 960 (*Sastra*)

submetallica Gressitt & Kimoto, 1963a: 413 (*Mimastracella*) **A:** CH (FUJ GUA HUN TAI)
tibialis Jacoby, 1900b: 131 **A:** ID (AP) **ORR**
violacea Weise, 1922b: 66 (*Sastra*) **A:** CH (FUJ GUA GUI GUX HUN JIX SCH) **ORR**

genus *Swargia* Maulik, 1936: 275 type species: *Swargia nila* Maulik, 1936
brevicollis S.-H. Chen & Jiang, 1981: 466, 485 (*Geinella*, ssp.) **A:** CH (XIZ)
limbaticollis S.-H. Chen & Jiang, 1981: 466, 485 (*Geinella*, ssp.) **A:** CH (XIZ)
nila Maulik, 1936: 276 A: CH (XIZ)

genus *Theone* Gistel, 1857b: 585 type species: *Galleruca silphoides* Dalman, 1823
Leptosonyx Weise, 1885b: 315 type species: *Galleruca silphoides* Dalman, 1823
margelanica margelanica Kraatz, 1882a: 117 (*Galeruca*) **A:** KI TD UZ
margelanica rugulosa Jakob, 1958a: 34 (*Leptosonyx*) **A:** KI
octocostata afghanistanica Mandl, 1968: 13 **A:** AF IN
octocostata beludshistanica Mandl, 1970: 156 (ssp.) **A:** PA
octocostata lopatini Mandl, 1970: 160 (ssp.) **A:** TD
octocostata octocostata Weise, 1912: 92 (*Leptosonyx*) **A:** ID (KA) PA "Pamir"
octocostata ornata Jakob, 1958a: 34 (*Leptosonyx*) **A:** TD (Pamir)
octocostata ovata Jakob, 1958a: 35 (*Leptosonyx*) **A:** CH (HEB XIZ)
octocostata woehrli Mandl, 1970: 159 (ssp.) **A:** AF PA TD
silphoides artemisiae Jacobson, 1896a: 555 (*Leptosonyx*) **A:** TD
silphoides costipennis Kirsch, 1880: 164 (*Adimonia*) **E:** RU (ST) **A:** KI TM UZ
bucharica Mandl, 1968: 15 (ssp.)
filicornis Jakob, 1958a: 33 (*Leptosonyx*)
silphoides kuldshensis Mandl, 1968: 23 (ssp.) **A:** CH (HEB SCH XIN XIZ) KZ MG
silphoides scrobiculata Mandl, 1968: 23 (ssp.) **A:** KZ
silphoides silphoides C. R. Sahlberg, 1823a: 62 [= 1823c: 62, = 1829: 27] (*Galleruca*) **E:** RU (ST) UK **A:** AB IN KZ
silphoides Dalman, 1823: 77 (*Galleruca*) [HN]

genus *Xanthogaleruca* Laboissière, 1934b: 67 type species: *Chrysomela luteola* O. F. Müller, 1766
aenescens Fairmaire, 1878b: 140 (*Galeruca*) **A:** CH (BEI CHQ GAN HEB HEI HEN HUN JIA JIL LIA NIN NMO SHA SHN SHX TIA ZHE) MG NC RU (FE) SC **[# 739]**
lusca Maulik, 1936: 234 (*Diorhabda*) **A:** ID (HP KA UT) PA
luteola O. F. Müller, 1766: 187 (*Chrysomela*) **E:** AL AU BE BH BU BY CR CZ DE FR GE GR HU IT KR LA LT LU MD ME NL NM PL PT RO RU (CT ST) SB SK SL SP SV

SZ TR UK **N:** AG MO TU **A:** AB AF AR GG CH (GAN NIN NMO SHA) CY ID (KA) IN IQ IS KI KZ LE PA SY TD TM TR **AUSi NARi NTRi** **[# 740, # 1145]**

bicarinata Ogloblin, 1936: 108, 391 (*Galerucella*, var.) [IN] **[# 698]**

nigra Csiki, 1953: 133 (*Galerucella*, ab.) [IN]

obscuridorsis Roubal, 1926: 247 (*Galerucella*, ab.) [IN]

ulmi Geoffroy, 1785: 103 (*Galeruca*)

xanthomelaena Schrank, 1781: 78 (*Chrysomela*)

maculicollis Motschulsky, 1854a: 49 (*Galleruca*) **A:** CH (BEI FUJ GAN GUA GUI GUX HEB HEI HEN HKG HUN JIA JIL JIX LIA NIN NMO SHA SHN SHX TAI ZHE) JA NC RU (FE "Southeast Siberia") SC

vageplicata Fairmaire, 1888b: 154 (*Galerucella*)

vittula Ogloblin, 1936: 106, 391 (*Galerucella*, ab.) [IN]

nigromarginata Jacoby, 1886a: 743 (*Galerucella*) **A:** CH (HAI HKG) JA

orientalis Ogloblin, 1936: 103, 390 (*Galerucella*) **A:** CH (BEI FUJ HEB LIA SHN SHX)

seminigra Jacoby, 1886a: 744 (*Galerucella*) **A:** CH (SHX) JA

subaenea Ogloblin, 1936: 102, 389 (*Galerucella*) **A:** CH (SCH YUN)

subcoerulescens Weise, 1884a: 159 (*Galerucella*) **A:** TR

yuae C.-F. Lee & Bezděk, 2021a: 9 **A:** CH (TAI)

genus *Zangastra* S.-H. Chen & Jiang, 1981: 460, 483 type species: *Zangastra tuberosa* S.-H. Chen & Jiang, 1981

angusta Jiang, 1988a: 392, 395 **A:** CH (XIZ)

nepalensis Takizawa, 1988a: 12 **A:** NP

nitidicollis S.-H. Chen & Jiang, 1981: 462, 484 **A:** CH (XIZ) ID (SD) NP

pallidicollis S.-H. Chen & Jiang, 1981: 462, 484 **A:** CH (XIZ)

picea Jiang, 1988a: 393, 395 **A:** CH (XIZ)

sichuanica Lopatin, 2007: 183 **A:** CH (SCH)

tuberosa S.-H. Chen & Jiang, 1981: 461, 483 **A:** CH (XIZ)

tribe Hylaspini Chapuis, 1875

genus *Agelasa* Motschulsky, 1861a: 25 type species: *Agelasa nigriceps* Motschulsky, 1861

nigriceps Motschulsky, 1861a: 25 **A:** CH (HEI HUN JIL) JA NC RU (ES FE) SC **[# 699]**

genus *Agelastica* Chevrolat, 1836: 381 type species: *Chrysomela alni* Linnaeus, 1758

Scholastica Gistel, 1848a: 402 [RN] type species: *Chrysomela alni* Linnaeus, 1758

alni alni Linnaeus, 1758: 369 (*Chrysomela*) **E:** AL AU BE BH BU BY CR CZ DE EN FI FR GB GE GR HU IT KR LA LS LT LU ME NL NM NR PL RO RU (CT NT ST) SB SK SL SP SV SZ TR UK **A:** AB CH (XIN) GG IN KZ RU (WS) TR **NARi**

atra Tenenbaum, 1927: 151 (ab.) [IN]

nigroviolacea Goeze, 1777: 326 (*Chrysomela*)

violacea Geoffroy, 1762: 232 (*Cryptocephalus*) [UN]
violacea Laicharting, 1781: 193 (*Adimonia*)
alni glabra Fischer von Waldheim, 1842: 20 (*Adimonia*) **A:** CH (CHQ GAN HEB QIN SCH SHX XIN) ID (KA) IN KZ MG TD TM
orientalis Baly, 1878h: 379
sillemi Laboissière, 1935a: 325 (ssp.)
coerulea Baly, 1874: 188 **A:** CH (GAN JIL) JA NC RU (ES FE) SC

genus *Aplosonyx* Chevrolat, 1836: 375 type species: *Galleruca albicornis* Wiedemann, 1821
Berecyntha Baly, 1865g: 98 type species: *Berecyntha tibialis* Baly, 1865
Caritheca Baly, 1877a: 226 type species: *Caritheca quadripustulata* Baly, 1877
Haplonyx Jacobson, 1896a: 555 [UE] type species: *Galleruca albicornis* Wiedemann, 1821
ancora Laboissière, 1935b: 110 [HN] **A:** CH (FUJ GUA GUX HAI YUN) **ORR** [# 692]
ancorella Feng, Yang, Liu & Z.-Q. Li, 2023: 213 **A:** CH (YUN)
chalybeus Hope, 1831: 28 (*Galleruca*) **A:** BT CH (XIZ YUN) ID (AP SD) NP **ORR**
elongata Baly, 1864a: 624
jeanvoinei Laboissière, 1935b: 109 (var.) [IN] [# 698]
cinctus S.-H. Chen, 1964: 203, 209 **A:** CH (YUN)
duvivieri Jacoby, 1900b: 130 (*Haplosonyx*) **A:** CH (YUN) ID (SD) **ORR**
emeishanicus Lopatin, 2005d: 877 (*Gallerucida*) **A:** CH (SCH)
metallicus S.-H. Chen, 2008: 65
flavipennis S.-H. Chen, 1964: 203, 209 **A:** CH (GUX XIZ YUN) [# 701]
fulvescens S.-H. Chen, 1964: 204, 210 (ssp.) **A:** CH (FUJ HAI)
gansuicus S.-H. Chen, 1942a: 38 (*Galerucida*) **A:** CH (GAN GUI HUB SCH)
montanus L. N. Medvedev, 2010a: 207 **A:** CH (YUN) **ORR**
nigriceps Yang, 1995a: 91, 93 **A:** CH (CHQ GUI HUB SCH)
nigricornis Feng, Yang, Liu & Z.-Q. Li, 2023: 216 **A:** CH (SCH)
omeiensis S.-H. Chen, 1942a: 40 (ssp.) **A:** CH (SCH)
orientalis Jacoby, 1892b: 962 (*Haplosonyx*) **A:** CH (FUJ GUA GUX YUN) ID ("India") NP **ORR** [# 702]
tonkinensis Laboissière, 1922a: 102 (*Sphenoraia*)
varipes Jacoby, 1892b: 964 (*Haplosonyx*)
ornatus Jacoby, 1892b: 963 (*Haplosonyx*) **A:** CH (YUN) **ORR**
pictus S.-H. Chen, 1942a: 39 **A:** CH (GAN HUB SCH SHA)
robinsoni Jacoby, 1905b: vi (*Haplosonyx*) **A:** CH (YUN) **ORR**
rufipennis Duvivier, 1892b: 439 (*Haplosonyx*) **A:** CH (SHG YUN) ID (SD) **ORR**
rubrus Maulik, 1936: 620
scutellatus Baly, 1879a: 452 (*Haplosonyx*) **A:** ID (AP SD UT) NP PA **ORR** **[Appendix]**
sublaevicollis Jacoby, 1889b: 218 (*Haplosonyx*) **A:** CH (YUN) **ORR**
tianpingshanensis Yang, 1995a: 91, 93 **A:** CH (GAN GUI HUB HUN SHA)

wudangensis Feng, Yang, Liu & Z.-Q. Li, 2023: 218 **A:** CH (HUB)
yunlongensis Jiang, 1992: 664, 673 **A:** CH (YUN)

genus *Arthrotidea* S.-H. Chen, 1942a: 45 type species: *Arthrotidea ruficollis* S.-H. Chen, 1942
cheni Yang, 1996: 30 **A:** CH (GAN)
luoxuensis Yang, 1996: 30 **A:** CH (YUN)
nepalensis Kimoto, 1970c: 418 (*Dercetina*) **A:** BT CH (GUI XIZ) ID (SD) NP
rubrica S.-H. Chen & Jiang, 1981: 474, 487 **A:** CH (XIZ) ID (SD)
ruficollis S.-H. Chen, 1942a: 44 **A:** CH (CHQ FUJ GAN GUI HUB HUN JIA SCH SHA XIZ YUN ZHE)
scutellaris Jiang, 1988a: 394, 396 **A:** CH (XIZ)
sikkima L. N. Medvedev, 2009b: 407 **A:** ID (SD)

genus *Arthrotus* Motschulsky, 1857: 38 type species: *Arthrotus niger* Motschulsky, 1857
Anastena Maulik, 1936: 296 type species: *Astena nigromaculata* Jacoby, 1896
Anicera Jacoby, 1884c: 207 type species: *Anicera bimaculata* Jacoby, 1884
Cerotrus Jacoby, 1884b: 33 type species: *Cerotrus melanocephalus* Jacoby, 1884
Dercestra Chûjô, 1962b: 163 type species: *Dercestra abdominalis* Chûjô, 1962
Taphinella Jacoby, 1889b: 224 type species: *Taphinella nigripennis* Jacoby, 1889
abdominalis Chûjô, 1962b: 166 (*Dercestra*) **A:** CH (TAI)
bipartitus Jacoby, 1890d: 196 (*Antipha*) **A:** CH (HUB ZHE)
brownelli Gressitt & Kimoto, 1963a: 693 **A:** CH (SCH YUN)
chinensis Baly, 1879b: 115 (*Antipha*) **A:** CH (CHQ FUJ GUI HAI HEN HUB HUN SCH SHA ZHE) **[# 705]**
caeruleiceps Fairmaire, 1889b: 80 (*Cneorane*)
coeruleus S.-H. Chen, 1942a: 43 (*Proegmena*) **A:** CH (SCH YUN)
cyaneus Chûjô, 1966b: 8 (*Anastena*) **A:** BT CH (SCH) ID (SD) NP
hauseri Kimoto, 1967b: 72
viridicyaneus Kimoto, 1977: 387 (*Dercetina*)
daliensis Lopatin, 2009a: 5 **A:** CH (YUN)
elongatus Gressitt & Kimoto, 1963a: 694 **A:** CH (SCH)
flavocinctus Hope, 1831: 29 (*Galleruca*) **A:** CH (ANH CHQ FUJ GAN GUA GUI HEB HUB HUN JIX SCH YUN ZHE) ID (AP HP SD UT) NP **ORR** **[# 741]**
femoralis Weise, 1922b: 97 (*Dercetes*)
flavofasciatus Baly, 1879a: 456 (*Antipha*)
freyi Gressitt & Kimoto, 1963a: 695 **A:** CH (SCH SHA)
fulvus Chûjô, 1938c: 139 **A:** CH (TAI)
nakanei Kimoto, 1969: 65 (*Dercetina*)

gressitti Kimoto, 1969: 61 **A:** CH (TAI)
hirashimai Kimoto, 1969: 60 **A:** CH (TAI)
  taiwanus Takizawa, 1978a: 125 (*Proegmena*)
histrio Baly, 1879a: 456 (*Antipha*) **A:** ID (SD) **ORR**
kalimponganus Kimoto, 2004: 51 **A:** ID (SD)
liquidus Gressitt & Kimoto, 1963a: 696 **A:** CH (HUB YUN)
maai Gressitt & Kimoto, 1963a: 698 **A:** CH (FUJ GUA)
malaisei Bryant, 1954a: 421 (*Gallerucida*) **A:** CH (YUN) **ORR**
melanocephalus Jacoby, 1884b: 34 (*Cerotrus*) **A:** BT ID (SD) NP **ORR**
  antennalis Laboissière, 1932b: 134
  bimaculatus Jacoby, 1884c: 208 (*Anicera*)
  frontalis Laboissière, 1932b: 133
  phaseoli Laboissière, 1932a: 174
  pumilus Laboissière, 1932b: 134
  tarsalis Laboissière, 1932b: 134
micans S.-H. Chen, 1942a: 43 (*Proegmena*) **A:** CH (HUN SCH ZHE)
niger Motschulsky, 1857: 39 **A:** JA
  suzukii S. Matsumura, 1934: 63 (*Luperodes*)
  variabilis Baly, 1874: 183
nigricollis S.-H. Chen, 1942a: 41 (*Proegmena*) **A:** CH (GUI JIX)
nigripennis Jacoby, 1889b: 224 (*Taphinella*) **A:** CH (HAI) NP **ORR**
  nigromarginatus Jacoby, 1892b: 940 (*Cerotrus*)
nigrofasciatus Jacoby, 1890d: 196 (*Antipha*) **A:** CH (ANH CHQ FUJ GAN GUA GUI HUB HUN JIX SCH SHA ZHE)
ochreipennis Gressitt & Kimoto, 1963a: 700 **A:** CH (FUJ GUI HUN)
pallimembris S.-H. Chen & Jiang, 1976: 206, 219 **A:** CH (XIZ YUN)
purpureus Gressitt & Kimoto, 1963a: 701 **A:** CH (SCH)
saigusai Kimoto, 1969: 62 **A:** CH (TAI)
testaceus Gressitt & Kimoto, 1963a: 702 **A:** CH (CHQ FUJ HUB SCH TAI ZHE)
  shibatai Kimoto, 1984: 55
tricolor Chûjô, 1965b: 95 (*Dercetis*) **A:** CH (TAI)
yangi C.-F. Lee, 2022a: 201 **A:** CH (TAI)
yuae C.-F. Lee, 2022a: 175 **A:** CH (TAI)

genus *Capula* Jacobson, 1925b: 52 [= 1925d: 145] type species: *Capula metallica* Jacobson, 1925
apicalis S.-H. Chen, S.-Y. Wang & Jiang, 1986: 399, 400 **A:** CH (SCH)
caudata S.-H. Chen, S.-Y. Wang & Jiang, 1986: 398, 400 **A:** CH (SCH)
metallica Jacobson, 1925b: 52 [= 1925d: 147] **A:** CH (QIN SCH XIZ)

genus *Dercetina* Gressitt & Kimoto, 1963a: 704 [RN] type species: *Dercetis depressa* Clark, 1865

Antipha Baly, 1865f: 251 [HN] type species: *Antipha picipes* Baly, 1865

Dercetes Hincks, 1949: 611 [UN]

Dercetis Clark, 1865a: 146 [HN] type species: *Dercetis depressa* Clark, 1865

azumai Kimoto & Gressitt, 1966: 534 **A:** CH (TAI) JA (Ryukyus)

bifasciata Clark, 1865a: 147 (*Dercetis*) **A:** CH (GUA HAI JIX YUN) **ORR**

feae Jacoby, 1892b: 973 (*Antipha*)

bretinghami Baly, 1879b: 118 (*Antipha*) **A:** BT CH (HAI) ID (SD) NP **ORR**

bhutanensis Kimoto, 1977: 385

hainana Gressitt & Kimoto, 1963a: 707

inornata Jacoby, 1892b: 972 (*Antipha*)

laetifica Weise, 1922b: 97 (*Dercetes*)

pallida Allard, 1889e: cviii (*Antipha*)

pallida Laboissière, 1932b: 133 (*Arthrotus*) [HN]

carinipennis Gressitt & Kimoto, 1963a: 705 **A:** CH (FUJ GAN SHA ZHE)

chinensis Weise, 1889g: 626 (*Arthrotus*) **A:** CH (HUB JIA SHA YUN ZHE)

varipennis Jacoby, 1890e: 214 (*Antipha*)

collina Weise, 1924a: 144 (*Dercetes*) [RN] **A:** ID ("India") NP **ORR**

bifasciata Jacoby, 1904a: 399 (*Antipha*) [HN]

cyanipennis S.-H. Chen, 1942a: 42 (*Proegmena*) **A:** CH (HUB SCH SHA)

dimidiaticornis Jacoby, 1891a: 33 (*Antipha*) **A:** ID (KA) PA

flaviventris Jacoby, 1890e: 214 (*Antipha*) **A:** CH (GUA GUI HUB HUN SCH TAI XIZ YUN)

abdominalis Laboissière, 1934a: 8 (*Proegmena*)

fulvomaculata Takizawa, 1990a: 285 **A:** ID (SD) NP **ORR**

grandis Kimoto, 1970c: 417 **A:** BT NP

itoi Kimoto, 1969: 64 **A:** CH (TAI)

laetissima L. N. Medvedev, 2011c: 253 **A:** NP

major Kimoto, 1977: 386 **A:** BT ID (SD) NP

mandarensis Jacoby, 1900b: 134 (*Antipha*) **A:** BT ID (SD) NP **ORR**

miniaticollis Hope, 1831: 29 (*Galleruca*) **A:** BT ID (SD) NP

minor Gressitt & Kimoto, 1963a: 708 **A:** CH (FUJ GUA JIX ZHE)

picipes Baly, 1865f: 251 (*Antipha*) **A:** BT ID (SD) NP

indica Duvivier, 1891b: clv (*Antipha*)

punctipennis Kimoto, 1977: 382

posticata Baly, 1879a: 455 (*Antipha*) **A:** CH (HAI SCH YUN) ID (SD) NP **ORR**

shirozui Kimoto, 1969: 63 **A:** CH (TAI)

shona Maulik, 1936: 361 (*Dercetis*) **A:** BT ID (AP) NP PA **ORR**

taiwana Chûjô, 1938c: 140 (*Dercetis*) **A:** CH (GUA HUB TAI)
melania Chûjô, 1938c: 140 (ab.) [IN]
tsoui C.-F. Lee, 2021: 316 **A:** CH (TAI)
unifasciata Allard, 1889e: cvii (*Antipha*) **A:** ID (SD) **ORR**
variabilis Kimoto, 1977: 383 **A:** BT NP
viridipennis Duvivier, 1887: xlix (*Antipha*) **A:** CH (SHA YUN) NP **ORR**

genus *Dercetisoma* Maulik, 1936: 455 type species: *Antipha concolor* Jacoby, 1889
concolor Jacoby, 1889b: 222 (*Antipha*) **A:** BT CH (CHQ HUB SCH) ID (HP SD UT) NP PA **ORR**
brunneum Allard, 1889e: cviii (*Antipha*)
modestum Allard, 1889f: 306 (*Sutrea*)
puncticolle Jacoby, 1889b: 221 (*Antipha*)
nepalicum L. N. Medvedev & Sprecher-Uebersax, 1998: 35 **A:** NP
persimile Kimoto, 1977: 389 (*Arthrotus*) **A:** BT ID (SD) NP

genus *Doryida* Baly, 1865g: 97 type species: *Doryida mouhoti* Baly, 1865
flava Allard, 1890: xciii **A:** NP
fraterna Laboissière, 1931b: 142 **A:** CH (YUN)
mouhoti Baly, 1865g: 98 **A** : ID (SD) **ORR**

genus *Doryidomorpha* Laboissière, 1931b: 144 type species: *Doryidomorpha souyrisi* Laboissière, 1931
fulva Laboissière, 1931b: 146 **A:** CH (YUN)
nigripennis Laboissière, 1931b: 146 **A:** CH (SCH YUN)
souyrisi Laboissière, 1931b: 145 **A:** CH (YUN)
variabilis Laboissière, 1931b: 147 **A:** CH (YUN)

genus *Emathea* Baly, 1865k: 147 type species: *Emathea aeneipennis* Baly, 1865
punctata Allard, 1889f: 307 (*Sphaeroderma*) **A:** CH (GAN GUA GUX HAI SHA) **ORR**
amoena Weise, 1922b: 95 (*Dercetes*)
moseri Weise, 1922b: 98
subcaerulea Jacoby, 1891a: 33 (*Antipha*) **A:** CH (XIZ YUN) ID (AP) NP **ORR**
semicoerulea Jacoby, 1892b: 998 (*Antipha*, var.) [UN]
ulyssis Laboissière, 1932b: 138 (*Cyclatipha*)

genus *Furusawaia* Chûjô, 1962b: 105 type species: *Furusawaia yosonis* Chûjô, 1962
Yunnaniata Lopatin, 2009a: 8 type species: *Yunnaniata konstantinovi* Lopatin, 2009
continentalis Lopatin, 2008d: 839 **A:** CH (SCH YUN)

jungchani C.-F. Lee & Bezděk, 2021b: 126 **A:** CH (TAI)
konstantinovi Lopatin, 2009a: 10 (*Yunnaniata*) **A:** CH (SCH YUN) [# 717]
lui C.-F. Lee & Bezděk, 2021b: 131 **A:** CH (TAI)
tahsiangi C.-F. Lee & Bezděk, 2021b: 134 **A:** CH (TAI)
tsoui C.-F. Lee & Bezděk, 2021b: 138 **A:** CH (TAI)
yosonis Chûjô, 1962b: 108 **A:** CH (TAI)

genus *Gallerucida* Motschulsky, 1861a: 24 type species: *Gallerucida bifasciata* Motschulsky, 1860
Coptomesa Weise, 1912: 91 type species: *Galerucida maculata* Weise, 1912
Eustetha Baly, 1861b: 296 type species: *Eustetha flaviventris* Baly, 1861
Hylaspes Baly, 1865e: 436 type species: *Hylaspes longicornis* Baly, 1865
Melospila Baly, 1861b: 297 type species: *Melospila nigromaculata* Baly, 1861
Stethidea Baly, 1890a: 13 type species: *Doryida balyi* Duvivier, 1885
abdominalis Gressitt & Kimoto, 1963a: 719 **A:** CH (FUJ GUA GUX YUN)
aenea Laboissière, 1935b: 118 (*Galerucida*) **A:** CH (YUN)
aeneomicans Ogloblin, 1936: 361, 446 (*Galerucida*) **A:** CH (CHQ GAN SCH SHA XIZ)
aenescens Weise, 1889g: 626 (*Galerucida*) **A:** CH (GAN SCH)
apicalis Laboissière, 1935b: 124 (*Galerucida*) **A:** CH (HAI SCH YUN)
apicipennis Duvivier, 1885b: 395 (*Parastetha*) **A:** CH (GUX HAI) **ORR**
fulva Laboissière, 1935b: 129 (*Galerucida*)
apurva Maulik, 1936: 603 (*Hylaspes*) **A:** BT ID ("India") **ORR**
apurvella Yang, 1994b: 343, 347 **A:** CH (YUN)
asticha Yang, 1997: 898, 903 **A:** CH (CHQ SCH)
balyi Duvivier, 1885b: 394 (*Doryida*) **A:** CH (GUX) ID (AP SD) **ORR**
fulva Laboissière, 1931b: 135 (*Stethidea*)
maculata Laboissière, 1931b: 133 (*Stethidea*) [HN]
ornata Laboissière, 1932b: 132 (*Stethidea*) [RN]
basalis S.-H. Chen, 1992: 133, 136 **A:** CH (YUN)
bifasciata Motschulsky, 1861a: 24 **A:** CH (ANH CHQ FUJ GAN GUI GUX HEB HEI HEN HUB HUN JIA JIL JIX LIA SCH SHA SHX TAI TIA YUN ZHE) NC RU (ES FE) JA SC
consociata Baly, 1874: 185 (*Melospila*)
nigrita Chûjô, 1935d: 168 (*Galerucida*)
nigrofasciata Baly, 1879a: 453
nigromaculata Baly, 1861b: 297 (*Melospila*)
bimaculata Laboissière, 1935b: 122 (*Galerucida*) **A:** CH (YUN)
confusa Ogloblin, 1936: 354, 444 (*Galerucida*) **A:** CH (SCH)
cupreata Jacoby, 1890d: 193 (*Sphenoraia*) **A:** CH (HUB)

duporti Laboissière, 1931b: 138 (*Hylaspes*) **A:** CH (YUN) **ORR**
erythroptera Yang, 1992b: 584, 589 **A:** CH (GUI HUN)
facialis Laboissière, 1931b: 136 (*Hylaspes*) **A:** CH (YUN)
flava Ogloblin, 1936: 351, 444 (*Galerucida*) **A:** RU (FE)
flavipennis Solsky, 1871: 257 **A:** CH (FUJ HEI JIL LIA ZHE) JA NC RU (ES FE) SC
 melanocephala Jacoby, 1886a: 748 (*Sphenoraia*)
flaviventris Baly, 1861b: 296 (*Eustetha*) **A:** CH (ANH FUJ GUX HAI HKG HUN JIA JIX SCH TAI ZHE)
 chujoi C.-F. Lee, 2014a: 143 (*Sphenoraia*)
fortispina S.-Y. Xu & Yang, 2022: 42 **A:** CH (GUX)
furvofovea Yang, 1992b: 585, 589 **A:** CH (FUJ HUN)
gebieni Weise, 1922b: 92 (*Galerucida*) **A:** CH (FUJ HKG TAI)
gloriosa Baly, 1861b: 296 (*Eustetha*) **A:** CH (ANH CHQ FUJ GAN GUA GUI HEB HEI HEN HUB HUN JIA JIL JIX LIA SCH SHA ZHE) NC RU (ES FE) SC **ORR** **[# 725]**
 jacobsoni Ogloblin, 1936: 364, 446 (*Galerucida*)
 seriata Fairmaire, 1878b: 136 (*Eustetha*)
heilongjiangana Yang, 1994b: 344, 348 **A:** CH (HEI)
indica Harold, 1880a: 146 (*Galerucida*) **A:** ID (HP SD) NP **ORR**
laboissierei Bezděk & Beenen, 2010: 74 [RN] **A:** CH (FUJ YUN) **ORR**
 tenuicornis Laboissière, 1931b: 140 (*Hylaspes*) [HN]
 laboissierei Yang & Nie, 2015: 173 [RN, HN]
levifasciata S.-Y. Xu & Nie, 2022: 43 **A:** CH (GAN)
lewisi Jacoby, 1886a: 747 (*Galerucida*) **A:** JA
limbata Baly, 1879a: 462 (*Eustetha*) **A:** CH (XIZ) ID (AP) **ORR**
limbatella S.-H. Chen, 1992: 134, 137 **A:** CH (HEB NMO SHA SHX)
longicornis Baly, 1865e: 436 (*Hylaspes*) **A:** BT ID (SD) NP **ORR**
 assamensis Baly, 1879a: 460 (*Hylaspes*)
 dohrni Duvivier, 1885a: 246 (*Hylaspes*)
lutea Gressitt & Kimoto, 1963a: 724 **A:** CH (GUA HUB HUN TAI) SC
magica Harold, 1880a: 145 (*Galerucida*) **A:** "Himalayas"
magnipunctata Yang, 1992d: 411, 412 **A:** CH (ZHE)
moseri Weise, 1922b: 93 (*Galerucida*) **A:** CH (FUJ GUA GUX GUI HUB HUN JIX YUN) **ORR**
nigra H.-J. Wang, Li & Yang, 2000: 118, 120 (*Sphenoraia*) **A:** CH (GAN)
nigricornis Laboissière, 1940a: 24 (*Galerucida*) **A:** ID (SD) NP
 binotata Takizawa, 1988b: 541
 nigricornis Duvivier, 1892b: 442 (*Parasthetha*, var.) [HN]
nigrofoveolata Fairmaire, 1889b: 80 (*Eustetha*) **A:** CH (CHQ FUJ GAN HUB SCH SHA XIZ YUN ZHE)

nigropicta Fairmaire, 1888c: 40 (*Galerucida*) **A:** CH (CHQ GAN GUI HUB SCH XIZ YUN)

fulvicollis Laboissière, 1935b: 119 (*Galerucida*, ssp.)

nigropunctata Fairmaire, 1889b: 79 (*Eustetha*)

nigropunctatoides Mader, 1938: 58 (*Galerucida*) **A:** CH (YUN)

confluens Mader, 1938: 58 (*Galerucida*, ab.) [IN]

juncta Mader, 1938: 58 (*Galerucida*, ab.) [IN]

semijuncta Mader, 1938: 58 (*Galerucida*, ab.) [IN]

nigrovittata S.-Y. Xu & Yang, 2022: 45 **A:** CH (YUN)

nothornata Yang, 1994b: 344, 348 **A:** CH (SCH YUN)

octodecimpunctata S.-Y. Xu & Yang, 2022: 46 **A:** CH (YUN)

ornatipennis Duvivier, 1885b: 397 (*Hylaspes*) **A:** CH (GUI GUX SCH XIZ YUN ZHE) **ORR**

aeneicollis Mader, 1938: 57 (*Galerucida*, ab.) [IN]

annulipennis Fairmaire, 1889b: 79 (*Eustetha*)

decolora Laboissière, 1935b: 120 (*Galerucida*, var.) [IN] **[# 698]**

inornata Mader, 1938: 57 (*Galerucida*, ab.) [IN]

varians Allard, 1891: 233 (*Eusthetha*)

violacea Laboissière, 1935b: 120 (*Galerucida*, var.) [IN] **[# 698]**

oshimana Kimoto & Gressitt, 1966: 536 **A:** JA (Ryukyus)

pallida Laboissière, 1935b: 123 (*Galerucida*) **A:** CH (SCH SHA)

parva S.-H. Chen, 1992: 135, 137 **A:** CH (YUN)

pectoralis Laboissière, 1935b: 123 (*Galerucida*) **A:** CH (GUI YUN)

humeralis Laboissière, 1935b: 124 (*Galerucida*, var.) [IN] **[# 698]**

piceusfasciata S.-Y. Xu & Yang, 2022: 48 **A:** CH (SCH)

podontioides S.-H. Chen, 1942a: 37 (*Galerucida*) **A:** CH (GUX)

posticalis Laboissière, 1935b: 123 (*Galerucida*) **A:** CH (YUN)

postifusca Yang, 1994b: 345, 349 **A:** CH (FUJ ZHE)

puncticollis Fairmaire, 1888c: 42 (*Eustetha*) **A:** CH (CHQ SCH)

reflecta Laboissière, 1935b: 126 (*Galerucida*) **A:** CH (FUJ GUI)

reflexa Gressitt & Kimoto, 1963a: 729 **A:** CH (SCH)

rubrozonata Fairmaire, 1889b: 75 (*Galerucida*) **A:** CH (SCH YUN)

atronotata Fairmaire, 1889b: 76 (*Galerucida*, var.)

atrozonata Weise, 1924a: 141 (*Galerucida*, ab.) [IN]

rufipectoralis S.-Y. Xu & Nie, 2022: 50 **A:** CH (YUN)

rufometallica Gressitt & Kimoto, 1963a: 731 **A:** CH (XIZ YUN) **[# 726]**

sauteri Chûjô, 1938c: 141 **A:** CH (TAI)

quadraticollis Takizawa, 1978a: 127

serricornis Fairmaire, 1888c: 41 (*Eustetha*) **A:** CH (SCH SHA XIZ)

shirozui Kimoto, 1969: 67 **A:** CH (TAI)
singularis Harold, 1880a: 146 (*Galerucida*) **A:** BT CH (FUJ GUA GUI GUX HAI HKG SCH TAI YUN) ID (AP SD) "Himalayas" **ORR**
tonkinensis Laboissière, 1935b: 120 (*Galerucida*)
sinica Yang, 1994a: 203, 205 **A:** CH (GUI HUN ZHE)
solenocephala Yang, 1994b: 346, 349 **A:** CH (SCH)
speciosa Laboisière, 1935b: 127 (*Galerucida*) **A:** CH (YUN)
spectabilis Laboissière, 1935b: 125 (*Galerucida*) **A:** CH (SCH XIZ YUN)
submetallica Gressitt & Kimoto, 1963a: 733 **A:** CH (FUJ GUI HEN HUN NMO ZHE) MG
tenuefasciata Fairmaire, 1888c: 40 (*Galerucida*) **A:** CH (SCH XIZ YUN)
potanini Ogloblin, 1936: 358, 445 (*Galerucida*)
thoracica Jacoby, 1888: 348 (*Eustetha*) **A:** CH (FUJ HUB JIX TAI)
tibialis Laboissière, 1931b: 138 (*Hylaspes*) **A:** CH (GUX XIZ YUN)
tienmushana Gressitt & Kimoto, 1963a: 735 **A:** CH (ZHE)
tricincta Laboissière, 1935b: 116 (*Galerucida*) **A:** CH (YUN)
tricolor Gressitt & Kimoto, 1963a: 736 **A:** CH (YUN)
trinotata Gressitt & Kimoto, 1963a: 738 **A:** CH (GUI HUN SCH)
variolosa Laboissière, 1938: 4 (*Galerucida*) **A:** CH (FUJ JIA JIX ZHE)

genus *Haemodoryida* S.-H. Chen, 1942a: 35 type species: *Haemodoryida sanguinea* S.-H. Chen, 1942
sanguinea S.-H. Chen, 1942a: 35 **A:** CH (SCH)

genus *Hamushia* Chûjô, 1956c: 9 type species: *Galerucida eburata* Harold, 1879
eburata Harold, 1879b: 368 (*Galerucida*) **A:** JA
konishii Chûjô, 1956c: 12 **A:** JA

genus *Himaplosonyx* S.-H. Chen, 1976: 209, 220 type species: *Himaplosonyx apterus* S.-H. Chen, 1976
apterus S.-H. Chen, 1976: 209, 220 **A:** CH (XIZ)

genus *Hylaspoides* Duvivier, 1892b: 443 type species: *Hylaspoides magnifica* Duvivier, 1892
magnifica Duvivier, 1892b: 444 **A:** ID (SD)

genus *Laphris* Baly, 1864d: 231 type species: *Laphris emarginata* Baly, 1864
Neohylaspes Chûjô, 1962b: 159 type species: *Neohylaspes rufofulva* Chûjô, 1962
apophysata Yang, 1993c: 363, 368 **A:** CH (ZHE)

collaris Yang, 1992b: 584, 589 [= 1993c: 363, 368] **A:** CH (HUN ZHE)
emarginata emarginata Baly, 1864d: 231 **A:** CH (ANH FUJ GUA GUI GUX HUB HUN JIX SCH ZHE)
emarginata rufofulva Chûjô, 1962b: 163 (*Neohylaspes*) **A:** CH (TAI)
moganshana Yang, 1992d: 410, 412 **A:** CH (ZHE)
nigripennis Yang, 2004: 76, 77 **A:** CH (XIZ)
sexplagiata Laboissière, 1935b: 113 **A:** CH (GUX HAI SCH) **ORR**
maxima Gressitt & Kimoto, 1963a: 726 (*Gallerucida*)
tricuspidata Yang, 1993c: 364, 369 **A:** CH (CHQ GUX SCH)

genus *Leptarthra* Baly, 1861a: 202 type species: *Leptarthra abdominalis* Baly, 1861
abdominalis Baly, 1861a: 203 **A:** BT ID (KA SD) NP
fasciata Jacoby, 1894: 314 **A:** ID ("Northern India") NP **ORR**
jayarami Vazirani, 1970: 111 (*Merista*) **A:** CH (XIZ) ID (AP)
pici Laboissière, 1935b: 112 **A:** CH (YUN) **ORR**

genus *Meristata* E. Strand, 1935: 297 [RN] type species: *Galleruca sexmaculata* Kollar & L. Redtenbacher, 1844
Merista Chapuis, 1875: 228 [HN] type species: *Galleruca sexmaculata* Kollar & L. Redtenbacher, 1844
dohrnii Baly, 1861a: 203 (*Leptarthra*) **A:** BT CH (XIZ) ID (AP SD) NP **ORR**
cardoni Duvivier, 1892b: 440 (*Merista*)
rufipennis Harold, 1880a: 144 (*Merista*)
elongata Jacoby, 1898a: 190 (*Merista*) **A:** CH (XIZ)
fallax Harold, 1880a: 143 (*Merista*) **A:** BT CH (XIZ) ID (AP SD) NP PA **ORR**
fraternalis fraternalis Baly, 1879a: 455 (*Leptarthra*) **A:** CH (YUN) ID (AP) **ORR**
fraternalis yunnanensis Laboissière, 1922a: 101 (*Merista*) **A:** CH (YUN) **ORR**
yunnanensis Maulik, 1936: 146 (*Merista*, var.) [HN]
pulunini occidentalis L. N. Medvedev, 1999c: 186 (ssp.) **A:** NP
pulunini pulunini Bryant, 1952: 607 (*Merista*) **A:** CH (XIZ) NP
quadrifasciata Hope, 1831: 28 (*Galleruca*) **A:** BT CH (XIZ) ID (HP KA SD UT) NP PA **ORR**
interrupta Kollar & L. Redtenbacher, 1844: 553 (*Galleruca*)
sexmaculata grossepunctata Laboissière, 1933a: 963 (*Merista*, ssp.) **A:** ID (SD)
sexmaculata sexmaculata Kollar & L. Redtenbacher, 1844: 555 (*Galleruca*) **A:** BT CH (XIZ YUN) ID (AP KA SD UT) NP PA **ORR**
spilota Hope, 1831: 28 (*Galleruca*) **A:** BT CH (XIZ YUN) ID (AP SD UT) NP PA **ORR**
flaviventris Harold, 1880a: 143 (*Merista*)
trifasciata Hope, 1831: 28 (*Galleruca*) [HN]
variabilis Harold, 1880a: 142 (*Merista*)

genus *Meristoides* Laboissière, 1929b: 289 type species: *Merista touzalini* Laboissière, 1922

subgenus *Meristoides* Laboissière, 1929b: 289 type species: *Merista touzalini* Laboissière, 1922

grandipennis Fairmaire, 1889b: 77 (*Leptarthra*) **A:** CH (GAN HUB SCH TAI XIZ) **ORR**

intermedius Jacoby, 1890d: 194 (*Leptarthra*)

keani Laboissière, 1929b: 292 **A:** CH (YUN) **ORR**

oberthueri Jacoby, 1883b: 404 (*Merista*) **A:** CH (SCH XIZ YUN)

touzalini Laboissière, 1922a: 101 (*Merista*)

obscurus Lopatin, 2011b: 385 **A:** CH (SCH)

subgenus *Sichuanella* Lopatin, 2008b: 138 [RN] type species: *Meristoides vigintiguttatus* Ogloblin, 1936

Sichuania Lopatin, 2002f: 881 [HN] type species: *Meristoides vigintiguttatus* Ogloblin, 1936

rugosus Lopatin, 2011b: 385 **A:** CH (SCH)

vigintiguttatus Ogloblin, 1936: 337, 439 **A:** CH (SCH)

genus *Miltina* Chapuis, 1875: 172 type species: *Miltina dilatata* Chapuis, 1875

Liroegala L. N. Medvedev, 2015c: 325 type species: *Liroegala pectinicornis* L. N. Medvedev, 2015

dilatata Chapuis, 1875: 173 **A:** BT CH (GUI HAI JIX SCH XIZ YUN) ID (AP SD) NP **ORR**

balyi Jacoby, 1889b: 206 (*Bonesia*)

pectinicornis L. N. Medvedev, 2015c: 325 (*Liroegala*) **A:** CH (YUN)

genus *Morphosphaera* Baly, 1861b: 298 type species: *Morphosphaera maculicollis* Baly, 1861

albipennis Allard, 1889d: lxvii **A:** CH (GUI GUX YUN) ID (AP) NP **ORR** [# 732]

brunnea Maulik, 1936: 321

simplex Weise, 1922b: 91 (*Galerucida*)

sodalis S.-H. Chen, 1935d: 778

bimaculata Chûjô, 1938c: 136 **A:** CH (TAI)

caerulea Schönfeldt, 1890: 173 (*Adorium*, var.) **A:** CH (HKG) JA (Ryukyus)

caerulea Jacoby, 1896a: 8 [HN]

chrysomeloides Bates, 1866: 355 (*Adorium*) **A:** CH (TAI)

formosa Laboissière, 1930: 367

collaris Laboissière, 1930: 362 **A:** CH (GUX SCH) **ORR**

coomani Laboissière, 1930: 363 **A:** BT CH (GUI SCH XIZ YUN ZHE) ID (AP SD) NP **ORR**

margaritacea Laboissière, 1930: 365

prava Maulik, 1936: 320

viridipennis Laboissière, 1930: 366

japonica Hornstedt, 1787: 1 (*Chrysomela*) **A:** CH (CHQ FUJ GUI GUX HUB HUN JIL JIX SCH SHN TAI YUN ZHE) ID (SD) JA NP PA RU (ES FE) SC **ORR**
maculicollis Baly, 1861b: 298 **A:** CH (XIZ) NP **ORR**
gracilicornis S.-H. Chen, 1963: 453, 457
quadrinotata S.-H. Chen, 1976: 207, 219
montivaga Maulik, 1936: 319 **A:** ID (SD) **ORR**
sumatrana Jacoby, 1886d: 60 **A:** BT CH (FUJ GAN GUA GUI HUB HUN JIX SCH XIZ YUN) ID (UT) NP SC **ORR**
cavaleriei Laboissière, 1930: 363
cincticollis Laboissière, 1930: 364
gingkoae Gressitt & Kimoto, 1963a: 557
marginata Laboissière, 1930: 366
metallescens Gressitt & Kimoto, 1963a: 559
purpurea Laboissière, 1930: 362

genus *Nepalogaleruca* Kimoto, 1970c: 413 type species: *Nepalogaleruca elegans* Kimoto, 1970
angustilineata Kimoto & Takizawa, 1972: 216 (ssp.) **A:** NP
conformis S.-H. Chen & Jiang, 1987: 54, 58 **A:** CH (XIZ)
elegans Kimoto, 1970c: 413 **A:** CH (XIZ) NP
hartmanni L. N. Medvedev, 2003a: 320 **A:** NP
laeta L. N. Medvedev, 1990b: 15 **A:** NP
nigriventris S.-H. Chen & Jiang, 1987: 55, 58 **A:** CH (XIZ)
schmidti L. N. Medvedev & Sprecher-Uebersax, 1997: 232 **A:** NP

genus *Paraspitiella* S.-H. Chen & Jiang, 1981: 470, 486 type species: *Paraspitiella nigromaculata* S.-H. Chen & Jiang, 1981
maculata Bryant, 1954a: 414 (*Merista*) **A:** CH (XIZ) **ORR**
nigrinota Yang, 2004: 75, 76
nigromaculata S.-H. Chen & Jiang, 1981: 471, 487 **A:** CH (XIZ)

genus *Proegmena* Weise, 1889g: 628 type species: *Proegmena pallidipennis* Weise, 1889
bipunctata S.-H. Chen, 1942a: 41 **A:** CH (GAN)
impressicollis Jacoby, 1891b: 38 (*Antipha*) **A:** CH (GAN SCH XIZ)
aeneipennis Ogloblin, 1936: 327, 437
fossicollis Fairmaire, 1889b: 80 (*Cneorane*)
nepalensis Chûjô, 1966b: 24 **A:** NP
pallidipennis Weise, 1889g: 630 **A:** CH (CHQ FUJ GAN HUB JIA NIN SCH SHA XIZ ZHE)

crux S.-H. Chen, 1942a: 40
elongata Jacoby, 1890d: 197 (*Antipha*)
smaragdina Gressitt & Kimoto, 1963a: 690 **A:** CH (HKG SCH)

genus *Pseudeustetha* Jacoby, 1900a: 311 type species: *Pseudeustetha quadriplagiata* Jacoby, 1900
Avinasa Maulik, 1936: 456 type species: *Antipha hirsuta* Jacoby, 1891
hirsuta Jacoby, 1891a: 32 (*Antipha*) **A:** CH (YUN) **ORR**
pubescens Jacoby, 1892b: 971 (*Anthipha*)

genus *Sermylassa* Reitter, 1913c: 138 [RN] type species: *Chrysomela halensis* Linnaeus, 1767
Sermyla Chapuis, 1875: 224 [HN] type species: *Chrysomela halensis* Linnaeus, 1767
halensis Linnaeus, 1767: 589 (*Chrysomela*) **E:** AL AU BE BH BU BY CR CZ DE FR GB GE GR HU IT LA LS LT LU MD NL NM PL RU (CT ST) SB SK SL SP SV SZ UK **A:** AB AR GG IN KZ RU (SS WS) TR [# 737]
cuprina Weise, 1886a: 662 (*Sermyla*, var.)
nigricornis Fabricius, 1775: 119 (*Crioceris*)
picea Laboissière, 1911: 303 (*Sermyla*, var.)
viridis Geoffroy, 1785: 104 (*Galeruca*)

genus *Sermyloides* Jacoby, 1884b: 64 type species: *Sermyloides basalis* Jacoby, 1884
Praeochralea Duvivier, 1885a: 245 type species: *Praeochralea antennalis* Duvivier, 1885
baishanzuia Yang, 1995c: 261, 263 **A:** CH (ZHE)
biconcavus Yang, 1991b: 299, 303 **A:** CH (GUI)
bimaculatus Gressitt & Kimoto, 1963a: 640 **A:** CH (GUA)
bipunctatus Lei, Yang & Nie, 2020: 415 **A:** CH (CHQ)
biunciatus Yang, 1991b: 300, 303 **A:** CH (YUN)
coomani Laboissière, 1936: 257 **A:** CH (GUX SCH YUN) **ORR**
cribellatus Yang, 1991b: 300, 303 **A:** CH (SCH)
cuspidatus Yang, 1991b: 300, 304 **A:** CH (YUN)
decoratus S.-H. Chen, 1942a: 55 **A:** CH (GUX HAI)
inornatus S.-H. Chen, 1942a: 53 **A:** CH (HAI)
lii Yang & W.-Z. Li, 2002: 308, 309 **A:** CH (GUI)
nigripennis Gressitt & Kimoto, 1963a: 642 **A:** CH (FUJ GUX JIX)
pilifer Yang, 1991d: 350, 351 **A:** CH (GUX)
pilosus Yang, 1991b: 301, 304 **A:** CH (GUI)
semiornatus S.-H. Chen, 1942a: 54 **A:** CH (FUJ GUX HUB SCH SHA YUN) **ORR**
sexmaculatus Yang, 1991b: 301, 304 **A:** CH (GUX) **ORR**

shaanxiensis Nie & Yang, 2017: 161, 266 **A:** CH (SHA)
sichuanus L. N. Medvedev, 2013b: 295 **A:** CH (SCH)
sulcatus Yang, 1991b: 301, 305 **A:** CH (YUN)
umbonatus Yang, 1991b: 302, 305 **A:** CH (YUN)
varicolor S.-H. Chen, 1942a: 51 **A:** CH (FUJ GUX QIN SCH XIZ YUN)
wangi Yang, 1993e: 456, 457 **A:** CH (CHQ GUI HUB HUN)
wuyiensis Lei, Yang & Nie, 2020: 416 A: CH (FUJ)
yunnanensis Yang, 1991b: 302, 305 **A:** CH (YUN)

genus *Shamshera* Maulik, 1936: 452 type species: *Galleruca bennettii* Hope, 1831
bennettii Hope, 1831: 29 (*Galleruca*) **A:** ID ("India") NP **ORR**

genus *Sphenoraia* Clark, 1865b: 262 type species: *Sphenoraia nigripennis* Clark, 1865 **[# 688]**
Neosermylassa Chûjô, 1956c: 14 type species: *Sermylassa japonica* Chûjô, 1956 (= *Sphenoraia intermedia* Jacoby, 1886)
Sphenoraioides Laboissière, 1935b: 131 type species: *Galleruca fulgida* Kollar & L. Redtenbacher, 1844 (= *Eumolpus rutilans* Hope, 1831)
anjiensis Yang & W.-Z. Li, 1998: 132, 135 **A:** CH (ZHE)
berberii Jiang, 1992: 665, 673 **A:** CH (YUN)
decemmaculata Feng, Yang & Liu, 2022: 75 **A:** CH (SCH YUN)
duvivieri Laboissière, 1926a: 53 (*Galerucida*) [RN] **A:** CH (GUA GUI GUX HUN YUN) ID ("India") **ORR**
amala Maulik, 1936: 549 (*Gallerucida*)
indica Duvivier, 1887: xlviii [HN]
flavicollis Clark, 1865b: 262 **A:** ID (UT) **ORR**
flavomarginata Feng, Yang & Li, 2022: 77 **A:** CH (SCH)
haizhuensis Yang, 2021: 247 **A:** CH (GUA)
intermedia Jacoby, 1886a: 747 **A:** JA SC
japonica Chûjô, 1956c: 15 (*Sermylassa*)
lopatini Bezděk & Beenen, 2010: 74 [RN] **A:** NP
cyanella Lopatin, 2005c: 232 [HN]
micans Fairmaire, 1888c: 42 (*Eustetha*) **A:** CH (CHQ FUJ GUA GUI GUX HEN HUB HUN SCH TAI XIZ ZHE) **ORR**
coerulescens Weise, 1922b: 91 (*Galerucida*, ab.) [IN]
cyanella Laboissière, 1935b: 132 (var.) [IN] **[# 698]**
multimaculata Kimoto, 1982a: 21 **A:** BT
nebulosa Gyllenhal, 1808: 292 (*Galleruca*) **A:** CH (GUA GUX HAI HKG YUN) ID (SD) **ORR**

nigripennis Clark, 1865b: 262 **A:** ID (AP HP SD UT) NP PA **ORR** [# 686]
bombayana Harold, 1880a: 147 (*Galerucida*)
bicolor Hope, 1831: 29 (*Galleruca*) [HN] [# 686]
hopei Beenen, 2005a: 18 [RN] [# 686]
nigromaculata Jiang, 1992: 665, 674 **A:** CH (SCH YUN)
punctipennis Jiang, 1992: 666, 674 **A:** CH (XIZ)
rutilans Hope, 1831: 30 (*Eumolpus*) **A:** BT ID (AP HP KA SD UT) NP PA **ORR**
cyanea Allard, 1890: xcii
fulgida Kollar & L. Redtenbacher, 1844: 554 (*Galleruca*)
mutabilis Hope, 1831: 30 (*Chrysomela*)
trifasciata Kimoto & Takizawa, 1972: 221 **A:** NP
yajiangensis Jiang, 1992: 667, 674 **A:** CH (SCH)

genus *Sphenorella* L. N. Medvedev & Sprecher-Uebersax, 1998: 35 type species: *Sphenoraia maculata* Kimoto & Takizawa, 1972
maculata Kimoto & Takizawa, 1972: 220 (*Sphenoraia*) **A:** NP
apicalis Kimoto & Takizawa, 1983: 93 (*Sphenoraia*)

genus *Spitiella* Laboissière, 1931b: 150 type species: *Spitiella auriculata* Laboissière, 1931 (= *Leptarthra collaris* Baly, 1878)
collaris Baly, 1878h: 382 (*Leptarthra*) **A:** BT CH (TAI XIZ) ID (AP HP KA SD) NP PA [# 1145]
auriculata Laboissière, 1931b: 151
maculipennis Kimoto, 1977: 390 **A:** BT

genus *Yunaspes* S.-H. Chen, 1976: 210, 221 type species: *Yunaspes nigritarsis* S.-H. Chen, 1976
medogensis Jiang, 1988a: 392, 395 **A:** CH (XIZ) [# 697]
nigritarsis S.-H. Chen, 1976: 210, 221 **A:** CH (YUN)

tribe Luperini Gistel, 1848

subtribe Aulacophorina Chapuis, 1875

genus *Agetocera* Hope, 1841: 170 type species: *Aegelocerus mirabilis* Hope, 1831 [# 694]
Aegelocerus Hope, 1831: 29 [UN] [# 694]
abdominalis Jiang, 1992: 653, 671 **A:** CH (YUN)
biclava L.-J. Zhang & Yang 2005a: 119 **A:** CH (GUX)
carinicornis S.-H. Chen, 1964: 205, 210 **A:** CH (YUN) **ORR**
chapana Laboissière, 1929a: 260 **A:** CH (QIN) **ORR**
choui C.-F. Lee, Bezděk & Staines, 2010: 9 **A:** CH (TAI)

cyanipennis Yang, 2001: 130 **A:** CH (YUN)
deformicornis Laboissière, 1927: 45 **A:** CH (CHQ GAN GUI HAI HUB HUN JIX SCH YUN ZHE) **ORR**
discedens Weise, 1922b: 64 **A:** CH (TAI)
femoralis S.-H. Chen, 1942a: 24 **A:** CH (SCH XIZ YUN)
filicornis Laboissière, 1927: 49 **A:** CH (CHQ FUJ GAN GUA GUI GUX HUB HUN JIX SCH SHA XIZ YUN ZHE) **ORR**
flaviventris Jacoby, 1880: 788 **A:** BT ID (SD) NP **ORR**
heinigi Beenen, 2019: 6 **A:** CH (GUX)
hopei Baly, 1865e: 438 **A:** BT CH (GUX NMO XIZ YUN) ID (AP SD UT) NP PA **ORR**
pulchella Chapuis, 1875: pl. 125
huatungensis C.-F. Lee, Bezděk & Staines, 2010: 15 **A:** CH (TAI)
lobicornis Baly, 1865e: 437 **A:** ID (AP SD) NP **ORR**
mirabilis Hope, 1831: 29 (*Aegelocerus*) **A:** BT CH (ANH GUA GUI GUX HAI HKG JIA TAI XIZ YUN ZHE) ID (SD) NP **ORR**
heterocera L. Redtenbacher, 1868: 206 (*Aplosonyx*)
nigripennis Laboissière, 1927: 48 **A:** CH (YUN) **ORR**
parva S.-H. Chen, 1964: 204, 210 **A:** CH (ANH FUJ HUB SHA ZHE)
similis S.-H. Chen, 1997: 387, 391 **A:** CH (GUX HUN YUN)
taiwana Chûjô, 1962b: 59 **A:** CH (TAI)
yuae C.-F. Lee, Bezděk & Staines, 2010: 9 **A:** CH (TAI)
yunnana S.-H. Chen, 1964: 204, 210 **A:** CH (GUX YUN) **ORR**

genus *Asbecesta* Harold, 1877c: 110 type species: *Asbecesta cyanipennis* Harold, 1877
cyanipennis Harold, 1877c: 110 **N:** EG **A:** YE **AFR**
aeneipennis Baly, 1878b: 206 (*Aulacophora*)
rugosa Jacoby, 1895a: 186
viridipennis Chapuis, 1879: 21 (*Malacosoma*)
viridipennis Jacoby, 1886b: 929 (*Diacantha*) [HN]
signata Kirsch, 1866: 284 (*Rhaphidopalpa*) **E:** IT (Sicily) **N:** EG **A:** SA YE **AFR**
allardi Weise, 1910c: 39 (ab.) [IN]
senegalensis Allard, 1887: cc

genus *Aulacophora* Chevrolat, 1836: 378 type species: *Galeruca quadraria* Olivier, 1808
Acutipalpa Rosenhauer, 1856: 327 type species: *Galleruca foveicollis* P. H. Lucas, 1847
Cerania Weise, 1892i: 396 type species: *Aulacophora cornuta* Baly, 1879
Ceratia Chapuis, 1877a: c type species: *Aulacophora marginalis* Chapuis, 1876
Orthaulaca Weise, 1892i: 393 type species: *Galeruca similis* Olivier, 1808
Pachypalpa Weise, 1892i: 392 type species: *Galleruca luteicornis* Fabricius, 1801

Raphidopalpa Chevrolat, 1836: 378 type species: *Crioceris abdominalis* Fabricius, 1781
Rhaphidopalpa Rosenhauer, 1856: 325 type species: *Galleruca foveicollis* P. H. Lucas, 1847
Sphaerarthra Weise, 1892i: 396 type species: *Aulacophora cyanoptera* Boisduval, 1835
Triaplatys Fairmaire, 1877b: 186 type species: *Triaplatys quadripartita* Fairmaire, 1877

analis Weber, 1801: 55 (*Galleruca*) **A:** CH (TAI: Lanyu Island) **ORR**
quadraria Olivier, 1808: 626 (*Galeruca*)

apicipes Jacoby, 1896d: 138 **A:** CH (GUX YUN) **ORR**
nigripalpis S.-H. Chen & Kung, 1959: 378, 386
tenuicincta Jacoby, 1897b: 406

bhamoensis Jacoby, 1892b: 944 **A:** NP **ORR**

bhimtalensis Gangola, 1969: 55 **A:** ID (UT)

bicolor Weber, 1801: 56 (*Galleruca*) **A:** CH (GUA GUX HAI HUB JIX SCH TAI YUN) ID (AP SD) JA (Ryukyus) "Eastern Himalayas" **ORR**
haemorrhoa Fabricius, 1803: 53 (*Galleruca*) [RN]
hemichroa Gistel, 1857b: 540
semiopaca Jacoby, 1886d: 51
sexnotata Chapuis, 1877a: c
sexpunctata Olivier, 1808: 627 (*Galeruca*)

calva Anand & Cox, 1986: 88 **A:** OM PA YE **ORR**

carinicauda S.-H. Chen & Kung, 1959: 379, 386 **A:** CH (FUJ GUA HAI HUN YUN) NP **ORR**

coomani Laboissière, 1929a: 257 **A:** CH (FUJ GAN GUA GUI HUN XIZ YUN) **ORR**

cornuta Baly, 1879a: 445 **A:** CH (YUN) ID ("India") **AUS ORR**
bicornuta Allard, 1889a: 309
duvivieri Baly, 1886b: 24
robusta Duvivier, 1884a: 124

cruenta Fabricius, 1781: 130 (*Chrysomela*) **A:** CH (HAI) ID (SD) **ORR**
biguttata Allard, 1889a: 313
gestroi Jacoby, 1892b: 943
lata Baly, 1886b: 8
nigripeta Duvivier, 1885b: 389
quadripunctata Allard, 1889a: 314
tonkinensis Laboissière, 1929a: 255

excavata Baly, 1886b: 18 **A:** ID (UT) **ORR**

foveicollis P. H. Lucas, 1847a: pl. 44 [= 1849: 542] (*Galleruca*) **E:** AL CR BH FR (Corsica) GR IT ME PT RU (ST) SP **N:** EG **A:** AF CY ID (AP SD UT) IN IS NP OM PA SA SY TR YE **AFR ORR**
africana Weise, 1903b: 36 (*Rhaphidopalpa*)
nigriventris L. Redtenbacher, 1849: 9 (*Galleruca*)

frontalis Baly, 1888b: 181 **A:** CH (GUA HAI TAI YUN) **ORR**
impressa Fabricius, 1801a: 485 (*Galleruca*) **A:** NP PA **ORR**
bidentata Fabricius, 1803: 53 (*Galleruca*) [RN]
limbatipennis Jacoby, 1903a: 117 (*Orthaulaca*)
indica Gmelin, 1790: 1720 (*Cryptocephalus*) **A:** AF BT CH (ANH CHQ FUJ GAN GUA GUI GUX HAI HEB HEN HKG HUB HUN JIA JIX LIA NIN SCH SHA SHG SHN SHX TAI XIZ YUN ZHE) ID (HP KA SD UT) JA NC NP PA RU (ES) SC **AUS ORR PAC**
bengalensis Weise, 1892i: 394 (*Rhaphidopalpa*)
ceramensis Weise, 1892i: 394 (*Rhaphidopalpa*)
chinensis Weise, 1892i: 395 (*Rhaphidopalpa*)
femoralis Motschulsky, 1857: 37 (*Raphidopalpa*)
flavipes Jacoby, 1883a: 202 (*Rhaphidopalpa*)
niasiensis Weise, 1892i: 394 (*Rhaphidopalpa*)
pubescens Allard, 1889a: 306 (*Rhaphidopalpa*)
similis Olivier, 1808: 624 (*Galeruca*)
testacea Fabricius, 1787: 87 (*Crioceris*)
jacobyi Weise, 1924a: 10 (*Ceratia*) [RN] **A:** CH (HAI YUN) ID ("India") **ORR**
denticornis Jacoby, 1896d: 137 [HN]
kotoensis Chûjô, 1962b: 79 **A:** CH (TAI)
lewisii Baly, 1886b: 24 **A:** AF BT CH (ANH CHQ FUJ GAN GUA GUI GUX HAI HKG HUB HUN JIA JIX LIA NIN SCH SHA TAI XIZ YUN ZHE) ID (AP SD UT) JA NP PA **ORR**
cattigarensis Weise, 1892i: 397 (*Orthaulaca*)
intermedia Jacoby, 1892b: 942
loochooensis Chûjô, 1957c: 2 **A:** JA (Ryukyus)
melanocephala Jacoby, 1892b: 941 **A:** CH (YUN) **ORR**
naseemi Abdullah & Qureshi, 1968a: 411 **A:** PA
nigripennis nigripennis Motschulsky, 1857: 38 **A:** CH (ANH CHQ FUJ GAN GUA GUI GUX HAI HEB HEI HKG HUB HUN JIA JIX NIN SCH SHA SHN SHX TAI XIZ YUN ZHE) JA RU (ES FE) SC **ORR**
atripennis Hope, 1843a: 66 [= 1843b: 64] (*Galleruca*) [HN]
nigripennis nitidipennis Chûjô, 1935a: 82 **A:** JA
nigripennis ohnoi Bezděk & Beenen, 2010: 74 (ssp.) [RN] **A:** JA (Izu Islands)
insularis Ohno, 1963: 86, 88 (ssp.) [HN]
opacipennis Chûjô, 1962b: 98 **A:** CH (TAI)
palliata Schaller, 1783: 279 (*Chrysomela*) **A:** CH (GUA HAI HKG TAI YUN) ID (SD) **ORR**
tibialis Chapuis, 1877a: xcix **A:** CH (CHQ FUJ GUA GUI GUX HAI HUB HUN JIA SCH TAI XIZ YUN ZHE) ID (AP HP SD UT) NP PA **ORR**
almora Maulik, 1936: 170

dohrni Jacoby, 1900a: 285
ritsemae Duvivier, 1884a: 121
semifusca Jacoby, 1892b: 942
simplex Chûjô, 1962b: 78
terminata Jacoby, 1900a: 284
yunnanensis S.-H. Chen & Kung, 1959: 383, 387 **A:** CH (FUJ GUA GUI HAI HUB HUN SCH YUN) **ORR**

genus *Bacteriaspis* Weise, 1905g: 328 type species: *Aulacophora scutellata* Baly, 1878
arabica L. N. Medvedev, 2012e: 417 **A:** YE

genus *Diacantha* Chevrolat, 1836: 378 type species: *Galeruca unifasciata* Olivier, 1808
Corynopalpa Chevrolat, 1836: 375 type species: *Adorium fasciatum* Olivier, 1807
Idacantha Fairmaire, 1869: 252 type species: *Galeruca unifasciata* Olivier, 1808
Hyperacantha Chapuis, 1879: 18 type species: *Hyperacantha fenestrata* Chapuis, 1879
dubia Gahan, 1896: 460 **A:** YE

genus *Lamprocopa* Hincks, 1949: 615 [RN] type species: *Galeruca delata* Erichson, 1843
Copa Weise, 1892i: 396 [HN] type species: *Galeruca delata* Erichson, 1843
delata Erichson, 1843: 265 (*Galeruca*) **A:** YE **AFR**

genus *Leptaulaca* Weise, 1902e: 142 type species: *Leptaulaca basalis* Weise, 1902
vinula Erichson, 1842b: 10 [= 1859: 85] (*Rhaphidopalpa*) **A:** YE **AFR**
festiva Gerstaecker, 1862: 343 (*Galleruca*)
fissicollis J. Thomson, 1858: 218 (*Galleruca*)
formosa Chapuis, 1879: 16 (*Aulacophora*)
mombonensis Weise, 1902e: 143 (var.)
nigrotransversa J. Thomson, 1858: 217 (*Galleruca*)
nubila Laboissière, 1921: 58 (var.)
oculata Karsch, 1882: 398 (*Aulacophora*)
obscurata Weise, 1902e: 144 (var.)
particularis Weise, 1902e: 143 (var.)
tristis Laboissière, 1921: 58 (var.)

genus *Mimagitocera* Maulik, 1936: 453 type species: *Agetocera flava* Jacoby, 1904
flava Jacoby, 1904a: 394 (*Agetocera*) **A:** CH (XIZ) ID ("India") NP **ORR**

genus *Paragetocera* Laboissière, 1929a: 262 type species: *Paragetocera involuta* Laboissière, 1929
dilatipennis L.-J. Zhang & Yang, 2004: 291 **A:** CH (HUN SHA)

fasciata Gressitt & Kimoto, 1963a: 492 **A:** CH (YUN) **ORR**
flavipes S.-H. Chen, 1942a: 27 **A:** CH (CHQ GAN GUI HUB HUN SCH SHA SHX YUN ZHE)
involuta Laboissière, 1929a: 263 **A:** CH (GAN GUI HEN HUB SCH SHA TAI XIZ YUN XIZ ZHE)
costata Chûjô, 1962b: 90 (*Aulacophora*)
nigricollis L.-J. Zhang & Yang, 2004: 294 **A:** CH (GAN SHA)
nigrimarginalis Jiang, 1992: 653, 672 **A:** CH (YUN)
pallida S.-H. Chen, 1942a: 25 **A:** CH (HAI SCH)
parvula metasternalis S.-H. Chen, 1942a: 27 (ssp.) **A:** CH (SCH)
parvula parvula Laboissière, 1929a: 261 (*Agetocera*) **A:** CH (GAN GUI HEN HUB HUN SCH SHA XIZ YUN ZHE)
tibialis S.-H. Chen, 1942a: 28 **A:** CH (GUI HUB SCH SHA)
violaceipennis L.-J. Zhang & Yang, 2004: 299 **A:** CH (SCH ZHE)
yunnanica Jiang, 1992: 654, 672 **A:** CH (YUN)

genus *Paridea* Baly, 1886b: 26 type species: *Paridea thoracica* Baly, 1886
subgenus *Paridea* Baly, 1886b: 26 type species: *Paridea thoracica* Baly, 1886
Carapaula Chûjô, 1962b: 198 type species: *Aulacophora quadriplagiata* Baly, 1874 (= *Paridea oculata* Laboissière, 1930)
Neosastra Abdullah & Qureshi, 1968a: 413 type species: *Neosastra murreeiensis* Abdullah & Qureshi, 1968
approximata Duvivier, 1892b: 430 **A:** ID (SD)
bengalica L. N. Medvedev & Samoderzhenkov, 1998: 63 **A:** ID (SD)
brachycornuta Yang, 1993b: 196, 199 **A:** CH (FUJ)
brevis Gressitt & Kimoto, 1963a: 510 **A:** CH (HAI YUN)
circumdata Laboissière, 1930: 345 **A:** CH (YUN) **ORR**
cornuta basalis Laboissière, 1930: 344 **A:** CH (GUX YUN) **ORR**
cornuta cornuta Jacoby, 1892b: 954 **A:** CH (GUX YUN) **ORR**
costallifera Yang, 1991a: 283, 292 **A:** CH (SCH)
costata Chûjô, 1935d: 164 (*Paraulaca*) **A:** CH (GAN GUI HEB HUN JIA JIX SCH SHA TAI ZHE) **[# 734]**
cyanea Yang, 1992b: 564, 588 **A:** CH (GUI HUN)
cyanipennis Chûjô, 1935d: 164 (*Paraulaca*) **A:** CH (TAI)
darjeelingensis Takizawa, 1990a: 286 **A:** ID (SD)
dohertyi Maulik, 1936: 505 **A:** CH (YUN) ID (AP) **ORR**
assama Maulik, 1936: 506 (var.)
epipleuralis S.-H. Chen, 1942a: 34 **A:** CH (GAN SHA SHX)
euryptera Yang, 1991a: 283, 292 **A:** CH (XIZ)
fasciata Laboissière, 1933a: 970 **A:** CH (XIZ) ID (SD) NP

foveipennis Jacoby, 1892b: 956 **A**: CH (HAI SHA XIZ YUN) **ORR**
multituberculata L. N. Medvedev & Samoderzhenkov, 1989: 458
phymatodea Yang, 1991a: 285, 293
fujiana Yang, 1991a: 283, 292 **A**: CH (FUJ GUX)
fusca Yang, 1991a: 284, 293 **A**: CH (YUN)
hirtipes S.-H. Chen & Jiang, 1981: 478, 488 **A**: CH (XIZ) ID ("India") NP
livida Duvivier, 1892b: 432 **A**: BT ID (SD) NP **ORR**
luteofasciata Laboissière, 1930: 343 **A**: CH (YUN) **ORR**
mimica L. N. Medvedev & Samoderzhenkov, 1998: 64 **A**: NP
monticola Gressitt & Kimoto, 1963a: 512 **A**: CH (GAN GUI HUB SCH XIZ YUN)
nepalica L. N. Medvedev & Samoderzhenkov, 1998: 63 **A**: NP
nigra Yang, 1992b: 565, 588 **A**: CH (GUI HUN)
nigrimaculata Yang, 1991a: 285, 293 **A**: CH (XIZ)
nigripennis Jacoby, 1892b: 955 **A**: CH (YUN) **ORR**
octomaculata Baly, 1886b: 17 (*Aulacophora*) **A**: BT CH (SCH XIZ YUN) ID (AP HP SD UT) NP PA
bimaculata Laboissière, 1933a: 968
murreeiensis Abdullah & Qureshi, 1968a: 414 (*Neosastra*)
patkaiensis Maulik, 1936: 505 (var.)
oculata Laboissière, 1930: 346 **A**: CH (GUI XIZ YUN) ID (AP SD) SC
fasciata Maulik, 1936: 508 (var.)
quadriguttata S.-H. Chen & Jiang, 1988: 342 [RN]
quadriplagiata Jacoby, 1894: 327 (*Aulacophora*) [HN]
quinqueplagiata Maulik, 1936: 507 (var.)
pallida Bryant, 1954a: 419 **A**: BT NP **ORR**
perplexa Baly, 1879a: 447 (*Aulacophora*) **A**: CH (FUJ GAN GUX JIX YUN) ID (SD) NP **ORR**
harmandi Laboissière, 1930: 340
plauta Yang, 1991a: 286, 294 **A**: CH (CHQ SCH)
quadrimaculata Kimoto, 2004: 58 **A**: NP
quadriplagiata Baly, 1874: 186 (*Aulacophora*) **A**: CH (ANH FUJ GUA GUI GUX HUB HUN JIX SCH YUN ZHE) ID ("India") JA
verticalis Laboissière, 1930: 343
ruficollis Jacoby, 1892b: 953 **A**: BT ID (SD) NP **ORR**
sancta Yang, 1991a: 286, 294 **A**: CH (YUN)
sauteri Chûjô, 1935d: 166 (*Paraulaca*) **A**: CH (TAI)
sichuana Yang, 1991a: 287, 294 **A**: CH (SCH)
sikkimia Laboissière, 1933a: 969 **A**: CH (XIZ) ID (SD)
sinensis Laboissière, 1930: 342 **A**: CH (CHQ FUJ GAN GUI HUB HUN JIX SCH SHA YUN ZHE)

subviridis Laboissière, 1930: 341 **A:** CH (GUX HAI YUN) **ORR**

taiwana Chûjô, 1935d: 167 (*Paraulaca*) **A:** CH (TAI)

terminata Yang, 1991a: 288, 294 **A:** CH (YUN)

testacea Gressitt & Kimoto, 1963a: 515 **A:** CH (FUJ TAI ZHE)

flavipennis Chûjô, 1935d: 165 (*Paraulaca*) [HN]

formosana Yang, 1991a: 272 [RN]

tetraspilota Hope, 1831: 29 (*Galleruca*) **A:** CH (TAI XIZ YUN) ID (SD) NP **ORR**

balyi Jacoby, 1898a: 190

thoracica Baly, 1886b: 27

tuberculata Gressitt & Kimoto, 1963a: 516 **A:** CH (FUJ)

unifasciata Jacoby, 1892b: 957 **A:** CH (YUN) ID (SD) NP **ORR**

yunnana Yang, 1991a: 289, 295 **A:** CH (YUN)

subgenus *Paraulaca* Baly, 1888b: 168 type species: *Rhaphidopalpa angulicollis* Motschulsky, 1854

Aeropa Weise, 1889g: 621 type species: *Aeropa maculata* Weise, 1889

Semacia Fairmaire, 1889b: 82 type species: *Semacia biplagiata* Fairmaire, 1889

Semacianella Laboissière, 1930: 337 type species: *Semacianella coomani* Laboissière, 1930

angulicollis Motschulsky, 1854a: 50 (*Raphidopalpa*) **A:** CH (FUJ GAN GUI GUX HAI HEB HEI HUB HUN JIA JIL JIX SHA SCH ZHE) JA RU (FE) SC

nigrimarginata Yang, 1991a: 279, 191

nipponensis Laboissière, 1930: 335 (*Semacia*)

avicauda Laboissière, 1930: 331 (*Semacia*) **A:** CH (SCH SHA XIZ)

biplagiata Fairmaire, 1889b: 82 (*Semacia*) **A:** CH (FUJ GAN GUI HUB HUN SCH XIZ)

maculata Weise, 1889g: 622 (*Aeropa*)

crenata Yang, 1993b: 197, 199 **A:** CH (GUI HUN)

flavipennis Laboissière, 1930: 334 (*Semacia*) **A:** CH (FUJ GUI HUN YUN)

flavipoda Yang, 1991a: 277, 290 **A:** CH (XIZ)

glyphea Yang, 1993b: 197, 200 **A:** CH (YUN)

grandifolia Yang, 1991a: 278, 290 **A:** CH (YUN)

houjayi C.-F. Lee & Bezděk, 2014: 87 **A:** CH (TAI)

kaoi C.-F. Lee & Bezděk, 2014: 89 **A:** CH (TAI)

lateralis L. N. Medvedev & Samoderzhenkov, 1989: 456 **A:** CH (GUX YUN) NP **ORR**

biconvexa Yang, 1991a: 275, 290

libita Yang, 1991a: 278, 291 **A:** CH (CHQ FUJ HUB HUN ZHE)

nigricaudata Yang, 1991a: 279, 291 **A:** CH (YUN)

nigrocephala Laboissière, 1930: 332 (*Semacia*) **A:** CH (FUJ GUA YUN)

pectoralis Laboissière, 1930: 333 (*Semacia*) **A:** CH (GUX) **ORR**

recava Yang, 1991a: 281, 292 **A:** CH (YUN)

sexmaculata Laboissière, 1930: 336 (*Semacia*) **A:** CH (BEI FUJ GUA GUI GUX HAI HEB HUB HUN JIA JIX SHG TAI ZHE)

transversofasciata Laboissière, 1930: 335 (*Semacia*) **A:** CH (CHQ FUJ GAN GUI HUB HUN JIA SCH YUN ZHE)

genus *Paridea*, species unassigned to subgenera

fulva Kimoto, 1977: 356 **A:** BT

bhutanensis Kimoto, 1977: 357 **A:** BT

genus *Pseudocophora* Jacoby, 1884b: 69 type species: *Galleruca buquetii* Guérin-Méneville, 1838

bicolor Jacoby, 1887a: 111 **A:** CH (FUJ GAN GUA GUX HAI HUN JIX YUN ZHE) NP PA **ORR**

nitens Allard, 1887: cc

carinata Yang, 1991c: 308, 311 **A:** CH (YUN)

cochleata Yang, 1991c: 309, 312 **A:** CH (YUN)

femoralis Laboissière, 1940a: 15 **A:** CH (YUN) **ORR**

flava Allard, 1887: cc **A:** CH (HAI YUN) ID (SD) NP **ORR**

flaveola Baly, 1888b: 173

pectoralis Baly, 1888b: 174 **A:** CH (GUA GUI GUX HUN SCH XIZ YUN) ID (AP SD) NP **ORR**

uniplagiata yunnana S.-H. Chen, 1976: 207, 219 (ssp.) **A:** CH (YUN XIZ) **ORR**

subtribe Diabroticina Chapuis, 1875

genus *Diabrotica* Chevrolat, 1836: 380 type species: *Crioceris fucata* Fabricius, 1787

virgifera virgifera LeConte, 1868: 59 Ei: AL AU BE BH BU BY CR CZ FR GB GE GR HU IT KO MD ME NL PL RO RU (CT ST) SB SK SL SP SZ UK **NAR NTR**

filicornis Horn, 1893: 94

subtribe Luperina Gistel, 1848

genus *Abdullahius* Abdullah & Qureshi, 1968b: 423 type species: *Abdullahius shameemae* Abdullah & Qureshi, 1968

shameemae Abdullah & Qureshi, 1968b: 424 **A:** PA

genus *Afromaculepta* Hasenkamp & Wagner, 2000: 4 type species: *Monolepta frontalis* Chevrolat, 1836

decemmaculata Jacoby, 1886e: 127 (*Monolepta*) **A:** YE **AFR**

contaminata Weise, 1914: 265 (*Monolepta*)

estcourtiana Jacoby, 1899d: 372 (*Monolepta*)

nodieri Laboissière, 1920: 322 (*Monolepta*)

senegalensis Bryant, 1948: 61 (*Monolepta*)

genus *Agelopsis* Jacoby, 1896c: 301 type species: *Agelopsis coeruleus* Jacoby, 1896

Liroetinus Lopatin, 2004a: 621 type species: *Liroetis belousovi* Lopatin, 2004

Zizonia S.-H. Chen, 1976: 209, 220 type species: *Zizonia tibetana* S.-H. Chen, 1976

aeneoviridis Lopatin, 2004a: 618 (*Liroetis*) **A:** CH (SCH)

belousovi Lopatin, 2004a: 620 (*Liroetis*) **A:** CH (SCH YUN)

caeruleus Jacoby, 1896c: 301 **A:** ID (SD) NP **ORR**

konstantinovi Bezděk, 2020a: 236 **A:** CH (SCH SHA)

malaisei Bezděk, 2020a: 237 **A:** CH (YUN) **ORR**

spinipes Ogloblin, 1936: 212, 406 (*Liroetis*) **A:** CH (GAN SCH)

tibetanus S.-H. Chen, 1976: 209, 220 (*Zizonia*) **A:** CH (XIZ)

traxlerorum Bezděk, 2020a: 240 **A:** CH (YUN)

genus *Apteromicrus* S.-H. Chen & Jiang, 1981: 475, 488 type species: *Apteromicrus flavipes* S.-H. Chen & Jiang, 1981

flavipes S.-H. Chen & Jiang, 1981: 476, 488 **A:** CH (XIZ)

genus *Ashrafia* Abdullah & Qureshi, 1968c: 425 type species: *Ashrafia anwarullahi* Abdullah & Qureshi, 1968

anwarullahi Abdullah & Qureshi, 1968c: 426 **A:** PA

genus *Atrachya* Chevrolat, 1836: 377 type species: *Galleruca menetriesii* Faldermann, 1835

Cnecodes Motschulsky, 1858: 99 type species: *Cnecodes bisignatus* Mostchulsky, 1858

Iphidea Baly, 1865i: 127 type species: *Iphidea discrepens* Baly, 1865 (= *Galleruca menetriesii* Faldermann, 1835)

bipartita Jacoby, 1890c: 163 (*Luperodes*) **A:** CH (FUJ GUI GUX HUB HUN SCH SHA ZHE)

caudata S.-H. Chen, 1942a: 47 (*Luperodes*) **A:** CH (HUB SCH)

flavomaculata Chûjô, 1935b: 206 (*Luperodes*) **A:** JA (Ryukyus)

fokiensis Weise, 1922b: 104 (*Monolepta*) **A:** CH (FUJ)

haemodera S.-H. Chen, 1942a: 48 (*Luperodes*) **A:** CH (GUA GUX)

haemoptera S.-H. Chen, 1942a: 49 (*Luperodes*) **A:** CH (FUJ GUA GUI GUX HUN SCH XIZ)

kusamai Takizawa, 1985b: 109 **A:** ID (SD) NP

menetriesii Faldermann, 1835: 439 (*Galleruca*) **A:** CH (CHQ FUJ GAN GUA GUI GUX HEB HEI HUB HUN JIA JIL JIX LIA NIN NMO QIN SCH SHA SHX XIZ YUN ZHE) JA NC RU (ES FE) SC

discrepens Baly, 1865i: 127 (*Iphidea*)

insularis Weise, 1922b: 81 (*Luperodes*, var.)

nigripennis Motschulsky, 1860b: 232 (*Luperodes*)
praeusta Motschulsky, 1860b: 232 (*Luperodes*)
pedestris Gressitt & Kimoto, 1963a: 589 **A:** CH (YUN)
rubripennis Gressitt & Kimoto, 1963a: 590 **A:** CH (CHQ GUI HUN SCH SHA ZHE)
tricolor Gressitt & Kimoto, 1963a: 591 **A:** CH (FUJ HUN) **ORR**

genus *Brachyphora* Jacoby, 1890d: 195 type species: *Brachyphora nigrovittata* Jacoby, 1890
nigrovittata Jacoby, 1890d: 195 **A:** CH (CHQ FUJ GUA GUI GUX HUB HUN JIA JIX SCH SHA SHX XIZ ZHE) **ORR**
vittata L. N. Medvedev, 1999c: 183 **A:** NP

genus *Calomicrus* Dillwyn, 1829: 52 type species: *Crioceris circumfusa* Marsham, 1802
albanicus Csiki, 1940b: 273 (*Luperus*) **E:** AL NM
macedonicus Tomov, 1975: 188 (*Luperus*)
algiricus Weise, 1924a: 113 (*Luperus*) [RN] **N:** AG
bimaculatus Pic, 1907i: 170 (*Luperus*) [HN]
angorensis Pic, 1912d: 34 (*Luperus*) **A:** TR
annapurnae L. N. Medvedev & Sprecher-Uebersax, 1999: 367 **A:** NP PA
apicalis Demaison, 1891: cxciv **A:** CY SY TR
atrocephalus Reitter, 1895a: 88 (*Luperus*)
peyroni Pic, 1899a: 209 (*Luperus*)
arabicus Lopatin & Nesterova, 2006: 537 **A:** AE OM
atroviridis Lopatin, 2006c: 601 **A:** CH (SCH)
bispiniger bispiniger Israelson, 1969: 161 (*Luperus*) **N:** CI (La Palma)
bispiniger longicornis Israelson, 1980: 193 (ssp.) **N:** CI (Gomera)
brunneus L. N. Medvedev, 1992d: 15 **A:** NP
carbunculus Peyerimhoff, 1949: 294 (*Luperus*) **N:** MO
circumfusus Marsham, 1802: 227 (*Crioceris*) **E:** AU BE BH BU BY CR CZ FR GB GE HU IT LU MD NL NM PL PO PT SK SL SP SB SL SZ UK **A:** TR
bedeli Weise, 1924a: 114 (*Luperus*, ab.) [IN]
bipunctatus Laboissière, 1912a: 329 (*Luperus*, var.)
brassicae Panzer, 1794a: 18 (*Altica*)
luteovittatus Pic, 1934h: 27 (*Luperus*, var.) [IN] [# 698]
nigripennis Bedel, 1892a: 590 (*Luperus*) [HN]
nigrofasciatus Weise, 1886a: 600 (*Luperus*)
spartii Koch, 1803a: 76 (*Haltica*)
talensis Pic, 1912e: 43 (*Luperus*, var.)
collaris Kimoto, 1984: 54 **A:** CH (TAI)

cous Weise, 1889b: 64 (*Luperus*) E: GR (Kos)
darvazicus Lopatin, 1975b: 49 (*Luperus*) **A:** TD
descarpentriesi Codina Padilla, 1963c: 367 (*Luperus*) **N:** AG
deserticola Ogloblin, 1936: 251, 420 (*Luperus*) **A:** TM
doramasensis Vela & García Becerra, 1996: 147 **N:** CI
fallax Joannis, 1865: 123 (*Luperus*) **N:** AG TU [# 706]
notaticollis Pic, 1912e: 43 (*Luperus*, var.)
foveolatus Rosenhauer, 1856: 328 **E:** SP
ghilarovi Lopatin, 1988c: 66 **A:** TD
gularis gularis Gredler, 1857: 157 (*Luperus*) E: AU FR IT SL
gularis ruficeps J. Müller, 1948: 98 (*Luperus*, ssp.) E: CR
gussakovskyi Ogloblin, 1936: 252, 421 (*Luperus*) **A:** TD
hartmanni L. N. Medvedev, 1999c: 182 **A:** NP
hissaricus Ogloblin, 1936: 253, 421 (*Luperus*) **A:** TD
ibericus Vela, 2018: 5 **E:** PT SP
impressithorax Pic, 1898f: 311 (*Luperus*) E: SP
jungchangi C.-F. Lee & Beenen, 2012: 123 **A:** CH (TAI)
kaszabi Lopatin, 1963d: 357 (*Luperus*) **A:** AF IN
kocheri Codina Padilla, 1961b: 75 (*Luperus*) **N:** MO
koenigi Jacobson, 1897c: 216 (*Luperus*) **A:** TR
kasikoporanus Pic, 1912e: 42 (*Luperus*)
kurosawai Kimoto, 1969: 44 **A:** CH (TAI)
lethierryi Guillebeau, 1891: 293 (*Luperus*) **N:** AG
fallax Fairmaire, 1875a: 496 [HN]
lividus Joannis, 1865: 125 (*Luperus*) **A:** IN IS JO LE SY TR
heydeni Weise, 1900b: 287 (*Luperus*)
punctatissimus Fairmaire, 1884d: 176 (*Luperus*)
trabzonus Lopatin & Nesterova, 2013: 95
lopatini Bezděk, 2012b: 397 [RN] **A:** CH (YUN)
minutissimus Lopatin, 2009a: 11 [HN]
mainlingus S.-H. Chen & Jiang, 1981: 475, 487 **A:** CH (XIZ)
malkini Warchałowski, 1991a: 49 **A:** TR
mercurini Laboissière, 1917: 159 (*Luperus*) **N:** MO
minutissimus Kimoto, 1996: 36 **A:** CH (TAI)
moralesi Codina Padilla, 1963c: 371 (*Luperus*) **N:** MO
xauensis Codina Padilla, 1963c: 374 (var.) [IN]
nigritarsis Joannis, 1865: 122 (*Luperus*) **N:** AG MO TU
dissonus Codina Padilla, 1963c: 378 (var.) [IN]
nigrosuturalis L. N. Medvedev, 2013a: 37 **A:** CH (YUN)
nuristanicus L. N. Medvedev, 1985a: 375 **A:** AF PA
ochraceus Lopatin, 2002e: 86 **A:** CH (SCH)

ophthalmicus Ogloblin, 1936: 254, 422 (*Luperus*) **A:** AF IN OM SA
palii Lopatin, 1965b: 12 (*Luperus*) **A:** KI
pardoi Codina Padilla, 1961b: 70 (*Luperus*) **N:** MO
pallidipennis Codina Padilla, 1963c: 377 (var.) [IN]
patanicus Lopatin, 1966c: 368 (*Luperus*) **A:** AF
populi Lopatin, 1963d: 358 (*Luperus*) **A:** AF TD
porrectus Normand, 1937: 128 (*Luperus*) **N:** AG TU
prujai Codina Padilla, 1963c: 375 (*Luperus*) **N:** MO
quadrilineatus L. N. Medvedev, 2013a: 37 **A:** CH (YUN)
quercus Pic, 1895b: cxxx (*Luperus*) **N:** AG MO
rottenbergi Ragusa, 1874: 234 **E:** IT (Sicily)
setulosus Weise, 1886a: 592 (*Luperus*) **E:** SP **N:** AG MO TU
subglabratus Codina Padilla, 1963a: 73 (var.) [IN]
sordidus Kiesenwetter, 1873: 24 (*Luperus*) **E:** SP
sugonjaevi Lopatin, 1983a: 94 (*Luperus*) **A:** TM
suturalis Joannis, 1865: 119 (*Luperus*) **E:** SP PT **N:** MO
intermedius Codina Padilla, 1963c: 377 (ab.) [IN]
tripolitanus Pic, 1939e: 285 (*Luperus*, var.) **N:** AG LB TU
turcicus L. N. Medvedev, 1975a: 14 (*Luperus*) **A:** TR
vanharteni Lopatin, 2001b: 18 **A:** YE
velai Bezděk, 2013a: 368 **A:** IN
warchalowskii Lopatin, 2005d: 877 **A:** CH (GAN SCH)
wilcoxi Lopatin, 1984a: 90 **A:** IN
wollastoni Paiva, 1861: 210 **N:** CI
yushunicus S.-H. Chen & Jiang, 1981: 475, 488 **A:** CH (QIN)

genus *Cassena* Weise, 1892i: 389 type species: *Cassena celebensis* Weise, 1892
subgenus *Cassena* Weise, 1892i: 389 type species: *Cassena celebensis* Weise, 1892
Euphyma Baly, 1879a: 457 [HN] type species: *Euphyma collaris* Baly, 1879
Solenia Jacoby, 1886d: 87 [RN] type species: *Euphyma collaris* Baly, 1879
Solephyma Maulik, 1936: 329 [RN] type species: *Euphyma collaris* Baly, 1879
Taphinellina Maulik, 1936: 299 type species: *Taphinellina bengalensis* Maulik, 1936
alticoides Gressitt & Kimoto, 1963a: 658 (*Solephyma*) **A:** CH (HAI)
apicalis Bryant, 1954a: 423 (*Solephyma*) **A:** CH (FUJ GUA GUI HUB HUN ZHE) **ORR**
terminalis Gressitt & Kimoto, 1963a: 662 (*Solephyma*)
bicolor Gressitt & Kimoto, 1963a: 660 (*Solephyma*) **A:** CH (GUA HUB) [# 707]
collaris collaris Baly, 1879a: 457 (*Euphyma*) **A:** BT CH (GUA) NP “India” **ORR**
malayana L. N. Medvedev, 2019a: 167 (*Cneorane*)
tinkhami Gressitt & Kimoto, 1963a: 663 (*Solephyma*)
tonkinensis Weise, 1922b: 128

sasajii Kimoto, 1969: 55 **A:** CH (TAI)
tricolor Gressitt & Kimoto, 1963a: 665 (*Solephyma*) **A:** CH (GUA HAI HKG YUN)
subgenus *Nepalocassena* L. N. Medvedev, 2009c: 223 type species: *Cassena antennata* Takizawa, 1988
antennata Takizawa, 1988b: 546 **A:** NP
takizawai L. N. Medvedev, 2009c: 223 **A:** NP

genus *Cerophysa* Chevrolat, 1836: 379 type species: *Galleruca nodicornis* Wiedemann, 1823
Luperomorphella Chûjô, 1964: 306 type species: *Luperomorpha albomaculata* Chûjô, 1964
Oedicerus Kollar & L. Redtenbacher, 1844: 556 type species: *Oedicerus cyanipennis* Kollar & L. Redtenbacher, 1844
Ozomena Chevrolat, 1845: 5 type species: *Galeruca nodicornis* Wiedemann, 1823
Taumaceroides Lopatin, 2009a: 6 type species: *Taumaceroides sinicus* Lopatin, 2009
andrewesi Jacoby, 1904a: 396 **A:** NP **ORR**
biplagiata Duvivier, 1885b: 393 **A:** CH (FUJ GUA GUX HAI HKG SCH YUN ZHE) **ORR**
collaris S.-H. Chen, 1942a: 32 (ssp.)
ruficollis Weise, 1922b: 69 (ab.) [IN]
chujoi Kimoto, 1966c: 33 (*Hoplosaenidea*) [RN] **A:** CH (TAI)
taiwana Chûjô, 1965b: 93 (*Cerophysa*, ssp.) [HN]
coomani Laboissière, 1930: 351 **A:** CH (YUN) **ORR**
cyanea Kimoto, 1977: 358 (*Oedicerus*) **A:** BT **ORR**
cyanipennis Kollar & L. Redtenbacher, 1844: 557 (*Oedicerus*) **A:** ID (KA) NP PA "India"
darjeelingensis Takizawa & Basu, 1987: 280 **A:** ID (SD) NP
erberi Beenen, 2005b: 384 **A:** PA
gracilicornis Gressitt & Kimoto, 1963a: 523 (*Taumacera*) **A:** CH (YUN)
mandarensis Jacoby, 1904a: 397 **A:** NP **ORR**
mauliki Lopatin, 1962d: 1813 (*Hoplasomedia*) **A:** AF PA
pakistanica Kimoto, 2004: 59 (*Taumacera*)
miyatakei Kimoto & Gressitt, 1966: 533 (*Hoplosaenidea*) **A:** JA (Ryukyus)
monstrosa Jacoby, 1892b: 948 **A:** NP **ORR**
nigripennis Kimoto, 1966c: 34 (*Hoplosaenidea*) **A:** CH (TAI)
laboissierei Beenen, 2008: 65 [= 2010a: 74] [RN]
pulchella Laboissière, 1930: 348 **A:** CH (GUA GUX YUN) **ORR**
sinica Lopatin, 2009a: 7 (*Taumaceroides*) **A:** CH (YUN)
taiwana Chûjô, 1935d: 171 (*Aenidea*) **A:** CH (TAI)
tibialis Jacoby, 1886a: 750 (*Aenidea*) **A:** CH (TAI) JA
zhenzhuristi Ogloblin, 1936: 172, 399 **A:** CH (GUA GUI JIA SHN ZHE)

genus *Cerophysella* Laboissière, 1930: 352 type species: *Cerophysella tonkinensis* Laboissière, 1930

Choganía Laboissière, 1930: 355 type species: *Chogania duporti* Laboissière, 1930 (= *Phyllectrus viridipennis* Allard, 1889)

basalis Baly, 1874: 181 (*Aenidea*) **A:** CH (GUA GUX HAI HKG JIX TAI YUN) JA **ORR**

ceylanica Allard, 1889d: lxxxii (*Phyllectrus*)

tonkinensis Laboissière, 1930: 353

plagiata Laboissière, 1930: 354 **A:** CH (YUN) **ORR**

viridipennis Allard, 1889d: lxxxiii (*Phyllectrus*) **A:** CH (SCH YUN) NP **ORR**

duporti Laboissière, 1930: 356 (*Chogania*)

yunnana L. N. Medvedev, 2015c: 325 **A:** CH (YUN)

genus *Charaea* Baly, 1878h: 376 type species: *Charaea flaviventre* Baly, 1878 (= *Calomicrus balyi* L. N. Medvedev & Sprecher-Uebersax, 1998)

Cneorides Jacoby, 1896c: 281 type species: *Cneorides flaviventris* Jacoby, 1896 (= *Charaea jacobyi* Bezděk & Viswajyothi, 2019)

abdominale Jacoby, 1890c: 167 (*Cneorane*) **A:** CH (HUB)

aeneofuscum Weise, 1889g: 612 (*Luperus*) **A:** CH (GAN SCH XIZ YUN)

akkoae Chûjô, 1954c: 53 (*Calomicrus*) **A:** JA

amamiense Nakane & Kimoto, 1961a: 19 (*Calomicrus*) **A:** JA (Ryukyus)

balyi L. N. Medvedev & Sprecher-Uebersax, 1998: 32 (*Calomicrus*) [RN] **A:** NP PA

flaviventre Baly, 1878h: 376 (*Charaea*) [HN]

chinense Gressitt & Kimoto, 1963a: 600 (*Euluperus*) **A:** CH (YUN)

chujoi Nakane, 1958: 309 (*Calomicrus*) **A:** CH (CHQ FUJ GAN GUI HEN HUB HUN SCH ZHE) JA

concolor Ogloblin, 1936: 269, 424 (*Luperus*) **A:** CH (GAN SCH XIZ)

costatum Kimoto, 1996: 36 (*Exosoma*) **A:** CH (TAI)

cyaneum Jacoby, 1886a: 748 (*Monolepta*) **A:** JA

japonicum Fleischer, 1916b: 223 (*Luperus*)

diadematum Ogloblin, 1936: 268, 423 (*Luperus*) **A:** CH (XIZ) NC

flaviventre Motschulsky, 1861a: 26 (*Calomicrus*) **A:** CH (ANH CHQ FUJ GAN GUA GUI GUX HEI HUB HUN JIA JIL JIX SCH SHA TAI ZHE) JA NC RU (ES FE) SC

hainanicum Gressitt & Kimoto, 1963a: 571 (*Calomicrus*) **A:** CH (HAI)

haruoi Bezděk & C.-F. Lee, 2014: 31 **A:** CH (TAI)

houaphanicum Bezděk, 2017: 478 **A:** CH (GUA) **ORR**

houjayi Bezděk & C.-F. Lee, 2014: 35 **A:** CH (TAI)

hummeli Laboissière, 1934a: 2 (*Exosoma*) **A:** CH (GAN)

iniquum Weise, 1889g: 617 (*Luperus*) **A:** BT CH (CHQ GAN SCH XIZ) ID (KA SD UT) JA NP [# 708]

aconiticola Chûjô, 1959: 10 (ssp.)

kelloggi Gressitt & Kimoto, 1963a: 572 (*Calomicrus*) **A:** CH (FUJ GUA GUI HAI HKG TAI) **ORR** [# 709]
lineatum Weise, 1889g: 612 (*Luperus*) **A:** CH (SCH)
maatsingi Gressitt & Kimoto, 1963a: 564 (*Calomicrus*) **A:** CH (FUJ)
maxbarclayi Bezděk & C.-F. Lee, 2014: 30 **A:** CH (TAI)
mimicum L. N. Medvedev, 1998f: 172 (*Calomicrus*) **A:** CH (TAI)
minutum Joannis, 1865: 136 (*Luperus*) **A:** CH (CHQ HEI JIL SCH) MG NC RU (ES FE SS) SC
miyamotoi Kimoto, 1969: 42 (*Calomicrus*) **A:** CH (TAI)
takizawai L. N. Medvedev, 1998f: 172 (*Calomicrus*)
nigriventre Ogloblin, 1936: 268, 424 (*Luperus*) **A:** CH (GAN HEI HEN JIL SCH SHA SHN SHX)
nobyi Chûjô, 1954c: 55 (*Calomicrus*) **A:** JA
parvicolle Weise, 1889g: 614 (*Luperus*, var.) **A:** CH (GAN SCH XIZ)
pratti Jacoby, 1890c: 164 (*Luperus*) **A:** CH (HUB)
pseudoakkoae L. N. Medvedev, 1998f: 172 (*Exosoma*) **A:** CH (TAI)
pseudominutum Beenen & Warchałowski, 2010: 58 **A:** CH (HUB) NC RU (FE)
punctatolineatum Laboissière, 1934a: 4 (*Luperus*) **A:** CH (GAN SCH XIZ)
sasajii Kimoto, 1969: 46 (*Exosoma*) **A:** CH (TAI)
sensarmai Takizawa, 1986: 96 (*Taphinellina*) **A:** ID (UT)
shirozui Kimoto, 1969: 43 (*Calomicrus*) **A:** CH (TAI)
sikanga Gressitt & Kimoto, 1963a: 567 (*Exosoma*) **A:** CH (CHQ SCH)
aenescens Weise, 1889g: 613 (*Luperus*) [HN]
grahami Wilcox, 1973: 449 (*Taphinellina*) [RN]
taiwanum Kimoto, 1969: 45 (*Exosoma*) **A:** CH (TAI)
takagii Takizawa, 1988a: 15 (*Calomicrus*) **A:** ID (SD) NP **ORR**
tomokunii Takizawa, 1990a: 290 (*Exosoma*) **A:** ID (SD UT) NP [# 690, # 710]
yasudai Chûjô, 1966b: 11 (*Cynorta*) **A:** BT ID (SD) NP **ORR**
yunnanum Lopatin, 2009a: 10 (*Calomicrus*) **A:** CH (YUN)

genus *Chikatunolepta* Bezděk, 2020b: 12 type species: *Calomicrus buettikeri* L. N. Medvedev, 1996
buettikeri L. N. Medvedev, 1996: 251 (*Calomicrus*) **A:** SA YE
emir Lopatin, 2006a: 261 (*Calomicrus*) **A:** AE OM
millingeni Pic, 1915e: 18 (*Luperus*) **N:** AG **A:** IS OM SA YE
artificiosus Peyerimhoff, 1931: 111 (*Lyperodes*)
friedmani Lopatin, 2002b: 379 (*Calomicrus*)

genus *Chinochya* C.-F. Lee, 2020: 147 type species: *Monolepta sublata* Gressitt & Kimoto, 1963

sublata Gressitt & Kimoto, 1963a: 635 (*Monolepta*) **A:** CH (FUJ SCH YUN ZHE)

unifasciata Takizawa, 1978a: 132 (*Atrachya*) **A:** CH (TAI)

genus *Clerotilia* Jacoby, 1886a: 751 type species: *Clerotilia flavomarginata* Jacoby, 1886

bicolor Gressitt & Kimoto, 1963a: 480 **A:** CH (YUN)

capitata S.-H. Chen, 1942a: 23 (*Clerotillia*) **A:** CH (GUI HUN SCH)

cheni C.-F. Lee, 2018: 358 **A:** CH (TAI)

depressa Gressitt & Kimoto, 1963a: 481 **A:** CH (GUI)

flavomarginata Jacoby, 1886a: 751 **A:** CH (ZHE) JA

formosana Chûjô, 1963b: 391 (*Hoplasoma*) **A:** CH (TAI)

shirozui Kimoto, 1969: 57 (*Acroxena*)

fulvicollis L. N. Medvedev, 2019a: 167 **A:** CH (SCH)

jungchani C.-F. Lee, 2018: 365 **A:** CH (TAI)

meihuai C.-F. Lee, 2018: 368 **A:** CH (TAI)

sichuanica Lopatin, 2002e: 85 **A:** CH (SCH)

terminata S.-H. Chen, 1942a: 24 (*Clerotillia*) **A:** CH (GUX SHA)

tsoui C.-F. Lee, 2018: 370 **A:** CH (TAI)

wulaiensis C.-F. Lee, 2018: 373 **A:** CH (TAI)

yuae C.-F. Lee, 2018: 375 **A:** CH (TAI)

genus *Cneorane* Baly, 1865g: 97 type species: *Cneorane fulvicollis* Baly, 1865 (= *Galleruca rubricollis* Hope, 1831)

apicicornis Jacoby, 1890c: 166 **A:** CH (HUB)

bhutana L. N. Medvedev, 2011e: 363 **A:** BT

birmanica Jacoby, 1900b: 132 **A:** CH (GUA SCH YUN) NP **ORR**

braeti Duvivier, 1892b: 435 **A:** ID (SD) NP **ORR** [# 711]

cariosipennis Fairmaire, 1888c: 45 **A:** CH (GUA GUI GUX HAI HUB SCH SHA XIZ YUN) **ORR**

nigripennis Laboissière, 1922a: 102

coeruleipes S.-H. Chen & Jiang, 1976: 206, 219 **A:** CH (XIZ)

crassicornis Fairmaire, 1889b: 81 **A:** CH (FUJ SCH SHA XIZ YUN) **ORR**

cribratissima Fairmaire, 1888c: 44 **A:** CH (FUJ GAN GUI HUB SCH)

dilaticornis S.-H. Chen, 1964: 201, 208 **A:** CH (YUN)

dohertyi Maulik, 1936: 341 **A:** BT ID (AP) NP **ORR**

elegans Baly, 1874: 182 **A:** CH (ANH BEI FUJ GAN GUA GUI GUX HEB HEI HUB HUN JIA JIL JIX LIA SCH SHA SHX TAI ZHE) JA NC RU (ES FE) SC

cyanipennis Chûjô, 1938c: 135

rufipes Weise, 1889g: 620

episcopalis Fairmaire, 1889b: 81 **A:** CH (GUI)
feae Jacoby, 1892b: 947 **A:** NP **ORR**
rubyana Maulik, 1936: 343
fulvicornis Jacoby, 1889b: 209 **A:** CH (SCH YUN) **ORR**
hirsuta Kimoto & Takizawa, 1972: 219 **A:** NP
intermedia Fairmaire, 1889b: 81 **A:** CH (FUJ GUA GUI GUX HUB SCH SHA XIZ YUN)
manipurana Maulik, 1936: 343 **A:** NP **ORR**
medvedevi Beenen, 2008: 65 [= 2010a: 74] [RN] **A:** NP
nigripennis L. N. Medvedev, 2004d: 203 [HN]
minuta L. N. Medvedev, 1992d: 14 **A:** NP
orientalis Jacoby, 1892b: 946 **A:** CH (GUA GUI GUX HAI HUB SCH YUN) ID (SD) NP **ORR** [# 712]
rubricollis Hope, 1831: 29 (*Galleruca*) **A:** BT ID (AP SD) NP **ORR**
fulvicollis Baly, 1865g: 97
rufocaerulea Fairmaire, 1889a: 374 **A:** CH (HUB) **ORR**
rugulipennis Baly, 1886a: 27 (*Cneoraus*) **A:** BT CH (CHQ FUJ GUA GUI HAI HUB HUN SCH SHA TAI XIZ YUN) ID (AP HP SD UT) NP PA **ORR**
delatouchii Fairmaire, 1888c: 45
femoralis Jacoby, 1888: 350
formosana Weise, 1922b: 72
subcaerulescens Fairmaire, 1888c: 44 **A:** CH (FUJ GUA JIX SCH YUN) **ORR**
tibialis Chûjô, 1966b: 10 **A:** ID (SD) NP
variabilis Kimoto, 1977: 361 **A:** BT
varipes Jacoby, 1896c: 299 **A:** BT CH (XIZ) ID (SD UT) NP PA
bicolor Kollar & L. Redtenbacher, 1844: 554 (*Galleruca*) [HN]
violaceipennis Allard, 1889d: lxx **A:** CH (CHQ ANH FUJ GAN GUA GUI GUX HEB HEI HUB HUN JIA JIL JIX LIA NIN SCH SHA SHX TAI XIZ ZHE) MG NC RU (ES) SC
fokiensis Weise, 1922b: 71

genus *Cneoranidea* S.-H. Chen, 1942a: 31 type species: *Cneoranidea signatipes* S.-H. Chen, 1942
Neocrane Chûjô, 1962b: 207 type species: *Neocrane bicoloripes* Chûjô, 1962 (= *Cneoranidea signatipes* S.-H. Chen, 1942)
bicoloripes Chûjô, 1962b: 210 (*Neocrane*) **A:** CH (TAI)
coryli S.-H. Chen & Jiang, 1984: 83, 89 **A:** CH (ANH FUJ HUB HUN)
flammea Yang, 1993a: 337, 346 **A:** CH (GUI HUN)
hirta Yang, 1991e: 200, 205 **A:** CH (SCH)
melanocephala Yang, 1997: 877, 903 **A:** CH (HUB)

miroshnikovi L. N. Medvedev, 2005c: 288 (*Haplomela*) **A:** CH (SCH)
parasinica L.-J. Zhang & Yang, 2005b: 51 **A:** CH (FUJ ZHE) [# 713]
signatipes S.-H. Chen, 1942a: 31 **A:** CH (ANH FUJ GAN GUA GUI GUX HUB HUN JIX SCH ZHE)
sinica Yang, 1991e: 201, 205 **A:** CH (FUJ GUI HUN JIX ZHE)
wuyiensis Yang & J.-S. Wang, 1998: 261, 264 **A:** CH (FUJ)

genus *Cneorella* L. N. Medvedev & Dang, 1981: 629 type species: *Cneorella chapaensis* L. N. Medvedev & Dang, 1981
eurycollis Chûjô, 1965b: 93 (*Dercetis*) **A:** CH (TAI)
spuria Gressitt & Kimoto, 1963a: 575 (*Calomicrus*) **A:** CH (FUJ GUA HAI JIX)
suisapana Gressitt & Kimoto, 1963a: 577 (*Calomicrus*) **A:** CH (GUX HUB) **ORR**

genus *Doryscus* Jacoby, 1887a: 115 type species: *Doryscus testaceus* Jacoby, 1887
chujoi Takizawa, 1978a: 130 **A:** CH (TAI) SC **ORR**
indochinensis C.-F. Lee, 2017: 9 **A:** CH (YUN) **ORR**
kubani C.-F. Lee, 2017: 14 **A:** CH (YUN)
nepalensis C.-F. Lee, 2017: 21 **A:** BT ID (AP SD) NP
niger L. N. Medvedev & Sprecher-Uebersax, 1998: 34 **A:** NP
varians Gressitt & Kimoto, 1963a: 674 (*Trichobalya*) **A:** CH (FUJ YUN) **ORR**
  marginicollis Jiang, 1992: 649, 671
  nigricollis Jiang, 1992: 649, 671

genus *Epaenidea* Gressitt & Kimoto, 1963a: 684 type species: *Epaenidea subvirida* Gressitt & Kimoto, 1963
elegans Kimoto & Gressitt, 1966: 535 **A:** JA (Ryukyus)
subvirida Gressitt & Kimoto, 1963a: 684 **A:** CH (HAI)

genus *Epiluperodes* Kimoto & Gressitt, 1966: 531 type species: *Epiluperodes ryukyuana* Kimoto & Gressitt, 1966
ryukyuana Kimoto & Gressitt, 1966: 531 **A:** JA (Ryukyus)

genus *Erganoides* Jacoby, 1903a: 125 type species: *Erganoides flavicollis* Jacoby, 1903
capito Weise, 1889g: 610 (*Luperus*) **A:** CH (GAN HUB SCH SHA)
  mundulus Weise, 1889g: 610 (*Luperus*, var.)
collaris Kimoto, 1977: 370 **A:** BT
discalis Gressitt & Kimoto, 1963a: 593 **A:** CH (SCH)
ictericus Weise, 1889g: 609 (*Luperus*) **A:** CH (GAN SCH) [# 691]
  unicolor Ogloblin, 1936: 256, 423 (*Luperus*, ab.) [IN]

occipitalis Laboissière, 1940b: 2 **A:** CH (HUB ZHE)
pallens Beenen, 2010c: 259 **A:** CH (SCH)
punctulatus Weise, 1889g: 614 (*Luperus*) **A:** CH (GAN HUB SCH XIZ)
hedini Laboissière, 1934a: 5
similis S.-H. Chen, 1942a: 46 **A:** CH (GAN SCH XIZ) [# 714]
suturalis Gressitt & Kimoto, 1963a: 595 **A:** CH (YUN)
tibialis S.-H. Chen, 1942a: 46 **A:** CH (FUJ GUX HAI)
tsoui C.-F. Lee & Beenen, 2010: 261 **A:** CH (TAI)
variabilis Gressitt & Kimoto, 1963a: 596 **A:** BT CH (ANH GUA GUI HAI SCH YUN)

genus *Euliroetis* Ogloblin, 1936: 197, 403 type species: *Aenidea ornata* Baly, 1874
abdominalis Baly, 1874: 180 (*Aenidea*) **A:** JA
lameyi Laboissière, 1929a: 278 (*Liroetis*) **A:** CH (FUJ HUN ZHE) RU (FE) **ORR**
obscuripes Ogloblin, 1936: 200, 405 (var.) [IN] [# 698]
melanocephala Bowditch, 1925: 246 (*Hoplosoma*) **A:** CH (FUJ GUA GUX HUN JIX ZHE)
nigrinota Gressitt & Kimoto, 1963a: 503 **A:** CH (FUJ ZHE)
nigripes Baly, 1874: 182 (*Aenidea*) **A:** JA
ornata Baly, 1874: 180 (*Aenidea*) **A:** CH (FUJ GUA GUI GUX HEI HUN JIA JIL LIA SCH SHA XIZ ZHE) JA RU (ES FE) SC
abdominalis Laboissière, 1929a: 278 (*Liroetis*) [HN]
ornata Jacoby, 1888: 349 (*Phyllobrotica*) [HN]
simulonigrinota Yang, 1992b: 563, 588 **A:** CH (HUN)
suturalis Laboissière, 1929a: 279 (*Liroetis*) **A:** CH (CHQ FUJ GAN GUI HUB HUN JIA SCH SHA XIZ YUN)

genus *Euluperus* Weise, 1886a: 662 type species: *Crioceris xanthopus* Duftschmid, 1825
azureus Fairmaire, 1884d: 176 (*Luperus*) **A:** IS LE SY TR
hermonensis Lopatin, 1997b: 376
cyaneus Joannis, 1865: 129 (*Luperus*) **E:** AL BH CR GR ME NM SB
costalis Joannis, 1865: 134 (*Luperus*)
espanoli Codina Padilla, 1963c: 368 (*Luperus*) **E:** SP
major Weise, 1886a: 665 (var.) **E:** BU HU NM RO SB **A:** TR
nairicus Lopatin, 1990d: 51 (*Calomicrus*) **A:** AR IN TR
napolovi L. N. Medvedev, 2012d: 504 (*Sceloluperus*) **A:** KZ
pinicola Duftschmid, 1825: 234 (*Crioceris*) **E:** AL AU BE BH BU BY CR CZ EN FI FR GE GR HU IT LA LS LT LU ME NL PL RO RU (CT NT) SK SL SP SV SZ UK **A:** IN TR [# 715]
concolor Apfelbeck, 1912: 259 (*Luperus*, var.)
silvestris Weise, 1886a: 601 (*Luperus*, var.)

pseudoaltaicus L. N. Medvedev, 1975a: 18 (*Luperus*) **A:** GG IN [# 716]
xanthopus virescens Weise, 1886a: 665 (var.) **A:** AB
fuscipes Weise, 1886a: 665 (var.)
xanthopus xanthopus Duftschmid, 1825: 234 (*Crioceris*) **E:** AU BU FR HU KR MD NM PL RU (ST) SB SK UK **A:** AB GG KZ TR
chevrolati Joannis, 1865: 128 (*Luperus*)
fossulatus Joannis, 1865: 131 (*Luperus*)
tritici Zabarinskiy, 1891: 59 (*Phyllobrotica*)
violaceus Harold, 1876b: 174 (*Luperus*) [RN]

genus *Eumelepta* Jacoby, 1892b: 988 type species: *Eumelepta biplagiata* Jacoby, 1892
biplagiata Jacoby, 1892b: 988 **A:** CH (YUN) NP **ORR**

genus *Exosoma* Jacoby, 1903b: 25 [RN] type species: *Chrysomela lusitanica* Linnaeus, 1767
Malacodora Bedel, 1905: 236 [RN] type species: *Chrysomela lusitanica* Linnaeus, 1767
Malacoptera Hope, 1841: 168 type species: *Chrysomela lusitanica* Linnaeus, 1767 **[Appendix]**
Malacosoma Chevrolat, 1836: 379 [HN] type species: *Chrysomela lusitanica* Linnaeus, 1767
albanicum Meschnigg, 1934: 68 (*Malacosoma*) **E:** AL
collare Hummel, 1825: 69 (*Luperus*) **E:** BY RU (CT ST) UK **A:** AB GG KZ SY RU (WS)
cyanopterum Krynicki, 1832: 166 (*Galleruca*)
lepidum Gebler, 1848: 14 (*Malacosoma*)
luteicolle Gebler, 1830: 219 (*Galleruca*)
russicum Pic, 1915e: 18 (var.)
triumphans Faldermann, 1837: 335 (*Malocosoma*)
flavipes Heyden, 1878: 336 (*Malacosoma*) **A:** AB AR IN TR
gaudionis Reiche, 1862b: 545 (*Malacosoma*) **E:** AL BU GR NM **A:** TR
grossepunctatum Roubal, 1931b: 454 (var.)
lusitanicum Linnaeus, 1767: 1066 (*Chrysomela*) **E:** FR IT PT SP SZ **N:** AG MO TU **A:** IN [# 674]
abdominale Schoenherr, 1808: 272 (*Crioceris*) [RN]
haemorrhoa Thunberg, 1827: 7 (*Crioceris*)
hosti Laboissière, 1912c: 20 (var.)
nigripes Olivier, 1792: 588 (*Galeruca*)
testaceum Fabricius, 1775: 116 (*Cistela*)
theryi Guillebeau, 1897a: 166 (*Malacosoma*)
neglectum Mohr, 1968c: 241 **A:** SY TR
thoracicum L. Redtenbacher, 1843: 989 (*Galleruca*) **A:** CY IN IS JO SY TR

genus *Falsoexosoma* Pic, 1926e: 10 type species: *Exosoma cyanipenne* Reitter, 1902
cyanipenne Reitter, 1902c: 217 (*Malacosoma*) **A:** AB IN TM
diversipenne Pic, 1915e: 18 (var.)

genus *Fleutiauxia* Laboissière, 1933b: 53 type species: *Fleutiauxia cyanipennis* Laboisièrre, 1933
armata Baly, 1874: 179 (*Aenidea*) **A:** CH (CHQ GAN HEI HEN HUB HUN JIL LIA SCH SHA ZHE) JA NC RU (ES FE) SC
koltzei Heyden, 1895: 204 (*Aenidea*)
bicavifrons Gressitt & Kimoto, 1963a: 527 **A:** CH (HUB)
chinensis Maulik, 1933: 564 (*Platyxantha*) **A:** CH (GUA GUI HEN HUN JIA SCH ZHE)
mutifrons Gressitt & Kimoto, 1963a: 528
flavida costata Yang & W.-Z. Li, 1998: 129, 132 (ssp.) **A:** CH (ZHE)
flavida flavida Yang, 1993d: 221, 225 **A:** CH (CHQ GAN SHA)
fuscialata Yang, 1993d: 222, 226 **A:** CH (YUN)
glossophylla Yang, 1997: 875, 903 **A:** CH (CHQ HUB SCH SHA)
rufipennis Yang, 1993d: 222, 226 **A:** CH (SCH)
septentrionalis Weise, 1922b: 82 (*Prasyptera*) **A:** CH (FUJ GUA HAI)
nigricornis Laboissière, 1933b: 55
yuae Yang, 1993d: 223, 227 **A:** CH (HEB)

genus *Galerudolphia* Hincks, 1949: 615 [RN] type species: *Rudolphia pallida* Jacoby, 1899
Rudolphia Jacoby, 1899a: 534 [HN] type species: *Rudolphia pallida* Jacoby, 1899
arabica L. N. Medvedev, 1996: 252 (*Monolepta*) **A:** SA YE

genus *Haplosomoides* Duvivier, 1890: xxxiv type species: *Rhaphidopalpa serena* Boheman, 1859
Hoplasomedia Maulik, 1936: 493 type species: *Hoplasomedia chinmatra* Maulik, 1936
abdominalis Kimoto, 1984: 48 **A:** CH (TAI)
annamita annamita Allard, 1889a: 330 (*Haplosoma*) **A:** BT CH (FUJ GUA GUX HKG JIA SCH XIZ YUN ZHE) ID (SD) NP **ORR**
egena Weise, 1922b: 74 (*Hoplasomoides*)
annamita occidentalis Gressitt & Kimoto, 1963a: 519 (ssp.) **A:** CH (SCH XIZ)
appendiculata Laboissière, 1930: 328 **A:** CH (HAI GUX XIZ) ID (AP) **ORR**
hainana Jiang, 1988b: 180, 182
sarata Maulik, 1936: 496 (*Hoplasomedia*)
arunachalensis Basu, 1985: 206 (*Hoplasomedia*) **A:** ID (AP)
brushei Jiang, 1988b: 178, 182 **A:** CH (YUN)

changi C.-F. Lee, Bezděk & Staines, 2011: 125 **A:** CH (TAI)
chengi C.-F. Lee, Bezděk & Staines, 2011: 119 **A:** CH (TAI)
costata Baly, 1878j: 415 (*Mimastra*) **A:** CH (CHQ FUJ GAN GUA GUI GUX HAI HKG HUB HUN JIX SCH SHX TAI XIZ ZHE) **ORR**
flava Laboissière, 1930: 326 **A:** CH (ZHE) **ORR**
krishila Maulik, 1936: 497 (*Hoplasomedia*) **A:** ID (SD) NP **ORR**
laticornis Laboissière, 1930: 329 **A:** CH (YUN) **ORR**
chinmatra Maulik, 1936: 494 (*Hoplasomedia*)
medvedevi Bezděk & L.-J. Zhang, 2007: 190 [RN] **A:** CH (YUN)
nigricollis L. N. Medvedev, 2000e: 28 [HN]
miyamotoi Kimoto, 1965: 371 **A:** JA (Ryukyus)
pusilla Laboissière, 1930: 327 **A:** CH (SCH YUN) ID (SD) **ORR**
krisha Maulik, 1936: 496 (*Hoplasomedia*)
rasha Maulik, 1936: 497 (*Hoplasomedia*) **A:** BT ID (AP) NP **ORR**
nirada Maulik, 1936: 498 (*Hoplasomedia*)
ryukyuensis C.-F. Lee, Bezděk & Staines, 2011: 125 **A:** JA (Ryukyus)
shirozui Kimoto, 1969: 37 (*Phyllobrotica*) **A:** CH (TAI)
tsoui C.-F. Lee, Bezděk & Staines, 2011: 135 **A:** CH (TAI)
ustulata Laboissière, 1938: 2 **A:** CH (HUN ZHE)
verticalis Jiang, 1988b: 178, 182 **A:** CH (GUI HUB SCH)
occipitalis L. N. Medvedev, 2000e: 29

genus *Hesperomorpha* Ogloblin, 1936: 298, 429 type species: *Luperus hirsutus* Jacoby, 1886
antennalis Chûjô, 1936b: 87 (*Hespera*) **A:** CH (TAI)
atra S.-H. Chen, 1942a: 63 (*Luperus*) **A:** CH (HEB)
collaris Kimoto, 1996: 34 **A:** CH (TAI)
hirsuta Jacoby, 1886a: 742 (*Luperus*) **A:** CH (HUB JIX) JA NP **ORR**
potanini Ogloblin, 1936: 300, 430 **A:** CH (SCH SHA XIZ)
taiwana Kimoto, 1996: 35 **A:** CH (TAI)
viridimicans Beenen, 2019: 8 **A:** CH (YUN)

genus *Hesperopenna* L. N. Medvedev & Dang, 1981: 634 type species: *Hesperopenna flava* L. N. Medvedev & Dang, 1981 (= *Hesperopenna medvedevi* Bezděk, 2013)
Levnma Özdikmen, 2008: 644 [RN] type species: *Martinella merkli* L. N. Medvedev, 2000 (= *Liroetiella tibialis* Kimoto, 1989)
Liroetiella Kimoto, 1989a: 106 type species: *Liroetiella tibialis* Kimoto, 1989
Martinella L. N. Medvedev, 2000f: 166 [HN] type species: *Martinella merkli* L. N. Medvedev, 2000 (= *Liroetiella tibialis* Kimoto, 1989)

arnoldi Bezděk, 2013b: 729 **A:** CH (YUN) **ORR**
brancuccii L. N. Medvedev, 2007c: 295 (*Liroetis*) **A:** NP **ORR**
helferi Bezděk, 2013b: 734
fulva Kimoto, 1977: 368 (*Calomicrus*) **A:** BT
thailandica Kimoto, 1989a: 100 (*Luperus*) **A:** CH (YUN) **ORR**
panfilovi L. N. Medvedev, 2012c: 253 (*Luperus*)

genus *Hoplasoma* Jacoby, 1884d: 233 type species: *Hoplasoma apicale* Jacoby, 1884 **[# 744]**
Paraulacophora Csiki, 1953: 131 type species: *Aulacophora pannonica* Csiki, 1953 (= *Galleruca unicolor* Illiger, 1800)
Haplomela S.-H. Chen, 1942a: 28 type species: *Haplomela semiopaca* S.-H. Chen, 1942
carinatum Kimoto, 1977: 355 **A:** BT ID (SD) NP
costatipenne Jacoby, 1896c: 273 (*Haplosoma*) **A:** ID (SD) NP **ORR**
furcatum L. N. Medvedev, 2000b: 126 **A:** ID (SD) NP **ORR**
majorinum Laboissière, 1929a: 258 **A:** CH (FUJ GAN GUA GUI GUX SCH TAI XIZ YUN ZHE) ID (AP SD) **ORR**
cheni L. N. Medvedev, 2004c: 330 (*Haplomela*)
semiopaca S.-H. Chen, 1942a: 28 (*Haplomela*)
minus Gressitt & Kimoto, 1963a: 497 **A:** CH (SCH)
biclavatum Jiang, 1988b: 177, 181 (*Haplosomoides*)
nigricolle Jiang, 1988b: 177, 181 (*Haplosomoides*) **A:** CH (SCH)
sexmaculatum Hope, 1831: 29 (*Auchenia*) **A:** BT CH (SCH XIZ YUN) ID (HP KA SD UT) NP PA **ORR**
bifasciatum Allard, 1889a: 327 (*Haplosoma*)
bistrimaculatum L. N. Medvedev, 2015a: 477 (*Mimastra*)
quadrinotatum Gressitt & Kimoto, 1963a: 540 (*Mimastra*)
sexmaculatum Jacoby, 1899c: 82 (*Haplosoma*)
unicolor Illiger, 1800: 135 (*Galleruca*) **A:** BT CH (FUJ GUA GUX HAI HKG YUN) ID (AP HP SD UT) NP **ORR**
abdominale Jacoby, 1896c: 272 (*Haplosoma*)
corniculatum Allard, 1889a: 328 (*Haplosoma*)
pannonicum Csiki, 1953: 132 (*Aulacophora*)

genus *Hoplosaenidea* Laboissière, 1933b: 59 type species: *Hoplosaenidea touzalini* Laboissière, 1933
Cynorita Laboissière, 1940b: 28 type species: *Cynorta citrina* Jacoby, 1894
Cynorta Baly, 1865f: 249 [HN] type species: *Cynorta porrecta* Baly, 1865
Cynortana E. Strand, 1942: 392 [RN] type species: *Cynorta porrecta* Baly, 1865
Diaphaenidea Laboissière, 1933b: 61 type species: *Diaphaenidea aerosa* Laboissière, 1933
Micraenidea Laboissière, 1933b: 64 type species: *Micraenidea coomani* Laboissière, 1933

aerosa Laboissière, 1933b: 62 (*Diaphaenidea*) **A:** CH (FUJ GUI JIX SHN YUN) **ORR**
bicolor Gressitt & Kimoto, 1963a: 522 (*Taumacera*) **A:** CH (SHA)
fragilis Gressitt & Kimoto, 1963a: 669 (*Haplosaenidea*) **A:** CH (GUA)
fulva Kimoto, 1977: 377 **A:** BT NP
gorbunovi L. N. Medvedev, 2011c: 251 **A:** NP
gressitti Beenen, 2008: 65 [= 2010a: 74] [RN] **A:** CH (GUA)
    cavifrons Gressitt & Kimoto, 1963a: 667 (*Haplosaenidea*) [HN]
nepalica L. N. Medvedev, 2011c: 250 **A:** NP
nitida Gressitt & Kimoto, 1963a: 671 (*Haplosaenidea*) **A:** CH (HAI)
persimilis Kimoto, 1977: 378 **A:** BT
pulchella Laboissière, 1933b: 65 (*Micraenidea*) **A:** CH (YUN)
salicis Gressitt & Kimoto, 1963a: 672 **A:** CH (HUB)
touzalini Laboissière, 1933b: 60 **A:** CH (YUN)
victori Bezděk & Beenen, 2010: 74 [RN] **A:** CH (YUN)
    apicalis Laboissière, 1933b: 61 [HN]

genus *Hyphaenia* Baly, 1865f: 410 type species: *Luperus pilicornis* Motschulsky, 1858
    Trichocerastes Motschulsky, 1866b: 413 type species: *Trichocerastes flavofemoratus* Motschulsky, 1866
aenea Laboissière, 1936: 243 **A:** CH (YUN)
apicalis Takizawa, 1988b: 551 **A:** NP
convexicollis L. N. Medvedev & Sprecher-Uebersax, 1997: 233 **A:** ID (SD) NP
cyanescens Laboissière, 1936: 242 **A:** CH (FUJ YUN) **ORR**
kabaki Lopatin, 2013: 775 **A:** CH (SHA)
konstantinovi Lopatin, 2004c: 190 **A:** CH (YUN)
lutescens Lopatin, 2013: 773 **A:** CH (CHQ)
medvedevi Lopatin, 2006b: 211 **A:** CH (YUN)
patrikeevi L. N. Medvedev & Romantsov, 2013: 138 **A:** CH (SCH)
potanini Lopatin, 2013: 773 **A:** CH (SCH)
volkovitshi Lopatin, 2009a: 4 **A:** CH (YUN)
yasudai Takizawa, 1985b: 100 **A:** ID (SD) NP

genus *Japonitata* E. Strand, 1935: 294 [RN] type species: *Phyllobrotica nigrita* Jacoby, 1886
    Japonia Weise, 1922b: 70 [HN] type species: *Phyllobrotica nigrita* Jacoby, 1886
abdominalis Jiang, 1989: 221, 224 **A:** CH (XIZ)
antennata S.-H. Chen & Jiang, 1986a: 74, 79 **A:** CH (SCH)
bipartita S.-H. Chen & Jiang, 1986a: 73, 79 **A:** CH (CHQ FUJ SHA)
biramosa S.-H. Chen & Jiang, 1986a: 75, 79 **A:** CH (YUN)
concaviuscula Yang, 1997: 870, 901 **A:** CH (CHQ SCH)
confragosa Yang & W.-Z. Li, 1997b: 871, 902 **A:** CH (CHQ SCH ZHE)

cordiformis S.-H. Chen & Jiang, 1986a: 77, 79 **A:** CH (YUN)
costata Allard, 1889e: cxv (*Xenarthra*) **A:** CH (XIZ) ID (HP UT) NP PA
tricostata S.-H. Chen & Jiang, 1981: 479, 489
diformis L. N. Medvedev & Sprecher-Uebersax, 1998: 31 **A:** NP
eberti Kimoto, 1970c: 417 (*Paridea*) **A:** BT CH (XIZ) ID (HP UT) NP
flavoterminata Yang, 1992b: 562, 588 **A:** CH (GUI HUN)
hongpingana Jiang, 1989: 222, 224 **A:** CH (HUB)
houjayi C.-F. Lee, 2022c: 174 **A:** CH (TAI)
jungchani C.-F. Lee, 2022c: 178 **A:** CH (TAI)
litocephala Yang & W.-Z. Li, 1998: 130, 134 **A:** CH (ZHE)
lunata S.-H. Chen & Jiang, 1986a: 78, 79 **A:** CH (SCH YUN)
mirabilis L. N. Medvedev & Sprecher-Uebersax, 1999: 368 **A:** NP
nigricans Yang & W.-Z. Li, 1998: 130, 134 **A:** CH (ZHE)
nigriceps S.-H. Chen & Jiang, 1986a: 73, 79 **A:** CH (SCH)
nigrita Jacoby, 1886a: 742 (*Phyllobrotica*) **A:** JA
ochracea Kimoto, 2004: 54 **A:** ID (SD)
pallipes S.-H. Chen & Jiang, 1986a: 75, 79 **A:** CH (GUX)
picea S.-H. Chen & Jiang, 1986a: 76, 79 **A:** CH (YUN)
ruficollis Jiang, 1989: 221, 224 **A:** CH (XIZ)
striata Yang & W.-Z. Li, 1997b: 871, 902 **A:** CH (CHQ SCH)
unicostata Laboissière, 1929a: 274 (*Japonia*) **A:** CH (FUJ YUN)

genus *Jolibrotica* C.-F. Lee & Bezděk, 2015: 77 type species: *Luperus sauteri* Chûjô, 1935
chujoi Kimoto, 1969: 38 (*Phyllobrotica*) [RN] **A:** CH (TAI)
sauteri Chûjô, 1963b: 395 (*Phyllobrotica*) [HN]
sauteri Chûjô, 1935d: 162 (*Luperus*) **A:** CH (GUX TAI)

genus *Kanarella* Jacoby, 1896c: 279 type species: *Kanarella unicolor* Jacoby, 1896
Cneoranella Maulik, 1936: 334 type species: *Cneorane pallida* Jacoby, 1887
unicolor Jacoby, 1896c: 280 **A:** BT CH (GUA GUX HAI YUN) ID (SD) NP **ORR**

genus *Khasia* Jacoby, 1899c: 83 type species: *Khasia kraatzi* Jacoby, 1899
Harmandinia Laboissière, 1933a: 964 type species: *Harmandinia paradoxa* Laboissière, 1933
kraatzi Jacoby, 1899c: 83 **A:** ID (SD) NP PA **ORR**
paradoxa Laboissière, 1933a: 966 (*Harmandinia*) **A:** ID (SD)

genus *Lanolepta* Kimoto, 1991a: 15 type species: *Lanolepta fulva* Kimoto, 1991
fulva Kimoto, 1991a: 16 **A:** CH (TAI)

genus *Leptomona* Bechyné, 1958c: 601 type species: *Chrysomela erythrocephala* Olivier, 1791

erythrocephala Olivier, 1791a: 704 (*Chrysomela*) **E:** FR IT RUi (ST) PT SP **N:** AG MO TU **[Appendix]**

terrestris Rosenhauer, 1856: 331 (*Monolepta*)

verticalis Reitter, 1886: 72 (*Monolepta*)

fulvicollis Jacoby, 1886a: 749 (*Monolepta*) **A:** JA

heydeni Joannis, 1865: 156 (*Monolepta*) **N:** EG **A:** YE

russica Gmelin, 1790: 1674 (*Chrysomela*) **E:** RU (ST) **A:** AB KI KZ TD UZ

angustula Weise, 1889d: 128 (*Monolepta*)

dubia Motschulsky, 1838: 179 (*Luperus*)

subseriata Weise, 1887h: 191 (*Monolepta*) **A:** RU (ES FE)

genus *Liroetis* Weise, 1889g: 607 type species: *Liroetis aeneipennis* Weise, 1889

Pseudoliroetis Laboissière, 1929a: 280 type species: *Liroetis fulvipennis* Jacoby, 1890

Siemssenius Weise, 1922b: 73 type species: *Siemssenius modestus* Weise, 1922

Zangia S.-H. Chen, 1976: 208, 220 type species: *Zangia latispina* S.-H. Chen, 1976

aeneipennis Weise, 1889g: 608 **A:** CH (CHQ FUJ GAN GUI HEN HUB HUN NIN SCH SHA ZHE)

tiemushannis Jiang, 1988c: 186, 195

alticola Jiang, 1988c: 194, 198 **A:** CH (YUN)

apicalis Gressitt & Kimoto, 1963a: 533 **A:** CH (SCH)

aurantiacus Bezděk, 2021b: 546 **A:** CH (YUN) **ORR**

cheni C.-F. Lee, 2016: 368 (*Siemssenius*) **A:** CH (TAI)

coeruleipennis Weise, 1889g: 609 **A:** JA

coeruleus Jiang, 1990b: 143, 144 (*Zangia*) **A:** CH (SCH)

elongatus Kimoto, 1977: 359 (*Pseudoliroetis*) **A:** BT

ephippiatus Laboissière, 1930: 346 (*Cneorane*) **A:** CH (YUN)

postmaculatus Lopatin, 2004c: 191

signatus Jiang, 1990b: 142, 144 (*Zangia*)

flavipennis Bryant, 1954b: 547 **A:** CH (YUN) **ORR**

leycesteriae Jiang, 1988c: 189, 196

fulvipennis Jacoby, 1890e: 215 (*Liroetis*) **A:** CH (CHQ FUJ GAN GUI GUX HUB HUN JIA JIX SCH SHG ZHE)

grandis S.-H. Chen & Jiang, 1986b: 199, 200 **A:** CH (SCH)

humeralis Jiang, 1988c: 192, 197 **A:** CH (YUN)

jungchani C.-F. Lee, 2016: 375 (*Siemssenius*) **A:** CH (TAI)

latispinus S.-H. Chen, 1976: 208, 220 (*Zangia*) **A:** CH (XIZ)

leechi Jacoby, 1890e: 215 **A:** CH (CHQ FUJ GAN GUI HUB SHA YUN ZHE)

verticalis Jiang, 1988c: 186, 195

liui C.-F. Lee 2016: 376 (*Siemssenius*) **A:** CH (TAI)
lonicernis Jiang, 1988c: 190, 196 **A:** CH (YUN)
medvedevi Bezděk, 2021b: 555 [RN] **A:** ID (SD) NP
nigricollis L. N. Medvedev, 2009b: 407 [HN]
metallipennis Chûjô, 1962b: 181 (*Pseudoliroetis*) **A:** CH (TAI)
modestus Weise, 1922b: 73 (*Siemssenius*) **A:** CH (FUJ ZHE)
nepalensis Chûjô, 1966b: 15 **A:** BT ID (SD) NP
bhutanus L. N. Medvedev, 2009b: 408 (ssp.)
nigriceps Laboissière, 1929a: 282 (*Pseudoliroetis*) **A:** CH (GUI SCH)
nigricollis Jiang, 1990b: 143, 144 (*Zangia*) **A:** CH (SCH)
unicolor L.-J. Zhang, W.-Z. Li & Yang, 2008: 23
nigropictus Fairmaire, 1889b: 76 (*Leptarthra*) **A:** CH (SCH)
obliquevirgatus Lopatin, 2013: 771 **A:** CH (SCH)
octopunctatus Weise, 1889g: 619 (*Mimastra*) **A:** CH (GAN QIN SCH XIZ)
pallidulus Jiang, 1990b: 141, 144 (*Zangia*) **A:** CH (XIZ)
paragrandis Jiang, 1988c: 192, 198 **A:** CH (SCH XIZ)
prominensis Jiang, 1988c: 187, 196 **A:** CH (SCH)
reitteri Pic, 1934g: 87 (*Merista*) **A:** CH (SCH YUN)
trifasciatus Jiang, 1992: 659, 672 (*Pseudoliroetis*)
rufipennis Chûjô, 1962b: 178 (*Pseudoliroetis*) **A:** CH (TAI)
sichuanensis Jiang, 1988c: 188, 196 **A:** CH (GAN GUI SCH SHA)
sulcipennis L.-J. Zhang & Yang, 2008: 127 (*Siemssenius*) **A:** CH (SCH YUN)
suwai Takizawa, 1988a: 14 (*Mimastra*) **A:** NP
tibetanus Jiang, 1988c: 191, 197 **A:** CH (XIZ) ID (AP) NP
tibialis Jiang, 1988c: 190, 197 **A:** CH (YUN)
tsoui C.-F. Lee, 2016: 380 (*Siemssenius*) **A:** CH (TAI)
violaceipennis L.-J. Zhang, W.-Z. Li & Yang, 2008: 24 **A:** CH (SCH) **ORR**
yuae C.-F. Lee, 2016: 381 (*Siemssenius*) **A:** CH (TAI)
yulongnis Jiang, 1988c: 193, 198 **A:** CH (YUN)
zhongdianicus Jiang, 1988c: 189, 196 **A:** CH (FUJ HUN YUN ZHE)

genus *Luperogala* L. N. Medvedev & Samoderzhenkov, 1989: 454 type species: *Luperogala mirabilis* L. N. Medvedev & Samoderzhenkov, 1989
Liroetina L. N. Medvedev, 2013b: 293 type species: *Liroetina viridipennis* L. N. Medvedev, 2013
viridipennis Kimoto, 1989a: 83 (*Liroetis*) **A:** ID (SD) **ORR**
viridipennis Medvedev, 2013b: 294 (*Liroetina*) [HN]

genus *Luperus* Geoffroy, 1762: 230 type species: *Chrysomela flavipes* Linnaeus, 1767 [ICZN (1984b), Opinion 1273]
Lyperus Billberg, 1820: 57 type species: *Chrysomela flavipes* Linnaeus, 1767
Trichelytron Apfelbeck, 1912: 259 type species: *Luperus apollinis* Apfelbeck, 1912 [# 685]

abdominalis Rosenhauer, 1856: 329 E: SP
abeillei Guillebeau, 1891: 301 A: SY
acutipennis Fairmaire, 1867: 415 N: MO
aetolicus Kiesenwetter, 1861: 393 E: AL BH BU CR GR ME NM SB
alpicus Desbrochers des Loges, 1898: 46 E: FR IT SP
antennalis L. N. Medvedev, 1998d: 613 A: RU (FE)
apollinis Apfelbeck, 1912: 259 E: GR [# 685]
armeniacus Kiesenwetter, 1878: 337 E: BU KR RU (ST) UK A: AB GG TR
biraghii Ragusa, 1871: 315 E: IT
aetnensis Rottenberg, 1872a: 244
paganettii Pic, 1912d: 34 (ssp.)
calabricus Laboissière, 1912c: 40 E: IT
caucasicus Weise, 1880a: 486 (var.) E: RU (ST) A: AB GG
circassicus L. N. Medvedev, 1962a: 386
mixtus Weise, 1886a: 599 (var.)
cavicollis S.-H. Chen, 1942a: 62 A: CH (GUX)
cyanipennis Küster, 1848b: 98 E: BH BU CR KR MD ME NM UK A: GG
dmitrii Bezděk & Beenen, 2010: 74 [RN] A: CH (GAN NMO XIN) MG RU (WS)
anthracinus Ogloblin, 1936: 289, 426 [HN]
ogloblini Aslam, 1967: 128 [RN, HN]
discolor Faldermann, 1837: 340 A: AB GG TR "Caucasus"
diversepunctatus Pic, 1912d: 34 E: SP
doeberli Vela & Bastazo, 2017: 241 A: TR
ehnbergi Jacobson, 1901b: 135 A: KI KZ
fiorii Weise, 1895b: 212 E: IT (Abruzzo)
flaviceps Apfelbeck, 1912: 261 E: AL BH GR [# 728]
flavilabris Ogloblin, 1941: 23 A: KI TD UZ
flavimanus Weise, 1889g: 618 A: CH (JIA LIA SCH)
flavipennis P. H. Lucas, 1847a: pl. 44 [= 1849: 543] N: AG MO
flavipes flavipes Linnaeus, 1767: 601 (*Chrysomela*) E: AL AU BE BH BU BY CR CZ DE EN FI FR GB GE GR HU IT KR LA LS LT LU ME NL NM NR PL RU (CT NT ST) SB SK SL SP SV SZ UK A: AB GG IN KZ MG RU (FE WS) TR
carniolicus Kiesenwetter, 1861: 393
hypsophilus Apfelbeck, 1912: 261 (var.)
megalophthalmus Joannis, 1865: 147
ochropus Gmelin, 1790: 1723 (*Cryptocephalus*)
purpurea Voet, 1806: 37 (*Crioceris*) [UN]
flavipes obscuricornis Ogloblin, 1936: 294, 426 (ssp.) A: RU (FE)
flavus Rosenhauer, 1856: 330 E: PT SP [# 729]
floralis Faldermann, 1837: 341 A: AB IN TR

graecus Weise, 1886a: 598 **E:** AL BU CR GR ME NM SB TR **A:** TR

balcanicus Apfelbeck, 1912: 260

kiesenwetteri Joannis, 1865: 146 **E:** RU (ST) **A:** AB IN KZ "Transcaucasia"

kiritshenkoi Ogloblin, 1936: 294, 427 **E:** KR

klimenkoi Romantsov, 2004: 623 **A:** KZ

kusanagii Chûjô, 1941b: 165 **A:** SC

laricis Motschulsky, 1859: 574 **A:** JA MG NC RU (ES FE)

murakamii Kimoto, 1965: 379 (ssp.)

leonardii Fogato, 1980: 46 **E:** IT

longicornis Fabricius, 1781: 72 (*Ptinus*) **E:** AL AU BE BH BU BY CR CZ DE EN FI FR GB GE GR HU IT KR LA LS LT LU NL NR PL RU (CT NT ST) SB SK SL SP SV SZ UK **A:** AB GG MG RU (ES FE WS) TR

bicolor Herbst, 1786: 165 (*Crioceris*)

diniensis Bellier, 1870: xxvii

fulvipes Gmelin, 1790: 1725 (*Cryptocephalus*)

subviridis Laboissière, 1912c: 38 (var.)

tenebricosus Ogloblin, 1936: 291, 426 (var.) [IN] **[# 698]**

luperus Sulzer, 1776: 35 (*Chrysomela*) **E:** AU BE BH BY CR CZ FR GE HU IT KR LT LU ME NL PL RU (CT NT ST) SB SK SP SZ UK **A:** CH ("China") GG MG RU (SS WS)

betulinus Geoffroy, 1785: 90

dispar L. Redtenbacher, 1874: 492 (*Chrysomela*)

erytromelas Gmelin, 1790: 1725 (*Cryptocephalus*)

maerkelii Motschulsky, 1850: 368

niger Goeze, 1777: 321

nigricans Gmelin, 1790: 1725 (*Cryptocephalus*)

rufipes Goeze, 1777: 322

tristis Schrank, 1837: 7 (*Chrysomela*)

ulmarius Geoffroy, 1785: 89

maculicornis Desbrochers des Loges, 1873: 431 (var.) **E:** FR (Corsica)

laevis Kiesenwetter, 1873: 30

margaritae Lopatin, 1990d: 53 **A:** AR

maroccanus Peyerimhoff, 1923: 79 (*Lyperus*) **N:** AG MO

mauritanicus Pic, 1898f: 312 **N:** AG MO

nigripes Kiesenwetter, 1861: 392 **E:** AU FR IT SZ

garieli Aubé, 1862: 74

geniculatus Joannis, 1865: 151

maritimus J. Daniel, 1903b: 255 (ab.) [IN]

orientalis Faldermann, 1837: 336 **E:** RU (ST) **A:** AB GG IN TM "Caucasus"

amoenus Faldermann, 1837: 339

oxypterus Faldermann, 1837: 338

perlucidus Iablokoff-Khnzorian, 1956: 137 **A:** AR IN TR
pygmaeus Joannis, 1865: 140 **E:** IT
pyrenaeus Germar, 1823: 597 **E:** FR (Pyrénées) IT (Prealpi Veneto e Trentine) SP (Pirineos)
arvernus Laboissière, 1913: 35 (var.)
quittardi Pic, 1902b: 27 (var.)
ragusai Laboissière, 1919a: 116 **E:** IT
rectangulus Weise, 1898: 214 **A:** TR
revelieri Perris, 1864: 301 **E:** FR (Corsica) IT (Sardinia)
rhilensis Weise, 1900b: 287 **E:** BU (Rila Mts.)
rugifrons Weise, 1886a: 595 **E:** AL BU GR HU SB
saxonicus Gmelin, 1790: 1723 (*Cryptocephalus*) **E:** CZ GE SK
rufipes Fabricius, 1787: 89 (*Crioceris*) [HN]
semiflavus Ogloblin, 1936: 294, 427 **A:** CH (JIL LIA SCH TAI) MG NC RU (ES) SC [# 730]
mandzhuricus Ogloblin, 1936: 295, 428
seminiger Pic, 1903h: 147 (*Eugalera*) **A:** CH ("Manchuria")
sichuanus L. N. Medvedev, 2012c: 253 **A:** CH (SCH)
silfverbergi Lopatin, 1984a: 90 **A:** AR IN
similis Iablokoff-Khnzorian, 1970: 72 **A:** RU (SS)
spurius Ogloblin, 1936: 297, 429 (ssp.) **A:** RU (FE: Sakhalin) JA
sulphuripes Graells, 1858a: 95 [= 1858b: 133] (*Lupersus*) **E:** PT SP
turkestanicus Weise, 1892d: 140 **A:** KI
variabilis Kimoto, 2004: 54 **A:** PA
viridipennis Germar, 1823: 597 **E:** AL AU BU CR CZ FR HU IT LS GE ME PL RU (NT ST) SK SL SZ UK **A:** AB GG RU (WS) TR
coerulescens Duftschmid, 1825: 233 (*Crioceris*)
gredleri J. Müller, 1948: 98 (ssp.)
pallidicollis Guillebeau, 1891: 305 (var.)
pallidipennis Pic, 1896g: 87 (var.)
rhenanus Sterlin, 1890: 251
varicollis Weise, 1886a: 609 (var.)
winkleri J. Müller, 1948: 98 (ssp.)
vitalei Ragusa, 1923: 15 **E:** IT (Sicily)
xanthopoda Schrank, 1781: 93 (*Chrysomela*) **E:** AL AU BH BU BY CR CZ FR GE GR HU IT KR MD ME NM PL RO RU (CT ST) SB SK SL SZ UK **A:** AB GG KI KZ MG TR
rufipes Scopoli, 1763: 73 (*Coccinella*) [QS]
yunnanus L. N. Medvedev, 2012c: 252 **A:** CH (YUN)

genus *Macrima* Baly, 1878h: 377 type species: *Macrima armata* Baly, 1878

Glechonis Weise, 1889g: 630 type species: *Glechonis rubripennis* Weise, 1889

Sepharia Fairmaire, 1889b: 78 type species: *Sepharia rubricata* Fairmaire, 1889

armata Baly, 1878h: 377 **A:** CH (XIZ) ID (HP KA SD) NP PA

frontalis Jacoby, 1890f: 254 (*Sepharia*)

aurantiaca Laboissière, 1936: 250 (*Sepharia*) **A:** BT CH (XIZ YUN) ID (SD) NP

bifida Yang, 1992c: 258, 261 **A:** CH (SCH)

cornuta Laboissière, 1936: 250 (*Sepharia*) **A:** BT CH (GAN NIN SCH XIZ YUN)

ferrugina Jiang, 1990a: 138, 139 **A:** CH (XIZ)

hartmanni L. N. Medvedev, 2009d: 411 **A:** NP

pallida Laboissière, 1936: 249 (*Sepharia*) **A:** BT CH (XIZ) ID (HP SD UT) NP

roseofulva L. N. Medvedev, 2011c: 251 **A:** NP

rubricata Fairmaire, 1889b: 78 (*Sepharia*) **A:** CH (GAN SCH XIZ YUN)

rubripennis Weise, 1889g: 632 (*Glechonis*)

straminea Ogloblin, 1936: 324, 436 (*Sepharia*) **A:** CH (SCH XIZ YUN)

yunnanensis Laboissière, 1936: 249 (*Sepharia*) **A:** CH (GUI XIZ YUN)

genus *Madurasia* Jacoby, 1896c: 280 type species: *Madurasia obscurella* Jacoby, 1896

Neorudolphia Laboissière, 1926b: 190 type species: *Neorudolphia bedfordi* Laboissière, 1926 (= *Madurasia obscurella* Jacoby, 1896)

undulatovittata Motschulsky, 1866b: 417 (*Teinodactila*) **A:** ID (UT) NP YE **AFR ORR**

bedfordi Laboissière, 1926b: 191 (*Neorudolphia*)

obscurella Jacoby, 1896c: 281

genus *Marseulia* Joannis, 1865: 160 type species: *Agelastica dilativentris* Reiche & Saulcy, 1858

aenea Beenen, 2019: 10 **A:** IQ

dilativentris Reiche & Saulcy, 1858b: 42 (*Agelastica*) **A:** IS LE SY

reuteri Beenen, 2019: 9 **A:** IQ

genus *Medythia* Jacoby, 1887b: 242 type species: *Medythia quadrimaculata* Jacoby, 1887

Paraluperodes Ogloblin, 1936: 310, 431 type species: *Cnecodes suturalis* Motschulsky, 1858

nigrobilineata Motschulsky, 1861a: 26 (*Cnecodes*) Ei: RU (ST) **A:** CH (ANH FUJ GAN GUI GUX HEB HEI HUB HUN JIA JIL JIX SCH SHA SHN SHX TAI XIZ YUN ZHE) GGi ID (UT) JA NC NP PA RU (ES FE) SC **ORR**

quaterna Fairmaire, 1880c: 317 (*Luperus*) **A:** OM YE **AFR**

lineata Karsch, 1882: 400 (*Monolepta*)

siamensis Kimoto, 1989a: 104 **A:** CH (HUB) **ORR**

suturalis Motschulsky, 1858: 100 (*Cnecodes*) **A:** CH (ANH FUJ GAN GUA GUX HAI HKG JIX SCH TAI) ID (UT) JA (Ryukyus) NP PA **ORR**

abbreviata Weise, 1922b: 81 (*Luperodes*, ab.) [IN]
vittata Weise, 1922b: 80 (*Luperodes*, ab.) [IN]
yemenica L. N. Medvedev, 2012e: 417 **A:** YE

genus *Mimastra* Baly, 1865f: 253 type species: *Mimastra arcuata* Baly, 1865
Anthraxantha Fairmaire, 1878b: 137 type species: *Anthraxantha davidis* Fairmaire, 1878
Brachita Allard, 1889e: ciii type species: *Brachita terminata* Allard, 1889
Neoatysa Abdullah & Qureshi, 1968a: 412 type species: *Neoatysa shahidi* Abdullah & Qureshi, 1968
aptera Takizawa, 1990a: 287 **A:** ID (SD)
arcuata Baly, 1865f: 253 **A:** CH (XIZ YUN) ID (AP SD) NP **ORR** **[Appendix]**
brancuccii L. N. Medvedev, 2011c: 249 [DA] **A:** NP **[# 693]**
chennelli Baly, 1879a: 450 **A:** BT CH (FUJ GUA HKG HUN JIX SCH SHA YUN ZHE) ID (HP UT) NP PA **ORR**
uncitarsis Laboissière, 1940b: 4
cyanura Hope, 1831: 29 (*Auchenia*) **A:** ID (HP) NP
davidis Fairmaire, 1878b: 137 (*Anthraxantha*) **A:** BT CH (CHQ FUJ GAN GUA GUI GUX HUB HUN JIA JIX SCH SHA XIZ YUN ZHE) ID (KA) NP **ORR**
limbata Baly, 1879a: 449
fortipunctata Maulik, 1936: 542 **A:** BT ID ("India") NP
bhutanica Kimoto, 1982a: 14
gracilicornis Jacoby, 1889b: 210 **A:** CH (YUN) ID (SD) NP **ORR**
gracilis Baly, 1878h: 378 **A:** AF BT CH (XIN XIZ) ID (KA SD UT) NP PA SA
nainitalensis Gangola, 1969: 58 (*Hoplosomedia*)
grahami Gressitt & Kimoto, 1963a: 537 **A:** CH (GUI HUN SCH SHA)
guerryi Laboissière, 1929a: 283 **A:** CH (CHQ SCH SHX YUN) ID (AP)
quadrivittata Mader, 1938: 59
hsuehleeae Bezděk & C.-F. Lee, 2011: 719 **A:** CH (TAI)
jacobyi Bezděk, 2010a: 38 [RN] **A:** CH (YUN) **ORR**
longicornis Jacoby, 1892b: 944 [HN]
kremitovskyi Bezděk, 2009: 833 **A:** CH (SHA YUN)
kumatai Kimoto & Takizawa, 1972: 218 (*Trichomimastra*) **A:** NP ID (SD) **ORR**
fulva Kimoto & Takizawa, 1983: 87
latimana Allard, 1889f: 308 **A:** CH (YUN) **ORR**
longicornis Allard, 1889a: 330 (*Haplosoma*) **A:** CH (SHG) ID ("India")
lunata Kollar & L. Redtenbacher, 1844: 556 (*Phyllobrotica*) **A:** CH (FUJ GUI HUN SCH) ID (KA UT) NP
apicalis Baly 1886a: 28
maai Gressitt & Kimoto, 1963a: 539 **A:** CH (FUJ GUX) **ORR**
malvi S.-H. Chen, 1942a: 30 **A:** CH (GUI HUN SCH SHA)

modesta Fairmaire, 1889b: 82 **A:** CH (SCH)
nitida Maulik, 1936: 540 **A:** ID (SD) NP **ORR**
oblonga Gyllenhal, 1808: 295 (*Galleruca*) **A:** CH (FUJ GUI GUX HKG HUN JIX SHG ZHE) **ORR**
procerula L.-J. Zhang & Yang, 2006: 206 **A:** CH (YUN)
pygidialis Laboissière, 1929a: 285 **A:** CH (GUX) **ORR**
quadripartita Baly, 1879a: 448 **A:** CH (YUN) ID (AP SD) PA **ORR**
riedeli Bezděk, 2009: 834 **A:** ID (UT) NP
scutellata Jacoby, 1904a: 395 **A:** NP **ORR**
shahidi Abdullah & Qureshi, 1968a: 412 (*Neoatysa*) **A:** PA
soreli Baly, 1878j: 414 **A:** CH (GUI HUN SCH) **ORR**

genus *Monolepta* Chevrolat, 1836: 383 type species: *Crioceris bioculata* Fabricius, 1781
Aemulaphthona Scherer, 1969: 89 type species: *Aphthona ochracea* Weise, 1922
Chimporia Laboissière, 1931c: 413 type species: *Chimporia monardi* Laboissière, 1931
Damais Jacoby, 1903a: 118 type species: *Damais humeralis* Jacoby, 1903
afghana Lopatin, 1963d: 359 **A:** AF
aglaonemae Gressitt & Kimoto, 1963a: 607 **A:** CH (FUJ GUA HAI)
albipunctata Lei, S.-Y. Xu, Yang & Nie, 2021: 44 **A:** CH (GUX)
albomaculata Maulik, 1936: 399 [RN] **A:** BT NP **ORR**
alboplagiata Jacoby, 1892b: 983 [HN]
alnivora S.-H. Chen, 1976: 205, 218 **A:** CH (XIZ)
alticola Lei, S.-Y. Xu, Yang & Nie, 2021: 40, 46 **A:** CH (YUN)
analis Weise, 1915a: 177 [RN] **A:** BT ID ("India")
humeralis Jacoby, 1903a: 125 (*Candezea*) [HN]
annamita Laboissière, 1935c: 143 [RN] **A:** CH (TAI) **ORR**
ferruginea Allard, 1889f: 311 (*Luperodes*) [HN]
arundinariae Gressitt & Kimoto, 1963a: 609 **A:** CH (CHQ FUJ GUI GUX HEN HUB HUN JIA JIX SCH)
asahinai Chûjô, 1962b: 116 **A:** CH (TAI)
arvensis Bryant, 1957b: 362 **A:** SA YE
babai Kimoto, 1996: 40 **A:** CH (TAI)
bacboensis L. N. Medvedev, 2012d: 506 **A:** CH (YUN) **ORR**
basifasciata L. N. Medvedev, 2015c: 326 **A:** NP
bhutanensis Kimoto, 1977: 374 **A:** BT NP
bicavipennis S.-H. Chen, 1942a: 55 **A:** CH (ANH FUJ GAN GUI GUX HEN HUB HUN JIX SHA SHX TAI YUN ZHE)
bimaculicollis Jacoby, 1904a: 405 **A:** NP **ORR**
bimaculipennis Kimoto, 1977: 373 **A:** BT NP

birmanensis Jacoby, 1892b: 982 **A:** ID (SD) **ORR**
bivittata Lei, S.-Y. Xu, Yang & Nie, 2021: 43, 48 **A:** CH (ZHE)
braeti Duvivier, 1892b: 437 (*Luperodes*) **A:** ID (SD) NP
brittoni Gressitt & Kimoto, 1963a: 611 **A:** CH (HAI)
capitata S.-H. Chen, 1942a: 58 **A:** CH (CHQ FUJ GUI HUB HUN SCH SHA)
cardoni Jacoby, 1900b: 137 **A:** ID (SD) NP **ORR**
carsteni Schlich & Wagner, 2010: 317 **A:** SA YE
cavipennis Baly, 1879a: 459 **A:** CH (FUJ GUA GUX HAI HKG YUN) ID ("India") **ORR**
  albofasciata Allard, 1889f: 310 (*Luperodes*)
  fasciata Allard, 1889a: 318 (*Diacantha*)
  siamensis Weise, 1915a: 177 [RN]
  trifasciata Jacoby, 1900b: 139 (*Candezea*) [HN]
cheni Beenen, 2008: 65 [= 2010a: 74] [RN] **A:** CH (XIZ)
  straminea S.-H. Chen, 1976: 206, 218 [HN]
chinkinyui Kimoto, 1996: 38 **A:** CH (TAI)
chujoi Nakane & Kimoto, 1961a: 20 **A:** JA (Ryukyus)
conformis Weise, 1922b: 105 **A:** ID (SD) NP **ORR**
dehradunensis Takizawa, 1986: 99 **A:** ID (UT)
dichroa Harold, 1877a: 366 **A:** JA RU (FE) SC
  apicipennis Jacoby, 1886a: 749 (var.)
  kurilensis L. N. Medvedev, 1966: 42
discalis Gressitt & Kimoto, 1963a: 612 **A:** CH (GAN GUI YUN ZHE)
eoa Ogloblin, 1936: 319, 435 **A:** RU (FE)
epistomalis Laboissière, 1934a: 9 **A:** CH (GAN GUI HUN SHA ZHE)
erythrocephala Baly, 1878h: 380 (*Luperodes*) **A:** BT CH (SCH XIN) ID (HP KA SD UT) NP PA **ORR**
  fulvifrons Jacoby, 1896c: 291
eunicia Maulik, 1936: 395 **A:** CH (GUI SCH) **ORR**
flavovittata S.-H. Chen, 1942a: 61 **A:** CH (FUJ GAN HEB HUB SHA) **ORR**
gantokensis Kimoto, 2004: 56 **A:** ID (SD)
gestroi Jacoby, 1892b: 983 **A:** ID (UT) **ORR**
godavariensis Takizawa, 1988b: 547 **A:** NP
gracilipes Chûjô, 1938c: 145 **A:** CH (TAI)
himalayaensis himalayaensis Kimoto, 1970c: 419 **A:** ID (SD UT) NP
himalayaensis postmaculata Kimoto, 1977: 375 (ssp.) **A:** BT
hiurai Kimoto, 1965: 392 **A:** JA
hlavaci Bezděk, 2012d: 417 **A:** YE (Socotra)
hongkongensis Kimoto, 1967a: 61 **A:** CH (HKG)
horni Chûjô, 1938c: 146 **A:** CH (TAI)

hupehensis Gressitt & Kimoto, 1963a: 615 **A:** CH (HEB HUB)
impressipennis Ogloblin, 1936: 321, 436 **A:** IN **AFR**
indicola Takizawa, 1986: 100 **A:** ID (UT)
indochinensis L. N. Medvedev, 1999c: 184 **A:** CH (YUN) **ORR**
kabakovi L. N. Medvedev, 1985a: 375 **A:** AF
kmenti Bezděk, 2012d: 420 **A:** YE (Socotra)
kuroheri Kimoto, 1966c: 31 **A:** CH (TAI)
kurosawai Chûjô & Ohno, 1961: 10 **A:** JA
kwangtunga Gressitt & Kimoto, 1963a: 616 **A:** CH (FUJ GUA GUX)
labiata Jacoby, 1900b: 138 (*Candezea*) **A:** ID (UT) NP **ORR**
labiatella Kimoto, 2004: 56 **A:** ID (SD) **ORR**
laterimarginata L. N. Medvedev, 1999c: 183 **A:** NP
lauta Gressitt & Kimoto, 1963a: 617 **A:** CH (GUI HAI) **ORR**
leechi Jacoby, 1890e: 216 **A:** CH (FUJ GUA GUI HUB TAI YUN ZHE) ID (SD) NP **ORR**
lepida Reiche & Saulcy, 1858b: 44 **N:** EG **A:** IS JO OM SA SI SY YE **AFR**
lesagei Takizawa, 1988b: 548 **A:** NP
liui Gressitt & Kimoto, 1963a: 618 **A:** CH (GUI YUN)
longicornis Jacoby, 1890d: 194 (*Taphinella*) **A:** CH (HUB)
longitarsoides Chûjô, 1938c: 147 **A:** CH (CHQ FUJ GUA GUI GUX HAI HUB HUN JIX SCH TAI ZHE)
lunata Gressitt & Kimoto, 1963a: 620 **A:** CH (HAI) **ORR**
maana Gressitt & Kimoto, 1963a: 621 **A:** CH (FUJ TAI ZHE)
mandibularis Chûjô, 1962b: 120 **A:** CH (TAI)
martensi L. N. Medvedev, 1992d: 16 **A:** NP
meihuai C.-F. Lee, Tian & Staines, 2010: 306 **A:** CH (TAI)
mengsongensis Lei, S.-Y. Xu, Yang & Nie, 2021: 43, 50 **A:** CH (YUN)
meridionalis Gressitt & Kimoto, 1963a: 622 **A:** CH (GUA HAI ZHE)
minor Chûjô, 1938c: 149 **A:** CH (TAI) JA
  seminigra Chûjô, 1962b: 118
minutissima S.-H. Chen, 1942a: 60 **A:** CH (GUX)
miyamotoi Kimoto, 1965: 393 **A:** JA (Ryukyus)
mordelloides S.-H. Chen, 1942a: 56 **A:** CH (GUI GUX HUB JIX SCH XIZ)
nakanei Kimoto, 1969: 52 **A:** CH (TAI)
nepalensis Kimoto, 1970c: 420 **A:** ID (SD) NP
nigripennis Motschulsky, 1866b: 418 (*Teinodactila*) **A:** NP **ORR**
nigrobasalis Jacoby, 1903a: 127 **A:** BT NP **ORR**
  humeralis Jacoby, 1903a: 119 (*Damais*)
nigrosinuata nigrosinuata Pic, 1921e: 14 **N:** EG
nigrosinuata rectelineata Pic, 1921e: 14 (var.) **N:** EG
nojiriensis Nakane, 1963a: 20 **A:** JA RU (FE)
  tsushimanum Kimoto, 1965: 393

occifluvis Gressitt & Kimoto, 1963a: 624 **A:** CH (GUA GUX)
ogloblini Papp, 1946b: 7 [RN] **A:** NC RU (ES FE)
    beeneniana Warchałowski, 2004a: 113 [RN]
    nebulosa Ogloblin, 1936: 320, 435 [HN]
ongi C.-F. Lee & Staines, 2010: 538 **A:** CH (TAI)
orientalis Jacoby, 1889b: 227 **A:** ID (AP UT) NP **ORR**
    konbirensis Duvivier, 1891a: xlvii
ovatula S.-H. Chen, 1942a: 60 **A:** CH (FUJ GUI GUX HUB JIX)
pahari Basu, 1996: 711 **A:** ID (SD)
pallidula Baly, 1874: 187 (*Luperodes*) **A:** CH (ANH CHQ FUJ GAN GUA GUI GUX HAI HEN HUB HUN JIX SCH XIZ YUN ZHE) JA NC NP RU (ES FE) SC **ORR**
palliparva Gressitt & Kimoto, 1963a: 626 **A:** CH (GUI HAI HKG JIX YUN)
paradichroa J.-K. Li & Z.-G. Wang, 1993: 166 **A:** CH (ANH)
parenthetica Gressitt & Kimoto, 1963a: 627 **A:** CH (HUB SCH)
parvezi Aslam, 1967: 128 [RN] **A:** CH (TAI)
    luperoides Chûjô, 1962b: 126 [HN]
postfasciata Gressitt & Kimoto, 1963a: 628 **A:** CH (ANH FUJ GAN GUI HUB HUN)
quadricavata S.-H. Chen, 1976: 205, 218 **A:** CH (XIZ) NP
quadriguttata Motschulsky, 1860b: 233 (*Luperodes*) **A:** CH (GAN GUX HEI NIN YUN) JA MG NC RU (ES FE) SC
rubripennis Lei, S.-Y. Xu, Yang & Nie, 2021: 39, 52 **A:** CH (FUJ HUN SCH)
rufa Takizawa, 1988b: 549 **A:** ID ("India") NP
rufofulva Chûjô, 1938c: 149 **A:** CH (SCH TAI YUN)
sakishimana Kimoto & Gressitt, 1966: 529 **A:** JA (Ryukyus)
sasajii Kimoto, 1969: 53 **A:** CH (TAI)
satoi Kimoto & Takizawa, 1983: 89 **A:** NP ID (SD)
saudica L. N. Medvedev, 1996: 253 **A:** OM SA YE
sauteri Chûjô, 1935d: 173 **A:** CH (TAI)
schawalleri L. N. Medvedev, 1990b: 21 **A:** NP
schereri Gressitt & Kimoto, 1963a: 630 **A:** CH (GUI HUB)
selmani Gressitt & Kimoto, 1963a: 631 **A:** CH (GAN GUI HUB HUN YUN ZHE) **ORR**
semenovi Ogloblin, 1936: 318, 434 **A:** CH (GUI SCH) RU (ES FE)
semiluperina Kimoto, 1989a: 148 **A:** NP **ORR**
septentrionalis L. N. Medvedev & Yavorskaya, 2008: 284 **A:** RU (FE)
severini Jacoby, 1896c: 276 (*Luperus*) **A:** CH (YUN) NP **ORR**
sexlineata Chûjô, 1938c: 150 **A:** BT CH (FUJ GAN GUA GUX HAI HEB JIL SHA SHX TAI YUN) ID (KA) NP PA **ORR**
    duvivieri Jacoby, 1904a: 404 [HN]
    lineata Weise, 1915a: 177 [HN]
    madrasensis Wilcox, 1973: 562 [RN]
shaowuensis Gressitt & Kimoto, 1963a: 632 **A:** CH (FUJ GUA HUB HUN JIX YUN)

shirozui Kimoto, 1965: 390 **A:** JA NC SC
signata Olivier, 1808: 665 (*Galeruca*) **A:** BT CH (CHQ FUJ GAN GUA GUI GUX HAI HEB HEI HEN HKG HUB HUN JIL JIX LIA NIN NMO SCH SHA SHX TAI XIZ YUN ZHE) ID (AP HP SD UT) JA MG NC NP PA RU (FE) SC **AUS ORR**
belmonti Piton, 1943: 138 (ab.) [IN]
biarcuata Weise, 1889g: 632
bipunctata Piton, 1943: 138 (ab.) [IN]
coalita Piton, 1943: 138 (ab.) [IN]
dorsalis Motschulsky, 1866b: 415 (*Luperodes*)
elegantula Boheman, 1859: 183
hieroglyphica Motschulsky, 1858: 104 (*Luperodes*)
lutea Piton, 1943: 138 (ab.) [IN]
neglecta C. R. Sahlberg, 1823b: 72 [= 1823c: 72, = 1829: 29] (*Crioceris*)
pici Piton, 1943: 138 (ab.) [IN]
picturata Jacoby, 1896c: 292
quadriguttata Fairmaire, 1887d: 333 (*Luperodes*)
quadripustulata Motschulsky, 1858: 105 (*Luperodes*)
sexmaculata Piton, 1943: 138 (ab.) [IN]
simplex Weise, 1913c: 229 (ab.) [IN]
yunnanesis Piton, 1943: 138 (ab.) [IN]
simlensis Kimoto, 1967b: 73 **A:** ID (HP) NP
subapicalis Gressitt & Kimoto, 1963a: 634 **A:** BT CH (CHQ FUJ GAN GUI HUB HUN SHA) **ORR**
subflavipennis Kimoto, 1989a: 154 **A:** CH (GAN GUI HUB SHA) **ORR**
subrubra S.-H. Chen, 1942a: 57 **A:** CH (FUJ GUI HUN SHA)
suchomeli Bezděk, 2012d: 412 **A:** YE (Socotra)
surkheta L. N. Medvedev, 2000e: 13 **A:** NP
syriaca Weise, 1924a: 116 (*Luperus*) [RN] **A:** IS SY TR
anatolica Bezděk, 1998: 150
brevicollis Weise, 1898: 214 (*Luperus*) [HN]
volkovitshi Lopatin & Nesterova, 2013: 95 (*Calomicrus*)
takizawai Kimoto, 1996: 39 **A:** CH (TAI)
tatemizo Kimoto, 2004: 58 **A:** BT NP
tricolorata Kimoto, 1977: 372 **A:** BT
trifasciata Jacoby, 1896c: 295 **A:** NP **ORR**
tuai Basu, 1996: 713 **A:** ID (SD)
tuberculipyga L. N. Medvedev, 2007c: 297 **A:** NP
tumpai Basu, 1996: 713 **A:** ID (SD)
weigeli L. N. Medvedev, 2012d: 506 **A:** CH (YUN)

wilcoxi Gressitt & Kimoto, 1965: 801 [RN] **A:** CH (CHQ GUI SCH)
antennalis Gressitt & Kimoto, 1963a: 608 [HN]
xanthodera S.-H. Chen, 1942a: 58 **A:** CH (FUJ GAN GUI HUB HUN SCH SHA TAI XIZ YUN) **ORR**
yama Gressitt & Kimoto, 1965: 802 [RN] **A:** CH (CHQ GAN GUI HAI HEN HUB JIX SCH SHA YUN ZHE)
monticola Gressitt & Kimoto, 1963a: 623 [HN]
yaosanica S.-H. Chen, 1942a: 59 **A:** CH (FUJ GUX HEN HUB JIX ZHE)
yunnanica Gressitt & Kimoto, 1963a: 637 **A:** CH (CHQ FUJ GUI HUN SCH YUN)
zonalis Gressitt & Kimoto, 1963a: 638 **A:** CH (YUN) **ORR**

genus *Munina* S.-H. Chen, 1976: 210, 221 type species: *Munina donacioides* S.-H. Chen, 1976 (= *Antipha blanchardi* Allard, 1891)
blanchardi Allard, 1891: 234 (*Antipha*) **A:** CH (YUN) **ORR**
donacioides S.-H. Chen, 1976: 210, 221
flavida Yang & Yao, 1997: 388, 391

genus *Neobarombiella* Bolz & Wagner, 2012: 6 type species: *Monolepta flavilabris* Weise, 1903
socotrana Bolz & Wagner, 2014: 278 **A:** YE (Socotra)

genus *Neochya* C.-F. Lee, 2020: 121 type species: *Atrachya hirashimai* Kimoto, 1969
chengi C.-F. Lee, 2020: 124 **A:** CH (TAI)
hirashimai Kimoto, 1969: 54 (*Atrachya*) **A:** CH (TAI)
nitidissima Chûjô, 1935d: 161 (*Luperodes*) **A:** CH (TAI)
bicoloripennis Chûjô, 1938c: 138 (*Luperodes*)
saramao Chûjô, 1962b: 230 (*Luperodes*)
tsoui C.-F. Lee, 2020: 140 **A:** CH (TAI)

genus *Nepalolepta* L. N. Medvedev, 1992d: 17 type species: *Nepalolepta carinata* L. N. Medvedev, 1992
carinata L. N. Medvedev, 1992d: 17 **A:** NP

genus *Nymphius* Weise, 1900c: 135 [= 1900d: 164] type species: *Luperus forcipifer* Weise, 1900
ensifer Guillebeau, 1891: 297 (*Luperus*) **A:** IS SY TR
forcipifer Weise, 1900c: 135 [= 1900d: 164] (*Luperus*) **A:** TR
gianassoi Bezděk, 2008: 80 **A:** IN
lydius Weise, 1886a: 594 (*Luperus*) **E:** BU **A:** AR IN TR

pravei Jacobson, 1899a: 141 (*Lyperus*) E: KR KZ RU (CT ST) UK
stylifer kadleci Bezděk, 2008: 83 (ssp.) **A:** TR
stylifer ogloblini Bogachev, 1947: 16 [= 1948: 131] (*Luperus*) **A:** AR IN TR
stylifer stylifer Weise, 1900a: 380 (*Luperus*) E: RU (ST) **A:** AB AR GG IN TR

genus *Ochralepta* Beenen, nom. nov. [RN] type species: *Ochralea nigricornis* Clark, 1865 (= *Galeruca nigripes* Olivier, 1808) **[# 689]**
Ochralea Clark, 1865a: 144 [HN] type species: *Ochralea nigricornis* Clark, 1865 (= *Galeruca nigripes* Olivier, 1808) [# 689]
nigripes Olivier, 1808: 648 (*Galeruca*) **A:** CH (GUX HAI YUN) ID (SD) **ORR**
erythromelas Weise, 1922b: 107 (*Monolepta*) [RN]
nigricornis Clark, 1865a: 144 (*Ochralea*)
pectoralis Harold, 1880a: 149 (*Ochralea*)

genus *Paleosepharia* Laboissière, 1936: 251 type species: *Paleosepharia truncata* Laboissière, 1936
amiana Chûjô, 1962b: 136 (*Monolepta*) **A:** CH (TAI)
basipennis Gressitt & Kimoto, 1963a: 645 **A:** CH (HAI GUX)
basituberculata S.-H. Chen & Jiang, 1984: 85, 90 **A:** CH (YUN)
castanoceps S.-H. Chen & Jiang, 1984: 87, 92 **A:** CH (YUN)
caudata S.-H. Chen & Jiang, 1984: 86, 91 **A:** CH (YUN)
centromaculata L. N. Medvedev, 2013b: 296 **A:** ID (UT)
excavata Chûjô, 1938c: 144 (*Monolepta*) **A:** CH (FUJ GAN GUA GUI GUX HUB HUN JIA JIX SCH SHA TAI YUN ZHE)
polychroma Laboissière, 1938: 8
fasciata Gressitt & Kimoto, 1963a: 646 **A:** CH (GUA)
formosana Chûjô, 1935d: 172 (*Monolepta*) **A:** CH (TAI)
fulvicornis S.-H. Chen, 1942a: 51 **A:** CH (CHQ FUJ GUA GUI GUX HAI HUB HUN SCH YUN ZHE) **ORR**
fusiformis S.-H. Chen & Jiang, 1984: 86, 91 **A:** CH (YUN)
gongshana S.-H. Chen & Jiang, 1986b: 198, 199 **A:** CH (YUN)
humeralis S.-H. Chen & Jiang, 1984: 87, 91 **A:** CH (YUN)
jiangae Beenen, 2008: 65 [= 2010a: 74] [RN] **A:** CH (XIZ YUN) ID (SD)
costata Jiang, 1990a: 137, 139 [HN]
j-signata S.-H. Chen & Jiang, 1984: 84, 89 **A:** CH (YUN)
kolthoffi Laboissière, 1938: 8 **A:** CH (ANH CHQ GUI HUB JIA SHA ZHE)
lingulata S.-H. Chen & Jiang, 1984: 85, 90 **A:** CH (YUN) **ORR**
liquidambara Gressitt & Kimoto, 1963a: 647 **A:** CH (ANH CHQ FUJ GAN GUA GUI GUX HUB HUN JIA JIX SCH YUN ZHE)
nantouensis Kimoto, 1996: 38 (*Monolepta*) **A:** CH (TAI)

orbiculata S.-H. Chen & Jiang, 1984: 86, 91 **A:** CH (YUN)
pallens S.-H. Chen, 1998: 123 **A:** CH (FUJ)
pokharensis Kimoto, 1982a: 17 (*Monolepta*) **A:** CH (XIZ) NP ID (SD)
costata Takizawa & Basu, 1987: 278
posticata S.-H. Chen, 1942a: 50 **A:** CH (HEB HUB SCH SHA SHX)
quercicola S.-H. Chen & Jiang, 1984: 84, 89 **A:** CH (LIA)
subnigra Gressitt & Kimoto, 1963a: 649 **A:** CH (JIX)
tibialis S.-H. Chen & Jiang, 1984: 88, 92 **A:** CH (GUX)
tomokunii Kimoto & Takizawa, 1983: 90 **A:** NP
truncatipennis S.-H. Chen & Jiang, 1984: 86, 90 **A:** CH (JIX)
verticalis S.-H. Chen & Jiang, 1984: 88, 92 **A:** CH (YUN)
yasumatsui Kimoto, 1969: 51 (*Monolepta*) **A:** CH (TAI)

genus *Palpoxena* Baly, 1861a: 203 type species: *Palpoxena laeta* Baly, 1861
Aenidea Baly, 1874: 179 [RN] type species: *Palpoxena laeta* Baly, 1861
konbirensis Weise, 1924a: 157 (*Platyxantha*) [RN] **A:** CH (HUB) ID ("India") NP **ORR**
indica Duvivier, 1892b: 445 (*Platyxantha*)
nasuta Westwood, 1838b: 177 (*Luperus*) **A:** ID (UT) NP **ORR**
bengalensis Jacoby, 1900b: 133 (*Aenidea*)
pallida Jacoby, 1896c: 286 (*Aenidea*) **A:** PA **ORR**
rufofulva Jacoby, 1896c: 287 (*Aenidea*) **A:** ID (HP UT) NP **ORR**

genus *Paraplotes* Laboissière, 1933b: 51 type species: *Paraplotes rugosa* Laboissière, 1933
Shensia S.-H. Chen, 1964: 202, 209 type species: *Shensia parvula* S.-H. Chen, 1964
antennalis S.-H. Chen, 1942a: 33 **A:** CH (SHA SHX)
parvula S.-H. Chen, 1964: 202, 209 (*Shensia*)
cheni C.-F. Lee, 2015a: 224 **A:** CH (TAI)
clavicornis Gressitt & Kimoto, 1963a: 682 **A:** CH (FUJ JIX) [# 733]
clavata Yang & Wu, 1998a: 182, 183 (*Japonitata*)
indica Takizawa & Basu, 1987: 279 **A:** ID (SD) NP
jengi C.-F. Lee, 2015a: 227 **A:** CH (TAI)
meihuai C.-F. Lee, 2015a: 229 **A:** CH (TAI)
nepalensis L. N. Medvedev, 1998c: 34 **A:** NP
rugatipennis S.-H. Chen & Jiang, 1986a: 76, 79 (*Japonitata*) **A:** CH (FUJ) **ORR**
semifulva Jiang, 1989: 222, 225 (*Japonitata*) **A:** CH (YUN)
tahsiangi C.-F. Lee, 2015a: 229 **A:** CH (TAI)
taiwana Chûjô, 1963a: 25 **A:** CH (TAI)
tatakaensis C.-F. Lee, 2015a: 236 **A:** CH (TAI)

tsoui C.-F. Lee, 2015a: 237 **A:** CH (TAI)
tsuenensis C.-F. Lee, 2015a: 239 **A:** CH (TAI)
yaoi C.-F. Lee, 2015a: 239 **A:** CH (TAI)
yuae C.-F. Lee, 2015a: 242 **A:** CH (TAI)

genus *Parexosoma* Laboissière, 1933a: 967 type species: *Malacosoma flaviventre* Baly, 1878
Bijukta Maulik, 1936: 321 type species: *Malacosoma flaviventre* Baly, 1878
beeneni Döberl, 1995a: 51 **A:** CH (SCH)
cuprescens Chûjô, 1966b: 23 (*Perexosoma*) **A:** BT NP
flaviventre Baly, 1878h: 379 (*Malacosoma*) **A:** BT CH (SCH) ID (SD) NP PA **ORR**
balyi Duvivier, 1884b: cccxiv (*Malacosoma*) [RN]
nigripenne Jiang, 1990a: 137, 139 **A:** CH (XIZ)
sikanga Gressitt & Kimoto, 1963a: 551 (*Cneorane*) **A:** CH (SCH)

genus *Phyllobrotica* Chevrolat, 1836: 381 type species: *Chrysomela quadrimaculata* Linnaeus, 1758
adusta adusta Creutzer, 1799: 121 (*Crioceris*) **E:** AL AU BH BU CR CZ GR HU MD NM RO SB SK TR UK **A:** TR
adusta litoralis J. Müller, 1924a: 78 (ssp.) **E:** CR ("Dalmatia")
binotata Ogloblin, 1936: 194, 402 **A:** AR TR
aslani Warchałowski, 1998b: 93
elegans Kraatz, 1866: 285 **E:** BU KR MD RU (ST) TR UK **A:** AB AR GG IS LE SY TR
nigropunctata Pic, 1894c: 72 (*Luperus*)
trimaculata Ballion, 1890: 33
frontalis Weise, 1886a: 587 **A:** AR IN SY TR
conjuncta Pic, 1904e: 58 (var.)
humeralis Kraatz, 1891: 124 (*Phyllopertha*)
komiyai Takizawa, 1985b: 107 **A:** ID (SD) NP **ORR**
lorestanica Romantsov, 2019: 20 **A:** IN
malinka Bezděk, 2010b: 564 **A:** IN TR
quadrimaculata Linnaeus, 1758: 376 (*Chrysomela*) **E:** AU BE BH BU BY CZ DE EN FI FR GB GE GR HU IT KR LA LT LU MD NL NR PL RU (CT NT ST) SB SK SL SV SZ UK **A:** CH (JIA) GG KZ RU (FE SS WS)
bimaculata Panzer, 1795a: 169 (*Crioceris*)
melanogaster Gmelin, 1790: 1720 (*Cryptocephalus*)
munda Weise, 1886a: 585 (var.)
nigroscutata Trella, 1932: 227 (ab.) [IN]
signata Mannerheim, 1825: 38 (*Galleruca*) **A:** CH (GAN HEB HEI JIL LIA NIN NMO SHN SHX) KZ MG RU (ES FE SS WS) SC
bisignata Gebler, 1830: 219 (*Galleruca*)

parasuturalis Gressitt & Kimoto, 1963a: 524 (*Taumacera*)
sibirica Joannis, 1865: 113
spinicoxa Laboissière, 1929a: 275 **A:** CH (YUN)

genus *Pseudergana* Beenen, 2019: 11 type species: *Luperus biplagiatus* Jacoby, 1890
biplagiata Jacoby, 1890c: 165 (*Luperus*) **A:** CH (HUB YUN)

genus *Pseudespera* S.-H. Chen, S.-Y. Wang & Jiang, 1985: 372, 375 type species: *Pseudespera femoralis* S.-H. Chen, S.-Y. Wang & Jiang, 1985
femoralis S.-H. Chen, S.-Y. Wang & Jiang, 1985: 374, 376 **A:** CH (HUB) **ORR**
paulowniae Jiang, 1992: 648, 670 **A:** CH (SCH)
sericea S.-H. Chen, S.-Y. Wang & Jiang, 1985: 373, 375 **A:** CH (SCH YUN)
shennongjiana S.-H. Chen, S.-Y. Wang & Jiang, 1985: 374, 376 **A:** CH (HUB)
sodalis S.-H. Chen, S.-Y. Wang & Jiang, 1985: 373, 376 **A:** CH (SCH)
subfemoralis Jiang, 1992: 648, 670 **A:** CH (YUN)

genus *Pseudosepharia* Laboissière, 1936: 253 type species: *Sepharia dilatipennis* Fairmaire, 1889
dilatipennis Fairmaire, 1889b: 78 (*Sepharia*) **A:** CH (CHQ GAN SHA SCH XIZ)
nigriceps Jiang, 1990c: 455, 456 **A:** CH (GAN HUB SHA)
pallinotata Jiang, 1992: 664, 673 **A:** CH (YUN)
rufula Jiang, 1992: 663, 672 **A:** CH (YUN)

genus *Pseudotheopea* C.-F. Lee & Bezděk, 2020: 76 type species: *Theopea sauteri* Chûjô, 1935a
aeneipennis Gressitt & Kimoto, 1963a: 675 (*Theopea*) **A:** CH (FUJ GUA HAI JIX)
aureoviridis Chûjô, 1935a: 85 (*Theopea*) **A:** JA (Ryukyus)
azurea Gressitt & Kimoto, 1963a: 677 (*Theopea*) **A:** CH (FUJ GUA HAI)
cheni C.-F. Lee & Bezděk, 2018: 340 (*Theopea*) **A:** CH (TAI)
coerulea Gressitt & Kimoto, 1963a: 679 (*Theopea*) **A:** CH (FUJ)
collaris Kimoto, 1989b: 75 (*Theopea*) **A:** CH (TAI)
hainanensis C.-F. Lee & Bezděk, 2018: 361 (*Theopea*) **A:** CH (HAI)
irregularis Takizawa, 1978a: 129 (*Theopea*) **A:** CH (TAI)
kanmiyai Kimoto, 1984: 53 (*Theopea*) **A:** CH (TAI)
laosensis C.-F. Lee & Bezděk, 2018: 363 (*Theopea*) **A:** CH (GUA GUX YUN) **ORR**
sauteri Chûjô, 1935d: 169 (*Theopea*) **A:** CH (TAI)
minuta Chûjô, 1935d: 171 (var.)
smaragdina Gressitt & Kimoto, 1963a: 680 (*Theopea*) **A:** CH (FUJ GUA HAI HKG)
sufangae C.-F. Lee & Bezděk, 2020: 107 **A:** CH (TAI)

genus *Scelolyperus* Crotch, 1874: 79 type species: *Scelolyperus tejonicus* Crotch, 1874

Eugalera Brancsik, 1899: 103 type species: *Eugalera reitteri* Brancsik, 1899 (= *Luperus altaicus* Mannerheim, 1825)

Tuomueria S.-H. Chen & Jiang, 1986c: 55, 58 type species: *Tuomueria tibialis* S.-H. Chen & Jiang, 1986

Tuomuria S.-H. Chen & Jiang, 1985: 105, 107 type species: *Tuomuria tibialis* S.-H. Chen & Jiang, 1985

altaicus Mannerheim, 1825: 38 (*Luperus*) **A:** CH (XIN) KZ MG RU (ES FE SS WS)

reitteri Brancsik, 1899: 104 (*Eugalera*)

caelesticus Beenen & Bezděk, 2007: 88 **A:** KI

eous Ogloblin, 1936: 274, 425 (*Luperus*, ssp.) **A:** NC RU (FE)

grandis Jacobson, 1894a: 106 (*Luperus*) **A:** CH (XIN) KI KZ

perreus Bezděk, 2015: 32 **A:** TR

sericeus Jacobson, 1894a: 105 (*Luperus*, var.) **A:** KZ KI UZ

jucundus Weise, 1900b: 286 (*Luperus*)

reitteri Laboissière, 1912c: 31 (*Luperus*, var.) [HN]

tibialis S.-H. Chen & Jiang, 1985: 106, 107 (*Tuomuria*) **A:** CH (XIN) KI KZ

kroliki Borowiec, 2005b: 380

tibialis S.-H. Chen & Jiang, 1986c: 56, 58 (*Tuomueria*) [HN]

genus *Shaira* Maulik, 1936: 277 type species: *Shaira maculata* Maulik, 1936

atra S.-H. Chen, Jiang & S.-Y. Wang, 1987: 63, 68 **A:** CH (YUN)

fulvicollis S.-H. Chen, Jiang & S.-Y. Wang, 1987: 64, 69 **A:** CH (SCH)

hemipteroides Lopatin, 2006c: 600 **A:** CH (SCH)

quadriguttata S.-H. Chen, Jiang & S.-Y. Wang, 1987: 63, 69 **A:** CH (YUN)

tenuipes S.-H. Chen & Jiang, 1981: 468, 486 (*Geinella*) **A:** CH (XIZ)

genus *Shairella* Chûjô, 1962b: 101 type species: *Shairella aeneipennis* Chûjô, 1962

aeneipennis Chûjô, 1962b: 105 **A:** CH (TAI)

itorum Kimoto, 1984: 56 (*Khasia*)

borowieci C.-F. Lee & C.-L. Huang, 2023: 412 **A:** CH (YUN)

caerulea Kimoto, 1996: 33 (*Japonitata*) **A:** CH (TAI)

cheni C.-F. Lee & Beenen, 2017a: 494 **A:** CH (TAI)

chungi C.-F. Lee & Beenen, 2017a: 496 **A:** CH (TAI)

guoi C.-F. Lee & Beenen, 2017a: 499 **A:** CH (TAI)

hsiehae C.-F. Lee & C.-L. Huang, 2023: 415 **A:** CH (TAI)

motienensis C.-F. Lee & Beenen, 2017a: 501 **A:** CH (TAI)

quadricostata Kimoto, 1996: 34 (*Japonitata*) **A:** CH (TAI)

tsoui C.-F. Lee & Beenen, 2017a: 503 **A:** CH (TAI)

genus *Sikkimia* Duvivier, 1891b: cliv type species: *Sikkimia antennata* Duvivier, 1891

Taiwanolepta Kimoto, 1989b: 73 type species: *Taiwanolepta babai* Kimoto, 1989

Vietocerus Lopatin, 2003a: 103 type species: *Vietocerus kabakovi* Lopatin, 2003

Yunomela S.-H. Chen, 1964: 201, 208 type species: *Yunomela rufa* S.-H. Chen, 1964

antennata Duvivier, 1891b: clv **A:** ID (SD)

babai Kimoto, 1989b: 74 (*Taiwanolepta*) **A:** CH (TAI)

meihuai C.-F. Lee & Bezděk, 2016: 88 **A:** CH (TAI)

rufa S.-H. Chen, 1964: 202, 208 (*Yunomela*) **A:** CH (YUN) **ORR**

sufangae C.-F. Lee & Bezděk, 2016: 93 **A:** CH (TAI)

tsoui C.-F. Lee & Bezděk, 2016: 96 **A:** CH (TAI)

yuae C.-F. Lee & Bezděk, 2016: 100 **A:** CH (TAI)

genus *Sinoluperoides* Kimoto, 1989a: 191 type species: *Sinoluperoides maculatus* Kimoto, 1989

marginalis Kimoto, 1989a: 195 **A:** NP **ORR**

genus *Sinoluperus* Gressitt & Kimoto, 1963a: 583 type species: *Sinoluperus subcostatus* Gressitt & Kimoto, 1963

subcostatus Gressitt & Kimoto, 1963a: 584 **A:** CH (GUA HAI HKG JIX SCH ZHE) **ORR**

wuyiensis Yang & Wu, 1998b: 262, 264 **A:** CH (FUJ)

genus *Strobiderus* Jacoby, 1884b: 61 type species: *Strobiderus excavatus* Jacoby, 1884 [# 745]

Syoplia Jacoby, 1886d: 85 type species: *Syoplia javanensis* Jacoby, 1886

fulvus Kimoto, 1977: 379 **A:** BT NP **ORR**

guiganus Yang, 1992a: 187, 189 **A:** CH (GUX)

nigriceps Laboissière, 1936: 258 **A:** CH (CHQ FUJ GUI HUN YUN) **ORR** [# 738]

tonkinensis Laboissière, 1936: 259

xianganus Yang, 1992b: 581, 589 **A:** CH (GUI HUN)

genus *Taiwanaenidea* Kimoto, 1984: 50 type species: *Taiwanaenidea strigosa* Kimoto, 1984

cheni C.-F. Lee & Beenen, 2015: 155 **A:** CH (TAI)

collaris Kimoto, 1984: 52 **A:** CH (TAI)

jungchangi C.-F. Lee & Beenen, 2015: 162 **A:** CH (TAI)

strigosa Kimoto, 1984: 51 **A:** CH (TAI)

genus *Taiwanoshaira* C.-F. Lee & Beenen, 2020: 132 type species: *Taiwanoshaira tsoui* Lee & Beenen, 2020

chujoi Kimoto, 1982b: 151 (*Shaira*) **A:** CH (TAI)
taipingshanensis C.-F. Lee & Beenen, 2020: 139 **A:** CH (TAI)
tsoui C.-F. Lee & Beenen, 2020: 142 **A:** CH (TAI)

genus *Taumacera* Thunberg, 1814: 48 type species: *Taumacera deusta* Thunberg, 1814
Acroxena Baly, 1879a: 462 type species: *Acroxena nasuta* Baly, 1879
Doridea Baly, 1864d: 236 type species: *Doridea insignis* Baly, 1864
Kinabalua Mohamedsaid, 1997: 132 type species: *Kinabalua antennata* Mohamedsaid, 1997
Metellus Jacoby, 1886d: 63 [RN] type species: *Neocharis fulvicollis* Jacoby, 1881
Nacrea Baly, 1886a: 29 type species: *Nacrea maculata* Baly, 1886
Neocharis Jacoby, 1881: 448 [HN] type species: *Neocharis fulvicollis* Jacoby, 1881
Neochrolea Jacoby, 1887a: 117 type species: *Neochrolea cavifrons* Jacoby, 1887
Paraenidea Laboissière, 1933b: 66 type species: *Paraenidea azurea* Laboissière, 1933
Platyxantha Baly, 1864d: 233 type species: *Platyxantha apicalis* Baly, 1864
Platyxanthoides Laboissière, 1933b: 71 type species: *Platyxanthoides variceps* Laboissière, 1933
Thaumacera Harold, 1876c: 3560 [UE] type species: *Taumacera deusta* Thunberg, 1814
Xenarthra Baly, 1861b: 298 type species: *Xenarthra cervicornis* Baly, 1861
aureipennis Laboissière, 1933b: 69 (*Paraenidea*) **A:** CH (YUN)
indica Jacoby, 1889b: 231 (*Dorydea*) **A:** CH (YUN) **ORR**
azurea Laboissière, 1933b: 67 (*Paraenidea*)
hoabinhia Laboissière, 1933b: 69 (*Paraenidea*, var.)
yunnana L. N. Medvedev, 2015c: 327 (*Palpoxena*)
insularis Gressitt & Kimoto, 1963b: 926 (*Paraenidea*) **A:** CH (HAI)
magenta Gressitt & Kimoto, 1963b: 927 (*Paraenidea*) **A:** CH (HAI)
martensi L. N. Medvedev, 1990b: 20 (*Acroxena*) **A:** NP **ORR**
nasuta Baly, 1879a: 463 (*Acroxena*) **A:** NP **ORR**
occipitalis Laboissière, 1933b: 70 (*Paraenidea*) **A:** CH (YUN) **ORR**
paradoxa Laboissière, 1936: 240 **A:** NP **ORR**
variceps Laboissière, 1933b: 71 (*Platyxanthoides*) **A:** CH (YUN) **ORR**
viridis Hope, 1831: 29 (*Auchenia*) **A:** ID (AP) NP **ORR**

genus *Theopea* Baly, 1864d: 237 type species: *Crioceris impressa* Fabricius, 1801
bicolor Kimoto, 1989a: 199 **A:** CH (YUN) **ORR**
bicoloroides C.-F. Lee & Bezděk, 2020: 71 **A:** CH (YUN) **ORR**

genus *Trichobalya* Weise, 1924a: 152 [RN] type species: *Trichidea bowringii* Baly, 1890
Trichidea Baly, 1890a: 13 [HN] type species: *Trichidea bowringii* Baly, 1890
apicalis Kimoto, 1982a: 23 **A:** BT ID (SD) NP
bowringii Baly, 1890a: 13 (*Trichidea*) **A:** CH (GUA HAI HKG YUN) ID (SD) **ORR**
tonkinensis Laboissière, 1936: 260

melanocephala Jacoby, 1889b: 220 (*Cynorta*) **A:** CH (YUN) **ORR**

gularis Laboissière, 1936: 261

mouhoti Baly, 1890a: 14 (*Trichidea*)

viridipennis Kimoto, 1977: 380 **A:** BT ID (SD) NP

genus *Trichomimastra* Weise, 1922b: 75 type species: *Mimastra seminigra* Weise, 1922

atriceps Lopatin, 1979c: 439 **A:** ID (KA)

attenuata albida Gressitt & Kimoto, 1963a: 545 (ssp.) **A:** CH (SCH XIZ)

attenuata attenuata Gressitt & Kimoto, 1963a: 543 **A:** CH (CHQ FUJ SCH ZHE)

gracilipes Gressitt & Kimoto, 1963a: 545 **A:** CH (GUI HUB SCH SHA XIZ)

hirsuta Jacoby, 1892b: 945 (*Mimastra*) **A:** CH (HAI) ID ("India") NP **ORR**

membranacea Weise, 1922b: 77 (*Mimastra*)

piceipennis Laboissière, 1929a: 287 (*Mimastra*)

jejuna Weise, 1889g: 616 (*Luperus*) **A:** CH (GAN SCH XIZ)

kaganica L. N. Medvedev, 2013f: 778 **A:** PA

nigroapicalis L. N. Medvedev, 2011c: 250 **A:** NP

pakistanica L. N. Medvedev, 2013f: 777 **A:** PA

pectoralis Laboissière, 1929a: 286 (*Mimastra*) **A:** CH (HAI) **ORR**

pellucida Ogloblin, 1936: 182, 400 (*Mimastra*) **A:** CH (SCH XIZ YUN)

genus *Trichosepharia* Laboissière, 1936: 254 type species: *Trichosepharia pubescens* Laboissière, 1936

pubescens Laboissière, 1936: 255 **A:** CH (YUN) **ORR**

basalis Laboissière, 1936: 255 (var.)

genus *Tschitscherinula* Jacobson, 1908: 619 type species: *Tschitscherinula paradoxocara* Jacobson, 1908

paradoxocara Jacobson, 1908: 621 **A:** TM TZ UZ

genus *Tsouchya* C.-F. Lee, 2020: 142 type species: *Atrachya mediofasciata* Kimoto, 1976

mediofasciata Kimoto, 1976: 6 (*Atrachya*) **A:** CH (TAI)

tsoui C.-F. Lee, 2009: 23 (*Monolepta*)

genus *Vietoluperus* L. N. Medvedev & Dang, 1981: 632 type species: *Vietoluperus alleculoides* L. N. Medvedev & Dang, 1981

alleculoides L. N. Medvedev & Dang, 1981: 632 **A:** ID (SD) **ORR**

bengalicus L. N. Medvedev, 2009b: 408

genus *Xingeina* S.-H. Chen, Jiang & S.-Y. Wang, 1987: 61, 67 type species: *Xingeina vittata* S.-H. Chen, Jiang & S.-Y. Wang, 1987

femoralis S.-H. Chen, Jiang & S.-Y. Wang, 1987: 63, 68 **A:** CH (SCH)
nigra S.-H. Chen, Jiang & S.-Y. Wang, 1987: 62, 68 **A:** CH (SCH)
nigrolucens Lopatin, 2006c: 599 **A:** CH (SCH)
vittata S.-H. Chen, Jiang & S.-Y. Wang, 1987: 61, 68 **A:** CH (YUN)

tribe Metacyclini Chapuis, 1875

genus *Pseudoides* Jacoby, 1892b: 966 type species: *Pseudoides bivitta* Jacoby, 1892 (= *Calomicrus flavovittis* Motschulsky, 1858)
flavovittis Motschulsky, 1858: 101 (*Calomicrus*) **A:** BT CH (YUN) ID (SD) NP **ORR**
bivittata Jacoby, 1892b: 966
marginalis Chûjô, 1966b: 12 (*Exosoma*) **A:** NP

tribe Oidini Laboissière, 1921 (1875)

genus *Oides* Weber, 1801: 53 type species: *Chrysomela bipunctata* Fabricius, 1781 (= *Oides andrewesi* Jacoby, 1900)
Adorium Fabricius, 1801a: 409 type species: *Chrysomela bipunctata* Fabricius, 1781 (= *Oides andrewesi* Jacoby, 1900)
Boisduvalia Montrouzier, 1856: 72 type species: *Galeruca sexlineata* Montrouzier, 1856
Botanoctona Fairmaire, 1877b: 185 type species: *Botanoctona pallidocincta* Fairmaire, 1877
Callipepla Dejean, 1836: 375 [HN] type species: *Adorium posticum* Boisduval, 1835
Eparchista Gistel, 1848b: xi [RN] type species: *Adorium flavum* Olivier, 1807
Ochralea Chevrolat, 1836: 375 type species: *Adorium flavum* Olivier, 1807
Rhombopalpa Chevrolat, 1836: 375 type species: *Adorium decempunctatum* Billberg, 1808
bowringii Baly, 1864a: 623 (*Adorium*) **A:** CH (CHQ FUJ GAN GUA GUI GUX HAI HKG HUB HUN JIX SCH SHA SHX XIZ YUN ZHE) JAi **ORR**
elegans Laboissière, 1919b: 161
tonkinensis Laboissière, 1929a: 252
coccinelloides Gahan, 1891: 458 **A:** CH (XIZ YUN) ID (SD) **ORR**
decempunctata Billberg, 1808: 230 (*Adorium*) **A:** CH (ANH BEI CHQ FUJ GAN GUA GUI GUX HAI HEB HEN HKG HUB HUN JIA JIL JIX NIN NMO SCH SHA SHG SHN SHX TAI TIA YUN ZHE) NC RU (FE) SC **ORR**
decemmaculata Laboissière, 1927: 39
gigantea Roubal, 1929a: 96 (*Solanophila*)
epipleuralis Laboissière, 1929a: 254 **A:** CH (TAI)
flava Olivier, 1807: 611 (*Adorium*) **A:** CH (HAI) ID (AP HP) NP **ORR**
abdominalis Duvivier, 1884b: cxxxiii
albicans Duvivier, 1884c: 237
bakeri Laboissière, 1925: 609
coomani Laboissière, 1927: 42
elliptica Duvivier, 1884b: cxxxvii

flavida Duvivier, 1884b: cxxxvii
flavipes Duvivier, 1884c: 239
inornata Baly, 1879a: 444
marcida Duvivier, 1884b: cxxxviii
ovalis Laboissière, 1925: 610
philippinensis Boheman, 1859: 175 (*Adorium*)
rhabdoscelis Weise, 1922b: 57
subtilissima Duvivier, 1884b: cxxxvi
sumatrensis Blanchard, 1853: 337 (*Adorium*)
geiseri C.-F. Lee & Beenen, 2017b: 61 **A:** ID (SD) NP **ORR**
laticlava Fairmaire, 1889b: 74 (*Adorium*) **A:** CH (ANH FUJ GAN GUA GUI GUX HAI HEN HKG HUB HUN JIA JIX SCH SHA SHN TAI YUN ZHE)
leucomelaena Weise, 1922b: 58 **A:** CH (ANH CHQ FUJ GUA GUI GUX HAI HUB YUN ZHE) **ORR**
disjuncta Pic, 1928g: 29 (var.)
subsinuata Pic, 1928g: 29 (var.)
livida Fabricius, 1801a: 410 (*Adorium*) **A:** BT CH (CHQ FUJ GUA GUI GUX HUN SCH YUN XIZ ZHE) ID (AP SD UT) NP **ORR**
analis Schoenherr, 1808: 231 (*Adorium*) [RN]
diardi Guérin-Méneville, 1838: 148 (*Adorium*)
livida Weber, 1801: 53 [HN]
nigripes Jacoby, 1891a: 34
pallidicornis Jacoby, 1900a: 284
pectoralis Clark, 1865a: 144 (*Rhombopala*)
maculata Olivier, 1807: 611 (*Adorium*) **A:** ID (AP SD UT) NP **ORR**
subhemisphaerica Guérin-Méneville, 1838: 146 (*Adorium*)
palleata Fabricius, 1801a: 410 (*Adorium*) **A:** CH (GUA GUI GUX HAI TAI YUN) ID (AP SD) NP **ORR**
bipunctata Fabricius, 1781a: 127 (*Chrysomela*) [HN]
indosinensis Laboissière, 1927: 37
laboissierei Pic, 1927k: 23 (var.)
scutellata Hope, 1831: 28 (*Adorium*) **A:** BT CH (XIZ) ID (HP SD UT) NP PA **ORR**
gyironga S.-H. Chen & Jiang, 1981: 468, 486
semipunctata Duvivier, 1884b: cxxxiii **A:** CH (YUN) ID (AP SD) NP **ORR**
quadrimaculata Jacoby, 1900b: 126
tarsata Baly, 1865e: 435 (*Adorium*) **A:** CH (ANH CHQ FUJ GAN GUA GUI GUX HAI HEB HEN HUB HUN JIA JIX SCH SHA XIZ ZHE) **ORR**
chinensis Weise, 1922b: 57
indica Baly, 1879a: 443
sordida Baly, 1865e: 435 (*Adorium*)

thibettana Jacoby, 1900b: 128 **A:** CH (SCH XIZ) **ORR**
ustulaticia Laboissière, 1927: 41 **A:** CH (GUI YUN)

supertribe Galerucitae, genera unassigned to tribes

genus *Demarchus* Jacoby, 1887a: 101 type species: *Demarchus pubipennis* Jacoby, 1887
hsui C.-F. Lee & J.-C. Chen, 2023: 5 **A:** CH (TAI)
nigriceps S.-H. Chen & S.-Y. Wang, 1988: 350, 353 **A:** CH (XIZ)
pubipennis Jacoby, 1887: 101 **A:** PA **ORR**

genus *Hespera* Weise, 1889g: 638 type species: *Hespera sericea* Weise, 1889
Allomorpha Jacoby, 1892b: 934 type species: *Allomorpha sericea* Jacoby, 1892 (= *Hespera lomasa* Maulik, 1926)
abdominalis S.-Y. Wang, 1997: 915, 929 **A:** CH (CHQ HUB)
aenea S.-H. Chen & S.-Y. Wang, 1984b: 315, 320 **A:** CH (YUN)
aeneocuprea S.-H. Chen & S.-Y. Wang, 1986b: 290, 301 **A:** CH (YUN)
aeneonigra S.-H. Chen & S.-Y. Wang, 1986b: 291, 302 **A:** CH (YUN)
angusticollis S.-H. Chen & S.-Y. Wang, 1986b: 291, 302 **A:** CH (XIZ)
aterrima Kung & S.-H. Chen, 1954b: 155, 162 **A:** CH (HUB SCH ZHE)
auricuprea S.-H. Chen & S.-Y. Wang, 1986b: 292, 302 **A:** CH (YUN)
aurisericea S.-H. Chen & P.-Y. Yu, 1976: 211, 222 **A:** CH (XIZ)
bengalensis Scherer, 1969: 40 **A:** ID (SD)
bhutanensis Scherer, 1979a: 128 **A:** BT NP
bipilosa S.-H. Chen & S.-Y. Wang, 1984b: 315, 320 **A:** CH (YUN)
biswasi Basu, 1985: 209 **A:** ID (AP)
brachyelytra S.-H. Chen & S.-Y. Wang, 1984b: 315, 320 **A:** CH (YUN)
brevipilosa Kung & S.-H. Chen, 1954b: 155, 161 **A:** CH (GUI SCH) **ORR**
byrsa Gressitt & Kimoto, 1963a: 804 **A:** CH (SCH XIZ)
cavaleriei S.-H. Chen, 1932b: 194 **A:** CH (CHQ FUJ GUI HUB SCH SHA YUN) **ORR**
chagyabana S.-H. Chen & S.-Y. Wang, 1981a: 499, 508 **A:** CH (SCH XIZ YUN)
coeruleipennis S.-H. Chen & S.-Y. Wang, 1984b: 316, 321 **A:** CH (YUN)
crassicornis S.-H. Chen, 1932b: 194 **A:** CH (SCH YUN)
cyanea Maulik, 1926: 140 **A:** CH (SCH XIZ YUN) ID (SD) NP **ORR**
dakshina Maulik, 1926: 145 **A:** NP **ORR**
dasytoides L. N. Medvedev, 1992d: 22 **A:** NP
elegans L. N. Medvedev, 1993c: 367 **A:** CH (YUN)
excavata S.-H. Chen & S.-Y. Wang, 1986b: 292, 303 **A:** CH (SCH)
flavipes S.-H. Chen, 1939a: 39 **A:** ID (SD) **ORR**
flaviventris L. N. Medvedev, 1997a: 13 **A:** NP
flavodorsata S.-H. Chen & S.-Y. Wang, 1984b: 314, 319 **A:** CH (SCH YUN)

fulvicollis Weise, 1924b: 23 **A:** YE **AFR**
fulvicornis Kung & S.-H. Chen, 1954b: 157, 162 **A:** CH (SCH SHA SHX)
fulvimembris Kung & S.-H. Chen, 1954b: 157, 162 **A:** NP **ORR**
fulvipes S.-H. Chen & S.-Y. Wang, 1986b: 293, 303 **A:** CH (YUN)
gerhardi Döberl, 2001: 382 [RN] **A:** BT
longicornis Scherer, 1979a: 129 [HN]
glabriceps S.-H. Chen & S.-Y. Wang, 1984b: 316, 321 **A:** CH (YUN)
gracilicornis S.-H. Chen & S.-Y. Wang, 1986b: 293, 303 **A:** CH (XIZ YUN)
gyirongana S.-H. Chen & S.-Y. Wang, 1986b: 294, 304 **A:** CH (XIZ)
krishna Maulik, 1926: 144 **A:** BT CH (CHQ GAN HUB SCH XIZ YUN ZHE) ID (SD UT) NP **ORR**
nepalensis Chûjô, 1966b: 33
kumaonensis Scherer, 1969: 47 **A:** ID (UT) NP
lijiangana S.-H. Chen & S.-Y. Wang, 1986b: 294, 304 **A:** CH (YUN)
lomasa Maulik, 1926: 142 [RN] **A:** CH (FUJ GUA GUI HAI HKG HUB JIA SCH SHA TAI XIZ YUN ZHE) JA (Ryukyus) NP **ORR**
albopilosa Chûjô, 1957c: 7 (ssp.)
auripilosa Chûjô, 1936b: 90
formosana Chûjô, 1936b: 90
insulana Chûjô, 1936b: 89
loochooana Chûjô, 1957c: 6 (ssp.)
rufotibialis Chûjô, 1936b: 88
sericea Jacoby, 1892b: 934 (*Allomorpha*) [HN]
longicornis S.-H. Chen, 1932b: 196 **A:** CH (YUN)
medogana S.-H. Chen & S.-Y. Wang, 1986b: 294, 304 **A:** CH (XIZ)
melanoptera S.-H. Chen & S.-Y. Wang, 1986b: 295, 305 **A:** CH (XIZ)
melanosoma S.-H. Chen & S.-Y. Wang, 1984b: 317, 321 **A:** CH (XIZ YUN)
metallica Scherer, 1969: 41 **A:** ID (SD) NP
mimica L. N. Medvedev, 1993c: 367 **A:** BT NP
montana Chûjô, 1936b: 87 **A:** CH (TAI)
naini Scherer, 1969: 45 **A:** ID (UT) NP
nepalica L. N. Medvedev, 2003a: 321 **A:** NP
nigriceps L. N. Medvedev, 1992d: 22 **A:** NP
nigripes Maulik, 1926: 143 **A:** ID (UT) NP **ORR**
nitidicollis S.-H. Chen & S.-Y. Wang, 1984b: 318, 322 **A:** CH (YUN)
nitididorsata S.-H. Chen & S.-Y. Wang, 1986b: 295, 305 **A:** CH (YUN)
obesula Kung & S.-H. Chen, 1954b: 158, 163 **A:** CH (FUJ) **ORR**
oculata Gressitt & Kimoto, 1963a: 807 **A:** CH (FUJ HUB)
pubiceps Chen & S.-Y. Wang, 1986b: 296, 305 **A:** CH (XIZ)

punctata S.-H. Chen, 1932b: 196 **A:** NP **ORR**
puncticeps S.-H. Chen & S.-Y. Wang, 1984b: 316, 321 **A:** CH (YUN)
ruficollis S.-H. Chen, 1932b: 195 **A:** CH (YUN) **ORR**
rufipes Maulik, 1926: 139 **A:** ID (UT) NP **ORR**
sangzhiensis S.-Y. Wang, 1993: 319, 328 **A:** CH (HUB HUN)
schawalleri L. N. Medvedev, 1990b: 26 **A:** NP
semicyanea S.-H. Chen, 1939a: 38 **A:** ID (SD)
sericea Weise, 1889g: 639 **A:** BT CH (CHQ FUJ GAN GUI GUX HUB HUN SCH SHA XIZ YUN ZHE) ID (HP SD) NP **ORR**
sinensis S.-H. Chen, 1934a: 12 **A:** CH (SCH "North-east China")
sparsa S.-Y. Wang, 1992c: 701, 741 **A:** CH (YUN)
strigiceps Kimoto, 2001: 57 **A:** NP
tenebrosa Gressitt & Kimoto, 1963a: 809 **A:** CH (YUN)
tibetana S.-H. Chen & P.-Y. Yu, 1976: 212, 222 **A:** CH (XIZ)
univestis S.-H. Chen & S.-Y. Wang, 1986b: 296, 306 **A:** CH (SCH)
varicolor S.-H. Chen & S.-Y. Wang, 1986b: 297, 306 **A:** CH (SCH)
violaceipennis Scherer, 1969: 43 **A:** ID (SD) NP
viridis S.-H. Chen, 1936d: 597 **A:** CH (SCH XIZ YUN) ID (SD) NP **ORR**

genus *Luperomorpha* Weise, 1887h: 202 type species: *Luperomorpha trivialis* Weise, 1887 (= *Phyllotreta funesta* Baly, 1874)
Docemasia Jacoby, 1900a: 283 type species: *Docemasia caerulea* Jacoby, 1900
Luperocnemus Fairmaire, 1888c: 43 type species: *Luperocnemus xanthoderus* Fairmaire, 1888
Pushtunaltica Lopatin, 1962d: 1814 type species: *Pushtunaltica klapperichi* Lopatin, 1962
aeneipennis S.-H. Chen, 1934h: 346 **A:** ID (SD) **ORR**
albofasciata Duvivier, 1892b: 428 **A:** CH (HAI XIZ YUN) ID (SD UT) **ORR**
amamiana Chûjô & Ohno, 1965: 8 **A:** JA (Ryukyus)
antennata S.-H. Chen, 1941c: 192 **A:** CH (SCH)
arabica Doguet, 1979a: 308 **A:** OM SA YE [# 727]
basantapurica Kimoto, 2001: 52 (*Aphthona*) **A:** NP
biondii Döberl, 2012: 439 **A:** YE (Socotra)
birmanica Jacoby, 1892b: 920 (*Aphthona*) **A:** BT CH (FUJ GUX HAI HUB SCH TAI YUN) JA (Ryukyus) NP **ORR**
bipustulata S.-H. Chen, 1933f: 52 (var.)
boja Gressitt & Kimoto, 1963a: 862 **A:** CH (HUB SCH)
bryanti Bezděk & Konstantinov, **nom. nov.** [RN] **A:** NP **ORR** [# 746]
caerulea Bryant, 1939: 18 [HN]
cheni S.-Y. Wang & S.-Q. Ge, 2010: 197, 199 **A:** CH (SCH)
clypeata S.-Y. Wang, 1992c: 707, 743 **A:** CH (SCH)

collaris Baly, 1874: 198 (*Aphthona*) **A:** CH (GAN GUX JIA SCH TAI ZHE) JA
japonica Chûjô & Ohno, 1965: 6
costipennis S.-Y. Wang, 1998: 185 [= 2002b: 684, 693] **A:** CH (FUJ ZHE)
deorali Bezděk & Konstantinov, **nom. nov.** [RN] **A:** NP [# 747]
nepalensis L. N. Medvedev, 1990b: 28 [HN]
dilatata S.-Y. Wang, 1992c: 707, 744 **A:** CH (YUN)
fasciata L. N. Medvedev, 2015c: 329 **A:** CH (YUN)
funesta Baly, 1874: 196 (*Phyllotreta*) **A:** CH (HEB SHA) JA RU (ES FE) SC
trivialis Weise, 1887h: 204
glabricollis S.-Y. Wang & S.-Q. Ge, 2010: 197, 199 **A:** CH (GUA)
guangxiana S.-Y. Wang & S.-Q. Ge, 2010: 197, 200 **A:** CH (GUA)
hainana S.-Y. Wang & S.-Q. Ge, 2010: 197, 200 **A:** CH (HAI)
hidakai Kimoto, 1966a: 603 **A:** JA (Ryukyus)
hirsuta L. N. Medvedev, 1997a: 14 **A:** NP
josifovi Gruev, 1994: 82 **A:** NC
klapperichi Lopatin, 1962d: 1815 (*Pushtunaltica*) **A:** AF
kurosawai Kimoto, 1970d: 304 **A:** CH (TAI)
lushuinensis S.-Y. Wang, 1992c: 708, 744 **A:** CH (YUN)
maculata S.-Y. Wang, 1992c: 708, 744 **A:** CH (SCH)
medvedevi Bezděk & Konstantinov, **nom. nov.** [RN] **A:** NP [# 748]
schmidti L. N. Medvedev, 2009d: 412 [HN]
metallica S.-H. Chen, 1935a: 78 **A:** BT CH (XIZ) NP ID (SD)
naini Scherer, 1969: 53 **A:** ID (UT)
nepalensis L. N. Medvedev, 1984: 51 **A:** NP
nepalica Kimoto, 2001: 60 **A:** NP
nigra S.-H. Chen, 1933a: 246 (var.) **A:** CH (GAN GUX HEN) RU (FE)
nigripennis Duvivier, 1892b: 428 **A:** ID (HP KA) NP PA **ORR**
flaviventris Jacoby, 1896c: 257 (*Phyllotreta*)
weisei Jacoby, 1898a: 189
nobilis Weise, 1889g: 640 **A:** CH (GAN GUI SHA SCH TAI) JA (Ryukyus)
pedicelis S.-Y. Wang & S.-Q. Ge, 2010: 198, 201 **A:** CH (YUN)
pryeri Baly, 1874: 198 (*Aphthona*) **A:** JA (Kyushu, Shikoku) NC SC
collaris Chûjô, 1940b: 117 (*Phyllotreta*, var.) [IN] [# 698]
rubra S.-H. Chen, 1933a: 246 **A:** CH (GUA HAI HKG) **ORR**
saigusai Kimoto, 1970d: 303 **A:** CH (TAI)
sakishimana Kimoto & Gressitt, 1966: 557 **A:** JA (Ishigaki, Iriomote)
sasajii Kimoto, 1970d: 304 **A:** CH (TAI)
schmidti L. N. Medvedev, 2004d: 204 **A:** NP
sibirica Csiki, 1916: 127 (*Luperus*) **A:** CH (HEB JIA JIL NMO SHA) MG RU (ES FE)

schultzi Csiki, 1916: 127 (*Luperus*, ab.) [IN]
suturalis S.-H. Chen, 1938a: 4
similimetallica S.-Y. Wang & S.-Q. Ge, 2010: 198, 201 **A**: CH (YUN)
tenebrosa Jacoby, 1886a: 731 (*Phyllotreta*) **A**: CH (TAI) JA
tokejii Chûjô & Ohno, 1965: 9 **A**: JA
irisae Kimoto, 1966a: 603
viridipennis Scherer, 1969: 50 **A**: ID (UT)
viridis S.-Y. Wang, 1992c: 709, 745 **A**: CH (GAN SCH)
xanthodera Fairmaire, 1888c: 43 (*Luperocnemus*) Ei: AU BE CZ FR GB GE HU IT ME NL PL PT RU (ST) SP SZ **A**: CH (CHQ FUJ GAN GUA GUI HUB HUN JIA SCH SHA SHN YUN ZHE) NC SC
similis Chûjô, 1938d: 166
yunnanensis Kung & Chen, 1954a: 94, 95 **A**: CH (YUN)

genus *Mandarella* Duvivier, 1892b: 433 type species: *Mandarella nagpurensis* Duvivier, 1892
Paramesopa L. N. Medvedev, 1984: 49 type species: *Paramesopa violacea* L. N. Medvedev, 1984
Stenoluperus Ogloblin, 1936: 247, 419 type species: *Luperus potanini* Weise, 1889
bhutanensis Kimoto, 1977: 366 (*Stenoluperus*) **A**: BT ID (SD)
bicarinata Weise, 1889g: 615 (*Luperus*) **A**: JA (Honshu)
birmanica Bryant, 1954a: 419 **A**: BT NP **ORR**
cyanea Baly, 1874: 184 (*Arthrotus*) **A**: JA RU (FE)
cyanipennis S.-Y. Wang, 1992c: 703, 741 (*Stenoluperus*) **A**: CH (SCH)
darjeelingensis Takizawa, 1985b: 108 (*Stenoluperus*) **A**: ID (SD)
flavimembris S.-H. Chen, 1942a: 66 (*Stenoluperus*) **A**: CH (GAN)
flavipes S.-H. Chen, 1942a: 65 (*Stenoluperus*) **A**: CH (SCH XIZ)
flaviventris S.-H. Chen, 1942a: 67 (*Stenoluperus*) **A**: CH (CHQ FUJ HUN JIA SCH TAI XIZ YUN ZHE) **ORR**
esakii Kimoto, 1969: 40 (*Stenoluperus*)
itoi Kimoto, 1991b: 116 (*Stenoluperus*) [HN]
matsumurai Takizawa, 1978a: 128 (*Stenoluperus*)
taiwanensis Medvedev, 2012b: 427 [RN]
itoi Chûjô, 1966b: 16 **A**: NP
lemoides Weise, 1889g: 616 (*Luperus*) **A**: CH (SCH XIZ)
major Kimoto, 1977: 367 (*Stenoluperus*) **A**: BT
minor Kimoto, 1977: 364 (*Stenoluperus*) **A**: BT ID ("India") NP
moorii Baly, 1874: 188 (*Luperus*) **A**: JA SC
nagpurensis Duvivier, 1892b: 434 **A**: NP **ORR**
nepalica Medvedev, 2012b: 424 [RN] **A**: NP
flavipes L. N. Medvedev, 1990b: 28 (*Paramesopa*) [HN]

nigra S.-Y. Wang, 1992c: 704, 742 (*Stenoluperus*) **A:** CH (YUN)
nigricornis L. N. Medvedev, 2000e: 17 (*Stenoluperus*) **A:** NP
emotoi Kimoto, 2001: 47 (*Stenoluperus*)
thudamensis Kimoto, 2001: 49 (*Stenoluperus*)
nigrimembris S.-H. Chen, 1942a: 66 (*Stenoluperus*) **A:** CH (GAN SCH XIZ)
nipponensis Laboissière, 1913: 24 (*Luperus*) [RN] **A:** CH (CHQ FUJ GAN HUN NIN SCH SHA-XIZ YUN ZHE) JA NC RU (ES FE) SC
jacobyi Weise, 1924a: 119 (*Luperus*) [RN]
longicornis Jacoby, 1886a: 742 (*Luperus*) [HN]
pallipes Gressitt & Kimoto, 1963a: 581 (*Stenoluperus*) **A:** CH (HUB)
parva Gressitt & Kimoto, 1963a: 582 (*Stenoluperus*) **A:** CH (HUB)
piceae S.-Y. Wang, 1992c: 705, 742 (*Stenoluperus*) **A:** CH (SCH)
potanini Weise, 1889g: 615 (*Luperus*) **A:** CH (CHQ GAN SCH XIZ)
pulchella Lopatin, 2005d: 879 (*Luperus*) **A:** CH (GAN)
punctata Takizawa, 1988b: 542 (*Stenoluperus*) **A:** ID (SD) NP
puncticeps S.-Y. Wang, 1992c: 706, 743 (*Stenoluperus*) **A:** CH (YUN)
puncticollis S.-Y. Wang, 1997: 917, 929 (*Stenoluperus*) **A:** CH (CHQ HUB)
smetanai Takizawa, 1988b: 544 (*Stenoluperus*) **A:** ID (SD) NP
tibialis S.-H. Chen, 1942a: 63 (*Stenoluperus*) **A:** CH (SCH XIZ)
tsoui C.-F. Lee, Tsai, Konstantinov & Yeh, 2016: 34 **A:** CH (TAI)
uenoi Kimoto, 1969: 39 (*Luperus*) **A:** CH (TAI)
kimotoi Döberl, 2001: 383 (*Stenoluperus*) [RN]
kimotoi Medvedev, 2012b: 427 [RN]
minor Kimoto, 1991b: 117 (*Stenoluperus*) [HN]
taiwana Kimoto, 1991a: 14 (*Stenoluperus*)
verticalis Kimoto, 1977: 365 (*Stenoluperus*) **A:** BT NP
violacea L. N. Medvedev, 1984: 49 (*Paramesopa*) **A:** ID (UT) NP

genus *Nonarthra* Baly, 1862b: 455 type species: *Nonarthra variabilis* Baly, 1862
Enneamera Harold, 1875b: 185 [RN] type species: *Nonarthra variabilis* Baly, 1862
amamiana Chûjô, 1957a: 18 **A:** JA (Ryukyus)
flavomarginata Chûjô, 1957c: 7 (ssp.)
bimaculata Kimoto, 1970b: 206 **A:** CH (TAI)
birmanica Jacoby, 1892b: 935 (*Enneamera*) **A:** BT NP **ORR**
dakshina Maulik, 1926: 121
dhumala Maulik, 1926: 121
chengi C.-F. Lee, 2014c: 252 **A:** CH (TAI)
coreana Chûjô, 1935f: 359 **A:** CH (GUX SCH ZHE) NC SC
cyanea alticola S.-Y. Wang, 1992c: 676, 730 (ssp.) **A:** CH (SCH)

cyanea cyanea Baly, 1874: 210 **A:** CH (AHN CHQ FUJ GAN GUA GUI HEB HUB JIA SCH SHA XIZ ZHE) JA NC RU (FE) SC **NTRi ORR**

fulva Baly, 1874: 211

nigricollis Weise, 1889g: 641

formosensis Chûjô, 1935f: 360 **A:** CH (TAI)

nigripennis S.-Y. Wang, 1992c: 676, 731 **A:** CH (YUN)

nigripes S.-Y. Wang, 1992c: 676, 731 **A:** CH (SCH)

pallidicornis S.-H. Chen & S.-Y. Wang, 1980: 1, 15 **A:** CH (FUJ) RU (FE)

postfasciata Fairmaire, 1889a: 73 (*Amphimela*) **A:** CH (CHQ FUJ GAN SCH SHA XIZ ZHE)

nigriceps Weise, 1889g: 642

pulchra S.-H. Chen, 1934g: 65 **A:** CH (AHN GUI JIA ZHE)

tibialis Jacoby, 1886a: 740 (*Enneamera*) **A:** CH ("South China") JA **ORR**

tsoui C.-F. Lee, 2014c: 258 **A:** CH (TAI)

variabilis Baly, 1862b: 456 **A:** AF BT CH (CHQ FUJ GAN GUI HAI HKG XIZ ZHE) ID (SD UT) NP **ORR**

albofasciata Duvivier, 1892b: 424

apicalis Jacoby, 1889b: 200 (*Enneamera*)

ceylonensis Jacoby, 1887a: 84 (*Enneamera*)

scutellata Jacoby, 1900b: 126 (*Enneamera*)

viridiceps S.-H. Chen, 1939b: 56 **A:** CH (GAN HEB)

supertribe Alticitae Newman, 1834 [# 749]

genus *Acrocrypta* Baly, 1862b: 457 type species: *Acrocrypta mouhoti* Baly, 1862

Colpodes Chevrolat, 1836: 394 [HN] type species: *Altica rotundata* Olivier, 1808

Imolia Jacoby, 1884b: 32 type species: *Imolia nigrofasciata* Jacoby, 1884

Pseudocrypta L. N. Medvedev, 1997c: 78 type species: *Pseudocrypta nigripennis* Medvedev, 1997

Pusio Gistel, 1848b: xi [RN, HN] type species: *Altica rotundata* Olivier, 1808

Sebaethiella L. N. Medvedev, 1994: 16 type species: *Sebaethiella mindanaica* Medvedev, 1994

assamensis Jacoby, 1893a: 151 **A:** CH (GUA GUI SCH YUN) **ORR**

convexa Gressitt & Kimoto, 1963a: 795 (*Colpodes*) **A:** CH (HAI)

intermedia Jacoby, 1892b: 925 (*Sphaerometopa*) **A:** CH (YUN) **ORR**

gracilicornis S.-H. Chen & S.-Y. Wang, 1980: 1, 15 **A:** CH (GUX)

purpurea Baly, 1877a: 224 Ei: NL **ORR**

violaceicuprea S.-Y. Wang, 2007: 463, 464 **A:** CH (YUN)

genus *Aeschrocnemis* Weise, 1888b: 855 type species: *Batophila graeca* Allard, 1884

subgenus *Aeschrocnemis* Weise, 1888b: 855 type species: *Batophila graeca* Allard, 1884

Derocrepisomus Pic, 1911d: 109 type species: *Crepidodera serbica* Kutschera, 1860

byzantica Nadein, 2011: 268 **A:** TR
caucasica Weise, 1886a: 690 (*Derocrepis*, var.) **E:** RU (ST) **A:** GG
graeca Allard, 1884b: 248 (*Batophila*) **E:** GR (Peloponnese, Ionian Is.)
obscurithorax Pic, 1909f: 145 (var.)
hellenica Doguet & Bergeal, 2000: 131 (*Derocrepis*, ssp.) **E:** GR
jailensis Heikertinger, 1915: 328 [= 1922: 328, = 1925b: 165] (*Derocrepis*, ssp.) **E:** KR
laterufa Pic, 1909k: 178 (*Derocrepis*, var.) **E:** RU (ST) **A:** GG
obscuripes Pic, 1911d: 109 (var.)
merditensis Heikertinger, 1915: 327 [= 1922: 327, = 1925b: 164] (*Derocrepis*, ssp.) **E:** AL
osmanica Nadein, 2011: 270 **A:** TR
peloponnesiaca Heikertinger, 1910d: 52 (*Derocrepis*, var.) **E:** GR (Peloponnese)
rhilensis Gruev, 1974a: 124 (*Derocrepis*, ssp.) **E:** BU (Rila Mts.) GR
rhodopensis Gruev, 1973a: 98 [= 1974a: 121] (*Derocrepis*, ssp.) **E:** BU (Rodopi Mts.)
serbica Kutschera, 1860b: 74 (*Haltica*) **E:** BU CR GR KR NM RO SB TR **A:** AB TR
bicolor Weise, 1886a: 690 (*Derocrepis*, var.)
strangulata Allard, 1860c: 61 (*Crepidodera*)
slavica Gruev, 1979: 134 (*Derocrepis*, ssp.) **E:** BU GR KO NM SB
subgenus *Hirticnemis* Nadein, 2011: 271 type species: *Derocrepis pubipennis* Reitter, 1892
anatolica Heikertinger, 1915: 332 [= 1922: 332, = 1925b: 171] (*Derocrepis*, ssp.) **A:** TR
caria Nadein, 2011: 279 **A:** TR
curda Nadein, 2011: 280 **A:** TR
delagrangei Pic, 1903d: 124 **A:** TR
iconiensis Nadein, 2011: 278 **A:** TR
ossetica Heikertinger, 1915: 329 [= 1922: 329, = 1925b: 167] (*Derocrepis*, ssp.) **E:** RU (ST: Caucasus) **A:** AB AR GG
pubipennis Reitter, 1892a: 154 (*Derocrepis*) **E:** RU (ST) **A:** AB (Arax Valley) AR (Arax Valley) GG TR
shirvana Nadein, 2011: 276 **A:** AB
subgenus *Nudicnemis* Nadein, 2011: 282 type species: *Derocrepis whiteheadi* Warchałowski, 1998
antiocha Nadein, 2011: 285 **A:** TR
turcica Nadein & Gök, 2009: 193 **A:** TR
whiteheadi Warchałowski, 1998d: 361 (*Derocrepis*) **A:** TR

genus *Agasicles* Jacoby, 1905c: 400 type species: *Systena connexa* Boheman, 1859
hygrophila Selman & Vogt, 1971: 1018 **Ai:** CH (ANH CHQ FUJ GUA GUI GUX HAI HKG HUB HUN JIA JIX SCH SHG TAI YUN ZHE) **AUSi NARi NTR ORRi**

genus *Altica* Geoffroy, 1762: 244 [conserved name, ICZN (1994), Opinion 1754] type species: *Chrysomela oleracea* Linnaeus, 1758

Adimus Billberg, 1820: 57 type species: *Galleruca lausoniae* Fabricius, 1798 (= *Galeruca caerulea* Olivier, 1792)
Graptodera Chevrolat, 1836: 388 type species: *Chrysomela oleracea* Linnaeus, 1758
Haltica Illiger, 1801a: 138 [UE] type species: *Chrysomela oleracea* Linnaeus, 1758
Halticella Germain, 1903: 60 type species: *Halticella atrocyanea* Germain, 1903
Halticoidea Germain, 1903: 57 type species: *Halticoidea costifera* Germain, 1903
Neoclitena Abdullah & Qureshi, 1968a: 413 type species: *Neoclitena simplex* Abdullah & Qureshi, 1968
Pantryptes Gistel, 1856: 378 type species: *Chrysomela oleracea* Linnaeus, 1758
Rybakowia Jacobson, 1892b: 465 type species: *Rybakowia pyriformis* Jacobson, 1892

aenea Olivier, 1808: 646 (*Galeruca*) **A**: CH (TAI) ID (UT) NP **AUS ORR**
australis Blackburn, 1889: 1493 (*Haltica*)
bicolora Jacoby, 1905a: 482 (*Haltica*)
foveicollis Jacoby, 1889b: 190 (*Haltica*) [QS]
ignea Blackburn, 1889: 1494 (*Haltica*)
jussiaeae Gressitt, 1955: 34
nepalensis Chûjô, 1966b: 29 [QS]
subaurichalcea Weise, 1922d: 156 (*Haltica*) [QS]

aenescens Weise, 1888b: 837 (*Haltica*, var.) **E**: AU BE BY CZ DE EN FI FR GE IT KR LA LT LU NL PL RU (CT NT ST) SK SL SZ UK **A**: RU (WS)
pseudolythri Heikertinger, 1940: 576 [RN]

alticola S.-Y. Wang, 1992c: 724, 752 **A**: CH (SCH)

ampelophaga ampelophaga Guérin-Méneville, 1858: 415 (*Graptodera*) **E**: AZ CR FR GE GR (North) IT KO KR MA ME PT SB SP SZ **N**: AG MO TU **A**: AB IN TR
crassa Allard, 1889c: 43 (*Graptodera*) [QS]
hispana Allard, 1869: 477 (*Graptodera*)

ampelophaga koreana Ogloblin, 1925: 93 (*Haltica*) **A**: CH (BEI HEB) NC SC

ancyrensis Weise, 1897c: 67 (*Haltica*) **E**: RU (ST) **A**: AR IS TR

balassogloi Jacobson, 1892b: 463 (*Haltica*) **A**: AF IN KI KZ MG PA TD UZ
lopatini Paliy, 1968b: 28 (*Haltica*)
suvorovi Ogloblin, 1921: 27 (*Haltica*)

bicarinata Kutschera, 1860a: 14 (*Haltica*) **E**: GR MD NM **N**: EG **A**: CY IN IS JO LE SA SI SY TR
ciliciensis Weise, 1889g: 636 (*Haltica*)

bicosta Shukla, 1960: 78 (*Haltica*) **A**: ID (HP)

birmanensis Jacoby, 1896c: 254 (*Haltica*) **A**: CH (TAI) ID (HP SD) JA **AUS ORR**
indica Shukla, 1960: 80 (*Haltica*)

bisulcata Weise, 1887h: 195 (*Haltica*) **A**: CH (HEI) MG RU (FE) *brevicollis* Foudras, 1860b: 296 [= 1861: 40] **E:** AU BE BH BU BY CR CZ DE EN FR GB GE GR (North) HU IT KR LA LT LU ME NL NM NR PL PT RO RU (CT NT ST) SB SK SL SP SV SZ TR UK WS **A:** AB AR IN KI KZ RU (ES FE SS) TR

azurescens Weise, 1888b: 833 (*Haltica*, var.)

coryletorum Král, 1964b: 127 (ssp.)

coryli Allard, 1860c: 77 (*Graptodera*)

breviuscula Weise, 1900b: 293 (*Haltica*) **A:** AB IN

bulgharensis Král, 1969b: 79 **A:** TR

caerulea Olivier, 1792: 590 (*Galeruca*) **A:** BT CH (FUJ GUA GUX HAI TAI ZHE) ID (KA SD) NC NP PA **AUS ORR**

brevicosta Weise, 1922b: 110 (*Haltica*)

elongata Jacoby, 1884b: 28 (*Haltica*)

lausoniae Fabricius, 1798: 95 (*Galleruca*)

caerulescens Baly, 1874: 190 (*Graptodera*) **A:** CH (BEI CHQ FUJ GUA HUB HUN JIA JIX SCH SHA ZHE TAI) JA NC RU (FE) SC **ORR**

carduorum Guérin-Méneville, 1858: 415 (*Graptodera*) **E:** AL AU BH BU BY CR FI FR GB GE GR HU IT KR LA MD ME NM RO RU (CT ST) SK SL SP SV SZ TR UK **A:** AB AR GG IN KI KZ RU (ES FE SS WS) TD TR **NAR**i

basalis Allard, 1859c: clxviii (*Graptodera*)

cirsii Israelson, 1956: 130 (*Haltica*)

carinthiaca Weise, 1888b: 840 (*Haltica*) **E:** AU BE BH BY EN FI FR GE IT LA LU NL NR PL RU (CT NT ST) SV SZ UK **A:** AR CH (XIN) GG KI RU (ES FE SS WS)

daghestanica Shapiro, 1970: 656 (*Haltica*)

issykensis Ogloblin, 1921: 31 (*Haltica*)

chamaenerii Håk. Lindberg, 1926: 68 (*Haltica*) **E:** EN FI LA LT NR RU (NT) SV

chingfui Suenaga & C.-F. Lee, 2023: 52 **A:** CH (TAI)

circaeae Ohno, 1960c: 89 **A:** JA RU (FE) SC

cirsicola Ohno, 1960c: 81 **A:** CH (BEI CHQ FUJ GAN HEB HEI HUB HUN JIA JIL SCH SHN TAI XIN YUN ZHE) JA NC RU (FE) SC **NAR**i

cornivorax Král, 1969b: 72 **E:** AU BU CZ HU MD PL RU (ST) SK SZ TR UK (Poltava)

deserticola Weise, 1889g: 635 (*Haltica*) **E:** RU (CT ST) UK **A:** AB AF AR CH (NMO XIN) IN IQ IS JO KI KZ MG RU (WS) SY TM TR

chotanica Ogloblin, 1925: 92 (*Haltica*, ssp.)

eichleri Ogloblin, 1925: 92 (*Haltica*, ssp.)

glycyrrhizae Ogloblin, 1925: 91 (*Haltica*)

difficilis Ogloblin, 1921: 33 (*Haltica*) **A:** KI

elaeagnusae Y. Zhang, S.-Y. Wang & Yang, 2006: 855 **A:** CH (XIN)

engstromi J. R. Sahlberg, 1893: 19, 167 (*Haltica*) **E:** EN FI LA LT NR RU (CT NT) SV **A:** KZ RU (ES FE SS) *ericeti* Allard, 1859c: clxvi (*Graptodera*) **E:** FR GB GE NL PT SP

filipendulae Chashchina, 2006: 858 **E:** RU (CT)

foveigera Harold, 1877c: 107 (*Haltica*) **A:** YE **AFR**

fragariae Nakane, 1955: 38 (*Haltica*) **A:** CH (BEI FUJ GUI HEB HEI HUB HUN JIL NMO SHN ZHE) JA NC RU (FE) SC

fruticola Weise, 1888b: 842 (*Haltica*) **E:** BH RO SP (North) UK (Carpathians)

globicollis Weise, 1889g: 635 (*Haltica*) **E:** RU (ST: Caucasus) TR **A:** AR IN TR

graeca Král, 1966: 164 **E:** GR **A:** IN TR

hampei Allard, 1867: 499 (*Graptodera*) **E:** KR RU (ST) **A:** AR GG TR

khnzoriani Král, 1969b: 81

helianthemi Allard, 1859c: clxvi (*Graptodera*) **E:** AU BE BH BU BY CR CZ EN FR GB GE GR HU IT KR LA LS LT LU NL PL RO RU (CT ST) SK SL SP SZ UK **A:** GG KZ RU (ES SS WS)

cognata Kutschera, 1860a: 17 (*Haltica*)

montana Foudras, 1860b: 295 [= 1861: 39]

potentillae Allard, 1859c: clxvii (*Graptodera*)

pusilla Duftschmid, 1825: 253 (*Haltica*) [HN]

himalayensis S.-H. Chen, 1936a: 80 (*Haltica*) **A:** BT CH (XIZ) ID (HP SD UT) NP PA **ORR**

himensis Shukla, 1960: 79 (*Haltica*)

iberica Weise, 1892a: 373 (*Haltica*) **E:** SP **A:** AG MO

constantini Doguet, 1972: 134

impressicollis Reiche, 1862a: 298 (*Graptodera*) **E:** AL AU BH BU CR CZ FR (Corsica) GE GR HU IT KR LA NM PL RO RU (ST) SB SK SZ TR UK **A:** AB IN IS SY TR

lacunosa Weise, 1900c: 136 [= 1900d: 164] (*Haltica*)

inconspicua Král, 1966: 167 **E:** SP

ivlievi L. N. Medvedev, 1968: 85 (*Haltica*) **A:** RU (FE)

japonica Ohno, 1960c: 92 (var.) **A:** CH (AHN TAI ZHE) JA NC SC

fukutai Nakane, 1963a: 20

jarmilae Král, 1979: 98 **E:** BU GR **A:** GG IS TR

kimotoi Suenaga, 2020: 194 **A:** JA

komiyai Suenaga, 2020: 195 **A:** JA

kozlovi Ogloblin, 1921: 25 (*Haltica*) **A:** CH (GAN)

kurosawai Ohno, 1960c: 87 **A:** JA (Honshu)

latericosta latericosta Jacoby, 1886a: 726 (*Haltica*) **A:** CH (GAN SCH "West China") JA (Hokkaido)

latericosta subcostata Ohno, 1960c: 80 (ssp.) **A:** CH (JIL) JA NC SC

lencorana Konstantinov, 1992b: 41 **A:** AB TR

talyshana Konstantinov, 1995: 196

longicollis Allard, 1860c: 83 (*Graptodera*) **E:** BE BY DE EN FI FR GB GE LA LU NL NR PL RU (NT) SP (North) SV SZ **A:** TR

britteni Sharp, 1914: 261 (*Haltica*)

sandini Kemner, 1919: 160 (*Haltica*)

lubischevi Paliy, 1963: 35 (*Haltica*) **A:** CH (XIZ) KI KZ TD TM UZ

aeneola Lopatin, 1956c: 46 (*Haltica*) [HN]

lythri Aubé, 1843: 8 **E:** AL AU BE BH BU BY CR CZ DE EN FI FR GB GE GR HU IT KO KR LA LU NL NM PL PT RO RU (CT NT ST) SB SK SL SP SV SZ UK **A:** AB IN RU (ES FE) TR

epilobii Allard, 1859c: clxvii (*Graptodera*)

malvernensis Jacoby, 1903b: 1 (*Haltica*) **A:** YE **AFR**

mohri Lopatin, 1990b: 603 **A:** IN

nipponica Ohno, 1960c: 80 **A:** JA (Honshu)

nitens Weise, 1924b: 21 (*Haltica*) **A:** YE **AFR**

nuristanica Lopatin, 1963d: 368 (*Haltica*) **A:** AF

oleracea Linnaeus, 1758: 372 (*Chrysomela*) **E:** AL AN AU BE BH BU BY CR CZ DE EN FI FR GB GE GR HU IT KO KR LA LS LT LU MD ME NL NM NR PL RO RU (CT NT ST) SB SK SL SP SV SZ TR UK **A:** AB AR CH (CHQ GAN GUX HEB HEI JIL LIA SCH XIN YUN) GG IN IS JA KI KZ MG NC RU (ES FE SS WS) SC TR **AUS NAR**

alchimillae Paliy, 1968b: 29 (*Haltica*)

breddini Mohr, 1958: 52 (*Haltica*, ssp.)

geminata Ogloblin, 1925: 94 (*Haltica*, ssp.)

lugubris Weise, 1888b: 845 (*Haltica*, var.)

nigra Fowler, 1890: 359 (*Haltica*, var.)

nobilis Weise, 1888b: 845 (*Haltica*, var.)

opacifrons Har. Lindberg, 1938: 45 (*Haltica*)

sicula Allard, 1859c: clxvii (*Graptodera*)

splendens Mulsant & Rey, 1874: 244

tenebrosa G. Rossi, 1882: 213 (*Haltica*, var.)

tyrolensis Siede, 1998: 88 (forma) [IN]

ytenensis Sharp, 1914: 263 (*Haltica*)

palustris Weise, 1888b: 844 (*Haltica*) **E:** AU BE BU BY CZ DE EN FI FR GB GE HU IT KR LA LS LT LU NL NR PL PT RU (CT NT ST) SK SP SV SZ UK **N:** AG TU **A:** AB AF AR GG IN KI KZ RU (SS) TD TR UZ

pamiranica Weise, 1889g: 635 (*Haltica*) **A:** CH (GAN HEI QIN SHX XIN) KI KZ TD

tianshanica Ogloblin, 1921: 30 (*Haltica*)

plotnikovi Paliy, 1963: 36 (*Haltica*) **A:** KI

pontica Ogloblin, 1925: 94 (*Haltica*, ssp.) **A:** TR

pyriformis Jacobson, 1892b: 466 (*Rybakowia*) **A**: CH (NMO XIZ) MG
pyritosa Erichson, 1843: 266 (*Haltica*) **A**: SA YE **AFR**
quercetorum Foudras, 1860b: 293 [= 1861: 37] **E**: AU BE BH BU BY CR CZ FI FR GE GR HU IT KR LA LT LU MD NL NR PL RO RU (CT ST) SB SK SL SP SV SZ TR UK **A**: AB AR GG IN TR
saliceti Weise, 1888b: 835 (*Haltica*)
rosicola Král, 1969b: 82 **E**: RU (ST) **A**: AR
sajanica Csiki, 1901a: 119 (*Haltica*) **A**: CH (NMO QIN XIN) MG RU (ES FE SS)
convexipennis Ogloblin, 1921: 28 (*Haltica*)
mongolica Král, 1973: 121
transbaicalica Ogloblin, 1925: 95 (*Haltica*)
sanguisorbae Ohno, 1960c: 83 **A**: JA SC
semenowi Jacobson, 1892b: 462 (*Haltica*) **A**: CH (XIZ) KI KZ TD UZ
alaiensis Paliy, 1968b: 26 (*Haltica*)
lazulina Ogloblin, 1921: 23 (*Haltica*)
susamyrica Paliy, 1968b: 27 (*Haltica*)
sibirica Csiki, 1901a: 119 (*Haltica*) **A**: RU (WS)
talassicola Lopatin, 1976a: 115 **A**: KZ KI
tamaricis franzi Král, 1966: 166 (ssp.) **E**: SP (North)
tamaricis tamaricis Schrank, 1785: 318 **E**: AL AU BE BH BU BY CR CZ DE EN FI FR GE GR HU IT KR LA LS LT MD NL NM PL PT RO RU (CT NT ST) SB SK SL SP (North) SV SZ TR UK **A**: AB AF AR GG IN IQ KI KZ MG RU (ES FE SS WS) TR UZ
articulata Beck, 1817: 17 (*Galleruca*)
azurea Sturm, 1843: 282 (*Graptodera*) [RN]
consobrina Duftschmid, 1825: 252 (*Haltica*)
erucae Fabricius, 1792b: 28 (*Galleruca*)
hippophaes Aubé, 1843: 8
nigra Tenenbaum, 1927: 152 (*Haltica*, ab.) [IN]
tristis Ogloblin, 1921: 20 (*Hermaeophaga*) **A**: RU (ES FE)
nigricornis Ogloblin, 1921: 21 (*Hermaeophaga*)
tsharynensis Ogloblin, 1921: 24 (*Haltica*) **E**: RU (ST) **A**: AB AR GG IN KZ
stichai Král, 1965a: 162
viridicyanea Baly, 1874: 191 (*Graptodera*) **A**: CH (BEI FUJ GAN GUA GUI GUX HEI HKG HUB JIA JIL SCH SHA SHN YUN ZHE) JA RU (FE) NC SC **ORR**
viridula Weise, 1889g: 636 (*Haltica*) **E**: RU (ST) **A**: AB AR GG IN
armeniaca Iablokoff-Khnzorian, 1954: 59 (*Haltica*)
avellanae Král, 1965a: 164
weisei Jacobson, 1892b: 463 (*Haltica*) [RN] **A**: CH (BEI GAN HEB HEI JIA JIL LIA NMO SCH SHX XIN XIZ) MG NC RU (ES FE)

laeviuscula Weise, 1889g: 634 (*Haltica*) [HN]
yunnana S.-Y. Wang, 1992c: 726, 753 **A:** CH (YUN)
zangana S.-H. Chen & S.-Y. Wang, 1981a: 504, 508 **A:** CH (SCH XIZ)

genus *Altica*, nomen dubium
simplex Abdullah & Qureshi, 1968a: 413 (*Neoclitena*) **A:** PA

genus *Amphimela* Chapuis, 1875: 34 type species: *Amphimela mouhoti* Chapuis, 1875
Cercyonia Weise, 1902d: 304 type species: *Cercyonia variabilis* Weise, 1902
Clitea Baly, 1877g: 287 type species: *Clitea picta* Baly, 1877
Diboloides Jacoby, 1897c: 553 type species: *Diboloides bicolor* Jacoby, 1897
Halticella Jacoby, 1899d: 357 [HN] type species: *Halticella flavopustulata* Jacoby, 1899
Halticorthaea Csiki, 1940a: 418 [RN] type species: *Halticella flavopustulata* Jacoby, 1899
Throscoryssa Maulik, 1928: 159 type species: *Throscoryssa citri* Maulik, 1928
Xanthocycla Baly, 1875a: 29 type species: *Xanthocycla chapuisii* Baly, 1875
fulva S.-H. Chen, 1933e: 383 (*Clitea*) **A:** CH (CHQ FUJ GUX) **ORR**
indica Jacoby, 1896c: 269 (*Mantura*) **A:** NP **ORR**
mauliki Prathapan, 2017: 279 [RN] **A:** CH (GUA) **ORR**
citri Maulik, 1928: 160 (*Throscoryssa*) [HN]
metallica S.-H. Chen, 1933e: 383 (*Clitea*) **A:** CH (FUJ GUA HAI HUN JIA SCH TAI) JA (Ryukyus) **ORR**
citri Chûjô, 1958c: 10 (*Clitea*) [HN]
mouhoti Chapuis, 1875: 36 **A:** ID (SD) **ORR**
picta Baly, 1877g: 287 (*Clitea*) **A:** NP PA **ORR**
inornata S.-H. Chen, 1933e: 383 (*Clitea*, var.)
raydahensis El Torkey & Al Dhafer, 2015: 435 **A:** SA
rufula S.-H. Chen & Zia, 1966a: 74, 75 (*Clitea*) **A:** CH (GUX)
subgeminata Kimoto, 2001: 51 **A:** NP

genus *Angulaphthona* Bechyné, 1960b: 74 type species: *Aphthona heteromorpha* Bechyné, 1955
latipennis Pic, 1921e: 15 (*Aphthona*) **A:** SA YE **N:** EG **AFR**
schereri Gruev, 1981c: 56 **A:** SA YE **AFR**

genus *Anthobiodes* Weise, 1887f: 402 type species: *Anthobiodes simoni* Weise, 1887 (= *Thyamis angusta* Allard, 1876)
Wittmeriana L. N. Medvedev, 1975a: 15 type species: *Wittmeriana turcica* L. N. Medvedev, 1975
angustus Allard, 1876: 24 (*Thyamis*) **A:** IS
simoni Weise, 1887f: 402

heydeni Allard, 1870b: 172 [= 1875: 439] (*Aphthona*) **E:** SP (Ronda)
turcicus L. N. Medvedev, 1975a: 15 (*Wittmeriana*) **E:** GR (Rhodes) **A:** TR (Antalya)

genus *Aphthona* Chevrolat, 1836: 391 type species: *Haltica cyparissiae* Koch, 1803
Aphthonaria L. N. Medvedev, 1990b: 37 type species: *Aphthonaria martensi* L. N. Medvedev, 1990
Aphthonotarsa L. N. Medvedev, 1984: 55 type species: *Aphthonotarsa brunnea* L. N. Medvedev, 1984 (= *Aphtona bombayensis* Scherer, 1969)
Asialtica Scherer, 1969: 123 type species: *Aphthona indica* Jacoby, 1900
Bhutajana Scherer, 1979a: 132 type species: *Bhutajana metallica* Scherer, 1979
Ectonia Weise, 1922b: 119 type species: *Ectonia laeta* Weise, 1922
abdominalis Duftschmid, 1825: 262 (*Haltica*) **E:** AU BH BU BY CR CZ FR GE GR HU IT KO KR LA LT MD ME NR PL RO RU (CT NT ST) SK SL SP SV SZ UK **A:** AB AF AR CH (JIA HEB HEI SCH SHA SHX) GG JA IN KZ MG NC RU (ES WS) SC TR **ORR**
albinea C. G. Thomson, 1866: 182 (*Teinodactyla*) [HN]
finitima Ogloblin, 1927: 300
foudrasi Jacoby, 1886a: 729
licentana S.-H. Chen, 1939b: 71
sajanica Ogloblin, 1927: 302
aeneomicans Allard, 1875: 439 **E:** FR IT (Gran Sasso) MD RU (CT) SB SL SP **A:** CY TR
metallica Weise, 1889j: 415
alanyensis Fritzlar, 2004: 107 **A:** TR
albertinae Allard, 1866: 230 **E:** FR (Pyrénées) GR PT SP
allardi Brisout de Barneville, 1866: 424
albescens Motschulsky, 1866b: 417 (*Teinodactila*) **A:** CH (GUX TAI) JA (Ryukyus) NP **ORR**
opaca Allard, 1889f: 305
ihai Chûjô, 1958c: 14 (*Longitarsus*)
alcina Leonardi, 1975a: 10 **E:** FR (Corsica) IT (Sardinia)
alexander Berti & Rapilly, 1973: 885 **A:** IN
zarudnyi L. N. Medvedev, 1973b: 883
almorensis Konstantinov & Lingafelter, 2002: 42 **A:** NP **ORR**
amamiana Ohno, 1962d: 66 **A:** JA (Ryukyus)
andrewesi Jacoby, 1896c: 256 **A:** ID (HP UT) NP **ORR**
mimica L. N. Medvedev, 1997e: 260
punctata Shukla, 1960: 76
antennata Scherer, 1969: 75 **A:** ID (SD)
archeri Bryant, 1941b: 411 **A:** ID (SD UT)

atrocaerulea Stephens, 1831: 299 (*Haltica*) E: AU BE BH BY CR CZ DE FR GB GE HU IT (Toscana) KR LA LT LU NL NR PL RO RU (ST) SB SK SL SP SV SZ UK **N**: AG MO **A**: IN TR

cyanella L. Redtenbacher, 1848: 531 (*Haltica*)

puncticollis Allard, 1866: 240

atrovirens Förster, 1849: 37 (*Haltica*) E: AU BE BU CR CZ FR GE GR HU IR IT KR LU ME NL PL RO RU (ST) SK SL SZ UK **A**: TR

aeneipennis Weise, 1893a: 1138 (var.)

tantilla Foudras, 1860b: 357 [= 1861: 118]

beauprei Pic, 1915a: 2 **N**: AG

beckeri Jacobson, 1896b: 108 **E**: AU KR PL RU (CT ST) SK UK **A**: GG KZ MG RU (ES FE SS WS)

brunneomicans Heikertinger, 1912d: 47

daurica L. N. Medvedev, 1973b: 883

bengalica Konstantinov & Lingafelter, 2002: 53 **A**: CH (YUN) ID (SD)

bergeali Fritzlar, 2001: 205 **A**: TR

bergealoides Fritzlar, 2004: 116 **A**: TR

bhutana L. N. Medvedev, 2009b: 408 **A**: BT

biokovensis Penecke, 1907: 16 E: AL BH CR

bombayensis Scherer, 1969: 78 **A**: NP **ORR**

brunnea L. N. Medvedev, 1984: 56 (*Aphthonotarsa*)

bonvouloiri Allard, 1862: 333 E: BU GR (Rhodes) IT TR **N**: EG **A**: IN IS LE SY TR

tauricola Pic, 1910c: 26

brevis S.-H. Chen, 1939b: 73 **A**: CH (SCH)

cangshanensis Konstantinov & Lingafelter, 2002: 60 **A**: CH (YUN)

carbonaria Rosenhauer, 1856: 337 (*Haltica*) **E**: FR (South) GR (Kérkyra Isl.) IT MA PT SP **N**: AG CY MO TU **A**: TR

hispana Allard, 1866: 232

janthina Allard, 1866: 237

leonhardi Heikertinger, 1944: 91 (ssp.)

pallidimembris Codina Padilla, 1963d: 46 (var.) [IN]

subrufescens Pic, 1909h: 156

chalchica L. N. Medvedev, 1980a: 319 **A**: CH (NMO) MG

chayuana S.-H. Chen & P.-Y. Yu, 1976: 211, 221 (ssp.) **A**: CH (XIZ)

chekanovskii Konstantinov, 1998b: 76 **A**: RU (ES)

chinchihi S.-H. Chen, 1939a: 49 **A**: CH (GUX NMO SHN) NC

famula Warchałowski, 1969: 227

chinensis Baly, 1877g: 295 A: CH (GUA) NC

hananoi Chûjô, 1942c: 61

chrysomelina Heikertinger, 1944: 79 **A**: KI UZ

irisha Paliy, 1970b: 6

constantini Doguet & Petitpierre, 1986: 146 E: SP (East)
cookei Gerstaecker, 1871: 85 (*Haltica*) **A**: OM YE **AFR**
coreana Heikertinger, 1944: 99 **A**: NC
crassicornis Lopatin, 1990b: 601 **A**: TR
crassipes Wollaston, 1860a: 3 (*Haltica*) **N**: CI
heringi Heikertinger, 1944: 48 (var.) [IN] [# 771]
crypta Scherer, 1969: 73 **A**: ID (SD) NP
cryptomorpha Konstantinov, 1998a: 143 **A**: CH (YUN)
cyanipennis Motschulsky, 1866b: 419 **A**: CH (YUN) ID (UT) NP
flavicollis S.-Y. Wang, 1992c: 712, 746
himalayana S.-H. Chen, 1934h: 371
cyparissiae Koch, 1803a: 80 (*Haltica*) E: AL AN AU BE BH BU BY CZ FR GB GE HU IT KR LS LU MD ME NL NM PL RO RU (CT) SB SK SL SP SZ UK **A**: TR **NAR**i
tabida Duftschmid, 1825: 256 (*Haltica*)
czwalinai Weise, 1888b: 911 E: AU CZ GE KR PL RO RU (CT ST) SK UK **A**: AB ES IN KI KZ MG **NAR**i
asiatica Heikertinger, 1911a: 10 (var.)
coerulescens Weise, 1889l: 420 (var.)
delicatula Foudras, 1860b: 373 [= 1861: 117] E: FR GE GR IT (North) SB SZ
graeca Doguet, 1988b: 303 (ssp.)
depressa Allard, 1859a: ci E: FR IT (Liguria) PT SP **N**: AG MO TU
aenea Allard, 1866: 233
biskrensis Pic, 1910c: 26
diminuta Abeille de Perrin, 1909: 180
ouarsenisiensis Pic, 1910c: 26 (var.)
subimpressa Mulsant & Rey, 1874: 255
subparallela Normand, 1937: 132 (var.) [IN] [# 771]
dhrubi Konstantinov & Sprecher-Uebersax, 2005: 184 **A**: NP
dhunche Konstantinov & Sprecher-Uebersax, 2005: 188 **A**: NP
dobangensis Kimoto, 2001: 52 **A**: NP
dohertyi Jacoby, 1894: 287 **A**: NP **ORR**
dongchuanica Konstantinov & Lingafelter, 2002: 80 **A**: CH (YUN)
erichsoni Zetterstedt, 1837: 222 (*Haltica*) E: BY EN FI GE KR LA LT NR PL RU (CT NT) SV **A**: MG RU (ES FE WS)
curvifrons Bach, 1858: 144 (*Haltica*)
nigritarsis Motschulsky, 1845b: 381
erythropoda S.-H. Chen, 1939b: 68 **A**: CH (SHA SHX) SC
espagnoli Král, 1965b: 131 E: SP (Cataluña)
euphorbiae Schrank, 1781: 83 (*Chrysomela*) E: AL AU BE BH BU BY CR CZ DE EN FI FR GB GE GR HU IT KO KR LA LS LT LU MA MD ME NL NM NR PL PT RO RU

(CT NT ST) SB SK SL SP SV SZ TR UK **N:** AG LB MO TU **A:** AB CY GG IN IS JO KZ LE RU (ES WS) SY TR

atratula Allard, 1859a: ci

hilaris Stephens, 1831: 316 (*Thyamis*)

orientalis Roubal, 1936b: 198 (ssp.) [HN]

pieninensis Roubal, 1940: 575 [RN]

punctigera Mulsant & Rey, 1874: 252

subaptera Mulsant & Rey, 1874: 256

virescens Foudras, 1860b: 375 [= 1861: 119]

flava Guillebeau, 1894: cxciii **E:** AL AU BH BU CR CZ GR HU IT NM PL RO SB SK SL TR UK **A:** AB TR **NARi**

flaviceps Allard, 1859a: c **E:** BH BU CR CZ FR GR HU IT KR MA MD ME NM RU (ST) SK SL SP SZ TR UK **N:** AG EG TU **A:** AB AF AR CY GG IN IQ IS JO KZ RU (FE) SI SY TD TM TR UZ

flavipes Allard, 1866: 221 [incorrect subsequent spelling]

pallida Boieldieu, 1859: 478 (*Haltica*) [HN]

straminea Foudras, 1860b: 363 [= 1861: 107]

formosana S.-H. Chen, 1934e: 185 **A:** CH (TAI) JA (Kyushu, Shikoku, Ryukyus) **AUS**

yakuana Nakane, 1958: 313 (ssp.)

franzi Heikertinger, 1944: 53 **E:** AU BU BY HU IT (South) KR MD ME RO RU (CT ST) SK UK **A:** AR IN IS RU (ES WS) SY TR

fuentei Reitter, 1901e: 202 **E:** GR IT (South) SP **N:** AG EG MO TU **A:** IN IS SI TR **AFR**

bavula Bechyné, 1960a: 11

camayennensis Bechyné, 1955b: 516

furthi L. N. Medvedev, 1997a: 14 **A:** NP

gardneri Bryant, 1941b: 412 **A:** ID (SD UT) NP

gerhardi Konstantinov & Prathapan, 2008: 385 [RN] **A:** NP

nepalensis Scherer, 1989: 42 (*Bhutajana*) [HN]

glebi Konstantinov, 2006: 179 **A:** IN UZ

gracilipes Ogloblin, 1927: 283 **A:** KI KZ

gracilis Faldermann, 1837: 344 **E:** KR MD RO RU (CT ST) UK **A:** AB AF AR GG IN IQ IS KZ MG RU (ES SS WS) TD TR TM UZ

armeniaca Weise, 1889h: 260

flavofemorata Heikertinger, 1911a: 8 (var.)

picipes Weise, 1888b: 894

gressitti Konstantinov, 1998b: 110 **A:** CH (FUJ HUB)

gruevi Konstantinov, 1998b: 111 **A:** CH (SHA)

habashanica Konstantinov & Lingafelter, 2002: 96 **A:** CH (YUN)

hakonensis Takizawa, 2012: 321 **A:** JA

hammarstroemi Jacobson, 1901b: 142 **A:** KI KZ RU (ES FE SS) TD

fischeri Pic, 1910c: 26
issykensis Ogloblin, 1921: 36
harputensis Özdikmen, Bal & Bolu, 2021: 28 **A:** TR
hauseri Heikertinger, 1911a: 6 **A:** AF IN TM UZ
claripes Rapilly, 1978: 335
herbigrada Curtis, 1837: 630 (*Altica*) **E:** AL AU BE BH BU CR CZ FR GB GE GR HU IT KR LS LU MD ME PL RO SB SK SL SP SZ UK **N:** AG MO TU
campanulae W. Redtenbacher, 1842: 28 (*Haltica*)
dimidiata Mulsant & Rey, 1874: 258
fossulata Allard, 1859a: ci
gibbula Weise, 1892c: 379 (var.)
laevicollis Mulsant & Rey, 1874: 257
heveli Konstantinov & Lingafelter, 2002: 98 **A:** PA
hissarica Lopatin, 1975b: 50 **A:** AF CH (GAN HEI QIN SHX XIN) ID (KA) IN KI KZ TD
howenchuni angustata S.-Y. Wang, 1992c: 711, 745 (ssp.) **A:** CH (YUN)
howenchuni howenchuni S.-H. Chen, 1934g: 63 (*Aphthonaltica*) **A:** CH (CHQ FUJ GAN GUI SCH SHA)
howenchuni nigripes S.-Y. Wang, 1992c: 712, 745 (ssp.) **A:** CH (SCH XIZ)
illigeri Bedel, 1898: 195 [RN] **E:** CR FR GE IT PT SL SP **N:** AG CI (Gran Canaria) MO TU
laevigata Illiger, 1807a: 61 (*Haltica*) [HN]
indica Jacoby, 1900b: 120 **A:** ID (SD) NP **ORR**
indochinensis S.-H. Chen, 1934h: 366 **A:** ID (SD UT) NP **ORR**
malaisei Bryant, 1939: 13
violaceomicans S.-H. Chen, 1936a: 83
interstitialis Weise, 1887h: 202 **A:** CH (FUJ HEB HUB NIN QIN SHX XIZ ZHE) JA (Honshu) MG RU (ES FE)
tibetana Heikertinger, 1911a: 5 (var.)
suvorovi Ogloblin, 1921: 37
suturanigra S.-H. Chen, 1939b: 71
yuasai Ohno, 1962d: 78
jacobsoni Ogloblin, 1927: 285 **A:** AF IN KI KZ TD TM UZ
jakuta Ogloblin, 1927: 296 **A:** MG RU (ES FE)
jalalabadensis Jind. Madar & Jiř. Madar, 1970: 206 **A:** AF
jordanica Mohr, 1962: 316 **A:** IS JO SY
juliana Springer, 1953: 510 **E:** IT (Alpi) SL
kanaraensis Jacoby, 1896c: 255 **A:** AF ID (HP UT) NP PA **ORR**
atriventris Maulik, 1926: 368
imitatrix Lopatin, 1963d: 362

kashmirensis Konstantinov & Lingafelter, 2002: 111 **A:** ID (KA)
kaszabi Král, 1967: 184 **A:** MG
amygdali L. N. Medvedev, 1974: 182 (*Longitarsus*)
kocheri Král, 1969c: 77 **N:** MO
konstantinovi Lopatin, 1998a: 132 **E:** KR RU (ST) **A:** TR
kubani Konstantinov & Lingafelter, 2002: 113 **A:** CH (YUN)
kuntzei Roubal, 1931a: 116 **E:** BH CR GR RU (CT) UK **A:** CY IN IS JO LE SY TR
kurosawai Ohno, 1962d: 81 **A:** JA (Honshu)
lacertosa Rosenhauer, 1847: 60 **E:** AU BU CR CZ FR HU KR MD PL RO RU (ST) SB SK TR UK **A:** AB **NARi**
divaricata L. Redtenbacher, 1848: 531 (*Haltica*)
laeta Weise, 1922b: 120 (*Ectonia*) **A:** CH ("South China") **ORR**
laetoides Chûjô, 1963b: 397 **A:** CH (TAI)
lantangi Konstantinov & Sprecher-Uebersax, 2005: 196 **A:** NP
lopatini Konstantinov, 1998b: 138 **A:** KZ
lubischevi Konstantinov, 1998b: 140 **E:** KR
lutescens Gyllenhal, 1813: 546 (*Haltica*) **E:** AL AU BE BH BU BY CR CZ DE EN FI FR GB GE GR HU IT KR LA LS LT LU MD ME NL NM NR PL PT RO RU (CT NT ST) SB SK SL SP SV SZ TR UK **A:** AB AE AR GG CY IN IS KZ MG RU (ES SS WS) TR YE [# 773]
piceoconcolor Döberl, 1994a: 179 (forma) [IN]
praeclara Weise, 1906c: 434 (var.)
sternalis Mulsant & Rey, 1874: 233 (*Thyamis*)
maculata Allard, 1876: 23 **E:** GR (Rhodes) RU (ST: Caucasus) **A:** AR CY IN IQ IS JO KZ LE SY TD TR
suturella Weise, 1888b: 890
maghrebina Doguet & Bergeal, 1989: 50 **N:** AG MO
maldesi Doguet & Petitpierre, 1986: 145 **E:** SP
mariki Konstantinov & Sprecher-Uebersax, 2005: 198 **A:** NP
marshalli Jacoby, 1906: 11 **A:** YE **AFR**
martensi L. N. Medvedev, 1990b: 38 (*Aphthonaria*) **A:** NP
medvedevi Konstantinov & Lingafelter, 2002: 121 [RN] **A:** NP
ovatipennis L. N. Medvedev, 1992d: 24 [HN]
melancholica Weise, 1888b: 893 **E:** GB PT SP
melanopoda S.-H. Chen, 1939b: 69 **A:** CH (GAN SCH)
metallica Scherer, 1979a: 133 (*Bhutajana*) **A:** BT
microcephala Weise, 1888b: 896 **A:** IS
modesta Weise, 1887h: 200 **A:** CH (AHN XIZ) NC RU (FE) SC
mohri Warchałowski, 1973b: 666 **A:** IN TM
mude Konstantinov & Sprecher-Uebersax, 2005: 206 **A:** NP

neoerythropoda Konstantinov & Lingafelter, 2002: 126 **A:** CH (SCH)

nepalensis L. N. Medvedev, 1984: 52 **A:** NP

nigriceps W. Redtenbacher, 1842: 27 (*Haltica*) **E:** AU BH BU BY CR CZ FR GB GR IT KR ME NM PT RO RU (ST) SB SK SL SP TR UK **N:** AG MO TU **A:** AB AR GG IN TR

decorata Kutschera, 1861: 240 (*Haltica*)

sicula Foudras, 1860b: 365 [= 1861: 109]

nigrilabris Duvivier, 1892b: 426 **A:** ID (KA UT) NP PA **ORR**

recticollis Jacoby, 1898a: 188 (*Longitarsus*)

rubida S.-H. Chen, 1936a: 83

nigriscutis Foudras, 1860b: 357 [= 1861: 101] **E:** AL BH BU BY CR CZ GE GR HU IT (North) KR MD NM PL RO RU (CT ST) SB SK UK **A:** AB AR GG IN KZ RU (ES WS) TR **NARi**

nigriventris Allard, 1860d: 390

nigroscutellata Reitter, 1904: 82

nigrita Motschulsky, 1866b: 418 **A:** TM **ORR**

niisatoi Takizawa, 2017: 287 **A:** JA

nonstriata Goeze, 1777: 312 (*Chrysomela*) **E:** AL AU BE BH BU BY CR CZ DE EN FI FR GB GE GR HU IR IT KR LA LS LT LU MA MD ME NL NM NR PL PT RO RU (CT NT ST) SB SK SL SP (North-East and Balearic Islands) SV SZ TR UK **N:** CI (La Palma) **A:** AB AR GG IN KZ RU (SS WS) TR

aenescens Weise, 1888b: 912 (var.)

asthmatica Weise, 1888b: 912 (var.)

coerulea Geoffroy, 1785: 100 (*Altica*)

pseudacori Marsham, 1802: 196 (*Chrysomela*)

thoracica Ogloblin, 1921: 35

occitana Doguet, 1988a: 147 **E:** AN FR (South) PT SP

okinawaensis Konstantinov & Lingafelter, 2002: 133 **A:** JA (Ryukyus)

olegi Konstantinov, 2006: 180 **A:** AB

omeishanica Konstantinov & Lingafelter, 2002: 134 **A:** CH (SCH)

ovata Foudras, 1860b: 372 [= 1861: 116] **E:** AU BH BU CR CZ FR GE GR HU IT KR PL RO RU (CT ST) SB SK SL SP (North) SZ TR UK **A:** TR

ovatoidea Gruev, 1981d: 148 **A:** CH (SHA)

pallida Bach, 1855: 141 (*Haltica*) **E:** AU CR CZ EN FR GE HU IT (North) KO KR LA LS LT MD NR PL PT RO RU (CT ST) SB SK SV SZ UK **A:** TR

geranii Weise, 1892c: 379 (var.)

parnassicola Heikertinger, 1944: 92 **E:** BU GR NM (Ohrid)

gruszkorum Warchałowski, 1974: 515

paropaca Konstantinov & Lingafelter, 2002: 141 **A:** ID (UT)

perminuta Baly, 1875b: 213 [RN] **A:** CH (TAI) JA RU (FE) SC **ORR**

pygmaea Baly, 1874: 198 [HN]

perrisi Allard, 1869: 477 E: FR (Corsica) IT
silvana Peyerimhoff, 1915a: 41 (ssp.)
phalchoki Konstantinov & Sprecher-Uebersax, 2005: 210 **A**: NP
piceipes Scherer, 1969: 125 (*Asialtica*) **A**: BT CH (YUN) ID (SD) NP
piciventris S.-H. Chen, 1939b: 72 **A**: CH (NMO)
placida Kutschera, 1864f: 472 (*Haltica*) [RN] **E**: AU BU CZ GR HU IT (North) KO KR MD PL RO RU (CT ST) SB SK UK
pallida Kutschera, 1861: 241 (*Haltica*) [HN]
potanini Konstantinov, 1998b: 174 **A**: CH (SCH) [# 783]
poupillieri Allard, 1860d: 401 E: SP (South) **N**: AG MO TU
moreli Pic, 1910c: 26
operosa Peyerimhoff, 1919: 229 (ssp.)
promissa Ogloblin, 1927: 289 **A**: AF KZ KI TD TM UZ
przhevalskii Konstantinov, 1998b: 179 **A**: KZ
pseudocrypta Konstantinov, 1998a: 144 **A**: NP
punctiventris Mulsant & Rey, 1874: 250 **E**: FR IT (Sardinia, Sicily) PT SP **N**: AG MO
pusilla Laboissière, 1942: 20 A: OM YE (Socotra) **AFR**
pygmaea Kutschera, 1861: 246 (*Haltica*) **E**: AL AU BE BH BU BY CR CZ FI (South) FR GE GR HU IT KO KR LA LS LT LU MA MD ME NL NM PL RO RU (CT ST) SB SK SL SZ TR UK **A**: AB AR CY GG IN IS JO LE SY TR YE **N**: EG LB
atrocoerulea C. G. Thomson, 1866: 202
fulvipes J. R. Sahlberg, 1913a: 78
nigella Kutschera, 1861: 247 (*Haltica*)
orientalis Mulsant & Rey, 1874: 253
substriata L. N. Medvedev, 1962a: 386
rawalpindica Lopatin, 2002d: 153 **A**: PA
reinecki Lopatin, 1963d: 363 **A**: AF (North-East)
reitteri Allard, 1884b: 248 E: RU (ST) **A**: AR GG
georgica Ogloblin, 1927: 298
renhwai S.-H. Chen, 1939a: 50 **A**: CH (GAN GUX)
rhodiensis Heikertinger, 1944: 67 E: GR (Rhodes) **A**: TR
rufosanguinea S.-H. Chen, 1939a: 49 **A**: CH (CHQ GUI GUX SCH)
rugipennis Ogloblin, 1927: 290 E: CZ HU KR RU (ST) SK UK **A**: AB AR GG KZ TR UZ
nitidior Warchałowski, 1995b: 464 [RN]
orientalis L. N. Medvedev, 1962a: 388 (ssp.) [HN]
praenuntiata Král, 1964a: 34
viridis Iablokoff-Khnzorian, 1964: 168
russica Konstantinov, Volkovitsh & Cristofaro, 2001: 33 **E**: RU (ST)
sandrae Baselga & Novoa, 2002: 2 **E**: SP (Galicia)
sardea Allard, 1867: 490 E: IT (Sardinia)
brunnipes Allard, 1870a: 50

sarmatica Ogloblin, 1928: 28 **E:** MD KR RU (ST) UK
schereri Konstantinov, 1998a: 145 **A:** NP
semicyanea Allard, 1859a: ci **E:** AL AU BH BU CR CZ FR (South-East) GR HU IT KR MD ME RO RU (ST) SB SK SL UK **A:** AB AR GG ID (HP) IN IS KI KZ MG RU (ES SS) SY TD TR TM UZ
alata Jacobson, 1901a: 93 (var.) [IN] [# 771]
atra Csiki, 1901a: 120 (var.)
crassipes Jacobson, 1901a: 92 [HN]
grossa Iablokoff-Khnzorian, 1960: 72
heptapotamica Ogloblin, 1921: 34
melanopeza Jacobson, 1901b: 143 [RN]
pachypus Jacobson, 1924c: 242 [RN]
peregrina Weise, 1906e: 577 [RN]
phrygia Weise, 1902a: 204 (var.)
picipes Weise, 1892e: 208 (var.) [HN]
punctatissima Weise, 1888b: 893
sundmani Jacobson, 1901b: 143
tenebrosa Ogloblin, 1927: 294 (ssp.)
viridula Allard, 1867: 491
semiviridis Jacoby, 1886a: 730 **A:** JA RU (FE)
sicelidis Weise, 1888b: 892 **E:** FR (Savoie) IT
obscuripes Heikertinger, 1912h: 387 (ab.) [IN]
signatifrons Wollaston, 1868: 150 (*Haltica*) **N:** MO **A:** AE OM SA YE **AFR**
asirica Doguet, 1979a: 311
dubia Laboissière, 1942: 22
silinica S.-H. Chen, 1939b: 70 **A:** CH (HEB)
simlaensis Konstantinov & Lingafelter, 2002: 170 **A:** ID (HP) NP **ORR**
singalilaensis Konstantinov & Lingafelter, 2002: 171 **A:** ID (SD) NP
sinuatipennis Konstantinov & Lingafelter, 2002: 173 **A:** CH (YUN) **ORR**
socotrana Döberl, 2012: 430 **A:** YE (Socotra)
splendida Weise, 1889g: 639 **A:** CH (CHQ FUJ HEB HUB GAN SCH SHA TAI XIZ YUN ZHE)
strigosa Baly, 1874: 197 **A:** CH (CHQ FUJ GUI HAI HKG TAI YUN ZHE) JA NP SC **ORR**
brancuccii L. N. Medvedev, 1997a: 15
glabrata Jacoby, 1896b: 457 (*Allomorpha*)
wallacei Baly, 1877d: 178
stussineri Weise, 1888b: 918 **E:** AU BH CR FR IT (North) PL RO SL SP
pyrenaea Heikertinger, 1944: 94
subovata Allard, 1859a: ci **E:** SP (Andalucía) **N:** AG MO TU
syriaca Heikertinger, 1944: 93 **A:** IQ IS SY TR

taiwana Takizawa, 1979: 346 **A:** CH (TAI)
taniae Konstantinov, 2006: 179 **A:** AF KZ KI UZ
tatrica Čížek, 2019: 183 **E:** SK (High Tatra Mts.)
testaceicornis Weise, 1894h: 152 **E:** RU (ST) **A:** GG
thikana Bryant, 1940: 44 **A:** YE **AFR**
tolli Ogloblin, 1927: 292 **E:** RU (CT) **A:** CH (HEB NIN NMO SHX) KZ RU (ES FE)
  seriata S.-H. Chen, 1939b: 68
  mongolica L. N. Medvedev, 1976a: 234
tonkinea S.-H. Chen, 1934h: 367 **A:** CH (CHQ) **ORR**
trivialis Weise, 1887h: 201 **A:** NC RU (ES FE) SC
  merkli Gruev, 1994: 80
ussuriensis Romantsov, 2021: 165 **A:** RU (FE)
valachica Heikertinger, 1944: 72 **E:** RO **A:** TR
variolosa Foudras, 1860b: 364 [= 1861: 108] **E:** FR IT SP SZ **N:** MO
varipes Jacoby, 1890c: 161 **A:** CH (FUJ GAN HUB HUN SCH SHA TAI XIZ YUN ZHE) JA (Ryukyus) NC **ORR**
vaulogeri Pic, 1910c: 26 **N:** AG LB MO TU
  cyrenaica Heikertinger, 1944: 80 (var.) [IN] [# 771]
venustula Kutschera, 1861: 244 (*Haltica*) **E:** AL AU BE BH BU BY CR CZ EN FR GE GR HU IR IT KO KR LA LS LT LU MD ME NL NM PT RO RU (CT ST) SB SK SL SP SZ TR UK **A:** TR
  attica Weise, 1890b: 113
violacea Koch, 1803a: 56 (*Haltica*) **E:** AU BY CR CZ EN FR GE HU IT KR LA LT LU NL PL RO RU (CT ST) SK SL SP SV SZ UK **A:** AR GG RU (ES) TR
  ciscaucasica L. N. Medvedev, 1962a: 387
  discolor Weise, 1893a: 1136 (var.)
  pseudacori L. Redtenbacher, 1858: 938 (*Haltica*) [HN]
  sculptipennis Ogloblin, 1927: 292 (ab.) [IN]
  sublaevis Boheman, 1851: 233
  tbilisiensis Král, 1969a: 67
wagneri Heikertinger, 1909a: 13 **E:** FR (Corsica) IT
warchalowskii Fritzlar, 2001: 201 **A:** TR (Antalya)
weiseana Konstantinov, 1998b: 228 **A:** IN TM UZ
wittmeri Doguet, 1979a: 311 **A:** AE SA YE
wrasei Medvedev, 2014: 240 **A:** CH (SHA)
yangsoensis S.-H. Chen, 1939a: 48 **A:** CH (FUJ GUX)
yaosanica S.-H. Chen, 1939a: 42 (*Aphthonaltica*) **A:** CH (CHQ FUJ YUN ZHE)
yunnanica Konstantinov & Lingafelter, 2002: 199 **A:** CH (YUN)
yunnomontana Konstantinov & Lingafelter, 2002: 201 **A:** CH (YUN)

genus *Aphthonaltica* Heikertinger, 1924a: 39 type species: *Graptodera angustata* Baly, 1874

Burmaltica Scherer, 1969: 122 type species: *Haltica kambaitiensis* Bryant, 1939

angustata Baly, 1874: 191 (*Graptodera*) **A:** JA

genus *Aphthonella* Jacoby, 1889b: 204 type species: *Aphthonella bhamoensis* Jacoby, 1889

nigripennis S.-H. Chen & S.-Y. Wang, 1980: 8, 21 **A:** CH (YUN)

nigronitida S.-H. Chen & S.-Y. Wang, 1980: 8, 21 **A:** CH (YUN)

genus *Aphthonoides* Jacoby, 1885a: 59 type species: *Aphthonoides beccarii* Jacoby, 1885

armipes Bryant, 1939: 19 **A:** CH (XIZ) **ORR**

beccarii Jacoby, 1885a: 59 **A:** CH (CHQ FUJ GAN HAI HUB SHA TAI ZHE) JA (Kyushu) **ORR**

beroni Gruev, 1985a: 37 **A:** NP

besucheti Döberl, 1991b: 617 **A:** NP

bhutanensis Döberl, 2005: 68 **A:** BT

carinipennis Scherer, 1989: 33 **A:** NP

castaneus S.-Y. Wang, 1992c: 686, 735 **A:** CH (TAI)

himalayensis L. N. Medvedev, 1984: 49 **A:** NP

latipennis S.-H. Chen & S.-Y. Wang, 1980: 9, 22 **A:** CH (GUX)

lesagei Kimoto, 1996: 42 **A:** CH (TAI)

lopatini Döberl, 2005: 70 **A:** CH (YUN)

piceus Scherer, 1989: 34 **A:** ID (SD) NP

pubipennis S.-Y. Wang, 1992c: 686, 735 **A:** CH (YUN)

rotundipennis Scherer, 1969: 33 **A:** ID (SD) NP

rugiceps S.-Y. Wang, 1992c: 687, 735 **A:** CH (YUN) **ORR**

sagaris Gressitt & Kimoto, 1963a: 797 **A:** CH (HAI)

schereri Döberl, 1991b: 618 **A:** NP

sprecherae Döberl, 2005: 72 **A:** CH (SCH)

taiwanicus Döberl, 2005: 72 **A:** CH (TAI)

tuberifrons S.-Y. Wang, 2004b: 78, 81 **A:** CH (XIZ)

genus *Apteropeda* Chevrolat, 1836: 393 type species: *Altica ciliata* Olivier, 1808 (= *Chrysomela orbiculata* Marsham, 1802)

globosa Illiger, 1794: 602 (*Altica*) **E:** AU BE CZ DE FR GB GE IR IT (North) LU NL PL PT RO SL SK SP (North) SZ UK (Carpathians)

conglomerata Illiger, 1807b: 164 (*Haltica*)

crassina Gistel, 1857b: 592

globus Duftschmid, 1825: 288 (*Haltica*)
majuscula Foudras, 1860b: 281 [= 1861: 25]
nigroaenea Weidenbach, 1859: 85
orbiculata Marsham, 1802: 200 (*Chrysomela*) **E:** AU BE BH BU CR CZ DE FR GB GE HU IR IT LU NL NM PL PT RO SK SL SP SZ UK (Carpathians)
aurichalcea Weise, 1893a: 1045 (var.)
ciliata Olivier, 1808: 710 (*Altica*)
coerulans Weise, 1893a: 1045 (var.)
graminis Koch, 1803a: 47 (*Haltica*)
hederae Illiger, 1807b: 164 (*Haltica*)
ovulum Illiger, 1807a: 65 (*Haltica*) **E:** FR PT SP **N:** AG MO TU
splendida Allard, 1860e: 577 **E:** AN AU CZ DE FR GB GE IR IT (North) PL SK SP SZ UK

genus *Argopistes* Motschulsky, 1860b: 236 type species: *Argopistes biplagiatus* Motschulsky, 1860
Sphaerophyma Baly, 1878c: 478 type species: *Sphaerophyma simoni* Baly, 1878
atricollis S.-H. Chen, 1934h: 317 **A:** BT NP ID (HP SD UT)
biplagiatus Motschulsky, 1860b: 236 **A:** CH (GAN SCH) JA RU (FE) SC (Dagelet Is., Dockdo Is.)
flavitaris Motschulsky, 1860b: 237
limbatus Motschulsky, 1860b: 237
suturalis Motschulsky, 1860b: 237
undecimmaculatus Jacoby, 1886a: 738
coccinelliformis Csiki, 1940a: 524 [RN] **A:** CH (GUA HKG TAI ZHE) JA **ORR**
coccinelloides Baly, 1874: 202 [HN]
rufus S.-H. Chen, 1934b: 72 (var.)
flavus S.-H. Chen, 1934h: 317 **A:** NP **ORR**
hoenei Maulik, 1934: 25 **A:** CH (GUX ZHE)
ryukyuensis Shigetoh & Suenaga, 2022: 4 **A:** JA (Ryukyus)
sinensis S.-H. Chen, 1939a: 51 **A:** CH (GUX)
tsekooni S.-H. Chen, 1934h: 316 **A:** CH (JIA ZHE) JA (Honshu, Kyushu) NC SC (Ullung-do, Dockdo Is.)
udege Konstantinov, 1994: 527 **A:** RU (FE)
unicolor Jacoby, 1886a: 738 **A:** JA (Kyushu) RU (ES FE)

genus *Argopistoides* Jacoby, 1892b: 931 type species: *Argopistoides septempunctata* Jacoby, 1892
Torodera Weise, 1902e: 163 type species: *Torodera octomaculata* Weise, 1902

septempunctata Jacoby, 1892b: 932 **A**: CH (FUJ GUA YUN) **ORR**
sexpunctata Scherer, 1987: 69 (*Torodera*) **A**: BT ID (SD) NP

genus *Argopus* Fischer von Waldheim, 1824: 182 type species: *Rhyssodes bicolor* Fischer von Waldheim, 1823
Dicherosis Foudras, 1860b: 348 [= 1861: 92] type species: *Haltica hemisphaerica* Duftschmid, 1825 (= *Argopus ahrensii Germar*, 1817)
ahrensii Germar, 1817: 206 (*Haltica*) **E**: AL AU BH BY CR CZ GE GR HU IT LA MD ME NM PL RO RU (CT ST) SB SK SL SZ UK
hemisphaericus Duftschmid, 1825: 253 (*Haltica*)
balyi Harold, 1878: 88 **A**: JA SC
bicolor Fischer von Waldheim, 1823: pl. XLVII (*Rhyssodes*) [= 1824: 183] **E**: BU CR HU KR PL RO RU (ST) SB SK UK **A**: AB
discolor Bielz, 1852: 16
punctatus Schaufuss, 1877: 423
bidentatus S.-Y. Wang, 1992c: 693, 739 **A**: CH (SCH SHA)
brevis Allard, 1860b: cclx **E**: FR GR (North Sporades) SP **N**: AG MO TU
chiui Kimoto, 1970d: 294 **A**: CH (TAI)
circumaedeagus Özdikmen, Kılıç, Coral Şahin & Bal, 2019: 448 **A**: TR
clarki Jacoby, 1886a: 734 **A**: JA (Honshu, Izu Isl.)
hasegawai Ohno, 1967: 49
clematidis Rapilly, 1978: 339 **A**: IN
clypeatus Baly, 1874: 206 **A**: JA
formosanus Chûjô, 1936c: 112 **A**: CH (TAI)
fortunei Baly, 1877d: 181 **A**: CH (GUA JIX YUN ZHE) **ORR**
foveolatus S.-Y. Wang & S.-Q. Ge, 2009b: 900, 904 **A**: CH (HEI)
frontoclypeatus Gressitt & Kimoto, 1963a: 816 **A**: CH (SCH)
intermedius Weise, 1887h: 208 **A**: RU (FE)
koreanus Chûjô, 1941b: 168 **A**: NC SC
melanocephalus Gressitt & Kimoto, 1963a: 818 **A**: CH (FUJ ZHE)
miyakei Kimoto, 1991b: 120 **A**: CH (TAI)
nigrifrons S.-H. Chen, 1933a: 237 **A**: CH (FUJ GUA HUB ZHE)
nigripennis Jacoby, 1886a: 734 **A**: JA (Honshu)
nigripes Weise, 1889g: 642 **A**: CH (GAN SCH SHA) SC
nigritarsis Gebler, 1823: 125 (*Chrysomela*) **E**: BY HU KZ LA NT PL RU (CT ST) SK UK **A**: CH (FUJ GAN HEB HUB JIX NIN SCH SHA SHX TAI ZHE) JA KZ MG NC RU (ES FE SS WS) SC
punctipennis punctipennis Motschulsky, 1866a: 176 (*Dicherosis*) **A**: JA RU (FE)
amagianus Ohno, 1967: 42 (ssp.)
japonicus S.-H. Chen, 1936d: 596 [RN]

orientalis Baly, 1874: 206
unicolor Jacoby, 1886a: 736 (*Sphaeroderma*) [HN]
univestis Heikertinger, 1940: 514 [RN]
punctipennis substriatus Weise, 1887h: 206 **A:** NC RU (FE) SC
riedeli Warchałowski, 1969: 228
similibidentatus S.-Y. Wang & S.-Q. Ge, 2009b: 901, 904 **A:** CH (SHA ZHE)
similis S.-H. Chen, 1939b: 64 **A:** CH (JIX ZHE)
unicolor Motschulsky, 1860b: 235 **A:** JA NC RU (FE) SC

genus *Arrhenocoela* Foudras, 1860b: 334 [= 1861: 78] type species: *Chrysomela lineata* P. Rossi, 1790
lineata P. Rossi, 1790: 88 (*Chrysomela*) **E:** AL CR FR GR IT ME PT SP **A:** AG MO TU
geminata Fabricius, 1801a: 498 (*Galleruca*)

genus *Asiophrida* L. N. Medvedev, 2000c: 180 type species: *Haltica marmorea* Wiedemann, 1819
subgenus *Asiophrida* L. N. Medvedev, 2000c: 180 type species: *Haltica marmorea* Wiedemann, 1819
marmorea Wiedemann, 1819: 181 (*Haltica*) **A:** NP **ORR**
scaphoides Baly, 1865e: 430 (*Podontia*) **A:** CH (CHQ FUJ GAN GUI HKG HUB JIA SCH SHA TAI XIZ YUN ZHE) NP
binduta Maulik, 1926: 233 (*Ophrida*)
spectabilis Baly, 1862b: 452 (*Podontia*) **A:** CH (CHQ FUJ GAN GUI HUB JIA SCH SHA TAI XIZ YUN ZHE) NC SC
rufoflava Fairmaire, 1889a: 73 (*Podontia*)
xanthospilota Baly, 1881: 54 (*Blepharida*) **A:** CH (GAN HEB SCH)

genus *Asutoshana* L. N. Medvedev, 2011c: 253 type species: *Asutoshana ornata* L. N. Medvedev, 2011
ornata L. N. Medvedev, 2011c: 254 **A:** NP

genus *Baoshanaltica* Konstantinov & Ruan, 2017b: 104 type species: *Baoshanaltica minuta* Konstantinov & Ruan, 2017
minuta Konstantinov & Ruan, 2017b: 108 **A:** CH (YUN)

genus *Batophila* Foudras, 1860a: 378 [= 1860b: 266] type species: *Galeruca rubi* Paykull, 1799
acutangula Heikertinger, 1921b: 91 **A:** CH (FUJ HUB JIA TAI ZHE) JA NC RU (FE) SC
hira Nakane, 1958: 313 (forma) [IN] [# 770]
kamikochiana Nakane, 1958: 313 (ssp.)

omogo Nakane, 1958: 313 (ssp.)
yakuensis Nakane, 1958: 313 (ssp.)
yangweii S.-H. Chen, 1933a: 250
aerata Marsham, 1802: 204 (*Chrysomela*) **E:** FR GB GE GR (South) IT SL SP SZ **N:** AG MO TU
bertolinii Gredler, 1866: 438
brunnicornis Stephens, 1831: 301 (*Haltica*) [QS]
thymi Abeille de Perrin, 1907: lxxx (var.)
alticola Döberl, 2003b: 149 **A:** NP
angustata S.-Y. Wang, 1992c: 720, 750 **A:** CH (YUN)
beroni Gruev, 1990a: 74 **A:** NP
costata Scherer, 1989: 49 **A:** ID (SD) NP
costipennis S.-Y. Wang, 1997: 925, 930 **A:** CH (CHQ HUB)
dogueti Döberl, 1994b: 149 **A:** AB IN
fallax Weise, 1888b: 854 (var.) **E:** HU KR NM PL RO RU (CT ST) TR UK **A:** AB GG IN TR
femorata Scherer, 1989: 50 **A:** NP
fragariae S.-Y. Wang, 1992c: 720, 750 **A:** CH (YUN)
impressa S.-Y. Wang, 1992c: 721, 751 **A:** CH (YUN)
latissima Chûjô, 1957a: 17 **A:** JA (Ryukyus)
moesica Heikertinger, 1948: 53 **E:** BU RO
nepalica Döberl, 2003b: 150 **A:** NP
olexai Král, 1964a: 37 **E:** BU GR RU (ST) **A:** AB GG IN TR (Giresun)
potentillae S.-Y. Wang, 1992c: 721, 751 **A:** CH (YUN)
punctifrons S.-Y. Wang, 1992c: 722, 751 **A:** CH (SCH)
pyrenaea Allard, 1866: 271 **E:** FR (Pyrenees) SP (Pyrenees)
rubi Paykull, 1799a: 112 (*Galeruca*) **E:** AL AU BE BH BU BY CR CZ DE EN FI FR GB GE HU IR IT (North) KO KR LA LS LT LU MD ME NL NM NR PL RO RU (CT NT ST) SB SK SL SV SZ UK **A:** RU (WS)
striatula Marsham, 1802: 205 (*Chrysomela*)
sinensis S.-H. Chen, 1939b: 62 **A:** CH (GAN SHX)
subcostata S.-H. Chen, 1939b: 63 **A:** CH (GAN HUN)
subplana S.-H. Chen, 1939b: 63 **A:** CH (GAN HEB)
taiwanica Döberl, 2010a: 65 **A:** CH (TAI)

genus *Benedictus* Scherer, 1969: 99 type species: *Benedictus elisabethae* Scherer, 1969
Himalalta L. N. Medvedev, 1990b: 41 type species: *Himalalta brevicornis* L. N. Medvedev, 1990 (= *Benedictus leoi* Scherer, 1989)
bagmati Sprecher-Uebersax, Konstantinov, Prathapan & Döberl, 2009: 476 **A:** NP

belousovi Sprecher-Uebersax, Konstantinov, Prathapan & Döberl, 2009: 478 **A:** CH (SCH)
cangshanicus Sprecher-Uebersax, Konstantinov, Prathapan & Döberl, 2009: 479 **A:** CH (YUN)
chilalla Sprecher-Uebersax, Konstantinov, Prathapan & Döberl, 2009: 483 **A:** BT
dochula Sprecher-Uebersax, Konstantinov, Prathapan & Döberl, 2009: 484 **A:** BT
elisabethae Scherer, 1969: 100 **A:** ID (SD) NP
flavicalli Scherer, 1989: 38 **A:** ID (SD) NP
fuanensis Ruan & Konstantinov, 2023: 151 **A:** CH (FUJ GUA)
ha Sprecher-Uebersax, Konstantinov, Prathapan & Döberl, 2009: 487 **A:** BT
kabaki Sprecher-Uebersax, Konstantinov, Prathapan & Döberl, 2009: 490 **A:** CH (SCH)
kurbatovi Sprecher-Uebersax, Konstantinov, Prathapan & Döberl, 2009: 490 **A:** CH (SCH)
lauribina Sprecher-Uebersax, Konstantinov, Prathapan & Döberl, 2009: 490 **A:** NP
leoi Scherer, 1989: 39 **A:** ID (SD) NP
  brevicornis L. N. Medvedev, 1990b: 42 (*Himalalta*)
medvedevi Döberl, 1991b: 623 **A:** NP
minutus L. N. Medvedev, 1984: 53 **A:** NP
  striatus L. N. Medvedev, 1990b: 42 (*Himalalta*)
nigrinus Sprecher-Uebersax, Konstantinov, Prathapan & Döberl, 2009: 498 **A:** CH (GAN SCH)
nobding Sprecher-Uebersax, Konstantinov, Prathapan & Döberl, 2009: 501 **A:** BT
quadrimaculatus Ruan & Konstantinov, 2023: 156 **A:** CH (YUN)
sagittalis Damaška & Aston, 2019: 152 **A:** CH (HKG)
shivalayanicus Sprecher-Uebersax, Konstantinov, Prathapan & Döberl, 2009: 502 **A:** NP
sichuanensis Sprecher-Uebersax, Konstantinov, Prathapan & Döberl, 2009: 503 **A:** CH (SCH SHA)
thumsila Sprecher-Uebersax, Konstantinov, Prathapan & Döberl, 2009: 504 **A:** BT
tibetanus Sprecher-Uebersax, Konstantinov, Prathapan & Döberl, 2009: 508 **A:** CH (SCH)
wangi Ruan & Konstantinov, 2023: 159 **A:** CH (XIZ)
yatongla Sprecher-Uebersax, Konstantinov, Prathapan & Döberl, 2009: 511 **A:** BT

genus *Bezdekaltica* Döberl, 2012: 434 type species: *Bezdekaltica socotrana* Döberl, 2012
socotrana Döberl, 2012: 435 **A:** YE (Socotra)

genus *Bhamoina* Bechyné, 1958a: 91 type species: *Sphaeroderma acutangulum* Jacoby, 1892 (= *Eucycla varipes* Jacoby, 1884)
nigripes L. N. Medvedev, 2019a: 167 **A:** CH (SCH)

varipes Jacoby, 1884c: 210 (*Eucycla*) **A:** CH (GUX XIZ) ID (UT) NP **ORR**
acutangulum Jacoby, 1892b: 927 (*Sphaeroderma*)
variipennis Jacoby, 1892b: 928 (*Sphaeroderma*)

genus *Bikasha* Maulik, 1931: 256 type species: *Bikasha tenuipunctata* Maulik, 1931
Aphthonomorpha S.-H. Chen, 1934h: 356 type species: *Crepidodera collaris* Baly, 1877
Manobidia S.-H. Chen, 1934h: 358 type species: *Manobidia antennata* S.-H. Chen, 1934
antennata S.-H. Chen, 1934h: 359 (*Manobidia*) **A:** CH (GUX YUN) **ORR**
atra L. N. Medvedev, 1992d: 28 (*Manobidia*) **A:** NP **ORR**
collaris Baly, 1877d: 161 (*Crepidodera*) **A:** CH (CHQ FUJ GUA GUI HKG HUB HUN JIA JIX TAI) JA (Shikoku, Kyushu) SC **ORR**
intermedia S.-H. Chen, 1934h: 360 (*Manobidia*) **A:** CH (GUA HKG) NP **ORR**
major Kimoto, 2001: 63 (*Manobidia*) **A:** NP
minuta S.-H. Chen, 1934h: 357 (*Aphthonomorpha*) **A:** NP **ORR**
nepalica L. N. Medvedev, 2007c: 300 (*Aphthonomorpha*) **A:** NP
nipponica Chûjô, 1959: 11 (*Manobidia*) **A:** CH (FUJ HUN HUB JIA TAI YUN) JA (Shikoku, Kyushu)
simplicithorax S.-H. Chen, 1934h: 360 (*Manobidia*) **A:** CH (GUA GUX HAI HKG TAI YUN) JA (Kyushu, Ryukyus) **ORR**
fulva Chûjô, 1957a: 16 (*Aphthonomorpha*)
violaceipennis L. N. Medvedev, 1992d: 27 (*Manobidia*) **A:** NP

genus *Bimala* Maulik, 1926: 195 type species: *Erystus indicus* Jacoby, 1896
indica Jacoby, 1896c: 265 (*Erystus*) **A:** ID (UT) **ORR**

genus *Blepharidina* Bechyné, 1969: 1725 type species: *Blepharida guttulata* Baly, 1881
subgenus *Afroblepharida* Biondi & D'Alessandro, 2017: 133 type species: *Blepharida antinorii* Chapuis, 1879
socotrana Biondi, Ianella & D'Alessandro, 2019: 502 **A:** YE (Socotra)

genus *Burumoseria* Csiki, 1939: 13 [RN] type species: *Moseria partita* Weise, 1922
Moseria Weise, 1922b: 110 [HN] type species: *Moseria partita* Weise, 1922
yuae C.-F. Lee & Konstantinov, 2015: 328 **A:** CH (TAI)

genus *Calotheca* Heyden, 1887a: 98 [RN] type species: *Eutheca haroldi* Baly, 1878
Blepharidella Weise, 1909: 220 type species: *Blepharida sjostedti* Weise, 1910
Blepharidula Weise, 1916a: 39 [RN] type species: *Eutheca haroldi* Baly, 1878
Eutheca Baly, 1878b: 204 [HN] type species: *Eutheca haroldi* Baly, 1878
sacra Weise, 1897b: 81 (*Podontia*) **N:** EG **A:** IS SA **AFR**
arabica L. N. Medvedev, 1996: 254 (*Blepharida*)

genus *Cangshanaltica* Konstantinov, Chamorro, Prathapan, Ge & Yang, 2013: 6 type species: *Cangshanaltica nigra* Konstantinov, Chamorro, Prathapan, Ge & Yang, 2013

castanea Gruev, 1985b: 126 (*Taizonia*) **A:** CH (ZHE)

gruevi Nadein, 2013b: 396 (*Ivalia*) [RN]

fuanensis Ruan, Konstantinov & Damaška, 2022: 29 **A:** CH (FUJ)

fuanensis Ruan, Konstantinov & Damaška, 2020: 4 [UN]

guangxiensis Damaška, Ruan & Fikáček, 2022: 54 **A:** CH (GUX)

marginata Damaška, Ruan & Fikáček, 2022: 54 **A:** CH (YUN)

nigra Konstantinov, Chamorro, Prathapan, Ge & Yang, 2013: 16 **A:** CH (YUN)

ogatai Takemoto & Suenaga, 2021: 146 **A:** JA

sprynari Damaška & Aston, 2019: 153 **A:** CH (HKG)

genus *Chaetocnema* Stephens, 1831: 325 type species: *Chrysomela concinna* Marsham, 1802 **[# 751, Appendix]**

Biondiana Özdikmen, 2021: 51 type species: *Chaetocnema zulu* Biondi & Nardis, 2000 [# 751]

Brinckaltica Bechyné, 1959: 237 type species: *Chaetocnema subaterrima* Jacoby, 1900

Carcharodis Weise, 1910a: 434 type species: *Carcharodis virens* Weise, 1910

Confinoides Özdikmen, 2021: 52 type species: *Chaetocnema confinis* Crotch, 1873 [# 751]

Dalessandroiana Özdikmen, 2021: 53 type species: *Chaetocnema audisiana* Biondi, 2000 [# 751]

Exorhina Weise, 1886a: 750 type species: *Haltica chlorophana* Duftschmid, 1825

Gahanioides Özdikmen, 2021: 54 type species: *Chaetocnema gahani* Jacoby, 1897 [# 751]

Hortensoides Özdikmen, 2021: 56 type species: *Altica hortensis* Geoffroy, 1785 [# 751]

Hydropus Motschulsky, 1860b: 235 type species: *Udorpes splendens* Motschulsky, 1845

Longicornoides Özdikmen, 2021: 60 type species: *Chaetocnema longicornis* Jacoby, 1895 [# 751]

Majoroides Özdikmen, 2021: 61 type species: *Plectroscelis major* Jacquelin du Val, 1853 [# 751]

Nigricoides Özdikmen, 2021: 62 type species: *Chaetocnema nigrica* Motschulsky, 1858 [# 751]

Nitidoides Özdikmen, 2021: 64 type species: *Phyllotreta nitida* Broun, 1880 [# 751]

Plectroscelis Chevrolat, 1836: 393 type species: *Altica dentipes* sensu Olivier (1808) (= *Haltica chlorophana* Duftschmid, 1825) [# 760]

Pseudochaetocnema Özdikmen, 2021: 66 type species: *Haltica conducta* Motschulsky, 1838 [# 751]

Tlanoma Motschulsky, 1845a: 108 type species: *Haltica dentipes* Koch, 1803 (= *Chrysomela concinna* Marsham, 1802)

Udorpes Motschulsky, 1845a: 107 type species: *Udorpes splendens* Motschulsky, 1845

Ydorpes Motschulsky, 1845a: [unnumb. page] type species: *Udorpes splendens* Motschulsky, 1845

aerosa Letzner, 1847: 84 (*Haltica*) E: AU BE BH BY CR CZ EN FI FR GB GE HU IT KO LA LU ME NL NR PL RO RU (CT NT ST) SK SP SV SZ UK **A:** IS RU (SS WS) TR

laevilinea J. R. Sahlberg, 1903: 34 (var.)

punctatissima Graells, 1858: 96 [= 1858b: 134] (*Plectroscelis*)

afghana Gruev, 1988b: 155 **A:** AE AF PA **ORR** [# 774]
alticola Maulik, 1926: 217 **A:** CH (XIN XIZ) ID (UT) NP
angustula Rosenhauer, 1847: 62 (*Plectroscelis*) **E:** AN AU FR GE IT (North) SL SP SZ
aeneicolor Pic, 1915b: 5 (var.)
alpicola Weise, 1886a: 767 (var.)
arenacea Allard, 1860e: 569 (*Plectroscelis*) **E:** AL AU BU CR CZ FR GR (Ionian Is.) HU IT (South) KR PT RO RU (ST) SB SK SP SZ TR UK **N:** AG MO **A:** AB AR IN TR
arida Foudras, 1860a: 234 [= 1860b: 122] **E:** AU BE BH BU BY CR CZ EN FR GB GE GR HU IT KO KR LA LS LT LU NL NM PL PT RO SB SK SL SP SZ UK **A:** IN TR **N:** AG MO TU
aestiva Weise, 1888b: 775 (var.)
surcoufi Pic, 1915h: 41
aridula Gyllenhal, 1827: 663 (*Haltica*) **E:** AL AU BE BH BU BY CR CY CZ DE EN FI FR GE GR HU IT KR KZ LA LS LT LU MA MD ME NL NM NR PL PT RO RU (CT NT ST) SB SK SL SP SV SZ TR UK **N:** AG **A:** AB IN KI KZ RU (ES SS WS) TR **ORR**i
weisei Gerhardt, 1906: 238 (var.)
medvedevi Paliy, 1968a: 15
babai Kimoto, 1991a: 17 **A:** CH (TAI)
balanomorpha Boieldieu, 1859: 481 (*Plectroscelis*) **E:** FR (Pyrénées) SP
baoshanica Ruan, Konstantinov & Yang, 2019: 29 **A:** CH (YUN)
batophiloides Abeille de Perrin, 1909: 180 **N:** AG **A:** IS JO **AFR**
belka Konstantinov, Baselga, Grebennikov, Prena & Lingafelter, 2011: 73 **A:** CH (GAN)
bella Baly, 1876d: 595 (*Plectroscelis*) **A:** CH (FUJ GUX HAI HUB JIA SCH SHA TAI YUN XIN XIZ ZHE) JA (Ryukyus) NP PA **ORR**
formosensis Chûjô, 1935h: 471
nepalensis Scherer, 1969: 159
placida Jacoby, 1896b: 441
shanensis Bryant, 1939: 16
belli Jacoby, 1904a: 392 **A:** BT NP PA **ORR**
bergeali Konstantinov, Baselga, Grebennikov, Prena & Lingafelter, 2011: 79 **E:** FR
bicolorata Kimoto, 1971b: 3 **A:** JA (Honshu) SC
bilunulata Demaison, 1902: 24 **N:** EG **AFR**
diluta Laboissière, 1942: 85
praescutellaris Pic, 1911a: 10 (var.)
vincenti Reitter, 1906a: 37
sylvia Bechyné, 1964: 157
bretinghami Baly, 1877d: 170 **A:** BT ID (SD) IS NP PA **ORR**
breviuscula Faldermann, 1837: 349 (*Plectroscelis*) **E:** BU BY EN GR HU KR MD RO RU (CT ST) SK TR UK **A:** AB AF AR CH (HEI HUB) GG IN IQ JO KZ MG NC RU (SS WS) SY TD TM TR UZ
turhala İriboz, 1934: 84

cheni Ruan, Konstantinov & Yang, 2014a: 21 **A:** CH (HUN JIX SCH YUN)

chlorophana Duftschmid, 1825: 286 (*Haltica*) **E:** AL AU BE BU CR CZ FR GE GR HU IT KR LU NM PT RO RU (ST) SB SK SL SP TR UK **N:** AG MO TU **A:** AB IN IQ IS SY TR

amoena Weise, 1886a: 756 (var.)

bicolora Roubal, 1948b: 125 (ab.) [IN]

kolbei Weise, 1886a: 756 (var.)

laeta Weise, 1886a: 756 (var.)

syriaca Demaison, 1896: 13

chujoi Takizawa, 2021a: 156 **A:** JA

cognata Baly, 1877d: 168 **A:** BT ID (SD UT) NP PA YE **ORR**

squarrosa Baly, 1877d: 169

compressa Letzner, 1847: 85 (*Haltica*) **E:** AU BU BY CZ GE GR HU KR PL RO RU (CT incl. Kaliningrad Region ST) SB SK SL SZ UK **A:** KZ RU (WS)

tarda Bach, 1858: 160 [HN]

tarda Foudras, 1860a: 224 [= 1860b: 112] [HN]

concinna Marsham, 1802: 196 (*Chrysomela*) **E:** AL AU BE BH BU BY CR CZ DE EN FI FR GB GE GR HU IT KO KR LA LS LT LU MD ME NL NM NR PL RO RU (CT NT ST) SB SK SL SP SV SZ TR UK **N:** AG MO **A:** AB AR GG IS JA KZ MG NC RU (ES FE SS WS) SC TR TM **NARi**

dentipes Koch, 1803a: 38 (*Haltica*)

lewisii Chûjô 1942a: 31

semirufescens Pic, 1909h: 155 (var.)

concinnicollis Baly, 1874: 208 (*Plectroscelis*) **A:** CH (FUJ GAN HAI HKG SHA TAI XIZ ZHE) JA NC NP RU (FE) SC **ORR**

kaibarensis Jind. Madar, 1960: 48 (ssp.)

philoxena Baly, 1876d: 595 (*Plectroscelis*)

concinnipennis Baly, 1877d: 170 **A:** CH (HAI HKG YUN) ID (SD) JA NP PA **ORR**

conducta Motschulsky, 1838: 180 (*Haltica*) **E:** AL AR AU BH BU CR CZ FR GE GR HU IT KR MD ME NM PT RO RU (CT ST) SB SK SL SP SZ TR UK **N:** AG EG MO TU **A:** AB IN IS KI KZ SY TD TR TM **AFR**

ahngeri Jacobson, 1901b: 139

foudrasi Bauduer, 1874a: 198 (*Plectroscelis*) [RN]

confinis Crotch, 1873: 75 Ai: CH (FUJ GUA GUX TAI) JA (Ryukyus) **AFRi AUS NAR NTR ORRi**

etiennei Jolivet, 1979: 641

flavicornis LeConte, 1878: 418

confusa Boheman, 1851: 234 (*Plectroscelis*) **E:** AU BE BH BY CR CZ DE FR GB GE GR HU IT LA LU NL PL SB SK SL SP SV SZ UK **A:** IN RU (WS) TR

constricta Ruan, Konstantinov & Yang, 2014a: 24 **A:** CH (ANH CHQ FUJ GUI GUX JIA JIX SCH YUN ZHE) [# 784] *costulata* Motschulsky, 1860b: 234 (*Plectroscelis*) **A:** CH (GAN GUA HEI JIL) KI MG NC RU (FE ES) SC

sonkulica Paliy, 1968a: 19

coyei Allard, 1864: 659 (*Plectroscelis*) **E:** AL BU CR GR (North) KR RO RU (ST) SB TR UK **A:** AB AR CY IN IQ IS JO SY TR

delagrangei Pic, 1909e: 138

metallica Weise, 1886a: 751 (var.)

cupreata S.-H. Chen, 1934h: 249 **A:** NP **ORR**

cylindrica Baly, 1874: 208 (*Plectroscelis*) **A:** CH (GAN GUX HUB JIA ZHE) JA (Honshu, Kyushu) SC

delarouzeei Brisout de Barneville, 1884: 89 (*Plectroscelis*) **N:** AG MO TU **A:** IS JO TR

depressa Boieldieu, 1859: 482 (*Plectroscelis*) **E:** BE CR FR IT LU SP **N:** AG TU

chrysicollis Foudras, 1860a: 217 [= 1860b: 105] (*Plectroscelis*)

gibbifrons Pic, 1909h: 155

deqinensis Ruan, Konstantinov & Yang, 2014b: 70 **A:** CH (GUI YUN) **ORR**

duvivieri Jacoby, 1892b: 939 **A:** CH (XIZ) **ORR**

eastafghanica Konstantinov, Baselga, Grebennikov, Prena & Lingafelter, 2011: 140 **A:** AF NP PA

excavata L. N. Medvedev, 1997a: 21 **A:** ID (SD) NP

fortecostata S.-H. Chen, 1939a: 33 **A:** CH (CHQ FUJ GUX HUB HUN JIX SCH SHA YUN ZHE) JA **ORR**

franzi Konstantinov, Baselga, Grebennikov, Prena & Lingafelter, 2011: 143 **E:** NM

fusiformis S.-H. Chen & S.-Y. Wang, 1980: 4, 17 **A:** CH (YUN)

ganganensis Bechyné, 1955b: 548 **N:** AG **AFR**

gottwaldi Král, 1969a: 69 **A:** IN KZ

gracilis Motschulsky, 1858: 107 (*Tlanoma*) **A:** BT CH (YUN) NP PA **AUS ORR**

indica Weise, 1916a: 39 [RN]

minuta Jacoby, 1896c: 271 [HN]

grandis Pic, 1909e: 139 (var.) **A:** KI KZ MG RU TD TM UZ

arisi Pic, 1915h: 42 (var.)

granulicollis Jacoby, 1896b: 439 **A:** CH (TAI) **ORR**

flavipennis Medvedev, 1997c: 68

granulosa Baly, 1874: 207 (*Plectroscelis*) **A:** CH (TAI) JA SC

hainanensis S.-H. Chen, 1933a: 228 **A:** CH (FUJ GUX HAI HKG) NP **ORR**

assamensis Scherer, 1969: 161

kwangsiensis S.-H. Chen, 1939a: 32

heptapotamica Lyubishchev, 1963: 858 **E:** RU (ST) **A:** GG KI KZ (Alma Ata) UZ

hiromui Takizawa, 2021a: 163 **A:** JA

hongkongensis Ruan, Konstantinov & Yang, 2019: 65 **A:** CH (HKG) **ORR**

hortensis Geoffroy, 1785: 98 (*Altica*) **E:** AL AU AZ BE BH BU BY CR CZ DE EN FI FR GB GE GR HU IT KO KR LA LS LT LU MA MD ME NL NM NR PL PT RO RU (CT

NT ST) SB SK SL SP SV SZ TR UK **N:** AG MO MR TU **A:** AB AF AR CH (GAN JIA) GG IN IQ CY IS KZ MG RU (ES FE SS WS) SY TD TR TM UZ YE **NARi**

aridella Paykull, 1799a: 111 (*Galeruca*)

brenskei Pic, 1910f: 305 (var.)

convexa Motschulsky, 1860b: 234 (*Plectroscelis*)

granosa Motschulsky, 1860b: 234 (*Plectroscelis*) [QS]

scabricollis Allard, 1860e: 569 (*Plectroscelis*)

igori Konstantinov, Baselga, Grebennikov, Prena & Lingafelter, 2011: 169 **A:** AB AF AR KZ TD TR UZ

imitatrix Gruev, 1990c: 89 **E:** BU (Vitosha Mt.) **A:** TR

ingenua Baly, 1876d: 594 (*Plectroscelis*) **A:** AF CH (CHQ FUJ GAN HEI HEB HEN HUB HUN JIA JIL JIX NIN NMO SHA SHN SHX TIA YUN ZHE) JA MG NC PA RU (FE) SC **ORR**

aurifrons Jacoby, 1886a: 733

fulvipes Jacoby, 1886a: 732

japonica Jacoby, 1886a: 732

ogloblini Paliy, 1970a: 197

jelineki Lopatin, 1990b: 605 **A:** IN (South-East)

jinxiuensis Ruan, Konstantinov & Yang, 2019: 67 **A:** CH (GUX)

kabakovi Lopatin, 1995a: 102 **A:** AF

kanmiyai Kimoto, 1974a: 148 **A:** JA (Kyushu)

kerimii Fairmaire, 1875b: 539 (*Plectroscelis*) **A:** AG LB MO TU

sycophanta Abeille de Perrin, 1896: 54

kimotoi Gruev, 1980a: 35 **A:** JA NC RU (FE)

kingpinensis Ruan, Konstantinov & Yang, 2014a: 27 **A:** CH (GUX JIX YUN)

klapperichi Lopatin, 1963d: 370 **A:** AF (East)

koreana Chûjô, 1942a: 33 **A:** CH (HEI) JA (Hokkaido, Honshu) NC SC

kumaonensis Scherer, 1969: 158 **A:** CH (YUN) ID (UT) NP **ORR**

latapronota Ruan, Konstantinov & Yang, 2019: 164 **A:** PA **ORR**

latipennis Pic, 1911b: 98 **N:** EG **AFR**

leonhardi Heikertinger, 1951: 63 **E:** BH (Bielashnica Mt., Ilidzha Mt.)

ljuba Bechyné, 1955b: 551 **A:** AE OM YE **AFR** [# 775]

segueia Bechyné, 1955b: 550

szumowskii Bechyné, 1955b: 551

ljudmilae Lopatin, 1961b: 145 **E:** RU (ST) **A:** KZ (South) KI RU (ES) TD (South)

lubischevi Konstantinov, Baselga, Grebennikov, Prena & Lingafelter, 2011: 209 **A:** AB IN KZ

major Jacquelin du Val, 1853: 717 (*Plectroscelis*) **E:** AL AU BU CZ FR GR HU KR MD RO RU (CT ST) SB SK SL SP TR UK **N:** AG **A:** AB AF AR CY IN IQ IS JA (Honshu) KZ SY TD TR

perrisii Bauduer, 1874a: 197 (*Plectroscelis*)

malayana Baly, 1877g: 312 **A:** CH (TAI YUN) **ORR**

birmanica Jacoby, 1892b: 939

mandschurica Heikertinger, 1951: 50 (ssp.) **A:** CH (HEI JIL) JA (Honshu) RU (FE)

mannerheimii Gyllenhal, 1827: 664 (*Haltica*) **E:** AL AU BE BU BY CR CZ EN FI FR GE GR (Kefallonia) HU IT (North) KR LA LS LT LU MD NL PL RO RU (CT NT ST) SB SK SL SV SZ TR UK **A:** KI KZ MG RU (ES FE WS)

fallax Weise, 1888b: 771 (var.)

melonae S.-H. Chen, 1934h: 252 **A:** CH (XIN XIZ) ID (SD) NP **ORR**

himalayana L. N. Medvedev, 1993c: 372

midimpunctata Ruan, Konstantinov & Yang, 2019: 170 **A:** CH (HAI) **ORR**

modesta Gressitt & Kimoto, 1963a: 780 **A:** CH (HUB SCH) **ORR**

modiglianii Jacoby, 1896b: 438 **A:** BT CH (HAI) NP **ORR**

vietnamica S.-H. Chen & S.-Y. Wang, 1980: 4, 18

montenegrina Heikertinger, 1912d: 44 (ssp.) **E:** AL BH BU CR GR IT (South) ME NM RO RU (CT ST) SB UK **A:** AF AR IN KI KZ TD TR TM UZ

ndesei Laboissière, 1942: 91 **A:** SA **AFR**

cephalica Bechyné, 1960a: 27

nebulosa Weise, 1886a: 753 **E:** KR KZ RU (CT ST) UK **A:** IN KI KZ MG PA RU (WS)

nigrica Motschulsky, 1858: 106 (*Tlanoma*) **A:** AF BT CH (CHQ FUJ GUI GUX HAI TAI XIZ ZHE) JA (Yakushima, Ryukyus) NP PA **AUS ORR**

basalis Baly, 1877g: 310

geniculata Jacoby, 1896c: 270

gestroi Jacoby, 1889c: 283

loriae Jacoby, 1905a: 489

nitens Baly, 1877g: 312

parvula Baly, 1877g: 310

nocticolor Rapilly, 1978: 331 **A:** IN

obesa Boieldieu, 1859: 480 (*Plectroscelis*) **E:** AL AU BE BH BU BY CR CZ FR GE GR HU IT KR LA MD ME NM PL PT RO RU (CT ST) SB SK SL SP SZ TR UK **N:** AG MO TU **A:** AB AR CH (XIZ) GG IN IQ KZ MG RU (ES) TR

corcyrica Pic, 1909e: 139 (var.)

jurassica Pic, 1915h: 42

meridionalis Foudras, 1860a: 228 [= 1860b: 116]

unicolor Weise, 1888b: 769 (var.)

oblonga Lopatin, 1990b: 606 **A:** IN (South)

orientalis Bauduer, 1874a: 198 (*Plectroscelis*) **E:** BU GR NM RO RU (ST: Caucasus) TR **A:** GG IN IQ IS JO SY TR TM

paganettii Heikertinger, 1913c: 143 **E:** PT SP (Astorga, Balearic Islands) **N:** AG MO

parafusiformis Ruan, Konstantinov & Yang, 2019: 86 **A:** BT IN NP **ORR**

paragreenica Ruan, Konstantinov & Yang, 2019: 178 **A:** NP PA **ORR** *pelagica* Caillol, 1921: 515 E: FR IT PT SP **N**: AG

picipes Stephens, 1831: 327 E: AU BE BU BY CZ DE EN FI FR GB GE HU IT LA LS LT LU MD NL NR PL RO RU (CT NT ST) SK SL SP SV SZ UK **A**: AB AR CH (BEI GAN HEB HEI LIA NIN NMO QIN SHA SHN SHX TIA) GG IN KI KZ MG NC RU (ES FE WS) TR

chalceola Jacoby, 1886a: 731

heikertingeri Lyubishchev, 1963: 858

laevicollis C. G. Thomson, 1866: 229 (*Plectroscelis*)

nitidicollis Jacobson, 1901a: 91 (var.) [IN] [# 771]

polita Abeille de Perrin, 1907: lxxix (*Plectroscelis*) **N**: AG

procerula Rosenhauer, 1856: 344 (*Plectroscelis*) E: AL AU BU CR CZ FR GE GR HU IT ME NM PL RU (ST) SL SK SP SZ TR UK **N**: AG MO TU **A**: AB TR

aenescens Weise, 1886a: 764 (var.)

psylloides Pic, 1909e: 139 **A**: AF CH (XIZ) IN IQ KZ KI PA TD TM UZ

altisocia S.-H. Chen & S.-Y. Wang, 1981a: 493, 506

pulla Chapuis, 1879: 14 **A**: SA YE **AFR**

vadoni Bechyné, 1948c: 9

zeae Bryant, 1926: 407

puncticollis Motschulsky, 1858: 107 (*Tlanoma*) **A**: CH (CHQ FUJ GUA GUI GUX HAI HKG HUB HUN JIA JIX SCH TAI XIN XIZ YUN ZHE) ID (SD) JA NC NP SC **ORR**

discreta Baly, 1876d: 596 (*Plectroscelis*)

granulifrons Baly, 1876d: 596 (*Plectroscelis*)

kanika Maulik, 1926: 216

punctifrons Abeille de Perrin, 1907: lxxix (*Plectroscelis*) E: IT **N**: AG TU

pusaensis Maulik, 1926: 205 **A**: PA **ORR**

resplendens Warchałowski, 1973a: 48 **A**: BT ID (SD) NP **ORR**

rufofemorata Pic, 1915h: 42 E: PT SP

castillana Bergeal & Doguet, 2005: 30

rahlensis Shukla, 1960: 77 **A**: ID (HP)

sahlbergii Gyllenhal, 1827: 662 (*Haltica*) E: AL AU BE BU BY CZ DE EN FI FR GB GE GR HU IR IT (North) KO LA LS LT LU ME NL NM NR PL RO RU (CT NT ST) SB SK SL SV SZ UK **A**: KZ MG RU (ES FE) TR [# 1085]

cyanescens Weise, 1888b: 777 (var.)

fairmairei Boieldieu, 1852: pl. 11 (*Plectroscelis*)

salixis Ruan, Konstantinov & Yang, 2014b: 64 **A**: CH (GAN SCH SHA)

scheffleri Kutschera, 1864d: 315 (*Plectroscelis*) E: AU BU CR FR GE GR HU IT KR MD RO RU (CT ST) SB SL SP SZ UK **N**: AG MO TU **A**: AB IN IQ IS TR UZ

longula Weise, 1890b: 111

tunisea Pic, 1909j: 162

schlaeflii Stierlin, 1866: 31 (*Plectroscelis*) [JE] E: RU (ST: Caucasus) **A:** AF CH (XIN) IN IQ TM **AFR**

impunctifrons Pic, 1909e: 138

persica Baly, 1877d: 167

schaeflini Stierlin, 1866: 31 (*Plectroscelis*) [incorrect original spelling]

tarda Motschulsky, 1845a: 107 (*Cardiapus*) [QS]

villiersi Bechyné, 1955b: 544

semicoerulea Koch, 1803a: 40 (*Haltica*) **E:** AL AU BH BU BY CR CZ EN FR GE GR HU IT KR LA LT LU MD ME NM PL RO RU (CT NT ST) SB SK SL SP SZ UK **A:** AB CH (HEI) GG KI RU (ES SS WS) TR

femoralis Weise, 1886a: 758 (var.)

meridionalis Allard, 1859b: cv (*Plectroscelis*)

saliceti Weise, 1886a: 758 (var.)

saltitans Stephens, 1831: 327 [QS]

septentrionalis Kimoto, 1963: 18 **A:** JA (Hokkaido)

shabalini Paliy, 1968a: 17 **A:** KI

shanxiensis S.-H. Chen & S.-Y. Wang, 1980: 4, 18 **A:** CH (SHX)

simplicifrons Baly, 1876d: 594 (*Plectroscelis*) **A:** CH (JIA XIN XIZ YUN) NP PA **ORR**

singala Maulik, 1926: 212 **A:** ID (UT) NP **ORR**

sinuata Weise, 1889g: 633 **A:** CH (GAN NMO) KI KZ MG RU (ES) TD TM UZ

splendens Motschulsky, 1845a: 108 (*Udorpes*) **A:** CH ("North China") KZ MG RU (ES FE)

aenea Motschulsky, 1860b: 235 (*Hydropus*) [HN]

daurica Heyden, 1882b: 212 [RN]

sticta Maulik, 1926: 213 **A:** ID (UT) **ORR**

subbasalis Ruan, Konstantinov & Yang, 2019: 104 **A:** CH (YUN)

subcoerulea Kutschera, 1864e: 346 (*Plectroscelis*) **E:** AL AU BE BU BY CR CZ DE EN FI FR GB GE GR HU IT KR LA LT LU NL NM PL RO RU (CT NT) SK SL SV SZ UK **A:** KI TR

christinae Heikertinger, 1910b: 369

punctatula Mulsant & Rey, 1874: 222

sulcicollis S.-H. Chen & S.-Y. Wang, 1980: 5, 18 **A:** CH (YUN)

sumatrana Jacoby, 1896b: 440 **A:** CH (YUN) **ORR**

syowaensis Takizawa, 2021a: 161 **A:** JA

taiwanensis Chûjô, 1965b: 99 (ssp.) **A:** CH (TAI)

tarsalis Wollaston, 1860a: 11 **N:** AG CI TU **A:** AE OM SA **AFR** **[# 776]**

tbilisiensis Konstantinov, Baselga, Grebennikov, Prena & Lingafelter, 2011: 323 **A:** GG

tibialis Illiger, 1807a: 64 (*Haltica*) **E:** AL AN AU BH BU BY CR CZ FI FR GE GR HU IT KO KR LA MA MD ME NM PL PT RO RU (CT ST) SB SK SL SP SZ TR UK **N:** AG EG LB MO TU **A:** AB AF AR CH (GAN SHA) CY GG IN IQ IS JO KI KZ MG RU (ES SS) SA SI SY TD TM TR UZ

caesaraugustana Fuente, 1909b: 138
obscuripes Pic, 1909e: 138
pumila Allard, 1859b: cv (*Plectroscelis*)
tonkinensis S.-H. Chen, 1934h: 250 **A:** CH (FUJ HAI JIA JIX GUX TAI) **ORR**
transbaicalica Heikertinger, 1951: 41 (ssp.) **A:** CH (HEI) MG RU (ES FE)
tristis Allard, 1889f: 307 **A:** CH (GUX YUN XIZ) **ORR**
fulvocostata S.-H. Chen, 1933a: 229 (var.)
ussuriensis Heikertinger, 1951: 69 **A:** CH (HEI) RU (ES FE)
wallacei Baly, 1877d: 171 **A:** CH (TAI) **ORR**
subcostata Jacoby, 1889b: 203
vietnamica L. N. Medvedev, 2009e: 50 [HN]
warchalowskii Döberl, 2009a: 22 [RN] **A:** BT CH (TAI) PA **ORR**
punctatula Warchałowski, 1973a: 48 [HN]
westwoodi Baly, 1877g: 311 **A:** NP **ORR**
harita Maulik, 1926: 211
wollastoni Baly, 1877d: 167 **Ei:** IT (Sardinia) **A:** AE CY IS SA TR YE **AFR**
dunbrodensis Jacoby, 1906: 17
fraterna Harold, 1879c: 232
monomorpha Bechyné, 1964: 156
yaosanica S.-H. Chen, 1939a: 31 **A:** CH (FUJ GUX YUN) **ORR**
yiei Kimoto, 1970b: 208 **A:** CH (TAI) **ORR**
ebenina Warchałowski, 1973a: 47
yulongensis Ruan, Konstantinov & Yang, 2014b: 67 **A:** CH (YUN)
yunnanica Heikertinger, 1951: 73 (ssp.) **A:** CH (YUN)
zangana S.-H. Chen & S.-Y. Wang, 1981a: 493, 506 **A:** CH (XIZ)

genus *Chalaenaria* L. N. Medvedev, 2003a: 321 type species: *Chalaenaria viridis* L. N. Medvedev, 2003
viridis L. N. Medvedev, 2003a: 321 **A:** NP

genus *Chilocoristes* Weise, 1895e: 336 type species: *Argopistes bistripunctata* Duvivier, 1892
funestus Weise, 1910c: 41 **A:** CH (FUJ GUX TAI YUN) **ORR**
pallidus Baly, 1877a: 224 (*Acrocrypta*) **A:** CH (HAI) **ORR**
schawalleri Medvedev, 2016: 261 **A:** NP
smilacis Gressitt & Kimoto, 1963a: 814 **A:** CH (HKG) **ORR**

genus *Clavicornaltica* Scherer, 1974: 58 type species: *Clavicornaltica besucheti* Scherer, 1974
dali Konstantinov & Duckett, 2005: 54 **A:** CH (YUN)
doeberli Damaška & Aston, 2019: 155 **A:** CH (HKG)

himalayensis L. N. Medvedev, 1984: 56 **A:** BT ID (SD) NP
loebli Scherer, 1974: 65 **A:** CH (TAI) **ORR**
longipenis Damaška & Aston, 2019: 156 **A:** CH (HKG)
longsheng Konstantinov & Duckett, 2005: 61 **A:** CH (GUX)
mizusawai Suenaga & Yoshida, 2016: 3 **A:** CH (TAI)
nakanoi Suenaga, 2023: 454 **A:** JA (Ryukyus)
pusilla Scherer, 1974: 66 **A:** CH (TAI) **ORR**
sakishimana Suenaga & Yoshida, 2016: 6 **A:** CH (TAI) JA (Ryukyus)
schereri Basu & Sengupta, 1982: 273 **A:** ID (SD)
takimotoi LeSage, 1998: 244 **A:** CH (TAI)
takizawai Döberl, 2009b: 414 **A:** NP
tokushigei Suenaga, 2023: 451 **A:** JA (Ryukyus)

genus *Cleonica* Jacoby, 1887b: 233 type species: *Cleonica quadriplagiata* Jacoby, 1887
unicolor L. N. Medvedev, 2004d: 204 **A:** NP

genus *Crepidodera* Chevrolat, 1836: 391 type species: *Chrysomela nitidula* Linnaeus, 1758
Chalcoides Foudras, 1860b: 312 [= 1861: 56] [HN] type species: *Chrysomela nitidula* Linnaeus, 1758
Foudrasia Gozis, 1881: 180 [RN] type species: *Chrysomela nitidula* Linnaeus, 1758
aurata Marsham, 1802: 195 (*Chrysomela*) **E:** AL AU BE BH BU BY CR CZ DE EN FI FR GB GE GR HU IR IT KO KR LA LS LT LU MD ME NL NM NR PL RO RU (CT NT ST) SB SK SL SP SV SZ TR UK **N:** MO **A:** AB AR CH (GAN JIA SHA) GG IN KZ MG RU (ES FE SS WS) TR **ORR**
pulchella Stephens, 1835: 423 (*Haltica*)
subunicolor Pic, 1918d: 18 (*Chalcoides*, var.) [IN] [# 771]
versicolor Kutschera, 1860b: 77 (*Haltica*)
aurea Geoffroy, 1785: 100 (*Altica*) **E:** AU BE BH BU BY CR CZ DE EN FR GB GE GR HU IT KO KR LA LS LT LU NL NM PL RO RU (CT NT ST) SB SK SL SP SZ TR UK **A:** AR GG KZ (East) KI LE RU (ES SS) TR
cyanea Marsham, 1802: 196 (*Chrysomela*)
gaudens Stephens, 1831: 306 (*Haltica*)
laeta Weise, 1886a: 718 (*Chalcoides*, var.)
purpurascens Pic, 1918d: 18 (*Chalcoides*, var.) [IN] [# 771]
splendens Weise, 1886a: 718 (*Chalcoides*, var.)
syriaca Pic, 1918d: 18 (*Chalcoides*, ssp.)
aureola Foudras, 1860b: 320 (*Chalcoides*) [= 1861: 64] **E:** FR (South) PT SP **N:** AG MO TU
formosa Guillebeau, 1897b: 227 (*Chalcoides*)
fulvipennis Pic, 1918d: 18 (*Chalcoides*)

fulvicornis Fabricius, 1792b: 30 (*Galleruca*) **E:** AU BE BH BU BY CR CZ DE EN FI FR GB GE GR HU IT KO KR LA LS LT LU NL NR PL PT RO RU (CT NT ST) SB SK SL SP SV SZ UK **A:** AB CH (GAN XIN) GG IN KZ LE MG RU (ES FE SS WS) TR

jucunda Weise, 1886a: 719 (*Chalcoides*, var.)

lapponica Heikertinger, 1950: 112 (*Chalcoides*, ssp.)

picicornis Weise, 1886a: 719 (*Chalcoides*, var.)

smaragdina Foudras, 1860b: 319 [= 1861: 63] (*Chalcoides*)

gemmata Abeille de Perrin, 1895a: cdii (*Chalcoides*) **A:** AG MO TU

semicyanea Pic, 1918d: 19 (*Chalcoides*, var.) [IN] [# 771]

semiviridis Pic, 1918d: 19 (*Chalcoides*, var.) [IN] [# 771]

hupehensis Döberl, 2001: 381 [RN] **A:** CH (GAN HUB)

viridis Gressitt & Kimoto, 1963a: 775 [HN]

japonica Baly, 1877d: 160 **A:** JA (Hokkaido, Honshu) RU (FE)

lamina Bedel, 1901: 398 (*Chalcoides*) **E:** AU BH BU CR CZ FR GE HU IT (North) KR MD ME NR PL RO RU (CT ST) SB SK SL SP SZ TR UK **A:** GG TR

aeruginosa Weise, 1886a: 716 (*Chalcoides*, var.)

cuprea Weise, 1886a: 716 (*Chalcoides*, var.)

metallica Duftschmid, 1825: 273 (*Haltica*) [QS]

sapphirina Weise, 1886a: 716 (*Chalcoides*, var.)

nigricoxis Allard, 1878: 340 **E:** AL AU BU GR ME NM RO RU (ST: Caucasus) SB **A:** NC TR

nitidula Linnaeus, 1758: 373 (*Chrysomela*) **E:** AU BE BH BU BY CZ DE EN FI FR GB GE HU IT KR LA LT LU ME NL NM NR PL RO RU (CT ST) SK SL SV SZ UK **A:** KZ RU (ES SS WS) TR

obscuripes Bechyně, 1956b: 581 **A:** MG RU (ES FE SS) **[# 761]**

obscuripes Heikertinger, 1912g: 104 (*Chalcoides*, ab.) [IN]

ovata L. N. Medvedev, 1993c: 370 **A:** BT

peloponnesiaca Heikertinger, 1910d: 54 (*Chalcoides*, var.) **E:** GR

picipes Weise, 1887h: 192 (*Chalcoides*) **A:** CH (CHQ GAN HUB JIL SCH SHA "Manchuria") NC RU (ES FE) SC

mroczkowskii Warchałowski, 1969: 230 (*Chalcoides*)

plutus Latreille, 1804b: 7 (*Altica*) **E:** AL AU BE BH BU BY CR CZ DE FR GB GE GR HU IT KO KR LA LU MD ME NL NM PL PT RO RU (CT NT ST) SB SK SL SP SV SZ UK **A:** AB CH (CHQ GAN HEB HEI NIN SHA XIZ) GG IN KI KZ MG NC RU (ES FE SS WS) SC TD TR **AFR**i

chloris Foudras, 1860b: 318 [= 1861: 62] (*Chalcoides*)

foudrasi Weise, 1886a: 722 (*Chalcoides*, var.)

melanocera Everts, 1919a: 161 (ab.) [IN]

nigrofemorata Heikertinger, 1912g: 106 (ab.) [IN]

punctatithorax Pic, 1918d: 18 (*Chalcoides*, var.) [IN] [# 771]

unicolor Westhoff, 1882: 279 (var.) [QS]

sahalinensis Konstantinov, 1996: 30 **A:** RU (FE)
sinensis S.-H. Chen, 1935d: 777 [= 1938a: 2] (*Chalcoides*) **A:** CH (JIA)
ussuriensis Konstantinov, 1996: 34 **A:** RU (FE)
wittmeri Doguet, 1976: 92 (*Chalcoides*) **A:** IN
yahiroi Suzuki, 2006: 189 **A:** JA

genus *Crepidodera*, nomina dubia
caerulescens Gmelin, 1790: 1698 (*Chrysomela*)
chrysocollis Scopoli, 1763: 64 (*Buprestis*)
cyanopicta Gmelin, 1790: 1698 (*Chrysomela*)
helxines Linnaeus, 1758: 373 (*Chrysomela*)
libanicola Pic, 1918d: 18 (*Chalcoides*, var.) [IN] [# 771]
violaceopunctata DeGeer, 1775: 344 (*Chrysomela*)
viridiaurata DeGeer, 1775: 345 (*Chrysomela*)
viridicaerulea DeGeer, 1775: 346 (*Chrysomela*)

genus *Crepidodera*, excluded taxon
teotiai Kacker, 1969: 55 **ORR** [# 766]

genus *Crepidoderoides* Chûjô, 1942a: 36 type species: *Crepidoderoides choi* Chûjô, 1942
choi Chûjô, 1942a: 37 **A:** NC
suturalis L. N. Medvedev, 1993c: 369 **A:** CH (SCH)

genus *Crepidosoma* S.-H. Chen, 1939b: 57 type species: *Crepidodera incerta* S.-H. Chen, 1935
cribricolle S.-H. Chen, 1941c: 194 **A:** CH (FUJ SCH)
incertum S.-H. Chen, 1935d: 775 (*Crepidodera*) **A:** CH (YUN)

genus *Cyrsylus* Jacoby, 1891d: 306 type species: *Cyrsylus recticollis* Jacoby, 1891
pulcherrimus Allard, 1867: 489 (*Aphthona*) **N:** AG

genus *Derocrepis* Weise, 1886a: 686 type species: *Chrysomela rufipes* Linnaeus, 1758
rufipes Linnaeus, 1758: 373 (*Chrysomela*) **E:** AU BE BH BU BY CR CZ DE EN FR GB GE HU IR IT (North) KO KR LA LS LT LU MD ME NL PL RO RU (CT NT ST) SB SK SL SP SV SZ UK **A:** AB IN RU (ES SS WS) TR
caeruleostriata DeGeer, 1775: 343 (*Chrysomela*)
latithorax Pic, 1911d: 109 (var.)
obscura Weise, 1886a: 688 (var.)
ruficornis Fabricius, 1792b: 32 (*Galleruca*)

sodalis Kutschera, 1860b: 73 (*Haltica*) E: FR (Alps) IT (Alps, Apennini) SZ (Alps)
scutellaris Pic, 1910c: 27 (var.)

genus *Dibolia* Latreille, 1829: 155 type species: *Haltica occultans* Koch, 1803
subgenus *Dibolia* Latreille, 1829: 155 type species: *Haltica occultans* Koch, 1803
alpestris Mohr, 1981: 438 E: AU (West Alps) IT
chevrolati Allard, 1862: 338 E: GR (Rhodes) IT (Sicily) SP (Andalucía) **N:** AG EG MO TU **A:** IS SY
cryptocephala Koch, 1803a: 22 (*Haltica*) E: AL AU BE BH BU BY CR CZ EN FR GE GR HU IT KR LA LT LU MD ME NM PL RO RU (CT ST) SB SK SL SP SZ UK **A:** AB AR GG IN KZ TR
cynoglossi Koch, 1803a: 20 (*Haltica*) E: AL AU BE BH BU BY CZ DE EN FR GB GE HU IT KR LA MD NL PL RO RU (CT ST) SB SK SL SP SZ UK **A:** AB AR TR
paludina Foudras, 1860a: 202 [= 1860b: 90]
depressiuscula Letzner, 1847: 83 (*Haltica*) E: AU BE BU BY CR CZ DE EN FI FR GE GR HU IT KR LA LT LU MD ME NM PL RO RU (CT ST) SB SK SL SZ UK **A:** IS JO MG KZ RU (ES SS) TR
laevicollis Foudras, 1860a: 201 [= 1860b: 89]
marrubii Hering, 1924: 36
djelfana Doguet, 1975: 202 **A:** MO TU **N:** AG
dogueti Mohr, 1981: 463 E: SP
foersteri Bach, 1858: 163 E: AL AU BE BH BU BY CR CZ FR GE GR HU IT KR NM PL RO RU (CT ST) SB SK SL SP SZ UK **A:** AR
buglossi Foudras, 1860a: 198 [= 1860b: 86]
foudrasi Mulsant & Rey, 1874: 219
iranica Mohr, 1981: 418 **A:** IN (North)
japonica S.-H. Chen, 1933d: 187 **A:** JA (Honshu)
kralii Mohr, 1981: 437 **A:** IN (North-West) TR
magnifica Har. Lindberg, 1950: 16 **N:** CI (Gran Canaria)
maura Allard, 1861: 795 E: IT (Sicily) SP (Valencia) **N:** AG MO TU
mesasiatica Lopatin, 1965c: 40 **A:** TD UZ
numidica Doguet, 1972: 133 **N:** AG TU **A:** TR
obtusa Wollaston, 1864: 417 **N:** CI (Fuerteventura)
occultans Koch, 1803a: 23 (*Haltica*) E: AL AU BE BH BU BY CR CZ DE EN FR GE GR HU IT KR LA LU MA ME NL NM PL PT RO RU (CT ST) SB SK SL SP SV SZ UK **N:** AG CI MO **A:** IN TR
menthae Hering, 1924: 37
nigripes Boheman, 1853: 95 (*Psylliodes*)
ordosana S.-H. Chen, 1939b: 59 **A:** CH (NMO)

oudai Čížek, 2007: 15 **E:** HU SK

pelleti Allard, 1861: 788 **E:** FR (South) SP **N:** AG MO

dimidiata Fuente, 1910b: 447

ifranensis Hoffmann, 1953: 188

vaulogeri Pic, 1909f: 146

peyerimhoffi Doguet, 1975: 200 **E:** PT SP (Balearic Islands) **N:** AG MO

phoenicia Allard, 1866: 424 **E:** "Balkan Peninsula" **A:** GG IS JO LE TR

cyanescens Weise, 1893a: 1033 (var.)

orientalis Weise, 1893a: 1033

potanini Weise, 1889g: 643 **A:** CH (GAN HEI NMO) RU (FE)

rufofemorata Reitter, 1896: 267 **A:** CY IS JO SY TR

marginicollis Abeille de Perrin, 1907: lxxx

notatipes Pic, 1903d: 124

rugulosa L. Redtenbacher, 1848: 541 **E:** AL AN AR AU BE BH BU CR CZ FR GE GR HU IT KO KR LU ME NM PL RO RU (CT ST) SB SK SL SP SZ UK **A:** AB GG IN TR

russica Weise, 1893a: 1029 **E:** BY CR KR RU (ST) SL UK

sinensis S.-H. Chen, 1939b: 59 **A:** CH (SHX)

tibialis S.-Y. Wang, 1992c: 685, 734 **A:** CH (SCH XIZ)

timida Illiger, 1807a: 71 (*Haltica*) **E:** AL AU BH BU CR CZ FR GE GR HU IT KR NM PT RO SB SK SL SP SZ UK **N:** AG MO TU **A:** TR

eryngii Bach, 1858: 163

oranensis Pic, 1909f: 146 (var.)

tshatcalica tshatcalica Paliy, 1968a: 20 **A:** KI TM

turkmenica Iablokoff-Khnzorian, 1978: 122

tshatcalica corpulenta Mohr, 1981: 425 (ssp.) **A:** KI

tyrrhenica Mohr, 1981: 462 **E:** GR (Makedonia) IT MA

velox Weise, 1889g: 643 **A:** CH (JIL NMO SCH)

veyreti Doguet, 1975: 209 **E:** FR (South)

weisei Mohr, 1981: 432 **A:** GG (Mesket Mt.)

zaitzevi L. N. Medvedev, 1980a: 320 **A:** CH (LIA) MG RU (ES)

subgenus *Eudibolia* Iablokoff-Khnzorian, 1968: 270 type species: *Dibolia schillingii* Letzner, 1847 (= *Petalopus metallica* Motschulsky, 1845)

Petalopus Motschulsky, 1845a: 107 [HN] type species: *Petalopus metallicus* Motschulsky, 1845

carpathica Weise, 1893a: 1030 **E:** BU CR GR HU KR ME RO RU (CT) SK UK **A:** AB AR GG IN IS KZ RU (WS) TR

transcaucasica Iablokoff-Khnzorian, 1968: 268

femoralis erythrogaster Allard, 1862: 337 **E:** IT GR (Crete, Rhodes) SP (South) **N:** AG MO TU

femoralis femoralis L. Redtenbacher, 1848: 540 [NP] **E:** AL AU BH BU CR CZ FR GE GR HU IT KR LU MD ME NM PL RO RU (CT ST) SB SK SL SP SZ UK **A:** TR [# 764]

aerata Dalman, 1823: 81 (*Haltica*) [NO] [# 764]
aurichalcea Förster, 1849: 37
peregrina Weise, 1893a: 1032 (var.)
theresae Pic, 1909f: 145
metallica Motschulsky, 1845a: 107 (*Petalopus*) [ICZN (2015b), Opinion 2357] **E**: AU BE BU BY CR CZ GE GR HU IT (North) KR MD NM PL RO RU (CT ST) SB SK SL SZ UK **A**: AR GG KZ RU (WS) TR
ciliciensis Heikertinger, 1911c: 119 (var.)
punctillata Foudras, 1860a: 196 [= 1860b: 84]
schillingii Letzner, 1847: 82 (*Haltica*) [ICZN (2015b), Opinion 2357]
tricolor Reitter, 1898b: 127 **A**: AR TR
subgenus *Pseudodibolia* Iablokoff-Khnzorian, 1968: 269 type species: *Dibolia zangezurica* Iablokoff-Khnzorian, 1968
zangezurica Iablokoff-Khnzorian, 1968: 266 **A**: AR IN TR

genus *Disonycha* Chevrolat, 1836: 390 type species: *Crioceris collata* Fabricius, 1801
politula Horn, 1889: 211 Ai: JA **NAR**

genus *Epitrix* Foudras, 1860a: 147 [= 1860b: 35] type species: *Epitrix atropae* Foudras, 1860
Epithrix Bedel, 1897: 179 [UE] type species: *Epitrix atropae* Foudras, 1860
Euplecnema Jacoby, 1906: 22 type species: *Euplecnema nigrita* Jacoby, 1906
abeillei Bauduer, 1874b: 211 (*Crepidodera*) **N**: EG **A**: AB AF CH (GAN NIN XIN) IN IQ IS JO KZ LE MG SI SY TR TM UZ YE [# 777]
judaea Allard, 1876: 21 (*Crepidodera*)
testaceipes Pic, 1909g: 226 (*Epithrix*, var.)
allardii Wollaston, 1860a: 1 (*Haltica*) **N**: CI (Tenerife, Gran Canaria)
atropae Foudras, 1860b: 311 [= 1861: 55] **E**: AL AU BE BH BU CR CZ FR GB GE HU IT KR LU ME NL NM PL RO RU (ST) SB SK SL SP SZ UK **N**: AG **A**: AB AR GG IN TR
nigritula Weise, 1886a: 711 (var.)
quadrimaculata Weise, 1886a: 711 (var.)
bipunctata Medvedev, 2016: 259 (*Epithrix*) **A**: CH (SCH)
caucasica Heikertinger, 1950: 120 (*Epithrix*) **E**: KR RU (ST) **A**: AB GG IN KZ TR TM "Himalaya"
cucumeris Harris, 1851: 100 (*Haltica*) Ei: AZ PT **Ni**: MR **AUSi NAR NTR**
azorica Gruev, 1981a: 1
dieckmanni Mohr, 1968b: 58 (*Epithrix*) **A**: AE IN IS JO OM SA TR TM **[Appendix]**
paliji Gruev, 1975c: 173 (*Epithrix*)
ermischi Mohr, 1968b: 59 (*Epithrix*) **A**: IN TM UZ
fasciata Blatchley, 1918: 56 Ei: AZ **NAR**
parvula Fabricius, 1801a: 468 (*Crioceris*) [HN]

hirtipennis Melsheimer, 1847: 165 (*Crepidodera*) Ei: AL AZ BU FR GR HU IT NM PT RU (ST) SL SP Ai: GG IN JA SY TR **NAR NTR PAC**i
intermedia Foudras, 1860b: 311 [= 1861: 55] E: AL BU FR GR HU IT (North) KO KR NM RO RU (ST) SB SK SL SP UK **A:** AB GG TR
krali Döberl, 2000: 9 **A:** IN JO TD
muehlei Döberl, 2000: 10 **A:** SA YE
ogloblini Iablokoff-Khnzorian, 1961: 151 (*Epithrix*) **A:** AR
papa Orlova-Bienkowskaya, 2015: 825 Ei: PT SP **NAR?**
priesneri Heikertinger, 1950: 120 (*Epithrix*) **N:** EG **A:** AE IN OM SA SI YE
pubescens Koch, 1803a: 37 (*Haltica*) E: AL AU AZ BE BH BU BY CR CZ DE EN FI FR GB GE GR HU IT KO KR LA LS LT LU MD ME NL NM PL PT RO RU (CT ST) SB SK SL SP SV SZ TR UK **A:** AB CY GG IN IS KI KZ RU (ES WS) TR **NAR**i
ferruginea Weise, 1886a: 710 (var.)
lenkorana Pic, 1903e: 131
suturalis Bedel, 1897: 179 (*Epithrix*, var.)
setosella Fairmaire, 1888c: 45 (*Crepidodera*) **A:** CH (FUJ GUX HEB JIA SHA) GG RU (FE)
wuorentausi Kontkanen, 1950: 121 (*Epithrix*)
warchalowskii Mohr, 1968b: 60 (*Epithrix*) **A:** IN
yunnana L. N. Medvedev, 2012d: 507 (*Epithrix*) **A:** CH (YUN)

genus *Eudolia* Jacoby, 1885a: 69 type species: *Eudolia sumatrana* Jacoby, 1885
himalayensis Maulik, 1926: 199 **A:** ID (SD UT)
nepalica L. N. Medvedev, 1990b: 39 **A:** NP
nila Maulik, 1926: 200 **A:** ID (AP) **ORR**
rufipes S.-Y. Wang, 2004b: 78, 81 **A:** CH (XIZ)
vietnamica L. N. Medvedev, 2007b: 78 **A:** CH (YUN) **ORR**

genus *Eudoliamorpha* Scherer, 1989: 46 type species: *Eudoliamorpha darjeelingensis* Scherer, 1989
darjeelingensis Scherer, 1989: 47 **A:** ID (SD)

genus *Euphitrea* Baly, 1875a: 27 type species: *Euphitrea wallacei* Baly, 1875
Euphymasia Jacoby, 1900a: 310 type species: *Euphymasia dohrni* Jacoby, 1900
Neorthaea Maulik, 1926: 259 [RN] type species: *Orthaea viridipennis* Jacoby, 1889
Orthaea Jacoby, 1889b: 201 [HN] type species: *Orthaea viridipennis* Jacoby, 1889
antennata Y. Zhang & Yang, 2006: 846 **A:** CH (YUN)
bhutanica L. N. Medvedev, 1992f: 386 **A:** BT
cheni Y. Zhang & Yang, 2006: 849 **A:** CH (YUN)
chinensis L. N. Medvedev, 1999a: 150 **A:** CH (SCH)

coerulea S.-H. Chen, 1933b: 95 (*Neorthaea*) **A:** CH (GUA) **ORR**
cribripennis S.-H. Chen & S.-Y. Wang, 1980: 7, 19 **A:** CH (YUN)
excavata S.-Y. Wang & Yang, 2008a: 46 **A:** CH (XIZ)
flavicornis S.-H. Chen, 1933b: 90 (*Neorthaea*) **A:** CH (TAI)
flavipes S.-H. Chen, 1933b: 95 (*Neorthaea*) **A:** CH (CHQ FUJ GUA GUX HUB)
foveicollis Jacoby, 1893a: 149 **A:** ID (SD) NP
gressitti Chûjô, 1965a: 55 (*Neorthaea*) **A:** CH (HAI HKG)
hainana Y. Zhang & Yang, 2006: 851 **A:** CH (HAI)
indica Jacoby, 1904a: 391 **A:** PA **ORR**
laboissierei S.-H. Chen, 1933b: 94 (*Neorthaea*) **A:** NP **ORR**
laevicollis S.-Y. Wang, 2004c: 380, 384 **A:** CH (GUX)
laticostata S.-H. Chen & S.-Y. Wang, 1980: 6, 19 **A:** CH (YUN)
longicornis S.-Y. Wang, 2004c: 379, 384 **A:** CH (GUX)
mandibula S.-Y. Wang & Y. Zhang, 2006: 853 **A:** CH (XIZ)
micans Baly, 1875a: 28 **A:** CH (GUA GUI GUX HKG HUB) **ORR**
  assamensis Baly, 1879a: 443
nisotroides S.-H. Chen, 1933b: 92 (*Neorthaea*) **A:** CH (FUJ GAN GUI SHA TAI) JA (Ryukyus)
omeia Y. Zhang & Yang, 2006: 856 **A:** CH (HUN SCH)
parva Döberl, 2011: 272 **A:** CH (JIX)
piceicollis S.-H. Chen, 1933f: 54 (*Neorthaea*) **A:** CH (CHQ FUJ GUA GUX HUB SCH)
rufipes S.-H. Chen & S.-Y. Wang, 1980: 7, 20 **A:** CH (YUN)
rufomarginata S.-Y. Wang, 1992c: 685, 734 **A:** CH (YUN)
signata Weise, 1922b: 117 (*Sphaeroderma*) **A:** CH (FUJ GUX) **ORR**
  suturalis S.-H. Chen, 1933b: 92 (*Neorthaea*)
subglobosa Hope, 1831: 31 (*Coccinella*) **A:** ID ("North India") NP
subregularis S.-H. Chen & S.-Y. Wang, 1980: 7, 20 **A:** CH (YUN)
taiwana Kimoto, 1991a: 18 **A:** CH (TAI)
xia S.-H. Chen & S.-Y. Wang, 1981a: 495, 507 **A:** CH (XIZ)

genus *Eutrea* Baly, 1875a: 24 type species: *Eutrea bowringii* Baly, 1875
bowringii Baly, 1875a: 25 **A:** CH (HKG)

genus *Gansuapteris* Ruan & Konstantinov, 2017: 112 type species: *Gansuapteris fulvus* Ruan & Konstantinov, 2017
fulvus Ruan & Konstantinov, 2017: 115 **A:** CH (GAN)

genus *Glaucosphaera* Maulik, 1926: 183 type species: *Amphimela cyanea* Duvivier, 1892
cyanea Duvivier, 1892b: 420 (*Amphimela*) **A:** CH (TAI XIZ) ID (SD) NP **ORR**
  nepalensis Chûjô, 1966b: 31 (*Hemipyxis*)

genus *Haemaltica* S.-H. Chen, 1933e: 386 type species: *Haemaltica tonkinensis* S.-H. Chen, 1933

nigripes S.-H. Chen & S.-Y. Wang, 1980: 5, 18 **A:** CH (GAN YUN)

parva S.-H. Chen & S.-Y. Wang, 1980: 5, 18 **A:** CH (YUN)

genus *Halticorcus* Lea, 1917: 319 type species: *Halticorcus platycerii* Lea, 1917

Eucycla Baly, 1876c: 439 [HN] type species: *Eucycla quadripustulata* Baly, 1876

Eucyclomela S.-H. Chen, 1934c: 73 type species: *Eucyclomela clypealis* S.-H. Chen, 1934

Schenklingia Csiki & Heikertinger, 1940: 516 [RN] type species: *Eucycla quadripustulata* Baly, 1876

bhaumiki Basu & Sengupta, 1982: 267 (*Schenklingia*) **A:** ID (SD)

bhutanicus Medvedev, 2016: 262 **A:** BT

duodecimmaculatus S.-H. Chen, 1934c: 75 (*Eucycla*) **A:** JA (Kyushu) "Korea"

himalayensis Basu & Sengupta, 1982: 265 (*Schenklingia*) **A:** ID (HP UT) NP

hiranoi Takizawa, 1982: 59 (*Schenklingia*) **A:** JA (Kyushu, Ryukyus)

kasuga Nakane, 1963a: 20 (*Schenklingia*) **A:** JA (Honshu)

miwai Chûjô, 1937b: 53 (*Eucycla*) **A:** CH (TAI)

ornatipennis S.-H. Chen, 1933a: 243 (*Eucycla*) **A:** CH (HUN TAI YUN) SC

saigusai Kimoto, 1970d: 293 (*Schenklingia*) **A:** CH (TAI)

sasajii Kimoto, 1970d: 292 (*Schenklingia*) **A:** CH (TAI)

sauteri sauteri S.-H. Chen, 1934e: 181 (*Eucycla*) **A:** CH (TAI) JA (Ryukyus)

sauteri insularis Nakane, 1984: 598 (*Schenklingia*, ssp.) **A:** JA (Yakushima)

smetanai Döberl, 2010a: 66 (*Schenklingia*) **A:** CH (TAI)

yunnanus L. N. Medvedev, 2015c: 330 **A:** CH (YUN)

genus *Hemipyxis* Chevrolat, 1836: 387 type species: *Altica troglodytes* Olivier, 1808 (= *Haltica fulvipennis* Illiger, 1807)

Epiotis Solsky, 1871: 259 type species: *Oedionychis plagioderoides* Motschulsky, 1861

Sebaethe Baly, 1864e: 438 type species: *Haltica badia* Erichson, 1834

abdominalis Döberl, 2007a: 85 **A:** CH (YUN)

albofasciata S.-H. Chen, 1933g: 278 (*Sebaethe*) **A:** CH (YUN) **ORR**

amurensis Weise, 1887h: 196 (*Sebaethe*) **A:** CH (BEI HEI JIA SCH "Manchuria") RU (FE)

apicicostata Kimoto, 1989b: 76 **A:** CH (TAI)

balyi balyi Bates, 1866: 355 (*Sebaethe*) **A:** CH (FUJ TAI)

balyi yaeyamana Kimoto, 1978a: 18 (ssp.) **A:** JA (Ryukyus)

bipustulata Jacoby, 1894: 291 (*Sebaethe*) **A:** CH (YUN) ID (SD) **ORR**

bouvieri S.-H. Chen, 1933g: 276 (*Sebaethe*) **A:** CH (SCH YUN) **ORR**

brevicollis Jacoby, 1903a: 106 (*Sebaethe*) **A:** ID (AP) **ORR**

caenotes Maulik, 1926: 401 (*Sebaethe*) **A:** CH (GUX) NP **ORR**

castaneipennis Scherer, 1969: 193 (*Sebaethe*) **A:** BT ID (SD UT) NP
changi Kimoto, 1970b: 219 **A:** CH (TAI) JA (Ryukyus)
chinensis Weise, 1912: 95 (*Sebaethe*) **A:** CH (GAN YUN) **ORR**
cinctipennis cinctipennis Weise, 1890d: 174 (*Sebaethe*) **A:** CH (TAI) JA (Kyushu, Ryukyus)
flavolimbata Jacoby, 1896a: 7 (*Sebaethe*)
cinctipennis okinawana Nakane & Kimoto, 1961c: 108 (ssp.) **A:** JA (Ryukyus)
dichroa Weise, 1922b: 113 (*Sebaethe*) **A:** CH (FUJ GUA HAI JIA)
difficilis Döberl, 2007a: 93 **A:** CH (YUN) **ORR**
elongata Jacoby, 1892b: 922 (*Sebaethe*) **A:** BT CH (GUI) ID (SD) NP **ORR**
foveifrons S.-H. Chen, 1933a: 245 (*Sebaethe*)
erberi Döberl, 2007a: 87 **A:** CH (SCH)
flaviabdominalis Chûjô, 1965b: 100 (ssp.) **A:** CH (TAI) JA (Ryukyus)
flavipennis Baly, 1874: 194 (*Sebaethe*) **A:** CH (FUJ ZHE) JA SC **ORR**
flavipes Kimoto, 1978b: 70 **A:** CH (TAI)
formosana Chûjô, 1937a: 46 (*Sebaethe*) **A:** CH (TAI)
fortepunctata S.-H. Chen, 1936d: 594 (*Sebaethe*) **A:** ID (SD)
foveolata foveolata Chûjô, 1958c: 17 (*Sebaethe*) **A:** JA (Ryukyus)
foveolata miyakoana Kimoto, 1978a: 18 (ssp.) **A:** JA (Ryukyus)
fulvipennis Illiger, 1807b: 156 (*Haltica*) **A:** CH (XIZ) **ORR**
fulvoculata Takizawa, 1979: 342 **A:** CH (TAI) **ORR**
gansuensis S.-Y. Wang, 2006: 228 **A:** CH (GAN)
gibbosula Gressitt & Kimoto, 1963a: 840 **A:** CH (CHQ FUJ GAN HUB)
glabricollis S.-Y. Wang, 1992c: 694, 740 **A:** (SCH YUN)
bocaki L. N. Medvedev, 1993c: 365 (*Sebaethe*)
indica S.-H. Chen, 1933h: 455 (*Sebaethe*) **A:** ID (SD) NP **ORR**
intermedia Jacoby, 1903a: 105 (*Sebaethe*) **A:** NP **ORR**
ioscopa Maulik, 1926: 390 (*Sebaethe*) **A:** NP **ORR**
jeanneli S.-H. Chen, 1933g: 277 (*Sebaethe*) **A:** CH (GUX) **ORR**
faceta Gressitt & Kimoto, 1963a: 839
kachinensis Döberl, 2011: 274 **A:** CH (YUN) **ORR**
kangdingana S.-Y. Wang, 1992c: 694, 739 **A:** CH (SCH XIZ)
kimotoi Döberl, 2007a: 94 **A:** CH (YUN) **ORR**
kiangsuana S.-H. Chen, 1934h: 308 [= 1938a: 3] (*Sebaethe*) **A:** CH (CHQ FUJ GAN HUB HUN JIA SCH ZHE)
limbata Gressitt & Kimoto, 1963a: 842 **A:** CH (FUJ GAN GUA JIA)
lineata Kimoto, 1978b: 69 **A:** CH (TAI)
liukueiana Kimoto, 1996: 43 **A:** CH (TAI)
lusca Fabricius, 1801a: 456 (*Crioceris*) **A:** CH (CHQ ZHE)
margitae Döberl, 2007a: 95 **A:** CH (SCH SHA YUN)
monticola Scherer, 1969: 195 (*Sebaethe*) **A:** ID (SD)

montivaga Maulik, 1926: 399 (*Sebaethe*) **A:** CH (ZHE) NP **ORR**
moseri Weise, 1922b: 114 (*Sebaethe*) **A:** CH (CHQ FUJ GUA GUI HUB HUN JIA SCH SHN ZHE) NP **ORR**
fleutiauxi S.-H. Chen, 1933g: 276 (*Sebaethe*)
iobaphes Maulik, 1926: 388 (*Sebaethe*)
neelys Maulik, 1926: 401 (*Sebaethe*) **A:** NP **ORR**
nigricornis Baly, 1877d: 164 (*Sebaethe*) **A:** CH (FUJ GUA JIX TAI) NP **ORR**
nila Maulik, 1926: 386 (*Sebaethe*) **A:** CH (SCH XIZ YUN) **ORR**
ornata L. N. Medvedev, 1993c: 365 **A:** CH (YUN)
parva S.-Y. Wang, 1992c: 695, 740 **A:** CH (SCH YUN)
patkaia Maulik, 1926: 402 (*Sebaethe*) **A:** BT ID (SD) NP **ORR**
persimilis Kimoto, 1996: 44 **A:** CH (TAI)
piceolimbatus Medvedev, 2013b: 296 **A:** CH (YUN)
pingala Maulik, 1926: 402 (*Sebaethe*) **A:** NP **ORR**
plagioderoides Motschulsky, 1861a: 27 (*Oedionychis*) **A:** CH (CHQ FUJ GAN HEB SHA ZHE) JA NC RU (FE) SC
privigna Gressitt & Kimoto, 1963a: 845 **A:** CH (FUJ GUA HUB JIA SCH XIZ ZHE)
pseudoprivigna Döberl, 2007a: 98 **A:** CH (FUJ GUA) **ORR**
pyrobapta Maulik, 1926: 387 (*Sebaethe*) **A:** CH (XIZ YUN) **ORR**
quadrimaculata Jacoby, 1892b: 922 (*Sebaethe*) **A:** CH (TAI) ID (SD) **ORR**
quadripustulata Baly, 1876a: 80 (*Sebaethe*) **A:** CH (FUJ TAI) JA (Ryukyus) **ORR**
rarashana Kimoto, 1996: 44 **A:** CH (TAI)
semicaerulea Bryant, 1939: 10 (*Sebaethe*) **A:** CH (SCH) **ORR**
schereri Döberl, 2007a: 91 **A:** ID (SD) NP **ORR**
shirakii Nakane & Kimoto, 1961c: 108 **A:** JA (Ryukyus)
similis Gressitt & Kimoto, 1963a: 848 **A:** CH (GUA)
sulphurea Jacoby, 1898b: 372 (*Sebaethe*) **A:** CH (GUA) **ORR**
taihorinensis Kimoto, 1966c: 36 **A:** CH (TAI)
takarai Kimoto & Gressitt, 1966: 551 **A:** JA (Ryukyus: Iriomote)
takedai Kimoto, 1976: 7 **A:** CH (TAI)
tendomarginalis Gressitt & Kimoto, 1963a: 848 **A:** CH (FUJ)
tonkinensis S.-H. Chen, 1933g: 277 (*Sebaethe*) **A:** CH (FUJ GUI GUX HUN SCH) **ORR**
troglodytes Olivier, 1808: 700 (*Altica*) **A:** CH (GAN "South China") ID (SD UT) NP PA **ORR**
bicolor Weise, 1922b: 114 (*Sebaethe*)
fulvipennis Baly, 1877d: 164 (*Sebaethe*)
pallidipennis Baly, 1879a: 442 (*Sebaethe*)
variabilis Jacoby, 1885a: 48 (*Sebaethe*) **A:** CH (FUJ GAN GUX SHA) **ORR**
wangi Döberl, 2007a: 87 **A:** CH (FUJ HUB SCH YUN)
yasumatsui Kimoto, 1970b: 220 **A:** CH (TAI)

yui C.-F. Lee & Staines, 2009: 65 **A:** CH (TAI)
yunnanica S.-H. Chen, 1933h: 455 (*Sebaethe*) **A:** CH (HEI SCH SHX XIZ YUN)

genus *Hermaeophaga* Foudras, 1860b: 299 [= 1861: 43] type species: *Haltica cicatrix* Illiger, 1807
subgenus *Hermaeophaga* Foudras, 1860b: 299 [= 1861: 43] type species: *Haltica cicatrix* Illiger, 1807
Linozosta Allard, 1860c: 72 type species: *Galleruca mercurialis* Fabricius, 1792
belkadavi Konstantinov, 2009: 12 **A:** CH (SCH)
cicatrix Illiger, 1807a: 59 (*Haltica*) **E:** BE FR GE IT LU NL PT SL SP SZ **A:** MO
dali Konstantinov, 2009: 14 **A:** CH (YUN)
flavitarsa Döberl, 1991b: 619 **A:** NP
korotyaevi Konstantinov, 2009: 16 **A:** SC
mercurialis Fabricius, 1792b: 33 (*Galleruca*) **E:** AU BE BH BU BY CR CZ DE EN FR GB GE HU IT KO KR LA LS LT LU MD NL NM PL RO RU (CT ST) SB SK SL SV SZ UK **A:** JO TR
subgenus *Orthocrepis* Weise, 1888b: 850 type species: *Altica ruficollis* P. H. Lucas, 1847
Leiometopona Brancsik, 1899: 105 type species: *Leiometopona suturalis* Brancsik, 1899 (= *Altica ruficollis* P. H. Lucas, 1847)
adamsii Baly, 1874: 193 (*Hermaeophaga*) **A:** CH (FUJ) JA NC
asahinai Kimoto, 1971a: 77 (*Orthocrepis*) **A:** CH (TAI)
hanoiensis S.-H. Chen, 1934h: 379 (*Lactica*) **A:** CH (GAN GUX SCH SHA) ID (HP) NP **ORR**
kuluensis Scherer, 1969: 110 (*Orthocrepis*)
nigripes Kimoto, 1987: 97 (*Orthocrepis*) **A:** CH (TAI)
parlinoides S.-H. Chen, 1939a: 35 (*Lactica*) **A:** CH (GUX)
perraudieri Allard, 1889f: 304 (*Lactica*) **A:** CH (HAI SCH) **ORR** [# 768]
nisotroides S.-H. Chen, 1939a: 36 (*Lactica*)
ruficollis P. H. Lucas, 1847a: pl. 45 [= 1849: 546] (*Altica*) **E:** AL FR (Corsica) GR NM SP **N:** AG EG LB MO TU **A:** AE AF CY IN IQ IS JO KZ LE NP OM PA SA SI SY TD TR TM UZ YE (incl. Socotra) **AFR ORR**
bamakoensis Bechyné, 1955b: 527 (*Orthocrepis*, ssp.)
thaiensis Chujo, 1964: 303 (*Lactica*)
suturalis Brancsik, 1899: 105 (*Leiometopona*)
unicolor Jacoby, 1886e: 124 (*Lactica*)
taiwanensis Kimoto, 1971a: 76 (*Orthocrepis*) **A:** CH (TAI)

genus *Heyrovskya* Jind. Madar & Jiř. Madar, 1968: 138 type species: *Haltica paivana* Wollaston, 1860
convexior Har. Lindberg, 1950: 15 (*Aphthona*) **N:** CI SG

cylindrica Hoffmann, 1953: 187 (*Aphthona*) **N:** MO
gomerensis Gruev & Petitpierre, 1979: 126 **N:** CI (Gomera)
moralesi Jind. Madar & Jiř. Madar, 1965: 8 (*Aphthona*) **N:** CI (Tenerife, Gomera)
occidentalis Har. Lindberg, 1950: 13 [= 1953: 12] (*Aphthona*) **N:** CI
paivana Wollaston, 1860a: 2 (*Haltica*) **N:** CI (Tenerife)
plenifrons Wollaston, 1864: 408 (*Haltica*) **N:** CI
tristis Har. Lindberg, 1950: 13 [= 1953: 12] (*Aphthona*) **N:** CI (La Palma)
wachnitzae Jind. Madar & Jiř. Madar, 1968: 144 **N:** CI MO
oromii Gruev & Petitpierre, 1979: 127 (*Heyrovskia*)
wollastoni Har. Lindberg, 1950: 15 (*Aphthona*) **N:** CI
atroviridescens Har. Lindberg, 1950: 15 (*Aphthona*)

genus *Hippuriphila* Foudras, 1860b: 306 [= **1861: 50**] type species: *Chrysomela modeeri* Linnaeus, 1761
Parachalcoides Chûjô, 1959: 13 type species: *Parachalcoides babai* Chûjô, 1959: 14
Pseudocrepidosoma L. N. Medvedev, 1966: 43 type species: *Pseudocrepidosoma ohnoi* L. N. Medvedev, 1966
babai Chûjô, 1959: 14 (*Parachalcoides*) **A:** JA (Hokkaido, Honshu) RU (FE)
ohnoi L. N. Medvedev, 1966: 43 (*Pseudocrepidosoma*)
modeeri Linnaeus, 1761: 167 (*Chrysomela*) **E:** AL AU BE BH BU BY CR CZ DE EN FI FR GB GE HU IR IT KR LA LS LT LU MD NL PL NR RO RU (CT NT ST) SB SK SL SP SV SZ UK WS **A:** GG MG RU (ES WS) TR
bimaculata Weise, 1886a: 725 (var.)
chalybaea Weise, 1886a: 725 (var.)
chrysopygis Beck, 1817: 17 (*Haltica*)
laevicollis Hellén, 1933: 40
prescutellaris Pic, 1909h: 155 (var.)

genus *Hyphasis* **Harold, 1877b: 434** type species: *Oedionychis magica* Harold, 1877
Hyphasoma Jacoby, 1903a: 110 type species: *Hyphasoma inconspicuum* Jacoby, 1903
anaimalaiensis Scherer, 1969: 179 **A:** NP **ORR**
atricornis S.-H. Chen, 1936d: 595 [RN] **A:** BT CH (FUJ) ID (HP SD UT) NP **ORR**
nigricornis Baly, 1878d: 314 [HN]
discipennis Jacoby, 1903a: 112 (*Hyphasoma*) **A:** NP PA **ORR**
femoralis Jacoby, 1889b: 198 **A:** PA **ORR**
fuscifrons Weise, 1922b: 125 (*Hyphasoma*) **A:** CH (FUJ GUI GUX) **ORR**
piceicollis S.-H. Chen, 1933g: 275 (*Hyphasoma*)
fuscipennis Weise, 1922b: 115 (*Sebaethe*) **A:** BT NP **ORR**
grandis S.-Y. Wang, 1993: 318, 328 **A:** CH (GUI GUX HUB)
inconstans Jacoby, 1886a: 733 **A:** CH (CHQ FUJ GUA HAI HKG HUN JIA) JA (Kyushu) **ORR**

indica Baly, 1879a: 442 **A:** CH (GUX) ID (AP SD) NP **ORR**
distincta Jacoby, 1892b: 936
inornata Jacoby, 1892b: 937 **A:** NP **ORR**
limbatipennis Jacoby, 1889b: 197 **A:** BT ID (SD) NP **ORR**
magica Harold, 1877b: 433 (*Oedionychis*) **A:** BT CH (XIZ) ID (SD) NP PA **ORR**
parvula Jacoby, 1884b: 29 **A:** CH (FUJ HKG SHA) ID (SD) NP SC **ORR**
feae Jacoby, 1889b: 196
moseri Weise, 1922b: 124 (*Hyphasoma*)
tristis S.-H. Chen, 1934h: 295 **A:** CH (GUI) **ORR**
unicolor Jacoby, 1889b: 197 **A:** CH (HAI) NP **ORR**
fulvicornis Jacoby, 1905b: iv

genus *Ivalia* Jacoby, 1887a: 100 type species: *Ivalia viridipennis* Jacoby, 1887 (= *Mniophila ruficollis* Motschulsky, 1866)
Amphimeloides Jacoby, 1887a: 96 type species: *Amphimeloides dorsalis* Jacoby, 1887
Ancyloscelis Ogloblin, 1930: 100 type species: *Mniophila ruficollis* Motschulsky, 1866
Schereria L. N. Medvedev, 1984: 60 type species: *Chabria minima* Scherer, 1969
Taizonia S.-H. Chen, 1934e: 182 type species: *Taizonia bella* S.-H. Chen, 1934
andreevi Gruev, 1985a: 42 (*Taizonia*) **A:** NP
bella S.-H. Chen, 1934e: 183 (*Taizonia*) **A:** CH (TAI)
bistripunctata S.-H. Chen, 1933d: 188 (*Amphimeloides*) **A:** JA (Honshu, Kyushu)
doeberli L. N. Medvedev, 2003a: 321 (*Taizonia*) **A:** NP
excavata S.-Y. Wang, 1993: 321, 329 (*Taizonia*) **A:** CH (HKG HUN)
fulva L. N. Medvedev, 1993c: 366 (*Taizonia*) **A:** ID (SD) NP
hiranoi Takemoto & Suenaga, 2021: 142 **A:** JA
kodama Takemoto & Suenaga, 2021: 132 **A:** JA
loebli Döberl, 1991b: 633 (*Taizonia*) **A:** NP
maculata Gressitt & Kimoto, 1963a: 835 (*Taizonia*) **A:** CH (FUJ)
martensi L. N. Medvedev, 1984: 61 (*Schereria*) **A:** NP
minima Scherer, 1969: 217 (*Chabria*) **A:** ID (SD) NP
nakanoi Takemoto & Suenaga, 2021: 138 **A:** JA
nigrina Döberl, 2009b: 416 **A:** NP
ochracea Gressitt & Kimoto, 1963a: 836 (*Taizonia*) **A:** CH (JIA)
schereri L. N. Medvedev, 1990b: 34 (*Taizonia*) **A:** NP
sexmaculata Kimoto & Takizawa, 2002: 143 (*Amphimeloides*) **A:** NP
sichuana Medvedev, 2016: 264 **A:** CH (SCH)
uenoi Kimoto, 1970d: 299 (*Taizonia*) **A:** CH (TAI)
variabilis Takemoto & Suenaga, 2021: 135 **A:** JA (Ryukyus)

genus *Jacobyana* Maulik, 1926: 302 type species: *Sphaerophysa piceicollis* Jacoby, 1889
naini Scherer, 1969: 231 **A:** ID (UT)

nepalica L. N. Medvedev, 1990b: 25 **A:** NP
nigrofasciata S.-H. Chen, 1935a: 79 **A:** CH (GUX) ID (SD)

genus *Kashmirobia* Konstantinov & Prathapan, 2006: 62 type species: *Aphthona hugeli* Jacoby, 1900
hugeli Jacoby, 1900b: 121 (*Aphthona*) **A:** ID (KA) **ORR**

genus *Laboisierella* S.-H. Chen, 1933e: 383 type species: *Laboisierella cyanipennis* S.-H. Chen, 1933
Luperaria L. N. Medvedev, 1993c: 373 type species: *Luperaria elongata* L. N. Medvedev, 1993
elongata L. N. Medvedev, 1993c: 373 (*Luperaria*) **A:** CH (YUN)

genus *Lanka* Maulik, 1926: 304 type species: *Lanka brunnea* Maulik, 1926
Dentisterna L. N. Medvedev, 1993b: 53 type species: *Dentisterna bicolor* L. N. Medvedev, 1993
Horaia Chûjô, 1937b: 55 type species: *Horaia nigra* Chûjô, 1937
Neorthella L. N. Medvedev, 2010c: 292 type species: *Neorthella rufofulva* Medvedev, 2010
bicolor Chûjô, 1937b: 57 (*Horaia*) **A:** CH (TAI) ID ("India")
esakii Chûjô & Ohno, 1961: 5 (*Horaia*) **A:** JA (Honshu, Shikoku)
fulva Chûjô, 1937b: 58 (*Horaia*) **A:** CH (TAI) JA (Ryukyus)
laevigata S.-H. Chen & S.-Y. Wang, 1980: 14, 25 (*Horaia*) **A:** CH (YUN)
magnoliae Chûjô & Ohno, 1961: 2 (*Horaia*) **A:** JA (Honshu, Kyushu)
minor Chûjô, 1937b: 57 (*Horaia*) **A:** CH (TAI) JA (Ryukyus)
nagaii Nakane, 1985: 2 (*Sphaeroderma*) **A:** JA (Kyushu)
nigra Chûjô, 1937b: 56 (*Horaia*) **A:** CH (TAI)
puncticolla S.-Y. Wang & Ge, 2012a: 331, 334 **A:** CH (YUN)
regularia S.-Y. Wang & Ge, 2012a: 331, 335 **A:** CH (XIZ)

genus *Laotzeus* S.-H. Chen, 1933a: 249 type species: *Laotzeus gracilicornis* S.-H. Chen, 1933
bicolor S.-Y. Wang, 1992c: 703, 741 **A:** CH (HUN YUN)
gracilicornis S.-H. Chen, 1933a: 250 **A:** CH (GUI)
niger Chen & S.-Y. Wang, 1980: 9, 22 **A:** CH (SCH)

genus *Leptodibolia* S.-H. Chen, 1941c: 193 type species: *Leptodibolia cyanipennis* S.-H. Chen, 1941
cyanipennis S.-H. Chen, 1941c: 194 **A:** CH (SCH)

genus *Lesneana* S.-H. Chen, 1933a: 241 type species: *Lesneana rufopicea* S.-H. Chen, 1933
rufopicea S.-H. Chen, 1933a: 242 **A:** CH (YUN) **ORR**

genus *Letzuella* S.-H. Chen, 1933a: 247 type species: *Letzuella yonyonae* S.-H. Chen, 1933
chinensis Döberl, 2011: 276 **A:** CH (HUB SCH ZHE)
himalayana L. N. Medvedev & Sprecher-Uebersax, 1997: 234 (*Aphthona*) **A:** NP
sprecherae Döberl, 2001: 381 [RN]
viridis S.-H. Chen, 1933a: 248 **A:** CH (YUN)
yonyonae S.-H. Chen, 1933a: 248 **A:** CH (GUI)

genus *Lipromela* S.-H. Chen, 1933c: 144 type species: *Lipromela costata* S.-H. Chen, 1933 (= *Lema minutissima* Pic, 1923)
Lipraria L. N. Medvedev, 1990b: 38 type species: *Lipraria variipennis* L. N. Medvedev, 1990
decemmaculata S.-Y. Wang & S.-Q. Ge, 2009a: 767 **A:** CH (HAI)
formosana Ohno, 1960e: 197 **A:** CH (TAI)
minutissima Pic, 1923g: 19 (*Lema*) **A:** JA
costata S.-H. Chen, 1933c: 144
okinawana Chûjô, 1958c: 12 **A:** JA (Ryukyus)
pubipennis S.-H. Chen & S.-Y. Wang, 1980: 10, 23 **A:** CH (GUX)
variipennis L. N. Medvedev, 1990b: 38 (*Lipraria*) **A:** NP

genus *Lipromima* Heikertinger, 1924a: 41 type species: *Liprus minutus* Jacoby, 1885
confusa L. N. Medvedev, 1993c: 371 **A:** CH (YUN)
minuta Jacoby, 1886a: 725 (*Liprus*) **A:** CH (CHQ FUJ GAN HUB JIA SCH SHA TAI ZHE) JA SC
fulvipes Chûjô, 1935g: 399
bicolor Chûjô, 1935g: 399 (ab.) [IN]
quadrimaculata Chûjô, 1956c: 17 (var.)
ornata L. N. Medvedev, 1993c: 370 **A:** ID (SD)

genus *Lipromorpha* Chûjô & Kimoto, 1960: 9 type species: *Liprus difficilis* S.-H. Chen, 1934
alutacea Nadein, 2012: 47 **A:** CH (TAI)
aptera L. N. Medvedev, 1992d: 27 **A:** NP
assamensis Maulik, 1926: 131 (*Liprus*) **A:** NP **ORR**
costipennis S.-H. Chen & S.-Y. Wang, 1980: 4, 17 **A:** CH (FUJ)
cyanea S.-H. Chen & S.-Y. Wang, 1980: 3, 17 **A:** CH (YUN)
difficilis S.-H. Chen, 1934h: 264 (*Liprus*) **A:** CH (FUJ GUA HUB JIA TAI YUN) JA (Ryukyus) **ORR**
emarginata S.-H. Chen & S.-Y. Wang, 1980: 2, 16 **A:** CH (SCH)
inexpectata Komiya, 2006b: 375 **A:** JA (Okinoerabu-jima)
loochooana Chûjô, 1961: 89 (ssp.) **A:** JA (Ryukyus)

marginata S.-Y. Wang, 1992c: 682, 733 **A:** CH (YUN)
meishanica S.-H. Chen & S.-Y. Wang, 1980: 2, 16 **A:** CH (SCH)
melanoptera S.-H. Chen & S.-Y. Wang, 1980: 3, 16 **A:** CH (YUN)
montana Chûjô, 1935g: 396 (*Liprus*) **A:** CH (TAI)
piceiventris S.-H. Chen & S.-Y. Wang, 1980: 3, 16 **A:** CH (YUN)
sakishimana Komiya, 2006a: 201 **A:** JA (Ryukyus)
shirozui Kimoto, 1970b: 210 **A:** CH (TAI)
variabilis Scherer, 1969: 96 **A:** BT CH (XIZ) ID (SD) NP **ORR**

genus *Liprus* Motschulsky, 1861a: 26 type species: *Liprus punctatostriatus* Motschulsky, 1861
Asiorella L. N. Medvedev, 1990b: 30 type species: *Asiorella caraboides* L. N. Medvedev, 1990
Crepidomorpha Fleischer, 1916b: 222 type species: *Crepidodera carinulata* Fleischer, 1916 (= *Liprus punctatostriatus* Motschulsky, 1861)
Crepidomorpha Heikertinger, 1923b: 137 [HN] type species: *Crepidodera carinulata* Fleischer, 1916 (= *Liprus punctatostriatus* Motschulsky, 1861)
Hartmannia L. N. Medvedev, 2015c: 329 [HN] type species: *Hartmannia matthiasi* Medvedev, 2015 [# 755]
caraboides L. N. Medvedev, 1990b: 31 (*Asiorella*) **A:** NP
geminatus S.-H. Chen & S.-Y. Wang, 1980: 10, 22 **A:** CH (JIX)
matthiasi L. N. Medvedev, 2015c: 330 (*Hartmannia*) **A:** CH (YUN) [# 755, # 758]
nuchalis Gressitt & Kimoto, 1963a: 878 **A:** CH (FUJ HUN SCH TAI ZHE)
punctatostriatus Motschulsky, 1861a: 26 **A:** CH (JIL GAN SHA TAI) RU (FE) JA NC
carinulatus Fleischer, 1916b: 222 (*Crepidodera*)
japanensis Schönfeldt, 1887: 184 (*Crepidodera*) [RN]
japonicus Jacoby, 1886a: 723 (*Crepidodera*) [HN]
rufotestaceus Motschulsky, 1866a: 175 (*Diabrotica*)

genus *Loeblaltica* Scherer, 1989: 40 type species: *Loeblaltica decorata* Scherer, 1989
antennalis L. N. Medvedev, 1984: 53 (*Benedictus*) **A:** NP
decorata Scherer, 1989: 41 **A:** ID (SD) NP
gerhardi L. N. Medvedev, 1984: 55 (*Benedictus*) **A:** ID ("India") NP **ORR**

genus *Longitarsus* Latreille, 1829: 155 type species: *Chrysomela atricilla* Linnaeus, 1761
subgenus *Longitarsus* Latreille, 1829: 155 type species: *Chrysomela atricilla* Linnaeus, 1761
Inopelonia Broun, 1893: 1392 type species: *Phyllotreta testacea* Broun, 1880
Teinodactyla Chevrolat, 1836: 392 [RN] type species: *Chrysomela atricilla* Linnaeus, 1761
Thyamis Stephens, 1831: 307 type species: *Altica quadripustulata* Fabricius, 1775 (= *Chrysomela quadriguttata* Pontoppidan, 1763)

absynthii Kutschera, 1862c: 217 **E**: AU BH CR CZ EN FR (East) GB GE HU IT KR LA MD PL RO RU (CT ST) SK SL SP SV UK **A**: KZ MG RU (WS) TR UZ "Turkestan"
acuticollis Lopatin, 1976a: 115 **A**: KZ
adamsii Baly, 1874: 199 (*Thyamis*) **A**: JA (Hokkaido, Honshu)
aeneicollis Faldermann, 1837: 346 (*Teinodactyla*) **E**: AU BE BH BU BY CR EN FR GB GB GE GR HU IT KO KR LA LT LU MD ME NL NM PT RO RU (ST) SB SK SP SZ UK **N**: AG LB MO TU **A**: AB AF GG IN IQ IS KI KZ LE SY TD TR TM UZ
cribripennis Abeille de Perrin, 1909: 181 (*Thyamis*)
nigricollis Foudras, 1860: 273 [= 1860b: 161] (*Teinodactyla*)
rubenticollis Allard, 1860c: 118 (*Teinodactyla*)
suturalis Marsham, 1802: 201 [HN] (*Chrysomela*)
aeneus Kutschera, 1862b: 107 **E**: CR FR IT GR (Crete) MA PT SP **N**: AG CI EG LB MO MR SG TU **A**: JO SY YE **AUS**i
involucer Weise, 1888b: 932 (var.)
aeruginosus aeruginosus Foudras, 1860a: 315 [= 1860b: 203] (*Teinodactyla*) **E**: AL AN AU BE BY FR GB GE GR IT KR LU MA NL PT SP UK **A**: CY TR
aeruginosus algiricus Heikertinger, 1913e: 181 (var.) **N**: AG MO TU
afghanicus Lopatin, 1963d: 366 (ssp.) **A**: AF
agilis Rye, 1868: 133 (*Thyamis*) E: BE CZ FR GB GE NL
albineus Foudras, 1860a: 310 [= 1860b: 198] (*Teinodactyla*) **E**: BU CR FR GR HU IT KR NM PT RO RU (ST) SK SP UK **N**: AG EG MO TU **A**: AB AF CY GG IN IQ IS SI TD TR UZ **AUS**i
albus Allard, 1866: 405 (*Thyamis*) **N**: AG MO **A**: IS
alfierii alfierii Pic, 1923c: 7 **E**: GR (North Sporades) **N**: EG **A**: AF IN IS JO KI LE SI SY TR
klapperichi Mohr, 1962: 323
alfierii antineae Peyerimhoff, 1929: 144 (*Thyamis*) **N**: AG
alfierii furthi Gruev, 1982c: 471 (ssp.) **E**: BU GR (Crete) KR NM RU (CT ST) SB SP UK **A**: GG TR
allotrophus Furth, 1980: 85 **A**: IS JO TR
almorae Maulik, 1926: 342 **A**: CH (XIZ) ID (KA UT) NP **ORR**
krishna Maulik, 1926: 340
anacardius Allard, 1866: 324 (*Thyamis*) **E**: IT (Sicily) PT SP **N**: AG MO TU
andalusicus Gruev, 1973c: 138 **E**: SP (Andalucía)
angelikae Fritzlar, 2001: 211 **A**: TR (Antalya)
angusticollis S.-Y. Wang, 1992c: 714, 747 **A**: CH (SCH)
aphthonoides Weise, 1887h: 206 **E**: AU CZ FR GE HU IT (North) PL RU (CT) SK SL SZ **A**: AB CH (FUJ HEI) IN KZ MG NC RU (ES FE) TR **N**: TU
callidus Warchałowski, 1967a: 59

apicalis Beck, 1817: 18 (*Haltica*) **E:** AL AU BE BH BU BY CR CZ DE EN FI FR GE GR HU IT KR LA LS ME NR PL RO RU (CT ST) SB SK SL SP SV SZ UK **N:** CI **A:** RU (ES)

analis Duftschmid, 1825: 264 (*Haltica*)

fischeri Zetterstedt, 1837: 223 (*Haltica*)

praticola R. F. Sahlberg, 1834: 279 (*Haltica*)

quadrimaculatus Weise, 1888b: 948 (var.)

apricus Warchałowski, 1966: 104 **A:** ID (SD HP UT) NP **ORR**

arabicus Doguet, 1979a: 311 **A:** SA YE

aramaicus Leonardi, 1979: 296 **A:** CY IS JO TR

arisanus Chûjô, 1937c: 100 **A:** CH (TAI)

arnoldi Bergeal & Doguet, 1991: 221 **N:** AG

artvinus Gruev & Aslan, 1998: 164 **A:** TR (Artvin)

asperifoliarum Weise, 1887d: 334 **A:** AF IN KZ KI RU (ES) TM UZ

defectus Jacobson, 1893b: 246 (var.)

atlanticus Döberl, 2003: 243 **A:** MO

atricillus Linnaeus, 1761: 166 (*Chrysomela*) **E:** AL AU BE BH BU BY CR CZ DE EN FI FR GB GE GR HU IT KR LA LT LU MD ME NL NR PL PT RO RU (CT NT ST) SB SK SL SP SV SZ UK **N:** AG LB MO TU **A:** AB CY IN RU (ES FE SS) TR TM UZ

declivis Weise, 1893a: 968 (var.)

similis Weise, 1893a: 968 (var.)

suturalis Duftschmid, 1825: 261 (*Haltica*)

audisioi Biondi, 1992: 342 **A:** TR

australis Mulsant & Rey, 1874: 227 (*Thyamis*) **E:** CR FR GE GR IT PT SP (incl. Balearic Islands) SZ **N:** AG MO **A:** TR

caninae Buysson, 1908a: 1 [= 1908b: 20] (*Thyamis*)

gracilicornis Mulsant & Rey, 1874: 238 (*Thyamis*)

awadi Lopatin, 2001b: 18 **A:** YE

baeticus Leonardi, 1979: 292 **E:** SP **N:** AG MO **A:** TR

ballotae Marsham, 1802: 205 (*Chrysomela*) **E:** AL AU BE BH BU BY CR CZ EN FR GB GE GR HU IR IT KR LA LU MD NL NM PL PT RO RU (CT ST) SB SK SL SP SZ UK **N:** AG LB MO TU **A:** CY GG IN IS JO KZ MG RU (ES) SY TR

canescens Foudras, 1860a: 327 [= 1860b: 215] (*Teinodactyla*)

caligans Král, 1945: 31 (forma)

septorum Buysson, 1908b: 29 (*Thyamis*, var.)

barbarae Doguet & Bergeal, 2002: 372 **E:** GR **A:** TR

bedelii Uhagón, 1887: 399 (*Thyamis*) **E:** SP

belgaumensis Jacoby, 1896c: 260 **A:** CH (CHQ FUJ GUX) ID (SD UT) NP **ORR**

bergeali Doguet & Gruev, 1988: 262 **E:** FR SP

beroni Gruev, 1988b: 158 **A:** AF PA
bertii Leonardi, 1973: 467 **E:** AL AU BH BU CZ GR HU IT NM RU (ST) SB SK SL TR UK **A:** CY IN IS SY TR TM
bicoloriceps Chûjô, 1937c: 99 **A:** CH (TAI)
bimaculatus Baly, 1874: 200 (*Thyamis*) **A:** CH (FUJ GUA HUB TAI) JA **AUS ORR**
lewisiellus Chûjô, 1937c: 102
ligustrivorus Chûjô, 1958c: 15
maebarai Ohno, 1968a: 8 (forma) [IN]
nigrofasciatus Ohno, 1968a: 8 (forma) [IN]
birmanicus Jacoby, 1892b: 921 **A:** CH (GUX XIZ) ID (SD UT) NP **ORR**
boharti Kimoto, 1965: 453 **A:** JA (Ryukyus)
boraginicolus Ohno, 1968a: 16 **A:** JA
suturalis Ohno, 1968a: 19 (forma) [IN]
bourdonnei Doguet & Bergeal, 2007: 109 **N:** MO
brevicornis S.-H. Chen, 1939b: 82 **A:** CH ("North China")
brevipennis Wollaston, 1860a: 8 **N:** CI (Lanzarote, Tenerife)
brisouti Heikertinger, 1912i: 293 **E:** BE BH BU CR FR GE GR HU IT KR LU ME RO RU (CT ST) SK SL SP UK **A:** RU (WS) TR
brunneus Duftschmid, 1825: 260 (*Haltica*) **E:** AL AU BE BH BU BY CR CZ DE EN FI FR GB GE GR (Crete) HU IT KR LA LS LT LU MA MD NL NR PL PT RO RU (CT NT ST) SB SK SL SP SV SZ UK **A:** KI KZ NC RU (ES FE WS)
castaneus Duftschmid, 1825: 260 (*Haltica*)
laevicollis Weise, 1888b: 950 (var.)
picinus Weise, 1888b: 950 (var.)
robustus Weise, 1888b: 950 (var.)
buettikeri Doguet, 1985: 363 **A:** AE OM SA YE (incl. Socotra) [# 778]
bytinskii Furth, 1980: 91 **A:** AR IS TR
candidulus candidulus Foudras, 1860a: 319 [= 1860b: 207] (*Teinodactyla*) **E:** CR FR GR IT ME PT SP **N:** AG EG LB MO TU **A:** SI TR
breviusculus Mulsant & Rey, 1876: 253 (*Thyamis*)
latifrons Allard, 1860c: 130 (*Teinodactyla*)
candidulus thymelaearum Peyerimhoff, 1911: 212 (*Thyamis*, ssp.) **N:** AG
carolileonardi Farina, 2023: 18 E: BH
celticus Leonardi, 1975b: 206 **E:** AU CZ FR GE IT (North) RU (CT ST) SB SK SP SZ UK **A:** TR
cerinthes Schrank, 1798: 562 (*Altica*) **E:** AU CR CZ FR GE GR HU IT PT RO SK SL SP SZ UK (Carpathians) **N:** AG CI LB MO MR TU **A:** CY TR
nervosus Wollaston, 1854: 447
pectoralis Foudras, 1860a: 308 [= 1860b: 196] (*Teinodactyla*)

cervinus Baly, 1875b: 213 (*Thyamis*) [RN] **A:** JA (Kyushu, Tsushima) NC SC

inconspicuus Baly, 1874: 201 (*Thyamis*) [HN]

championi Maulik, 1926: 358 **A:** ID (UT) NP **ORR**

cheni Scherer, 1969: 62 **A:** ID (SD) NP

chitwanus L. N. Medvedev, 2000e: 19 **A:** NP

chujoi Csiki, 1939: 183 [RN] **A:** CH (TAI) JA (Honshu)

flavicornis Chûjô, 1937c: 97 [HN]

cinerariae Wollaston, 1854: 444 **N:** MR

consanguineus Wollaston, 1857: 132

cizeki Döberl, 2004: 15 **E:** SK

codinai Jind. Madar & Jiř. Madar, 1965: 9 **E:** CR FR IT PT SP **N:** AG MR MO TU

consobrinellus S.-H. Chen, 1939a: 46 **A:** CH (GUX) **ORR**

corpulentus Weise, 1887d: 333 **A:** KZ TD TM TR UZ

croaticus Gruev, 1975b: 139 **E:** CR

curtus Allard, 1861: 832 (*Teinodactyla*) [RN, NP] **E:** AL AU BE BH BU BY CR CZ DE EN FI FR GB GE GR HU IR IT KO KR LA LT LU ME NL NR PL PT RO RU (CT ST) SB SK SL SP SV SZ UK **A:** AF AR RU (ES) TD TR "Turkestan" **[# 763]**

arbustorum Gistel, 1857b: 521 (*Crepidodera*) [NO] [# 763]

ciliifer Méquignon, 1948: 80 (var.)

lichtneckerti Kaszab, 1962: 73 (ab.) [IN]

liliputanus Allard, 1866: 348 (*Thyamis*)

papaveris Allard, 1866: 394 (*Thyamis*)

pratensis Allard, 1860c: 137 (*Teinodactyla*) [HN]

subquadratus Allard, 1866: 414 (*Thyamis*) [QS]

viduus Allard, 1866: 340 (*Thyamis*)

cyanipennis Bryant, 1924b: 249 **A:** CH (GAN HUN SCH SHA TAI XIZ YUN ZHE) ID (HP SD) NP **ORR**

danieli Mohr, 1962: 321 **E:** SP

debernardii Leonardi, 1979: 303 **A:** IS JO

desertorum Heikertinger, 1913b: 69 **A:** AF KI KZ MG TD TM UZ

dlabolai Král, 1964a: 36 **A:** GG

dorsalis Fabricius, 1781: 135 (*Chrysomela*) **E:** BE CZ FR GB GE IT LU NL PT SP SZ **N:** AG MO [# 757]

dorsopictus S.-H. Chen, 1939a: 45 **A:** CH (BEI CHQ FUJ GUX HUN YUN) JA (Honshu)

echii Koch, 1803a: 52 (*Haltica*) **E:** AL AU BH BU CR CZ FR GE GR HU IT KR LT MA MD ME NM PL PT RO RU (CT ST) SB SK SL SP SZ UK **N:** AG CI MO MR TU **A:** AB AF IN SY TR **AFR AUSi**

excurvus Wollaston, 1857: 133

kulikovskii Jacobson, 1896a: 557

tibialis Duftschmid, 1825: 258 (*Haltica*)

eminatus Furth, 1980: 95 **A:** IS LE SY
eminus Warchałowski, 1967a: 62 **A:** AF IN IQ IS KI KZ SI TD TR UZ
erberi Mohr, 1984: 172 **A:** PA
exsoletus exsoletus Linnaeus, 1758: 373 (*Chrysomela*) **E:** AL AN AU BE BH BU BY CR CZ DE EN FR GB GE GR HU IT KR LA LS LT LU MA MD ME NL NR PL PT RO RU (CT ST) SB SK SL SP SV SZ UK **A:** AB AR CY GG IN SY TR TM
arctulus Weise, 1893a: 1000 (var.)
boppardiensis Bach, 1858: 152
exoletus Linnaeus, 1761: 167 (*Chrysomela*) [incorrect subsequent spelling]
femoralis Marsham, 1802: 201 (*Chrysomela*)
livens Mulsant & Rey, 1874: 235 [HN]
lividus Fauvel, 1888: 251 [RN]
exsoletus megaloleucus Allard, 1860c: 122 (*Teinodactyla*) **E:** SP (Menorca) **N:** AG TU
exsoletus rufulus Foudras, 1860a: 317 [= 1860b: 205] (*Teinodactyla*) **E:** BH CR FR GR IT **A:** CY IN TR
fallax Weise, 1888b: 929 (var.) **E:** BU CR FR GR IT KR MD RO RU (CT ST) TR UK **N:** AG EG LB MO **A:** AB AF AR IN IQ IS KI KZ SI TD TR TM UZ
cynoglossi Faliy, 1970c: 135 [HN]
dimidiatus Allard, 1866: 329 (*Thyamis*) [HN]
nitens Iablokoff-Khnzorian, 1968: 274
scrutator Weise, 1890b: 115 (var.)
ferrugineus Foudras, 1860a: 328 [= 1860b: 216] (*Teinodactyla*) **E:** AU BE BY CR CZ DE EN FR GB GE IT LA LT LU NL RU (CT ST) SB SK SL SP SV SZ UK **N:** AG MO TU **NAR**i
menthae Bedel, 1898: 189 (*Thyamis*)
menthaphagus Gentner, 1926: 152
waterhousei Kutschera, 1864c: 274
ferruginipennis Fuente, 1910b: 446 **E:** PT SP
finitimus Konstantinov, 1993: 42 **A:** KZ
flavicornis Stephens, 1831: 309 (*Thyamis*) **E:** BE FR GB MD PT SP SZ **N:** AG MO **AUS**i **NAR**i
rufescens Fowler, 1890: 352 (var.)
citrinus Fuente, 1910b: 446
formosanus Chûjô, 1937c: 101 **A:** CH (TAI) JA (Ryukyus)
foudrasi Weise, 1893a: 987 [NP] **E:** AL AN AU BH BU CR CZ FR GE GR HU IT KO KR MD ME NM PL RO RU (ST) SB SK SL SP SZ UK **N:** AG MO TU **A:** AR RU (ES FE) TR [# 765]
frontosus Normand, 1947a: 87 **E:** IT (Toscana) SP **N:** AG TU

fulgens Foudras, 1860a: 261 [= 1860b: 149] (*Teinodactyla*) **E:** AU BE BU BY CZ EN FI FR GE HU IT KR LA LU NL PL RO RU (CT ST) SB SK SL SP SZ NL UK **A:** AB GG RU (ES)

exhaustus Weise, 1888b: 952 (var.)

fuscorufus Döberl, 2011: 277 **A:** CH (YUN)

fusus S.-H. Chen, 1939b: 83 **A:** CH (GAN NIN SHX)

ganglbaueri ganglbaueri Heikertinger, 1912i: 292 [RN, NP] **E:** AU BE BH BU BY CR CZ DE EN FI FR GB GE GR (Ionian Is.) HU IR IT KR LA LS LT LU ME NL NM NR PL RO RU (CT ST) SB SK SL SP SV SZ UK **A:** MG RU (ES FE SS) TR **NARi** [# 765]

foudrasi Crotch, 1871: 36 (*Thyamis*) [RN, NO] [# 765]

senecionis Brisout de Barneville, 1873b: 6 [HN]

ganglbaueri balachowskyi Hoffmann, 1950: 192 (ssp.) **N:** AG MO

gavira Maulik, 1926: 349 **A:** ID (UT)

georgianus Allard, 1866: 362 (*Thyamis*) **A:** GG TR

gerhardi Lopatin, 1979c: 440 **A:** ID (KA)

gerhardschereri Gruev, 1996: 49 [RN] **A:** ID (SD) NP **ORR**

bengalensis Döberl, 2001: 382 [RN]

kimotoi Scherer, 1969: 65 [HN]

girardi Doguet, 1987: 55 **N:** AG

gloriae Doguet & Bergeal, 2002: 370 **N:** MO

godmani Baly, 1876d: 583 (*Thyamis*) **A:** CH (FUJ GAN HUB HUN JIA SCH SHG YUN) NC NP RU (FE) SC **ORR**

daovantieni Warchałowski, 1966: 106

subniger S.-H. Chen, 1939b: 79

gola Maulik, 1926: 353 **A:** ID (HP UT) NP PA

gomerensis Biondi, 1986b: 356 (ssp.) **N:** CI (La Gomera)

gracilis Kutschera, 1864c: 275 **E:** AU BE BU BY CR CZ DE EN FR GB GE GR HU IR IT KR LA LT LU MA MD ME NL PL RO RU (ST: Caucasus) SB SK SL SP SV SZ UK **N:** AG MO **A:** IS RU (WS) TR

nigrithorax Heikertinger, 1912c: 65 (ab.) [IN]

poweri Allard, 1866: 408 (*Thyamis*)

sagittifer Mohr, 1962: 320

grandis Rapilly, 1978: 338 **A:** IN

gressitti Scherer, 1969: 63 **A:** CH (SCH XIZ) ID (UT) NP

gruevi Leonardi & Mohr, 1974: 192 E FR (Alpi-Maritime, Cottian Alps) IT (North) SZ

hammondi Gruev, 1981d: 151 **A:** CH (HEI)

hartmanni L. N. Medvedev, 2004d: 204 **A:** NP

hedini S.-H. Chen, 1934a: 10 **A:** CH (FUJ HUB JIA SCH XIZ) ID ("North India") NP

heinigi Döberl, 2003: 243 **A:** KI KZ TD UZ

helvolus Kutschera, 1864c: 269 **E:** AU CR CZ FR GE HU IT SK SL SZ **A:** TR

hermonensis Furth, 1980: 99 **A:** IS TR
himalayensis S.-H. Chen, 1935a: 78 **A:** ID (SD) NP
hina Maulik, 1926: 347 **A:** ID (UT) NP
hissaricus Lopatin, 2000a: 204 **A:** AF TD
hoberlandti Lopatin, 1990b: 601 **A:** CH (HUB JIA SCH XIZ) ID ("North India") IN NP TD TM
hoggarensis Cobos, 1958: 353 **N:** AG
hohuanshanus Kimoto, 1970d: 309 **A:** CH (TAI)
holsaticus Linnaeus, 1758: 373 (*Chrysomela*) **E:** AU BE BH BU BY CR CZ DE EN FI FR GB GE HU IR IT LA LS LT LU ME NL NM NR PL RO RU (CT NT) SB SK SL SP SV SZ UK **A:** CH (FUJ SCH) JA NC RU (ES FE)
discoideus Weise, 1888b: 945 (var.)
haemorrhoidalis Jacoby, 1886a: 728
pulicarius Linnaeus, 1767: 595 (*Chrysomela*)
seriepunctatus Roubal, 1943: 58 (var.)
tsii S.-H. Chen, 1941c: 195
hopeianus S.-H. Chen, 1941c: 195 [RN] **A:** CH (HEB) RU (FE)
frontalis S.-H. Chen, 1939b: 78 [HN]
hsienweni S.-H. Chen, 1939a: 44 **A:** BT CH (GUX HUB) ID (UT) NP **ORR**
ibericus Leonardi & Mohr, 1974: 192 **E:** SP
idilphilus Biondi, 1984: 379 **N:** MO
impressus Döberl, 2009b: 418 **A:** NP
inconspicuus Wollaston, 1860a: 9 **N:** CI (La Gomera, Tenerife)
ellipsodes Wollaston, 1865: 371 (var.)
indigonaceus Lopatin, 1963d: 366 **A:** AF CH (TAI) ID (KA SD) NP **ORR**
warchalowskii Scherer, 1969: 60
ishigakiensis Kimoto, 1965: 456 **A:** JA (Ryukyus)
ishikawai Kimoto, 1970d: 308 **A:** CH (TAI)
isoplexidis Wollaston, 1854: 443 **N:** MR
masoni Wollaston, 1857: 131
jacobaeae G. R. Waterhouse, 1861: 94 (*Thyamis*) **E:** AU BE BU BY CR CZ DE EN FR GB GE GR HU IR IT KO KR LA LS LT LU MD ME NL NR PL RO RU (CT ST) SB SK SL SV SZ UK **A:** AB CH (XIZ) KI KZ MG RU (ES WS) TR **NARi AUSi**
laevis Duftschmid, 1825: 261 [QS]
jailensis Heikertinger, 1913d: 145 **E:** BU (Strandzha Mt.) KR TR **A:** TR
jandiensis Biondi, 1986b: 356 **N:** CI (Fuerteventura)
juncicola Foudras, 1860a: 301 [= 1860b: 189] (*Teinodactyla*) **E:** AL AU BE BH BU CR FR GR (Ionian Is., Crete) HU IT MA ME PT RO RU (ST) SK SL SP UK "Transcaspia" **N:** AG LB MO TU **A:** AB TR
fulviventris Weise, 1893a: 989 (var.)

karlheinzi Warchałowski, 1972: 315 **E:** KR **A:** IN IS SY TR
khnsoriani Paliy, 1970b: 8 **E:** RU (ST) **A:** KI
kippenbergi Warchałowski, 1998c: 351 **N:** MO
kleiniiperda Wollaston, 1860a: 4 **N:** CI
komiyai Ohno, 1968a: 44 **A:** JA (Honshu, Shikoku, Tsushima)
kopdagiensis Gruev & Aslan, 1998: 165 **A:** TR
kutscherai Rye, 1872a: 199 (*Thyamis*) [RN] **E:** AL AU BE BH BU BY CZ DE FI FR GB GE GR HU IR IT LA NL NR PL PT RO RU (CT NT ST) SK SL SP SV SZ UK **N:** AG **A:** AB CH (NMO) GG NC RU (SS WS) SC TR
atriceps Kutschera, 1864a: 38 [HN]
femoratus S.-H. Chen, 1939b: 80
laevicollis S.-Y. Wang, 1992c: 716, 747 **A:** CH (SCH YUN)
langtangensis Gruev, 1990b: 61 **A:** NP
languidus Kutschera, 1863a: 163 **E:** CR CZ GE HU IT (North-East) KZ (Transcaspia) MD RO RU (CT ST) SB SK SL UK
latens Warchałowski, 1999a: 291 **A:** TR (Trabzon)
lateripunctatus Rosenhauer, 1856: 339 **E:** AZ CR FR GR IT MA ME PT SP **N:** AG MO TU **A:** CY IS JO
azoricus Israelson, 1991: 1
biguttatus Foudras, 1860a: 284 [= 1860b: 172] (*Teinodactyla*)
signatus Reiche & Saulcy, 1858b: 49 (*Haltica*)
laureolae Biondi, 1989: 137 **E:** IT (South, Sicily)
ledouxi Doguet, 1979b: 52 **E:** RU (ST: Caucasus) **A:** GG TR
abchasicus Konstantinov, 1986: 81
leonardicarloi Konstantinov & Dorr, 2023: 46 **A:** TM
leonardii Doguet, 1973: 125 **E:** FR (Pyrénées) SP (Cordillera Cantabríca)
lewisii Baly, 1874: 199 (*Thyamis*) **E:** AU BH BU BY CZ EN FI FR GE GR HU IT LA NM PL RO RU (CT NT ST) SK SL SP SZ UK **A:** AF CH (FUJ HEB HEI SHA SHX TAI) GG IN JA KI MG NC PA RU (ES FE SS WS) SC TD TR UZ **ORR NAR**i
stramineus Weise, 1887h: 205
borodinensis Chûjô, 1940e: 363
linnaei Duftschmid, 1825: 265 (*Haltica*) **E:** AU BH BU CR CZ FR GE GR (Ionian Is.) HU IT KR PL PT RO RU (ST) SB SK SL SP SZ TR UK **A:** AB AR GG IN IS SY TR
amoenus Weise, 1888b: 929 (var.)
litanganus S.-Y. Wang, 1992c: 716, 748 **A:** CH (SCH)
lohita Maulik, 1926: 352 **A:** CH (HUN YUN) **ORR**
longipennis Kutschera, 1863b: 306 **E:** AU BE BU BY CR CZ GE GR HU IT KR LU MD ME NL PL RO RU (CT ST) SB SK SZ UK **N:** MO **A:** AB KZ RU (ES) TR "Turkestan"

longiseta Weise, 1889l: 420 **E:** AU BU BY CZ EN FI FR GB GE HU IT (North) KR LA LS LT NR PL RO RU (CT ST) SK SL SV SZ UK **A:** CH (GUX HEI) JA (Hokkaido, Honshu) NC RU (ES FE WS)
clarus Allen, 1967: 78
kwangsiensis S.-H. Chen, 1939a: 45
luridus conspiciabilis Faldermann, 1837: 345 (*Teinodactyla*) **E:** RU (ST)
luridus luridus Scopoli, 1763: 70 (*Chrysomela*) **E:** AL AN AU BE BH BU BY CR CZ DE EN FI FR GB GE GR HU IR IT KR LA LS LT LU MA MD ME NL NM NR PL PT RO RU (CT NT ST) SB SK SL SP SV SZ TR UK **A:** AB AR CY GG IN IQ IS KZ LE MG RU (ES FE SS WS) SY TR UZ **NARi**
amurensis Weise, 1887h: 205
bicolor L. N. Medvedev, 1992a: 589
cognatus Weise, 1888b: 957 (var.)
fuscescens Stephens, 1831: 315 (*Thyamis*)
nigricans Weise, 1888b: 957 (var.)
piceonitens Guillebeau, 1895b: ccclxxxviii
quadrisignatus Duftschmid, 1825: 259 (*Haltica*)
syriacus Allard, 1866: 379 (*Thyamis*)
luridus mauritanicus Peyerimhoff, 1915b: 221 (*Thyamis*, ssp.) **N:** AG TU
lycopi Foudras, 1860a: 305 [= 1860b: 193] (*Teinodactyla*) **E:** AL AU BE BH BU BY CR CZ DE EN FI FR GB GE GR HU IT KO KR LA LT LU MA MD ME NL NM PL PT RO RU (CT NT ST) SB SK SL SP SV SZ TR UK **N:** AG MO TU **A:** AB AR CY GG IN IQ IS JO KZ RU (SS WS) SA SY TD TR TM UZ YE **AFR**
limnophilus Abeille de Perrin, 1907: lxxxi
seriatus Kutschera, 1864b: 154 [QS]
malina Maulik, 1926: 349 **A:** ID (UT) NP **ORR**
manfredi Fritzlar, 2004: 120 **A:** TR
marguzoricus Konstantinov, 2000: 206 **A:** TD
medvedevi Shapiro, 1956: 152 **E:** AU BU CZ HU KR PL RO RU (CT ST) SB SK SV UK **A:** RU (WS) TR UZ
melanocephalus DeGeer, 1775: 348 (*Chrysomela*) **E:** AL AN AU BE BH BU BY CR CZ DE EN FI FR GB GE GR HU IT KO KR LA LS LT LU MD ME NL NM NR PL PT RO RU (CT NT ST) SB SK SL SP SV SZ TR UK **N:** MR **A:** AB AF GG IN IS KZ MG RU (ES WS) TD TR TD
atricapillus Duftschmid, 1825: 257 (*Haltica*)
crassicornis Foudras, 1860a: 325 [= 1860b: 213] (*Teinodactyla*)
nigrinus Weise, 1893a: 979 (var.)
pallidus Geoffroy, 1785: 101 (*Altica*)
piciceps Stephens, 1831: 309 (*Thyamis*)

melanoxanthus Lopatin, 1963d: 365 **A:** AF TM
membranaceus Foudras, 1860a: 334 [= 1860b: 222] (*Teinodactyla*) **E:** BE BH BU CR FR GB GE GR HU IT KR LU MA NL PT RO RU (ST) SP UK **A:** AB CY IN IS TR **N:** AG MO TU
teucrii Allard, 1860c: 139 (*Teinodactyla*)
meridionalis Weise, 1888b: 938 (var.) **A:** IS SY TR
messerschmidtiae Wollaston, 1860a: 6 **N:** CI **AFR**
stenocyphon Wollaston, 1868: 150
minimus Kutschera, 1864b: 144 **E:** AU BH BY CZ GE HU IT KR PL RU (CT ST) SB SK SZ UK **A:** RU (ES WS) TR
minusculus Foudras, 1860a: 266 [= 1860b: 154] (*Teinodactyla*) **E:** AU BE BH BU CR CZ EN FR GB GE GR HU IT KR ME PT RO RU (CT ST) SK SL SP SZ UK **A:** TR
mirei Doguet, 1979a: 313 **A:** IS SA **AFR**
montanus S.-Y. Wang, 1992c: 717, 748 **A:** CH (SCH)
monticola Kutschera, 1864a: 44 **E:** AL AU BU CZ FR GE HU IT MD PL RO RU (CT ST) SK SL SZ **A:** TD
moscovitus Allard, 1866: 415 (*Thyamis*) [QS]
morrisonus Chûjô, 1937c: 98 **A:** CH (TAI) JA
kimotoi Ohno, 1968a: 14
muchei Mohr, 1984: 171 **A:** UZ
muralis S.-H. Chen, 1939b: 85 **A:** CH (SHX)
murteus Warchałowski, 1972: 316 **A:** AF KZ
nakanei Kimoto, 1970d: 308 **A:** CH (TAI)
nanus Foudras, 1860a: 335 [= 1860b: 223] (*Teinodactyla*) **E:** AU BE BH BU BY CR CZ FR GE GR HU IT KR LS LU MD NL RU (CT ST) SB SK SL SP SZ UK **N:** AG **A:** AB GG IS TR
brevicollis Kutschera, 1864c: 271
cerinus Foudras, 1860a: 332 [= 331] [= 1860b: 219] (*Teinodactyla*) [QS]
vitreus Kutschera, 1864c: 270
nasturtii Fabricius, 1792b: 31 (*Galleruca*) **E:** AU BE BH BU BY CR CZ DE EN FI FR GB GE GR HU IT KR LA LS LT LU MD ME NL NR PL RO RU (CT NT ST) SB SK SL SP SV SZ UK **A:** AB CH (HEB HEI SHA SHX XIZ) KI KZ MG NC RU (ES FE WS) SC TR
auctumnalis Weise, 1893a: 965 (var.)
circumscriptus Bach, 1858: 150
nebulosus Allard, 1867: 495 (*Thyamis*) **E:** FR (Corsica) IT (Sardinia)
neckeri Döberl, 2011: 279 **A:** ID (HP UT)
nepalensis Gruev, 1988a: 101 **A:** NP PA
flavicollis L. N. Medvedev, 1993c: 368

niger Koch, 1803a: 57 (*Haltica*) **E:** AL AU BE BH BU BY CR CZ DE EN FR GE GR HU IT KR LA LT LU ME NL NM PL RO RU (ST) SB SK SL SP SV SZ UK **A:** GG TR

elongatus Bach, 1858: 147

peregii Kaszab, 1962: 74 (ab.) [IN]

nigerrimus Gyllenhal, 1827: 656 (*Haltica*) **E:** AU BE BY CZ DE EN FI FR GB GE IT (North) KR LA LT LU NL NR PL RU (CT ST) SL SV SZ UK **A:** RU (FE)

atricornis Stephens, 1835: 423 (*Thyamis*)

nigrilividus Furth, 1980: 105 **A:** IS

nigrocillus atriplaga Heikertinger, 1913a: 29 (forma) **N:** AG MO TU [# 772]

atriplaga J. R. Sahlberg, 1913a: 82 [HN]

nigrocillus nigrocillus Motschulsky, 1849: 146 (*Thyamis*) **E:** FR IT PT SP

subterlucens Foudras, 1860a: 280 [= 1860b: 168] (*Teinodactyla*)

nigrofasciatus nigrofasciatus Goeze, 1777: 312 (*Chrysomela*) **E:** AL AN AU BE BH BU BY CR CZ EN FR GB GE GR HU IR IT KR LA LS LU MD ME NL NM NR PL PT RO RU (CT ST) SB SK SL SP SZ TR UK **A:** AB AF AR CY ES IN IQ IS JO KI KZ SI SY TD TR TM UZ **N:** AG CI MO MR TU

distinguendus Rye, 1872b: 157 (*Thyamis*)

domesticus Weise, 1893a: 983 (var.)

fractus Wollaston, 1857: 133

lateralis Illiger, 1807a: 68 (*Haltica*)

longipes Bach, 1858: 153

marginatus Geoffroy, 1785: 99 (*Altica*)

patruelis Allard, 1866: 398 (*Thyamis*)

saltator Wollaston, 1854: 445

trilineolatus Foudras, 1860a: 290 [= 1860b: 178] (*Teinodactyla*) [UN]

unionis J. R. Sahlberg, 1913c: 257 (var.) [IN] [# 771]

nigrofasciatus rudipennis Allard, 1866: 358 (*Thyamis*) **E:** FR (Corsica)

nigrofasciatus secutorius Peyerimhoff, 1911: 211 (*Thyamis*, ssp.) **N:** AG

nimrodi Furth, 1980: 108 **E:** AU BU CZ IT (North) NM SK SL **A:** IS TR

nipponensis Csiki, 1940a: 576 [RN] **A:** JA (Hokkaido) RU (FE)

menthae Chûjô, 1938a: 15 [HN]

nitidiamiculus Kimoto, 1965: 458 **A:** JA RU (FE)

nitidus Jacoby, 1886a: 727 **A:** CH (CHQ GAN SCH SHA "North-east China") JA NC RU (ES FE)

nodulus S.-Y. Wang, 1992c: 718, 749 **A:** CH (YUN)

noricus Leonardi, 1976: 239 **E:** AU BH BU BY CZ EN GE HU IT (North) KR LA LT PL RO RU (CT ST) SK SL SZ UK **A:** AB AR GG TR

nubigena Wollaston, 1854: 447 **N:** CI (Gomera, Tenerife) MR *obliteratoides* Gruev, 1973c: 133 **E:** FR GB GE GR (Halkidiki) KR PT SK SP SZ **N:** MO **A:** AB CY GG TR

obliteratus Rosenhauer, 1847: 61 (*Teinodactyla*) **E:** AL AU BH BU BY CR CZ FR GB GE GR HU IT KO KR LT MD ME NL NM PL PT RO RU (CT ST) SB SK SL SP (Andalucía) SZ TR UK **A:** AB AR GG IN IS JO RU (SS) SI SY TR

consociatus Förster, 1849: 38

ochraceicornis Maulik, 1926: 353 **A:** NP **ORR**

ochroleucus lindbergi Jind. Madar & Jiř. Madar, 1963: 123 (ssp.) **N:** MR

ochroleucus ochroleucus Marsham, 1802: 202 (*Chrysomela*) **E:** AN AU BE BU BY CR CZ DE FI FR GB GE GR HU IT KO KR LA LU MA MD NL NM NR PL PT RO RU (ST: Caucasus) SB SK SL SP SV SZ UK **N:** AG CI IN MO TU **A:** AB CH (YUN) NP TR

albellus Duméril, 1816: 524 (*Altica*)

cognatus Wollaston, 1860a: 7

scutellata Gistel, 1857b: 536, 549 (*Thyamis*)

subnigra Roubal, 1943: 57 (ab.) [IN]

ohnoi Gruev, 1996: 52 [RN] **A:** JA

chujoi Ohno, 1968a: 35 [HN]

ordinatus Foudras, 1860a: 333 [= 1860b: 221] (*Teinodactyla*) **E:** FR GR IT PT SP SZ **N:** AG MR MO **AFR**

brunniceps Allard, 1860c: 138 (*Teinodactyla*)

tantulus Foudras, 1860a: 307 [= 1860b: 195] (*Teinodactyla*)

orientalis Jacoby, 1886a: 728 **A:** CH (SHA) JA (Honshu, Awashima Is.)

osimaensis Takizawa, 2015: 238 **A:** JA (Hokkaido)

ozbeki Aslan & Warchałowski, 2005: 222 **A:** TR

paitanus S.-H. Chen, 1939b: 84 **A:** CH (HEB)

palliatus Peyerimhoff, 1939: 36 **N:** AG MO

pallidicornis Kutschera, 1863a: 164 **E:** AU BH CZ FR HU IT (North) PL RO SK SL UK

hubenthali Wanka, 1917: 74

paludivagus Peyerimhoff, 1915a: 45 (*Thyamis*, ssp.) **E:** FR (Corsica) SP (Mallorca) **N:** AG

pardoi Doguet, 1974: 125 **E:** FR (Pyrénées-Orientales) SP (Sierra Nevada) **N:** MO

parvulus Paykull, 1799a: 102 (*Galeruca*) **E:** AL AU BE BH BU BY CR CZ DE EN FI FR GB GE GR HU IR IT KO KR LA LT LU MD ME NL NM NR PL PT RO RU (CT NT ST: Caucasus) SB SK SL SP SV SZ UK **N:** AG CI LB MO MR TU **A:** AB CY IS RU (ES) TR

concinnus Weise, 1888b: 942 (var.)

maderensis Allard, 1864: 659 (*Teinodactyla*)

pumilus Illiger, 1807b: 170 (*Haltica*)

pavidus Warchałowski, 1972: 313 **A:** PA

pekingensis Z.-L. Liang, Konstantinov & S.-Q. Ge, 2023: 113 **A:** CH (BEI) *pellucidus* Foudras, 1860a: 322 [= 1860b: 210] (*Teinodactyla*) E: AL AN AU BE BH BU BY CR CZ DE EN FR GB GE GR HU IT KO KR LA LT LU MA MD ME NL NM NR PL PT RO RU (CT ST) SB SK SL SP SV SZ UK N: AG MO TU A: AB AF CY GG ID (KA) IN IS JO KA KI KZ MG RU (ES WS) SI SY TR UZ **NARi**

ionius Mohr, 1962: 317

nigriventris Weise, 1893a: 1007 (var.)

paleaceus Mulsant & Rey, 1874: 236 (*Thyamis*)

testaceus Allard, 1860c: 127 (*Teinodactyla*)

persimilis Wollaston, 1860a: 4 N: CI

personatus Weise, 1893a: 964 (var.) E: AU BH BU CR CZ FR GE GR HU IT KR MD ME RO RU (ST) SB SK SL SZ UK A: CY IS SY TR N: AG

peyerimhoffi Abeille de Perrin, 1909: 181 (*Thyamis*) E: SP N: AG MO

piceorufus S.-H. Chen, 1939b: 84 A: CH (BEI FUJ HUB JIA ZHE) NC RU (FE) **ORR**

piceus Kimoto, 2001: 66 (*Tegyrius*) A: NP

picicollis Weise, 1900c: 137 [= 1900d: 165] E: BU KR RO UK A: AB AF GG IN IQ KI KZ TD TR TM UZ “Transcaucasus”

pinfanus S.-H. Chen, 1934h: 352 A: CH (CHQ GUI SCH XIZ YUN)

plantagomaritimus Dollman, 1912: 187 E: BE DE GB GE NL RU (NT) SV

perplexus Dollman, 1912: 188 (ab.) [IN]

postmaculatus L. N. Medvedev, 1993c: 368 (*Aphthona*) A: BT ID (UT)

pratensis Panzer, 1794a: 16 (*Altica*) E: AL AU BE BH BU BY CR CZ DE EN FI FR GB GE GR HU IR IT KO KR LA LS LT LU MA MD ME NL NM NR PL PT RO RU (CT NT ST) SB SK SL SP SV SZ TR UK N: AG CI (Tenerife) MO A: AB AR AF GG IN IS KI KZ RU (FE WS) TD TR TM UZ **AFR NARi**

bearei Kevan, 1967: 85

collaris Stephens, 1831: 313 (*Thyamis*)

funereus Mulsant & Rey, 1874: 232 (*Thyamis*)

medicaginis Allard, 1860c: 124 (*Teinodactyla*)

obsoletus Mulsant & Rey, 1874: 230 (*Thyamis*)

pusillus Gyllenhal, 1813: 549 (*Haltica*)

testaceus Melsheimer, 1847: 166 (*Thyamis*) [QS]

pubipennis S.-H. Chen & S.-Y. Wang, 1980: 14, 25 A: CH (YUN)

pulexoides S.-H. Chen, 1939a: 47 A: CH (FUJ GUX)

pulmonariae Weise, 1893a: 998 E: AU BE BH BU CR CZ DE GE HU KR LU NM PL RO RU (CT ST) SB SK SL UK A: AB AR GG IN TR

puncti Maulik, 1926: 350 A: ID (SD) NP **ORR**

punctiger J. R. Sahlberg, 1913a: 80 A: IS JO

quadraticollis Jacoby, 1886a: 729 A: JA **ORR**

quadriguttatus Pontoppidan, 1763: 670 (*Chrysomela*) **E:** AU BU BY CZ DE EN FR GB GE GR HU IT (North) KO KR LU MD ME NM PL RO RU (CT ST) SB SK SP SV SZ UK **A:** TR **NARi**

binotatus Weise, 1888b: 947 (var.)

cynoglossi Marsham, 1802: 205 (*Chrysomela*)

immaculatus Weise, 1888b: 947 (var.)

quadrimaculatus Koch, 1803b: 128 (*Haltica*)

quadrimaculatus Weidenbach, 1859: 58 [HN]

quadrinotatus Gmelin, 1790: 1697 (*Chrysomela*)

quadripunctatus Geoffroy, 1785: 100 (*Altica*)

quadripustulatus Fabricius, 1775: 114 (*Altica*)

quatuorpustulatus Olivier, 1789: 109 (*Altica*)

vittatus Weise, 1888b: 947 (var.)

rangoonensis Jacoby, 1892b: 920 **A:** AF CH (GAN GUX HEB JIA NIN NMO SCH SHX YUN) ID (UT) NP **ORR**

kophir Lopatin, 1966c: 368

madurensis Jacoby, 1896c: 258

manilensis Weise, 1913c: 233

puncticeps S.-H. Chen, 1939b: 81

suturellinus Csiki, 1939: 185 [RN]

suturellus Motschulsky, 1866b: 416 (*Teinodactila*) [HN]

ratshensis Iablokoff-Khnzorian, 1962: 119 **A:** GG

reconditus Warchałowski, 1967a: 58 **A:** IN

rectilineatus Foudras, 1860a: 250 [= 1860b: 138] (*Teinodactyla*) **E:** AU BH BU CR FR GE GR HU IT KR MD NM RU (ST) SP SZ UK **N:** AG TU **A:** AF CY GG IN IQ IS KI KZ SY TD TR TM UZ

kaufmanni Kaszab, 1962: 74 (ab.) [IN]

spilotus Weise, 1900b: 294

refugiensis Leonardi & Mohr, 1974: 192 **E:** IT (North)

reichei Allard, 1860c: 132 (*Teinodactyla*) **E:** AL AU BH BU CR CZ DE FR GB GE GR HU IT NL NR PL RO RU (ST: Caucasus) SB SK SL SP SV SZ TR UK **A:** AR IN TR

fusculus Kutschera, 1864c: 273

rohtangensis Shukla, 1960: 82 **A:** ID (HP)

rubellus Foudras, 1860a: 268 [= 1860b: 156] (*Teinodactyla*) **E:** AU BH CR CZ GE IT (North) KR PL RO SL UK

gravidulus Kutschera, 1863a: 166

rubiginosus Foudras, 1860a: 316 [= 1860b: 204] (*Teinodactyla*) **E:** AL AU BE BH BU BY CR (Adriatic Is.) CZ DE EN FI FR GB GE GR (Crete) HU IT KR LA LS LT LU MD ME NL PL RO RU (CT ST) SK SL SP SV SZ UK **A:** AB CH ("North China") IN RU (ES FE WS) TR "Turkestan" **NARi**

amurensis Weise, 1887h: 205
fumigatus Weise, 1893a: 1005 (var.)
profugus Weise, 1893a: 1005 (var.)
rufotestaceus S.-H. Chen, 1933a: 247 **A**: CH (CHQ FUJ GUX HUN SCH XIZ YUN ZHE) **ORR**
rugipunctatus S.-Y. Wang, 1992c: 719, 749 **A**: CH (YUN)
rugithorax S.-H. Chen, 1939b: 86 **A**: CH (HEB)
rutilus Illiger, 1807a: 67 (*Haltica*) **E**: FR GB IT NL PT SP **N**: AG MO TU **A**: YE
eryngii Gistel, 1831: 303 (*Haltica*)
sahlbergi Pic, 1907b: 112 [RN] **E**: RU (CT) **A**: KZ
weisei Jacobson, 1901b: 144 [HN]
salarius Lopatin & Kulenova, 1985: 48 **E**: RU (ST) **A**: KI KZ
salviae Gruev, 1975a: 89 **E**: AU BH BU BY CR CZ FR GE GR HU IT KR MD NM PL RO RU (CT ST) SB SK SL SP SZ **A**: AR GG TR
sari Maulik, 1926: 350 **A**: NP **ORR**
scrobipennis Heikertinger, 1913a: 31 **E**: BU CR (Ugljan Isl.) HU ME RU (ST) SK
scrophulariae Ohno, 1968a: 42 **A**: JA (Hokkaido, Honshu) **[# 769]**
okushirensis Gruev, 1996: 52 [RN]
scutellariae Ohno, 1968a: 20 **A**: JA (Hokkaido, Honshu)
scutellaris Mulsant & Rey, 1874: 231 (*Thyamis*) **E**: AL AN AU BH BU BY CR EN FI FR GE GR HU IT KR MD ME PL RO RU (CT ST) SB SK SL SP SZ UK **A**: AB CH (HEI NMO) GG IN JA MG NC RU (ES FE) SC TR
sencieri Allard, 1860c: 106 (*Teinodactyla*) **E**: PT SP **N**: AG MO TU
seticollis Mohr, 1962: 319 **E**: SP
caroli Bastazo & Garcia Raso, 1985: 94
shuteae Gruev, 1981d: 151 **A**: CH (HEI)
silbursaicus Lopatin, 2004b: 174 [RN] **A**: TD
seriepunctatus Lopatin, 1966b: 46 [HN]
sinensis S.-H. Chen, 1934a: 11 **A**: CH (GAN)
sjoestedti S.-H. Chen, 1934a: 10 **A**: CH (GAN)
solaris Gruev, 1977: 69 **E**: BU GR **A**: TR
springeri Leonardi, 1975b: 200 **E**: IT (South)
stragulatoides Gruev, 1990b: 62 **A**: NP
stragulatus dichrous Iablokoff-Khnzorian, 1962: 118 **A**: AR (Erevan) TD TR (Diyabakir)
stragulatus pallidicollis Wollaston, 1865: 369 (var.) **N**: CI
pseudodorsalis Har. Lindberg, 1950: 16
stragulatus punicus Peyerimhoff, 1915a: 45 (*Thyamis*, ssp.) **N**: AG (Aurès Mts.)
stragulatus stragulatus Foudras, 1860a: 282 [= 1860b: 170] (*Teinodactyla*) **E**: FR IT (incl. Sardinia, Sicily) MA SP **N**: AG EG LB MO TU **A**: IS JO TR
circumseptus Allard, 1860c: 105 (*Teinodactyla*)
saturatus Peyerimhoff, 1919: 231 (*Thyamis*, var.) [IN]

sellatus Weise, 1893b: 963 (var.)

strigicollis Wollaston, 1864: 412 **E:** AU BU CR CZ FR GB GE HU IT MD PT SK SL SP SZ UK **N:** AG CI MO TU **A:** GG TR **AFR**

bombycinus Mohr, 1962: 318

exiguus Bedel, 1898: 190 (*Thyamis*)

fowleri Allen, 1967: 75

submaculatus Kutschera, 1863a: 154 **E:** FI

subruber S.-H. Chen, 1939b: 79 **A:** CH (SHN)

substriatus Kutschera, 1864a: 43 **E:** AL AU BU CR CZ FR GE GR HU IT KR ME RO RU (ST) SB SK SL SP SV UK **A:** KZ TR

succineus Foudras, 1860a: 329 [= 330] [= 1860b: 218] (*Teinodactyla*) **E:** AL AN AU BE BH BU BY CR CZ DE EN FI FR GB GE GR HU IT KO KR LA LS LT LU MD ME NL NM NR PL PT RO RU (CT NT ST) SB SK SL SP SV SZ TR UK **N:** AG MO **A:** AB AF AR CH (BEI CHQ FUJ GAN SCH SHA) CY GG IN IS JA KI KZ MG NC NP RU (ES FE WS) SC TD TR UZ **NARi**

amiculus Baly, 1874: 201 (*Thyamis*)

arakii Chûjô, 1942a: 39

ovalis Gentner, 1926: 153

perfectus Weise, 1893a: 1008 (var.)

sundara Maulik, 1926: 356 **A:** NP **ORR**

suturatus Foudras, 1860a: 292 [= 1860b: 180] (*Teinodactyla*) **E:** FR IT (South, Sardinia) SP **A:** AG MO TU

scrophulariae Normand, 1937: 132 (var.) [IN] [# 771]

suturellus Duftschmid, 1825: 262 (*Haltica*) **E:** AL AU BE BH BU BY CR CZ DE EN FI FR GB GE GR HU IR IT KR LA LS LT LU MD ME NL NM NR PL RO RU (CT ST) SB SK SL SV SZ UK **A:** AB CH JA (Hokkaido, Honshu) MG NC RU (ES FE SS WS) TR

dimidiatus Stephens, 1835: 423 (*Thyamis*)

freminvillei Guillebeau, 1895c: ccclxxxix

fuscicollis Stephens, 1831: 312 (*Thyamis*)

macer Weise, 1893a: 970 (var.)

paludosus Weise, 1893a: 970 (var.)

senecionis Bach, 1858: 150 [HN]

senecionis Motschulsky, 1852: 599 (*Teinodactyla*)

thoracicus Stephens, 1831: 312 (*Thyamis*)

suzukii Ohno, 1968a: 8 **A:** JA (Hokkaido, Honshu)

symphyti Heikertinger, 1912c: 69 [conserved name, ICZN (1991), Opinion 1656] **E:** AU BE BY CR CZ EN FR GB GE HU IT (North) LS LT LU NL NM PL RO RU (CT) SK SL SZ UK **A:** GG RU (ES) UZ

luctator Weise, 1893a: 1010 (var.) [suppressed name, ICZN (1991), Opinion 1656]

obscurus Döberl, 1994a: 181 (forma) [IN]

szechuanicus S.-H. Chen, 1934a: 9 **A:** CH (SCH)

tabidus Fabricius, 1775: 115 (*Altica*) **E:** AL AN AU BE BH BU BY CR CZ DE EN FI FR GB GE GR HU IT KR LA LS LT LU MA MD ME NL NM NR PL PT RO RU (CT NT ST) SB SK SL SP SV SZ TR UK **N:** AG MO TU **A:** AB AR CY IN IS KI KZ MG RU (ES FE) TR TM

borealis Zetterstedt, 1837: 223 (*Haltica*)

dorsalis P. Rossi, 1792: 32 (*Chrysomela*) [HN]

elongatus Weise, 1893a: 986 (var.)

jaceae Panzer, 1806: 7 (*Altica*)

sisymbrii Fabricius, 1792b: 31 (*Galleruca*)

thapsi Marsham, 1802: 202 (*Chrysomela*)

verbasci Panzer, 1794a: 17 (*Altica*)

vulgaris Weise, 1893a: 985 (var.)

taiwanicus S.-H. Chen, 1934e: 184 **A:** CH (TAI)

taklechensis Bryant, 1941b: 408 **A:** ID (HP)

tarraconensis Leonardi, 1979: 301 **E:** SP

thamensis Gruev, 1990a: 73 **A:** NP

tibetanus S.-H. Chen, 1963: 454, 457 **A:** CH (XIZ)

tienshanicus Konstantinov, 1993: 41 **A:** UZ

tishechkini Konstantinov, 2000: 214 **A:** KZ

tokaranus Nakane & Kimoto, 1961b: 77 **A:** JA (Ryukyus)

tournefortiae L. N. Medvedev & Voronova, 1979: 132 **A:** MG

transbaicalicus Ogloblin, 1921: 38 **A:** MG RU (ES FE)

transversalis S.-H. Chen, 1935a: 77 **A:** ID (SD) **ORR**

trepidus Warchałowski, 1973b: 674 **E:** RU (ST) **A:** GG AF IN IQ TR

tristis Weise, 1888b: 952 (var.) **E:** BU CR (North) CZ GE HU KR RO RU (ST) SK SV UK

pannonicus Kaszab, 1962: 74

tsinicus S.-H. Chen, 1939b: 77 **A:** CH (CHQ GAN SCH SHA SHX)

tunetanus Csiki, 1940a: 575 [RN] **E:** SP **N:** TU

perforatus Peyerimhoff, 1925a: 16 (*Thyamis*) [HN]

ventricosus Foudras, 1860a: 259 [= 1860b: 137] (*Teinodactyla*) **E:** FR (Pyrenees) SP (Pyrenees)

gibbosus Foudras, 1860a: 267 [= 1860b: 155] (*Teinodactyla*)

subrotundus Allard, 1860a: 100 (*Teinodactyla*)

vilis Wollaston, 1864: 415 **E:** IT PT SP (North) **N:** AG CI (Gran Canaria, Tenerife) MO

violentoides Konstantinov, 2000: 215 **A:** AR

violentus Weise, 1893a: 1016 **E:** KR RU (CT ST) UK **A:** AB AF AR GG IN KI KZ MG RU (ES FE WS) TM TR UZ

waltherhorni Csiki, 1939: 185 [RN] **A:** CH (FUJ TAI) SC

horni S.-H. Chen, 1934e: 183 [HN]

wangi Döberl, 2001: 382 [RN] **A:** CH (YUN)
nigriceps S.-Y. Wang, 1992c: 718, 748 [HN]
xinjiangensis Z.-L. Liang, Konstantinov & S.-Q. Ge, 2023: 115 **A:** CH (XIN)
yangsoensis S.-H. Chen, 1939a: 47 **A:** CH (FUJ GUX ZHE) **ORR**
yoshiakii Ohno, 1968a: 51 **A:** JA (Honshu, Kyushu)
zangherii Warchałowski, 1968: 209 **E:** IT
zhamicus S.-H. Chen & S.-Y. Wang, 1981a: 503, 508 **A:** CH (XIZ)
subgenus *Testergus* Weise, 1893a: 1013 type species: *Longitarsus lederi* Weise, 1889
Truncatus Paliy, 1970b: 10 type species: *Longitarsus zeravshanicus* Paliy, 1970 (= *Longitarsus tmetopterus* Jacobson, 1893)
anatolicus Weise, 1900c: 138 [= 1900d: 166] **E:** BU **A:** AB IN IS TR "Palaestina"
anchusae Paykull, 1799a: 101 (*Galeruca*) **E:** AL AU BE BH BU BY CR CZ DE EN FR GB GE GR HU IT KO KR LA LT LU MD ME NL NM PL PT RO RU (CT ST) SB SK SL SP SV SZ TR UK **A:** AB AF AR CY GG IN IQ IS JO KI KZ RO RU (ES WS) SY TD TR
convexus Stephens, 1835: 423 (*Thyamis*)
gagathinus Duftschmid, 1825: 267 (*Haltica*)
punctatissimus Foudras, 1860a: 250 [= 1860b: 138] (*Teinodactyla*)
aubozaorum Biondi, 1997: 383 **A:** TR
behnei Gruev & Arnold, 1989: 175 **E:** BU
bonnairei Allard, 1866: 344 (*Thyamis*) **E:** FR (Corsica) IT (Sardinia)
borisi Konstantinov, 2005: 24 **A:** TD
brachypterus Weise, 1890b: 116 **E:** RU (ST: Caucasus)
bremondi Peyerimhoff, 1939: 36 **N:** MO
bulgaricus Gruev, 1973b: 117 **E:** BU RO (Baile Herculane)
corynthius corynthius Reiche & Saulcy, 1858b: 47 (*Haltica*) **E:** AL BH BU CR GR NM **A:** CY IN TR
cuprinus Kutschera, 1862b: 108
corynthius metallescens Foudras, 1860a: 245 [= 1860b: 133] (*Teinodactyla*) **E:** CR FR IT MA SP
mediterraneus Allard, 1866: 332 (*Thyamis*)
corynthius multipunctatus Allard, 1866: 331 (*Thyamis*) **N:** AG MO TU
pennatus Normand, 1937: 134 (ab.) [IN]
danilevskyi Konstantinov, 2005: 25 **A:** KZ
doeberli Biondi & D'Alessandro, 2017c: 166 **A:** YE (Socotra)
ellipticus Reitter, 1909a: 57 **A:** KZ
emarginatus Weise, 1890b: 115 **A:** IS JO SY
excisipennis Lopatin, 1968a: 114 **A:** TD (South)
fuscoaeneus fuscoaeneus L. Redtenbacher, 1848: 535 **E:** AU BE BU CZ FR GR HU IT (North) KR LU MD NL PT RO RU (CT ST) SB SK UK **A:** AB AR CY GG IS JO SY TR
lopatini Iablokoff-Khnzorian, 1960: 71

fuscoaeneus persicus Warchałowski, 1967b: 628 (ssp.) **A:** IN TM UZ
fuscoaeneus polonicus Warchałowski, 1967b: 628 (ssp.) E: PL
hittita Biondi, 1996: 262 **A:** AR TR [# 779]
iconiensis Weise, 1900c: 137 [= 1900d: 165] **A:** TR
angorensis Gruev & Kasap, 1985: 64
igori Konstantinov, 2005: 27 **A:** TD
imitator Lopatin, 1968a: 115 **A:** KI KZ TD UZ
josiphi Konstantinov, 1994: 529 [RN] **A:** KZ
philippi Konstantinov, 1993: 43 [HN]
lederi Weise, 1889a: 3 E: RU (ST: Caucasus) **A:** GG TR
maracandicus Lopatin, 1966b: 45 **A:** KZ UZ
mohri Lopatin, 1963d: 367 **A:** AF
nadiae Konstantinov, 2005: 28 **A:** KI
nemethi Hoffmann, 1953: 188 **N:** MO
onosmae Peyerimhoff, 1912: 295 (*Thyamis*) **N:** AG MO TU **A:** TR
onosmatis Peyerimhoff, 1915a: 42 (*Thyamis*) [UE]
petitpierrei Bastazo, 1998: 32 E: SP (Cádiz)
pinguis Weise, 1888b: 935 E: BU BY CZ GE GR HU IT KR PL RO RU (ST) SB SK SL SZ UK **A:** AR GG IS JO SY TR
alpigradus Iablokoff-Khnzorian, 1962: 117
primaeveris Lopatin, 1968a: 112 **A:** KZ KI TD (South-West) UZ
pubescens Weise, 1890b: 114 E: RU (ST) **A:** AR GG
radiatus Konstantinov, 1993: 43 **A:** UZ
saulicus Gruev & Döberl, 2005: 104 [RN] **A:** IS JO
morio J. R. Sahlberg, 1913a: 79 [HN]
scaphidioides Abeille de Perrin, 1896: 53 E: IT (Sicily) **N:** AG MO TU
vaulogeri Pic, 1911a: 10
sengloki Konstantinov, 1994: 528 **A:** TD
sogdianus Lopatin, 1956a: 72 **A:** TD
tmetopterus Jacobson, 1893b: 246 **A:** AF KZ KI TM UZ
nurataicus Paliy, 1970b: 14
zeravshanicus Paliy, 1970b: 13
truncatellus Weise, 1890b: 115 E: GR RU (ST) **A:** IS JO TR
tshikatunovi Lopatin, 1966b: 45 **A:** KI KZ
turcomanorum Lopatin, 1968a: 113 **A:** TM UZ
vestumnalis Paliy, 1970b: 12
velai Bastazo, 1998: 35 E: SP (Cádiz, Malaga)
weilli Doguet, 2013: 174 **N:** LB
weisei Guillebeau, 1895b: ccclxxxvii E: AL FR GE PL RU (CT ST) SP SZ **A:** AF IN NP RU (ES WS) TD TM TR
maassi Hubenthal, 1898: 92 (var.)

genus *Longitarsus*, nomen dubium
pallens Stephens, 1831: 313 (*Thyamis*) E: GB [# 752]

genus *Lypnea* Baly, 1876c: 446 type species: *Lypnea flava* Baly, 1876
pubipennis S.-Y. Wang & Yang, 2007: 952, 954 **A:** CH (YUN)

genus *Lythraria* Bedel, 1897: 179 type species: *Galeruca salicariae* Paykull, 1800
Paraphthonomorpha Ohno, 1960b: 69 type species: *Paraphthonomorpha komiyamai* Ohno, 1960
komiyamai Ohno, 1960b: 69 (*Paraphthonomorpha*) **A:** JA
salicariae Paykull, 1800: 453 (*Galeruca*) E: AU BE BH BU BY CR CZ DE EN FI FR GB GE GR HU IT KR LA LS LT LU NL NR PL RO RU (CT NT ST) SB SK SL SP SV SZ UK **A:** AR IN IQ JA KZ MG NC RU (ES FE WS) TR **NARi**
nitidicollis Motschulsky, 1860b: 233 (*Haltica*)
picicollis Weise, 1886a: 707 (*Ochrosis*, var.)
striatella Illiger, 1807b: 159 (*Haltica*)
unicolor Reitter, 1906b: 451 (*Ochrosis*, var.)

genus *Manobia* Jacoby, 1885a: 73 type species: *Manobia nigripennis* Jacoby, 1885
aoshimensis Nakane, 1985: 3 (*Sphaeroderma*) **A:** JA (Kyushu)
bhutanensis Scherer, 1979a: 134 **A:** BT NP
bimaculata Kimoto, 1991a: 19 **A:** CH (TAI)
cheni Scherer, 1969: 106 [RN] **A:** CH (XIZ YUN) **ORR**
castanea Chen, 1939a: 44 [HN]
coomani S.-H. Chen, 1934h: 385 **A:** CH (GUX) **ORR**
fasciata L. N. Medvedev, 1998b: 209 **A:** ID (SD)
formosana Chûjô, 1936a: 23 **A:** CH (TAI)
gressitti Nakane & Kimoto, 1961c: 106 **A:** JA (Ryukyus)
hayashii Kimoto, 1991b: 123 **A:** CH (TAI)
inhumeralis Kimoto, 1997: 410 [RN] **A:** CH (TAI)
humeralis Kimoto, 1996: 48 [HN]
krishna Scherer, 1969: 106 **A:** ID (SD) NP
lewisi Jacoby, 1886a: 741 **A:** CH (TAI) JA (Kyushu, Ryukyus)
merkli L. N. Medvedev, 1998b: 207 **A:** ID (SD) NP **ORR**
nigra Scherer, 1969: 121 (*Microcrepis*) **A:** CH (YUN) ID (SD) NP
nigrita Chûjô, 1936a: 21 **A:** CH (TAI)
parvula Baly, 1874: 195 (*Crepidodera*) **A:** CH (TAI) JA NC NP SC **ORR**
piceipennis S.-H. Chen, 1934h: 384 **A:** CH (GUA JIA) **ORR**
punctata Kimoto, 1997: 411 [RN] **A:** CH (TAI)
puncticollis Kimoto, 1971a: 75 [HN] (*Manobia*)

puncticollis S.-H. Chen, 1934h: 386 **A:** CH (GUX) **ORR**
shimai Kimoto, 2001: 62 **A:** NP
shirozui Kimoto, 1996: 48 **A:** CH (TAI)
sinensis Gressitt & Kimoto, 1963a: 881 **A:** CH (CHQ FUJ HUB ZHE)
violaceipennis L. N. Medvedev, 1992d: 27 **A:** NP
yunnana L. N. Medvedev, 2015c: 328 **A:** CH (YUN)

genus *Mantura* Stephens, 1831: 322 type species: *Chrysomela rustica* Linnaeus, 1767
subgenus *Mantura* Stephens, 1831: 322 type species: *Chrysomela rustica* Linnaeus, 1767
Balanomorpha Chevrolat, 1836: 393 type species: *Haltica chrysanthemi* Koch, 1803
Cardiapus Curtis, 1833: 435 type species: *Cardiapus mathewsii* Curtis, 1833
bicoloripes S.-H. Chen, 1939a: 37 **A:** CH (GUX)
chrysanthemi chrysanthemi Koch, 1803a: 45 (*Haltica*) **E:** AU BE BY CZ DE EN FI FR GB GE HU IR IT LA LT LU NL NR PL PT RO RU (CT NT) SB SK SP SV SZ UK **N:** AG CI MO **A:** TR **AFR NARi**
crotchi Allard, 1867: 501 (*Balanomorpha*, var.)
reclairei Everts, 1921b: 24
chrysanthemi diluta Abeille de Perrin, 1895a: cdiii **E:** IT (Sardinia) **N:** MR "North Africa"
clavareaui Heikertinger, 1912d: 45 **A:** JA RU (FE: Sakhalin) SC
fulvipes Jacoby 1886a: 720 **A:** CH (TAI) JA
henoni Pic, 1909e: 138 **N:** AG TU
subelongata Pic, 1909e: 138
sparsepunctata Pic, 1911a: 9 (var.)
horioni Heikertinger, 1940: 369 [RN] **E:** FR GE
japonica Jacoby, 1886a: 720 **A:** JA (Hokkaido) RU (FE)
judaea Heikertinger, 1951: 22 **A:** IS
lutea Allard, 1859b: cv (*Balanomorpha*) **E:** FR GR (Ionian Is.) IT PT SP **A:** AG MO TU
mathewsii Curtis, 1833: 435 (*Cardiapus*) **E:** AU BE BH BU CR (Adriatic Is.) FR GB GE IT LU ME NL SK SP SZ UK (Carpathians) **A:** AB TR
aeraria Foudras, 1860b: 275 [= 1861: 19] (*Balanomorpha*)
bicolor Weise, 1886a: 743 (var.)
dichroa Bedel, 1900: 283 [RN]
helianthemi Stephens, 1839: 301 (*Cardiapus*, var.)
mesasiatica Lopatin, 1965c: 39 **A:** KI TD
nepala L. N. Medvedev, 2004d: 205 **A:** NP
nitens Allard, 1867: 501 (*Balanomorpha*) **A:** AG MO
obtusata Gyllenhal, 1813: 579 (*Haltica*) **E:** BE BY DE EN FI FR GB GE LA LS NR SB SV SZ
ambigua Kutschera, 1862: 52 (*Haltica*)

pallidicornis Waltl, 1839: 225 (*Haltica*) **E:** AU BH BU CR CZ EN FR GE GR HU IT LA LS ME NL PL RO RU (ST) SB SK SL SZ UK

subobtusata Jansson, 1947: 118

rustica Linnaeus, 1767: 595 (*Chrysomela*) **E:** AU BE BH BU BY CZ DE EN FI FR GB GE GR HU IT KR LA LT LU ME NL NM NR PL RO RU (CT ST) SB SK SL SP SV SZ TR UK **A:** AF CH (GAN XIZ) IN KI KZ MG PA RU (ES FE) SC TD TR UZ

semiaenea Fabricius, 1792b: 30 (*Galleruca*)

suturalis Weise, 1886a: 746 (var.)

suturata Fairmaire, 1873c: 359 (*Balanomorpha*) **N:** AG EG LB MO TU **A:** IS JO

limbata Allard, 1876: 24 (*Balanomorpha*)

burlinii Codina Padilla, 1961a: 51

subgenus *Stenomantura* Heikertinger, 1910b: 370 type species: *Mantura cylindrica* Miller, 1880

cylindrica Miller, 1880: 2 **E:** BU CR (South) GR IT RU (ST: Caucasus) **A:** AB SY TR

genus *Maulika* Basu & Sengupta, 1980: 143 type species: *Maulika decemmaculata* Basu & Sengupta, 1980

bengalica Lopatin, 1984b: 198 **A:** NP

cornuta L. N. Medvedev, 1999a: 156 **A:** NP

decemmaculata Basu & Sengupta, 1980: 145 **A:** ID (AP)

depressa S.-Y. Wang, 1992c: 688, 736 (*Pentamesa*) **A:** CH (SCH)

emarginata S.-H. Chen & S.-Y. Wang, 1987b: 59, 62 (*Pentamesa*) **A:** CH (XIZ) NP

genus *Meishania* Chen & S.-Y. Wang, 1980: 13, 24 type species: *Meishania rufa* S.-H. Chen & S.-Y. Wang, 1980

bhutanensis Konstantinov, Ruan & Prathapan, 2018: 190 **A:** BT

cangshanensis Konstantinov, Ruan & Prathapan, 2018: 192 **A:** CH (YUN)

flavipennis Konstantinov, Ruan & Prathapan, 2018: 194 **A:** CH (YUN)

fulvotigera Konstantinov, Ruan & Prathapan, 2018: 196 **A:** CH (SCH)

rufa S.-H. Chen & S.-Y. Wang, 1980: 13, 25 **A:** CH (SCH)

sichuanica Konstantinov, Ruan & Prathapan, 2018: 197 **A:** CH (SCH)

genus *Mellipora* Chûjô, 1965b: 101 type species: *Mellipora shirozui* Chûjô, 1965

shirozui Chûjô, 1965b: 102 **A:** CH (TAI)

genus *Micrespera* S.-H. Chen & S.-Y. Wang, 1987a: 196, 200 type species: *Micrespera castanea* S.-H. Chen & S.-Y. Wang, 1987

castanea S.-H. Chen & S.-Y. Wang, 1987a: 196, 201 **A:** CH (YUN)

genus *Microcrepis* S.-H. Chen, 1933h: 449 type species: *Microcrepis politus* S.-H. Chen, 1933

Martensomela L. N. Medvedev, 1984: 61 type species: *Martensomela aptera* L. N. Medvedev, 1984

Nepalicrepis Scherer, 1969: 119 type species: *Nepalicrepis darjeelingensis* Scherer, 1969

apterus L. N. Medvedev, 1984: 62 (*Martensomela*) **A:** NP

besucheti Scherer, 1989: 45 (*Nepalicrepis*) **A:** ID (SD) NP

brunneus L. N. Medvedev, 1990b: 36 (*Nepalicrepis*) **A:** NP

castaneus Gruev, 1990b: 63 (*Nepalicrepis*) **A:** NP

himalayensis Döberl, 1991b: 624 (*Nepalicrepis*) **A:** NP

laevigatus S.-Y. Wang & Ge, 2012b: 337, 339 **A:** CH (YUN)

loebli Scherer, 1989: 44 (*Nepalicrepis*) **A:** NP

politus S.-H. Chen, 1933h: 449 **A:** ID (SD) NP

darjeelingensis Scherer, 1969: 119 (*Nepalicrepis*)

schereri Döberl, 1991b: 625 (*Nepalicrepis*) **A:** NP

smetanai Scherer, 1989: 43 (*Nepalicrepis*) **A:** NP

genus *Minota* Kutschera, 1859: 141 type species: *Haltica obesa* Waltl, 1839

subgenus *Cardax* Weise, 1893a: 1138 type species: *Cardax stussineri* Weise, 1893

stussineri Weise, 1893a: 1140 (*Cardax*) E: BH CR ME

subgenus *Minota* Kutschera, 1859: 141 type species: *Haltica obesa* Waltl, 1839

Hypnophila Foudras, 1860b: 282 [= 1861: 26] type species: *Balanomorpha caricis* Märkel, 1847 (= *Haltica obesa* Waltl, 1839)

alpina Biondi, 1986a: 53 E: IT SZ

carpathica Heikertinger, 1912e: 60 (ssp.) E: AU BU CR CZ GE HU IT PL RO SK SL SZ UK

chinensis Döberl, 2007b: 330 **A:** CH (SCH)

halmae Apfelbeck, 1906: 249 (*Hypnophila*) E: AU BH BU CR CZ HU IT RO SK SL

himalayensis Scherer, 1989: 51 **A:** ID (SD) NP

impuncticollis Allard, 1860e: 552 (*Balanomorpha*) E: AL AU BH FR IT ME SB SP SZ

minima Heikertinger, 1912e: 60 (var.) [# 772]

nivalis Apfelbeck, 1906: 249 (*Hypnophila*)

ovoides Allard, 1859b: cvi (*Apteropeda*)

medvedevi Döberl, 2007b: 330 **A:** CH (HUB)

nepalensis Döberl, 2007b: 331 **A:** NP

nigropicea Baly, 1874: 207 (*Apteropeda*) **A:** CH (FUJ) JA

japonica Chûjô, 1951b: 41 (*Apteropeda*)

obesa Waltl, 1839: 225 (*Haltica*) E: AU BE CZ GE IT LU NL PL SB SK SL SZ

caricis Märkel, 1847: 86 (*Balanomorpha*)

schereri L. N. Medvedev, 2004c: 335 **A:** ID (SD)
sichuanica S.-H. Chen & S.-Y. Wang, 1980: 7, 20 **A:** CH (SCH)

genus *Mniophila* Stephens, 1831: 330 type species: *Haltica muscorum* Koch, 1803
bosnica Apfelbeck, 1914: 446 **E:** AL AU BH BU CR CZ FR GB GE GG IT ME NM PL RO SL
caucasica Nadein, 2009: 122 **E:** RU (ST) **A:** AB GG
muscorum Koch, 1803a: 48 (*Haltica*) **E:** AL AU BE BH BU BY CR CZ DE FR GB GE HU IR IT KR LT LU ME NL NM NR PL RO RU (ST) SB SK SL SP (Pyrenees) SV SZ UK **A:** GG [# 767]
bryozoa Gistel, 1857b: 536 (*Teinodactyla*)
fortepunctata Horion, 1939: 142 (forma) [IN] [# 770]
monticola Grimmer, 1841: 48 (*Altica*)
wroblewskii Wankowicz, 1880: 170
taurica Nadein, 2009: 120 **E:** KR
transcaucasica Nadein, 2009: 125 **A:** AR GG
turcica L. N. Medvedev, 1970a: 317 (ssp.) **A:** TR

genus *Mniophilosoma* Wollaston, 1854: 453 type species: *Mniophilosoma laeve* Wollaston, 1854
laeve Wollaston, 1854: 454 **N:** MR
obscurum Gillerfors, 1986: 7 **E:** AZ

genus *Neocrepidodera* Heikertinger, 1911e: 35 type species: *Ochrosis sibirica* Pic, 1909
Asiorestia Jacobson, 1926a: 274 type species: *Asiorestia kozhantshikovi* Jacobson, 1926
Orestioides Hatch, 1935: 276 type species: *Crepidodera robusta* LeConte, 1874
acuminata Jacoby, 1886a: 722 (*Crepidodera*) **A:** JA (Honshu, Kyushu)
adelinae Binaghi, 1947: 47 (*Crepidodera*) **E:** IT (Alps)
adelinae Heikertinger, 1948: 60 (*Crepidodera*, ssp.) [HN]
albanica Mohr, 1965: 719 (*Crepidodera*, ssp.) **E:** AL NM
arunachalensis Basu, 1991: 125 (*Asiorestia*) **A:** ID (AP)
basalis K. Daniel, 1900: 140 (*Crepidodera*) **E:** IT (Alps) SZ
bolognai Biondi, 1982a: 230 (*Crepidodera*) **N:** MO
brevicollis J. Daniel, 1904b: 249 (*Crepidodera*) **E:** AU BY CZ DE FR GE IT LA PL RU (CT) SK SL SV SZ
carolinae Baselga & Novoa, 2005: 898 **N:** MO
cheni Gressitt & Kimoto, 1963a: 762 (*Asiorestia*) **A:** CH (HUB)
chinensis Gruev, 1981d: 153 (*Ochrosis*) **A:** CH (SHA)
convexa Gressitt & Kimoto, 1963a: 763 (*Asiorestia*) **A:** CH (FUJ ZHE)

corpulenta Kutschera, 1860c: 132 (*Haltica*) E: AL BH BU CR FR IT HU ME NM RO SB UK
crassicornis Faldermann, 1837: 343 (*Crepidodera*) E: AU BH BU CR CZ EN FR GR HU IT (North) KR MD NM PT RO RU (CT ST) SB SK SL SP SZ UK A: AB GG KZ TR
cyanescens coeruleicollis Pic, 1901a: 20 (*Crepidodera*) E: FR (Alps)
cyanescens concolor K. Daniel, 1900: 140 (*Crepidodera*) E: FR IT SZ [# 750]
agrostidis Gistel, 1857b: 517 (*Aphthona*) [# 750]
cyanescens cyanescens Duftschmid, 1825: 274 (*Haltica*) E: AU CZ IT PL RO SK SL UK
alpicola F. J. Schmidt, 1850: 184 (*Crepidodera*)
fugax Gistel, 1857b: 600 (*Haltica*)
sabauda Pic, 1904e: 57 (*Crepidodera*)
subelongata Pic, 1910g: 82 (*Crepidodera*, var.)
tatrica Reitter 1914b: 111 (*Crepidodera*)
cyanipennis Kutschera, 1860c: 135 (*Haltica*) E: AU FR IT RO SL SZ
inconcolor Pic, 1905e: 99 (*Crepidodera*)
danahina Basu, 1991: 127 (*Asiorestia*) **A:** ID (HP)
femorata Gyllenhal, 1813: 559 (*Haltica*) E: AU BE BH BU BY CR (Skrad) CZ DE EN FI FR GE LT HU IT (North) NR NT PL RO RU (CT NT) SK SL SP SV SZ UK (Carpathians) **A:** RU (ES WS)
aeneipennis Weise, 1886a: 694 (*Crepidodera*, var.)
femoralis Duftschmid, 1825: 271 (*Haltica*)
infuscipes Foudras, 1860b: 326 [= 1861: 70] (*Crepidodera*, var.)
kossmanni Gerhardt, 1904: 78 (*Crepidodera*, var.)
ferruginea Scopoli, 1763: 70 (*Chrysomela*) E: AL AN AU AZ BE BH BU BY CR CZ DE EN FI FR GB GE GR HU IR IT KO KR LA LT LU MD ME NL NM NR NT PL PT RO RU (CT NT ST) SB SK SL SP SV SZ TR UK **A:** AB GG IN RU (SS) TR **NAR**i
melanaria Everts, 1919b: 174 (ab.) [IN]
similis Stephens, 1831: 303 (*Haltica*)
fulva Kimoto, 1991b: 119 **A:** CH (TAI)
gruevi Kimoto, 1983b: 49 (*Asiorestia*) **A:** JA (Honshu, Kyushu)
himalayana L. N. Medvedev & Sprecher-Uebersax, 1997: 233 (*Asiorestia*) **A:** NP
hiranoi Takizawa, 2021a: 155 (*Asiorestia*) **A:** JA
hispanica J. Daniel, 1904b: 255 (*Crepidodera*, ssp.) E: FR (Camarague) PT (Mogaduro) SP
hummeli S.-H. Chen, 1934a: 7 (*Crepidodera*) **A:** CH (AHN FUJ HUB JIA SCH XIZ)
impressa impressa Fabricius, 1801a: 496 (*Galleruca*) E: AL BH BY CR FR GB GE GR (Ionian Is., Crete, Rhodes) IT MA ME PT SB SL SP SV SZ **N:** AG MO TU **A:** CY IS SY
punctulata Allard, 1859a: c (*Crepidodera*)
rufa Küster, 1848c: 91 (*Crepidodera*)

impressa obtusangula J. Daniel, 1904b: 249 (*Crepidodera*, var.) **E**: BU GR **A**: CY TR [# 772]

interpunctata Motschulsky, 1860a: 498 (*Crepidodera*) **E**: BY DE EN FI GE KR LA LT MD NR NT PL RU (CT) SV **A**: IN JA (Hokkaido) KZ MG NC RU (ES FE SS) SC

mitsuhashii S. Matsumura, 1911b: 143 (*Crepidodera*)

irrorata L. N. Medvedev, 1997a: 19 (*Asiorestia*) **A**: NP

komatsui Nakane, 1963a: 20 (*Asiorestia*) **A**: JA (Honshu)

konstantinovi Baselga, 2006: 58 **A**: CH (HEI) RU (FE)

kozhantshikovi Jacobson, 1926a: 275 (*Asiorestia*) **A**: RU (ES)

laevicollis Jacoby, 1886a: 722 (*Crepidodera*) **A**: CH (HUB JIA SHA) JA

ligurica J. Daniel, 1904b: 285 (*Crepidodera*, ssp.) **E**: IT (Alps) FR (Alps)

manobioides S.-H. Chen, 1939a: 34 (*Crepidodera*) **A**: CH (FUJ GUX)

melanopus Kutschera, 1860c: 130 (*Haltica*) **E**: AN FR (Pyrenees) SP (Pyrenees, Cordillera Cantabríca)

melanostoma L. Redtenbacher, 1848: 529 (*Haltica*) **E**: AL AU BH CR FR GE IT LS RO SK SL SP SZ UK

frigida Weise, 1886a: 696 (*Crepidodera*)

theresae Pic, 1903i: 155 (*Crepidodera*)

minima Gressitt & Kimoto, 1963a: 766 (*Asiorestia*) **A**: CH (HUB SCH)

motschulskii Konstantinov, 1991: 143 (*Asiorestia*) **E**: AU BE BU BY CZ DE EN FI FR GE HU LA LT LU NL NR NT PL RU (CT NT ST) SK SV UK **A**: AR GG IN KI KZ RU (ES FE WS) TD TR

naini Scherer, 1969: 134 (*Asiorestia*) **A**: ID (UT) NP

nepalica L. N. Medvedev, 1990b: 32 (*Asiorestia*) **A**: NP

nigritula Gyllenhal, 1813: 557 (*Haltica*) **E**: AU BH BU BY CR CZ EN FI GE HU IT (North) LA LT NM PL RO RU (CT NT) SB SK SL SV SZ TR UK

impressa Waltl, 1839: 226 (*Haltica*)

marani Roubal, 1939: 34 (*Orestia*, ssp.)

ovulum Duftschmid, 1825: 274 (*Haltica*)

nobilis J. Daniel, 1904b: 281 (*Crepidodera*) **E**: FR (Alps) IT (Alps) SZ

interstitialis J. Daniel, 1904b: 281 (*Crepidodera*, var.) [IN] [# 771]

norica Weise, 1890f: 24 (*Crepidodera*) **E**: AU BH CR IT (North) NM RO SL

globipennis K. Daniel, 1900: 140 (*Crepidodera*)

obirensis Ganglbauer, 1897c: 571 (*Crepidodera*) **E**: AU (East Alps) SL

obscuritarsis Motschulsky, 1860a: 498 (*Crepidodera*) **A**: CH (CHQ FUJ GUI SCH SHA "Manchuria" "North-East China") JA NC RU (ES FE) SC

lewisi Jacoby, 1886a: 721 (*Crepidodera*)

oculata Gressitt & Kimoto, 1963a: 767 (*Asiorestia*) **A**: CH (YUN)

ohkawai Takizawa, 2002: 47 **A**: JA (Honshu) RU (FE) SC

peirolerii melanothorax Biondi, 1982b: 55 (*Crepidodera*, ssp.) E: IT (Apennines)
peirolerii peirolerii Kutschera, 1860c: 131 (*Haltica*) E: AU BH BU FR GE IT (North) LS NM SL SZ
peregrina Harold, 1875b: 185 (*Crepidodera*) [RN] **N:** AG MO TU
laevigata Foudras, 1860b: 331 [= 1861: 75] (*Crepidodera*) [HN]
precaria Baselga & Novoa, 2005: 904 E: SP **N:** MO
puncticollis Reitter, 1879: 55 (*Orestia*) E: RO SK UK
picea Petri, 1907: 60 (*Crepidodera*)
recticollis Jacoby, 1886a: 721 (*Crepidodera*) **A:** JA RU (FE)
resina Gressitt & Kimoto, 1963a: 769 (*Asiorestia*) **A:** CH (FUJ ZHE) SC
rhaetica Kutschera, 1860c: 133 (*Haltica*) E: AU FR GE IT SL SZ
rufoconcolor J. Daniel, 1904b: 273 (*Crepidodera*, ab.) [IN]
satoi Takizawa, 2002: 49 **A:** JA (Honshu)
schenklingi Csiki, 1939: 298 [RN] **A:** ID (SD) NP
sublaevis S.-H. Chen, 1935a: 76 (*Crepidodera*) [HN]
sibirica Pic, 1909h: 155 (*Ochrosis*) **A:** FE JA (Hokkaido) NC SC
simplicipes Kutschera, 1860c: 137 (*Haltica*) E: AU (Steiermark, Kärten)
spectabilis Döberl, 1995b: 84 (*Asiorestia*) E: IT (Monte Rosa) SZ (Monte Rosa) **[# 762]**
spectabilis J. Daniel, 1904b: 273 (*Crepidodera*, var.) [IN] [# 771]
springeri Heikertinger, 1923a: 122 (*Crepidodera*) E: AL BH ME SB
sublaevis Motschulsky, 1860a: 498 (*Crepidodera*) **A:** AF ES FE JA (Hokkaido) KI KZ MG RU (SS) TD UZ
taiwana Kimoto, 1996: 41 (*Asiorestia*) **A:** CH (TAI)
takara Nakane, 1963a: 20 **A:** JA (Ryukyus)
thoracica L. N. Medvedev, 1990b: 31 (*Asiorestia*) **A:** NP
transsilvanica Fuss, 1864: 147 (*Haltica*) E: PL RO SK UK
rufipennis J. Daniel, 1904b: 278 (*Crepidodera*, ab.) [IN]
transversa Marsham, 1802: 203 (*Chrysomela*) E: AL AU BE BH BU BY CR CZ DE EN FR GB GE GR (Cyclades) HU IR IT KR LA LS LT LU MD ME NL NM PL PT RO RU (CT ST) SB SK SL SP SV SZ TR UK **A:** AB CY IN TR
marginicollis Küster, 1848c: 92 (*Crepidodera*)
melancholica J. Daniel, 1904b: 246 (*Crepidodera*, ab.) [IN]
nigra Schilsky, 1908: 603 (var.)
scutellaris Olivier, 1808: 699 (*Altica*) [QS]
wittmeri L. N. Medvedev, 1997a: 17 (*Asiorestia*) **A:** NP

genus *Nisotra* Baly, 1864e: 437 type species: *Haltica gemella* Erichson, 1834
apicefulva Bryant, 1941b: 410 **A:** NP **ORR**
cardoni naini Scherer, 1969: 146 (ssp.) **A:** ID (UT)

chrysomeloides Jacoby, 1885a: 36 **A:** CH (SCH XIZ YUN) NP **ORR**
dohertyi Maulik, 1926: 280 (*Podagrica*) **A:** CH (YUN) NP **ORR**
gemella Erichson, 1834: 275 (*Haltica*) **A:** BT CH (CHQ FUJ GUA GUI GUX HAI HKG JIA SCH TAI XIZ) ID (SD) NP PA **ORR**
bowringii Baly, 1876d: 584
hibisci Bryant, 1941a: 286 (*Podagrica*)
javana Motschulsky, 1866b: 421 (*Sphaeroderma*)
orbiculata Motschulsky, 1866b: 421 (*Sphaeroderma*)
madurensis Jacoby, 1896c: 267 **A:** CH (FUJ) PA
nigripennis Jacoby, 1903a: 109 (*Podagrica*)
nigripes Jacoby, 1894: 293 **A:** CH (YUN) NP **ORR**
semicoerulea Jacoby, 1903a: 109 **A:** PA **ORR**
unicolor Harold, 1879c: 231 **A:** NP **ORR**
xinjiangana Y. Zhang & Yang, 2007: 844 **A:** CH (XIN)

genus *Novascuta* Özdikmen, 2008: 645 [RN] type species: *Ascuta paradoxa* L. N. Medvedev, 1997
Ascuta L. N. Medvedev, 1997g: 1218 [HN] type species: *Ascuta paradoxa* L. N. Medvedev, 1997
Ascutella L. N. Medvedev, 2009a: 58 [RN] type species: *Ascuta paradoxa* L. N. Medvedev, 1997
paradoxa L. N. Medvedev, 1997g: 1219 (*Ascuta*) **A:** NP

genus *Novofoudrasia* Jacobson, 1901b: 139 [RN] type species: *Foudrasia rufiventris* Weise, 1900
Foudrasia Weise, 1900b: 289 [HN] type species: *Foudrasia rufiventris* Weise, 1900
Griva Maulik, 1926: 241 type species: *Pseudodera cyanipennis* Jacoby, 1900
curvata P.-Y. Yu, 1997b: 907, 928 (*Griva*) **A:** CH (HUB)
nigricollis S.-H. Chen, 1939b: 57 (*Griva*) **A:** CH (SCH)
regularis S.-H. Chen, 1934h: 258 (*Griva*) **A:** BT CH (CHQ GUA HUB SCH YUN) ID (SD) **ORR**
rufiventris Weise, 1900b: 291 (*Foudrasia*) **A:** AF KZ NP TD UZ

genus *Ochrosis* Foudras, 1860b: 302 [= 1861: 46] type species: *Haltica ventralis* Illiger, 1807
ventralis Illiger, 1807a: 58 (*Haltica*) **E:** AL AU BE BH BU BY CR CZ FR GB GE GR HU IR IT KR LU MA ME NL PL PT RO RU (ST: Caucasus) SK SP SZ TR UK **N:** AG CI EG MO MR TU **A:** CY GG IN IS LE SY TR **AFR**
abdominalis Küster, 1848c: 93 (*Crepidodera*)
corcyrea Pic, 1909f: 145
corsica Allard, 1866: 184 (*Crepidodera*)
krueperi Weise, 1886a: 706

lubrica Wollaston, 1864: 406 (*Haltica*)
nigriventris Bach, 1856: 245 (*Haltica*)
obscuricollis Heikertinger, 1911d: 28 (ab.) [IN]
pisana Allard, 1862: 308 (*Crepidodera*, var.)
rubicunda Perris, 1864: 302 (*Crepidodera*, var.)
viridana Reitter, 1906b: 451 (var.)

genus *Oedionychis* Latreille, 1829: 154 type species: *Galleruca marginella* Fabricius, 1801 (= *Chrysomela cincta* Fabricius, 1787) [# 753]
Chloephaga Weise, 1899a: 138 [HN] type species: *Oedionychis cretica* Jacoby, 1886 (= *Haltica sexmaculata* Illiger, 1807)
Haplosonyx Gistel, 1848b: xiv [RN] type species: *Galleruca marginella* Fabricius, 1801 (= *Chrysomela cincta* Fabricius, 1787)
Lithonoma Chevrolat, 1836: 384 type species: *Galleruca marginella* Fabricius, 1801 (= *Chrysomela cincta* Fabricius, 1781)
Pedema Chevrolat, 1836: 384 [UN]
cincta Fabricius, 1781: 132 (*Chrysomela*) **E:** FR PT SP
dorsalis Weise, 1886a: 677 (*Lithonoma*, var.)
lusitanica Weise, 1886a: 677 (*Lithonoma*, var.)
marginella Fabricius, 1801a: 496 (*Galleruca*) [HN]
marginula Fabricius, 1803: 54 (*Galleruca*) [RN]
favieri Fairmaire, 1874a: 388 (*Lithonoma*) **N:** MO
cyanea Tournier, 1875: 475 (*Lithonoma*)
limbata limbata Fabricius, 1798: 96 (*Galleruca*) **E:** SP **N:** AG MO
africana Clark, 1861: 272 (*Lithonoma*)
andalusica Rosenhauer, 1856: 333 (*Lithonoma*)
excellens Fabricius, 1803: 54 (*Galleruca*) [RN]
schrammi Kocher, 1959: 79 (var.) [IN] [# 771]
limbata heikertingeri Csiki, 1940a: 432 [RN] **N:** MO (Moyen Atlas)
cruciata Kocher, 1959: 67 (var.) [IN] [# 770]
lateralis Bedel, 1921: 61 (*Lithonoma*) [HN]

genus *Omeiana* S.-H. Chen, 1934h: 411 [RN] type species: *Omeia rufipes* S.-H. Chen, 1934
Omeia S.-H. Chen, 1934g: 63 [HN] type species: *Omeia rufipes* S.-H. Chen, 1934
rufipes S.-H. Chen, 1934g: 65 (*Omeia*) **A:** CH (CHQ SCH)

genus *Omeisphaera* S.-H. Chen & Zia, 1974: 479, 480 type species: *Omeisphaera anticata* S.-H. Chen & Zia, 1974
anticata S.-H. Chen & Zia, 1974: 480, 481 **A:** CH (FUJ HUN SCH YUN)
flavimaculata S.-Y. Wang, 2002a: 313, 316 **A:** CH (GUI)

genus *Ophrida* Chapuis, 1875: 31 type species: *Ophrida guttata* Chapuis, 1875

Blepharoides Jacoby, 1893a: 148 type species: *Blepharoides flavitarsis* Jacoby, 1893

parva S.-H. Chen & S.-Y. Wang, 1980: 6, 19 **A:** CH (FUJ)

genus *Orestia* Chevrolat, 1836: 440 type species: *Lycoperdina alpina* Germar, 1823

Pyladia Weise, 1886a: 730 type species: *Orestia bruleriei* Allard, 1876

alpina Germar, 1823: 622 (*Lycoperdina*) E: AL AU BH CR HU IT (North) ME NM RO SB SL

bosnica Apfelbeck, 1906: 248

chionophila Apfelbeck, 1916: 385

curtula Apfelbeck, 1906: 248

hampei Miller, 1868: 32 (*Haltica*)

hungarica Heikertinger, 1910c: 43 (var.) [# 772]

styriae Gistel, 1857b: 530

apennina Weise, 1886a: 733 E: IT (Apennines)

aubei Allard, 1860a: ccxlii E: BH GR ME PL RO SB SK SL (North) UK (Carpathians)

arcuata Miller, 1868: 32 (*Haltica*)

brandstetteri Kapp, 2007: 96 E: IT

bruleriei Allard, 1876: 23 **A:** IS JO LE

brevis Pic, 1909g: 226

bulgarica Heikertinger, 1910c: 43 E: BU SB

calabra calabra Heikertinger, 1910a: 361 E: IT (Calabria)

calabra corcyrea Heikertinger, 1924b: 120 (ssp.) E: GR (Kérkyra Isl.)

carnica Leonardi, 1974: 386 (ssp.) E: IT (North)

carniolica Weise, 1886a: 734 E: IT (North) SL

carpathica Reitter, 1879a: 55 E: CZ HU PL RO SK UK

caucasica Reitter, 1880: 487 E: RU (ST) **A:** GG

pseudoaubei Pic, 1910g: 83 (var.)

coiffaiti Doguet, 1990: 112 E: FR (Corsica)

delagrangei Pic, 1909h: 155 **A:** IS JO LE SY TR

elektra brunnea Halbherr, 1898: 4 (*Hypnophyla*) E: IT (North)

elektra elektra Gredler, 1868: 78 E: IT (Alps and Prealps of Lombardia and Veneto)

heikertingeri Leonardi, 1974: 387 (ssp.) E: IT (Lombardia, Piemonte) SZ

kraatzi Allard, 1862: 312 E: CR IT (South)

semijanthina Reitter, 1909c: 88

leprieuri Allard, 1860b: cclx **N:** AG

loebli Biondi, 1992: 346 **A:** TR

maura P. H. Lucas, 1846a: pl. 22 [= 1847b: 223] (*Cryptophagus*) **N:** AG MO TU

pommereaui Perris, 1869: 33

olympica Frivaldszky, 1884: 7 **A:** TR

merkli Weise, 1886a: 732

oselliana Leonardi, 1977: 124 **A:** TR (Ilgazh Dagh)
pandellei Allard, 1863: 121 E: FR (Pyrenees)
parallela Allard, 1876: 22 **A:** JO SY
paveli Frivaldszky, 1877: 229 (*Haltica*) E: AL NM RO SB UK
punctipennis P. H. Lucas, 1847a: pl. 45 [= 1849: 545] (*Altica*) E: FR (Corsica) PT SP **N:** AG MO TU
andalusica Allard, 1869: 476
schawalleri L. N. Medvedev, 2000e: 22 **A:** ID (HP KA) NP
sierrana Heyden, 1882a: 48 E: FR (Pyrenees) PT SP
parallela Reitter, 1909b: 103 (var.) [HN]
subparallela Pic, 1909h: 155 [RN]

genus *Orhespera* S.-H. Chen & S.-Y. Wang, 1984b: 313, 319 type species: *Orhespera glabricollis* S.-H. Chen & S.-Y. Wang, 1984
fulvohirsuta S.-H. Chen & S.-Y. Wang, 1987a: 198, 200 **A:** CH (YUN)
glabricollis S.-H. Chen & S.-Y. Wang, 1984b: 313, 319 (*Hespera*) **A:** CH (YUN)
impressicollis S.-H. Chen & S.-Y. Wang, 1987a: 198, 200 **A:** CH (YUN)

genus *Paraminota* Scherer, 1989: 52 type species: *Paraminota minima* Scherer, 1989
Chabriella L. N. Medvedev, 1990b: 34 type species: *Chabriella minuta* L. N. Medvedev, 1990
Schawalleria L. N. Medvedev, 1990b: 35 type species: *Schawalleria lamprosomoides* L. N. Medvedev, 1990
lamprosomoides L. N. Medvedev, 1990b: 35 (*Schawalleria*) **A:** NP
lauribina Konstantinov, 2002: 244 **A:** NP
minima Scherer, 1989: 53 **A:** NP
minuta L. N. Medvedev, 1990b: 35 (*Chabriella*) **A:** NP

genus *Paraminotella* Döberl & Konstantinov, 2003: 205 type species: *Paraminota nepalensis* Döberl, 1991
nepalensis Döberl, 1991b: 629 (*Paraminota*) **A:** NP
nigrita Döberl & Konstantinov, 2003: 214 **A:** NP

genus *Parargopus* S.-H. Chen, 1939b: 65 type species: *Parargopus sphaerodermoides* S.-H. Chen, 1939
sphaerodermoides S.-H. Chen, 1939b: 66 **A:** CH (HEB)

genus *Parategyrius* Kimoto & Gressitt, 1966: 552 type species: *Parategyrius unicolor* Kimoto & Gressitt, 1966
Lankanella Kimoto, 2000a: 26 type species: *Lankanella thailandica* Kimoto, 2000
Lankaphthona L. N. Medvedev, 2001: 162 type species: *Lankaphthona micheli* Medvedev, 2001

Longitarsella L. N. Medvedev, 2009e: 202 type species: *Thyamis binotata* Baly, 1876 sensu Medvedev (2009b) (= *Philotarsa laosica* Medvedev, 2009)
Philotarsa L. N. Medvedev, 2009e: 147 type species: *Philotarsa laosica* Medvedev, 2009
binotata Baly, 1876d: 583 (*Thyamis*) **A:** CH (FUJ JIA SHG)
nigronotata Jacoby, 1896c: 259 (*Longitarsus*) **A:** CH (YUN) **ORR**
unicolor Kimoto & Gressitt, 1966: 552 **A:** JA (Ryukyus)
yunnantarsella Ruan, Konstantinov & Prathapan, 2019: 50 (*Lankaphthona*) **A:** CH (YUN)

genus *Parathrylea* Duvivier, 1892b: 420 type species: *Parathrylea apicipennis* Duvivier, 1892
apicipennis Duvivier, 1892b: 421 **A:** CH (GUA XIZ) ID (SD UT) NP **ORR**
rectimarginata S.-Y. Wang, 1993: 320, 328 **A:** CH (HUN)
tuckuchensis Kimoto & Takizawa, 1973: 177 (*Acrocrypta*) **A:** NP

genus *Parazipangia* Ohno, 1965b: 101 type species: *Parazipangia okiana* Ohno, 1965
okiana Ohno, 1965b: 102 **A:** JA (Honshu)

genus *Penghou* Ruan, Konstantinov, Prathapan, Ge & Yang, 2015: 301 type species: *Penghou yulongshan* Ruan, Konstantinov, Prathapan, Ge & Yang, 2015
yulongshan Ruan, Konstantinov, Prathapan, Ge & Yang, 2015: 305 **A:** CH (YUN)

genus *Pentamesa* Harold, 1876a: 124 type species: *Pentamesa duodecimmaculata* Harold, 1876
anemoneae anemoneae S.-H. Chen & Zia, 1966b: 148, 150 **A:** CH (SCH QIN XIZ)
anemoneae canaliculata S.-Y. Wang, 1992c: 668, 736 (ssp.) **A:** CH (SCH)
cribellata Weise, 1895e: 335 **A:** ID (SD)
duodecimmaculata Harold, 1876a: 124 **A:** ID (HP SD UT) NP PA RU (ES) TD **ORR**
generosa Weise, 1895e: 332
duodecimpunctata Motschulsky, 1861b: 41 (*Colaphus*) **A:** MG RU (ES FE)
tredecimpunctata Jacobson, 1901a: 93
fulva S.-Y. Wang, 1992c: 688, 736 **A:** CH (SCH)
gonggana S.-Y. Wang, 1992c: 689, 737 **A:** CH (SCH)
guttata Weise, 1895e: 334 **A:** BT ID (SD) NP
trigrapha Maulik, 1926: 292
guttipennis S.-H. Chen & S.-Y. Wang, 1980: 11, 23 **A:** CH (SCH)
haroldi Baly, 1876c: 439 (*Argopus*) **A:** CH (XIZ) ID (KA SD UT) NP
inornata S.-H. Chen & S.-Y. Wang, 1980: 11, 23 **A:** CH (SCH)
kondarensis arnoldi Lopatin, 1996: 641 (ssp.) **A:** UZ

kondarensis kondarensis Lopatin, 1956a: 71 **A:** TD UZ
laevicollis S.-H. Chen, 1935a: 79 **A:** ID (SD)
nigrofasciata S.-H. Chen, 1933a: 238 **A:** CH (YUN) **ORR**
parva S.-H. Chen & S.-Y. Wang, 1981a: 497, 507 **A:** CH (XIZ)
scripta L. N. Medvedev, 1997e: 263 **A:** NP
sophiae An, 1988: 44 **A:** SC
subfasciata Weise, 1895e: 333 **A:** ID (HP KA UT) NP
trifasciata S.-H. Chen, 1935d: 773 **A:** CH (GAN HUB SCH SHA YUN)
xiangchengana S.-Y. Wang, 1992c: 690, 737 **A:** CH (SCH YUN)

genus *Philopona* Weise, 1903d: 216 type species: *Oedionychis vernicata* Gerstaecker, 1871
mouhoti Baly, 1878d: 316 (*Oedionychis*) **A:** CH (HAI) **ORR**
pseudomouhoti D.-Y. Ge, S.-Y. Wang & Yang, 2009: 28 **A:** CH (HAI)
vibex Erichson, 1834: 273 (*Haltica*) **A:** CH (CHQ FUJ GUA GUX HUB NMO SCH TAI ZHE) JA NC RU (ES FE) SC **ORR**
japonica Baly, 1874: 189 (*Aedionychis*)
nigra Chûjô, 1951b: 44 (var.)
signata Duvivier, 1892b: 429 (*Hyphasis*)
zangana S.-H. Chen & S.-Y. Wang, 1987b: 58, 62 **A:** CH (XIZ)

genus *Phygasia* Chevrolat, 1836: 387 [NP] type species: *Altica unicolor* Olivier, 1808 (= *Haltica silacea* Illiger, 1807)
Aldrisma Fairmaire, 1888b: 156 type species: *Aldrisma externecostata* Fairmaire, 1888 (= *Graptodera fulvipennis* Baly, 1874)
Isosoma Billberg, 1820: 56 [NO] type species: *Chrysomela concolor* Fabricius, 1781
Scallodera Harold, 1877a: 365 type species: *Graptodera fulvipennis* Baly, 1874
africana Chapuis, 1879: 16 (*Lactica*) **A:** OM **AFR**
carinipennis S.-H. Chen & S.-Y. Wang, 1980: 11, 23 **A:** CH (GUX YUN)
chengi C.-F. Lee, 2012: 28 **A:** CH (TAI)
cobosi Jind. Madar & Jiř. Madar, 1965: 10 **N:** AG
diancangana S.-Y. Wang, 1992c: 722, 752 **A:** CH (SCH YUN)
wittmeri L. N. Medvedev, 1995b: 485
diluta Chûjô, 1963b: 400 (ssp.) **A:** CH (TAI)
taiwanensis D.-Y. Ge, S.-Y. Wang & Yang, 2010: 325
dorsata Baly, 1878g: 231 **A:** CH (GUI GUX HKG SCH YUN) ID (SD) **ORR**
subfasciata S.-H. Chen, 1933a: 236 (var.)
eschatia Gressitt & Kimoto, 1963a: 883 **A:** CH (CHQ HUN SCH XIZ YUN)
foveolata S.-Y. Wang, 1992c: 723, 752 **A:** CH (GAN HEN HUB HUN SCH)

fulvipennis Baly, 1874: 193 (*Graptodera*) **A:** CH (BEI CHQ HEB HEI HUB HUN JIA JIL JIX LIA SCH YUN ZHE) JA NC RU (FE) SC
externecostata Fairmaire, 1888b: 156 (*Aldrisma*)
gracilicornis S.-Y. Wang & Yang, 2008b: 557 **A:** CH (GUA GUI SCH)
heikertingeri Peyerimhoff, 1929: 142 **N:** AG
hookeri Baly, 1876c: 445 **A:** BT CH (GUI XIZ YUN) ID (SD) NP
media S.-H. Chen & S.-Y. Wang, 1980: 12, 24 **A:** CH (YUN)
nigricollis S.-Y. Wang & Yang, 2008b: 558 **A:** CH (GUA HAI)
ornata Baly, 1876c: 445 **A:** CH (AHN CHQ FUJ GAN GUA GUI GUX HAI HEN HKG HUB HUN JIA JIX SCH YUN ZHE) **ORR**
bipustulata Jacoby, 1892b: 919 (*Lactica*)
pallidipennis S.-H. Chen & S.-Y. Wang, 1980: 12, 24 **A:** CH (YUN)
parva S.-Y. Wang & Yang, 2008b: 559 **A:** CH (SCH ZHE)
potanini Lopatin, 1995a: 103 **A:** CH (SCH)
potanini L. N. Medvedev, 1995b: 484 [HN]
pseudomedia S.-Y. Wang & Yang, 2008b: 561 **A:** CH (GUI)
pseudornata S.-Y. Wang & Yang, 2008b: 562 **A:** CH (HAI)
quadriplagiata Scherer, 1969: 114 **A:** ID (UT) NP
ruficollis S.-Y. Wang, 1993: 325, 329 **A:** CH (GAN HUB HUN SHA)
simidorsata S.-Y. Wang & Yang, 2008b: 562 **A:** CH (HAI)
silacea Illiger, 1807b: 121 (*Haltica*) **A:** CH (SHX) ID (UT) NP
indica Jacoby, 1898a: 187
unicolor Olivier, 1808: 699 (*Altica*)
sulphuripennis Jacoby, 1899d: 339 **A:** YE **AFR**
suturalis S.-Y. Wang & Yang, 2008b: 563 **A:** CH (GUA)
tricolora L. N. Medvedev, 1995b: 485 **A:** CH (SCH)
yunnana S.-Y. Wang & Yang, 2008b: 564 **A:** CH (YUN)

genus *Phyllotreta* Chevrolat, 1836: 391 type species: *Chrysomela nemorum* Linnaeus, 1758
Letzuana S.-H. Chen, 1934h: 340 type species: *Letzuana depressa* S.-H. Chen, 1934
Orchestris Kirby, 1837: 217 type species: *Chrysomela nemorum* Linnaeus, 1758
Tanygaster Blatchley, 1921: 26 type species: *Tanygaster ovalis* Blatchley, 1921
acutecarinata Heikertinger, 1941b: 81 E: AU GR (Thessaloniki) PL RU (ST) SK UK **A:** AF IN TR
andreevae Lopatin, 1992c: 264 **A:** TD
annae Konstantinov, 1992a: 265 **A:** KZ
aptera S.-Y. Wang, 1992c: 713, 746 **A:** CH (YUN)
araxicola Iablokoff-Khnzorian, 1968: 273 **A:** AB AR TR

armoraciae Koch, 1803a: 75 (*Haltica*) E: AU BU BY CR (North) CZ DE EN FI FR (East) GE HU IT (North) KR LA LT LU MD NL PL RO RU (CT ST) SB SK SL SV SZ UK **A:** AB KZ RU (ES SS WS) "Turkestan" **NARi**

biplagiata Chittenden, 1927: 16 (var.)

astrachanica Lopatin, 1977d: 32 E: AU BE BU BY CZ FR GB GE GR HU IT LA LU MD ME NL PL RU (CT ST) SB SK SL SP SV SZ UK **A:** AB CY GG IN KZ RU (WS) TR UZ

asturica Heikertinger, 1941a: 42 E: SP

atra Fabricius, 1775: 115 (*Altica*) E: AL AU BE BH BU BY CR CZ DE EN FI FR GB GE GR HU IT KO KR LA LT LU MD ME NL NM NR PL RO RU (CT NT ST) SB SK SL SP SV SZ TR UK **N:** AG MO **A:** AB AF AR GG IN JO KI KZ MG RU (ES FE SS WS) SY TD TR YE

aethiopissa Schrank, 1789: 69 (*Altica*)

aterrima Schrank, 1798: 561 (*Altica*)

melaena Illiger, 1807a: 60 (*Haltica*)

pulex Schrank, 1781a: 85 (*Chrysomela*)

austriaca aligera Heikertinger, 1912b: 164 **A:** CH (GAN) NC RU (ES FE)

austriaca austriaca Heikertinger, 1909a: 9 **E:** AU CZ HU RO RU (ST) SK UK **A:** RU (ES FE)

aygulae Özdikmen, Coral Şahin & Bal, 2017: 147 **A:** TR

bactriana Heikertinger, 1941b: 108 **A:** TD UZ "Turkestan"

balcanica Heikertinger, 1909b: 292 **E:** AL AU BH BU CR FR GR HU IT ME NM KR RO RU (ST) SB SL SP UK **A:** AB AF IN KI KZ TR UZ

adanensis Pic, 1910c: 25

anatolica Pic 1910f: 306 (var.)

banghaasi Heikertinger, 1941b: 96 **A:** KI KZ TM UZ

bartanga Lopatin, 1966b: 47 **A:** TD

basalis Brisout de Barneville, 1884: 90 **N:** AG

beschkentica Lopatin, 1961c: 103 **A:** IN TD

bilgeae Özdikmen & Coral Şahin, 2017: 175 **A:** TR

bolognai Biondi, 1992: 349 **A:** TR

buchtarmensis Lopatin, 1990c: 57 **A:** KZ

buettikeri Doguet, 1979a: 310 **N:** AG **A:** IS SA

buhseae Iablokoff-Khnzorian, 1978: 121 **A:** TM

bulgarica Gruev, 1977: 67 E: BU GR **A:** IS TR

caucasicola Heikertinger, 1941b: 82 **E:** RU (ST: Caucasus) **A:** AR IN IQ TR

cheiranthi Weise, 1903d: 218 **N:** EG **A:** AE OM SA YE (incl. Socotra) **ORR**

hargreavesi Bryant, 1930: 362

chotanica Duvivier, 1892b: 426 **A:** CH (TAI YUN XIZ) ID (KA SD UT) NP PA **ORR**

christinae Heikertinger, 1941a: 27 **E:** AU BH BU CR CZ FR GE HU IT (North) LU NL PL RO RU (CT) SB SK SL SZ UK
chujoe Jind. Madar, 1959: 3 **A:** JA (Hokkaido, Honshu) RU (FE)
brevistriata Kimoto, 1966a: 606
cleomica Furth, 1979: 21 **A:** IS SI
coiffaiti Doguet, 1979b: 50 **A:** LE
consobrina consobrina Curtis, 1837: 630 (*Altica*) **E:** BE CR FR GB GE IT LU MA NL PT SP SZ **N:** AG MO MR TU **AFR**
hintoni Donisthorpe, 1944: 227
consobrina springeri Wittmer, 1936: 16 **A:** IS JO LE SI
corrugata Reiche & Saulcy, 1858b: 46 (*Haltica*) **E:** BU FR GB GR IT KR MA RU (ST) SP UK **N:** AG EG LB MO TU **A:** AB AF AR CY IN IQ IS JO KZ SY TR TM UZ
beauprei Pic, 1909h: 156 (var.)
galloprovincialis Caillol, 1925: 540
discophora Iablokoff-Khnzorian, 1961: 152
bella Paliy, 1970b: 3
crassicornis Allard, 1866: 255 **E:** BE FR IT LU NL SP **N:** AG MO
distinctipes Pic, 1906c: 35
cruciferae Goeze, 1777: 312 (*Chrysomela*) **E:** AL AU BE BH BU BY CR CZ DE FR GB GE GR HU IT KO KR LA LS LT LU MA MD ME NL NM PL PT RO RU (CT ST) SB SK SL SP SV SZ TR UK **N:** AG EG MO TU **A:** AB AF CY GG ID ("India") IN IS JA (Kyushu) JO KI KZ MG PA RU (WS) SY TD TR **AFR NARi**
brassicae Geoffroy, 1785: 98 (*Altica*)
colorea Foudras, 1860a: 370 [= 1860b: 258]
columbiana Chittenden, 1927: 46
elongata Stephens, 1835: 423 (*Haltica*)
hortensis Olivier, 1789: 108 (*Altica*)
nigroaenea Stephens, 1831: 298 (*Haltica*)
obscurella Illiger, 1807b: 154 (*Haltica*)
poeciloceras Comolli, 1837: 48 (*Altica*)
punctipennis Weidenbach, 1859: 85 (*Haltica*)
cruralis Abeille de Perrin, 1895a: cdiii **E:** SP **N:** AG TU **A:** IS JO SI
cupreata S.-H. Chen & Kung, 1955: 321, 325 **A:** CH (GAN)
curvipes Heikertinger, 1941b: 108 **A:** KZ
dacica Heikertinger, 1941b: 84 **E:** BH BU CR KO SB RO UK **A:** AB TR
diademata Foudras, 1860a: 369 [= 1860b: 257] **E:** AL AU BE BH BU BY CR CZ FR GE GR HU IT KR LU MD ME NL NM PL RO RU (CT ST) SB SK SL SP SZ TR UK **A:** AB AF GG IN IQ SD SU TR "Turkestan"
dilatata C. G. Thomson, 1866: 192 **E:** AU BE BY CZ DE EN FR GE LA LT LU NL PL RU (CT NT ST) SB SK SV SZ UK **A:** RU (ES)

djurdjurensis Doguet, 1977: 185 **N:** AG
downesi Baly, 1877g: 300 **A:** CH (TAI) **ORR**
egridirensis Gruev & Kasap, 1985: 60 **A:** IN TR
erysimi erysimi Weise, 1900c: 138 [= 1900d: 166] **E:** BU GR KR MD ME NM RO RU (ST) SB TR UK **A:** AB AF GG IN IS KI KZ MG RU (ES FE) SY TD TR TM UZ
baicalica Heikertinger, 1941a: 37 (ssp.)
kutscherai Heikertinger, 1941a: 38 (ssp.)
erysimi iranella Lopatin, 1990b: 600 (ssp.) **A:** IN
erysimi tekensis Lopatin, 1992c: 261 (ssp.) **A:** TM
exclamationis Thunberg, 1784: 14 (*Chrysomela*) **E:** AU BE BH BU BY CR CZ DE EN FI FR GE HU IT LA LS LT LU ME NL NR PL RO RU (CT ST) SK SL SP SV SZ UK **A:** AB RU (ES FE WS)
brassicae Fabricius, 1787: 78 (*Chrysomela*) [HN]
quadriguttata Stephens, 1831: 299 (*Haltica*)
vibex Weise, 1888b: 866 (var.)
ezoensis Kimoto, 1993: 97 **A:** JA (Hokkaido) RU (FE)
fallaciosa Heikertinger, 1941b: 77 **E:** FR GR (Rhodes) IT SP **N:** AG MO **A:** CY IN IS JO SI TR YE [# 780]
farsetiarum Peyerimhoff, 1929: 143 **A:** AG **AFR**
flavifrons Jacoby, 1899a: 529 **A:** YE **AFR**
flavoguttata Kutschera, 1860d: 207 (*Haltica*) **E:** AL GR **A:** IS
flexuosa Illiger, 1794: 601 (*Altica*) **E:** AU BE BH BU BY CZ DE EN FI FR GB GE GR HU IR IT (North) KR LA LT LU NL NR PL RO RU (CT NT ST) SK SL SV SZ UK **A:** AB GG RU (ES FE)
fallax Allard, 1861: 834
fenestrata Weise, 1888b: 870 (var.)
sinuata Stephens, 1831: 297 (*Haltica*)
florieni Pic, 1910c: 25 **N:** EG **A:** AE IS SA
fornuseki Čížek, 2003: 63 **E:** CZ GR SK **A:** TR
foudrasi Brisout de Barneville, 1873a: 5 **E:** FR IT (North-West) PT SP **N:** AG MO TU
dentifera Guillebeau, 1885: 3
saadana Pic, 1910b: 18 (var.)
furthi Doguet, 1984: 258 **A:** AG TU
gafsana Normand, 1937: 129 **N:** TU
gallica Brisout de Barneville, 1891: clxxxv **E:** FR SP SZ
ganglbaueri Heikertinger, 1909b: 290 **E:** AU BH BU CR CZ FR (Alpes-Maritimes) IT KR ME PT RO SB SL SP SZ **A:** IN TR
gillerforsi Biondi, 1991: 34 **N:** CI (Gran Canaria)
gloriae Biondi, 1994: 111 **E:** SP
gurskii Lopatin, 1966b: 47 **A:** TD

hamata J. Park, 2013b: 121 **A:** SC
hebraea Heikertinger, 1941b: 106 **A:** IS
hemipoda Abeille de Perrin, 1909: 180 **E:** PT **N:** AG MO TU
hermonensis Furth, 1979: 23 **A:** IN IS
hispanica Pic, 1903c: 115 **E:** SP (South)
hochetlingeri Fleischer, 1917: 17 **E:** CR PL SK
extincta Roubal, 1923: 65 (ab.) [IN]
humilis Weise, 1887h: 198 **A:** CH (GAN HEI NIN SHN) MG NC RU (ES FE SS) SC
iberica Heikertinger, 1912b: 162 **E:** SP (North-West)
insularis Heikertinger, 1942: 145 (ssp.) **A:** CH (TAI)
judaea Pic, 1901b: 27 **E:** BU SK **A:** AR IN IS JO TR
koltzei Weise, 1887h: 197 **A:** NC RU (FE)
konevi Lopatin, 1985a: 48 **A:** KZ (Barsakelmes Isl. in Aral Sea)
krali Lopatin, 1990b: 598 **A:** IN
kurbatovi L. N. Medvedev, 2010d: 485 **A:** RU (FE)
lacerta Heikertinger, 1941b: 98 **E:** SP (North-West)
lativittata Kutschera, 1860e: 307 (*Haltica*) **E:** GR IT MA RU (ST) **A:** AB AF AR CH (XIN) CY GG IN IQ IS JO KI KZ LE OM SY TD TR TM UZ
bisbinotata Pic, 1909h: 156 (var.)
ininterrupta Pic, 1909h: 156 (var.)
iris Reitter, 1891: 35
orientalis Baly, 1877d: 178
rubrithorax Pic, 1900f: 80
ruficollis Weise, 1888b: 859
lijiangana S.-Y. Wang, 1992c: 713, 746 **A:** CH (YUN)
lopatini Konstantinov, 1992a: 266 [= 1992b: 38] **A:** AB
lorestanica Warchałowski, 1973b: 664 (ssp.) **A:** IN TR
lubischevi Lopatin, 1992c: 265 **A:** KI
maculicornis Pic, 1906c: 35 **A:** SY TR
ispartaensis Gök, 2005a: 2
maculosa Iablokoff-Khnzorian, 1968: 271 **E:** AR **A:** IN
mashonana Jacoby, 1897c: 554 **A:** SA YE **AFR**
usambarica Weise, 1903d: 217
melichari Heikertinger, 1941b: 98 **E:** SP
misella Jacobson, 1901b: 141 **E:** RU (CT) **A:** CH (HEI) MG RU (ES FE)
mollis Konstantinov, 1992a: 263 **A:** UZ
mongolica L. N. Medvedev, 1980a: 319 **A:** MG
muehlei Döberl, 2010c: 50 **A:** IN
nemorum Linnaeus, 1758: 373 (*Chrysomela*) **E:** AL AU BE BH BU BY CR CZ DE EN FI FR GB GE GR HU IT KR LA LS LT LU MD ME NL NM NR PL RO RU (CT NT ST)

SB SK SL SP SV SZ TR UK WS **A:** AB CY GG IN IS KZ MG RU (ES FE SS) SC TD TR UZ **AUSi**

theresae Pic, 1909i: 157 (var.)

kuennemanni Reitter, 1905b: 251

nigripes impennis Peyerimhoff, 1919: 228 (ssp.) **N:** AG

nigripes nigripes Fabricius, 1775: 113 (*Altica*) **E:** AL AN AU BE BH BU BY CR CZ DE EN FI (South) FR GB GE GR HU IT KO KR LA LT LU MD ME NL NM PL RO RU (CT NT ST) SB SK SL SP SV SZ TR UK **N:** AG CI EG MO TU **A:** AB AF CY GG IN IS KI KZ LE SI SY TD TR UZ

arabidis Hoffmann, 1953: 187 (var.)

cruciferarum Gmelin, 1790: 1699 (*Chrysomela*) [QS]

lens Thunberg, 1784: 13 (*Chrysomela*)

lepidii Koch, 1803a: 64 (*Haltica*)

nitidicollis Weise, 1888b: 863 **E:** IT (Sicily) **N:** AG LB MO TU

alutacea Normand, 1937: 131 (ab.) [IN]

distincta Monnot, 1914: 170

nodicornis Marsham, 1802: 204 (*Chrysomela*) **E:** AU BE BH BU CR CZ FR GB GE HU IT KR LU MD NL PL PT RO RU (CT ST) SB SK SL SP SZ UK **A:** AB ID (UT) TR

antennata Koch, 1803a: 67 (*Haltica*)

oncera Maulik, 1926: 378

ochripes Curtis, 1837b: 630 (*Altica*) **E:** AU BE BH BU BY CR CZ DE FI FR GB GE GR HU IT KR LA LT LU MD ME NL NR PL PT RO RU (CT NT ST) SB SK SL SP SV SZ TR UK **A:** AB GG IN JA (Honshu, Kyushu) RU (ES FE WS) TR

burdigalensis Pic, 1909k: 178 (var.)

caucasica Harold, 1879c: 231 [= 1880b: 486] [QS]

comanensis Pic, 1912c: 136 (var.)

cruciata Weise, 1888b: 867 (var.)

excisa L. Redtenbacher, 1848: 532 (*Haltica*)

eximia Guillebeau, 1895a: xxvii [QS]

horticola Richter, 1820: 8 (*Haltica*) [QS]

kerkhoveni Everts, 1919b: 175 (ab.) [IN]

ogloblini Shapiro, 1960: 188 **A:** UK

oltuensis Gruev & Aslan, 1998: 166 **A:** TR (Erzurum)

ozbeki Gruev & Aslan, 1998: 167 **A:** IN TR (Bayburt)

pallidipennis Reitter, 1891: 34 **E:** KR KZ RU (CT ST) UK **A:** AF CH (XIZ) IN KI KZ MG PA RU (ES SS WS) TD TM TR UZ

dilutipennis Reitter, 1891: 35

schreineri Jacobson, 1915a: 72

turcmenica Weise, 1900c: 138

panousei Hoffmann, 1953: 186 **A:** MO

paradoxa Lopatin, 1956c: 48 **E:** RU (ST) **A:** AF TD
parallela Boieldieu, 1859: 476 (*Haltica*) E: FR IT PT SP **N:** AG MO TU
humeralis Foudras, 1860a: 347 [= 1860b: 235]
parfentjevi Shapiro, 1958: 432 **A:** IN KZ TM UZ
peyerimhoffi Heikertinger, 1941b: 107 **A:** AE IS JO OM SA SI **[Appendix]**
pontoaegeica Gruev, 1982a: 99 **E:** BU GR NM **A:** TR
praticola Weise, 1887d: 333 **E:** KR RU (CT ST) **A:** AB AF CH ("North-west China") ID ("India") IN KI KZ MG RU (ES SS WS) TD TR UZ
fucata Weise, 1890b: 112
procera L. Redtenbacher, 1848: 530 (*Haltica*) E: AN AU BE BH BU CR CZ FR GB GE GR HU IT KO KR LU MA MD NL NM PT RO RU (CT ST) SB SK SL SP SV SZ UK **N:** AG CI EG MO MR SG TU **A:** AB CY GG IN IS JO SI TD TR TM UZ YE (incl. Socotra) **AFR** [# 781]
carreti Monnot, 1913: 233
procerior Peyerimhoff, 1941: 15 (ssp.)
subtilis Wollaston, 1854: 441 (*Haltica*)
pseudoexclamationis Konstantinov, 1992a: 262 **A:** GG
punctulata Marsham, 1802: 200 (*Chrysomela*) **E:** AU BE BH BU CR CZ FR GB GE GR HU IT KR LU MD ME NL NM PL PT RO RU (ST) SB SK SL SP (incl. Balearic Islands) SZ TR UK **N:** MO **A:** AB IS TR **NAR**i
aerea Allard, 1859a: c
nigroaenea Marsham, 1802: 197 (*Chrysomela*) [QS]
randoniae Peyerimhoff, 1920: 42 **E:** SP **N:** AG MO
rapillyi Doguet, 1984: 255 **N:** AG
rectilineata S.-H. Chen, 1939a: 50 **A:** CH (FUJ GAN GUX HEI HUB JIA NIN) JA (Kyushu) NC RU (FE) SC **ORR**
chinensis Heikertinger, 1941a: 27
reitteri Heikertinger, 1911f: 159 **E**i: KR RU (ST) **A:** KZ TRi UZ
rufitarsis Allard, 1859a: c **E:** IT (Sicily) SP **N:** AG CI MO TU **A:** IS JO SI **AFR**
rugifrons Küster, 1849b: 95 **E:** BE CR FR GE IT MA SP **N:** AG MO TU **A:** IS SY
biguttata Foudras, 1860a: 363 [= 1860b: 251]
bimaculata Allard, 1859a: c
sachalinensis L. N. Medvedev, 1973b: 883 **A:** RU (FE)
scheuchi Heikertinger, 1941b: 90 **E:** AU BU CZ GB GE HU KR SK SP UK **A:** MG
schuelkei Döberl, 2011: 281 **A:** CH (SHA)
shirahatai Jind. Madar, 1959: 5 (forma) **A:** JA (Honshu) [# 772]
sholaksori Konstantinov & Moseyko, 2019: 500 **E:** RU (ST) **A:** RU (WS)
sisymbrii Weise, 1888b: 860 **E:** RU (ST: Caucasus) **A:** AB AR GG IN SY TR
stehliki Jind. Madar & Jiř. Madar, 1969: 141 **A:** AF

striolata Fabricius, 1803: 38 (*Crioceris*) [RN] **E:** AL AU BE BH BU BY CR CZ DE EN FI FR GB GE GR HU IT KO KR LA LS LT LU MD ME NL NM NR PL PT RO RU (CT NT ST) SB SK SL SP SV SZ TR UK **A:** AB AR CH (AHN CHQ FUJ GAN GUA GUI GUX HAI HEI HKG HUB JIA NIN SHA SCH TAI XIZ YUN ZHE) GG ID (SD) IN JA KZ MG NC NP RU (FE SS) SC TR **AFRi AUS NARi ORR**

atrivitta Chittenden, 1927: 26 (var.)

discedens Weise, 1888b: 871 (var.)

lineolata Chittenden, 1927: 25 (var.)

monticola Weise, 1888b: 871 (var.)

sinuata L. Redtenbacher, 1848: 532 (*Haltica*)

sinuata Gistel, 1857b: 525, 564 [HN]

strigula Montrouzier, 1865: 202 (*Altica*)

vernicosa Chittenden, 1927: 25 (var.)

vittata Fabricius, 1801a: 469 (*Crioceris*) [HN]

subtilior Csiki, 1939: 66 [RN] **A:** ID (HP)

subtilis Weise, 1922b: 122 [HN]

talassicola Heikertinger, 1944: 124 **A:** AF KI KZ TD UZ

temperei Doguet, 1974: 121 **E:** AN FR SP

tenuimarginata Jacoby, 1899a: 528 **A:** AE IN YE **AFR**

tetrastigma Comolli, 1837: 47 (*Altica*) **E:** AU BE BH BU BY CR CZ DE EN FI FR GB GE HU IT (North) KR LA LS LT LU NL NR PL RO RU (CT NT) SK SL SP SV SZ UK **A:** RU (ES SS WS) TR

toelgi Heikertinger, 1941b: 105 **A:** TR

tomboi Lopatin, 1967d: 169 **A:** MG

brachyptera L. N. Medvedev, 1977a: 351

tunisea Pic, 1909f: 145 **N:** AG MO TU

ubsunurica L. N. Medvedev, 1980a: 319 **A:** MG

undulata Kutschera, 1860e: 301 (*Haltica*) [conserved name, ICZN (2015a), Opinion 2351] **E:** AL AU BE BH BU BY CR CZ DE EN FI FR GB GE GR HU IT KO KR LA LS LT LU MD ME NL NM NR PL PT RO RU (CT NT ST) SB SK SL SP SV SZ TR UK **N:** AG **A:** AB AR KI KZ MG RU (ES FE SS WS) TR TM UZ **AUSi NARi**

australis Blackburn, 1890b: 146

biinterrupta Pic, 1915f: 33 (var.)

bivittata Waterhouse, 1838: 133 (*Haltica*) [suppressed name, ICZN (2015a), Opinion 2351]

blackburni Bryant, 1925: 596

evanescens Döberl, 1994a: 182 (forma) [IN]

nigroconcolor Döberl, 1994a: 182 (forma) [IN]

vittigera Broun, 1893: 1392

ustulata iraqensis Gruev, 1998: 40 (ssp.) **A:** IQ

ustulata ustulata Lopatin, 1961b: 144 **A:** KZ TD UZ

variipennis aegyptiaca Pic, 1915a: 3 (var.) **N:** AG CI EG LB **A:** SA **AFR**

variipennis variipennis Boieldieu, 1859: 477 (*Haltica*) **E:** BH BU CR FR GR IT MA ME NM PT RO SL SP SZ TR **N:** AG CI (Tenerife) MO TU **A:** CY IN IQ IS SI TR

varians Foudras, 1860a: 360 [= 1860b: 248]

vilis Weise, 1888b: 861 **E:** BU CR GR IT (South) NM **A:** TR

seidlitzi Weise, 1888b: 861

vittula L. Redtenbacher, 1848: 532 (*Haltica*) **E:** AL AU BE BH BU BY CR CZ DE EN FI FR GB GE GR HU IR IT KO KR LA LT LU MD ME NL NM NR PL RO RU (CT NT ST) SB SK SL SP SV SZ TR UK **A:** AB AF CH (GAN HEI NIN) GG IN KI KZ MG NC RU (ES FE SS WS) TD TR **NAR**i

exigua Heikertinger, 1911f: 160

rivularis Motschulsky, 1849: 147

weiseana Jacobson, 1901b: 141 [RN] **E:** KR RU (CT ST) UK **A:** AB IN KZ TR

denticornis Weise, 1890b: 112 [HN]

yoffei Furth, 1979: 24 **A:** IS

yunnanica S.-H. Chen, 1933a: 251 **A:** CH (GAN SCH YUN)

zerchei Döberl, 1998: 41 **E:** GR (Peloponnese)

ziegleri Lohse, 1980: 155 **E:** AU (East Alps)

zimmermanni Crotch, 1873: 66 (*Orchestris*) **E:** FI RU (NT) SV **A:** RU (ES FE) **NAR**

sibirica Weise, 1887h: 200

genus *Podagrica* Chevrolat, 1836: 394 type species: *Altica fuscipes* Fabricius, 1775

Geochoreutes Gistel, 1856: 379 type species: *Altica fuscipes* Fabricius, 1775

aeneipennis L. N. Medvedev, 1990b: 39 **A:** NP

altica Shukla, 1960: 84 (*Podagarica* [sic]) **A:** ID (HP)

atlantica Heikertinger, 1951: 9 **N:** MO

audisioi Biondi, 1982a: 233 **N:** MO

decolorata Duvivier, 1892a: 60 **A:** YE **AFR**

uniforma Jacoby, 1899d: 354 (*Nisotra*)

fuscicornis Linnaeus, 1767: 595 (*Chrysomela*) **E:** AL AU BE BH BU BY CR CZ DE FR GB GE GR HU IT KO KR LU MD ME NL NM PL PT RO RU (CT ST) SB SK SL SP SZ TR UK **N:** AG CI LB MO TU **A:** AB IS LE SY TR **AFR**i

chrysomelina Waltl, 1835: 83 (*Haltica*, var.)

fulvipes Fabricius, 1801a: 463 (*Crioceris*)

meridionalis Weise, 1886a: 685 (var.)

fuscipes Fabricius, 1775: 114 (*Altica*) **E:** BE FR GB GE IR IT LU NL PT SP SZ

bicolor Goeze, 1777: 312 (*Chrysomela*)

malvae malvae Illiger, 1807a: 63 (*Haltica*) **E:** AL AU BH BU BY CR CZ FR GE GR HU IT KO KR LU MA ME NM PL PT RO RU (ST) SB SK SL SP SZ TR UK **A:** AB CY GG IN IQ IS JO LE SI SY TR

aenescens Weise, 1886a: 683 (var.)
angusticollis J. R. Sahlberg, 1913c: 251 (ab.) [IN]
cedricola J. R. Sahlberg, 1913c: 251 (var.) [IN] [# 771]
convexicollis J. R. Sahlberg, 1913c: 251 (var.) [IN] [# 771]
cyanella Reiche & Saulcy, 1858b: 52 (*Haltica*, var.)
nigerrima Weise, 1894h: 152 (var.)
saracena Reiche & Saulcy, 1858b: 52 (*Haltica*)
subaptera J. R. Sahlberg, 1913c: 251 (forma) [IN] [# 771]
tristicula Weise, 1886a: 684 [# 1121]
unicolor Marseul, 1868: 213

malvae semirufa Küster, 1847: 86 (*Crepidodera*) **E**: AL CR FR GR IT MA ME SP **N**: AG MO TU
intermedia Kutschera, 1860d: 197 (*Haltica*)
italica Allard, 1860e: 542
laevis Abeille de Perrin, 1895a: cdi
picicollis Heikertinger, 1912a: 18 (ab.) [IN]
rufa Allard, 1884b: 248

menetriesii Faldermann, 1837: 396 (*Triplax*) **E**: AL AU BH BU CR CZ FR GE GR HU IT KR MD ME NM PT RO RU (CT ST) SB SK SL SP TR UK **A**: AB CH (XIN) GG IQ KI KZ TR TM UZ
desertorum Brancsik, 1899: 104
discedens Boieldieu, 1859: 475 (*Haltica*)
luctuosa Demaison, 1902: 25 (var.)
minor J. R. Sahlberg, 1913b: 100 (forma) [IN] [# 770]
punctatostriata Motschulsky, 1845a: 110
robusta Ballion, 1878: 386
rudicollis Foudras, 1860b: 341 [= 1861: 85]

pallidicolor Pic, 1909h: 154 **N**: EG **A**: SA SI YE **AFR**

puncticollis Weise, 1902e: 166 **N**: EG **A**: OM SA SI YE **AFR**

genus *Podagricella* S.-H. Chen, 1933a: 234 type species: *Podagricella cyanipennis* S.-H. Chen, 1933

cyanipennis S.-H. Chen, 1933a: 235 **A**: CH (YUN)

genus *Podagricomela* Heikertinger, 1924a: 36 type species: *Podagricomela weisei* Heikertinger, 1924

cheni L. N. Medvedev, 2002b: 212 [RN] **A**: CH (YUN) **ORR**
geminata S.-H. Chen & Zia, 1966a: 73, 74

cuprea S.-Y. Wang, 1990a: 123, 125 **A**: CH (GAN SCH)

cyanea S.-H. Chen, 1939b: 58 **A**: CH (GAN JIA)

decempunctata L. N. Medvedev, 2002b: 210 **A**: ID (SD)

flavitibialis S.-Y. Wang, 1990a: 123, 125 **A:** CH (GAN SHA SHX)
grossa L. N. Medvedev, 2002b: 212 **A:** NP
metallica L. N. Medvedev, 1997a: 17 **A:** ID (SD)
nigricollis S.-H. Chen, 1933f: 58 **A:** CH (FUJ GAN GUA GUX HKG HUB HUN JIA SCH TAI YUN ZHE)
nigripes L. N. Medvedev, 1993c: 372 **A:** BT CH (XIZ) ID (SD) **ORR**
parva S.-H. Chen & Zia, 1966a: 73, 75 **A:** CH (YUN) **ORR**
shavrini L. N. Medvedev, 2013c: 86 **A:** ID (UT)
shirahatai Chûjô, 1957b: 4 (*Clitea*) **A:** CH (GAN HUN SCH SHA SHX)
striatipennis Jacoby, 1884b: 35 (*Sphaeroderma*) **A:** CH (HAI YUN) NP **ORR**
apicalis Kimoto, 2000a: 19 (*Amphimela*)
apicipennis Jacoby, 1905b: v (*Notomela*)
taiwana L. N. Medvedev, 2002b: 211 **A:** CH (TAI)
viridicyanea L. N. Medvedev & Sprecher-Uebersax, 1997: 235 **A:** NP
weisei Heikertinger, 1924a: 36 **A:** CH (CHQ FUJ GAN GUA GUX HEB HUN JIA SCH SHN SHX ZHE)

genus *Podontia* Dalman, 1824: 23 type species: *Galleruca grandis* Gröndal, 1808 (= *Chrysomela lutea* Olivier, 1791)
Paramerista Lopatin, 2011a: 143 type species: *Paramerista luteola* Lopatin, 2011
affinis affinis Gröndal, 1808: 289 (*Galleruca*) **A:** CH (GUI YUN) ID (SD UT) **ORR**
affinis indosinensis Scherer, 1978: 276 (ssp.) **A:** CH (HAI GUI GUX) ID (SD UT) **ORR**
congregata Baly, 1865j: 405 **A:** PA **ORR**
dalmani Baly, 1865j: 405 **A:** CH (HAI GUI GUX YUN) **ORR**
laosensis Scherer, 1969: 169 **A:** CH (YUN) **ORR**
lutea Olivier, 1791a: 692 (*Chrysomela*) **A:** CH (CHQ FUJ GAN GUA GUI GUX HUB HKG HUN JIA SCH SHA TAI XIZ YUN ZHE) PA **AUS ORR**
grandis Gröndal, 1808: 288 (*Gallerruca*)
luteola Lopatin, 2011a: 143 (*Paramerista*)
quatuordecimpunctata Linnaeus, 1767: 599 (*Chrysomela*) **A:** CH (XIZ YUN) ID (KA SD UT) JA NP PA **ORR**
quatuordecimpunctata Kollar & L. Redtenbacher, 1844: 559 (*Gonioctena*) [HN]
rufocastanea Baly, 1865j: 405 **A:** CH (YUN) **ORR**
pitalohita Maulik, 1926: 224

genus *Polyclada* Chevrolat, 1836: 375 type species: *Clytra pectinicornis* Olivier, 1791
benti Gahan, 1895: 291 **A:** OM SA YE **AFR**

genus *Primulavorus* Konstantinov & Ruan, 2017: 116 type species: *Primulavorus maculatus* Konstantinov & Ruan, 2017
maculatus Konstantinov & Ruan, 2017: 117 **A:** CH (YUN)

genus *Priobolia* S.-H. Chen & S.-Y. Wang, 1987b: 58, 61 type species: *Priobolia viridiaurata* S.-H. Chen & S.-Y. Wang, 1987
viridiaurata S.-H. Chen & S.-Y. Wang, 1987b: 58, 62 **A:** CH (XIZ)

genus *Pseudodera* Baly, 1861a: 200 type species: *Pseudodera xanthospila* Baly, 1861
fulva L. N. Medvedev, 1997e: 261 **A:** NP
himalayensis Scherer, 1969: 140 **A:** ID (SD) NP
inornata S.-H. Chen, 1933a: 233 **A:** CH (YUN)
laeta L. N. Medvedev, 1997a: 19 **A:** NP
leigongshanensis P.-Y. Yu, 1993: 316, 327 **A:** CH (GUI)
nepalica Medvedev, 2015a: 477 **A:** NP
orientalis Baly, 1877g: 286 **A:** ID (SD) NP **ORR**
xanthospila Baly, 1861a: 200 **A:** CH (FUJ GUA GUI HUB HUN JIA TAI YUN ZHE) JA
balyi Jacoby, 1891c: 62
bimaculata Jacoby, 1886a: 723 (*Crepidodera*)

genus *Pseudoliprus* Chûjô & Kimoto, 1960: 9 type species: *Liprus hirtus* Baly, 1874
bisulcatus S.-H. Chen & S.-Y. Wang, 1980: 1, 15 **A:** CH (SCH)
chinensis L. N. Medvedev, 1993c: 371 **A:** CH (SCH)
hirtus flaviceps Ohno, 1959: 38 (*Liprus*, ssp.) **A:** JA (Hokkaido, Honshu)
hirtus hirtus Baly, 1874: 194 (*Liprus*) **A:** JA
hirtus tosanus Ohno, 1966: 46 (ssp.) **A:** JA (Shikoku)
kimotonis Komiya, 2006c: 185 **A:** CH (TAI)
komiyai Ohno, 1966: 44 **A:** JA (Kyushu)
kurosawai Nakane, 1958: 50 (*Liprus*) **A:** CH (TAI) JA (Yakushima)
lalashanensis Komiya, 2006c: 189 **A:** CH (TAI)
nigritus nigritus Jacoby, 1886a: 724 (*Liprus*) **A:** JA (Honshu)
nigritus watanabei Ohno, 1959: 36 (*Liprus*, ssp.) **A:** JA (Honshu)
saigusai Kimoto, 1970b: 209 **A:** CH (TAI)
suturalis flavus Ohno, 1966: 46 (ssp.) **A:** JA (Honshu)
suturalis suturalis Jacoby, 1886a: 725 (*Liprus*) **A:** JA SC
maebarai Ohno, 1959: 40 (*Liprus*, ssp.)

genus *Psylliodes* Latreille, 1829: 154 type species: *Galeruca affinis* Paykull, 1799
[# 754, Appendix]
subgenus *Eupus* Wollaston, 1854: 452 type species: *Psylliodes tarsata* Wollaston, 1854
aemulans Har. Lindberg, 1953: 12 **N:** CI (Gran Canaria, Tenerife)
amplicollis Wollaston, 1865: 56 **N:** MR
stolida Wollaston, 1860a: 11 **N:** CI MR SG
tarsata Wollaston, 1854: 452 **N:** MR
wollastoni Nadein, 2007a: 312 **N:** MR

subgenus *Minicnema* Nadein, 2007a: 313 type species: *Psylliodes elliptica* Allard, 1861
belarbii Döberl, 1991a: 337 **N:** MO
elliptica Allard, 1862: 340 **A:** IS JO SY TR
nitidula Heikertinger, 1940: 561 [RN]
subgenus *Psylliodes* Latreille, 1829: 154 type species: *Galeruca affinis* Paykull, 1799 **[Appendix]**
Macronema Stephens, 1829b: 217 type species: *Chrysomela hyoscyami* Linnaeus, 1758
Macrocnema Stephens, 1831: 317 type species: *Chrysomela hyoscyami* Linnaeus, 1758
Macrocnema Weise, 1888b: 785 [HN] type species: *Haltica cucullata* Illiger, 1807
Psyllides Gistel, 1856: 379 [UE] type species: *Chrysomela chrysocephala* Linnaeus, 1758
Psyllomima Bedel, 1898: 200 [RN] type species: *Haltica cucullata* Illiger, 1807
abdominalis S.-Y. Wang, 1992c: 677, 731 **A:** CH (YUN)
aeneola Heikertinger, 1911b: 16 **A:** KZ
aerea Foudras, 1860a: 165 [= 1860b: 53] E: AU BE BU CZ FR GE HU IT RO SK SP UK (Carpathians) **A:** AR TR
araratica Iablokoff-Khnzorian, 1968: 265
austriaca Heikertinger, 1911b: 21 (var.)
affinis Paykull, 1799a: 109 (*Galeruca*) E: AU BE BH BU BY CR CZ DE EN FI FR GB GE GR HU IT KR LA LS LT LU MD ME NL NM NR PL PT RO RU (CT NT ST) SB SK SL SP SV SZ UK **N:** MO **A:** AB IN KZ RU (ES SS WS) TR **AFRi NARi**
exoleta Stephens, 1831: 321 (*Macrocnema*)
agropyri Paliy, 1961: 390 E: KR RU (CT ST) **A:** KZ
algirica Allard, 1860b: cclxi E: GR (Kefallonia) IT SP **N:** AG TU **NARi**
altimontana L. N. Medvedev, 2003a: 322 **A:** NP
amurensis Nadein, 2006: 862 **A:** RU (FE)
analogica Nadein, 2005: 136 **A:** KI (Central Tienshan)
anatolica Gök & Çilbiroğlu, 2004: 2 **A:** TR
angusticeps Israelson, 1980: 194 **N:** CI (Gomera, Tenerife)
arista Iablokoff-Khnzorian, 1962: 121 E: RU (ST) **A:** AB AR GG IN TR
kasyi Lopatin, 1967e: 439
astenica Nadein, 2005: 138 **A:** TD
attenuata Koch, 1803a: 34 (*Haltica*) E: AU BE BH BU BY CR CZ EN FI FR GB GE GR HU IR IT KO KR LA LT LU MD ME NL PL RO RU (CT NT ST) SB SK SL SP SV SZ UK **A:** AB AR CH (GAN GUI HEI LIA NIN SHA SHX TAI) GG JA KI KZ MG NC RU (ES FE SS WS) SC TR UZ
coerulea Král, 1947: 21 (forma)
japonica Jacoby, 1886a: 740
picicornis Stephens, 1831: 321 (*Macrocnema*)
taiwanica Chûjô, 1935f: 364
vicina L. Redtenbacher, 1848: 537

ausoniae Leonardi, 2013: 101 **E:** FR IT **A:** TR
baluchistana Lopatin, 1990b: 607 **A:** IN
biondii Leonardi, 2007: 180 **E:** IT
brettinghami Baly, 1862b: 457 **A:** CH (GUX SCH TAI YUN) ID (KA) JA NC NP PA **AUS ORR**
cucurbitae Gressitt, 1955: 42
difficilis Baly, 1874: 210
fulvipes Jacoby, 1896b: 455
gigantea Chûjô, 1935f: 367
howensis Lea, 1929: 239 (var.)
illigeri Perroud, 1864: 199
lubricata Blackburn, 1896: 78
norfolcensis Lea, 1929: 239 (var.)
novaecaledoniae Baly, 1876d: 600
sera Jacobson, 1922a: 527
solanae Broun, 1910: 302
vitiensis Bryant, 1925: 597
brisouti Bedel, 1898: 202 (var.) **E:** AN AU BU CR CZ FR GE GR HU IT KR ME NM NR PL RO RU (ST) SK SL SP SV SZ **A:** TR
burangana S.-H. Chen & S.-Y. Wang, 1981a: 492, 505 **A:** CH (XIZ)
callinota Faldermann, 1837: 348 **E:** RU (ST: Caucasus)
caneparii Leonardi, 2007: 176 **E:** IT
cantonensis Gruev, 1981b: 74 **A:** CH (GUA TAI)
cerenae Gök, Doguet & Çilbiroğlu, 2003: 201 **A:** TR
cereola Peyerimhoff, 1925c: 268 **N:** AG MO
cervinoi Baselga & Novoa, 2003: 689 **E:** SP
chalcomera Illiger, 1807a: 75 (*Haltica*) **E:** AL AU BE BH BU BY CR CZ DE EN FI FR GB GE GR HU IT KR LA LT LU MD ME NL NM PL PT RO RO RU (CT ST) SB SK SL SP SV SZ TR UK **N:** AG MO TU **A:** AB AR CH (HEI) CY GG IN IS KI KZ RU (ES FE SS WS) SY TR **NARi**
brunnipes Duftschmid, 1825: 280 (*Haltica*)
cardui Weise, 1888b: 821 (var.)
cyanescens Normand, 1937: 137 (ab.) [IN]
erythroceros Abeille de Perrin, 1895a: cdiii (var.)
unimaculata Curtis, 1834: 486 (*Macrocnema*)
chlorophana Erichson, 1842a: 237 **A:** CH (FUJ GUI TAI YUN) ID (AP) JA (Ryukyus) **ORR**
balyi Jacoby, 1884b: 30
chrysocephala chrysocephala Linnaeus, 1758: 372 (*Chrysomela*) **E:** AL AU AZ BE BH BU BY CR CZ DE EN FI FR GB GE GR HU IT KR LA LS LT LU MA MD ME NL NM

NR PL PT RO RU (ST) SB SK SL SP SV SZ TR UK **N:** AG MO MR TU **A:** AB AR CY GG IN IS JO LE SY TR **AFR NARi**

allardi Bach, 1867: 355
anglicana Gmelin, 1790: 1693 (*Chrysomela*)
anglica Fabricius, 1787: 77 (*Chrysomela*)
angulicollis Heikertinger, 1926: 119 (var.) [# 772]
brachybiotes Gistel, 1857b: 539
castanea Abeille de Perrin, 1907: lxxxi
catinensis Rottenberg, 1872a: 245
collaris Weise, 1888b: 803 (var.)
cyanoptera Illiger, 1807a: 77 (*Haltica*)
erythrocephala Linnaeus, 1758: 373 (*Chrysomela*)
eubiotes Gistel, 1857b: 525
kunzei Foudras, 1860a: 187 [= 1860b: 75]
longipennis Pic, 1903d: 125
nigricollis Marsham, 1802: 205 (*Chrysomela*)
nigrifrons Heikertinger, 1911b: 16 (ab.) [IN]
nucea Illiger, 1807a: 78 (*Haltica*)
peregrina Weise, 1888b: 803 (var.)
sicardi Weise, 1894c: 90
sordida Stephens, 1831: 320 (*Macrocnema*)

chrysocephala inops Peyerimhoff, 1915a: 37 **N:** AG

chujoe Jind. Madar, 1960: 48 **A:** CH (TAI) JA (Kyushu)

circumdata W. Redtenbacher, 1842: 27 (*Haltica*) **E:** AL BH BU BY CR FR GE GR HU IT KR ME NM PL RO RU (ST) SB SK SL SP UK **A:** AB GG IN TR

integra Weise, 1888b: 817 (var.)

coelestis Warchałowski, 2000c: 588 [RN] **E:** SP **N:** MO

coerulea Weise, 1888b: 789 [HN]

concolor Nadein, 2006: 867 **A:** GG

crambicola Lohse, 1954: 211 **E:** DE EN FI GE NR RU SV

cretica Weise, 1888b: 791 **E:** GR (Crete)

cucullata Illiger, 1807a: 73 (*Haltica*) **E:** AU BE BY CR CZ DE EN FI FR GB GE IT LA LT LU NL NR PL PT RO RU (CT NT) SK SP SV SZ UK **A:** CH (GAN XIN) KZ MG NC RU (ES FE SS WS) **NARi**

angustata Waltl, 1835: 82 (*Haltica*, var.) [QS]
augustalisi Pic, 1909f: 146 [QS]
gansuica Jacobson, 1922a: 526 (ssp.)
spergulae Gyllenhal, 1813: 571 (*Haltica*)
ussuriensis L. N. Medvedev, 1973b: 884 (ssp.)
vicina Boieldieu, 1859: 479 [HN]

cuprea Koch, 1803a: 28 (*Haltica*) E: AL AU BE BH BU BY CR CZ DE EN FR GB GE GR HU IT KR LA LS LT LU MA MD NL NM PL PT RO RU (CT NT ST) SB SK SL SP SV SZ TR UK **N:** AG EG MO TU **A:** AB AR CY GG IN IS JO KI KZ LE MG RU (ES) SI SY TM TR

cypricolor Allard, 1861: 811

foudrasi Bach, 1867: 355

herbacea Foudras, 1860a: 170 [= 1860b: 58]

laevata Foudras, 1860a: 158 [= 1860b: 46] [QS]

obscura Duftschmid, 1825: 281 (*Haltica*)

polita Gravenhorst, 1807: 143 (*Haltica*)

cupreata Duftschmid, 1825: 282 (*Haltica*) E: AU BY CR CZ EN FR GE HU KR LT NM PL RO RU (CT ST) SB SP UK **A:** AB AR KI KZ MG RU (ES SS WS) UZ

cyanescens Weise, 1887h: 194 **A:** RU (ES FE) "Korea"

danieli Weise, 1900b: 292 E: AU (Alps) CR IT (Alps) SL

forojuliensis Heikertinger, 1926: 138 (ssp.)

rambouseki Heikertinger, 1910a: 363

deplanata L. N. Medvedev, 1962a: 388 E: KR RU (ST) UK **A:** GG

dilutella Heikertinger, 1911b: 22 **A:** KI KZ TD UZ

diversicolor Nadein, 2006: 868 **A:** TR

kasnakensis Gök & Aslan, 2007: 372

dogueti Warchałowski, 1993b: 359 **A:** TR (Gümüsane)

drusei Furth, 1984: 52 **A:** IS TR

dulcamarae Koch, 1803a: 24 (*Haltica*) E: AL AU BE BH BU BY CR CZ DE EN FI (South) FR GB GE GR (Corfu) HU IR IT KR LA LT LU MD NL NT PL PT RO RU (CT ST) SB SK SL SP SV SZ UK **A:** AB KZ MG RU (ES WS) TR

erberi Döberl, 1995c: 1 **N:** MR

fageli Bechyné, 1957: 4 **N:** AG (ssp.)

feroniae Leonardi, 1978: 294 E: IT (South)

festae Dodero, 1925: 19 E: GR (Crete, Peloponnese) **A:** CY IS JO LE SY TR **N:** EG LB

fiorellae Leonardi, 1978: 290 E: IT (North)

frivaldszkyi Weise, 1888b: 786 E: AU HU PL RO SK UK

fusiformis Illiger, 1807a: 76 (*Haltica*) E: CR (Istria) FR (Corsica) IT PT SP SZ **N:** AG MO TU

luteipes Küster, 1849b: 96

gibbosa Allard, 1861: 820 E: CR FR IT NM SL SP SZ **N:** AG MO **A:** IS TR

sicula Stierlin, 1867: 228

glabra Duftschmid, 1825: 282 (*Haltica*) E: AU BH CR (North) CZ GE HU IT (North) RO SL

alpina L. Redtenbacher, 1848: 538

valligenus Gistel, 1857b: 599 (*Longitarsus*)

globosa Nadein, 2007a: 318 **A:** NP
gougeleti Allard, 1860b: cclx **E:** PT SP **N:** AG MO
grigorievi Jacobson, 1922a: 530 **A:** UZ
gruevi Döberl, 2009b: 419 **A:** NP
gyirongana S.-H. Chen & S.-Y. Wang, 1981a: 492, 505 **A:** CH (XIZ)
heikertingeri Jacobson, 1922a: 532 **E:** IT (Sicily) MA **N:** AG TU
hermonensis Furth, 1984: 47 **A:** IS
heydeni Weise, 1888b: 785 **E:** FR (Pyrenees) PT SP
hispana Heikertinger, 1911b: 19 **E:** FR PT SP
hospes Wollaston, 1854: 449 **E:** FR IT (Sardinia, Sicily) SP **N:** AG CI EG LB MO MR TU **A:** AE IS OM SA SI TR [# 782]
aridissima Peyerimhoff, 1941: 15 (ssp.)
leprieuri Pic, 1911a: 10 (var.)
huaxiensis S.-Y. Wang, 1992c: 679, 732 **A:** CH (SCH YUN)
hyoscyami Linnaeus, 1758: 372 (*Chrysomela*) **E:** AU BE BH BU BY CR CZ DE EN FI FR GB GE GR HU IT KO KR LA LT LU MD ME NR PL RO RU (CT NT ST) SB SK SP SV SZ UK **A:** AB AR CY GG IQ IN IS JO KI KZ RU (ES FE SS) SY TR UZ
coerulea Paykull, 1799a: 97 (*Galeruca*)
coerulescens Weise, 1888b: 821 (var.)
cupronitens Förster, 1849: 37
illyrica Leonardi & Gruev, 1993: 15 **E:** AU BH BU CR CZ RO SK UK (Carpathians) SB **A:** TR
infanda Nadein, 2005: 133 **A:** MG (South-west)
inflata Reiche & Saulcy, 1858b: 50 (*Haltica*) **E:** FR IT (Sardinia, Sicily) MA PT SP **N:** AG LB MO TU **A:** IN IQ IS JO LE SY TR
completa Normand, 1937: 137 (var.) [IN] [# 771]
ventricosa Rottenberg, 1872a: 246
instabilis Foudras, 1860a: 171 [= 1860b: 59] **E:** AU BE BH BU CR CZ FR GE GR HU IT KR NM PL RO RU (CT ST) SK SP SZ UK **N:** AG LB MO TU **A:** AR CY GG IS JO LE MG SY TR
lauticollis Allard, 1867: 468 [QS]
insularis Nadein, 2006: 869 **N:** CI
isatidis Heikertinger, 1912j: 125 [= 1913f: 207] (var.) **E:** AU BE BU CR CZ FI FR GE GR HU IT KR LS LU MD ME NM NR RO RU (CT ST) SB SK SV SZ UK **A:** AB AR GG IN KZ MG RU (ES WS) TR UZ
kiesenwetteri Kutschera, 1864f: 424 **E:** AL AU BH BU CR FR (Corsica) GR HU IT ME NM SL TR **A:** TR
latifrons Weise, 1888b: 799
konstantinovi Lopatin, 1997b: 376 **A:** KI
laevicollis Dufour, 1851: 354 (*Altica*) **E:** FR PT SP **N:** MO

laevifrons Kutschera, 1864f: 414 **E:** IT

laticollis Kutschera, 1864f: 388 **E:** BE BU CR CZ FR GB GE GR IT LU NL PL PT SK SL SP SZ **N:** AG MO MR TU **A:** TR **AFR**

rufofemorata Heikertinger, 1912f: 240 (ab.) [IN]

weberi Lohse, 1956: 88

laurisilvae Biondi, 1987: 94 **N:** CI (La Palma, Tenerife)

laxa Nadein, 2006: 864 **A:** RU (FE)

leonhardi Heikertinger, 1926: 129 **E:** IT (Sicily)

lethierryi Allard, 1861: 808 **E:** IT (Sicily) **A:** AG TU

sicana Mulsant & Rey, 1874: 217

libanicola Pic, 1903d: 124 **A:** LE

libertii Leonardi, 2013: 89 **A:** FR (incl. Corsica) IT (incl. Sardinia)

littoralis Biondi, 1997: 386 **A:** CY TR (Mugla)

longicollis Weise, 1900b: 292 **E:** RU (ST) **A:** AB AR GG TR

caucasica Heikertinger, 1916: 44

nivalis Iablokoff-Khnzorian, 1962: 120

oreophila Iablokoff-Khnzorian, 1957: 170

luridipennis Kutschera, 1864f: 393 **E:** GB (Lundy Is.)

luteola O. F. Müller, 1776: 84 (*Altica*) **E:** AU BE BH BU BY CR CZ EN FR GB GE GR HU IT LA LU MD NL PL RO RU (ST) SK SP SZ UK **N:** AG MO **A:** AB CY GG IN KZ LE SY TR

aberrans Roubal, 1916a: 14 (var.)

nana Duftschmid, 1825: 285 (*Haltica*)

pallida Stephens, 1831: 321 (*Macrocnema*)

propinqua L. Redtenbacher, 1848: 536

maculatipes Pic, 1924c: 22 **N:** AG EG **A:** AE IS OM SA SI

magnifica Gruev, 1975a: 92 **E:** BU (Strandzha Mts.) GR (North) **A:** TR

marcida Illiger, 1807b: 175 (*Haltica*) **E:** AZ BE BU CR DE EN FI FR GB GE GR IT KR LA LT NL NR PL PT SP RU (NT) SV UK **N:** AG MO TU **A:** CY GG IS LE TR

operosa Foudras, 1860a: 188 [= 1860b: 76]

maroccana Heikertinger, 1916: 36 **E:** FR (Corsica) IT (South, Sicily, Sardinia) SP (Huelva) **N:** AG CI (Tenerife) MO

metatarsalis Leonardi, 2007: 200 **N:** AG

milleri lindbergi Jind. Madar & Jiř. Madar, 1964: 118 **E:** SP (South) **N:** AG MO

milleri milleri Kutschera, 1864f: 390 **E:** GR IT SP **A:** CY TR

moricandiae Peyerimhoff, 1925a: 15 (ssp.) **N:** AG LB TU

napi Fabricius, 1792b: 29 (*Galleruca*) **E:** AL AU BE BH BU BY CR CZ DE EN FI FR GB GE GR HU IT KO KR LA LS LT LU MA MD ME NL NR PL PT RO RU (CT NT ST) SB SK SL SP SV SZ TR UK **N:** AG MO **A:** AB GG KZ RU (ES SS) TR **NARi**

asturiensis Weise, 1888b: 807 (var.)

ecalcarata L. Redtenbacher, 1848: 537
flavicornis Weise, 1883b: 219 (var.)
hignetti Donisthorpe, 1940: 120 (var.)
lunariae E. Jacquet, 1887: 2 (var.)
rapae Stephens, 1831: 319 (*Macrocnema*)
nigripennis Allard, 1861: 827 **N:** AG TU
bicolorata Abeille de Perrin, 1907: cxxxi
nitida L. N. Medvedev, 1973b: 884 **A:** RU (FE)
nyalamana S.-H. Chen & S.-Y. Wang, 1981a: 492, 506 **A:** CH (XIZ)
obscuroaenea Rosenhauer, 1856: 342 **E:** IT (South, Sicily) PT SP **N:** AG MO TU
obscurofasciata S.-H. Chen, 1933c: 143 **A:** CH (GAN NIN SHA)
olgae Nadein, 2007b: 656 **E:** SP
ozisiki Leonardi & Arnold, 1995: 306 **A:** AR TR
pallidicolor Pic, 1903d: 125 **E:** BU GR SP **N:** AG **A:** IQ IS LE SY TR
pallidicornis Heikertinger, 1921a: 44 **E:** RU (ST: Caucasus) **A:** AR TM TR
pallidipennis Rosenhauer, 1856: 340 **E:** FR IT MA PT SP **N:** AG MO TU **A:** LE SY
parodii Leonardi, 2007: 222 **E:** IT
persica Allard, 1867: 457 **E:** RU (ST) **A:** AB AF IN IQ IS JO KI KZ SA SY TD TR TM UZ YE (Socotra) "Transcaucasus"
petasata Foudras, 1860a: 176 [= 1860b: 64] **E:** FR (Pyrenees) SP (Pyrenees)
minima Allard, 1861: 824
peyerimhoffi Heikertinger, 1916: 33 **N:** EG **A:** AE IS SA SI
picina Marsham, 1802: 206 (*Chrysomela*) **E:** AU BE BH BU BY CR CZ DE EN FI FR GB GE HU IR IT (North) LA LT LU MD NL NR PL RO RU (CT NT ST) SK SL SP SV SZ TR UK **A:** RU (WS) **NARi**
apicalis Stephens, 1831: 321 (*Macrocnema*) [QS]
melanophthalma Duftschmid, 1825: 284 (*Haltica*)
picea L. Redtenbacher, 1848: 538
rufilabris Koch, 1803a: 33 (*Haltica*)
rufopicea Letzner, 1847: 83 (*Haltica*)
picipes L. Redtenbacher, 1848: 538 **E:** AU CR FR (Alpes Maritimes) IT (North) SL
ogloblini Jacobson, 1922a: 533
plana Maulik, 1926: 128 **A:** CH (GAN GUX) ID (UT) NC NP **ORR**
puncticollis Rosenhauer, 1856: 341 **E:** AL BU CR (South) FR GR (Crete) IT SP **N:** AG TU
crassicollis Fairmaire, 1857b: 641
dilatata Foudras, 1860a: 179 [= 1860b: 67]
punctifrons Baly, 1874: 209 **A:** CH (CHQ FUJ GAN GUI HUN JIA SHA SCH XIZ TAI YUN XIZ ZHE) JA RU (FE) SC **ORR**
pyrenaea Heikertinger, 1921a: 43 (ssp.) **E:** FR (Pyrénées, Basses-Alpes) SP (Avila)
pyritosa Kutschera, 1864f: 396 **E:** AU BH CR FR (South) GR IT ME PT SL SP **N:** AG CI MO MR **A:** TR **AFR**

rhaica Jacobson, 1922a: 526 **E:** RU (ST) UK **A:** KZ
ridenda Nadein, 2008: 348 **A:** TR
rubroaenea Heikertinger, 1916: 43 **E:** RU (ST) **A:** GG
ruffoi Leonardi, 1975c: 75 **E:** IT
ruficolor Doguet, 1993: 358 **E:** FR SP **N:** AG TU
rufitarsis Graells, 1858: 96 [= 1858b: 134] **E:** SP (Madrid)
sasakii Takizawa, 2015: 242 **A:** JA
saulcyi Allard, 1867: 469 **E:** KR RU (CT ST) UK **N:** EG **A:** AE CY IN IQ IS JO KZ MG SI SY TR
atriplicis Jacobson, 1922a: 529
schwarzi Weise, 1900b: 293 **E:** FR (Alps) IT (Alps) SZ
semicnemis Gruev, 1990b: 64 **A:** NP
shira Maulik, 1926: 128 **A:** NP **ORR**
solarii Leonardi, 1975c: 70 **E:** FR (Savoie) IT (Apennines)
springeri Leonardi, 1975c: 61 **E:** IT (South)
sturanyi Apfelbeck, 1906: 249 **E:** BH BU ME RO SB
subaenea styriaca Heikertinger, 1921a: 44 (ssp.) **E:** AU (East Alps) SL
subaenea subaenea Kutschera, 1864f: 407 **E:** AU CR BH GE HU ME PL RO SB SK UK
submontana Nadein, 2006: 866 **A:** RU (WS)
subrugosa Jacoby, 1886a: 739 **A:** CH (HUB SHA TAI ZHE) JA RU (FE) SC **ORR**
intermedia Jacoby, 1886a: 739
mitchyi Chûjô, 1951b: 44
taiwana Takizawa, 1979: 340 (ssp.) **A:** CH (TAI) JA (Ryukyus)
takizawai Gruev, 1990d: 129 **A:** NC
taurica Leonardi, 1971: 516 **A:** TR (Taurus Mts.)
tenebrosa Jacoby, 1896c: 269 **A:** ID (HP SD UT) NP PA **ORR**
tenuidentata Nadein, 2008: 347 **A:** IS
testaceoconcolor Heikertinger, 1926: 122 (forma) **A:** AB CY GG IS LE TR [# 772]
thlaspis Foudras, 1860a: 154 [= 1860b: 42] **E:** AN AU BE BU CR CZ FR GE GR (Karpathos) HU IT KR MD NM PL RO RU (ST) SB SK SL SP SZ UK **A:** AB GG IN TR
tibetana S.-H. Chen, 1976: 211, 221 **A:** CH (XIZ)
toelgi Heikertinger, 1914: 95 **E:** AU CR CZ FR GE GR HU IT (North) ME RO SK SL SP SZ UK (Carpathians) **A:** TR
tricolor Weise, 1888b: 806 (var.) **E:** AU BU BY CZ DE EN FI FR GB GE GR HU IT (North) KR LT MD NM NR PL RO RU (CT ST) SK SP SV SZ UK **N:** AG MO **A:** AB AF AR GG IN IS KI KZ LE RU (ES SS WS) SY TD TR TM UZ
autumnalis Kaszab, 1962: 73 (ab.) [IN]
lichtneckerti Kaszab, 1962: 73 (ab.) [IN]
nigricapilla Král, 1945: 31 (forma)
nigrifrons Heikertinger, 1914: 104 (ab.) [IN]

sophiae Heikertinger, 1914: 104 [RN]
tsinghaina S.-H. Chen & Zia, 1966b: 149, 150 **A:** CH (QIN)
umbratilis Wollaston, 1854: 450 **N:** MR
urbaniae Biondi & D'Alessandro, 2017b: 122 **E:** IT
valida Weise, 1889a: 2 **E:** RU (ST) **A:** GG TR
vehemens normandi Heikertinger, 1916: 34 **E:** AZ SP (Malaga) **N:** AG CI MO MR TU
vehemens vehemens Wollaston, 1854: 451 **N:** MR
azorica Jacobson, 1922a: 528
canarica Jacobson, 1922a: 531
storai Uyttenboogaart, 1935: 12 (ab.) [IN]
vindobonensis Heikertinger, 1914: 97 **E:** AU (East Alps) CR (Velebit Mt.) CZ FR (Pyrenees, Alps) GE IT (Liguria, Apennines) SP (Pyrenees) SZ (Jura) UK (Carpathians) **A:** TR (Konya)
viridana Motschulsky, 1858: 108 **A:** CH (CHQ FUJ GAN GUA HUN SCH SHA TAI YUN ZHE) JA NC NP RU (ES FE) SC **ORR**
angusticollis Baly, 1874: 209
formosana Chûjô, 1963b: 401
loochooana Chûjô, 1961: 90 (ssp.)
palleola Motschulsky, 1866b: 418
rishiriensis Chûjô, 1959: 15 (ssp.)
sinensis S.-H. Chen, 1934h: 240
wachsmanni Csiki, 1903: 40 **E:** AL BH CR IT (North) ME SL TR **A:** CY JO
wrasei Leonardi & Arnold, 1995: 300 **E:** BU GR (North) KR MD NM RU (ST) UK **A:** GG TR
wunderlei Döberl, 1998: 41 **E:** SP (Santander)
yalvacensis Gök, 2005b: 133 **A:** TR (Southwest)
yuae Nadein & C.-F. Lee, 2012: 44 **A:** CH (TAI)
subgenus *Psyllobactra* Lopatin, 1976a: 115 type species: *Psylliodes pubipennis* Lopatin, 1958
pubipennis Lopatin, 1958a: 55 **A:** TD (South)
subgenus *Semicnema* Weise, 1888b: 784 type species: *Psylliodes reitteri* Weise, 1888
elongata S.-Y. Wang, 1992c: 680, 732 **A:** CH (YUN)
longicornis S.-Y. Wang, 1992c: 681, 733 **A:** CH (YUN)
macella Weise, 1900b: 291 (*Semicnema*) **A:** KZ MG RU (ES FE SS)
roddi Jacobson, 1922a: 524
reitteri parallela Weise, 1890g: 366 **E:** RU (ST) **A:** CH (XIZ) JA (Hokkaido) KI KZ MG RU (ES FE) UZ "Turkestan"
reitteri reitteri Weise, 1888b: 792 **E:** AU BH BU CZ GE HU KR MD RU (CT ST) SK UK **A:** TR

genus *Psylliodes*, nomen dubium
elongata Gyllenhal, 1813: 570 (*Haltica*) E: SV

genus *Sangariola* Jacobson, 1922a: 522 type species: *Galleruca punctatostriata* Motschulsky, 1861
Allophyla Weise, 1889g: 624 [HN] type species: *Allophyla aurora* Weise, 1889 (= *Charidea fortunei* Baly, 1888)
Charidea Baly, 1888b: 157 [HN] type species: *Galeruca punctatostriata* Motschulsky, 1861
Lochmaeina L. N. Medvedev, 1956a: 157 type species: *Lochmaeina rosea* L. N. Medvedev, 1956
Lophallya Hincks, 1949: 616 [RN] type species: *Allophyla aurora* Weise, 1889 (= *Charidea fortunei* Baly, 1888)
costata Chûjô, 1935g: 394 **A**: CH (TAI)
fortunei Baly, 1888b: 158 (*Charidea*) **A**: CH (FUJ GAN GUA GUI HUN TAI YUN ZHE) SC
atriceps Pic, 1938b: 18 (*Charidea*, var.)
aurora Weise, 1889g: 626 (*Allophyla*)
bicostata Pic, 1928c: 32 (*Charidea*)
nigrosuturalis Pic, 1937b: 176 (*Charidea*)
obliterata Pic, 1937b: 176 (*Charidea*, var.)
pieli Pic, 1937b: 175 (*Charidea*)
unicoloripennis Chûjô, 1941b: 173 (var.) [IN] [# 771]
hirashimai Kimoto, 1970b: 213 **A**: CH (TAI)
punctatostriata Motschulsky, 1861a: 25 (*Galleruca*) **A**: CH (TAI "North China") JA RU (FE) SC
aequicostata Chûjô, 1938f: 10 (ssp.)
multicostata Jacoby, 1886a: 746 (*Galeruca*)
regularis Pic, 1928c: 31 (*Charidea*)
rosea L. N. Medvedev, 1956a: 157 (*Lochmaeina*)
yuae C.-F. Lee, 2014b: 192 **A**: CH (TAI)

genus *Sebaethoides* S.-H. Chen, 1934h: 298 type species: *Sebaethoides castanea* S.-H. Chen, 1934
castanea S.-H. Chen, 1934h: 299 **A**: ID (SD) **ORR**

genus *Sinaltica* S.-H. Chen, 1939b: 60 type species: *Sinaltica exigua* S.-H. Chen, 1939
exigua S.-H. Chen, 1939b: 61 **A**: CH (GAN SHX)

genus *Sinocrepis* S.-H. Chen, 1933a: 232 type species: *Sinocrepis micans* S.-H. Chen, 1933 (= *Crepidodera obscurofasciata* Jacoby, 1892)

fulva Kimoto, 1996: 46 **A:** CH (TAI)
obscurofasciata Jacoby, 1892b: 933 (*Crepidodera*) **A:** CH (CHQ GUI HUB JIA SHA TAI ZHE) **ORR**
micans S.-H. Chen, 1933a: 233

genus *Sinosphaera* Konstantinov & Ruan, 2017b: 112 type species: *Sinosphaera aptera* Konstantinov & Ruan, 2017
aptera Konstantinov & Ruan, 2017b: 114 **A:** CH (SCH)

genus *Sphaeraltica* Ohno, 1961a: 84 type species: *Graptodera flavicornis* Baly, 1874
Lesagealtica Döberl, 2009a: 22 [RN] type species: *Ochrosis nigripennis* Motschulsky, 1866
Ochrosoma Ogloblin, 1930: 104 [HN] type species: *Ochrosis nigripennis* Motschulsky, 1866
Ogloblinia Csiki, 1939: 280 [RN, HN] type species: *Ochrosis nigripennis* Motschulsky, 1866
Philaphthona L. N. Medvedev, 1994: 21 type species: *Philaphthona tenebrosa* L. N. Medvedev, 1994
affinis S.-H. Chen, 1939a: 43 (*Ochrosoma*) **A:** CH (FUJ GUX ZHE)
berberii Ohno, 1961a: 87 (*Sphaeraltica*) **A:** JA
flavicornis Baly, 1874: 192 (*Graptodera*) **A:** CH (TAI) JA
laurenti Döberl, 2009b: 417 **A:** NP
tarsalis L. N. Medvedev, 2012d: 507 (*Ogloblinia*) **A:** CH (YUN)

genus *Sphaeroderma* Stephens, 1831: 328 type species: *Altica testacea* Fabricius, 1775
Argosomus Wollaston, 1868: 152 type species: *Argosomus epilachnoides* Wollaston, 1868
Kimotoa Gruev, 1985b: 125 type species: *Argopus splendens* Gressitt & Kimoto, 1963
Musaka Bechyné, 1958a: 91 type species: *Sphaeroderma freyi* Bechyné, 1955
Sphaerochabria L. N. Medvedev, 1999c: 187 type species: *Sphaerochabria nepalica* L. N. Medvedev, 1999 **[# 756]**
Sphaerodermus Gistel, 1856: 379 [UE] type species: *Altica testacea* Fabricius, 1775
akebiae Ohno, 1964a: 57 **A:** JA
pseudoapicale Ohno, 1964a: 59 (forma) [IN]
alienum Weise, 1922b: 116 **A:** CH (FUJ)
alishanense Takizawa, 1979: 344 **A:** CH (TAI)
alternatum S.-H. Chen, 1939b: 66 **A:** CH (FUJ GAN SHA)
apicale Baly, 1874: 205 **A:** CH (CHQ FUJ GAN GUA GUI HUN JIA NIN SCH YUN ZHE) JA NC SC **ORR**
immaculithorax Nakane, 1958: 311 (var.) [IN] **[# 771]**
atrithorax S.-H. Chen, 1934h: 326 **A:** CH (YUN) **ORR**
atrum Jacoby, 1886a: 735 **A:** JA
balyi balyi Jacoby, 1886a: 735 **A:** CH (ZHE) JA NC RU (FE)
balyi hupehense Gressitt & Kimoto, 1963a: 824 (ssp.) **A:** CH (FUJ HUB)

bambusicola S.-Y. Wang, S.-Q. Ge & W.-Z. Li, 2010: 907, 910 **A:** CH (GUX YUN)
bicarinatum S.-Y. Wang, S.-Q. Ge & Cui, 2010: 906, 909 **A:** CH (YUN)
birmanicum Jacoby, 1892b: 927 **A:** NP **ORR**
brevicorne Jacoby, 1900b: 124 **A:** NP PA **ORR**
carinatum S.-Y. Wang, 1992c: 691, 737 **A:** CH (SCH SHA)
changi C.-F. Lee, 2023: 8 **A:** CH (TAI)
cheni L. N. Medvedev, 1973b: 884 **A:** CH (SCH)
chongi S.-H. Chen, 1935d: 772 **A:** CH (YUN)
chui Kimoto, 1970d: 297 **A:** CH (TAI)
confine S.-H. Chen, 1939a: 53 **A:** CH (GUX) **ORR**
darjeelingense Scherer, 1969: 210 **A:** ID (SD) NP
doeberli L. N. Medvedev, 1997a: 15 **A:** ID (SD) NP
flavitarse S.-Y. Wang & W.-Z. Li, 2004: 80, 81 **A:** CH (XIZ)
flavonotatum Chûjô, 1937a: 40 **A:** CH (TAI) JA (Ryukyus)
fraternale S.-H. Chen, 1939a: 53 **A:** CH (FUJ GUX HUB JIA) NC
fulvoapicale Kimoto & Gressitt, 1966: 546 **A:** JA (Ryukyus) **ORR**
fulvum Motschulsky, 1866b: 421 **A:** NP **ORR**
fuscicorne Baly, 1865b: 184 **A:** CH (HKG YUN) RU (FE)
gressitti Scherer, 1969: 205 **A:** ID (SD)
himalayense Scherer, 1969: 208 **A:** ID (SD) **ORR**
hsui C.-F. Lee, 2023: 4 **A:** CH (TAI)
inaizumii Takizawa, 2015: 244 **A:** JA (Honshu)
japanum Baly, 1874: 204 **A:** JA
jungchani C.-F. Lee, 2023: 16 **A:** CH (TAI)
kimotoi Scherer, 1969: 206 **A:** ID (SD) NP
kolibaci L. N. Medvedev, 2009d: 411 **A:** NP
kondoi Ohno, 1964c: 15 **A:** CH (TAI)
kuroashi Kimoto, 2000b: 153 [RN] **A:** CH (SCH) JA (Honshu) NP
  nigripes Kimoto, 1965: 444 [HN]
laterimarginatum L. N. Medvedev, 2015c: 328 **A:** CH (YUN)
leei Takizawa, 1980a: 9 **A:** NC SC
luteipenne Weise, 1922b: 118 **A:** NP **ORR**
maculatum S.-Y. Wang, 1992c: 691, 738 **A:** CH (YUN)
mandarense Jacoby, 1900b: 123 **A:** NP **ORR**
melli S.-H. Chen, 1933f: 56 **A:** CH (GUA)
monticola Scherer, 1969: 207 **A:** CH (XIZ) NP
morimotoi Chûjô & Ohno, 1964: 42 **A:** JA (Honshu, Kyushu)
nakanishii Kimoto, 2001: 66 **A:** NP
nepalense Bryant, 1952: 606 **A:** CH (CHQ SCH XIZ) NP
nepalicum L. N. Medvedev, 1999c: 187 **A:** NP (*Sphaerochabria*) [# 756, # 759]

nigriceps Scherer, 1969: 210 **A:** ID (SD)
nigricolle Jacoby, 1886a: 737 **A:** JA
abdominale Jacoby, 1886a: 736 [HN]
flaviventre Weise, 1916a: 40 [RN]
marginale Chûjô, 1937a: 36 (var.)
nigripes S.-H. Chen, 1934h: 329 **A:** JA (Honshu) NP **ORR**
nigroapicale Takizawa, 1979: 345 **A:** CH (TAI)
nigrocephalum S.-Y. Wang, 1992c: 692, 738 **A:** CH (SHA XIZ YUN)
nilum Gressitt & Kimoto, 1963a: 826 **A:** CH (HUB)
obscurum Ohno, 1964a: 45 **A:** JA (Honshu)
ohkuboi Chûjô, 1940b: 120 **A:** JA
orientale Jacoby, 1887a: 92 **A:** NP **ORR**
pallidipes Scherer, 1969: 208 **A:** ID (SD) NP
piceum Baly, 1876d: 582 **A:** CH (HUB JIA SCH SHG ZHE)
placidum Harold, 1877a: 364 **A:** JA
chujoi Ohno, 1964a: 51
pseudoplacidum Ohno, 1964a: 52 (forma) [IN]
postfasciatum S.-H. Chen, 1939b: 67 **A:** CH (GAN SCH)
postnigrum Chûjô, 1963a: 28 **A:** CH (TAI)
pseudapicale L. N. Medvedev, 1997a: 16 **A:** CH (YUN) NP **ORR**
pygmaeum Döberl, 2001: 383 [RN] **A:** NP
minutissimum L. N. Medvedev, 1997a: 15 [HN]
schawalleri L. N. Medvedev, 2009a: 58 [RN]
quadrimaculatum Chûjô, 1935b: 207 **A:** JA (Kyushu, Ryukyus)
resinulum Gressitt & Kimoto, 1963a: 828 **A:** CH (GUA HUB)
rubi Chûjô, 1937a: 42 **A:** CH (HUB TAI)
rubidum Graells, 1858: 97 [= 1858b: 135] (*Argopus*) **E:** AL AU BE BH BY CR CZ DE FR GB GE GR HU IT KR LA LS LT LU MA MD ME NL PL PT RO RU (ST) SB SK SL SP SV SZ TR UK **N:** AG CI MO MR TU **A:** AB CY IS JA (Honshu) LE SY TR
gyllenhali Capra, 1933: 165 (var.)
nitidum Portevin, 1934: 308
ocularium Allard, 1860d: 416
rufotestaceum Gressitt & Kimoto, 1963a: 829 **A:** CH (FUJ HUB)
schereri L. N. Medvedev, 1997e: 261 **A:** NP
seminigrum Jacoby, 1900a: 275 **A:** CH (GUX) NP **ORR**
separatum Baly, 1874: 205 **A:** CH (HUB FUJ) JA SC
seriatum Baly, 1874: 203 **A:** CH (FUJ HUN YUN ZHE) JA SC **ORR**
sexmaculatum L. N. Medvedev, 2007c: 300 **A:** NP
sheipaensis C.-F. Lee, 2023: 10 **A:** CH (TAI)
shirakii Csiki, 1940a: 509 [RN] **A:** CH (TAI)
tibiale Chûjô, 1937a: 41 [HN]

sinuatum Gressitt & Kimoto, 1963a: 831 **A:** CH (FUJ)
splendens Gressitt & Kimoto, 1963a: 820 (*Argopus*) **A:** CH (YUN)
subfurcatum S.-H. Chen, 1939a: 52 (*Argopus*) **A:** CH (GUA GUX)
tarsatum Baly, 1874: 203 **A:** JA RU (FE: Kuriles)
piceipenne Chûjô, 1937a: 37 (var.)
testaceum Fabricius, 1775: 114 (*Altica*) **E:** AL AU BE BH BU BY CR CZ DE EN FI FR GB GE GR HU IT KR LA LT LU MD ME NL NM NR PL RO RU (CT NT ST) SB SK SL SP SV SZ UK **A:** AB CY TR **NAR**i
cardui Gyllenhal, 1827: 659 (*Haltica*)
centaureae Stephens, 1831: 329
fulvum Geoffroy, 1785: 101 (*Altica*) [QS]
pacoviense Roubal, 1913b: 112 (ab.) [IN]
volutum Laicharting, 1781: 196 (*Altica*) [QS]
uenoi Takizawa, 2021b: 220 **A:** JA
unicolor Kimoto, 1965: 445 **A:** JA RU (ES FE)
varicolor Takizawa, 1979: 343 **A:** TAI
viride S.-Y. Wang, 1992c: 692, 738 **A:** CH (SCH)

genus *Taiwanohespera* Kimoto, 1970d: 300 type species: *Taiwanohespera sasajii* Kimoto, 1970
sasajii Kimoto, 1970d: 300 **A:** CH (TAI)
bicolor Kimoto, 1970d: 301

genus *Taiwanoliprus* Komiya, 2006c: 190 type species: *Taiwanoliprus endonis* Komiya, 2006
endonis Komiya, 2006c: 191 **A:** CH (TAI)
wenroni Komiya, 2006c: 192 **A:** CH (TAI)

genus *Taiwanorestia* Kimoto, 1991b: 122 type species: *Taiwanorestia nigra* Kimoto, 1991
nigra Kimoto, 1991b: 122 **A:** CH (TAI)

genus *Tebalia* Fairmaire, 1889b: 83 type species: *Tebalia caeruleata* Fairmaire, 1889
coeruleata Fairmaire, 1889b: 83 **A:** CH (XIZ)

genus *Trachytetra* Sharp, 1886b: 448 type species: *Phyllotreta rugulosa* Broun, 1880
Amydus S.-H. Chen, 1935a: 76 type species: *Amydus castaneus* S.-H. Chen, 1935
Monodaltica Bechyné, 1955b: 509 type species: *Monodaltica guineensis* Bechyné, 1955
Nesohaltica Maulik, 1929: 201 type species: *Nesohaltica nigra* Maulik, 1929
Trachyaphthona Heikertinger, 1924a: 34 type species: *Aphthona sordida* Baly, 1874
Typhodes Samuelson, 1984: 32 type species: *Typhodes aetherius* Samuelson, 1984
Zipangia Heikertinger, 1924a: 39 type species: *Haltica obscura* Jacoby, 1885

amamiana Ohno, 1961a: 83 (*Trachyaphthona*, ssp.) **A:** JA (Ryukyus)
aptera L. N. Medvedev, 1992d: 29 (*Zipangia*) **A:** NP
bicolora L. N. Medvedev, 1990b: 41 (*Zipangia*) **A:** NP
bidentata S.-H. Chen & S.-Y. Wang, 1980: 12, 24 (*Trachyaphthona*) **A:** CH (CHQ GAN GUI HUN SCH SHA YUN)
brevicornis Takizawa, 1979: 348 (*Trachyaphthona*) **A:** CH (TAI)
buddlejae S.-Y. Wang, 1990b: 127, 134 (*Trachyaphthona*) **A:** CH (YUN)
castanea S.-H. Chen, 1935a: 77 (*Amydus*) **A:** ID (SD) NP
gibbicollis Döberl, 1991b: 615 (*Amydus*)
nepalensis Scherer, 1989: 36 (*Amydus*)
cerambycina Scherer, 1969: 83 (*Zipangia*) **A:** ID (SD) NP
collaris Kimoto, 1996: 45 (*Trachyaphthona*) **A:** CH (TAI) JA (Ryukyus)
cyanea S.-H. Chen, 1939a: 41 (*Zipangia*) **A:** CH (GUX SCH)
formosana Takizawa, 1979: 347 (*Trachyaphthona*) **A:** CH (TAI)
fulva S.-Y. Wang, 1990b: 128, 134 (*Trachyaphthona*) **A:** CH (YUN)
fulvicornis Scherer, 1979b: 479 (*Zipangia*) **A:** ID (SD) NP
fusca Scherer, 1969: 86 (*Zipangia*) **A:** ID (UT)
hammondi Gruev, 1981d: 149 (*Zipangia*) **A:** CH (SHA)
hiunchulii Sprecher-Uebersax, 2000: 204 (*Trachyaphthona*) **A:** NP
imasakai Takizawa, 1990b: 6 (*Zipangia*) **A:** JA
infuscaticornis Scherer, 1979b: 480 (*Zipangia*) **A:** ID (SD)
latispina S.-Y. Wang, 1990b: 129, 135 (*Trachyaphthona*) **A:** CH (GAN SCH)
lewisi Jacoby, 1886a: 727 (*Haltica*) **A:** CH (FUJ) JA
micans Scherer, 1969: 84 (*Zipangia*) **A:** BT ID (SD) NP
montana Bryant, 1939: 14 (*Aphthona*) **A:** NP **ORR**
nepala Gruev, 1990b: 64 (*Zipangia*) **A:** NP
nigricornis Kimoto & Gressitt, 1966: 565 (*Zipangia*) **A:** JA (Ryukyus)
nigrocyanea Scherer, 1969: 83 (*Zipangia*) **A:** ID (SD)
nigrita Ohno, 1961a: 77 (*Trachyaphthona*) **A:** CH (TAI) JA (Kyushu, Ryukyus, Yakushima)
maebarai Ohno, 1961a: 79 (*Trachyaphthona*, ssp.)
picea Nakane & Kimoto, 1961b: 76 (*Zipangia*)
shibatai Chûjô, 1961: 88 (*Aphthona*)
nigrosterna S.-Y. Wang, 1990b: 129, 135 (*Trachyaphthona*) **A:** CH (SCH)
nishii Ohno, 1965a: 25 (*Trachyaphthona*) **A:** JA (Tanegashima)
obscura Jacoby, 1886a: 726 (*Haltica*) **A:** CH (FUJ JIA TAI ZHE) JA **ORR**
okinawana Takizawa, 1979: 350 (*Zipangia*) **A:** JA (Ryukyus)
ornata L. N. Medvedev, 2007d: 437 (*Trachyaphthona*) **A:** CH (YUN)
punctata Kimoto, 1996: 45 (*Trachyaphthona*) **A:** CH (TAI)
recticollis Takizawa, 1979: 348 (*Zipangia*) **A:** CH (TAI) JA (Ryukyus)

rugicollis S.-Y. Wang, 1997: 921, 930 (*Trachyaphthona*) **A:** CH (HUB)
sordida Baly, 1874: 197 (*Aphthona*) **A:** CH (FUJ HUB TAI) JA
subcostata L. N. Medvedev, 1990b: 40 (*Zipangia*) **A:** NP
suturalis S.-H. Chen, 1934h: 389 (*Zipangia*) **A:** CH (GAN "China-Vietnam border")
takizawai Kimoto, 1996: 49 (*Zipangia*) **A:** CH (TAI)

genus *Xuthea* Baly, 1865f: 248 type species: *Xuthea orientalis* Baly, 1865
subgenus *Paraxuthea* L. N. Medvedev, 1997d: 75 type species: *Xuthea leonardii* L. N. Medvedev, 1997
pallida L. N. Medvedev, 1997a: 20 **A:** NP
sinuata Gressitt & Kimoto, 1963a: 771 **A:** CH (SCH)
subgenus *Xuthea* Baly, 1865f: 248 type species: *Xuthea orientalis* Baly, 1865
bengalensis Sengupta & Basu, 1977: 199 **A:** ID (SD)
coerulea L. N. Medvedev, 1997a: 20 **A:** NP
elliptica Sengupta & Basu, 1977: 199 **A:** ID (SD)
laevicollis S.-H. Chen, 1933e: 387 **A:** ID (SD)
laticollis S.-H. Chen & S.-Y. Wang, 1981a: 495, 507 **A:** CH (XIZ)
nepalensis Scherer, 1983: 157 **A:** NP
orientalis Baly, 1865f: 249 **A:** BT CH (GAN SCH XIZ YUN) ID (AP SD) NP **ORR**
yunnanensis Heikertinger, 1948: 42 (ssp.) **A:** CH (TAI YUN) **ORR**
  geminalis S.-Y. Wang, 1992c: 683, 733

genus *Yemenaltica* Scherer, 1985: 86 type species: *Yemenaltica scorteccii* Scherer, 1985: 86
furthi Döberl, 2012: 444 **A:** YE (Socotra)
marginata L. N. Medvedev, 2012e: 418 **A:** YE
scorteccii Scherer, 1985: 86 **A:** AE OM YE

genus *Yetialtica* Döberl, 1991b: 626 type species: *Yetialtica besucheti* Döberl, 1991
besucheti Döberl, 1991b: 627 **A:** NP

genus *Yunohespera* S.-H. Chen & S.-Y. Wang, 1984b: 318, 322 type species: *Yunohespera sulcicollis* S.-H. Chen & S.-Y. Wang, 1984
sulcicollis S.-H. Chen & S.-Y. Wang, 1984b: 318, 322 **A:** CH (YUN)

genus *Yunotrichia* S.-H. Chen & S.-Y. Wang, 1980: 8, 21 type species: *Yunotrichia mediovittata* S.-H. Chen & S.-Y. Wang, 1980
mediovittata S.-H. Chen & S.-Y. Wang, 1980: 9, 22 **A:** CH (XIZ YUN)

genus *Zangaltica* S.-H. Chen & S.-Y. Wang, 1988: 348, 353 type species: *Zangaltica multicostata* S.-H. Chen & S.-Y. Wang, 1988
multicostata S.-H. Chen & S.-Y. Wang, 1988: 348, 353 **A:** CH (XIZ)

genus *Zipanginia* Ohno, 1962b: 23 type species: *Graptodera picipes* Baly, 1874
Yoshiakia Takizawa, 2009: 149 type species: *Yoshiakia iwatensis* Takizawa, 2009
bungoensis Takizawa, 2015: 247 **A:** JA (Kyushu)
iwatensis Takizawa, 2009: 151 (*Yoshiakia*) **A:** JA (Honshu)
leehsuehae C.-F. Lee, 2022b: 183 **A:** CH (TAI)
loochooensis Ohno, 1962b: 27 **A:** JA (Ryukyus)
miyatakei Kimoto, 1971b: 5 **A:** JA (Honshu)
picipes katoi Ohno, 1962b: 26 (ssp.) **A:** JA (Honshu)
picipes picipes Baly, 1874: 191 (*Graptodera*) **A:** JA
sakishimana Kimoto & Gressitt, 1966: 566 **A:** CH (TAI) JA (Ryukyus)
tosana Ohno, 1964c: 38 **A:** JA (Shikoku)
tuberosa Ohno, 1964c: 35 **A:** JA (Honshu)

supertribe Alticitae, nomina dubia
Altica brunnea Grimmer, 1841: 48 E: AU
Altica fossulata Grimmer, 1841: 48 E: AU

supertribe Serraticollitae B. White, 1942

genus *Aulacothorax* Boheman, 1858: 35 type species: *Aulacothorax exilis* Boheman, 1858
Epithrella L. N. Medvedev, 1994: 17 type species: *Epithrella philippina* L. N. Medvedev, 1994
Leptotrichaltica Heikertinger, 1925a: 68 [RN] type species: *Haltica recticollis* LeConte, 1861
Leptotrix Horn, 1889: 249 type species: *Haltica recticollis* LeConte, 1861
Livolia Jacoby, 1903b: 15 type species: *Livolia sulcicollis* Jacoby, 1903
Livoliella L. N. Medvedev, 1997b: 100 type species: *Livoliella luzonica* L. N. Medvedev, 1997
Micrepitrix Laboissière, 1933c: 205 type species: *Micrepitrix coomani* Laboissière, 1933
Orthaltica Crotch, 1873: 69 type species: *Crioceris copalina* Fabricius, 1801
Serraticollis B. White, 1942: 17 type species: *Serraticollis rhois* B. White, 1942
arabica L. N. Medvedev, 1997f: 323 (*Livolia*) **A:** SA
assamensis Scherer, 1971: 26 (*Livolia*) **A:** NP **ORR**
coomani Laboissière, 1933c: 208 (*Micrepitrix*) **A:** CH (HAI HKG TAI) **ORR**
laboissierei S.-H. Chen, 1935d: 777 (*Micrepitrix*) **A:** CH (FUJ JIA)
okinawana Kimoto & Gressitt, 1966: 543 (*Micrepitrix*) **A:** CH (TAI) JA (Ryukyus)
purba Basu & Sengupta, 1985: 47 (*Orthaltica*) **A:** ID (SD)
shirozui Chûjô, 1957c: 5 (*Epithrix*) **A:** JA (Ryukyus)

subfamily **Lamprosomatinae** Lacordaire, 1848

tribe Lamprosomatini Lacordaire, 1848

genus *Asisia* Bezděk, Löbl & Konstantinov, 2010: 80 [RN] type species: *Guggenheimia sumatrana* Monrós, 1956

Guggenheimia Monrós, 1956: 52 [HN] type species: *Guggenheimia sumatrana* Monrós, 1956

assamensis Jacoby, 1908: 281 (*Lamprosoma*) **A:** NP **ORR**

genus *Oomorphus* Curtis, 1831: 347 type species: *Byrrhus concolor* Sturm, 1807

subgenus *Oomorphus* Curtis, 1831: 347 type species: *Byrrhus concolor* Sturm, 1807

Delphastobia Casey, 1924: 171 type species: *Delphastobia mexicana* Casey, 1924

Laophilus Gistel, 1856: 381 type species: *Byrrhus concolor* Sturm, 1807

concolor Sturm, 1807: 109 (*Byrrhus*) **E:** AU BE BH BU CR CZ DE EN FR GB GE HU IT KR LA LU NL PL RO RU (CT ST) SB SK SL SP SV SZ UK **A:** AB GG IN

kolbei Scholz, 1926: 104 (*Lamprosoma*)

maritimus Stephens, 1829a: 159 (*Phalacrus*)

unicolor Brullé, 1837: 355

japanus Jacoby, 1885b: 197 **A:** JA

chujoi Ohno, 1961d: 48 (forma) [IN]

genus *Oomorphoides* Monrós, 1956: 54 type species: *Lamprosoma cupreatum* Baly, 1873

aeneus S.-H. Chen, 1932a: 108 (*Lamprosoma*) **A:** CH (FUJ HUB SCH)

alienus Bates, 1866: 353 (*Lamprosoma*) **A:** CH (TAI)

chujoi Takizawa, 1987: 48 **A:** NP

confucii Weise, 1889g: 594 (*Lamprosoma*) **A:** CH (GAN)

cupreatus Baly, 1873: 82 (*Lamprosoma*) **A:** JA NC SC

yuasai Ohno, 1961d: 49 (forma) [IN]

flavicornis Tan, 1964: 102, 104 **A:** CH (YUN)

formosensis Chûjô, 1934a: 245 (*Lamprosoma*) **A:** CH (TAI)

foveatus Tan, 1992b: 774, 811 **A:** CH (SCH)

loochooensis Chûjô, 1958c: 4 **A:** JA

yakushimensis Ohno, 1961d: 50 (ssp.)

martensi L. N. Medvedev, 1990b: 7 **A:** NP

metallicus Gressitt & Kimoto, 1961: 191 **A:** CH (HUB)

nepalensis Takizawa, 1987: 49 **A:** ID (SD) NP

nigrocaeruleus Baly, 1873: 83 (*Lamprosoma*) **A:** JA NC RU (FE) SC

okinawensis Chûjô, 1935a: 75 (*Lamprosoma*) **A:** JA

kurosawai Nakane, 1958: 304 (ssp.)

omeiensis Tan, 1964: 103, 104 **A:** CH (SCH)

pallidicornis Gressitt & Kimoto, 1961: 192 **A:** CH (GUA JIX HAI)

piliceps S.-H. Chen, 1940a: 204 (*Lamprosoma*) **A:** CH (FUJ GAN)

punctatus Tan, 1992b: 775, 812 **A**: CH (SCH)
purpurascens Tan, 1964: 103, 105 **A**: CH (SCH)
sakaii Takizawa, 1989a: 329 **A**: ID (SD)
sakishimanus Kimoto & Gressitt, 1966: 499 **A**: CH (TAI) JA (Ryukyu Isl.)
tonkinensis Chûjô, 1935c: 94 (*Lamprosoma*) **A**: CH (FUJ GUA HAI HUB SCH) **ORR**
violaceonigrus Chûjô, 1953b: 243 (*Lamprosoma*) **A**: CH (FUJ TAI)
yaosanicus S.-H. Chen, 1940a: 205 (*Lamprosoma*) **A**: CH (FUJ GUA GUX HAI JIX)

subfamily **Cryptocephalinae** Gyllenhal, 1813
tribe Clytrini Kirby, 1837
subtribe Acolastina Schöller, **subtrib. nov.** [# 785]

genus *Acolastus* Gerstaecker, 1855: 636 type species: *Acolastus callosus* Gerstaecker, 1855

subgenus *Acolastus* Gerstaecker, 1855: 636 type species: *Acolastus callosus* Gerstaecker, 1855

Falsopachybrachys Pic, 1947a: 5 type species: *Falsopachybrachys diversepubens* Pic, 1947
Pachylanka L. N. Medvedev, 1989: 6 type species: *Pachylanka denticollis* L. N. Medvedev, 1989
Serrinotus Tan, 1992b: 785, 818 type species: *Serrinotus albopilosus* Tan, 1992
Thelylankus Lopatin, 1987b: 778 type species: *Thelyterotarsus insularis* Lopatin, 1987
Thelyterotarsus Weise, 1882b: 246 type species: *Thelyterotarsus fausti* Weise, 1882

albopilosus Tan, 1992b: 785, 819 (*Serrinotus*) **A**: CH (SCH YUN)
altaicus L. N. Medvedev & Voronova, 1977: 315 (*Thelyterotarsus*) **A**: MG
anthracinus Lopatin, 1976a: 107 (*Thelyterotarsus*) **A**: KI
atasicus L. N. Medvedev & Voronova, 1977: 314 (*Thelyterotarsus*) **A**: MG
atraphaxidis Lopatin, 1960: 39 (*Thelyterotarsus*) **A**: TD TM UZ
badakshanicus Lopatin, 1967a: 218 (*Thelyterotarsus*) **A**: AF TD
baeckmanni Jacobson, 1917: 274 (*Thelyterotarsus*) **A**: KI KZ TM UZ
ogloblini Jankowski, 1932: 168 (*Thelyterotarsus*)
balchaschensis Lopatin & Kulenova, 1982: 44 (*Thelyterotarsus*) **A**: KZ
batangensis Tan, 1992b: 786, 819 (*Serrinotus*) **A**: CH (SCH YUN)
callosomarginatus Schöller, 2018: 410 **A**: YE
cribripennis Pic, 1924h: 30 (*Thelyterotarsus*) **A**: IS SI **N**: EG
darvazicus Lopatin, 1975b: 48 (*Thelyterotarsus*) **A**: TD
dzhungarus Lopatin, 1976a: 109 (*Thelyterotarsus*) **A**: KZ TD
fausti fausti Weise, 1882b: 247 (*Thelyterotarsus*) **A**: AB AR GG IN TM
armeniacus Pic, 1906c: 35 (*Thelyterotarsus*, var.)
georgicus Lopatin, 1986b: 28 (*Thelyterotarsus*)
nigronotatus Pic, 1906c: 35 (*Thelyterotarsus*, var.)
fausti transcaspicus Pic, 1909d: 131 (*Thelyterotarsus*, var.) **A**: TM UZ
bergi Jacobson, 1917: 274 (*Thelyterotarsus*)
margaritae Lopatin, 1997b: 372 (*Thelyterotarsus*)

furcatus Schöller, 2000: 26 **A:** CH (XIZ)
furthi Lopatin, 1994: 524 (*Thelyterotarsus*) **A:** IS SI
gobicus L. N. Medvedev & Voronova, 1977: 313 (*Thelyterotarsus*) **A:** MG
gurjevae gurjevae Lopatin, 1968c: 541 (*Thelyterotarsus*) **A:** UZ
gurjevae unicolor Lopatin, 2009c: 172 (ssp.) **A:** UZ [# 798]
hauseri Weise, 1887d: 331 (*Thelyterotarsus*) **A:** TD (Kyndyr-Tau)
  confluens Weise, 1887d: 332 (*Thelyterotarsus*, var.)
hoberlandti Lopatin, 1980: 622 (*Thelyterotarsus*) **A:** IN
iliensis Lopatin, 1967a: 218 (*Thelyterotarsus*) **A:** KZ
inopinatus Lopatin, 1992b: 397 (*Thelyterotarsus*) **A:** KZ
intermedius Lopatin, 1968c: 543 (*Thelyterotarsus*) **A:** TD
issykensis Lopatin, 1992b: 396 (*Thelyterotarsus*) **A:** KI
ivanovi Jacobson, 1925c: 12 (*Thelyterotarsus*) **A:** KI TD TM UZ
jacobsoni Lopatin, 1968c: 542 (*Thelyterotarsus*) **A:** TD
karakirgiza Jacobson, 1925c: 11 (*Thelyterotarsus*) **A:** TD UZ
karatavicus Lopatin, 1976a: 110 (*Thelyterotarsus*) **A:** KZ TD UZ
karateginus Lopatin, 1992b: 402 (*Thelyterotarsus*) **A:** KI TD
kaszabi Lopatin, 1968b: 216 (*Thelyterotarsus*) **A:** MG
khnzoriani Lopatin, 1976a: 110 (*Thelyterotarsus*) **A:** TM
korotyaevi Lopatin, 1992b: 397 (*Thelyterotarsus*) **A:** KZ TD
kryzhanovskii Lopatin, 1976a: 109 (*Thelyterotarsus*) **A:** TM
kuramensis Lopatin, 1997b: 374 (*Thelyterotarsus*) **A:** TD
limbatus Lopatin, 1992b: 395 (*Thelyterotarsus*) **A:** KI
loebli Schöller, 2018: 408 **A:** OM
lopatini L. N. Medvedev & Voronova, 1977: 315 (*Thelyterotarsus*) **A:** MG (west)
lygaeus Iablokoff-Khnzorian, 1965: 21 (*Thelyterotarsus*) **A:** KI
medvedevi Lopatin, 1977a: 171 (*Thelyterotarsus*) **A:** MG
miscellus Berti & Rapilly, 1973: 872 (*Thelyterotarsus*) **A:** IN
mitjaevi Lopatin, 2008c: 37 **A:** UZ
mogoltavicus Lopatin, 1992b: 401 (*Thelyterotarsus*) **A:** TD
mongolicus Lopatin, 1971a: 231 (*Thelyterotarsus*) **A:** MG
murinus Lopatin, 1961a: 203 (*Thelyterotarsus*) **A:** TD
nanus Lopatin, 1976a: 107 (*Thelyterotarsus*) **A:** KZ (Karatau Ridge, Kyzylkum desert)
nigrifrons Jacobson, 1917: 272 (*Thelyterotarsus*) **A:** TD (Hissar Ridge)
nuratavicus Lopatin, 1992b: 401 (*Thelyterotarsus*) **A:** UZ
ochraceus Lopatin, 1966c: 367 (*Thelyterotarsus*) **A:** AF
ophthalmicus Lopatin, 1997b: 372 (*Thelyterotarsus*) **A:** IN
paki Romantsov, 2017: 351 **A:** TD
pallidus montanus Lopatin, 1976a: 111 (*Thelyterotarsus*) **A:** TD (Badakshan)
pallidus pallidus Lopatin, 1956c: 43 (*Thelyterotarsus*) **A:** TD (south) UZ
pallidus tianshanicus Lopatin, 2009c: 173 (ssp.) **A:** KI [# 798]

postfasciatus Lopatin, 1975b: 49 (*Thelyterotarsus*) **A:** TD
przevalskii Lopatin, 1992b: 402 (*Thelyterotarsus*) **A:** KI
regeli Jacobson, 1898: 194 (*Thelyterotarsus*) **A:** CH (XIN) KZ
rufus Lopatin, 1980: 621 (*Thelyterotarsus*) **A:** IN
semenovi Jacobson, 1917: 272 (*Thelyterotarsus*) **A:** IN
seravschanicus Lopatin, 1960: 40 (*Thelyterotarsus*) **A:** TD UZ
shahristanus Lopatin, 1987a: 82 (*Thelyterotarsus*) **A:** TD
sogdianus Lopatin, 1992b: 400 (*Thelyterotarsus*) **A:** TD UZ
spitameni Lopatin, 1990d: 48 (*Thelyterotarsus*) **A:** TD
substriatus L. N. Medvedev, 1996: 241 (*Thelyterotarsus*) **A:** AE OM SA YE
syrdariensis Romantsov, 2003: 698 **A:** KZ
tadzhibajevi Lopatin, 1975b: 47 (*Thelyterotarsus*) **A:** TD
tadzhicus Lopatin, 1968c: 543 (*Thelyterotarsus*) **A:** TD UZ
tatianae Lopatin, 1983a: 92 (*Thelyterotarsus*) **A:** TD
tsaganicus L. N. Medvedev, 1980a: 316 (*Thelyterotarsus*) **A:** MG
turkestanicus Lopatin, 1965b: 8 (*Thelyterotarsus*) **A:** KI TD
velutinus Lopatin, 1967a: 219 (*Thelyterotarsus*) **A:** TD
volkovitshi Lopatin, 1986b: 27 (*Thelyterotarsus*) **A:** UZ
wittmeri Lopatin, 1979a: 299 (*Thelyterotarsus*) **A:** AE OM SA YE
x-signatus Lopatin, 1985b: 768 (*Thelyterotarsus*) **A:** IN TR **[# 799]**
zaissanicus Lopatin, 1967a: 217 (*Thelyterotarsus*) **A:** KZ
subgenus *Anodontelytrus* Jacobson, 1917: 268 type species: *Pachybrachys caroli* Marseul, 1884
brunneomaculatus Pic, 1896e: 204 (*Pachybrachis*, var.) **N:** AG TU
brunneonotatus Clavareau, 1913: 96 (ab.) [IN]
brunneonotatus Jacobson, 1917: 268 (ssp.)
buettikeri Lopatin, 1983b: 207 (*Thelyterotarsus*) **A:** JO SA
caroli Marseul, 1884: 188 (*Pachybrachys*) **N:** AG TU
leopardinus Chobaut, 1899: 389 (*Pachybrachis*)
denticulatus L. N. Medvedev, 1996: 241 (*Thelyterotarsus*) **A:** SA
granulatus Berti & Doguet, 1994: 152 (*Thelyterotarsus*) **N:** MO
mirandus Lopatin, 1980: 620 (*Thelyterotarsus*) **A:** IN
subgenus *Lapsionus* Ponting, 2020: 298 [RN] type species: *Pachybrachys theryi* Chobaut, 1896
Anopsilus Jacobson, 1917: 268 [HN] type species: *Pachybrachys theryi* Chobaut, 1896
afghanicus L. N. Medvedev, 1978c: 868 (*Thelyterotarsus*) **A:** AF
apicenotatus Chobaut, 1899: 388 (*Pachybrachis*) **N:** AG TU
deletus Chobaut, 1899: 389 (*Pachybrachis*, var.)
multinotatus Pic, 1906b: 27 (*Pachybrachis*, var.)
arabicus Lopatin, 1982a: 465 (*Thelyterotarsus*) **A:** AE OM **NTR**i
balchanus Lopatin, 2009c: 172 **A:** TM

bezdeki Schöller, 2020: 85 **A:** AE
costatus L. N. Medvedev & Sprecher-Uebersax, 1999: 361 (*Thelyterotarsus*) **A:** NP
glabratus Lopatin, 1985b: 765 (*Thelyterotarsus*) **A:** IN TR
gobustanus Lopatin, 1992b: 403 [= 1992e: 34] (*Thelyterotarsus*) **A:** AB
hebraeus J. R. Sahlberg, 1913a: 71 (*Pachybrachys*) **A:** IQ IS SI
sinaicus Pic, 1937c: 13 (*Thelyterotarsus*)
iranicus Lopatin, 1980: 621 (*Thelyterotarsus*) **A:** IN SY TR
jelineki Lopatin, 1985b: 766 (*Thelyterotarsus*) **A:** IN
klimenkoi Romantsov & Bukejs, 2009: 90 **A:** IN
latifrons Lopatin & Nesterova, 2007a: 308 **A:** AE OM
lugubris Berti & Rapilly, 1973: 871 (*Thelyterotarsus*) **A:** IN
mandli Lopatin, 1967e: 438 (*Thelyterotarsus*) **A:** AF IN
mesopotamicus Lopatin, 1996: 638 (*Thelyterotarsus*) **A:** IQ
minimus Jacobson, 1917: 270 (*Thelyterotarsus*) **A:** KZ TM (Karatau)
pici Lopatin, 1985b: 766 (*Thelyterotarsus*) **A:** IN
praevius Suffrian, 1857: 249 (*Pachybrachys*) **N:** AG EG MO **A:** IS SI **AFR** [# 800]
curtus Lopatin, 1994: 525 (*Thelyterotarsus*)
shirazicus Nesterova & Lopatin, 2009: 273 **A:** IN
similis Lopatin, 1976a: 111 (*Thelyterotarsus*) **A:** KZ TD
theryi Chobaut, 1896a: 151 (*Pachybrachis*) **N:** AG TU
cardui Pic, 1896e: 204 (*Pachybrachis*, var.)
tuniseus Pic, 1900c: 63 (*Pachybrachys*)
zarudnyi Jacobson, 1917: 268 (*Thelyterotarsus*) **A:** AF IN

subtribe Clytrina Kirby, 1837

genus *Aetheanta* L. N. Medvedev, 1988a: 476 type species: *Aetheanta fasciata* L. N. Medvedev, 1988 (= *Smaragdina higuchii* Kimoto & Takizawa, 1981)
higuchii Kimoto & Takizawa, 1981: 53 (*Smaragdina*) **A:** CH (XIZ) ID (SD) NP **ORR**
bothrionota Tan, 1987: 42 (*Smaragdina*)
fasciata L. N. Medvedev, 1988a: 476

genus *Aetheodactyla* Lacordaire, 1848: 188 type species: *Clythra lateralis* Lacordaire, 1848
Pseudoclytra Jacoby, 1908: 101 type species: *Tituboea andrewesi* Jacoby, 1895
dimidiatipennis Baly, 1878h: 371 (*Coptocephala*) **A:** ID (KA UT) NP PA
plagiata Duvivier, 1891a: xxix (*Clytra*) **A:** ID (UT) **ORR**

genus *Aetheomorpha* Lacordaire, 1848: 311 type species: *Clythra nematoides* Lacordaire, 1848
apicata L. N. Medvedev, 1988a: 474 **A:** ID (SD) NP
nigroapicalis Lopatin, 2005c: 231 (*Smaragdina*)

bacboensis L. N. Medvedev, 1992b: 72 **A:** CH (GUX YUN) **ORR**
bidentata F.-Y. Wang & Zhou, 2012: 1410 **A:** CH (SCH)
bilobata F.-Y. Wang & Zhou, 2012: 1413 **A:** CH (YUN)
brevelineata Pic, 1943c: 14 (*Diapromorpha*) **A:** ID (SD) **ORR**
cochinchinensis cochinchinensis Lefèvre, 1889a: 290 (*Gynandrophthalma*) **A:** NP **ORR**
coerulea Jacoby, 1892b: 882 (*Aspidolopha*) **A:** CH (HAI YUN) NP **ORR**
cyanea Pic, 1928g: 25 **A:** CH (YUN) **ORR**
decemnotata Jacoby, 1892b: 884 (*Gynandrophthalma*) **A:** CH (YUN) **ORR**
bavayi Pic, 1932c: 13 (*Cyaniris*, var.)
gressitti L. N. Medvedev & Regalin, 1998: 19 **A:** CH (YUN)
fallax Lacordaire, 1848: 314 (*Clythra*) **A:** PA **ORR**
furcata L. N. Medvedev & Kantner, 2002: 262 **A:** CH (YUN) **ORR**
fuscicornis Lacordaire, 1848: 386 (*Clythra*) **A:** NP **ORR**
hakka Gressitt & Kimoto, 1961: 86 **A:** CH (GUA SCH) [# 835]
laeta L. N. Medvedev, 1995a: 90 **A:** CH (YUN)
malayana Baly, 1865c: 51 (*Gynandrophthalma*) **A:** ID (HP SD) NP **ORR**
unijuncta Pic, 1927c: 29 (*Cyaniris*, var.)
nitidicollis Jacoby, 1908: 143 (*Aspidolopha*) **A:** NP **ORR**
obtusapicata F.-Y. Wang & Zhou, 2012: 1415 **A:** CH (YUN)
pictipennis Jacoby, 1892b: 883 (*Gynandrophthalma*) **A:** NP **ORR**
pseudonagaensis L. N. Medvedev, 1988a: 474 **A:** ID (SD)
punctistriata F.-Y. Wang & Zhou, 2012: 1418 **A:** CH (YUN)
sauteri Chûjô, 1934f: 286 (*Cyaniris*) **A:** CH (TAI)
seminigra pumilio Lacordaire, 1848: 313 (*Clythra*) **N:** EG **A:** AE IN OM SA YE **AFR** [# 801, # 833]
minuta Pic, 1950e: 203 (*Diapromorpha*)
substriata L. N. Medvedev, 2003a: 317 **A:** ID (SD) NP
suturata Jacoby, 1898a: 185 (*Clythra*) **A:** ID (HP) NP **ORR**
morimotoi Kimoto & Gressitt, 1981: 310 (*Clytra*)
multiplagiata Jacoby, 1903a: 87
taiwana Chûjô, 1952a: 41 **A:** CH (TAI YUN)
bowditchi Gressitt & Kimoto, 1961: 87
unifasciata Jacoby, 1908: 129 **A:** ID (SD) **ORR**
virgula Jacoby, 1908: 128 **A:** NP ID (SD UT) [# 835]
wangi F.-Y. Wang & Zhou, 2012: 1420 **A:** CH (YUN)
weigeli L. N. Medvedev, 2010f: 195 **A:** CH (YUN)
yunnana Pic, 1927c: 27 **A:** CH (YUN)

genus *Afrophthalma* L. N. Medvedev, 1978b: 57 type species: *Gynandrophthalma apicalis* Jacoby, 1891
antipaeformis L. N. Medvedev, 1993a: 131 **A:** OM SA YE

arabica Bryant, 1957b: 354 (*Gynandrophthalma*) **A:** OM SA YE
filiformis Lacordaire, 1848: 317 (*Clythra*) **A:** SA **AFR**
miopristoides Burgeon, 1942: 49 (*Gynandrophthalma*)

genus *Aspidolopha* Lacordaire, 1848: 252 type species: *Clythra melanophtalma* Lacordaire, 1848
buqueti egregia Boheman, 1859: 154 (*Clythra*) **A:** CH (YUN) ID (SD) NP **ORR**
humeralis Pic, 1928g: 25 (var.)
siamensis Pic, 1941b: 8
trinotata Pic, 1928g: 24
centromaculata L. N. Medvedev, 1988c: 46 **A:** ID (SD)
costata Jacoby, 1892b: 881 **A:** CH (YUN) **ORR**
himalayensis L. N. Medvedev, 1988a: 475 **A:** ID (SD)
medogensis Tan, 1988a: 322, 332 **A:** CH (XIZ)
melanophtalma Lacordaire, 1848: 254 (*Clythra*) **A:** CH (GUA GUX HAI YUN) ID (SD) NP **ORR**
instigmata Pic, 1933a: 3 (var.)
minuta Pic, 1933a: 3 (*Diapromorpha*) [HN]
pseudospilota L. N. Medvedev, 1988c: 47 **A:** ID (SD) NP **ORR**
quadrinotata Bryant, 1952: 603 **A:** ID (SD) NP
spilota Hope, 1831: 30 (*Cryptocephalus*) **A:** CH (YUN) ID (SD UT) NP **ORR**
rugosa Jacoby, 1891a: 32
thoracica Jacoby, 1892b: 879 **A:** CH (GUA GUI GUX HAI JIL NIN SCH YUN) **ORR** [# 835]
bisignata Pic 1927c: 26 (var.)
infasciata Pic, 1927c: 26 (var.)
lajoyei Pic, 1927c: 26
medioobliterata Pic, 1927m: 7 (var.)
obliterata Pic, 1927c: 26 (var.)
submaculata Pic, 1927m: 6
unifasciata Pic, 1927c: 26 **A:** CH (YUN)

genus *Ceratobasis* Lacordaire, 1848: 362 type species: *Cryptocephalus koenigii* Fabricius, 1775
koenigii Fabricius, 1775: 111 (*Cryptocephalus*) **A:** ID (SD UT) NP **ORR**
bimaculata Jacoby, 1908: 99 (*Miopristis*)
binotaticeps Pic, 1943c: 8
ceylonensis Jacoby, 1908: 173
discoidalis Pic, 1943c: 8
imbasalis Pic, 1943b: 16 (var.)
impolita Weise, 1913d: 219 (*Coptocephala*) [RN]

innotatithorax Pic, 1943b: 16
nair Lacordaire, 1848: 363 (*Clythra*)
obliterata Pic, 1943c: 8 (var.)
opacipennis Jacoby, 1903a: 86 (*Gynandrophthalma*)
paulofasciata Pic, 1943c: 8 (var.)
unireducta Pic, 1943c: 8 (var.)

genus *Cheilotoma* Chevrolat, 1836: 420 type species: *Chrysomela bucephala* Schaller, 1783 (= *Chrysomela musciformis* Goeze, 1777)
subgenus *Cheilotoma* Chevrolat, 1836: 420 type species: *Chrysomela bucephala* Schaller, 1783 (= *Chrysomela musciformis* Goeze, 1777)
Chilotoma Agassiz, 1846b: 78 [UE] type species: *Chrysomela bucephala* Schaller, 1783 (= *Chrysomela musciformis* Goeze, 1777)
beldei Kasap, 1984: 216 **A:** IN IS JO TR
ammanica Lopatin, 1995a: 100 (ssp.)
cankiriensis Özdikmen & Bal, 2016a: 305 **A:** TR
erythrostoma erythrostoma Faldermann, 1837: 376 **E:** BU CZ HU KR RO RU (ST) SK UK **A:** AB AR GG IN KZ TR [# 802]
erythostoma Faldermann, 1837: 376 [unused original spelling]
erythrostoma italica Leoni, 1906: 147 (*Chilotoma*) **E:** IT SP (Villamanreque de Tajo)
baudii Leoni, 1906: 147 (*Chilotoma*, var.)
leonii Depoli, 1932: 87 (*Chilotoma*, ab.) [IN]
fulvicollis J. R. Sahlberg, 1913a: 66 (*Chilotoma*) **A:** SY
musciformis apennina L. N. Medvedev, 2004b: 38 (ssp.) **E:** IT
hispanica L. N. Medvedev, 2004b: 38 (ssp.)
musciformis iranica L. N. Medvedev, 2004b: 38 (ssp.) **A:** IN
musciformis musciformis Goeze, 1777: 319 (*Chrysomela*) **E:** AU BE BH BU CR CZ FR GE GR HU IT (Venezia Giulia) LU MD ME NM PL RO RU (CT ST) SB SK SL UK **A:** AB GG IS KZ MG RU (SS WS) TR
bucephala Schaller, 1783: 276 (*Chrysomela*)
leosinii Leoni, 1906: 146 (*Chilotoma*, var.) [QS]
muscoides Geoffroy, 1785: 72 (*Melolontha*)
orientalis Jacobson, 1901b: 107 (*Chilotoma*, var.) [IN] [# 795]
siegae J. Müller, 1921: 71 (*Chilotoma*, ab.) [IN]
voriseki L. N. Medvedev & Kantner, 2003: 268 **A:** IQ TR (Nemrut Dagh) **[Appendix]**
subgenus *Exaesiognatha* Jacobson, 1923: 233 type species: *Exaesiognatha ivanovi* Jacobson, 1923
ivanovi ivanovi Jacobson, 1923: 234 (*Exaesiognatha*) **A:** TD UZ
ivanovi turcomanica L. N. Medvedev, 1971a: 692 (*Chilotoma*, ssp.) **A:** TM (Kopet-Dagh)

genus *Chilotomina* Reitter, 1913a: 84 type species: *Clythra nigritarsis* Lacordaire, 1848

apicitarsis Fairmaire, 1876b: 50 (*Gynandrophthalma*) **N:** AG

bergeali Warchałowski, 2000b: 575

korbi Weise, 1895c: 216 (*Gynandrophthalma*) E: SP

marocana Pic, 1936c: 26 (*Cyaniris*) **N:** MO

erberi Warchałowski, 2000b: 577

moroderi bacarescensis Cobos, 1961: 30 (*Gynandrophthalma*, ssp.) E: SP

moroderi moroderi Escalera, 1928: 432 (*Gynandrophthalma*) E: SP

moroderi Cobos, 1969: 73 (*Gynandrophthalma*) [HN]

nigritarsis Lacordaire, 1848: 297 (*Clythra*) E: FR SP [# 1086]

oberthuri Lefèvre, 1876a: 77 (*Gynandrophthalma*) E: PT SP

baetica Weise, 1881: 123 (*Gynandrophthalma*)

regalini Warchałowski, 2000b: 581 **N:** MO

genus *Clytra* Laicharting, 1781: 165 type species: *Chrysomela quadripunctata* Linnaeus, 1758

subgenus *Clytra* Laicharting, 1781: 165 type species: *Chrysomela quadripunctata* Linnaeus, 1758

Clythra Fabricius, 1798: 5 [incorrect subsequent spelling]

Clythraria Rafinesque, 1815: 116 [RN] type species: *Chrysomela quadripunctata* Linnaeus, 1758

Hirtoclytra L. N. Medvedev, 1961: 640 type species: *Clytra puberula* Weise, 1898

Melolontha Geoffroy, 1762: 195 [suppressed name, ICZN (1994), Opinion 1754] type species: *Chrysomela quadripunctata* Linnaeus, 1758

aliena Weise, 1897c: 64 (*Clythra*) E: GR **A:** AR TR

appendicina Lacordaire, 1848: 208 (*Clythra*) **E:** AU BH BU CR CZ GR HU IT KO KR MD ME NM PT RO RU (ST) SB SK SL SZ **A:** AB AF AR IN IS KZ TD TM TR [# 803]

latina J. Müller, 1951: 292 (ssp.)

pseudolaeviuscula L. N. Medvedev, 1961: 647 (var.) [IN]

arida Weise, 1889g: 563 (var.) E: NT (North Ural) **A:** CH (BEI GAN HEB HEI HEN HUB HUN JIA JIL LIA NIN NMO QIN SCH SHA SHN SHX) JA KZ MG NC RU (ES FE SS WS) SC

antistita Weise, 1889g: 563 (var.)

ehnbergi Jacobson, 1901a: 94 (var.) [IN] [# 795]

ehnbergi Jacobson, 1901b: 110 (var.) [IN] [# 795]

duodecimmaculata duodecimmaculata Fabricius, 1781: 139 (*Cryptocephalus*) [RN] **A:** CH (GUA GUX HAI YUN) **ORR**

bella Baly, 1865c: 48 (*Clythra*)

bicoloriceps Pic, 1933a: 3 [HN]

binhana Pic, 1927c: 31
cryptocephala Hornstedt, 1787: 7 (*Chrysomela*)
decemmaculata Fabricius, 1775: 106 (*Cryptocephalus*) [HN]
gigantea Jacoby, 1892b: 878 (*Diapromorpha*)
gigas Forsberg, 1821: 283 (*Clythra*)
inapicalis Pic, 1927c: 31 (var.)
theresae Pic, 1927c: 30
espanoli Daccordi & Petitpierre, 1977: 227 **E:** FR IT (Sardinia) SP **N:** MO [# 835]
guerryi Pic, 1927h: 187 **A:** CH (YUN)
hajeki L. N. Medvedev & Kantner, 2002: 260 **A:** IN
kadleci Bezděk, 2016: 154 **A:** YE
laeviuscula Ratzeburg, 1837: 202 (*Chrysomela*) **E:** AL AU BE BH BU CR CZ EN FR GB GE GR HU IT KR LA LU MD ME NL NM PL PT RO RU (CT ST) SB SK SL SP SZ UK **N:** MO **A:** AB AF AR CH (HEB HEI JIL NIN XIN) GG IN KI KZ RU (ES SS WS) TD TR
connexa Fricken, 1888: 325 (*Clythra*, var.)
fasciata Ratzeburg, 1837: 201 (*Chrysomela*) [HN]
quadripunctata puberula Weise, 1898: 181 **E:** AN FR (East Pyrenees) SP [# 835]
quadripunctata quadripunctata Linnaeus, 1758: 374 (*Chrysomela*) **E:** AL AN AU BE BH BU BY CR CZ DE EN FI FR GB GE GR HU IR IT KO KR LA LS LT LU MD ME NL NM NR PL PT RO RU (CT NT ST) SB SP SK SL SV SZ UK **A:** AB AF CH (HEI) GG IN KI KZ MG RU (ES FE SS WS) TD TM UZ TR [# 807, # 835]
dissimilis Weise, 1889g: 563 (var.)
messae J. Müller, 1921: 66 (*Clythra*)
quadrisignata Märkel, 1841: 223 (*Clythra*) [HN]
quadripunctata turfanica Lopatin, 1962c: 320 (ssp.) **A:** CH (XIN)
reductoapicalis Pic, 1932g: 35 **A:** CH ("Chine")
rubrimaculata Tan, 1992b: 780, 815 **A:** CH (SCH YUN)
subopacipennis Pic, 1943b: 15 **A:** ID (HP SD) PA **ORR** [# 1087]
inhumeralis Pic, 1943c: 14 (var.)
octonotata Pic, 1943b: 15 (var.)
neobengalensis Rizvi & Kamaluddin, 2011: 57 (*Oides*)
oblita Monrós, 1953: 49 [RN]
succincta Lacordaire, 1848: 201 (*Clythra*) [HN] [# 1087]
tsinensis Pic, 1927c: 31 **A:** CH (GUI SCH YUN)
subgenus *Clytraria* Semenov, 1903b: 173 type species: *Clytra jacobsoni* Semenov, 1903
Clytrella L. N. Medvedev, 1961: 639 type species: *Chrysomela atraphaxidis* Pallas, 1773
afghanica Reineck, 1937: 184 (ssp.) **A:** AF PA
incomperta Lopatin, 1966c: 365

atraphaxidis asiatica Chûjô, 1941c: 455 (var.) **A:** CH (BEI GAN HEB JIA JIL JIX LIA QIN SHA SHG SHN SHX YUN) NC RU (FE) SC [# 796]

atraphaxidis atraphaxidis Pallas, 1773: 725 (*Chrysomela*) **E:** AL BU FR GR IT KR MD PT RO RU (ST) SP TR UK **A:** AB AF AR CY GG IN IS KI KZ LE RU (WS) SY TD TM TR UZ [# 804]

angulata Weise, 1900b: 273 (var.)

ashabadensis Brancsik, 1899: 103 (*Clythra*, var.)

cobosi Tiberghien, 1969: 121 (ssp.)

conjuncta Heyden, 1892a: 368 (var.)

croceicollis Weise, 1900b: 273 (var.)

delagrangei Pic, 1896c: 62 (*Elythra* [sic!], var.)

flexuosa Weise, 1900b: 274 (var.)

lacordairei Weise, 1900b: 273 (var.)

maculifrons Zubkov, 1833: 338 (*Clythra*) [# 788]

marginicollis Weise, 1900b: 273 (var.)

milliati Pic, 1942b: 7 (var.)

nigromaculata Pic, 1897a: 84 (var.)

quinquemaculata Pic, 1920a: 7 (var.) [IN] [# 795]

sierrana J. Daniel, 1903b: 253 (var.)

vulgaris Weise, 1900b: 274 (var.)

atraphaxidis punctata Weise, 1890a: 477 (var.) **A:** CH (XIN) KZ MG RU (SS)

jacobsoni Semenov, 1903b: 173 [RN] **A:** AF IN TD UZ

bucharica Pic, 1915b: 5

notatithorax Pic, 1915b: 5 (var.)

opaca Jacobson, 1898: 191 [HN]

jelineki Lopatin, 1980: 614 **A:** IN

novempunctata Olivier, 1808: 852 **E:** AL BU GR IT KO NM RO RU (ST) SB TR **N:** EG **A:** AB AR CY GG IN IQ IS JO LE SY TM TR [# 806, # 1145]

aleppensis L. Redtenbacher, 1843: 989 (*Clythra*)

elata Weise, 1881: 117

elegans Faldermann, 1837: 368 (*Clythra*) [HN]

juncta Pic, 1920a: 7 (var.) [IN] [# 795]

persica Pic, 1920a: 7

signicollis Weise, 1900b: 272 (var.)

popovi L. N. Medvedev, 1961: 647 **A:** AF TD

rufina Solsky, 1881: 60 (*Coptocephala*) **A:** KZ TD TM UZ

defecta Jacobson, 1901b: 110 (var.) [IN] [# 795]

deficiens Weise, 1900b: 274 (var.) [HN]

tanae F.-Y. Wang & Zhou, 2011: 4 **A:** CH (QIN YUN)

truncatula F.-Y. Wang & Zhou, 2011: 6 **A:** CH (XIN)
valeriana iranica L. N. Medvedev, 1961: 647 (ssp.) **A:** IN TM
valeriana taurica L. N. Medvedev, 1961: 646 (ssp.) E: KR UK **A:** TR
valeriana valeriana Ménétriés, 1832: 237 (*Clythra*) E: BU GR NM RO RU (ST) SB TR UK **A:** AB AR CY GG IN IQ KZ SY TR
biinterrupta Reitter, 1913c: 87 (ab.) [IN] [# 789]
dissimilithorax Desbrochers des Loges, 1870a: 129 (*Clythra*)
drurei Pic, 1920a: 7 (var.) [IN] [# 795]
opacipennis Lopatin, 1962c: 319 (ssp.)
subjuncta Pic, 1920a: 7 (var.) [IN] [# 795]
tetrastigma F. J. Schmidt, 1841: 151 (*Clythra*) [HN]
yunnana L. N. Medvedev, 2008b: 280 **A:** CH (SCH YUN)
subgenus *Ovoclytra* L. N. Medvedev, 1961: 640 type species: *Clythra nigrocincta* Lacordaire, 1848
binominata Monrós, 1953: 49 [RN] E: GR (Halki, Rhodes, Megysti) **A:** JO TR
laticollis Weise, 1889b: 63 [HN]
bodemeyeri arabica L. N. Medvedev, 1961: 651 (ssp.) **A:** IQ [# 792]
bodemeyeri bodemeyeri Weise, 1900c: 134 [= 1900d: 163] **A:** IQ IS JO TR
deficiens Heyden, 1892a: 368 (var.) **A:** CY
cyprica Pic, 1918b: 11 (var.) [IN] [# 795]
graeca Pic, 1915b: 5 (var.)
multipunctata Pic, 1920a: 7 (var.) [IN] [# 795]
rotundata L. N. Medvedev, 1961: 650
subfasciata Melichar, 1912a: 35 (var.)
subinterrupta Pic, 1920a: 7 (var.) [IN] [# 795]
medvedevi Lopatin, 1966c: 364 **A:** AF PA [# 805]
nigrocincta bagdatensis Pic, 1920a: 7 (ssp.) **A:** IQ
nigrocincta nigrocincta Lacordaire, 1848: 200 (*Clythra*) **A:** GG IN IS JO SI SY TR
semireducta Pic, 1918c: 15 (var.)
unifasciata Ménétriés, 1836a: 151 (*Clythra*) [HN]
obesa Rapilly, 1982a: 18 **A:** IN
ovata borealis L. N. Medvedev & Kantner, 2002: 259 (ssp.) **A:** LE SY TR
ovata ovata Weise, 1881: 117 (var.) **A:** IQ IS JO SI TR [# 1122]
weisei Monrós, 1953: 49 [RN] **A:** IN IQ SY TR [# 808]
cingulata Weise, 1898: 181 [HN]

genus *Clytrasoma* Jacoby, 1908: 149 type species: *Clythra palliata* Fabricius, 1801
subgenus *Clytrasoma* Jacoby, 1908: 149 type species: *Clythra palliata* Fabricius, 1801
conforme Lacordaire, 1848: 194 (*Clythra*) **A:** ID (HP KA) "Himalaya" **ORR**
chinense Weise, 1922b: 42 (ab.) [IN, QS]
octomaculatum Pic, 1932g: 35 (var.)

dispar Kollar & L. Redtenbacher, 1844: 560 (*Clythra*) **A:** ID (KA)
trifasciatum Kollar & L. Redtenbacher, 1844: pl. 28 (*Clytra*)
palliatum Fabricius, 1801b: 30 (*Clythra*) **A:** CH (FUJ GUA GUI GUX HAI HEN HUB HUN JIA JIX SCH SHN TAI YUN ZHE) ID (SD) NP **ORR**
ceylonense Jacoby, 1908: 151
donckieri Pic 1932g: 35
gibbosum Vigors, 1825: 241 (*Clythra*)
marginatum Pic, 1932g: 35
quadrisignatum Pic, 1937a: 19
reductum Pic, 1932g: 35 (var.)
rufithorax Pic, 1932g: 35 (var.)
subobliteratum Pic, 1932g: 35 (var.)
transversum Illiger, 1801b: 182 (*Clytra*)

genus *Coptocephala* Chevrolat, 1836: 419 type species: *Chrysomela scopolina* Linnaeus, 1767
Ecphorastes Gistel, 1856: 380 type species: *Chrysomela scopolina* Linnaeus, 1767
aeneopicta Fairmaire, 1864: 646 (*Gynandrophtalma*) **N:** AG EG MO TU
biinterrupta Pic, 1918e: 23 (var.)
biscrensis Pic, 1918e: 23 (var.)
bistrijuncta Pic, 1905f: 156 (var.)
trimaculata Pic, 1918e: 23 (var.)
arcasi Báguena, 1960a: 25 **E:** BH CR IT (Venezia Giulia) ME SL
dalmatina Pic, 1918d: 17 (var.) [IN] [# 795]
inornata Fleischer, 1918b: 106 (ab.) [IN]
bechynei Cobos, 1956: 179 (*Gynandrophthalma*) **N:** MO
bistrinotata Fabricius, 1803: 30 [RN] (*Clythra*) **N:** AG MO TU [# 835]
biloba Báguena, 1960a: 27 (ab.) [IN, QS]
bistriolata Gistel, 1857a: 140 [RN] (*Clythra*)
divisa Báguena, 1960a: 27 (ab.) [IN, QS]
malacaensis Báguena, 1960a: 27 (ab.) [IN, QS]
sexnotata Fabricius, 1801b: 35 (*Clythra*) [HN]
trinotata Forsberg, 1821: 264 (*Clythra*) [RN]
brevicornis Lefèvre, 1872b: 351 (*Gynandrophthalma*) **E:** PT SP
extensa Báguena, 1960a: 27 (ab.) [IN]
chalybaea apicalis Lacordaire, 1848: 361 (*Clythra*) **E:** KR RU (CT ST) UK **A:** KZ RU (WS)
chalybaea chalybaea Germar, 1823: 546 (*Clytra*) **E:** AU BU CR CZ GR HU IT (Venezia Giulia) KR MD PT RU (CT ST) SB SK SL SP RO UK **A:** AB GG KZ
coptocephaloides Lacordaire, 1848: 310 (*Clythra*) **N:** EG **A:** IQ IS JO SI SY
furthi L. N. Medvedev, 1992e: 55
jaechi Warchałowski, 1991c: 247

crassipes crassipes Lefèvre, 1876a: 78 **E:** IT (Sardinia: isola di Sant'Antioco) **N:** AG MO TU

leprieuri Pic, 1897f: 198 (var.)

crassipes iranica L. N. Medvedev, 1971a: 693 (ssp.) **A:** AF IN OM PA SA YE **[Appendix]**

crassipes nepalica L. N. Medvedev, 1999c: 181 (ssp.) **A:** NP

crassipes vavilovi Lopatin, 1966c: 363 (*Smaragdina*) **A:** AF ID (HP)

destinoi Fairmaire, 1884d: 174 **E:** GR (Kos, Megysti, Rhodes, Sámos) TR **A:** AB AR CY GG IN IQ IS JO SY TR [# 809]

intermedia Reineck, 1908: 368

latenotata Pic, 1949a: 4 (var.)

tambei Pic, 1942b: 7 (var.)

dilatipes Pic, 1923d: 11 **N:** EG

dubia Baly, 1878h: 370 **A:** ID (HP) PA

fallaciosa Fairmaire, 1884d: 175 **A:** IN IS JO LE SY TR

judaea L. N. Medvedev, 1992e: 55

fossulata Lefèvre, 1873a: 372 **E:** IT

vitalei Pic, 1913d: 114 (var.)

gebleri Gebler, 1841b: 617 **E:** AL BU GR KR MD NM RO RU (CT ST) UK **A:** AB AR IN IQ KI KZ RU (SS) TR

amasiensis Weise, 1889e: 128 (var.)

inundata Reitter, 1905b: 251 (var.)

hellenica Warchałowski, 1991c: 246 **E:** AL GR

hissarica Lopatin, 1963c: 118 **A:** TD UZ

tadzhica L. N. Medvedev, 1965a: 13 (ssp.)

kabakovi L. N. Medvedev, 1978c: 865 [DA] **A:** AF

maidaquensis Lopatin, 2008a: 312 **A:** AE

massiliensis Pic, 1914b: 11 (var.) **E:** FR GE IT SP

barcinonensis Báguena, 1960a: 28 (ab.) [IN]

catalaunica Báguena, 1960a: 28 (ab.) [IN]

linnaeana Petitpierre & Alonso-Zarazaga, 2000: 201 [RN]

quadrimaculata Linnaeus, 1767: 596 (*Chrysomela*) [HN]

normandi Pic, 1914d: 18 **N:** TU

orientalis Baly, 1873: 81 **A:** CH (BEI GAN HEB HEI HUB JIL LIA NIN NMO QIN SHA SHN SHX XIN) JA MG NC RU (ES FE SS) SC

asiatica Chûjô, 1940d: 355

freija Reitter, 1900a: 165

panousei Kocher, 1959: 15 **N:** AG MO

peresi Vauloger de Beaupré, 1895: 194 (*Clytra*) **N:** AG EG TU

flavolimbata Pic, 1905c: 122

maculicollis Pic, 1919b: 14 (var.)

perrisi Desbrochers des Loges, 1870c: 79 (*Clythra*) **N:** AG MO TU
bleusei Pic, 1897a: 84
holoxantha Peyerimhoff, 1949: 294
impressiceps Pic, 1918d: 18 (var.) [IN] [# 795]
kerimii Fairmaire, 1875b: 537
rubriceps Pic, 1916b: 10 (var.) [IN] [# 795]
rubriceps Roubal, 1948a: 34 (forma)
schrammi Kocher, 1959: 70
sexstigma Pic, 1918d: 17
plagiocephala Fabricius, 1792b: 60 (*Cryptocephalus*) **E:** IT (Sicily) **N:** AG MO TU
bipunctata Forsberg, 1821: 277 (*Clythra*)
espanoli Pic, 1933c: 15 (var.) [IN] [# 795]
externepunctata Pic, 1895d: 89 (var.)
melanocephala Olivier, 1808: 854 (*Clytra*)
theryi Pic, 1918d: 17 (var.) [IN] [# 795]
tunisea Pic, 1901g: 80 (var.)
postmaculata Lopatin, 1980: 613 (ssp.) **A:** IN IQ
proxima Escalera, 1914: 513 **N:** MO
raffrayi Desbrochers des Loges, 1870a: 130 (*Clythra*) **E:** FR (Corsica) IT (Sardinia)
cyanocephala Lacordaire, 1848: 348 (*Clythra*) [HN]
fabianii Burlini, 1946: 78 (var.)
quinquenotata Lefèvre, 1873a: 364
rubicunda rossica L. N. Medvedev, 1977c: 34 (ssp.) **E:** BY KR PL RU (CT ST) UK **A:** AB GG KZ MG RU (SS WS)
rubicunda rubicunda Laicharting, 1781: 169 (*Clytra*) **E:** AL AU BE BU CR CZ FR GE GR HU IT ME NM PL PT RO SB SK SL SP (Pirineos de Lérida) SZ UK **A:** TR
bimaculata Polentz, 1939: 10 (forma) [IN] [# 795]
connexa Csiki, 1953: 122 (ab.) [IN]
didymata Roubal, 1948a: 34 (forma)
efasciata Csiki, 1953: 122 (ab.) [IN]
fuentei Vázquez, 1904: 374 (var.) [QS]
mediodisjuncta Fuente, 1910a: 181 (var.) [QS]
rubra Geoffroy, 1785: 72 (*Melolontha*)
subobliterata Pic, 1901h: 83 (var.)
tetradyma Küster, 1845a: 89
vittata Petri, 1926: 196 (ab.) [IN]
rungsi Pic, 1953e: 62 **N:** MO
kocheri Pic, 1953e: 62 (var.)
scopolina alticola Cobos, 1954a: 146 **E:** SP (Sierra Nevada)
scopolina floralis Olivier, 1791b: 37 (*Clytra*) **E:** PT SP

bisbimaculata Báguena, 1960a: 23 (ab.) [IN]
bistrimaculata Küster, 1848c: 100
humeralis Fuente, 1918: 46 (ab.) [IN]
incorrecta Báguena, 1960a: 22 (ab.) [IN]
inhumeralis Fuente, 1910b: 445 (var.)
majoricensis Báguena, 1960a: 24 (ab.) [IN]
nudista Báguena, 1960a: 23 (ab.) [IN]
peninsularis Báguena, 1960a: 24 (ab.) [IN]
scutellaris Fuente, 1913a: 474 (ab.) [IN]
subfasciata Pic, 1897g: 206 (var.)
scopolina kuesteri Kraatz, 1872b: 230 [RN] E: BH CR IT SB SL SZ [# 810]
andalusiaca Pic, 1918d: 17 (var.) [IN, QS] [# 795]
melanocephala Küster, 1847: 100 [HN]
scopolina nana Báguena, 1960a: 24 E: SP
scopolina punctata Weise, 1889i: 336 E: SP
soriana Báguena, 1960b: 143 (ab.) [IN]
scopolina scopolina Linnaeus, 1767: 597 (*Chrysomela*) E: AN BH CR FR IT SP (Pyrenees) SZ [# 811, # 835]
bipunctata Depoli, 1926b: 92 (forma) [IN, QS] [# 795]
destituta Depoli, 1926b: 92 (forma) [IN, QS] [# 795]
lateremaculata Depoli, 1926b: 92 (forma) [IN, QS] [# 795]
rufilabris J. Müller, 1948: 82 (ab.) [IN]
scutellaris Depoli, 1926b: 92 (forma) [IN, QS] [# 795]
separata Depoli, 1926b: 92 (forma) [IN, QS] [# 795]
signithorax Depoli, 1940: 307 (ab.) [IN]
sefrensis Pic, 1897c: 165 **N:** AG MO
metalliconotata Pic, 1933b: 11
rotroui Kocher, 1969a: 111
theryi Pic, 1936c: 26 (var.)
simillima Lodewyckx, 1995: 103 **E:** NM **A:** AR TR [# 812]
tani Regalin & L. N. Medvedev, 2010: 76 [RN] **A:** CH (HUN)
cyanea Tan, 1992a: 601, 606 [HN]
unicolor P. H. Lucas, 1845: 125 (*Clythra*) E: IT MA PT SP **N:** AG MO TU
aenescens Pic, 1932d: 23 (var.)
atra Pic, 1932d: 23
unifasciata australis L. N. Medvedev, 1965a: 12 (ssp.) **A:** AF IN KI KZ MG TD TM UZ [# 835]
deserta L. N. Medvedev, 1965a: 13 (ssp.)
unifasciata babylonica Lopatin, 1995a: 100 (ssp.) **A:** IQ
unifasciata unifasciata Scopoli, 1763: 66 (*Buprestis*) E: AL AN AU BE BH BU BY CR CZ EN FR GE GR HU IT KO KR LA LT MD ME NM PL PT RO RU (CT ST)

SB SK SL SP SZ TR UK **A:** AB AR GG IN KI KZ MG RU (ES SS WS) TD TM TR UZ [# 813, # 835]

conjuncta Csiki, 1953: 121 (ab.) [IN]

fallax Weise, 1887c: 56 (var.)

femoralis Küster, 1845a: 91

flavipes Csiki, 1953: 121 (ab.) [IN]

maculiceps Kraatz, 1872b: 223 (var.)

melanocephala Thunberg, 1787b: 46 (*Cryptocephalus*)

mihoki Kaszab, 1962: 31 (ab.) [IN]

nigripes Jacobson, 1915b: 160 (ab.) [IN]

phenax Jacobson, 1896b: 106 (var.) [QS]

reducta J. Müller, 1948: 80 (forma) [IN] [# 795]

ruficeps Weise, 1890g: 363 (var.) [QS]

triornata Roubal, 1948a: 33 (forma) [# 794]

xiangchengensis Tan, 1992b: 778, 814 **A:** CH (SCH)

zhaosuensis Tan, 1985: 103, 104 **A:** CH (XIN)

genus *Coptocephala*, nomen dubium

cyanocephala Olivier, 1791b: 35 (*Clytra*) **E:** FR (Corsica)

genus *Diapromorpha* Lacordaire, 1848: 227 type species: *Clytra trifasciata* Olivier, 1808

dejeanii dejeanii Lacordaire, 1848: 235 (*Clythra*) **A:** BT ID (AP HP SD) NP **ORR** [# 814, # 1035]

dejeanii yunnana L. N. Medvedev & Kantner, 2002: 261 (ssp.) **A:** CH (YUN)

pallens Fabricius, 1787: 81 (*Cryptocephalus*) **A:** BT CH (GUA GUX HAI SHA XIZ YUN) ID (AP SD UT) NP **ORR** [# 815, # 835]

elongata Pic, 1927m: 6

melanopus Lacordaire, 1848: 238 (*Clythra*)

pallidiceps Gressitt, 1942c: 326

pinguis Lacordaire, 1848: 234 (*Clythra*) **A:** CH (YUN) **ORR** [# 835]

postbinotata Pic, 1928g: 24 (*Ceratobasis*, var.)

vitalisi Pic, 1928g: 24 (*Ceratobasis*)

turcica turcica Fabricius, 1801b: 37 (*Clythra*) **A:** ID (SD UT) **ORR**

postsignata Pic, 1941b: 8 (var.)

genus *Epimela* Weise, 1903a: 27 type species: *Coptocephala ornata* Kollar & L. Redtenbacher, 1844

subgenus *Epimela* Weise, 1903a: 27 type species: *Coptocephala ornata* Kollar & L. Redtenbacher, 1844

nepalensis Takizawa, 1987: 47 **A:** NP

ornata Kollar & L. Redtenbacher, 1844: 560 (*Coptocephala*) **A:** ID (KA)

trimaculata Fabricius, 1798: 110 (*Clythra*) **A:** NP **ORR**

cicticollis Pic, 1941b: 8 (*Diapromorpha*)

dilutipes Jacoby, 1908: 147

insularis Weise, 1903a: 26

interrupta Pic, 1941b: 8 (var.)

viridescens Pic, 1941b: 8

tristis L. N. Medvedev, 2003b: 283 **A:** NP

viridicollis Jacoby, 1899b: 69 (*Pantocometis*) **A:** ID (HP KA SD UT) **ORR** **[# 816]**

genus *Exomis* Weise, 1889g: 577 type species: *Exomis peplopteroides* Weise, 1889

degenevei Pic, 1932g: 36 (*Clytra*) **A:** CH (GUX YUN) **[# 835]**

deqinensis F.-Y. Wang & Zhou, 2020: 353 **A:** CH (YUN)

huangi F.-Y. Wang & Zhou, 2020: 354 **A:** CH (HUB SCH)

oblonga Lopatin, 2009a: 3 (*Smaragdina*) **A:** CH (YUN)

peplopteroides Weise, 1889g: 577 **A:** CH (GAN QIN SCH SHA YUN XIZ) **[# 835]**

davidis Lefèvre, 1893: 113 (*Clytra*)

pubescens F.-Y. Wang & Zhou, 2020: 355 **A:** CH (XIZ)

pubipennis F.-Y. Wang & Zhou, 2020: 356 **A:** CH (YUN)

tanae F.-Y. Wang & Zhou, 2020: 357 **A:** CH (SCH)

viridis F.-Y. Wang & Zhou, 2020: 359 **A:** CH (QIN XIZ)

genus *Labidostomis* Chevrolat, 1836: 418 type species: *Chrysomela longimana* Linnaeus, 1761

subgenus *Aphobera* Warchałowski, 1985: 654 type species: *Labidostomis funerea* Fairmaire, 1891

funerea Fairmaire, 1891: ciii **A:** ID (KA)

subgenus *Chlorostola* Weise, 1881: 99 type species: *Cryptocephalus hordei* Fabricius, 1787

andreinii Escalera, 1931: 121 **N:** LB TU

ballsi Antoine, 1937: 244 **N:** MO

bolivari bolivari Escalera, 1931: 128 **N:** MO

bolivari garciaolmoi Cobos, 1954b: 37 E: SP

centromaculata centromaculata Gené, 1839: 81 E: IT (Sardinia)

damryi Weise, 1881: 95 (var.)

centromaculata orophila Bouyon, Doguet & Debreuil, 2010: 102 (ssp.) E: FR (Corsica)

embergeri Théry, 1930: 90 **N:** MO

guerini guerini Bassi, 1834: 472 (*Clythra*) E: IT **N:** AG TU

guerinii Bassi, 1834: 472 (*Clythra*) [unused original spelling]

trifoveolata Desbrochers des Loges, 1866: 42 (*Clythra*)

violacea Escalera, 1931: 115 (ab.) [IN]

guerini lejeunei Fairmaire, 1866a: 70 **N:** AG
lejeunii Fairmaire, 1866a: 70 [unused original spelling]
pelissieri Buquet, 1869: cv (*Clythra*)
violaceipennis Pic, 1932d: 23 (var.)
hordei Fabricius, 1787: 83 (*Cryptocephalus*) **E:** PT SP **N:** AG MO TU
atropurpurea Báguena, 1960b: 125 (ab.) [IN]
nevadensis J. Daniel, 1904a: 91 E: SP
rufomarginata Vauloger de Beaupré, 1895: 194 (*Clytra*) **N:** MO
cuprea Escalera, 1931: 115 (ab.) [IN]
reymondi Kocher, 1959: 7 (var.) [IN] [# 795]
syriaca Weise, 1881: 95 (var.) E: FR (Corsica) [# 1123]
lineata Pic, 1920a: 6 (var.) [IN] [# 795]
obliterata Pic, 1920a: 6 (var.) [IN] [# 795]
suturella Pic, 1920a: 6 (var.) [IN] [# 795]
subgenus *Labidostomis* Chevrolat, 1836: 418 type species: *Chrysomela longimana* Linnaeus, 1761
Labidostomus Gistel, 1848d: [10] [UE] type species: *Chrysomela longimana* Linnaeus, 1761
amurensis amurensis Heyden, 1884: 299 **A:** CH (HEI JIL) NC RU (ES FE) SC
fallacissima Jacobson, 1901b: 106 (var.) [IN] [# 795]
orientalis Chûjô, 1940d: 360
amurensis nigrilabris L. N. Medvedev, 1980a: 311 (ssp.) **A:** MG
arcuata Pic, 1920a: 7 **A:** CH (XIN) KI KZ
arisi Pic, 1920a: 7 (var.) [IN] [# 795]
arnoldii L. N. Medvedev, 1962a: 384 E: RU (ST) **A:** GG
asiatica Faldermann, 1837: 370 E: RU (ST) **A:** AB AR GG IN JO SY TR
atkaracalarica Özdikmen & Bal, 2016b: 517 **A:** TR
axillaris Lacordaire, 1848: 69 (*Clythra*) **E:** AU BH BU CR CZ FR GE GR HU IT KO KR MD NM PL RO RU (ST) SB SK SL SZ TR UK **A:** AB GG IN KZ RU (SS WS) TR [# 817, # 1088]
sareptana Kraatz, 1870c: 31 (var.)
balcanica Tomov, 1987: 25 E: CR
basanica J. R. Sahlberg, 1913a: 63 **A:** IQ IS JO SY TR
bcharrensis Bezděk, 2018b: 4 **A:** LE
beckeri Weise, 1881: 93 E: KR RO RU (ST) UK **A:** AB KZ TR WS
boreopersica Lopatin, 1997b: 363 **A:** IN
brevipennis Faldermann, 1837: 375 **A:** AB AR GG IN IQ JO SY TR [# 818, # 1089]
armeniaca Lacordaire, 1848: 57 (*Clythra*)
cavifrons Lefèvre, 1872a: 80 **N:** MO
centrisculpta centrisculpta Pic, 1920a: 6 **A:** CH (XIN) KI KZ TD UZ [# 835]
alaiensis Pic, 1920a: 7

centrisculpta kafiristanica Regalin & L. N. Medvedev, 2011: 56 [RN] **A:** AF

afghanica Lopatin & Nesterova, 2007b: 430 (ssp.) [HN]

cheni Lopatin, 1995a: 98 **A:** CH (XIN)

chinensis Lefèvre, 1887a: lv **A:** CH (BEI GAN HEB HEI JIL LIA NIN NMO SHA SHN SHX) MG NC RU (ES FE) SC

crebrecollis L. N. Medvedev, 1962b: 254 **A:** MG NC RU (ES FE)

cummingi Jacoby, 1908: 97 **A:** PA

curta Lopatin, 1961a: 201 **A:** KZ TD TM UZ [# 835]

cyanicornis Germar, 1822b: 7 (*Clytra*) **E:** AL AU BH BU CR CZ EN FR GE GR HU KO KR IT ME MD NM RO RU (CT ST) SB SK SL SP SZ UK **A:** KZ MG RU (ES SS WS) TR [# 835]

fulvipennis Lacordaire, 1848: 54 (*Clythra*, var.)

humeralis Rüschkamp, 1935: 67 (ab.) [IN]

daccordii Bezděk & Regalin, 2017: 331 **E:** GR

damavandensis Rapilly, 1984: 216 **A:** IN

decipiens Faldermann, 1837: 373 **A:** AB AR CY GG IN IQ IS JO SY TR

lineola L. Redtenbacher, 1843: 990

deserta L. N. Medvedev, 1971a: 691 (ssp.) **A:** KZ TD UZ [# 835]

diversifrons Lefèvre, 1872a: 90 **A:** CY IN IS JO LE SY TR

attenuata Pic, 1897g: 202

speculifrons Kraatz, 1872b: 200

elegans Lefèvre, 1876a: 76 **A:** AB AR IN TR

inhumeralis Pic, 1920a: 6 (var.) [IN] [# 795]

reitteri Weise, 1885a: 314

fedtschenkoi Lopatin, 1963c: 116 **A:** KI KZ TD UZ

graeca Tomov, 1990: 135 **E:** GR

harazensis Rapilly, 1984: 219 **A:** IN

hebraea Lacordaire, 1848: 55 (*Clythra*) **A:** IS JO LE SY TR

bijuncta Pic, 1912b: 17 (var.)

bisbijuncta Pic, 1912b: 17 (var.)

posticejuncta Pic, 1912b: 17 (var.)

humeralis D. H. Schneider, 1792: 192 (*Clytra*) **E:** AU BE BH BU BY CR CZ FR GE GR HU IT KO KR LA LT LU MD ME NM NR PL RO RU (CT ST) SB SK SP SV SZ UK **A:** GG RU (WS) TR

hybrida P. H. Lucas, 1845: 121 (*Clythra*) **N:** AG MO

atlasica Kocher, 1959: 7 (var.) [IN] [# 795]

deficiens Kocher, 1959: 7 (ab.) [IN]

roberti Pic, 1919d: 23

issykensis L. N. Medvedev, 1971a: 691 **A:** KI RU (SS)

kantneri Warchałowski, 2004b: 558 **A:** IN

karamanica Weise, 1900c: 134 [= 1900d: 163] E: GR (Rhodes) **A:** CY IN IS LE SY TR
cypria Lopatin, 1962c: 319
kaszabi L. N. Medvedev, 1962d: 333 **A:** TR
korbi Weise, 1902a: 203 **A:** TR
krimaea Lopatin & Nesterova, 2007b: 431 **E:** KR
leonardii Bezděk & Regalin, 2017: 333 **E:** GR (Kos) **A:** TR
lepida Lefèvre, 1872a: 94 **E:** BY CZ EN LA LT RU (CT ST) SK UK **A:** RU (ES SS WS)
limbata Lacordaire, 1848: 56 (*Clythra*) **A:** IS SY [# 793]
heinzi Lopatin, 1993a: 169
laeta L. N. Medvedev, 1992e: 51
lipskyi Lopatin, 1961a: 203 **A:** TD
longimana Linnaeus, 1761: 170 (*Chrysomela*) **E:** AL AN AU BE BH BU BY CR CZ DE EN FR GE GR HU IT KO KR LA LT LU MD ME NL NM NR PL RO RU (CT NT ST) SB SK SL SP SV SZ TR UK **A:** AB AR ES GG IN KZ RU (ES SS WS) TR
caucasica L. N. Medvedev, 1962a: 385 (ssp.)
dalmatina L. N. Medvedev, 1990a: 128 (ssp.) [# 1124]
glycyrrhizae Becker, 1880: 148
obscura Trella, 1930: 198 (ab.) [IN]
pallida Geoffroy, 1785: 72 (*Melolontha*)
uralensis Lacordaire, 1848: 73 (*Clythra*)
lucida Germar, 1823: 548 (*Clytra*) **E:** AN AU BH FR IT RU (CT ST) SB SP SZ **A:** GG KZ RU (SS WS) TR [# 819, # 835]
notata Gebler, 1830: 198 (*Clythra*) [QS]
luristanica Bezděk, **sp. nov.** **A:** IN [# 786]
luristanica Pic, 1920a: 6 (var.) [IN] [# 795]
lusitanica Germar, 1823: 549 (*Clytra*) **E:** FR IT (Sardinia) PT SP **N:** AG TU [# 835]
bigemina Suffrian, 1851b: 200 (*Clythra*)
carpetana Báguena, 1958: 23 (ab.) [IN]
croceopennis Motschulsky, 1849: 148
lacordairei Reiche & Saulcy, 1858b: 26 [RN]
melanostigma J. Daniel, 1903b: 253 (ab.) [IN]
meridionalis Lacordaire, 1848: 38 (*Clythra*)
montana Cobos, 1954a: 143 (ssp.)
pseudohumeralis L. N. Medvedev, 1962d: 334
pyrenaea Kraatz, 1872b: 197
quadrimaculata Motschulsky, 1849: 149
semideficiens Pic, 1906a: 19 (var.)
tibialis Lacordaire, 1848: 36 (*Clythra*) [HN]
maculipennis Lefèvre, 1870: xlii **A:** AR GG IN LE RU (ST) SY TR
kindermanni Kraatz, 1872b: 201

mairei Peyerimhoff, 1922: 62 **N:** MO

subinterrupta Pic, 1932d: 22 (var.)

martensi L. N. Medvedev, 1983: 134 **A:** IN

medvedevi Warchałowski, 1985: 699 [RN] **A:** AB AR IN TM TR

montana L. N. Medvedev, 1970d: 1000 [= 1971a: 686] [HN]

mesopotamica Heyden, 1886d: 279 **A:** IN JO SY TR

escherichi Weise, 1897c: 63

frontalis Weise, 1900c: 132 [= 1900d: 162]

iconiensis Weise, 1900c: 133 [= 1900d: 162]

metallica dzhungarica L. N. Medvedev, 1980a: 312 (ssp.) **A:** CH (XIN) MG

metallica glasunowi Jacobson, 1893b: 238 **A:** AF IN KZ TD

metallica imitatrix Ogloblin & L. N. Medvedev, 1971: 689 **A:** RU (ES WS)

metallica indica Jacoby, 1895b: 254 **A:** ID (HP KA) PA

metallica metallica Lefèvre, 1872a: 83 **E:** KR RU (ST) **A:** IN KZ TR UZ [# 835]

nitidicollis Lefèvre, 1872a: 85

metallica mongolica L. N. Medvedev, 1980a: 312 (ssp.) **A:** MG RU (SS)

metallica ornatipennis L. N. Medvedev, 1971a: 689 (ssp.) **A:** KZ UZ

metallica steppensis L. N. Medvedev, 1971a: 688 (ssp.) **A:** KZ

nuristanica afghanica L. N. Medvedev, 1978c: 861 (ssp.) **A:** AF (Kabul) PA

nuristanica bamiana L. N. Medvedev, 1978c: 861 (ssp.) **A:** AF (Bamian)

nuristanica nuristanica Reineck, 1937: 183 **A:** AF

oertzeni Weise, 1889b: 62 **E:** BU GR NM TR **A:** AR GG IN TR

heteromalla Lefèvre, 1889b: cxix

villosula Weise, 1900b: 270

ornata L. N. Medvedev, 1962b: 256 **A:** KI KZ UZ

signatipennis Lopatin, 1963c: 117

pachysoma L. N. Medvedev, 1965b: 429 (ssp.) **E:** KR RU (ST) UK **A:** AB GG

pallidipennis Gebler, 1830: 199 (*Clythra*) **E:** AL AU BH BU CR CZ FR GR HU IT KR ME NM RO RU (CT ST) SB SK SL SP SZ TR UK **A:** AB AR CH (XIN) GG IN KI KZ RU (SS WS) TR

chalybeicornis Brullé, 1833: 268 (*Clythra*)

distinguenda Rosenhauer, 1847: 63

pilicollis Lacordaire, 1848: 52 (*Clythra*)

peregrina Weise, 1900b: 271 **E:** RU (ST) **A:** AB AR GG IN TR

propinqua Faldermann, 1837: 372 **E:** AL BU GR KR NM RO RU (ST) SB TR **A:** AB AR GG IN IQ JO SY TR [# 835]

binotata Waltl, 1838: 472 (*Clythra*)

stevenii Lacordaire, 1848: 48 (*Clythra*)

regalini Bezděk, 2018a: 513 **A:** TR

rubripennis P. H. Lucas, 1845: 120 (*Clythra*) **N:** AG MO TU
lucaniformis Lefèvre, 1873a: 382
mogadorica Escalera, 1914: 507
rufa Waltl, 1838: 472 (*Clythra*) **E:** BU GR NM TR **A:** IN IS TR
rufa Lacordaire, 1848: 43 (*Clythra*) [HN]
rugicollis Lefèvre, 1872a: 86 **A:** IN KI KZ TD TM UZ [# 835]
schneideri Weise, 1900b: 272 **A:** AF KI TD UZ [# 835]
polygoni Lopatin, 1961a: 201
senicula Kraatz, 1870c: 29 **E:** RU (ST) **A:** AB AF CH (XIN) IN KI KZ MG TD TM UZ
albida Jacobson, 1894e: 271
nitida Ballion, 1878: 373
turcomanica Lopatin, 1954: 96
shensiensis Gressitt & Kimoto, 1961: 82 **A:** CH (SHA)
shirazica Lopatin, 1979b: 586 **A:** IN
sibirica Germar, 1823: 545 (*Clytra*) **E:** RU (ST) **A:** CH (NMO) KZ MG RU (ES SS WS)
altaica Gebler, 1848b: 42
stenostoma Weise, 1900b: 270 **A:** AF KZ TM TD UZ
subfasciata Weise, 1885a: 313 **A:** AB AR IN TM TR
trigemina Weise, 1898: 180
sulcicollis Lacordaire, 1848: 49 (*Clythra*) **E:** TR **A:** AR IN TR
taxicornis maroccana Lefèvre, 1873a: 380 **N:** MO
taxicornis taxicornis Fabricius, 1792b: 56 (*Cryptocephalus*) **E:** CR FR IT MA PT SP **N:** AG LB MO TU
latina Ulrich, 1924: 154 (forma) [IN] [# 795]
sardoa Jacobson, 1906a: 30
similis D. H. Schneider, 1792: 191 (*Clytra*)
testaceipes Pic, 1904f: 93 **A:** IQ SY TR
delagrangei Pic, 1904f: 94
robusta Reineck, 1911: 336 (*Macrolenes*)
tjutschewi Jacobson, 1901a: 75 **A:** MG NC RU (ES FE SS WS)
transitoria Jacobson, 1901a: 94 (var.) **A:** KZ MG RU (ES SS WS) [# 796]
transitoria Jacobson, 1901b: 106 (var.) [IN] [# 795]
tridentata Linnaeus, 1758: 374 (*Chrysomela*) **E:** AU BE BH BU BY CR CZ DE EN FI FR GB GE HU IT KR LA LT MD NL NR PL RO RU (CT NT ST) SB SK SL SP SV SZ UK **A:** CH (HEI) KZ MG RU (ES FE SS WS) [# 820, # 835]
kluczyckii Nowicki, 1872: 3 (*Clythra*)
leithneri L. Redtenbacher, 1873: 448
tymphristica Tomov, 1990: 135 **E:** GR
urticarum indigirensis L. N. Medvedev, 1978e: 116 (ssp.) **A:** RU (ES)

urticarum urticarum Frivaldszky, 1892: 119 **A:** CH (BEI GAN HEB HEI HUB HUN JIL LIA NIN NMO QIN SHA SHN SHX) KZ MG NC RU (ES FE SS WS)
bipunctata Mannerheim, 1825: 40 (*Clythra*) [HN]
mannerheimi Monrós, 1953: 49 [RN]
ujgur Lopatin & Nesterova, 2007b: 432 **A:** CH (XIN)
warchalowskii Kantner, 2007: 560 **A:** CH (HEB SHA)
yakovlevi Guskova, 2006: 304 **A:** MG
subgenus *Wellschmiedia* Warchałowski, 1985: 655 type species: *Clythra ghilianii* Lacordaire, 1848
ghilianii Lacordaire, 1848: 77 (*Clythra*) E: PT SP
atricollis Pic, 1922d: 27 (*Cyaniris*)

genus *Labidostomis*, nomina dubia
bisignata Faldermann, 1835: 444 (*Clythra*) **A:** "China borealis"
caerulescens Scopoli, 1763: 65 (*Buprestis*) E: SL
hispanica Lacordaire, 1848: 74 (*Clythra*) E: SP

genus *Labidostomis*, unavailable name
rubimacula X.-P. Li, 2013: 29, 66 **A:** CH (HEI) [# 797]

genus *Lachnaia* Chevrolat, 1836: 418 type species: *Chrysomela variolosa* Linnaeus, 1767
subgenus *Barathraea* Lacordaire, 1848: 874 type species: *Clytra cerealis* Olivier, 1808
Barathroea Lacordaire, 1848: 164 [unused original spelling] type species: *Clytra cerealis* Olivier, 1808
cerealis Olivier, 1808: 844 (*Clytra*) **N:** AG EG LB TU **A:** IS JO LE SI SY [# 1090]
padillai Tomov, 1982b: 108 **N:** TU
straminipennis P. H. Lucas, 1845: 122 (*Clythra*) **N:** AG MO
octomaculata Pic, 1895e: 243 (*Barathraea*)
octopunctata Pic, 1897g: 204 [incorrect subsequent spelling]
pardoi Cobos, 1956: 174 (*Lachnaea*)
separata Pic, 1897g: 204 (*Lachnea*)
subgenus *Lachnaia* Chevrolat, 1836: 418 type species: *Chrysomela variolosa* Linnaeus, 1767
Bucedestes Gistel, 1856: 380 type species: *Cryptocephalus longipes* Fabricius, 1775 (= *Buprestis sexpunctata* Scopoli, 1763)
Lachnaea Agassiz, 1846b: 199 [UE] type species: *Chrysomela variolosa* Linnaeus, 1767
Vervex Gistel, 1848b: xi [RN] type species: *Chrysomela variolosa* Linnaeus, 1767
caprai Grasso, 1958: 59 (*Lachnaea*) E: IT (Sicily)
cylindrica Lacordaire, 1848: 186 (*Clythra*) E: CR FR GR (Crete) IT PT SP **N:** AG MO TU

fuchsi Papp, 1951b: 83 **N:** AG
gallaeca Baselga & Ruiz-García, 2007: 40 **E:** PT SP
hirta Fabricius, 1801b: 30 (*Clythra*) **E:** PT SP **N:** MO
hirtipes Allard, 1865: 383 (*Clythra*)
suffriani Kraatz, 1872b: 210 (*Lachnaea*)
italica italica Weise, 1881: 108 (*Lachnaea*) **E:** CR FR IT SL SZ
italica occidentalis Grasso, 1963: 72 (*Lachnaea*, ssp.) **E:** FR (Corsica) IT (Sardinia)
lucidipennis Peyerimhoff, 1926: 95 (*Lachnaea*) **N:** MO
orientalis Weise, 1881: 111 (*Lachnaea*) **E:** AL GR NM
paradoxa Olivier, 1808: 844 (*Clytra*) **E:** FR (East Pyrenees) IT (Sicily) PT SP **A:** IS **N:** AG EG LB MO TU [# 835]
bijuncta Kocher, 1953c: 86 (*Lachnaea*, ab.) [IN]
bistigmata Pic, 1912k: 90 (*Lachnaea*, var.)
jurjurensis Pic, 1913e: 129 (*Lachnaea*, var.)
kabyliana Pic, 1912k: 90 (*Lachnaea*, var.)
vicina Lacordaire, 1848: 173 (*Clythra*)
peyerimhoffi Kocher, 1953c: 86 (*Lachnaea*, var.) **N:** MO [# 796]
pseudobarathraea K. Daniel & J. Daniel, 1898: 80 (*Lachnaea*) **E:** SP
inesperata Báguena, 1958: 21 (*Lachnaea*, ab.) [IN]
insidiosa Cobos, 1954a: 144 (*Lachnaea*)
pubescens Dufour, 1820: 307 (*Clytra*) **E:** FR PT SP
macrodactyla Lacordaire, 1848: 177 (*Clythra*)
palmata Lacordaire, 1848: 175 (*Clythra*)
subfasciata Pic, 1897f: 197 (*Lachnaea*, var.)
puncticollis Chevrolat, 1840: 17 **E:** PT SP **N:** AG LB MO TU
curtipennis Pic, 1936e: 214 (*Lachnaea*)
instigmata Normand, 1937: 120 (*Lachnaea*, ab.) [IN]
uniustigmata Pic, 1898a: 74 (*Lachnaea*, var.)
sexpunctata Scopoli, 1763: 67 (*Buprestis*) **E:** AL AU BE BH BU CR CZ GE GR HU IT KO MD ME NL NM PL RO RU (ST) SB SK SL SZ TR UK **A:** IN LE IS SY TR [# 835]
brachialis Küster, 1845b: 81
glabricollis Suffrian, 1851b: 210 (*Clythra*) [QS]
longipes Fabricius, 1775: 105 (*Cryptocephalus*)
pontica Weise, 1884a: 157 (*Lachnaea*, var.)
taurinensis O. F. Müller, 1766: 187 (*Chrysomela*) [QS]
tristigma Lacordaire, 1848: 183 (*Clythra*) **E:** FR PT SP **N:** AG TU [# 1091]
variolosa Linnaeus, 1767: 591 (*Chrysomela*) **E:** PT SP **N:** AG MO
lentisci Fabricius, 1792b: 57 (*Cryptocephalus*)
zoiai Regalin, 1997: 72 **E:** GR

genus *Lachnaia*, nomina dubia
rugicollis Normand, 1937: 121 (*Lachnaea*) **N:** TU
uniformis Normand, 1937: 122 (*Lachnaea*, ab.) [IN]
tripunctata D. H. Schneider, 1792: 191 (*Clytra*) E: SP

genus *Macrolenes* Chevrolat, 1836: 419 type species: *Cryptocephalus ruficollis* Fabricius, 1792 (= *Clytra dentipes* Olivier, 1808)
bellieri Reiche, 1861a: 735 (*Clythra*) E: IT (Sicily)
dentipes Olivier, 1808: 857 (*Clytra*) E: AL BH BU CR FR GR IT ME NM PT SL SP TR **N:** AG MO TU **A:** IS JO TR [# 821, # 835]
binotaticollis Pic, 1916a: 6 (var.) [IN] [# 795]
confluens Weise, 1893a: 1118 (var.)
crassimana Forsberg, 1821: 276 (*Clythra*)
gallica Schulz, 1909: 323 (var.)
humeralis Vitale, 1933: 66 (*Miopristis*, ab.) [IN]
immaculata Schulz, 1909: 323 (var.)
latemaculata Pic, 1897c: 165 (var.)
oculatothorax Vitale, 1933: 66 (*Miopristis*, ab.) [IN]
picturata Normand, 1950a: 91 (*Cyaniris*) [HN] [# 790]
quadrimaculata Schulz, 1909: 323 (var.)
ruficollis Fabricius, 1792b: 61 (*Cryptocephalus*) [HN]
salicaria Ménétriés, 1836a: 151 (*Clythra*)
transversa Depoli, 1915: 110 (ab.) [IN]
trimaculata Schulz, 1909: 323 (var.)
tunisea Warchałowski, 2000c: 587 [RN] [# 790]
zoufali Reitter, 1918: 44 (*Coptocephala*)

genus *Melitonoma* Chevrolat, 1836: 419 type species: *Cryptocephalus pallens* Fabricius, 1787 [# 787]
sobrina Lacordaire, 1848: 377 (*Clythra*) **N:** TU **AFR**
scita Weise, 1909: 164 (var.) [# 1125]

genus *Merilia* Lacordaire, 1848: 125 type species: *Cryptocephalus lunulatus* Fabricius, 1781
bipartita Tan & S.-Y. Wang, 1981: 447, 454 **A:** CH (XIZ)

genus *Miochira* Lacordaire, 1848: 315 type species: *Clythra gracilis* Lacordaire, 1848
gracilis Lacordaire, 1848: 316 (*Clythra*) **A:** AF BT CH (YUN) ID (HP KA SD UT) NP **ORR** [# 822]

affinis Jacoby, 1889b: 158 (*Melitonoma*)
argyrica Weise, 1913d: 220 (*Gynandrophthalma*) [RN]
cardoni Jacoby, 1897d: 424 (*Clythra*)
indica Jacoby, 1889b: 158 (*Melitonoma*)
pectoralis Lefèvre, 1893: 113 (*Aetheomorpha*)
nepalica L. N. Medvedev, 1998a: 27 **A:** NP
signaticollis signaticollis Kollar & L. Redtenbacher, 1844: 561 (*Coptocephala*) **A:** AF CH (XIZ) ID (HP KA SD UT) NP PA **ORR** [# 835]
montana Jacoby, 1895b: 255 (*Clythra*)
nuristanica Lopatin, 1966c: 363
signaticollis thibetana Pic, 1927c: 29 (*Cyaniris*) **A:** CH (XIZ)
bistrimaculata Pic, 1927c: 29 (*Cyaniris*, var.)
disjuncta Pic 1927c: 29 (*Cyaniris*, var.)
subcincta Pic, 1927c: 29 (*Cyaniris*, var.)
trinotaticollis Pic, 1927c: 29 (*Cyaniris*, var.)
variegata Lefèvre, 1890b: 191 (*Diapromorpha*) **A:** CH (YUN) **ORR**

genus *Miopristis* Lacordaire, 1848: 25 type species: *Clythra lepida* Lacordaire, 1848
humeralis Papp, 1951b: 81 **N:** AG

genus *Otiocephala* Lefèvre, 1872b: 317 type species: *Clythra opaca* Rosenhauer, 1856
antoinei Cobos, 1957: 178 (*Gynandrophthalma*) **N:** MO
atlantica Cobos, 1957: 179 (*Gynandrophthalma*) **N:** MO
grossepunctata Cobos, 1957: 187 (*Gynandrophthalma*) **N:** MO
forcipifera P. H. Lucas, 1845: 122 (*Clythra*) **N:** AG MO TU
freyi Cobos, 1957: 172 (*Gynandrophthalma*) **N:** MO TU
kocheri Cobos, 1957: 180 (*Gynandrophthalma*) **N:** MO
maroccana Cobos, 1957: 181 (*Gynandrophthalma*) **N:** MO
opaca Rosenhauer, 1856: 308 (*Clythra*) **E:** SP **N:** AG EG MO
andalusica Heyden, 1870: 165 (*Calyptorhina*)
carnerii Pic 1920e: 21 (*Cyaniris*, var.) [QS]
bisbipunctata Desbrochers des Loges, 1870c: 81 (*Clythra*) [QS]
latecyanescens Pic, 1946a: 3 (var.) [QS]
rufolimbata Pic, 1897b: 135 [QS]
ornatissima Escalera, 1914: 511 **N:** MO
otini Cobos, 1969: 69 (*Gynandrophthalma*) **N:** MO
pardoi Cobos, 1957: 171 (*Gynandrophthalma*) **N:** MO
proxima obsoletesculpta Cobos, 1957: 174 (*Gynandrophthalma*, ssp.) **N:** MO
proxima proxima Cobos, 1957: 174 (*Gynandrophthalma*) **N:** MO

quadrinotata atlasica Cobos, 1957: 186 (*Gynandrophthalma*) **N:** MO
quadrinotata quadrinotata Fabricius, 1787: 79 (*Cryptocephalus*) **N:** AG MO TU
rotroui Pic, 1934h: 26 (*Cyaniris*) **N:** MO
tripolitana Cobos, 1957: 177 (*Gynandrophthalma*) **N:** LB TU
warioni Lefèvre, 1873c: 4 **N:** AG LB MO

genus *Otiothraea* Warchałowski, 1990: 793 type species: *Otiothraea ghardaiensis* Warchałowski, 1990
avilai Vela & Bastazo, 1993: 33 **E:** SP
filabrensis Cobos, 1957: 187 (*Gynandrophthalma*) **E:** SP
ghardaiensis Warchałowski, 1990: 795 **N:** AG
riffensis Romantsov, 2011: 145 **N:** MO
rotroui Kocher, 1961: 238 (*Chilotoma*) **N:** MO

genus *Paraclytra* L. N. Medvedev, 1971d: 655 type species: *Clythra notata* Klug, 1835 (= *Clythra signata* Lacordaire, 1848)
cervenkai Bezděk & Kantner, 2010: 4 **A:** OM YE
crocata Lacordaire, 1848: 226 (*Clythra*) **A:** SA YE
melanaria L. N. Medvedev, 1979a: 295 (var.) [IN]
rufitarsis Lacordaire, 1848: 226 (*Clythra*) **N:** EG **A:** IS SA SY YE **AFR** [# 791]
aethiopica L. N. Medvedev, 1971b: 23 (*Clytra*, ssp.) [# 791]
plicatipennis Pic, 1939d: 31 (*Clytra*) [# 791]
sennariensis L. N. Medvedev, 1971b: 23 (*Clytra*, ssp.) [# 791, # 1126]

genus *Physauchenia* Lacordaire, 1848: 367 type species: *Clythra pallens* Lacordaire, 1848
cheni Pic, 1943c: 8 (*Coptocephala*) **A:** CH ("China")
pallens Lacordaire, 1848: 368 (*Clythra*) **A:** CH (FUJ GUA GUI GUX HAI HEN HKG HUB HUN JIA JIX SCH SHA TAI YUN ZHE) NC SC **ORR**
bifasciata Jacoby, 1888: 341 (*Coptocephala*)
binotata Pic, 1927f: 7 (var.)
fokienensis Pic, 1927f: 7 (var.)
formosana Pic, 1928c: 35 (*Coptocephala*)
octonotata Chûjô, 1952b: 89 (*Coptocephala*, var.) [IN] [# 795]
uninotata Pic, 1927f: 7 (var.)

genus *Physosmaragdina* L. N. Medvedev, 1971a: 694 type species: *Clythra nigrifrons* Hope, 1843
atriceps Pic, 1927c: 28 (*Cyaniris*) **A:** CH (YUN)
rufithorax Pic, 1927c: 28 (*Cyaniris*)
crassipes Duvivier, 1891a: xxxi (*Clytra*) **A:** ID (UT) NP **ORR** [# 835]

motschulskyi L. N. Medvedev, 1992f: 382 (*Smaragdina*) **A:** NP

nigrifrons Hope 1843a: 64 [= 1843b: 62] (*Clythra*) **A:** CH (ANH BEI CHQ FUJ GAN GUA GUI GUX HEB HEN HKG HUB HUN JIA JIX LIA SCH SHA SHN SHX TAI TIA YUN ZHE) JA NC SC **ORR** [# 823]

atripes Pic, 1927f: 7 (*Physauchenia*)

atrobasalis Pic, 1932c: 13 (*Cyaniris*, var.)

basidisjuncta Pic, 1934b: 20 (*Cyaniris*, var.)

coreana Kolbe, 1886: 226 (*Clythra*)

formosana Chûjô, 1934f: 284 (*Cyaniris*, var.)

immaculata Heyden, 1887b: 295 (*Coptocephala*, var.)

japonica Baly, 1873: 79 (*Clythra*)

kiangsuna Pic 1938b: 17 (*Cyaniris*, var.)

kiotensis Pic, 1927f: 7 (*Physauchenia*)

kolthoffi Pic, 1938b: 17 (*Cyaniris*)

latereducta Pic, 1934b: 20 (*Cyaniris*, var.)

mandarina Weise, 1889g: 579 (*Gynandrophthalma*, var.)

submarginata Pic, 1927f: 7 (*Physauchenia*)

subsignata Fairmaire, 1888c: 36 (*Gynandrophthalma*)

tonkinensis Lefèvre, 1891: ccliv [= 1893: 114] (*Damia*) **A:** ID (HP SD UT) NP **ORR**

brancuccii L. N. Medvedev, 1988a: 472 (*Smaragdina*)

mungphuensis Jacoby, 1908: 109 (*Gynandrophthalma*)

genus *Saudiclytra* L. N. Medvedev, El Torkey & Al Dhafer, 2014: 7 type species: *Saudiclytra spinifemorata* L. N. Medvedev, El Torkey & Al Dhafer, 2014 (= *Antipa wittmeri* L. N. Medvedev, 1979)

wittmeri L. N. Medvedev, 1979a: 296 (*Antipa*) **A:** OM SA

spinifemorata L. N. Medvedev, El Torkey & Al Dhafer, 2014: 7

genus *Smaragdina* Chevrolat, 1836: 420 type species: *Clythra menetriesii* Ménétriés, 1832 (= *Clythra unipunctata* Lacordaire, 1848)

Calyptorhina Lacordaire, 1848: 81 type species: *Clythra chloris* Lacordaire, 1848

Carmentis Gistel, 1848a: 404 [= 1848b: 123] [RN] type species: *Clythra menetriesii* Ménétriés, 1832 (= *Clythra unipunctata* Lacordaire, 1848)

Cyaniris Chevrolat, 1836: 420 [HN] type species: *Cryptocephalus collaris* Fabricius, 1781

Medvedevella Özdikmen, 2008b: 643 [RN] type species: *Gynandrophthalma macilenta* Weise, 1887

Monrosia L. N. Medvedev, 1971a: 694 [HN] type species: *Cryptocephalus cyaneus* Fabricius, 1775 (= *Buprestis salicina* Scopoli, 1763)

Nanosmaragdina Lopatin & Kulenova, 1986: 42 type species: *Gynandrophthalma macilenta* Weise, 1887

Necyomantes Gistel, 1848a: 404 [= 1848b: 123] [RN] type species: *Cryptocephalus collaris* Fabricius, 1781

Saulopalpa Weise, 1881: 123 type species: *Clythra viridana* Lacordaire, 1848

Smaragdinella L. N. Medvedev, 1971a: 693 [HN] type species: *Gynandrophthalma macilenta* Weise, 1887

aeneoviridis Lopatin, 1975b: 45 **A:** TD

aethiops Lopatin, 2004c: 188 **A:** CH (YUN)

affinis affinis Illiger, 1794: 611 (*Clytra*) E: AU BE BH BU BY CR CZ EN FI FR GB GE HU IT LA LU MD ME NL NM PL RO RU (CT NT ST) SB SK SL SP SZ UK **A:** AB GG IS TR

noesskei Schatzmayr, 1928: 47 (*Gynandrophthalma*, ssp.)

trimaculata Porta, 1934: 257 (*Gynandrophthalma*, ab.) [IN]

affinis manicata Lefèvre, 1872b: 341 (*Gynandrophthalma*) E: PT SP **[# 1127]**

lucidipes Pic, 1897g: 206 (*Chilotoma*, var.)

algirica Weise, 1894b: 84 (*Gynandrophthalma*) **N:** AG MO TU

apiciflava Chûjô, 1952a: 49 (*Gynandrophthalma*) **A:** CH (TAI)

aurita aurita Linnaeus, 1767: 596 (*Chrysomela*) E: AL AU BE BH BU CR CZ DE FR GE GR HU IT KO LU MD ME NL NM PL RO RU (CT ST) SB SK SL SP SV SZ TR UK **A:** TR

alpigena Gistel, 1857b: 520 (*Gynandrophthalma*)

bicolor Grimmer, 1841: 46 (*Clythra*) [HN]

dissimilis Lefèvre, 1872b: 339 (*Gynandrophthalma*, var.)

aurita auritoides Achard, 1923b: 93 (*Cyaniris*) E: RU (ST) **A:** AB AR GG IN JO **[# 835]**

pseudocyanea Ogloblin, 1948: 480

aurita hammarstroemi Jacobson, 1901b: 108 (*Gynandrophthalma*, var.) **A:** CH (GAN HEB HEI JIL LIA NIN NMO SHA SHN) MG NC RU (ES FE SS WS) SC **[# 796, # 835]**

aurita nigrocyanea Motschulsky, 1866a: 177 (*Gynandrophthalma*) **A:** JA RU (FE)

geniculata Motschulsky, 1866a: 177 (*Cheilotoma*)

biornata angorensis Lopatin, 2002c: 89 (*Calyptorhina*, ssp.) **A:** TR (Ankara)

biornata biornata Lefèvre, 1872b: 322 (*Calyptorhina*) **A:** AB AR GG IN TR

schelkownikowi Jacobson, 1894e: 272 (*Cyaniris*)

blackwelderi Gressitt & Kimoto, 1961: 97 **A:** CH (SCH SHA XIZ)

boreosinica Gressitt & Kimoto, 1961: 97 **A:** CH (BEI GAN HEB SHA)

centromaculata L. N. Medvedev, 1995a: 90 **A:** CH (YUN)

chloris caucasica L. N. Medvedev, 1971a: 693 (ssp.) E: RU (ST) **A:** GG

chloris chloris Lacordaire, 1848: 83 (*Clythra*) E: BU RO SB **A:** JO TR

banatica Weise, 1881: 123 (*Gynandrophthalma*, var.)

clavareaui Jacobson, 1906c: 311 [RN] (*Gynandrophthalma*) E: PT SP

alcarriensis Escalera, 1928: 431 (*Gynandrophthalma*)

thoracica Lacordaire, 1848: 299 [HN] (*Clythra*)

clypealis L. N. Medvedev, 1992e: 54 (*Calyptorrhina*) **A:** IS JO

cobosi Codina Padilla, 1963d: 43 (*Gynandrophthalma*) E: SP *collaris collaris* Fabricius, 1781: 143 (*Cryptocephalus*) **A:** CH (GAN HEB) MG RU (ES SS WS)

gotoi Chûjô, 1957b: 1 (*Gynandrophthalma*)

collaris lateralis Gebler, 1848: 43 (*Cyaniris*) **A:** KZ RU (ES SS)

compressipennis Pic, 1927c: 28 (*Cyaniris*) **A:** CH (GUI GUX SCH YUN)

obscuricolor Pic, 1927c: 28 (*Cyaniris*)

concolor amabilis Lacordaire, 1848: 292 (*Clythra*) **E:** PT SP

concolor concolor Fabricius, 1792b: 69 (*Cryptocephalus*) **E:** AN FR IT PT SP SZ [# 824, # 834]

aenea Germar, 1823: 547 (*Clytra*)

costata Tan & S.-Y. Wang, 1981: 448, 455 **A:** CH (XIZ)

cribripennis Tan, 1988a: 322, 332 **A:** CH (XIZ)

discolor discolor Solsky, 1881: 58 (*Gynandrophthalma*) **A:** KI KZ TD UZ [# 825]

discolor rufilabris Weise, 1887d: 326 (*Gynandrophthalma*) **A:** KZ UZ

discolor viridiceps Weise, 1892d: 133 (*Gynandrophthalma*) **A:** KI KZ TD UZ

ignorata Lopatin, 1954: 97 (*Cyaniris*)

diversiceps Pic, 1941a: 15 (*Cyaniris*) **A:** CH (YUN) **ORR**

diversipes Letzner, 1840: 114 (*Clythra*) **E:** AU CZ FR GE IT SP SZ [# 826]

divisa Jacoby, 1889b: 156 (*Gynandrophthalma*) **A:** CH (HAI YUN) ID (SD) NP **ORR**

indica Jacoby, 1895b: 263 (*Gynandrophthalma*)

djebellina Lefèvre, 1872b: 349 (*Gynandrophthalma*) **A:** IS JO LE SY TR

dohertyi Jacoby, 1908: 107 (*Gynandrophthalma*) **A:** ID (SD UT) NP **ORR**

duvivieri Jacoby, 1908: 155 (*Clytra*) **A:** ID (HP) **ORR**

emarginata L. N. Medvedev, 1995a: 88 **A:** CH (YUN)

fabrei Lefèvre, 1883b: 164 (*Gynandrophthalma*) **A:** NP **ORR**

ferulae Gené, 1839: 82 **E:** FR (Corsica) IT (Sardinia)

flavicollis Charpentier, 1825: 236 (*Clytra*) **E:** AU BE BH BU BY CR CZ EN FI FR GE GR HU IT KR LA LS LT NL PL RO RU (CT NT ST) SB SK SL SZ UK **A:** IN RU (WS) TR [# 827, # 835]

amasina Pic, 1897f: 197 (*Gynandrophthalma*)

picticollis Weise, 1889e: 128 (*Gynandrophthalma*, var.)

flavicoxis L. N. Medvedev, 1992f: 382 **A:** CH (SCH)

flavifrons Gressitt & Kimoto, 1961: 103 **A:** CH (SCH XIZ) [# 835]

flavilabris Breit, 1917: 66 (*Cyaniris*) **A:** CH (QIN XIZ)

flavobasalis Jacoby, 1908: 121 (*Gynandrophthalma*) **A:** NP **ORR**

fulveola Jacoby, 1890a: 86 (*Gynandrophthalma*) **A:** CH (HUB TAI)

divisoides Chûjô, 1952a: 48 (*Gynandrophthalma*)

fulvitarsis L. N. Medvedev, 1992f: 380 **A:** ID (SD) NP

gerhardi Bezděk & Beenen, 2020: 320 [RN] **A:** CH (SCH)

schereri Lopatin, 2006c: 593 [HN]

golda Jacobson, 1925c: 9 (*Calyptorrhina*) **A:** CH (NMO) JA NC RU (FE) SC
kusanagii Chûjô, 1940d: 359 (*Cyaniris*)
graeca Kraatz, 1872b: 217 (*Gynandrophthalma*) E: BU GR KO NM SB TR **A:** JO TR [# 828, # 835]
brucki Pic, 1897a: 85 (*Gynandrophthalma*) [RN]
graeca Lefèvre, 1872b: 346 (*Gynandrophthalma*) [HN]
gratiosa P. H. Lucas, 1845: 124 (*Clythra*) E: SP **N:** AG MO
limbifera Escalera, 1928: 432 (*Gynandrophthalma*, var.)
guillebeaui Pic, 1927c: 27 (*Cyaniris*) **A:** CH (GUX YUN)
haskourensis Warchałowski, 2000c: 587 [RN] **N:** MO
viridipennis Cobos, 1969: 70 (*Gynandrophthalma*) [HN]
hejingensis Duan, F.-Y. Wang & Zhou, 2022: 4 **A:** CH (XIN)
hypocrita Lacordaire, 1848: 288 (*Clythra*) E: BU KR RO SB TR UK **A:** GG JO TR [# 1092]
ihai Chûjô, 1958c: 2 (*Gynandrophthalma*) **A:** JA (Ryukyus)
impressicollis Tan, 1992b: 779, 814 **A:** CH (YUN)
insulana L. N. Medvedev, 1992b: 73 **A:** CH (GUX) **ORR**
judaica Lefèvre, 1872b: 342 (*Gynandrophthalma*) **A:** IS JO LE SY TR
berchmansi Reitter, 1911: 235 (*Gynandrophthalma*)
kimotoi Lopatin, 2003b: 301 **A:** CH (HAI) **ORR**
kuromon Kimoto, 1984: 43 **A:** CH (TAI)
kurosuji Kimoto, 1984: 42 **A:** CH (TAI)
labilis labilis Weise, 1889g: 579 (*Gynandrophthalma*) **A:** CH (GAN NIN NMO SHA SHX)
labilis sahlbergi Jacobson, 1901b: 109 (*Gynandrophthalma*) **A:** CH (HEI JIL LIA SHN) MG NC RU (ES FE) SC
coreana Chûjô, 1936d: 7 (*Cyaniris*) [HN]
discalis Chûjô, 1941c: 454 (*Calyptorrhina*) [RN]
discithorax Pic, 1902f: 49 (*Gynandrophthalma*)
laboissierei Pic, 1928c: 34 (*Cyaniris*) **A:** CH (GUX) **ORR**
binhana Pic, 1928c: 34 (*Cyaniris*, var.) [IN] [# 795]
reducta L. N. Medvedev, 1988b: 29 (ab.) [IN]
laevicollis Jacoby, 1890a: 86 (*Gynandrophthalma*) **A:** CH (FUJ HUB HUN JIA JIX SCH SHA ZHE)
apicenotata Pic, 1951b: 10 [= 1955a: 3] (*Cyaniris*, var.)
fukiensis Pic, 1955a: 3 (*Cyaniris*, var.)
kuatunensis Pic, 1954b: 56 (*Cyaniris*, var.)
marginata Pic, 1938b: 17 (*Cyaniris*)
laosensis Kimoto & Gressitt, 1981: 320 **A:** CH (YUN) **ORR**
laticollis Duvivier, 1892b: 399 (*Gynandrophthalma*) **A:** ID (SD) NP

levi Lopatin, 2004d: 31 [RN] **A:** CH (YUN)
minuta L. N. Medvedev, 1995a: 88 [HN]
limbata Steven, 1806: 157 (*Clytra*) **E:** BU GR NM RO RU (ST) SB TR **A:** AB AR CY GG IN IS JO LE SY TR
aurichalcea Marcu, 1938: 336 (*Gynandrophthalma*, ab.) [IN]
cincta Lefèvre, 1876a: 77 (*Gynandrophthalma*)
dorsalis Olivier, 1808: 866 (*Clytra*)
macilenta Weise, 1887d: 326 (*Gynandrophthalma*) **A:** KZ TD UZ
maculicollis Chûjô, 1952b: 73 (*Gynandrophthalma*) **A:** CH (TAI)
magnipunctata Duan, F.-Y. Wang & Zhou, 2022: 7 **A:** CH (YUN)
mandzhura Jacobson, 1925c: 10 (*Calyptorrhina*) **A:** CH (GAN HEB HEI JIA JIL LIA NMO SHA SHN SHX) MG RU (FE) SC
mangkamensis Tan & S.-Y. Wang, 1981: 448, 455 **A:** CH (XIZ)
micheli L. N. Medvedev, 1995a: 89 **A:** CH (SCH)
minuta Jacoby, 1908: 115 (*Gynandrophthalma*) **A:** ID (SD) **ORR**
minutissima Lopatin, 1967e: 437 (*Antipa*) **A:** AF ID (KA UT) NP PA
miyakei Kimoto, 1976: 1 **A:** CH (TAI)
miyatakei Kimoto, 1976: 3 **A:** CH (TAI)
moutoni Pic, 1897a: 86 (*Gynandrophthalma*) [RN] **A:** CH (HUN JIA SCH)
mandarina Lefèvre, 1892: xciv (*Gynandrophthalma*) [HN]
muehlei Kippenberg, 2018: 217 **A:** IN
nigripennis Chûjô, 1952b: 77 (*Gynandrophthalma*) **A:** CH (TAI)
nigriscutis L. N. Medvedev, 1970b: 187 **A:** ID (SD) NP **ORR**
nigroguttata Lopatin, 2002f: 874 **A:** CH (SCH)
nigrosignata Pic, 1954d: 53 (*Cyaniris*) **A:** CH (FUJ)
nigrosternum Erber & L. N. Medvedev, 1999: 6 **A:** CH (GAN SCH YUN)
nigroviolacea Lopatin, 2004d: 29 **A:** CH (YUN)
nipponensis Chûjô, 1951b: 33 (*Gynandrophthalma*) **A:** CH (TAI ZHE) JA SC
flavimana Chûjô, 1952b: 76 (*Gynandrophthalma*)
nomurai Kimoto, 1976: 2 **A:** CH (TAI)
obscuripes Weise, 1887h: 165 (*Gynandrophthalma*) **A:** CH (HEI JIL) NC RU (ES FE) SC
occidentalis L. N. Medvedev, 2010e: 275 **A:** ID (KA) PA
octomaculata Chûjô, 1952b: 81 (*Gynandrophthalma*) **A:** CH (TAI)
melania Chûjô, 1952b: 82 (*Gynandrophthalma*, var.) [IN] [# 795]
oculata L. N. Medvedev, 1988a: 471 **A:** CH (YUN) **ORR**
pakistanica L. N. Medvedev, 2010e: 279 **A:** PA
pallescens Pic, 1895c: 81 (*Gynandrophtalma*) **N:** AG TU **A:** IS
persica Pic, 1911c: 107 (*Gynandrophthalma*) **A:** IN
mirabilis Romantsov, 2012: 64
potanini L. N. Medvedev, 1970b: 187 **A:** CH (SCH) RU (FE)

punctatissima Weise, 1892d: 133 (*Gynandrophthalma*) **A:** KI KZ
quadratomaculata Jacoby, 1896a: 5 (*Gynandrophthalma*) **A:** JA (Ryukyus)
quadrimaculata Lopatin, 2009a: 2 **A:** CH (YUN)
rapillyi Lopatin, 2002c: 87 (*Calyptorrhina*) **A:** IN
regalini L. N. Medvedev & Kantner, 2002: 264 **A:** ID (SD) **ORR**
reyi Brisout de Barneville, 1866: 422 (*Cheilotoma*) **E:** FR (Pyrénées-Atlantiques) PT SP
rufimana Lacordaire, 1848: 299 (*Clythra*) **E:** PT SP (Andalucía) **N:** AG MO
salicina Scopoli, 1763: 65 (*Buprestis*) **E:** AL AU BE BH BU BY CR CZ DE EN FI FR GB GE GR HU IT KO KR LA LT LU MD ME NL NM PL RO RU (CT NT ST) SB SK SL SP SV SZ TR UK **A:** AB AR CH (XIN) GG IN IS KZ RU (SS WS) TR **[# 829]**
corporaali Trella, 1936: 97 (ab.) [IN]
cyanea Fabricius, 1775: 109 (*Cryptocephalus*)
nigripennis Trella, 1930: 198 (ab.) [IN]
notaticollis Roubal, 1922: 116 (ab.) [IN]
sapphirina Gmelin, 1790: 1705 (*Cryptocephalus*)
transsylvanica Frivaldszky, 1883: 14 (*Gynandrophthalma*)
saudica L. N. Medvedev, 1993a: 132 **A:** SA YE
scalaris Pic, 1927c: 27 (*Cyaniris*) **A:** CH (YUN)
scutellaris Lefèvre, 1872b: 335 (*Gynandrophthalma*) **A:** IN IS IQ JO LE SY TR
furthi Erber & L. N. Medvedev, 1999: 3
jordanica L. N. Medvedev & Katbeh-Bader, 2002: 255
latemaculata Pic, 1897g: 205 (*Gynandrophthalma*, var.)
semiaurantiaca Fairmaire, 1888b: 150 (*Gynandrophthalma*) **A:** CH (BEI HEB HEI HEN HUB JIA JIL LIA NIN QIN SHA SHN) JA NC RU (FE) SC
fleischeri Papp, 1946b: 7 (*Cyaniris*) [RN]
garretai Achard, 1921a: 61 (*Gynandrophthalma*) [RN]
japonica Fleischer, 1916b: 223 (*Gynandrophthalma*) [HN]
semiviridis Pic, 1922d: 27 (*Cyaniris*) **A:** CH (XIZ)
sikkimia Jacoby, 1903a: 86 (*Gynandrophthalma*) **A:** BT ID (SD) NP **ORR**
stenroosi Jacobson, 1901b: 108 (*Gynandrophthalma*) **A:** KI KZ UZ **[# 835]**
subacuminata Pic, 1927c: 28 (*Cyaniris*) **A:** CH (YUN)
symmetria Tan, 1988a: 323, 333 **A:** BT CH (XIZ) NP
schmidti L. N. Medvedev, 2003a: 318 (*Aetheomorpha*)
szechuana L. N. Medvedev, 1995a: 87 **A:** CH (SCH)
tani Lopatin, 2004d: 30 **A:** CH (YUN)
thibetana L. N. Medvedev & Romantsov, 2012: 75 **A:** CH (XIZ)
thoracica thoracica Fischer von Waldheim, 1842: 20 (*Coptocephala*) **A:** CH (XIN) KI KZ RU (WS) TD **[# 835]**
bijuncta Pic, 1906e: 42 (*Coptocephala*, var.)
fischeri Heyden, 1882b: 199 (*Clythra*) [RN]

hilaris Kraatz, 1879h: 288 (*Coptocephala*)
latejuncta Pic, 1914d: 18 (*Cyaniris*, var.)
subjuncta Pic, 1914c: 139 (*Cyaniris*, var.)
undecimnotata Weise, 1892d: 132 (*Gynandrophthalma*)
thoracica dzhungarica L. N. Medvedev, 1980a: 312 (ssp.) **A:** MG
tianmuensis F.-Y. Wang & Zhou, 2013: 252 **A:** CH (ZHE)
tibialis Brullé, 1833: 268 (*Clythra*) **E:** AL BU GR NM RO TR UK **A:** CY IN IS JO LE SY TR
bioculata Lacordaire, 1848: 289 (*Clythra*)
flavipes L. N. Medvedev, 1992e: 54 (ssp.)
hellenica Marseul, 1868: 205 (*Clythra*)
hungarica Weise, 1895d: 224 (*Gynandrophthalma*, var.)
kraatzi Clavareau, 1906: 44 (*Gynandrophthalma*, var.) [RN]
liturata Weise, 1881: 128 (*Gynandrophthalma*, var.)
tonkinensis Jacoby, 1895a: 172 (*Gynandrophthalma*) [RN] **A:** ID (SD) **ORR**
apicalis Lefèvre, 1893: 114 (*Gynandrophthalma*) [HN]
unipunctata Olivier, 1808: 870 (*Clytra*) **E:** RU (ST) **N:** EG **A:** AB AR GG IN IQ IS JO SI SY TM TR **[# 830, # 835]**
aegyptiaca Lefèvre, 1872b: 337 (*Gynandrophthalma*, var.) [# 1128]
menetriesii Ménétriés, 1832: 236 (*Clythra*)
venusta Lefèvre, 1872b: 337 (*Gynandrophthalma*, var.) [# 1129]
variabilis Chûjô, 1952b: 74 (*Gynandrophthalma*) **A:** CH (TAI)
vaulogeri Pic, 1894g: cclxxxv (*Gynandrophthalma*) **A:** CY IN IS LE SY TR
laeviceps Abeille de Perrin, 1895b: cdv (*Gynandrophthalma*)
uyguni Warchałowski, 1993a: 129
virgata Lopatin, 2004c: 187 **A:** CH (YUN)
viridana heinzi Erber & L. N. Medvedev, 1999: 3 **A:** IS JO
viridana marioutensis Wittmer, 1936: 16 (*Cyaniris*, ssp.) **N:** EG **A:** SI
viridana viridana Lacordaire, 1848: 291 (*Clythra*) **A:** IS JO LE SI SY TR "Mesopotamia"
azurea Reiche & Saulcy, 1858b: 26 (*Clythra*)
viridipennis Pic, 1937a: 19 (*Cyaniris*) **A:** CH ("Chine")
viridis comans Jacobson, 1925c: 8 (*Calyptorrhina*) **A:** AB IN KI TM
viridis viridis Kraatz, 1882b: 338 (*Gynandrophthalma*) **A:** AF IN KI KZ TD TM UZ
russowi Heyden, 1889b: 330 (*Gynandrophthalma*)
volkovitshi Lopatin, 2004d: 29 **A:** CH (YUN)
xanthaspis Germar, 1823: 547 (*Clytra*) **E:** AL AU BH BU CR CZ GE GR HU IT KO MD ME NM RO RU (CT ST) SB SK SL TR UK **A:** AB AR GG IN IS JO RU (WS) SY TR [# 835]
bicoloripes Pic, 1922d: 27 (*Cyaniris*)
collaris D. H. Schneider, 1794: 613 (*Clytra*) [HN]
thoracica Küster, 1846b: 99 (*Cyaniris*) [HN]

yajiangensis F.-Y. Wang & Zhou, 2013: 252 **A**: CH (SCH)
yangae F.-Y. Wang & Zhou, 2013: 255 **A**: CH (SCH)
yunnana L. N. Medvedev, 1995a: 87 **A**: CH (YUN)
zhangi F.-Y. Wang & Zhou, 2013: 259 **A**: CH (SCH YUN)

genus *Tituboea* Lacordaire, 1848: 141 type species: *Cryptocephalus sexmaculatus* Fabricius, 1781
affinis L. N. Medvedev, 1957b: 330 **A**: IN
sulcata L. N. Medvedev, 1962c: 622 (*Antipa*, var.) [IN]
afghana L. N. Medvedev, 1978c: 864 (*Antipa*) **A**: AF
ahwasa Lopatin, 1985b: 761 (*Antipa*) **A**: IN
arabica Olivier, 1808: 860 (*Clytra*) **N**: EG **A**: IN IQ IS JO SA SI SY TR YE
bisbinotata Pic, 1912k: 90 (*Titubaea*, var.)
mokattamensis Pic, 1912h: 73 (*Titubaea*)
notaticeps Pic, 1912h: 73 (*Titubaea*, var.)
robustior Pic, 1912h: 73 (*Titubaea*, var.)
subabbreviata Pic, 1912h: 73 (*Titubaea*)
sericea Olivier, 1808: 865 (*Clytra*) [QS]
atriceps Pic, 1924c: 22 (*Titubaea*, var.) **N**: EG **A**: IQ IS
decemmaculata Pic, 1937c: 13 (*Antipa*, var.)
israelita L. N. Medvedev, 1992e: 54 (*Antipa*)
biguttata Olivier, 1791b: 34 (*Clytra*) **E**: FR GR IT PT SP **N**: AG MO TU **AFR**
attenuata Fairmaire, 1875b: 536 (*Clythra*)
bicoloripes Pic, 1942c: 78 [= 1942e: 10] (*Antipa*)
dispar P. H. Lucas, 1845: 123 (*Clythra*)
flavipennis Pic, 1950a: 16 (*Antipa*, var.)
legionis Kocher, 1959: 12 (*Antipa*, var.) [IN] [# 795]
maculicollis Brullé, 1833: 267 (*Clythra*)
nigrithorax Porta, 1934: 254 (*Clytra*, ab.) [IN]
novempunctata Dufour, 1820: 309 (*Clytra*) [HN]
rectangula Bonelli, 1812: 169 (*Clythra*)
sexpunctata Olivier, 1791b: 34 (*Clytra*)
sexpunctata Forsberg, 1821: 280 (*Clythra*) [HN]
umbellatarum Olivier, 1808: 847 (*Clytra*) [QS]
vagenotata Pic, 1939a: 18 (*Antipa*, var.)
carmelica Lopatin & Chikatunov, 2001: 437 (*Antipa*) **A**: IS
chobauti Pic, 1896h: 142 (*Gynandrophthalma*) **N**: AG EG MO **A**: IS JO SI
semijuncta Pic, 1918d: 18 (*Titubaea*, var.) [IN] [# 795]
cingulata Lefèvre, 1883a: 150 (*Camptolenes*) **A**: SA YE **AFR**
donceeli Pic, 1929l: 14 (*Antina* [sic!], var.)
pubipennis L. N. Medvedev, 1962c: 622 (*Antipa*)

daccordii Lopatin, 1982c: 257 (*Antipa*) **A:** AF PA
demoflysi Normand, 1947b: 84 (*Coptocephala*) **N:** TU
elegans Weise, 1898: 182 **N:** MO
elegantissima Lopatin, 1966c: 361 (*Antipa*) **A:** AF
fasciata Lefèvre, 1872a: 129 (var.) **N:** AG MO
binotaticollis Pic, 1914a: 1 (*Titubaea*, var.)
infasciata Kocher, 1953a: 117 (*Antipa*, var.) [HN]
obliterata Pic, 1897c: 164 (var.)
semiconjuncta Pic, 1914a: 1 (*Titubaea*, var.)
filitarsis Lacordaire, 1848: 160 (*Clythra*) **N:** EG LB **A:** OM SA SI YE **AFR**
fairmairei Lefèvre, 1891: ccxlviii (*Camptolenes*)
friedmani Bezděk, 2018b: 11 **A:** IS
harteni Bezděk, 2018b: 15 **A:** AE OM **[Appendix]**
hebraica Lopatin, 1996: 637 (*Antipa*) **A:** IS SI
heptneri L. N. Medvedev, 1957b: 329 **A:** AF IN
illigeri Lacordaire, 1848: 142 (*Clythra*) **N:** AG LB TU
leprieuri Pic, 1895d: 89 (*Titubaea*, var.)
nigricollis Heyden, 1886b: 49 (var.)
semiundulata Pic, 1939a: 18 (*Antipa*, var.)
uninotata Pic, 1939a: 18 (*Antipa*, ab.) [IN]
vaulogeri Pic, 1939a: 18 (*Antipa*, ab.) [IN]
iranica L. N. Medvedev, 1957b: 326 **A:** IN
unicolor L. N. Medvedev, 1957b: 327 (ab.) [IN]
kermanica L. N. Medvedev, 1962c: 619 (*Antipa*) **A:** IN
lacordairei Pic, 1929l: 14 (*Antina* [sic!], var.) **A:** IS JO OM SA SI SY YE
laticollis Olivier, 1808: 846 (*Clytra*) **N:** AG EG MO TU
algerica Desbrochers des Loges, 1870a: 128 (*Clythra*)
subjuncta Pic, 1903b: 114 (*Titubaea*, var.)
lefevrei Pic, 1894e: 221 (*Melitonoma*) **A:** IS JO **N:** AG MO TU
chikatunovi Lopatin, 1995a: 100 (*Antipa*)
palaestina Pic, 1929l: 14 (*Antina* [sic!], var.)
macropus Illiger, 1800: 128 (*Clytra*) **E:** AL AU BH BU CR CZ GR HU IT KR MD ME NM RO RU (CT ST) SB SK SL SZ TR UK **A:** AB AR GG IN IS JO KZ LE SY TM TR UZ [# 831, # 835]
armeniaca Pic, 1918d: 18 (*Titubaea*, var.) [IN] [# 795]
binaeva Reitter, 1913c: 86 (ab.) [IN]
ciliciensis Pic, 1904e: 57 (*Titubaea*)
coalita Forsberg, 1821: 280 (*Clythra*)
grandipes Forsberg, 1821: 262 (*Clythra*) [RN]
mecheriensis Pic, 1895a: 13 (*Titubaea*, var.) **N:** AG MO TU
hypomelaena Bedel, 1921: 61 (var.) [# 796]

janczyki Cobos, 1956: 169 (*Macrolenes*)
kocheri Pic, 1949d: 16 (*Titubaea*, var.)
urikkana Tomov, 1983: 64 (*Antipa*)
medvedevi Lopatin, 1982c: 255 (*Antipa*) **A:** AF
mirzayani Lopatin, 1985b: 763 (*Antipa*) **A:** IN
nevoi Lopatin & Chikatunov, 2001: 437 (*Antipa*) **A:** IS **N:** LB
nigriventris Lefèvre, 1872a: 136 **A:** AF IN KI KZ TD TM UZ [# 832]
infasciata Pic, 1912h: 73 (*Titubaea*, var.)
nuristanica Lopatin, 1966c: 361 (*Antipa*) **A:** AF
obliquata Lacordaire, 1848: 216 (*Clythra*) **A:** YE **AFR**
octopunctata Fabricius, 1787: 79 (*Cryptocephalus*) **N:** AG MO TU
quadrisignata Forsberg, 1821: 283 (*Clythra*)
siciliensis Pic, 1912h: 74 (*Titubaea*, var.)
tetrapunctata Vitale, 1933: 66 (*Antipa*, ab.) [IN, QS]
unipunctata Pic, 1897c: 165 [= 1897e: 204] (var.)
octosignata Fabricius, 1801b: 36 (*Clythra*) **N:** AG MO
ogloblini L. N. Medvedev, 1962c: 623 (*Antipa*) **A:** YE
immaculata L. N. Medvedev, 1978b: 57 (*Antipa*, ab.) [IN]
ohbayashii Kimoto, 1984: 40 **A:** CH (TAI)
olivieri Lacordaire, 1848: 159 (*Clythra*) **N:** EG LB TU
femoralis L. N. Medvedev, 1962d: 335
ornaticollis L. N. Medvedev, 1957b: 328 **A:** IN IQ
paykulli Lacordaire, 1848: 161 (*Clythra*) **N:** AG MO [# 1093]
hirsutula Kocher, 1959: 70 (*Antipa*)
paykullii Lacordaire, 1848: 161 (*Clythra*) [unused original spelling]
testaceipes Pic, 1939a: 18 (*Antipa*)
trinotaticollis Kocher, 1959: 11 (ab.) [IN]
piffli Lopatin, 1967b: 323 (*Antipa*) **A:** PA
pindai Bezděk, 2011: 259 **A:** AE OM
puncticollis L. N. Medvedev, 1962c: 622 (*Antipa*) **A:** AF IN PA TM
maculicollis L. N. Medvedev, 1962c: 622 (*Antipa*, var.) [IN]
purcharti Bezděk, 2012c: 396 **A:** YE (Socotra)
pusilla Lopatin, 2001a: 41 **A:** IN
radeki Bezděk, 2018b: 19 **A:** OM YE
rufopyga Lopatin, 1985b: 761 (*Antipa*) **A:** IN
rugulosa Weise, 1898: 183 **A:** AB AR IN
saadensis Pic, 1894e: 221 (*Melitonoma*) **N:** AG MO TU **AFR**
adrarensis Pic, 1942c: 78 (*Coptocephala*)
reymondi Kocher, 1956: 127 (*Antipa*)
sefrensis Pic, 1897a: 82 (*Melitonoma*, var.)

subfasciata Kocher, 1956: 128 (*Antipa*, ab.) [IN]

volatica Normand, 1950a: 92 (*Coptocephala*)

saudica L. N. Medvedev, 2012e: 416 **A:** SA

sexmaculata Fabricius, 1781: 138 (*Cryptocephalus*) **E:** FR IT PT SP **N:** AG MO TU **A:** IN IS LE SY TR [# 835]

akbesiana Pic, 1897a: 82 (var.)

fulvipes Lefèvre, 1872a: 123 (var.)

humeralifer Pic, 1900h: 91 (*Titubaea*, var.)

parviceps Lacordaire, 1848: 148 (*Clythra*)

quadrexterna Báguena, 1960b: 139 (*Antipa*, ab.) [IN]

quadrinterna Báguena, 1960b: 140 (*Antipa*, ab.) [IN]

silensis Weise, 1894a: 70 **A:** AF IN KZ TD TM UZ

testaceiventris Pic, 1913l: 186 (*Titubaea*) **A:** IQ SY

tredecimpunctata Desbrochers des Loges, 1870a: 128 (*Clythra*) **N:** AG EG MO TU **A:** IS JO SI

peyerimhoffi Pic, 1902f: 48 (*Titubaea*)

zarudnii Lopatin, 2001a: 40 **A:** IN

genus *Tituboea*, nomina dubia

decemguttata Walker, 1871: 18 (*Titubaea*) **N:** EG

fasciaticeps Pic, 1929j: 95 (*Antipa*) **N:** LB

minor Fairmaire, 1894b: 312 (*Titubaea*) **N:** AG EG LB

tribe Cryptocephalini Gyllenhal, 1813

subtribe Cryptocephalina Gyllenhal, 1813

genus *Cryptocephalus* Geoffroy, 1762: 231 [conserved name, ICZN (1994), Opinion 1754] type species: *Chrysomela sericea* Linnaeus, 1758

subgenus *Bertiellus* Lopatin, 1977e: 72 type species: *Cryptocephalus* (*Asiopus*) *umarovi* Lopatin, 1970

furcatus Schöller, **nom. nov.** [RN] **A:** AF [# 837]

bifurcatus L. N. Medvedev, 1978c: 866 [HN]

lopatini L. N. Medvedev, 1978c: 866 **A:** AF

pushtunicus Lopatin, 1982c: 259 **A:** AF

umarovi Lopatin, 1970: 201 **A:** AF

subgenus *Burlinius* Lopatin, 1965a: 455 type species: *Chrysomela fulva* Goeze, 1777

alnicola A. Costa, 1884: 49 E: FR (Corsica) IT (Sardinia)

lostiae Burlini, 1942a: 13 (ab.) [IN]

aphthonoides S.-H. Chen, 1942c: 118 **A:** CH (GAN YUN) [# 862]

appositus Lopatin, 1958b: 33 **A:** KI

araxicola Iablokoff-Khnzorian, 1968: 262 **A:** AB AR IN TM *rastegari* Rapilly, 1980: 83

xanthus Iablokoff-Khnzorian, 1968: 264

ayvazi Gök & Sassi, 2002: 155 **A:** TR

baowenzhengi Duan & Zhou, 2021: 8 **A:** CH (SCH SHA)

bidorsalis Marseul, 1874: 218 **A:** IS LE SI

morfini Pic, 1914a: 2 (var.)

bilineatus Linnaeus, 1767: 597 (*Chrysomela*) **E:** AU BE BH BU BY CR CZ DE EN FI FR GB GE HU IT LA LS LT LU MD NL NR PL PT RO RU (CT NT ST) SB SK SL SP SV SZ UK **A:** AR CH (HEI NMO QIN XIN) GG IN JA (Hokkaido, Honshu) KZ MG NC RU (ES FE incl. Sakhalin SS WS) SC TR **[# 1036]**

acciculatus Tamanuki, 1927: 79 (ab.) [IN]

armeniacus Faldermann, 1837: 382 (*Pachybrachis*)

bakonyensis Kaszab, 1962: 47 (ab.) [IN]

bisbilineatus Pic, 1904e: 57 (var.)

csikii Kaszab & Székessy, 1953: 197 (ab.) [IN]

ehmanni Csiki, 1953: 127 (ab.) [IN]

gesztelensis Csiki, 1953: 127 (ab.) [IN]

moestus Weise, 1882b: 230 (var.) [HN]

partitus Jacoby, 1885b: 200 [typo 100]

pseudoarmeniacus Burlini, 1956b: 184 (ab.) [IN]

quadrus Schrank, 1798: 551 [HN]

szekessyi Csiki, 1953: 127 (ab.) [IN]

ujhelyii Csiki, 1953: 127 (ab.) [IN]

weisei Portevin, 1934: 345 (ab.) [IN]

biondii Sassi & Regalin, 1998: 136 **E:** FR (Corsica) IT (incl. Capraia Isl., Sardinia)

brevisignaticollis Pic, 1914a: 1 **A:** JO LE SY

carpathicus Frivaldszky, 1883: 15 **E:** CZ MD PL RO SK UK

chafarinensis Petitpierre, 2007: 572 **N:** SP (Islas Chafarinas: Isla del Congreso)

chrysopus Gmelin, 1790: 1713 **E:** AU BE BH BU BY CR CZ FR GE GR HU IT KO KR LT MD ME NM PL RO RU (CT ST) SB SK SL SP SZ TR UK **A:** AB GG RU (SS WS) TR **[# 863]**

biguttatus Schaller, 1783: 278 (*Chrysomela*) [HN]

haemorrhoidalis D. H. Schneider, 1792: 202 [HN]

huebneri Fabricius, 1792b: 65

pillichi Kaszab, 1962: 50 (ab.) [IN]

cognatus A. Costa, 1886: 28 **E:** FR (Corsica) IT (Sardinia)

confusus Suffrian, 1854b: 140 **A:** CH (BEI GUA HEB HEI HUB JIA JIL LIA NMO SHX ZHE) JA (Honshu, Kyushu, Shikoku) MG NC RU (ES FE incl. Sakhalin SS WS) SC

discretus Baly, 1873: 97

rectipennis Jacoby, 1890a: 87

connexus Olivier, 1808: 836 **E:** AL AU BH BU CR CZ FR (incl. Corsica) GR HU IT (incl. Sicily) KO KR MD ME NM PL RO RU (CT ST) SB SK SL SZ TR UK **N:** AG **A:** AB AR CY GG IN IS JO KZ SY TM TR

ahngeri Jacobson, 1901b: 114 (var.) [IN] [# 856]

amoenus Charpentier, 1825: 242

arenarius Weise, 1882b: 235 (var.) [HN]

bisbireductellus Kaszab, 1962: 51 (ab.) [IN]

fasciatus Petri, 1912: 261 (var.) [HN]

heterepistaticus Kaszab, 1962: 53 (ab.) [IN]

kaszabanus Burlini 1963a: 158 (ab.) [IN]

lichtneckerti Kaszab, 1962: 53 (ab.) [IN]

petrii Burlini, 1956b: 214 (ab.) [IN]

reductevittatus Kaszab, 1962: 51 (ab.) [IN]

sajoi Kaszab, 1962: 53 (ab.) [IN]

subconnexus Weise, 1882b: 235 (var.)

corynetes Warchałowski, 1999b: 556 **N:** MO

crenatus Wollaston, 1854: 456 **N:** MR (incl. Porto Santo)

brunneobinotatus Pic, 1938a: 1 (var.)

madeirensis Pic, 1938a: 1 (var.)

czwalinae Weise, 1882b: 216 **E:** IT

latenigriceps Pic, 1953b: 4 (ab.) [IN]

luigionii Pic, 1953b: 4 (ab.) [IN]

subiacosus Pic, 1953b: 4 (ab.) [IN]

toscanensis Pic, 1953b: 4 (ab.) [IN]

daccordii Biondi, 1995: 39 **E:** IT

dahdah Marseul, 1868: 209 **A:** IS JO LE SY

darjilingensis Jacoby, 1908: 263 **A:** ID (SD) NP **ORR**

dilutellus Jacobson, 1901b: 114 **A:** AF KI KZ MG TD TM UZ [# 854]

fedtschenkoi Jacobson, 1901b: 114 (var.) [IN] [# 856]

discicollis Fairmaire, 1867: 413 **N:** AG MO

bisignatipennis Pic, 1900i: 96 (var.)

broudeli Pic, 1900i: 96 (var.)

favarcqi Pic, 1902g: 72 (var.)

kabylianus Pic, 1902g: 72 (var.)

paulopunctatus Pic, 1907j: 179 (var.)

theresae Pic, 1900i: 96 (var.)

ejimai Takizawa, 1990b: 3 **A:** JA (Kyushu)

elegantulus Gravenhorst, 1807: 152 **E:** AL AU BH BU BY CR CZ DE FR GB GE GR HU IT KO KR LT MD ME NL NM PL RO RU (CT NT ST) SB SK SL SP SZ TR UK **A:** AB AR CH (BEI GAN HEI) GG IN KZ MG NC RU (ES FE SS WS) SC TR

deubeli Kaszab, 1962: 47 (ab.) [IN]

elongatulus Olivier, 1808: 835

horvathi Csiki, 1953: 126 (ab.) [IN]

inadumbratus Pic, 1904e: 57 (var.)

jucundus Faldermann, 1837: 393 [typo 396]

kuatunensis Pic, 1954d: 56

negrei Burlini, 1967: 131 (morpha) [IN]

tesselatus Germar, 1813: 128 [# 1037]

exiguus amiculus Baly, 1873: 98 **A:** CH (ANH BEI GAN HEI HEB JIL NMO SCH SHA SHN SHX) JA (Hokkaido, Honshu, Kyushu, Shikoku) MG NC RU (ES FE incl. Kuriles Isl., Sakhalin WS) SC (incl. Jeju-do Isl.) **[# 864]**

adocetus Jacobson, 1901b: 114 (ssp.)

kiyosatonus Kimoto, 1964a: 153

exiguus exiguus D. H. Schneider, 1792: 204 **E:** AU BE BU BY CR CZ EN FI FR GB GE HU IT LA LS LT NL NR PL RO RU (NT) SK SL SV SZ UK

wasastjernii Gyllenhal, 1827: 669

exiguus variceps Weise, 1884b: 161 **E:** BU GR TR **A:** AB AR GG IN TR **[# 865]**

areshanus Pic, 1914b: 12 (var.)

karsantianus Pic, 1914b: 13

kuchtai Breit, 1918: 50 (ab.) [IN]

fausti Weise, 1882b: 236 **E:** RU (ST) **A:** AB TR

ozbeki Aslan & Warchałowski, 1998: 343

flabellatus Schöller, 2009b: 107 [RN] **A:** IN

orientalis Rapilly, 1980: 84 (ssp.) [HN]

rapillyi Warchałowski, 2010c: 583 [RN]

flavolimbatus Pic, 1920c: 25 **A:** CH (YUN) RU (FE)

flavoscutellaris L. N. Medvedev, 1973b: 879 **A:** NC RU (FE incl. Sakhalin)

frontalis Marsham, 1802: 211 **E:** AU BE BU BY CZ DE EN (incl. Hiiumaa) FI FR GB GE HU IT LA LT NR PL RO RU (CT NT) SK SV SZ UK **A:** AB JA (Hokkaido) RU (ES FE incl. Sakhalin WS)

marginellus Gravenhorst, 1807: 148 [HN]

fulvus fulvus Goeze, 1777: 321 (*Chrysomela*) **E:** AL AN AU BE BH BU BY CR CZ DE EN FI FR (incl. Corsica) GB GE GR HU IT (incl. Elba, Pantelleria, Sardinia, Sicily) LA LT LU MA MD ME NL NM NR PL PT RO RU (CT NT ST) SK SL SP (incl. Balearic Islands: Ibiza, Formentera) SV (incl. Gotland) SZ TR UK **A:** AB AR GG KZ RU (WS) TD TR

apricus Weise, 1882b: 235 (unspecified infrasubspecific entity) [IN]

desgozisi Pic, 1908j: 94 (var.) [IN] [# 856]

minutus Fabricius, 1792b: 70 [HN] [# 1094]

ochraceus Stephens, 1831: 362

ochraceus Curtis, 1840: 281 [HN]
signaticollis Suffrian, 1848: 68
succinctus Weise, 1882b: 234 (unspecified infrasubspecific entity) [IN]
weiseanus Breit, 1918a: 52 (ab.) [IN]
vittatus Gozis, 1907: 166 (var.) [HN]
fulvus fuscolineatus Chûjô, 1940f: 385 **A:** CH (BEI GAN HEB HEI SHA) JA (Honshu, Kyushu, Shikoku, Tsuno Isl.) NC RU (FE incl. Kuriles Isl., Sakhalin) SC
fulvus schatzmayri Burlini, 1969a: 114 (ssp.) **E:** GR **A:** IS JO SA TR
gangaridus Lopatin, 1979c: 431 **A:** ID (KA) NP [# 866]
glabrelegantulus Duan & Zhou, 2021: 12 **A:** CH (SCH)
gounellei Pic, 1922a: 19 (var.) **N:** CI (Fuerteventura, Gomera, Gran Canaria, Hierro, La Palma, Lanzarote, Tenerife)
nubigena Franz, 1982: 4
palmensis Franz, 1982: 4
gussakovskii Lopatin, 1952: 28 **A:** RU (FE) SC
pseudopopuli Schöller, 2011: 26
hartmanni L. N. Medvedev, 2003a: 318 **A:** NP
heinigi Warchałowski, 1999b: 569 **A:** SY
hohuanshanus Kimoto, 1996: 29 **A:** CH (TAI)
honshuensis Takizawa, 2015: 233 **A:** JA (Honshu)
ivanovi Pic, 1953g: 13 **A:** UZ
invisus Lopatin, 1963c: 119 **A:** KI KZ TD UZ
jocularius Normand, 1947a: 143 **E:** SP **N:** AG MO TU
baguenai Burlini, 1961: 95 (ssp.)
mogadoricus Pic, 1950c: 2
kabaki Lopatin, 2002e: 84 **A:** CH (XIN)
kabakovi L. N. Medvedev, 1978c: 867 **A:** AF
katranus Lopatin & Chikatunov, 1997: 102 **A:** IS JO
kimotoi Nakane, 1963a: 19 **A:** JA (Honshu)
kurentzovi L. N. Medvedev, 1966: 39 **A:** RU (FE: Kuriles Is.)
labiatus Linnaeus, 1761: 169 (*Chrysomela*) **E:** AL AU BE BH BU BY CR CZ DE EN FI FR GB GE GR HU IR IT (incl. Sardinia, Sicily) KO KR LA LS LT LU MD ME NL NM NR PL RO RU (CT NT ST) SB SK SL SP SV (incl. Gotland) SZ UK **A:** AB GG IN KZ MG RU (ES FE incl. Sakhalin SS) TR [# 867]
digrammus Suffrian, 1848: 88
exilis Stephens, 1831: 361 [HN]
longicornes C. G. Thomson, 1866: 311
niger Geoffroy, 1785: 93
ocularis Heyden, 1863: 46 (var.)
parisinus Gmelin, 1790: 1711

laghouatensis Pic, 1903b: 115 (var.) **N:** AG MO

borowieci Warchałowski, 1999b: 548

lederi Weise, 1889h: 259 **A:** AB AF AR GG IN IQ TM TR

oomorphus Jacobson, 1896a: 530 [RN]

ovulum Jacobson, 1894e: 275 [HN]

limbifer Seidlitz, 1867: 190 **E:** SP (Sierra Nevada)

longchiensis Duan & Zhou, 2021: 4 **A:** CH (SCH)

lostiai Burlini, 1951: 46 **E:** FR (Corsica) IT (Sardinia)

lostianus Burlini, 1956b: 221 [RN]

luridicollis cisti Pic, 1905d: 147 **N:** AG TU

luridicollis douaranus Pic, 1934d: 21 (var.) **N:** EG MO

codinai Burlini, 1959a: 196 (var.)

luridicollis luridicollis Suffrian, 1868: 175 [RN] **E:** FR PT SP (incl. Balearic Islands: Mallorca, Menorca)

maculicollis Mulsant & Rey, 1861a: 343 [= 1861b: 90] [HN]

reyi Tappes, 1872: 253 [RN]

macellus Suffrian, 1860: 53 **E:** AL AN AU BE BH BU CR FR (incl. Corsica) GE GR (incl. Crete) HU IT (incl. Pantelleria, Sardinia, Sicily) MA MD ME NL NM PL RU (ST) SL SP (incl. Balearic Islands: Mallorca) SZ TR UK **A:** AB AR CY GG IN IS JO SY TR **N:** AG EG TU **[# 868]**

iranicus L. N. Medvedev, 1975a: 17 (ssp.)

pseudoconnexus J. Müller, 1948: 91 (ab.) [IN]

zantensis Pic, 1953g: 14

macilentus Lopatin, 2001c: 92 **A:** AR IN

maghribanus Aslam, 1967: 127 [RN] **N:** MO

nigroscutus Pic, 1950a: 16 [HN]

majoricensis Fuente, 1918: 44 **E:** SP (Balearic Islands: Cabrera, Dragonera, Formentera, Mallorca, Menorca)

mayeti Marseul, 1878: 82 **E:** FR SP

millari Pic, 1953c: 19 **N:** MO

moroderi Pic, 1914a: 2 (var.) **E:** FR SP

capucinus Suffrian, 1853a: 141 [HN]

cobosi Burlini, 1956b: 210 (ab.) [IN]

mystacatus Suffrian, 1848: 94 **E:** PT SP

larvatus Suffrian, 1848: 98

nebulo Weise, 1889g: 588 **A:** CH (GAN SHA)

nigellus Lopatin & Chikatunov, 2000a: 329 **A:** IS

nigrofasciatus Jacoby, 1885b: 200 **A:** CH (HEB HEI JIA JIL SCH SHA SHX YUN XIZ) JA (Hokkaido, Honshu, Kyushu, Shikoku) MG NC RU (FE incl. Sakhalin) SC

sagamensis Tomov, 1982a: 41 [QS]

turpis S.-H. Chen, 1942c: 121

nigrolimbatus Jacoby, 1890a: 87 **A**: CH (FUJ HUB) [# 848]
nigrorufus Gressitt, 1942c: 348 **A**: CH (HAI)
nitidicollis Wollaston, 1864: 397 **N**: CI (Gomera, Gran Canaria, Hierro, La Palma, Tenerife)
validicornis Har. Lindberg, 1953: 11
notogrammus Suffrian, 1854b: 143 **A**: AF ID (KA) NP PA **ORR**
ocellatus Drapiez, 1819d: 201 **E**: AL AN AU BE BH BU BY CR CZ EN FR (incl. Corsica) GB GE GR HU IT (incl. Sardinia, Sicily) KO KR KZ LA LS LT LU MD ME NL NM PL RO RU (CT NT ST) SB SK SL SP SZ TR UK **A**: AB AR CY GG IN IQ KZ LE RU (ES SS WS) SY TR
corsicus Pic, 1909h: 153
geminus Gyllenhal, 1827: 670
hummleri Pic, 1909k: 179
montanellus J. Müller, 1948: 88 (var.) [IN] [# 856]
nigrifrons Bedel, 1899: 243 (var.)
ochropezus Suffrian, 1853a: 144
ochroleucus Fairmaire, 1859a: 63 **E**: BE BU FR (incl. Corsica) GE GR HU IT (incl. Sicily) LT NL PL SL UK
fallax Suffrian, 1868: 174
oranensis Weise, 1882b: 244 **N**: AG
discolugubris Pic, 1905e: 95 (var.)
petiti Pic, 1905d: 147 (var.)
semiinterruptus Pic, 1905e: 96 (var.)
signifer Weise, 1886c: 208 (var.)
pallidoapicalis Pic, 1917a: 10 **A**: CH (LIA YUN) RU (FE)
pallidocinctus Fairmaire, 1867: 412 **N**: AG MO
latecinctus Pic, 1908j: 93 (var.) [IN] [# 856]
rifanus Burlini, 1962: 328 (morpha) [IN]
pallifrons Gyllenhal, 1813: 625 **E**: AU BE BY CZ DE EN FI FR GB GE HU LA LT NL NR PL RU (CT NT ST) SK SV SZ UK **A**: KZ MG RU (ES FE SS WS)
epipleuralis Weise, 1898: 186 (var.) [HN]
furcifrons Mannerheim, 1844: 201
irkutensis Clavareau, 1913: 171 (ab.) [IN]
pallidifrons Clavareau, 1913: 171 [UE]
petulans Weise, 1889g: 587 **A**: CH (GAN SHA)
piceoverticalis Pic, 1939a: 18 **N**: TU
planifrons Weise, 1882b: 239 **E**: AU BH BU BY CR HU IT KR MD RO RU (CT ST) SK SL UK **A**: AR IN RU (WS)
plantaris Suffrian, 1868: 174 **E**: IT (Sicily) MA

politus politus Suffrian, 1853a: 143 **E:** FR IT (incl. Sardinia, Sicily) SP (incl. Balearic Islands: Mallorca)

raphaelensis Gautier des Cottes, 1861: 193

politus sindonicus Marseul, 1874: 236 **N:** AG MO TU **[# 869]**

polymorphus parallelus Jacobson, 1896a: 534 **A:** CH (XIZ) KZ NP TD (Vakhsh and Khozratishoh Ridge)

norensis Pic, 1911c: 107

polymorphus polymorphus Solsky, 1881: 78 **A:** AF KI KZ TD UZ

dilectus Lopatin, 1956b: 162 (ab.) [IN]

disjunctipennis Pic, 1907c: 3 (var.)

laevifrons Weise, 1894a: 68

quadrinaevus Pic, 1907c: 3 (var.)

polymorphus schmidti Jacobson, 1894a: 100 **A:** KI (Issyk-kul)

populi Suffrian, 1848: 76 **E:** AL AU BE BH BU BY CR CZ FR GE HU IT KR MD NM PL RO RU (CT ST) SK SZ TR UK **A:** AB AR GG IN KZ TR

brachialis Mulsant & Rey, 1859: 45

henningsi Burlini, 1956b: 232 (ab.) [IN]

potanini Lopatin, 2001c: 91 **A:** CH (SCH XIZ)

presuturalis Pic, 1907j: 179 (var.) **E:** FR (Corsica) IT (incl. Pantelleria, Sardinia) **N:** AG MO TU

blanduloides Normand, 1947a: 142

pseudofulvus L. N. Medvedev, 1973b: 880 **A:** JA (Hokkaido, Honshu)

hakonensis Takizawa, 1982: 55

pseudosindonicus Burlini, 1960: 137 **N:** MO

pulchellus equiseti A. Costa, 1886: 27 **E:** FR (Corsica) IT (Sardinia)

pallidus Krausse, 1911: 145 (var.)

pulchellus pulchellus Suffrian, 1848: 71 **E:** FR IT (Sicily) PT SP **N:** AG MO TU

blandulus Harold, 1872: 254 [RN]

fuenteanus Roubal, 1913a: 22 (ab.) [IN]

henoni Pic, 1914b: 12 (var.)

inconnexus Fuente, 1910b: 445 (var.)

mimicus Pic, 1905e: 96 (var.)

pullatus Marseul, 1868: 218 [RN] **A:** LE

aquilus Schöller, 2009b: 106 [RN]

betmeriensis Pic, 1914a: 1 (var.)

pullus Marseul, 1868: 208 [HN]

puncticollis Wollaston, 1864: 398 **N:** CI (Gomera, Gran Canaria, Hierro, La Palma, Tenerife)

punctiger Paykull, 1799a: 146 **E:** AU BE BH BY CZ DE EN FI FR GB GE HU IT LA LT NL NR PL RU (CT NT) SK SV SZ **A:** KZ RU (ES SS WS)

reticulaticollis Roubal, 1929b: 56 (var.)

pusillus Fabricius, 1777: 221 **E:** AL AU BE BH BU BY CR CZ DE EN FI FR GB GE GR HU IR IT LA LS LT LU MD ME NL NM NR PL PT RO RU (CT NT ST) SK SL SP SV SZ TR UK **A:** GG KZ RU (ES WS) TR

arcuatetestaceus Pic, 1926e: 10 (var.) [IN] [# 856]

baudyi Pic, 1945a: 2 (var.)

colini Pic, 1945a: 1 (var.)

ernesti Pic, 1926e: 10 (var.) [IN] [# 856]

exilis Gmelin, 1790: 1686 [RN]

flavissimus Roubal 1915: 170 (ab.) [IN]

immaculatus Westhoff, 1882: 266 (var.)

minutus Herbst, 1783: 57 (*Chrysomela*)

marginellus Donovan, 1801: [25], pl. CCCXXXV (*Chrysomela*) [HN]

marginellus Marsham, 1802: 211 [HN]

marshami Weise, 1882b: 243 (var.)

muelleri Schleicher, 1924: 58 (ab.) [IN]

multinigronotatus Pic, 1926e: 10 (var.) [IN] [# 856]

nigripennis Stephens, 1831: 359

parisiensis Pic, 1926e: 10 (var.) [IN] [# 856]

philiberti Pic, 1945a: 2 (var.)

planchezi Pic, 1926e: 10 (var.) [IN] [# 856]

sonchi Schrank, 1785: 319

testouti Pic, 1945a: 2 (var.)

verticalis Boheman, 1853: 97

viduus Weise, 1889f: 218 (var.)

pusus Schöller, 2009b: 106 [RN] **A:** CH (SCH YUN) **[# 870]**

nanus Tan, 1992b: 791, 825 [HN]

pygmaeus pygmaeus Fabricius, 1792b: 70 **E:** AU BE BH CR CZ FR GE LU NL PT SB SL SP SZ **N:** AG MO

amaenus Drapiez, 1819c: 48

atlanticus Burlini, 1970: 38 (ssp.) [# 839]

lugubris Demaison, 1905: 286

otini Burlini, 1970: 39 (morpha) [IN]

pygmaeus vittula Suffrian, 1848: 63 **E:** AL AU BH BU CR CZ FR GE GR (incl. Crete) HU IT (incl. Sicily) KO KR LS MD ME NM PL RO RU (CT ST) SB SK SL SP SZ TR UK **A:** AB AR GG IN KZ SY TR **[# 871, # 1038]**

mediterraneus J. Müller, 1948: 90 (ssp.)

orientalis Weise, 1882b: 233 (var.) [HN]

pazsiczkyi Kaszab, 1962: 51 (ab.) [IN]

picticollis J. Müller, 1948: 90 (ssp.)

querceti Suffrian, 1848: 96 **E**: AL AU BH BU CR CZ DE EN (incl. Saaremaa) FI FR GB GE GR HU IT KO LA LT ME NR PL RO RU (CT NT) SB SK SL SV SZ **[# 872, # 1039]**

paganensis Pic, 1914b: 13 (var.)

subocularis Reitter, 1913c: 100 (ab.) [IN]

rabatensis Pic, 1953g: 13 **N**: MO

ragusanus Roubal, 1912a: 1 **E**: IT (incl. Sicily)

pictus Ragusa, 1882: 251 (var.) [HN]

sicilianus Clavareau, 1913: 163 (ab.) [IN]

sicilianus Burlini, 1956b: 124 (ssp.)

rufipes Goeze, 1777: 321 (*Chrysomela*) **E**: AU BE BH CR CZ DE FR GB GE IT LT NL PL PT RO RU (NT) SL SP SZ TR UK **N**: AG MO TU

abdominalis Weise, 1886c: 208

bimaculatus Čepelák, 1928: 123 (ab.) [IN]

cepelaki Kaszab, 1962: 77 (ab.) [IN]

dormeyeri Wagner, 1919: 73 (ab.) [IN]

gracilis Fabricius, 1792b: 70

hispanus Seidlitz, 1891a: 688 [= 1891b: 766] (var.)

limbaticeps Pic, 1920a: 6 (var.) [IN] **[# 856]**

mauritianus Burlini, 1956b: 249 (ab.) [IN]

prescutellaris Pic, 1913l: 187 (var.) [HN]

pseudoquadrinaevus Pic, 1913l: 187

sanguinicollis Frölich, 1792: 130

septimaniensis Pic, 1905g: 163

signatipennis Pic, 1913l: 187

subflavoapicalis Pic, 1955d: 6 (var.) [IN] **[# 856]**

timhaditus Pic, 1949d: 16 (var.)

vitticollis Weise, 1891b: 149 (var.)

sagraensis Pic, 1953g: 13 **E**: SP

saintpierrei Tappes, 1869: 13 **N**: AG MO

saliceti Zebe, 1855: 28 **E**: AU CZ EN (incl. Hiiumaa) FI FR GE HU IT ME PL RO SK SL

resinaceus Pic, 1903i: 154 (var.)

schaeferi Burlini, 1967: 131 (morpha) [IN]

saucius Truqui, 1852: 67 (var.) **E**: CR FR (incl. Corsica) IT (Sardinia, Sicily) SP (incl. Balearic Islands: Mallorca, Menorca) **N**: AG MO TU

alboscutellatus Suffrian, 1853a: 118

discus Truqui, 1852: 67 (var.)

escalerai Pic, 1908j: 93 (var.) [IN] **[# 856]**

immaculatissimus Pic, 1900i: 96 (var.)

inlateralis Escalera, 1914: 516 (var.)

nigridorsum Chevrolat, 1860: 453
scutellaris Truqui, 1852: 65 [HN]
tenietensis Pic, 1900i: 95 (var.)
ulmineus Truqui, 1852: 67 (var.)
scapularis Suffrian, 1848: 99 **E:** CR FR (Corsica) IT (incl. Sicily) SL SZ [# 873]
diversus Burlini, 1956b: 195 (ab.) [IN]
scarificollis Lopatin, 1984a: 85 **A:** IN
scutemaculatus Tan, 1992b: 794, 827 **A:** CH (YUN)
shabalinae Lopatin, 1968c: 547 **A:** KI
shaowuanus Gressitt & Kimoto, 1961: 163 **A:** CH (FUJ GAN HUB SCH)
sichuanicus Lopatin, 1999a: 89 **A:** CH (SCH)
strigosus Germar, 1823: 560 **E:** AU BH BU CR CZ FR (Alps) GR HU IT NM PL RO SB SK SL SZ **A:** TR [# 1040]
betulinus Gistel, 1857b: 522
caducus Gistel, 1857b: 544
galeazii Pic, 1900e: 69 (var.)
solarii Burlini, 1967: 131 (morpha) [IN]
sultani Pic, 1920e: 22 **A:** CY TR
tauricus Romantsov & Moseyko, 2022: 619 **E:** KR
theryi Pic, 1950a: 15 **N:** MO
tomurensis Duan & Zhou, 2021: 10 **A:** CH (XIN)
tramuntanae Petitpierre, 1993: 299 **E:** SP (Balearic Islands: Mallorca)
trapezicollis Har. Lindberg, 1953: 12 **N:** CI (Gran Canaria, La Palma)
tschimganensis narzykulovi Lopatin, 1958b: 31 **A:** TD
puzanovi Lopatin, 1958b: 32
tschimganensis tschimganensis Weise, 1894a: 68 **A:** KI KZ TD TR UZ
tshorumae Tomov, 1984: 376 **A:** TR
vividus Lopatin, 1997b: 369 **A:** CH (HEB SHA)
volubilisus Pic, 1950a: 16 **N:** MO
yangweii S.-H. Chen, 1942c: 122 **A:** CH (FUJ GAN HUB JIX ZHE)
subgenus *Cryptocephalus* Geoffroy, 1762: 231 [conserved name, ICZN (1994), Opinion 1754] type species: *Chrysomela sericea* Linnaeus, 1758
Anteriscus Weise, 1906a: 39 type species: *Cryptocephalus ertli* Weise, 1906
Cerodens Burlini, 1969b: 539 [RN] type species: *Cryptocephalus kocheri* Burlini, 1953 (= *Cryptocephalus emiliae* Burlini, 1956)
Ceropachys Burlini, 1953: 75 [HN] type species: *Cryptocephalus kocheri* Burlini, 1953 (= *Cryptocephalus emiliae* Burlini, 1956)
Chrysocryptocephalus Steinhausen, 2007: 31 type species: *Chrysomela sericea* Linnaeus, 1758
Pertuisanus Bourdonné & Doguet, 2017: 120 type species: *Cryptocephalus tristigma* Charpentier, 1825

abdeldayemi Bezděk, El-Torkey & Al Dhafer, 2017: 70 **A:** SA YE
acupunctatus Fairmaire, 1873a: pl. 16 [= 1873c: 354] **N:** AG MO TU
acupictus Marseul, 1874: 172
acutesternalis L. N. Medvedev, 1978c: 865 **A:** AF
aeneoblitus Takizawa, 1975b: 427 **A:** JA (Honshu, Shikoku, Kyushu) SC
ajeschae Schöller, 2006: 149 **A:** SA YE [# 874]
albidiceps Pic, 1913j: 171 **A:** RU (ES)
albolineatus Suffrian, 1847b: 55 **E:** AU FR (Alps) IT (Alps) SL SZ [# 875]
bischoffi Tappes, 1869: 8
perrieri Fairmaire, 1862b: 649
somi Pic, 1932e: 25 (var.)
suffrianii Suffrian, 1849a: 18
tamaninii Burlini, 1956b: 63 (ab.) [IN]
alborzensis Rapilly, 1980: 77 **A:** AB IN
almorensis Bryant, 1954c: 449 **A:** ID (UT)
alpigradus alpigradus Lopatin, 1982b: 416 **A:** NP
alpigradus kharikolensis Lopatin, 1982b: 418 (ssp.) **A:** NP
analis Olivier, 1808: 786 **A:** NP **ORR**
anastasei Pic, 1924h: 30 **A:** SI **N:** EG
angustelimbatus Reitter, 1913c: 91 **E:** RO RU (ST) **A:** "Kaukasus"
anticus Suffrian, 1848: 37 **E:** AL AU BE BH BU BY CR CZ FR GE GR HU IT KR LT ME MD NM PL RO RU (CT ST) SB SK SL SP SZ TR UK **A:** AB CH (XIN) GG IN JO KI KZ RU (SS WS) SY TR UZ [# 876, # 1041]
babadjanidesi Pic, 1924a: 18 (var.)
balkanicus Burlini, 1962: 327 (morpha) [IN]
ballioni Clavareau, 1913: 130 (ab.) [IN]
lateniger Pic, 1913j: 171 (var.)
nigerrimus Ballion, 1878: 376 [HN]
octacosmus Bedel, 1891: 131 [RN]
omissus Weise, 1882b: 227 (var.)
oneratus Weise, 1882b: 227 (var.)
praeceptor Kaszab, 1962: 50 (ab.) [IN]
pusztae Kaszab, 1962: 50 (ab.) [IN]
sexpustulatus P. Rossi, 1790: 97 [HN]
vastus Kaszab, 1962: 49 (ab.) [IN]
aquitanus Sassi, 2001a: 129 **E:** AN FR SP
andorrensis Burlini, 1956b: 124 (ab.) [IN]
arbensis arbensis Pic, 1900p: 67 (var.) **N:** AG MO TU
antoinei Kocher, 1959: 22 (ab.) [IN]
bedeli Pic, 1900p: 68
marocanus Pic, 1953d: 62

mimeuri Pic, 1949a: 16 (var.)
normandi Pic, 1900p: 68
siccensis Normand, 1938: 368
submerus Normand, 1938: 369 (ab.) [IN]
tangerianus Pic, 1908d: 13 (ab.) [IN]
arbensis mizanensis Pic, 1900p: 68 (var.) **N:** AG
asturiensis Heyden, 1870: 166 E: FR SP
atrifrons Abeille de Perrin, 1901: 70 E: FR IT (Alps)
ochsi Burlini, 1956: 170 (ab.) [IN]
aureolus Suffrian, 1847b: 132 E: AN AU BE BY CZ DE EN FI FR GB (incl. Hebrides Isl.) GE IT KZ LA LS LT LU MD NL PL RU (CT NT) SB SK SP SV SZ UK **A:** KZ RU (ES WS) **[# 1042]**
coerulescens Schilsky, 1888: 322 (var.) [HN]
discolor Gerhardt, 1909: 420 (ab.) [IN]
subcaeruleicolor Pic, 1912a: 3 (ab.) [IN]
azurescens Escalera, 1914: 514 **N:** MO
malhommei Kocher, 1959: 23 (ab.) [IN]
pseudobaeticus Burlini, 1959a: 196 (var.)
telueticus Escalera, 1914: 515
vaucheri Pic, 1950c: 2 (var.)
baenai Vela & Bastazo, 2012: 32 E: PT SP
baeticus Suffrian, 1847b: 38 E: SP **N:** AG MO
bahadur Lopatin, 1984c: 331 **A:** NP **ORR**
bahilloi López-Colón, 2004: 96 E: SP (central)
baillyi Pic, 1920c: 27 **A:** CH (YUN)
bameuli Duhaldeborde, 1999: 124 E: AL AU BE BU BY CR CZ DE EN FR GE GR HU IT (incl. Sardinia) KR LA LT MD NM PL RO RU (CT ST) SB SK SL SP SV SZ TR UK **A:** AB AR KZ RU (ES SS WS) TR [# 877]
dispar Weise, 1882b: 214 (var.) [HN]
cordifrons J. Müller, 1951: 338 (ab.) [IN]
variifrons Clavareau, 1913: 149 (ab.) [IN]
barii Burlini, 1948: 226 E: IT
barkeri Jacoby, 1901b: 235 **A:** SA **AFR**
baroniurbanii Lopatin, 1982b: 414 **A:** BT ID (SD) NP **ORR**
bhutanensis Lopatin, 1975a: 368 **A:** BT NP
bhutanus L. N. Medvedev & Sprecher-Uebersax, 1997: 216
bicolor Eschscholtz, 1818: 466 E: BH BU CR FR KR NM RO RU (ST) UK **A:** AB GG IN TR
concinnus Suffrian, 1853a: 129
bicoloripennis Chûjô, 1934e: 518 **A:** CH (FUJ GUA TAI)
breveflavomaculatus Pic, 1954d: 54

biguttatus Scopoli, 1763: 65 (*Buprestis*) E: AU BE BH BU BY CR CZ DE EN FI FR (incl. Corsica) GB GE HU IT LA LS LT LU MD NL PL RO RU (CT NT) SB SK SL SP SZ UK A: AB IN KZ RU (ES SS WS) TR

bipustulatus Fabricius, 1775: 821

carniolicus Clavareau, 1913: 133 (ab.) [IN]

ferrugatus Schrank, 1776: 64 (*Chrysomela*)

ferruginosus Gmelin, 1790: 1711

ornatus Roubal, 1910: 374 (ab.) [IN]

podae Laicharting 1781: 179

punctatus Gravenhorst, 1807: 151 [HN]

voetii Lichtenstein, 1796: 62 [UN]

biguttulatus Gebler, 1841b: 618 A: MG RU (ES FE SS WS)

biguttulus Suffrian, 1848: 24 E: KR

rufilabris Suffrian, 1853a: 128 **[# 1130]**

biledjekensis Pic, 1909h: 153 (var.) E: BU A: AR GG IN SY TR **[# 878]**

bimaculatus Fabricius, 1781: 141 E: AU FR IT (incl. Sicily) SZ **[# 1043]**

agraphus Gozis, 1886: 34 (forma)

bisbipustulatus Suffrian, 1853a: 93 [HN]

blockii P. Rossi, 1794: 91

humeralifer Pic, 1900e: 68 (var.)

lissopterus Gozis, 1886: 34 (forma)

pominii Burlini, 1948: 221

binotatithorax Pic, 1920c: 26 A: CH (YUN) **ORR**

biordopunctatus Duan, F.-Y. Wang & Zhou, 2021b: 6 A: CH (YUN)

bipunctatus bipunctatus Linnaeus, 1758: 374 (*Chrysomela*) E: AL AU BE BH BU BY CR CZ DE EN FI FR GB GE GR HU IR IT (incl. Elba) KO KR LA LS LT LU MD ME NL NM NR PL PT RO RU (CT NT ST) SB SK SL SP SV (incl. Gotland) SZ TR UK A: AB AR GG IN KZ RU (ES SS WS) TR **[# 1095]**

agnani Pic, 1908e: 58 (var.)

bilituralus Donovan, 1794: [71], pl. XCIX (*Chrysomela*)

bimacula Csiki, 1953: 126 (ab.) [IN]

bimaculatus Trella, 1930: 192 (ab.) [IN]

canigouensis Pic, 1951a: 9 (var.)

clericus Seidlitz, 1891a: 684 [= 1891b: 763] (var.)

dispar Paykull, 1799a: 142

immaculatus Sharp, 1917: 78 (var.)

immaculipennis Sharp, 1917: 78 (var.)

intemeratus Gozis, 1907: 156 (var.)

limbatus Laicharting 1781: 181

lineolus Fabricius, 1792b: 59

nebulosonotatus Pic, 1911j: 186 (var.)
oblongoguttatus Harrer, 1784: 55
quadrinotatus Herrich-Schäffer, 1835c: 22
rubrostriatus Goeze, 1777: 320 (*Chrysomela*)
sanguinolentus Scopoli, 1763: 66 (*Buprestis*)
submaculatus Pic, 1919e: 20 (var.)
suturatus Trella, 1930: 192 (ab.) [IN]
thomsoni Weise, 1881: 167 (var.)
vibex Gmelin, 1790: 1711

bipunctatus cautus Weise, 1893a: 1119 (var.) **A:** CH (FUJ GUA HEB HEI JIA JIL SHN) NC RU (ES FE) SC [# 879]
immaculatipennis Pic, 1900c: 62 (var.) [HN]
katabosi Chûjô, 1940f: 383 (ssp.)
subimmaculatus Pic, 1900c: 62 (var.)

bissexguttatus Suffrian, 1854b: 18 **A:** CH (HKG)
bissexsignatus Suffrian, 1854b: 56 **A:** ID (SD) NP **ORR**
bodungeni bodungeni Jacobson, 1905: 111 **A:** KZ UZ
bodungeni tadzhikus Jacobson, 1925c: 14 **A:** TD UZ
boreoindicus boreoindicus Lopatin, 1995c: 458 **A:** NP
boreoindicus matiltanus Lopatin, 1995c: 460 (ssp.) **A:** ID (KA) PA
brahminus Jacoby, 1908: 223 **A:** NP **ORR**
brancuccii Lopatin, 1984c: 333 **A:** ID (SD UT) NP [# 880]
brevebilineatus Pic, 1922e: 28 **A:** ID (SD) NP **ORR**
binhanus Pic, 1927m: 4 (var.)
internefasciatus Pic, 1927m: 4 (var.)
internelineatus Pic, 1927m: 4 (var.)
monticola Bryant, 1954c: 452 [HN]
subarcuatus Pic, 1927m: 4 (var.)

brevesignatus Pic, 1922c: 10 **A:** CH (YUN)
brevesignatus Pic, 1943d: 15 [HN]

brittoni Bryant, 1957b: 354 (*Anteriscus*) **A:** YE
brunneopunctatus Pic, 1922f: 9 **A:** CH (SCH YUN) **ORR**
bryanti Gressitt, 1942c: 340 **A:** CH (HAI HKG)
buettikeri Lopatin, 1979a: 301 **A:** SA YE
caerulans Marseul, 1874: 152 [RN] **A:** CH (JIL) JA (Hokkaido, Honshu, Shikoku) MG NC RU (ES FE SS WS) SC (incl. Jeju-do Isl.)
approximans S.-H. Chen, 1942c: 113
coerulescens Suffrian, 1847b: 170 [HN]
kraatzi Chûjô, 1935e: 220

caerulescens caerulescens C. R. Sahlberg, 1839: 261 **E:** BE BY CZ EN FI FR (incl. Corsica) GE IT KR LA LT NL PL RO RU (CT NT) SP SV SZ UK **A:** AB AR GG IN IS KZ MG RU (WS FE incl. Sakhalin) SY TM TR **[# 846, # 881]**

androgyne Marseul, 1874: 184 **[# 846]**

distractus Burlini, 1967: 131 (morpha) [IN]

flavilabris C. G. Thomson, 1866: 304 [HN]

larozei Pic, 1908a: 14 (ab.) [IN]

matsumurai Takizawa, 1971: 137

prehumeralis Pic, 1908a: 14 (ab.) [IN]

quadripunctata Polentz, 1938: 59 (forma) [IN] **[# 856]**

similis L. N. Medvedev, 1956c: 250 [HN]

caerulescens pelleti Marseul, 1874: 183 **E:** FR (incl. Corsica) IT (incl. Sardinia) PT SP **[# 882]**

cantabricus Franz, 1958: 125 **E:** PT SP

castilianus Weise, 1894e: 93 **E:** SP

celtibericus Suffrian, 1848: 53 **E:** FR PT SP **N:** AG TU

apiceprolungatus Pic, 1900e: 69 (var.)

cheni Pic, 1935e: 20 **A:** CH ("Chine")

chikatunovi Lopatin, 2008a: 313 **A:** AE OM

chinensis Jacoby, 1888: 341 **A:** CH (FUJ GUX JIA JIX TAI ZHE)

subfenestratus Pic, 1954d: 54

chitwanus L. N. Medvedev, 2009d: 409 **A:** NP

cicatricosus cicatricosus P. H. Lucas, 1845: 125 **E:** SP **N:** AG MO

cicatricosus punctonotatus Pic, 1896d: 94 (var.) **N:** AG MO TU

subcicatricosus Pic, 1908a: 13 (ab.) [IN]

subcicatricosus Kocher, 1958: 70 (ssp.)

cilicius Duhaldeborde, 2018: 261 **A:** TR

columnarius Tan & Pu, 1991: 253, 255 (*Melixanthus*) **A:** CH (SCH)

concolor Suffrian, 1847b: 127 **E:** BU **A:** AB AR GG IN (north) TR

persicus Rapilly, 1980: 75 (ssp.)

unicolor Faldermann, 1837: 386 [HN]

conspicuus Weise, 1922a: 44 (*Anteriscus*) **A:** CH (FUJ)

convergens Sassi, 2001b: 136 **E:** AN AU FR IT (Alps) SP SZ

cordiger Linnaeus, 1758: 375 (*Chrysomela*) **E:** AU BE BH BU BY CR CZ DE EN FI FR GE GR HU IT LA LT LU MD ME NM NR PL RO RU (CT NT ST) SB SK SL SV SZ UK **A:** KZ MG RU (ES SS WS) TR

cirsii Geoffroy, 1785: 92

completus Jacobson, 1901b: 111 (var.) [IN] **[# 856]**

magistrettii Burlini, 1948: 229 (ab.) [IN]

nigripes J. Müller, 1948: 85 (ab.) [IN]
talaris Csiki, 1953: 123 (ab.) [IN]
tihanyensis Székessy, 1943: 396 (ab.) [IN]
vlascanus Pic, 1910e: 66 (var.)
crassus Olivier, 1792: 620 **E**: FR (incl. Corsica) IT (incl. Lampedusa, Sardinia) PT SP **N**: AG MO TU **A**: IS SI SY TR
anticemaculatus Chobaut, 1902: 165 (var.)
gravidus Herrich-Schäffer, 1836: 23 [# 1044]
gravidus P. H. Lucas, 1845: 126 [HN]
japhaaensis Pic, 1924a: 18 (var.)
posticemaculatus Pic, 1898c: 139 (var.)
prescutellaris Pic, 1901d: 38 (var.)
sefrensis Pic, 1900e: 69 (var.)
susicus Kocher, 1959: 25 (var.) [IN] [# 856]
tenebricosus Weise, 1882b: 226 (var.)
creticus Suffrian, 1847b: 178 **E**: GR (Crete)
cribratus Suffrian, 1847b: 90 **E**: RU (ST) TR **A**: AB GG IN SY TM TR
notaticoxis Pic, 1924a: 17 (var.)
syriacus Pic, 1908j: 92 (var.) [IN] [# 856]
trapezensis Tappes, 1872: 266
cristula Dufour, 1843: 113 **E**: FR SP
cristatus Suffrian, 1849a: 19 [HN]
cruciger Hellén, 1922: 103 **E**: FI RU (NT) SV
cruciger Aurivillius, 1917: 65 (ab.) [IN]
crucipennis Suffrian, 1854b: 64 **A**: CH (HAI HUB JIX SCH YUN) **ORR**
birmanicus Jacoby, 1889b: 159
lajoyei Pic, 1920c: 26
peguensis Weise, 1910c: 28 (ab.) [IN]
violaceocinctus Pic, 1943d: 16
curtipennis Pic, 1920c: 25 **A**: CH (YUN)
curtissimus Pic, 1907c: 1 **A**: IN KZ TD TM UZ
curvilinea Olivier, 1808: 812 **E**: IT (Sardinia, Lampedusa) **N**: AG EG LB MO TU
carnerii Pic, 1924h: 30 (var.)
dahlii P. H. Lucas, 1845: 126
inops Weise, 1882b: 204 (var.)
octopunctatus Dalman, 1808: 368 [HN]
ornatus Herrich-Schäffer, 1835c: 21 [HN]
cynarae Suffrian, 1847b: 27 **E**: FR PT SP **N**: MO
duodecimplagiatus Fairmaire, 1859c: cli

duodecimplagiatus Gozis, 1907: 75 (var.) [HN]
fucatus Burlini, 1956b: 142 (ab.) [IN]
plagiatus Gozis, 1907: 75 (var.)
dalianus Lopatin, 2004c: 188 **A**: CH (YUN)
decastictus Fairmaire, 1888c: 36 **A**: CH (SCH YUN)
arcuatomaculatus Pic, 1915c: 11 (var.)
arcuatovittatus Pic, 1915c: 11 (var.)
subconnexus Pic, 1920c: 25 (var.) [HN]
uniarcuatus Pic, 1920c: 25 (var.)
decemmaculatus Linnaeus, 1758: 375 (*Chrysomela*) E: AU BE BH BY CR CZ DE EN FI FR GB GE HU IR KR LA LT LU NL NR PL RO RU (CT NT ST) SB SK SV SZ UK **A**: AB RU (WS)
banatensis Pic, 1917c: 22 (var.)
barbareae Linnaeus, 1761: 169 (*Chrysomela*)
betulaenanae Schilling, 1837: 99 [UN]
bothnicus Linnaeus, 1758: 375 (*Chrysomela*)
champenoisi Pic, 1925b: 9 (var.) [IN] [# 856]
croissandeaui Pic, 1939g: 30 (ab.) [IN]
decempunctatus Linnaeus, 1761: 170 (*Chrysomela*)
externesignatus Pic, 1925b: 9 (var.) [IN] [# 856]
frenatus Fabricius, 1792b: 60 [HN]
gallicus Pic, 1917c: 22 (var.)
gardanus Pic, 1927l: 11 (var.) [IN] [# 856]
germanicus Pic, 1927l: 11 (var.) [IN] [# 856]
goulettei Pic, 1927l: 11 (var.) [IN] [# 856]
lapponicus Pic, 1939g: 30 (var.) [IN] [# 856]
leveillei Pic, 1939g: 30 (ab.) [IN]
marionensis Pic, 1927l: 11 (var.) [IN] [# 856]
miscerus Pic, 1927l: 11 (var.) [IN] [# 856]
moestus Weise, 1882b: 208 (var.)
morvandicus Pic, 1925b: 9 (var.) [IN] [# 856]
ornatus Herbst, 1783: 63
prixi Pic, 1927l: 11 (var.) [IN] [# 856]
scenicus Weise, 1882b: 207 (var.)
solutus Weise, 1882b: 207 (var.)
teutonicus Pic, 1927l: 11 (var.) [IN] [# 856]
deficiens Suffrian, 1854b: 146 **A**: BT ID (SD UT) NP **ORR**
dimidiatipennis Jacoby, 1895b: 269 **A**: CH (XIZ) ID (SD UT) NP **ORR** [# 883]
manipurensis Jacoby, 1908: 225

dinae Lopatin & Chikatunov, 2000a: 329 **A:** IS

discoderus Fairmaire, 1889b: 69 **A:** CH (GUA GUI HUB JIX SCH TAI YUN)

discoidalis Jacoby, 1890a: 89

semicaerulescens Pic, 1922e: 28 (var.)

senensis Pic, 1922f: 9

suboblongus Pic, 1922e: 28

suboblongus Pic, 1943c: 9 [HN]

yunnanus Pic, 1917a: 10

dislocatus Suffrian, 1857: 176 **A:** YE **AFR**

distinctenotatus Pic, 1901a: 19 **A:** IS LE SI

libanensis Pic, 1901j: 38

distinguendus J. H. Schneider, 1792: 209 **E:** AU BE BY CZ DE EN FI FR GE LA LT NR PL RO RU (CT NT) SK SV SZ UK **A:** AB IN KZ MG RU (ES SS WS)

alni Ménétriés, 1837: 181

bihumeralis Pic, 1908a: 13 (var.) [HN]

ochthephilus Clavareau, 1913: 146 (var.) [RN]

ongudajensis Jacobson, 1901a: 95

pomeranicus Burlini, 1962: 328 (morpha) [IN]

subdistinguendus Pic, 1908a: 13 (ab.) [IN]

dogueti Bourdonné, 1994: 199 **N:** AG MO TU

dumonti Peyerimhoff, 1924: 62 **N:** AG EG TU **A:** AE IS SI

assioutensis Pic, 1927o: 2

margaritae Lopatin, 1997b: 368

duplicatus Suffrian, 1847b: 126 **E:** BU GR RO RU (ST) TR **A:** AB AR GG IN IS JO SY TR [# 884]

durius Suffrian, 1860: 35 **A:** ID (AP) **ORR**

eberti Kimoto, 2004: 47 **A:** NP

egenus Tan, 1992b: 786, 820 **A:** CH (SCH)

egerickxi Tappes, 1884: CCLVI **A:** AE IS JO SA

elatus Fabricius, 1801b: 41 **N:** MO

bremondi Pic, 1953b: 2 (ab.) [IN]

molossus Fairmaire, 1881: 217

elongatus Germar, 1823: 557 **E:** AU BH CZ HU MD RO RU (CT ST) SK SL SV (Öland) UK **A:** KZ RU (ES SS WS) [# 1045]

emiliae Burlini, 1956b: 178 [RN] **N:** MO

anchorarius Burlini, 1970: 40 (morpha) [IN]

kocheri Burlini, 1953: 73 [HN]

pseudorugicollis Burlini, 1953: 75 (ab.) [IN]

subpersonatus Kocher, 1959: 26 (ab.) [IN]

ensifer Hope, 1831: 30 **A:** ID (SD UT) NP **[# 885]**
rasuanus Bryant, 1952: 604
eous Lopatin, 1952: 28 **A:** CH (HEN JIL) MG RU (FE) **[# 886]**
erberi Lopatin, 1991b: 405 **A:** NP
eridani Sassi, 2001a: 126 **E:** FR IT
espanoli Burlini, 1965: 17 E: SP
etruscus Heyden, 1892a: 368 **E:** FR IT SP **[# 847, # 1131]**
evae Lopatin, 2002e: 84 **A:** NP
exsulans Suffrian, 1854b: 149 **A:** BT CH (XIN XIZ) ID (AP SD) NP **ORR**
falzonii Burlini, 1956b: 110 (ssp.) **E:** IT
fatuus Suffrian, 1860: 43 **A:** ID (KA) **ORR** **[# 887]**
festivus Jacoby, 1890a: 88 **A:** CH (FUJ GUX HUB JIA JIX SHA TAI ZHE)
flavopygidialis Pic, 1922c: 11
flavopygidialis Pic, 1943c: 9 [HN]
flavicaudis Tan, 1992b: 787, 821 **A:** CH (YUN)
flavipes Fabricius, 1781: 146 **E:** AL AU BE BH BU BY CR CZ EN FR GE GR HU IT (incl. Sicily) KR LA LS LT LU MD ME NL NM PL RO RU (CT NT ST) SB SK SL SP SZ TR UK **A:** AR GG IN IS KZ MG RU (ES FE WS) **[# 1046, # 1096]**
glaucocephalus Schaller, 1783: 277 (*Chrysomela*)
nigrescens Gradl, 1881: 307 (var.)
obscuripes Weise, 1893a: 1120 (var.)
parenthesis Schrank, 1782: 153 (*Chrysomela*)
pumilionis Apfelbeck, 1912: 244 (var.)
virago Burlini, 1967: 131 (morpha) [IN]
wydleri Faldermann, 1837: 388
fokiensis Weise, 1922a: 44 **A:** CH (FUJ)
formosanus Chûjô, 1934f: 287 **A:** CH (TAI)
foveatus L. N. Medvedev, 2003a: 318 **A:** NP
freidbergi Lopatin & Chikatunov, 1997: 100 **A:** IS
frenatus Laicharting, 1781: 183 **E:** AL AU BH BU BY CR CZ FR GE HU IT LT PL RO SB SK SL SZ UK **A:** GG RU (WS)
argentorati Pic, 1917c: 22 (var.)
barbarorum Pic, 1917b: 15 (var.)
basicinctus Csiki, 1953: 126 (ab.) [IN]
binotatipennis Pic, 1917c: 23 (var.)
discojunctus Pic, 1917b: 15 (var.)
fabricii Weise, 1882b: 209 (var.)
flavescens D. H. Schneider, 1792: 196 [# 1047]
nigricans Trella, 1932: 227, 229 (ab.) [IN]

notatithorax Pic, 1909l: 186 (var.)
pisanus Burlini, 1967: 131 (morpha) [IN]
provocatoris Pic, 1917c: 22 (var.)
seminiger Weise, 1882b: 209 (var.)
servillorum Pic, 1917b: 15 (var.)
servorum Pic, 1917c: 23 (var.)
trilineatus Fabricius, 1798: 105
unicolor Eichler, 1924: 66 (var.) [HN]
vastatorum Pic, 1917c: 23 (var.)
zoltani Csiki, 1953: 126 (ab.) [IN]
freyi Gressitt & Kimoto, 1961: 150 **A:** CH (YUN)
furvicornis Tan, 1992b: 788, 821 **A:** CH (SCH)
gandolphei Pic, 1902f: 49 (var.) **N:** AG MO
berakensis Pic, 1911f: 130 (var.)
diversepunctatus Pic, 1937c: 15
geisthardti Schöller, 2021a: 17 **A:** OM
gestroi Jacoby, 1892b: 890 **A:** CH (FUJ GUX SCH SHX XIZ YUN) ID (AP) NP **ORR**
talianus Pic, 1915g: 12
globicollis Suffrian, 1847b: 141 **E:** FR IT (Alps) SP SZ **N:** AG [# 888]
aurantiacus Tiberghien, 1971: 191 (ab.) [IN]
cyaneomicans Heyden, 1892a: 368 (var.)
purpureomicans Pic, 1900e: 68 (var.)
semipurpureus Pic, 1914b: 12 (var.)
simulatus Burlini, 1958: 122 (ab.) [IN]
spectabilis Tiberghien, 1971: 191 (ab.) [IN]
gorbunovi L. N. Medvedev, 1999c: 182 **A:** NP
grahami Gressitt & Kimoto, 1961: 151 **A:** CH (SCH)
gridellii Burlini, 1950: 54 **E:** BH BU CR CZ HU IT SB SK SL
grohmanni Suffrian, 1848: 22 **E:** IT (incl. Sicily)
cruciatus Ragusa, 1922: 125 (ab.) [IN]
guttifer Suffrian, 1854b: 22 **A:** CH (YUN) ID (HP KA SD UT) NP **ORR**
kashmirensis Jacoby, 1908: 261
hainanicus Gressitt, 1942c: 343 **A:** CH (HAI)
halleri Costessèque, 2008: 53 **E:** FR
halyzioides Weise, 1889g: 585 **A:** CH (SCH SHA)
hamifasciatus Chûjô, 1934e: 516 **A:** CH (TAI)
kesi Chûjô, 1954d: 213 [RN]
minutissimus Chûjô, 1934e: 514 [HN]
hani Duan, F.-Y. Wang & Zhou, 2021b: 8 **A:** CH (GAN HUB SHA SHX)

haroldi Kraatz, 1879b: 130 **A:** CH (northeastern China) RU (FE)

helobius Tan, 1992b: 788, 822 **A:** (SCH)

hennigi Sassi, 2011: 124 **E:** IT (Piedmontese Western Alps)

heraldicus Suffrian, 1854b: 145 **A:** ID (UT) NP **ORR**

poultoni Jacoby, 1908: 230

herbstii Suffrian, 1854b: 25 **A:** ID (UT) **ORR**

hiro Chûjô, 1954d: 200 **A:** CH (TAI)

hirticollis Suffrian, 1847b: 35 **E:** IT (Egadi islands, Levano Marettimo, Sicily)

hongshanus Gressitt, 1942c: 343 **A:** CH (JIX)

hopei L. N. Medvedev & Sprecher-Uebersax, 1997: 214 **A:** ID (SD UT) NP

hyacinthinus Suffrian, 1860: 46 **A:** CH (HEB HEI JIA JIX LIA NIN SCH SHA ZHE) JA (Hokkaido, Honshu, Kyushu, Shikoku) NC RU (FE) SC **[# 889]**

approximatus Baly, 1873: 93

fortunatus Baly, 1873: 94

hypochaeridis Linnaeus, 1758: 370 (*Chrysomela*) **E:** AL AU BE BH BU BY CR CZ DE EN FI FR GB GE HU IT KO LA LS LT LU NL NM NR PL RU (CT NT) SB SK SL SV SZ UK

anteviridis Burlini, 1958: 123 (ab.) [IN]

auratus Gerhardt, 1909: 420 (ab.) [IN]

austerus Burlini, 1958: 123 (ab.) [IN]

auricollis Clavareau, 1913: 154 (ab.) [IN]

bicolor Gerhardt, 1909: 420 (ab.) [IN, HN]

burlinianus Kaszab, 1962: 46 (ab.) [IN]

cyanescens Weise, 1893a: 1119 (var.) [HN]

fiorii Della Beffa, 1909: 200 (*Chryptocephalus*, ab.) [IN]

frigidus Jacobson, 1901b: 113 (var.) [IN] **[# 856]**

iridescens Trella, 1930: 192 (ab.) [IN]

nigricans Burlini, 1962: 328 (morpha) [IN]

nigrinus Burlini, 1963a: 157 (morpha) [IN]

purpurascens Jacobson, 1897c: 431 (var.)

rugulipennis Suffrian, 1853a: 111

similis Stephens, 1834: 358

suturalis Depoli, 1926a: 15 (forma) [IN] **[# 856]**

syngenesiae Scopoli, 1763: 63 (*Buprestis*)

violaceus Gerhardt, 1909: 420 (ab.) [IN, HN]

viridicollis Depoli, 1926a: 15 (forma) [IN] **[# 856]**

ilicis Olivier, 1808: 801 **E:** AL FR (Corsica) GR IT (Aeolian and Egadi isl., Sicily) TR **A:** JO SY TR **[# 1048]**

binotatus Gozis, 1907: 72 (var.) [HN] **[# 1132]**

humerinotatus Burlini, 1956b: 75 (ab.) [IN]
obliteratipennis Pic, 1900e: 67 (var.)
oblitteratipennis Pic, 1900c: 62 (var.)
prinkipensis Pic, 1908a: 13 (ab.) [IN]
pseudoprinkipensis Burlini, 1962: 327 (morpha) [IN]
siculus Herrich-Schäffer, 1835c: 23
imperialis Laicharting, 1781: 179 **E:** AL AU BH BU CR CZ FR GE GR HU IT KR MD ME NM PL RO RU (ST) SB SK SL SP SZ UK **A:** GG IN TR [# 890]
bistripunctatus Germar, 1823: 557
inexternus Pic, 1910d: 58 (var.)
posticeinstitutus Pic, 1900g: 88 (var.)
inaequalis Tan, 1988a: 321, 331 (*Melixanthus*) **A:** CH (XIZ)
incisodentatus Duan, F.-Y. Wang & Zhou, 2021b: 11 **A:** CH (SCH YUN)
incisus Tan, 1992b: 789, 822 **A:** CH (YUN)
indicus Suffrian, 1854b: 58 **A:** NP **ORR**
infirmior Kraatz, 1876: 138 **E:** FR IT PT SP
posticeobliteratus Pic, 1900e: 68 (var.)
infraflavus Pic, 1922f: 9 **A:** CH (HAI) **ORR**
diversepleuris Pic, 1922f: 9 (var.)
instabilis Baly, 1873: 91 **A:** JA (Honshu, Kyushu, Shikoku)
kiotosinus Pic, 1908j: 92 (var.) [IN] [# 856]
multiconnexus Pic, 1908j: 92 (var.) [IN] [# 856]
jani Medvedev, 2011d: 336 **A:** CH (YUN)
jansoni Baly, 1877e: 218 **A:** CH (BEI HEI SHN) MG [# 891]
janthinus Germar, 1823: 555 **E:** AL AU BH BU BY CR CZ EN FR (incl. Corsica) GE HU IT (incl. Sardinia) LT MD NM PL RO RU (CT ST) SB SK SL SZ TR UK **A:** AB AR CH (HEI JIL) GG IN IS JA (Hokkaido, Honshu) MG RU (ES FE incl. Sakhalin SS WS) TR [# 1049]
convexissimus Pic, 1908j: 93
insulcaticeps Pic, 1911j: 186
megalosomus Faldermann, 1837: 390
melanarius Burlini, 1967: 130 (morpha) [IN]
rusticus Faldermann, 1837: 392 [typo 395]
satanas Kaszab, 1962: 46 (ab.) [IN]
weymarni Gressitt & Kimoto, 1961: 168
juniperi Pic, 1905d: 147 **N:** AG
kanoi Chûjô, 1954d: 212 **A:** CH (TAI)
kermanicus Lopatin, 1979b: 587 **A:** IN IQ
kiritschenkiellus Lopatin, 1961c: 99 **A:** UZ

kiyoyamai Kimoto, 1974b: 22 **A:** CH (TAI)
kopetzi L. N. Medvedev, 2003a: 318 **A:** ID (HP) NP
krutovskyi gebleri Jacobson, 1924c: 242 [RN] **A:** KZ MG RU (ES SS WS)
niger Gebler, 1830: 207 [HN]
krutovskyi krutovskyi Jacobson, 1901a: 76 **A:** JA (Hokkaido) MG RU (ES FE incl. Sakhalin SS WS)
karafutonis Kôno & Tamanuki, 1926: 293
triangulifer Jacobson, 1926a: 273 (ab.) [IN]
kubani Medvedev, 2011d: 338 **A:** CH (YUN)
kulibini kulibini Gebler, 1832: 71 **A:** CH (GUA HEI JIL LIA NIN) MG NC RU (ES FE) SC
kulibini nasutulus Weise, 1889g: 584 **A:** CH (GAN HUB JIL QIN SHA SHN)
lacosus Pic, 1922f: 9 **A:** CH (YUN) **ORR**
lactineus Tan, 1992b: 789, 823 **A:** CH (SCH) [# 840]
laetus Fabricius, 1792b: 54 **E:** AU BU BY CZ HU KR MD PL RO RU (CT ST) SK UK **A:** GG KZ MG RU (WS south)
basifasciatus Kaszab, 1962: 45 (ab.) [IN]
balatonicus Kaszab, 1962: 45 (ab.) [IN]
berladensis Pic, 1915d: 13 (var.)
ernoi Kaszab, 1962: 44 (ab.) [IN]
innotatus Csiki, 1953: 124 (ab.) [IN]
micronotatus Pic, 1908a: 13 (ab.) [IN]
mixtus D. H. Schneider, 1792: 212
octosignatus Csiki, 1953: 124 (ab.) [IN]
salisburgensis Moll, 1784a: 43 (*Chrysomela*) [HN]
steveni Adams, 1817: 313
lanpingensis Tan, 1992b: 790, 823 **A:** CH (YUN) [# 841]
latimargo L. N. Medvedev, 1971c: 129 **A:** MG RU (ES SS)
lefevrei Jacoby, 1895b: 268 **A:** NP **ORR**
leonhardi Breit, 1918a: 42 **E:** CR IT SL
leucographus Weise, 1922a: 45 (*Anteriscus*) **A:** CH (FUJ)
lentiginosus Tan, 1992b: 790, 824 **A:** CH (SCH)
licenti S.-H. Chen, 1942c: 115 **A:** CH (GAN HEB HEI SHX)
limarius Lopatin, 1984a: 83 **A:** IN
limbatipennis Jacoby, 1885b: 199 **A:** CH (FUJ JIX NIN SHA SHN) JA (Honshu, Kyushu) RU (FE) SC
klapperichi Pic, 1954d: 54
yanoi Chûjô, 1938d: 161
lineellus Suffrian, 1849b: 290 **E:** SP **N:** MO
lingnanensis Gressitt, 1942c: 344 **A:** CH (GUA HAI)
lofgrenae Gressitt & Kimoto, 1961: 154 **A:** CH (GUA HKG)

longiscapus Tan & Pu, 1991: 254, 256 (*Melixanthus*) **A:** CH (YUN)
loochooensis loochooensis Chûjô, 1935a: 72 **A:** JA (Aka Isl., Amami-Ōshima, Ishigaki Isl., Kume-jima Isl., Niijima Isl., Okinawa main Isl., Taketomi Isl.)
loochooensis miyakoensis Kimoto, 1974c: 281 (ssp.) **A:** JA (Miyako Isl.)
lopatianus L. N. Medvedev & Sprecher-Uebersax, 1997: 213 **A:** ID (SD UT) NP
ludingensis Tan, 1992b: 791, 825 **A:** CH (SCH)
luridipennis luridipennis Suffrian, 1854b: 99 **A:** CH (HUB JIA JIL SHG) MG NC RU (ES FE incl. Sakhalin SS WS)
bodoi Pic, 1913a: 102 (var.)
multicolor Pic, 1922c: 10
multicolor Pic, 1943d: 15 [HN]
quindecimnotatus Suffrian, 1854b: 101
luridipennis signaticeps Baly, 1873: 91 **A:** JA (Hokkaido, Honshu, Kyushu, Shikoku) MG NC RU (FE) SC
chujoi Nakane, 1963a: 19
distinctior Pic, 1908j: 92 (ssp.)
kapfereri Pic, 1913a: 102 (var.)
latepallescens Pic, 1908a: 14 (ab.) [IN]
pallescens Kraatz, 1879b: 133
serinicolor Pic, 1907c: 3
sterenanus Pic, 1924a: 17 (var.)
subpallescens Pic, 1908a: 14 (ab.) [IN]
lushanensis Gressitt, 1942c: 345 **A:** CH (SCH)
luteosignatus Pic, 1922c: 10 **A:** CH (BEI FUJ GUA GUX HAI HKG JIA JIX NIN SCH TAI ZHE) ORR [# 892]
auripennis Chûjô, 1934e: 515
luteodecemmaculatus Pic, 1943c: 16
multifenestratus Pic, 1928c: 33
variifasciatus Chûjô, 1934e: 522
luteosuturalis Pic, 1928c: 34 **A:** CH (YUN)
biquinquemaculatus Pic, 1928c: 34 (var.) [IN] [# 856]
fouanus Pic, 1928c: 34 (var.) [IN] [# 856]
humeromaculatus Pic, 1928c: 34 (var.) [IN] [# 856]
makii Chûjô, 1934e: 523 **A:** CH (TAI)
mannerheimii Gebler, 1825: 56 **A:** CH (GAN HEB HEI NIN SHX) JA ("Japan") MG NC RU (ES FE SS WS)
bicolor Gebler, 1825: 56 [UN]
immaculatus Bodemeyer, 1916: 112 (var.) [HN]
infrajunctus Pic, 1924a: 17 (var.)
medioflavus Pic, 1904a: 27 (var.) [IN] [# 856]

medioniger Pic, 1904a: 27 (var.) [IN] [# 856]
ussuriensis Weise, 1898: 184 (var.)
marginatus Fabricius, 1781: 140 **E:** AU BE BH BU CR CZ FI FR GE HU IT NL PL RO RU (NT ST) SK SL SZ UK
crucifer Burlini, 1946: 78 (ab.) [IN]
devaulxi Méquignon, 1948: 79 (var.) [IN] [# 856]
discoidalis Burlini, 1946: 78 (ab.) [IN]
palliatus Suffrian, 1861a: 91
phaleratus Schaller, 1783: 276 (*Chrysomela*)
pseudobicolor Costessèque, 2008: 53 (ab.) [IN]
pseudoterminatus Burlini, 1948: 228 (ab.) [IN]
sabaudus Méquignon, 1948: 79 (var.) [IN] [# 856]
semidophilus Gistel, 1857b: 522
terminatus Germar, 1823: 555
viridiflavus Marquet, 1864: 373
wellschmiedi Burlini, 1961: 95 (morpha) [IN]
marginellus Olivier, 1792: 616 **E:** FR GE SZ
inexpectus Fairmaire, 1861: xi
mariae Mulsant & Rey, 1852a: 5 [= 1852b: 145] **E:** FR IT SP **N:** AG
doctoris Pic, 1900e: 69 (var.)
jacqueti Pic, 1900e: 69 (var.)
pyrenaeus Weise, 1882b: 224 (var.)
signatus Olivier, 1792: 619 [HN]
mashonanus Jacoby, 1901b: 234 **A:** AE YE **AFR**
maytreiae Lopatin, 1979c: 434 **A:** ID (SD) NP
melanoxanthus chotanensis Jacobson, 1896a: 538 **A:** CH (XIN)
melanoxanthus melanoxanthus Solsky, 1876: 297 [= 1881: 73] **A:** AF KI KZ TD UZ
grombczewskii Jacobson, 1896a: 540
melanoxanthus solskii Jacobson, 1896a: 542 (var.) **A:** KZ TM
flaviventris Lopatin, 1952: 28 (ssp.)
melanoxanthus variolosus Jacobson, 1894a: 99 **A:** KI
melaphaeus Schöller, Smetana & Lopatin, 2010: 80 [RN] **A:** CH (GAN)
bisbicruciatus S.-H. Chen, 1942c: 115 [HN]
messutati Kippenberg, 2011: 43 **E:** BU GR **A:** TR [# 893]
mitchyi Chûjô, 1954d: 214 **A:** CH (TAI)
modestus Suffrian, 1848: 107 **E:** RU (ST) **A:** AB KZ RU (WS)
rugulifrons Jacobson, 1896a: 532 (var.)
moniliformis Lopatin, 1980: 619 **A:** IN
monilis Weise, 1890a: 478 **A:** CH (HEI) KZ MG

moraei Linnaeus, 1758: 374 (*Chrysomela*) **E:** AL AU BE BH BU BY CR CZ DE EN FI FR (incl. Corsica) GB GE GR HU IR IT (incl. Sicily) KO KR LA LS LT LU MD ME NL NM NR PL PT RO RU (CT NT ST) SB SK SL SP (incl. Balearic Islands: Ibiza) SV (incl. Gotland) SZ TR UK **A:** AB AR GG IN IQ JO RU (ES SS WS) TR **NARi** [# 1097]

alleoni Portevin, 1934: 214 (ab.) [IN]

arquatus Weise, 1882b: 225 (var.)

bifasciatus Townson, 1797: 462 [HN]

bivittatus Gyllenhal, 1813: 619 (var.)

bivittiger Pic, 1909d: 131 (var.)

cruciatus Gozis, 1907: 151 (var.) [# 1133]

flavoguttatus Schrank, 1781: 90 (*Chrysomela*)

hispanicus Burlini, 1967: 131 (morpha) [IN]

greineri Burlini, 1956b: 157 (ab.) [IN]

marseuli Csiki, 1953: 126 (ab.) [IN]

messai Burlini, 1967: 131 (morpha) [IN]

nigellus Burlini, 1956b: 157 (ab.) [IN]

nigellus Burlini, 1967: 131 (morpha) [IN]

quadriguttatus O. F. Müller, 1776: 58 [NO] [# 845]

quadrimaculatus Geoffroy, 1785: 91

vittiger Marseul, 1874: 165 (var.) [HN]

winkleri Pic, 1913f: 138 (var.)

moutoni Pic, 1922c: 10 **A:** CH (FUJ GUX JIX SHG TAI ZHE)

flavobinotatus Chûjô, 1934e: 517

montoni Pic, 1943d: 15

moya Chûjô, 1954d: 237 **A:** CH (TAI)

muelleri Burlini, 1948: 224 **E:** FR SP (incl. Balearic Islands: Mallorca)

muellerianus Burlini, 1956b: 143 [RN]

multiplex liothorax Solsky, 1871: 250 **A:** CH (HEI JIL QIN XIZ) RU (ES FE)

multiplex multiplex Suffrian, 1860: 39 **A:** CH (GAN GUX HUN JIA NIN SHA ZHE) JA NC NP RU (ES FE) SC [# 894]

jacobsoni Clavareau, 1913: 157 [RN]

parvicollis Jacobson, 1896a: 535 [HN]

savioi Pic, 1928c: 33 (var.) [IN] [# 856]

nepalensis Bryant, 1952: 605 **A:** NP

ngae Gressitt, 1942: 347 **A:** CH (GUA GUX) **ORR**

nigrobasalis Bryant, 1946: 613 **A:** YE **AFR**

gabbari Schöller, 2005: 1556

nigroflavus Lopatin, 2004a: 614 **A:** CH (SCH)

nigroflavusiventerus Duan, F.-Y. Wang & Zhou, 2021b: 13 **A:** CH (YUN)

nigronotaticollis Chûjô, 1934e: 519 **A:** CH (TAI)
nigronotatus Bryant, 1954c: 453 **A:** ID (SD UT) NP **ORR**
balajensis Kimoto & Takizawa, 1973: 174
nigropygius Pic, 1913l: 186 **A:** RU (ES)
nilgiriensis Jacoby, 1903a: 90 **A:** NP **ORR**
nitidissimus Chûjô, 1934f: 290 **A:** CH (TAI)
nitidulus Fabricius, 1787: 84 **E:** AU BE BH BU BY CR CZ DE EN FI FR GB GE HU IT LA LS LT LU MD NL NR PL RO RU (CT NT ST) SK SL SP SV SZ UK **A:** JA (Hokkaido) RU (ES FE incl. Kuriles Isl., Sakhalin SS WS)
aurithorax Kaszab, 1962: 46 (ab.) [IN]
borealis L. N. Medvedev, 1966: 39 (ssp.)
iturupensis L. N. Medvedev, 1966: 39 (forma) [UN]
ochrostoma Harold, 1872: 254 [RN]
perfectecinctus Burlini, 1962: 327 (morpha) [IN]
sikotanensis L. N. Medvedev, 1966: 39 (forma) [UN]
splendidulus Csiki, 1953: 125 (ab.) [IN]
stigmula Burlini, 1956b: 126 (ab.) [IN]
nitidus Linnaeus, 1758: 374 (*Chrysomela*) **E:** AN AU BE BH BU BY CR CZ DE EN FI FR GE GR HU IT KO LA LS LT LU MD ME NL NM NR PL RO RU (CT NT) SB SK SL SP SV SZ UK **A:** AB RU (ES SS WS) **[# 895]**
assimilis Herbst, 1783: 63
berchtesgadensis Schrank, 1785: 319
flavifrons Fabricius, 1792b: 65
nitens Linnaeus, 1761: 169 (*Chrysomela*) [# 1050]
punctatus Goeze, 1777: 320 (*Chrysomela*)
romanus Weise, 1900b: 278 (var.)
rufimanus Eschscholtz, 1818: 467
nitobei Chûjô, 1954d: 227 **A:** CH (TAI)
nitobei Chûjô, 1934e: 521 (ab.) [IN]
nobilis Kraatz, 1879b: 132 **A:** CH (HEN) JA (Honshu, Kyushu, Shikoku) NC RU (FE) SC **[# 896]**
numidicus Bourdonné, 1994: 192 **N:** AG MO TU
obliteratus Suffrian, 1854b: 116 **A:** ID (UT) PA **ORR**
ochroloma Gebler, 1830: 208 **A:** CH (GAN HEB HEI JIL SHX) MG NC RU (ES FE incl. Sakhalin SS WS) SC
inlineolatus Pic, 1908a: 13 (ab.) [IN]
octoguttatus Linnaeus, 1767: 597 (*Chrysomela*) **E:** FR GE PT SP **N:** MO
bisbiconjunctus Pic, 1900e: 69 (var.)
koyi Suffrian, 1848: 46

marisimilis Burlini, 1958: 123 (ab.) [IN]
martinezi Weise, 1882b: 227 (var.)
ponferradanus Pic, 1924a: 18 (var.)
subinterruptus Pic, 1900e: 69 (var.)

octomaculatus P. Rossi, 1790: 96 **E**: AU BU CR CZ FR GE GR HU IT MD PL RO RU (CT ST) SB SK SL SP SZ TR UK **A**: AB GG IN TR
biroi Csiki, 1952: 124 (ab.) [IN]
bisquinquenotatus Kaszab, 1962: 44 (ab.) [IN]
bisquinquepunctatus Harrer, 1791: 133
bokori Csiki, 1952: 124 (ab.) [IN]
cumaniensis Pic, 1924a: 17 (var.)
derivatus Pic, 1924a: 17 (var.)
duodecimpunctatus Fabricius, 1792b: 67
ehumeralis Kaszab, 1962: 41 (ab.) [IN]
fodori Csiki, 1952: 124 (ab.) [IN]
hajossi Csiki, 1952: 124 (ab.) [IN]
horvathianus Kaszab, 1962: 41 (ab.) [IN]
quinquepunctatus Harrer, 1784: 55 [HN]
octonotatus D. H. Schneider, 1792: 210
pilisensis Kaszab, 1962: 43 (ab.) [IN]
stramineus Suffrian, 1847b: 153
testaceus A. Villa & G. B. Villa, 1833: 36
volviciensis Pic, 1924a: 17 (var.)

octopunctatus Scopoli, 1763: 67 (*Buprestis*) **E**: AN AU BE BH BU BY CR CZ EN FI FR GE HU IT KO LA LT MD ME NL NM PL RO RU (CT NT ST) SB SK SL SP SV SZ UK **A**: CH (SCH) GG KZ RU (ES SS WS) TR [# 1145]
anticejunctus Csiki, 1953: 122 (ab.) [IN]
apotmetus Jacobson, 1901a: 77 (ssp.)
bicrux Jacobson, 1901a: 78 (var.) [IN] [# 855]
bijunctus Csiki, 1953: 122 (ab.) [IN]
bipunctatus Trella, 1930: 192 (ab.) [IN]
bucegianus Marcu, 1957: 544 (ab.) [IN]
colmarensis Pic, 1918e: 23 (var.)
cordigeroides Pic, 1918e: 24 (var.)
decempunctatus Trella, 1930: 192 (ab.) [IN]
diversenotatus Pic, 1907h: 154 (var.)
gorgenyensis Csiki, 1942: 379 (ab.) [IN]
hongyuanensis Tan, 1992b: 792, 826 (ssp.)
impunctatus Trella, 1930: 192 (ab.) [IN]

internotatus Csiki, 1953: 122 (ab.) [IN]

jaszoensis Kaszab, 1962: 39 (ab.) [IN]

latedeletus Pic, 1903g: 140 (var.)

mauritii Pic, 1927e: 7 (var.)

multiinterruptus Pic, 1918e: 23 (var.)

postiunctus Trella, 1930: 192 (ab.) [IN]

potaissiae Csiki, 1953: 122

praetermissus Burlini, 1948: 228 (ab.) [IN]

quadripunctatus Trella, 1930: 192 (ab.) [IN]

reductus Csiki, 1953: 122 (ab.) [IN]

sanguinarius Pic, 1918e: 24 (var.)

scalaris Kaszab, 1962: 38 (ab.) [IN]

sexpunctatus Trella, 1930: 192 (ab.) [IN]

subhumeralis Pic, 1918e: 23 (var.)

subinterruptus Pic, 1939g: 30 (ab.) [IN]

variabilis D. H. Schneider, 1791: 95

viturati Pic, 1903m: 140 (var.)

ohnoi Kimoto, 1983b: 46 **A:** CH (TAI) JA (Honshu)

omanicus L. N. Medvedev, 1996: 238 **A:** AE OM

opacus Tappes, 1872: 264 **A:** RU (ES)

oppositus Jacoby, 1908: 249 **A:** ID (HP SD) NP

orotshena Jacobson, 1926a: 272 **E:** RU (NT) **A:** RU (ES FE)

ovulum Suffrian, 1854b: 30 **A:** NP **ORR**

oxysternus Jacobson, 1896a: 531 **A:** CH ("North-East") MG RU (FE)

pallidipennis kalimpongi Medvedev, 2011d: 344 (ssp.) **A:** ID (SD)

pallidipes nakatae Gressitt & Kimoto, 1961: 159 (ssp.) **A:** CH (YUN)

pallidipes pallidipes Pic, 1927j: 184 **A:** CH (YUN)

paphlagonius Sassi & Kismali, 2000: 97 **A:** TR

paradisiacus Weise, 1900b: 276 **E:** TR

parvulus O. F. Müller, 1776: 58 **E:** AL AU BE BH BU BY CR CZ DE EN FI FR (incl. Corsica) GB GE GR HU IR IT (incl. Sardinia) LA LT MD NL NM NR PL RO RU (CT NT) SB SK SL SP SV (incl. Gotland) SZ UK **A:** AB AR CH (FUJ HEB HEI JIA JIL LIA SCH) GG IN IS JA (Hokkaido, Honshu, Kyushu, Shikoku, Kurile Isl.) KZ LE MG NC RU (ES FE incl. Sakhalin SS WS) SC SY TR [# 897]

amatus Baly, 1873: 96 [HN]

betulae Grimmer, 1841: 46

burlinii Tiberghien, 1971: 191 (ab.) [IN]

caeruleus Olivier, 1791a: 718 (*Chrysomela*)

consalanus Baly, 1874: 217 [RN]

devillei J. Müller, 1948: 85 (var.) [IN] [# 856]

flavilabris Fabricius, 1787: 84
fulcratus Germar, 1823: 556
inurbanus Harold, 1874b: 152 [RN]
klettei Gerhardt, 1909: 420 (ab.) [IN]
livens Gmelin 1790: 1711
obliquostriatus Motschulsky, 1866a: 176
nigrocoeruleus Goeze, 1777: 320 (*Chrysomela*)
permodestus Baly, 1873: 95
violaceus Geoffroy, 1785: 92 [HN]

pedatus Tan, 1992b: 793, 826 **A:** CH (SCH)

peliopterus Solsky, 1871: 251 **A:** CH (HEB HEI JIL NIN SHN SHX) JA (Honshu, Kyushu, Shikoku) RU (FE WS) SC
bistrinotatus S.-H. Chen, 1942c: 116 (ssp.)
bivittatus Gebler, 1830: 202 [HN]
japanus Baly, 1873: 92
obliteratithorax Pic, 1904a: 27 (var.) [IN] [# 855]
posticedepictus Pic, 1908j: 92 (var.) [IN] [# 856]
quadrilineatus Weise, 1887h: 168 (var.)
raddei Kraatz, 1879b: 131
raddensis Pic, 1904a: 27 (var.) [IN] [# 855]
rosinae Pic, 1904a: 27 (ssp.)

perelegans insulanus Chûjô, 1935a: 72 (var.) **A:** JA (Ryukyu Archipelago)
insularis Chûjô, 1954: 223 [incorrect subsequent spelling]

perelegans perelegans Baly, 1873: 88 **A:** CH (TAI) JA (Hachijo Isl., Hokkaido, Honshu, Kyushu, Okinawa Isl., Shikoku, Shimokoshiki Isl., Shodo Isl., Yaku Isl.)
ater Kimoto 1974c: 279 [UN]
gotoi Nakane, 1963b: 225 (forma) [IN]
kuro Nakane, 1963b: 225 (forma) [IN]
kurobane Nakane, 1963b: 225 (forma) [IN]
lividus Nakane, 1963b: 225 (forma) [IN]
usumon Nakane, 1963b: 225 (forma) [IN]
takahashii Chûjô, 1935a: 73
takasagonus Chûjô, 1954d: 225 (ssp.)
weberi Bodemeyer, 1916: 112

perelegans yonaguniensis Kimoto, 1974c: 281 (ssp.) **A:** JA (Ryukyu Archipelago, Yonaguni Isl.)

perrisi Tappes, 1869: 9 **N:** AG MO
jurjurensis Pic, 1900c: 62 (var.)
warioni Pic, 1909c: 106 (var.)

petraeus Suffrian, 1854b: 77 **N:** AG EG MO TU **A:** OM SA SI YE [# 1051]

arabs Suffrian, 1854b: 79
fulgurans Fairmaire, 1873a: pl. 16 [= 1873c: 353]
fulguritus Marseul, 1874: 49
mezounanus Pic, 1914i: 78 (var.)
pexicollis Suffrian, 1847b: 51 **E**: FR PT SP SZ **N**: AG
depictus Weise, 1891b: 148 (var.)
santanderi Burlini, 1956b: 73 (ab.) [IN]
peyroni Marseul, 1874: 169 **A**: IS JO LE SY TR **N**: EG **[# 898]**
weisei Reitter, 1886: 70
piceolimbatus Pic, 1920c: 25 **A**: CH (SCH)
seminigrolimbatus Pic, 1920c: 26
pieli pieli Pic, 1928c: 33 **A**: CH (FUJ GUA HUB HUN JIX ZHE)
szechuanus Pic, 1943d: 16
pieli simulator Gressitt, 1942c: 349 **A**: CH (YUN)
pistaciae Suffrian, 1853a: 124 **E**: AL BH BU CR FR (incl. Corsica) GR IT (incl. Elba, Sicily) ME NM RO SL SZ TR **A**: GG JO SY TR **[# 849, # 899, # 1134]**
deceptor Baudi di Selve, 1889: 198 (var.)
luctuosus J. Müller, 1948: 88 (ab.) [IN]
turcicus Weise, 1882b: 215 [# 849, # 1134]
pokharensis L. N. Medvedev & Sprecher-Uebersax, 1999: 360 **A**: NP
pominorum Burlini, 1956b: 65 **E**: FR PT SP
povolnyi Lopatin, 1970: 200 **A**: AF
praticola Weise, 1889a: 1 **E**: RO RU (ST) TR **A**: AB AR GG IN TR
laetior Roubal, 1911: 102 (var.)
ruguliventris Franz, 1949b: 185 (ssp.)
primarius Harold, 1872: 254 [RN] **E**: BE CZ FR GB GE IT PT SP SZ
decemmaculatus Geoffroy, 1785: 93 [HN]
grenieri Pic, 1912b: 17 (var.)
guedeli Pic, 1911b: 98 (var.)
imperialis Fabricius, 1798: 106 [HN] [# 1098]
manueli Tappes, 1874: 284
modanensis Pic, 1912b: 17 (var.)
nigrocunjunctus Pic, 1900c: 62 (var.)
vivesi Burlini, 1962: 327 (morpha) [IN]
przewalskii Lopatin, 1997b: 365 **A**: CH (SCH)
pseudocautus L. N. Medvedev, 1973a: 126 [= 1973b: 878] **A**: RU (ES)
pseudoexsulans Lopatin, 1982b: 420 **A**: BT
pseudoincisus Medvedev, 2011d: 338 **A**: CH (YUN)
pseudosplendens Lopatin, 2008d: 832 **A**: CH (GAN)

pseudovirens Medvedev, 2011d: 339 **A:** CH (SCH)
pubescens L. N. Medvedev, 1996: 240 **A:** OM
pulvillatus Suffrian, 1854b: 24 **A:** NP
pustulipes Ménétriés, 1837: 181 **A:** CH (BEI HEB HEI HEN JIA JIX QIN SCH XIZ) MG NC RU (ES FE SS WS) **[# 900]**
biquadrinotatus Pic, 1902c: 26 (var.)
discopallidior Pic, 1907c: 3 (var.)
kukunorensis Pic, 1907c: 2
muscifer Fairmaire, 1888b: 152
rufotibialis Pic, 1908j: 92 (ssp.)
transversalis Suffrian, 1853a: 98 [= 1854a: 87] **[# 1052]**
putjatae Jacobson, 1896a: 536 **A:** CH (HEI) MG RU (ES FE)
quadriguttatus C. F. W. Richter, 1820: 12 [NP] **E:** AU BH BU CR CZ GE GR HU IT MD PL RO RU (ST) SB SK SL SP UK **A:** AB AR GG IN KZ RU (ES SS WS) TR **[# 845, # 1053]**
bisbipustulatus Herrich-Schäffer, 1830: 10
corinthius Pic, 1914b: 12
kowarzii Gradl, 1881: 307 (var.)
latemaculatus Csiki, 1953: 126 (ab.) [IN]
maurus Stierlin, 1896a: 533 (var.) **[# 1135]**
nigribuccis Gebler, 1830: 208
noesskei Apfelbeck, 1912: 243 (var.)
nuntius Weise, 1898: 189 (var.)
quadripustulatus Gyllenhal, 1813: 613 **E:** AL AU BE BH BU BY CR CZ DE EN FI FR GE HU IT LA LS LT ME NM NR PL RO RU (CT NT) SB SK SL SV SZ UK **A:** GG RU (ES) **[# 901, # 1054]**
aethiops Weise, 1882b: 212 (var.)
apicalis Marseul, 1874: 161 (var.) [HN] **[# 1136]**
bisignatus Marseul, 1874: 161 (var.) **[# 1137]**
bosnicus Apfelbeck, 1912: 243 (var.)
liddesiensis Pic, 1903i: 155 (var.)
rhaeticus Stierlin, 1863b: 64
trismarginatus Burlini, 1967: 130 (morpha) [IN]
quinquepunctatus Scopoli, 1763: 67 (*Buprestis*) **E:** AU BE BH BU BY CZ FR GE HU IT LS MD PL RO RU (CT NT) SB SK SL SZ UK **A:** RU (WS) TR
anticecruciatus Csiki, 1953: 122 (ab.) [IN]
bifasciatus Trella, 1930: 192 (ab.) [IN]
biharicus Kaszab, 1862: 35 (ab.) [IN]
erratus Pic, 1919a: 2 (var.)

erratus Pic, 1939g: 30 (ab.) [IN]
interruptus Suffrian, 1847b: 88
ravouxi Pic, 1911g: 139 (var.)
rubellus Weise, 1881: 154 (var.)
signatus Laicharting, 1781: 174
tyrolensis Pic, 1919a: 2 (var.)
tyrolensis Pic, 1939g: 30 (ab.) [IN]
unifasciatus Trella, 1930: 192 (ab.) [IN]
vallorbensis Pic, 1939g: 30 (ab.) [IN]
rainwaterae Gressitt & Kimoto, 1961: 162 **A:** CH (YUN)
ramburii Suffrian, 1847b: 183 **E:** FR PT SP
ardoisi Fuente, 1913b: 476 (ab.) [IN]
pulcher Burlini, 1967: 130 (morpha) [IN]
regalis angaricus Franz, 1949b: 189 **A:** CH (XIZ)
regalis cyanescens Weise, 1887h: 166 (var.) [RN] **A:** CH (HEB HEI JIL NIN SHX) RU (ES FE)
cuprescens Jacobson, 1901b: 113 (var.) [IN] **[# 856]**
dauricus Weise, 1881: 178 (var.) [HN]
discoviridis Pic, 1909h: 153 (var.)
regalis mutatus Kraatz, 1879d: 262 (var.) **A:** NC RU (ES)
flavoposticatus Heyden, 1887d: 262 (var.)
regalis regalis Gebler, 1830: 208 **A:** CH (GAN HEI HUB JIA LIA NIN QIN SCH SHA SHN SHX) JA MG NC RU (ES FE SS WS) SC **[# 902]**
bohnhofi Pic, 1919a: 2 (var.)
hieracii Weise, 1889g: 583
luteocinctus Pic, 1919a: 2 (var.)
pilosus Baly, 1873: 90
tournieri Pic, 1909h: 153 (var.)
regalis sutschanus Machatschke, 1959: 751 (ssp.) **A:** RU (FE)
flavomarginatus Machatschke, 1959: 752 (forma) [IN]
renatae Sassi, 2001a: 119 **E:** IT (incl. Elba) NM
savonae Burlini, 1948: 227 (ab.) [IN]
reichei Marseul, 1874: 247 **N:** AG
mekalianus Pic, 1905d: 147
zizyphi Pic, 1914b: 13 (var.)
rosti Pic, 1902c: 25 **A:** RU (FE)
rubripes Hope, 1831: 30 **A:** NP
rufofasciatus Solsky, 1881: 69 **A:** KZ TM UZ
meyeri Reitter, 1897a: 241 [RN]

plustschewskyi Jacobson, 1894b: 155
rickmersi Weise, 1900b: 275
transversalis Reitter, 1895b: 162 [HN]
rufoflavus Lopatin, 1984a: 82 **A:** IN
rugicollis rugicollis Olivier, 1792: 611 **E:** CR FR (incl. Corsica) GR IT (incl. Sardinia, Sicily) PT SP **N:** AG LB MO TU **[# 1055, # 1099]**
biprolongatus Pic, 1936f: 30 (ab.) [IN]
exclamationis Weise, 1891b: 148 (var.)
humeralis Fabricius, 1801b: 43
invirgatus Pic, 1908a: 13 (ab.) [IN]
mekalisius Pic, 1900e: 68 (var.)
octonotatus Vitale, 1933: 66 (ab.) [IN]
perroudi Pic, 1905b: 107 (var.)
pseudovasquezi Tiberghien, 1971: 191 (ab.) [IN]
seynei Burlini, 1956b: 98 (ab.) [IN]
subverrucosus Pic, 1908a: 13 (ab.) [IN]
vasquezi Pic, 1911e: 114 (var.)
verrucosus Weise, 1881: 178 (unspecified infrasubspecific entity) [IN] [# 1099]
wagneri Küster, 1845a: 95
x-signatus Pic, 1911e: 114 (var.)
rugicollis sexnotatus Fabricius, 1792b: 53 **E:** GR (Rhodes) **A:** JO TR
ottomanus Medvedev, 2015b: 392 (ssp.)
virgatus Suffrian, 1847b: 43
ruralis Weise, 1887h: 170 **A:** RU (ES FE)
ruri Chûjô, 1934e: 514 **A:** CH (TAI)
sacchii Jacoby, 1899a: 523 **N:** EG **A:** YE **AFR**
sagittimaculatus Tan, 1992b: 793, 826 **A:** CH (YUN)
samniticus Leonardi & Sassi, 2001: 45 **E:** IT
saudiensis Lopatin, 1983b: 205 **A:** SA YE
sauteri Chûjô, 1934f: 289 **A:** CH (TAI)
schawalleri Medvedev, 2011d: 342 **A:** CH (YUN)
securus Weise, 1913c: 219 [RN] **A:** CH (SHA SHN SHX)
aethiops Weise, 1889g: 586 [HN]
seductus Lopatin, 1995c: 458 **A:** BT NP
sehestedii Fabricius, 1798: 107 **A:** NP **ORR**
semiargenteus Reitter, 1894c: 247 **A:** AF IN KZ TD TM UZ
mujunkumensis Lebedev, 1931: 79 (ssp.)
subniger Lebedev, 1931: 79 (ab.) [IN]
senegalensis Suffrian, 1857: 162 **A:** SA YE **N:** EG **AFR**
elkhalidii Lopatin, 1983b: 206

bisbiininterruptus Pic, 1912a: 2 (var.)
kerenensis Weise, 1902d: 302 (var.)
melanostictus Fairmaire, 1874a: 389
mokattamensis Pic, 1912a: 2 (var.)
steindachneri Reineck, 1915b: 410 (ab.) [IN]

senguptai Lopatin, 1979c: 433 **A:** ID (AP SD) **ORR**

sericeus Linnaeus, 1758: 374 (*Chrysomela*) E: AL AU BE BH BU BY CR CZ DE EN FR GE GR HU IT KO KR LA LS LT LU MD ME NL NM NR PL RO RU (CT NT ST) SB SK SL SP SV SZ TR UK **A:** AB AR GG IN KZ MG RU (ES FE SS WS) TR **[# 903, # 1056]**
bicolor Depoli, 1926b: 94 (forma) [IN] [# 856]
bicoloratus Kaszab, 1962: 77 (ab.) [IN]
bidens C. G. Thomson, 1868: 274
funebris Burlini, 1962: 327 (morpha) [IN]
mirandus Burlini, 1967: 130 (morpha) [IN]
parellinus Clavareau, 1913: 185 (ab.) [IN]
pratomus Portevin, 1934: 202 (ab.) [IN]
purpurascens Reitter, 1913c: 90 (ab.) [IN]
purpureicolor Pic, 1914b: 11 (var.)
robustus Suffrian, 1853a: 113
semilimbaticollis Pic, 1941d: 15 (var.)
semiviridescens Pic, 1914b: 12 (var.)

sexmaculatus Olivier, 1792: 611 **E:** FR IT PT SP SZ [# 852, # 1057]
obliteratifer Pic, 1900e: 67 (var.) [# 852]

sexpunctatus kiangsiensis S.-H. Chen, 1942c: 117 (ssp.) **A:** CH (JIA JIX)

sexpunctatus sexpunctatus Linnaeus, 1758: 375 (*Chrysomela*) E: AN AU BE BH BU BY CR CZ DE EN FI FR GB GE HU IT LA LS LT LU MD NL NR PL RO RU (CT NT ST) SB SK SL SP SV (incl. Gotland) SZ UK **A:** CH (JIA JIX NIN) GG IN JA (Hokkaido, Honshu, Kuriles Isl., Shikoku) NC RU (ES FE incl. Sakhalin SS WS) SC TR [# 1058]
fortinsi Pic, 1919a: 2 (var.)
fortinsi Pic, 1939g: 30 (ab.) [IN]
gyllenhali Weise, 1881: 152 (var.)
limbosus Geoffroy, 1785: 94
pedemontanus Pic, 1919a: 2 (var.)
pedemontanus Pic, 1939g: 30 (ab.) [IN]
simplarius Heyden, 1884: 284 (var.)
subnigrofasciatus Pic, 1918e: 23 (var.)
thoracicus Weise, 1881: 151 (var.)
trijunctus Trella, 1930: 192 (ab.) [IN]

sexpustulatus Villers, 1789: 158 (*Chrysomela*) E: FR IT PT SB SP
centrimaculatus Suffrian, 1849a: 20
evorae Burlini, 1961: 96 (morpha) [IN]
fari Burlini, 1961: 96 (morpha) [IN]
gaditanus Marseul, 1874: 206 (var.)
graellsi Weise, 1882: 228 (unspecified infrasubspecific entity) [IN]
hirtifrons Graëlls, 1858a: 98 [= 1858b: 136]
limbatipennis Pic, 1909l: 186 (var.) [HN]
nolousiensis Pic, 1908a: 13 (ab.) [IN]
quadrisignatus Suffrian, 1847b: 192
rossii Suffrian, 1848: 48 [# 1059]
schaeferi Pic, 1950b: 5 (var.)
subancorarius Pic, 1910h: 91 (var.)
sexsignatus Fabricius, 1801b: 41 **A**: ID (UT) NP PA **ORR** [# 904]
shansiensis S.-H. Chen, 1942c: 114 **A**: CH (SHX)
signatifrons Stierlin, 1896a: 533 (var.) **E**: AL AU BE BH BU CR CZ FR GE GR HU IT LS LU NM PL RO SB SK SL SZ TR UK **A**: TR [# 905, # 1138]
sikhimensis Jacoby, 1908: 203 **A**: BT ID (SD) NP **ORR**
simillimus Schöller, 2009b: 106 [RN] **A**: CH (YUN)
similis Tan, 1992b: 794, 827 [HN]
sinensis Weise, 1889g: 584 **A**: CH (SCH YUN)
sinuatolineatus Pic, 1920c: 25 (var.) **A**: CH (SCH YUN)
delkeskampi Gressitt & Kimoto, 1961: 147
piniosensis Pic, 1922b: 20
zhongdianensis Tan, 1992b: 795, 828
slocumi Gressitt, 1942c: 350 **A**: CH (SCH)
socotrensis Schöller, 2014b: 270 **A**: YE (Socotra)
solingensis Gressitt & Kimoto, 1961: 164 **A**: CH (YUN)
solivagus Leonardi & Sassi, 2001: 25 **E**: BH BU BY CR EN (incl. Saaremaa) GR HU LA LT MD NM PL RO RU (CT NT ST) SK UK **A**: CH (HEI) KZ RU (ES SS WS) TR UZ [# 906]
solus Chûjô, 1934e: 515 **A**: CH (TAI)
sonani Chûjô, 1934e: 518 **A**: CH (HUB JIX SCH TAI YUN) [# 907]
spilothorax Jacobson, 1924c: 241 [RN] **E**: GR **N**: EG **AFR**
inscitus Warchałowski, 1999b: 578 [RN]
maculicollis Suffrian, 1857: 222 [HN]
spilotus Hope, 1831: 30 **A**: NP
splendens Kraatz, 1879b: 134 **A**: CH ("North-East") MG NC RU (ES FE incl. Sakhalin SS WS)
flavoirroratus Jacobson, 1901a: 97 (var.) [# 857]
zubovskyi Jacobson, 1894e: 274

sprecherae Lopatin, 2005c: 232 [RN] **A:** NP
micheli L. N. Medvedev & Sprecher-Uebersax, 1997: 215 [HN]
suavis Duvivier, 1892b: 401 **A:** CH (FUJ GUA GUX HAI) ID (SD UT) NP
sublineelus Pic, 1915g: 12 **A:** CH (GUI YUN)
subnepalensis Lopatin, 1982b: 423 **A:** NP
subruber Rapilly, 1980: 78 **A:** IN TR
subsimulans S.-H. Chen, 1942c: 123 **A:** CH (HAI) **ORR**
subunicolor Gressitt, 1942c: 352 **A:** CH (GUA HAI)
sulphureus Olivier, 1808: 806 **E:** AZ FR (incl. Corsica) IT (incl. Sardinia) PT SP **N:** AG MO
holoxanthus Fairmaire, 1859a: 64
surdus Rapilly, 1980: 81 **A:** AR GG IN JO SY TR
swinhoei Bates, 1866: 354 **A:** CH (FUJ JIA JIX NIN TAI ZHE)
tainanensis Schöller, 2010a: 80 [RN] **A:** CH (TAI)
fainanensis Pic, 1943c: 8 [HN]
taiwanus Chûjô, 1934e: 520 **A:** CH (TAI)
innotaticollis Chûjô, 1954d: 207 (var.) [IN] [# 856]
tataricus Gebler, 1841a: 375 **E:** RU (ST) **A:** KI KZ MG RU (ES SS) TD UZ
dorsiger Breit, 1918a: 36
ephippiatus Jacobson, 1901b: 111 (var.) [IN] [# 856]
expictus Jacobson, 1925a: 50 (ab.) [IN]
hectastigmus Solsky, 1881: 71
montivagus Lopatin, 1954: 98
rufus Kraatz, 1882a: 116
tenuelimbatus Lopatin, 1995c: 456 **A:** ID (SD)
teruelensis Pic, 1908i: 91 [JE] **E:** SP
terolensis Pic, 1908g: 82 [incorrect original spelling]
tetradecaspilotus Baly, 1873: 89 **A:** CH (FUJ GUA GUX HUB JIA JIX LIA TAI ZHE) JA (Honshu, Kyushu, Shikoku) RU (FE) SC (incl. Jeju-do Isl.)
duodecatus Fairmaire, 1889b: 70
fukienensis Pic, 1954d: 54
kononovi L. N. Medvedev, 1973a: 127
tetraspilus Suffrian, 1852: 652 **E:** FR IT SP
bensai Burlini, 1956b: 121 (ab.) [IN]
fongondi Burlini, 1956b: 122 (ab.) [IN]
lepidus Mulsant & Rey, 1852a: 8 [= 1852b: 148]
tetrathyrus Solsky, 1871: 248 **A:** KI RU (FE) SC
amurensis Heyden, 1884: 299
therondi Franz, 1949b: 185 (ssp.) **E:** FR
thibetanus Pic, 1917a: 10 **A:** CH (SHA XIZ YUN)
tibialis Brisout de Barneville, 1866: 421 **E:** FR IT PT SP

fauconneti Pic, 1911e: 114 (var.)
lapicidinarum Pic, 1907f: 139 (var.)
ligeris Pic, 1901: 67 (var.)
poeciloceras Heyden, 1870: 168
transcaucasicus Jacobson, 1898: 193 **E:** AL AU BH BU CR CZ GR HU IT LS MD ME NM PL RO RU (ST) SB SK SL SZ TR **A:** AB AR GG IN JO TR
aureicollis Depoli, 1940: 307 (ab.) [IN]
coeruleotinctus Depoli, 1926b: 93 (forma) [IN] [# 856]
illyricus Franz, 1949b: 174 (ssp.)
lydiae Depoli, 1926a: 14 (forma) [IN] [# 855]
monticola Breit, 1918a: 38 (var.)
purpureus Csiki, 1953: 125 (ab.) [IN]
transiens Franz, 1949b: 184 (ssp.) **E:** AL AU BH CR CZ FR HU IT ME RO SK SL SZ
desiderii Kaszab, 1962: 46 (ab.) [IN]
triangularis boreooccidentalis L. N. Medvedev & Romantsov, 2012: 76 (ssp.) **A:** ID (HP) PA
triangularis triangularis Hope, 1831: 30 **A:** AF BT CH (XIZ YUN) ID (KA SD UT) NP PA **ORR**
ater Bryant, 1954c: 449
bifurcatus Pic, 1917a: 11 (var.)
dodecaspilus Suffrian, 1854b: 61
interjectus Baly, 1878h: 372
pygidiomaculatus Pic, 1928g: 25
postunifasciatus Pic, 1917a: 11 (var.)
sannio Kollar & L. Redtenbacher, 1844: 561
triangulomimus Medvedev, 2011d: 344 **A:** NP
tricinctus Kollar & L. Redtenbacher, 1844: 562 **A:** CH (XIZ YUN) ID (KA SD UT) NP [# 908]
lallieri Pic, 1920c: 28 (var.)
tricoloratus Jacobson, 1896b: 532 **A:** CH (SCH YUN) RU (FE)
monrosi Gressitt & Kimoto, 1961: 156
trifasciatus Fabricius, 1787: 81 **A:** CH (FUJ GUA GUX HAI HKG HUN JIX SCH TAI ZHE) JA (Kyushu) NP **ORR**
fainanensis Pic, 1928c: 34
orientalis Hope, 1831: 30
trimaculatus P. Rossi, 1790: 96 **E:** AL AU BE BH BU CR FR GR HU IT KR ME NM RO SL SP SZ **N:** EG **A:** AB AR GG IN IS JO LE RU (WS) SY TR [# 909]
posticefasciatus Pic, 1900e: 69 (var.)
salicis Fabricius, 1792b: 53
tristigma Charpentier, 1825: 236 **E:** FR PT SP (South) **N:** AG MO TU
turangae Lopatin, 1961a: 204 **A:** TD

umbrosus Lopatin, 1956b: 161 (ssp.) **A:** IN

ferrugineus Ogloblin, 1956: 893

unifasciatus Jacoby, 1889b: 161 **A:** CH (YUN) **ORR**

vahlii Fabricius, 1798: 108 **A:** AF **ORR**

mephistopheles Duvivier, 1891a: xxxvi

parasiticus Suffrian, 1854b: 42

piceitarsis Jacoby, 1903a: 90

spengleri Fabricius, 1798: 109

vanharteni Schöller, 2006: 152 **A:** YE

variegatus Fabricius, 1782: 497 **E:** AU CZ FR GE IT LS PL SK SL SV SZ **[# 1060]**

axillaris Charpentier, 1825: 239

raccolanae Burlini, 1967: 130 (morpha) [IN]

strasseri Weise, 1889k: 416 (var.)

vidali Peyerimhoff, 1939: 33 **N:** MO

pardoi Pic, 1953f: 10

violaceus scaffaiolus Burlini, 1961: 94 (ssp.) **E:** FR IT PT SP

violaceus violaceus Laicharting, 1781: 172 **E:** AL AU BE BH BU CR CZ GE GR HU IT KO LS LT LU MD ME NM PL RO RU (CT ST) SB SK SL SV SZ TR UK **A:** AB GG TR **[# 910, # 1100]**

cyaneus Olivier, 1792: 607

fuscatus Gmelin, 1790: 1713

fuscipes Geoffroy, 1785: 94

haneli Kaszab, 1962: 77 (ab.) [IN]

morelianus Pic, 1953f: 10 (ab.) [IN]

sepulchralis Csiki, 1953: 125 (ab.) [IN]

unicolor Latreille, 1804a: 363 [HN]

viridicollis Hänel, 1934a: 90 (ab.) [IN]

virens Suffrian, 1847b: 125 **E:** AU BH BU BY CR CZ GE GR HU IT KR MD ME PL RO RU (CT ST) SB SK SL SZ UK **A:** AB IN KZ MG RU (ES SS WS) TR

vittatus Fabricius, 1775: 110 **E:** AL AU BE BU BY CR CZ DE EN FR GE GR HU IT KR LA LS LT LU NL PL PT RO SB SK SP SV SZ UK

bolati Geoffroy, 1785: 91

digoniensis Pic, 1900e: 69 (var.)

desbrochersi Pic, 1898c: 139 (var.)

interruptus Everts, 1921a: 307 (ab.) [IN]

lineellus Gerhardt, 1909: 420 (ab.) [IN]

lisbonensis, 1924a: 18 (var.)

lotharingus Pic, 1900c: 63

negligens Weise, 1882b: 229 (var.)

optatus Weise, 1898: 185 (var.)

patens Csiki, 1953: 126 (ab.) [IN]
pici Portevin, 1934: 213 (ab.) [IN]
pici Burlini, 1956b: 248 (ab.) [IN]
quadratus Olivier, 1792: 615
quadrum Fabricius, 1777: 221
reductevittatus Pic, 1913j: 171 (var.)
reynosensis Pic, 1924a: 17 (var.)
schrammi Pic, 1907b: 112 (var.)
suturalis Gmelin, 1790: 1713
vittipennis Suffrian, 1854b: 31 **A:** AF PA **ORR**
weigeli L. N. Medvedev, 2015c: 319 **A:** CH (YUN)
yemenicus Lopatin, 1999a: 88 **A:** YE
zambanellus Marseul, 1874: 130 **E:** BH CR IT ME NM
intrusus Weise, 1881: 183 (var.) [# 1139]
viridiaureus Depoli, 1926b: 94 (forma) [IN] [# 856]
mozambanellus Weise, 1906e: 543 (ab.) [IN]
zejensis Mikhailov, 2000: 129 **A:** RU (FE WS)
zinicus Lopatin, 2004a: 614 **A:** CH (SCH)
zoiai Sassi, 2001a: 132 **E:** FR IT
subgenus *Disopus* Chevrolat, 1836: 425 type species: *Chrysomela pini* Linnaeus, 1758
Disops Gistel, 1856: 381 [UE] type species: *Chrysomela pini* Linnaeus, 1758
Taxaris Gistel, 1848a: 404 [= 1848b: 123] [RN] type species: *Chrysomela pini* Linnaeus, 1758
pini difformis Jacoby, 1885b: 201 **A:** JA (Hokkaido, Honshu) NC RU (ES FE WS)
pini guilhemi Schaefer, 1958: 69 (ssp.) **E:** FR
pini pini Linnaeus, 1758: 375 (*Chrysomela*) **E:** AL AU BE BH BU BY CR CZ DE EN FI FR GE HU IT LA LT LU MD NL NM NR PL PT RO RU (CT NT ST) SB SK SL SP SV (incl. Gotland) SZ UK [# 1061]
apiceflavus J. Müller, 1948: 86 (ab.) [IN]
abietis Suffrian, 1847b: 160 [# 1062]
dorsalis Marsham, 1802: 210
laevigatus Suffrian, 1847b: 156
obsoletus Schrank, 1789: 69
simoni Weise, 1882b: 205 **E:** SP
tardus Weise, 1888c: 368 **E:** FR IT SZ
subgenus *Homalopus* Chevrolat, 1837: 446 type species: *Cryptocephalus loreyi* Solier, 1837
Cryptodontus Burlini, 1969b: 536 [HN] type species: *Cryptocephalus informis* Suffrian, 1847
Heterichnus Warchałowski, 1991b: 76 [RN] type species: *Cryptocephalus macrodactylus* Gebler, 1830

Heterodactylus L. N. Medvedev, 1963a: 38 [HN] type species: *Cryptocephalus macrodactylus* Gebler, 1830
Sezerus Özdikmen, 2006: 155 [RN] type species: *Cryptocephalus informis* Suffrian, 1847
carinthiacus Suffrian, 1848: 1 E: AU CR FR IT SL SZ
abietinus Gautier des Cottes, 1861: 194
anticeincisus Pic, 1908j: 92 (var.) [IN] [# 856]
atroscutellaris Pic, 1912h: 74 (var.)
balmae Burlini, 1956b: 43 (ab.) [IN]
ravizzai Burlini, 1967: 130 (morpha) [IN]
boldettii Porta, 1949: 280 (ab.) [IN]
mondolei Burlini, 1956b: 43 (ab.) [IN]
naldii Burlini, 1956b: 44 (ab.) [IN]
pochoni Kippenberg, 1975: 64 (ab.) [IN]
tignensis Pic, 1908f: 66 (var.)
coryli Linnaeus, 1758: 375 (*Chrysomela*) E: AU BE BH BU BY CR CZ DE EN FI FR GB GE GR HU IR IT KZ LA LS LT LU MD NL NM NR PL RO RU (CT NT ST) SB SK SL SV (incl. Gotland) SZ TR UK A: AB CH (SHX) GG IN KZ MG NC RU (ES FE incl. Sakhalin SS WS) SC **[# 911, # 936, # 1101]**
benoiti Pic, 1900e: 69 (var.)
brevecinctus Pic, 1953a: 1 (var.)
chermesinus Geoffroy, 1785: 94
combinatus Kaszab, 1962: 33 (ab.) [IN]
curtii Mader, 1927: 196 (ab.) [IN]
dorsosignatus Reitter, 1913c: 93 (ab.) [IN]
franzi Burlini, 1948: 228 (ab.) [IN]
motasi Panin, 1943: 181
panzeri Schreckenstein, 1801: 41 [RN]
rubroscutellatus Burlini, 1967: 130 (morpha) [IN]
semiconnexus Pic, 1900e: 69 (var.)
stilleri Kaszab, 1962: 34 (ab.) [IN]
temesiensis Marseul, 1874: 55 (var.) [# 1140]
vitis Fabricius, 1775: 108 [# 936]
weiseianus Pic, 1953a: 2 (ab.) [IN]
cunctatus Clavareau, 1913: 143 [RN] A: CH (BEI HEB HEI SHN) MG
dichrous Fairmaire, 1888b: 151 [HN]
danieli Clavareau, 1913: 144 [RN] E: FR SP
championi J. Daniel, 1904a: 92 [HN]
erosus Seidlitz, 1867: 189 E: SP
excisus Seidlitz, 1872: 254 [RN]
floribundus Suffrian, 1866: 207 E: FR SP
hustachei Pic, 1909a: 99 (var.)

gallegoi Lencina & Petitpierre, 2011: 284 **A:** SP
hengduanensis Romantsov & Schöller, 2023: 706 **A:** CH (SCH YUN)
informis Suffrian, 1847b: 66 E: FR IT
lividimanus Suffrian, 1852: 651 E: AN PT SP [# 912]
albaracinus Pic, 1908a: 12 (var.)
albuferanus Pic, 1908g: 83 (var.)
iberus Pic, 1908g: 83 (var.)
inapicalis Pic, 1908g: 83 (var.)
noctifer Pic, 1900e: 69 (var.)
uniobliteratus Pic, 1913j: 171 (var.)
loebli Sassi, 1997: 57 **A:** TR
loreyi Solier, 1837: 687 E: FR IT SP SZ [# 1063]
gabilloti Pic, 1908a: 12 (ab.) [IN]
major Comolli, 1837: 49
mediooculatus Pic, 1931c: 10 (var.)
posticeoculatus Pic, 1900e: 68 (var.)
lusitanicus Suffrian, 1847b: 98 E: PT SP
macrodactylus levipes Lopatin, 1999b: 888 (ssp.) **A:** KI (Naryn-tau)
macrodactylus macrocyaneus Schöller, **nom. nov.** [RN] **A:** CH (XIN) [# 838]
cyaneus Ballion, 1878: 374 [HN]
macrodactylus macrodactylus Gebler, 1830: 206 **A:** KI KZ MG UZ
alaicus L. N. Medvedev, 1963a: 44 (forma) [UN]
apicimaculatus Jacobson, 1901b: 111 (var.) [IN] [# 856]
bicoloratus Breit, 1918a: 45 (ab.) [IN]
correspondens Jacobson, 1901b: 111 (var.) [IN] [# 856]
dilutipes Jacobson, 1896a: 544 (var.)
dissimilis Breit, 1918a: 45 (ab.) [IN]
externemarginatus Breit, 1918a: 47 (ab.) [IN]
humerosus Breit, 1918a: 47 (ab.) [IN]
obscuritarsis Breit, 1918a: 45 (ab.) [IN]
occidentalis L. N. Medvedev, 1963a: 44 (forma) [UN]
picipes Breit, 1918a: 47 (ab.) [IN]
reductus Breit, 1918a: 48 (ab.) [IN]
scapulitarsis Jacobson, 1894a: 98
viridis Breit, 1918a: 48 (ab.) [IN]
macrodactylus persimilis Lopatin, 1956b: 160 **A:** KI (Kavak-tau, Sussamyr-tau)
macrodactylus prosdocetus Jacobson, 1898: 192 **A:** KI (Ferganskiy Range)
superbus Lopatin, 1953: 309
podager Seidlitz, 1867: 190 E: FR SP
alberensis Pic, 1950b: 5 (var.)

prusias Suffrian, 1853a: 95 **E:** BU GR NM TR **A:** AB AR GG IN JO SY TR
delagrangei Pic, 1898b: 93
taurusiensis Pic, 1898b: 93 (var.)
pseudolusitanicus W. Arnold, 1938: 474 **E:** SP
gilblas Demaux, 1964: 178
quadripunctatus Olivier, 1808: 800 **E:** AN FR IT SP SZ **[# 913]**
arvernicus Pic, 1905g: 163 (var.)
bisbiobliteratus Pic, 1913m: 190 (var.)
bisinuatus Pic, 1913m: 191 (var.)
martini Pic, 1908j: 92 (var.) [IN] [# 856]
mediojunctus Pic, 1913m: 190 (var.)
pelissieri Pic, 1900e: 69 (var.)
quittardi Pic, 1900e: 70 (var.)
stierlini Weise, 1881: 147 (var.)
suturalifer Pic, 1900e: 69 (var.)
tiberghieni Burlini, 1967: 130 (morpha) [IN]
vesubiensis Pic, 1900e: 69 (var.)
siedei Warchałowski, 2001a: 81 **A:** CH (SHA)
sinuatus Harold, 1872: 254 [RN] **E:** FR **[# 1141]**
fasciatus Herrich-Schäffer, 1835a: 13 [HN]
mariei Pic, 1930c: 6 (var.)
stragula P. Rossi, 1794: 90 **E:** FR IT
anguissolae Luigioni, 1920: 28 (ab.) [IN]
princeps Rottenberg, 1872b: 248
tarsalis submontanus Lopatin, 1999b: 889 (ssp.) **A:** TD (Gasimajlik Range)
tarsalis tarsalis Weise, 1887d: 328 **A:** CH (QIN) KI KZ TD TM UZ
aksuensis Lopatin, 1952: 27
banghaasi Pic, 1908j: 93 (var.) [IN] [# 856]
bisbicruciatus Pic, 1907c: 1 (var.)
kaschgarensis Pic, 1908j: 93 (var.) [IN] [# 856]
mouzafferi Pic, 1907c: 1 (var.)
nigrescens Jacobson, 1896a: 543 (var.) [HN]
nigripes Jacobson, 1896a: 543 (var.) [HN]
subexquisitus Pic, 1908j: 94 (var.) [IN] [# 856]
tricolor P. Rossi, 1792: 33 **E:** IT SP SZ
bisbinotatithorax Pic, 1914i: 78 (var.)
emilianus Pic, 1914i: 78 (var.)
florentinus Olivier, 1808: 792
wittmerorum Lopatin, 1995c: 455 **A:** ID (KA) PA
subgenus *Lamellosus* Tomov, 1979: 43 type species: *Cryptocephalus angorensis* Pic, 1908
angorensis Pic, 1908a: 14 **A:** TR

justinii Pic, 1910e: 66 (var.)

laevicollis Gebler, 1830: 205 **E:** AU BU CZ GE HU KR MD RO RU (CT ST) SK TR UK **A:** AB KZ RU (SS WS) TR [# 1064]

nigrofemoratus Burlini, 1962: 328 (morpha) [IN]

pseudoviennensis Burlini, 1962: 328 (morpha) [IN]

rossicus Weise, 1900b: 277

viennensis Weise, 1881: 159 (var.)

subgenus *Protophysus* Chevrolat, 1836: 422 type species: *Cryptocephalus lobatus* Fabricius, 1792 (= *Cryptocephalus schaefferi* Schrank, 1789)

Proctophysus L. Redtenbacher, 1845: 118 [UE] type species: *Cryptocephalus lobatus* Fabricius, 1792 (= *Cryptocephalus schaefferi* Schrank, 1789)

cyanipes Suffrian, 1847b: 146 **E:** FR IT SZ

lantosquensis Pic, 1908a: 13 (ab.) [IN]

euchirus Kraatz, 1879h: 288 **A:** AR IN

obliteratus L. N. Medvedev, 1956d: 897 (var.) [HN]

schaefferi moehringi Weise, 1884a: 158 **E:** GR TR **A:** CY JO SY TR

bodemeyeri Weise, 1900c: 135 [= 1900d: 164]

schaefferi schaefferi Schrank, 1789: 69 **E:** AU BH BU CR CZ FR GE GR HU IT MD NM RO RU (CT) SB SK SP SZ UK **A:** AB GG IN KZ MG RU (ES SS WS)

aubryi Tiberghien, 1971: 192 (ab.) [IN]

gynaecoides Roubal, 1942: 33 (var.)

haemorrhoidalis Olivier, 1792: 608

hungaricus Csiki, 1901b: 103 (var.)

lobatus Fabricius, 1792b: 63 [# 1065]

notatus D. H. Schneider, 1792: 213

pulchrus Tiberghien, 1971: 192 (ab.) [IN]

sabaudus Pic, 1908a: 13 (ab.) [IN]

unicolor Olivier, 1792: 608

villosulus Suffrian, 1847b: 148 **E:** AU BH BU CR GE GR HU IT (Alps) NM RO SK

inapicipennis Pic, 1914i: 78 (var.)

wehnckei Weise, 1881: 179 **A:** SY TR

subgenus *Sopidus* Jacobson, 1901b: 112 type species: *Cryptocephalus gamma* Herrich-Schäffer, 1835 [# 836]

Aryana Berti & Rapilly, 1973: 867 [HN] type species: *Cryptocephalus davatchii* Berti & Rapilly, 1973 [# 836]

Asionus Lopatin, 1988a: 8 [RN] type species: *Cryptocephalus flavicollis* Fabricius, 1781 [# 836]

Asiopus Lopatin, 1965a: 452 [HN] type species: *Cryptocephalus flavicollis* Fabricius, 1781 [# 836]

altaicus Harold, 1872: 254 [RN] **A:** CH (HEB NIN NMO SHX XIN) MG RU (ES FE SS)

agnus Weise, 1898: 184

clavaceus Csiki, 1916: 127
trivittatus Gebler, 1830: 209 [HN]
amasiensis Weise, 1894d: 91 **A:** TR
apicalis Gebler, 1830: 201 **E:** AU BH BU CR HU IT KR KZ MD RO RU (CT ST) SB SK TR UK **A:** AB GG IN KZ RU (ES SS WS) TR
csepelensis Csiki, 1953: 125 (ab.) [IN]
delicatissimus Roubal, 1940a: 284 (ab.) [IN]
eburatus Weise, 1881: 176 (var.)
flavoguttatus Olivier, 1792: 623 [HN]
holtzi Pic, 1909d: 130 (var.)
hortobagianus Pic, 1909d: 130 (var.)
limbipennis Csiki, 1953: 125 (ab.) [IN]
marginatipennis Csiki, 1953: 125 (ab.) [IN]
quadriguttatus Gebler, 1848: 54 [HN, RN]
rimskyi Jacobson, 1907: LXIV (ab.) [IN]
arnoldii arnoldii L. N. Medvedev, 1956d: 895 **A:** KZ TD UZ
arnoldii sogdianus Lopatin, 1963c: 121 **A:** TD UZ
astracanicus Suffrian, 1867b: 310 **E:** RU (ST) **A:** AB AF CH (HEI NIN XIN) GG IN KZ MG TD TM UZ
tamaricis Solsky, 1867: 183
atriplicis Lopatin, 1967a: 214 **E:** RU (ST) **A:** AB KZ
augustalisi augustalisi Pic, 1913j: 170 **A:** IN IQ IS LE SY
alepensis Pic, 1916g: 23 (var.)
euphraticus Pic, 1913j: 170 (var.)
augustalisi dissipatus Schöller, 2009b: 107 [RN] **A:** IN
iranicus Berti & Rapilly, 1979: 260 [HN]
balassogloi Jacobson, 1896a: 537 **A:** KZ TD UZ
beckeri Morawitz, 1860: 299 **E:** KZ RU (ST) **A:** KZ
suffriani Dohrn, 1859: 426 [HN]
bidentatus L. N. Medvedev, 1956d: 897 **A:** KI UZ
bidentulus bidentulus Suffrian, 1854b: 132 **A:** MG RU (ES SS) **[# 914]**
bidentulus klementzi Ogloblin, 1956: 890 **A:** MG RU (ES SS WS)
bidentulus subdesertus L. N. Medvedev, 1980a: 316 (ssp.) **A:** MG
bitaeniatus Solsky, 1876: 298 **A:** TD TM UZ
adustus Weise, 1890g: 364
bivulneratus bivulneratus Faldermann, 1835: 445 **A:** MG RU (ES SS) **[# 1066]**
ourgana Pic, 1909e: 137 (var.)
bivulneratus gobiensis L. N. Medvedev, 1980a: 313 (ssp.) **A:** MG
bohemius Drapiez, 1819b: 296 **E:** AU CZ HU KR MD RO RU (CT ST) SB UK **A:** KZ MG RU (SS) **[# 915]**
boehmii Germar, 1823: 558

borochorensis Pic, 1907c: 2 **A:** CH (XIN) KZ (south east)
kulenovae Lopatin, 1967a: 214
coronatus Suffrian, 1847b: 53 **E:** KR RU (CT ST) UK **A:** KI KZ UZ
cupreatus S.-H. Chen, 1942c: 113 **A:** CH (GAN HEB NIN)
curda Jacobson, 1897c: 215 **A:** AB AR GG IN TR
araxidis Weise, 1898: 186
davatchii Berti & Rapilly, 1973: 867 **A:** IN
diadochus Lopatin, 1963c: 121 **A:** KI TD [# 916]
dilectus Weise, 1894a: 67 **A:** KI KZ TD UZ
dsungaricus Pic, 1907c: 1 **A:** CH (XIN)
ergenensis ergenensis Morawitz, 1863: 174 **E:** RU (ST) **A:** AB GG KZ
beckeri Suffrian, 1864: 265 [HN]
georgicus L. N. Medvedev & Seperteladze, 1963: 699 (ssp.)
ergenensis kalbensis Mikhailov, 2021a: 83 **A:** KZ
fasciatointerruptus Berti & Rapilly, 1979: 259 **A:** IN
flavicollis Fabricius, 1781: 140 **E:** KR RU (CT ST) UK **A:** KI KZ MG RU (SS WS)
eusenicus Weise, 1900b: 277 (var.)
kelecsenyi Pic, 1909l: 186 (var.)
montanus L. N. Medvedev, 1956d: 895 (var.)
nigritarsis Suffrian, 1847b: 107
ottoi Pic, 1908a: 13 (var.)
uralskensis Pic, 1908a: 13 (ab.) [IN]
flexuosus Krynicki, 1834: 173 **E:** BU KR RO RU (CT ST) UK **A:** AB AR GG KI KZ RU (SS WS) TR
ataenis Pic, 1903e: 131 (var.)
mesmini Pic, 1908a: 13 (ab.) [IN]
floralis Krynicki, 1834: 172 **E:** KR RU (ST)
aipetrensis L. N. Medvedev, 1956d: 895 (var.)
fulmenifer Reitter, 1889d: 376 **E:** KR **A:** AB AR GG IN TR
bucharicus Pic, 1907c: 1 (var.)
phaleratus Tappes, 1872: 256 [HN, RN]
tigrinus Berti & Rapilly, 1979: 246 [RN]
undatus Suffrian, 1847b: 60 [HN]
gamma Herrich-Schäffer, 1835b: 1 [NP] **E:** BU HU KR MD RO RU (CT ST) SK UK **A:** AB GG KI KZ RU (ES WS) [# 844]
sesquistriatus Krynicki, 1832: 174 [NO] [# 844]
maculipes Zubkov, 1833: 338 [NO] [# 844]
semilugens Dudich, 1924: 23 (ssp.)
sesquilineatus Gebler, 1841b: 618 [RN]
ypsilon Suffrian, 1847b: 111 [# 1067]

gansuicus S.-H. Chen, 1942c: 110 **A:** CH (GAN)
globulus Lopatin, 1953: 308 **A:** KI KZ MG TD UZ
gloriosus Mulsant & Wachanru, 1853b: 127 [= 1853c: 159] **A:** TR (Anatolia)
halophilus Gebler, 1830: 204 **E:** RU (ST) **A:** KZ RU (SS)
skopini Lopatin, 1961c: 100
hamatus Suffrian, 1854b: 134 **A:** MG RU (ES FE SS)
micropyga Weise, 1898: 184
hammadae Lopatin, 1961c: 98 **A:** TD
hauseri Weise, 1887d: 327 **A:** KI (Tchatkal Ridge)
perfectus Weise, 1887d: 327 (var.)
heydeni heydeni Weise, 1886b: 193 **A:** KI KZ (south east) TD UZ
biarcuatus Weise, 1886b: 193 (var.)
heydeni personatus Weise, 1892d: 136 **A:** KI
alaiensis Pic, 1907c: 2
hirtipennis Faldermann, 1835: 446 **A:** CH (HEB HEI NIN) JA (Hokkaido) MG NC RU (ES FE incl. Sakhalin SS WS) SC [# 1068]
ainu Chûjô, 1959: 6
egregius Weise, 1887h: 169
flavomixtus Jacobson, 1901b: 113 (var.) [IN] [# 856]
yamadai Chûjô, 1940f: 394
semiviridis Suffrian, 1854b: 120 [# 1069]
sibiricus Marseul, 1874: 135
impressipygus Ogloblin, 1956: 892 **E:** KZ RU (ST) **A:** KZ
ingamma Pic, 1908g: 83 (var.) **E:** SP
irenae Lopatin, 1958a: 53 **A:** TD
iskanderi Lopatin, 1961a: 205 **A:** TD
jaxarticus Lopatin, 1986c: 65 [RN] **A:** KZ
simulator Lopatin, 1971b: 80 [HN]
jirofti Schöller, 2014a: 651 **A:** IN
karakalensis L. N. Medvedev, 1955: 1177 **A:** AB IN KZ TD TM UZ
karatavicus Lopatin, 1971b: 79 **A:** KZ
kerzhneri Lopatin, 1968c: 547 **E:** RU (ST) **A:** KZ UZ
khajami Lopatin, 1980: 617 **A:** IN
klarae Lopatin, 1990c: 49 **A:** KZ
kokanda Jacobson, 1924c: 242 [RN] **A:** KI KZ TD UZ
kiritshenkoi Jacobson, 1925c: 12
negligens Weise, 1892d: 137 [HN]
koltzei Weise, 1887h: 171 **A:** CH (BEI FUJ GAN HEB HEI HUB JIA JIL NIN NMO SHA SHX ZHE) MG NC RU (ES FE) SC [# 917]
conjunctus S.-H. Chen, 1942c: 112 (ssp.)
flavopictus Jacoby, 1890a: 88

nigricaudis S.-H. Chen, 1942c: 112 (ssp.)
ussuriskensis Pic, 1922d: 28 (var.)
korotyaevi korotyaevi L. N. Medvedev, 1980a: 314 **A:** MG RU (SS)
korotyaevi shamoensis L. N. Medvedev, 1980a: 315 (ssp.) **A:** MG (south)
kozlovi Lopatin, 1977c: 306 **A:** CH (XIN) MG
lateralis Suffrian, 1853a: 108 **E:** KR RU (ST) UK **A:** KZ RU (WS)
aeneus Stierlin, 1863a: 499 [HN]
lemniscatus Suffrian, 1854b: 115 **A:** CH (BEI HEB HEI JIA JIL SHN XIN) MG RU (FE)
confucius Clavareau, 1913: 140 [RN]
discoideus Fairmaire, 1888b: 151 [HN]
hoffmanni Pic, 1938f: 333 [HN]
pici Papp, 1946b: 7 [RN]
limbellus limbellus Suffrian, 1847b: 119 **A:** MG RU (ES SS)
obscurior Jacobson, 1901b: 112 (var.) [IN] [# 856]
sublimbellus Pic, 1902c: 26 (var.)
limbellus semenowi Weise, 1889g: 580 **A:** CH (GAN HEI JIL NMO SHA SHN SHX) JA (Hokkaido, Honshu) MG NC RU (ES SS FE) SC
tentator Weise, 1889g: 581
limoniastri Pic, 1894a: 106 **N:** AG EG TU
alfierii Pic, 1912i: 83 (var.)
biscrensis Pic, 1912b: 17 (var.)
mroczkowskii Warchałowski, 1997: 273
linnavuorii Lopatin, 1986a: 109 **A:** IQ
lucidus Rapilly, 1980: 80 **A:** IN
lukjanovitschi Lopatin, 1971b: 79 **A:** KZ
malaysaryensis Kolov, 2017: 899 **A:** KZ
manchuricus Gressit & Kimoto, 1961: 140 **A:** CH (HEI JIL NIN XIN) KZ MG RU (ES FE SS WS)
crux Gebler, 1848: 47 [HN]
dariganicus L. N. Medvedev, 1980a: 314 (ssp.)
equestris Gebler, 1830: 204 [HN]
incompletus Jacobson, 1901b: 112 (var.) [IN] [# 856]
medvedevi Lopatin, 1953: 309 **A:** AF TD
lopatini Iablokoff-Khnzorian, 1983: 94 [HN]
mniszechi Tappes, 1869: 11 **A:** RU (SS)
ogloblini Lopatin, 1953: 307 **A:** KI KZ UZ
oxianus Lopatin, 1975b: 46 **A:** TD
pavlovskii Lopatin, 1956b: 158 **A:** KI KZ UZ
imitator Ogloblin & L. N. Medvedev, 1956d: 896
pilosellus Suffrian, 1854b: 111 **A:** CH (BEI HEB HEI JIL NIN SHN) MG RU (ES)
nigroscriptus Fairmaire, 1888b: 152

placidus Weise, 1892d: 135 **A:** KI
prasolovi Lopatin, 1992d: 72 **A:** KZ
pseudolateralis L. N. Medvedev, 1973b: 877 **A:** KZ
pseudoreitteri Tomov, 1976: 83 **A:** TR
punctatissimus Suffrian, 1857: 184 **N:** EG **A:** SI
signatithorax Pic, 1916g: 23 (var.)
quatuordecimmaculatus D. H. Schneider, 1792: 195 **E:** AL AU BU CZ HU MD ME NM RO RU (CT) SB SK UK **A:** TR
coloratus Fabricius, 1798: 108
flavicollis Townson, 1797: 462 [HN]
gyorffyi Csiki, 1953: 123 (ab.) [IN]
guranyii Csiki, 1953: 124 (ab.) [IN]
kaszabi Csiki, 1953: 124 (ab.) [IN]
kuthyi Csiki, 1953: 123 (ab.) [IN]
magyaranus Pic, 1909l: 186 (var.)
mimipennis Pic, 1909l: 186 (var.)
pilleri Schrank, 1798: 555
speiseri Kaszab, 1962: 39 (ab.) [IN]
stredae Csiki, 1953: 123 (ab.) [IN]
reitteri Weise, 1881: 172 **E:** BH CR
croaticus Csiki, 1953: 123 (ab.) [IN]
fiumensis Pic, 1908j: 93 (var.) [IN] [# 856]
suboctopunctatus Kaszab, 1962: 39 (ab.) [IN]
richteri L. N. Medvedev, 1957b: 331 **A:** AB AR IN
rubi glasunowi Jacobson, 1894e: 272 **A:** IN KZ TM
rubi properus Weise, 1900b: 278 **A:** KZ TD UZ
rubi reductosignatus Pic, 1907c: 1 **A:** CH (JIL XIN)
rubi rubi Ménétriés, 1832: 238 **A:** AB AR GG
subpallidus Pic, 1914b: 12 (var.)
sarafschanensis betpakdalensis Lopatin, 1976a: 107 (ssp.) **A:** KZ (Betpak-Dala)
sarafschanensis duvivieri Weise, 1892d: 134, 137 **A:** KI KZ (Karatau Ridge)
ferghanensis Pic, 1908j: 93
solskyanus Jacobson, 1901b: 112 [IN] [# 856]
solskyi Weise, 1900b: 275 (var.) [HN]
stenroosi Jacobson, 1901b: 112 (var.) [IN] [# 856]
talassicus Lopatin, 1976a: 106 (ssp.)
sarafschanensis grumi Lopatin, 1976a: 107 (ssp.) **A:** KZ (east) MG
sarafschanensis iliensis Weise, 1900b: 274 (var.) **A:** CH (XIN) KZ (south east) [# 918]
sarafschanensis narynensis Lopatin, 1976a: 106 (ssp.) **A:** KI

sarafschanensis sarafschanensis Solsky, 1881: 76 **A:** TD (north)
sarafschanensis sugatensis Lopatin, 1999b: 890 (ssp.) **A:** KZ (Sugaty Ridge)
sarafschanensis tiro Weise, 1900b: 274 (var.) **A:** KI RU (WS)
sarbazicus Lopatin, 1980: 615 **A:** IN
sareptanus Morawitz, 1863: 173 **E:** KZ RU (ST) **A:** KZ
saryarkensis L. N. Medvedev & Kulenova, 1980: 119 **A:** KZ
sinaita moricei Pic, 1908a: 14 (var.) **A:** IS JO SI
sinaita sinaita Suffrian, 1854b: 75 **A:** SI **N:** EG
spurius Lopatin, 1956b: 159 **A:** AB AF IN TM **[# 919]**
stackelbergi Lopatin, 1971b: 77 **A:** KZ MG
stchukini Faldermann, 1835: 447 **A:** CH (GAN HEB HEI JIL NIN SHX) MG RU (ES FE SS WS) **[# 920]**
 chaffanjoni Pic, 1909e: 137 (var.)
 congener Ménétriés, 1837: 181
 dauricus Marseul, 1874: 92 (var.)
 externopunctatus Jacobson, 1901b: 112 (var.) [IN] [# 856]
 subapicalis Pic, 1914b: 12
 sublimbalis S.-H. Chen, 1942c: 113 (ssp.)
 valens S.-H. Chen, 1942c: 114
subdeserticola Berti & Rapilly, 1979: 251 **A:** YE
subtigrinus Berti & Rapilly, 1979: 250 **A:** IN
tappesi Marseul, 1868: 206 **A:** LE SY TR
 bidens Suffrian, 1870: 24 [HN]
 disconiger Pic, 1904e: 57 (var.)
 missisiensis Pic, 1908j: 92 (var.) [IN] [# 856]
terminassianae Lopatin, 1967a: 212 **A:** CH (XIN) KZ (south-east)
undulatus Suffrian, 1854b: 74 **A:** AB AF IN KI KZ MG TD TM UZ
verae L. N. Medvedev, 1956c: 249 **A:** TM
vitellinus Lopatin, 1980: 615 **A:** IN
vouauxi Achard, 1921a: 61 [RN] **A:** CH (GAN QIN)
 thibetanus Breit, 1918a: 35 [HN]
volkovitshi Lopatin, 1976a: 106 **A:** AB AR IN TR
wittmeri Lopatin, 1977f: 116 **A:** IN
yaosanicus S.-H. Chen, 1942c: 120 **A:** CH (GUX)
zaitzevi Lopatin, 1977d: 31 **A:** AB GG
zarudnianus L. N. Medvedev, 1971c: 131 **A:** IN
zarudnyi Lopatin, 1953: 308 **A:** IN
 kharazii Berti & Rapilly, 1979: 242
zinovjevi L. N. Medvedev, 1973b: 878 **A:** CH (SHX) MG (north) RU (FE)

genus *Cryptocephalus*, nomina dubia
albopectoralis Pic, 1908j: 93 (var.) [IN] **A:** RU (FE) **[# 856]**
insignis Paykull, 1799b: 116 [specimen monstrosum] E: SV
cremeus Tan, 1988a: 320, 331 **A:** CH (XIZ)
striatus Gravenhorst, 1807: 151 [distribution not given]
transversus Olivier, 1808: 786 [distribution not given]

genus *Cryptocephalus*, unavailable names
apressus X.-P. Li, 2013: 38 **A:** CH (HEI) **[# 853]**
heilongjiangensis X.-P. Li, 2013: 47 [RN] **[# 853]**
maculatus X.-P. Li, 2013: 44 **A:** CH (HEI) **[# 853]**

genus *Diachus* LeConte, 1881: 196 type species: *Cryptocephalus auratus* Fabricius, 1801
Fabricianus Weise, 1895a: 58 type species: *Cryptocephalus auratus* Fabricius, 1801
auratus Fabricius, 1801b: 57 (*Cryptocephalus*) **Ai:** CH (TAI) JA (Ryukyu Archipelago) **AFRi AUSi NAR NTR ORRi** **[# 1102]**
aeneus Olivier, 1808: 916 (*Eumolpus*)
chalconatus Mannerheim, 1843: 312 (*Cryptocephalus*)
hisamatsui Takizawa, 1975a: 56 (*Melixanthus*)
levis Haldemann 1849: 257 (*Cryptocephalus*)
smaragdinus Melsheimer, 1847: 174 (*Cryptocephalus*) [UN]
viridis Melsheimer, 1847: 174 (*Monachus*)

genus *Jaxartiolus* Jacobson, 1922a: 517 type species: *Jaxartiolus baeckmannianus* Jacobson, 1922
baeckmannianus Jacobson, 1922a: 519 **A:** KZ (south) TD UZ
kozlovi Lopatin, 1997b: 365 **A:** CH (GAN)
reductosignatus Lopatin, 1963b: 855 **A:** KI KZ (south-east)

genus *Melixanthus* Suffrian, 1854b: 8 type species: *Melixanthus intermedius* Suffrian, 1854
Suffrianius Weise, 1895a: 58 type species: *Cryptocephalus pumilio* Suffrian, 1854
adamsi Baly, 1877e: 216 **A:** CH (GUA) **ORR**
assamensis Jacoby, 1908: 195 (*Cryptocephalus*) **A:** ID (AP) **ORR**
atricillus Suffrian, 1854b: 80 (*Cryptocephalus*) **A:** CH (HAI) ID (HP KA) PA **ORR**
kuluensis Lopatin, 1979c: 436
lutescens Suffrian, 1854b: 82 (*Cryptocephalus*)
minimus Gressitt, 1942c: 353
atriventris Pic, 1927a: 6 **A:** NP **ORR**
babai Chûjô, 1959: 8 (*Cryptocephalus*) **A:** JA (Honshu)

bimaculicollis Baly, 1865c: 65 **A:** CH (FUJ SCH) **ORR**

coomani Pic, 1926b: 11

moupinensis Gressitt, 1942c: 346

pendleburyi Bryant, 1954c: 450

brunnicollis Suffrian, 1857: 189 (*Cryptocephalus*) **N:** EG **A:** IS OM

aegyptiacus Pic, 1900i: 95 (var.)

innesi Pic, 1900i: 95 (var.)

letourneuxi Pic, 1916g: 23 (*Cryptocephalus*, var.)

luxorensis Pic, 1916g: 22 (*Cryptocephalus*, var.)

multimaculatus Pic, 1916g: 22 (*Cryptocephalus*, var.)

formosensis Chûjô, 1934f: 291 **A:** CH (TAI)

innotaticollis Pic, 1930a: 612 **A:** CH (FUJ) **ORR**

kashmirensis Lopatin, 1995c: 460 **A:** ID (KA)

luridus Motschulsky, 1866b: 411 (*Monachus*) **A:** CH (YUN) NP **ORR**

milleri Bryant, 1941a: 283

pallidipennis Pic, 1922f: 8

placidus Baly, 1877e: 217

luteofasciatus Pic, 1926b: 11 **A:** CH **ORR**

melanocephalus Suffrian, 1857: 186 (*Cryptocephalus*) **E:** BU **N:** EG LB **A:** AE AF AR IN IS JO OM SA SI TM YE (Socotra) **[# 833, # 921]**

atropectoralis Pic, 1916g: 23 (*Cryptocephalus*, var.)

granularis Suffrian, 1857: 188 (*Cryptocephalus*)

jordanicus Lopatin, 1979a: 301

mucidus Harold, 1872: 254 (*Cryptocephalus*) [RN]

menglaensis Duan, F.-Y. Wang & Zhou, 2021a: 113 **A:** CH (YUN) **ORR**

nuristanicus Lopatin, 1982c: 260 **A:** AF

pumilio Suffrian, 1854b: 139 (*Cryptocephalus*) **A:** CH (SCH QIN) JA (Hokkaido, Honshu,) MG RU (ES FE incl. Sakhalin, South Kuriles)

manturiformis Pic, 1920c: 27

yajiangensis Tan, 1992b: 796, 828

puncticollis Lopatin, 2005d: 873 **A:** CH (SCH)

scitulus Baly, 1873: 89 (*Cryptocephalus*) **A:** JA (Hokkaido, Honshu, Kyushu, Shikoku)

similibimaculicollis Duan, F.-Y. Wang & Zhou, 2021a: 116 **A:** CH (YUN)

tubu Chûjô, 1954d: 199 (*Cryptocephalus*) **A:** CH (TAI)

subtribe Monachulina Leng, 1920

genus *Adiscus* Gistel, 1857b: 604 type species: *Phaedon nigromaculatum* Kollar & L. Redtenbacher, 1844

Atropidius Chapuis, 1874: 175 type species: *Atropidius improbus* Chapuis, 1874

Dioryctus Suffrian, 1860: 3 type species: *Dioryctus porcatus* Suffrian, 1860

Falsodioryctus Pic, 1955c: 21 type species: *Falsodioryctus sinensis* Pic, 1955 (= *Dioryctus nigripennis* Jacoby, 1890)

annulatus Pic, 1922c: 14 (*Dioryctus*) **A:** CH (FUJ GUA GUX YUN)

alternatus S.-H. Chen, 1941c: 190 (*Dioryctus*)

apicalis Chûjô, 1954d: 159 (*Dioryctus*) **A:** CH (TAI)

arisanus Chûjô, 1954d: 162 (*Dioryctus*) **A:** CH (TAI)

atripennis S.-H. Chen & Pu, 1980: 109, 113 **A:** CH (FUJ)

bicoloriceps Pic, 1932f: 208 (*Dioryctus*) **A:** CH (FUJ TAI)

interruptus Pic, 1954d: 56 (*Dioryctus*, var.)

bimaculicollis S.-H. Chen & Fu, 1980: 110, 114 **A:** CH (YUN)

bodhisatva Gressitt, 1942c: 332 (*Dioryctus*) **A:** CH (SCH YUN)

chekianga Gressitt & Kimoto, 1961: 116 **A:** CH (ZHE)

crassicornis Tan, 1992b: 781, 816 **A:** CH (YUN)

cyaneus Tan, 1992b: 782, 816 **A:** CH (SCH YUN)

daofuensis Duan & Zhou, 2022: 7 **A:** CH (GAN HEN SCH)

exilis Weise, 1922a: 43 (*Dioryctus*) **A:** CH (GUA GUI SCH XIZ YUN)

fracticeps Gressitt, 1942c: 333 (*Dioryctus*) **A:** CH (HKG: Lantau Isl.)

glabrous Tan, 1988a: 319, 330 **A:** CH (XIZ)

grandipalpus Tan, 1992a: 602, 606 **A:** CH (HUN)

hauseri Weise, 1895e: 327 (*Atropidius*) **A:** BT ID (SD) NP [# 858]

humeralis Pic, 1922c: 14 (*Dioryctus*) **A:** CH (HUB SCH YUN) **ORR**

auricacumen Gressitt, 1942c: 331 (*Dioryctus*)

inornatus S.-H. Chen & Pu, 1980: 111, 114 **A:** CH (YUN)

issikii Chûjô, 1954d: 149 (*Dioryctus*) **A:** CH (TAI)

klapperichi Pic, 1955c: 22 (*Dioryctus*) **A:** CH (FUJ)

kweiyangensis Gressitt, 1942c: 334 (*Dioryctus*) **A:** CH (GUI HUB HUN SCH TAI)

monticolus Chûjô, 1954d: 161 (*Dioryctus*)

lewisii Baly, 1873: 87 (*Dioryctus*) **A:** JA (Honshu, Kyushu, Shikoku)

ogloblini Roubal, 1929a: 97 (*Dioryctus*)

testaceipes Pic, 1922c: 14 (*Dioryctus*)

longicornis Chûjô, 1954d: 169 (*Dioryctus*) **A:** CH (TAI)

nigrobasalis Chûjô, 1954d: 170 (*Dioryctus*, var.) [IN] [# 856]

lushuiensis Tan, 1992b: 782, 817 **A:** CH (YUN)

maculatus Weise, 1912: 80 (*Dioryctus*) **A:** CH (YUN) **ORR**

multimaculatus Pic, 1935e: 19 (*Dioryctus*)

matudai Chûjô, 1954d: 153 (*Dioryctus*) **A:** CH (TAI)

mediofulvus Chûjô, 1954d: 166 (*Dioryctus*) **A:** CH (TAI)

melanopterus Chûjô, 1954d: 164 (*Dioryctus*) **A:** CH (TAI)

mouhoti Baly, 1877b: 36 (*Dioryctus*) **A:** CH (GUA GUX HAI YUN) **ORR**

kabakovi L. N. Medvedev & Samoderzhenkov, 1987: 23

laetus Weise, 1904b: 161 (*Dioryctus*)
rubripennis Pic, 1926b: 12 (*Dioryctus*)
niger S.-H. Chen, 1941c: 189 (*Dioryctus*) **A:** CH (GAN GUI HUB SCH SHA) [# 859]
nigricollis Gressitt, 1942c: 334 (*Dioryctus*) **A:** CH (SCH)
nigripennis Jacoby, 1890a: 89 (*Dioryctus*) **A:** CH (FUJ GUI HUB ZHE) ID (SD) JA (Iriomote Isl., Ishigaki Isl.) **ORR**
major Pic, 1926b: 12 (*Dioryctus*)
sinensis Pic, 1955c: 21 (*Falsodioryctus*)
nigromaculatus Kollar & L. Redtenbacher, 1844: 562 (*Phaedon*) **A:** ID (KA UT) NP
angulatus L. N. Medvedev & Sprecher-Uebersax, 1997: 220
nigroplagiatus Jacoby, 1903a: 92 (*Atropidius*) **A:** ID (SD) NP **ORR**
ningshanensis Duan & Zhou, 2022: 10 **A:** CH (SHA)
obscurus L. N. Medvedev & Sprecher-Uebersax, 1997: 221 **A:** NP
occipitalis S.-H. Chen & Pu, 1980: 109, 113 **A:** CH (FUJ)
osawai Kimoto, 1987: 95 **A:** CH (TAI)
pectoralis Pic, 1926b: 12 (*Dioryctus*) **A:** CH (YUN)
pubiventris L. N. Medvedev, 2008a: 202 **A:** CH (GUI GUX JIX) **ORR**
punctithorax L. N. Medvedev, 2008a: 200 **A:** CH (ANH)
rufonigrus Chûjô, 1954d: 151 (*Dioryctus*) **A:** CH (TAI)
melanopus Chûjô, 1954d: 158 (*Dioryctus*, var.) [IN] [# 856]
nigrescens Chûjô, 1954d: 153 (*Dioryctus*, var.) [IN] [# 856]
nigrofasciatus Chûjô, 1954d: 156 (*Dioryctus*)
nigromaculatus Chûjô, 1954d: 154 (*Dioryctus*) [HN]
sauteri Chûjô, 1938e: 294 (*Dioryctus*) **A:** CH (TAI)
speciosus Tan, 1988a: 319, 330 **A:** CH (XIZ)
sungkangensis Kimoto, 1996: 28 **A:** CH (TAI)
taiwanus Chûjô, 1954d: 148 (*Dioryctus*) **A:** CH (TAI)
tanae Duan & Zhou, 2022: 13 **A:** CH (XIZ)
tibialis S.-H. Chen & Pu, 1980: 111, 114 **A:** CH (FUJ YUN)
transversalis Tan, 1992b: 783, 817 **A:** CH (YUN)
variabilis Jacoby, 1890b: 114 (*Dioryctus*) **A:** CH (FUJ GAN GUI HEN HUB SCH SHA XIN YUN) **ORR** [# 860]
weigeli L. N. Medvedev, 2019b: 282 **A:** CH (YUN) **ORR**
weisei L. N. Medvedev, 2008a: 203 **A:** ID (SD) **ORR**
wittmeri L. N. Medvedev, 1992f: 383 **A:** NP
yunnanus L. N. Medvedev, 2008a: 205 **A:** CH (YUN)

genus *Coenobius* Suffrian, 1857: 61 type species: *Coenobius triangulum* Suffrian, 1857
Inclica Walker, 1859a: 53 type species: *Inclica solida* Walker, 1859
azureus Schöller, 1999: 74 [RN] **A:** ID (SD)
cyaneus Lopatin, 1979c: 437 [HN]

babai Chûjô, 1963a: 21 **A:** CH (TAI)
baronii Lopatin, 1979c: 438 **A:** NP **ORR**
birmanicus Jacoby, 1892b: 889 **A:** ID (SD) NP **ORR**
brevioculatus Chûjô, 1954d: 173 **A:** CH (FUJ TAI)
caeruleipennis S.-H. Chen & Pu, 1980: 112, 115 **A:** CH (YUN)
castaneus S.-H. Chen & Pu, 1980: 112, 115 **A:** CH (FUJ)
chalceus Jacoby, 1908: 185 **A:** BT **ORR**
choanus Schöller, 1999: 75 [RN] **A:** CH (TAI)
collaris Kimoto, 1996: 27 [HN]
chujoi Aslam, 1967: 127 [RN] **A:** CH (TAI)
longicornis Chûjô, 1954d: 181 [HN]
distantis Tan, 1992b: 784, 818 **A:** CH (SCH)
endoi Chûjô, 1954d: 177 **A:** CH (TAI)
flavescens Chûjô, 1954d: 185 **A:** CH (TAI)
fulvipes Baly, 1877e: 213 **A:** BT CH (FUJ) ID (SD) NP PA **ORR**
bhaktai Lopatin, 1995c: 463
fulvicornis Jacoby, 1908: 188
lateralis Weise, 1903a: 34
manipurensis Jacoby, 1908: 184
pakistanicus Lopatin, 1995c: 461
pallipes Jacoby, 1908: 189
wittmeri Lopatin, 1995c: 462
indicus Weise, 1913d: 219 [RN] **A:** NP **ORR**
aterrimus Jacoby, 1908: 187 [HN]
insularis Weise, 1903a: 34 **A:** CH (FUJ) **ORR**
kaohsiungensis Kimoto & Chu, 1996: 21 [RN] **A:** CH (TAI)
babai Kimoto, 1991a: 13 [HN]
formosus Schöller, 1999: 74 [RN]
luteomaculatus Pic, 1915i: 48 **A:** CH ("Chine")
manturiformis Pic, 1943c: 16 **A:** JA (Honshu)
nigrocastaneus Chûjô, 1954d: 175 **A:** CH (TAI) JA (Iriomote Isl., Ishigaki Isl.)
obscuripennis Chûjô, 1935a: 74 **A:** CH (FUJ GUA JIX TAI) JA (Honshu, Kyushu, Ryukyu Archipelago, Shikoku) SC
flavicornis Gressitt, 1942c: 336
nigropiceus Chûjô, 1954d: 184
piceipes Gressitt, 1942c: 336 **A:** CH (GUI HUB JIX TAI YUN ZHE) JA (Honshu, Kyushu, Shikoku) **[# 861]**
piceus Baly, 1873: 86 **A:** CH (FUJ GUI TAI YUN) JA (Honshu, Kyushu, Shikoku) **ORR**
chinensis Baly, 1877e: 214

ruficollis L. N. Medvedev, 2000f: 162 **A**: CH (YUN) **ORR**
sculptatus Pic, 1932c: 12 **A**: CH (YUN) **ORR**
sikkimensis Lopatin, 1995c: 462 **A**: ID (SD)
sulcicollis Baly, 1873: 86 **A**: CH (FUJ GUA) JA (Honshu, Kyushu, Shikoku)
taiwanus Chûjô, 1954d: 178 **A**: CH (TAI)
variegatus Jacoby, 1908: 190 **A**: NP **ORR**
yosionis Chûjô, 1954d: 182 **A**: CH (TAI)

genus *Isnus* Weise, 1898: 216 type species: *Isnus niger* Weise, 1898
biseriatus Chapuis, 1877b: 341 (*Coenobius*) **A**: SA YE **AFR**
arabicus Lopatin, 1979a: 302 (*Coenobius*)
yunnanus Schöller, 2021b: 141 **A**: CH (YUN)

subtribe Stylosomina Chapuis, 1874

genus *Stylosomus* Suffrian, 1848: 146 type species: *Cryptocephalus tamarisci* Herrich-Schäffer, 1836 **[Appendix]**
subgenus *Microsomus* Burlini, 1957b: 62 type species: *Stylosomus bechynei* Burlini, 1957 (= *Stylosomus major* Breit, 1918)
costatus Lopatin, 1962b: 50 **A**: KI UZ
cylindricus Morawitz, 1860: 301 **E**: KR KZ RU (ST) UK **A**: KZ RU (SS WS)
major Breit, 1918b: 118 **A**: KZ MG TD TM
bechynei Burlini, 1957b: 62
hirsutus Lopatin, 1956c: 43
weberi Reitter, 1905a: 94 **A**: AF KI KZ TD TM UZ
amaenus Pic, 1909h: 154 (var.) [RN]
amoenus J. R. Sahlberg, 1913a: 73 [HN]
sahlbergi Winkler, 1929: 1245 [RN]
sublineatus J. R. Sahlberg, 1913a: 75
subgenus *Stylomicrus* Petitpierre & Alonso-Zarazaga, 2000: 247 [RN] type species: *Stylosomus rugithorax* Abeille de Perrin, 1877
Microstilus Pic, 1950d: 94 [incorrect subsequent spelling] [# 851]
Microstylus Rey, 1883c: 316 [HN] type species: *Stylosomus rugithorax* Abeille de Perrin, 1877
ater Lopatin, 1962b: 51 **A**: KZ UZ
ericeti Suffrian, 1852: 653 **E**: SP
bituberculatus Desbrochers des Loges, 1870b: 169
constrictus Schaufuss, 1862b: 92 (*Stylossomus*)
hartmanni L. N. Medvedev, 2003a: 320 **A**: NP
ilicicola Suffrian, 1848: 151 **E**: FR IT PT SP **N**: MO
marocanus Pic, 1950d: 94 **N**: MO

minutissimus Germar, 1823: 561 (*Cryptocephalus*) E: BH CR FR (incl. Corsica) GR IT (incl. Pantelleria, Sardinia, Sicily) ME PT SB SP SZ **N:** MO TU **[# 931]**

depilis Abeille de Perrin, 1877: xlix

lutetianus Sainte-Claire Deville, 1911: 377 [RN]

terminalis Fuente, 1907: 319 (var.)

nepalicus L. N. Medvedev, 2003c: 319 **A:** NP

rugithorax Abeille de Perrin, 1877: xlix E: FR IT SP

nigrisura Gozis, 1908: 47 (var.)

tadzhikus Lopatin, 1965c: 38 **A:** TD

subgenus *Stylosomus* Suffrian, 1848: 146 type species: *Cryptocephalus tamarisci* Herrich-Schäffer, 1836 **[Appendix]**

arnoldi Warchałowski, 2006: 497 **N:** TU

bipartitus Fairmaire, 1867: 414 **N:** AG TU **[# 932]**

adumbraticeps Roubal, 1929b: 56 (ab.) [IN]

biplagiatus Wollaston, 1864: 399 **N:** CI (Fuerteventura)

biskrensis Rey, 1883c: 320 **N:** AG

nigronotatus Pic, 1909h: 154 (var.)

trifasciatus Pic, 1909h: 154 (var.) [HN]

cheni Lopatin, 1962b: 49 **A:** KI

corsicus Rey, 1883c: 319 E: FR (Corsica) IT (Sardinia, Sicily)

erythrocephalus Suffrian, 1853a: 151 [= 1854b: 157] **A:** IQ

fausti Reitter, 1891a: 34 **A:** KI KZ TM UZ

flavus caucasicus J. Müller, 1948: 85 (ssp.) **A:** AR GG

flavus flavus Marseul, 1874: 295 E: AL AU BU CR GR KR NM RO RU (ST) TR UK **A:** AB CY GG KZ MG RU (SS WS) TR UZ **[# 933]**

obscuritarsis Pic, 1914b: 11

lateniger Pic, 1952: 13 **A:** CH (GAN)

macer Weise, 1882b: 271 **N:** AG

nigrifrons Pic, 1909h: 154 (var.) **A:** AF IN TM UZ

nigrifrons Fleischer, 1909b: 245 [HN]

niloticus Suffrian, 1857: 251 **N:** EG **A:** SA SI

ferrantei Pic, 1912g: 75 (var.)

multinotatus Pic, 1909h: 154 (var.)

tournieri Pic, 1909h: 154 (var.)

pubescens Pic, 1913k: 179 **A:** IQ

libertinus Pic, 1918e: 23 (var.)

subelongatus Pic, 1913k: 179 **A:** CY IS JO SY TR

notaticeps Pic, 1914b: 11 (var.)

sidonius Pic, 1913k: 179 (var.)

submetallicus S.-H. Chen, 1941b: 20 **A:** CH (GAN HEB QIN SCH SHA) MG RU (ES SS)
sinensis Lopatin, 1956b: 157
sulcicollis Champion, 1925: 269 **A:** ID (SD UT)
tamarisci Herrich-Schäffer, 1836: 24 (*Cryptocephalus*) **E:** BU CR GR FR IT (incl. Sardinia, Sicily) NM SP (incl. Balearic Islands: Ibiza) **N:** AG LB MO TU **A:** IS SI SY YE
cruciatus Weise, 1882b: 272 (var.)
leprieuri Pic, 1924c: 23 (var.)
nigripennis Roubal, 1922: 116 (ab.) [IN]
oberthuri Pic, 1909h: 154 (var.)
pallidicolor Pic, 1909h: 154 (var.)
rufonotatus Pic, 1924c: 22 (var.)
tamaricis Suffrian, 1848: 148 [UE]
xantholus Rey, 1885: 274
vestitus S.-H. Chen, 1941b: 19 **A:** CH (HEB SCH)
warchalowskii Schöller, 2007: 567 **A:** AF
x-signum Pic, 1899b: 261 **E:** SP (incl. Balearic Islands: Mallorca) **N:** AG MO **A:** IS [# **934**]
andalusiacus Pic, 1909h: 154 (var.)
externemaculatus Pic, 1899b: 261 (var.)

tribe Cryptocephalini, nomen dubium
Buprestis lunatus Scopoli, 1763: 66 **E:** SL [# 850]

tribe Pachybrachini Chapuis, 1874

genus *Pachybrachis* Chevrolat, 1836: 420 type species: *Cryptocephalus hieroglyphicus* Laicharting, 1781
subgenus *Chloropachys* Rey, 1883a: 264 type species: *Pachybrachys azureus* Suffrian, 1848
azureus Suffrian, 1848: 113 (*Pachybrachys*) **E:** FR PT SP **N:** MO
aeruginosus Weise, 1891b: 147 (*Pachybrachys*, forma)
bifoveolatus Rey, 1883a: 265 (var.)
notatipes Pic, 1919: 14 (*Pachybrachys*, var.)
verticalis Rey, 1883a: 265 (var.)
viridissimus Suffrian, 1848: 115 (*Pachybrachys*)
elegans Graells, 1851: 153 **E:** PT SP **N:** AG MO
pici Chobaut, 1896b: 389 **N:** AG MO
regius Schaufuss, 1862a: 312 (*Pachybrachys*) **E:** PT SP
prasinus Fairmaire, 1873c: 355 (*Pachybrachys*)
vicarius Burlini, 1968: 99 (morpha) [IN]

viedmai Burlini, 1968: 100 E: PT SP

subgenus *Pachybrachis* Chevrolat, 1836: 420 type species: *Cryptocephalus hieroglyphicus* Laicharting, 1781

Homoeostigmus Jacobson, 1917: 267 type species: *Pachybrachys capreus* Weise, 1887

Pachybrachia Gistel, 1856: 380 [UE] type species: *Cryptocephalus hieroglyphicus* Laicharting, 1781

Pachybrachys L. Redtenbacher, 1848: 563 [incorrect subsequent spelling]

Pachystylus Rey, 1883a: 263 type species: *Pachybrachis testaceus* Perris, 1866

abditus Lopatin, 1991a: 23 **A**: KI [# 842]

absinthii Lopatin, 1990c: 46 (*Pachybrachys*) **A**: KZ

adaliensis Weise, 1886e: 24 (*Pachybrachys*) **A**: TR

afghanensis Lopatin, 1966c: 366 (*Pachybrachys*) **A**: AF

albicans Weise, 1882b: 248 (*Pachybrachys*) E: RU (ST) **A**: AB AR GG IN TR [# 922]

alcantarensis De La Rosa, Tamurejo, Villanueva & Rodríguez, 2012: 88 **A**: SP

altimontanus Lopatin, 1968c: 545 (*Pachybrachys*) **A**: KI

amurensis L. N. Medvedev, 1973b: 880 **A**: MG NC RU (ES FE incl. Sakhalin SS) SC

mohri Lopatin, 1974b: 177

anatolicus Lopatin, 1985b: 768 **A**: TR

anoguttatus Suffrian, 1866: 208 (*Pachybrachys*) E: IT (Sardinia) SP (Balearic Islands: Mallorca, Menorca)

antigae Weise, 1900b: 281 (*Pachybrachys*) E: SP

aragonicus Tempère & Rapilly, 1981: 313 E: FR SP

arnoldii Lopatin, 1967a: 216 (*Pachybrachys*) **A**: KZ MG

parvissimus Lopatin, 1975c: 198

assiettae Burlini, 1968: 95 (ssp.) E: FR IT

atraphaxidis Lopatin, 1968c: 546 (*Pachybrachys*) **A**: KZ

auliensis Breit, 1921: 72 (*Pachybrachys*) **A**: CH (XIN) KZ TM

baeticus Weise, 1882b: 269 (*Pachybrachys*) E: SP

moroderi Pic, 1913f: 139 (*Pachybrachys*)

subinterruptus Pic, 1914i: 78 (var.)

bodemeyeri Weise, 1906d: 472 **A**: TR

anteconnexus Burlini, 1968: 86 (morpha) [IN]

humeralis Burlini, 1968: 86 (morpha) [IN]

boreopersicus Lopatin, 1991a: 26 **A**: IN

borowieci Lopatin & Nesterova, 2010: 85 **A**: IQ

burlinii Daccordi & Ruffo, 1971: 42 E: IT (Pontine Isl.)

canigouensis Lambelet, 2005: 29 E: FR (East Pyrenees)

capreus capreus Weise, 1887d: 329 (*Pachybrachys*) **A**: IN KZ TM

capreus dzhungaricus L. N. Medvedev & Rybakova, 1980: 44 (*Pachybrachys*, ssp.) **A**: MG

caraganae Lopatin, 1977g: 154 (*Pachybrachys*) **A**: MG

changaicus L. N. Medvedev & Rybakova, 1980: 44 (*Pachybrachys*, ssp.)

carpathicus Rey, 1883a: 271 (*Pachybrachys*) **E:** MD RO
catalonicus Burlini, 1968: 55 **E:** SP
cedri Pic, 1898c: 139 (*Pachybrachys*) **N:** AG
chitralensis Schöller, 2009a: 94 **A:** PA
cinctus Suffrian, 1848: 123 (*Pachybrachys*) **E:** FR (Corsica) IT (Sardinia)
coquereli Rey, 1883b: 304 (*Pachybrachys*) **N:** AG TU
pallidior Pic, 1900a: 5 (*Pachybrachys*, var.)
cordatus Sassi & Schöller, 2003: 144 **A:** JO SY TR
creticus Weise, 1886e: 21 (*Pachybrachys*) **E:** GR (Crete)
theresae Pic, 1907c: 4 (var.)
cribricollis Pic, 1907c: 4 **E:** RU (ST) **A:** KZ UZ
curtipennis Breit, 1921: 69 (*Pachybrachys*) **A:** TM UZ
dimorphus L. N. Medvedev, 1978c: 868 **A:** AF
distictopygus Jacobson, 1901b: 115 (*Pachybrachys*) **A:** CH (HEI) MG RU (ES FE SS) SC [# 923]
kaszabi Lopatin, 1966a: 237 (*Pachybrachys*)
eruditus Baly, 1873: 98 (*Pachybrachys*) **A:** JA (Hokkaido, Honshu, Kyushu, Shikoku)
doenitzi Harold, 1877a: 361 (*Pachybrachys*)
nigrescens Nakane, 1963a: 19 (ab.) [IN]
excisus Weise, 1897c: 66 (*Pachybrachys*) **E:** BU GR TR **A:** SY TR [# 924]
exclusus Rey, 1883b: 291 (*Pachybrachys*) **E:** CR FR IT (incl. Sicily)
etruscus Weise, 1886e: 23 (*Pachybrachys*)
guilhelmi Pic, 1955d: 6 (*Pachybrachus*, var.) [IN] [# 856]
liguricus Burlini, 1968: 73 (morpha) [IN]
fallaciosus Burlini, 1968: 56 (ssp.) **E:** FR
alpinus Rapilly, 1982b: 53
fimbriolatus Suffrian, 1848: 142 (*Pachybrachys*) **E:** AL AU BH BU BY CR CZ FR GE GR HU IT (incl. Sicily) KR MD ME PL RO RU (CT ST) SB SK SL SP SZ UK **A:** AB GG RU (WS) TR
dorsalis Grimmer, 1841: 47 (*Cryptocephalus*) [HN]
mulsantii Perris, 1852: 186 (*Pachybrachys*)
transmontanus Gistel, 1857b: 522 (*Cryptocephalus*)
tristis Olivier, 1808: 810 (*Cryptocephalus*) [HN]
varanensis Pic, 1913f: 138 (*Pachybrachys*, var.)
weisei Burlini, 1968: 93 (morpha) [IN]
flexuosus Weise, 1882b: 254 (*Pachybrachys*) **E:** AL AU BH CR FR GR HU IT (incl. Sicily) SL
dalmatinus Pic, 1908b: 45 (*Pachybrachys*)
oertzeni Weise, 1886e: 22 (*Pachybrachys*)
stramineus Burlini, 1968: 45 (morpha) [IN]
viennensis Weise, 1882b: 256 (*Pachybrachys*, var.)

fraudator Lopatin, 1991a: 24 **A:** KZ
fraudolentus J. Müller, 1955: 67 E: IT
freidbergi Lopatin, 1994: 525 **A:** IS SI
freyi Burlini, 1957a: 25 (*Pachybrachys*) E: GR (Ionian Is.: Cephalonia)
fulvipes albarracinus Wagner, 1927: 107 (ssp.) E: SP
fulvipes fulvipes Suffrian, 1848: 145 (*Pachybrachys*) E: SP
albolimbatus Weise, 1891b: 147 (*Pachybrachys*, var.)
ghilarovi Lopatin, 1974a: 659 (*Pachybrachys*) **A:** KZ
gilloti Pic, 1901j: 39 **A:** LE
glycirrhizae Olivier, 1808: 838 (*Cryptocephalus*) **A:** AB AF AR IN IQ IS JO LE SA SI SY TD TM TR UZ
pygidialis L. N. Medvedev, 2012e: 416
transcaspicus Breit, 1921: 65 (*Pachybrachys*, var.)
gobicus Lopatin, 1977c: 304 (*Pachybrachys*) **A:** MG
hauseri Pic, 1907c: 3 **A:** CH (HEI QIN XIZ)
hieroglyphicus Laicharting, 1781: 182 (*Cryptocephalus*) E: AL AU BE BH BU BY CR CZ EN FI FR GE GR HU IT (incl. Sicily) LA LU MD ME NL NM PL RO RU (CT NT) SB SK SL SP SV SZ TR UK **A:** RU (ES FE incl. Sakhalin SS WS) TR **[# 925]**
apicalis Rey, 1883a: 272 (*Pachybrachys*)
atomarius Gebler, 1830: 201 (*Cryptocephalus*)
cuencanus Burlini, 1968: 49 (morpha) [IN]
decempunctatus Olivier, 1792: 617
disconotatus Pic, 1908c: 47 (*Pachybrachys*, var.)
doctoris Pic, 1910a: 2 (*Pachybrachys*, var.)
histrio Fabricius, 1781: 145 (*Cryptocephalus*)
ictericus Weise, 1882b: 249 (*Pachybrachys*, var.)
imitans Burlini, 1968: 49 (morpha) [IN]
italicus Weise, 1882b: 249 (*Pachybrachys*, var.)
laichartingii Moll, 1784b: 195 (*Chrysomela*)
niger Burlini, 1968: 49 (morpha) [IN]
nyctophantus Gozis, 1908: 40 (var.)
piceus Suffrian, 1848: 116 (*Pachybrachys*) [QS]
posticinus Rey, 1883a: 270 (*Pachybrachys*, var.)
pseudoscriptus Wagner, 1927: 108
pseudosuturalis Burlini, 1968: 50 (morpha) [IN]
stellaris Gozis, 1908: 40 (var.)
subapicalis Pic, 1919b: 14 (*Pachybrachys*, var.)
tristis Laicharting, 1781: 184 (*Cryptocephalus*)
hippophaes Suffrian, 1848: 125 (*Pachybrachys*) E: AU CR FR GE IT (incl. Sardinia, Sicily) LS RO SL SP SZ **[# 1070]**
galeazii Pic, 1910a: 2 (*Pachybrachys*, var.)

lignosus J. Müller, 1951: 351 (ab.) [IN]
pratsensis Pic, 1955d: 6 (*Pachybrachus*, var.) [IN] [# 856]
holerorum Montagna & Sassi, 2013: 248 **E:** IT
instabilis Weise, 1887d: 330 (*Pachybrachys*) **E:** TR **A:** KI KZ TR
issykensis gussakovskii Lopatin, 1968c: 546 (*Pachybrachys*, ssp.) **A:** TD
issykensis issykensis Jacobson, 1901b: 116 (*Pachybrachys*, ssp.) **A:** KI KZ TD UZ
jacobsoni Lopatin, 1968c: 544 (*Pachybrachys*) **A:** AF IN KI KZ UZ
heptapotamicus Lopatin, 1997b: 370
jastschenkoi Lopatin, 1995d: 21 **A:** CH (XIN)
jordanicus Lopatin, 1984b: 197 **A:** IS JO SY [# 926]
kaplini Lopatin, 1986b: 25 **A:** TM
karamani Weise, 1893b: 348 (*Pachybrachys*) **E:** CR IT
ogloblini Roubal, 1927a: 74 (*Pachybrachys*)
kazakhstanicus Lopatin, 1974a: 661 (*Pachybrachys*) **A:** KZ
kirghizicus kirghizicus Lopatin, 1974a: 660 (*Pachybrachys*) **A:** KI
kirghizicus salsolae Lopatin, 1974a: 660 (*Pachybrachys*, ssp.) **A:** KI
koktumensis Lopatin, 1991a: 25 **A:** KI KZ
korbi Weise, 1891b: 145 (*Pachybrachys*) **E:** SP
korotjaevi Lopatin, 1995d: 20 **A:** RU (SS)
kraatzi Weise, 1882b: 255 (*Pachybrachys*) **E:** PT SP
obscuricolor Fuente, 1909a: 308
kuramensis Lopatin, 1974a: 660 (*Pachybrachys*) **A:** TD
laetificus Marseul, 1874: 282 **A:** GG UZ
laevigatus Breit, 1921: 71 (*Pachybrachys*) **A:** KZ
lallemanti Marseul, 1874: 272 **N:** AG MO
obscuriceps Pic, 1905d: 148 (*Pachybrachys*, var.)
semiinterruptus Pic, 1905d: 148 (*Pachybrachys*, var.)
laticollis Suffrian, 1860: 61 (*Pachybrachys*) **N:** EG **A:** IN IS SY TR
callosus J. R. Sahlberg, 1913a: 68 (*Pachybrachys*)
misellus Weise, 1900b: 280 (*Pachybrachys*)
latipes Lopatin, 1971a: 230 (*Pachybrachys*) **A:** MG
lecerfi Pic, 1947a: 6 (*Pachybrachys*) **N:** MO
lencinai Daccordi & Petitpierre, 2013: 60 **A:** SP
leonardii Sassi & Schöller, 2003: 141 **E:** GR (Rhodes, Kos) **A:** TR
libanicola Rey, 1883b: 304 (*Pachybrachys*) **A:** IS LE SY
akbesianus Pic, 1907c: 4
limbatus Ménétriés, 1836: 151 (*Cryptocephalus*) **E:** AL BH BU CR GR (incl. Crete) HU ME NM RO SB TR **A:** IS JO SY TR
ater Weise, 1882b: 264 (*Pachybrachys*, var.) [# 1142]
maculatus Suffrian, 1848: 139 (*Pachybrachys*)
lindbergi Burlini, 1963b: 148 **E:** PT SP

lineatus Weise, 1889g: 591 (*Pachybrachys*) **A:** CH (SCH) MG
lineolatus Suffrian, 1848: 122 (*Pachybrachys*) **E:** PT SP
lopatini L. N. Medvedev & Rybakova, 1980: 43 (*Pachybrachys*) **A:** CH (BEI) MG NC RU (ES FE SS) SC
mardinensis Weise, 1900b: 279 (*Pachybrachys*) **A:** IN IQ IS LE SY TR
seidenstueckeri Kippenberg, 1974: 47
marki Lopatin, 1997b: 371 **A:** TD
marocanus Pic, 1945b: 11 (*Pachybrachys*) **N:** MO
mendax mendax Suffrian, 1860: 60 (*Pachybrachys*) **E:** KR MD RU (ST) TR UK **A:** AB AF AR CH (JIL) GG IN KZ MG RU (ES SS) TR **[# 927]**
dissecticeps Roubal, 1916b: 239 (*Pachybrachys*)
kislovodskensis Roubal, 1916b: 239 (*Pachybrachys*, ab.) [IN]
probus Weise, 1882b: 267 (*Pachybrachys*)
mendax sibiricus Lopatin, 1962c: 320 (*Pachybrachys*, ssp.) **A:** RU (WS)
merkensis Lopatin, 1968c: 546 (*Pachybrachys*, ssp.) **A:** KI KZ UZ **[# 928]**
mitjaevi Lopatin & Kulenova, 1982: 43 **A:** KZ
mogol Lopatin, 1986b: 26 **A:** TD
mongolensis L. N. Medvedev & Rybakova, 1980: 46 (*Pachybrachys*) **A:** MG
nigropunctatus Suffrian, 1854b: 152 (*Pachybrachys*) **A:** AB AF AR IN IQ IS KZ SY TD TM TR **[# 1071]**
paulonotatus Pic, 1908b: 45 (*Pachybrachys*)
nitidicollis Weise, 1894a: 69 (*Pachybrachys*) **A:** KI KZ TD TM UZ
noticollis Rey, 1883a: 273 (*Pachybrachys*) **A:** SY
ochropygus Solsky, 1871: 253 (*Pachybrachys*) **A:** CH (ANH GAN HEB HEI HUB JIL LIA NIN) MG NC RU (ES FE) SC
sexsignatus Weise, 1889g: 589 (*Pachybrachys*)
osellai Daccordi & Ruffo, 1975: 428 (*Pachybrachys*) **E:** IT (Egadi Isl.: Levanzo and Marettimo)
paganus Lopatin, 1990d: 48 **A:** KZ
pallidulus Suffrian, 1852: 652 (*Pachybrachys*) **E:** AL BH BU CR FR (incl. Corsica) GE ME NM RO SB SL SP SZ
angustifrons Rey, 1883a: 278 (*Pachybrachys*, var.)
bimaculatus Rey, 1883a: 275 (*Pachybrachys*, var.)
gallicus Weise, 1882b: 253 (*Pachybrachys*, var.)
gavoyi Pic, 1955d: 6 (*Pachybrachus*, ssp.)
gavoyi Burlini, 1968: 54 (morpha) [IN]
leveillei Pic, 1913f: 138 (*Pachybrachys*, var.)
molloensis Pic, 1955d: 6 (*Pachybrachus*, var.) [IN] **[# 856]**
quadrimaculatus Rey, 1883a: 275 (*Pachybrachys*, var.)
reductus Rey, 1883a: 276 (*Pachybrachys*, var.)

subtilis Rey, 1883a: 275 (*Pachybrachys*, var.)
suturalis Weise, 1882b: 252 (*Pachybrachys*)
parallelus Schöller, 2021b: 138 **A:** CH (SCH)
parthicus Romantsov, 2012: 67 **A:** IN [# 843]
pentheri Ganglbauer, 1905: 280 **A:** TR
petitpierrei Daccordi, 1976c: 595 E: SP
peyroni Pic, 1914i: 78 **A:** "Syrie"
picus Weise, 1882b: 264 (*Pachybrachys*) E: AU BE BH CR CZ FR GE IT RO SK SL SP SZ
pistor Breit, 1921: 71 (*Pachybrachys*) **A:** KI MG TM UZ
planifrons Wagner, 1927: 106 E: FR SP
avignonensis Burlini, 1968: 66 (morpha) [IN]
perplexus Burlini, 1968: 66 (morpha) [IN]
potanini L. N. Medvedev & Rybakova, 1980: 48 (*Pachybrachys*) **A:** CH (SCH)
pradensis Marseul, 1874: 270 E: FR IT SP
incallidus Pic, 1897f: 196 (*Pachybrachys*)
schaeferi Pic, 1948: 14 (var.)
teruelensis Pic, 1913f: 138 (*Pachybrachys*, var.)
pteromelas Graëlls, 1858a: 99 [= 1858b: 137] E: SP
pubipennis Lopatin, 1977d: 30 (*Pachybrachys*) **A:** AR IN (north)
pudicus Lopatin, 1984a: 81 **A:** IN
rapillyi Lopatin, 1984a: 79 **A:** IN
rondanus Burlini, 1968: 37 E: PT SP
rufescens Lopatin, 1974a: 659 (*Pachybrachys*) **A:** KZ UZ
ruffoi ruffoi Burlini, 1956a: 4 (*Pachybrachys*) E: IT (incl. Sicily)
ruffoi sibillinus Burlini, 1959b: 201 (ssp.) E: IT (Monti Sibillini)
rugifer Abeille de Perrin, 1905: 281 E: FR PT SP **N:** MO
diversepunctatus Pic, 1907c: 3 (var.)
salfii Burlini, 1956a: 1 (*Pachybrachys*) E: FR IT
solarii Burlini, 1968: 62 (morpha) [IN]
salsuginosus Schöller, 2010b: 150 **A:** IN
sassii Montagna, 2011: 53 E: IT (Giglio Isl.)
saudicus Lopatin, 1979a: 300 (*Pachybrachys*) **A:** SA YE
schuelkei Schöller, 2021b: 136 **A:** CH (SCH)
scripticollis Faldermann, 1837: 381 **A:** AB AF AR IN IQ IS JO KI LE MG SY TM TR **[# 929]**
arisi Pic, 1908b: 45 (*Pachybrachys*, var.) [IN] [# 856]
israelita Tappes, 1872: 267 (*Pachybrachys*)
letourneuxi Pic, 1900a: 5 (*Pachybrachys*, var.)
persa Suffrian, 1860: 67 (*Pachybrachys*)

scriptidorsum Marseul, 1874: 261 **E:** BU BY MD RO RU (ST) UK **A:** AB AR CH (GAN HEB HEI JIL NIN SHA SHN XIN) GG KZ MG NC RU (ES FE SS WS) SC SY TR [# 930]

beckeri Pic, 1910a: 2 (*Pachybrachys*, var.)

chinensis Weise, 1913d: 219 (*Pachybrachys*, ab.) [IN]

chinensis Gressitt & Kimoto 1961: 111 (*Pachybrachys*, ssp.)

lugubris Weise, 1882b: 248 (*Pachybrachys*, var.)

thoracicus Weise, 1889g: 564 (*Pachybrachys*, var.) [HN]

scriptus Herrich-Schäffer, 1838c: 4 (*Cryptocephalus*) **E:** FR (Corsica) IT (incl. Elba, Sardinia, Sicily)

erycinus Weise, 1882b: 254 (*Pachybrachys*, var.) [# 1143]

hellwigi Weise, 1882b: 254 (*Pachybrachys*, var.) [# 1144]

marchii Pic, 1923a: 28 (*Pachybrachys*)

semidesertus Lopatin, 1994: 526 **E:** KZ

sericans Suffrian, 1860: 57 (*Pachybrachys*) **A:** KZ MG RU (ES)

unicolor Lopatin, 1962c: 320 (*Pachybrachys*, ab.) [IN]

siculus Weise, 1891b: 145 (*Pachybrachys*) **E:** IT (Pantelleria, Sicily) MA

imitator Burlini, 1968: 59 (morpha) [IN]

madoniensis Burlini, 1963c: 134 (morpha) [IN]

sidonius Pic, 1914i: 78 **A:** LE

simius Marseul, 1874: 283 **E:** IT (Lampedusa, Sicily) MA PT SP **N:** AG MO TU

externenotatus Pic, 1910a: 2 (*Pachybrachys*, var.)

favarcqi Pic, 1907k: 186 (*Pachybrachys*, var.) [IN] [# 856]

lepineyi Pic, 1955f: 41 (*Pachybrachys*) [RN]

marocanus Pic, 1955b: 206 (*Pachybrachys*) [HN]

multipunctatus Pic, 1919d: 22 (var.)

normandi Pic, 1919d: 22

obscurior Pic, 1898c: 140 (*Pachybrachys*, var.)

pallidicolor Pic, 1900a: 5 (*Pachybrachys*, var.)

siculus Pic, 1913f: 139 (*Pachybrachys*, var.) [HN]

sinkianensis Lopatin, 1995d: 21 **A:** CH (XIN)

sinuatus Mulsant & Rey, 1859: 47 (*Pachybrachys*) **E:** AL AU BH BU CR CZ FR GE GR HU IT LS LU MD ME NM NL PL RO SB SK SL SZ TR UK **A:** GG RU (WS) TR

haliciensis Miller, 1868: 29 (*Pachybrachys*)

rufimanus Weise, 1882b: 258 (*Pachybrachys*, var.)

skopini Lopatin, 1967a: 217 (*Pachybrachys*) **A:** KZ

suffrianii Schaufuss, 1862a: 312 (*Pachybrachys*) **E:** FR PT SP

danieli Burlini, 1968: 94

sulcithorax Pic, 1896e: 204 **N:** AG

tekensis Lopatin, 1983a: 91 **A:** TM
terminalis Suffrian, 1849b: 291 (*Pachybrachys*) **E:** PT SP
martini Pic, 1909l: 186 (var.)
tesselatus tauricus Suffrian, 1848: 137 (*Pachybrachys*) **E:** KR **A:** AB AR GG IN IS JO LE SY TR
orientalis Weise, 1894g: 144 (*Pachybrachys*, var.)
syriacus Weise, 1900b: 281 (*Pachybrachys*, var.)
tesselatus tesselatus Olivier, 1792: 618 (*Cryptocephalus*) **E:** AL AU BE BH BU CR CZ FR GE GR HU IT (incl. Sicily) LU MD ME NL NM PL RU (CT ST) SB SK SL SP SZ TR UK
aethiops Weise, 1882b: 262 (*Pachybrachys*, var.)
bisignatus L. Redtenbacher, 1848: 563 (*Pachybrachys*)
melanopygus J. Müller, 1948: 84 (var.) [IN] [# 856]
picturatus Rey, 1883b: 291 (*Pachybrachys*, var.)
semilunatus Marcu, 1957: 544 (ab.) [IN]
testaceus Perris, 1866: 510 (*Pachybrachys*) **E:** FR (Corsica) IT (Sardinia, Sicily)
paganettii Pic, 1935d: 12 (*Pachybrachys*, var.)
riguus Marseul, 1874: 287
tibetanus Gressitt & Kimoto, 1961: 112 (*Pachybrachys*) **A:** CH (XIZ)
transbaicalicus L. N. Medvedev, 1974: 181 (*Pachybrachis*) **A:** MG RU (ES)
tuvensis saluki Lopatin, 1995d: 21 (ssp.) **A:** RU (SS)
tuvensis tuvensis L. N. Medvedev, 1974: 179 (*Pachybrachis*) **A:** MG RU (ES SS)
velarum Warchałowski, 1998a: 85 **A:** TR
vermicularis Suffrian, 1854b: 156 (*Pachybrachys*) **E:** RU (ST) **A:** AB IN KZ TR
astragali Stierlin, 1863a: 500 (*Pachybrachys*)
pallidiventris Jacobson, 1894b: 158 (*Pachybrachys*, var.)
warchalowskii Lopatin & Nesterova, 2010: 83 **A:** TR

tribe Fulcidacini Jacobson, 1924

genus *Chlamisus* Rafinesque, 1815: 116 [RN] type species: *Chlamys tuberosa* Knoch, 1801 **[# 935]**
Arthrochlamys Ihering, 1904: 642 [RN] type species: *Chlamys tuberosa* Knoch, 1801 **[# 935]**
Boloschesis Jacobson, 1924c: 239 [RN] type species: *Chlamys tuberosa* Knoch, 1801 **[# 935]**
Chlamys Knoch, 1801: 122 [HN] type species: *Chlamys tuberosa* Knoch, 1801 **[# 935]**
aegyptiacus Desbrochers des Loges, 1898: 47 (*Chlamis*) **N:** EG **A:** SA
aegyptiacus Achard, 1913: 183 (*Chlamys*) [HN]
angularis Gressitt, 1942c: 356 (*Chlamys*) **A:** CH (GUA)
angulisternus Chûjô, 1940c: 273 (*Exema*) **A:** CH (TAI)
brancuccii L. N. Medvedev & Sprecher-Uebersax, 1997: 226 **A:** NP

capitatus Bowditch, 1913: 18 (*Chlamys*) **A:** CH (FUJ GUX HAI HKG SCH TAI YUN) **ORR**

latiusculus Chûjô, 1940c: 284 (*Exema*)

castaneus S.-H. Chen, 1940a: 201 (*Chlamys*) **A:** CH (FUJ "South China") **ORR**

championi Bryant, 1957a: 215 **A:** ID (UT)

cheni Gressitt, 1942c: 359 (*Chlamys*) **A:** CH (SCH)

chinensis Baly, 1878i: 345 (*Chlamys*) **A:** CH (FUJ HKG HAI)

chujoi Gressitt, 1942c: 366 (*Chlamys*) [RN] **A:** CH (TAI)

sexcostatus Chûjô, 1940c: 279 (*Exema*) [HN]

clermonti Achard, 1919: 36 (*Chlamys*) **A:** CH (GUX)

collaris L. N. Medvedev & Sprecher-Uebersax, 1997: 223 **A:** NP

consimilis Chûjô, 1942b: 86 (*Exema*) **A:** JA

corollinotatus Su & Zhou, 2017: 13 **A:** CH (GUX)

diminutus Gressitt, 1942c: 359 (*Chlamys*) **A:** CH (FUJ GUA HAI HKG HUB JIA TAI ZHE) JA (Kyushu) SC

ferrugineus Gressitt, 1946: 89 **A:** CH (FUJ JIX SCH TAI)

flavomaculatus Tan & Zhou, 1995: 232 **A:** CH (FUJ GUX)

flavosignatus Chûjô, 1940c: 275 (*Exema*) **A:** CH (TAI)

floralis Bryant, 1957a: 219 **A:** ID (UT)

formosanus Bates, 1866: 353 (*Chlamys*) **A:** CH (TAI)

fulvipes Baly, 1878i: 345 (*Chlamys*) **A:** ID ("Northern India") NP

fulvitarsis Achard, 1919: 38 (*Chlamys*) **A:** CH (FUJ GUX ZHE)

geniculatus Jacoby, 1896a: 6 (*Chlamys*) **A:** CH (TAI) JA (Ryukyus)

goldenae Gressitt & Kimoto, 1961: 177 **A:** CH (HUB)

granularis L. N. Medvedev & Sprecher-Uebersax, 1997: 225 **A:** BT ID (SD)

himalayanus L. N. Medvedev & Sprecher-Uebersax, 1997: 227 **A:** ID (SD)

holzschuhi Lopatin, 1995c: 465 **A:** NP

ibomune Chûjô, 1956a: 158 [RN] **A:** CH (TAI)

granulicollis Chûjô, 1940c: 278 (*Exema*) [HN]

indicus Guérin-Méneville, 1840a: 41 (*Chlamys*) **A:** NP **ORR**

interjectus Baly, 1873: 84 (*Chlamys*) **A:** JA

cirsicolus Chûjô, 1942b: 93 (*Exema*)

japonicus Jacoby, 1885b: 198 (*Chlamys*) **A:** CH (HUB JIX TAI) JA

kanaraensis Jacoby, 1908: 276 **A:** "Himalaya" **ORR**

laticollis Chûjô, 1942b: 82 (*Exema*) **A:** JA

lewisii Baly, 1873: 83 (*Chlamys*) **A:** CH (TAI) JA

lycocephalus Su & Zhou, 2017: 17 **A:** CH (YUN)

maculiceps Gressitt, 1942c: 361 (*Chlamys*) **A:** CH (FUJ GUA HUN SCH)

martialis Gressitt, 1942c: 362 (*Chlamys*) **A:** CH (GUA JIA)

metasequoiae Gressitt & Kimoto, 1961: 182 **A:** CH (HUB YUN)

montanus Gressitt, 1942c: 363 (*Chlamys*) **A:** CH (SCH)
monticolus Chûjô, 1940c: 282 (*Exema*) **A:** CH (TAI)
mosaicus Tan, 1992b: 776, 812 **A:** CH (YUN)
munemizo Chûjô, 1956a: 166 [RN] **A:** CH (FUJ HUB TAI)
sulcicollis Chûjô, 1940c: 290 (*Exema*) [HN]
hupehensis Gressitt & Kimoto, 1961: 179
nanshanchiensis Kimoto, 1969: 7 **A:** CH (TAI)
nepalensis Lopatin, 1995c: 464 **A:** NP
nigripes Chûjô, 1940c: 281 (*Exema*) **A:** CH (TAI)
orientalis Jacoby, 1895b: 269 (*Chlamys*) **A:** NP **ORR**
pallidiceps Gressitt, 1946: 92 **A:** CH (HAI)
pallidicornis Gressitt, 1942c: 365 (*Chlamys*) **A:** CH (GUA) **ORR**
palliditarsis S.-H. Chen, 1940a: 196 (*Chlamys*) **A:** CH (FUJ GUA GUI GUX HAI SCH YUN) **ORR**
pashokensis Maulik, 1920: 19 (*Chlamys*) **A:** ID (SD)
piceifrons Gressitt, 1942c: 366 (*Chlamys*) **A:** CH (HAI)
prominens Gressitt, 1946: 93 **A:** CH (FUJ TAI ZHE)
prosternalis L. N. Medvedev & Sprecher-Uebersax, 1997: 224 **A:** BT
pubiceps Chûjô, 1940f: 396 (*Exema*) **A:** CH (BEI FUJ HEB HEI SHN) NC RU (FE) SC
purpureocupreus S.-H. Chen, 1976: 217, 224 **A:** CH (XIZ)
reticulicollis Gressitt, 1942c: 367 (*Chlamys*) **A:** CH (GUA JIA)
robustus Chûjô, 1940c: 276 (*Exema*) **A:** CH (TAI)
rufescens Gressitt, 1946: 94 **A:** CH (GUA)
ruficeps S.-H. Chen, 1940a: 198 (*Chlamys*) **A:** CH (CHQ FUJ GUX HUB SCH) **ORR**
rufulus S.-H. Chen, 1940a: 200 (*Chlamys*) **A:** CH (FUJ GUA GUX JIX TAI YUN)
rugiceps S.-H. Chen, 1940a: 198 (*Chlamys*) **A:** CH (GUX)
rugosus Zhou & Tan, 1995: 228 **A:** CH (FUJ)
rusticus Gressitt, 1942c: 368 (*Chlamys*) **A:** CH (GUA HKG)
semirufus S.-H. Chen, 1940a: 199 (*Chlamys*) **A:** CH (FUJ GUA GUX JIX YUN) **ORR**
setosus Bowditch, 1913: 20 (*Chlamys*) **A:** CH (YUN) **ORR**
sexcarinatus Gressitt, 1942c: 370 (*Chlamys*) **A:** CH (GAN GUA JIX)
shirozui Chûjô, 1956a: 155 **A:** CH (HKG TAI)
siciencheni L. N. Medvedev, 2013e: 421 **A:** CH (GUI SCH YUN) **ORR**
speciosus Tan & Zhou, 1995: 233 **A:** CH (FUJ)
spilotus Baly, 1873: 85 (*Chlamys*) **A:** CH (FUJ JIA ZHE) JA SC
stercoralis Gressitt, 1942c: 371 (*Chlamys*) **A:** CH (FUJ GUA GUI GUX HAI SCH XIZ YUN) NP **ORR**
indicus Jacoby, 1901a: 163 (*Chlamys*) [HN]
subferrugineus Gressitt, 1942c: 373 (*Chlamys*) **A:** CH (GUA HAI)
subruficeps Tan, 1992b: 777, 812 **A:** CH (SCH)

suisapanus Gressitt & Kimoto, 1961: 186 **A:** CH (HUB)
sungkangensis Kimoto, 1969: 8 **A:** CH (TAI)
superciliosus Gressitt, 1946: 96 **A:** CH (GUA HAI YUN)
tuberculithorax Gressitt, 1942c: 374 (*Chlamys*) **A:** CH (HAI YUN) **ORR**
uniformis Gressitt, 1946: 97 **A:** CH (GUA)
varipennatus Su & Zhou, 2017: 21 **A:** CH (YUN)
velutinomaculatus Gressitt, 1946: 98 **A:** CH (FUJ HAI JIX)
wuyiensis Zhou & Tan, 1995: 230 **A:** CH (FUJ)
y-costatus Chûjô, 1940c: 288 (*Exema*) **A:** CH (TAI)
yakushimanus Ohno, 1961c: 1 **A:** JA (Shikoku, Ryukyus)
yunnanus Bowditch, 1913: 19 (*Chlamys*) **A:** CH (HAI XIZ YUN) **ORR**
  aterrimus Gressitt, 1942c: 357 (*Chlamys*)
zhamensis Tan, 1981b: 445, 454 **A:** CH (XIZ)

genus *Hymetes* Lacordaire, 1848: 861 type species: *Hymetes javana* Lacordaire, 1848
indica Chapuis, 1874: 205 **A:** ID ("Inde")

subfamily **Eumolpinae** Hope, 1841

tribe Adoxini Baly, 1863 [# 936, # 938]

genus *Acrothinium* Marshall, 1864: 47 type species: *Chrysochus gaschkevitchii* Motschulsky, 1861
cupricolle Jacoby, 1888: 342 **A:** CH (FUJ JIX ZHE)
cyaneum S.-H. Chen, 1940c: 504 **A:** CH (HAI)
gaschkevitchii gaschkevitchii Motschulsky, 1861a: 23 (*Chrysochus*) **A:** CH (FUJ JIX TAI ZHE) JA SC
  miyakensis Ohno, 1968b: 35 (forma) [IN] [# 944]
  obscurum Ohno, 1968b: 35 (forma) [IN] [# 944]
gaschkevitchii matsuii Nakane, 1956: 170 (ssp.) **A:** JA (Okinoerabu)
gaschkevitchii okinawense Nakane, 1985: 2 (ssp.) **A:** JA (Okinawa)
gaschkevitchii shirakii Nakane, 1956: 170 (ssp.) **A:** JA (Amami-Oshima, Okinawa)
gaschkevitchii tokaraense Nakane, 1956: 170 (ssp.) **A:** JA (Tokara)

genus *Aloria* Anonymous, 1940: 283 type species: *Aloria purpurea* Bryant, 1939 [# 939]
  Aloria Bryant, 1939: 9 [UN] [# 939]
  Enneaoria Tan, 1981a: 51, 53 type species: *Enneaoria yunnanensis* Tan, 1981
caerulea Jacoby, 1908: 400 (*Pseudaoria*) **A:** ID (SD) **ORR**
yunnanensis Tan, 1981a: 51, 53 **A:** CH (YUN)

genus *Andosia* Weise, 1896a: 79 type species: *Andosia reitteri* Weise, 1896
reitteri Weise, 1896a: 79 **A:** KZ UZ TD

genus *Anidania* Reitter, 1889e: 127 type species: *Anidania rubripes* Reitter, 1889 (= *Pseudocolaspis luctuosa* Solsky, 1881)
luctuosa Solsky, 1881: 67 (*Pseudocolaspis*) **A:** AF KZ TD TM UZ
rubripes Reitter, 1889e: 128

genus *Aoria* Baly, 1863b: 149 type species: *Adoxus nigripes* Baly, 1860
subgenus *Aoria* Baly, 1863b: 149 type species: *Adoxus nigripes* Baly, 1860
Pseudaoriana Pic, 1930d: 3 type species: *Pseudaoria lemoulti* Pic, 1930 (= *Aoria rufotestacea* Fairmaire, 1889)
annulipes Pic, 1935c: 8 **A:** CH ("Chine")
diversipes Pic, 1935c: 8 (var.)
pectoralis Pic, 1935c: 8
antennata S.-H. Chen, 1940c: 515 **A:** CH (FUJ JIX ZHE)
atra Pic, 1923g: 17 **A:** CH (YUN) **ORR**
bowringii Baly, 1860b: 27 (*Adoxus*) **A:** CH (GUA GUI GUX HAI HUB JIA JIX SHG TAI YUN) ID (SD) NP **ORR** [# 1145]
quinquemaculata Bates, 1866: 354
larvata Gressitt & Kimoto, 1961: 258
carinata Tan, 1993a: 360, 365 **A:** CH (GUI)
costata Tan, 1992a: 595, 605 **A:** CH (HUN)
cyanea S.-H. Chen, 1940c: 514 **A:** CH (GAN SHA)
cuprea L. N. Medvedev, 2012a: 48 **A:** CH (YUN)
fulva L. N. Medvedev, 2012a: 51 **A:** CH (SHA)
gracilicornis S.-H. Chen, 1940c: 515 **A:** CH (SHN)
heinzi L. N. Medvedev, 2012a: 48 **A:** CH (SCH)
marginipennis L. N. Medvedev, 2012a: 48 **A:** CH (SCH)
martensi L. N. Medvedev, 2012a: 51 **A:** CH (SHA)
nepalica L. N. Medvedev & Sprecher-Uebersax, 1997: 228 **A:** NP
nigripennis Gressitt & Kimoto, 1961: 259 **A:** CH (CHQ FUJ GUA JIX ZHE) [# 1145]
nigripes Baly, 1860b: 28 (*Adoxus*) **A:** CH (CHQ FUJ GUA GUI GUX HAI HEB HKG HUB HUN JIA JIX TAI YUN XIZ) ID (SD) NP **ORR**
bicolor Pic, 1935c: 8 (var.)
fulvifrons Jacoby, 1892b: 905 [QS]
fuscula Motschulsky, 1866b: 408 (*Eumolpus*)
mouhoti Baly, 1878f: 247
nigrita Jacoby, 1892b: 904 [QS]

pusilla Lefèvre, 1889a: 291
rufula Motschulsky, 1866b: 408 (*Eumolpus*)
taiwana Chûjô, 1956b: 119 (ssp.)
nigromarginata L. N. Medvedev, 2012a: 50 **A:** CH (GUA)
panfilovi L. N. Medvedev, 2012a: 50 **A:** CH (SCH)
rufotestacea Fairmaire, 1889b: 70 **A:** CH (FUJ GUI HEB HUB JIA LIA SCH SHA SHX ZHE) NC SC
chinensis Jacoby, 1890b: 114
lemoulti Pic, 1930d: 3 (*Pseudaoria*)
rufa Pic, 1928i: 378
testacea Pic, 1935c: 8
scutellaris rufipennis Pic, 1923g: 17 (var.) **A:** CH (SCH) **ORR**
scutellaris scutellaris Pic, 1923g: 17 (var.) **A:** CH (FUJ SCH YUN) SC **ORR**
thibetana Pic, 1928i: 378 **A:** CH (XIZ)
subgenus *Osnaparis* Fairmaire, 1889b: 72 type species: *Osnaparis nucea* Fairmaire, 1889
lushuiensis Tan, 1992b: 766, 805 **A:** CH (YUN)
montana Tan, 1992b: 767, 805 **A:** CH (SCH)
nucea Fairmaire, 1889b: 72 (*Osnaparis*) **A:** CH (CHQ FUJ GUX HUB JIX SCH SHA TAI YUN) JA **ORR**
taliana Pic, 1935c: 8
subgenus *Pseudaoria* Jacoby, 1908: 399 type species: *Pseudaoria burmanica* Jacoby, 1908
floccosa Tan, 1992b: 768, 806 (*Pseudaoria*) **A:** CH (SCH YUN)
irregularis Tan, 1992b: 769, 807 (*Pseudaoria*) **A:** CH (SCH)
petri Warchałowski, 2010a: 338 (*Pseudaoria*) **A:** CH (SCH)
rufina Gressitt & Kimoto, 1961: 261 (*Pseudaoria*) **A:** CH (SCH)
yunnana Tan, 1992b: 769, 807 (*Pseudaoria*) **A:** CH (YUN)

genus *Aulexis* Baly, 1863b: 153 type species: *Aulexis nigricollis* Baly, 1863
abbreviata Gressitt, 1942a: 279 (*Syneta*) **A:** CH (GUA)
atripennis Pic, 1923g: 17 **A:** CH (FUJ YUN)
brevidentata Gressitt, 1942a: 279 (*Syneta*) **A:** CH (HAI)
carinata Pic, 1935c: 8 **A:** CH (GUA) **ORR**
magniscapa Gressitt, 1942a: 281 (*Syneta*)
chengana S.-H. Chen, 1940c: 518 **A:** CH (SCH)
cinnamomi S.-H. Chen & S.-Y. Wang, 1976: 216, 224 **A:** CH (YUN)
gressitti Moseyko, **nom. nov.** [RN] **A:** CH (HAI) [# 945]
gracilicornis Gressitt & Kimoto, 1961: 264 [HN] [# 945]
hochii S.-H. Chen, 1940c: 519 **A:** CH (HAI) NP **ORR**
hainana Gressitt, 1942a: 280 (*Syneta*)

jiangkouensis Tan, 1993a: 361, 366 **A:** CH (GUI)
nepalensis L. N. Medvedev & Sprecher-Uebersax, 1997: 230 **A:** ID (SD) NP
nigripennis Jacoby, 1908: 411 **A:** NP **ORR**
obscura Gressitt, 1945a: 142 **A:** CH (YUN)
shaowuensis Gressitt & Kimoto, 1961: 265 **A:** CH (FUJ)
sichuanensis Tan, 1992b: 772, 810 **A:** CH (SCH)
sinensis S.-H. Chen, 1934f: 186 **A:** CH (FUJ GUI GUX)
tuberculata Tan, 1993a: 361, 366 **A:** CH (GUI)
unicolor Gressitt, 1942a: 282 (*Syneta*) **A:** CH (GUA)
ventralis Gressitt, 1942a: 282 (*Syneta*) **A:** CH (SCH)

genus *Bromius* Chevrolat, 1836: 412 type species: *Chrysomela obscura* Linnaeus, 1758 [ICZN (2012a), Opinion 2298]
Adoxus Kirby, 1837: 209 type species: *Cryptocephalus vitis* Fabricius, 1775 sensu Kirby (1837) (= *Chrysomela obscura* Linnaeus, 1758) **[# 936]**
Chartonia Gistel, 1856: 381 type species: *Chrysomela obscura* Linnaeus, 1758
Eumolpus Illiger, 1798: 498 [rejected name, ICZN (2012a), Opinion 2298]
obscurus Linnaeus, 1758: 375 (*Chrysomela*) **E:** AU AN BE BH BU BY CR CZ DE EN FI FR GR HU IT KR KZ LA LS LT LU MD ME NM NR PL RO RU (CT NT ST) SB SK SL SP SV SZ TR UK **A:** AB AR CH (GAN GUI HEB HEI HUN JIA SCH SHA SHX TAI XIN XIZ) GG IN JA KI KZ MG NC RU (ES FE SS WS) SC TD TR UZ **NAR** **[# 936, # 955]**
cochlearius Say, 1824b: 296 (*Eumolpus*)
concinnus Weise, 1898: 190 (*Adoxus*, var.)
epilobii Weise, 1882b: 295 (*Adoxus*, var.)
japonicus Ohno, 1960d: 241 (*Adoxus*, ssp.)
lewisi Weise, 1898: 190 (*Adoxus*, var.)
nigroquadratus DeGeer, 1775: 336 (*Chrysomela*)
villosulus Schrank, 1781: 95 (*Chrysomela*)
weisei Heyden, 1883b: 53 (*Adoxus*, var.)

genus *Callipta* Lefèvre, 1885b: cxxv [RN] type species: *Calliope fausti* Weise, 1882
Arachnipta L. N. Medvedev, 1957c: 1335 type species: *Callipta smirnovi* L. N. Medvedev, 1957 (= *Callipta fausti turcomana* L. N. Medvedev, 1957)
Calliope Weise, 1882b: 279 [HN] type species: *Calliope fausti* Weise, 1882
borealis Lopatin, 1976a: 112 **A:** CH (XIN) UZ
fausti badghyza Lopatin, 1997a: 170 (ssp.) **A:** TM (South Turkmenistan: Badkhyz)
fausti balchana Lopatin, 1997a: 170 (ssp.) **A:** TM (West Turkmenistan)
fausti fausti Weise, 1882b: 280 (*Calliope*) **E:** RU (ST) **A:** AB

fausti murgabica Lopatin, 1997a: 170 (ssp.) **A:** TM (South Turkmenistan: Murgab River)
fausti palvanica Lopatin, 1997a: 171 (ssp.) **A:** TM (North Turkmenistan: Kaplankyr)
fausti turcomana L. N. Medvedev, 1957c: 1334 (ssp.) **A:** TM (Kopet-Dag mountain ridge)
smirnovi L. N. Medvedev, 1957c: 1335
iranica Lopatin, 1997a: 168 **A:** IN
oberthurii Fairmaire, 1877a: 98 (*Pseudocolaspis*) **N:** EG

genus *Caspiana* Lopatin, 1978: 53 type species: *Caspiana armata* Lopatin, 1978
armata Lopatin, 1978: 53 **A:** TM

genus *Colaspidea* Laporte, 1833: 21 type species: *Chrysomela aeruginea* Fabricius, 1792 (= *Chrysomela metallica* P. Rossi, 1790)
Dia Dejean, 1836: 411 [RN] type species: *Chrysomela aeruginea* Fabricius, 1792 (= *Chrysomela metallica* P. Rossi, 1790)
Plestya Gistel, 1848a: 404 [= 1848b: 123] [RN] type species: *Chrysomela aeruginea* Fabricius, 1792 (= *Chrysomela metallica* P. Rossi, 1790)
algarvensis Zoia, 2014: 9 **E:** PT SP
confinis Zoia, 2014: 14 **N:** AG
dogueti Zoia, 2014: 12 **N:** AG
globosa Küster, 1848a: 92 (*Pachnephorus*) **E:** FR PT SP
globata L. Redtenbacher, 1858: 895 (*Dia*)
abbreviata Desbrochers des Loges, 1871: 358
graeca Zoia, 2014: 5 **E:** GR
grossa Fairmaire, 1866a: 73 **N:** MO
incerta Zoia, 2014: 11 **N:** AG
inflata Lefèvre, 1876b: 17 **N:** AG
juengeri Doguet, 1988b: 302 (ssp.) **E:** GR (Corfu) IT (Sicily)
maghrebina Zoia, 2014: 15 **N:** AG
maura Zoia, 2014: 14 **N:** AG MO
metallica P. Rossi, 1790: 84 (*Chrysomela*) **E:** IT
aeruginea Fabricius, 1792a: 330 (*Chrysomela*)
sphaeroides Fairmaire, 1862a: 590 (*Dia*)
oblonga Blanchard, 1845c: iv (*Dia*) **E:** AL CR IT (incl. Sardinia, Sicily) FR (Corsica) GR (incl. Corfu, Crete) **N:** AG MO TU
albanica Schatzmayr, 1923: 8 (ssp.)
nitida P. H. Lucas, 1846: pl. 46 [= 1847c: 515]
nitida L. Redtenbacher, 1858: 895 (*Dia*) [HN]
ovulum Fairmaire, 1866a: 72 **N:** AG
pallidipes Zoia, 2014: 13 **E:** SP **N:** MO
proxima Fairmaire, 1862a: 591 (*Dia*) **E:** CR FR IT SP **N:** AG TU

genus *Colaspina* Weise, 1893a: 1121 type species: *Dia saportae* Grenier, 1863
saportae Grenier, 1863: 119 (*Dia*) E: FR SP

genus *Damasus* Chapuis, 1874: 321 type species: *Damasus albicans* Chapuis, 1874
albicans Chapuis, 1874: 322 **A:** SY TR

genus *Demotina* Baly, 1863b: 158 type species: *Demotina bowringii* Baly, 1863
albomaculata Tan, 1992b: 764, 803 **A:** CH (SCH)
alni Chûjô, 1956b: 97 **A:** CH (TAI)
atra Pic, 1923g: 17 (*Pseudometaxis*) **A:** CH (YUN) **ORR**
aurosquama Chûjô, 1961: 84 **A:** JA
bicoloriceps Tan, 1992b: 763, 802 **A:** CH (YUN)
bipunctata Jacoby, 1885b: 204 **A:** JA
 inornata Nakane, 1958: 306
bowringii Baly, 1863b: 159 **A:** CH (FUJ GUA HAI HKG JIX) **ORR**
decorata Baly, 1874: 163 **A:** JA SC
decoratella Chûjô, 1956b: 96 **A:** CH (TAI)
elegans Chûjô & Shirôzu, 1955: 239 **A:** JA
 futamon Nakane, 1958: 306 (var.) [IN] [# 942]
fasciata Baly, 1874: 162 **A:** CH (JIX TAI) JA SC
 taiwana Chûjô, 1956b: 94
 pseudoimasakai J. Park, 2013: 116 [# 950]
fasciculata Baly, 1874: 162 **A:** CH (FUJ GUA HAI JIX) JA SC
flavicornis Tan & Zhou, 1997: 190 **A:** CH (FUJ YUN)
gansuica S.-H. Chen, 1940c: 511 **A:** CH (GAN)
imasakai Isono, 1990: 376 **A:** JA
inaequalis Pic, 1927d: 109 (*Pseudometaxis*) **A:** CH (SCH YUN)
incostata Takizawa, 1978b: 597 (ssp.) **A:** CH (TAI)
inhirsuta Pic, 1923g: 16 (*Pseudometaxis*) **A:** CH ("Chine") **ORR**
 robusta Pic, 1929i: 141 (*Pseudometaxis*)
minuta Jacoby, 1908: 428 **A:** NP **ORR**
major Chûjô, 1958c: 6 (ssp.) **A:** JA (Ryukyus)
modesta Baly, 1874: 164 **A:** CH (TAI) JA NP SC **NAR**
montana Chûjô, 1956b: 99 **A:** CH (TAI)
 punctata Takizawa, 1978b: 597
multinotata Pic, 1929e: 45 **A:** CH (YUN)
piceonotata Pic, 1929e: 45 **A:** CH (SHA YUN)
postica S.-H. Chen, 1935c: 353 **A:** CH (GUI)
pustulosa S.-H. Chen, 1935c: 355 **A:** CH (GUI)
sasakawai Nakane & Kimoto, 1959: 67 **A:** JA (Ryukyus)

serriventris Isono, 1990: 377 **A:** CH (TAI) JA NP
squamosa Isono, 1990: 379 **A:** JA
thei S.-H. Chen, 1940c: 510 **A:** CH (GUX YUN)
tuberosa S.-H. Chen, 1935c: 354 **A:** CH (FUJ GUI JIX ZHE) JA **ORR**
japana Ohno, 1960a: 65
vernalis Isono, 1990: 381 **A:** JA

genus *Dermestops* Jacobson, 1898: 195 type species: *Dermestops ahngeri* Jacobson, 1898
Allecumolpus L. N. Medvedev, 1956b: 242 type species: *Allecumolpus smirnovi* L. N. Medvedev, 1956 (= *Dermestops ahngeri* Jacobson, 1898)
ahngeri Jacobson, 1898: 195 **A:** TM
smirnovi L. N. Medvedev, 1956b: 242 (*Allecumolpus*)

genus *Erythraella* Zoia, 2012: 474 type species: *Erythraella bicuspidata* Zoia, 2012
bicuspidata Zoia, 2012: 475 **A:** YE (Socotra)

genus *Eryxia* Baly, 1865h: 437 type species: *Eryxia baikii* Baly, 1865
subgenus *Azerberyxia* Romantsov & Moseyko, 2020: 140 type species: *Eryxia serratotibialis* Romantsov & Moseyko, 2020
serratotibialis Romantsov & Moseyko, 2020: 140 **A:** AB
subgenus *Eryxia* Baly, 1865h: 437 type species: *Eryxia baikii* Baly, 1865
annobioides Escalera, 1914: 520 **N:** MO
gracilipes Lefèvre, 1890a: lvii **A:** YE

genus *Fidia* Motschulsky, 1861a: 22 [ICZN (2009), Opinion 2227] type species: *Fidia atra* Motschulsky, 1861
subgenus *Fidia* Motschulsky, 1861a: 22 [ICZN (2009), Opinion 2227] type species: *Fidia atra* Motschulsky, 1861
Endoxus Baly, 1861b: 285 [incorrect subsequent spelling of *Adoxus*]
Leprotes Baly, 1863b: 158 type species: *Adoxus gracilicornis* Baly, 1861
Lypesthes Baly, 1863b: 152 [ICZN (2009), Opinion 2227] type species: *Fidia atra* Motschulsky, 1861
Talmonus Fairmaire, 1889b: 71 type species: *Talmonus farinosus* Fairmaire, 1889 (= *Fidia atra* Motschulsky, 1861)
albida Pic, 1923g: 16 (*Lypesthes*) **A:** NP **ORR**
atra Motschulsky, 1861a: 22 **A:** CH (CHQ FUJ GUA GUI GUX HUB JIX SCH SHA TIA YUN ZHE) JA NC RU (FE) SC
farinosa Fairmaire, 1889b: 71 (*Talmonus*)

fulvipes Chûjô, 1954b: 106 (*Lypesthes*, ssp.)
pulverulenta Jacoby, 1885b: 203 (*Leprotes*)
testaceipes Pic, 1928g: 26 (*Leprotes*)
tibialis Ohno, 1958b: 174 (*Lypesthes*, forma) [IN] [# 941]
basalis S.-H. Chen, 1940c: 517 (*Lypesthes*) [RN] **A:** CH (YUN)
atra Pic, 1936a: 15 (*Leprotes*) [HN]
bicoloripes Pic, 1936a: 15 (*Leprotes*) **N:** AG
brancuccii L. N. Medvedev, 1993c: 362 (*Hyperaxis*) **A:** NP
fulva Baly, 1878f: 250 (*Leprotes*) **A:** CH (TAI XIN) JA SC
anger Chûjô, 1935b: 204 (*Lypesthes*)
taiwana Chûjô, 1956b: 105 (*Lypesthes*)
gracilicornis Baly, 1861b: 285 (*Endoxus*) **A:** CH (HKG TAI)
sauteri Chûjô, 1938b: 30 (*Lypesthes*)
indica Jacoby, 1908: 412 (*Lypesthes*) **A:** NP **ORR**
itoi Chûjô, 1954b: 103 (*Lypesthes*) **A:** JA
japonica Ohno, 1958b: 179 (*Lypesthes*) **A:** JA SC
babai Chûjô, 1958a: 4 (*Lypesthes*)
kiiensis Ohno, 1958b: 177 (*Lypesthes*) **A:** JA
immaculata Ohno, 1958b: 179 (*Lypesthes*, ab.) [IN]
lewisi Baly, 1878f: 251 (*Leprotes*) **A:** CH ("China") JA
perelegans Gressitt & Kimoto, 1961: 271 (*Lypesthes*) **A:** CH (GUA)
phoebicola Tan, 1983c: 130, 131 (*Hyperaxis*) **A:** CH (FUJ HKG) [# 1145]
vittata Zhou & Tan, 1997: 189 (*Lypesthes*) [# 951]
picea Gressitt & Kimoto, 1961: 271 (*Lypesthes*) **A:** CH (GUA)
shirozui Kimoto, 1969: 17 (*Lypesthes*) **A:** CH (TAI)
sinensis Gressitt & Kimoto, 1961: 272 (*Lypesthes*) **A:** CH (GUA)
subregularis Pic, 1923g: 17 (*Leprotes*) **A:** CH (GUI YUN)
farinosa Pic, 1923g: 17 (*Leprotes*) [HN]
sulcatifrons Gressitt & Kimoto, 1961: 273 (*Lypesthes*) **A:** CH (HAI)

genus *Heteraspis* Chevrolat, 1836: 413 type species: *Eumolpus vittatus* Olivier, 1808
Scelodonta Westwood, 1838a: 129 type species: *Scelodonta curculionoides* Westwood, 1838
Scelodontomorpha Pic, 1938c: 26 type species: *Scelodontomorpha tricostata* Pic, 1938 (= *Scelodonta costata* Jacoby, 1894)
dillwynii Stephens, 1831: 364 (*Eumolpus*) **A:** AF CH (GUA GUX HAI YUN) ID (SD) NP **ORR**
coomani Pic, 1926c: 145 (*Scelodonta*)
indica Duvivier, 1891a: xxxix (*Scelodonta*)
nitidula Baly, 1864b: 2 (*Scelodonta*)

strigicollis Motschulsky, 1866b: 409 (*Odontionopa*)
granulosa Baly, 1867: 158 (*Scelodonta*) **A:** CH (GUA GUX HKG) **ORR**
blaisiei Pic, 1926c: 144 (*Scelodonta*)
lewisii Baly, 1874: 165 (*Scelodonta*) **A:** CH (FUJ GUA GUI GUX HAI HEB HUB HUN JIA JIX SHA SHN TAI YUN ZHE) JA NC SC **ORR**
jeanvoinei Pic, 1941a: 13 (*Scelodonta*)
orientalis Lefèvre, 1887a: lvi (*Scelodonta*)
longicollis Jacoby, 1908: 387 (*Scelodonta*) **A:** ID (KA)
sauteri Chûjô, 1938b: 29 (*Scelodonta*) **A:** CH (TAI)
vicina Harold, 1877c: 106 (*Scelodonta*) **A:** OM SA YE **AFR**
vittata Olivier, 1808: 907 (*Eumolpus*) **A:** NP **ORR**

genus *Heterotrichus* Chapuis, 1874: 316 type species: *Heterotrichus balyi* Chapuis, 1874
balyi Chapuis, 1874: 317 **A:** CH (YUN) **ORR**
laosensis Pic, 1928g: 27 (*Trichochrysea*)
hirsutus Tan & S.-Y. Wang, 2005: 47, 382 (*Acrothinium*) **A:** CH (YUN)

genus *Hyperaxis* Harold, 1874c: 3377 [RN] type species: *Metaxis sellata* Baly, 1863
Metaxis Baly, 1863b: 157 [HN] type species: *Metaxis sellata* Baly, 1863
Pseudopiomera Jacoby, 1892b: 907 type species: *Pseudopiomera pallidicornis* Jacoby, 1892
nigra S.-H. Chen, 1935c: 348 (*Pseudometaxis*) **A:** CH (GUI)
nigrescens Chûjô, 1956b: 90 (*Pseudometaxis*) **A:** CH (TAI)
yaosanica S.-H. Chen, 1940c: 508 **A:** CH (FUJ GUI GUX HUB JIA JIX SCH) **ORR**
consors S.-H. Chen, 1940c: 507

genus *Lahejia* Gahan, 1896: 459 type species: *Lahejia cinerascens* Gahan, 1896
Pseudomalegia Jacoby, 1897c: 546 type species: *Pseudomalegia lefevrei* Jacoby, 1897
aenea S.-H. Chen, 1940c: 517 (*Microlypesthes*) **A:** CH (GUA HEB JIA SCH) RU (FE)
flavipes Chûjô, 1942c: 55 (*Malegia*)
cinerascens Gahan, 1896: 459 **A:** YE
donckieri Pic, 1904b: 36 (*Malegia*)
jacobsoni Sumakov, 1901a: 179 (*Malegia*) **A:** TM
weisei Reitter, 1912: 92 (*Malegia*)
turkestanica Reitter, 1890c: 198 (*Malegia*) **A:** KI KZ TD UZ
hirsuta Jacobson, 1894a: 101 (*Malegia*)
uralensis Reitter, 1912: 92 (*Malegia*, var.)

genus *Lepina* Baly, 1863b: 161 type species: *Lepina inconspicua* Baly, 1863
Demotinella Jacoby, 1908: 414 type species: *Demotinella pectoralis* Jacoby, 1908

inconspicua Baly, 1863b: 162 **A**: CH (YUN) **ORR**

genus *Macrocoma* Chapuis, 1874: 292 type species: *Macrocoma eriophora* Chapuis, 1874

Adoxinia Reitter, 1889e: 128 type species: *Adoxinia spinipes* Reitter, 1889

Eubrachis Baly, 1878f: 248 type species: *Pachnephorus cylindricus* Küster, 1846

Ischyromus Jacobson, 1893b: 241 type species: *Pseudocolaspis sarvadensis* Solsky, 1881

aeneonigra Fairmaire, 1873b: 392 (*Pseudocolaspis*) **N**: AG

aladina Daccordi & L. N. Medvedev, 1996: 244 **A**: SA

bertiae Moseyko, 2013b: 45 [RN] **A**: IN

impressa Berti & Rapilly, 1973: 875 (*Pseudocolaspis*) [HN]

bezdeki Zoia, 2012: 478 **A**: YE (Socotra)

bipartita Kocher, 1963: 212 **N**: MO

bolivari antiatlantis Kocher, 1967: 299 (ssp.) **N**: MO

bolivari bolivari Escalera, 1914: 519 (*Pseudocolaspis*) **N**: MO

rotroui Kocher, 1963: 211 [QS]

siruensis Kocher, 1969a: 113 (var.) [IN] [# 944]

smirnovi Kocher, 1969a: 112 (ab.) [IN]

brunnea Bryant, 1957b: 357 **A**: YE

brunnipes brunnipes Olivier, 1808: 913 (*Eumolpus*) **A**: IQ IS JO SI SY [# 960]

variabilis Schaufuss, 1871b: 202 (*Pseudocolaspis*)

brunnipes obscuricolor Pic, 1905h: 179 (*Pseudocolaspis*, var.) **A**: TR

budura Daccordi & L. N. Medvedev, 1996: 245 **A**: SA

buettikeriana Daccordi, 1979a: 305 **A**: IS SA SI YE

calliptoides Lopatin, 1996: 638 **A**: IQ

carbonaria Lefèvre, 1876b: 7 (*Pseudocolaspis*) **N**: AG

chrysites Gerstaecker, 1871: 81 (*Pseudocolaspis*) **A**: YE **AFR**

crassipes Lefèvre, 1876b: 3 (*Pseudocolaspis*) **N**: AG MO

cylindrica Küster, 1846a: 92 (*Pachnephorus*) **E**: SP

daccordii L. N. Medvedev, 1996: 246 **A**: YE

dakkai Kocher, 1963: 204 **N**: MO

delagrangei Pic, 1898b: 93 (*Pseudocolaspis*) **A**: IN SY TR

akbesiana Pic, 1898b: 94 (*Pseudocolaspis*, var.)

dimorpha Medvedev, 1956d: 897 (*Pseudocolaspis*) **A**: IQ

divisa Wollaston, 1864: 394 (*Pseudocolaspis*) **N**: CI (Fuerteventura, Lanzarote)

djurdjurensis Warchałowski, 2001c: 189 **N**: AG

doboszi Borowiec, 2005a: 374 **E**: TR **A**: TR

dubia Wollaston, 1864: 395 (*Pseudocolaspis*) **N**: CI (Fuerteventura, Lanzarote)

fortidens Berti & Rapilly, 1973: 874 (*Pseudocolaspis*) **A**: TR

haiensis Kocher, 1967: 296 **N:** MO
henoni babylonica Lopatin, 1986a: 110 (ssp.) **A:** IQ
henoni henoni Pic, 1894d: 210 (*Pseudocolaspis*) **N:** AG MO
reymondi Kocher, 1959: 30 (var.) [IN] [# 942]
sarroensis Kocher, 1959: 30 (var.) [IN] [# 942]
henoni occidentalis Escalera, 1914: 517 (*Pseudocolaspis*) **N:** AG MO
atlasica Kocher, 1959: 30 (var.) [# 943]
coerulescens Kocher, 1959: 30 (ab.) [IN]
cyanea Raffray, 1873: 384 (*Pseudocolaspis*) [QS]
debduensis Kocher, 1967: 293 [QS]
lindbergi Kocher, 1967: 292 (var.) [IN] [# 944]
mogadorensis Pic, 1912j: 220 (*Pseudocolaspis*, var.) [QS]
nigrita Kocher, 1959: 30 (var.) [IN] [# 942]
pardoi Kocher, 1967: 291 (var.) [IN] [# 944]
pici Escalera, 1914: 518 (*Pseudocolaspis*)
susica Kocher, 1967: 290 (var.) [IN] [# 944]
heydeni Lefèvre, 1876b: 5 (*Pseudocolaspis*) **N:** MO
coerulea Kocher, 1967: 298 (ab.) [IN]
himalayensis Jacoby, 1900b: 111 (*Pseudocolaspis*) **A:** CH (XIZ) ID (HP)
affinis Breit, 1913: 295 (*Ischyromus*)
hulai Zoia, 2012: 480 **A:** YE (Socotra)
indica afghana L. N. Medvedev, 1985a: 372 (ssp.) **A:** AF
indica indica Baly, 1878f: 249 (*Eubrachis*) **A:** CH (XIZ) ID (HP KA UT) NP **ORR**
banghaasi Breit, 1913: 296 (*Ischyromus*)
iranica Lopatin, 1984a: 88 **A:** IN
kabakovi L. N. Medvedev, 1985a: 372 **A:** AF
korbi Pic, 1901a: 19 (*Pseudocolaspis*) **A:** TR
korbi Weise, 1902a: 204 (*Pseudocolaspis*) [HN]
latifrons Har. Lindberg, 1953: 10 **N:** CI (Tenerife)
lefevrei Baly, 1878f: 260 (*Pseudocolaspis*) **A:** IN OM SA
decemlineata Pic, 1903a: 98 (*Pseudocolaspis*)
leprieuri leprieuri Lefèvre, 1876b: 6 (*Pseudocolaspis*) **N:** AG MO TU **A:** AE SI
alternata Kocher, 1969b: 6 (ab.) [IN]
melillensis Kocher, 1967: 301
pachydera Fairmaire, 1876c: 94 (*Pseudocolaspis*) [QS]
leprieuri majuscula Bechyné, 1957: 1 (ssp.) **A:** SI
marquardti Breit, 1913: 294 (*Ischyromus*) **A:** PA CH (XIZ)
mateui Kocher, 1959: 36 **N:** MO (West Sahara)
micula Lopatin, 1995b: 261 **A:** AF

millingeni Pic, 1898e: 154 (*Pseudocolaspis*) **A:** IN OM SA
mesopotamicus Pic, 1898e: 154 (*Pseudocolaspis*, var.)
minuta L. N. Medvedev, 1985a: 371 **A:** AF
niedobovae Zoia, 2012: 477 **A:** YE (Socotra)
obscuripes Wollaston, 1862b: 441 (*Pseudocolaspis*) **N:** CI (Gran Canaria)
oromiana Daccordi, 1978c: 156 **N:** CI (Alegranza) SG
pakistana L. N. Medvedev, 2000d: 261 **A:** PA
pelikani Zoia, 2017: 79 **N:** CI (Gran Canaria)
peyerimhoffi Kocher, 1959: 32 **N:** MO
robusta Lopatin, 1984a: 89 **A:** IN
rubripes rubripes Schaufuss, 1862a: 311 (*Pseudocolaspis*) **E:** BU GR RO TR **A:** AB AR CY GG IN JO SY TR
aenea Walker, 1871: 18 (*Pachnephorus*)
anatolica Pic, 1898b: 94 (*Pseudocolaspis*) [QS]
balcanica Apfelbeck, 1912: 246 (*Eubrachis*)
diversicolor Schaufuss, 1871b: 202 (*Pseudocolaspis*)
graeca Lefèvre, 1876b: 4 (*Pseudocolaspis*)
rubripes turcmena Lopatin, 1976a: 112 (ssp.) **A:** TM
rufotibialis Jacoby, 1908: 433 (*Eubrachis*) **A:** NP **ORR**
sacra Lopatin, 1983b: 208 (*Pseudocolaspis*) **A:** PA SA
saharica Kocher, 1959: 34 **N:** MO
sarvadensis Solsky, 1881: 65 (*Pseudocolaspis*) **A:** KZ TD TM UZ
saudica L. N. Medvedev, 1996: 247 **A:** IS SA SI
schereri Lopatin, 1995b: 260 **A:** AF
seriesericans Fairmaire, 1876c: 94 (*Pseudocolaspis*) **N:** AG
gossypiata Fairmaire, 1876c: 94 (*Pseudocolaspis*)
setosa mesatlantica Kocher, 1959: 39 (ssp.) **N:** MO
fauconnieri Kocher, 1959: 40 (var.) [IN] [# 941]
setosa setosa P. H. Lucas, 1846: pl. 46 [= 1847c: 518] (*Pseudocolaspis*) **N:** AG
vaucheri Pic, 1907d: 115 (*Pseudocolaspis*, ssp.)
spinipes Reitter, 1889e: 128 (*Adoxinia*) **A:** TD TM UZ
jacobsoni Reitter, 1894a: 16 (*Adoxinia*)
splendens Har. Lindberg, 1950: 13 **N:** CI (La Gomera, La Palma, Tenerife)
splendidula franzi Palm, 1976: 102 (ssp.) **N:** CI (El Hierro)
splendidula palmaensis Palm, 1977: 240 [RN] **N:** CI (La Palma)
occidentalis Palm, 1976: 102 [HN]
officiens Warchałowski, 2001e: 462 [RN]
splendidula splendidula Wollaston, 1862b: 442 (*Pseudocolaspis*) **N:** CI (Gran Canaria, Tenerife)
substriata Weise, 1904a: 100 (*Pseudocolaspis*) **A:** IN TR

vanharteni Lopatin, 2008a: 317 **A**: AE
zarudnii Lopatin, 1985b: 769 **A**: AE IN OM SA **[# 961]**
hormuziaca Warchałowski, 2001d: 345

genus *Malegia* Lefèvre, 1883c: 169 type species: *Malegia letourneuxi* Lefèvre, 1883
arabica Daccordi, 1979a: 306 **A**: SA
asiatica Pic, 1894f: 143 **A**: IQ IS JO
colchica Reitter, 1912: 92 **A**: AB AR IN TR
hauseri Pic, 1902a: 3 **A**: AF
letourneuxi Lefèvre, 1883c: 169 **N**: EG
olivacea Pic, 1941a: 13 [DA] **A**: CH ("Chine")

genus *Mecistes* Chapuis, 1874: 322 type species: *Mecistes tarsalis* Chapuis, 1874
seriatus Lefèvre, 1885d: 122 **A**: SA **AFR**
saudicus L. N. Medvedev, 1997f: 321

genus *Mesocolaspis* Jacoby, 1908: 437 type species: *Apolepis indica* Jacoby, 1904
spinulosa L. N. Medvedev & Zoia, 1996: 112 **A**: CH (YUN) **ORR**

genus *Pachnephoptrus* Reitter, 1892a: 153 type species: *Pachnephoptrus weisei* Reitter, 1892
Bromiodes Jacoby, 1895b: 286 type species: *Bromiodes indicus* Jacoby, 1895
Leprotella Jacoby, 1900b: 100 type species: *Leprotella cashmirensis* Jacoby, 1900
cashmirensis Jacoby, 1900b: 101 (*Leprotella*) **A**: ID (KA)
indicus Jacoby, 1895b: 287 (*Bromiodes*) **A**: ID (HP SD) NP
squamosus Bryant, 1923: 261 (*Bromiodes*) **A**: ID (HP) **ORR**
brunneus Papp, 1951a: 131 (*Bromiodes*, var.)
weisei Reitter, 1892a: 153 **A**: AB AR AF IN TM TR UZ

genus *Pachnephorus* Chevrolat, 1836: 414 type species: *Cryptocephalus arenarius* Panzer, 1797 (= *Cryptocephalus pilosus* P. Rossi, 1790)
subgenus *Pachnephorus* Chevrolat, 1836: 414 type species: *Cryptocephalus arenarius* Panzer, 1797 (= *Cryptocephalus pilosus* P. Rossi, 1790)
Euchthonius Gistel, 1856: 381 type species: *Cryptocephalus arenarius* Panzer, 1797 (= *Cryptocephalus pilosus* P. Rossi, 1790)
bruckii Fairmaire, 1862a: 589 E: FR IT SP
baeticus Weise, 1882b: 287
bistriatus Mulsant & Wachanru, 1852: 177 [= 1853a: 17] **E**: FR IT MA PT SP **N**: AG MO TU **A**: SI TR **ORR** **[Appendix]**
convexicollis Baly, 1867: 95 [QS]
impressus Rosenhauer, 1856: 310

brunneus L. N. Medvedev, 1957c: 1335 **A:** TM
canus Weise, 1882b: 285 **E:** AL BU GR IT RU (ST) TR UK **A:** TM TR
corinthius Fairmaire, 1862a: 590 **E:** AL GR IT **A:** TR [# 972]
edmondi Pic, 1901c: 35
graecus Pic, 1901c: 35 [QS]
curtus Pic, 1921d: 3 **A:** CH ("Chine")
cylindricus P. H. Lucas, 1846b: pl. 46 [= 1847c: 519] **E:** FR PT SP **N:** AG MO TU [# 972, **Appendix**]
hipponensis Desbrochers des Loges, 1871: 358
daccordii Zoia, 2007: 51 **A:** SA YE
fulvus Lopatin, 1976b: 69 **A:** UZ
hispidulus Fairmaire, 1866a: 72 **N:** AG MO
kaszabi Lopatin, 1962d: 1811 **A:** AF
laevicollis Fairmaire, 1862a: 590 **E:** IT SP **N:** AG MO
leprieuri Pic, 1898e: 154 (var.)
lateralis Reitter, 1901a: 54 **E:** KR RU (ST) UK **A:** GG TR [# 972, # 973]
lewisi Baly, 1878f: 257 **A:** CH (FUJ GUA GUX HAI HEB HUB JIA JIL JIX SCH TAI ZHE) ID (HP SD) NC NP **ORR**
bretinghami Baly, 1878f: 256
formosanus Chûjô, 1938b: 32
plagiatus Jacoby, 1892b: 913
variegatus Lefèvre, 1887a: lvii
pilosus P. Rossi, 1790: 100 (*Cryptocephalus*) **E:** AL AU BE BH BU BY CR CZ EN FI FR HU KO LA LS LT LU PL RO RU (CT NT ST) SB SK SP SZ TR UK **A:** AB GG IN RU (WS) TR
aeneus D. H. Schneider, 1792: 220 (*Cryptocephalus*)
arenarius Panzer, 1796b: 19 (*Cryptocephalus*)
carbonarius Grimmer, 1841: 47 (*Eumolpus*)
lepidopterus Küster, 1846a: 94
porosus Baly, 1878f: 257 **A:** CH (BEI FUJ GUX HEB HUB JIA JIX SCH TAI YUN ZHE) ID (HP) NC NP RU (FE) SC **ORR**
sauteri Chûjô, 1938b: 32
seriatus Lefèvre, 1887a: lvii
squamosus Chûjô, 1936d: 9
robustus Desbrochers des Loges, 1870a: 132 **E:** RU (ST) UK **A:** TR
micantipennis Reitter, 1901a: 54 (var.)
ruficornis Lefèvre, 1876b: 10 **A:** AR IN IQ JO SY TR
lefevrei Pic, 1894b: 65
syriacus Reitter, 1886: 71 **A:** IS
tessellatus Duftschmid, 1825: 217 (*Eumolpus*) **E:** AL AU BH BU BY CR CZ EN FR GE GR HU IT LA LS LT ME PL RO RU (CT NT ST) SB SK SP SZ TR UK **N:** CI **A:** AB

AF AR CH (HEI GAN NMO XIN) GG IN KI KZ MG PA RU (ES SS WS) TD TM TR UZ **NAR**

albovillosus Jacoby, 1882: 176 (*Myochrous*) [QS]

campestris Gistel, 1857a: 530

sabulosus Gebler, 1830: 210 (*Eumolpus*)

turcomanicus L. N. Medvedev, 1957c: 1335 **A:** TM UZ

villosus Duftschmid, 1825: 217 (*Eumolpus*) **E:** AL AU BH BU BY CR CZ GR HU IT KR ME NM RO RU (ST) SB SK SL TR UK **A:** AB CY GG IN IS TR

aspericollis Fairmaire, 1862a: 588

chalybaeus Weise, 1882b: 289 (var.)

rugaticollis Miller, 1862: 281

schlumbergerii Dufour, 1851: 354

silvanae Daccordi, 1977c: 83

yemenicus Lopatin, 2001b: 16 **A:** YE

subgenus *Pachnephoriscus* Lopatin, 1976a: 111 type species: *Pachnephorus jacobsoni* Lopatin, 1976

jacobsoni Lopatin, 1976a: 111 **A:** UZ

genus ***Parheminodes*** **S.-H. Chen, 1940b: 211** type species: *Parheminodes collaris* S.-H. Chen, 1940

collaris S.-H. Chen, 1940b: 211 **A:** CH (HAI YUN)

mouhoti Baly, 1864c: 222 (*Chrysochus*) **A:** CH (HAI) **ORR** [# 975]

adonis Pic, 1928g: 27 (*Chrysochus*)

siamensis Jacoby, 1895c: 67 (*Chrysochus*)

genus ***Parnops*** **Jacobson, 1894e: 275** type species: *Parnops glasunowi* Jacobson, 1894

atriceps Pic, 1903j: 163 **A:** CH (LIA XIN) [# 963, # 976]

glasunowi ferghanicus Lopatin, 1976a: 113 (ssp.) **A:** KI

glasunowi glasunowi Jacobson, 1894e: 277 **A:** CH (GAN HEB LIA NIN NMO SHA SHX XIN) IN TD TM UZ

ordossana Jacobson, 1910b: 54 **A:** CH (NMO)

vaillanti Pic, 1945c: 13 **A:** CH (XIN)

genus ***Pseudometaxis*** **Jacoby, 1900b: 101** type species: *Pseudometaxis serraticollis* Jacoby, 1900 (= *Pseudometaxis serratithorax* S.-H. Chen, 1940)

minutus Pic, 1923g: 16 **A:** CH (YUN)

angulosus Pic, 1929k: 147 (var.)

bicoloripes Pic, 1929i: 141

reductus Pic, 1929i: 141 (var.)

serraticollis Baly, 1867: 85 (*Demotina*) **A:** CH (SCH YUN) **ORR** *serratithorax* S.-H. Chen, 1940c: 509 [RN] **A:** CH (YUN) **ORR**

maculatus Chûjô, 1964: 279

serraticollis Jacoby, 1900b: 102 [HN]

submaculatus Pic, 1924b: 6 (*Leprotes*) **A:** CH (CHQ GUA GUI GUX HAI SCH YUN) **ORR**

nanus S.-H. Chen, 1935c: 350

genus *Trichochrysea* Baly, 1861a: 195 type species: *Trichochrysea mouhoti* Baly, 1861

Bromius Baly, 1865h: 439 [HN] type species: *Eumolpus hirtus* Fabricius, 1801

Heteraspibrachis Pic, 1907i: 170 type species: *Heteraspibrachis bipubescens* Pic, 1907

Heteraspis Chapuis, 1874: 284 [HN] type species: *Trichochrysea mouhoti* Baly, 1861

Lefevrella Jacobson, 1894e: 277 type species: *Heteraspis hauseri* Weise, 1890 (= *Heteraspis occidentalis* Weise, 1887)

aeneipennis Lefèvre, 1890b: 194 (*Heteraspis*) **A:** CH (GUI) **ORR**

gloriosa Lefèvre, 1893: 123 (*Heteraspis*)

amygdali amygdali Ogloblin, 1941: 20 (*Lefevrella*) **A:** TD

amygdali nuratavica Lopatin, 1976a: 112 (ssp.) **A:** UZ

annamita Lefèvre, 1878: 309 (*Heteraspis*) **A:** CH (GUA HKG) **ORR**

antennata Jacoby, 1908: 389 **A:** ID (SD) **ORR**

arnoldii L. N. Medvedev, 1957c: 1334 (*Lefevrella*) **A:** KI KZ

chejudoana Komiya, 1985: 19 **A:** SC

chihtuana Komiya, 1985: 11 **A:** CH (TAI)

clypeata Jacoby, 1889b: 172 (*Heteraspis*) **A:** CH (YUN) **ORR**

formosana Komiya, 1985: 14 **A:** CH (TAI)

fortipunctata Lopatin, 2005c: 231 **A:** NP

hebe Baly, 1864c: 220 (*Bromius*) **A:** CH (YUN) **ORR**

splendida Achard, 1921b: 171

hirta viridis Jacoby, 1892a: 86 (*Heteraspis*) **A:** CH (CHQ FUJ GUX HAI YUN) ID (SD) **ORR**

nitida Jacoby, 1892b: 909 (*Heteraspis*)

imperialis Baly, 1861b: 285 (*Callomorpha*) **A:** CH (CHQ FUJ GUX HAI HUN JIX SCH SHA YUN ZHE) **ORR**

reitteri Pic, 1938h: 158

japana Motschulsky, 1857: 37 (*Heteraspis*) **A:** CH (BEI CHQ FUJ GUA GUI GUX HAI HUB HUN JIA JIX NIN SCH TAI YUN ZHE) JA SC

lameyi Lefèvre, 1893: 123 (*Heteraspis*) **A:** CH (HAI) **ORR**

lesnei Berlioz, 1921: 332 (*Heteraspis*) **A:** CH (GUI ZHE)

mandarina Lefèvre, 1893: 125 (*Heteraspis*) **A:** CH (YUN) **ORR**

coomani Pic, 1926d: 11

marmorata Tan, 1984: 94, 96 **A:** CH (YUN)

mouhoti Baly, 1861a: 196 **A**: CH (YUN) **ORR**
superba Pic, 1934i: 6
nitidissima nitidissima Jacoby, 1888: 345 (*Heteraspis*) **A**: CH (FUJ GUX HAI HUN JIX SCH YUN ZHE) NP **ORR**
diversicolor Pic, 1939c: 344
nitidissima scutellaris Pic, 1926d: 11 **A**: NP **ORR**
donckieri Pic, 1926d: 11
occidentalis Weise, 1887d: 335 (*Heteraspis*) **A**: KZ UZ
bipubescens Pic, 1907i: 170 (*Heteraspibrachis*)
hauseri Weise, 1890g: 365 (*Heteraspis*)
orientalis Jacobson, 1909: pl. 55 (*Heteraspis*)
viridescens Pic, 1907i: 170 (*Heteraspibrachis*, var.)
okinawana meridiojaponica Komiya, 1985: 22 (ssp.) **A**: JA
okinawana okinawana Nakane, 1956: 171 (ssp.) **A**: JA (Ryukyus)
okinawana taiwana Komiya, 1985: 20 (ssp.) **A**: CH (TAI)
purpureonotata Pic, 1927n: 133 **A**: CH (YUN)
sakishimana Komiya, 1985: 16 **A**: JA (Ryukyus)
sericea Pic, 1926d: 12 **A**: CH (YUN)
similis S.-H. Chen, 1935c: 342 **A**: CH (GUX) **ORR**
sinensis S.-H. Chen, 1940c: 505 **A**: CH (GAN NIN SCH SHA)
tarsata Achard, 1921b: 171 **A**: CH (GUI) **ORR**
truncata L. N. Medvedev & Sprecher-Uebersax, 1999: 363 **A**: NP
undulata Pic, 1926d: 13 **A**: CH (XIZ)
vestita Baly, 1861a: 196 **A**: NP **ORR**
severini Jacoby, 1900b: 109
viridilabris Heller, 1923: 71 **A**: CH (SCH)

genus *Trichotheca* Baly, 1860b: 26 type species: *Trichotheca hirta* Baly, 1860
aeneopicta Bryant, 1924a: 300 **A**: ID (HP)
annularis Tan, 1981b: 444, 454 **A**: CH (XIZ)
apicalis Pic, 1928h: 264 **A**: CH (YUN)
attenuata Tan, 1992b: 770, 808 **A**: CH (SCH)
beesoni Bryant, 1941b: 407 **A**: ID (KA)
bicolor Tan, 1988a: 316, 328 **A**: CH (XIZ)
brunnea Tan, 1992b: 772, 810 (*Malegia*) **A**: CH (YUN)
dentata Tan, 1981b: 443, 454 **A**: CH (XIZ)
elongata Tan, 1988a: 316, 328 **A**: CH (XIZ)
flavinotata Tan, 1981b: 443, 453 **A**: CH (XIZ)
fulvopilosa S.-H. Chen & S.-Y. Wang, 1976: 214, 223 **A**: CH (XIZ)
fuscicornis S.-H. Chen, 1963: 451, 456 **A**: CH (XIZ)

hirta Baly, 1860b: 26 **A:** CH (HKG) ID (KA SD) NP PA **ORR**
basifemorata Duvivier, 1892b: 410
duvivieri Jacoby, 1904a: 386
fusca Jacoby, 1908: 402
scutellaris Pic, 1929k: 148 (var.)
sikkimensis Takizawa, 1987: 53
nodicollis S.-H. Chen & S.-Y. Wang, 1976: 214, 223 **A:** CH (XIZ)
nuristanica L. N. Medvedev, 1985a: 370 **A:** AF
parva S.-H. Chen & S.-Y. Wang, 1976: 213, 223 **A:** CH (XIZ)
rufofrontalis Tan, 1992b: 771, 808 **A:** CH (YUN)
similis Tan, 1992b: 771, 809 **A:** CH (SCH)
unicolor S.-H. Chen & S.-Y. Wang, 1976: 215, 223 **A:** CH (YUN)
variabilis Gressitt & Kimoto, 1961: 268 **A:** CH (XIZ) NP
ventralis S.-H. Chen, 1935d: 770 **A:** CH (SCH YUN)

genus *Trichoxantha* L. N. Medvedev, 1992f: 385 type species: *Trichoxantha nigripennis* L. N. Medvedev, 1992
nigripennis L. N. Medvedev, 1992f: 385 **A:** ID (SD) NP

genus *Xanthonia* Baly, 1863b: 151 type species: *Xanthonia stevensi* Baly, 1863
subgenus *Xanthonia* Baly, 1863b: 151 type species: *Xanthonia stevensi* Baly, 1863
Microlypesthes Pic, 1936a: 15 type species: *Microlypesthes coomani* Pic, 1936
collaris S.-H. Chen, 1940c: 512 **A:** CH (QIN SCH SHX XIZ YUN)
flavescens Tan, 1988a: 315, 327 **A:** CH (XIZ)
foveata Tan, 1992b: 765, 803 **A:** CH (YUN)
fulva Takizawa, 1987: 55 **A:** NP
glabrata Tan, 1992b: 765, 804 **A:** CH (YUN)
minuta Pic, 1929e: 45 (*Aoria*) **A:** CH (CHQ GUX SCH) **ORR**
nepalensis Takizawa, 1987: 56 **A:** NP
oblonga Takizawa & Basu, 1987: 275 **A:** ID (SD) NP
placida Baly, 1874: 161 **A:** JA SC
hachijoensis Ohno, 1960b: 66 (ssp.)
obscura Ohno, 1960b: 66 (forma) [IN] [# 941]
scutellata Ohno, 1960b: 66 (forma) [IN] [# 941]
signata S.-H. Chen, 1935c: 360 **A:** CH (CHQ GUI HUB SCH SHA YUN)
similis Tan, 1992b: 766, 804 **A:** CH (YUN)
sinica S.-H. Chen, 1935c: 361 **A:** CH (GUI SCH)
striatipennis Kimoto, 1969: 16 **A:** CH (TAI)
taiwana Chûjô, 1956b: 110 **A:** CH (TAI)
varipennis S.-H. Chen, 1940c: 512 **A:** CH (GUX)

subgenus *Xanthonioides* Chûjô, 1966b: 6 type species: *Xanthonia dorsalis* Chûjô, 1966
dorsalis Chûjô, 1966b: 5 **A:** NP

genus ***Xanthophorus*** **Jacoby, 1908: 404** type species: *Xanthonia flavopilosa* Jacoby, 1887
nepalicus L. N. Medvedev & Sprecher-Uebersax, 1999: 363 **A:** NP
seriatus Weise, 1922b: 49 **A:** ID (SD) NP

tribe Eumolpini Hope, 1841 [# **938**]

genus ***Abiromorphus*** **Pic, 1924b: 7** type species: *Abiromorphus anceyi* Pic, 1924
anceyi Pic, 1924b: 7 **A:** CH (JIA JIL HEB HEI SHG ZHE) NC RU (FE) SC [# **1145**]

genus ***Abirus*** **Chapuis, 1874: 310** type species: *Cryptocephalus aeneus* Wiedemann, 1821
elongatus Jacoby, 1908: 457 **A:** ID (SD)
fortunei Baly, 1861b: 283 (*Dermorhytis*) **A:** CH (CHQ FUJ GUA GUI GUX HKG HUB HUN JIA JIX SCH SHA SHN TAI YUN ZHE) JA (Ryukyus) NC **ORR** [# **1145**]
achardi Pic, 1923e: 12 (var.)
atricolor Pic, 1927k: 23
cuprescens Pic, 1946b: 14 (var.)
curtus Pic, 1946b: 14 (var.)
denticollis Lefèvre, 1893: 127
granosus Lefèvre, 1893: 128
guerryi Pic, 1927n: 133 (var.) [# **946**]
guerryi Pic, 1946b: 14 (var.) [HN]
harmandi Lefèvre, 1876c: 305
kiotensis Pic, 1944a: 8
latus Pic, 1944a: 8 [# **946**]
malleti Pic, 1944a: 8
recticollis Pic, 1927n: 133
semicyaneus Pic, 1944a: 8 (var.) [# **946**]
sinensis Pic, 1927n: 133
superbus Pic, 1927k: 23 (var.)
viridescens Pic, 1946b: 14 (var.)
yashiroi Yuasa, 1930: 294
laticornis Tan, 1982a: 84, 88 **A:** CH (YUN)
xishuangensis Tan, 1982a: 85, 89 **A:** CH (YUN)

genus ***Chrysochares*** **Morawitz, 1861: 160** type species: *Chrysomela asiatica* Pallas, 1771
asiaticus Pallas, 1771: 463 (*Chrysomela*) **E:** RU (ST) UK **A:** AB AF AR CH (NIN XIN) KZ MG TD TM TR UZ
aeneus Ballion, 1878: 377

coerulescens Jacobson, 1894b: 158 (var.)
ignitus Jacobson, 1894b: 158 (var.)
orientalis Lopatin, 1963a: 300 (ssp.)
sibiricus Frölich, 1792: 130 (*Cryptocephalus*)
violaceomixtus Jacobson, 1894b: 158 (var.)
virens Weise, 1890g: 366 (var.)
punctatus constricticollis Lopatin, 1963a: 300 **A:** AB GG
punctatus punctatus Gebler, 1844: 106 (*Chrysochus*) **A:** CH (NIN XIN) IN KZ MG TD TM UZ
aeneocupreus S.-H. Chen, 1935c: 276

genus *Chrysochus* Chevrolat, 1836: 413 type species: *Chrysomela praetiosa* Fabricius, 1792 (= *Chrysomela asclepiadea* Pallas, 1773) [ICZN (2012a), Opinion 2298]
Atymius Gistel, 1848a: 404 [= 1848b: 123] [RN] type species: *Chrysomela asclepiadea* Pallas, 1773
asclepiadeus Pallas, 1773: 725 (*Chrysomela*) **E:** AL AU BE BH BU BY CR CZ FR GR HU IT KO LT MD NM PL RO RU (ST) SB SK SL SP SZ UK **A:** AB AR GG IN KZ TR **[# 1145]**
asiaeminoris De Monte, 1948: 57 (ssp.)
cyaneus Olivier, 1792: 607 (*Cryptocephalus*)
praetiosus Fabricius, 1792a: 324 (*Chrysomela*)
brevefasciatus Pic, 1934c: 8 **A:** CH (SHG)
chinensis Baly, 1859d: 125 **A:** CH (FUJ GAN GUI HEB HEI HEN JIA JIL JIX LIA NIN NMO QIN SHA SHN SHX TIA ZHE) JA MG NC SC
cyclostoma Weise, 1889g: 593
singularis Lefèvre, 1884c: ccv
goniostoma Weise, 1889g: 592 **A:** CH (HEI NMO SHX) KZ MG RU (ES FE SS WS)
caerulescens Pic, 1927c: 25 (var.) **[# 947]**
globicollis Lefèvre, 1888: lxxxi **A:** CH (HEI) NC RU (FE)
curtus Pic, 1927g: 154
sikhima Jacoby, 1908: 507 **A:** ID (SD)

genus *Chrysolampra* Baly, 1859d: 126 type species: *Chrysolampra splendens* Baly, 1859
cyanea Lefèvre, 1884c: cxciv **A:** CH (GUI JIX SCH YUN)
hirta Tan, 1982a: 85, 89 **A:** CH (YUN)
longitarsis Tan, 1982a: 86, 90 **A:** CH (YUN)
monstrosa Tan, 1988a: 316, 329 **A:** CH (XIZ)
rugosa Tan, 1982a: 87, 91 **A:** CH (YUN)
smaragdula Boheman, 1859: 170 (*Colaspis*) **A:** CH (HKG)

splendens Baly, 1859d: 126 **A**: CH (ANH CHQ FUJ GUA GUI HUB HUN JIA JIX SCH ZHE) **ORR**

coerulea Pic, 1926b: 12

festiva Lefèvre, 1893: 116

testaceicornis Pic, 1907e: 135

tuberculata Pic, 1926b: 13

verrucosa Lefèvre, 1890b: 192

medvedevi Moseyko & Sprecher-Uebersax, 2010: 80 [RN] **A**: AF

tuberculata L. N. Medvedev, 2005c: 285 [HN]

genus *Colaspinella* Weise, 1893a: 1121 type species: *Colaspidea grandis* Frivaldszky, 1881 **[# 948]**

subgenus *Colaspinella* Weise, 1893a: 1121 type species: *Colaspidea grandis* Frivaldszky, 1881

Falsocolaphellus Pic, 1928f: 10 type species: *Falsocolaphellus syriacus* Pic, 1928

grandis Frivaldszky, 1881: 264 (*Colaspidea*) **E**: BU TR **A**: TR

syriaca Pic, 1928f: 10 (*Falsocolaphellus*) **A**: SY

subgenus *Weigelia* L. N. Medvedev, 2005a: 191 type species: *Weigelia fritzlari* L. N. Medvedev, 2005

fritzlari L. N. Medvedev, 2005a: 192 (*Weigelia*) **A**: TR

genus *Colaspoides* Laporte, 1833: 20 type species: *Colaspis limbata* Olivier, 1808

Amasia Chapuis, 1874: 313 type species: *Amasia spinipes* Chapuis, 1874 (= *Colaspoides varians* Baly, 1867)

Melina Chapuis, 1874: 345 [HN] type species: *Melina calceata* Chapuis, 1874 (= *Eumolpus tibialis* Germar, 1823)

Melinophora Lefèvre, 1885d: 157 [RN] type species: *Melina calceata* Chapuis, 1874 (= *Eumolpus tibialis* Germar, 1823)

Tailandia Chûjô, 1964: 281 type species: *Tailandia chakratongii* Chûjô, 1964

amamianus Komiya, 1988: 18 **A**: JA (Ryukyus)

bengalensis Duvivier, 1892b: 418 **A**: ID (SD) NP **ORR**

montanus Jacoby, 1900b: 114

cantonensis L. N. Medvedev, 2004a: 270 **A**: CH (GUA) **ORR**

chinensis Jacoby, 1888: 343 **A**: CH (FUJ GUA GUX JIA JIX ZHE) NC SC **ORR**

costalis L. N. Medvedev, 2004a: 284 **A**: CH (GUA)

crassifemur Tan & S.-Y. Wang, 1984: 56, 58 **A**: CH (YUN)

duvivieri L. N. Medvedev, 2015a: 476 **A**: ID (SD) NP

femoralis Lefèvre, 1885d: 158 **A**: CH (CHQ FUJ GUA GUI GUX HUB JIX MAC SCH SHA SHN) **ORR**

fulvus Chûjô, 1935a: 76 (*Chrysolampra*) **A:** JA (Ryukyus)
hainanensis Gressitt & Kimoto, 1961: 281 **A:** CH (HAI)
imasakai Komiya, 1991: 139 **A:** JA (Ryukyus)
yakuanus Komiya, 1991: 142 (ssp.)
jacobyi L. N. Medvedev, 2004a: 263 **A:** CH ("China")
japanus Chûjô, 1956c: 3 **A:** JA (Kyushu)
kasaharai Komiya, 1991: 143 **A:** JA (Ryukyus)
kimotoi L. N. Medvedev, 2004a: 269 **A:** CH (HAI)
martini Lefèvre, 1885a: lxvi **A:** CH (GUI YUN) **ORR**
medogensis Tan, 1988a: 317, 329 **A:** CH (XIZ)
miyatakei Kimoto, 1967a: 57 **A:** CH (HKG)
morimotoi Kimoto, 1967a: 58 **A:** CH (HKG)
nepalensis Kimoto, 2001: 28 **A:** NP
okinawanus Komiya, 1991: 142 (ssp.) **A:** JA (Ryukyus)
okumai Komiya, 1991: 146 **A:** CH (TAI)
opacus Jacoby, 1888: 343 **A:** CH (GUA GUI GUX HUN JIA JIX MAC SCH SHN YUN)
dentatus Jacoby, 1900b: 113 [RN]
spinipes Jacoby, 1888: 343 [HN]
paddis Aslam, 1967: 128 [RN] **A:** CH (JIX)
tarsalis S.-H. Chen, 1935c: 258 [HN]
pallidicornis Tan & Zhou, 1997: 192 **A:** CH (FUJ)
patrikeevi Moseyko & Romantsov, 2022: 149 **A:** CH (SCH)
persicariae Chûjô, 1956b: 163 **A:** CH (TAI)
pilicornis Lefèvre, 1889a: 298 **A:** CH (FUJ GUA GUX HAI HKG HUN JIX YUN) **ORR**
prasinus prasinus Lefèvre, 1890b: 202 **A:** CH (YUN) **ORR**
chlorolaema Berlioz, 1918: 316
pseudodiffinis L. N. Medvedev, 2004a: 293 **A:** CH (FUJ)
rufus Gressitt & Kimoto, 1961: 283 **A:** CH (JIX GUA) **ORR**
sauteri Chûjô, 1956b: 167 **A:** CH (TAI)
sublaevicollis Duvivier, 1892b: 417 **A:** ID (SD) NP **ORR**
glabricollis Jacoby, 1908: 471 (*Tricliona*) [RN]
laevicollis Jacoby, 1900b: 110 (*Tricliona*) [HN]
subovatus L. N. Medvedev, 2004a: 297 **A:** CH (YUN) **ORR**
subrugosus Jacoby, 1908: 515 **A:** ID (SD) NP **ORR**
brancuccii L. N. Medvedev, 2004a: 261
suginoi Komiya, 1988: 20 **A:** JA (Ryukyus)
taiwanus Chûjô, 1956b: 165 **A:** CH (TAI)
tenenbaumi Pic, 1942a: 9 **A:** CH (YUN) [# 970]
yunnanicus L. N. Medvedev, 2004a: 269 **A:** CH (YUN)

genus *Dermorhytis* Baly, 1861b: 282 type species: *Dermorhytis igneofasciata* Baly, 1861
foveata Tan, 1982a: 83, 88 **A:** CH (YUN)
yunnanensis Tan & S.-Y. Wang, 1984: 55, 57 **A:** CH (YUN)

genus *Glyptoscelis* Chevrolat, 1836: 414 type species: *Eumolpus hirtus* Olivier, 1808
sinicus Moseyko, 2021: 853 **A:** CH (HUN)

genus *Heminodes* Jacoby, 1895b: 287 type species: *Heminodes indicus* Jacoby, 1895
antennatus Jacoby, 1892b: 912 (*Abirus*) **A:** CH (YUN) **ORR**
unicolor Duvivier, 1891a: xxxvii (*Dermorhytis*) **A:** ID (SD) **ORR**

genus *Iphimoides* Jacoby, 1883a: 200 type species: *Iphimoides celebensis* Jacoby, 1883
subgenus *Clisitherella* S.-H. Chen, 1940b: 207 type species: *Clisitherella suturalis* S.-H. Chen, 1940 (= *Iphimoides suturalis* Pic, 1929)
suturalis Pic, 1929g: 32 **A:** CH (FUJ HUN) **ORR**
cheni L. N. Medvedev, 2001: 177 [RN]
fukienensis Tan, 1983c: 129, 131 (*Clisitherella*)
suturalis S.-H. Chen, 1940b: 208 (*Clisitherella*) [HN]
subgenus *Iphimoides* Jacoby, 1883a: 200 type species: *Iphimoides celebensis* Jacoby, 1883
pallidulus Jacoby, 1889b: 187 (*Colaspoides*) **A:** CH (GUA) **ORR**

genus *Lophea* Baly, 1865h: 441 type species: *Lophea melancholica* Baly, 1865
tonkinea Pic, 1927n: 133 **A:** CH (YUN) **ORR**

genus *Massiea* Lefèvre, 1893: 132 type species: *Massiea cyanipennis* Lefèvre, 1893
Abirellus Chûjô, 1956b: 133 type species: *Chrysochus cylindricus* Chûjô, 1938
Laoseumolpus Pic, 1935b: 3 type species: *Laoseumolpus bicolor* Pic, 1935 (= *Massiea cyanipennis* Lefèvre, 1893)
Tillopsis Berlioz, 1918: 312 type species: *Tillopsis angustata* Berlioz, 1918 (= *Massiea cyanipennis* Lefèvre, 1893)
chouioi S.-H. Chen, 1940c: 521 (*Chalcolema*) **A:** CH (SCH)
cinnamomi S.-H. Chen & S.-Y. Wang, 1976: 217, 224 (*Chalcolema*) **A:** CH (YUN)
costata S.-H. Chen & S.-Y. Wang, 1976: 217, 224 (*Chalcolema*) **A:** CH (YUN)
cribrata S.-H. Chen, 1940c: 522 (*Chalcolema*) **A:** CH (SCH CHQ)
cylindrica Chûjô, 1938b: 35 (*Chrysochus*) **A:** CH (TAI)
glabrata Tan, 1982a: 83, 87 (*Chalcolema*) **A:** CH (YUN)
gracilis S.-H. Chen, 1940c: 520 (*Chalcolema*) **A:** CH (YUN)
sangzhiensis Tan, 1993a: 361, 367 (*Chalcolema*) **A:** CH (HUN)
splendida Tan, 1992a: 597, 606 (*Chalcolema*) **A:** CH (HUN)

genus *Olorus* Chapuis, 1874: 311 type species: *Olorus femoralis* Chapuis, 1874

Autolampra Jacoby, 1908: 300 [RN] type species: *Nephus femoratus* Jacoby, 1889

Nephiusus Jacoby, 1892b: 995 [RN] type species: *Nephus femoratus* Jacoby, 1889

Nephus Jacoby, 1889b: 165 [HN] type species: *Nephus femoratus* Jacoby, 1889

dentipes Tan, 1984: 93, 95 **A:** CH (YUN)

genus *Platycorynus* Chevrolat, 1836: 413 type species: *Eumolpus compressicornis* Fabricius, 1801

Bathycolpus Marshall, 1864: 46 type species: *Eumolpus ignicollis* Hope, 1843

Corynodes Hope, 1841: 162 type species: *Eumolpus compressicornis* Fabricius, 1801 **[Appendix]**

Corynoeides Clark, 1865a: 139 type species: *Corynoeides tuberculata* Clark, 1865 (= *Corynodes monstrosus* Baly, 1867)

Erigenes Marshall, 1864: 45 type species: *Corynodes circumductus* Marshall, 1864

Eudora Laporte, 1840: 513 type species: *Eumolpus compressicornis* Fabricius, 1801

Eurycorynus Marshall, 1864: 36 type species: *Eumolpus chrysis* Olivier, 1808

Neolycaria Abdullah & Qureshi, 1969b: 118 type species: *Neolycaria ahmadi* Abdullah & Qureshi, 1969

Omodon Marshall, 1864: 44 type species: *Corynodes tuberculatus* Baly, 1864

Theumorus Marshall, 1864: 35 type species: *Corynodes amethystinus* Marshall, 1864

aemulus Lefèvre, 1889a: 297 (*Corynodes*) **A:** CH (GUA GUX HAI YUN) **ORR**

beauchenei Lefèvre, 1889a: 298 (*Corynodes*)

inlimbatus Pic, 1936a: 13 (*Corynodes*, var.)

affinis S.-H. Chen, 1934i: 500 (*Corynodes*) **A:** CH (GUI)

ahmadi Abdullah & Qureshi, 1969b: 118 (*Neolycaria*) **A:** PA

angularis Tan, 1982c: 285, 289 **A:** CH (GUI YUN)

argentipilus Tan, 1982c: 286, 290 **A:** CH (YUN)

asphodelus Marshall, 1864: 43 (*Corynodes*) **A:** NP **ORR**

bellus S.-H. Chen, 1940c: 524 (*Corynodes*) **A:** CH (HAI)

chalybeus Marshall, 1864: 44 (*Corynodes*) **A:** CH (HAI YUN) **ORR**

birmanicus Jacoby, 1892b: 917 (*Corynodes*)

dilaticollis Jacoby, 1892b: 916 (*Corynodes*)

paviei Lefèvre, 1890b: 198 (*Corynodes*)

rectesulcatus Pic, 1938d: 124 (*Corynodes*)

chrysochoides S.-H. Chen, 1940c: 523 **A:** CH (HAI)

coeruleicollis Pic, 1940b: 6 (*Corynodes*) **A:** CH ("Chine")

costipennis S.-H. Chen, 1940c: 522 (*Corynodes*) **A:** CH (HAI)

cupreoviridis Tan, 1992b: 774, 811 **A:** CH (SCH)

davidi Lefèvre, 1887a: lvi (*Corynodes*) **A:** CH (YUN)

deletus Lefèvre, 1890b: 199 (*Corynodes*) **A:** CH (GUA GUX HAI JIX YUN) **ORR**

dentatus Tan, 1982c: 286, 290
tonkineus Lefèvre, 1893: 130 (*Corynodes*)
viridimicans Gressitt & Kimoto, 1961: 295
farooqi Abdullah & Qureshi, 1969b: 118 (*Neolycaria*) **A:** PA
gibbosus S.-H. Chen, 1934i: 499 (*Corynodes*) **A:** CH (GUI YUN)
adonis Pic, 1936a: 13 (*Corynodes*)
grahami Gressitt & Kimoto, 1961: 291 **A:** CH (GAN SCH)
gratiosus Baly, 1864b: 7 **A:** CH (HAI) **ORR**
gibbifrons Lefèvre, 1885d: 148 (*Corynodes*)
ignicollis Hope, 1843a: 66 [= 1843b: 64] (*Eumolpus*) **A:** CH (ANH FUJ GUA HAI JIA JIX ZHE)
cyaneipes Pic, 1928g: 29 (*Acrothinium*)
thoracicus Baly, 1859d: 125 (*Chrysochus*)
impressicollis Jacoby, 1908: 499 (*Corynodes*) **A:** ID (SD)
japonicus japonicus Jacoby, 1896a: 7 (*Corynodes*) **A:** JA (Ryukyu: Amami-Oshima)
japonicus umebayashii Kimoto, 1974c: 281 (ssp.) **A:** JA (Ryukyu: Okinoerabu)
lateralis Hope, 1831: 30 (*Eumolpus*) **A:** ID (SD) NP **ORR**
nitens Hope, 1831: 30 (*Eumolpus*)
micans S.-H. Chen, 1934i: 494 (*Corynodes*) **A:** CH (GUA GUI GUX HAI JIX)
mouhoti Baly, 1864b: 7 **A:** CH (GUA HAI SCH YUN) ID (SD) **ORR**
bicavifrons S.-H. Chen, 1940c: 524 (*Corynodes*)
nasiri Abdullah & Qureshi, 1969b: 119 (*Anwarullahia*) **A:** PA
niger niger S.-H. Chen, 1940c: 525 (*Corynodes*) **A:** CH (HUB HUN SCH)
niger yunnanensis Tan, 1982c: 288, 292 (ssp.) **A:** CH (YUN)
parryi parryi Baly, 1864c: 223 **A:** CH (BEI CHQ FUJ GUA GUI GUX HUB HUN JIA JIX SCH ZHE) "Korea" **ORR**
jeanvoinei Pic, 1936a: 13 (*Corynodes*, var.)
peregrinus Herbst, 1783: 63 (*Cryptocephalus*) **A:** CH (CHQ GUI GUX HUB SCH XIZ YUN) ID (SD) NP PA **ORR**
bonneuili Pic, 1934i: 6 (*Corynodes*)
chapuisi Lefèvre, 1885d: 147 (*Corynodes*)
cyaneus Olivier, 1808: 899 (*Eumolpus*)
major Gressitt & Kimoto, 1961: 292
plebejus Weise, 1889g: 593 (*Corynodes*) **A:** CH (GUI HUB SCH SHA YUN ZHE) **ORR**
minutus Pic, 1929f: 35 (*Chrysochus*, var.) [QS]
yunnanus Pic, 1927c: 25 (*Chrysochus*)
purpureimicans Tan, 1982d: 394, 395 **A:** CH (YUN)
pyrophorus Parry, 1844a: 454 [= 1844b: 87] (*Eumolpus*) **A:** CH ("China") ID (SD) NP **ORR**
gloriosus Baly, 1859d: 125 (*Corynodes*)
roseus Tan, 1982c: 287, 291 **A:** CH (YUN)

rugipennis Jacoby, 1895c: 66 (*Corynodes*) **A:** CH (YUN) **ORR**
punctatus Tan, 1982c: 287, 291
sauteri Chûjô, 1938b: 34 (*Corynodes*) **A:** CH (TAI)
speciosus Lefèvre, 1891: cclxxvi (*Corynodes*) **A:** CH (GUI SCH YUN) ID (SD) NP **ORR**
florentini Lefèvre, 1893: 130 (*Corynodes*)
purpureipennis Pic, 1928g: 29 (*Corynodes*)
sulcus Tan, 1982c: 288, 292 **A:** CH (YUN)
undatus Olivier, 1792: 614 (*Cryptocephalus*) **A:** CH (FUJ GUA HAI HKG HUN MAC TAI) NP **ORR** [# 1145]
bifasciatus Olivier, 1808: 900 (*Eumolpus*)
cyaneofasciatus Pic, 1934i: 4 (*Corynodes*, var.)
flosculus Marshall, 1864: 37 (*Corynodes*)
groendalii Swartz, 1808: 235 (*Eumolpus*)
subconcolor Pic, 1934i: 4 (*Corynodes*, var.)
vicinus Pic, 1940a: 4 (*Corynodes*) **A:** CH ("Chine")

genus *Trichocolaspis* L. N. Medvedev, 2005c: 283 type species: *Trichocolaspis pubescens* L. N. Medvedev, 2005
pubescens L. N. Medvedev, 2005c: 283 **A:** CH (YUN)

tribe Euryopini Lefèvre, 1885 [# 938]

genus *Colasposoma* Laporte, 1833: 22 type species: *Colasposoma senegalense* Laporte, 1833
subgenus *Colasposoma* Laporte, 1833: 22 type species: *Colasposoma senegalense* Laporte, 1833
Acis Chevrolat, 1836: 411 [HN, RN] type species: *Colasposoma senegalense* Laporte, 1833
Anwarullahia Abdullah & Qureshi, 1969b: 119 type species: *Anwarullahia lahorensis* Abdullah & Qureshi, 1969 (= *Colasposoma ornatum* Jacoby, 1881)
Ballatro Gistel, 1848a: 404 [= 1848b: 123] [RN] type species: *Colasposoma senegalense* Laporte, 1833
Cheiriphyle Jacoby, 1901b: 241 type species: *Cheiriphyle metallica* Jacoby, 1901 (= *Colasposoma colmanti* Burgeon, 1941)
Chiriphyle Clavareau, 1914: 87 [incorrect subsequent spelling]
Dasychlorus Fairmaire, 1898a: 19 type species: *Dasychlorus passeti* Fairmaire, 1898 (= *Thysbe laticornis* J. Thomson, 1858)
Palesida Harold, 1874a: 23 type species: *Palesida chapuisi* Harold, 1874
Pseudomacetes Linell, 1896: 695 type species: *Pseudomacetes aeneus* Linell, 1896 (= *Colasposoma varicolor* Fairmaire, 1898)
Thysbe J. Thomson, 1858: 208 type species: *Thysbe aurichalcica* J. Thomson, 1858
apicipenne Tan, 1983a: 173, 175 **A:** CH (YUN)

atrocyaneum Zoia, 2012: 467 **A:** YE (Socotra)
austerum Zoia, 2012: 461 **A:** YE (Socotra)
brevepilosum brevepilosum Zoia, 2012: 469 **A:** YE (Socotra)
brevepilosum maritimum Zoia, 2012: 473 **A:** YE (Socotra)
brevepilosum orientale Zoia, 2012: 472 **A:** YE (Socotra)
dauricum Mannerheim, 1849: 247 (*Acis*) Ei: IT **A:** CH (ANH BEI GAN GUI HEB HEI HEN HUB HUN JIA LIA NIN NMO QIN SCH SHA SHN SHX TAI TIA XIN) JA KZ MG NC RU (ES FE WS) SC **[# 971]**
cyaneum Motschulsky, 1860b: 177
mongolicum Motschulsky, 1860b: 178
davidi Lefèvre, 1887a: lv **A:** CH (YUN)
densatum Fairmaire, 1888a: 350 **A:** YE **AFR**
downesii asperatum Lefèvre, 1885d: 104 **A:** CH (YUN) **ORR**
andrewesi Jacoby, 1895b: 272
multipunctatum Jacoby, 1889b: 177
planifrons Jacoby, 1900b: 116
prosternale Jacoby, 1896c: 303
serratulum Lefèvre, 1885d: 106
downesii downesii Baly, 1862a: 19 **A:** BT CH (XIZ) ID (SD UT) NP **ORR**
angulicolle Lefèvre, 1893: 127
aureovittatum Baly, 1864b: 14
confusum Tan & S.-Y. Wang, 1981: 440, 453
igneicolle Baly, 1865e: 428
insuturatum Pic, 1937a: 21
ornaticolle Jacoby, 1881: 444
purpureicolle Pic, 1937a: 22 (var.)
saigonense Pic, 1937a: 22
transversicolle Jacoby, 1889b: 176
viridiaeneum Baly, 1865e: 429
hajeki Zoia, 2012: 465 **A:** YE (Socotra)
laticolle Jacoby, 1900b: 115 **A:** NP **ORR**
ornatum Jacoby, 1881: 445 **A:** AF PA **ORR**
lahorense Abdullah & Qureshi, 1969b: 119 (*Anwarullahia*)
obscurofasciatum Jacoby, 1895b: 273
piffli Mandl, 1965: 25 (*Acrothinium*)
pretiosum Baly, 1860b: 36 **A:** CH (CHQ XIZ YUN) ID (SD) NP **ORR**
bonneuili Pic, 1943c: 12
coeruleatum Baly, 1879a: 440
purpureum Pic, 1937a: 21

rufipenne Pic, 1943c: 13
thibettanum Jacoby, 1891b: 35
purcharti Zoia, 2012: 463 **A:** YE (Socotra)
rufipes Jacoby, 1900b: 114 [DA] **A:** ID (SD)
semicostatum Jacoby, 1908: 443 **A:** ID (SD) NP
castetsi Pic, 1943c: 12
ramnadense Pic, 1943c: 12
tenenbaumi Pic, 1943c: 12 **A:** "Himalaya"
unicostatum Zoia, 2012: 462 **A:** YE (Socotra)
vicinale Tan, 1983a: 174, 176 **A:** CH (YUN)
villosulum Lefèvre, 1885d: 107 **A:** CH (YUN) **ORR**
albopilosum Pic, 1943c: 10
viridicoeruleum Motschulsky, 1860b: 178 **A:** CH (CHQ FUJ GUA GUI GUX HAI HKG HUN JIX MAC NIN SCH TAI XIZ YUN ZHE) ID (SD) JA **ORR** [# 1145]
affine Lefèvre, 1890b: 195
annamita Lefèvre, 1885d: 104
auripenne Motschulsky, 1860b: 179
auritum Papp, 1946a: 7 (ab.) [IN]
azureum Papp, 1946a: 7 (ab.) [IN]
azureolineatum Papp, 1946a: 7 (ab.) [IN]
brevenotatum Pic, 1937a: 22
bryanti Papp, 1946a: 7 (ab.) [IN]
capitatum Jacoby, 1889b: 178
chujoi Papp, 1946a: 7 (ab.) [IN]
cyaneum Papp, 1946a: 7 (ab.) [IN]
jeanvoinei Pic, 1937a: 21
metallicum Clark, 1865a: 142
mutabile Baly, 1867: 273
oberthuri Jacoby, 1896a: 6
perroudi Pic, 1937a: 22
pulcherrimum Baly, 1864b: 15
sauteri Papp, 1946a: 7 (ab.) [IN]
violaceum Papp, 1946a: 7 (ab.) [IN]
yunnanum Fairmaire, 1888c: 37 **A:** CH (YUN)
subgenus *Falsonerissus* Pic, 1951c: 16 type species: *Falsonerissus arabicus* Pic, 1951
Andosiomorpha Lopatin, 1981b: 623 type species: *Andosiomorpha argentata* Lopatin, 1981
Bezdekia Warchałowski, 2005: 303 type species: *Bezdekia tenebrosa* Warchałowski, 2005
Iranomolpus Lopatin, 1979b: 588 type species: *Iranomolpus badius* Lopatin, 1979
argentatum Lopatin, 1981b: 624 (*Andosiomorpha*) **A:** IN

badium Lopatin, 1979b: 589 (*Iranomolpus*) **A:** IN
coracinum Lopatin, 1996: 640 (*Eryxia*) **A:** AE IN OM [# 959]
grande grande Lefèvre, 1890a: lvii (*Eryxia*) **A:** AE SA YE
arabicum Pic, 1951c: 16 (*Falsonerissus*)
grande insulare Zoia, 2012: 455 **A:** YE (Socotra)
distinguendum Zoia, 2012: 457 **A:** YE (Socotra)
socotranum Gahan, 1903: 287 (*Eryxia*) **A:** YE (Socotra)
tenebrosum Warchałowski, 2005: 303 (*Bezdekia*) **A:** IN
villosum Zoia, 2012: 458 **A:** YE (Socotra)

genus *Euryope* Dalman, 1824: 17 type species: *Eumolpus ruber* Latreille, 1807 (= *Cryptocephalus subserricornis* Latreille, 1805) **[Appendix]**
Arachnosphoerus J. Thomson, 1856: 329 type species: *Arachnosphoerus megacephala* J. Thomson, 1856
Bechyneia Jolivet, 1950b: 36 type species: *Bechyneia spinosa* Jolivet, 1950 (= *Euryope monstrosa* Baly, 1862)
Ecranus Walker, 1871: 19 type species: *Ecranus nigripes* Walker, 1871 (= *Cryptocephalus subserricornis* Latreille, 1805)
subserricornis Latreille, 1805: xiv (*Cryptocephalus*) **A:** SA SI YE **AFR**
nigripes Walker, 1871: 19 (*Ecranus*)
quadrimaculatus Olivier, 1808: 905 (*Eumolpus*)
rubra Latreille, 1807: 56 (*Eumolpus*)

genus *Paracrothinium* S.-H. Chen, 1940b: 209 type species: *Paracrothinium cupricolle* S.-H. Chen, 1940
cupricolle S.-H. Chen, 1940b: 209 **A:** CH (GUX)
latum Pic, 1928i: 378 (*Iphimoides*) **A:** CH (HUN) **ORR**
rufum L. N. Medvedev, 1993c: 361 **A:** CH (SCH)

genus *Pathius* Aslam, 1967: 129 [RN] type species: *Taphius flavus* Jacoby, 1897
Taphius Jacoby, 1897c: 531 [HN] type species: *Taphius flavus* Jacoby, 1897
maculatus Bryant, 1957b: 358 (*Taphius*) **A:** YE

genus *Phascus* Lefèvre, 1884a: 91 type species: *Phascus maculatus* Lefèvre, 1884
pallidus Lefèvre, 1884a: 92 **A:** SA **AFR**

tribe Floricolini Moseyko, **trib. nov.** [# 937]

genus *Floricola* Gistel, 1848a: 404 [= 1848b: xi, 123] [RN] type species: *Colaspis ulema* Germar, 1813 [# 937]
Eupales Lefèvre, 1885d: 50 [RN] type species: *Colaspis ulema* Germar, 1813
Pales Chevrolat, 1836: 408 [HN] type species: *Colaspis ulema* Germar, 1813

ulema Germar, 1813: 125 (*Colaspis*) **E:** AL BH BU CR GR HU KO ME NM RO SB TR **A:** TR
viridis Pic, 1916a: 10 (*Eupales*, var.) [IN] [# 942]

genus *Rhodopaea* Gruev & Tomov, 1968: 553 type species: *Rhodopaea angelovi* Gruev & Tomov, 1968 [# 937]
Gruevia Warchałowski, 1974: 497 [RN] type species: *Rhodopaea angelovi* Gruev & Tomov, 1968
angelovi Gruev & Tomov, 1968: 553 **E:** BU
heinzi Kippenberg, 2019: 321 **A:** TR

tribe Typophorini Baly, 1865 [# 938]

genus *Aphilenia* Weise, 1889m: 129 type species: *Aphilenia interrupta* Weise, 1889
subgenus *Aphilenia* Weise, 1889m: 129 type species: *Aphilenia interrupta* Weise, 1889
astakhovi Moseyko, 2012: 101 E: RU (ST)
gobica Lopatin, 1970a: 256 (ssp.) **A:** CH (NMO) MG
interrupta Weise, 1889m: 130 **A:** KZ TD TR TM UZ
ornata Reitter, 1889e: 130 **A:** KZ TM UZ
parvula Weise, 1894a: 71 **A:** TM UZ
nigrescens Jacobson, 1901b: 117 (var.) [IN] [# 942]
subgenus *Pseudaphilenia* Lopatin, 1976a: 111 type species: *Aphilenia hauseri* Weise, 1894
lopatini Moseyko, 2013a: 108 **A:** KZ
mujunkumica Moseyko, 2013a: 109 **A:** KZ
unicolor Reitter 1889e: 130 (var.) **A:** KZ TM UZ
hauseri Weise, 1894a: 71

genus *Atomyria* Jacobson, 1894a: 102 type species: *Nodostoma sarafschanica* Solsky, 1881
kermanshahica Moseyko, 2020: 345 **A:** IN
sarafschanica Solsky, 1881: 62 (*Nodostoma*) **A:** KI KZ TD TM UZ

genus *Basilepta* Baly, 1860b: 23 type species: *Basilepta longipes* Baly, 1860
Falsoiphimoides Pic, 1935b: 2 type species: *Falsoiphimoides bicoloripes* Pic, 1935
Mimoparascela Pic, 1935b: 1 type species: *Mimoparascela viridis* Pic, 1935
Nodostoma Motschulsky, 1860b: 176 type species: *Nodostoma fulvipes* Motschulsky, 1860
abdominalis Jacoby, 1908: 337 (*Nodostoma*) **A:** NP **ORR**
amamiensis Chûjô, 1957a: 13 **A:** JA (Ryukyus)
angulicollis Duvivier, 1892b: 408 (*Nodostoma*) **A:** ID (SD)
apicipennis S.-H. Chen, 1935c: 315 **A:** CH (CHQ HUB JIX GUA SCH)
balyi Harold, 1877a: 361 (*Nodostoma*) **A:** JA RU (FE: Kuril, Sakhalin)
japonica Jacoby, 1885b: 205 (*Nodostoma*)
kurilensis L. N. Medvedev, 1966: 40 (ssp.)

otsukae S. Matsumura, 1911a: 61 (*Nodostoma*)
yezo Nakane, 1963b: 225 (ssp.)
beccarii Jacoby, 1884a: 214 (*Nodostoma*) **A:** NP **ORR**
bicollis Tan, 1988b: 161, 169 **A:** CH (YUN)
bicolorata S.-H. Chen, 1940c: 499 **A:** CH (GUX)
bidens Tan, 1988a: 311, 325 **A:** CH (XIZ)
binhana Pic, 1930d: 8 (*Nodostoma*) **A:** CH (GUA HAI) ID (SD) NP **ORR**
distincta S.-H. Chen, 1935c: 314
kumatai Kimoto & Takizawa, 1973: 175
bisulcata S.-H. Chen, 1976: 215, 223 **A:** CH (XIZ)
borodinensis Kimoto, 1979a: 41 **A:** JA (Ryukyus)
chalcea Jacoby, 1908: 305 (*Nodostoma*) **A:** CH (YUN) **ORR**
concinnicollis Baly, 1878h: 373 (*Nodostoma*) **A:** PA **ORR**
congregata Jacoby, 1908: 329 (*Nodostoma*) **A:** CH (HAI) **ORR**
consobrina S.-H. Chen, 1940c: 500 **A:** CH (FUJ GUX)
convexa Tan, 1988a: 313, 326 **A:** CH (XIZ)
cornuta Chûjô, 1956b: 23 **A:** CH (TAI)
crassipes S.-H. Chen, 1976: 215, 223 **A:** CH (XIZ)
cribricollis Motschulsky, 1860b: 176 (*Nodostoma*) **A:** CH (FUJ HEI HEN JIL JIX LIA SCH SHA TAI YUN) JA NC RU (FE) SC **[# 956, # 965, # 1145]**
atripes Motschulsky, 1861a: 23 (*Nodostoma*)
bicoloripes Pic, 1930d: 7 (*Nodostoma*) **[# 952]**
chinensis Lefèvre, 1877: 158 (*Nodostoma*) **[# 952]**
coerulescens Weise, 1889g: 597 (*Nodostoma*, var.)
guerryi Pic, 1930d: 7 (*Nodostoma*, var.) **[# 952]**
oberthuri Lefèvre, 1877: 158 (*Nodostoma*) **[# 952]**
cribrithorax S.-H. Chen, 1940c: 501 **A:** CH (SCH CHQ GAN)
cyanea Lefèvre, 1891: cclxvi (*Nodostoma*) **A:** ID (SD)
davidi Lefèvre, 1877: 157 (*Nodostoma*) **A:** CH (FUJ HAI TAI) JA (Ryukyus) SC
approximata Pic, 1933a: 8 (*Nodostoma*, var.)
atriventris Pic, 1933a: 7 (*Nodostoma*)
insulana Chûjô, 1935b: 204 (*Nodostoma*)
okinawensis Chûjô, 1935a: 77 (*Nodostoma*)
declivis Tan, 1988b: 161, 169 **A:** CH (SCH)
deqenensis Tan, 1988b: 162, 170 **A:** CH (YUN)
dhunchena Kimoto & Takizawa, 1981: 55 **A:** NP
djoui Gressitt & Kimoto, 1961: 220 **A:** CH (FUJ GUA GUI HUB JIX YUN)
elegans Chûjô, 1956b: 38 **A:** CH (TAI)
melanura Chûjô, 1956b: 40 (var.) [IN] **[# 942]**

elongata Tan, 1988b: 163, 171 **A:** CH (YUN)
fabrei Lefèvre, 1887b: clxvii (*Nodostoma*) **A:** ID (SD) NP PA **ORR**
flavicaudis Tan, 1988b: 163, 171 **A:** CH (YUN)
fukienensis Gressitt & Kimoto, 1961: 221 **A:** CH (FUJ)
fulvipes Motschulsky, 1860b: 176 (*Nodostoma*) **A:** CH (BEI CHQ FUJ GUI GUX HEB HEI HUB HUN JIA JIX LIA NMO NIN SCH SHA SHN SHX TAI YUN ZHE) JA MG NC RU (FE) SC **[# 965]**
aeneipennis Motschulsky, 1860b: 177 (*Nodostoma*)
picicollis Weise, 1889g: 597 (*Nodostoma*, var.)
rufotestacea Motschulsky, 1860b: 177 (*Nodostoma*)
fuscolimbata Tan, 1988b: 164, 172 **A:** CH (YUN)
gibbosa Jacoby, 1890b: 115 (*Nodostoma*) **A:** CH (HUB) **ORR**
granulosa Tan, 1988b: 164, 172 **A:** CH (YUN)
hirayamai Chûjô, 1935a: 77 (*Nodostoma*) **A:** CH (TAI) JA SC
hirticollis Baly, 1874: 167 (*Nodostoma*) **A:** JA
incerta Pic, 1929g: 33 (*Pagria*) **A:** CH (SCH) **ORR**
pallidicornis S.-H. Chen, 1935c: 312
indica Jacoby, 1892a: 86 (*Nodostoma*) **A:** ID (SD) **ORR**
issikii Chûjô, 1956b: 45 **A:** CH (CHQ TAI)
kaszabi Lopatin, 1962c: 321 **A:** CH (SHG) **[# 966]**
kaulbachi Bryant, 1939: 5 (*Nodostoma*) **A:** CH (XIZ) **ORR**
laeta L. N. Medvedev, 1992d: 7 **A:** NP
laevigata Tan, 1981b: 438, 452 **A:** CH (SCH XIZ)
latericosta Gressitt & Kimoto, 1961: 224 **A:** CH (XIZ)
leechi Jacoby, 1888: 344 (*Nodostoma*) **A:** CH (CHQ FUJ GUA GUI GUX HUB JIA JIX SCH YUN ZHE) **ORR**
longitarsalis Tan, 1992a: 591, 605 **A:** CH (HUN)
magnicollis Tan, 1988a: 312, 325 **A:** CH (XIZ)
makiharai Kimoto, 2001: 25 **A:** NP
manaliensis Shukla, 1960: 70 (*Nodostoma*) **A:** ID (HP)
manii Shukla, 1960: 72 (*Nodostoma*) **A:** ID (HP)
martini Lefèvre, 1885a: lxv (*Nodostoma*) **A:** CH (FUJ GUA GUX HAI HUN JIX TAI) **ORR**
atrovittata Pic, 1930b: 37 (*Nodostoma*, var.)
interrupta Pic, 1933a: 8 (*Nodostoma*, var.)
languei Pic, 1933a: 8 (*Nodostoma*, var.)
lineata Pic, 1933a: 8 (*Nodostoma*, var.)
sauteri Chûjô, 1938b: 26
unicoloripennis Chûjô, 1938b: 27 (ab.) [IN]

melanopus Lefèvre, 1893: 119 (*Nodostoma*) **A:** CH (FUJ GUA HUB HUN JIX TAI) **ORR**
minutipunctata Tan, 1988a: 313, 326 **A:** CH (XIZ)
modesta Jacoby, 1885b: 206 (*Nodostoma*) **A:** CH (FUJ TAI) JA
cupraria Chûjô, 1938b: 26 (ssp.)
gigantea Chûjô, 1938b: 25
nigricollis Medvedev, 2015a: 475 **A:** NP
nigripecta Gressitt & Kimoto, 1961: 226 **A:** CH (JIX)
nigrita L. N. Medvedev, 1997e: 257 **A:** NP
nyalamensis Tan, 1981b: 438, 452 **A:** CH (XIZ)
orientalis Jacoby, 1890b: 115 (*Nodostoma*) **A:** CH (HUB SCH)
ovalis S.-H. Chen, 1940c: 497 **A:** CH (HEB SCH ZHE)
pallidula Baly, 1874: 169 (*Nodostoma*) **A:** CH (FUJ GUA GUI HAI HKG HUB JIX SCH ZHE) JA NC SC
laeviuscula Weise, 1910c: 34 (*Nodostoma*)
pectoralis Pic, 1931b: 22 (*Nodostoma*) **A:** CH (YUN) [# 967]
pinguis S.-H. Chen, 1940c: 498 **A:** CH (HAI)
plagiosa Baly, 1878h: 373 (*Nodostoma*) **A:** CH (XIN) ID (SD) NP PA **ORR**
pretiosa Jacoby, 1908: 304 (*Nodostoma*) **A:** CH (XIZ) ID (SD) **ORR**
pubiventris Tan, 1988b: 165, 173 **A:** CH (SCH)
puncticollis Lefèvre, 1889a: 295 (*Nodostoma*) **A:** CH (FUJ GUA JIX YUN ZHE) ID (SD) NP **ORR**
atritarsis Pic, 1930d: 7 (*Nodostoma*)
atrithorax Pic, 1930d: 5 (*Nodostoma*)
birmanica Jacoby, 1892b: 898 (*Nodostoma*)
brunneopunctata Pic, 1933a: 8 (*Nodostoma*)
femoralis Pic, 1933a: 8 (*Nodostoma*)
hanoiensis Pic, 1933a: 8 (*Nodostoma*, var.)
limbata Lefèvre, 1893: 116 (*Nodostoma*)
nigriventris Lefèvre, 1893: 117 (*Nodostoma*)
notaticeps Pic, 1930d: 6 (*Nodostoma*, var.)
robusta Weise, 1922b: 48
punctifrons An, 1988: 43 **A:** SC
punctostriata Gressitt & Kimoto, 1961: 229 **A:** CH (JIX)
quadrifasciata Jacoby, 1908: 314 (*Nodostoma*) **A:** ID (SD) NP **ORR**
regularis Tan, 1981b: 439, 453 **A:** CH (XIZ)
remota Tan & S.-Y. Wang, 1984: 55, 57 **A:** CH (YUN)
riedeli Medvedev, 2015a: 475 **A:** ID (UT)
rubimaculata Tan, 1988a: 314, 327 **A:** CH (XIZ)

ruficollis Jacoby, 1885b: 205 (*Nodostoma*) **A:** CH (CHQ FUJ GUI GUX HUB SCH TAI YUN ZHE) JA **ORR**

kuromune Nakane, 1963b: 225 (ab.) [IN]

latipennis Pic, 1928i: 379 (*Nodostoma*)

melanicollis Chûjô, 1938b: 26 (ab.) [IN]

nigripennis Pic, 1928i: 379 (*Nodostoma*)

schawalleri L. N. Medvedev, 1992d: 6 **A:** NP

sculpturata Motschulsky, 1860b: 177 (*Nodostoma*) **A:** ID (SD) NP **ORR**

bhamoensis Jacoby, 1892b: 896 (*Nodostoma*)

scutellaris S.-H. Chen, 1963: 451, 456 **A:** CH (XIZ)

separata Jacoby, 1908: 324 (*Nodostoma*) **A:** ID (SD) NP

sikkimensis Jacoby, 1900b: 102 (*Nodostoma*) **A:** ID (SD) NP **ORR**

sinarum Weise, 1922b: 47 **A:** CH (CHQ FUJ GUI GUX HEB SHN)

splendens Hope, 1831: 30 (*Eumolpus*) **A:** ID (SD) NP **ORR**

femorata Jacoby, 1908: 333 (*Nodostoma*)

subcostata Jacoby, 1889b: 164 (*Nodostoma*) **A:** CH (YUN) ID (SD UT) NP **ORR**

cyanipennis Lefèvre, 1893: 120 (*Nodostoma*)

obscura Jacoby, 1908: 338 (*Nodostoma*)

sakaii Takizawa, 1987: 50

sublateralis Clavareau, 1914: 57 (*Nodostoma*) [RN] **A:** CH (TAI)

immaculatipennis Chûjô, 1938b: 28 (ab.) [IN]

lateralis S. Matsumura, 1911c: 140 (*Nodostoma*) [HN]

subpunctata Achard, 1914a: 71 (*Nodostoma*) [RN] **A:** CH (GAN)

puncticollis Weise, 1889g: 600 (*Nodostoma*) [HN]

subruficollis Tan, 1993a: 358, 365 **A:** CH (GUI)

subtuberosa Tan, 1988b: 167, 174 **A:** CH (YUN)

tricarinata Tan, 1988b: 167, 174 **A:** CH (YUN)

tristis Medvedev, 2015a: 476 **A:** NP

uenoi Nakane, 1958: 305 (ssp.) **A:** JA

variabilis Duvivier, 1892b: 406 (*Nodostoma*) **A:** ID (SD) NP **ORR**

abbreviata Jacoby, 1900b: 106 (*Nodostoma*)

mandarensis Jacoby, 1900b: 107 (*Nodostoma*)

varians Chûjô, 1956b: 51 **A:** CH (FUJ TAI)

bicolorata Chûjô, 1956b: 49 [HN]

consimilis Chûjô, 1956b: 47

dorunga Aslam, 1967: 128 [RN]

taiwana Chûjô, 1956b: 53

varicolor Jacoby, 1886a: 751 (*Nodostoma*) **A:** (GUA HAI SCH TAI) JA **ORR**

aeneomicans Chûjô, 1956b: 26

coomani Pic, 1931b: 23 (*Nodostoma*)
maebarai Chûjô, 1956c: 1
viridicollis Pic, 1929g: 34 (*Pagria*)
virendri Shukla, 1960: 73 (*Nodostoma*) **A:** ID (HP)
viridipennis Motschulsky, 1860b: 177 (*Nodostoma*) **A:** CH (HAI XIZ YUN) NP **ORR**
aeneipennis Baly, 1867: 235 (*Nodostoma*) [HN]
apicipes Jacoby, 1892a: 87 (*Nodostoma*)
frontalis Baly, 1867: 253 (*Nodostoma*)
haroldi Jacoby, 1908: 331 (*Nodostoma*) [RN]
occipitalis Jacoby, 1908: 335 (*Nodostoma*)
weisei Jacoby, 1900b: 105 (*Nodostoma*) **A:** PA **ORR**
weixiensis Tan, 1988b: 167, 175 **A:** CH (YUN)
yangsoensis S.-H. Chen, 1940c: 500 **A:** CH (FUJ GUX)
yimnae Gressitt & Kimoto, 1961: 231 **A:** CH (GUA)

genus *Bathseba* Motschulsky, 1866b: 412 type species: *Bathseba ferruginosa* Motschulsky, 1866
Tricliona Lefèvre, 1885c: cxlvii type species: *Tricliona fasciata* Lefèvre, 1885
bifasciata Jacoby, 1895b: 284 (*Tricliona*) **A:** BT ID (SD) NP **ORR**
consobrina S.-H. Chen, 1935c: 291 (*Tricliona*) **A:** CH (YUN) **ORR**
microdentata L. N. Medvedev & Sprecher-Uebersax, 1999: 364 (*Tricliona*) **A:** NP
puncticeps Duvivier, 1891a: xli **A:** NP (*Tricliona*) **ORR**
apicata Jacoby, 1895b: 285

genus *Bedelia* Lefèvre, 1875: 10 type species: *Bedelia insignis* Lefèvre, 1875
insignis Lefèvre, 1875: 10 E: RU (ST) **A:** AB AR AF CH (XIN) GG IN KI KZ TD TM TR UZ
angustata Lefèvre, 1875: 10
kokanica Solsky, 1881: 64 (*Nodostoma*)
kaschgarica Lopatin, 1962c: 320 **A:** CH (XIN)
viridicoerulea Reitter, 1901c: 84 **A:** KI TD UZ

genus *Callisina* Baly, 1860b: 30 type species: *Callisina fasciata* Baly, 1860
brunnea Jacoby, 1908: 362 **A:** ID (SD)
quadripustulata Baly, 1864b: 11 **A:** CH (YUN) **ORR**
mouhoti Baly, 1864b: 11
laosensis Pic, 1928g: 28
rufa Tan, 1992b: 761, 802 **A:** CH (HKG MAC YUN) **[# 1145]**
rufipes Pic, 1928g: 28 **A:** CH (YUN)

genus *Chloropterus* Morawitz, 1861: 162 [RN] type species: *Heterocnemis versicolor* Morawitz, 1860

Heterocnemis Morawitz, 1860: 301 [HN]

Heterocnensis Morawitz, 1860: 301 [incorrect original spelling]

bimaculatus Raffray, 1873: 385 (*Nodostoma*) **N:** AG MO **A:** SI

fiorii Ruffo, 1965: 246 **N:** LB

grandis Weise, 1889g: 595 **A:** CH (XIN) KI KZ MG RU (WS) TD TM

inornatus S.-H. Chen, 1935c: 237 (*Aphilenia*) **A:** CH (XIN)

lefevrei arabicus Lopatin, 2008a: 319 (ssp.) **A:** AE OM [# 957]

lefevrei lefevrei Reitter, 1890c: 197 **E:** RU (ST) **A:** AB IN IQ IS KZ SA TD TM UZ

bipunctatus Reitter, 1897b: 225 (var.)

bucharicus Reitter, 1897b: 226 (var.)

nigrofasciatus Reitter, 1897b: 225

mateui Selman, 1969: 201 (*Atomyria*) **N:** AG

moldaviensis Pic, 1909k: 179 **E:** RO (Dobrogea)

obscuricolor Pic, 1910e: 66 (var.)

pallidicolor Pic, 1910e: 66 (var.)

rufithorax Pic, 1911i: 178 (var.)

sublimbatus Pic, 1910e: 66 (var.)

ornatus Lopatin, 1984a: 86 **A:** IN IQ SY PA [# 958]

pallidus Chobaut, 1898b: 308 **N:** AG

persicus Baly, 1878f: 259 (*Bedelia*) **A:** IN

politus Berti & Rapilly, 1973: 877 **A:** AE IN OM

stigmaticollis Fairmaire, 1875b: 537 **N:** TU

normandi Pic, 1901i: 91 (var.)

unguiculatus Lopatin, 1965c: 39 **A:** TD

versicolor Morawitz, 1860: 302 (*Heterocnensis*) **E:** KR RU (ST) UK **A:** AB KZ TM TR

immaculatus Pic, 1909k: 179 (var.)

infuscatus J. F. Sahlberg, 1913c: 246 (var.) [IN] [# 942]

genus *Chrysonopa* Jacoby, 1908: 353 type species: *Chrysonopa viridis* Jacoby, 1908

apicalis Takizawa, 1989a: 329 **A:** ID (SD)

longipes Jacoby, 1894: 275 (*Dermorrhytis*) **A:** NP **ORR**

nigroscutella Tan, 1988a: 310, 324 **A:** CH (XIZ)

rotundicollis Jacoby, 1900b: 104 (*Nodostoma*) **A:** NP **ORR**

tibetana Gressitt & Kimoto, 1961: 215 **A:** CH (XIZ)

genus *Chrysopida* Baly, 1861b: 288 type species: *Chrysopida adonis* Baly, 1861 (= *Colaspis attelaboides* Erichson, 1834)

murina Baly, 1867: 162 **A:** CH (TAI) **ORR**

curta Lefèvre, 1885d: 99
signifera Weise, 1922c: 479 (ab.) [IN]

genus *Cleoporus* Lefèvre, 1884b: 106 type species: *Cleoporus cruciatus* Lefèvre, 1884 (= *Typophorus quadripustulatus* Baly, 1859)
badius Lefèvre, 1889a: 294 **A:** CH (GUA GUX HAI HKG HUN) **ORR** **[# 1145]**
rugiceps S.-H. Chen, 1940c: 503
costatus Chûjô, 1956b: 154 **A:** CH (TAI)
cuprescens Baly, 1878h: 374 (*Paria*) [DA] **A:** PA **ORR** **[# 968]**
lineatus L. N. Medvedev & Sprecher-Uebersax, 1999: 363 **A:** NP
pallidipes Fairmaire, 1888c: 36 (*Stethotes*) **A:** CH (JIX)
robustus Baly, 1874: 166 (*Paria*) **A:** CH (FUJ HUB SHG YUN) JA SC
fulvipes Chûjô, 1938b: 34 (*Tricliona*) [QS]
niger Weise, 1922b: 52 [QS]
pygmaeus Weise, 1922b: 53 [QS]
suturalis S.-H. Chen, 1935c: 287 **A:** CH (GUA HAI) **ORR**
variabilis Baly, 1874: 166 (*Paria*) **A:** CH (BEI CHQ HEI JIL LIA TAI) JA NC RU (FE)

genus *Cleorina* Lefèvre, 1885d: 143 type species: *Cleorina aulica* Lefèvre, 1885 (= *Nodostoma aeneomicans* Baly, 1867)
Colposcelis Dejean, 1836: 408 [HN] type species: *Colaspis viridiaenea* Gyllenhal, 1808 **[# 949]**
assamensis Jacoby, 1908: 486 **A:** ID (SD) **ORR**
bevani Baly, 1877c: 40 (*Nodostoma*) **A:** ID (SD) **ORR**
chlorina Takizawa, 1988a: 11 **A:** ID (SD) NP
costata Tan & S.-Y. Wang, 1981: 434, 452 **A:** CH (XIZ)
fulva Jacoby, 1908: 489 **A:** NP **ORR**
hainana Gressitt & Kimoto, 1961: 203 **A:** CH (HAI) **ORR**
instriata Pic, 1937a: 23 **A:** CH (YUN) **[# 969]**
jacobyi Duvivier, 1892b: 415 **A:** ID (SD) NP **ORR**
janthina Lefèvre, 1885d: 144 **A:** CH (FUJ GUA GUX HAI HKG HUB JIX MAC SCH TAI XIZ YUN) ID (SD) NP **ORR** **[# 1145]**
bella Jacoby, 1892b: 895 (*Nodostoma*)
lefevrei Jacoby, 1890b: 116
purpureicollis Pic, 1937a: 23 (var.)
violaceipennis Pic, 1937a: 23 (var.)
longicornia Tan, 1992b: 755, 800 **A:** CH (YUN)
metallica Shukla, 1960: 68 **A:** ID (HP)
nepalensis Chûjô, 1966b: 4 (*Basilepta*) **A:** ID (SD) NP
nitida Tan, 1981b: 433, 451 **A:** CH (XIZ)

nitidicollis Tan & S.-Y. Wang, 1981: 434, 451 **A:** CH (XIZ)
robusta Takizawa & Basu, 1987: 276 **A:** ID (SD)
splendida Tan, 1992b: 754, 799 **A:** CH (YUN) **[Appendix]**
strigosipleuris L. N. Medvedev, 2015c: 322 **A:** CH (YUN)
subnodosa Gressitt & Kimoto, 1961: 204 **A:** CH (HAI)
takizawai L. N. Medvedev & Sprecher-Uebersax, 1999: 364 [RN] **A:** ID (SD) NP
nepalensis Takizawa, 1985b: 97 [HN]
xizangensis Tan & S.-Y. Wang, 1981: 435, 452 **A:** CH (XIZ)

genus *Microeurydemus* Pic, 1938c: 35 type species: *Microeurydemus unimaculatus* Pic, 1938
africanus Jacoby, 1900c: 228 (*Pseudosyagrus*) **A:** SA **AFR**
flavescens Bryant, 1942b: 514 (*Syagrus*) **A:** OM SA YE **AFR**
semivittatus Jacoby, 1899a: 525 (*Eurydemus*) **A:** AE SA YE **AFR**
wraniki Lopatin, 1994: 526 **A:** YE

genus *Mireditha* Reitter, 1913c: 104 type species: *Nodostoma ovulum* Weise, 1889
ambigua S.-H. Chen & S.-Y. Wang, 1976: 216, 223 **A:** CH (XIZ)
cribrata Tan, 1981a: 52, 53 **A:** CH (QIN)
curtipennis Medvedev, 2011a: 191 **A:** CH (YUN)
flavomaculata Tan, 1992b: 756, 800 **A:** CH (SCH)
intermedia Tan, 1992b: 757, 800 **A:** CH (SCH)
nigra S.-H. Chen, 1940c: 496 **A:** CH (SHX)
ovula Weise, 1889g: 598 (*Nodostoma*) **A:** CH (GAN QIN)
vittata Tan, 1992b: 757, 801 **A:** CH (SCH)

genus *Nodina* Motschulsky, 1858: 108 type species: *Nodina pusilla* Motschulsky, 1858
aeneicollis Jacoby, 1895b: 271 **A:** NP **ORR**
alpicola Weise, 1889g: 600 **A:** CH (SCH)
bicoloripes L. N. Medvedev, 2015c: 320 **A:** CH (YUN)
chalcosoma Baly, 1874: 170 **A:** CH (FUJ GUA GUX HKG TAI ZHE) JA (Ryukyus)
ao Nakane, 1963b: 226 (ab.) [IN]
rufofulva Chûjô, 1951b: 37 (var.)
chinensis Weise, 1922b: 49 **A:** CH (CHQ FUJ GUA GUX HEB HKG JIA JIX SHA ZHE)
coerulea Chen, 1940c: 494 **A:** CH (YUN) **ORR**
cyanea S.-H. Chen, 1940c: 495 **A:** CH (ZHE)
dhadinga L. N. Medvedev, 1992d: 6 **A:** NP
hongshana Gressitt & Kimoto, 1961: 207 **A:** CH (JIX)
issikii Chûjô, 1956b: 12 **A:** CH (TAI) JA (Ryukyus)

kraussi Kimoto & Gressitt, 1966: 502 **A:** JA (Ryukyus)
liui Gressitt & Kimoto, 1961: 208 **A:** CH (YUN)
martensi L. N. Medvedev, 1992d: 6 **A:** NP
mengi L. N. Medvedev, 2015c: 320 **A:** CH (YUN)
meridiosinica Gressitt & Kimoto, 1961: 209 **A:** CH (GUA HAI JIX)
morimotoi Kimoto & Gressitt, 1966: 503 **A:** JA
nepalensis Takizawa, 1987: 52 **A:** ID (SD) NP **ORR**
obliquocostata L. N. Medvedev, 2011c: 249 **A:** NP **[# 940]**
parva Gressitt & Kimoto, 1961: 211 **A:** CH (HAI)
parvula Jacoby, 1892b: 902 **A:** NP **ORR**
pilifrons S.-H. Chen, 1940c: 496 **A:** CH (FUJ GUA GUX HAI JIX YUN)
punctostriolata Fairmaire, 1888c: 37 (*Lamprosoma*) **A:** CH (FUJ GUA GUX HAI HUN JIX YUN ZHE) **ORR**
chinensis Bryant, 1924b: 248 [HN]
fokienica S.-H. Chen, 1935c: 300 [RN]
metallica Bryant, 1937: 98 [RN]
minutula Fairmaire, 1888c: 37 (*Lamprosoma*)
pusilla Motschulsky, 1858: 108 **A:** ID (SD) **ORR**
robusta Jacoby, 1892b: 903 **A:** ID (SD UT) NP **ORR**
crassipes Jacoby, 1908: 293
indica Jacoby, 1895b: 271
major Kimoto & Gressitt, 1982: 21
sauteri Chûjô, 1956b: 16 **A:** CH (TAI)
striopunctata Tan, 1988a: 310, 324 **A:** CH (XIZ)
taliana S.-H. Chen, 1940c: 493 **A:** CH (YUN)
tibialis S.-H. Chen, 1940c: 493 **A:** CH (FUJ GUA GUI GUX HUB JIX SCH YUN ZHE)
tricarinata Gressitt & Kimoto, 1961: 213 **A:** CH (GUA HAI JIX)

genus *Pagria* Lefèvre, 1884a: 92 type species: *Pagria suturalis* Lefèvre, 1884
Aphthonesthis Weise, 1895e: 329 type species: *Aphthonesthis concinna* Weise, 1895
Odontionopa Motschulsky, 1866b: 408 [HN] type species: *Odontionopa aenea* Motschulsky, 1866 (= *Rhynchites restituens* Walker, 1859)
bengalensis Moseyko & L. N. Medvedev, 2005: 121 **A:** ID (SD)
consimilis Baly, 1874: 168 (*Nodostoma*) **A:** JA RU (FE) SC
anceyi Pic, 1929g: 35 (var.)
grata Baly, 1867: 224 (*Nodostoma*) **A:** CH (SCH TAI YUN) ID (SD) JA NC NP SC **ORR** **[# 974]**
aeneicollis Lefèvre, 1890b: 181
bicolor Weise, 1922c: 451 (ab.) [IN]
flavopustulata Baly, 1874: 169 (*Nodostoma*)

fulvipes Baly, 1867: 228 (*Nodostoma*) [HN]
grandis Chûjô, 1938b: 28 (ssp.)
innotata Pic, 1929g: 35 (var.)
kanaraensis Jacoby, 1895b: 279 (*Nodostoma*)
nodieri Pic, 1949c: 55
vignaphila Bryant, 1942a: 32
xanthopus Harold, 1874b: 152 (*Nodostoma*) [RN]
ingibbosa Pic, 1929g: 35 A: CH (GUA JIA SHG) ID (SD) JA NP **ORR**
lineolata Pic, 1949c: 55
signata Motschulsky, 1858: 110 (*Metachroma*) A: CH (YUN) **ORR** [# 974]
bipunctata Lefèvre, 1891: cclxvi
diversepunctata Pic, 1950c: 3
rufithorax Pic, 1929g: 35 (var.)
ussuriensis Moseyko & L. N. Medvedev, 2005: 120 A: CH (HEI JIL LIA) JA NC RU (FE) SC [# 962, # 974]
ussuriensis Roubal, 1929a: 98 (ab.) [IN]

genus *Parascela* Baly, 1878f: 252 type species: *Pseudocolaspis cribrata* Schaufuss, 1871
Pseudoparascela Pic, 1935b: 1 type species: *Pseudoparascela jeanvoinei* Pic, 1935 (= *Pseudocolaspis cribrata* Schaufuss, 1871)
cribrata Schaufuss, 1871a: 200 (*Pseudocolaspis*) A: CH (FUJ GUA HKG JIA JIX SCH TAI YUN ZHE) JA **ORR** [# 1145]
hirta S.-H. Chen, 1935d: 769 (*Basilepta*)
jeanvoinei Pic, 1935b: 1 (*Pseudoparascela*)
filimonovi Romantsov & Moseyko, 2019: 329 A: CH (YUN) **ORR**
hirsuta Jacoby, 1908: 343 (*Nodostoma*) A: CH (XIZ YUN) **ORR**
castanea Tan, 1983b: 392, 934
rugipennis Tan, 1988b: 166, 173 (*Basilepta*) A: CH (YUN)
tuberosa Tan & S.-Y. Wang, 1983: 393, 394 [HN]

genus *Pseudostonopa* Jacoby, 1903a: 98 type sepcies *Pseudostonopa bicolor* Jacoby, 1903
bicolor Jacoby, 1903a: 99 A: NP **ORR**
castanea Jacoby, 1908: 352 A: ID (SD)
fulviventris Kimoto, 2004: 49 A: NP
picea Jacoby, 1908: 352 A: ID (SD) NP **ORR**

genus *Rhyparida* Baly, 1861b: 286 type species: *Rhyparida dimidiata* Baly, 1861
Marsaeus Clark, 1864: 252 type species: *Cryptocephalus didimus* Fabricius, 1775
bilineata L. N. Medvedev, 2010b: 291 A: ID (SD)

bimaculata Jacoby, 1889b: 174 **A:** CH (YUN) NP **ORR** [# 964]

immaculata Jacoby, 1908: 379

dentipes S.-H. Chen, 1935c: 311 (*Basilepta*) **A:** CH (ZHE)

formosana Aslam, 1967: 129 [RN] **A:** CH (TAI)

basileptoides Chûjô, 1956b: 69 [HN]

himalayana L. N. Medvedev, 2010b: 291 **A:** ID (SD)

kotoensis Chûjô, 1956b: 67 **A:** CH (TAI)

lineolata Weise, 1913c: 220 (*Phytorus*) **A:** JA (Ogasawara) **ORR**

phytorella Moseyko & L. N. Medvedev, 2005: 123 [RN]

simplex Lefèvre, 1885d: 135 (*Phytorus*) [HN]

sakisimensis Yuasa, 1930: 293 **A:** CH (TAI) JA (Ryukyus)

thailandica L. N. Medvedev, 2001: 174 **A:** CH (YUN) **ORR**

subfamily Eumolpinae, tribe insertae sedis

genus *Chalcolema* Jacoby, 1890a: 84 type species: *Chalcolema foveicollis* Jacoby, 1890

foveicollis Jacoby, 1890a: 85 **A:** CH (HUB)

subfamily Eumolpinae, excluded taxa

genus *Graphops* LeConte, 1884: 26 type species: *Heteraspis nebulosa* LeConte, 1859

Phortus Weise, 1899a: 137 type species: *Phortus creticus* Weise, 1899

curtipennis Melsheimer, 1847: 169 (*Eumolpus*) **NAR** [# 953]

creticus Weise, 1899a: 138 (*Phortus*)

genus *Weiselina* Reitter, 1913c: 104 type species: *Nodostoma lenkorana* Reitter, 1890

lenkorana Reitter, 1890a: 164 **AFR** [# 954]

subfamily **Synetinae** LeConte & Horn, 1883

genus *Syneta* Dejean, 1835: 359 [= 1836: 361] type species: *Crioceris betulae* Fabricius, 1792

adamsi Baly, 1877f: 378 **A:** CH (BEI HEB HEI JIL LIA NMO SHX) JA NC RU (FE) SC

major Nakane, 1963a: 19

betulae amurensis Pic, 1901d: 19 (var.) **A:** MG RU (ES FE)

ussurica Brancsik, 1914: 61

betulae betulae Fabricius, 1792b: 5 (*Crioceris*) [NP] **E:** BY EN FI KZ LA NR RU (CT NT) SV **A:** KZ RU (SS WS)

betulina Thunberg, 1787: 47 (*Crioceris*) [NO]

brevitibialis Kimoto, 1971a: 2 **A:** JA

family Chrysomelidae, nomen dubium

Eumolpus badgeri Stephens, 1832: 408 **E:** GB

Appendix

In the final stage of proofreading, we received additional information, which we consider important to be included in the Catalogue. However, it was no longer possible to incorporate it to the respective chapters in the New Acts hence we present it in a form of an appendix

Lema cyanella kuwayamae Monrós, 1960. The name proposed by Monrós (1960) as a new substitute name for *L. cyanella orientalis* Kuwyama, which is a nomen nudum that appeared in Medvedev (1958) and hence is unavailable as already stated by White (1981).

Craspeda pseudolurida pappi Bourdonné, 2008 is unavailable name because it was proposed as a new substitute name for infrasubspecific taxon *Chrysomela lurida* ab. *lineata* Papp, 1946.

Malacoptera Hope, 1841 (type species: *Chrysomela lusitanica* Linnaeus, 1767) is an available name which threatens the well-established name *Exosoma* Jacoby, 1903. Solving this problem requires more time and will therefore be published separately after the publication of this Catalogue.

Cleorina splendida Tan, 1992 from Yunnan is a junior primary homonym of *C. splendida* Bryant, 1950 from Papua New Guinea. However, *C. splendida* Tan is probably a junior synonym of *C. janthina* Lefèvre, 1885, therefore, the new name is not proposed until a revision of *Cleorina* species will be done.

Type species. At the final stage of proof reading of the Catalogue, Bouchard et al. (2024) published the paper revising family-group names in Coleoptera including their type genera and type species, but not all respective implications could be included. Some of the overlooked older fixations pose a threat to nomenclatural stability and require an application to the Commission to conserve the present usage. This concerns Palaearctic genera *Amblycerus* Thunberg, 1815, *Chaetocnema* Stephens, 1831, *Phratora* Chevrolat, 1836, and *Stylosomus* Suffrian, 1847. Further nomenclatural issues were found for *Acanthoscelides* Schilsky, 1905, *Caryopemon* Jekel, 1855, and *Corynodes* Hope, 1841. Finally, older type species fixations not posing a threat to nomenclatural stability were discovered for *Euryope* Dalman, 1824 and *Psylliodes* Latreille, 1829.

Aplosonyx scutellatus Baly, 1879. INDIA: Uttarakhand: Ranikhet, Kumaon, viii.1920, 2 spec., H. G. Champion leg., J. Bezděk det. (NHMUK). New species for Uttarakhand (India).

Cheilotoma voriseki L. N. Medvedev & Kantner, 2003. IRAQ: NE Duhok, Sarsing, 37°01'24"N 43°13'19"E, 1000–1200 m, 21.v.2019, 1 ♂, Reuter leg., J. Bezděk det. (collection of M. Schöller, Berlin, Germany). New species for Iraq.

Chrysolina (*Chalcoidea*) *sanguineocincta* (Crotch, 1872). SYRIA: steppe between Homs and Palmyra 500–600 m, 13.iv.1978, 1 ♂ 1 ♀, Heinz leg., H. Kippenberg det. (collection of H. Kippenberg, Herzogenaurach, Germany). New species for Syria.

Coptocephala crassipes iranica L. N. Medvedev, 1971. OMAN: Dhofar, 12 km N of Rakhyut, dam, 900 m, 18.ix.2023, 12 ♂♂ 16 ♀♀, M. Halada leg., J. Bezděk det. (collection of J. Bezděk, Brno, Czech Republic). Confirmed occurrence in Oman.

Epitrix dieckmanni Mohr, 1968. OMAN: Al-Batina Gov., Wadi Rajma, near Rajmi, 24°40'36"N 56°08'47"E, 14.iii.2023, 1 ♂ 2 spec., O. Konvička leg., J. Bezděk det. (collection of O. Konvička, Zlín, Czech Republic); Al-Buraimi Gov., Mahzá, 28 km NE of Ash Shiyah, 24°39'18"N 56°03'52"E, 13.iii.2023, 4 spec., O. Konvička leg., J. Bezděk det. (collection of O. Konvička, Zlín, Czech Republic). New species for Oman.

Leptomona erythrocephala (Olivier, 1791). TUNISIA: Tabarka, 36°57'16"N 8°45'29"E, 7.vii.2008, 1 spec., M. Mantič leg., J. Bezděk det. (collection of M. Mantič, Hlučín, Czech Republic); Ain Draham, 36°46'38"N 8°41'33"E, 10.vii.2008, 1 spec., M. Mantič leg., J. Bezděk det. (collection of M. Mantič, Hlučín, Czech Republic). New species for Tunisia.

Mimastra arcuata Baly, 1865. INDIA: Arunachal Pradesh: Etalin vicinity, 28°36′56″N 95°53′21″E, 700 m, 12.–25.v.2012, 1 ♀, O. Šauša leg., J. Bezděk det. (NHMUK). New species for India: Arunachal Pradesh.

Oreina (Allorina) caerulea (Olivier, 1791). NORTH MACEDONIA: Strezimir, Korab Mts., Golem Korab Mt., 2100 m, 27.vi.2023, 3 ♂♂, Z. Martinová leg., J. Bezděk det. (collection of J. Háva, Únětice u Prahy, Czech Republic). New species for North Macedonia.

Pachnephorus bistriatus Mulsant & Wachanru, 1852. TUNISIA: Tabarka, 36°57'16"N 8°45'29"E, 7.vii.2008, 15 spec., M. Mantič leg., J. Bezděk det. (collection of M. Mantič, Hlučín, Czech Republic); Tabarka, 26.–28.iv.1997, 1 spec., T. Kopecký leg., J. Bezděk det. (collection of J. Bezděk, Brno, Czech Republic). New species for Tunisia.

Pachnephorus cylindricus P. H. Lucas, 1846. PORTUGAL: Faro Reg., Castro Marim, 37°13'06"N 7°26'39"E, 24.iii.2009, 28 spec., M. Mantič leg., J. Bezděk det. (collection of M. Mantič, Hlučín, Czech Republic). TUNISIA: Tabarka, 36°57'16"N 8°45'29"E, 7.vii.2008, 9 spec., M. Mantič leg., J. Bezděk det. (collection of M. Mantič, Hlučín, Czech Republic); Ain Draham, 23.–26.iv.1997, 4 spec., J. Mertlik leg., J. Bezděk det. (collection of J. Bezděk, Brno, Czech Republic). New species for Portugal and Tunisia.

Phyllotreta peyerimhoffi Heikertinger, 1941. OMAN: Al-Buraimi Gov., Mahzá, 28 km NE of Ash Shiyah, 24°39'18"N 56°03'52"E, 13.iii.2023, 6 spec., O. Konvička leg., J. Bezděk det. (collections of O. Konvička, Zlín, Czech Republic, and J. Bezděk, Brno, Czech Republic); Al-Buraimi Gov., Mahzá, 25 km NE of Ash Shuwayhah, 24°37'25"N 56°04'14"E, 14.iii.2023, 2 spec., O. Konvička leg., J. Bezděk det. (collection of O. Konvička, Zlín, Czech Republic). New species for Oman.

Tituboea harteni Bezděk, 2018. OMAN: Al-Buraimi Gov., Mahzá, 28 km NE of Ash Shiyah, 24°39'18"N 56°03'52"E, 13.iii.2023, 4 ♂♂ 6 ♀♀, O. Konvička leg., J. Bezděk det. (collections of O. Konvička, Zlín, Czech Republic, and J. Bezděk, Brno, Czech Republic). New species for Oman.

Index to Family-group Names

Acanthoscelidina Bridwell, 1946 180
Acolastina Schöller, **subtrib. nov.** 582
Adoxini Baly, 1863 688
Alticitae Newman, 1834 468
Amblycerina Bridwell, 1932 176
Amblycerini Bridwell, 1932 176
Anisoderini Chapuis, 1875 233
Aspidomorphina Chapuis, 1875 239
Aulacophorina Chapuis, 1875 413

Basiprionotini Gressitt, 1952 (1914) 234
Bothryonopini Chapuis, 1875 236
Bruchina Latreille, 1802 198
Bruchinae Latreille, 1802 176
Bruchini Latreille, 1802 180

Callispini Chapuis, 1875 237
Caryedontina Bridwell, 1929 203
Caryopemina Bridwell, 1929 205
Cassidina Gyllenhal, 1813 241
Cassidinae Gyllenhal, 1813 233
Cassidini Gyllenhal, 1813 239
CHRYSOMELIDAE Latreille, 1802 175
Chrysomelina Latreille, 1802 273
Chrysomelinae Latreille, 1802 273
Chrysomelini Latreille, 1802 273
Clytrina Kirby, 1837 585
Clytrini Kirby, 1837 582
Coelaenomenoderini Weise, 1911 257
Criocerinae Latreille, 1804 216
Criocerini Latreille, 1804 216
Cryptocephalina Gyllenhal, 1813 619
Cryptocephalinae Gyllenhal, 1813 582
Cryptocephalini Gyllenhal, 1813 619
Cryptonychini Chapuis, 1875 258

Decarthrocerini Laboisièrre, 1937 378
Diabroticina Chapuis, 1875 421
Donaciinae Kirby, 1837 206
Donaciini Kirby, 1837 206
Doryphorina Motschulsky, 1860 288

Entomoscelina Reitter, 1913 343
Eumolpinae Hope, 1841 688
Eumolpini Hope, 1841 706
Euryopini Lefèvre, 1885 713

Floricolini Moseyko, **trib. nov.** 716
Fulcidacini Jacobson, 1924 685

Galerucinae Latreille, 1802 378
Galerucini Latreille, 1802 378
Galerucitae Latreille, 1802 378
Gonioctenina Motschulsky, 1860 350
Gonophorini Chapuis, 1875 258

Haemoniini S.-H. Chen, 1941 212
Hispini Gyllenhal, 1813 261
Hylaspini Chapuis, 1875 398

Kytorhinini Bridwell, 1932 202

Lamprosomatinae Lacordaire, 1848 581
Lamprosomatini Lacordaire, 1848 581
Lemini Gyllenhal, 1813 225
Leptispini Fairmaire, 1868 270
Luperina Gistel, 1848 421
Luperini Gistel, 1848 413

Megacerina Bridwell, 1946 202
MEGALOPODIDAE Latreille, 1802 170
Megalopodinae Latreille, 1802 173
Megalopodini Latreille, 1802 173
Metacyclini Chapuis, 1875 460
Monachulina Leng, 1920 671

Notosacanthini Gressitt, 1952 (1929) 271

Oidini Laboissière, 1921 (1875) 460
Oncocephalini Chapuis, 1875 272
ORSODACNIDAE C. G. Thomson, 1859 169
Orsodacninae C. G. Thomson, 1859 169

Pachybrachini Chapuis, 1874 677
Pachymerina Bridwell, 1929 205
Pachymerini Bridwell, 1929 203
Plateumarini Böving, 1922 213

Rhaebini Blanchard, 1845 205

Sagrinae Leach, 1815 175
Sclerophaedonina Steinhausen, 1996 367
Serraticollitae B. White, 1942 580
Spermophagina Borowiec, 1987 176
Stylosomina Chapuis, 1874 675
Synetinae LeConte & Horn, 1883 728

Temnaspidina Rodríguez-Mirón, 2021 173
Timarchini Motschulsky, 1860 369
Typophorini Baly, 1865 717

Zeugophorinae Böving & Craighead, 1931 170

Index to Genus-group Names

Following abbreviations are used: g. – genus; sbg. – subgenus; syn. – synonym; un. – unavailable name; i. o. s. – incorrect original spelling; i. s. s. – incorrect subsequent spelling; u. o. s. – unused original spelling.

Abdullahius Abdullah & Qureshi, 1968 (g.) 421
Abirellus Chûjô, 1956 (syn.) 710
Abiromorphus Pic, 1924 (g.) 706
Abirus Chapuis, 1874 (g.) 706
Acacicola Lea, 1903 (syn.) 366
Acanthobruchidius Borowiec, 1980 (g.) 180
Acanthobruchus Yus Ramos, 1978 (syn.) 198
Acanthodon Weise, 1880 (syn.) 362
Acanthoscelides Schilsky, 1905 (g.) 180
Acis Chevrolat, 1836 (syn.) 713
Acmenychus Weise, 1905 (g.) 261
Acolastus Gerstaecker, 1855 (g.) 582
Acolastus Gerstaecker, 1855 (sbg.) 582
Acrocrypta Baly, 1862 (g.) 468
Acrothinium Marshall, 1864 (g.) 688
Acroxena Baly, 1879 (syn.) 458
Acutipalpa Rosenhauer, 1856 (syn.) 414
Adimonia Laicharting, 1781 (syn.) 383
Adimus Billberg, 1820 (syn.) 470
Adiscus Gistel, 1857 (g.) 671
Adorium Fabricius, 1801 (syn.) 460
Adoxinia Reitter, 1889 (syn.) 697
Adoxus Kirby, 1837 (syn.) 691
Aegelocerus Hope, 1831 (un.) 413
Aemulaphthona Scherer, 1969 (syn.) 446
Aenidea Baly, 1874 (syn.) 453
Aeropa Weise, 1889 (syn.) 420
Aeschrocnemis Weise, 1888 (g.) 468
Aeschrocnemis Weise, 1888 (sbg.) 468
Aetheanta L. N. Medvedev, 1988 (g.) 585
Aetheodactyla Lacordaire, 1848 (g.) 585
Aetheomorpha Lacordaire, 1848 (g.) 585
Afroblepharida Biondi & D'Alessandro, 2017 (sbg.) 492
Afromaculepta Hasenkamp & Wagner, 2000 (g.) 421
Afrophthalma L. N. Medvedev, 1978 (g.) 586
Agasicles Jacoby, 1905 (g.) 469
Agasta Hope, 1841 (g.) 273
Agelasa Motschulsky, 1861 (g.) 398
Agelastica Chevrolat, 1836 (g.) 398
Agelopsis Jacoby, 1896 (g.) 422
Agetocera Hope, 1841 (g.) 413
Agonia Weise, 1904 (syn.) 258
Agoniella Weise, 1911 (g.) 258
Agonita E. Strand, 1942 (g.) 258
Agrosteella L. N. Medvedev, 1987 (g.) 288
Agrosteomela Gistel, 1857 (g.) 288
Agrostithassa Jacobson, 1921 (syn.) 286
Aldrisma Fairmaire, 1888 (syn.) 551
Algarobius Bridwell, 1946 (g.) 180
Alitene Gistel, 1856 (syn.) 278
Allecumolpus L. N. Medvedev, 1956 (syn.) 694
Alledoya Hincks, 1950 (syn.) 241
Allochrysolina Bechyné, 1950 (sbg.) 290
Allohypericia Bechyné, 1950 (sbg.) 290
Allomorpha Jacoby, 1892 (syn.) 462
Allophyla Weise, 1889 (syn.) 573
Allorina Weise, 1902 (sbg.) 331
Allorinula Weise, 1916 (syn.) 331
Alonsina Warchałowski, 1995 (syn.) 286
Aloria Anonymous, 1940 (g.) 688
Aloria Bryant, 1939 (un.) 688
Alpaeixena Motschulsky, 1860 (syn.) 332
Altailina Mikhailov, 2000 (syn.) 317
Altica Geoffroy, 1762 (g.) 469
Amasia Chapuis, 1874 (syn.) 708
Amblis Gistel, 1848 (syn.) 363
Amblispa Baly, 1859 (g.) 237
Amblycerus Thunberg, 1815 (g.) 176
Ambrostoma Motschulsky, 1845 (g.) 288
Ambrostoma Motschulsky, 1845 (sbg.) 288
Amphimela Chapuis, 1875 (g.) 475
Amphimeloides Jacoby, 1887 (syn.) 515
Amydus S.-H. Chen, 1935 (syn.) 577
Anadimonia Ogloblin, 1936 (g.) 378
Anastena Maulik, 1936 (syn.) 400
Ancyloscelis Ogloblin, 1930 (syn.) 515
Andosia Weise, 1896 (g.) 689
Andosiomorpha Lopatin, 1981 (syn.) 715

Angulaphthona Bechyné, 1960 (g.) 475
Anicera Jacoby, 1884 (syn.) 400
Anidania Reitter, 1889 (g.) 689
Anisodera Baly, 1859 (syn.) 233
Anisodera Chevrolat, 1836 (g.) 233
Anisoderopsis Maulik, 1916 (syn.) 234
Anodontelytrus Jacobson, 1917 (sbg.) 584
Anopachys Motschulsky, 1860 (sbg.) 291
Anopsilus Jacobson, 1917 (syn.) 584
Anteriscus Weise, 1906 (syn.) 629
Anthobiodes Weise, 1887 (g.) 475
Anthraxantha Fairmaire, 1878 (syn.) 445
Anthurga Gistel, 1848 (syn.) 216
Antipha Baly, 1865 (syn.) 402
Anwarullahia Abdullah & Qureshi, 1969 (syn.) 713
Aoria Baly, 1863 (g.) 689
Aoria Baly, 1863 (sbg.) 689
Apaksha Maulik, 1926 (syn.) 347
Apelma Billberg, 1820 (syn.) 212
Aphilenia Weise, 1889 (g.) 717
Aphilenia Weise, 1889 (sbg.) 717
Aphobera Warchałowski, 1985 (sbg.) 598
Aphthona Chevrolat, 1836 (g.) 476
Aphthonaltica Heikertinger, 1924 (g.) 486
Aphthonaria L. N. Medvedev, 1990 (syn.) 476
Aphthonella Jacoby, 1889 (g.) 486
Aphthonesthis Weise, 1895 (syn.) 726
Aphthonoides Jacoby, 1885 (g.) 486
Aphthonomorpha S.-H. Chen, 1934 (syn.) 492
Aphthonotarsa L. N. Medvedev, 1984 (syn.) 476
Aplosonyx Chevrolat, 1836 (g.) 399
Apophylana L. N. Medvedev, 2019 (syn.) 378
Apophylia J. Thomson, 1858 (g.) 378
Apterocuris Jacobson, 1901 (g.) 367
Apterogaleruca Chûjô, 1962 (g.) 380
Apteromicrus S.-H. Chen & Jiang, 1981 (g.) 422
Apteropeda Chevrolat, 1836 (g.) 486
Apterosoma Motschulsky, 1861 (sbg.) 292
Arachnipta L. N. Medvedev, 1957 (syn.) 691
Arachnosphoerus J. Thomson, 1856 (syn.) 716
Arctolina Kontkanen, 1959 (sbg.) 292
Argopistes Motschulsky, 1860 (g.) 487
Argopistoides Jacoby, 1892 (g.) 487
Argopus Fischer von Waldheim, 1824 (g.) 488
Argosomus Wollaston, 1868 (syn.) 574
Arima Chapuis, 1875 (g.) 380
Arrhenocoela Foudras, 1860 (g.) 489
Arthrochlamys Ihering, 1904 (syn.) 685
Arthrotidea S.-H. Chen, 1942 (g.) 400
Arthrotus Motschulsky, 1857 (g.) 400
Arundonacia Kippenberg, 2015 (syn.) 206
Aryana Berti & Rapilly, 1973 (syn.) 663
Asamangulia Maulik, 1915 (g.) 261
Asbecesta Harold, 1877 (g.) 414
Ascuta L. N. Medvedev, 1997 (syn.) 546
Ascutella L. N. Medvedev, 2009 (syn.) 546
Ashrafia Abdullah & Qureshi, 1968 (g.) 422
Asialtica Scherer, 1969 (syn.) 476
Asionus Lopatin, 1988 (syn.) 663
Asiophrida L. N. Medvedev, 2000 (g.) 489
Asiophrida L. N. Medvedev, 2000 (sbg.) 489
Asiopus Lopatin, 1965 (syn.) 663
Asiorella L. N. Medvedev, 1990 (syn.) 518
Asiorestia Jacobson, 1926 (syn.) 542
Asiparopsis S.-H. Chen, 1934 (sbg.) 366
Asiphytodecta S.-H. Chen, 1935 (sbg.) 350
Asisia Bezděk, Löbl & Konstantinov, 2010 (g.) 581
Askevoldia Kippenberg, 1994 (syn.) 206
Asparagocrioceris Bogachev & Antsupova, 1980 (un.) 216
Aspidimorpha Hope, 1841 (i. o. s.) 239
Aspidolopha Lacordaire, 1848 (g.) 587
Aspidomorpha Hope, 1841 (g.) 239
Aspidomorpha Hope, 1841 (sbg.) 239
Asutoshana L. N. Medvedev, 2011 (g.) 489
Atactolema Heinze, 1927 (syn.) 225
Atlasiana Bourdonné & Doguet, 1991 (sbg.) 293
Atomyria Jacobson, 1894 (g.) 717
Atrachya Chevrolat, 1836 (g.) 422
Atropidius Chapuis, 1874 (syn.) 671
Atymius Gistel, 1848 (syn.) 707
Atysa Baly, 1864 (g.) 381
Auchenia Thunberg, 1792 (syn.) 171
Aulacophora Chevrolat, 1836 (g.) 414
Aulacothorax Boheman, 1858 (g.) 580
Aulexis Baly, 1863 (g.) 690
Austrolema Oke, 1932 (syn.) 171

Austrosternoplatys Jacobson, 1921 (sbg.) 287
Autolampra Jacoby, 1908 (syn.) 711
Avinasa Maulik, 1936 (syn.) 411
Azerberyxia Romantsov & Moseyko, 2020 (sbg.) 694

Bacteriaspis Weise, 1905 (g.) 417
Balanomorpha Chevrolat, 1836 (syn.) 539
Ballatro Gistel, 1848 (syn.) 713
Baoshanaltica Konstantinov & Ruan, 2017 (g.) 489
Barathraea Lacordaire, 1848 (sbg.) 604
Barathroea Lacordaire, 1848 (u. o. s.) 604
Basilepta Baly, 1860 (g.) 717
Basiprionota Chevrolat, 1836 (g.) 234
Bathseba Motschulsky, 1866 (g.) 722
Bathycolpus Marshall, 1864 (syn.) 711
Batophila Foudras, 1860 (g.) 489
Bechynea L. N. Medvedev, 1966 (syn.) 298
Bechyneia Jolivet, 1950 (syn.) 716
Bechynia Bourdonné, 1977 (sbg.) 294
Bedelia Lefèvre, 1875 (g.) 722
Beenenia Bezděk, 2012 (g.) 381
Belarima Reitter, 1913 (g.) 381
Benedictus Scherer, 1969 (g.) 490
Bequaertinia Laboissière, 1922 (syn.) 378
Berecyntha Baly, 1865 (syn.) 399
Bertiellus Lopatin, 1977 (sbg.) 619
Betacassida Steinhausen, 2002 (syn.) 241
Bezdekaltica Döberl, 2012 (g.) 491
Bezdekia Warchałowski, 2005 (syn.) 715
Bhamoina Bechyné, 1958 (g.) 491
Bhutajana Scherer, 1979 (syn.) 476
Bienkowskia L. N. Medvedev, 2009 (syn.) 310
Bijukta Maulik, 1936 (syn.) 454
Bikasha Maulik, 1931 (g.) 492
Bimala Maulik, 1926 (g.) 492
Biondiana Özdikmen, 2021 (syn.) 493
Bittotaenia Motschulsky, 1860 (sbg.) 294
Blepharidella Weise, 1909 (syn.) 492
Blepharidina Bechyné, 1969 (g.) 492
Blepharidula Weise, 1916 (syn.) 492
Blepharoides Jacoby, 1893 (syn.) 548
Boisduvalia Montrouzier, 1856 (syn.) 460
Boliographa Motschulsky, 1860 (syn.) 289
Boloschesis Jacobson, 1924 (syn.) 685
Borowiecius Anton, 1994 (g.) 181
Botanoctona Fairmaire, 1877 (syn.) 460
Bothryonopa Guérin-Méneville, 1840 (g.) 236
Bourdonneana Kippenberg, 2010 (syn.) 290
Brachita Allard, 1889 (syn.) 445
Brachyphora Jacoby, 1890 (g.) 423
Brachyphytodecta Bechyné, 1948 (sbg.) 351
Bradyceris Chûjô, 1951 (sbg.) 220
Bradylema Weise, 1901 (syn.) 229
Bradylemoides Heinze, 1930 (syn.) 229
Brevidonacia Kippenberg, 2015 (syn.) 206
Brinckaltica Bechyné, 1959 (syn.) 493
Bromiodes Jacoby, 1895 (syn.) 700
Bromius Baly, 1865 (syn.) 703
Bromius Chevrolat, 1836 (g.) 691
Bronthispa Weise, 1911 (syn.) 258
Brontispa Sharp, 1904 (g.) 258
Bruchidius Schilsky, 1905 (g.) 181
Bruchoides Yus Ramos, 1978 (syn.) 198
Bruchomima Achard, 1916 (syn.) 171
Bruchus Linnaeus, 1767 (g.) 198
Bucedestes Gistel, 1856 (syn.) 604
Burlinius Lopatin, 1965 (sbg.) 619
Burmaltica Scherer, 1969 (syn.) 486
Burumoseria Csiki, 1939 (g.) 492
Butiobruchus Prevett, 1966 (sbg.) 205
Byrrhiformis J. R. Sahlberg, 1913 (syn.) 312

Calamobia Gistel, 1856 (syn.) 212
Calligrapha Chevrolat, 1836 (g.) 289
Calligrapha Chevrolat, 1836 (sbg.) 289
Calliope Weise, 1882 (syn.) 691
Callipepla Dejean, 1836 (syn.) 460
Callipta Lefèvre, 1885 (g.) 691
Callisina Baly, 1860 (g.) 722
Callispa Baly, 1859 (g.) 237
Callispa Baly, 1859 (sbg.) 237
Callopistria Chevrolat, 1836 (syn.) 381
Callosobruchus Pic, 1902 (g.) 192
Calomicrus Dillwyn, 1829 (g.) 423
Calopepla Hope, 1841 (syn.) 235
Calotheca Heyden, 1887 (g.) 492
Calyptorhina Lacordaire, 1848 (syn.) 609
Cangshanaltica Konstantinov, Chamorro, Prathapan, Ge & Yang, 2013 (g.) 493
Capula Jacobson, 1925 (g.) 401
Carapaula Chûjô, 1962 (syn.) 418
Carcharodis Weise, 1910 (syn.) 493

Cardax Weise, 1893 (sbg.) 541
Cardiapus Curtis, 1833 (syn.) 539
Caritheca Baly, 1877 (syn.) 399
Carmentis Gistel, 1848 (syn.) 609
Caryedes Hummel, 1827 (g.) 193
Caryedon Schoenherr, 1823 (g.) 203
Caryobruchus Bridwell, 1929 (g.) 205
Caryopemon Jekel, 1855 (g.) 205
Caspiana Lopatin, 1978 (g.) 692
Cassena Weise, 1892 (g.) 425
Cassena Weise, 1892 (sbg.) 425
Cassida Linnaeus, 1758 (g.) 241
Cassidispa Gestro, 1899 (g.) 261
Cassidula Weise, 1889 (syn.) 241
Cassidulella E. Strand, 1928 (syn.) 241
Caudatochrysa Bechyné, 1950 (syn.) 292
Cecchiniola Jacobson, 1908 (sbg.) 294
Centoptera Motschulsky, 1860 (sbg.) 294
Centroscelis Chevrolat, 1836 (g.) 350
Centroscelis Chevrolat, 1836 (sbg.) 350
Cerania Weise, 1892 (syn.) 414
Ceratia Chapuis, 1877 (syn.) 414
Ceratobasis Lacordaire, 1848 (g.) 587
Cercidiestes Bridwell, 1946 (syn.) 195
Cercyonia Weise, 1902 (syn.) 475
Cercyonops Jacobson, 1901 (sbg.) 352
Cerodens Burlini, 1969 (syn.) 629
Ceropachys Burlini, 1953 (syn.) 629
Cerophysa Chevrolat, 1836 (g.) 426
Cerophysella Laboissière, 1930 (g.) 427
Cerotrus Jacoby, 1884 (syn.) 400
Chabriella L. N. Medvedev, 1990 (syn.) 549
Chaeridiona Baly, 1869 (g.) 272
Chaeroceta Mroczkowski, 1990 (syn.) 281
Chaetocera Weise, 1884 (syn.) 281
Chaetoceroides E. Strand, 1935 (sbg.) 281
Chaetocnema Stephens, 1831 (g.) 493
Chalaenaria L. N. Medvedev, 2003 (g.) 501
Chalcoidea Motschulsky, 1860 (sbg.) 295
Chalcoides Foudras, 1860 (syn.) 502
Chalcolema Jacoby, 1890 (g.) 728
Chapalia Laboissière, 1929 (syn.) 393
Charaea Baly, 1878 (g.) 427
Charidea Baly, 1888 (syn.) 573
Chartonia Gistel, 1856 (syn.) 691
Cheilotoma Chevrolat, 1836 (g.) 588
Cheilotoma Chevrolat, 1836 (sbg.) 588
Cheiriphyle Jacoby, 1901 (syn.) 713
Chikatunolepta Bezděk, 2020 (g.) 428
Chilocoristes Weise, 1895 (g.) 501
Chilotoma Agassiz, 1846 (syn.) 588
Chilotomina Reitter, 1913 (g.) 589
Chimporia Laboissière, 1931 (syn.) 446
Chinochya C.-F. Lee, 2020 (g.) 429
Chiridopsis Spaeth, 1922 (g.) 253
Chiridula Weise, 1889 (g.) 254
Chiriphyle Clavareau, 1914 (i. s. s.) 713
Chlamisus Rafinesque, 1815 (g.) 685
Chlamys Knoch, 1801 (syn.) 685
Chloemela Gistel, 1856 (syn.) 299
Chloephaga Weise, 1899 (syn.) 547
Chloocassis Spaeth, 1952 (syn.) 257
Chloropachys Rey, 1883 (sbg.) 677
Chloropterus Morawitz, 1861 (g.) 723
Chlorostola Weise, 1881 (sbg.) 598
Choeridiona Gemminger & Harold, 1876 (syn.) 272
Chogania Laboissière, 1930 (syn.) 427
Chondromela Weise, 1915 (syn.) 367
Chrysispa Weise, 1897 (g.) 261
Chrysochares Morawitz, 1861 (g.) 706
Chrysochloa Hope, 1841 (sbg.) 332
Chrysochus Chevrolat, 1836 (g.) 707
Chrysocrosita Bechyné, 1950 (sbg.) 298
Chrysocryptocephalus Steinhausen, 2007 (syn.) 629
Chrysolampra Baly, 1859 (g.) 707
Chrysolina Motschulsky, 1860 (g.) 289
Chrysolina Motschulsky, 1860 (sbg.) 299
Chrysolinopsis Bechyné, 1950 (sbg.) 300
Chrysomela Linnaeus, 1758 (g.) 273
Chrysomorpha Motschulsky, 1860 (sbg.) 300
Chrysonopa Jacoby, 1908 (g.) 723
Chrysonotum J. R. Sahlberg, 1913 (syn.) 319
Chrysophtharta Weise, 1901 (syn.) 366
Chrysopida Baly, 1861 (g.) 723
Chujoa Gressitt & Kimoto, 1963 (g.) 381
Chujoita Monrós, 1960 (sbg.) 220
Cirrispa Uhmann, 1936 (syn.) 266
Clavicornaltica Scherer, 1974 (g.) 501
Cleonica Jacoby, 1887 (g.) 502
Cleoporus Lefèvre, 1884 (g.) 724
Cleorina Lefèvre, 1885 (g.) 724
Clerotilia Jacoby, 1886 (g.) 429
Clisitherella S.-H. Chen, 1940 (sbg.) 710

Clitea Baly, 1877 (syn.) 475
Clitena Baly, 1864 (g.) 381
Clitenella Laboissière, 1927 (g.) 381
Clythra Fabricius, 1798 (i. s. s.) 589
Clythraria Rafinesque, 1815 (syn.) 589
Clythraxeloma Kraatz, 1879 (g.) 173
Clytra Laicharting, 1781 (g.) 589
Clytra Laicharting, 1781 (sbg.) 589
Clytraria Semenov, 1903 (sbg.) 590
Clytrasoma Jacoby, 1908 (g.) 592
Clytrasoma Jacoby, 1908 (sbg.) 592
Clytraxeloma Clavareau, 1913 (syn.) 173
Clytrella L. N. Medvedev, 1961 (syn.) 590
Cnecodes Motschulsky, 1858 (syn.) 422
Cneorane Baly, 1865 (g.) 429
Cneoranella Maulik, 1936 (syn.) 438
Cneoranidea S.-H. Chen, 1942 (g.) 430
Cneorella L. N. Medvedev & Dang, 1981 (g.) 431
Cneorides Jacoby, 1896 (syn.) 427
Cobosorina David, 1953 (syn.) 318
Coenobius Suffrian, 1857 (g.) 673
Colaphellus Weise, 1916 (g.) 367
Colaphodes Motschulsky, 1860 (sbg.) 301
Colaphomega Reitter, 1913 (sbg.) 368
Colaphoptera Motschulsky, 1860 (sbg.) 302
Colaphosoma Motschulsky, 1860 (sbg.) 303
Colaphus Dejean, 1836 (syn.) 368
Colaphus L. Redtenbacher, 1845 (syn.) 367
Colaspidea Laporte, 1833 (g.) 692
Colaspidema Laporte, 1833 (g.) 368
Colaspidema Laporte, 1833 (sbg.) 368
Colaspina Weise, 1893 (g.) 693
Colaspinella Weise, 1893 (g.) 708
Colaspinella Weise, 1893 (sbg.) 708
Colaspoides Laporte, 1833 (g.) 708
Colasposoma Laporte, 1833 (g.) 713
Colasposoma Laporte, 1833 (sbg.) 713
Colobaspis Fairmaire, 1894 (syn.) 173
Colpodes Chevrolat, 1836 (syn.) 468
Colposcelis Dejean, 1836 (syn.) 724
Conchyloctenia Spaeth, 1902 (g.) 240
Confinoides Özdikmen, 2021 (syn.) 493
Conicobruchus Decelle, 1951 (g.) 194
Conradsia Pic, 1936 (syn.) 231
Copa Weise, 1892 (syn.) 417
Coptocephala Chevrolat, 1836 (g.) 593
Coptomesa Weise, 1912 (syn.) 404
Cornutobruchus Decelle, 1975 (syn.) 194
Corynodes Hope, 1841 (syn.) 711
Corynoeides Clark, 1865 (syn.) 711
Corynopalpa Chevrolat, 1836 (syn.) 417
Craspeda Motschulsky, 1860 (syn.) 295
Craspedomela Jacobson, 1926 (sbg.) 345
Craspedonta Chevrolat, 1836 (g.) 235
Crassodonacia Kippenberg, 2015 (syn.) 206
Crepidaspis Spaeth, 1912 (syn.) 241
Crepidodera Chevrolat, 1836 (g.) 502
Crepidoderoides Chûjô, 1942 (g.) 504
Crepidomorpha Fleischer, 1916 (syn.) 518
Crepidomorpha Heikertinger, 1923 (syn.) 518
Crepidosoma S.-H. Chen, 1939 (g.) 504
Crioceris Geoffroy, 1762 (g.) 216
Crosita Motschulsky, 1860 (g.) 327
Crositops Marseul, 1883 (sbg.) 303
Cryptocephalus Geoffroy, 1762 (g.) 619
Cryptocephalus Geoffroy, 1762 (sbg.) 629
Cryptodontus Burlini, 1969 (syn.) 659
Cyaniris Chevrolat, 1836 (syn.) 609
Cyclocassida S.-H. Chen & Zia, 1961 (syn.) 241
Cynorita Laboissière, 1940 (syn.) 436
Cynorta Baly, 1865 (syn.) 436
Cynortana E. Strand, 1942 (syn.) 436
Cyphogaster Goecke, 1934 (syn.) 206
Cyrsylus Jacoby, 1891 (g.) 504
Cyrtochrysolina Kippenberg, 2012 (sbg.) 303
Cyrtochrysolina Kippenberg, 1994 (un.) 303
Cyrtocnemis Motschulsky, 1868 (syn.) 324
Cyrtonastes Fairmaire, 1874 (g.) 343
Cyrtonastes Fairmaire, 1874 (sbg.) 343
Cyrtonocassis S.-H. Chen & Zia, 1961 (syn.) 241
Cyrtonus Latreille, 1829 (g.) 328
Cyrtonussa E. Strand, 1935 (syn.) 328
Cystocnemis Motschulsky, 1860 (g.) 343
Cystocnemis Motschulsky, 1860 (sbg.) 343

Dactylispa Weise, 1897 (g.) 261
Dactylispa Weise, 1897 (sbg.) 261
Dalessandroiana Özdikmen, 2021 (syn.) 493
Damais Jacoby, 1903 (syn.) 446
Damasus Chapuis, 1874 (g.) 693
Dasychlorus Fairmaire, 1898 (syn.) 713
Decarthrocera Laboissière, 1937 (g.) 378

Decellebruchus Borowiec, 1987 (g.) 194
Decoomanius Laboissière, 1927 (syn.) 393
Deloyala L. Redtenbacher, 1858 (syn.) 241
Delphastobia Casey, 1924 (syn.) 581
Demarchus Jacoby, 1887 (g.) 462
Demotina Baly, 1863 (g.) 693
Demotinella Jacoby, 1908 (syn.) 696
Dentisterna L. N. Medvedev, 1993 (syn.) 516
Dercestra Chûjô, 1962 (syn.) 400
Dercetes Hincks, 1949 (un.) 402
Dercetina Gressitt & Kimoto, 1963 (g.) 402
Dercetis Clark, 1865 (syn.) 402
Dercetisoma Maulik, 1936 (g.) 403
Dermestops Jacobson, 1898 (g.) 694
Dermorhytis Baly, 1861 (g.) 710
Derocrepis Weise, 1886 (g.) 504
Derocrepisomus Pic, 1911 (syn.) 468
Dia Dejean, 1836 (syn.) 692
Diabrotica Chevrolat, 1836 (g.) 421
Diacantha Chevrolat, 1836 (g.) 417
Diachalcoidea Bechyné, 1955 (sbg.) 303
Diachus LeConte, 1881 (g.) 670
Dianaspis S.-H. Chen & Zia, 1984 (sbg.) 240
Diaphaenidea Laboissière, 1933 (syn.) 436
Diapromorpha Lacordaire, 1848 (g.) 597
Dibolia Latreille, 1829 (g.) 505
Dibolia Latreille, 1829 (sbg.) 505
Diboloides Jacoby, 1897 (syn.) 475
Dicherosis Foudras, 1860 (syn.) 488
Dichirispa Gestro, 1890 (sbg.) 268
Dicladispa Gestro, 1897 (g.) 266
Diorhabda Weise, 1883 (g.) 382
Dioryctus Suffrian, 1860 (syn.) 671
Disonycha Chevrolat, 1836 (g.) 507
Disops Gistel, 1856 (syn.) 659
Disopus Chevrolat, 1836 (sbg.) 659
Distolaca Baly, 1859 (syn.) 260
Diversivascula Özdikmen & Bal, 2021 (syn.) 241
Dlochrysa Motschulsky, 1860 (syn.) 336
Docemasia Jacoby, 1900 (syn.) 464
Doeberlia Warchałowski, 2007 (syn.) 324
Dolichocassida Günther, 1958 (syn.) 241
Donacia Fabricius, 1775 (g.) 206
Donaciasta Fairmaire, 1901 (g.) 212
Donaciella Reitter, 1920 (syn.) 206
Donacilla Fairmaire, 1901 (syn.) 212
Donaciocrioceris Pic, 1936 (syn.) 212
Donaciomima L. N. Medvedev, 1973 (syn.) 206
Donacocia Gistel, 1856 (syn.) 206
Doridea Baly, 1864 (syn.) 458
Doryida Baly, 1865 (g.) 403
Doryidomorpha Laboissière, 1931 (g.) 403
Doryscus Jacoby, 1887 (g.) 431
Doryxena Baly, 1861 (g.) 382
Doryxenoides Laboissière, 1927 (g.) 382
Downesia Baly, 1859 (g.) 259

Ecphorastes Gistel, 1856 (syn.) 593
Ecranus Walker, 1871 (syn.) 716
Ectonia Weise, 1922 (syn.) 476
Eleia Gistel, 1848 (syn.) 273
Emarhopa Weise, 1886 (sbg.) 383
Emathea Baly, 1865 (g.) 403
Emmetrus Motschulsky, 1860 (syn.) 279
Endoxus Baly, 1861 (i. s. s.) 694
Enneamera Harold, 1875 (syn.) 467
Enneaoria Tan, 1981 (syn.) 688
Enoplolema Heinze, 1943 (syn.) 229
Entomomela Jacobson, 1926 (sbg.) 344
Entomoscelis Chevrolat, 1836 (g.) 344
Epaenidea Gressitt & Kimoto, 1963 (g.) 431
Eparchista Gistel, 1848 (syn.) 460
Epiluperodes Kimoto & Gressitt, 1966 (g.) 431
Epimela Weise, 1903 (g.) 597
Epimela Weise, 1903 (sbg.) 597
Epiotis Solsky, 1871 (syn.) 510
Epistictia Boheman, 1850 (syn.) 236
Epistictina Hincks, 1950 (g.) 236
Epithrella L. N. Medvedev, 1994 (syn.) 580
Epithrix Bedel, 1897 (syn.) 507
Epitrix Foudras, 1860 (g.) 507
Eremocassis Spaeth, 1926 (syn.) 241
Eremosis Gozis, 1882 (syn.) 286
Erganoides Jacoby, 1903 (g.) 431
Erigenes Marshall, 1864 (syn.) 711
Eriosarda Jacoby, 1903 (syn.) 391
Eriosardella Chûjô, 1935 (syn.) 396
Ernobia Gistel, 1856 (syn.) 273
Erythraella Zoia, 2012 (g.) 694
Erythrochrysa Bechyné, 1950 (sbg.) 304
Eryxia Baly, 1865 (g.) 694
Eryxia Baly, 1865 (sbg.) 694
Estigmena Hope, 1841 (g.) 233

Eubrachis Baly, 1878 (syn.) 697
Euchrysolina Bechyné, 1950 (sbg.) 304
Euchthonius Gistel, 1856 (syn.) 700
Eucycla Baly, 1876 (syn.) 510
Eucyclomela S.-H. Chen, 1934 (syn.) 510
Eudibolia Iablokoff-Khnzorian, 1968 (sbg.) 506
Eudolia Jacoby, 1885 (g.) 508
Eudoliamorpha Scherer, 1989 (g.) 508
Eudora Laporte, 1840 (syn.) 711
Eugalera Brancsik, 1899 (syn.) 456
Euliroetis Ogloblin, 1936 (g.) 432
Euluperus Weise, 1886 (g.) 432
Eumela Baly, 1875 (syn.) 330
Eumelaella E. Strand, 1935 (syn.) 330
Eumelepta Jacoby, 1892 (g.) 433
Eumolpus Illiger, 1798 (syn.) 691
Eupales Lefèvre, 1885 (syn.) 716
Euphitrea Baly, 1875 (g.) 508
Euphyma Baly, 1879 (syn.) 425
Euphymasia Jacoby, 1900 (syn.) 508
Euplateumaris Iablokoff-Khnzorian, 1966 (syn.) 213
Euplecnema Jacoby, 1906 (syn.) 507
Eupus Wollaston, 1854 (sbg.) 563
Eurycorynus Marshall, 1864 (syn.) 711
Euryope Dalman, 1824 (g.) 716
Euspermophagus Zacher, 1930 (syn.) 176
Eustetha Baly, 1861 (syn.) 404
Eutheca Baly, 1878 (syn.) 492
Eutrea Baly, 1875 (g.) 509
Evaspistes Gistel, 1856: 381 (syn.) 241
Exaesiognatha Jacobson, 1923 (sbg.) 588
Exiguipenna Jolivet, 1951 (sbg.) 276
Exomis Weise, 1889 (g.) 598
Exorhina Weise, 1886 (syn.) 493
Exosoma Jacoby, 1903 (g.) 433

Fabricianus Weise, 1895 (syn.) 670
Falsobruchus Pic, 1913 (syn.) 193
Falsocassidula Pic, 1939 (syn.) 256
Falsocolaphellus Pic, 1928 (syn.) 708
Falsodioryctus Pic, 1955 (syn.) 672
Falsoexosoma Pic, 1926 (g.) 434
Falsoiphimoides Pic, 1935 (syn.) 717
Falsonerissus Pic, 1951 (sbg.) 715
Falsopachybrachys Pic, 1947 (syn.) 582
Falsoplatyxantha Pic, 1927 (syn.) 381
Farsogaleruca Lopatin, 1981 (syn.) 395
Fassatia Havelka, 1955 (sbg.) 383
Fasta Petitpierre & Alonso Zarazaga, 2019 (g.) 330
Fastuolina Warchałowski, 1991 (un.) 330
Fastuolina Warchałowski, 1993 (syn.) 336
Festa Gistel, 1848 (syn.) 271
Fidia Motschulsky, 1861 (g.) 694
Fidia Motschulsky, 1861 (sbg.) 694
Flavodonacia Kippenberg, 2015 (syn.) 206
Fleutiauxia Laboissière, 1933 (g.) 434
Floricola Gistel, 1848 (g.) 716
Formosogalerucella Pic, 1928 (syn.) 381
Foudrasia Gozis, 1881 (syn.) 502
Foudrasia Weise, 1900 (syn.) 546
Frigidorina Kühnelt, 1984 (sbg.) 336
Furusawaia Chûjô, 1962 (g.) 403

Gahanioides Özdikmen, 2021 (syn.) 493
Galemira Beenen, 2003 (syn.) 386
Galenaria L. N. Medvedev, 2007 (g.) 383
Galerima Reitter, 1903 (sbg.) 383
Galerotoma Reitter, 1903 (sbg.) 383
Galeruca Geoffroy, 1762 (g.) 383
Galeruca Geoffroy, 1762 (sbg.) 383
Galerucella Crotch, 1873 (g.) 387
Galerucella Crotch, 1873 (sbg.) 387
Galerucesthis Weise, 1897 (syn.) 378
Galerudolphia Hincks, 1949 (g.) 434
Galerupipla Maulik, 1936 (syn.) 395
Gallerucida Motschulsky, 1861 (g.) 404
Gansuapteris Ruan & Konstantinov, 2017 (g.) 509
Gasterophysia Gistel, 1856 (syn.) 276
Gastroeidea Hope, 1841 (syn.) 276
Gastroidea Agassiz, 1846 (syn.) 276
Gastrolina Baly, 1859 (g.) 276
Gastrolinoides Chûjô & Kimoto, 1960 (g.) 276
Gastrophysa Chevrolat, 1836 (g.) 276
Gastrophysa Chevrolat, 1836 (sbg.) 276
Geina Jacobson, 1925 (syn.) 389
Geinella E. Strand, 1935 (g.) 389
Geinula Ogloblin, 1936 (g.) 389
Gemellata J. R. Sahlberg, 1913 (syn.) 312
Geochoreutes Gistel, 1856 (syn.) 560
Glaucosphaera Maulik, 1926 (g.) 509
Glechonis Weise, 1889 (syn.) 444

Glyphocassis Spaeth, 1914 (g.) 254
Glyptolus Jacoby, 1884 (syn.) 378
Glyptoscelis Chevrolat, 1836 (g.) 710
Gnathomela Jacobson, 1895 (sbg.) 305
Goniactis Gistel, 1856 (syn.) 352
Gonioctena Chevrolat, 1836 (g.) 350
Gonioctena Chevrolat, 1836 (sbg.) 352
Goniomena Motschulsky, 1860 (sbg.) 360
Gonophora Baly, 1859 (syn.) 260
Gonophora Chevrolat, 1836 (g.) 260
Gracilema Chûjô, 1964 (sbg.) 233
Graphops LeConte, 1884 (g.) 728
Graptodera Chevrolat, 1836 (syn.) 470
Griva Maulik, 1926 (syn.) 546
Gruevia Warchałowski, 1974 (syn.) 717
Guggenheimia Monrós, 1956 (syn.) 581
Gymnota Gistel, 1848 (syn.) 273

Haemaltica S.-H. Chen, 1933 (g.) 510
Haemodoryida S.-H. Chen, 1942 (g.) 407
Haemonia Dejean, 1821 (syn.) 212
Haltica Illiger, 1801 (syn.) 470
Halticella Germain, 1903 (syn.) 470
Halticella Jacoby, 1899 (syn.) 475
Halticoidea Germain, 1903 (syn.) 470
Halticorcus Lea, 1917 (g.) 510
Halticorthaea Csiki, 1940 (syn.) 475
Hamushia Chûjô, 1956 (g.) 407
Hanoia Fairmaire, 1889 (syn.) 259
Haplomela S.-H. Chen, 1942 (syn.) 436
Haplonyx Jacobson, 1896 (syn.) 399
Haplosomoides Duvivier, 1890 (g.) 434
Haplosonyx Gistel, 1848 (syn.) 547
Hapsidolema Heinze, 1927 (syn.) 231
Haptoscelis Weise, 1886 (sbg.) 386
Harmandinia Laboissière, 1933 (syn.) 438
Hartmannia L. N. Medvedev, 2015 (syn.) 518
Hebdomecosta Spaeth, 1915 (syn.) 254
Heliostola Motschulsky, 1860 (sbg.) 305
Helodes Paykull, 1799 (syn.) 286
Heminodes Jacoby, 1895 (g.) 710
Hemiphaedon Jacobson, 1901 (sbg.) 279
Hemipyxis Chevrolat, 1836 (g.) 510
Hermaeophaga Foudras, 1860 (g.) 513
Hermaeophaga Foudras, 1860 (sbg.) 513
Hespera Weise, 1889 (g.) 462
Hesperomorpha Ogloblin, 1936 (g.) 435
Hesperopenna L. N. Medvedev & Dang, 1981 (g.) 435
Heteraspibrachis Pic, 1907 (syn.) 703
Heteraspis Chapuis, 1874 (syn.) 703
Heteraspis Chevrolat, 1836 (g.) 695
Heterichnus Warchałowski, 1991 (syn.) 659
Heterocnemis Morawitz, 1860 (syn.) 723
Heterocnensis Morawitz, 1860 (i. o. s.) 723
Heterodactylus L. N. Medvedev, 1963 (syn.) 660
Heterotrichus Chapuis, 1874 (g.) 696
Heyrovskya Jind. Madar & Jiř. Madar, 1968 (g.) 513
Himalalta L. N. Medvedev, 1990 (syn.) 490
Himaplosonyx S.-H. Chen, 1976 (g.) 407
Hippuriphila Foudras, 1860 (g.) 514
Hirticnemis Nadein, 2011 (sbg.) 469
Hirtigaleruca Chûjô, 1962 (g.) 389
Hirtoclytra L. N. Medvedev, 1961 (syn.) 589
Hispa Linnaeus, 1767 (g.) 267
Hispella Chapuis, 1875 (syn.) 267
Hispellinus Weise, 1897 (g.) 267
Hispopria Baly, 1859 (syn.) 236
Homalopus Chevrolat, 1837 (sbg.) 659
Homoeostigmus Jacobson, 1917 (syn.) 678
Hoplasoma Jacoby, 1884 (g.) 436
Hoplasomedia Maulik, 1936 (syn.) 434
Hoplionota Hope, 1841 (syn.) 271
Hoplosaenidea Laboissière, 1933 (g.) 436
Hoplosoma Motschulsky, 1860 (syn.) 311
Hoplostines Blackburn, 1890 (syn.) 393
Horaia Chûjô, 1937 (syn.) 516
Horridobruchus Borowiec, 1984 (g.) 194
Hortensoides Özdikmen, 2021 (syn.) 493
Humba S.-H. Chen, 1934 (g.) 330
Hydrogaleruca Laboissière, 1922 (syn.) 387
Hydropus Motschulsky, 1860 (syn.) 493
Hydrothassa C. G. Thomson, 1859 (sbg.) 286
Hylaspes Baly, 1865 (syn.) 404
Hylaspoides Duvivier, 1892 (g.) 407
Hymetes Lacordaire, 1848 (g.) 688
Hyperacantha Chapuis, 1879 (syn.) 417
Hyperaxis Harold, 1874 (g.) 696
Hypericia Bedel, 1899 (sbg.) 306
Hyphaenia Baly, 1865 (g.) 437
Hyphasis Harold, 1877 (g.) 514
Hyphasoma Jacoby, 1903 (syn.) 514

Hypnophila Foudras, 1860 (syn.) 541
Hypocassida Weise, 1893 (g.) 254
Hypochalcoidea Bourdonné, 2012 (sbg.) 308
Hyporeina Bourdonné & Doguet, 1986 (syn.) 342

Idacantha Fairmaire, 1869 (syn.) 417
Imolia Jacoby, 1884 (syn.) 468
Incisophthalma Heinze, 1929 (syn.) 231
Inclica Walker, 1859 (syn.) 673
Inopelonia Broun, 1893 (syn.) 518
Intricatorina Kühnelt, 1984 (sbg.) 336
Iphidea Baly, 1865 (syn.) 422
Iphimoides Jacoby 1883 (g.) 710
Iphimoides Jacoby, 1883 (sbg.) 710
Iranomolpus Lopatin, 1979 (syn.) 715
Ischyromus Jacobson, 1893 (syn.) 697
Ischyronota Weise, 1891 (g.) 255
Isnus Weise, 1898 (g.) 675
Isosoma Billberg, 1820 (syn.) 551
Issikia Chûjô, 1961 (syn.) 391
Ivalia Jacoby, 1887 (g.) 515

Jacobsonia L. N. Medvedev, 1970 (syn.) 310
Jacobyana Maulik, 1926 (g.) 515
Jambhala Würmli, 1975 (g.) 268
Japonia Weise, 1922 (syn.) 437
Japonitata E. Strand, 1935 (g.) 437
Javeta Baly, 1859 (g.) 257
Jaxartiolus Jacobson, 1922 (g.) 670
Jeanclaudia Mikhailov, 2009 (sbg.) 308
Jolibrotica C.-F. Lee & Bezděk, 2015 (g.) 438
Juliusina Reitter, 1920 (syn.) 213

Kanarella Jacoby 1896 (g.) 438
Kari Maulik, 1923 (syn.) 255
Kashmirobia Konstantinov & Prathapan, 2006 (g.) 516
Khasia Jacoby, 1899 (g.) 438
Kimotoa Gruev, 1985 (syn.) 574
Kinabalua Mohamedsaid, 1997 (syn.) 458
Kingsolverius Borowiec, 1987 (g.) 194
Klitispa Uhmann, 1940 (g.) 260
Kytorhinoides Legalov, 2011 (sbg.) 202
Kytorhinus Fischer von Waldheim, 1809 (g.) 202
Kytorhinus Fischer von Waldheim, 1809 (sbg.) 202

Labidostomis Chevrolat, 1836 (g.) 598
Labidostomis Chevrolat, 1836 (sbg.) 599
Labidostomus Gistel, 1848 (syn.) 599
Laboisierella S.-H. Chen, 1933 (g.) 516
Laccoptera Boheman, 1854 (g.) 240
Laccopteroidea Spaeth, 1952 (sbg.) 240
Lachnaea Agassiz, 1846 (syn.) 604
Lachnaia Chevrolat, 1836 (g.) 604
Lachnaia Chevrolat, 1836 (sbg.) 604
Lahejia Gahan, 1896 (g.) 696
Lamellosus Tomov, 1979 (sbg.) 662
Lamprocopa Hincks, 1949 (g.) 417
Lanka Maulik, 1926 (g.) 516
Lankanella Kimoto, 2000 (syn.) 549
Lankaphthona L. N. Medvedev, 2001 (syn.) 549
Lanolepta Kimoto, 1991 (g.) 438
Laophilus Gistel, 1856 (syn.) 581
Laoseumolpus Pic, 1935 (syn.) 710
Laotzeus S.-H. Chen, 1933 (g.) 516
Laphris Baly, 1864 (g.) 407
Lapsionus Ponting, 2020 (sbg.) 584
Lasiocassis Gressitt, 1952 (syn.) 241
Lasiochila Weise, 1916 (g.) 233
Lefevrella Jacobson, 1894 (syn.) 703
Leiometopona Brancsik, 1899 (syn.) 513
Lema Fabricius, 1798 (g.) 225
Lema Fabricius, 1798 (sbg.) 225
Lepina Baly, 1863 (g.) 696
Leprotella Jacoby, 1900 (syn.) 700
Leprotes Baly, 1863 (syn.) 694
Leptarthra Baly, 1861 (g.) 408
Leptaulaca Weise, 1902 (g.) 417
Lepthispa Weise, 1911 (syn.) 270
Leptinotarsa Chevrolat, 1836 (g.) 331
Leptispa Baly, 1859 (g.) 270
Leptodibolia S.-H. Chen, 1941 (g.) 516
Leptomela Jacobson, 1926 (sbg.) 345
Leptomona Bechyné, 1958 (g.) 439
Leptomorpha Germar, 1842 (syn.) 270
Leptosonyx Weise, 1885 (syn.) 397
Leptotrichaltica Heikertinger, 1925 (syn.) 580
Leptotrix Horn, 1889 (syn.) 580
Lesagealtica Döberl, 2009 (syn.) 574
Lesneana S.-H. Chen, 1933 (g.) 516
Letzuana S.-H. Chen, 1934 (syn.) 552
Letzuella S.-H. Chen, 1933 (g.) 517

Levnma Özdikmen, 2008 (syn.) 435
Lilioceris Reitter, 1913 (g.) 220
Lilioceris Reitter, 1913 (sbg.) 220
Lina Latreille, 1829 (syn.) 273
Linaeidea Motschulsky, 1860 (sbg.) 285
Linamorpha Motschulsky, 1860 (syn.) 284
Linastica Motschulsky, 1860 (syn.) 276
Linozosta Allard, 1860 (syn.) 513
Lipraria L. N. Medvedev, 1990 (syn.) 517
Lipromela S.-H. Chen, 1933 (g.) 517
Lipromima Heikertinger, 1924 (g.) 517
Lipromorpha Chûjô & Kimoto, 1960 (g.) 517
Liprus Motschulsky, 1861 (g.) 518
Liroegala L. N. Medvedev, 2015 (syn.) 409
Liroetiella Kimoto, 1989 (syn.) 435
Liroetina L. N. Medvedev, 2013 (syn.) 440
Liroetinus Lopatin, 2004 (syn.) 422
Liroetis Weise, 1889 (g.) 439
Lissochila Weise, 1911 (syn.) 233
Lithocrosita L. N. Medvedev, 1982 (syn.) 298
Lithonoma Chevrolat, 1836 (syn.) 547
Lithoptera Motschulsky, 1860 (syn.) 308
Lithopteroides E. Strand, 1935 (sbg.) 308
Livolia Jacoby, 1903 (syn.) 580
Livoliella L. N. Medvedev, 1997 (syn.) 580
Lobacantha Kirby, 1837 (syn.) 268
Lochmaea Weise, 1883 (g.) 389
Lochmaeata E. Strand, 1935 (syn.) 389
Lochmaeina L. N. Medvedev, 1956 (syn.) 573
Loeblaltica Scherer, 1989 (g.) 518
Longebruchus Pic, 1953 (syn.) 196
Longiampulla Özdikmen & Bal, 2021 (syn.) 241
Longicornoides Özdikmen, 2021 (syn.) 493
Longitarsella L. N. Medvedev, 2009 (syn.) 550
Longitarsus Latreille, 1829 (g.) 518
Longitarsus Latreille, 1829 (sbg.) 518
Lopatinica Kippenberg, 2012 (sbg.) 309
Lophallya Hincks, 1949 (syn.) 573
Lophea Baly, 1865 (g.) 710
Lordicassis Reitter, 1926 (syn.) 241
Lordiconia Reitter, 1926 (syn.) 241
Luperaria L. N. Medvedev, 1993 (syn.) 516
Luperocnemus Fairmaire, 1888 (syn.) 464
Luperogala L. N. Medvedev & Samoderzhenkov, 1989 (g.) 440
Luperomorpha Weise, 1887 (g.) 464
Luperomorphella Chûjô, 1964 (syn.) 426
Luperus Geoffroy, 1762 (g.) 440
Lyperus Billberg, 1820 (syn.) 440
Lypesthes Baly, 1863 (syn.) 694
Lypnea Baly, 1876 (g.) 538
Lythraria Bedel, 1897 (g.) 538

Machomena Dubois, 1887 (sbg.) 362
Macrima Baly, 1878 (g.) 444
Macrispa Baly, 1859 (g.) 236
Macrocnema Stephens, 1831 (syn.) 564
Macrocnema Weise, 1888 (syn.) 564
Macrocoma Chapuis, 1874 (g.) 697
Macrolenes Chevrolat, 1836 (g.) 606
Macrolina Motschulsky, 1860 (syn.) 273
Macromela Chûjô, 1958 (sbg.) 285
Macromonycha Spaeth, 1911 (g.) 255
Macronema Stephens, 1829 (syn.) 564
Macroplea Samouelle, 1819 (g.) 212
Macrozeugophora Achard, 1914 (syn.) 171
Madurasia Jacoby, 1896 (g.) 444
Maenadochrysa Bechyné, 1950 (sbg.) 309
Majoroides Özdikmen, 2021 (syn.) 493
Malacodora Bedel, 1905 (syn.) 433
Malacoptera Hope, 1841 (syn.) 433
Malacosoma Chevrolat, 1836 (syn.) 433
Malaxia Fairmaire, 1878 (syn.) 378
Malaxioides Fairmaire, 1888 (syn.) 378
Malegia Lefèvre, 1883 (g.) 700
Mandarella Duvivier, 1892 (g.) 466
Manipuria Jacoby, 1908 (g.) 224
Manobia Jacoby, 1885 (g.) 538
Manobidia S.-H. Chen, 1934 (syn.) 492
Mantura Stephens, 1831 (g.) 539
Mantura Stephens, 1831 (sbg.) 539
Marsaeus Clark, 1864 (syn.) 727
Marseulia Joannis, 1865 (g.) 444
Martensomela L. N. Medvedev, 1984 (syn.) 541
Martinella L. N. Medvedev, 2000 (syn.) 435
Massiea Lefèvre, 1893 (g.) 710
Maulika Basu & Sengupta, 1980 (g.) 540
Mecistes Chapuis, 1874 (g.) 700
Mecoprosopus Chûjô, 1951 (g.) 224
Medvedevella Özdikmen, 2008 (syn.) 609
Medvedevlevna Özdikmen, 2008 (sbg.) 310
Medythia Jacoby, 1887 (g.) 444
Megabruchidius Borowiec, 1984 (g.) 194

Megacerus Fåhraeus, 1839 (g.) 202
Megacerus Fåhraeus, 1839 (sbg.) 202
Meishania Chen & S.-Y. Wang, 1980 (g.) 540
Melasoma Dillwyn, 1829 (syn.) 273
Melasomida Schaeffer, 1920 (syn.) 284
Melasomoptera Bechyné, 1950 (sbg.) 311
Melina Chapuis, 1874 (syn.) 708
Melinophora Lefèvre, 1885 (syn.) 708
Melitonoma Chevrolat, 1836 (g.) 606
Melixanthus Suffrian, 1854 (g.) 670
Mellipora Chûjô, 1965 (g.) 540
Melolontha Geoffroy, 1762 (syn.) 589
Melospila Baly, 1861 (syn.) 404
Menippus Clark, 1864 (g.) 391
Menthastriella Bechyné, 1950 (syn.) 319
Mergodonacia Kippenberg, 2015 (syn.) 206
Merilia Lacordaire, 1848 (g.) 606
Merista Chapuis, 1875 (syn.) 408
Meristata E. Strand, 1935 (g.) 408
Meristoides Laboissière, 1929 (g.) 409
Meristoides Laboissière, 1929 (sbg.) 409
Merobruchus Bridwell, 1946 (g.) 195
Mesocolaspis Jacoby, 1908 (g.) 700
Mesodonta Baly, 1865 (syn.) 381
Metallographa Motschulsky, 1860 (syn.) 289
Metallotimarcha Motschulsky, 1860 (sbg.) 369
Metaxis Baly, 1863 (syn.) 696
Metellus Jacoby, 1886 (syn.) 458
Micraenidea Laboissière, 1933 (syn.) 436
Micrepitrix Laboissière, 1933 (syn.) 580
Micrespera S.-H. Chen & S.-Y. Wang, 1987 (g.) 540
Micrispa Gestro, 1897 (g.) 260
Microcrepis S.-H. Chen, 1933 (g.) 541
Microdera Stephens, 1839 (syn.) 273
Microeurydemus Pic, 1938 (g.) 725
Microlema Pic, 1932 (sbg.) 229
Microlina Lopatin, 1977 (un.) 273
Microlypesthes Pic, 1936 (syn.) 705
Microsomus Burlini, 1957 (sbg.) 675
Microstilus Pic, 1950 (i. s. s.) 675
Microstylus Rey, 1883 (syn.) 675
Miltina Chapuis, 1875 (g.) 409
Mimagitocera Maulik, 1936 (g.) 417
Mimastra Baly 1865 (g.) 445
Mimastracella Jacoby, 1903 (g.) 391
Mimoparascela Pic, 1935 (syn.) 717
Mimophaedon Bourdonné, 1997 (syn.) 293
Mimosestes Bridwell, 1946 (g.) 195
Minckia E. Strand, 1935 (sbg.) 311
Minicnema Nadein, 2007 (sbg.) 564
Minota Kutschera, 1859 (g.) 541
Minota Kutschera, 1859 (sbg.) 541
Miochira Lacordaire, 1848 (g.) 606
Mionycha Weise, 1891 (syn.) 241
Mionychella Spaeth, 1952 (syn.) 241
Miopristis Lacordaire, 1848 (g.) 607
Mireditha Reitter, 1913 (g.) 725
Mniophila Stephens, 1831 (g.) 542
Mniophilosoma Wollaston, 1854 (g.) 542
Monax Gistel, 1848 (syn.) 234
Monochirus Chapuis, 1875 (syn.) 267
Monodaltica Bechyné, 1955 (syn.) 577
Monohispa Weise, 1897 (g.) 268
Monolepta Chevrolat, 1836 (g.) 446
Monoxia LeConte, 1865 (g.) 391
Monrosia L. N. Medvedev, 1971 (syn.) 609
Morphosphaera Baly, 1861 (g.) 409
Moseria Weise, 1922 (syn.) 492
Munina S.-H. Chen, 1976 (g.) 451
Musaka Bechyné, 1958 (syn.) 574
Myocoryna Stål, 1859 (syn.) 331

Nabathaea Spaeth, 1911 (g.) 256
Nacrea Baly, 1886 (syn.) 458
Nanosmaragdina Lopatin & Kulenova, 1986 (syn.) 609
Natocyrstes Berti & Daccordi, 1974 (un.) 343
Natocyrstes Kippenberg, 2010 (sbg.) 343
Necyomantes Gistel, 1848 (syn.) 609
Neoatysa Abdullah & Qureshi, 1968 (syn.) 445
Neobarombiella Bolz & Wagner, 2012 (g.) 451
Neocharis Jacoby, 1881 (syn.) 458
Neochrolea Jacoby, 1887 (syn.) 458
Neochya C.-F. Lee, 2020 (g.) 451
Neoclitena Abdullah & Qureshi, 1968 (syn.) 470
Neocrane Chûjô, 1962 (syn.) 430
Neocrepidodera Heikertinger, 1911 (g.) 542
Neodownesia Gressitt, 1953 (g.) 236
Neogalerucella Chûjô, 1962 (sbg.) 388
Neohaemonia Székessy, 1941 (g.) 213
Neohylaspes Chûjô, 1962 (syn.) 407

Neolycaria Abdullah & Qureshi, 1969 (syn.) 711
Neonabathaea Borowiec, 1994 (syn.) 257
Neophaedon Jacobson, 1901 (g.) 278
Neopotanina Abdullah & Qureshi, 1969 (syn.) 308
Neorthaea Maulik, 1926 (syn.) 508
Neorthella L. N. Medvedev, 2010 (syn.) 516
Neorudolphia Laboissière, 1926 (syn.) 444
Neosastra Abdullah & Qureshi, 1968 (syn.) 418
Neosermylassa Chûjô, 1956 (syn.) 412
Nepalicrepis Scherer, 1969 (syn.) 541
Nepalocassena L. N. Medvedev, 2009 (sbg.) 426
Nepalogaleruca Kimoto, 1970 (g.) 410
Nepalolepta L. N. Medvedev, 1992 (g.) 451
Nephiusus Jacoby, 1892 (syn.) 711
Nephus Jacoby, 1889 (syn.) 711
Nepius J. Thomson, 1858 (syn.) 272
Nesohaltica Maulik, 1929 (syn.) 577
Nigricoides Özdikmen, 2021 (syn.) 493
Nilgiraspis Spaeth, 1932 (g.) 241
Niliosoma Motschulsky, 1860 (syn.) 366
Nisotra Baly, 1864 (g.) 545
Nitidoides Özdikmen, 2021 (syn.) 493
Nodina Motschulsky, 1858 (g.) 725
Nodostoma Motschulsky, 1860 (syn.) 717
Nonarthra Baly, 1862 (g.) 467
Notocantha Agassiz, 1846 (syn.) 271
Notosacantha Chevrolat, 1836 (g.) 271
Novascuta Özdikmen, 2008 (g.) 546
Novofoudrasia Jacobson, 1901 (g.) 546
Nudicnemis Nadein, 2011 (sbg.) 469
Nyctidromas Semenov, 1896 (g.) 391
Nyctiphantus Semenov, 1902 (syn.) 391
Nymphius Weise, 1900 (g.) 451

Ochralea Clark, 1865 (syn.) 452
Ochralea Chevrolat, 1836 (syn.) 460
Ochralepta Beenen, **nom. nov.** (g.) 452
Ochrosis Foudras, 1860 (g.) 546
Ochrosoma Ogloblin, 1930 (syn.) 574
Octodonta Chapuis, 1875 (g.) 258
Octodonta Chapuis, 1875 (sbg.) 258
Odontionopa Motschulsky, 1866 (syn.) 726
Odontionycha Weise, 1891 (syn.) 241
Odontoedon S.-Q. Ge & Daccordi, 2013 (g.) 278
Oedicerus Kollar & L. Redtenbacher, 1844 (syn.) 426
Oedionychis Latreille, 1829 (g.) 547
Ogloblinia Csiki, 1939 (syn.) 574
Oides Weber, 1801 (g.) 460
Olcomela Jacobson, 1926 (sbg.) 345
Olorus Chapuis, 1874 (g.) 711
Omeia S.-H. Chen, 1934 (syn.) 547
Omeiana S.-H. Chen, 1934 (g.) 547
Omeisphaera S.-H. Chen & Zia, 1974 (g.) 547
Omodon Marshall, 1864 (syn.) 711
Onchocephala Guérin-Méneville, 1844 (syn.) 272
Oncocephala Agassiz, 1846 (g.) 272
Onychocassis Spaeth, 1926 (syn.) 242
Oocassida Weise, 1897 (g.) 256
Oomorphoides Monrós, 1956 (g.) 581
Oomorphus Curtis, 1831 (g.) 581
Oomorphus Curtis, 1831 (sbg.) 581
Ophraella Wilcox, 1965 (g.) 392
Ophrida Chapuis, 1875 (g.) 548
Orchestris Kirby, 1837 (syn.) 552
Oreina Chevrolat, 1836 (g.) 331
Oreina Chevrolat, 1836 (sbg.) 336
Oreomela Jacobson, 1895 (g.) 345
Oreomela Jacobson, 1895 (sbg.) 345
Oreomelina Lopatin, 2004 (syn.) 348
Oreothassa Jacobson, 1901 (g.) 279
Orestia Chevrolat, 1836 (g.) 548
Orestioides Hatch, 1935 (syn.) 542
Orhespera S.-H. Chen & S.-Y. Wang, 1984 (g.) 549
Orina Agassiz, 1846 (syn.) 336
Orsodacna Latreille, 1804 (i. s. s.) 169
Orsodacne Latreille, 1802 (g.) 169
Orthaea Jacoby, 1889 (syn.) 508
Orthaltica Crotch, 1873 (syn.) 580
Orthaulaca Weise, 1892 (syn.) 414
Orthocrepis Weise, 1888 (sbg.) 513
Ortholema Heinze, 1943 (g.) 231
Orthosagra Crowson, 1946 (syn.) 175
Osnaparis Fairmaire, 1889 (sbg.) 690
Otiocephala Lefèvre, 1872 (g.) 607
Otiothraea Warchałowski, 1990 (g.) 608
Oulema Gozis, 1886 (g.) 231
Oulema Gozis, 1886 (sbg.) 231
Ovoclytra L. N. Medvedev, 1961 (sbg.) 592
Ovomorpha Motschulsky, 1860 (sbg.) 318
Ovosoma Motschulsky, 1860 (sbg.) 312

Ovostoma Motschulsky, 1860 (sbg.) 313
Ovotispa L. N. Medvedev, 1992 (g.) 271
Oxylepis Desbrochers des Loges, 1891 (syn.) 256
Oxylepus Desbrochers des Loges, 1884 (g.) 256
Ozomena Chevrolat, 1845 (syn.) 426

Pachnephoptrus Reitter, 1892 (g.) 700
Pachnephoriscus Lopatin, 1976 (sbg.) 702
Pachnephorus Chevrolat, 1836 (g.) 400
Pachnephorus Chevrolat, 1836 (sbg.) 400
Pachybrachia Gistel, 1856 (syn.) 678
Pachybrachis Chevrolat, 1836 (g.) 677
Pachybrachis Chevrolat, 1836 (sbg.) 678
Pachybrachys L. Redtenbacher, 1848 (i. s. s.) 678
Pachygnatha Weise, 1892 (syn.) 305
Pachylanka L. N. Medvedev, 1989 (syn.) 582
Pachylina L. N. Medvedev & Chernov, 1969 (syn.) 273
Pachymeroides Prevett, 1966 (syn.) 205
Pachymerus Thunberg, 1805 (g.) 205
Pachymerus Thunberg, 1805 (sbg.) 205
Pachypalpa Weise, 1892 (syn.) 414
Pachystylus Rey, 1883 (syn.) 678
Pagria Lefèvre, 1884 (g.) 726
Palaeosticta Bechyné, 1952 (sbg.) 314
Paleoacanthoscelides Borowiec, 1985 (g.) 195
Paleobruchidius Egorov, 1989 (g.) 195
Paleosepharia Laboissière, 1936 (g.) 452
Pales Chevrolat, 1836 (syn.) 716
Palesida Harold, 1874 (syn.) 713
Pallasia Weise, 1886 (syn.) 392
Pallasiola Jacobson, 1925 (g.) 392
Palpoxena Baly, 1861 (g.) 453
Pantryptes Gistel, 1856 (syn.) 470
Papuleptura Gressitt, 1959 (syn.) 171
Parachalcoides Chûjô, 1959 (syn.) 514
Paraclytra L. N. Medvedev, 1971 (g.) 608
Paracrosita Daccordi, 1982 (sbg.) 314
Paracrothinium S.-H. Chen, 1940 (g.) 716
Paradiachalcoidea Daccordi, 1978 (syn.) 303
Paradownesia Gestro, 1899 (syn.) 270
Paraenidea Laboissière, 1933 (syn.) 458
Paragetocera Laboissière, 1929 (g.) 417
Paraheliostola L. N. Medvedev, 1992 (syn.) 324
Paralina Baly, 1860 (syn.) 288
Parallelispa Fairmaire, 1885 (syn.) 270
Paraluperodes Ogloblin, 1936 (syn.) 444
Parambrostoma S.-H. Chen, 1936 (sbg.) 289
Paramerista Lopatin, 2011 (syn.) 562
Paramesopa L. N. Medvedev, 1984 (syn.) 466
Paraminota Scherer, 1989 (g.) 549
Paraminotella Döberl & Konstantinov, 2003 (g.) 549
Paraphaedon Sharp, 1910 (sbg.) 279
Paraphthonomorpha Ohno, 1960 (syn.) 538
Paraplagiomorpha Daccordi, 1987 (syn.) 284
Paraplotes Laboissière, 1933 (g.) 453
Parargopus S.-H. Chen, 1939 (g.) 549
Parascela Baly, 1878 (g.) 727
Paraspitiella S.-H. Chen & Jiang, 1981 (g.) 410
Parategyrius Kimoto & Gressitt, 1966 (g.) 549
Parathrylea Duvivier, 1892 (g.) 550
Paraulaca Baly, 1888 (sbg.) 420
Paraulacophora Csiki, 1953 (syn.) 436
Paraxuthea L. N. Medvedev, 1997 (sbg.) 579
Parazipangia Ohno, 1965 (g.) 550
Parexosoma Laboissière, 1933 (g.) 454
Parheminodes S.-H. Chen, 1940 (g.) 702
Paridea Baly, 1886 (g.) 418
Paridea Baly, 1886 (sbg.) 418
Parkaniola Bechyné, 1950 (syn.) 312
Parnops Jacobson, 1894 (g.) 702
Paropsides Motschulsky, 1860 (g.) 366
Paropsisterna Motschulsky, 1860 (g.) 366
Pathius Aslam, 1967 (g.) 716
Pedema Chevrolat, 1836 (un.) 547
Pedrillia Westwood, 1864 (syn.) 171
Pedrillimorpha Papp, 1946 (un.) 171
Pedrilliomorpha Pic, 1917 (syn.) 171
Pedrilonga Papp, 1946 (un.) 171
Pedrinella Papp, 1946 (un.) 171
Peltoschema Reitter, 1880 (g.) 366
Penghou Ruan, Konstantinov, Prathapan, Ge & Yang, 2015 (g.) 550
Pentamesa Harold, 1876 (g.) 550
Penthobruchus Kingsolver, 1973 (g.) 196
Periclitena Weise, 1902 (g.) 392
Pertuisanus Bourdonné & Doguet, 2017 (syn.) 629

Petalopus Motschulsky, 1845 (syn.) 506
Petauristes Latreille, 1829 (sbg.) 229
Pezocrosita Jacobson, 1901 (sbg.) 314
Phaedon Latreille, 1829 (g.) 279
Phaedon Latreille, 1829 (sbg.) 279
Phaedonia Weise, 1898 (g.) 281
Phascus Lefèvre, 1884 (g.) 716
Philaphthona L. N. Medvedev, 1994 (syn.) 574
Philopona Weise, 1903 (g.) 551
Philotarsa L. N. Medvedev, 2009 (syn.) 550
Phola Weise, 1890 (g.) 348
Phortus Weise, 1899 (syn.) 728
Phratora Chevrolat, 1836 (g.) 281
Phratora Chevrolat, 1836 (sbg.) 282
Phygasia Chevrolat, 1836 (g.) 551
Phyllis Gistel, 1848 (syn.) 289
Phyllobrotica Chevrolat, 1836 (g.) 454
Phyllodecta Kirby, 1837 (syn.) 282
Phyllophila Stål, 1857 (syn.) 348
Phyllotreta Chevrolat, 1836 (g.) 552
Physauchenia Lacordaire, 1848 (g.) 608
Physosmaragdina L. N. Medvedev, 1971 (g.) 608
Phytodecta Kirby, 1837 (syn.) 352
Phytodectella Cantonnet, 1968 (syn.) 364
Pierryvettia Bechyné, 1950 (sbg.) 315
Pilemostoma Desbrochers des Loges, 1891 (g.) 256
Plagiodera Chevrolat, 1836 (g.) 284
Plagiomorpha Motschulsky, 1860 (syn.) 285
Plagioschema Daccordi, 1987 (syn.) 285
Plagiosterna Motschulsky, 1860 (g.) 284
Planispa Chûjô, 1937 (syn.) 258
Plateumaris C. G. Thomson, 1859 (g.) 213
Plateumaroides Iablokoff-Khnzorian, 1962 (syn.) 206
Platycorynus Chevrolat, 1836 (g.) 711
Platyphytodecta Bechyné, 1948 (syn.) 352
Platypria Guérin-Méneville, 1840 (g.) 268
Platypria Guérin-Méneville, 1840 (sbg.) 268
Platypriella S.-H. Chen & Tan, 1961 (sbg.) 264
Platyxantha Baly, 1864 (syn.) 458
Platyxanthoides Laboissière, 1933 (syn.) 458
Plectroscelis Chevrolat, 1836 (syn.) 493
Plestya Gistel, 1848 (syn.) 692
Pleurosticha Motschulsky, 1860 (sbg.) 316
Podagrica Chevrolat, 1836 (g.) 560
Podagricella S.-H. Chen, 1933 (g.) 561
Podagricomela Heikertinger, 1924 (g.) 561
Podontia Dalman, 1824 (g.) 562
Poebia Gistel, 1856 (syn.) 279
Polyclada Chevrolat, 1836 (g.) 562
Polyconia Weise, 1905 (g.) 269
Polygramma Chevrolat, 1836 (syn.) 331
Polyspila Hope, 1841 (syn.) 289
Potaninia Weise, 1889 (g.) 348
Praeochralea Duvivier, 1885 (syn.) 411
Prasocuris Latreille, 1802 (g.) 286
Prasocuris Latreille, 1802 (sbg.) 286
Primulavorus Konstantinov & Ruan, 2017 (g.) 562
Priobolia S.-H. Chen & S.-Y. Wang, 1987 (g.) 563
Prionispa Chapuis, 1875 (g.) 272
Prioptera Hope, 1841 (syn.) 234
Proctophysus L. Redtenbacher, 1845 (syn.) 663
Prodonacia S.-H. Chen, 1966 (syn.) 212
Proegmena Weise, 1889 (g.) 410
Prophyllis Reitter, 1913 (syn.) 395
Prosagra Crowson, 1946 (syn.) 175
Protobruchus Zampetti & Toma, 2020 (syn.) 192
Protocaryopemon Borowiec, 1987 (syn.) 205
Protophysus Chevrolat, 1836 (sbg.) 663
Protorina Weise, 1894 (sbg.) 341
Pseudadimonia Duvivier, 1891 (g.) 392
Pseudaoria Jacoby, 1908 (sbg.) 690
Pseudaoriana Pic, 1930 (syn.) 689
Pseudaphilenia Lopatin, 1976 (sbg.) 717
Pseudergana Beenen, 2019 (g.) 455
Pseudespera S.-H. Chen, S.-Y. Wang & Jiang, 1985 (g.) 455
Pseudeustetha Jacoby, 1900 (g.) 411
Pseudocassida Desbrochers des Loges, 1891 (syn.) 242
Pseudocassis Steinhausen, 2002 (syn.) 242
Pseudoclytra Jacoby, 1908 (syn.) 585
Pseudocophora Jacoby, 1884 (g.) 421
Pseudocrepidosoma L. N. Medvedev, 1966 (syn.) 514
Pseudocrosita Lopatin, 1999 (sbg.) 317
Pseudocrypta L. N. Medvedev, 1997 (syn.) 468

Pseudodera Baly, 1861 (g.) 563
Pseudodibolia Iablokoff-Khnzorian, 1968 (sbg.) 507
Pseudodonacia Reitter, 1920 (syn.) 206
Pseudochaetocnema Özdikmen, 2021 (syn.) 493
Pseudoides Jacoby, 1892 (g.) 460
Pseudolema Pic, 1928 (syn.) 229
Pseudolina Jacoby, 1896 (sbg.) 347
Pseudolina Schaeffer, 1920 (syn.) 285
Pseudoliprus Chûjô & Kimoto, 1960 (g.) 563
Pseudoliroetis Laboissière, 1929 (syn.) 439
Pseudolithoptera L. N. Medvedev, 1970 (sbg.) 317
Pseudomacetes Linell, 1896 (syn.) 713
Pseudomalegia Jacoby, 1897 (syn.) 696
Pseudometaxis Jacoby, 1900 (g.) 702
Pseudopachymerina Zacher, 1952 (g.) 196
Pseudoparascela Pic, 1935 (syn.) 727
Pseudoparopsis Blackburn, 1899 (syn.) 284
Pseudopiomera Jacoby, 1892 (syn.) 696
Pseudosepharia Laboissière, 1936 (g.) 455
Pseudostonopa Jacoby, 1903 (g.) 727
Pseudotaeniochrysea Daccordi, 1980 (syn.) 321
Pseudotheopea C.-F. Lee & Bezděk, 2020 (g.) 455
Psyllides Gistel, 1856 (syn.) 564
Psylliodes Latreille, 1829 (g.) 563
Psylliodes Latreille, 1829 (sbg.) 564
Psyllobactra Lopatin, 1976 (sbg.) 572
Psyllomima Bedel, 1898 (syn.) 564
Pterophthinus Gressitt & Kimoto, 1963 (g.) 393
Pushtunaltica Lopatin, 1962 (syn.) 464
Pusio Gistel, 1848 (syn.) 468
Pygobruchidius Pic, 1951 (g.) 196
Pygobruchus Sharp, 1886 (sbg.) 203
Pyladia Weise, 1886 (syn.) 548
Pyrgo Weise, 1901 (syn.) 366
Pyrgoides Aslam, 1967 (un.) 366
Pyrgoides Kelly & Reid 1999 (syn.) 366
Pyrrhalta Joannis, 1865 (g.) 393

Quasilema Monrós, 1951 (sbg.) 230

Radymna Reitter, 1913 (g.) 395
Raphidopalpa Chevrolat, 1836 (syn.) 415
Reliquacassida Özdikmen & Bal, 2021 (syn.) 242
Rhabdotilla Jacobson, 1911 (sbg.) 386
Rhadinosa Weise, 1905 (g.) 269
Rhaebus Fischer von Waldheim, 1823 (g.) 205
Rhaphidopalpa Rosenhauer, 1856 (syn.) 415
Rhaphomela Jacobson, 1926 (sbg.) 347
Rhodopaea Gruev & Tomov, 1968 (g.) 717
Rhombopalpa Chevrolat, 1836 (syn.) 460
Rhoptrispa S.-H. Chen & Tan, 1964 (g.) 269
Rhyparida Baly, 1861 (g.) 727
Rhyssoloma Wollaston, 1854 (sbg.) 317
Rhytidocassis Spaeth, 1941 (g.) 257
Romalorina Weise, 1906 (syn.) 336
Rudolphia Jacoby, 1899 (syn.) 434
Rybakowia Jacobson, 1892 (syn.) 470

Sagra Fabricius, 1792 (g.) 175
Sagrania Rafinesque, 1815 (syn.) 175
Sagrina Crowson, 1946 (syn.) 175
Sagrinola Monrós & Bechyné, 1956 (syn.) 175
Salviabruchus Decelle, 1983 (g.) 197
Sangariola Jacobson, 1922 (g.) 573
Sastra Baly, 1865 (g.) 396
Sastracella Jacoby, 1900 (g.) 396
Sastroides Jacoby, 1884 (g.) 396
Saudiclytra L. N. Medvedev, El Torkey & Al Dhafer, 2014 (g.) 609
Saulopalpa Weise, 1881 (syn.) 610
Scallodera Harold, 1877 (syn.) 551
Scelodonta Westwood, 1838 (syn.) 695
Scelodontomorpha Pic, 1938 (syn.) 695
Scelolyperus Crotch, 1874 (g.) 456
Schawalleria L. N. Medvedev, 1990 (syn.) 549
Schenklingia Csiki & Heikertinger, 1940 (syn.) 510
Schereria L. N. Medvedev, 1984 (syn.) 515
Scholastica Gistel, 1848 (syn.) 398
Sclerophaedon Weise, 1882 (g.) 368
Sclerophaedon Weise, 1882 (sbg.) 368
Sebaethe Baly, 1864 (syn.) 510
Sebaethiella L. N. Medvedev, 1994 (syn.) 468
Sebaethoides S.-H. Chen, 1934 (g.) 573
Semacia Fairmaire, 1889 (syn.) 420
Semacianella Laboissière, 1930 (syn.) 420

Semenowia Weise, 1889 (sbg.) 317
Semicnema Weise, 1888 (sbg.) 572
Seminabathea Borowiec, 1994 (g.) 257
Sennius Bridwell, 1946 (g.) 197
Sepharia Fairmaire, 1889 (syn.) 444
Sermyla Chapuis, 1875 (syn.) 411
Sermylassa Reitter, 1913 (g.) 411
Sermyloides Jacoby, 1884 (g.) 411
Serraticollis B. White, 1942 (syn.) 580
Serratobruchidius Yus Ramos, 2013 (syn.) 181
Serrinotus Tan, 1992 (syn.) 582
Sezerus Özdikmen, 2006 (syn.) 660
Shaira Maulik, 1936 (g.) 456
Shairella Chûjô, 1962 (g.) 456
Shamshera Maulik, 1936 (g.) 412
Shensia S.-H. Chen, 1964 (syn.) 453
Sibiriella L. N. Medvedev, 1999 (sbg.) 317
Sichuanella Lopatin, 2008 (sbg.) 409
Sichuania Lopatin, 2002 (syn.) 409
Siemssenius Weise, 1922 (syn.) 439
Sikkimia Duvivier, 1891 (g.) 457
Sinagonia S.-H. Chen & Tan, 1962 (g.) 260
Sinaltica S.-H. Chen, 1939 (g.) 573
Sindiola Spaeth, 1903 (sbg.) 241
Sindiolina Świętojańska, 2001 (sbg.) 241
Sinispa Uhmann, 1938 (g.) 269
Sinocrepis S.-H. Chen, 1933 (g.) 573
Sinoluperoides Kimoto, 1989 (g.) 457
Sinoluperus Gressitt & Kimoto, 1963 (g.) 457
Sinomela S.-H. Chen, 1935 (sbg.) 362
Sinosphaera Konstantinov & Ruan, 2017 (g.) 574
Smaragdina Chevrolat, 1836 (g.) 609
Smaragdinella L. N. Medvedev, 1971 (syn.) 610
Smaragdonia Kippenberg, 2015 (syn.) 206
Solenia Jacoby, 1886 (syn.) 425
Solephyma Maulik, 1936 (syn.) 425
Sominella Jacobson, 1908 (g.) 212
Sopidus Jacobson, 1901 (sbg.) 663
Sotaiana Kippenberg, 2015 (syn.) 206
Sparteus Bridwell, 1946 (syn.) 181
Spartiophila Küster, 1848 (syn.) 363
Spartomena Reitter, 1913 (sbg.) 363
Spartophila Chevrolat, 1836 (sbg.) 363
Spartoxena Motschulsky, 1860 (sbg.) 364
Specularius Bridwell, 1938 (g.) 197
Spermatophagus Gistel, 1856 (syn.) 177
Spermophagus Schoenherr, 1833 (g.) 176
Sphaeraltica Ohno, 1961 (g.) 574
Sphaerarthra Weise, 1892 (syn.) 415
Sphaerochabria L. N. Medvedev, 1999 (syn.) 574
Sphaerochrysolina Kippenberg, 2010 (sbg.) 317
Sphaerochrysolina Kippenberg, 1994 (un.) 317
Sphaeroderma Stephens, 1831 (g.) 574
Sphaerodermus Gistel, 1856 (syn.) 574
Sphaeromela Bedel, 1899 (sbg.) 317
Sphaerophyma Baly, 1878 (syn.) 487
Sphenoraia Clark, 1865 (g.) 412
Sphenoraioides Laboissière, 1935 (syn.) 412
Sphenorella L. N. Medvedev & Sprecher-Uebersax, 1998 (g.) 413
Spitiella Laboissière, 1931 (g.) 413
Stator Bridwell, 1946 (g.) 198
Stenoluperus Ogloblin, 1936 (syn.) 466
Stenomantura Heikertinger, 1910 (sbg.) 540
Stenoprioptera Spaeth, 1914 (syn.) 234
Sternoplatys Motschulsky, 1860 (g.) 287
Sternoplatys Motschulsky, 1860 (sbg.) 287
Sterromela Weise, 1915 (syn.) 366
Stethidea Baly, 1890 (syn.) 404
Stichoptera Motschulsky, 1860 (sbg.) 318
Stichosoma Motschulsky, 1860 (syn.) 299
Strickerus R. Lucas, 1920 (syn.) 273
Strobiderus Jacoby, 1884 (g.) 457
Stylomicrus Petitpierre & Alonso-Zarazaga, 2000 (sbg.) 675
Stylosomus Suffrian, 1848 (g.) 675
Stylosomus Suffrian, 1848 (sbg.) 676
Suffrianius Weise, 1895 (syn.) 670
Suinzona S.-H. Chen, 1931 (g.) 348
Sulcatolema Pic, 1928 (syn.) 225
Sulcicollis J. R. Sahlberg, 1913 (syn.) 311
Sulcobruchus Chûjô, 1937 (g.) 198
Swargia Maulik, 1936 (g.) 397
Synerga Weise, 1900 (sbg.) 319
Syneta Dejean, 1835 (g.) 728
Syoplia Jacoby, 1886 (syn.) 457

Taeniochrysea Bechyné, 1950 (sbg.) 321
Taeniosticha Motschulsky, 1860 (sbg.) 321
Tailandia Chûjô, 1964 (syn.) 708

Taipinus Lopatin, 2007 (g.) 349
Taiwanaenidea Kimoto, 1984 (g.) 457
Taiwania Spaeth, 1913 (syn.) 242
Taiwanohespera Kimoto, 1970 (g.) 577
Taiwanolepta Kimoto, 1989 (syn.) 457
Taiwanoliprus Komiya, 2006 (g.) 577
Taiwanorestia Kimoto, 1991 (g.) 577
Taiwanoshaira C.-F. Lee & Beenen, 2020 (g.) 457
Taizonia S.-H. Chen, 1934 (syn.) 515
Talmonus Fairmaire, 1889 (syn.) 694
Tantraedon Daccordi & L. N. Medvedev, 2000 (sbg.) 369
Tanygaster Blatchley, 1921 (syn.) 552
Taphinella Jacoby, 1889 (syn.) 400
Taphinellina Maulik, 1936 (syn.) 425
Taphius Jacoby, 1897 (syn.) 716
Taraxis LeConte, 1850 (syn.) 171
Taumacera Thunberg, 1814 (g.) 458
Taumaceroides Lopatin, 2009 (syn.) 426
Taxaris Gistel, 1848 (syn.) 659
Tebalia Fairmaire, 1889 (g.) 577
Teinodactyla Chevrolat, 1836 (syn.) 518
Temnaspis Lacordaire, 1845 (g.) 173
Testergus Weise, 1893 (sbg.) 536
Thaumacera Harold, 1876 (syn.) 458
Thelylankus Lopatin, 1987 (syn.) 582
Thelyterotarsus Weise, 1882 (syn.) 582
Theone Gistel, 1857 (g.) 397
Theopea Baly, 1864 (g.) 458
Theumorus Marshall, 1864 (syn.) 711
Thlaspida Weise, 1899 (g.) 257
Thlaspidosoma Spaeth, 1901 (g.) 257
Thlibocoryna C. v. Riley, 1875 (syn.) 331
Threnosoma Motschulsky, 1860 (sbg.) 322
Throscoryssa Maulik, 1928 (syn.) 475
Thyamis Stephens, 1831 (syn.) 518
Thysbe J. Thomson, 1858 (syn.) 713
Tillopsis Berlioz, 1918 (syn.) 710
Timarcha Samouelle, 1819 (g.) 369
Timarcha Samouelle, 1819 (sbg.) 369
Timarchida Ganglbauer, 1897 (sbg.) 323
Timarchomela Achard, 1922 (sbg.) 324
Timarchoptera Motschulsky, 1860 (sbg.) 324
Timarchostoma Motschulsky, 1860 (syn.) 369
Tinosagra Weise, 1905 (syn.) 175
Tituboea Lacordaire, 1848 (g.) 616
Tlanoma Motschulsky, 1845 (syn.) 493
Torodera Weise, 1902 (syn.) 487
Trachyaphthona Heikertinger, 1924 (syn.) 577
Trachymela Weise, 1908 (g.) 367
Trachytetra Sharp, 1886 (g.) 577
Triaplatarthris Fairmaire, 1878 (syn.) 381
Triaplatys Fairmaire, 1877 (syn.) 415
Trichelytron Apfelbeck, 1912 (syn.) 440
Trichidea Baly, 1890 (syn.) 458
Trichobalya Weise, 1924 (g.) 458
Trichocerastes Motschulsky, 1866 (syn.) 437
Trichocerophysa Gressitt & Kimoto, 1963 (syn.) 378
Trichochrysea Baly, 1861 (g.) 703
Trichocolaspis L. N. Medvedev, 2005 (g.) 713
Tricholochmaea Laboissière, 1933 (syn.) 393
Trichomimastra Weise, 1922 (g.) 459
Trichonotolema Heinze, 1927 (syn.) 225
Trichosepharia Laboissière, 1936 (g.) 459
Trichotheca Baly, 1860 (g.) 704
Trichoxantha L. N. Medvedev, 1992 (g.) 705
Tricliona Lefèvre, 1885 (syn.) 722
Triplispa Weise, 1897 (sbg.) 264
Truncatus Paliy, 1970 (syn.) 536
Tschitscherinula Jacobson, 1908 (g.) 459
Tsouchya C.-F. Lee, 2020 (g.) 459
Tuberculobruchus Decelle, 1951 (syn.) 196
Tumidocnema Cho, 2016 (un.) 365
Tuomueria S.-H. Chen & Jiang, 1986 (syn.) 456
Tuomuria S.-H. Chen & Jiang, 1985 (syn.) 456
Tylocentra Reitter, 1926 (syn.) 242
Typhodes Samuelson, 1984 (syn.) 577

Udorpes Motschulsky, 1845 (syn.) 493
Ulema Bedel, 1891 (syn.) 231
Upseleatlasia Bourdonné, 2012 (sbg.) 324

Vervex Gistel, 1848 (syn.) 604
Vietocerus Lopatin, 2003 (syn.) 457
Vietoluperus L. N. Medvedev & Dang, 1981 (g.) 459
Virgulatorina Kühnelt, 1984 (sbg.) 342
Vittatochrysa Lopatin, 1977 (sbg.) 324
Volosatik Bieńkowski, 2022 (sbg.) 324

Wallacea Baly, 1859 (g.) 237
Wallaceana Maulik, 1928 (syn.) 237
Weigelia L. N. Medvedev, 2005 (sbg.) 708
Weiselina Reitter, 1913 (g.) 728
Wellschmiedia Warchałowski, 1985 (sbg.) 604
Wittmeriana L. N. Medvedev, 1975 (syn.) 475

Xanthocycla Baly, 1875 (syn.) 475
Xanthogaleruca Laboissière, 1934 (g.) 397
Xanthogramma Weise, 1923 (syn.) 366
Xanthonia Baly, 1863 (g.) 705
Xanthonia Baly, 1863 (sbg.) 705
Xanthonioides Chûjô, 1966 (sbg.) 706
Xanthophorus Jacoby, 1908 (g.) 706
Xenarthra Baly, 1861 (syn.) 458
Xenolina Jacoby, 1905 (sbg.) 285
Xenomela Weise, 1884 (g.) 349
Xingeina S.-H. Chen, Jiang & S.-Y. Wang, 1987 (g.) 459
Xoidolema Heinze, 1931 (syn.) 231
Xuthea Baly, 1865 (g.) 579
Xuthea Baly, 1865 (sbg.) 579

Ydorpes Motschulsky, 1845 (syn.) 493
Yemenaltica Scherer, 1985 (g.) 579
Yetialtica Döberl, 1991 (g.) 579
Yoshiakia Takizawa, 2009 (syn.) 580
Yulongedon Daccordi & S.-Q. Ge, 2012 (g.) 288
Yunaspes S.-H. Chen, 1976 (g.) 413
Yunnaedon Daccordi & L. N. Medvedev, 2000 (g.) 350
Yunnaniata Lopatin, 2009 (syn.) 403
Yunocassis S.-H. Chen & Zia, 1961 (syn.) 242
Yunohespera S.-H. Chen & S.-Y. Wang, 1984 (g.) 579
Yunomela S.-H. Chen, 1964 (syn.) 457
Yunotrichia S.-H. Chen & S.-Y. Wang, 1980 (g.) 579

Zabrotes Horn, 1885 (g.) 179
Zangaltica S.-H. Chen & S.-Y. Wang, 1988 (g.) 580
Zangastra S.-H. Chen & Jiang, 1981 (g.) 398
Zangia S.-H. Chen, 1976 (syn.) 439
Zeugophora Kunze, 1818 (g.) 170
Zeugotaenia Motschulsky, 1860 (sbg.) 325
Zipangia Heikertinger, 1924 (syn.) 577
Zipanginia Ohno, 1962 (g.) 580
Zizonia S.-H. Chen, 1976 (syn.) 422
Zygogramma Chevrolat, 1836 (g.) 342
Zygospila Achard, 1923 (sbg.) 343